3. Tree diagrams
   n = number of paths of tree

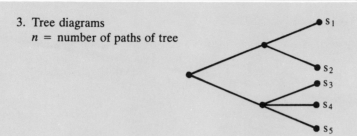

4. Multiplication principle

$$n = n_1 \cdot n_2 \cdot n_3 \cdots n_k$$

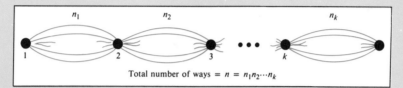

Total number of ways $= n = n_1 n_2 \cdots n_k$

Permutations and Combinations

Number of permutations $= P(n,r) = \dfrac{n!}{(n-r)!}$

Number of combinations $= \dbinom{n}{r} = \dfrac{n!}{r!(n-r)!}$

Rules of Probability

Let S be a finite sample space with n outcomes. Denote the probability that a subset or event A of the sample space occurs by $P(A)$. The following formulas allow one to find some important probabilities.

1. The probability of an event A with k outcomes in an equally likely sample space is given by

$$P(A) = \frac{k}{n}$$

2. The probability of A or B is given by

$$P(A \cup B) = P(A) + P(B) - P(A \cap B)$$

3. The probability of A or B when A and B are disjoint events is given by

$$P(A \cup B) = P(A) + P(B)$$

4. The conditional probability of A given B is given by

$$P(A \mid B) = \frac{P(A \cap B)}{P(B)}$$

5. The probability of A and B is given by

$$P(A \cap B) = P(A)P(B \mid A)$$
$$= P(B)P(A \mid B)$$

6. The probability of A and B when A and B are independent events is given by

$$P(A \cap B) = P(A) \cdot P(B)$$

APPLIED MATHEMATICS

for Management, Life Sciences, and Social Sciences

STANLEY J. FARLOW

University of Maine

GARY M. HAGGARD

Bucknell University

The Random House/Birkhäuser
Mathematics Series

RANDOM HOUSE · New York

First Edition

987654321

Library of Congress Cataloging-in-Publication Data

Farlow, Stanley J., 1937–
    Applied mathematics for management, life sciences, and social sciences/
Stanley J. Farlow and Gary M. Haggard.
        p. cm.
    Includes index.
    ISBN 0-394-35160-6
    1. Mathematics—1961—   I. Haggard, Gary.   II. Title.
QA37.2.F37 1987
    510—dc19   87-20999
                                                                    CIP

Preface

The primary goal of this text is to provide a survey of the essential quantitative ideas and mathematical techniques used in decision making in a diversity of disciplines. There is an emphasis on developing understanding and comprehension of the mathematical methods as well as establishing sound technical proficiency. The level of the presentation is easily accessible to most students, and there is a strong degree of reliance on intuition—more so than on overly formal and abstract mathematical theory. We have tried to write a text that is useful to students and helpful for instructors. In order to fulfill our goal we have used a very broad and rich selection of topics, features, and motivational items in conjunction with proven pedagogical techniques in the teaching of mathematics.

Applied Mathematics is designed for use in a two-semester or three-quarter survey course in finite mathematics and calculus taken primarily by students majoring in business, economics, life sciences, and social sciences. Part I covers the major topics of finite mathematics and Part II presents an introduction to the essential topics of the calculus. The flexible organization and rich selection of topics make the book easily adaptable to a wide variety of courses. The only prerequisite for studying the material is two years of high-school algebra or its equivalent.

Pedagogical Features

Emphasis and Writing Style We have written the text in a lively and motivational style. We use real-world examples, historical comments, business profiles, and intuitive presentations to *explain* to the student the intelligent use of quantitative concepts and techniques. Our basic approach is to present the mathematics in a more *humanistic* manner and thereby enhance its use as a genuine aid to decision making by nonmathematicians.

Format Major concepts and definitions are highlighted with a box and in a second color so they may be easily found and referred to throughout the book. All motivational interest material is also set off in special boxes.

Strong Visual Program We have included more than 600 figures and numerous photographs to convey a strong visual sense of the mathematics for ease of learning and to provide a realistic context to the applications. We have tried to provide helpful captions to all figures and photographs, either reinforcing an idea or providing additional explanation.

Realistic Applications We have put forth much effort to provide realistic applications in the examples and exercise sets. We selected many that will appeal to *all* students in the course, regardless of their major area of study. All the applications were chosen and developed for their pedagogical appeal and effectiveness in helping to teach the mathematics.

Worked Examples The book contains over 400 worked examples, each carefully chosen to illustrate a particular concept or technique. We collected these over the many years we have been teaching the material to our own students.

Exercises Effective exercises are at the heart of any mathematics textbook. The text contains over 3800 well chosen and interesting exercises designed to reinforce understanding as well as develop technical skills. They are carefully graded as to the level of difficulty and include many challenging applied problems.

Interviews We have interviewed people at several major American companies to illustrate the importance of various finite mathematics topics to their successful operation.

Historical Comments There are many historical comments and profiles of key historical figures in mathematics. These should be of interest to both students and instructors.

Chapter Epilogues Each chapter ends with a brief prose epilogue that relates the material to larger contemporary society. Some epilogues are taken from newspaper articles or magazines; others simply provide an educational diversion.

End-of-Chapter Review Material At the end of each chapter is a list of key terms, a varied and extensive chapter review exercise set, and a brief practice test.

Section Purposes All sections in a chapter begin with a brief listing of the important concepts and skills. These provide the students with a brief introduction to the overall purpose of the section. They also allow greater flexibility for an instructor who wishes to omit or reorder some sections.

Chapter Organization

The charts on the facing page illustrate how the chapters are related and how the book can be used in a variety of courses.

Algebra Review Material The algebra reviews, "Finite Mathematics Preliminaries" and "Calculus Preliminaries," are intended for students whose background may be weak in some topics in algebra. Each review section presents only the relevant material needed to review either Part I or Part II; some instructors may wish to combine the two sections for a more intensive, unified review at the beginning of the course.

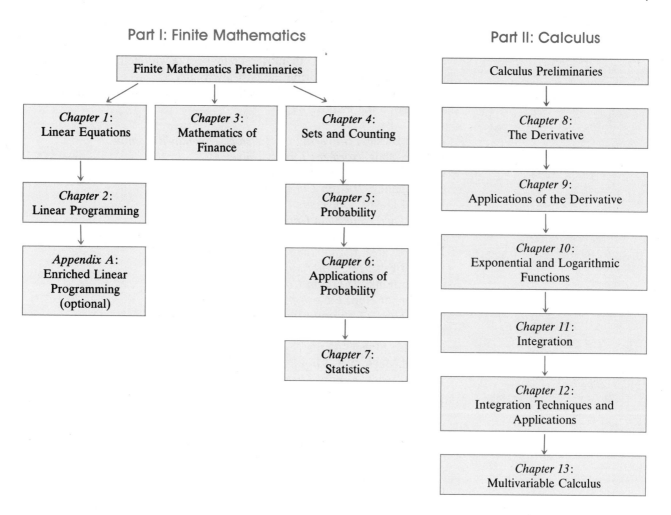

An instructor may omit the material, cover all or part of the material in class, or assign portions for students to work on their own. Practice tests have been included to aid instructors in the placement of students or assessing the amount of review needed.

Student and Instructor Aids

Student Solutions Manual   This manual is available to students at a nominal cost. It contains detailed solutions to all odd-numbered exercises in the book.

Instructor Aids   An instructor's solutions manual, available to adopters, contains detailed solutions to all even-numbered exercises in the book. A printed test bank provides several versions of tests for each chapter in the book. Also available is a set of transparency masters for classroom use of selected figures from the text.

Acknowledgments

We would like sincerely to thank the many people who helped us at various stages with this project during the past few years.

The following people offered excellent advice, suggestions, and ideas as they reviewed the manuscript:

Ronald Barnes, *University of Houston; Downtown Campus*
William Blair, *Northern Illinois University*
Gordon Brown, *University of Colorado*
Card Chaney, *Husson College*
Richard Davitt, *University of Louisville*
Duane Deal, *Ball State University*
Bruce Edwards, *University of Florida*
Joseph Evans, *Middle Tennessee State University*
Howard Frisinger, *Colorado State University*
Carolyn Funk, *Thornton Community College*
Christopher Hee, *Eastern Michigan University*
Myron Hood, *California Polytechnic State University*
Evan Houston, *University of North Carolina, Charlotte*
Lawrence Maher, *North Texas State University*
Michael Mays, *West Virginia University*
Thomas Miles, *Central Michigan University*
Robert Moreland, *Texas Tech University*
David Ponick, *University of Wisconsin, Eau Claire*
Robert Pruitt, *San Jose State University*
Douglas Schafer, *University of North Carolina, Charlotte*
Cynthia Siegel, *University of Missouri, St. Louis*
William Soule, *University of Maine*
Louis Talman, *Metropolitan State University*

We especially appreciate the help of Ronald Barnes, David Finkel (Bucknell University), and Gloria Langer with checking the accuracy of the examples and answers to the exercises. They did an excellent and thorough job for the text.

Finally, we want to thank all the people at Random House who have contributed to the project and worked so hard to support us. The editorial staff was always helpful and supportive through every phase: Wayne Yuhasz, Senior Editor; Alexa Barnes and Anne Wightman, Developmental Editors; and Louise Bush, Editorial Assistant. The Random House production staff provided professional and efficient support in producing a very attractive book: Margaret Pinette, Project Manager; her assistant, Pamela Niebauer; Michael Weinstein, Production Manager; and his assistant, Susan Brown. Also, we appreciate the assistance of Mary Stuart Lang with research and choice of photographs.

Any errors that appear are the responsibility of the authors. We would appreciate having these brought to our attention. We would also appreciate comments from students and instructors.

Contents

Part I
Finite Mathematics

Finite
Mathematics
Preliminaries

The purpose of this preliminary section is to review some basic mathematical topics that will be used throughout the book. The reader can either use this material to review topics when needed or study these topics carefully before beginning the new material in the book. We begin by reviewing the most fundamental mathematical structure, the real number system.

The Real Numbers

One of the distinguishing features of modern business and science is the importance that is placed on quantification. It seems that almost everything is measured, weighed, or timed. We calculate speeds of objects, the size of the economy, scientific measurements of many kinds, and on and on. We even compute baseball statistics. In making these measurements we use whole numbers, positive numbers, negative numbers, fractional numbers, and irrational numbers. Of course, if we were interested only in making simple measurements, such as counting something or computing an average, we could probably get by with simple numbers, such as integers or fractions. However, for performing sophisticated quantitative analyses, as is often done in the modern business world, we must use the properties of the **real number system**.

When we refer to numbers in this book, unless otherwise stated, we are normally referring to real numbers. Real numbers are often represented as points on a line, and the entire line is called the **real number line**. (See Figure 1.)

Every real number is associated with a point on the real number line and vice versa. The number 0 is generally labeled, while the **positive numbers** are drawn to the right and the **negative numbers** to the left. We can think of the real number line as two **infinite rulers** back to back, a positive one going infinitely far to the right and a negative one going infinitely far to the left. (See Figure 2.)

There are various kinds of real numbers within the real number system. (See Figure 3.) Table 1 lists the basic kinds of real numbers.

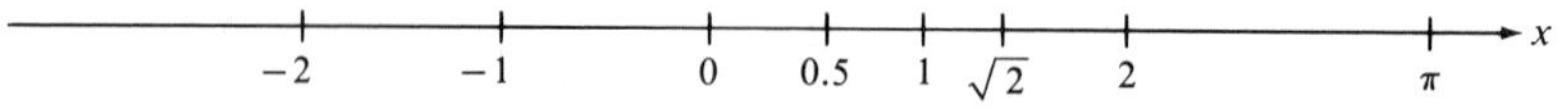

Figure 1
The real number line and typical numbers

Figure 2
Real number line illustrated by back-to-back infinite rulers

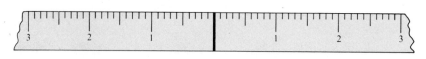

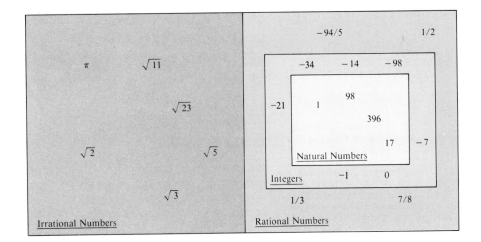

Figure 3
The family of real numbers

TABLE 1
Types of Real Numbers

Type of Number	Definition	Examples
Natural numbers	1, 2, 3, …	7, 11, 25
Integers	…, −2, −1, 0, 1, 2, …	−3, 0, 5
Rational numbers	Fractions, any number that can be written as a/b, where a and b are integers with $b \neq 0$	$\frac{4}{3}, 5.34, \frac{3}{25}, 5, 0, \frac{5}{1}$
Irrational numbers	Numbers that cannot be written as fractions	$\sqrt{2}, \pi, e$

Properties of the Real Numbers

Table 2 lists several useful properties of the real numbers that are used often. We generally use these properties so matter-of-factly that we often do not even realize that we are using them.

TABLE 2
Properties of Addition and Multiplication

Property	Addition	Multiplication
Commutative Properties	$a + b = b + a$	$ab = ba$
Associative Properties	$a + (b + c) = (a + b) + c$	$a(bc) = (ab)c$
Identities	$0 + a = a$	$1 \cdot a = a$
Distributive Properties	$a(b + c) = ab + ac$	
	$(b + c)a = ba + ca$	

The following example illustrates the above properties. The reader should determine which rule is being used in each equation.

Example

(a) $2 + 3 = 3 + 2$

(b) $3 \cdot 4 = 4 \cdot 3$

(c) $x^2 + (y^2 + z^2) = (x^2 + y^2) + z^2$

(d) $(2 \cdot 4) \cdot 3 = 2 \cdot (4 \cdot 3)$

(e) $(-1) + 0 = -1$

(f) $1 \cdot 5 = 5$

(g) $4(x^2 + y^2) = 4x^2 + 4y^2$

(h) $(2 + 5)x = 2x + 5x$ ☐

HISTORICAL NOTE

The first recorded appearance in print of plus and minus signs was in an arithmetic text published in Leipzig in 1489 by the German Renaissance mathematician Johann Widmann (ca. 1460–1520).

The signs replaced the "Italian" p and m notation that was previously used for plus and minus. Imagine writing

$$2p2m1 = 3$$

Exponents

Most people just do not appreciate the power of exponents. Do you realize that by using the three numbers 2, 3, and 4 only once each, you can construct a number that is 100 billion times larger than the distance (in miles) from the earth to the sun? It's true. The number 3^{42}, which uses the digits 2, 3, and 4 only once and represents the number 3 times itself 42 times, is a number that is 100 billion times larger than the 93,000,000 miles to the sun. In science and modern business, one often encounters repeated multiplication of some number like $6 \cdot 6 \cdot 6 \cdot 6$ or $10 \cdot 10 \cdot 10 \cdot 10 \cdot 10 \cdot 10$. These products can be written more compactly and efficiently as

$$6^4 = 6 \cdot 6 \cdot 6 \cdot 6$$
$$10^6 = 10 \cdot 10 \cdot 10 \cdot 10 \cdot 10 \cdot 10$$

The expression on the left is called **exponential notation**.

Definition of a^n

If a is a real number and if n is a positive integer, then we define

$$a^n = a \cdot a \cdot \cdots \cdot a$$

$$n \text{ factors}$$

The number a is called the **base**, and n is called the **exponent**. We also write the negative exponent as the **reciprocal**:

$$a^{-n} = \frac{1}{a^n} \qquad (a \neq 0)$$

We also define $a^0 = 1$ for $a \neq 0$, while the expression 0^0 is left **undefined**. There are some serious logical problems in trying to define it.

Example

The following equations show numbers represented by exponential notation.

(a) $2^{10} = 2 \cdot 2 \cdot 2 \cdot 2 \cdot 2 \cdot 2 \cdot 2 \cdot 2 \cdot 2 \cdot 2 = 1024$

(b) $5^{-1} = \dfrac{1}{5}$

(c) $(55{,}456{,}930{,}346{,}339)^0 = 1$

(d) $x^4 = x \cdot x \cdot x \cdot x$

(e) $4^{-3} = \dfrac{1}{4 \cdot 4 \cdot 4} = \dfrac{1}{64}$ □

To use exponents effectively, you should be familiar with the following rules.

Properties of Exponentials

If a and b are real numbers and m and n are integers, then the following rules of exponents hold.

General Rule	Example
$a^m a^n = a^{m+n}$	$a^2 a^{-4} = a^{2+(-4)} = a^{-2}$
$(a^m)^n = a^{mn}$	$(a^2)^{-4} = a^{(2)(-4)} = a^{-8}$
$(ab)^m = a^m b^m$	$(ab)^2 = a^2 b^2$
$\left(\dfrac{a}{b}\right)^m = \dfrac{a^m}{b^m}$	$\left(\dfrac{a}{b}\right)^3 = \dfrac{a^3}{b^3}$
$\dfrac{a^m}{a^n} = a^{m-n}$	$\dfrac{a^2}{a^3} = a^{2-3} = a^{-1} = \dfrac{1}{a}$

Example

The following equations illustrate the rules of exponents. The reader should identify which rule is being applied in each equation.

(a) $3^7 3^{-4} = 3^3$

(b) $(2^2)^{-1} = \dfrac{1}{2^2}$

(c) $(2 \cdot 3)^2 = 2^2 \cdot 3^2$

(d) $\left(\dfrac{7}{3}\right)^2 = \dfrac{7^2}{3^2}$

(e) $(3r^2)^4 = 3^4 \cdot (r^2)^4 = 3^4 r^8$

(f) $5^{-1} + 3^{-1} = \dfrac{1}{5} + \dfrac{1}{3}$

(g) $\dfrac{2^{-4} \cdot 2^3}{2^2 \cdot 2^6} = \dfrac{2^{-1}}{2^8} = 2^{-9}$

(h) $(3x)^2 = 3^2 x^2$

(i) $(3x^2)(5x^4) = 15x^6$

(j) $-3^2 = -(3^2) = -9$

(k) $-3^{-2} = -(3^{-2}) = -\dfrac{1}{3^2}$ □

Problems 1

Evaluate the expressions in Problems 1–20. Use your calculator for Problems 15–20.

1. 5^3
2. 4^{-2}
3. $-(1)^{-4}$
4. $(1/5)^{-2}$
5. $(2/5)^{-1}$
6. 6°
7. 0°
8. $3^{-1} + 4^2$
9. $-(2)^{-4}$
10. $-(-3)^{-1}$
11. -2^{-4}
12. $-(2)^{-4}$
13. -1°
14. $(1/8)^2$
15. 2^{10}
16. $(-2)^{10}$ (error message)
17. $(2/7)^2$
18. $5^{-1} + 3^{-4}$
19. 1.01^{50}
20. 1.05^{75}

Simplify the expressions in Problems 21–30 so that they have only positive exponents. Assume that all variables are positive real numbers.

21. $\dfrac{5^2}{5^4}$

22. $\dfrac{3^{-4}}{3^{-5}}$

23. $\dfrac{10^5 \cdot 10^{-5} \cdot 10^6}{10^5 \cdot 10^{-7}}$

24. $\dfrac{x^4 \cdot y^3}{x^3 \cdot y^2}$

25. $\dfrac{a^5 \cdot b^{-3}}{a^4 \cdot b^{-7}}$

26. $\left(\dfrac{3x^{-2}}{y^4}\right)^{-3}$

27. $\dfrac{(4m^{-1})^{-1}}{2m^2}$

28. $\dfrac{5^{-2}x^2 y^{-1}}{5x^{-1}y^{-3}}$

29. $\dfrac{x^{-1}}{y^{-1}}$

30. $\dfrac{x^{-1}}{y}$

31. **Hmmmmmmm** Which is the larger number, 2 or $(2,465,543,076,345)^0$?

32. **English to Algebra** Write algebraically the following statement: The product of powers of the same quantity is equal to the power of that same quantity with exponent the sum of the exponents of the two factors.

33. **Simple Puzzler** What is the largest number that can be written by using three 2's and the operations of addition, subtraction, multiplication, division, and exponentiation? Some examples are $2 + 2 + 2$, $2 \cdot 2 \cdot 2$, $2^2 + 2$, $(2 + 2)/2$, and 222. Can you get a number as large as 4,194,304? What would be the answer if exponents were not allowed?

34. **Puzzler** What is the largest number that can be written by using each of the numbers 3, 4, and 5 only once?

"WE HAVE REASON TO BELIEVE BINGLEMAN IS AN IRRATIONAL NUMBER HIMSELF."

Roots

It is useful to introduce the concept of **fractional exponents** into mathematics. A beginner in mathematics often wonders how it is possible to raise a number to a fractional power. For instance, it is easy to understand that 2^3 means to multiply 2 times itself the appropriate number of times, but what is the meaning of $2^{1/3}$?

We begin by saying that r is a **square root** of a number a if $r^2 = a$. In general, we say that r is an **nth root** of a number a if $r^n = a$. In this book we are interested in real roots (in contrast to complex roots), and so when we say "roots," we mean real roots. The interesting thing is that there are *always* 0, 1, or 2 distinct nth roots of a number for any positive integer n. Table 3 illustrates the nature of nth roots.

2.√ 9 ist 3.ptingen in einer su
Exempl von communicanten
item √ 20 ßû √ 45 item √ 1
facit √ 125.   fa:
μ⅔it.√ 12−½ßû √ 40−½it.√ 1
−⅔ fa: √ 98   fa
Exempl von irracionaln
facit √ Des collects 12−+ √ 14
1 √ 13 facit √ Des collects 17−+

HISTORICAL NOTE

The radical notation $\sqrt{}$ was introduced in 1525 by the German algebraist Christoff Rudolff (ca. 1500–1545) in his book on algebra entitled *Die Coss*. The symbol was chosen because it looks like a lowercase *r*, which stood for *radix*, the Latin word for the square root. Rudolff is also remembered because he was one of the first mathematicians to introduce decimal fractions into texts.

TABLE 3
Nature of *n*th Roots of a
Real Number

	a Positive	*a* Negative
n Even	Two roots (9 has two square roots, 3 and -3)	No roots (-2 has no square root)
n Odd	One root (8 has one cube root, 2)	One root (-8 has one cube root, -2)

For completeness, we should mention that the only *n*th root of zero is zero.

It is convenient to introduce the **fractional notation** $a^{1/n}$ or $\sqrt[n]{a}$ to stand for the *n*th root of a number *a*. If there are two *n*th roots, then by notational agreement, $a^{1/n}$ is taken as the positive of the two roots. The following example illustrates the ideas.

Example

(a) $4^{1/2} = \sqrt{4} = +2$ ⟵ *positive root*

(b) $8^{1/3} = \sqrt[3]{8} = +2$ ⟵ *only root*

(c) $(-8)^{1/3} = \sqrt[3]{-8} = -2$ ⟵ *only root*

(d) $16^{1/4} = \sqrt[4]{16} = +2$ ⟵ *positive root*

(e) $(4/9)^{1/2} = \sqrt{4/9} = +2/3$ ⟵ *positive root*

(f) $(-2)^{1/2} = \sqrt{-2}$ ⟵ *no root*

(g) $(-3)^{1/2} = \sqrt{-3}$ ⟵ *no root* □

Problems II

Evaluate the roots in Problems 1–10. Use a calculator if necessary.

1. $8^{1/3}$

2. $16^{1/4}$

3. $125^{1/3}$

4. $1024^{1/10}$

5. $\sqrt[3]{27}$

6. $\sqrt[3]{64}$

7. $\sqrt[3]{-125}$

8. $\sqrt[5]{2}$

9. $\sqrt[9]{512}$

10. $\sqrt[8]{256}$

11. **Hmmmmm** The next time you owe someone half a dollar, say that since 25 cents $= \frac{1}{4}$ dollar you will just take the square root of these equal quantities and obtain the new equal quantities: 5 cents $= \frac{1}{2}$ dollar. Then give the person a nickel. What is wrong with your argument?

Simplest Form for a Radical Square Root

We say that an expression involving a square root $\sqrt{}$ is in **simplest form** when the following conditions are satisfied.

- No factor within a radical should be raised to a power greater than one. For example, $\sqrt{x^2}$ and $\sqrt{x^3}$ are not in simplest form.
- No square root should be in the denominator. For example,

$$\frac{1}{\sqrt{x}} \quad \text{should be rewritten as} \quad \frac{\sqrt{x}}{x}$$

- No fraction should be within a square root. For example,

$$\sqrt{\frac{5}{2}} \quad \text{should be rewritten as} \quad \frac{\sqrt{5}}{\sqrt{2}}$$

The following properties are useful when writing expressions involving square roots in simplest form.

Properties of the Square Root

If a and b are nonnegative real numbers, then the following rules hold.

- $\sqrt{a^2} = a$
- $\sqrt{ab} = \sqrt{a}\sqrt{b}$

- $\sqrt{\dfrac{a}{b}} = \dfrac{\sqrt{a}}{\sqrt{b}} \quad b \neq 0$

Example

The following equations illustrate how the properties of the square root can be used to simplify expressions. Assume that all variables are nonnegative.

(a) $\sqrt{20} = \sqrt{5 \cdot 4} = \sqrt{5} \cdot \sqrt{4} = 2\sqrt{5}$ (b) $\sqrt{\dfrac{25}{4}} = \dfrac{\sqrt{25}}{\sqrt{4}} = \dfrac{5}{2}$

(c) $\sqrt{25z^5} = \sqrt{25z^4 \cdot z} = \sqrt{25z^4} \cdot \sqrt{z} = 5z^2\sqrt{z}$

(d) $\sqrt{25x^3} = \sqrt{25x^2 \cdot x} = \sqrt{25x^2} \cdot \sqrt{x} = 5x\sqrt{x}$

(e) $\dfrac{2xy}{\sqrt{y}} = \dfrac{2x\sqrt{y}\sqrt{y}}{\sqrt{y}} = 2x\sqrt{y}$ □

Problems III

Reduce the expressions in Problems 1–10 to simplest form. Assume that all variables are nonnegative.

1. $\sqrt{25x^3}$

2. $\sqrt{16x^5}$

3. $\dfrac{3x}{\sqrt{x}}$

4. $\dfrac{\sqrt{25x^2y^2}}{\sqrt{x}}$

5. $\dfrac{1}{\sqrt{xy}}$

6. $\sqrt{16x^2y^2}$

7. $\sqrt{z^7}$

8. $\sqrt{x^2y^3z^5}$

9. $\sqrt{u^2v^4}$

10. $\sqrt{2x^2y^5}$

11. **Hmmmmm**   What is wrong with the following argument?

$$1 = \sqrt{1} = \sqrt{(-1)(-1)} = \sqrt{-1}\sqrt{-1} = [(-1)^{1/2}]^2 = -1$$

Logarithms

Roughly 400 years ago, the Scottish laird John Napier looked at two rows of numbers, somewhat similar to the rows of numbers below, and made a discovery that literally changed the mathematical world.

0	1	2	3	4	5	6	7	8	9	10
1	2	4	8	16	32	64	128	256	512	1024

This was a time before computers, calculators, and even slide rules. It was a time when all computations were done by hand. What Napier saw in the above numbers was an ingenious way to multiply numbers by *adding*. Since addition is far easier and quicker to perform, he discovered what might be called a sixteenth century computer.

To understand Napier's discovery, let us suppose that we wish to multiply the numbers 16 and 64. If we carry out this multiplication, we will see that the product is 1024. However, there is a far easier way to find this product using the above table. (Do you see what it is?) What Napier saw was that if the two numbers directly above 16 and 64 (namely, 4 and 6) were added, then the sum (10) is the number directly above the product of 16 and 64. In fact, this scheme will work for the product of any two numbers in the bottom row (4 times 128 is the number below the sum $2 + 7 = 9$, or 512). Of course, Napier still had a few bugs to work out. In particular, how could he multiply numbers that were not in the bottom row? These problems were all dealt with, and after 20 years of refinement, Napier published his results in 1614 in a paper entitled "A Description of the Marvelous Rule of Logarithms."

What Napier discovered in essence was that it is possible to convert multiplication to addition by adding the exponents or logarithms of two numbers, instead of multiplying the numbers themselves. This brings us to the following definition of one of the most important discoveries in all of computational mathematics.

Definition of the Logarithm

The **logarithm** of a positive number x to a base b, denoted

$$y = \log_b x$$

where b is any positive number such that $b \neq 1$, is the exponent that we must apply to b in order to get the number x. In other words, $b^y = x$ means the same as $y = \log_b x$. If the base b is 10, the logarithm is called the **common logarithm**, and we simply write

$$y = \log_{10} x = \log x$$

Later we will see that it is convenient to use a logarithm with base $b = 2.718\ldots$. This number is called "e" in mathematics, and the logarithm with base $e = 2.718\ldots$ is called the **natural logarithm**. It is denoted by

$$y = \log_e x = \ln x$$

Later, when we study the mathematics of finance, we will call the number $e = 2.718\ldots$ the **banker's constant**, for reasons that will become apparent.

The following logarithms are found using the definition of the logarithm.

Logarithmic Form		Exponential Form
(a) $\log 1000 = 3$	means the same as	$10^3 = 1000$
(b) $\log 100 = 2$	means the same as	$10^2 = 100$
(c) $\log(1/10) = -1$	means the same as	$10^{-1} = 1/10$
(d) $\log_2 16 = 4$	means the same as	$2^4 = 16$
(e) $\log_2 8 = 3$	means the same as	$2^3 = 8$
(f) $\log_2(1/8) = -3$	means the same as	$2^{-3} = 1/8$
(g) $\ln 2 = 0.693\ldots$	means the same as	$e^{0.693\ldots} = 2$
(h) $\ln 2.5 = 0.916\ldots$	means the same as	$e^{0.916\ldots} = 2.5$

The logarithms in parts (g) and (h) can be approximated either by using the tables in the back of this book or by using a calculator. Since logarithms were introduced about 400 years ago, they have been computed mostly by means of tables. Over the past 20 years, however, most people have used pocket calculators to find them. Pocket calculators having both common ($\log x$) and natural ($\ln x$) logarithm keys can be purchased for under \$10.   □

Solving Equations Using Logarithms

In the chapter on the mathematics of finance we will solve equations in which the **unknown variable** x occurs in the exponent. Specifically, we will solve equations of the form

$$e^x = b$$

where e is $2.718\ldots$ and b is a given constant. To solve for x in this equation, we simply rewrite the equation in logarithmic form, getting

$$x = \ln b$$

We can then approximate x by using a calculator or a table of logarithms. For example, to solve the equation

$$e^x = 20$$

we write it in logarithmic form:

$$x = \ln 20$$

and then evaluate this logarithm using a calculator. This gives $x \cong 2.996$.

In the mathematics of finance it is important to solve equations such as

$$100e^{0.05t} = 50{,}000$$

for the variable t. To solve this equation, we first divide each side by 100, getting

$$e^{0.05t} = 500$$

We then rewrite this equation in logarithmic form as

$$0.05t = \ln(500)$$

Dividing each side of the equation by 0.05, we get

$$t = \frac{1}{0.05} \ln(500) \cong \frac{6.215}{0.05} = 124.3$$

Often, the solution of such an equation will represent the number of years required for a given amount of money deposited in a bank to grow to a given amount. □

Problems IV

Solve the following equations for the indicated variable.

1. $e^x = 5$ **2.** $e^{2x} = 10$ **5.** $5e^t = 20$ **6.** $50e^{0.08t} = 1000$
3. $e^{5x} = 50$ **4.** $e^{-x} = 0.5$ **7.** $50e^{-0.2t} = 5$

Linear Equations in One Unknown

People solve linear equations every day whether they know it or not. If you see apples advertised at three for 75 cents, you quickly calculate that each apple costs 25 cents. By finding the cost of a single apple, you have essentially solved the linear equation

$$3x = 75$$

Other examples of linear equations are

$$2x + 1 = 0$$
$$-3w + 4 = 7$$
$$2p - 6 = -3$$

The letter in the equation is called the **unknown**, and the convention is that it is denoted by one of the letters in the latter part of the alphabet, such as u, v, w, x, y, and z.

General Linear Equation

The **general linear equation** is an equation of the form

$$ax + b = c$$

where a, b, and c are any real numbers, with the exception that $a \neq 0$.

To solve a linear equation for the unknown variable, we isolate the variable on one side of the equation using various properties of the real numbers. We illustrate these ideas with the example that follows.

Example

Solve the following equation for x:

$$\frac{1}{5}x = 4 - x$$

Solution   Multiply each side of the equation by 5 to get

$$x = 20 - 5x$$

Now add $5x$ to each side of the equation, getting

$$x + 5x = 20 - 5x + 5x$$

or

$$6x = 20$$

Finally, if we multiply each side of the equation by 1/6, we get the value of x:

$$x = \frac{10}{3}$$

The reader should substitute this value back into the equation to see that it is a **solution**.   □

HISTORICAL NOTE

The English mathematician Robert Recorde (1510–1558) was the first person to use the modern symbolism for the equals sign in an algebra book, *The Whetstone on Witte*, published in 1557. Recorde used the pair of parallel line segments, =, "bicuase noe 2 thynges can be moare equalle."

Problems V

Solve the linear equations.

1. $7 - 5x = x - 19$
2. $2(x - 3) = 3(x + 1)$
3. $15x - 10 = 2x + 1$
4. $9x + (1 - x) = -x$
5. $2x + 3(x + 1) = 5$
6. $5x - 5 = 7$
7. $-x + 1 = 5$
8. $5x + 2x = 0$
9. $7x - 2(x - 1) = 3x$
10. $1 - x = x - 1$
11. $2x - 1 = 1 - (2x - 1)$
12. $1 - [1 - (1 - x)] = 0$
13. $1 + [1 + (1 + x)] = 0$
14. $1 - [1 - (1 - x)] = 1 + [1 + (1 + x)]$ (a real good one)

Linear Inequalities

Inequalities express relations between two quantities that are not equal to one another. Since there are more instances of inequality than of equality in the world, it makes sense that inequalities should be an important tool for expressing relations. Most beginning students find the study of inequalities more difficult than that of equalities, since the rules of operation for inequalities are more general than the rules of operation for equalities. We will see in this section that they can also be quite useful.

Definition of the Inequality

If a and b are real numbers, then

- $a < b$ means that $b - a$ is positive,
- $a \leq b$ means that $b - a$ is positive or zero.

Geometrically, $a < b$ means that a is to the *left* of b on the real number line.

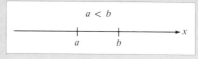

Likewise, we say that $a > b$ when $a - b$ is positive and that $a \geq b$ when $a - b$ is positive or zero.

Example

(a) $2 < 3$ since $3 - 2$ is positive

(b) $2 \leq 3$ since $3 - 2$ is positive or zero

(c) $-5 < -2$ since $-2 - (-5)$ is positive

(d) $-1 < 0$ since $0 - (-1)$ is positive

(e) $-5 < -3$ since $-3 - (-5)$ is positive

(f) $3 \leq 3$ since $3 - 3$ is positive or zero

(g) $4 > 3$ since $4 - 3$ is positive

(h) $-2 > -3$ since $-2 - (-3)$ is positive   □

To solve problems involving inequalities, it is useful to know the rules they satisfy.

Six Properties of Inequalities

For real numbers a, b, and c the following properties hold.

1. If $a < b$ and $b < c$, then $a < c$.
2. If $a < b$, then $a + c < b + c$.
3. If $a < b$ and $c < d$, then $a + c < b + d$.
4. If $a < b$ and c is positive, then $ac < bc$.
5. If $a < b$ and c is negative, then $ac > bc$.
6. If $a < b$ and both a and b have the same sign (both positive or both negative), then $1/a > 1/b$.

One of the major uses of inequalities is to describe different regions on the real number line. In particular, they can describe the following **intervals**.

Important Intervals on the Real Line

We say that:

- $a < x < b$ when $a < x$ and $x < b$
- $a < x \leq b$ when $a < x$ and $x \leq b$
- $a \leq x < b$ when $a \leq x$ and $x < b$
- $a \leq x \leq b$ when $a \leq x$ and $x \leq b$

By means of the above **double inequalities**, we can define the following intervals.

Open Interval

The open interval (a, b) consists of the real numbers x that satisfy $a < x < b$.

Closed Interval

The closed interval $[a, b]$ consists of the real numbers x that satisfy $a \leq x \leq b$.

Half-Open Interval

The half-open interval $(a, b]$ consists of the real numbers x that satisfy $a < x \leq b$. The half-open interval $[a, b)$ consists of the real numbers x that satisfy $a \leq x < b$.

Semi-Infinite Intervals

Other important intervals of real numbers are the semi-infinite intervals:

$$(a, \infty) = \{\text{all numbers greater than } a\}$$
$$[a, \infty) = \{\text{all numbers greater than or equal to } a\}$$
$$(-\infty, a) = \{\text{all numbers less than } a\}$$
$$(-\infty, a] = \{\text{all numbers less than or equal to } a\}$$

Infinite Interval

The infinite interval $(-\infty, \infty)$ consists of all the real numbers.

Some typical intervals are shown in Figure 4.

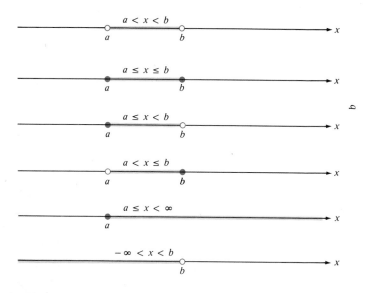

Figure 4
Typical intervals

It is often important to find the numbers that satisfy a given inequality. The following example illustrates how inequalities can be solved.

Example

Find all possible numbers that satisfy the inequality

$$x + 1 < 2x - 1$$

Solution   We are looking for all real numbers x for which this inequality is a true statement. Starting with

$$x + 1 < 2x - 1$$

add $-x$ to each side, getting

$$1 < x - 1 \qquad \text{(Property 2)}$$

Then add 1 to each side, getting

$$2 < x \qquad \text{(Property 2)}$$

Hence the solution consists of all real numbers greater than 2, or $(2, \infty)$. You should check a few of these values for yourself.   □

Can You Find the Flaw?

Can you find the flaw in the following use of inequalities? We will show that all positive numbers are negative.

We begin by selecting any two numbers a and b, where $a < b$. Hence

$$a(a - b) < b(a - b) \qquad \text{(just multiply each side by } a - b\text{)}$$

Hence $a^2 - ab < ab - b^2$ (just algebra)

Hence $a^2 - 2ab + b^2 < 0$ (just add $a^2 - ab$ to each side)

Hence $(a - b)^2 < 0$ (just algebra)

But this last inequality says that the (obviously) positive number $(a - b)^2$ is negative. What went wrong?

Problems VI

Solve the following inequalities.

1. $x - 1 < 2x + 3$
2. $x + 4 < -x + 4$
3. $2x + 5 \leq 4$
4. $5x + 6 \leq 5x + 7$
5. $8x + 5 \leq 0$

"BUT GERSHON, YOU CAN'T CALL IT GERSHON'S EQUATION IF EVERYONE HAS KNOWN IT FOR AGES."

Cartesian Coordinate System

One of the greatest mathematical developments of all time was the merging of the major mathematical disciplines of algebra and geometry into what is now called *analytical geometry*. To understand how the ideas of algebra (equations) can be merged with the ideas of geometry (lines and planes), we begin by drawing two perpendicular lines, one horizontal (*x*-axis) and one vertical (*y*-axis), as shown in Figure 5.

The point where the two lines intersect is called the **origin** and is labeled $(0, 0)$. All other points are labeled accordingly by a pair of numbers (x, y) called the **coordinates** of the point. The coordinates locate the position of the point relative to both axes. For example, to find the point labeled $(3, 4)$, we go 3 units to the right on the *x*-axis and 4 units upward parallel to the *y*-axis. To find the

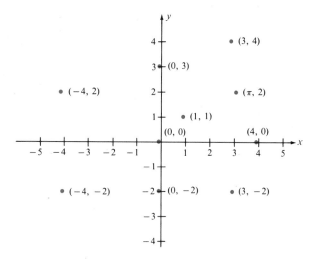

Figure 5
The cartesian coordinate system

point $(-1, -4)$, we move 1 unit to the left on the *x*-axis and 4 units downward parallel to the *y*-axis. For the point $(3, 4)$, we would say that 3 is the **x-coordinate** and 4 is the **y-coordinate**.

We can easily find the distance from one point to another in this system by applying probably the most famous theorem in mathematics, the **Pythagorean theorem**. This geometric theorem was first proved by the Greek mathematician Pythagoras in about 500 B.C., according to mathematical tradition. The theorem says that the sum of squares of the lengths of the two sides of a right triangle (a triangle that has one 90 degree angle) is the square of the length of the hypotenuse (the third side, the one opposite the right angle). Figure 6 illustrates this idea.

Using the Pythagorean theorem, we can find the distance D between any two points in the plane.

Distance between Points in the Plane

The distance D between the points (a, b) and (c, d) in the *xy*-plane is given by

$$D = \sqrt{(a - c)^2 + (b - d)^2}$$

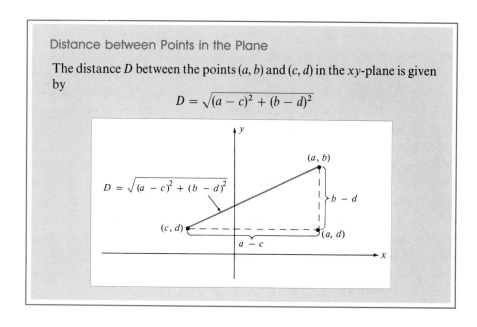

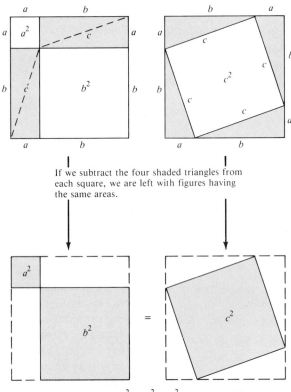

Two squares each subdivided differently having the same area:

If we subtract the four shaded triangles from each square, we are left with figures having the same areas.

Figure 6
The Pythagorean theorem

Hence we have $a^2 + b^2 = c^2$, where a and b are the lengths of the legs of a right triangle and c is the length of the hypoteneuse.

Example

Find the distance between the points $(1, 3)$ and $(-2, 2)$.

Solution
$$D = \sqrt{(1 + 2)^2 + (3 - 2)^2}$$

HISTORICAL NOTE

The time was ripe in the early 1600s for the brilliant French mathematician René Descartes (1596–1650) and others to found the subject of *analytical geometry*. The merging of the two great mathematical disciplines of the time, algebra and geometry, into a single unifying body of knowledge was one of the great mathematical developments of all time. Analytical geometry enabled mathematicians to describe curves (geometry), such as the trajectories of the planets, by means of equations (algebra). This new way of looking at the universe was responsible in great part for the "invention" of the calculus, which would come 50 years later.

It is interesting to note here that Descartes was also responsible for popularizing the exponential notation x^n that we studied earlier.

Problems VII

In Problems 1–5, connect the following points with straight line segments to find the mystery figures.

$$A = (0, 0) \qquad E = (2, 4)$$
$$B = (4, 0) \qquad F = (4, 3)$$
$$C = (2, 2) \qquad G = (2, 6)$$
$$D = (0, 3)$$

1. Connect A to C, C to B, C to E, D to E, E to F, and E to G.
2. Connect A to D, D to E, E to F, F to B, and B to A.
3. Connect A to F and D to B.
4. Connect A to G, G to B, and B to A.
5. Connect D to A, A to C, C to B, and B to F.

Find the mystery figures in Problems 6–9 by connecting the points in each list (in order) with straight line segments.

6. (3, 5), (6, −2), (0, −2), (3, 5)   (You've seen this before.)
7. (0, 0), (0, 4), (5, 4), (5, 0), (0, 0)   (You've seen this before.)
8. (3, −1), (−5, −1), (−2, 4), (6, 4), (3, −1)   (What is this called?)
9. (−2, 0), (5, −3), (6, 2), (1, 4), (−2, 0)   (What is this called?)

10. **Hmmmmmmmm**   Connect the following points in the order they are given with straight line segments to find the initials of one of the major universities in the United States.

   First initial: (0, 4), (0, 0), (4, 0), (4, 4)
   Second initial: (6, 0), (6, 4), (8, 2), (10, 4), (10, 0)

In Problems 11–14, locate the following points in the xy-plane and find the distance between them.

11. (0, 0), (1, 1)   12. (2, 3), (3, 5)
13. (−1, −1), (3, −3)   14. (1, 0), (2, −3)

15. What is the total distance around the perimeter of the triangle that has corner points (1, 1), (2, 3), and (4, −1)?

The Straight Line

An equation of the form

$$ax + by = c$$

where a and b are not both zero is called a **linear equation** in the two variables x and y. Examples are

$$2x + 3y = 3$$
$$x - y = 0$$
$$-3x + y = -2$$
$$y = 5$$

A solution of a linear equation in two variables is an **ordered pair** of real numbers (x, y), or point, that satisfies the equation. For example, the pair $(1, 3)$ is a solution of the equation $3x + y = 6$ because $3(1) + 3 = 6$. The **solution set** of a linear equation is the set of all solutions of a linear equation. By the **graph of an equation**, we mean a plot of all its solutions (x, y) in the xy-plane. We will not prove it here, but it can be shown that the graph of a linear equation is a straight line. To graph a linear equation, it is sufficient to plot just two solutions (x, y) in the xy-plane and then draw the straight line that passes through those two points because any straight line is completely determined by any pair of points on it.

Example

Draw the graph of $2x + y = 2$.

Solution It is usually convenient to first set x to zero and solve for y and then set y to zero and solve for x. Doing this, we get

Setting $x = 0$ and Solving for y	Setting $y = 0$ and Solving for x
$2(0) + y = 2$	$2x + 0 = 2$
$\qquad\;\; y = 2$	$\qquad 2x = 2$
So $(0, 2)$ is a point on the graph	$\qquad\;\; x = 1$
(the y-intercept).	So $(1, 0)$ is a point on the graph
	(the x-intercept).

Hence we have

x	y	
0	2	← y-intercept
1	0	← x-intercept

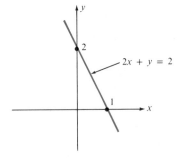

Figure 7
Line passing through $(0, 2)$ and $(1, 0)$

These values where the graph in Figure 7 crosses the x- and y-axes are called the x- and y-intercepts, respectively. □

Example —————————

Draw the graph of the line $x - 3y = 6$.

Solution After finding the x- and y-intercepts, we can easily graph the line. (See Figure 8.)

x	y	
0	−2	← y-intercept
6	0	← x-intercept

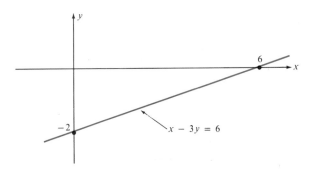

Figure 8
Line passing through $(0, 2)$ and $(6, 0)$

□

Slope of a Straight Line

An important characteristic of a line is its "steepness." To measure the steepness of a line, we introduce the concept of the **slope** of the straight line that passes through two given points.

Definition of Slope of a Straight Line

The slope of the nonvertical straight line that passes through the points (x_1, y_1) and (x_2, y_2) is given by

$$m = \frac{\text{Rise}}{\text{Run}} = \frac{y_2 - y_1}{x_2 - x_1} \qquad (x_1 \neq x_2)$$

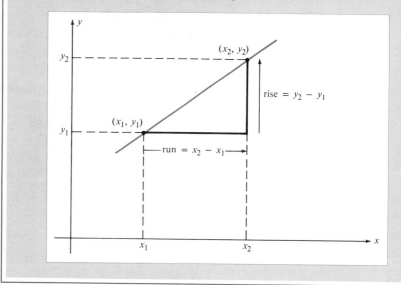

Example

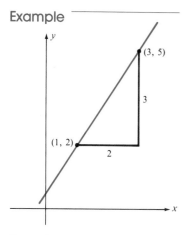

Figure 9
Line passing through (1, 2) and (3, 5)

Find the slope of the line that passes through each pair of points

(a) (1, 2) and (3, 5)

(b) (2, −1) and (2, 3)

Solution

(a) If we say that $(x_1, y_1) = (1, 2)$ and $(x_2, y_2) = (3, 5)$, we have

$$m = \frac{y_2 - y_1}{x_2 - x_1} = \frac{5 - 2}{3 - 1} = 1.5$$

The reader should know that it does not make any difference how we identify (x_1, y_1) and (x_2, y_2) with the two points. For example if we let $(x_1, y_1) = (3, 5)$ and $(x_2, y_2) = (1, 2)$, we have the same slope

$$m = \frac{2 - 5}{1 - 3} = \frac{-3}{-2} = 1.5$$

(See Figure 9.)

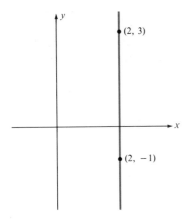

Figure 10
Vertical line x = 2

(b) If we say that $(x_1, y_1) = (2, -1)$ and $(x_2, y_2) = (2, 3)$, we have

$$m = \frac{y_2 - y_1}{x_2 - x_1} = \frac{3 - (-1)}{2 - 2} = \frac{4}{0} \longleftarrow \textit{Not defined!}$$

(See Figure 10.)   □

This last example illustrates the fact that **vertical lines** do not have slopes. Table 4 illustrates the general nature of the slope of a line.

"DOES THIS APPLY ALWAYS, SOMETIMES, OR NEVER?"

Special Forms of Lines

Slope-Intercept Form of a Line

The graph of the equation

$$y = mx + b$$

where m and b are constants is a nonvertical straight line. This form of a straight line is called the **slope-intercept form** of a line.

Note that the graph of the above equation crosses the y-axis when $x = 0$, or at the point $(0, b)$. The constant b is thus called the **y-intercept** of the line. To determine the significance of m, we consider $x = 0$ and $x = 1$. For these

TABLE 4
Slope of a Line

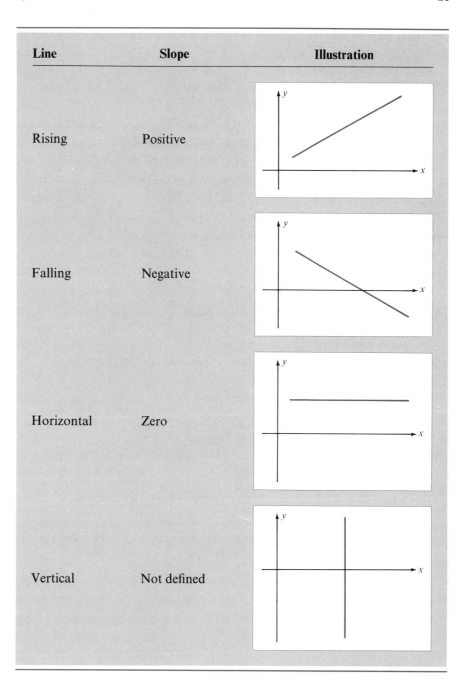

Line	Slope	Illustration
Rising	Positive	
Falling	Negative	
Horizontal	Zero	
Vertical	Not defined	

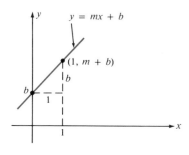

Figure 11
Constructing a line with slope m

values of x, the line passes through the points $(0, b)$ and $(1, m + b)$. Hence the slope of this line is given by

$$\text{Slope} = \frac{(m + b) - b}{1 - 0} = m$$

In other words, m represents the slope of the line. (See Figure 11.)

Example

Find the slope and y-intercept of the following lines:

(a) $y = 2x + 3$ (b) $y = -x - 2$

(c) $y = 5$

Solution

(a) The slope of the line $y = 2x + 3$ is 2, and the y-intercept is 3. (See Figure 12.)

(b) The slope of the line $y = -x - 2$ is -1, and the y-intercept is -2. (See Figure 13.)

(c) The slope of the line $y = 5$ is 0, and the y-intercept is 5. The line is horizontal. (See Figure 14.)

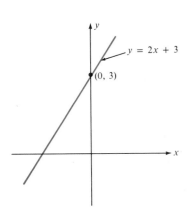

Figure 12
Line with slope 2 and y-intercept 3

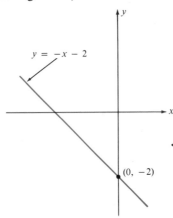

Figure 13
Line with slope -1 and y-intercept -2

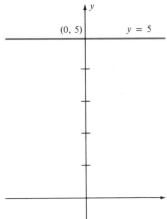

Figure 14
Horizontal line $y = 5$

Point-Slope Form of a Line

Imagine a straight line that passes through a fixed point (x_1, y_1) and has a given slope m. (See Figure 15.) The interesting thing about this line is that every *other* point (x, y) on this line satisfies the equation

$$\frac{y - y_1}{x - x_1} = m$$

This gives us the general equation for the **point-slope form** of a straight line.

> ### Point-Slope Form of a Straight Line
>
> The point-slope form of a straight line with slope m that passes through the point (x_1, y_1) is given by the equation
>
> $$y - y_1 = m(x - x_1)$$

The point-slope form of a straight line is useful because it allows us to find the equation of a line when we know either the coordinates of a point on the line and the slope or just the coordinates of two points on the line.

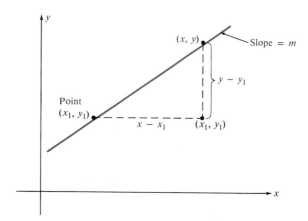

Figure 15
Point-slope form of a line

Example

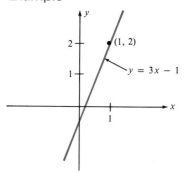

Figure 16
*Line passing through (1, 3)
with slope 3*

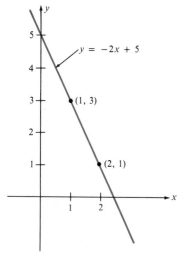

Figure 17
*Line passing through (1, 3)
and (2, 1)*

(a) Find an equation of the line that passes through the point $(1, 2)$ and has slope 3.
(b) Find an equation of the line that passes through the two points $(1, 3)$ and $(2, 1)$.

Solution

(a) If we let $(x_1, y_1) = (1, 2)$ and $m = 3$, we have

$$y - y_1 = m(x - x_1)$$

or

$$y - 2 = 3(x - 1)$$

We can rewrite this equation in slope-intercept form:

$$y = 3x - 1$$

which means that the line has a y-intercept of -1. (See Figure 16.)

(b) If we let $(x_1, y_1) = (1, 3)$ and $(x_2, y_2) = (2, 1)$, we first find the slope $m = (1 - 3)/(2 - 1) = -2$. Hence the point-slope formula is

$$y - y_1 = m(x - x_1)$$

Letting $(x_1, y_1) = (1, 3)$, we get

$$y - 3 = -2(x - 1)$$

We rewrite this equation, too, in the slope-intercept form:

$$y = -2x + 5$$

(See Figure 17.)

  Note: It is important to observe that in the above point-slope formula, we could have just as well substituted the second point $(x_2, y_2) = (2, 1)$, and we would have gotten the same equation. To convince you of this fact, we make this substitution, getting

$$y - 1 = -2(x - 2)$$

or

$$y = -2x + 5$$ □

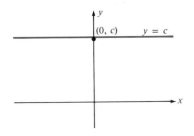

Figure 18
Horizontal line $y = c$

Horizontal and Vertical Lines

The simplest line equations are those for horizontal and vertical lines. Since a horizontal line is satisfied by all points (x, y) that have the same y-coordinate, the equation of the horizontal line whose y-coordinate is always c is given by

$$y = c \qquad \text{(horizontal line)}$$

Note that the slope of a horizontal line is zero and, conversely, that all lines with slope zero are horizontal. (See Figure 18.)

On the other hand, a vertical line consists of points (x, y) whose x-coordinate is a constant. (See Figure 19.) That means an equation of the form

$$x = c$$

As we noted earlier, vertical lines do not have a slope.

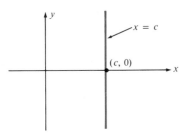

Figure 19
Vertical line $x = c$

Straight Lines in Modern Business (Break-Even Analysis)

Consider a manufacturing plant such as a General Electric plant that produces refrigerators. Suppose the daily cost to operate the plant has been broken into two kinds of costs: **fixed cost** (cost that stays the same no matter how many refrigerators are produced) and **variable cost** (cost that depends on the number of refrigerators produced). Every manufacturing plant has fixed costs, which include such costs as those for building maintenance, electricity, and heating. An important problem in production analysis is the *break-even problem.* Suppose the fixed daily cost at this plant is $50,000 and the cost to produce a single refrigerator is $250. If the company sells the refrigerators for $400 each, how many refrigerators must be produced for the company to break even at this plant for one day?

The pool deck on Royal Caribbean Cruise Line's Song of Norway. *The cruise industry is keenly aware of break-even analysis.*

Let

$$x = \text{Number of refrigerators produced per day}$$
$$C = \text{Daily cost of producing } x \text{ refrigerators}$$
$$R = \text{Daily revenue (return) on sales of } x \text{ refrigerators}$$
$$P = \text{Daily profit when } x \text{ refrigerators are sold}$$

The company's revenue R and total cost C when x refrigerators are produced (assuming that it sells all it produces) are given by

$$R = \$400x$$
$$C = \text{Fixed costs} + \text{Variable costs} = \$50,\!000 + \$250x$$

The company's profit P when x refrigerators are produced (we assume that the company sells all that it produces) is simply the revenue minus the cost. That is,

$$P = R - C = 400x - (50,\!000 + 250x) = 150x - 50,\!000$$

Note that when no refrigerators are produced, the daily profit is $-\$50,\!000$, which means that the company suffers a daily loss equal to the fixed costs. As the number of refrigerators produced increases, however, the profit will increase at a rate of \$150 per refrigerator. After a certain number of refrigerators are produced, the company's revenue will finally equal the fixed cost, and the profit will be zero. This gives the number of refrigerators that must be produced in a given day for the company to *break even*. Only when more units are produced is it worthwhile for the company to be in business. In this problem this occurs when

$$150x - 50,\!000 = 0$$

or
$$x = 333.33$$

In other words, the company must produce roughly 333 or 334 refrigerators at this plant each day just to break even. We will pick 334 as the break-even point, since when 333 refrigerators are produced per day, the profit is slightly negative, but when 334 refrigerators are produced, the profit becomes slightly positive.

In other words, when production reaches 334 or more refrigerators per day, the company starts to turn a profit. For instance, if the company produces 750 refrigerators per day, then its daily profit from this plant will be

$$P = 150x - 50,\!000 = 150(750) - 50,\!000 = \$62,\!500$$

The **profit line** $P = 150x - 50,\!000$ is graphed in Figure 20.

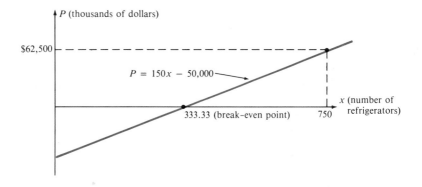

Figure 20
Profit line showing break-even point

Problems VIII

Graph the lines in Problems 1–5 and identify the slopes and y-intercepts (if such exist).

1. $y = 2x + 1$
2. $y = -x + 2$
3. $y - 1 = x + 1$
4. $y = 5$
5. $x = 2$

In Problems 6–10, find the equation of the straight line in slope-intercept form with the given slope that passes through the given point. What is the y-intercept of each line? Sketch the graph of each line.

6. $(x_1, y_1) = (0, 0)$, slope 1
7. $(x_1, y_1) = (1, 2)$, slope 2
8. $(x_1, y_1) = (0, -1)$, slope -2
9. $(x_1, y_1) = (-1, 2)$, slope 3
10. $(x_1, y_1) = (2, 0)$, slope -1

Find the slope-intercept form of the straight line that passes through the points given in Exercises 11–15. What are the slope and y-intercepts of each line? Sketch the graph of each line.

11. $(0, 0), (1, 1)$
12. $(0, 1), (1, 0)$
13. $(-1, 1), (1, -1)$
14. $(2, 3), (4, 7)$
15. $(0, 4), (1, 3)$

Mathematics in Business Management

16. **Break-even Analysis** A major publishing company like Random House that produces college textbooks generally has large fixed costs but relatively small variable costs. Suppose Random House has a fixed cost of $200,000 (production costs, artwork, editorial costs, ...) to produce a Finite Mathematics textbook and a variable cost of $8 for each book produced. If the company sells the books for $20 each to college bookstores, how many books must be sold (or produced) for the company to break even? Find the company's loss or profit if it sells
    (a) 10,000 books
    (b) 15,000 books
    (c) 25,000 books

For Problems 17–21, use the following fixed and variable costs in the different industries to determine the break-even point for each. Sketch the graphs of the revenue and cost functions.

17. Record company (Warner Brothers, Columbia, ...)
    Fixed costs = $500,000 (one-time cost)
    Variable costs = $0.50 per album
    Revenue = $5.00 per album
18. Computer company (IBM, Leading Edge, Apple, ...)
    Fixed costs = $100,000 (daily costs)
    Variable costs = $250 per computer
    Revenue = $450 per computer

19. Food products company (General Foods, Kelloggs, ...)
    Fixed costs = $25,000 (weekly costs)
    Variable costs = $0.25 per box of cereal
    Revenue = $0.90 per box of cereal
20. Automobile company (Ford, General Motors, ...)
    Fixed costs = $250,000 (weekly costs)
    Variable costs = $5000 per car
    Revenue = $7000 per car
21. Newspaper publisher (*Boston Globe, Denver Post,* ...)
    Fixed costs = $100,000 (daily costs)
    Variable costs = $0.15 per newspaper
    Revenue = $0.35 per paper
22. **Break-Even Analysis in the Airline Industry** Suppose the fixed cost to an airline company to make a particular flight from New York to Los Angeles is $50,000.
    (a) If the company makes a profit of $200 per passenger, how many passengers must be on this flight for the company to break even?
    (b) If the airplane on this flight has a capacity of 400 passengers, what percentage of the plane must be filled for the company to break even?
    (c) What expenses would be included in the $50,000 fixed cost?
23. **Break-Even Analysis in the Movie Industry** Suppose the fixed cost to a movie theater owner to show a movie is $200.
    (a) If the movie theater owner makes a profit of $4 per moviegoer, how many tickets must be sold for the owner to break even?
    (b) Do the above fixed cost and profit per moviegoer appear to reflect the true values in the movie industry? How would a theater owner find these values?

Mathematics in the Natural Sciences

24. **Celsius to Fahrenheit** Suppose you are in Canada, where the temperature is measured in Celsius instead of Fahrenheit. The relationship between the temperature in Celsius (C) and the temperature in Fahrenheit (F) is the linear equation

$$F = \frac{9}{5}C + 32$$

Sketch the graph that relates these two temperatures. Let C vary from -20 and $+40$. (Use C for the x-axis and F for the y-axis.) If tomorrow's weather is predicted to be 25 degrees Celsius, what kind of clothes will you wear outside?

25. **Sulfur Dioxide Pollution** A major by-product of burning fossil fuels is the poisonous chemical sulfur dioxide. In an experiment conducted in Oslo, Norway, it was shown that

the number N of deaths per week in the world is approximately a linear function of the mean concentration C of sulfur dioxide in the air (measured in parts per million of a gram per cubic meter). The equation found was

$$N = 94 + 0.031C \qquad 50 < C < 700$$

(a) Sketch the graph of this function.
(b) Find the predicted number of deaths per week that will result if the earth's atmosphere contains 100 parts per million of a gram of sulfur dioxide.

Mathematics in the Social Sciences

26. Gasoline Efficiency An experiment conducted by the Federal Highway Administration showed that (over a given range of speeds) the relationship between the fuel efficiency N of an automobile (in miles per gallon) and the speed S of the automobile (in miles per hour) is a linear one. Suppose that for a certain automobile the relationship is

$$N = -0.5S + 40$$

(a) Sketch the graph of this function.
(b) What is the predicted mileage for this type of automobile when the speed is 50 miles per hour?
(c) How fast must you drive if you expect to get 20 miles per gallon?

27. Is There an Equation That Predicts World Records? The past world records for the men's mile run are plotted in Figure 21. Statisticians have ways of finding the "best line" that will approximate a given set of observations. A linear approximation to the points in Figure 21 has been found to be

$$T = -0.0064Y + 16.6$$

where T is the world record in minutes and Y is the calendar year in which the record was set. Using this linear model as a guide, what will be the world record for the men's mile run in the year 1995? Be sure to convert the time to minutes and seconds.

Interesting Problem

28. How Much Farther Did You Run? The next time you are jogging around a quarter-mile track (or any oval track) and you are running on the outside, with a friend on the inside, just say, "Did you know that the extra distance I'm running does not depend on the length of the track,

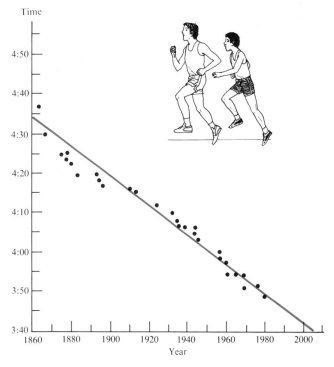

Figure 21

World record times for the mile run (men's)

but is a linear function of the distance we are apart." If you want to be more impressive, tell your friend that the extra distance D the outside runner runs is

$$D = 2\pi x$$

where $\pi = 3.14159\ldots$ and x is the distance between the runners. For example, if you are three feet farther outside than your friend is, you will run

$$D = 2\pi(3)$$
$$= 18.8$$

feet farther on each lap than your running partner. Sketch the graph of the distance D as a function of x for $0 < x < 8$. How much farther do you run per lap if you are five feet further outside?

The interesting thing about this formula is that it holds for any oval track with circular ends (which is the shape of most tracks).

Functions

The world is filled with relationships between quantities. If one picks up a newspaper, it is impossible not to see graphs, tables, and diagrams. Mathematics allows us to discuss relationships between phenomena in a precise way, using the notion of a **function**.

A Function

A function is a rule that assigns to each element of a given set, called the **domain set**, an element in another set, called the **range set**.

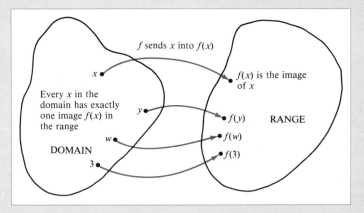

We often think of functions as black boxes into which we enter a value of x (element of the domain), and out comes a corresponding value of y (element of the range) as determined by the black box.

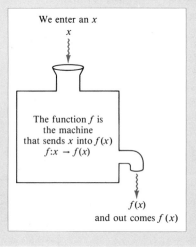

A typical function is the rule that assigns to each real number x its square x^2. The domain of this function is the set of real numbers, and the range is the set of nonnegative real numbers $[0, \infty)$. We can denote some values of this function by writing

$$3 \longrightarrow 9$$
$$-1 \longrightarrow 1$$
$$-5 \longrightarrow 25$$
$$2.5 \longrightarrow 6.25$$

to indicate the value corresponding to a given element of the domain. We often denote

$$x \longrightarrow x^2$$

to represent the value of the function for all real numbers x. We generally denote functions by letters, such as f, g, or h, and illustrate the rule of correspondence by an arrow, such as

$$f: x \longrightarrow y$$

The value y above is called the *image* of the function f at the point x; it is also written as $f(x)$. For example, if f is the function that assigns to each real number x its square x^2, then we could also write

$$f: x \longrightarrow x^2 \qquad (f \text{ "sends" } x \text{ into } x^2)$$

or $\qquad\qquad f(x) = x^2 \qquad$ (image of f at x, read "f of x")

Confusion of the Meaning of f and $f(x)$

There is often a great deal of confusion between the symbols f and $f(x)$. A function (the rule) is denoted by the letter f, while the value of a function at a point x of its domain is denoted by $f(x)$. For example, the square root function, which takes the positive square root of a positive number, is denoted by

$$f: x \longrightarrow \sqrt{x}$$

but the value of f at a particular real number x is $f(x) = \sqrt{x}$.

Example

If

$$f(x) = 2x + 1 \qquad g(x) = \frac{1}{x+1} \qquad h(x) = \sqrt{x+1}$$

then

(a) $f(2) = 2 \cdot 2 + 1 = 5$

(b) $f(0) = 2 \cdot 0 + 1 = 1$

(c) $f(-1) = 2 \cdot (-1) + 1 = -1$

(d) $f(\text{BLAH}) = 2 \cdot \text{BLAH} + 1 \qquad$ (BLAH stands for any real number)

(e) $g(0) = \dfrac{1}{0+1} = 1$

(f) $g(3) = \dfrac{1}{3+1} = 0.25$

(g) $g(x^2) = \dfrac{1}{x^2+1}$

(h) $g(\text{BLAH}) = \dfrac{1}{\text{BLAH}+1} \qquad$ (BLAH is any real number $\neq -1$)

(i) $h(0) = \sqrt{0+1} = 1$

(j) $h(1) = \sqrt{1+1} = \sqrt{2}$

(k) $h(t+1) = \sqrt{(t+1)+1} = \sqrt{t+2}$

(l) $h(\text{BLAH}) = \sqrt{\text{BLAH}+1} \qquad$ (BLAH is any real number ≥ -1) $\qquad\qquad\square$

Problems IX

For Problems 1–30, find the indicated values given the functions defined by

$$f(x) = x - 2 \qquad g(x) = \frac{1}{x} \qquad h(x) = x^2$$

1. $f(1)$

2. $f(-1)$

3. $f(0)$

4. $f(x + 1)$

5. $f(x^2)$

6. $f(x^2 + 1)$

7. $f(x^2) + 1$

8. $f(2x) + f(x^2)$

9. $f(2x) + 2x$

10. $f(-x) - f(x)$

11. $g(1)$

12. $g(-1)$

13. $g(0)$

14. $g(x + 1)$

15. $g(x^2)$

16. $g(x^2 + 1)$

17. $g(x^2) + 1$

18. $g(2x) + g(x^2)$

19. $g(2x) + 2x$

20. $g(-x) - g(x)$

21. $h(1)$

22. $h(-1)$

23. $h(0)$

24. $h(x + 1)$

25. $h(x^2)$

26. $h(x^2 + 1)$

27. $h(x^2) + 1$

28. $h(2x) + h(x^2)$

29. $h(2x) + 2x$

30. $h(-x) - h(x)$

Practice Test

The following test can be used by students to determine whether they need additional work in the Finite Mathematics Preliminaries. Before beginning Chapter 1, all students should spend some time and take this test. There are 11 well-chosen questions. Experience has shown that students should use the following guidelines:

Correct Answers	Review Strategy
11	Excellent, no review necessary
10–9	Good, just a little brush-up needed
8–7	Average, review the weak sections
6–5	Advisable to review, quick review
4–0	Necessary to review, detailed review

It would also be useful for students to talk with their professors about their weaknesses.

1. Find the numerical values of the following expressions:
   (a) 2^4
   (b) 3^{-2}
   (c) $(2/3)^2$
   (d) $-(-2)^{-4}$
   (e) $(1,244,043)^0$

2. Find the following numerical values of the expressions (if they exist):
   (a) $\sqrt{16}$
   (b) $\sqrt[3]{8}$
   (c) $\sqrt{-25}$
   (d) $\sqrt[3]{-125}$
   (e) $(4/9)^{1/2}$

3. Reduce the following expressions to simplest form:
   (a) $\sqrt{36x^2}$

   (b) $\dfrac{3x}{\sqrt{x}}$

4. Solve for x in the following equation:

$$10e^{2x} = 100$$

by writing x in terms of the natural logarithm of some constant.

5. Solve for x in the following equation:

$$5 - 2x = x + 10$$

6. Solve the following inequality:

$$x + 1 < 2x - 5$$

7. Find the distance between the points $(1, 2)$ and $(3, 5)$.

8. Graph the line described by the following equation:

$$y = 3x - 1$$

and identify the slope and y-intercept.

9. Find the equation of the straight line that passes through the point $(1, 3)$ with slope 2.

10. Find the equation of the straight line that passes through the points $(1, 1)$ and $(2, 0)$.

11. If $f(x) = x^2$ and $g(x) = 2x + 1$, find
   (a) $f[g(x)]$
   (b) $g[f(x)]$

1

Systems of
Linear Equations
and Matrices

Although simultaneous linear equations have been in existence and solved as early as 200 B.C. in the Chinese book of mathematics *Chui-chang suan-shu*, they have recently taken on new importance because of computers. Instead of solving systems of two or three equations with two or three unknowns, today we routinely solve systems of a thousand equations with a thousand unknowns. This allows us to solve a wide variety of problems that had not been imagined just a few years ago. In this chapter we will show how systems of equations can be used to describe phenomena in a technological society. We will then solve these systems using the Gauss-Jordan and Gaussian elimination methods.

The matrix, one of the most useful concepts in mathematics, was introduced only in the nineteenth century. We will present the arithmetic operations of matrices and show how matrices are used to describe some common phenomena.

1.1 Introduction to Linear Systems

The role of mathematics in business, economics, and the life sciences is to express phenomena in these disciplines in a precise language. This is done by means of **mathematical models**, which are an attempt to describe some part of the real world in mathematical terms. Mathematical models are not as complex as reality; if they were, there would be no reason for their use. Fortunately, we can often construct models that are much simpler than reality but complex enough to be used to predict and explain phenomena with a high degree of accuracy. One advantage of using a mathematical model to solve a problem is that the entire arsenal of mathematical results about the model can be brought to bear on the problem. Once the solution to the mathematical problem is found, this information can then be translated back into a real-world setting.

One special area of mathematics that has been particularly useful in solving problems in business, economics, and the life sciences is the theory of linear systems of equations. The following example illustrates how a linear system of equations can describe a problem in the health sciences.

The Dietician's Problem

A dietician is planning a meal around two foods. Each ounce of Food I contains 5% of the daily requirements of carbohydrates and 10% of the daily requirements of protein. On the other hand, each ounce of Food II contains 15% of the daily requirements for carbohydrates and 5% of the daily requirements for protein. This information is summarized in Table 1.

Mathematics is being used on a daily basis by hospitals that use automated dietary planning systems. The patients may not know it, but their nutritional requirements are being monitored by a computer.

TABLE 1

Percent of Carbohydrates and Protein in Foods 1 and II

	Food I	Food II	Percent Required
Carbohydrates	5% carbohydrates per ounce of food I	15% carbohydrates per ounce of food II	100%
Protein	10% protein per ounce of food I	5% protein per ounce of food II	100%

To formalize this problem, we define variables x and y as follows:

$$x = \text{number of ounces of Food I served}$$
$$y = \text{number of ounces of Food II served}$$

For the meal to contain exactly 100% of the daily requirements for both carbohydrates and protein, the values of x and y must satisfy the following two conditions:

$$5x + 15y = 100 \qquad \text{(carbohydrate requirement)}$$
$$10x + 5y = 100 \qquad \text{(protein requirement)}$$

Mathematically, we are looking for the quantities x and y that simultaneously satisfy both of these equations. In the next section we will see how to find these values. Equations written in this form are called **systems of linear equations**. We will discover that

$$x = 8 \text{ ounces}$$
$$y = 4 \text{ ounces}$$

are the only values for x and y that simultaneously satisfy both equations. This means that the only serving of Food I and Food II that exactly meets the daily requirements for carbohydrates and protein will consist of 8 ounces of Food I and 4 ounces of Food II.

There are only two variables in this example, but the ideas used in solving this system are the same as those used to solve systems with several variables. The limitations on problem size have become a function more of skill in formulating complex problems than of skill in solving the equations, since sophisticated computer software and high-speed computers have made it possible to solve systems of equations with thousands of variables.

The characteristics that make linear systems so useful include the following:

- Many real-world problems can be described accurately by systems of linear equations.
- There is a wealth of mathematical knowledge about linear systems that can be applied.
- A growing number of computer packages have been developed that are tailored to solve linear systems that arise in many application areas.

Two special topics that are closely related to systems of linear equations, which will be studied in this book, are the Leontief input-output model and linear programming. The development of these two theories was deemed so important that in 1973 the Nobel Prize in Economics was awarded to Professor Wassily Leontief of Harvard University for his pioneering work in input-output analysis. (Interesting articles about the work of Leontief can be found in *Scientific American*, April 1965, and *Newsweek*, October 29, 1973.) Professor George Dantzig of Stanford University has also been awarded honors for his development of the simplex method for the solution of the linear programming problem.

1.2 Linear Systems with Two and Three Equations

PURPOSE

We introduce the basic ideas of linear systems with two and three unknowns. We show

- how to solve 2×2 systems by the method of substitution and the add-subtract method of elimination and
- how to interpret 2×2 and 3×3 linear systems geometrically.

Two Equations in Two Unknowns

A system of two linear equations in two unknowns x and y consists of two equations of the form

$$ax + by = c$$
$$dx + ey = f$$

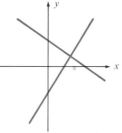 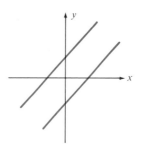 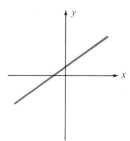

One solution when the lines are nonparallel

(a)

No solutions when the lines are parallel and distinct

(b)

Infinite number of solutions when the lines are the same

(c)

Figure 1
Three possibilities for 2 × 2 linear systems

where a, b, c, d, e, and f are real numbers. We refer to such systems as 2×2 systems, indicating that there are two equations and two unknowns.

A solution (x, y) of a 2×2 system is a pair of real numbers x and y that simultaneously satisfy both equations. Finding a solution of a 2×2 system corresponds geometrically to finding points of intersection of two lines. (See Figure 1.)

If the two equations describe two distinct lines that are not parallel, then the system will have exactly one solution. If the two equations describe two distinct lines that are parallel, then the system will not have any solutions. Finally, if the two equations describe the same line, then there will be an infinite number of solutions (every point on the line).

Example 1

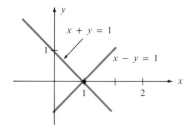

Figure 2
Two nonparallel lines always intersect at one point

One Solution of a Linear System   Determine the number of solutions of the linear system

$$x + y = 1$$
$$x - y = 1$$

Solution   Writing each of the equations in slope-intercept form, we have

$$y = -x + 1$$
$$y = x - 1$$

The lines described by these equations are not parallel, since they have the distinct slopes -1 and 1, respectively. Hence the linear system has exactly one solution. (See Figure 2.)   □

Example 2

No Solution of a Linear System   Determine the number of solutions of the linear system

$$x + y = 5$$
$$x + y = 1$$

Solution   Writing the equations in slope-intercept form, we have

$$y = -x + 5$$
$$y = -x + 1$$

Since the slopes of these two equations are both -1 and the y-intercepts are different, 5 and 1, we conclude that the lines are parallel but distinct. Hence the system does not have any solutions. (See Figure 3.)

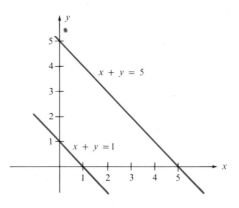

Figure 3
Linear system with no solution

Example 3

Infinite Number of Solutions   Determine the solutions of the linear system

$$x + y = 1$$
$$3x + 3y = 3$$

Solution   Writing the equations in slope-intercept form, we have

$$y = -x + 1$$
$$y = -x + 1$$

Since the two equations are identical, they describe the same line. Hence the linear system has an infinite number of solutions, all the points on the line. (See Figure 4.)

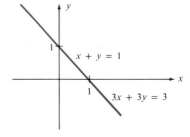

Figure 4
Linear system with an infinite number of solutions

Graphing straight lines tells us whether a 2×2 linear system has no solutions, one solution, or an infinite number of solutions. However, graphs are rarely accurate enough to provide us with the exact coordinates of the solution. We now present a method that lets us find the exact solution when it exists.

Method of Substitution for Solving 2 × 2 Systems

When a linear system has a single solution, we can find this solution by the **method of substitution**. Although the method works for larger systems, it works best when applied to systems with only a few variables. We apply the method to the 2×2 system here.

Method of Substitution for 2 × 2 System

$$ax + by = c \quad \text{(Eq. 1)}$$
$$dx + ey = f \quad \text{(Eq. 2)}$$

- **Step 1.** Solve for x in terms of y in Eq. 1.
- **Step 2.** Substitute the expression for x found in Step 1 into Eq. 2, giving an equation in y.
- **Step 3.** Solve the equation found in Step 2 for y.
- **Step 4.** Substitute the value of y found in Step 3 into one of the original equations and solve for x.

The substitution method need not be followed verbatim. It is possible to begin by solving for x in the second equation, provided that term is present, or even by solving for y in one of the two equations.

Example 4

Method of Substitution Suppose the total cost of one ticket to a Yankees game and three tickets to a Red Sox game (Figure 5) is $27. Suppose, too, that the ticket to the Yankees game costs $1 less than the ticket to the Red Sox game. What is the cost of each ticket?

Solution We let

$$x = \text{cost of a Red Sox ticket}$$
$$y = \text{cost of a Yankee ticket}$$

The given restrictions on x and y say that

$$3x + y = 27 \quad \text{(Eq. 1)}$$
$$x - y = \ 1 \quad \text{(Eq. 2)}$$

Writing these equations in slope-intercept form, we can see that the slopes of these two equations are not the same. Hence the linear system has exactly one solution.

We now use the method of substitution to find the solution.

- **Step 1** (solve for y in Eq. 1). It is more convenient to solve for y in Eq. 1. Solving for y gives

$$y = 27 - 3x$$

- **Step 2** (substitute the value of y into Eq. 2). Substituting the above value of y into Eq. 2, we have

$$x - (27 - 3x) = 1$$

- **Step 3** (solve for x in the equation found in Step 2). Solving the above equation for x, we get

$$x = 7$$

Figure 5
Yankee and Red Sox tickets

- **Step 4** (substitute the value of x into one of the original equations). Substituting $x = 7$ into the original Eq. 2, we find

$$7 - y = 1$$

Solving for y gives

$$y = 6$$

Hence the solution to the above 2×2 system is

$$x = 7$$
$$y = 6$$

We conclude that the cost for a ticket to a Red Sox game is \$7.00, while the cost for a ticket to a Yankees game is \$6.00.

Check:

$$3(7) + (6) = 27 \qquad \text{(Eq. 1)}$$
$$(7) - (6) = \ \ 1 \qquad \text{(Eq. 2)} \qquad \qquad \square$$

The Add-Subtract Method of Elimination for 2×2 Systems

Another useful method for solving systems of equations with only a few variables is the **add-subtract method**. The method becomes bogged down in a maze of equations when it is applied to systems with more than two or three variables, but it is convenient for small systems. We describe the method for 2×2 systems.

The Add-Subtract Method for 2×2 Systems

$$ax + by = c \qquad \text{(Eq. 1)}$$
$$dx + ey = f \qquad \text{(Eq. 2)}$$

- **Step 1.** Multiply Eq. 1 by d and Eq. 2 by a. Subtract the resulting Eq. 2 from the resulting Eq. 1, giving an equation in the unknown y. Doing this, we get

$$\begin{aligned} d(ax + by = c) \\ -a(dx + ey = f) \\ \hline dby - aey = dc - af \end{aligned}$$

- **Step 2.** Solve the equation found in Step 1 for y. We get

$$y = \frac{af - cd}{ae - bd}$$

- **Step 3.** Substitute the value of y found in Step 2 into one of the original equations and solve for x. Doing this gives

$$x = \frac{ce - bf}{ae - bd}$$

Again, it is not necessary that the above rules be followed verbatim. Sometimes x can be eliminated by simply multiplying one of the two equations by a constant and then subtracting the resulting equations. Sometimes it is even possible simply to add or subtract the equations as given to eliminate one of the variables. It is also possible to eliminate the y variable from the two equations by multiplying Eq. 1 by e and Eq. 2 by b and then subtracting one of the resulting equations from the other. The variable that is selected for elimination depends to a great extent on the relative simplicity of the arithmetic operations.

Example 5

Add-Subtract Method of Elimination   Use the add-subtract method to solve the following system of equations:

$$3x + 5y = 21 \quad \text{(Eq. 1)}$$
$$5x + 9y = 37 \quad \text{(Eq. 2)}$$

Solution

- **Step 1** (multiply each equation by the appropriate constant). To eliminate x, we multiply Eq. 1 by 5 and Eq. 2 by 3 and subtract. Doing this, we get

$$5(3x + 5y = 21)$$
$$\underline{-3(5x + 9y = 37)}$$
$$25y - 27y = 105 - 111$$

  or

$$-2y = -6$$

- **Step 2** (solve for y).   Solving the equation above for y gives

$$y = 3$$

- **Step 3** (substitute back into original equations).   Substituting $y = 3$ into Eq. 1, we get

$$3x + 5(3) = 21$$

  or

$$3x = 6$$

  Solving for x, we find

$$x = 2$$

Hence $x = 2$, $y = 3$ is the solution of the linear system.
   Check:

$$3(2) + 5(3) = 21 \quad \text{(Eq. 1)}$$
$$5(2) + 9(3) = 37 \quad \text{(Eq. 2)}$$ □

Market and Equilibrium Analysis (Supply and Demand)

The number of items consumers are willing to buy depends on many things. One of the major factors that affects sales is the price of the item. Suppose a market study is performed in San Francisco with the goal of determining how the price of an item affects the number of items sold. Six stores are carefully selected on the basis of their size and sales records to participate in the study. Each store sells the product at a different price. The results of the study are shown in Table 2.

Mathematics can be used to effectively analyze the relationship between the supply and demand of a commodity.

TABLE 2
Changing Demand with Price

Store	Price of Item, p	Number of Items Purchased, q (weekly)
Store 1	$80	0
Store 2	$68	60
Store 3	$56	120
Store 4	$44	180
Store 5	$32	240
Store 6	$20	300

Table 2 illustrates a basic principle in economics: the **law of consumer demand**. This law states that as the price of a commodity rises, the amount demanded by consumers will fall. The mathematical relationship between the amount q of a given commodity demanded by consumers and the price p of the commodity is called the **demand curve**:

$$q = D(p)$$

For the observations in Table 2, the relationship between the demand q and price p can be described mathematically by the **linear demand curve**

$$q = D(p) = -5p + 400$$

(See Figure 6.)

Just as the consumer has a law of demand, there is a law of supply for the producers of a commodity. The **law of supply** states that as the price rises, the

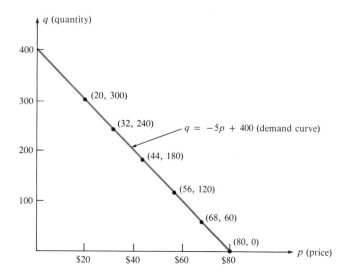

Figure 6
A linear demand curve

amount of a commodity that will be supplied to the marketplace also increases. The reason for this phenomenon is that more firms will be willing and able to produce the commodity profitably at a higher unit price. The functional relationship that describes the quantity q supplied to the marketplace at a price p is called the **supply curve**:

$$q = S(p)$$

Producers of a commodity prefer high prices, while consumers prefer low prices. The resulting price in the marketplace is usually some compromise between these two opposing interests. Consider the supply and demand curves for a certain commodity as shown in Figure 7.

If the price of this commodity is $10, then it is seen in Figure 7 that producers are supplying more items than consumers are buying. In other words, a price of $10 will give rise to a surplus in the commodity. The net result is that the price will come down. On the other hand, if the price of the commodity is low, say $8, then consumers demand more items than are being produced.

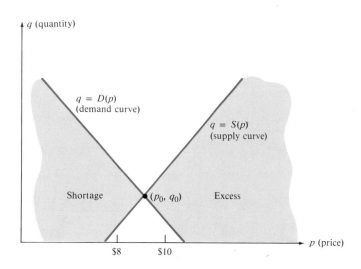

Figure 7
Supply and demand curves

The net result is that there is a shortage of the commodity, and prices can be increased by the producers without fear that items will go unsold.

The net result of both commodity surplus and shortage is that the price will stabilize at an **equilibrium price** p_0, at which time the suppliers will produce an **equilibrium amount** q_0. At this amount the suppliers will fill the need of the consumers with no surplus. The point (p_0, q_0) is called the **equilibrium point**.

Example 6

Supply and Demand   A consumer demand table and producer supply table for a commodity are shown in Table 3. Draw supply and demand curves and use these curves to estimate the equilibrium price for this commodity.

TABLE 3
Supply and Demand Tables

Price, p	Consumer Demand, $q = D(p)$	Producer Supply, $q = S(p)$
\$5	20	10
\$6	19	12
\$7	18	14
\$8	17	16
\$9	16	18
\$10	15	20

Solution   The values in Table 3 have been plotted and are shown in Figure 8.

To find the demand curve $D(p)$, note from Figure 8(a) that the six observations (p, q), relating consumer demand and price, all lie on a straight line. Hence we conclude that the demand curve is a linear function that has the form

$$q = D(p) = ap + b$$

Therefore any two observations (p, q) relating the demand q with the price p will determine $D(p)$. We select the points $(5, 20)$ and $(10, 15)$ from Figure 8(a). Since the slope of the straight line passing through two points $(a, f(a))$ and $(b, f(b))$ is

$$m = \frac{f(b) - f(a)}{b - a}$$

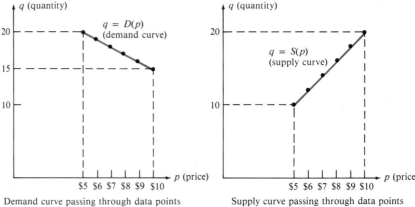

Figure 8
Finding the equations for the supply and demand curves

Demand curve passing through data points   (a)

Supply curve passing through data points   (b)

we can use the point-slope formula

$$y - f(a) = m(x - a)$$

for the straight line passing through $(a, f(a))$. Hence we have

$$q - 20 = \frac{15 - 20}{10 - 5}(p - 5)$$

Simplifying, we get

$$q = 25 - p \quad \text{(demand curve)}$$

Similarly, we can find the supply curve by finding the line passing through any two of the points (p, q) in Figure 8(b). Selecting $(5, 10)$ and $(10, 20)$, we have

$$q - 10 = \frac{20 - 10}{10 - 5}(p - 5)$$

Simplifying, we get

$$q = 2p \quad \text{(supply curve)}$$

To find the equilibrium point (p_0, q_0), we rewrite the above supply and demand curves as

$$p + q = 25 \quad \text{(demand curve)}$$
$$2p - q = 0 \quad \text{(supply curve)}$$

and solve for p and q. To solve this 2×2 system, we use the add-subtract method of elimination. For this system we add the two equations to eliminate q. Hence we have

$$3p = 25$$

Solving for p gives

$$p = \frac{25}{3} \cong 8.33$$

Substituting this value into the supply curve, we find

$$q = 2(8.33) = 16.66$$

Conclusion: We conclude that the equilibrium price for this commodity will be \$8.33 for each item and that the number of items of the commodity demanded by the consumers (each week) at this price will be 16.66 (around 16 or 17). (See Figure 9.) □

Three Equations in Three Unknowns

The general linear system for three equations in three unknowns is a system of equations of the form

$$a_{11}x + a_{12}y + a_{13}z = b_1$$
$$a_{21}x + a_{22}y + a_{23}z = b_2$$
$$a_{31}x + a_{32}y + a_{33}z = b_3$$

where a_{ij} and b_i for $i, j = 1, 2, 3$ are real numbers. A **solution** of a 3×3 system is a triple (x, y, z) of three real numbers x, y, and z that simultaneously satisfy

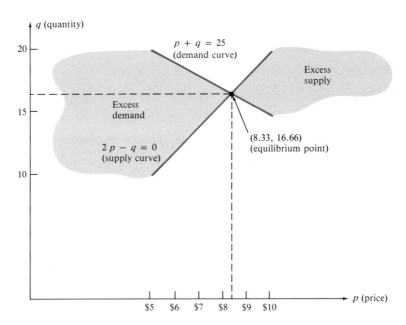

Figure 9
Equilibrium price of $8.33 when the supply is equal to the demand

each of the three equations. Although the method of substitution and the add-subtract method can be used to solve linear systems in three variables, a better method for solving systems of this size and larger will be studied in the next section. In this section we concentrate on the nature of solutions of 3×3 systems and leave the actual solving for later.

Solutions of 3×3 Systems

We have seen that an equation of the form

$$ax + by = c$$

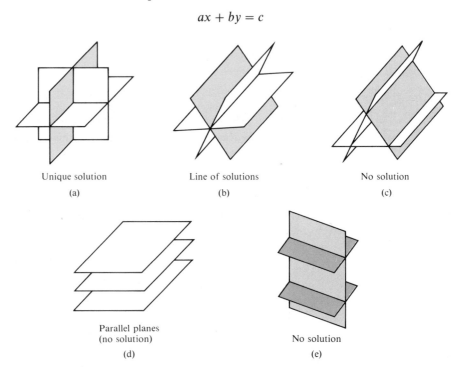

Figure 10
Possible solutions for 3×3 linear systems

describes a straight line in the xy-plane. Similarly, an equation in three variables of the form

$$ax + by + cz = d$$

represents a plane in three-dimensional space. Hence a solution (x, y, z) to a 3×3 system of equations represents a point that lies on each of the planes described by the three equations. If we can visualize the possible ways in which three planes can intersect, we will gain some insight into the nature of the solutions for 3×3 systems. Figure 10 illustrates several of the many ways in which three planes can intersect.

An interesting fact about the solutions of 3×3 systems (and about linear systems in general), as Figure 10 indicates, is that the system will have either no solutions, one solution, or an infinite number of solutions.

Problems

1. For the system of equations

$$x + y = 9$$
$$x - 2y = 0$$

substitute the following pairs of numbers (x, y) into the equations to determine which are solutions.
(a) $(x, y) = (-2, -3)$
(b) $(x, y) = (1, 4)$
(c) $(x, y) = (6, 3)$

2. For the following two equations

$$3j + 4k = -1$$
$$5j + 2k = 3$$

substitute the following pairs of numbers into the equations to determine which are solutions.
(a) $(j, k) = (2, 1)$
(b) $(j, k) = (1, 2)$
(c) $(j, k) = (1, -1)$

3. For the following system of three equations in three unknowns

$$3x + 4y + 2z = 6$$
$$x + y - 2z = 9$$
$$3x + 3y - 6z = 27$$

substitute the following values of x, y, z into the equations to determine which triples (x, y, z) are solutions.
(a) $(x, y, z) = (1, 4, -2)$
(b) $(x, y, z) = (-6, 15, 0)$
(c) $(x, y, z) = (40, -29, 1)$

4. For the following system of three equations in three unknowns

$$2x + 3y - z = 4$$
$$x - 2y + 3z = 2$$
$$-x + y + z = 1$$

substitute the following values of x, y, z into the equations to determine which triples (x, y, z) are solutions.
(a) $(x, y, z) = (2, 1, 3)$
(b) $(x, y, z) = (1, 1, 1)$
(c) $(x, y, z) = (3, 3, 1)$

For Problems 5–8, approximate the solutions of the following systems by graphing on graph paper the lines described by the equations and finding the intersection. Check your graphical solution by substituting the values of x and y into the equations.

5. $3x - y = 3$
$2x + y = 2$
6. $x + 3y = 3$
$2x + 6y = 9$
7. $2x - y = 2$
$4x + 2y = 4$
8. $3x + 2y = 1$
$2x - y = 2$

For Problems 9–12, solve the following systems of equations both by the substitution method and by the add-subtract method of elimination.

9. $5x + 4y = 18$
$-3x + y = -4$
10. $x + y = 10$
$2x - y = 0$
11. $3x + 7y = 4$
$-x + y = 10$
12. $x - y = 0$
$x - 2y = -1$

For Problems 13–16, determine the solutions of the following systems of equations. Be sure to find *all solutions*.

13. $4x - 6y = 10$
$2x - 3y = 5$
14. $x + y = 3$
$x - y = 1$
15. $3x + 3y = 3$
$x + y = 3$
16. $2x - y = 1$
$x - 3y = 3$

17. Conditions for Solutions For the system of two equations in two unknowns

$$ax + by = r$$
$$cx + dy = s$$

find relations involving the values $a, b, c, d, r,$ and s such that the system of equations has
(a) one solution,
(b) an infinite number of solutions,
(c) no solution.

18. Line Passing Through Points   Find that line

$$ax + by = 5$$

that passes through the points $(-2, 1)$ and $(-1, -2)$.

19. Conditions for No Solution   For what values of the constant k does the system

$$3a - 3b = k$$
$$a - b = 2$$

fail to have a solution?

20. Simultaneous Equations   Recently, Oregon played Oregon State in basketball, and the two teams scored a total of 176 points. In that game, Oregon State beat Oregon by 18 points. What was the final score?

21. Finding Simultaneous Equations   A collection of nickels, dimes, and quarters is worth $10.00. There are 52 coins in all. If there are three times as many nickels as quarters, write the three linear equations that one must solve to find the number of nickels, dimes, and quarters.

22. Finding Simultaneous Equations   A certain three-digit number is equal to 46 times the sum of its digits. The difference between the number and the number obtained by reversing the digits is 495. The sum of the units and hundreds digits is one more than the tens digit. Write the three equations that one must solve to find the number.

Linear Systems in Business Management

23. Supply and Demand   Supply and demand for a given commodity are known to be linear functions and satisfy the following conditions:

Demand	Supply	Price
30	10	$5
15	20	$10

(a) What is the demand curve for this commodity?
(b) What is the supply curve for this commodity?
(c) Draw graphs of the supply and demand curves.
(d) What are the equilibrium price and equilibrium quantity?
(e) When the price of the commodity is $8, is there a surplus or shortage of the commodity? What is the value of this shortage or surplus?
(f) When the price of the commodity is $12, is there a surplus or shortage of the commodity? What is the value of this shortage or surplus?

24. Economic Equilibrium   Find the equilibrium point for the supply and demand curves

$$q = S(p) = 2p - 4 \quad \text{(supply curve)}$$
$$q = D(p) = -p + 10 \quad \text{(demand curve)}$$

Draw graphs of these curves. What is the lowest price at which the supplier will produce the commodity? What is the highest price at which the consumer will buy any of the commodity? What is the surplus of the commodity when the price is $6? What is the surplus when the price is $8?

25. Supply and Demand   Find the linear demand curve $q = mp + b$ that passes through the points $(1, 5)$ and $(2, 3)$. Draw the graph of the curve. At what price will the consumer stop buying the commodity?

26. Supply and Demand   For the supply and demand curves

$$q = p - 1 \quad \text{(supply curve)}$$
$$q = -p + 1 \quad \text{(demand curve)}$$

find the following values:
(a) the price at which supply equals demand,
(b) the price at which the supplier stops supplying the commodity,
(c) the price at which the buyer stops buying the commodity,
(d) the range of prices at which supply is greater than demand,
(e) the range of prices at which supply is less than demand.

27. Supply and Demand   Answer parts (a)–(e) in Problem 26 for the following supply and demand curves:

$$q = 2p - 8 \quad \text{(supply curve)}$$
$$q = -2p + 10 \quad \text{(demand curve)}$$

28. Supply and Demand   Table 4 lists the monthly supply and demand for a given product in the city of Chicago.

TABLE 4
Supply and Demand Table for Problem 28

Demand	Supply	Price
500,000	200,000	$1.00
400,000	210,000	$1.25
290,000	240,000	$1.50
200,000	300,000	$1.75
50,000	500,000	$2.00

(a) What is the surplus of this commodity when the price is $2.00?
(b) What is the shortage of this commodity when the price is $1.00?

(c) Plot the supply and demand table as a function of price p.

(d) Do the supply and demand functions appear to be linear from the plotted points?

(e) From the above numbers, what is your approximation of the equilibrium price?

29. **Investment**   An individual invested a total of $50,000 in stocks and bonds. The stock investment earned 8%, and the bond investment earned 15%. The total return on these investments was $5000. How much money was invested in stocks and how much in bonds?

30. **Blending of Coffee**   A company sells Jamaican and Colombian coffee. This company sells the Jamaican coffee at $4.00 per pound and the Colombian coffee at $3.00 per pound. The company has an offer to sell 5000 pounds of a blend of these coffees at $3.20 per pound. How much of each coffee should be added to this blend so that the company's revenues remain constant?

31. **The Age Riddle**   Professor Bigelow as asked how old his daughters were. He replied, "Their combined ages are 4 times the differences of their ages. Five years ago, one was twice as old as the other." How old are Professor Bigelow's daughters?

32. **The Income Riddle**   Ms. Smith and Ms. Jones were complaining about their incomes. Ms. Smith said, "If you give me $100,000 of your income, my income will be twice your income." Ms. Jones then replied, "Well, that's nothing, if you give me $100,000 of your income, our incomes will be the same." How much does each person make?

33. **Dietician's Problem**   A given Food I contains 1 mg of vitamin A per ounce of the food and 3 mg of vitamin B per ounce of the food. A Food II contains 2 mg of vitamin A per ounce of the food and 1 mg of vitamin B per ounce of the food. How many ounces of each food must one eat to consume 50 mg of vitamin A and 70 mg of vitamin B? *Hint:* It is convenient to present the above information in the form in Table 5.

TABLE 5
Tabulated Information for Problem 33

	Food I	Food II	Required
Vitamin A	1 mg	2 mg	50 mg
Vitamin B	3 mg	1 mg	70 mg

The Game Dept—The ST Game

For Problems 34–37, suppose you have a collection of cardboard squares and triangles. On each square is drawn one cat and four dogs as in Figure 11.

On each triangle is drawn one cat and one dog as in Figure 12.

Figure 11
Square with one cat and four dogs

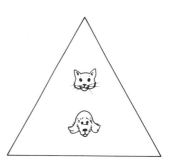

Figure 12
Triangle with one cat and one dog

How many squares and triangles should you select so the total number of cats and dogs selected are the numbers given? For example, to select two cats and five dogs, you should pick one square and one triangle.

This problem is essentially the same as the dietician's problem in Problem 33. The square and triangle play the role of the two foods, and the cats and dogs play the role of vitamin A and B. (See Table 6.)

TABLE 6
Tabulated Information for Problems 34–37

	Triangle	Square	Required Cats and Dogs
Cats	1 cat	1 cat	*
Dogs	1 dog	4 dogs	*

34. Number of cats = 9
    Number of dogs = 27
35. Number of cats = 5
    Number of dogs = 17
36. Number of cats = 20
    Number of dogs = 50
37. Number of cats = 27
    Number of dogs = 81

1.3 Elementary Operations and the Gauss-Jordan Method

PURPOSE

We show
- how elementary operations performed on a system of equations will change the system to a new system with the same solutions and
- how to solve a system of equations by the Gauss-Jordan method by repeated use of elementary operations.

Introduction

In the previous section we studied two methods for solving systems of equations, the method of substitution and the add-subtract method of elimination. Although these methods work for systems that are larger than 2×2, they are computationally complex, even for systems with three variables. A method that is more suited for solving larger systems of equations is the **Gauss-Jordan method**. This method solves a system of equations by systematically transforming it, one step at a time, into a system of equations whose solutions can be found by inspection. Each step of this sequence of transformations is carried out by an appropriate elementary operation. In cases in which the original system of equations does not have a solution, the method will come to a halt and will provide the necessary information to draw this conclusion.

To understand the Gauss-Jordan method, we need to understand the idea of **equivalent systems of equations**. Two systems of equations are equivalent if they have the same solutions.

Example 1

Equivalent Systems of Equations   Show that the two systems of equations A and B below are equivalent.

$$\text{A:} \quad \begin{matrix} 2x + y = 1 \\ x - y = 2 \end{matrix} \qquad \text{B:} \quad \begin{matrix} 4x + 2y = 2 \\ y = -1 \end{matrix}$$

Solution   From the graphs of these equations in Figure 13 we see that each system has the same solution $x = 1$, $y = -1$. Hence A and B are equivalent

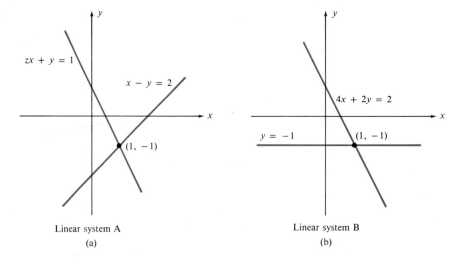

Figure 13
Equivalent systems may not consist of the same equations, but they have the same solutions

Linear system A
(a)

Linear system B
(b)

systems of equations. Observe that the equations are not the same, only that each system has the same solution.   □

Elementary Operations on Linear Systems

Three operations can be performed on a system of equations that will change or transform one system of equations into an equivalent system of equations.

Elementary Operations

The following three operations, called **elementary operations**, performed on a system of equations will result in an equivalent system of equations. By calling the ith equation in a linear system E_i, the three operations can be represented easily as shown below.

- **Operation 1** (interchange equations).   We can *interchange* any two equations of a system of equations. The notation

$$E_i \longleftrightarrow E_j$$

  means to interchange equations i and j.

- **Operation 2** (multiply an equation by a constant).   We can *multiply* both sides of an equation by any nonzero real number. The notation

$$cE_i \longrightarrow E_i$$

  means to replace the ith equation by the ith equation multiplied by c.

- **Operation 3** (add a multiple of one equation to another equation).   We can *add* to any equation a constant multiple of another equation. The notation

$$E_i + cE_j \longrightarrow E_i$$

  means to replace the ith equation by the sum of the ith equation and the jth equation multiplied by the constant c.

Example 2

$E_2 \longleftrightarrow E_3$ The linear system

$$
\begin{aligned}
x + 3y - z &= 2 \quad &(E_1) \\
2x + y + z &= 4 \quad &(E_2) \\
7x + 5y - z &= 5 \quad &(E_3)
\end{aligned}
$$

has the same solution as the system

$$
\begin{aligned}
x + 3y - z &= 2 \quad &(E_1) \\
7x + 5y - z &= 5 \quad &(E_3) \\
2x + y + z &= 4 \quad &(E_2)
\end{aligned}
$$

Note that we have simply interchanged E_2 and E_3.   □

Example 3 ————————— $2E_1 \longrightarrow E_1$ The linear system

$$3x + 2y - z = 4 \qquad (E_1)$$
$$x - y + 2z = 5 \qquad (E_2)$$
$$3x + 3y + z = 6 \qquad (E_3)$$

has the same solution as the system

$$6x + 4y - 2z = 8 \qquad (2E_1)$$
$$x - y + 2z = 5 \qquad (E_2)$$
$$3x + 3y + z = 6 \qquad (E_3)$$

Here we have simply replaced E_1 in the original system of equations by 2 times E_1.   □

Example 4 ————————— $E_2 + 2E_1 \longrightarrow E_2$ The linear system

$$2x + y - z = 4 \qquad (E_1)$$
$$3x + y + z = 2 \qquad (E_2)$$
$$x + y + z = 1 \qquad (E_3)$$

has the same solution as the linear system

$$2x + y - z = 4 \qquad (E_1)$$
$$7x + 3y - z = 10 \qquad (E_2 + 2E_1)$$
$$x + y + z = 1 \qquad (E_3)$$

The second equation in the original system was replaced by the sum of that equation and 2 times the first equation. That is,

$$
\begin{array}{ll}
3x + y + z = 2 & (E_2) \\
4x + 2y - 2z = 8 & +(2E_1) \\
\hline
7x + 3y - z = 10 & (\text{new } E_2)
\end{array}
$$
     □

We now are in position to develop a method for solving systems of linear equations that can be either carried out by hand or programmed on a computer.

The Gauss-Jordan Method

The three elementary operations just described lead to an efficient method for solving systems of linear equations. The method is the Gauss-Jordan method, named after two nineteenth century mathematicians whose work provided the theoretical basis for the procedure.

To understand how the method works, consider the following two systems of equations:

Original System	Transformed System
$2x + 5y = 45$	$x + 0y = 10$
$x - 3y = -5$	$0x + y = 5$

The goal is to solve the original system. It can be shown that this system can be transformed into the system on the right using the three elementary operations; hence both systems have the same solution. Since the transformed system clearly has the solution $x = 10$ and $y = 5$, we can conclude that the original system also has this solution.

We now show how to carry out the steps of the Gauss-Jordan method. We illustrate the method for a general 3×3 system of equations.

Gauss-Jordan Method for Solving the 3 × 3 Linear System

Start with the following general 3×3 linear system:

$$
\begin{aligned}
a_{11}x + a_{12}y + a_{13}z &= b_1 \quad (E_1) \\
a_{21}x + a_{22}y + a_{23}z &= b_2 \quad (E_2) \\
a_{31}x + a_{32}y + a_{33}z &= b_3 \quad (E_3)
\end{aligned}
$$

- **Step 1** (transform coefficients of x).   Use elementary operations to make the coefficient of x equal to 1 in the first equation and 0 in the second and third equations. That is, use elementary operations so that the three coefficients of x:

$$
\text{Coefficients of } x = \begin{bmatrix} a_{11} \\ a_{21} \\ a_{31} \end{bmatrix} \quad \begin{matrix} (E_1) \\ (E_2) \\ (E_3) \end{matrix}
$$

are transformed into

$$
\text{New coefficients of } x = \begin{bmatrix} 1 \\ 0 \\ 0 \end{bmatrix} \quad \begin{matrix} (E_1) \\ (E_2) \\ (E_3) \end{matrix}
$$

- **Step 2** (transform coefficients of y).   Starting with the equations obtained in Step 1, use elementary operations to make the coefficient of y equal to 1 in the second equation and 0 in the first and third equations. After this step the coefficients of y will be

$$
\text{Transformed coefficients of } y = \begin{bmatrix} 0 \\ 1 \\ 0 \end{bmatrix} \quad \begin{matrix} (E_1) \\ (E_2) \\ (E_3) \end{matrix}
$$

  It should be noted that by transforming the coefficients of y, the coefficients of x have not been changed.

- **Step 3** (transform coefficients of z).   Starting with the equations obtained in Step 2, use elementary operations to make the coefficient of z equal to 1 in the third equation and 0 in the first and second equations. After this step the three coefficients of z will be

$$
\text{Transformed coefficients of } z = \begin{bmatrix} 0 \\ 0 \\ 1 \end{bmatrix} \quad \begin{matrix} (E_1) \\ (E_2) \\ (E_3) \end{matrix}
$$

  It should be noted that by transforming the coefficients of z, the coefficients of x and y have not been changed. The solution of the transformed system is now obvious.

Solving linear systems with a different number of variables uses this procedure with obvious modifications.

Example 5

Gauss-Jordan Method Solve the linear system

$$3x + 12y = 42 \qquad (E_1)$$
$$6x + 26y = 90 \qquad (E_2)$$

Solution

- **Step 1** (transform coefficients of x). We first transform the equations so that the coefficient of x in equation E_1 is 1. We do this by multiplying the equation E_1 by 1/3, which is the elementary operation

$$(1/3)E_1 \longrightarrow E_1$$

This gives us the following equivalent system of equations:

$$x + 4y = 14 \qquad (E_1)$$
$$6x + 26y = 90 \qquad (E_2)$$

We now want the coefficient of x in the equation E_2 to be 0. A 0 can be placed in this location if we replace E_2 by the sum of E_2 plus -6 times E_1. That is, we perform the elementary operation

$$E_1 + (-6)E_2 \longrightarrow E_2$$

which gives us the new linear system

$$x + 4y = 14 \qquad (E_1)$$
$$2y = 6 \qquad (E_2)$$

- **Step 2** (transform coefficients of y). We now transform the above system of equations so that the coefficient of y in E_2 is equal to 1. We do this by multiplying E_2 by 1/2. In other words, we perform the operation

$$(1/2)E_2 \longrightarrow E_2$$

This gives the equivalent system

$$x + 4y = 14 \qquad (E_1)$$
$$0x + y = 3 \qquad (E_2)$$

Finally, to eliminate the y term in E_1 (change the 4 to 0), we replace E_1 by the sum of E_1 plus -4 times E_2. That is, we perform the elementary operation

$$E_1 + (-4)E_2 \longrightarrow E_1$$

This gives the final linear system

$$x + 0y = 2$$
$$0x + y = 3$$

Since this system was found by applying elementary operations to the original system of equations, it has the same set of solutions as the original system of equations. But this final system can be solved by inspection, and its solution clearly is

$$x = 2$$
$$y = 3$$

This completes the Gauss-Jordan process. These values can be checked to see that they satisfy both of the original equations. □

HISTORICAL NOTE

Carl Friedrich Gauss (1777–1855), pictured at left, was born into a poor working-class family in Brunswick, Germany. At the time of his death he was proclaimed the most brilliant mathematician the world had ever known. He made important contributions to number theory, astronomy, statistics, physics, and geometry. His work with systems of equations and the method of solution that we study here represents only a small portion of his life's work.

Camille Jordan (1838–1922) was born into a well-to-do French family that had a reputation for scholarship. At the age of 17, Jordan entered France's prestigious *Ecole Polytechnique* and became an engineer, pursuing the study of mathematics only in his spare time.

Example 6

Gauss-Jordan Method   Solve the linear system

$$2x + 8y + 6z = 20$$
$$4x + 2y - 2z = -2$$
$$3x - y + z = 11$$

by the Gauss-Jordan method.

Solution

- **Step 1** (transform the coefficients of x).   The following operations will transform the linear system to an equivalent system having the proper coefficients of x.

$$(1/2)E_1 \longrightarrow E_1$$

want 1

$$
\begin{array}{lll}
(1/2)[2x + 8y + 6z = 20] & x + 4y + 3z = 10 & (E_1) \\
4x + 2y - 2z = -2 \longrightarrow & 4x + 2y - 2z = -2 & (E_2) \\
3x - y + z = 11 & 3x - y + z = 11 & (E_3)
\end{array}
$$

$$E_2 + (-4)E_1 \longrightarrow E_2$$

want 0

$$
\begin{array}{lll}
(-4)[\, x + 4y + 3z = 10] & x + 4y + 3z = 10 & (E_1) \\
\longrightarrow 4x + 2y - 2z = -2 \longrightarrow & -14y - 14z = -42 & (E_2) \\
3x - y + z = 11 & 3x - y + z = 11 & (E_3)
\end{array}
$$

$$E_3 + (-3)E_1 \longrightarrow E_3$$

want 0

$$
\begin{array}{lll}
(-3)[\, x + 4y + 3z = 10] & x + 4y + 3z = 10 & (E_1) \\
-14y - 14z = -42 \longrightarrow & -14y - 14z = -42 & (E_2) \\
\longrightarrow 3x - y + z = 11 & -13y - 8z = -19 & (E_3)
\end{array}
$$

- **Step 2** (transform the coefficients of y).

$$(-1/14)E_2 \longrightarrow E_2$$

(want 1)

$$
\begin{aligned}
x + 4y + 3z &= 10 \\
(-1/14)[\quad -14y - 14z &= -42] \\
-13y - 8z &= -19
\end{aligned}
\longrightarrow
\begin{aligned}
x + 4y + 3z &= 10 \quad (E_1) \\
y + z &= 3 \quad (E_2) \\
-13y - 8z &= -19 \quad (E_3)
\end{aligned}
$$

$$E_1 + (-4)E_2 \longrightarrow E_1$$

(want 0)

$$
\begin{aligned}
x + 4y + 3z &= 10 \\
(-4)[\quad y + z &= 3] \\
-13y - 8z &= -19
\end{aligned}
\qquad
\begin{aligned}
x \quad\quad - z &= -2 \quad (E_1) \\
y + z &= 3 \quad (E_2) \\
-13y - 8z &= -19 \quad (E_3)
\end{aligned}
$$

$$E_3 + 13E_2 \longrightarrow E_3$$

$$
\begin{aligned}
x \quad - z &= -2 \\
(13)[\quad y + z &= 3] \\
-13y - 8z &= -19
\end{aligned}
\longrightarrow
\begin{aligned}
x \quad - z &= -2 \quad (E_1) \\
y + z &= 3 \quad (E_2) \\
5z &= 20 \quad (E_3)
\end{aligned}
$$

(want 0)

- **Step 3** (transform the coefficients of z).

$$(1/5)E_3 \longrightarrow E_3$$

$$
\begin{aligned}
x \quad - z &= -2 \\
y + z &= 3 \\
(1/5)[\quad 5z &= 20]
\end{aligned}
\longrightarrow
\begin{aligned}
x \quad - z &= -2 \quad (E_1) \\
y + z &= 3 \quad (E_2) \\
z &= 4 \quad (E_3)
\end{aligned}
$$

(want 1)

$$E_1 + E_3 \longrightarrow E_1$$

(want 0)

$$
\begin{aligned}
x \quad - z &= -2 \\
y + z &= 3 \\
(1)[\quad z &= 4]
\end{aligned}
\longrightarrow
\begin{aligned}
x \quad\quad &= 2 \quad (E_1) \\
y + z &= 3 \quad (E_2) \\
z &= 4 \quad (E_3)
\end{aligned}
$$

$$E_2 + (-1)E_3 \longrightarrow E_2$$

(want 0)

$$
\begin{aligned}
x \quad\quad &= 2 \\
y + z &= 3 \\
(-1)[\quad z &= 4]
\end{aligned}
\longrightarrow
\begin{aligned}
x \quad\quad &= 2 \quad (E_1) \\
y \quad &= -1 \quad (E_2) \\
z &= 4 \quad (E_3)
\end{aligned}
$$

The solution can now be read directly from the right-hand sides of the above equations as

$$
\begin{aligned}
x &= 2 \\
y &= -1 \\
z &= 4
\end{aligned}
$$

Check:

$$2(2) + 8(-1) + 6(4) = 20$$
$$4(2) + 2(-1) - 2(4) = -2$$
$$3(2) - (-1) + (4) = 11$$

The RST Game (More Than a Game)

The following RST game may seem at first glance to be a recreational diversion and nothing more. On closer inspection, however, the game illustrates clearly many of the important phenomena studied in this book. By a thorough understanding of this game, the reader will be better equipped to understand many problems involving linear systems in this chapter and linear programming in the next chapter. We will see how the RST game can be solved by solving a system of linear equations using the Gauss-Jordan method.

Rules for the RST Game

You are given a large collection of cardboard rectangles, squares, and triangles. Drawn inside these figures are the initials of three major universities in the United States: BU, UA, and UL. In particular we have the following figures.

Rectangle: On each rectangle (Figure 14) we draw two BU's, three UA's, and one UL.

Square: On each square (Figure 15) we draw three BU's, one UA, and four UL's.

Triangle: On each triangle (Figure 16) we draw one BU, one UA, and two UL's.

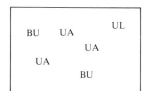

Figure 14
Rectangle for the RST game

Figure 15
Square for the RST game

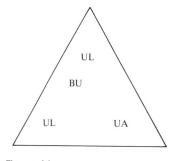

Figure 16
Triangle for the RST game

The goal of the game is to find a collection (if any) of rectangles, squares, and triangles that should be placed in a pot so that the total number of BU's in the pot is 15, the total number of UA's is 16, and the total number of UL's is 19. You may attempt to find the solution by experimentation before seeing how mathematics can be used to solve the problem.

To determine the correct combination of rectangles, squares, and triangles, we denote

$$r = \text{number of rectangles chosen}$$
$$s = \text{number of squares chosen}$$
$$t = \text{number of triangles chosen}$$

To find the solution, first organize the above information into Table 7.

TABLE 7 ————————————
Distribution of BU's, UA's, and UL's in Figures 14, 15, and 16

	Rectangle	Square	Triangle	Total Number of Initials
BU	BU BU	BU BU BU	BU	15
UA	UA UA UA	UA	UA	16
UL	UL	UL UL UL UL	UL UL	19
	BU BU UA UA UA UL	BU BU BU UA UL UL UL UL	BU UA UL UL	

If we now place *r* rectangles, *s* squares, and *t* triangles in the pot, then the numbers of BU's, UA's, and UL's that will be in the pot are

$$\text{Total number of BU's} = 2r + 3s + t$$
$$\text{Total number of UA's} = 3r + s + t$$
$$\text{Total number of UL's} = r + 4s + 2t$$

Hence to have 15 BU's, 16 UA's, and 19 UL's in the pot, the above total number of rectangles, squares, and triangles must meet the following conditions:

$$2r + 3s + t = 15 \quad \text{(BU constraint)}$$
$$3r + s + t = 16 \quad \text{(UA constraint)}$$
$$r + 4s + 2t = 19 \quad \text{(UL constraint)}$$

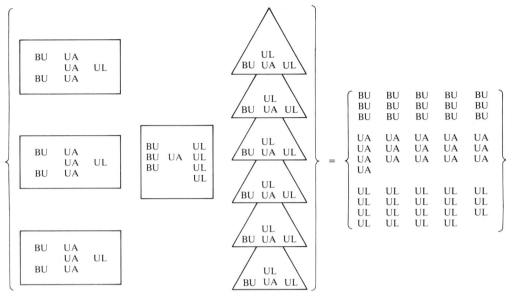

Figure 17
Solution of the RST game

This 3 × 3 system can be solved by the Gauss-Jordan method (Problem 34 at the end of this section) and has the following solution:

$$r = 3 \qquad \text{(number of rectangles)}$$
$$s = 1 \qquad \text{(number of squares)}$$
$$t = 6 \qquad \text{(number of triangles)}$$

This combination of figures is illustrated in Figure 17.

We now show how the RST game is similar to a problem that on the surface is quite different.

The RST Game and the Dietician's Problem

Three foods contain given amounts of vitamins A, B, and C as shown in Table 8.

TABLE 8
Vitamin Amounts in
Three Foods

	Food I (oz)	Food II (oz)	Food III (oz)	Vitamins Required
Vitamin A	2 mg/oz	3 mg/oz	1 mg/oz	15 mg
Vitamin B	3 mg/oz	1 mg/oz	1 mg/oz	16 mg
Vitamin C	1 mg/oz	4 mg/oz	2 mg/oz	19 mg

The numbers in Table 8 refer to the amounts of vitamins in the three different kinds of foods. For example, the number 2 in the upper left-hand corner means that there are 2 milligrams (mg) of vitamin A in each ounce (oz) of Food I. A dietician is faced with the problem of determining the amounts of Foods I, II, and III a person should ingest to ensure that the total intake of vitamins A, B, and C is 15, 16, and 19 mg, respectively.

In this problem the three vitamins are analogous to the three universities, and the three foods are analogous to the three figures in the RST game. It is interesting to note the similarities between Tables 7 and 8.

To solve the dietician's problem, we let

$$x = \text{amount of Food I ingested}$$
$$y = \text{amount of Food II ingested}$$
$$z = \text{amount of Food III ingested}$$

For 15 mg of vitamin A, 16 mg of vitamin B, and 19 mg of vitamin C to be ingested, the variables x, y, and z must satisfy the following equations:

$$
\begin{array}{ll}
2x + 3y + \ z = 15 & \text{(vitamin A requirement)} \\
3x + \ y + \ z = 16 & \text{(vitamin B requirement)} \\
x + 4y + 2z = 19 & \text{(vitamin C requirement)}
\end{array}
$$

Solving this 3×3 system, we will find

$$x = 3$$
$$y = 1$$
$$z = 6$$

In other words, the dietician determines that a meal consisting of three ounces of Food I, one ounce of Food II, and six ounces of Food III will exactly meet the nutritional requirements of the three vitamins.

In addition to the dietician's problem, the RST game is analogous to several other problems. For example, a furniture manufacturer who builds tables, chairs, and cabinets may make these products primarily from the three basic resources of wood, plastic, and labor. If the amount of each of these resources used in the production of these products is known, and if the manufacturer has a given amount of these resources available, then it is possible to solve a 3×3 linear system to determine the number of products the manufacturer can produce.

Problems

For Problems 1–4, perform the indicated elementary operations. The second operation should be performed on the system as transformed by the first operation.

1. $E_1: 2x - y = 5 \quad (1/2)E_1 \longrightarrow E_1$
$E_2: \ x + 2y = 3 \quad (1/2)E_2 \longrightarrow E_2$

2. $E_1: \ x - 2y = \ \ 3$
$E_2: 3x + \ y = -1 \quad E_2 + (-3)E_1 \longrightarrow E_2$

3. $E_1: \ x + 2y = -1$
$E_2: 2x - 5y = \ \ 3 \quad E_2 + (-2)E_1 \longrightarrow E_2$

4. $E_1: \ \ \ x + 2y + 3z = 4$
$E_2: \ 2x + 3y + \ z = 6 \quad E_2 + (-2)E_1 \longrightarrow E_2$
$E_3: -x + \ y + 3z = 5 \quad (-1)E_3 \longrightarrow E_3$

For Problems 5–7, perform the indicated elementary operations. For each operation, use the system as it is given or as it is transformed by earlier operations.

5. $E_1: 2x - \ y = 3 \quad (1/2)E_1 \longrightarrow E_1$
$E_2: 4x - 3y = 2 \quad E_2 + (-4)E_1 \longrightarrow E_2$

6. $E_1: 3x + 2y - \ z = 3 \quad (1/3)E_1 \longrightarrow E_1$
$E_2: \ x - \ y + 2z = 4 \quad E_2 + (-1)E_1 \longrightarrow E_2$
$E_3: 2x + 2y - 2z = 7 \quad E_3 + (-2)E_1 \longrightarrow E_3$

7. $E_1: 3x + 2y + \ z = \ \ 2 \quad E_1 + E_2 \longrightarrow E_1$
$E_2: \ x - \ y - \ z = -1 \quad E_3 + 2E_1 \longrightarrow E_3$
$E_3: 2x + 3y + 2z = \ \ 4 \quad (1/2)E_3 \longrightarrow E_3$

For Problems 8–18, solve the following systems of equations using elementary row operations to transform the equations to a simpler form.

8. $5x + 2y = 1$
$-4x + 3y = 10$

9. $x - y = 3$
$2x + 3y = 4$

10. $x + y = 1$
$x - y = 1$

11. $x = 0$
$x + 2y = 1$

12. $y = 0$
$x - 3y = 1$

13. $x + y + z = 1$
$x + y - z = 0$
$x - y + 2z = 4$

14. $x + z = 4$
$y - 2z = 2$
$x + y = 0$

15. $-2x + y + z = 3$
$x - y + z = 2$
$2x + 2y + z = 4$

16. $2x - y + 3z = 15$
$x + y - z = 1$
$3x + 2y - z = 8$

17. $x + 2z = 3$
$x - y = 4$
$2y + z = 3$

18. $x - 3y - z = -4$
$2x - 4y - z = -7$
$x - y + z = -2$

Linear Systems in Business Management

For Problems 19–22, use Table 9.

TABLE 9
Table of Values for Problems 19–22

	Growth Stocks	Dividend-Producing Stocks	Tax-Free Bonds
Expected Appreciation per Year	12%	8%	5%
Potential Loss per Year	10%	8%	3%

19. Portfolio Analysis Determine a strategy for investing $250,000 in a portfolio of the three types of investments in Table 9. The annual appreciation desired is 7% with a potential loss of no more than 5%.

20. Portfolio Analysis Determine a strategy for investing $50,000 in a portfolio of the three types of investments in Table 9. The annual appreciation desired is 6% with a potential loss of no more than 5%.

21. Portfolio Analysis Determine a strategy for investing $1,000,000 in a portfolio of the three types of investments in Table 9. The annual appreciation desired is 9% with a potential loss of no more than 7%.

22. Portfolio Analysis Determine a strategy for investing $500,000 in a portfolio of the three types of investments in Table 9. The annual appreciation desired is 8% with a potential loss of no more than 7%.

23. Blending Problem Fast Foods, Inc. sells three specials. The number of hamburgers, french fry orders, and sodas in each special is summarized in Table 10.

TABLE 10
Information for Problem 23

	Special 1	Special 2	Special 3
Hamburger	1	0	1
French Fries	2	1	0
Soda	1	2	0

The day's receipts indicate the following sales:

300 hamburgers

500 french fries

400 sodas

How many units of each special were sold?

24. Blending Problem Repeat Problem 23 but replace Table 10 with Table 11.

TABLE 11
Information for Problem 24

	Special 1	Special 2	Special 3
Hamburger	1	0	1
French Fries	2	1	1
Soda	2	2	1

The daily sales were 300 hamburgers, 600 french fries, and 500 sodas.

25. Gold and Silver Gold and silver can be extracted from two types of ore, I and II. Each 100 pounds of ore I yields 10 ounces of gold and 20 ounces of silver. Each 100 pounds of ore II yields 15 ounces of gold and 10 ounces of silver. How many pounds of ores I and II are required to produce 175 ounces of gold and 250 ounces of silver?

26. Lead, Copper, and Zinc Lead, copper, and zinc are extracted from three types of ores. Each 100 pounds of ore I yields 20 ounces of lead, 10 ounces of copper, and 15 ounces of zinc. Each 100 pounds of ore II yields 15 ounces of lead, 20 ounces of copper, and 10 ounces of zinc. Each 100 pounds of ore III yields 10 ounces of lead, 5 ounces of copper, and 10 ounces of zinc. How many pounds of each ore are required to produce 85,000 ounces of lead, 55,000 ounces of copper, and 70,000 ounces of zinc?

27. **Dietician's Problem**   Protein and carbohydrates can be obtained from two foods, Food I and Food II. Each ounce of Food I contains 5 grams of protein and 10 grams of carbohydrates. Each ounce of Food II contains 10 grams of protein and 5 grams of carbohydrates. How much of each food should a person consume to ingest 150 grams of protein and 150 grams of carbohydrates?

28. **Dietician's Problem**   Protein, carbohydrates, and fats can be obtained from three foods. Each ounce of Food I contains 5 grams of protein, 10 grams of carbohydrates, and 40 grams of fat. Each ounce of Food II contains 10 grams of protein, 5 grams of carbohydrates, and 30 grams of fat. Each ounce of Food III contains 15 grams of protein, 15 grams of carbohydrates, and 80 grams of fat. How many ounces of each of the three foods are required to yield 300 grams of protein, 300 grams of carbohydrates, and 1500 grams of fat?

More RST Games (Business Problems in Disguise)

For Problems 29–33, suppose you have a collection of rectangles, squares, and triangles. On each of these figures are drawn cats, dogs, and mice as follows.

Rectangle: On each rectangle is drawn one cat, two dogs, and one mouse. (See Figure 18.)

Figure 18
Rectangle for Problems 29–33

Square: On each square is drawn two cats, three dogs, and four mice. (See Figure 19.)

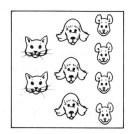

Figure 19
Square for Problems 29–33

Triangle: On each triangle is drawn one cat, one dog, and two mice. (See Figure 20.)

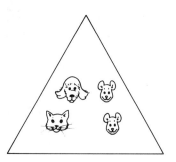

Figure 20
Triangle for Problems 29–33

The information above is summarized in Table 12.

TABLE 12
Number of Cats, Dogs, and Mice in Each Rectangle, Square, and Triangle

	Rectangle	**Square**	**Triangle**
Cat	1	2	1
Dog	2	3	1
Mouse	1	4	2

Select the combination of rectangles, squares, and triangles so that the total number of animals selected is given by the following values. Some of these problems have an infinite number of solutions, and some have no solutions. Note how this game is a restatement of many problems in business management.

29. Number of cats $= 17$
    Number of dogs $= 25$
    Number of mice $= 34$

30. Number of cats $= 4$
    Number of dogs $= 6$
    Number of mice $= 8$

31. Number of cats $= 3$
    Number of dogs $= 5$
    Number of mice $= 6$

32. Number of cats $= 40$
    Number of dogs $= 60$
    Number of mice $= 80$

33. Number of cats $= 18$
    Number of dogs $= 30$
    Number of mice $= 44$

34. Solve the RST game discussed in this section by solving the linear system

$$2r + 3s + t = 15$$
$$3r + s + t = 16$$
$$r + 4s + 2t = 19$$

Linear Systems in the Social Sciences

Problems 35–40 are concerned with the following traffic flow problem. Suppose the city of Cleveland wishes to improve its

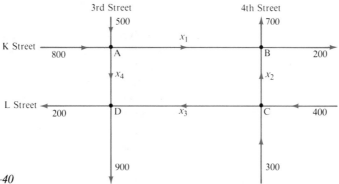

Figure 21
Traffic flow for Problems 35–40

traffic signals in order to increase traffic flow. Figure 21 illustrates the flow of cars (per hour) at a typical intersection (note that all streets are one-way).

From Figure 21 we can see that 500 cars per hour come down 3rd Street to intersection A and 800 cars per hour enter intersection A on K Street. Among these cars, x_1 of them leave intersection A on K Street, and x_4 leave intersection A on 3rd Street. Since the number of cars entering and leaving intersection A is the same, the following equation must hold:

$$\text{Intersection A:} \quad x_1 + x_4 = 500 + 800$$

or

$$x_1 + x_4 = 1300$$

35. **Finding the Traffic Model**   Carry out a similar analysis at the other intersections B, C, and D to find the four linear equations that must be satisfied by the variables x_1, x_2, x_3, and x_4. (We have already found one equation that corresponds to intersection A.)

36. **Solving the Traffic Model**   Solve the system of equations found in Problem 35 by the Gauss-Jordan method. *Hint:* You should arrive at a row of 0's and hence an infinite number of solutions. What is the interpretation of each solution (x_1, x_2, x_3, x_4)?

37. **Maximum Traffic Flow on 3rd Street**   The linear system in Problem 36 has an infinite number of solutions. They are

$$x_1 = 1300 - x_4$$
$$x_2 = -400 + x_4$$
$$x_3 = 1100 - x_4$$

What is the largest value that x_4 can be such that x_1, x_2, and x_3 are all nonnegative? What is the interpretation of this value?

38. **Maximum Traffic Flow on K Street**   Solve the equations in Problem 37 for x_2, x_3, and x_4 in terms of x_1. What is the largest value that x_1 can be such that x_2, x_3, and x_4 are all nonnegative? What is the interpretation of this value?

39. **Maximum Traffic Flow on 4th Street**   Solve the equations in Problem 37 for x_1, x_3, and x_4 in terms of x_2. What is the largest value that x_2 can be such that x_1, x_3, and x_4 are all nonnegative? What is the interpretation of this value?

40. **Maximum Traffic Flow on L Street**   Solve the equations in Problem 37 for x_1, x_2, and x_4 in terms of x_3. What is the largest value that x_3 can be such that x_1, x_2, and x_4 are all nonnegative? What is the interpretation of this value?

1.4   The Gauss-Jordan and Gaussian Elimination Methods

PURPOSE

We introduce the concept of the augmented matrix of a linear system of equations and show how the Gauss-Jordan method can be streamlined. We also determine when a system of equations has an infinite number of solutions or no solutions.

We also introduce the Gaussian elimination method for solving linear systems.

$$x + 2y - z = 3$$
$$x + y + z = 6$$
$$3x + 5y + 4z = 7$$

That's just dead weight!

$$\begin{bmatrix} 1 & 2 & -1 & 3 \\ 1 & 1 & 1 & 6 \\ 3 & 5 & 4 & 7 \end{bmatrix}$$

Introduction

The systems of equations solved in the previous section involved two and three unknowns. Much of the difficulty one faces when solving larger systems arises from trying to keep track of all the intermediate systems as we go along. The names $x, y, z, \ldots$ of the variables are not essential for carrying out any of the computations. Therefore instead of working with the equations, we work only with the coefficients and the numbers on the right-hand side of the equations. This brings us to the important topics of the coefficient and augmented matrices of a linear system of equations.

Coefficient and Augmented Matrices

Consider a typical linear system

$$x + y + 2z = 4$$
$$3x - y + 6z = 0$$
$$4x + 2y - 9z = 15$$

To specify a system of equations, it is not necessary that we actually write the names of the variables. The above system of equations would be the same if we replaced the three variables x, y, and z by u, v, and w. We can just as well specify the above system by the following 3×4 array of three rows and four columns of numbers:

	Column 1	Column 2	Column 3	Column 4
Row 1	1	1	2	4
Row 2	3	-1	6	0
Row 3	4	2	-9	15
	Coefficient Matrix			Right-Hand Side

Any rectangular array of numbers enclosed in brackets such as the one above is called a **matrix**. This particular matrix is called the **augmented matrix** associated with the previous linear system. The 3×3 array of three rows and three columns to the left of the vertical bar is called the **coefficient matrix** of the linear system. The numbers in any matrix are called the **elements** of the matrix. An element in position (i, j) is the element in row i and column j. We call this element the ij-element.

Example 1 ──────────────

Augmented and Coefficient Matrices Find the augmented and coefficient matrices for the following linear system:

$$x + 2y - z = 7$$
$$-x + y + 4z = 9$$
$$4x + 5y - 7z = -4$$

Solution The augmented matrix of the above linear system is

$$\text{Augmented matrix} = \begin{bmatrix} 1 & 2 & -1 & 7 \\ -1 & 1 & 4 & 9 \\ 4 & 5 & -7 & -4 \end{bmatrix}$$

The matrix to the left of the vertical bar is the coefficient matrix:

$$\text{Coefficient matrix} = \begin{bmatrix} 1 & 2 & -1 \\ -1 & 1 & 4 \\ 4 & 5 & -7 \end{bmatrix}$$

□

When solving systems of equations, it is possible to work only with the augmented matrix. The advantage of this approach lies in the fact that we do not need to carry along the names of the variables. When solving linear systems using computers, we work only with the augmented matrix.

The elementary operations of the previous section can be restated in terms of elementary row operations on augmented matrices.

Elementary Row Operations on Augmented Matrices

The following operations applied to an augmented matrix of a linear system will produce an augmented matrix of an equivalent linear system. If the ith row of the augmented matrix is denoted by R_i, we illustrate these operations symbolically.

- **Row Operation 1** (interchange rows). We can interchange any two rows of the augmented matrix. The notation

$$R_i \longleftrightarrow R_j$$

means to interchange rows i and j.

- **Row Operation 2** (multiply a row times a constant). We can multiply all entries of a row by a nonzero real number. The notation

$$cR_i \longrightarrow R_i$$

means to replace each element in the ith row by its value multiplied by c.

- **Row Operation 3** (add to a row a multiple of another row). We can add to any row a constant multiple of another row. The notation

$$R_i + cR_j \longrightarrow R_i$$

means to replace the ith row by the sum of the ith row and the jth row times c.

On the next page we state the Gauss-Jordan method for solving linear systems in terms of the augmented matrix.

Computer programmers often find it convenient to describe procedures such as the Gauss-Jordan method by means of **flow diagrams**. Flow diagrams give a visual description of the method. A flow diagram for the Gauss-Jordan method is described in Figure 22 on page 67.

Gauss-Jordan Method Applied to Augmented Matrices

Starting with the augmented matrix for a linear system, perform the elementary row operations as indicated in the following steps. For simplicity we illustrate the method for the general 3×3 system:

$$
\begin{aligned}
a_{11}x + a_{12}y + a_{13}z &= b_1 \\
a_{21}x + a_{22}y + a_{23}z &= b_2 \\
a_{31}x + a_{32}y + a_{33}z &= b_3
\end{aligned}
\longleftrightarrow
\left[\begin{array}{ccc|c}
a_{11} & a_{12} & a_{13} & b_1 \\
a_{21} & a_{22} & a_{23} & b_2 \\
a_{31} & a_{32} & a_{33} & b_3
\end{array}\right]
$$

Original Linear System   *Original Augmented Matrix*

- **Step 1** (transform column 1).   Transform the original augmented matrix to a new augmented matrix in which column 1 has a 1 in the top position and 0's in all lower positions (the asterisks in the following augmented matrices represent real numbers whose values depend on the specific problem being solved).

$$
\left[\begin{array}{ccc|c}
a_{11} & a_{12} & a_{13} & b_1 \\
a_{21} & a_{22} & a_{23} & b_2 \\
a_{31} & a_{32} & a_{33} & b_3
\end{array}\right]
\longrightarrow
\left[\begin{array}{ccc|c}
1 & * & * & * \\
0 & * & * & * \\
0 & * & * & *
\end{array}\right]
$$

Original Augmented Matrix   *Transformed Augmented Matrix*

- **Step 2** (transform column 2).   Transform the augmented matrix found in Step 1 to another augmented matrix in which column 2 has a 1 in the second position from the top and 0's in the other positions.

$$
\left[\begin{array}{ccc|c}
1 & * & * & * \\
0 & * & * & * \\
0 & * & * & *
\end{array}\right]
\longrightarrow
\left[\begin{array}{ccc|c}
1 & 0 & * & * \\
0 & 1 & * & * \\
0 & 0 & * & *
\end{array}\right]
$$

Augmented Matrix
from Step 1

- **Steps 3, 4, ...** (transform columns $3, 4, \ldots$).   Continue in this way until all columns of the coefficient matrix have been transformed. The rightmost column of the coefficient matrix will have a 1 in the bottom position and 0's in all positions above it.

$$
\left[\begin{array}{ccc|c}
1 & 0 & * & * \\
0 & 1 & * & * \\
0 & 0 & * & *
\end{array}\right]
\longrightarrow
\left[\begin{array}{ccc|c}
1 & 0 & 0 & r_1 \\
0 & 1 & 0 & r_2 \\
0 & 0 & 1 & r_3
\end{array}\right]
$$

Final Transformed Matrix

- **Last Step** (read solution from column at right).   The final transformed matrix represents the linear system

$$
\begin{aligned}
x + 0y + 0z &= r_1 \\
0x + y + 0z &= r_2 \\
0x + 0y + z &= r_3
\end{aligned}
$$

Hence the solution $(x, y, z) = (r_1, r_2, r_3)$ is located in the column to the right of the vertical bar.

After each individual step of the process is performed, any row of the augmented matrix with all elements zero should be immediately interchanged with the bottom row that contains at least one nonzero element.

The array of numbers to the left of the vertical bar in the final transformed matrix (with 1's down the "diagonal" and 0's elsewhere) is called the **identity matrix**. In other words, the Gauss-Jordan method can be interpreted as a procedure whereby the coefficient matrix of a linear system is transformed by means of elementary row operations into the identity matrix. In so doing, the right-hand side of the linear system is transformed into the solution.

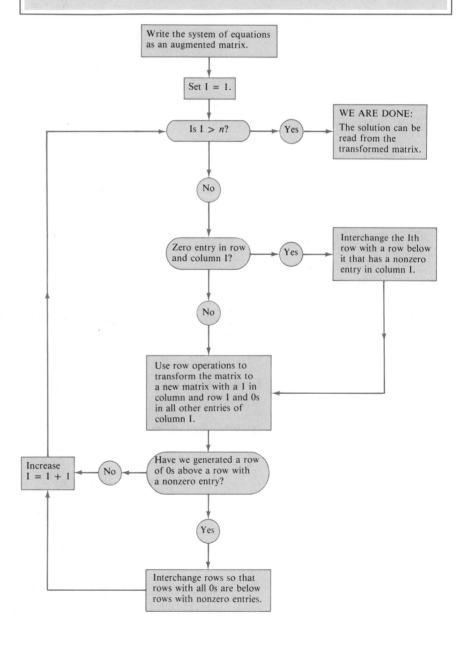

Figure 22
Flow diagram of the Gauss-Jordan method for solving n simultaneous equations

Example 2

A 2×2 System with a Unique Solution   Solve the following 2×2 system of equations:

$$x + y = 5$$
$$-x + y = -1$$

Solution   We start with the augmented matrix

$$\left[\begin{array}{cc|c} 1 & 1 & 5 \\ -1 & 1 & -1 \end{array}\right] \begin{array}{c} R_1 \\ R_2 \end{array}$$

- **Step 1** (transform column 1).   We transform column 1 so that the top element is 1 and the bottom element is 0. Since the top element is already 1, we perform the following operation.

$$R_2 + R_1 \longrightarrow R_2$$

$$(1)\left[\begin{array}{cc|c} 1 & 1 & 5 \\ \underrightarrow{-1} & 1 & -1 \end{array}\right] \longrightarrow \left[\begin{array}{cc|c} 1 & 1 & 5 \\ 0 & 2 & 4 \end{array}\right] \begin{array}{c} R_1 \\ R_2 \end{array}$$

(want 0)

Original Augmented Matrix

- **Step 2** (transform column 2).   We transform column 2 so that the bottom element is 1 and the top element is 0.

$$(1/2)R_2 \longrightarrow R_2$$

$$(1/2)\left[\begin{array}{cc|c} 1 & 1 & 5 \\ 0 & 2 & 4 \end{array}\right] \longrightarrow \left[\begin{array}{cc|c} 1 & 1 & 5 \\ 0 & 1 & 2 \end{array}\right] \begin{array}{c} R_1 \\ R_2 \end{array}$$

(want 1)

$$R_1 + (-1)R_2 \longrightarrow R_1$$

(want 0)

$$(-1)\left[\begin{array}{cc|c} 1 & 1 & 5 \\ 0 & 1 & 2 \end{array}\right] \longrightarrow \left[\begin{array}{cc|c} 1 & 0 & 3 \\ 0 & 1 & 2 \end{array}\right] \begin{array}{c} R_1 \\ R_2 \end{array}$$

Final Augmented Matrix

The final augmented matrix above represents the linear system

$$x + 0y = 3$$
$$0x + y = 2$$

Hence the solution of the original system of equations is $x = 3$ and $y = 2$.   □

Example 3

A 3×3 System Using the Gauss-Jordan Method   Solve the following system of equations:

$$x + y + z = 6$$
$$2x + y + 6z = 22$$
$$3x + 6y + z = 18$$

Solution   The augmented matrix is

$$\left[\begin{array}{ccc|c} 1 & 1 & 1 & 6 \\ 2 & 1 & 6 & 22 \\ 3 & 6 & 1 & 18 \end{array}\right] \begin{array}{c} R_1 \\ R_2 \\ R_3 \end{array}$$

- **Step 1** (transform column 1).   Transform the augmented matrix so that column 1 will have a 1 in the top position and 0's in the bottom two positions. Since the top element is already 1, it is necessary only to make the bottom two entries 0. We do this by performing the following two row operations.

$$R_2 + (-2)R_1 \longrightarrow R_2$$

$$(-2)\begin{bmatrix} 1 & 1 & 1 & | & 6 \\ 2 & 1 & 6 & | & 22 \\ 3 & 6 & 1 & | & 18 \end{bmatrix} \longrightarrow \begin{bmatrix} 1 & 1 & 1 & | & 6 \\ 0 & -1 & 4 & | & 10 \\ 3 & 6 & 1 & | & 18 \end{bmatrix} \begin{matrix} R_1 \\ R_2 \\ R_3 \end{matrix}$$

(want 0)

$$R_3 + (-3)R_1 \longrightarrow R_3$$

$$(-3)\begin{bmatrix} 1 & 1 & 1 & | & 6 \\ 0 & -1 & 4 & | & 10 \\ 3 & 6 & 1 & | & 18 \end{bmatrix} \longrightarrow \begin{bmatrix} 1 & 1 & 1 & | & 6 \\ 0 & -1 & 4 & | & 10 \\ 0 & 3 & -2 & | & 0 \end{bmatrix} \begin{matrix} R_1 \\ R_2 \\ R_3 \end{matrix}$$

(want 0)

- **Step 2** (transform column 2).

$$(-1)R_2 \longrightarrow R_2$$

$$(-1)\begin{bmatrix} 1 & 1 & 1 & | & 6 \\ 0 & -1 & 4 & | & 10 \\ 0 & 3 & -2 & | & 0 \end{bmatrix} \longrightarrow \begin{bmatrix} 1 & 1 & 1 & | & 6 \\ 0 & 1 & -4 & | & -10 \\ 0 & 3 & -2 & | & 0 \end{bmatrix} \begin{matrix} R_1 \\ R_2 \\ R_3 \end{matrix}$$

(want 1)

$$R_1 + (-1)R_2 \longrightarrow R_1$$

$$(-1)\begin{bmatrix} 1 & 1 & 1 & | & 6 \\ 0 & 1 & -4 & | & -10 \\ 0 & 3 & -2 & | & 0 \end{bmatrix} \longrightarrow \begin{bmatrix} 1 & 0 & 5 & | & 16 \\ 0 & 1 & -4 & | & -10 \\ 0 & 3 & -2 & | & 0 \end{bmatrix} \begin{matrix} R_1 \\ R_2 \\ R_3 \end{matrix}$$

(want 0)

$$R_3 + (-3)R_2 \longrightarrow R_3$$

$$(-3)\begin{bmatrix} 1 & 0 & 5 & | & 16 \\ 0 & 1 & -4 & | & -10 \\ 0 & 3 & -2 & | & 0 \end{bmatrix} \longrightarrow \begin{bmatrix} 1 & 0 & 5 & | & 16 \\ 0 & 1 & -4 & | & -10 \\ 0 & 0 & 10 & | & 30 \end{bmatrix} \begin{matrix} R_1 \\ R_2 \\ R_3 \end{matrix}$$

(want 0)

- **Step 3** (transform column 3).

$$(1/10)R_3 \longrightarrow R_3$$

$$(1/10)\begin{bmatrix} 1 & 0 & 5 & | & 16 \\ 0 & 1 & -4 & | & -10 \\ 0 & 0 & 10 & | & 30 \end{bmatrix} \longrightarrow \begin{bmatrix} 1 & 0 & 5 & | & 16 \\ 0 & 1 & -4 & | & -10 \\ 0 & 0 & 1 & | & 3 \end{bmatrix} \begin{matrix} R_1 \\ R_2 \\ R_3 \end{matrix}$$

(want 1)

$$R_1 + (-5)R_3 \longrightarrow R_1$$

$$(-5)\begin{bmatrix} 1 & 0 & 5 & | & 16 \\ 0 & 1 & -4 & | & -10 \\ 0 & 0 & 1 & | & 3 \end{bmatrix} \longrightarrow \begin{bmatrix} 1 & 0 & 0 & | & 1 \\ 0 & 1 & -4 & | & -10 \\ 0 & 0 & 1 & | & 3 \end{bmatrix} \begin{matrix} R_1 \\ R_2 \\ R_3 \end{matrix}$$

(want 0)

$$R_2 + 4R_3 \longrightarrow R_2$$

$$(4)\begin{bmatrix} 1 & 0 & 0 & | & 1 \\ 0 & 1 & -4 & | & -10 \\ 0 & 0 & 1 & | & 3 \end{bmatrix} \longrightarrow \begin{bmatrix} 1 & 0 & 0 & | & 1 \\ 0 & 1 & 0 & | & 2 \\ 0 & 0 & 1 & | & 3 \end{bmatrix} \begin{matrix} R_1 \\ R_2 \\ R_3 \end{matrix}$$

(want 0)

Final Transformed Matrix

Writing the system of equations represented by the final transformed matrix above gives

$$x + 0y + 0z = 1$$
$$0x + y + 0z = 2$$
$$0x + 0y + z = 3$$

Hence the values of x, y, and z are easily determined. The solution is

$$x = 1$$
$$y = 2$$
$$z = 3$$
□

Summary of the Gauss-Jordan Method

The Gauss-Jordan method is a procedure that uses a sequence of transformations on a system of equations to derive an equivalent system with a simpler form. In Example 3 we made a series of transformations that gave

$$\begin{bmatrix} 1 & 1 & 1 & | & 6 \\ 2 & 1 & 6 & | & 22 \\ 3 & 6 & 1 & | & 18 \end{bmatrix} \xrightarrow[\text{operations}]{\text{elementary}} \begin{bmatrix} 1 & 0 & 0 & | & 1 \\ 0 & 1 & 0 & | & 2 \\ 0 & 0 & 1 & | & 3 \end{bmatrix}$$

Original Augmented Matrix　　　　　*Identity Solution Matrix*

In the above example the solution can be read from the column on the far right. In general, we can say the Gauss-Jordan method transforms the coefficient matrix of the original linear system into the identity matrix for those systems of equations that have a unique solution.

Linear Systems with No Solution or an Infinite Number of Solutions

The following examples illustrate how the Gauss-Jordan method recognizes linear systems that have no solutions or an infinite number of solutions.

Example 4

No Solutions　Solve the following systems of equations:

$$x + y + z = 1$$
$$x + 2y + z = 4$$
$$x + y + z = 2$$

Solution　Writing the augmented matrix, we have

$$\begin{bmatrix} 1 & 1 & 1 & | & 1 \\ 1 & 2 & 1 & | & 4 \\ 1 & 1 & 1 & | & 2 \end{bmatrix} \begin{matrix} R_1 \\ R_2 \\ R_3 \end{matrix}$$

• **Step 1** (transform column 1).　We begin by transforming column 1.

$$R_2 + (-1)R_1 \longrightarrow R_2$$

$$(-1)\begin{bmatrix} 1 & 1 & 1 & | & 1 \\ 1 & 2 & 1 & | & 4 \\ 1 & 1 & 1 & | & 2 \end{bmatrix} \longrightarrow \begin{bmatrix} 1 & 1 & 1 & | & 1 \\ 0 & 1 & 0 & | & 3 \\ 1 & 1 & 1 & | & 2 \end{bmatrix} \begin{matrix} R_1 \\ R_2 \\ R_3 \end{matrix}$$

want 0

$$R_3 + (-1)R_1 \longrightarrow R_3$$

$$(-1)\begin{bmatrix} 1 & 1 & 1 & | & 1 \\ 0 & 1 & 0 & | & 3 \\ 1 & 1 & 1 & | & 2 \end{bmatrix} \longrightarrow \begin{bmatrix} 1 & 1 & 1 & | & 1 \\ 0 & 1 & 0 & | & 3 \\ 0 & 0 & 0 & | & 1 \end{bmatrix} \begin{matrix} R_1 \\ R_2 \\ R_3 \end{matrix}$$

want 0

At this stage we stop the process because the bottom row has the form

$$\begin{bmatrix} 0 & 0 & 0 & | & 1 \end{bmatrix}$$

The equation represented by this row is

$$0x + 0y + 0z = 1$$

or

$$0 = 1$$

This, of course, cannot be. Whenever the Gauss-Jordan method generates a row that has all 0's to the left of the vertical bar in the augmented matrix and a nonzero value to the right of the bar, the system of equations does not have a solution.   □

Example 5 —————— Infinite Number of Solutions   Solve the following system of equations:

$$\begin{aligned} x + y + z &= 1 \\ 2x + 7y - 3z &= 7 \\ 3x + 3y + 3z &= 3 \end{aligned}$$

Solution   The augmented matrix is

$$\begin{bmatrix} 1 & 1 & 1 & | & 1 \\ 2 & 7 & -3 & | & 7 \\ 3 & 3 & 3 & | & 3 \end{bmatrix} \begin{matrix} R_1 \\ R_2 \\ R_3 \end{matrix}$$

We begin by transforming column one.

- **Step 1** (transform column 1). Since the top element in column 1 is already 1, we concentrate on making the entries below it 0.

$$R_2 + (-2)R_1 \longrightarrow R_2$$

$$(-2)\begin{bmatrix} 1 & 1 & 1 & | & 1 \\ 2 & 7 & -3 & | & 7 \\ 3 & 3 & 3 & | & 3 \end{bmatrix} \longrightarrow \begin{bmatrix} 1 & 1 & 1 & | & 1 \\ 0 & 5 & -5 & | & 5 \\ 3 & 3 & 3 & | & 3 \end{bmatrix} \begin{matrix} R_1 \\ R_2 \\ R_3 \end{matrix}$$

want 0

$$R_3 + (-3)R_1 \longrightarrow R_3$$

$$(-3)\begin{bmatrix} 1 & 1 & 1 & | & 1 \\ 0 & 5 & -5 & | & 5 \\ 3 & 3 & 3 & | & 3 \end{bmatrix} \longrightarrow \begin{bmatrix} 1 & 1 & 1 & | & 1 \\ 0 & 5 & -5 & | & 5 \\ 0 & 0 & 0 & | & 0 \end{bmatrix} \begin{matrix} R_1 \\ R_2 \\ R_3 \end{matrix}$$

want 0

At this stage, observe that all the elements in the bottom row are 0's. If we write the equation this row represents, we get

$$0x + 0y + 0z = 0$$

or

$$0 = 0$$

This is true, of course, but it says nothing about the system of equations. Hence we are left with only two equations in which to find the unknowns x, y, and z. This situation arises when we have an infinite number of solutions. To find all the solutions, we continue the process by transforming column 2 so that the coefficient of y in R_2 is 1 and the coefficients of y in R_1 and R_3 are both 0's. Doing this, we get

$$(1/5)R_2 \longrightarrow R_2$$

$$\left[\begin{array}{ccc|c} 1 & 1 & 1 & 1 \\ 0 & 5 & -5 & 5 \\ 0 & 0 & 0 & 0 \end{array}\right] \longrightarrow \left[\begin{array}{ccc|c} 1 & 1 & 1 & 1 \\ 0 & 1 & -1 & 1 \\ 0 & 0 & 0 & 0 \end{array}\right] \begin{array}{c} R_1 \\ R_2 \\ R_3 \end{array}$$

$$R_1 + (-1)R_2 \longrightarrow R_1$$

$$\left[\begin{array}{ccc|c} 1 & 1 & 1 & 1 \\ 0 & 1 & -1 & 1 \\ 0 & 0 & 0 & 0 \end{array}\right] \longrightarrow \left[\begin{array}{ccc|c} 1 & 0 & 2 & 0 \\ 0 & 1 & -1 & 1 \\ 0 & 0 & 0 & 0 \end{array}\right] \begin{array}{c} R_1 \\ R_2 \\ R_3 \end{array}$$

We have now transformed the first two columns into the desired form. However, the third column cannot be transformed, since the coefficient of z in R_3 is 0. Hence it is best at this stage to write the equations that represent the last augmented matrix (neglecting the bottom row of 0's, which says $0 = 0$). Doing this, we have

$$\begin{aligned} x \quad\quad + 2z &= 0 \\ y - z &= 1 \end{aligned}$$

We now solve for the first two variables, x and y, in terms of the third variable z, getting

$$\begin{aligned} x &= -2z \\ y &= z + 1 \end{aligned}$$

Since z can be any real number, we conclude that the solution set consists of all values of x, y, and z of the form

$$\begin{aligned} x &= -2z \\ y &= z + 1 \\ z &= \text{any real number} \end{aligned}$$

Some solutions among the infinite number of solutions would be the ones shown in Table 13.

The set of all solutions of this system of equations consists of the points (x, y, z) lying on the straight line shown in Figure 23. □

TABLE 13
Solutions for the System of
Equations in Example 5

z	$x = -2z$	$y = z + 1$
0	0	1
1	-2	2
2	-4	3
3	-6	4
4	-8	5

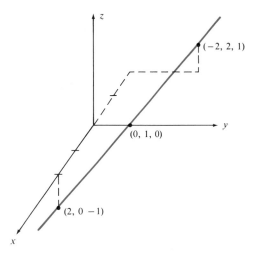

Figure 23
A line of solutions

Gaussian Elimination

Gauss's original method for solving systems of equations varies from the Gauss-Jordan method previously discussed. The original method is called **Gaussian elimination.** It is similar to the Gauss-Jordan method except that Gaussian elimination transforms the original coefficient matrix into a new coefficient matrix with 1's on the diagonal of the matrix and 0's below the diagonal and any value in the positions above the diagonal. A matrix of this form is called an **upper triangular matrix.** For example, to solve the 3×3 system of equations

$$
\begin{aligned}
-y + z &= 3 \\
x - y - z &= 0 \\
-x \quad\ - z &= -3
\end{aligned}
$$

we begin by writing the augmented matrix:

$$
\left[
\begin{array}{ccc|c}
0 & -1 & 1 & 3 \\
1 & -1 & -1 & 0 \\
-1 & 0 & -1 & -3
\end{array}
\right]
$$

To transform the coefficient matrix to upper triangular form, we carry out the following row operations.

- **Step 1** (transform column 1).

$$R_1 \longleftrightarrow R_2$$

(want 1) $\begin{bmatrix} 0 & -1 & 1 & | & 3 \\ 1 & -1 & -1 & | & 0 \\ -1 & 0 & -1 & | & -3 \end{bmatrix} \longrightarrow \begin{bmatrix} 1 & -1 & -1 & | & 0 \\ 0 & -1 & 1 & | & 3 \\ -1 & 0 & -1 & | & -3 \end{bmatrix} \begin{matrix} R_1 \\ R_2 \\ R_3 \end{matrix}$

$$R_3 + (1)R_1 \longrightarrow R_3$$

$(1)\begin{bmatrix} 1 & -1 & -1 & | & 0 \\ 0 & -1 & 1 & | & 3 \\ -1 & 0 & -1 & | & -3 \end{bmatrix} \longrightarrow \begin{bmatrix} 1 & -1 & -1 & | & 0 \\ 0 & -1 & 1 & | & 3 \\ 0 & -1 & -2 & | & -3 \end{bmatrix} \begin{matrix} R_1 \\ R_2 \\ R_3 \end{matrix}$

(want 0)

- **Step 2** (transform column 2).

$$(-1)R_2 \longrightarrow R_2$$

(want 1) $(-1)\begin{bmatrix} 1 & -1 & -1 & | & 0 \\ 0 & -1 & 1 & | & 3 \\ 0 & -1 & -2 & | & -3 \end{bmatrix} \longrightarrow \begin{bmatrix} 1 & -1 & -1 & | & 0 \\ 0 & 1 & -1 & | & -3 \\ 0 & -1 & -2 & | & -3 \end{bmatrix} \begin{matrix} R_1 \\ R_2 \\ R_3 \end{matrix}$

$$R_3 + (1)R_2 \longrightarrow R_3$$

$(1)\begin{bmatrix} 1 & -1 & -1 & | & 0 \\ 0 & 1 & -1 & | & -3 \\ 0 & -1 & -2 & | & -3 \end{bmatrix} \longrightarrow \begin{bmatrix} 1 & -1 & -1 & | & 0 \\ 0 & 1 & -1 & | & -3 \\ 0 & 0 & -3 & | & -6 \end{bmatrix} \begin{matrix} R_1 \\ R_2 \\ R_3 \end{matrix}$

(want 0)

- **Step 3** (transform column 3).

$$(-1/3)R_3 \longrightarrow R_3$$

$(-1/3)\begin{bmatrix} 1 & -1 & -1 & | & 0 \\ 0 & 1 & -1 & | & -3 \\ 0 & 0 & -3 & | & -6 \end{bmatrix} \longrightarrow \begin{bmatrix} 1 & -1 & -1 & | & 0 \\ 0 & 1 & -1 & | & -3 \\ 0 & 0 & 1 & | & 2 \end{bmatrix}$

(want 1)

Note that the above transformed coefficient matrix has 1's down the main diagonal and 0's below the diagonal. To solve this new system of equations, we use a process called **back-substitution**. To understand back-substitution, it is best to write the equations represented by the above augmented matrix. Doing this, we have

$$\begin{aligned} x - y - z &= 0 \\ y - z &= -3 \\ z &= 2 \end{aligned}$$

From these equations we see that the last equation gives the value of z, namely,

$$z = 2$$

Working backward, we substitute this value into the second from the last equation and solve for y. This gives

$$y = z - 3 = 2 - 3 = -1$$

Finally, we solve for x in the first equation, getting

$$x = y + z = -1 + 2 = 1$$

Hence we have the solution

$$x = 1$$
$$y = -1$$
$$z = 2$$

Comparison of Gaussian Elimination and the Gauss-Jordan Method

One advantage of Gaussian elimination over the Gauss-Jordan method is that it does not use as many arithmetic operations. For large problems this could be a critical factor. If you were to keep a tally of the number of additions and multiplications performed by each of the two methods, you would find that the Gauss-Jordan method would require more. Table 14 shows the total number of arithmetic steps (either an addition or a multiplication) used to solve a general system with a given number of unknowns using each of the two methods.

TABLE 14

Comparison of the Gauss-Jordan Method and Gaussian Elimination

Number of Unknowns	Number of Steps in Gaussian Elimination	Number of Steps in the Gauss-Jordan Method
2	9	10
3	28	33
10	805	1090
100	681,550	1,009,900
1000	668,165,500	1,009,999,000

If one assumes that a modern computer can perform 100,000 arithmetic operations a second, Gaussian elimination for a 1000×1000 system would take approximately 1.8 hours, while the Gauss-Jordan elimination would take approximately 2.8 hours. This difference indicates why computer scientists are continually looking for better ways to handle problems that contain many variables.

The Gauss-Jordan method has advantages over Gaussian elimination as well. Later, when we study matrix inverses, we will see that the Gauss-Jordan method is well suited for finding the inverse of a matrix.

This section closes with an example from the airline industry, which will illustrate the Gaussian elimination method.

Example 6

Gaussian Elimination   Campus Express offers college students summer charter trips to Europe. The company flies three kinds of airplanes: the Airbus 100, the Airbus 200, and the Airbus 300. Each plane is outfitted with tourist, economy, and first-class seats. The number of each kind of seat in the three types of planes is shown in Table 15.

TABLE 15
Number of Each Type of
Seat in the Different Planes

	Airbus 100	Airbus 200	Airbus 300
Tourist	50	75	40
Economy	30	45	25
First Class	32	50	30

The company lists the following number of reservations for its July flight to France:

Category	Number of Reservations
Tourist class	305
Economy class	185
First class	206

How many planes of each kind should the company fly to fill all of the seats?

Solution   Let

$$x = \text{number of Airbus 100's}$$
$$y = \text{number of Airbus 200's}$$
$$z = \text{number of Airbus 300's}$$

A simple analysis will conclude that the variables x, y, and z must satisfy the following constraints:

$$50x + 75y + 40z = 305 \quad \text{(tourist constraint)}$$
$$30x + 45y + 25z = 185 \quad \text{(economy constraint)}$$
$$32x + 50y + 30z = 206 \quad \text{(first class constraint)}$$

The final step is to solve this 3×3 linear system for x, y, and z. To do this, we use Gaussian elimination.

We start with the augmented matrix

$$\begin{bmatrix} 50 & 75 & 40 & | & 305 \\ 30 & 45 & 25 & | & 185 \\ 32 & 50 & 30 & | & 206 \end{bmatrix}$$

· **Step 1** (transform column 1).

$$(1/50)R_1 \longrightarrow R_1$$

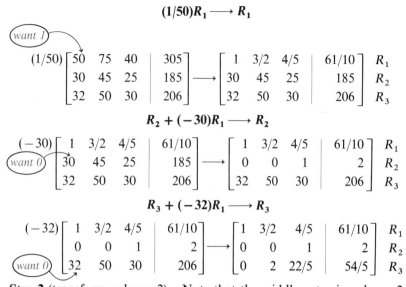

· **Step 2** (transform column 2). Note that the middle entry in column 2 of the previous matrix is 0. Hence we cannot hope to make this entry a 1 by multiplying the second row by the reciprocal of this entry. We can, however, interchange rows 2 and 3 so that the middle element in column 2 is nonzero. Doing this, we have

$$R_2 \longleftrightarrow R_3$$

$$\begin{bmatrix} 1 & 3/2 & 4/5 & 61/10 \\ 0 & 0 & 1 & 2 \\ 0 & 2 & 22/5 & 54/5 \end{bmatrix} \longrightarrow \begin{bmatrix} 1 & 3/2 & 4/5 & 61/10 \\ 0 & 2 & 22/5 & 54/5 \\ 0 & 0 & 1 & 2 \end{bmatrix} \begin{matrix} R_1 \\ R_2 \\ R_3 \end{matrix}$$

want nonzero

We now perform the last step.

$$(1/2)R_2 \longrightarrow R_2$$

want 1

$$(1/2)\begin{bmatrix} 1 & 3/2 & 4/5 & 61/10 \\ 0 & 2 & 22/5 & 54/5 \\ 0 & 0 & 1 & 2 \end{bmatrix} \longrightarrow \begin{bmatrix} 1 & 3/2 & 4/5 & 61/10 \\ 0 & 1 & 11/5 & 27/5 \\ 0 & 0 & 1 & 2 \end{bmatrix} \begin{matrix} R_1 \\ R_2 \\ R_3 \end{matrix}$$

The coefficient matrix has now been transformed to upper triangular form. To find the solution, we back-substitute. Doing this, we get

$$z = 2$$
$$y = 27/5 - (11/5)z$$
$$= 27/5 - (11/5)(2)$$
$$= 1$$
$$x = 61/10 - (4/5)z - (3/2)y$$
$$= 61/10 - (4/5)(2) - (3/2)(1)$$
$$= 3$$

Hence the company should use

$$x = 3 \text{ Airbus } 100\text{'s}$$
$$y = 1 \text{ Airbus } 200\text{'s}$$
$$z = 2 \text{ Airbus } 300\text{'s}$$

☐

Suspicion

You may suspect that the authors set this problem up so that the answer came out as integers. In a realistic problem, x, y, and z would not generally be integers. If the answer had turned out, say, 3.2 Airbus 100's, 1.5 Airbus 200's, and 2.2 Airbus 300's, then we would have to round up to the next integer at least one of the numbers to provide enough flights to carry all the passengers.

Problems

For Problems 1–9, write the linear system represented by each augmented matrix.

1. $\begin{bmatrix} 1 & 3 & | & 4 \\ 0 & 1 & | & 1 \end{bmatrix}$

2. $\begin{bmatrix} 1 & 0 & | & 2 \\ 0 & 1 & | & 4 \end{bmatrix}$

3. $\begin{bmatrix} 1 & 3 & | & 4 \\ 0 & 1 & | & 2 \end{bmatrix}$

4. $\begin{bmatrix} 1 & 2 & | & 3 \\ 0 & 1 & | & 4 \\ 0 & 0 & | & 1 \end{bmatrix}$

5. $\begin{bmatrix} 1 & 0 & 0 & | & 2 \\ 0 & 0 & 1 & | & 1 \\ 0 & 0 & 3 & | & 4 \end{bmatrix}$

6. $\begin{bmatrix} 1 & 2 & -3 & | & 1 \\ 0 & 1 & -4/3 & | & -1 \\ 0 & 0 & 1 & | & -2/5 \end{bmatrix}$

7. $\begin{bmatrix} 1 & -3/2 & 7/2 & | & 1 \\ 0 & 1 & -1/4 & | & -3/2 \\ 0 & 0 & 1 & | & 6 \end{bmatrix}$

8. $\begin{bmatrix} 1 & 0 & 0 & | & 3 \\ 0 & 1 & 0 & | & 5 \\ 0 & 0 & 1 & | & -1 \end{bmatrix}$

9. $\begin{bmatrix} 1 & 3 & 4 & 6 & | & 3 \\ 0 & 0 & 2 & 0 & | & 1 \\ 0 & 0 & 1 & 0 & | & 2 \\ 0 & 0 & 0 & 1 & | & 5 \end{bmatrix}$

For Problems 10–14, the following augmented matrices have been found by carrying out the steps of the Gauss-Jordan method. Determine the solution to these systems of equations.

10. $\begin{bmatrix} 1 & 2 & | & 3 \\ 0 & 1 & | & 4 \\ 0 & 0 & | & 1 \end{bmatrix}$

11. $\begin{bmatrix} 1 & 0 & 0 & | & 3 \\ 0 & 1 & 0 & | & 2 \\ 0 & 0 & 1 & | & 1 \end{bmatrix}$

12. $\begin{bmatrix} 1 & 3 & 4 & | & 3 \\ 0 & 1 & 3 & | & 2 \\ 0 & 0 & 1 & | & 2 \end{bmatrix}$

13. $\begin{bmatrix} 1 & 0 & 0 & 1 & | & -3 \\ 0 & 1 & 0 & 0 & | & 2 \\ 0 & 0 & 1 & -1 & | & 3 \end{bmatrix}$

14. $\begin{bmatrix} 1 & 0 & 0 & 1 & | & 2 \\ 0 & 1 & 0 & 2 & | & 3 \end{bmatrix}$

For Problems 15–22, solve the system of equations by the Gauss-Jordan method.

15. $\begin{bmatrix} 1 & 3 & | & 4 \\ 2 & 5 & | & 7 \end{bmatrix}$

16. $\begin{bmatrix} -1 & 31 & | & 31 \\ 2 & 42 & | & 42 \end{bmatrix}$

17. $\begin{bmatrix} 3 & 1 & | & -1 \\ 2 & 5 & | & 0 \end{bmatrix}$

18. $\begin{bmatrix} 2 & 1 & 1 & | & 3 \\ 1 & 0 & 3 & | & 1 \\ 2 & 1 & 0 & | & 3 \end{bmatrix}$

19. $\begin{bmatrix} 1 & -1 & 1 & | & -1 \\ 2 & 1 & 3 & | & 1 \\ 2 & 2 & 3 & | & 4 \end{bmatrix}$

20. $\begin{bmatrix} 0 & 1 & 3 & | & 2 \\ 2 & 1 & 1 & | & -1 \\ 1 & 3 & 2 & | & 1 \end{bmatrix}$

21. $\begin{bmatrix} 1 & -1 & 1 & 1 & | & 1 \\ 0 & 2 & 1 & 2 & | & 2 \\ 3 & 1 & 5 & 3 & | & 3 \end{bmatrix}$

22. $\begin{bmatrix} 2 & -1 & 3 & 4 & | & 3 \\ 5 & 3 & 4 & 6 & | & 4 \\ 3 & 2 & 5 & 1 & | & 5 \end{bmatrix}$

For Problems 23–30, write the augmented matrix for the system and solve by the Gauss-Jordan method.

23. $3x + 2y = 3$
$8x + 4y = 8$

24. $3x + y = 6$
$3x - 4y = -24$

25. $\begin{aligned} 3x + 2y + z &= 6 \\ 9x + 4y + 5z &= 12 \\ x - 2y + 3z &= 6 \end{aligned}$

26. $\begin{aligned} 2x - y + z &= 3 \\ x \quad\;\; + 3z &= 4 \\ -x + y + 2z &= 1 \end{aligned}$

27. $\begin{aligned} x + 2y + z + 2t &= 3 \\ 2x + 5y + 4z + 4t &= 8 \end{aligned}$

28. $\begin{aligned} 2x + 5y - 3z &= 6 \\ -3x + 2y + 4z &= -3 \\ 5x + 4y - 2z &= 7 \end{aligned}$

29. $\begin{aligned} -2x + 4y + 2z &= 6 \\ -x + 2y + z &= 3 \\ -x + 7y + 3z &= 7 \end{aligned}$

30. $\begin{aligned} x + y + z &= 3 \\ 2x - y \quad\;\; &= -1 \\ x \quad\;\; - z &= -3 \end{aligned}$

For Problems 31–38, solve the systems of equations using the Gauss-Jordan method. These systems of equations have an infinite number of solutions.

31. $\begin{aligned} x - y &= 3 \\ 2x - 2y &= 6 \end{aligned}$

32. $\begin{aligned} x_1 + x_2 + x_3 &= 1 \\ 5x_1 + 2x_2 - 2x_3 &= 3 \end{aligned}$

33. $\begin{aligned} u + 2v - w &= 0 \\ v - 3w &= 1 \end{aligned}$

34. $\begin{aligned} x + y - z &= 3 \\ 2x - y \quad\;\; &= 4 \\ -x - 4y + 3z &= -5 \end{aligned}$

35. $\begin{aligned} 2x - y + 3z &= 1 \\ x + y + z &= 4 \\ 2x - 4y + 4z &= -6 \end{aligned}$

36. $\begin{aligned} x - 2y + z + 3w + t &= 1 \\ 2x + 2y - z + 6w + 2t &= 2 \\ 3x + 4y + 2z - 9w - 3t &= 3 \end{aligned}$

37. $\begin{aligned} 2a + b + c + d &= 5 \\ -a + 2b - 2c + 3d &= 2 \\ 3a + 4b \quad\quad + 5d &= 12 \end{aligned}$

38. $\begin{aligned} x + y + z + t &= 2 \\ 2x \quad\; + 3z - t &= 4 \\ x - y \quad\; + t &= 5 \\ 3x - y + 3z \quad\; &= 9 \end{aligned}$

For Problems 39–44, solve the systems of equations by Gaussian elimination.

39. $\begin{aligned} x + y &= 7 \\ x - y &= 1 \end{aligned}$

40. $\begin{aligned} 3x - y &= 2 \\ x + 2y &= 3 \end{aligned}$

41. $\begin{aligned} x + 3y - 2z &= 2 \\ 3x - y - 2z &= 0 \\ 2x + 2y + z &= 5 \end{aligned}$

42. $\begin{aligned} 2x + 5y + 8z &= 15 \\ 3x - 4y - 3z &= -4 \\ 4x - 6y - 2z &= -4 \end{aligned}$

43. $\begin{aligned} x + y + z &= 1 \\ x + y + 2z &= 2 \end{aligned}$

44. $\begin{aligned} x \quad\; + z &= 1 \\ 2x + 2y - 5z &= 0 \end{aligned}$

Linear Systems in Business Management

The management of the Royal Caribbean cruise company plans on building a new luxury cruise ship. The plans are for

Royal Caribbean Cruise Line's new Song of America *at her home port in Miami. The billion-dollar cruise industry operates on a slim margin of profit, necessitating as efficient an operation as possible.*

the ship to contain single rooms, double rooms, and four-person suites. The staff necessary to maintain these three types of rooms is determined on the basis of one attendant for ten single rooms, five double rooms, or two four-person suites. For each of the configurations in Problems 45–49, determine

$$x = \text{number of single rooms}$$

$$y = \text{number of double rooms}$$

$$z = \text{number of four-person suites}$$

required in the ship. *Hint:* Each of Problems 45–49 uses the basic array in Table 16.

45. Number of rooms $= 50$
Staff size $\quad\quad\;\; = 10$
Number of guests $= 100$

46. Number of rooms $= 100$
Staff size $\quad\quad\;\; = 20$
Number of guests $= 200$

TABLE 16
Basic Array for Problems 45–49

	Single Room	Double Room	Four-Person Suite	Requirements
Room	1	1	1	*
Staff	1/10	1/5	1/2	*
Guests	1	2	4	*

47. Number of rooms = 300
Staff size   = 50
Number of guests = 500

48. Number of rooms = 200
Staff size   = 70
Number of guests = 700

49. Number of rooms = 40
Staff size   = 40
Number of guests = 120

50. Pharmaceutical Problem   Joe Smith takes a daily prescription composed of 8 units of A, 14 units of B, and 16 units of C. The generic drug capsules that contain these ingredients have the units of each of these components shown in Table 17.
How many capsules of each generic drug should Joe take daily?

51. Pharmaceutical Problem   Mary Chan takes a daily prescription composed of 13 units of A, 22 units of B, and 20 units of C. How many capsules of each generic drug described in Problem 50 should she take daily?

TABLE 17
Array for Problems 50 and 51

	Generic I	Generic II	Generic III	Requirements
A	2	1	3	8 units
B	3	2	5	14 units
C	2	4	2	16 units

52. Production Problem   A large paper company in Oregon has three plants in Eugene, Corvallis, and Monroe supplying newsprint to three daily newspapers: the *New York Star Ledger*, the *Washington Register*, and the *Atlanta Tribune*. Table 18 gives the percentage of output of each plant that is produced for each of the three newspapers as well as the weekly demands of each newspaper (for example, the 20% in the upper left-hand corner means that 20% of the paper produced at the Eugene plant is produced for the *New York Star Ledger*). What should be the weekly production in tons of each plant to meet the given demands?

53. Production Problem   The *New York Star Ledger* is having a strike and has canceled all orders of newsprint. Resolve Problem 52 with the new demands: *New York Star Ledger* 50 tons, *Washington Register* 50 tons, and *Atlanta Tribune* 55 tons.

TABLE 18
Array for Problems 52 and 53

	Eugene	Corvallis	Monroe	Demands
New York	20%	30%	50%	120 tons
Washington	30%	40%	30%	130 tons
Atlanta	30%	30%	40%	130 tons

1.5

Introduction to Matrices

PURPOSE

> We define an $m \times n$ matrix and show how to add, subtract, and multiply matrices. We also show how to represent systems of equations using matrix notation.

The Concept of a Matrix

In the previous section we used the augmented matrix to represent systems of equations when using the Gauss-Jordan method. There, however, we thought of matrices only as a compact way to write systems of equations by deleting the names of the variables. In this section we will show how matrices can be interpreted as "generalized numbers" in the sense that they can be added, subtracted, and multiplied. We begin with some basic definitions.

Definition of a Matrix

An $m \times n$ matrix is a rectangular array of mn numbers arranged in m rows and n columns:

$$A = \begin{bmatrix} a_{11} & a_{12} & a_{13} & \cdots & a_{1n} \\ a_{21} & a_{22} & a_{23} & \cdots & a_{2n} \\ \vdots & \vdots & & & \\ a_{m1} & a_{m2} & a_{m3} & \cdots & a_{mn} \end{bmatrix}$$

Matrices are generally denoted by capital letters, and the **elements** or entries in the matrix are denoted by lowercase letters. For example, the number appearing in the ith row and jth column of a matrix A would be called a_{ij}. Matrices are also classified according to their dimension, which simply indicates the number of rows and the number of columns in the matrix. For instance, a matrix having m rows and n columns is said to have dimension m by n and is referred to as an **$m \times n$ matrix**.

Example 1 ———————— Sample Matrices

(a) The matrix

$$\begin{bmatrix} 2 & 4 & 9 \\ 0 & -7 & 0 \\ 1 & 3 & 4 \end{bmatrix}$$

has dimension 3 by 3 and would be called a 3×3 matrix. Matrices with the same number of rows and columns are called **square matrices**.

(b) The matrix

$$\begin{bmatrix} 3 & 4 \\ 1 & 0 \\ 0 & 1 \end{bmatrix}$$

is a 3×2 matrix, since it has three rows and two columns.

(c) The matrix

$$\begin{bmatrix} 1 & 4 & 3 \end{bmatrix}$$

is a 1×3 matrix. Matrices with only one row are called **row matrices** or **row vectors.**

(d) The matrix

$$\begin{bmatrix} 4 \\ 1 \\ -3 \end{bmatrix}$$

is a 3×1 matrix. Matrices with only one column are called **column matrices** or **column vectors**. We would call this matrix a 3 by 1 column matrix.   □

Equality of Matrices

Two matrices with the same number of rows and columns are equal when all corresponding elements are the same. By this definition, the two matrices

$$\begin{bmatrix} 2 & 0 \\ 1 & 4 \end{bmatrix} \text{ and } \begin{bmatrix} 2 & 0 \\ 1 & 5 \end{bmatrix}$$

are not equal (although they do have the same dimension and three out of the four elements are the same). From the definition of equality of matrices, the statement

$$\begin{bmatrix} x \\ y \\ z \end{bmatrix} = \begin{bmatrix} 4 \\ 5 \\ 7 \end{bmatrix}$$

implies that $x = 4$, $y = 5$, and $z = 7$.

Example 2

Equality of Matrices Which pairs of the three matrices A, B, and C are equal?

$$A = \begin{bmatrix} 3 & 5 & 8 \\ 4 & 0 & -2 \\ 9 & 1 & 7 \end{bmatrix} \quad B = \begin{bmatrix} 3 & 5 & 8 \\ 4 & 0 & -2 \\ 9 & 1 & 7 \end{bmatrix} \quad C = \begin{bmatrix} 3 & 5 & 9 \\ 4 & 1 & 0 \\ 3 & 0 & 5 \end{bmatrix}$$

Solution First observe that all three matrices have the same dimension. Using the definition of equality, we have

- $A = B$, since all corresponding entries are the same.
- $A \neq C$, since at least one corresponding pair differs ($a_{13} = 8 \neq 9 = c_{13}$).
- $B \neq C$, since at least one corresponding pair differs ($b_{33} \neq c_{33}$). □

We now show how matrices can be used in a business setting.

Matrices in a Business Setting (Takach Shoe Co.)

The Takach Shoe Company stocks three brands of basketball sneakers—the Inflight, the Scrapper, and the Celtic—at each of its four retail outlets. The inventory levels at each store on the first of June are listed in Table 19.

TABLE 19
Present Inventory of Shoes

	Inflight	Scrapper	Celtic
Chicago	75	90	60
Cleveland	80	100	75
Denver	60	50	120
Portland	95	200	150

The numbers in this table can be interpreted as elements in an array or matrix, called INVENTORY.

$$\text{INVENTORY} = \begin{bmatrix} 75 & 90 & 60 \\ 80 & 100 & 75 \\ 60 & 50 & 120 \\ 95 & 200 & 150 \end{bmatrix} \begin{array}{l} \text{Chicago} \\ \text{Cleveland} \\ \text{Denver} \\ \text{Portland} \end{array}$$

(columns: Inflight Scrapper Celtic)

The company determines new inventory levels at the end of the month. These totals will depend on the number of shoes they started the month with, the number of shoes ordered during that month, and the number of shoes sold during that period. Suppose sales and the number of sneakers ordered during this month are given in the following 4×3 matrices:

$$\text{ORDERS} = \begin{bmatrix} 25 & 45 & 50 \\ 60 & 50 & 40 \\ 25 & 40 & 80 \\ 75 & 90 & 110 \end{bmatrix} \begin{array}{l} \text{Chicago} \\ \text{Cleveland} \\ \text{Denver} \\ \text{Portland} \end{array}$$

(columns: Inflight Scrapper Celtic)

$$\text{SALES} = \begin{bmatrix} 40 & 25 & 70 \\ 75 & 60 & 35 \\ 35 & 20 & 60 \\ 75 & 90 & 110 \end{bmatrix} \begin{array}{l} \text{Chicago} \\ \text{Cleveland} \\ \text{Denver} \\ \text{Portland} \end{array}$$

(columns: Inflight Scrapper Celtic)

...omputerized systems monitor the inventories ...most large companies.

For example, the Chicago outlet sold 40 Inflights in June, and Denver ordered 40 Scrappers. If we wanted to find the new inventory levels for each type of shoe at each outlet at the beginning of the next month (July), we would compute the following new matrix:

$$\begin{array}{l} \text{NEW} \\ \text{INVENTORY} \end{array} = \begin{bmatrix} 75 + 25 - 40 & 90 + 45 - 25 & 60 + 50 - 70 \\ 80 + 60 - 75 & 100 + 50 - 60 & 75 + 40 - 35 \\ 60 + 25 - 35 & 50 + 40 - 20 & 120 + 80 - 60 \\ 95 + 75 - 75 & 200 + 90 - 90 & 150 + 110 - 110 \end{bmatrix}$$

(columns: Inflight Scrapper Celtic)

old inventory   new orders   sales

$$= \begin{bmatrix} 60 & 110 & 40 \\ 65 & 90 & 80 \\ 50 & 70 & 140 \\ 95 & 200 & 150 \end{bmatrix} \begin{array}{l} \text{Chicago} \\ \text{Cleveland} \\ \text{Denver} \\ \text{Portland} \end{array}$$

(columns: Inflight Scrapper Celtic)

For large retail companies like Sears, Roebuck and Co., J. C. Penney, and K-Mart, which have hundreds of outlets and thousands of products, the computation of exact inventory levels at any time is an important management tool. Management decisions are often based on knowing exact inventory levels of every product in every region of the country.

In the example above, we computed the new inventory level, which are elements of the matrix NEW INVENTORY, knowing the three matrices

INVENTORY   (old inventory levels)

SALES   (monthly sales)

ORDERS   (new orders)

In the language of matrix arithmetic, what we did was add and subtract matrices. This brings us to the definitions of these important operations on matrices.

Matrix Addition and Subtraction

The **sum** of two $m \times n$ matrices A and B, denoted $A + B$, is the $m \times n$ matrix that consists of entries formed by taking the sum of the corresponding entries of A and B. If a_{ij} and b_{ij} are the ijth entries of the matrices A and B, respectively, then the ijth entry of the sum is $a_{ij} + b_{ij}$.

The **difference** of two $m \times n$ matrices A and B, denoted $A - B$, is the $m \times n$ matrix that consists of entries formed by taking the difference between corresponding entries of A and B. If a_{ij} and b_{ij} are the ijth entries of the matrices A and B, respectively, then the ijth entry of the difference is $a_{ij} - b_{ij}$.

Note that matrices must have the same dimension in order to be added or subtracted.

In the language of matrix arithmetic we can now rephrase the above inventory problem by stating that the new inventory level, given by the matrix NEW INVENTORY, can be computed from the matrix formula:

NEW INVENTORY =   INVENTORY +   ORDER   −   SALES

$$
= \begin{bmatrix} 75 & 90 & 60 \\ 80 & 100 & 75 \\ 60 & 50 & 120 \\ 95 & 200 & 150 \end{bmatrix} + \begin{bmatrix} 25 & 45 & 50 \\ 60 & 50 & 40 \\ 25 & 40 & 80 \\ 75 & 90 & 110 \end{bmatrix} - \begin{bmatrix} 40 & 25 & 70 \\ 75 & 60 & 35 \\ 35 & 20 & 60 \\ 75 & 90 & 110 \end{bmatrix}
$$

$$
= \begin{bmatrix} 60 & 110 & 40 \\ 65 & 90 & 80 \\ 50 & 70 & 140 \\ 95 & 200 & 150 \end{bmatrix}
$$

Example 3

Addition and Subtraction of Matrices

(a) $\begin{bmatrix} 2 & 1 \\ 0 & 2 \\ 3 & 4 \end{bmatrix} + \begin{bmatrix} -1 & 0 \\ 4 & 7 \\ 3 & 9 \end{bmatrix} = \begin{bmatrix} 1 & 1 \\ 4 & 9 \\ 6 & 13 \end{bmatrix}$

(b) $\begin{bmatrix} 2 & 1 \\ 0 & 3 \\ 0 & 0 \end{bmatrix} - \begin{bmatrix} 4 & 7 \\ 3 & -3 \\ 0 & 1 \end{bmatrix} = \begin{bmatrix} -2 & -6 \\ -3 & 6 \\ 0 & -1 \end{bmatrix}$

(c) $\begin{bmatrix} 1 & 2 \\ 0 & 4 \end{bmatrix} + \begin{bmatrix} 2 & 3 \end{bmatrix}$ (not the same dimension, cannot add)

(d) $\begin{bmatrix} 3 & 7 \end{bmatrix} + \begin{bmatrix} 2 & 7 \end{bmatrix} = \begin{bmatrix} 5 & 14 \end{bmatrix}$

(e) $\begin{bmatrix} 1 \\ 0 \\ 1 \end{bmatrix} + \begin{bmatrix} 5 \\ 6 \\ 4 \end{bmatrix} = \begin{bmatrix} 6 \\ 6 \\ 5 \end{bmatrix}$

□

Before moving to other matrix operations we state some useful rules for matrix addition.

Rules For Matrix Addition

Let A, B, and C be any $m \times n$ matrices. The following three properties always hold:

1. $A + B = B + A$ (commutative rule)
2. $(A + B) + C = A + (B + C)$ (associative rule)
3. $A + 0_{mn} = A$ where 0_{mn} is the $m \times n$ matrix with all elements zero, called the **zero matrix**.

To motivate the matrix operation of multiplication of a matrix times a constant, let us return to the Takach Shoe Company example. At the end of each month, suppose the management sets sales goals for the coming month. Suppose, too, that next month will be the height of the tourist season, which is the best sales period of the year. The goal of the company is to increase sales by 20% over the previous month. This means that next month's goal is determined by multiplying by 1.2 each entry of last month's sales, as represented by the matrix SALES. If we define the matrix GOALS, whose elements are next month's sales goals, then we can represent this by the matrix equation

GOALS = 1.2 SALES

$$
= 1.2 \begin{array}{c}
\begin{array}{ccc} \text{Inflight} & \text{Scrapper} & \text{Celtic} \end{array} \\
\begin{bmatrix} 40 & 25 & 70 \\ 75 & 60 & 35 \\ 35 & 20 & 60 \\ 75 & 90 & 110 \end{bmatrix}
\begin{array}{l} \text{Chicago} \\ \text{Cleveland} \\ \text{Denver} \\ \text{Portland} \end{array}
\end{array}
$$

$$
= \begin{array}{c}
\begin{array}{ccc} \text{Inflight} & \text{Scrapper} & \text{Celtic} \end{array} \\
\begin{bmatrix} 48 & 30 & 84 \\ 90 & 72 & 42 \\ 42 & 24 & 72 \\ 90 & 108 & 132 \end{bmatrix}
\begin{array}{l} \text{Chicago} \\ \text{Cleveland} \\ \text{Denver} \\ \text{Portland} \end{array}
\end{array}
$$

We see from the matrix above that the company would like to sell 108 Scrappers in Portland next month.

This motivates the following definition.

Matrix Multiplication by a Constant

Let A be a matrix and k a real number. Define $k \cdot A$ as the matrix whose entries are formed by multiplying each entry of A by k.

Example 4

Multiplication by a Constant

(a) $2 \begin{bmatrix} 2 & 4 \\ 1 & 0 \\ 3 & 3 \end{bmatrix} = \begin{bmatrix} 2 \cdot 2 & 2 \cdot 4 \\ 2 \cdot 1 & 2 \cdot 0 \\ 2 \cdot 3 & 2 \cdot 3 \end{bmatrix} = \begin{bmatrix} 4 & 8 \\ 2 & 0 \\ 6 & 6 \end{bmatrix}$

(b) $3 \begin{bmatrix} 3 & 6 \\ 6 & 2 \\ 3 & 6 \end{bmatrix} = \begin{bmatrix} 3 \cdot 3 & 3 \cdot 6 \\ 3 \cdot 6 & 3 \cdot 2 \\ 3 \cdot 3 & 3 \cdot 6 \end{bmatrix} = \begin{bmatrix} 9 & 18 \\ 18 & 6 \\ 9 & 18 \end{bmatrix}$ ☐

Matrix Multiplication

We now show how to multiply two matrices. This operation is not as intuitive as the matrix operations of addition and subtraction, but it does arise naturally in many problems. We will see instances of its use later in this section and in the problem set. Before defining matrix multiplication, however, we first define the dot product of a row matrix times a column matrix, which plays a central role in the operation of the multiplication of matrices.

Dot Product of a Row and Column Matrix

Let A be a $1 \times n$ row matrix and B an $n \times 1$ column matrix. The **dot product** of A and B, denoted by $A \cdot B$, is the real number computed by

$$\begin{bmatrix} a_{11} & a_{12} & \cdots & a_{1n} \end{bmatrix} \begin{bmatrix} b_{11} \\ b_{21} \\ \vdots \\ b_{n1} \end{bmatrix}$$

$$= a_{11}b_{11} + a_{12}b_{21} + \cdots + a_{1n}b_{n1} \qquad \text{(a real number)}$$

Example 5

Computation of a Dot Product   Let A be the 1×3 row matrix and B the 3×1 column matrix given as follows:

$$A = \begin{matrix} \begin{bmatrix} 3 & 6 & 8 \end{bmatrix} \\ a_{11} \quad a_{12} \quad a_{13} \end{matrix} \qquad B = \begin{matrix} \begin{bmatrix} 9 \\ 1 \\ 2 \end{bmatrix} \begin{matrix} b_{11} \\ b_{21} \\ b_{31} \end{matrix} \end{matrix}$$

Find the dot product $A \cdot B$.

Solution By direct computation we have

$$A \cdot B = a_{11}b_{11} + a_{12}b_{21} + a_{13}b_{31}$$
$$= (3)(9) + (6)(1) + (8)(2)$$
$$= 49 \qquad \square$$

Multiplication of Matrices

Two matrices are compatible if the number of *columns* of A is the same as the number of *rows* of B. It is only when two matrices A and B are compatible that we can define the matrix product AB.

Definition of Matrix Multiplication

Let A be an $m \times p$ matrix and B an $p \times n$ matrix. The product of A times B, denoted AB, is a matrix C of dimension $m \times n$ whose ijth entry c_{ij} is the dot product of the ith row of A times the jth column of B.

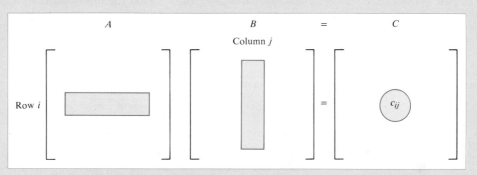

Stated in algebraic language, the ijth element c_{ij} of the product $C = AB$ is determined by the formula

$$c_{ij} = a_{i1}b_{1j} + a_{i2}b_{2j} + a_{i3}b_{3j} + \cdots + a_{ip}b_{pj}$$

for $i = 1, 2, \ldots, m$ and $j = 1, 2, \ldots, n$.

Example 6

Matrix Multiplication Find the product AB of the matrices

$$A = \begin{bmatrix} 2 & 1 & 7 \\ 3 & 4 & 9 \\ 1 & 0 & 0 \end{bmatrix} \qquad B = \begin{bmatrix} 1 & 3 \\ 4 & 5 \\ 5 & 8 \end{bmatrix}$$

Solution Observe that the matrices are compatible, since A has three columns and B has three rows. The product matrix, which we will call C, will be a 3×2 matrix.

The entry c_{11} in the product matrix is found by taking the dot product of the first row of A and the first column of B. All other entries c_{ij} are found by taking the dot product of the ith row of A and the jth column of B. For example, the element c_{32} is the dot product of the third row of A and the second column of B. The product $C = AB$ is computed below:

$$C = \begin{bmatrix} 2 & 1 & 7 \\ 3 & 4 & 9 \\ 1 & 0 & 0 \end{bmatrix} \begin{bmatrix} 1 & 3 \\ 4 & 5 \\ 5 & 8 \end{bmatrix} = \begin{bmatrix} [2 \ 1 \ 7]\begin{bmatrix}1\\4\\5\end{bmatrix} & [2 \ 1 \ 7]\begin{bmatrix}3\\5\\8\end{bmatrix} \\ [3 \ 4 \ 9]\begin{bmatrix}1\\4\\5\end{bmatrix} & [3 \ 4 \ 9]\begin{bmatrix}3\\5\\8\end{bmatrix} \\ [1 \ 0 \ 0]\begin{bmatrix}1\\4\\5\end{bmatrix} & [1 \ 0 \ 0]\begin{bmatrix}3\\5\\8\end{bmatrix} \end{bmatrix} = \begin{bmatrix} 41 & 67 \\ 64 & 101 \\ 1 & 3 \end{bmatrix}$$

$$A \qquad B \qquad\qquad\qquad\qquad\qquad\qquad\qquad\qquad (3 \times 2)$$
$$(3 \times 3) \quad (3 \times 2)$$

compatible

└─dimension of ─┘
product matrix   □

Example 7

Matrix Multiplication   Find the following matrix products:

(a) $\begin{bmatrix} 2 & 3 \\ 4 & 8 \end{bmatrix}\begin{bmatrix} 1 & 3 \\ 2 & 7 \end{bmatrix}$ (b) $\begin{bmatrix} 1 & 0 & 3 \\ 2 & 1 & 0 \end{bmatrix}\begin{bmatrix} 3 & 5 \\ 4 & 0 \\ 2 & 9 \end{bmatrix}$ (c) $\begin{bmatrix} 2 \\ 5 \\ 1 \end{bmatrix}[4 \ 2 \ 8]$

(d) $\begin{bmatrix} 2 & 4 \\ 1 & -2 \\ 3 & 5 \end{bmatrix}\begin{bmatrix} 3 & 8 & 0 \\ 1 & 2 & 4 \end{bmatrix}$ (e) $\begin{bmatrix} 2 & 4 & 5 \\ 0 & 0 & 2 \end{bmatrix}\begin{bmatrix} 3 & 4 \\ 4 & 2 \end{bmatrix}$

Solution   Using the definition of matrix multiplication, we find the following products:

(a) $\begin{bmatrix} 2 & 3 \\ 4 & 8 \end{bmatrix}\begin{bmatrix} 1 & 3 \\ 2 & 7 \end{bmatrix} = \begin{bmatrix} (2)(1)+(3)(2) & (2)(3)+(3)(7) \\ (4)(1)+(8)(2) & (4)(3)+(8)(7) \end{bmatrix}$

$$= \begin{bmatrix} 8 & 27 \\ 20 & 68 \end{bmatrix}$$

Dimension check: $(2 \times 2)(2 \times 2) \longrightarrow (2 \times 2)$

(b) $\begin{bmatrix} 1 & 0 & 3 \\ 2 & 1 & 0 \end{bmatrix}\begin{bmatrix} 3 & 5 \\ 4 & 0 \\ 2 & 9 \end{bmatrix} = \begin{bmatrix} 9 & 32 \\ 10 & 10 \end{bmatrix}$

Dimension check: $(2 \times 3)(3 \times 2) \longrightarrow (2 \times 2)$

(c) $\begin{bmatrix} 2 \\ 5 \\ 1 \end{bmatrix}[4 \ 2 \ 8] = \begin{bmatrix} 8 & 4 & 16 \\ 20 & 10 & 40 \\ 4 & 2 & 8 \end{bmatrix}$

Dimension check: $(3 \times 1)(1 \times 3) \longrightarrow (3 \times 3)$

(d) $\begin{bmatrix} 2 & 4 \\ 1 & -2 \\ 3 & 5 \end{bmatrix} \begin{bmatrix} 3 & 8 & 0 \\ 1 & 2 & 4 \end{bmatrix} = \begin{bmatrix} 10 & 24 & 16 \\ 1 & 4 & -8 \\ 14 & 34 & 20 \end{bmatrix}$

Dimension check: $(3 \times 2)(2 \times 3) \longrightarrow (3 \times 3)$

(e) $\begin{bmatrix} 2 & 4 & 5 \\ 0 & 0 & 2 \end{bmatrix} \begin{bmatrix} 3 & 4 \\ 4 & 2 \end{bmatrix}$ not compatible

Dimension check: $(2 \times 3)(2 \times 2)$ is not compatible.   □

An interesting fact about the product AB of two matrices A and B is that AB is not always equal to the product BA. That is, matrix multiplication is not commutative. Matrix products do, however, satisfy many of the same types of properties as the products of real numbers. We list a few of them here.

Properties of Matrix Multiplication

If all the products and sums are defined for the following matrices A, B, and C, then the following four rules of arithmetic for matrices hold:

1. $A(BC) = (AB)C$ (associative law)
2. $A(B + C) = AB + AC$ (left distributive law)
3. $(A + B)C = AC + BC$ (right distributive law)
4. $AI_n = I_nA = A$ where I_n is the $n \times n$ matrix, called the identity matrix, that has 1's down the diagonal and 0's elsewhere; that is,

$$I_n = \begin{bmatrix} 1 & 0 & 0 & \cdots & & 0 & 0 \\ 0 & 1 & 0 & & & & 0 \\ 0 & 0 & 1 & \cdots & 0 & & 0 \\ \vdots & & & 1 & & & \\ & & 0 & \cdots & 1 & & \vdots \\ 0 & & & & & 1 & 0 \\ 0 & 0 & 0 & \cdots & & 0 & 1 \end{bmatrix}$$

We now see how the concept of the matrix product arises in a business setting.

Matrix Multiplication and Tennis Rackets

A sports equipment company has four major outlets in New York, Chicago, Dallas, and Los Angeles. Its current inventory in tennis rackets is given by Table 20.

TABLE 20
Current Inventory of Tennis
Rackets at Four Outlets

Outlet	Type		
	Forehander	Overhander	Backhander
New York	60	110	40
Chicago	65	90	80
Dallas	50	70	140
Los Angeles	95	210	130

The cost of each racket is listed as

Forehander	$41.80
Overhander	$51.20
Backhander	$46.70

The current inventory and racket costs can be denoted by the two matrices COST and INVENTORY.

$$COST = \begin{bmatrix} 41.80 \\ 51.20 \\ 46.70 \end{bmatrix} \begin{matrix} Forehander \\ Overhander \\ Backhander \end{matrix}$$

$$INVENTORY = \begin{bmatrix} 60 & 110 & 40 \\ 65 & 90 & 80 \\ 50 & 70 & 140 \\ 95 & 210 & 130 \end{bmatrix} \begin{matrix} New York \\ Chicago \\ Dallas \\ Los Angeles \end{matrix}$$

with columns Forehander, Overhander, Backhander.

With a little thought you can see that the total value of the inventories at each of the four locations are the values in the matrix product VALUE, determined by

$$VALUE = INVENTORY \cdot COST$$

$$= \begin{bmatrix} 60 & 110 & 40 \\ 65 & 90 & 80 \\ 50 & 70 & 140 \\ 95 & 210 & 130 \end{bmatrix} \begin{bmatrix} 41.80 \\ 51.20 \\ 46.70 \end{bmatrix}$$

$$= \begin{bmatrix} (60)(41.80) + (110)(51.20) + (40)(46.70) \\ (65)(41.80) + (90)(51.20) + (80)(46.70) \\ (50)(41.80) + (70)(51.20) + (140)(46.70) \\ (95)(41.80) + (210)(51.20) + (130)(46.70) \end{bmatrix}$$

$$= \begin{bmatrix} \$10,008.00 \\ \$11,061.00 \\ \$12,212.00 \\ \$20,794.00 \end{bmatrix} \begin{matrix} New York \\ Chicago \\ Dallas \\ Los Angeles \end{matrix}$$

We now give the weekly sales at each location for each kind of racket in Table 21.

TABLE 21
Number of Rackets Sold at
Each Outlet

Outlet	Forehander	Overhander	Backhander
New York	60	35	90
Chicago	100	70	40
Dallas	50	30	70
Los Angeles	80	100	140

To find the total cost of goods sold at each of the four outlets, we multiply the above sales array SALES formed from Table 21 times PRICE, getting

$$\text{TOTAL SALES} = \text{SALES} \cdot \text{PRICE}$$

$$= \begin{bmatrix} 60 & 35 & 90 \\ 100 & 70 & 40 \\ 50 & 30 & 70 \\ 80 & 100 & 140 \end{bmatrix} \begin{bmatrix} \$41.80 \\ \$51.20 \\ \$46.70 \end{bmatrix}$$

$$= \begin{bmatrix} (60)(41.80) + (35)(51.20) + (90)(46.70) \\ (100)(41.80) + (70)(51.20) + (40)(46.70) \\ (50)(41.80) + (30)(51.20) + (70)(46.70) \\ (80)(41.80) + (100)(51.20) + (140)(46.70) \end{bmatrix}$$

$$= \begin{bmatrix} \$8{,}503.00 \\ \$9{,}632.00 \\ \$6{,}895.00 \\ \$15{,}002.00 \end{bmatrix} \begin{matrix} \text{New York} \\ \text{Chicago} \\ \text{Dallas} \\ \text{Los Angeles} \end{matrix}$$

Matrix Representation of Systems of Equations

We close this section by showing how to write a system of equations as a single matrix equation. This is a convenient and compact way for writing linear systems. We illustrate this idea for three equations and three unknowns. Consider the following linear system of equations with three unknowns:

$$a_{11}x_1 + a_{12}x_2 + a_{13}x_3 = b_1$$
$$a_{21}x_1 + a_{22}x_2 + a_{23}x_3 = b_2$$
$$a_{31}x_1 + a_{32}x_2 + a_{33}x_3 = b_3$$

We start by writing the 3×1 column matrix

$$\begin{bmatrix} a_{11}x_1 + a_{12}x_2 + a_{13}x_3 \\ a_{21}x_1 + a_{22}x_2 + a_{23}x_3 \\ a_{31}x_1 + a_{32}x_2 + a_{33}x_3 \end{bmatrix}$$

as the product of two matrices:

$$\underset{(3 \times 1)}{\begin{bmatrix} a_{11}x_1 + a_{12}x_2 + a_{13}x_3 \\ a_{21}x_1 + a_{22}x_2 + a_{23}x_3 \\ a_{31}x_1 + a_{32}x_2 + a_{33}x_3 \end{bmatrix}} = \underset{(3 \times 3)}{\begin{bmatrix} a_{11} & a_{12} & a_{13} \\ a_{21} & a_{22} & a_{23} \\ a_{31} & a_{32} & a_{33} \end{bmatrix}} \underset{(3 \times 1)}{\begin{bmatrix} x_1 \\ x_2 \\ x_3 \end{bmatrix}}$$

Hence the linear system

$$a_{11}x_1 + a_{12}x_2 + a_{13}x_3 = b_1$$
$$a_{21}x_1 + a_{22}x_2 + a_{23}x_3 = b_2$$
$$a_{31}x_1 + a_{32}x_2 + a_{33}x_3 = b_3$$

is the same as the matrix equation

$$\begin{bmatrix} a_{11} & a_{12} & a_{13} \\ a_{21} & a_{22} & a_{23} \\ a_{31} & a_{32} & a_{33} \end{bmatrix} \begin{bmatrix} x_1 \\ x_2 \\ x_3 \end{bmatrix} = \begin{bmatrix} b_1 \\ b_2 \\ b_3 \end{bmatrix}$$

or

$$Ax = b$$

where

$$A = \begin{bmatrix} a_{11} & a_{12} & a_{13} \\ a_{21} & a_{22} & a_{23} \\ a_{31} & a_{32} & a_{33} \end{bmatrix} \qquad x = \begin{bmatrix} x_1 \\ x_2 \\ x_3 \end{bmatrix} \qquad b = \begin{bmatrix} b_1 \\ b_2 \\ b_3 \end{bmatrix}$$

Example 8

Matrix Representations The following systems of equations are written in both algebraic and matrix form:

Algebraic Form	Matrix Form

(a) $\quad 2x + 3y = 4$
    $\quad\quad x - y = 7$
$$\begin{bmatrix} 2 & 3 \\ 1 & -1 \end{bmatrix} \begin{bmatrix} x \\ y \end{bmatrix} = \begin{bmatrix} 4 \\ 7 \end{bmatrix}$$

(b) $\quad 7x_1 + 5x_2 = 7$
    $\quad 5x_1 + x_2 = 0$
$$\begin{bmatrix} 7 & 5 \\ 5 & 1 \end{bmatrix} \begin{bmatrix} x_1 \\ x_2 \end{bmatrix} = \begin{bmatrix} 7 \\ 0 \end{bmatrix}$$

(c) $\quad 4u + 4v - 9w = 0$
    $\quad 2u - 9v = 4$
    $\quad\quad v + w = -3$
$$\begin{bmatrix} 4 & 4 & -9 \\ 2 & -9 & 0 \\ 0 & 1 & 1 \end{bmatrix} \begin{bmatrix} u \\ v \\ w \end{bmatrix} = \begin{bmatrix} 0 \\ 4 \\ -3 \end{bmatrix}$$

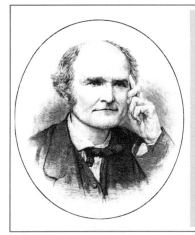

HISTORICAL NOTE

Matrices had their origins in England in the nineteenth century with the work of the brilliant Arthur Cayley (1821–1895). He invented the subject in 1858 and spent the rest of his life working on related subjects. Originally, matrices were thought to have no practical value and were viewed by many as only interesting games. Not until 1925, 30 years after Cayley's death, did physicists find a use for matrices in quantum mechanics. Since then, engineers, biologists, economists, computer scientists, and many others have used matrices to great advantage. Some people have gone so far as to say that matrices may be the most important single discovery in mathematics in the past 100 years.

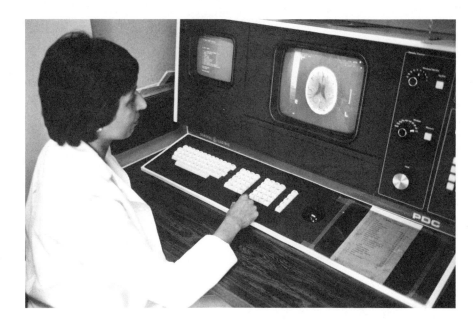

A patient undergoing diagnostic imaging for a brain tumor is probably unaware that a computer is multiplying thousands of matrices in order to construct the image of the tumor.

Problems

1. Given the matrices A, B, and C,

$$A = \begin{bmatrix} -2 & 1 \\ 2 & -1 \end{bmatrix} \qquad B = \begin{bmatrix} 2 & 1 & 3 \\ 1 & 4 & 2 \end{bmatrix}$$

$$C = \begin{bmatrix} -1 & 2 & -2 \\ 3 & 1 & 2 \end{bmatrix}$$

  (a) Determine the dimension of each matrix.
  (b) Calculate the sum of any pair of matrices for which the operation is defined.
  (c) Calculate $3A$, $-2B$, $2A - B$, and $-C$.

2. Write out the 3×4 matrix A whose elements are

$$a_{ij} = \begin{cases} 3i + 2j & i > j \\ i - j & i = j \\ i + j & i < j \end{cases}$$

Solve for the variables in the matrix equations in Problems 3–6.

3. $\begin{bmatrix} y & 2z \\ x & 3w \end{bmatrix} = \begin{bmatrix} 4 & 6 \\ 0 & 7 \end{bmatrix}$

4. $\begin{bmatrix} 6 & 2 \\ x & 7 \end{bmatrix} = \begin{bmatrix} 6 & 2 \\ 6 & 7 \end{bmatrix}$

5. $\begin{bmatrix} x \\ y \end{bmatrix} + \begin{bmatrix} 3 \\ 5 \end{bmatrix} = \begin{bmatrix} 1 \\ 2 \end{bmatrix}$

6. $\begin{bmatrix} 2 & -1 & 2 \\ 3 & 0 & 4 \end{bmatrix} + \begin{bmatrix} 4 & 1 & 3 \\ 8 & -3 & -2 \end{bmatrix} = \begin{bmatrix} a & b & c \\ d & e & f \end{bmatrix}$

Solve for x, y, and z in Problems 7–8.

7. $\begin{bmatrix} 3 & x \\ 2 & y \end{bmatrix} + \begin{bmatrix} x & -y \\ 3 & 5 \end{bmatrix} = \begin{bmatrix} 3 & 1 \\ 5 & 4 \end{bmatrix}$

8. $\begin{bmatrix} 3 \\ 1 \\ x \end{bmatrix} + \begin{bmatrix} 2 \\ y \\ 5 \end{bmatrix} = \begin{bmatrix} z \\ 4 \\ 2 \end{bmatrix}$

9. Given the matrices

$$A = \begin{bmatrix} 1 & 0 & 2 & -1 \\ 4 & 1 & 3 & 0 \\ 0 & -1 & 3 & 8 \\ 1 & 1 & 2 & 2 \end{bmatrix}$$

$$B = \begin{bmatrix} 2 & -4 & 1 & 3 \\ 4 & 0 & 4 & 5 \\ 5 & 0 & 0 & 3 \\ 9 & 4 & 1 & 8 \end{bmatrix}$$

find
  (a) $A + B$ (b) $A - B$ (c) $3A + 4B$

10. Given the matrices

$$A = \begin{bmatrix} 1 & 2 \\ 3 & 0 \\ 6 & 1 \\ 0 & 1 \end{bmatrix} \qquad B = \begin{bmatrix} 1 & 1 \\ 4 & 4 \\ 0 & 0 \\ 6 & 6 \end{bmatrix} \qquad C = \begin{bmatrix} 9 & 4 \\ 3 & -1 \\ 1 & 1 \\ 9 & 2 \end{bmatrix}$$

calculate
(a) $A + B$ (b) $A - B$
(c) $2A$ (d) $-3B$
(e) $2A + 3B$ (f) $A + (B - C)$
(g) $(A + B) + C$ (h) $(2A - B) + 3C$

11. Given the dimensions of the following matrices, determine whether the indicated multiplication can be performed. If matrix multiplication is defined, determine the dimension of the resulting matrix

$$A: 2 \times 3 \quad B: 3 \times 1 \quad C: 2 \times 5$$
$$D: 4 \times 3 \quad E: 3 \times 2 \quad F: 2 \times 3$$

(a) AE (b) FA
(c) EC (d) FBA
(e) FC (f) EA
(g) DB (h) AEF

For Problems 12–15, determine the matrix product.

12. $\begin{bmatrix} 1 & -1 \\ 3 & 2 \end{bmatrix}\begin{bmatrix} -2 & 1 \\ 3 & 4 \end{bmatrix}$

13. $\begin{bmatrix} 1 & -1 & 2 \end{bmatrix}\begin{bmatrix} 2 & 3 \\ 0 & 0 \\ 1 & 3 \end{bmatrix}$

14. $\begin{bmatrix} -2 & 4 \\ 1 & 5 \end{bmatrix}\begin{bmatrix} 3 & 1 & 4 \\ 0 & 2 & 3 \end{bmatrix}$

15. $\begin{bmatrix} 4 & 4 & 4 \\ 0 & 0 & 1 \\ 1 & 2 & 3 \end{bmatrix}\begin{bmatrix} 1 & 1 & 1 \\ 2 & 4 & 3 \\ 2 & 5 & 2 \end{bmatrix}$

16. Given the matrices

$$A = \begin{bmatrix} 1 & 2 & -2 \\ 3 & 1 & 4 \end{bmatrix}$$

$$B = \begin{bmatrix} 1 & 3 \\ 1 & 4 \\ 0 & -1 \end{bmatrix}$$

$$C = \begin{bmatrix} 2 & 4 \\ 1 & 0 \\ 3 & 4 \end{bmatrix}$$

find
(a) $A(B + C)$ (b) AB (c) AC

17. Show that $A(B + C) = AB + AC$ for the following matrices:

$$A = \begin{bmatrix} -1 & 1 \\ 6 & 4 \\ 2 & 3 \end{bmatrix} \quad B = \begin{bmatrix} 1 & 1 \\ 2 & 2 \end{bmatrix} \quad C = \begin{bmatrix} 1 & 0 \\ 0 & 1 \end{bmatrix}$$

18. Use the matrix

$$A = \begin{bmatrix} 0 & 1 & 0 \\ 0 & 0 & 1 \\ 1 & 2 & -1 \end{bmatrix}$$

to calculate

$$A^3 + A^2 - 2A$$

where $A^2 = A \cdot A$ and $A^3 = A \cdot A^2$.

19. Use the matrix

$$A = \begin{bmatrix} 0 & 1 & 0 \\ -1 & 4 & 5 \\ 1 & -3 & 3 \end{bmatrix}$$

to calculate $(A - I)^3$ where I_3 is the 3×3 identity matrix:

$$I_3 = \begin{bmatrix} 1 & 0 & 0 \\ 0 & 1 & 0 \\ 0 & 0 & 1 \end{bmatrix}$$

20. Use the matrix

$$A = \begin{bmatrix} 2 & -3 & -5 \\ -1 & 4 & 5 \\ 1 & -3 & -4 \end{bmatrix}$$

to calculate A^2 and A^3. Can you determine A^n for $n > 3$?

21. Prove the following matrix identities for arbitrary 2×2 matrices A, B, and C:

$$A = \begin{bmatrix} a_{11} & a_{12} \\ a_{21} & a_{22} \end{bmatrix} \quad B = \begin{bmatrix} b_{11} & b_{12} \\ b_{21} & b_{22} \end{bmatrix} \quad C = \begin{bmatrix} c_{11} & c_{12} \\ c_{21} & c_{22} \end{bmatrix}$$

(a) $A(BC) = (AB)C$
(b) $A(B + C) = AB + AC$
(c) $(A + B)C = AC + BC$
(d) $k(A + B) = kA + kB$ (k any real numbers)
(e) $(r + s)A = rA + sA$ (r, s any real numbers)

For Problems 22–25, represent the systems of equations in matrix form.

22. $2x - y = 3$
    $3x + 4y = 0$

23. $3w + 3x + 2y + z = 1$
    $w - 4x \quad\quad - z = 0$
    $x + 3y \quad\quad = 2$
    $6w + x \quad\quad + z = 3$

24. $x + y = 3$
    $x - y = 4$
    $2x + y = 0$
    $x + y = 1$

25. $u + 4v + w = 1$
    $u - v + 3w = 0$

26. For the 2×2 matrices A and B:

$$A = \begin{bmatrix} a & b \\ b & a \end{bmatrix} \quad B = \begin{bmatrix} c & d \\ d & c \end{bmatrix}$$

prove that $AB = BA$. Although $AB \neq BA$ in general, in any individual instance it still may be true that $AB = BA$.

Matrices in Business Management

27. **Land Values** The total assets of a company are based on the table of values in Table 22.

TABLE 22
Table of Values for Problem 27

Item	Value/Unit
Land	7 million dollars
Labor	3 million dollars
Capital	14 million dollars
Raw materials	4 million dollars

Using matrix operations, calculate the total monetary value of a company with the assets in Table 23. Also calculate the value of each asset for each of the next two years if the value of each unit is projected to increase at a rate of 8% a year.

TABLE 23
Assets of the Company in Problem 27

Item	Units
Land	4
Labor	2
Capital	3
Raw materials	6

28. Total Cost Ajan Inc. manufactures three products used in the production of cars. Table 24 gives the number of units of each input for each product. Table 25 gives the cost per unit for the inputs.

Using matrix operations, calculate the total cost of the inputs for 50 units of product 1, 75 units of product 2, and 100 units of product 3.

29. Building Construction A building contractor accepts summer orders for 135 houses, 3 condominiums, and 1 high-rise building. The construction materials (in appropriate units) that go into each of these buildings are listed in Table 26. How much of each raw material will be needed for all of the summer contracts?

TABLE 26
Construction Materials for Buildings in Problem 29

	House	Condominium	High-Rise
Lumber	10	400	500
Glass	5	150	1000
Steel	0	50	2000
Concrete	0	100	1000
Labor	20	1000	5000

Matrices in the Social Sciences

30. Voting Problem Suppose the electorate of the city of Buffalo, New York, has the following percentages of residents registered in Party X and Party Y in each of the

TABLE 24
Units of Input for the Three Products in Problem 28

	Aluminum	Plastic	Machine Time	Labor	Misc.
Product 1	3	2	1	1.5	0.2
Product 2	3.3	3	2	2.5	0.4
Product 3	2	2	1.2	2.3	0.1

TABLE 25
Cost per Unit for Problem 28

	Aluminum	Plastic	Machine Time	Labor	Misc.
Cost/Unit	$1.42	$0.50	$0.43	$3.85	$0.05

indicated age groups in Table 27. The population itself is split into different age groups as follows:

TABLE 27
Party Registration by Age Group in Buffalo
(Problem 30)

Age Group	Party X	Party Y
18–24	50%	50%
25–40	70%	30%
41 +	80%	20%

Age Group	Percentage
18–24	40%
25–40	40%
41 +	20%

For an issue to pass in a special election, a favorable vote of over 50% is required. Buffalo has roughly 250,000 citizens aged 18 or over. A vote was held on two issues, and the result by party is listed below.

	Party X Favorable Vote	Party Y Favorable Vote
Issue 1	50%	60%
Issue 2	60%	30%

Which of the two issues passed?

31. Computational Complexity   Associativity of matrix multiplication is stated by the property

$$(AB)C = A(BC)$$

In other words, it does not make any difference which product AB or BC is computed first when finding the product of three matrices ABC. There can, however, be quite a difference in the number of operations one must perform, depending on the order in which the operations are performed. Determine the number of multiplications needed for the following two computations of the multiplication of the matrices A, B, and C with the dimensions given:

Matrix	Dimension
A	1×50
B	50×20
C	20×100

(a) $(AB)C$ (b) $A(BC)$

Hint: The following example shows what is meant by the number of multiplications:

$$\begin{bmatrix} 1 & 2 \\ 3 & 4 \end{bmatrix}\begin{bmatrix} 5 & 6 \\ 7 & 8 \end{bmatrix} = \begin{bmatrix} 1\cdot 5 + 2\cdot 7 & 1\cdot 6 + 2\cdot 8 \\ 3\cdot 5 + 4\cdot 7 & 3\cdot 6 + 4\cdot 8 \end{bmatrix}$$

The number of individual multiplications performed here is eight, as shown by the entries in the matrix at the right.

32. The RST Problem Revisited   Suppose you have at your disposal a large collection of cardboard figures in the shapes of rectangles, squares, and triangles. On each figure is drawn cats, dogs, and mice. The distribution of the three animals on the three figures is given in Table 28. Suppose you select 15 rectangles, 10 squares, and 5 triangles from the collection. What is the total number of cats, dogs, and mice on the selected figures? Note that this problem can be solved by matrix multiplication, whereas the earlier versions of this problem were solved by solving a system of linear equations.

TABLE 28
Number of Different Animals Drawn on Each Figure
for Problems 32–33

	Rectangle	Square	Triangle
Cats	3	3	0
Dogs	4	5	3
Mice	1	2	3

33. The RST Problem Revisited   For Problem 32, what is the total number of cats, dogs, and mice on the selected figures if 20 rectangles, 10 squares, and 2 triangles are selected?

Matrices in the Life Sciences

34. Aquaculture   A fish of species I consumes 20 grams of food A and 30 grams of food B in one day. A fish of species II consumes 30 grams of food A and 45 grams of food B in one day. How many grams of the two foods will 1000 fish of species I and 2000 fish of species II consume in one day?

35. Aquaculture   A scientist trying to grow lobsters in a controlled environment mixes two grains in varying amounts to make the ideal lobster food. The scientist makes three mixes according to the following mixtures:

$$M = \begin{matrix} & \text{Mix I} & \text{Mix II} & \text{Mix III} \\ \begin{bmatrix} 45 \text{ oz} & 30 \text{ oz} & 15 \text{ oz} \\ 15 \text{ oz} & 30 \text{ oz} & 45 \text{ oz} \end{bmatrix} & & & \end{matrix} \begin{matrix} \text{Grain A} \\ \text{Grain B} \end{matrix}$$

The nutritional value of each of the two grains is given by the values in the 3×2 matrix

$$N = \begin{matrix} \text{Grain A} & \text{Grain B} \\ \begin{bmatrix} 5 \text{ gm} & 10 \text{ gm} \\ 50 \text{ gm} & 30 \text{ gm} \\ 10 \text{ gm} & 5 \text{ gm} \end{bmatrix} & \end{matrix} \begin{matrix} \text{Protein} \\ \text{Carbohydrates} \\ \text{Fat} \end{matrix}$$

(a) What is the amount of carbohydrates in mix I?
(b) What is the amount of protein in mix II?
(c) What is the amount of fat in mix III?
(d) What is the interpretation of the elements in the matrix NM?

36. Diabetic Requirements   Four diabetic patients in a hospital each require injections of three kinds of insulin each day. The 4×3 matrix shown in Table 29 gives the amounts of the three kinds of insulin each patient requires each day. If patient 1 stays 10 days, patient 2 stays 5 days, patient 3 stays 20 days, and patient 4 stays 10 days, how much of each type of insulin is required?

TABLE 29
Amounts of Insulin for the Patients in Problem 36

	Patient 1	Patient 2	Patient 3	Patient 4
Semi-lente insulin	10 g	40 g	30 g	20 g
Lente-insulin	20 g	0 g	10 g	10 g
Ultra-insulin	30 g	0 g	0 g	20 g

Matrices in Food Chains

Ecologists find it useful to study the following matrices C and H. The carnivore matrix C gives the fraction of a given kind of herbivore (animal that eats mostly plant matter) that is consumed by a given kind of carnivore (animal that eats mostly animal matter) in a given time period. For example, the average leopard will eat 0.1 tapir in every time period.

On the other hand, the herbivore matrix H gives the fraction of a given kind of plant that is consumed by a given kind of herbivore in a given time period. For example, the average tapir will eat 0.1 unit of berries in every time period.

$$C = \begin{matrix} & & \text{Kind of Carnivore} \\ & & \text{Leopard} \quad \text{Crocodile} \quad \text{Boa} \quad \text{Piranha} \\ \begin{matrix} \text{Tapir} \\ \text{Sloth} \\ \text{Monkey} \end{matrix} & \begin{bmatrix} 0.1 & 0 & 0.1 & 0 \\ 0.3 & 0.1 & 0.2 & 0 \\ 0.3 & 0.2 & 0.3 & 0.5 \end{bmatrix} \end{matrix} \begin{matrix} \text{Kind of} \\ \text{Herbivore} \end{matrix}$$

$$H = \begin{matrix} & & \text{Kind of Herbivore} \\ & & \text{Tapir} \quad \text{Sloth} \quad \text{Monkey} \\ \begin{matrix} \text{Berries} \\ \text{Nuts} \\ \text{Roots} \\ \text{Leaves} \\ \text{Bark} \end{matrix} & \begin{bmatrix} 0.1 & 0.2 & 0.3 \\ 0.2 & 0.3 & 0.2 \\ 0.2 & 0.1 & 0.1 \\ 0.1 & 0 & 0.1 \\ 0.3 & 0 & 0 \end{bmatrix} \end{matrix} \begin{matrix} \text{Kind of} \\ \text{Plant} \end{matrix}$$

The product HC gives the fraction of each plant that is consumed by each kind of carnivore.

37. Compute the matrix product HC and interpret the results for the above carnivores, herbivores, and plants.

38. Find the carnivore that consumes the most of each type of plant.

Estimating Future Populations Using Matrices

The following 3×3 matrix, known as the *Leslie matrix*, illustrates how matrices can be used to predict future populations (populations of countries, universities, states, biological populations, and so on). Suppose the following Leslie matrix has been found on the basis of a study of birth and survival rates. The numbers 0.33, 0.67, and 0.10 in the first row of the matrix represent the average number of daughters a female in age groups 0–14, 15–29, and 30–44 will have during the next 15 years. The two numbers .99 and .98 represent the probabilities that females in age groups 0–14 and 15–29, respectively, will survive to be in age groups 15–29 and 30–44 in 15 years. Note that any female who is alive in 15 years will automatically be in the next age group. Females who are currently in age group 30–44 are not of interest to this study 15 years from now.

$$L = \begin{matrix} & & \text{Age} & \text{Age} & \text{Age} \\ & & \text{0–14} & \text{15–29} & \text{30–44} \\ \begin{matrix} \text{Age 0–14} \\ \text{Age 15–29} \\ \text{Age 30–44} \end{matrix} & \begin{bmatrix} 0.33 & 0.67 & 0.10 \\ 0.99 & 0 & 0 \\ 0 & 0.98 & 0 \end{bmatrix} \end{matrix}$$

If the Leslie matrix L is multiplied by the 3×1 column matrix P_1,

$$P_1 = \begin{bmatrix} 16{,}000 \\ 18{,}000 \\ 15{,}000 \end{bmatrix} \begin{matrix} \text{Age } 0–14 \\ \text{Age } 15–29 \\ \text{Age } 30–44 \end{matrix}$$

representing the current populations in each of the age groups, then the matrix product LP_1 will be a 3×1 column matrix that will give the population of females in each age category 15 years from now. By repeated multiplication

$$P_2 = LP_1$$
$$P_3 = LP_2$$
$$P_4 = LP_3$$
$$\vdots \qquad \vdots$$

it is possible (as long as current birth and survival rates hold) to estimate the population in 30, 45, 60, . . . years.

39. If the current population of females in each of the three age groups is given by P_1, estimate the population P_2 of females in each of the three age groups in 15 years.

40. If the current population of females in each of the three age groups is given by P_1, estimate the population P_3 of females in each of the three age groups in 30 years.

41. If the current population of females in each of the three age groups is given by P_1, estimate the population P_4 of females in each of the three age groups in 45 years.

42. Why is the above mathematical model restricted to females?

43. What is the general Leslie matrix if there are four age groups of females and the birth and survival rates during the next time period (15 years in the above example) are given by b_1–b_4 and s_1–s_3, respectively.

Modeling Disease Transmission by Matrix Multiplication

Suppose Mary, Ed, and Sue have contracted a contagious virus. A second group of five persons is questioned to determine who has been in contact with the three infected persons. A third group of four persons is then questioned to determine contacts with any of the five persons in the second group. The following matrices A and B describe the contacts between the first and second groups and the second and third groups, respectively.

$$A = \begin{array}{c} \\ \text{Mary} \\ \text{Ed} \\ \text{Sue} \end{array} \begin{array}{c} \overset{\text{Group 2}}{\overbrace{\begin{array}{ccccc} \text{Peter} & \text{Sam} & \text{Rose} & \text{Sally} & \text{Ann} \end{array}}} \\ \left[\begin{array}{ccccc} 1 & 0 & 1 & 0 & 1 \\ 0 & 1 & 1 & 1 & 1 \\ 0 & 0 & 0 & 1 & 0 \end{array}\right] \end{array} \begin{array}{c} \\ \\ \text{Group 1} \end{array}$$

$$B = \begin{array}{c} \\ \text{Peter} \\ \text{Sam} \\ \text{Rose} \\ \text{Sally} \\ \text{Ann} \end{array} \begin{array}{c} \overset{\text{Group 3}}{\overbrace{\begin{array}{cccc} \text{Henry} & \text{Gary} & \text{Jerry} & \text{Carol} \end{array}}} \\ \left[\begin{array}{cccc} 1 & 1 & 0 & 1 \\ 1 & 1 & 0 & 0 \\ 0 & 0 & 1 & 0 \\ 0 & 0 & 1 & 0 \\ 1 & 0 & 0 & 1 \end{array}\right] \end{array} \begin{array}{c} \\ \\ \text{Group 2} \\ \\ \\ \end{array}$$

An entry of 1 in the above matrices indicates a contact between two individuals; a 0 indicates no contact. For instance, Mary has had contact with Peter, Rose, and Ann but no contact with Sam or Sally. Ann has had contact with Henry and Carol in Group 3. The elements in the matrix product AB represent the second-order contacts between individuals in Group 1 and Group 3.

44. Compute the second-order contacts between individuals in Group 1 and Group 3 by computing the matrix product AB. Interpret the entries in this matrix.

45. How many second-order contacts are there between Mary and Carol? What are the different ways that Carol can contract the disease from Mary?

46. How many second-order contacts are there between Ed and Jerry? What are the different ways that Jerry can contract the disease from Ed?

1.6   Matrix Inverses

We show that some square matrices A have an inverse A^{-1} that satisfies the matrix equation

$$AA^{-1} = I$$

where I is the identity matrix of the same dimension as A. We also learn

- how to find the inverse of a matrix by use of the Gauss-Jordan method and
- how to use the inverse of a matrix to solve linear systems of equations.

Identity Matrix

For all real numbers we know that

$$1 \cdot a = a \cdot 1 = a$$

The number 1 is called the **identity** for multiplication. Since we have defined addition, subtraction, and multiplication for matrices, we are tempted to think of a matrix as a kind of "generalized number." It is then natural to ask whether we can find a matrix "1" that satisfies an analogous matrix equation

$$1 \cdot A = A \cdot 1 = A$$

The answer to this question is that for any square $n \times n$ matrix A there exists an $n \times n$ matrix I_n, called the **identity matrix**, that has 1's down the main diagonal of the matrix and 0's elsewhere and that satisfies

$$AI_n = I_nA = A$$

For example, given a 3×3 matrix

$$A = \begin{bmatrix} 4 & 1 & 8 \\ 6 & -5 & 9 \\ 0 & 4 & 2 \end{bmatrix}$$

the 3×3 identity matrix is

$$I_3 = \begin{bmatrix} 1 & 0 & 0 \\ 0 & 1 & 0 \\ 0 & 0 & 1 \end{bmatrix}$$

Clearly, we have

$$\overset{A}{\begin{bmatrix} 4 & 1 & 8 \\ 6 & -5 & 9 \\ 0 & 4 & 2 \end{bmatrix}} \overset{I_3}{\begin{bmatrix} 1 & 0 & 0 \\ 0 & 1 & 0 \\ 0 & 0 & 1 \end{bmatrix}} = \overset{I_3}{\begin{bmatrix} 1 & 0 & 0 \\ 0 & 1 & 0 \\ 0 & 0 & 1 \end{bmatrix}} \overset{A}{\begin{bmatrix} 4 & 1 & 8 \\ 6 & -5 & 9 \\ 0 & 4 & 2 \end{bmatrix}} = \overset{A}{\begin{bmatrix} 4 & 1 & 8 \\ 6 & -5 & 9 \\ 0 & 4 & 2 \end{bmatrix}}$$

In the future we will usually denote the identity matrix by the symbol I and drop the designation of the dimension.

The Matrix Inverse

For all real numbers a, except zero, we know that

$$\frac{1}{a} a = 1$$

The number $1/a$ is called the **multiplicative inverse** or simply the **inverse** of the number a. It makes sense to ask whether we can find a **matrix inverse** $1/A$ for a matrix A that satisfies the analogous equation

$$\frac{1}{A} A = 1$$

For some square matrices A there exists a matrix A^{-1}, called the **inverse of A**, that satisfies the equations

$$A^{-1}A = AA^{-1} = I$$

The inverse of the matrix A, when it exists, will be a square matrix and have the same dimension as A. We do not define inverse matrices for rectangular matrices, since, for example, the inverse matrix A^{-1} of a 3×2 matrix A would have to be a 2×3 matrix for both AA^{-1} and $A^{-1}A$ to exist. However, AA^{-1} is not equal to $A^{-1}A$, since AA^{-1} is a 3×3 matrix and $A^{-1}A$ is a 2×2 matrix.

Example 1

Matrix Inverse   Verify that the following matrices A and B satisfy the equations $AB = BA = I$. Hence we conclude that $B = A^{-1}$.

$$A = \begin{bmatrix} 2 & -1 \\ 0 & 1 \end{bmatrix} \qquad B = \begin{bmatrix} 1/2 & 1/2 \\ 0 & 1 \end{bmatrix}$$

Solution   By direct computation we have

$$\overset{A}{\begin{bmatrix} 2 & -1 \\ 0 & 1 \end{bmatrix}} \overset{B}{\begin{bmatrix} 1/2 & 1/2 \\ 0 & 1 \end{bmatrix}} = \overset{I}{\begin{bmatrix} 1 & 0 \\ 0 & 1 \end{bmatrix}}$$

and

$$\overset{B}{\begin{bmatrix} 1/2 & 1/2 \\ 0 & 1 \end{bmatrix}} \overset{A}{\begin{bmatrix} 2 & -1 \\ 0 & 1 \end{bmatrix}} = \overset{I}{\begin{bmatrix} 1 & 0 \\ 0 & 1 \end{bmatrix}}$$

Therefore B is the inverse of A.   □

It can be proved that if one of the conditions $AA^{-1} = I$ or $A^{-1}A = I$ holds, then the other does too. Hence to prove that A^{-1} is the inverse of A, it is necessary only to verify *one* of the above conditions.

Before learning how to find the inverse of a matrix we will learn how the inverse can be useful for solving systems of linear equations.

Solving $AX = B$ by Calculating $X = A^{-1}B$

Consider the linear system

$$\begin{aligned} x - y &= 2 \\ 2x + y &= 3 \end{aligned}$$

Let

$$A = \begin{bmatrix} 1 & -1 \\ 2 & 1 \end{bmatrix} \qquad X = \begin{bmatrix} x \\ y \end{bmatrix} \qquad B = \begin{bmatrix} 2 \\ 3 \end{bmatrix}$$

We can write the above system of equations in matrix form:

$$\begin{bmatrix} 1 & -1 \\ 2 & 1 \end{bmatrix} \begin{bmatrix} x \\ y \end{bmatrix} = \begin{bmatrix} 2 \\ 3 \end{bmatrix}$$

or simply as

$$AX = B$$

We assume now that the inverse of the coefficient matrix A is known. In this case it is

$$A^{-1} = \begin{bmatrix} 1/3 & 1/3 \\ -2/3 & 1/3 \end{bmatrix}$$

To solve for X in the equation $AX = B$, we multiply the equation by A^{-1}, which gives

$$A^{-1}(AX) = A^{-1}B$$

or

$$(A^{-1}A)X = A^{-1}B$$

But since $A^{-1}A = I$, the above equation is simply

$$IX = A^{-1}B$$

or

$$X = A^{-1}B$$

If the reader is bothered by matrix notation, we can resolve the system above by writing out the matrices. Start again with the matrix equation

$$\overset{A}{\begin{bmatrix} 1 & -1 \\ 2 & 1 \end{bmatrix}} \overset{X}{\begin{bmatrix} x \\ y \end{bmatrix}} = \overset{B}{\begin{bmatrix} 2 \\ 3 \end{bmatrix}}$$

Multiply each side of the equation by A^{-1}, getting

$$\overset{A^{-1}}{\begin{bmatrix} 1/3 & 1/3 \\ -2/3 & 1/3 \end{bmatrix}} \overset{A}{\begin{bmatrix} 1 & -1 \\ 2 & 1 \end{bmatrix}} \overset{X}{\begin{bmatrix} x \\ y \end{bmatrix}} = \overset{A^{-1}}{\begin{bmatrix} 1/3 & 1/3 \\ -2/3 & 1/3 \end{bmatrix}} \overset{B}{\begin{bmatrix} 2 \\ 3 \end{bmatrix}}$$

But the product of A^{-1} times A is I; hence

$$\overset{I}{\begin{bmatrix} 1 & 0 \\ 0 & 1 \end{bmatrix}} \overset{X}{\begin{bmatrix} x \\ y \end{bmatrix}} = \overset{A^{-1}}{\begin{bmatrix} 1/3 & 1/3 \\ -2/3 & 1/3 \end{bmatrix}} \overset{B}{\begin{bmatrix} 2 \\ 3 \end{bmatrix}}$$

Since $IX = X$, we have

$$\overset{X}{\begin{bmatrix} x \\ y \end{bmatrix}} = \overset{A^{-1}}{\begin{bmatrix} 1/3 & 1/3 \\ -2/3 & 1/3 \end{bmatrix}} \overset{B}{\begin{bmatrix} 2 \\ 3 \end{bmatrix}}$$

$$= \begin{bmatrix} 5/3 \\ -1/3 \end{bmatrix}$$

In other words the solution is

$$x = 5/3$$
$$y = -1/3$$

The above strategy of multiplying each side of a matrix equation by an appropriate matrix is similar to the method used in elementary algebra for solving equations such as

$$3x = 2$$

There we multiplied each side of the equation by 1/3 (the multiplicative inverse of 3) and found the solution

$$x = (1/3)2 = 2/3$$

For systems of equations $AX = B$ the procedure is similar, except that now both sides of the equation are multiplied by A^{-1}.

The major goal of this section is to explain a procedure for finding the inverse of a matrix A, when it exists. We begin by describing the procedure for 2×2 matrices.

Finding the Inverse of a 2 × 2 Matrix

Start with a typical 2×2 matrix:

$$A = \begin{bmatrix} 1 & 2 \\ -1 & 1 \end{bmatrix}$$

We know that the unknown inverse A^{-1} is also a 2×2 matrix, which we denote as

$$A^{-1} = \begin{bmatrix} a & b \\ c & d \end{bmatrix}$$

Since A^{-1} satisfies $AA^{-1} = I$, we have

$$\overset{A}{\begin{bmatrix} 1 & 2 \\ -1 & 1 \end{bmatrix}} \overset{A^{-1}}{\begin{bmatrix} a & b \\ c & d \end{bmatrix}} = \overset{I}{\begin{bmatrix} 1 & 0 \\ 0 & 1 \end{bmatrix}}$$

By multiplying the two matrices A and A^{-1} we get

$$\begin{bmatrix} a + 2c & b + 2d \\ -a + c & -b + d \end{bmatrix} = \begin{bmatrix} 1 & 0 \\ 0 & 1 \end{bmatrix}$$

Now, by setting the corresponding elements of the matrices equal to each other we have

$$a + 2c = 1$$
$$b + 2d = 0$$
$$-a + c = 0$$
$$-b + d = 1$$

These four equations in a, b, c, and d can be separated into two different 2×2 systems of equations, one involving the unknowns a and c and the other involving the unknowns b and d.

System I	System II
$a + 2c = 1$	$b + 2d = 0$
$-a + c = 0$	$-b + d = 1$

Writing the augmented matrix for each system, we have

System I   System II

$$\begin{bmatrix} 1 & 2 & | & 1 \\ -1 & 1 & | & 0 \end{bmatrix} \qquad \begin{bmatrix} 1 & 2 & | & 0 \\ -1 & 1 & | & 1 \end{bmatrix}$$

The goal now is to solve these systems for a, b, c, and d. When doing this, note that the same sequence of elementary row operations would be used for each system, since the two coefficient matrices are the same. Therefore it makes sense to combine the two augmented matrices above into one larger matrix of the form

$$
\begin{array}{cc}
A & I \\
\end{array}
$$

$$
\left[\begin{array}{cc|cc}
1 & 2 & 1 & 0 \\
-1 & 1 & 0 & 1
\end{array}\right]
$$

We are really solving two separate systems of equations at one time; and to do this, we simply apply the elementary row operations to this new matrix and transform the matrix A into the identity matrix I. The solution of one system of equations with values a and c will then be read from the first column to the right of the vertical bar, and the values of the other unknowns b and d will be read from the second column to the right of the vertical bar. But these values are the two columns of A^{-1}. In other words, the inverse A^{-1} will be located to the right of the vertical bar after the matrix A is transformed into the identity matrix I.

If we were to carry out these elementary row operations, we would find

$$
\begin{array}{cc}
I & A^{-1} \\
\end{array}
$$

$$
\left[\begin{array}{cc|cc}
1 & 0 & 1/3 & -2/3 \\
0 & 1 & 1/3 & 1/3
\end{array}\right]
$$

Hence the inverse of A is

$$
A^{-1} = \left[\begin{array}{cc}
1/3 & -2/3 \\
1/3 & 1/3
\end{array}\right]
$$

The above example is formalized in the following method for finding the inverse of a matrix A.

Finding the Inverse of a Matrix

To find the inverse A^{-1} of a square matrix A, whenever it exists, we carry out the following two steps.

- **Step 1.**   Write the matrix A and the identity matrix I of the same dimension alongside each other in the single matrix $[A \mid I]$.
- **Step 2.**   Transform the matrix A into the identity matrix I by means of elementary row operations. After this has been carried out, the matrix to the right of the newly formed identity matrix will be A^{-1}. In other words, the elementary row operations will effect the transformation

$$
[A \mid I] \longrightarrow [I \mid A^{-1}]
$$

When the matrix A cannot be transformed into the identity matrix by elementary row operations, the matrix does not have an inverse. This occurs when one or more rows to the left of the vertical bar are transformed into a row consisting of all 0's.

Example 2 _____   Inverse of a 2 × 2 Matrix   Find the inverse of the matrix A:

$$
A = \left[\begin{array}{cc}
1 & 1 \\
4 & 1
\end{array}\right]
$$

Solution

- **Step 1** (construct $[A \mid I]$). We start by writing the matrix

$$[A \mid I] = \begin{bmatrix} 1 & 1 & | & 1 & 0 \\ 4 & 1 & | & 0 & 1 \end{bmatrix}$$

- **Step 2** (transform A into I, using row operations). We perform the following elementary row operations.

$$R_2 + (-4)R_1 \longrightarrow R_2$$

$$\begin{array}{c} (-4) \\ \text{(want 0)} \end{array} \begin{bmatrix} 1 & 1 & | & 1 & 0 \\ 4 & 1 & | & 0 & 1 \end{bmatrix} \longrightarrow \begin{bmatrix} 1 & 1 & | & 1 & 0 \\ 0 & -3 & | & -4 & 1 \end{bmatrix} \begin{array}{c} R_1 \\ R_2 \end{array}$$

Starting Matrix $[A \mid I]$

$$(-1/3)R_2 \longrightarrow R_2$$

$$(-1/3) \begin{bmatrix} 1 & 1 & | & 1 & 0 \\ 0 & -3 & | & -4 & 1 \end{bmatrix} \begin{array}{c} \\ \text{(want 1)} \end{array} \begin{bmatrix} 1 & 1 & | & 1 & 0 \\ 0 & 1 & | & 4/3 & -1/3 \end{bmatrix} \begin{array}{c} R_1 \\ R_2 \end{array}$$

$$R_1 + (-1)R_2 \longrightarrow R_1$$

$$\begin{array}{c} \text{(want 0)} \\ (-1) \end{array} \begin{bmatrix} 1 & 1 & | & 1 & 0 \\ 0 & 1 & | & 4/3 & -1/3 \end{bmatrix} \longrightarrow \begin{bmatrix} 1 & 0 & | & -1/3 & 1/3 \\ 0 & 1 & | & 4/3 & -1/3 \end{bmatrix} \begin{array}{c} R_1 \\ R_2 \end{array}$$

Final Matrix $[I \quad A^{-1}]$

The matrix to the right of the vertical bar is the inverse of A. That is,

$$A^{-1} = \begin{bmatrix} -1/3 & 1/3 \\ 4/3 & -1/3 \end{bmatrix}$$

This result is checked by computing

$$\overset{A^{-1}}{\begin{bmatrix} -1/3 & 1/3 \\ 4/3 & -1/3 \end{bmatrix}} \overset{A}{\begin{bmatrix} 1 & 1 \\ 4 & 1 \end{bmatrix}} = \overset{I}{\begin{bmatrix} 1 & 0 \\ 0 & 1 \end{bmatrix}}$$

□

General Formula for the 2 × 2 Inverse

By using the steps above, it can be shown that the matrix inverse, if it exists, for the general 2 × 2 matrix

$$A = \begin{bmatrix} a & b \\ c & d \end{bmatrix}$$

is

$$A^{-1} = \frac{1}{(ad - bc)} \begin{bmatrix} d & -b \\ -c & a \end{bmatrix}$$

The matrix A does not have an inverse when $ad - bc = 0$.

Example 3 ——————— Using the Formula for 2×2 Inverses Find the inverse of the following matrix:

$$A = \begin{bmatrix} -1 & 3 \\ 2 & -4 \end{bmatrix}$$

Solution Since

$$\begin{array}{cc} a = -1 & b = 3 \\ c = 2 & d = -4 \end{array}$$

we have

$$ad - bc = (-1)(-4) - (3)(2)$$
$$= -2$$

Hence the inverse is

$$A^{-1} = \frac{1}{-2}\begin{bmatrix} -4 & -3 \\ -2 & -1 \end{bmatrix}$$

$$= \begin{bmatrix} 2 & 3/2 \\ 1 & 1/2 \end{bmatrix}$$

Check:

$$\overset{A}{\begin{bmatrix} -1 & 3 \\ 2 & -4 \end{bmatrix}} \overset{A^{-1}}{\begin{bmatrix} 2 & 3/2 \\ 1 & 1/2 \end{bmatrix}} = \overset{I}{\begin{bmatrix} 1 & 0 \\ 0 & 1 \end{bmatrix}}$$

☐

Example 4 ——————— Solving a System of Equations Using a Matrix Inverse Solve the linear system

$$\begin{array}{rcl} x + y + 2z & = & 3 \\ 2x + 3y + 2z & = & 4 \\ x + y + 3z & = & 5 \end{array}$$

Solution Denoting the above system by $AX = B$, we will compute $X = A^{-1}B$. We first find the inverse of A by writing

$$[A \mid I] = \begin{bmatrix} 1 & 1 & 2 & | & 1 & 0 & 0 \\ 2 & 3 & 2 & | & 0 & 1 & 0 \\ 1 & 1 & 3 & | & 0 & 0 & 1 \end{bmatrix}$$

We now transform the coefficient matrix A located to the left of the vertical bar into the identity matrix. The inverse A^{-1} will then be located to the right of the vertical bar. Without writing down the individual steps, the transformed matrix is

$$[I \mid A^{-1}] = \begin{bmatrix} 1 & 0 & 0 & | & 7 & -1 & -4 \\ 0 & 1 & 0 & | & -4 & 1 & 2 \\ 0 & 0 & 1 & | & -1 & 0 & 1 \end{bmatrix}$$

Hence the inverse of A is

$$A^{-1} = \begin{bmatrix} 7 & -1 & -4 \\ -4 & 1 & 2 \\ -1 & 0 & 1 \end{bmatrix}$$

Now that A^{-1} is known, the solution can be found by matrix multiplication:

$$\begin{array}{ccc} X & A^{-1} & B \end{array}$$

$$\begin{bmatrix} x \\ y \\ z \end{bmatrix} = \begin{bmatrix} 7 & -1 & -4 \\ -4 & 1 & 2 \\ -1 & 0 & 1 \end{bmatrix}\begin{bmatrix} 3 \\ 4 \\ 5 \end{bmatrix}$$

$$= \begin{bmatrix} -3 \\ 2 \\ 2 \end{bmatrix}$$

In other words, the solution is

$$\begin{aligned} x &= -3 \\ y &= 2 \\ z &= 2 \end{aligned}$$

You can check these values to see that they are a solution of the system of equations.   □

Summary for Solving Systems of Equations by Matrix Inversion

To solve the linear system $AX = B$, carry out the following two steps.

  Step 1. Find the inverse of A by transforming

$$\begin{bmatrix} A & | & I \end{bmatrix} \longrightarrow \begin{bmatrix} I & | & A^{-1} \end{bmatrix}$$

  Step 2. Compute the matrix product

$$X = A^{-1}B$$

There are systems of equations $AX = B$ whose coefficient matrix A does not have an inverse. If the matrix $\begin{bmatrix} A & | & I \end{bmatrix}$ cannot be transformed to $\begin{bmatrix} I & | & A^{-1} \end{bmatrix}$ by means of a sequence of elementary row operations, then A does not have an inverse. This happens when, at some stage when the elementary row operations are being performed, one or more rows of the current coefficient matrix consist of all 0's to the left of the vertical bar. When the matrix A does not have an inverse, the linear system $AX = B$ has either no solutions or an infinite number of solutions. When either of these situations arises, computation of this infinite set of solutions (when they exist) of the linear system $AX = B$ should proceed by using the Gauss-Jordan method discussed in Section 1.5.

Example 5

Recognizing When a Matrix Does Not Have an Inverse Solve the following linear system:

$$\begin{aligned} 3x + 2y + 2z &= 4 \\ x - y + z &= 3 \\ 2x + 3y + z &= 7 \end{aligned}$$

Solution We begin by trying to transform the matrix A in the matrix $[A \mid I]$ into the identity matrix I. We write

$$[A \mid I] = \begin{bmatrix} 3 & 2 & 2 & 1 & 0 & 0 \\ 1 & -1 & 1 & 0 & 1 & 0 \\ 2 & 3 & 1 & 0 & 0 & 1 \end{bmatrix}$$

Without carrying out the individual row operations we will arrive at a matrix where the bottom row of the coefficient matrix consists of all 0's:

$$\begin{bmatrix} 1 & 2/3 & 2/3 & 1/3 & 1 & 0 \\ 0 & 1 & -1/5 & 1/5 & -3/5 & 0 \\ 0 & 0 & 0 & -1 & 1 & 1 \end{bmatrix}$$

At this point we see that the coefficient matrix cannot be transformed into the identity matrix. The fact that there is a nonzero entry in the bottom row to the *right* of the vertical bar means that the original system of equations has no solution. In the case of the -1 to the right of the vertical bar, this corresponds to the equation

$$0x + 0y + 0z = -1$$

If each of the elements in the bottom row to the right of the vertical bar had been 0's, this would have meant that the original system has an infinite number of solutions. □

Matrix Inverses and Production Scheduling

We close this section by presenting an example that shows how the matrix inverse comes about in the business world. This example is particularly interesting, since it shows how several systems of equations can be solved at one time once the inverse of the coefficient matrix has been found.

Example 6

Production Scheduling The Outdoor Suppliers Co. recently had the problem of determining production schedules for a four-week period. The company manufactures three products: life rafts, tents, and boat covers. Each of these products is made from the raw materials canvas, rubber, and nylon. The specific requirements for each of the products are listed in Table 30.

TABLE 30
Material Requirements for
Three Products

	Life Raft	**Tent**	**Boat Cover**
Canvas	1 unit	3 units	3 units
Rubber	1 unit	4 units	3 units
Nylon	1 unit	3 units	4 units

For instance, the 1 in the upper left-hand corner of the table means that each life raft uses 1 unit of canvas in its production. Over the next four weeks the company has estimated that it will have available the quantity of raw materials listed in Table 31.

A linear system at work on an assembly line.

TABLE 31
Estimate of the Next Four
Weeks Deliveries of Raw
Materials

	Week 1	Week 2	Week 3	Week 4
Canvas	70 units	50 units	80 units	50 units
Rubber	75 units	55 units	85 units	50 units
Nylon	80 units	60 units	90 units	50 units

Solution　To find the production schedule for the next four weeks, we let

$$x_i = \text{number of rafts to produce in week } i$$
$$y_i = \text{number of tents to produce in week } i$$
$$z_i = \text{number of covers to produce in week } i$$

for $i = 1, 2, 3, 4$. For the company to use all the material available during the four weeks, the variables x_i, y_i, and z_i must satisfy the following equations:

$$\text{Week 1:} \begin{cases} x_1 + 3y_1 + 3z_1 = 70 & \text{Canvas} \\ x_1 + 4y_1 + 3z_1 = 75 & \text{Rubber} \\ x_1 + 3y_1 + 4z_1 = 80 & \text{Nylon} \end{cases}$$

$$\text{Week 2:} \begin{cases} x_2 + 3y_2 + 3z_2 = 50 & \text{Canvas} \\ x_2 + 4y_2 + 3z_2 = 55 & \text{Rubber} \\ x_2 + 3y_2 + 4z_2 = 60 & \text{Nylon} \end{cases}$$

$$\text{Week 3:} \begin{cases} x_3 + 3y_3 + 3z_3 = 80 & \text{Canvas} \\ x_3 + 4y_3 + 3z_3 = 85 & \text{Rubber} \\ x_3 + 3y_3 + 4z_3 = 90 & \text{Nylon} \end{cases}$$

$$\text{Week 4:} \begin{cases} x_4 + 3y_4 + 3z_4 = 50 & \text{Canvas} \\ x_4 + 4y_4 + 3z_4 = 50 & \text{Rubber} \\ x_4 + 3y_4 + 4z_4 = 50 & \text{Nylon} \end{cases}$$

Each of the four systems of equations above has the same coefficient matrix. Hence we can find the solutions of all four systems of equations by finding the inverse of this common coefficient matrix and multiplying it times the respective right-hand side.

To start finding the inverse of the coefficient matrix, we write

$$[A \mid I] = \begin{bmatrix} 1 & 3 & 3 & | & 1 & 0 & 0 \\ 1 & 4 & 3 & | & 0 & 1 & 0 \\ 1 & 3 & 4 & | & 0 & 0 & 1 \end{bmatrix}$$

We now use elementary row operations to transform the matrix A into the identity matrix I.

Transform column 1:

$$R_2 + (-1)R_1 \longrightarrow R_2$$

$$\begin{matrix} (-1) \\ \\ \\ \end{matrix} \begin{bmatrix} 1 & 3 & 3 & | & 1 & 0 & 0 \\ 1 & 4 & 3 & | & 0 & 1 & 0 \\ 1 & 3 & 4 & | & 0 & 0 & 1 \end{bmatrix} \longrightarrow \begin{bmatrix} 1 & 3 & 3 & | & 1 & 0 & 0 \\ 0 & 1 & 0 & | & -1 & 1 & 0 \\ 1 & 3 & 4 & | & 0 & 0 & 1 \end{bmatrix} \begin{matrix} R_1 \\ R_2 \\ R_3 \end{matrix}$$

(want 0)

Original Matrix $[A \mid I]$

$$R_3 + (-1)R_1 \longrightarrow R_3$$

$$
(-1)\begin{bmatrix} 1 & 3 & 3 & | & 1 & 0 & 0 \\ 0 & 1 & 0 & | & -1 & 1 & 0 \\ \text{(want 0)} \ 1 & 3 & 4 & | & 0 & 0 & 1 \end{bmatrix} \longrightarrow \begin{bmatrix} 1 & 3 & 3 & | & 1 & 0 & 0 \\ 0 & 1 & 0 & | & -1 & 1 & 0 \\ 0 & 0 & 1 & | & -1 & 0 & 1 \end{bmatrix} \begin{matrix} R_1 \\ R_2 \\ R_3 \end{matrix}
$$

After transforming columns 2 and 3 we arrive at the final matrix

$$
[I \mid A^{-1}] = \begin{bmatrix} 1 & 0 & 0 & | & 7 & -3 & -3 \\ 0 & 1 & 0 & | & -1 & 1 & 0 \\ 0 & 0 & 1 & | & -1 & 0 & 1 \end{bmatrix}
$$

Hence the inverse of A is

$$
A^{-1} = \begin{bmatrix} 7 & -3 & -3 \\ -1 & 1 & 0 \\ -1 & 0 & 1 \end{bmatrix}
$$

Now that we have the inverse of the common coefficient matrix, we can multiply it times each of the right-hand sides of each system of equations to find the production schedules for each of the four weeks. Doing this, we get

$$
\text{Week 1} \quad \begin{bmatrix} x_1 \\ y_1 \\ z_1 \end{bmatrix} = \begin{bmatrix} 7 & -3 & -3 \\ -1 & 1 & 0 \\ -1 & 0 & 1 \end{bmatrix} \begin{bmatrix} 70 \\ 75 \\ 80 \end{bmatrix} = \begin{bmatrix} 25 \\ 5 \\ 10 \end{bmatrix}
$$

$$
\text{Week 2} \quad \begin{bmatrix} x_2 \\ y_2 \\ z_2 \end{bmatrix} = \begin{bmatrix} 7 & -3 & -3 \\ -1 & 1 & 0 \\ -1 & 0 & 1 \end{bmatrix} \begin{bmatrix} 50 \\ 55 \\ 60 \end{bmatrix} = \begin{bmatrix} 5 \\ 5 \\ 10 \end{bmatrix}
$$

$$
\text{Week 3} \quad \begin{bmatrix} x_3 \\ y_3 \\ z_3 \end{bmatrix} = \begin{bmatrix} 7 & -3 & -3 \\ -1 & 1 & 0 \\ -1 & 0 & 1 \end{bmatrix} \begin{bmatrix} 80 \\ 85 \\ 90 \end{bmatrix} = \begin{bmatrix} 35 \\ 5 \\ 10 \end{bmatrix}
$$

$$
\text{Week 4} \quad \begin{bmatrix} x_4 \\ y_4 \\ z_4 \end{bmatrix} = \begin{bmatrix} 7 & -3 & -3 \\ -1 & 1 & 0 \\ -1 & 0 & 1 \end{bmatrix} \begin{bmatrix} 50 \\ 50 \\ 50 \end{bmatrix} = \begin{bmatrix} 50 \\ 0 \\ 0 \end{bmatrix}
$$

These results are summarized in Table 32.

TABLE 32
Production Schedule for the
Next Four Weeks

	Week 1	Week 2	Week 3	Week 4
Rafts	25	5	35	50
Tents	5	5	5	0
Boat Covers	10	10	10	0

Comments on the Solution: The array in Table 32 is simply the matrix product $A^{-1}B$, where A^{-1} is the inverse of the coefficient matrix (which we just found) times the 3×4 matrix of raw materials given in Table 31. You should carry out this multiplication to verify this result.

It is useful to note that we have just solved four systems of equations. However, since the coefficient matrix was the same for each system, we needed to find only one inverse. If we had solved the four systems separately by the Gauss-Jordan method rather than by finding the inverse, it would have taken considerably more time and effort. This is one advantage of the matrix inverse method for solving linear systems.   □

Problems

Find the inverse for the matrices 1–15 if they exist. Verify your answers.

1. $\begin{bmatrix} 1 & 2 \\ 3 & 4 \end{bmatrix}$

2. $\begin{bmatrix} 1 & 2 \\ 4 & 3 \end{bmatrix}$

3. $\begin{bmatrix} 1 & 3 \\ 2 & 4 \end{bmatrix}$

4. $\begin{bmatrix} 1 & 4 \\ 2 & 3 \end{bmatrix}$

5. $\begin{bmatrix} 2 & 3 \\ 3 & 5 \end{bmatrix}$

6. $\begin{bmatrix} 1 & 1 & 3 \\ 1 & 3 & 2 \\ 3 & 2 & 1 \end{bmatrix}$

7. $\begin{bmatrix} 1 & 2 & 3 \\ 1 & 3 & 4 \\ 1 & 4 & 3 \end{bmatrix}$

8. $\begin{bmatrix} 1 & 2 & 3 \\ 3 & 5 & 6 \\ 2 & 4 & 5 \end{bmatrix}$

9. $\begin{bmatrix} 1 & 3 & 2 \\ 3 & 6 & 5 \\ 2 & 5 & 4 \end{bmatrix}$

10. $\begin{bmatrix} 0 & 0 & 1 \\ 1 & 0 & 0 \\ 0 & 1 & 0 \end{bmatrix}$

11. $\begin{bmatrix} 6 & 7 & 2 \\ 4 & 2 & 1 \\ 6 & 1 & 1 \end{bmatrix}$

12. $\begin{bmatrix} 3 & 2 & 1 \\ 3 & 1 & 3 \\ 4 & 1 & 2 \end{bmatrix}$

13. $\begin{bmatrix} 2 & 1 & 3 \\ 3 & 1 & 3 \\ 2 & 1 & 4 \end{bmatrix}$

14. $\begin{bmatrix} 0 & 1 & 3 \\ 1 & 1 & 1 \\ 2 & -1 & 1 \end{bmatrix}$

15. $\begin{bmatrix} 1 & 2 & 3 \\ 0 & 1 & 3 \\ 0 & 0 & 1 \end{bmatrix}$

16. Verify that the inverse of the matrix

$$A = \begin{bmatrix} 1 & 1 & 2 \\ 2 & 3 & 2 \\ 1 & 1 & 3 \end{bmatrix}$$

is

$$A^{-1} = \begin{bmatrix} 7 & -1 & -4 \\ -4 & 1 & 2 \\ -1 & 0 & 1 \end{bmatrix}$$

For Problems 17–21, use the formula

$$\begin{bmatrix} a & b \\ c & d \end{bmatrix}^{-1} = \frac{1}{(ad - bc)} \begin{bmatrix} d & -b \\ -c & a \end{bmatrix}$$

to find the inverse of the following 2 × 2 matrices. What condition is necessary for a 2 × 2 matrix to have an inverse?

17. $\begin{bmatrix} 1 & 1 \\ 3 & 2 \end{bmatrix}$

18. $\begin{bmatrix} 2 & 6 \\ 3 & 3 \end{bmatrix}$

19. $\begin{bmatrix} 4 & 2 \\ 2 & 3 \end{bmatrix}$

20. $\begin{bmatrix} 1 & 0 \\ 0 & 1 \end{bmatrix}$

21. $\begin{bmatrix} 5 & 5 \\ 1 & 0 \end{bmatrix}$

22. Verify that the formula for the inverse of a 2 × 2 matrix stated before Problem 17 is valid.

For Problems 23–26, use the inverse of the coefficient matrix to solve the system of equations.

23. $2x - 7y = 4$
$3x + 9y = -3$

24. $x + z = 1$
$2x + y = 3$
$x - y + z = 4$

25. $x + y = 2$
$x - y = 2$

26. $x + y + z = 1$
$x + z = 1$
$y + z = 1$

For Problems 27–29, find the inverse of

$$A = \begin{bmatrix} 1 & 2 & 2 \\ 1 & 3 & 2 \\ 1 & 2 & 3 \end{bmatrix}$$

to solve the systems of equations.

27. $x + 2y + 2z = 5$
$x + 3y + 2z = 1$
$x + 2y + 3z = 2$

28. $x + 2y + 2z = 3$
$x + 3y + 2z = 4$
$x + 2y + 3z = 1$

29. $x + 2y + 2z = 1$
$x + 3y + 2z = 4$
$x + 2y + 3z = 3$

Matrices in Business Management

30. Production Scheduling   The parts requirements for three products are summarized in Table 33.

TABLE 33
Parts Requirements for Problem 30

	Product 1	Product 2	Product 3
Molded Plastic Parts	2	1	3
Fasteners	3	2	2
Units of Paint	1	2	4

For the next two weeks the parts available will be the following:

	Week 1	**Week 2**
Molded Plastic Parts	50	80
Fasteners	60	100
Units of Paint	60	100

How many units of each product should be produced each week?

31. Refining of Metals   Gold and silver are extracted from the two grades of ore at a mine in Climax, Colorado. From each 100 pounds of the high-grade ore, 10 grams of silver and 4 grams of gold are extracted. From each 100 pounds of the low-grade ore, 5 grams of silver and 1 gram of gold are extracted. How many pounds of the high- and low-grade ores must be mined so that 1600 gram of gold and 5000 grams of silver are extracted?

Encoding and Decoding Secret Messages

Matrices and their inverses can be used to encode and decode secret messages. Suppose you assign to the letters of the alphabet the following numbers:

a	b	c	d	e	f	g	h	i	j	k	l	m	n	o
1	2	3	4	5	6	7	8	9	10	11	12	13	14	15

p	q	r	s	t	u	v	w	x	y	z	blank
16	17	18	19	20	21	22	23	24	25	26	27

Suppose you wish to send the secret message "not now" by a matrix encoding/decoding scheme. You start by assigning the appropriate number to each letter as follows:

n	o	t	blank	n	o	w
14	15	20	27	14	15	23

Then break the sequence of numbers into groups of a given length, and multiply each of these groups by a given encoding matrix. For instance, if we pick the encoding matrix as the 2×2 matrix

$$\text{Encoding matrix} = \begin{bmatrix} 1 & 1 \\ 2 & 4 \end{bmatrix}$$

then we would break the original sequence into groups of two and compute the following matrix products:

$$\begin{bmatrix} 1 & 1 \\ 2 & 4 \end{bmatrix}\begin{bmatrix} 14 \\ 15 \end{bmatrix} = \begin{bmatrix} 29 \\ 88 \end{bmatrix}$$

$$\begin{bmatrix} 1 & 1 \\ 2 & 4 \end{bmatrix}\begin{bmatrix} 20 \\ 27 \end{bmatrix} = \begin{bmatrix} 47 \\ 148 \end{bmatrix}$$

$$\begin{bmatrix} 1 & 1 \\ 2 & 4 \end{bmatrix}\begin{bmatrix} 14 \\ 15 \end{bmatrix} = \begin{bmatrix} 29 \\ 88 \end{bmatrix}$$

$$\begin{bmatrix} 1 & 1 \\ 2 & 4 \end{bmatrix}\begin{bmatrix} 23 \\ 27 \end{bmatrix} = \begin{bmatrix} 50 \\ 154 \end{bmatrix}$$

That is, the message sent would be 29 88 47 148 29 88 50 154. To decode the message, the receiver would have to know the encoding matrix and multiply in a similar fashion the received message times the inverse matrix of the encoding matrix.

32. Use the above encoding matrix to encode the message "Meet me at the Casbah."

33. Use the above encoding matrix to encode the message "Purdue beat Notre Dame."

34. Find the inverse of the above encoding matrix and decode the message 29 88 47 148 29 88 50 154.

35. Find the inverse of the above encoding matrix. The authors have a message for the reader: 17 52.

36. The encoding matrix is

$$\text{Encoding matrix} = \begin{bmatrix} 4 & 5 \\ 2 & 5 \end{bmatrix}$$

Find the decoding matrix.

37. Encode the message "Iowa beat Purdue" using the encoding matrix in Problem 36.

38. If the encoding matrix is

$$\text{Encoding matrix} = \begin{bmatrix} 1 & 1 & 1 \\ 1 & 1 & 0 \\ 1 & 0 & 0 \end{bmatrix}$$

find the decoding matrix.

39. Encode the message "I got an A in mathematics" using the encoding matrix in Problem 38.

40. You have just received a secret message on your wrist radio. It says: 37 28 9 45 36 9 42 36 6 62 36 27. The sender used the encoding matrix in Problem 38. What does the message say?

INTERVIEW WITH MR. RON TARVIN
Senior Research Associate
Cincinnati Milacron

Cincinnati Milacron is a world leader in the design, manufacture, and implementation of robots.

TOPIC: Robots and Matrices

Random House: Why would a world leader in the field of robotics be interested in matrices?

Tarvin: Most people don't realize it, but every time a robot loads a machine tool or welds two pieces of metal, the computer that drives the robot is multiplying a half dozen 3×3 matrices.

Random House: Why is it necessary that the robot, or computer as you say, multiply these matrices?

Tarvin: The matrices represent rotations of the joints of the robot. The multiplications relate these joint rotations to the robot's location in space.

Random House: What about other matrix operations? For example, does the robot ever find the inverse of a matrix when applying sealant to a car?

Tarvin: Yes. The computer must calculate the inverse of a matrix to determine the joint rotations that keep the robot moving along the desired sealant path.

Random House: Then you would say that matrices are very important in the operation of industrial robots?

Tarvin: Without matrices, the task of achieving sophisticated robot motion would be extremely difficult.

1.7 The Leontief Input-Output Model

PURPOSE

We introduce one of the most important applications of matrices and systems of equations to economics, the Leontief input-output model. We show how the input-output model is constructed by presenting a simplified three-sector economy of the electric, coal, and steel industries.

We also show how input-output analysis can be used to solve a problem in shop scheduling.

Introduction

Suppose the world price of oil is going up. We would expect income to decline and unemployment to increase in many segments of the economy as a result of this action. For example in the 1979 oil price rise the automobile industry, rubber industry, chemical industry, and many others suffered heavily. Suppose now to offset this increase in oil prices, the United States Congress authorizes a large expenditure in military spending, which will stimulate many of the affected industries. What effect will these actions have on income levels or unemployment levels? The problem is complicated, since two conflicting economic forces are at work and the U.S. economy is composed of many extremely interrelated sectors. Layoffs in one industry may cause layoffs in several other industries. On the other hand, an economic recovery in one sector of the economy will help spur a recovery in other areas.

In 1973 the Nobel Prize in Economics was awarded to Professor Wassily Leontief of Harvard University for his development of **input-output analysis**, a body of knowledge that is useful for studying interrelationships between different sectors of the economy. We illustrate the ideas of Leontief's methodology by presenting a simple three-sector economy.

A Three-Sector Economy

Consider a simple model of an economy in which only three goods are produced: electricity, coal, and steel. These three industries are called **sectors** in a three-sector economy. Any industry that does not produce electricity, coal, or steel, is said to be outside of the economy. It is possible to study economies that have hundreds of sectors, but we include only three for simplicity. In Leontief's original model the entire U.S. economy was broken into 81 different sectors.

Suppose now that industries other than the electric, coal, and steel industries make demands on the outputs of these three sectors. Such demands by outside businesses, industries, governments, and other parties are called **outside** or **external demands**. As an example, suppose that other parties are offering to buy electricity worth $1,156,500, coal worth $685,000, and steel worth $1,155,000. The question to answer is: How much should each sector produce to meet these outside or external demands? We are tempted to say that each sector should produce exactly the amounts that equal the external demands. However, the situation is more complicated, since the three industries themselves use their own outputs. For example, it takes steel to produce steel (for construction of plants, equipment, and so on). While each sector produces goods for external use, it also may demand as input the outputs of other sectors. These demands are called **internal demands**. The dynamics of a three-sector economy are illustrated in Figure 24.

To analyze these internal demands, let us assume that for each dollar of electricity produced, electric companies use $0.15 worth of electricity, $0.50 worth of coal, and $0.10 worth of steel. Also, for each dollar's worth of coal produced, the coal companies use electricity valued at $0.15 and steel valued at $0.03. Finally, for each dollar's worth of steel produced, steel companies use electricity valued at $0.09, coal valued at $0.10, and steel valued at $0.20. This information can be summarized in the matrix shown in Table 34, called the **input-output** or **technological matrix**.

Each column of the input-output matrix gives the values in dollars of the three row inputs needed to produce a dollar's worth of the column output. For example, the column labeled "Coal" should be interpreted as saying that the coal sector uses 15 cents' worth of electricity, no coal, and 3 cents' worth of steel for every dollar's worth of coal produced.

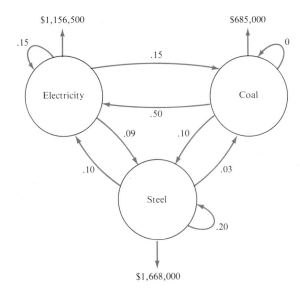

Figure 24
Dynamics of a three-sector economy

TABLE 34
Input-Output Matrix for a
Three-Sector Economy

Items Consumed (input)	Items Produced (output)		
	Electricity	*Coal*	*Steel*
Electricity	$0.15	$0.15	$0.09
Coal	$0.50	$0.00	$0.10
Steel	$0.10	$0.03	$0.20

The major goal of input-output analysis is described below.

The Input-Output Problem

Given the input-output or technological matrix for an economy and the external demands, determine the output of each sector that will meet both the external and internal demands of the economy.

In the context of the electricity, coal, and steel economy we must determine the outputs (stated in dollar values for convenience) of electricity, coal, and steel that will ensure that all external (or consumer) demands are met.

To solve this problem, we make the fundamental observation of input-output analysis that the output of each sector is equal to the external demand on these products plus the amount of these products consumed within the economy:

TOTAL OUTPUT = EXTERNAL DEMAND
+ INTERNAL CONSUMPTION

The goal of input-output analysis is to determine the output of each sector that meets the demands outside of the economy and at the same time also meets the demands of the other sectors within the economy.

We begin by calling the outputs of the three sectors

$$x_1 = \text{dollar value of the output of electricity}$$
$$x_2 = \text{dollar value of the output of coal}$$
$$x_3 = \text{dollar value of the output of steel}$$

We now construct the input-output matrix, which tells how the sectors of the economy interact:

$$
\begin{array}{ccc}
\text{Electricity} & \text{Coal} & \text{Steel}
\end{array}
$$
$$
\begin{bmatrix}
0.15 & 0.15 & 0.09 \\
0.50 & 0 & 0.10 \\
0.10 & 0.03 & 0.20
\end{bmatrix}
\begin{array}{l}
\text{Electricity} \\
\text{Coal} \\
\text{Steel}
\end{array}
$$

Using the input-output matrix, we can find the dollar amounts of electricity, coal, and steel that are consumed within the economy to produce x_1, x_2, and x_3 dollar amounts of these products. We have

$$\text{Electricity consumed} = 0.15x_1 + 0.15x_2 + 0.09x_3$$
$$\text{Coal consumed} = 0.50x_1 + 0.10x_3$$
$$\text{Steel consumed} = 0.10x_1 + 0.03x_2 + 0.20x_3$$

We can now substitute the required values into the fundamental result of input-output analysis to get

Total Output	External Demand	Internal Consumption

$$x_1 = 1{,}156{,}500 + 0.15x_1 + 0.15x_2 + 0.09x_3$$
$$x_2 = 685{,}000 + 0.50x_1 + 0.10x_3$$
$$x_3 = 1{,}668{,}000 + 0.10x_1 + 0.03x_2 + 0.20x_3$$

The three equations above constitute three linear equations in the three unknowns x_1, x_2, and x_3. To solve for the unknowns, we rewrite the equations as

$$0.85x_1 - 0.15x_2 - 0.09x_3 = 1{,}156{,}500$$
$$-0.50x_1 + x_2 - 0.10x_3 = 685{,}000$$
$$-0.10x_1 - 0.03x_2 + 0.80x_3 = 1{,}668{,}000$$

Solving for x_1, x_2, and x_3 by the Gauss-Jordan method, we find

$$x_1 = \$1{,}950{,}000$$
$$x_2 = \$1{,}900{,}000$$
$$x_3 = \$2{,}400{,}000$$

The conclusion is that the three sectors must produce $1,950,000 worth of electricity, $1,900,000 worth of coal, and $2,400,000 worth of steel to meet both external and internal demands.

Conclusion of Input-Output Model

The total output required of each sector is considerably more than the external demand of the corresponding product. The difference between the total output of each product and the external demand of that product is the amount (dollar value) of that product used within the economy. In this example they have the values

$$\text{Electricity consumed} = \text{Total Output} - \text{External Demand}$$
$$= \$1,950,000 - \$1,156,500$$
$$= \$793,500$$
$$\text{Coal consumed} = \text{Total Output} - \text{External Demand}$$
$$= \$1,900,000 - \$685,000$$
$$= \$1,215,000$$
$$\text{Steel consumed} = \text{Total Output} - \text{External Demand}$$
$$= \$2,400,000 - \$1,668,000$$
$$= \$732,000$$

The solution is illustrated in Figure 25.

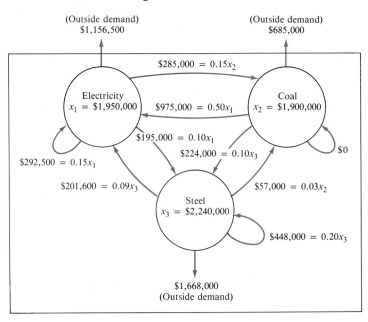

Figure 25
Flow of goods in dollars for a three-sector economy

The above analysis of this problem can be carried out by using matrix algebra. We summarize the results here.

Leontief Analysis of Three-Sector Economy with External Demand

Let the input-output matrix A and the external demand D be given as follows:

$$A = \begin{bmatrix} a_{11} & a_{12} & a_{13} \\ a_{21} & a_{22} & a_{23} \\ a_{31} & a_{32} & a_{33} \end{bmatrix} \qquad D = \begin{bmatrix} d_1 \\ d_2 \\ d_3 \end{bmatrix}$$

The total output of each sector X,

$$X = \begin{bmatrix} x_1 \\ x_2 \\ x_3 \end{bmatrix}$$

satisfies the linear system of equations:

$$\begin{array}{ccccc}
\text{Total} & & \text{External} & & \text{Internal} \\
\text{Output} & & \text{Demand} & & \text{Consumption} \\
X & = & D & + & AX
\end{array}$$

To solve for X, rewrite the above matrix equation as

$$X - AX = D$$

Factoring the left-hand side as the matrix product,

$$(I - A)X = D$$

we can solve the 3×3 linear system above either by the Gauss-Jordan method, by Gaussian elimination, or by finding the matrix inverse $(I - A)^{-1}$ and computing

$$X = (I - A)^{-1}D$$

Example 1

Two-Sector Economy   Find the output X of a two-sector economy with the following input-output matrix A and external demand D:

$$A = \begin{bmatrix} 0.3 & 0.4 \\ 0.5 & 0.3 \end{bmatrix} \qquad D = \begin{bmatrix} 150 \\ 100 \end{bmatrix}$$

Solution

- **Step 1** (compute $I - A$).   Computing the difference $I - A$, we find

$$I - A = \begin{bmatrix} 1 & 0 \\ 0 & 1 \end{bmatrix} - \begin{bmatrix} 0.3 & 0.4 \\ 0.5 & 0.3 \end{bmatrix}$$

$$= \begin{bmatrix} 0.7 & -0.4 \\ -0.5 & 0.7 \end{bmatrix}$$

- **Step 2** (find the inverse $(I - A)^{-1}$).   Finding the inverse of $(I - A)$, we have

$$(I - A)^{-1} = \begin{bmatrix} 0.7 & -0.4 \\ -0.5 & 0.7 \end{bmatrix}^{-1}$$

$$= \begin{bmatrix} 70/29 & 40/29 \\ 50/29 & 70/29 \end{bmatrix}$$

- **Step 3** (find $(I - A)^{-1}D$).   Computing $X = (I - A)^{-1}D$, we get

$$X = \begin{bmatrix} 70/29 & 40/29 \\ 50/29 & 70/29 \end{bmatrix} \begin{bmatrix} 150 \\ 100 \end{bmatrix}$$

$$= \begin{bmatrix} 500 \\ 500 \end{bmatrix}$$

Conclusion: In other words, if both sectors of this economy produce $500 worth of their respective products, then the external demands of $150 and $100 for these products will be met. Each sector must produce considerably more than the consumer (external) demand, owing to the high internal consumption of these products. □

HISTORICAL NOTE

Wassily W. Leontief (1906–   ) was born in St. Petersburg in Russia (now Leningrad) and began his studies at what is now the University of Leningrad at the age of 15. By the time he was 22, Leontief had received his Ph.D. in economics from the University of Berlin. His original economic theories were resisted by most economists of the time. It was not until World War II and afterward that detailed input-output tables for the U.S. economy were developed. At that time, input-output analysis was recognized as having a solid theoretical base and a tremendous potential for analyzing complex problems in economics.

Shop Scheduling Using Input-Output Analysis

Input-output analysis can be used in a variety of ways that have nothing to do with economic systems. An important area of business management is *shop scheduling.* Although the name may not sound very glamorous, it is a critical problem for many companies. Shop scheduling involves making sure the right amount of raw materials are available to produce a finished product. The difference between efficient and inefficient inventory control may represent millions of dollars in profit to some companies.

Suppose a manufacturing company makes three finished products, called P1, P2, and P3. Assume that these products are built from subassemblies A1, A2, A3, and A4, which the company buys from subcontractors. Suppose further that each subassembly is built from other subassemblies B1, B2, and B3, which the company also buys from subcontractors. The above rules for assembly are described by the diagram in Figure 26.

A typical problem in shop scheduling is the following. Suppose a market forecast determines a market for 30 P1's, 50 P2's, and 75 P3's. Someone must

Figure 26
Component diagram of a shop scheduling problem

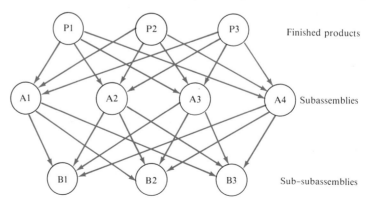

determine how many subassemblies should be ordered from subcontractors. The problem can be very complicated when dozens of subassemblies and subcontractors are involved. Many companies order subassemblies from hundreds of subcontractors. If care is not taken in these matters, a firm may find that it faces a critical deadline without sufficient inventory to produce the desired products.

We now present an example that will show how input-output analysis can be used to solve problems of this kind.

Example 2

Shop Scheduling Cleveland Prefab Metal Builders, Inc. makes and sells prefabricated metal buildings. A major bulwark in their product line is the Warehouse, a utility shed that sells for $499. Buyers can purchase the entire building (which is shipped unassembled) or one of a variety of replacement components. A component analysis of the Warehouse has been performed to determine its individual parts. The results are shown in Figure 27.

The numbers in Figure 27 signify

3 walls are used in each utility shed
20 bolts are used in each utility shed
1 roof is used in each utility shed
1 door is used in each utility shed
1 window is used in each wall
35 bolts are used in each wall
4 braces are used in each wall
10 bolts are used in each roof
8 braces are used in each roof
4 bolts are used in each door
2 braces are used in each door

Figure 27 seems to say that to make a wall, all that is needed is 1 window, 35 bolts, and 4 braces. Naturally, it takes more than this to make a wall. It is

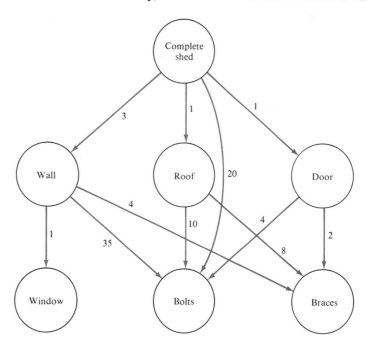

Figure 27
Component analysis of the Warehouse

just that the only components listed are those that the company sells. Suppose now that the company receives orders (external demands) for the following quantities:

100 complete sheds
60 walls
25 roofs
500 windows
200 doors
10,000 bolts
2000 braces

How many sheds, walls, roofs, windows, doors, bolts, and braces are required to meet the above orders?

Solution   We begin by constructing the component matrix A shown in Table 35. The columns of this matrix give the ingredients of each item. For example, the second column, labeled "Wall," says that each wall requires one window, 35 bolts, and 4 braces (at least those are the components of a wall that this company sells).

TABLE 35
Component Matrix *A* for the
Warehouse

	Shed	Wall	Roof	Door	Window	Bolts	Braces
Shed	0	0	0	0	0	0	0
Wall	3	0	0	0	0	0	0
Roof	1	0	0	0	0	0	0
Door	1	0	0	0	0	0	0
Window	0	1	0	0	0	0	0
Bolts	20	35	10	4	0	0	0
Braces	0	4	8	2	0	0	0

This problem can be interpreted as a seven-sector economy problem in disguise if we make the following analogies:

Input-Output Analysis   *Shop Scheduling*

input-output matrix $A \longleftrightarrow$ component matrix A

external demands $D \longleftrightarrow$ product demands D

output of sectors $X \longleftrightarrow$ number of components X

Hence we can find the number X of each component from the input-output equation for total output:

$$X = (I - A)^{-1}D$$

where

$$D = \begin{bmatrix} 100 \\ 60 \\ 25 \\ 500 \\ 200 \\ 10{,}000 \\ 2000 \end{bmatrix} \begin{matrix} \text{Sheds} \\ \text{Walls} \\ \text{Roofs} \\ \text{Windows} \\ \text{Doors} \\ \text{Bolts} \\ \text{Braces} \end{matrix} \qquad X = \begin{bmatrix} x_1 \\ x_2 \\ x_3 \\ x_4 \\ x_5 \\ x_6 \\ x_7 \end{bmatrix} \begin{matrix} \text{Sheds required} \\ \text{Walls required} \\ \text{Roofs required} \\ \text{Windows required} \\ \text{Doors required} \\ \text{Bolts required} \\ \text{Braces required} \end{matrix}$$

Substituting the values above into the equation for X, we get

$$X = \begin{bmatrix} 100 \\ 360 \\ 125 \\ 600 \\ 560 \\ 28,250 \\ 5640 \end{bmatrix}$$

The interpretation is that to meet a demand of

$$D = \begin{bmatrix} 100 \\ 60 \\ 25 \\ 500 \\ 200 \\ 10,000 \\ 2000 \end{bmatrix} \begin{array}{l} \text{Sheds} \\ \text{Walls} \\ \text{Roofs} \\ \text{Windows} \\ \text{Doors} \\ \text{Bolts} \\ \text{Braces} \end{array}$$

the company must produce or subcontract the following number of components:

$$\begin{aligned} x_1 &= 100 \text{ Complete sheds} \\ x_2 &= 360 \text{ Walls} \\ x_3 &= 125 \text{ Roofs} \\ x_4 &= 600 \text{ Windows} \\ x_5 &= 560 \text{ Doors} \\ x_6 &= 28{,}250 \text{ Bolts} \\ x_7 &= 5640 \text{ Braces} \end{aligned}$$

Conclusion: Although the above shed problem has only a few variables, it illustrates the general idea of shop scheduling. One of the recent trends in the automobile industry (and many other industries) is **zero inventory production**. That is, by sophisticated mathematical and computer ordering and scheduling systems, automobile companies keep the weekly inventory of raw material at a minimum. This means that profits can be increased, but it also means that the exact number of raw materials must be known for the upcoming weekly production schedule. Here is where shop scheduling becomes important.   ☐

Problems

For Problems 1–6, consider an economy with two sectors, Agriculture and Manufacturing. The interaction between them is given by Table 36.

1. Draw a diagram similar to Figure 24 illustrating the interaction between the two sectors.
2. In order that Agriculture produce a dollar's worth of output, how many dollars worth of the output of Agriculture and Manufacturing are used? In order that Manu-

facturing produce a dollar's worth of output, how many dollars' worth of output of Agriculture and Manufacturing are used?
3. Find the input-output matrix A.
4. Find the inverse matrix $(I - A)^{-1}$.
5. Find the total outputs that each sector must produce to meet both external and internal demands.
6. Draw a diagram similar to Figure 25 illustrating the flow of goods in dollars between the sectors.

TABLE 36
Interaction of Two-Sector Economy for Problems 1–6

	Agriculture	Manufacturing	External Demands
Agriculture	$0.06	$0.03	200
Manufacturing	$0.02	$0.05	300

For Problems 7–12, the input-output matrix in a two-sector economy is given by the following table:

	Sector 1	Sector 2
Sector 1	0.05	0.04
Sector 2	0.06	0.08

7. In order that Sector 1 produce a dollar's worth of output, how many dollars' worth of the output of Sectors 1 and 2 are used? In order that Sector 2 produce a dollar's worth of output, how many dollars' worth of output of Sectors 1 and 2 are used?
8. Find the inverse matrix $(I - A)^{-1}$.
9. Find the total outputs of each sector so that each sector meets both external and internal demands if the external demand of Sector 1 is $100 and the external demand of Sector 2 is $200.
10. Draw a diagram similar to Figure 25 illustrating the flow of goods in dollars.
11. Determine the new production levels if the external demand of Sector 1 increases by 50% and the external demand of Sector 2 increases by 25%.
12. Suppose the production levels of the two products are $150 worth of Product 1 and $250 worth of Product 2. How many dollars' worth of Products 1 and 2 are available for external (consumer) demand or consumption? *Hint:* In this problem, X is given and D is the unknown. Since we know A, simply compute $D = X - AX$.

For Problems 13–17, use the following input-output matrix for a three-sector economy:

	Sector 1	Sector 2	Sector 3
Sector 1	1/3	1/2	0
Sector 2	1/3	1/3	1/6
Sector 3	1/4	0	1/6

13. Determine $I - A$.

14. Verify that
$$\begin{bmatrix} 2.6 & 2.0 & 0.4 \\ 1.5 & 2.6 & 0.5 \\ 0.8 & 0.6 & 1.3 \end{bmatrix}$$
is a close approximation of $(I - A)^{-1}$.
15. **Production Levels** Calculate the production levels needed for each sector if the total demand is given by
$$D = \begin{bmatrix} 250 \\ 300 \\ 500 \end{bmatrix} \begin{matrix} \text{(dollars' worth of Sector 1)} \\ \text{(dollars' worth of Sector 2)} \\ \text{(dollars' worth of Sector 3)} \end{matrix}$$
16. **New Production Levels** Calculate the production levels needed for each sector if the external demand for each sector increases by 25% from the demands in Problem 15.
17. **Outside Consumption** How many dollars' worth of the output of each sector will be available for external (or consumer) use if the production levels are as follows:

700 dollars' worth of Sector 1
800 dollars' worth of Sector 2
750 dollars' worth of Sector 3

Hint: In this problem the production levels X are given, A was given earlier, and the problem is to find $D = X - AX$.

18. **Production Levels** The input-output matrix for a two-sector economy is given by the following matrix:
$$A = \begin{bmatrix} 0.25 & 0.40 \\ 0.14 & 0.12 \end{bmatrix}$$

For each of the external demand vectors given below, find the production levels for each sector.

	External Demands				
Sector 1	10	15	20	20	40
Sector 2	20	20	20	40	30

19. **Production Levels** The input-output matrix for a three-sector economy is given by the following matrix:

$$A = \begin{bmatrix} 11/41 & 19/240 & 1/185 \\ 5/41 & 89/240 & 40/185 \\ 5/41 & 37/240 & 37/185 \end{bmatrix}$$

For each of the external demands given below, find the total amount that must be produced by each sector.

	External Demands		
Sector 1	10	50	90
Sector 2	106	125	300
Sector 3	106	130	280

As an aid in the computations, $(I - A)^{-1}$ is given by

$$(I - A)^{-1} = \begin{bmatrix} 1.41 & 0.19 & 0.06 \\ 0.36 & 1.70 & 0.46 \\ 0.28 & 0.35 & 1.34 \end{bmatrix}$$

For Problems 20–22, use the input-output matrix for a five-sector economy given as follows:

	Metal	Machinery	Fuel	Agriculture	Labor	External Demands
Metal	0.10	0.20	0.30	0.02	0.10	100
Machinery	0.65	0.12	0.10	0.01	0.20	200
Fuel	0.10	0.18	0.10	0.07	0.10	50
Agriculture	0.05	0.37	0.10	0.08	0.55	650
Labor	0.10	0.13	0.40	0.82	0.05	1000

20. Input-Output Matrix Column by Column Which sector is the most important to the fuel sector? Explain.

21. Input-Output Matrix Column by Column For each dollar's worth of machinery produced, what are the labor costs?

22. Input-Output Matrix Column by Column For each dollar's worth of metals produced, what are the labor costs?

Figure 28 describes some of the major components of a microcomputer. A U.S. company buys computer components from suppliers in Korea and assembles the computers in the United States. The U.S. company sells both the assembled computer and the components in the United States. Assume that the company receives orders for 500 computers, 2500 memory chips, and 4000 disk drives. For each dollar earned on the sale of a computer it costs the computer company $0.25 for computer chips and $0.10 for disk drives.

23. What is the component matrix A that corresponds to the diagram in Figure 28?

24. What is the inverse matrix $(I - A)^{-1}$?

25. How many computers, memory chips, and disk drives must the company have in stock to meet the given orders?

For Problems 26–29, the component matrix A in Table 37 describes some of the major components of a bicycle:

TABLE 37
Component Matrix A for Problems 26–29

	Bicycle	Wheels	Spokes
Bicycle	0	0	0
Wheels	2	0	0
Spokes	0	50	0

26. Draw the diagram for this matrix.
27. Compute $(I - A)$.
28. Compute $(I - A)^{-1}$.
29. How many bicycles, wheels, and spokes are needed to supply an order for 50 bicycles, 75 wheels, and 1000 spokes? You should be able to answer this problem using

common sense. Does the given answer make sense?

A company manufacturers an air-conditioning control system, which it sells to builders of large buildings. The company also sells a given component of this air-conditioning system

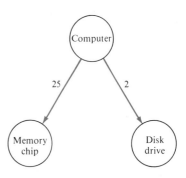

Figure 28
Components of a microcomputer

to other companies or to builders. The component matrix is given by the following 2 × 2 matrix:

	Control System	**Component**
Control System	0	0
Component	10	0

Problems 30–32 refer to the component matrix.

30. What is the inverse matrix $(I - A)^{-1}$?
31. Draw a diagram for this component matrix.
32. How many complete control systems and components will be needed to meet the expected sales demand of 400 control systems and 350 components?

Epilogue: Summary of Linear Systems

Matrices and systems of linear equations are two of the most important topics in all of mathematics. If every computer in the world suddenly stopped solving systems of equations and manipulating matrices, the modern world would be greatly disrupted. Airplane and hotel reservation systems would falter, communication networks would be in shambles, modern business would be in chaos, the Pentagon would be in a state of alert, and other operations too numerous to mention would be at a standstill. It is evident that any person who is competing in the modern business world should have an understanding of these contemporary tools.

Beginning with the first discussion in this chapter on solving small systems of equations by substitution and the add-subtract methods, there was a careful development of increasingly powerful methods for solving linear systems. A list of the tools and techniques learned in this chapter will help to focus attention on how they fit together.

- **Elementary Operations and Augmented Matrices.** After an explanation of how a system of equations could be transformed by elementary operations into an equivalent system, it became clear that the more compact and convenient

notation of augmented matrices would make the Gauss-Jordan method more efficient.

- **Matrix Operations and Inverses.** Matrices were found to be useful for compactly representing systems of equations as well as describing a wide variety of important problems. The interpretation of certain matrix operations often gives meaningful answers to a wide variety of problems. Although matrix division was not defined, some matrices had an inverse, which acted like a divisor. The existence of an inverse made the solution to many systems of equations easier. The ability to solve several systems with the same coefficient matrix is useful in scheduling problems.

- **Input-Output Analysis.** Input-output analysis is a powerful tool in macroeconomics. This method of economic analysis depends only on the mathematical tools learned in this chapter. A clear understanding of how this technique can be used to solve small problems, in which the calculations can be carried out by hand, provides a basis for insight into larger economic systems.

Key Ideas for Review

Gauss-Jordan method, 52
input-output matrix, 113
internal demand, 113
matrices, 81
   addition, 84
   augmented, 64

identity, 89
inverse, 99
multiplication, 87
subtraction, 84
method of substitution, 38
row operations on matrices, 65

RS game, 49
RST game, 57
technological matrix, 113

Review Exercises

1. Determine whether any of the pairs of values for x and y are solutions for the following system of equations:

$$2x + y = 4$$
$$x - 2y = 2$$

  (a) $x = 1$ (b) $x = 2$ (c) $x = 0$
      $y = 2$ $y = 0$ $y = 4$

2. A system of three equations in three unknowns has as its solution:

$$x = z + 1$$
$$y = 2z - 3$$
$$z = \text{arbitrary}$$

What is the solution for the system when $z = 1$? When $z = -1$?

Solve Problems 3–5 by using the method of substitution.

3. $3x + y = 3$
   $3x - 2y = 1$

4. $x + 2y = 0$
   $2x - 3y = 1$

5. $2x + y = 2$
   $x + 3y = 3$

Solve Problems 6–8 by finding the intersection of the two lines described by the equations.

6. $2x + y = 3$
   $x + y = 2$

7. $x + 3y = 5$
   $2x - y = 2$

8. $x - y = 3$
   $2x - 2y = 6$

9. Solve the following system of equations for x and y in terms of z:

$$3x - y + 2z = 3$$
$$x + 2y + z = 1$$

10. Curlew beat Emmetsburg by two goals. The two teams together scored 12 goals. What was the final score of the game?

11. Solve the following system of equations by using the method of substitution:

$$x + 2y \quad\quad = 3$$
$$x \quad\quad - z = 2$$
$$x + y + z = 4$$

For Problems 12–15, write the resulting system of equations after performing the indicated elementary operations. Start with the following system of equations:

$$2x + 3y - z = 4 \quad E_1$$
$$4x - 2y + 3z = 5 \quad E_2$$
$$x + y - 3z = 8 \quad E_3$$

12. $4E_2 \longrightarrow E_2$
14. $(1/2)E_1 \longrightarrow E_1$

13. $E_3 + (-2)E_1 \longrightarrow E_3$
15. $E_1 + 2E_2 \longrightarrow E_1$

Use elementary operations to solve the systems of equations in Problems 16–18.

16. $2x + y = 3$
   $x + 3y = 4$

17. $x - y + z = 3$
   $2x \quad + 3z = 5$
   $y + 4z = 2$

18. $x - 2y - 3z = 4$
   $2x + 3y - z = 3$
   $3x - y + 2x = -5$

For Problems 19–21, perform the indicated elementary row operations on the given matrices. A second or third operation is to be applied to the result of earlier operations.

   Row Operation   *Augmented Matrix*

19. $R_2 \longleftrightarrow R_3$ $\begin{bmatrix} 1 & 0 & 0 & 1 & | & 1 \\ 0 & 0 & 1 & 2 & | & 1 \\ 0 & 1 & 0 & 3 & | & 2 \end{bmatrix}$

20. $(1/3)R_1 \longrightarrow R_1$ $\begin{bmatrix} 3 & 1 & | & 2 \\ 4 & 0 & | & 3 \end{bmatrix}$
     $R_2 + (-4)R_1 \longrightarrow R_2$

21. $(1/2)R_1 \longrightarrow R_1$ $\begin{bmatrix} 2 & -1 & 3 & | & 2 \\ 3 & 2 & 0 & | & 1 \\ 4 & 3 & -1 & | & 1 \end{bmatrix}$
     $R_2 + (-3)R_1 \longrightarrow R_2$
     $R_3 + (-4)R_1 \longrightarrow R_3$

For Problems 22–25, determine the solution of the system of equations from the augmented matrices. Call the variables x, y, and z.

22. $\begin{bmatrix} 0 & 1 & 3 & | & 2 \\ 1 & 0 & 4 & | & 1 \\ 0 & 0 & 0 & | & 0 \end{bmatrix}$

23. $\begin{bmatrix} 1 & 0 & 0 & | & -3 \\ 0 & 1 & 0 & | & -2 \\ 0 & 0 & 0 & | & 0 \end{bmatrix}$

24. $\begin{bmatrix} 1 & 0 & 1 & | & 3 \\ 0 & 1 & 3 & | & 2 \\ 0 & 0 & 0 & | & 1 \end{bmatrix}$ **25.** $\begin{bmatrix} 1 & 0 & 1 & | & 1 \\ 0 & 0 & 0 & | & 0 \\ 0 & 0 & 0 & | & 0 \end{bmatrix}$

For Problems 26–29, complete the transformation of the matrix to final form by using elementary row operations. What is the solution of the system of equations?

26. $\begin{bmatrix} 1 & 3 & | & 1 \\ 0 & 2 & | & 4 \end{bmatrix}$ **27.** $\begin{bmatrix} 1 & 2 & 0 & | & 3 \\ 0 & 3 & 1 & | & 1 \\ 0 & 1 & 0 & | & 4 \end{bmatrix}$

28. $\begin{bmatrix} 2 & 1 & | & 2 \\ 3 & 4 & | & 3 \end{bmatrix}$ **29.** $\begin{bmatrix} 1 & 2 & 1 & | & 0 \\ 0 & 3 & 3 & | & 0 \\ 0 & -1 & 2 & | & 3 \end{bmatrix}$

For Problems 30–33, solve the system of equations using Gaussian elimination.

30. $2x + 3y = -1$
  $x - 2y = 3$

31. $x + y = 0$
  $2x - 4y = 1$

32. $2x - y + 2z = 3$
  $x + 2y = 1$
  $3y - 2z = 2$

33. $3x + 2y - z = 1$
  $x + z = 2$
  $x + y + z = 5$

34. Solve the following system of equations using the Gauss-Jordan method:

$$3x + 2y - z = 1$$
$$x + z = 2$$
$$x + y + z = 5$$

For Problems 35–45, use the matrices given to evaluate the given matrix expressions (if possible).

$$A = \begin{bmatrix} 1 & 2 \\ 2 & 1 \\ 3 & 2 \end{bmatrix} \quad B = \begin{bmatrix} 1 & -1 & 0 \\ 2 & 1 & 0 \\ 1 & 1 & -1 \end{bmatrix} \quad D = \begin{bmatrix} 3 & 1 & 5 \\ 2 & 1 & 4 \\ 1 & 2 & 4 \end{bmatrix}$$

$$C = \begin{bmatrix} 3 & 1 & 5 \end{bmatrix}$$

35. $B + D$ **36.** $3A$
37. $-2C$ **38.** BA
39. DB **40.** $2A + B - C$
41. $CD - DC$ **42.** $2B - D$
43. D^2 (D times itself)   **44.** A^2 (A times itself)
45. $B^2 + D^2$

For Problems 46–48, find the coefficient matrix for the system of equations and express the system of equations in terms of matrix operations.

46. $2x - 3y = 3$
  $x - 2y = 4$

47. $x + 2y - 3z = 5$
  $4x - 3y + z = 2$
  $-x + 2y - 4z = 6$

48. $x + z = 1$
  $ y + z + t = 2$
  $x + t - u = 3$
  $ y - z + u = 4$

Problems 49–51 refer to the following problem description. The Fast Freeze Ice Cream Co. sells sundaes, shakes, cones, and freezies. Table 38 shows how many units of each item each of its outlets sold last week.

TABLE 38
Information for Problems 49–51

	Sundaes	Shakes	Cones	Freezies
Fruit St.	150	275	800	900
Fourth St.	275	350	700	50
View Ave.	500	450	650	300
Campus Ave.	200	400	300	2500

The cost of each item is given below:

	Cost
Sundaes	$1.25
Shakes	$0.95
Cones	$0.75
Freezies	$0.50

49. What is the total sales for each store?
50. What was the total sales last week if this week saw a 50% sales increase in each item at each location?
51. What was the sales amount for just sundaes and cones at the Fruit St. and View Ave. stores?

52. Verify that the following matrices A and B are inverses:

$$A = \begin{bmatrix} 2 & 1 \\ 3 & 2 \end{bmatrix} \quad B = \begin{bmatrix} 2 & -1 \\ -3 & 2 \end{bmatrix}$$

53. Verify that the following matrices A and B are inverses:

$$A = \begin{bmatrix} 0 & 1 & -1 \\ 1 & 0 & 1 \\ 2 & 1 & 0 \end{bmatrix} \quad B = \begin{bmatrix} -1 & -1 & 1 \\ 2 & 2 & -1 \\ 1 & 2 & -1 \end{bmatrix}$$

For Problems 54–57, find the inverse of the given matrix, if it exists.

54. $\begin{bmatrix} 1 & 3 \\ 2 & 4 \end{bmatrix}$ **55.** $\begin{bmatrix} 3 & 2 \\ 1 & 1 \end{bmatrix}$

56. $\begin{bmatrix} 2 & 1 & 2 \\ 3 & 2 & 4 \\ 1 & 1 & 2 \end{bmatrix}$ **57.** $\begin{bmatrix} 2 & 1 & 2 \\ 3 & 2 & 4 \\ 1 & -1 & 1 \end{bmatrix}$

For Problems 58–60, find the inverse of the system of equations shown below and then use the inverse to solve the system of equations with the indicated values for a, b, and c.

$$x \qquad + 2z = a$$
$$5x + 3y - z = b$$
$$4x \qquad + 6z = c$$

58. $a = 6, b = 9, c = 4$ **59.** $a = 2, b = 3, c = 6$
60. $a = 4, b = 6, c = 6$

61. The input-output matrix of a two-sector economy is given by

$$A = \begin{bmatrix} 1/2 & 1/6 \\ 1/3 & 1/6 \end{bmatrix}$$

The external demands are given by

$$D = \begin{bmatrix} 100 \\ 300 \end{bmatrix}$$

What will the total output be for each sector of the economy?

62. The input-output matrix of a two-sector economy is given by

$$A = \begin{bmatrix} 0.3 & 0.4 \\ 0.5 & 0.3 \end{bmatrix}$$

The external demands are given by

$$D = \begin{bmatrix} 500 \\ 100 \end{bmatrix}$$

What should be the total output for each sector of the economy?

Chapter Test

For Problems 1–3, use the following system of equations:

$$x + y = 4$$
$$4x + y = 4$$

1. Solve by Gaussian elimination.
2. Solve by the Gauss-Jordan method.
3. Solve by finding the inverse of the coefficient matrix and multiplying it by the right-hand-side matrix.

4. Perform the indicated row operations on the following matrices. The second and third operation should be applied to the results of earlier operations.

Row Operations	Matrix
$(1/3)R_1 \longrightarrow R_1$	$\begin{bmatrix} 3 & 6 & 9 & 1 \\ 2 & 3 & 5 & 2 \\ 4 & 1 & 0 & 0 \end{bmatrix}$
$R_2 + (-2)R_1 \longrightarrow R_2$	
$R_3 + (-4)R_1 \longrightarrow R_3$	

5. What are the solutions (if any) of the systems of equations represented by the following augmented matrices? Call the variables x and y for two equations and x, y, and z for three equations.

(a) $\begin{bmatrix} 1 & 0 & | & 3 \\ 0 & 0 & | & 0 \end{bmatrix}$ (b) $\begin{bmatrix} 1 & 0 & 0 & | & 3 \\ 0 & 1 & 0 & | & 4 \\ 0 & 0 & 1 & | & 0 \end{bmatrix}$

(c) $\begin{bmatrix} 1 & 0 & 0 & | & 1 \\ 0 & 1 & 0 & | & 0 \\ 0 & 0 & 0 & | & 1 \end{bmatrix}$ (d) $\begin{bmatrix} 1 & 0 & 0 & | & 0 \\ 0 & 1 & 0 & | & 1 \\ 0 & 0 & 0 & | & 0 \end{bmatrix}$

6. The input-output matrix for a two-sector economy is given by

$$A = \begin{bmatrix} 0.25 & 0.40 \\ 0.10 & 0.10 \end{bmatrix}$$

If the external demand for the output of each sector is

$$D = \begin{bmatrix} 10 \\ 20 \end{bmatrix}$$

find the production level of each sector.

2

Linear Programming

The general formulation of the linear programming problem and its solution, found in the middle part of this century, is one of the most important mathematical discoveries of the past 100 years. In this chapter we present the linear programming problem in two variables and show how it can be solved by graphical methods. The core of the chapter is the simplex method, which is a step-by-step technique for solving the linear programming problem.

We close by presenting the important related topics of duality and postoptimal analysis.

2.1 Systems of Linear Inequalities

PURPOSE

We introduce the concept of a system of linear inequalities, which is used in linear programming. In particular, we learn

- how to interpret a system of linear inequalities and
- how to solve a system of linear inequalities.

Introduction

Most students are probably more familiar with systems of linear equations than with systems of linear inequalities. However, many problems in business and science are better described by inequalities than by equalities. For example, a nutritionist planning a diet must make sure that certain minimum requirements of vitamins and minerals are met. A system of linear inequalities is found that represents conditions on the amounts of foods that the diet must contain. A manufacturing problem may be solved by using certain inequalities that represent production constraints such as the availability of materials and/or labor. To study optimization problems using linear programming, you need a clear understanding of linear systems of inequalities.

A Single Inequality

We begin by studying a single **linear inequality** in two variables x and y. Inequalities of this type come in four kinds.

Linear Inequalities in Two Variables

Let a, b, and c be real numbers. The four linear inequalities in two variables x and y are

1. $ax + by \leq c$ (less than or equal to)

2. $ax + by < c$ (less than)

3. $ax + by \geq c$ (greater than or equal to)

4. $ax + by > c$ (greater than)

For inequalities of these kinds the goal is to find all pairs (x, y) for which the inequality holds. Pairs (x, y) for which the inequality holds are called solutions of the inequality or **feasible points**. Points that do not satisfy the inequality are called **infeasible points**.

To understand the geometric interpretation of the solutions of a linear inequality, remember from the Finite Mathematics Preliminaries that a linear equation of the form

$$ax + by = c$$

defines points on a straight line in the xy-plane. If the equality is replaced by an inequality, the inequality defines points (x, y) that lie on the line and/or on one side of the line.

Geometric Interpretation of Linear Inequalities

The straight line

$$ax + by = c$$

separates the xy-plane into two separate sides. (Neither side includes the line itself.) Points (x, y) on one side of the line are solutions of the strict inequality

$$ax + by < c$$

and the points on the other side are solutions of

$$ax + by > c$$

The two sides of the line *and* the line itself are defined by the nonstrict inequalities

$$ax + by \leq c$$
$$ax + by \geq c$$

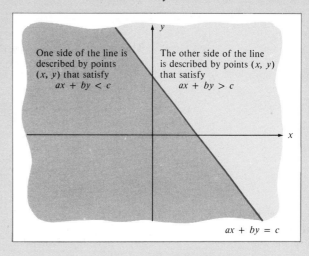

Solving a Linear Inequality

- **Step 1** (draw the boundary line).   Replace the inequality by an equality and graph the resulting straight line. Draw a dotted line if the inequality is of the form < or >; draw a solid line if the inequality is of the form $\leq$ or $\geq$. This line is the **boundary line** of the feasible set.

- **Step 2** (determine the correct side of the line to shade).   Pick any point that is not on the boundary line as a **test point**. If the point satisfies the inequality, then shade the side of the line that includes the point; otherwise, shade the points on the other side of the line. The origin is often a good test point, provided that it does not lie on the line. (See Figure 1.)

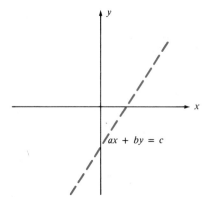

Figure 1
Two-step process for solving a linear inequality

Step 1. Draw dotted line for strict inequality   Step 2. Test point satisfies the inequality

---

Example 1 _____   Solving a Nonstrict Inequality   Solve the following linear inequality

$$2x + y \leq 3$$

and shade the set of feasible points in the xy-plane.

Solution

- **Step 1** (draw the boundary line).   Replacing the inequality by an equality, we have

$$2x + y = 3$$

This equation describes the boundary line of the set of feasible points shown in Figure 2(a).

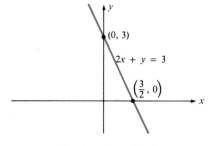

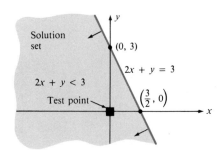

Figure 2
Solution consists of points on the line and points below and to the left of the line

(a) First draw the straight line   (b) Then pick a test point

· **Step 2** (shade the correct side of the boundary line). To determine the correct side of the boundary line, select a point that is not on the line and determine whether this test point satisfies the inequality. In this problem the origin satisfies the inequality ($0 + 0 \leq 3$ is true), and so the feasible set is shaded *below and to the left of the line* as indicated in Figure 2(b).   □

Example 2 ——————

Solution of an Inequality   Find the solution set of the strict inequality

$$x + y > 1$$

Solution

· **Step 1** (draw the boundary line).   Replacing the inequality by an equality, we have

$$x + y = 1$$

We draw this boundary line as a dotted line (since we have a strict inequality) as shown in Figure 3(a).

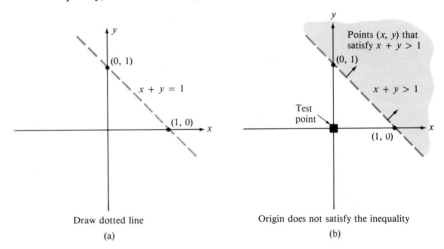

Figure 3
Solution of the inequality
$x + y > 1$

Draw dotted line   Origin does not satisfy the inequality
(a)   (b)

· **Step 2** (shade the correct side of the boundary line).   Since the origin $(0, 0)$ does not satisfy the inequality ($0 + 0 > 1$ is false), we shade the side of the line *above and to the right* as shown in Figure 3(b).   □

Systems of Linear Inequalities

For a **system of linear inequalities** in two variables a pair (x, y) is a solution if it simultaneously satisfies each linear inequality of the system. Geometrically, we seek to find those points in the plane that simultaneously lie on the correct side of several lines. We illustrate these ideas by means of examples.

Example 3 ——————

Two Inequalities   Find the points in the plane that satisfy the following system of inequalities:

$$x + y \leq 4$$
$$x - y \leq 0$$

Solution

- **Step 1** (replace inequalities by equalities). Replacing inequalities by equalities, we have

$$x + y = 4$$
$$x - y = 0$$

We draw each of these lines as a solid line, since each inequality is a nonstrict inequality. See Figure 4(a).

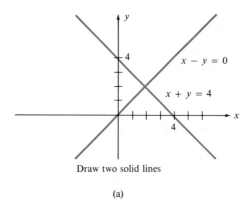

Draw two solid lines

(a)

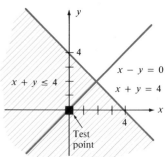

Shade the solution of one of
the inequalities

(b)

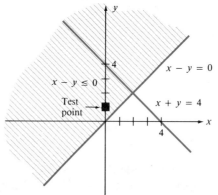

Shade the solution of the other
inequality

(c)

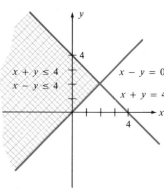

Shade the intersection of the
regions found in (b) and (c)

(d)

Figure 4
*Four steps in solving a
system of two inequalities*

- **Step 2** (determine the correct sides of the lines). Pick a test point for each inequality and shade the correct sides. Since the origin satisfies the first inequality

$$x + y \leq 4$$

we shade the side of the line

$$x + y = 4$$

that includes the origin. See Figure 4(b). For the second inequality

$$x - y \leq 0$$

we pick the test point (0, 1). Since it satisfies the inequality, we shade the region above and to the left of the line

$$x - y = 0$$

as shown in Figure 4(c). The *intersection* of the two shaded areas, illustrated in Figure 4(d), is the solution of this system of inequalities.   □

Example 4

Three Inequalities   Shade the set of feasible points for the system of three inequalities

$$x + y \leq 4$$
$$2x + y \geq 4$$
$$x - y \leq 0$$

Solution

• **Step 1** (replace inequalities by equalities).   Replacing the inequalities by equalities, we have

$$x + y = 4$$
$$2x + y = 4$$
$$x - y = 0$$

Draw the graphs of these three lines as *solid lines* as shown in Figure 5(a).

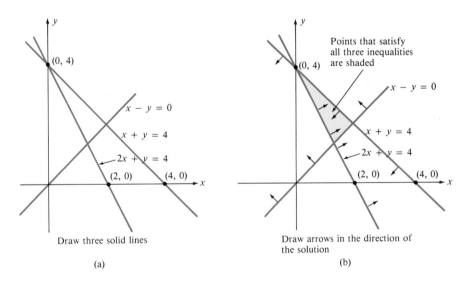

Figure 5
*Solution of three
simultaneous inequalities*

Draw three solid lines   Draw arrows in the direction of
                                          the solution

(a)   (b)

• **Step 2** (determine the correct side of the lines).   Select a test point for each of the three inequalities and shade the correct side of each of the three lines. After the three sides have been determined, shade the region common to all three. These steps are illustrated in Figure 5(b).   □

Inequalities and the American Diabetes Association

Inequalities and systems of inequalities are important in business management and science, since they allow us to analyze in a precise way relationships between different quantities. The following example is only one of many that we could present to illustrate how systems of inequalities arise in the real world.

The American Diabetes Association's meal plans (diets) are given in terms of **exchanges**. The six food exchanges represent the different nutrients that any healthy person should eat. The basic food exchanges are bread, meat, vegetables, fruit, fat, and milk. Within each exchange are a number of foods, each having the same number of carbohydrates, protein, and fat (and hence calories). For example, typical bread exchanges are 1 slice of bread, 5 white crackers, or 2 graham crackers. Typical meat exchanges are 1 oz chicken, 1/4 oz tuna, or 1 oz fish. By selecting different foods within a given exchange, a diabetic person can conveniently plan a wide variety of meals that have the same nutritional properties. Table 1 shows the amounts of carbohydrates, protein, fat, and calories in the bread and meat exchanges.

TABLE 1

Amounts of Carbohydrates, Protein, and Fat in the Bread and Meat Exchanges

Ingredients	Exchanges	
	Bread	*Meat*
Carbohydrates	15 g/exchange	0 g/exchange
Protein	2 g/exchange	7 g/exchange
Fat	0 g/exchange	5 g/exchange
Calories	60 calories	75 calories

Suppose you are a dietician and want to recommend to a patient a snack consisting of bread and meat exchanges. Suppose, too, that you would like the snack to contain at least

20 grams of carbohydrates

14 grams of protein

5 grams of fat

How many bread and meat exchanges should you include in the snack to meet these requirements?

To solve this problem, we begin by letting

x = number of bread exchanges in the snack

y = number of meat exchanges in the snack

If the snack consists of x exchanges of bread and y exchanges of meat, then the number of grams of carbohydrates, protein, and fat in the snack will be

Grams of carbohydrates in snack = $15x + 0y$

Grams of protein in snack = $2x + 7y$

Grams of fat in snack = $0x + 5y$

For the snack to contain at least the given amounts of carbohydrates, protein, and fat, the number of bread and meat exchanges x and y must satisfy the inequalities

$$15x + 0y \geq 20$$
$$2x + 7y \geq 14$$
$$0x + 5y \geq 5$$

In addition, since x and y represent the number of food exchanges, it is clear that they are nonnegative, and so the following inequalities must be satisfied:

$$x \geq 0$$
$$y \geq 0$$

Hence any solution (x, y) of this system of five inequalities will define a snack of x bread exchanges and y meat exchanges that will meet the requirements for carbohydrates, protein, and fat. The feasible set is shaded in Figure 6.

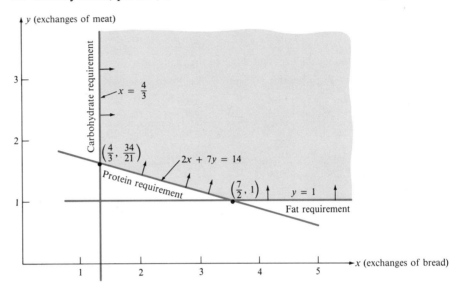

Figure 6
Feasible bread and meat exchanges that meet carbohydrate, protein, and fat requirements

Comments on the Feasible Set

There are an infinite number of points (x, y) in the feasible set in Figure 6. This means that there are an infinite number of possible snacks that satisfy the given requirements. The problem the dietician faces is: Which feasible snack should be chosen? Of course, a serving of 500 exchanges of bread and 500 exchanges of meat will obviously satisfy the three requirements, but it is not exactly what the dietician had in mind. Common sense tells us that the feasible point (4/3, 34/21), which means 1.33 bread exchanges and 1.62 meat exchanges, is a reasonable snack. A natural question to ask at this point is: What are the values of x and y that meet the requirements for carbohydrates, protein, and fat and at the same time have the smallest number of calories? This is the linear programming problem that will be studied in the next section.

Problems

For Problems 1–10, shade the points in the xy-plane that satisfy the following inequalities.

1. $x + y \leq 1$
2. $-x - y \geq -1$
3. $x + 2y \leq 2$
4. $2x + y \leq 2$
5. $x \geq 2$
6. $-x + y \geq 3$
7. $3x - 4y \leq 0$
8. $y \leq -2$
9. $5x + 15y \leq 50$
10. $x - y \leq 5$

For Problems 11–14, shade the points in the xy-plane that satisfy the following systems of linear inequalities.

11. The square inequalities
$$x \geq 0$$
$$y \geq 0$$
$$x \leq 1$$
$$y \leq 1$$

12. The triangle inequalities
$$x - y \geq 0$$
$$x + y \leq 1$$
$$y \geq 0$$

13. Mystery figure
$$y \leq 1$$
$$y - x \leq 0$$
$$x + y \leq 3$$
$$x + y \geq 1$$
$$y - x \geq -2$$
$$y \geq 0$$

14. House inequalities
$$y - x \leq 1$$
$$x + y \leq 3$$
$$x \leq 2$$
$$x \geq 0$$
$$y \geq 0$$

15. Can you find the inequalities that describe the diamond in Figure 7?

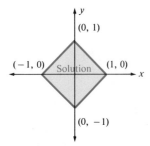

Figure 7
Diamond for Problem 15

For Problems 16–17, find the coordinates of the corner-points A, B, and C of the following solution sets. Corner-points occur at the intersection of the boundary lines.

16.
$$x + 2y \geq 2$$
$$2x + \quad y \geq 2$$
$$x \geq 0$$
$$y \geq 0$$
(Figure 8)

17.
$$x + y \leq 1$$
$$x \geq 0$$
$$y \geq 0$$
(Figure 9)

Linear inequalities are said to be written in **standardized form** if they are written in one of the following two forms:

$$y \leq \text{"everything else"}$$
$$y \geq \text{"everything else"}$$

For example, the inequality

$$-x + y \leq 5$$

can be rewritten in standardized form as

$$y \leq x + 5$$

It is clear from this standardized form that the solutions of the inequality consist of all points on and below the straight line

$$y = x + 5$$

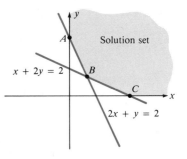

Figure 8

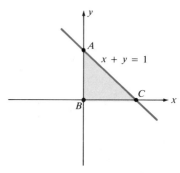

Figure 9

For Problems 18–20, rewrite the following inequalities in standardized form and shade the solution set:

18. $2x - y \leq -3$ **19.** $-x + y \geq 5$

20. $-3x + 4y \leq 0$
$$x - \quad y \geq 2$$
$$x \geq 0$$
$$y \geq 0$$

For Problems 21–24, determine which of the following test points satisfy the given inequalities.

21. $x + y \leq 1$, $(0, 0)$ **22.** $-2x + 3y \geq 4$, $(1, -1)$
23. $2x + y \leq 2$, $(1/2, 1/2)$ **24.** $x \leq 5$, $(5, 4)$

Linear Inequalities in Business Management and Science

25. Manufacturing A manufacturer of running shoes makes two basic kinds, jogging shoes and racing flats. Data collected from the company shows some of the major raw materials that make up these two products, as seen in Table 2.

Let

$x =$ the number of jogging shoes produced each day

$y =$ the number of racing flats produced each day

Write the three inequalities that x and y must satisfy so that the company does not use more than the maximum available resources. Illustrate the solutions of these inequalities in the xy-plane. Suppose the company makes a profit of \$18 for each pair of jogging shoes and \$25 for each pair of racing flats. Pick any feasible point and compute the profit for that point.

TABLE 2
Raw Materials for Running Shoes in Problem 25

	Jogging Shoes	Racing Flats	Maximum Resource Available per Day
Labor	0.15 hr	0.25 hr	800 hours
Machine	3 min	3 min	10,000 minutes
Nylon	10 sq. in.	6 sq. in.	30,000 square inches
Profits	$18.00	$25.00	

26. Patient Requirements A patient is required to take at least 100 units of Drug I and 150 units of Drug II each day. Suppose that drug companies do not manufacture tablets of these drugs in pure form but that there are medicines available that contain these drugs in varying amounts. Suppose the three medicines A, B, and C contain the mixes of these two drugs shown in Table 3. What inequalities describe the mixtures of medicines A, B, and C (in grams) the patient can take each day to meet the minimum daily requirements?

TABLE 3
Drug Information for Problem 26

	Medicine A	Medicine B	Medicine C
Drug I	7 units/g	9 units/g	5 units/g
Drug II	5 units/g	7 units/g	5 units/g

27. Mixing Problem What are the different meals of Corn Munchies and Wheat Munchies that will meet the minimum requirements for vitamins A, B_2, and C? The relevant information is given in Table 4.

TABLE 4
Vitamin Information for Problem 27

	Units of Vitamins per oz Corn Munchies	Units of Vitamins per oz Wheat Munchies	Minimum Requirements
A	1000	800	4500 Units
B_2	5	8	200 Units
C	10	10	500 Units

28. Manufacturing A furniture manufacturer makes chairs, tables, desks, and bookcases. The raw materials that go into these products are lumber, glue, labor, leather, and glass. Suppose the total amount of resources available per week is as follows:

- 20,000 board-feet mahogany lumber
- 4000 hours labor
- 2000 ounces glue
- 3000 square feet leather
- 500 square feet glass

Suppose the constraints on the number of chairs (C), tables (T), desks (D), and bookcases (B) are described in Figure 10. Use the inequalities in this figure to answer the following questions.

(a) What is the meaning of the numbers 5, 10, 3, and 4 in the first column of Figure 10?

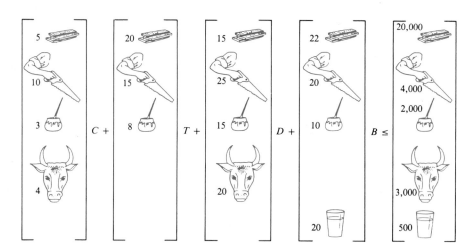

Figure 10
*Pictorial illustration of a system
of inequalities*

(b) What is the meaning of the numbers 20, 15, and 8 in the second column of Figure 10?

(c) What is the meaning of the numbers 15, 25, 15, and 20 in the third column of Figure 10?

(d) What is one set of feasible values of C, T, D, and B?

29. Mixing Problem A nut company has the following inventory of nuts:

- 5000 lb of peanuts
- 2000 lb of Brazil nuts
- 3000 lb of cashews
- 1000 lb of almonds

The company sells three different mixes of these four basic kinds of nuts and labels them Plain, Fancy, and Extra Fancy. Table 5 lists the contents of each of these three mixes per one-pound can. What system of inequalities must the number of cans of Plain, Fancy, and Extra Fancy mixes satisfy in order for the company to stay within its inventory?

TABLE 5
Nut Mixtures for Problem 29

	Plain (P)	Fancy (F)	Extra Fancy (E)
Peanuts	8 oz/can	4 oz/can	2 oz/can
Brazil nuts	4 oz/can	4 oz/can	8 oz/can
Cashews	2 oz/can	4 oz/can	2 oz/can
Almonds	2 oz/can	4 oz/can	4 oz/can

30. Manufacturing A manufacturing company is considering marketing three new products, which we call Products 1, 2, and 3. The products require many raw materials but only two that might limit production. The availability of these raw materials is given below.

The number of hours of use for these two raw materials needed to make each unit of the three products is listed in Table 6. Let P_1, P_2, and P_3 be the number of units of Products 1, 2, and 3 produced. What inequalities must these values satisfy so that they stay within the given inventory?

Raw Material	Available Each Week
Labor	500 hours
Lathe	100 hours

TABLE 6
Production Information for Problem 30

	Product 1 (P_1)	Product 2 (P_2)	Product 3 (P_3)
Labor hours	0.10 hr/unit	0.10 hr/unit	0.05 hr/unit
Lathe	0.05 hr/unit	0.10 hr/unit	0.10 hr/unit

31. Inequality Version of the ST Game   You have available a large collection of squares and triangles as shown in Figure 11. On each square is drawn three black dots and five color dots. On each triangle is drawn five black dots and one color dot. What combination of squares and triangles can be chosen so that the number of black dots is less than 40 and the number of color dots is less than 20? *Hint:* This problem involves inequalities, but you should realize that the number of squares and triangles must be positive integers.

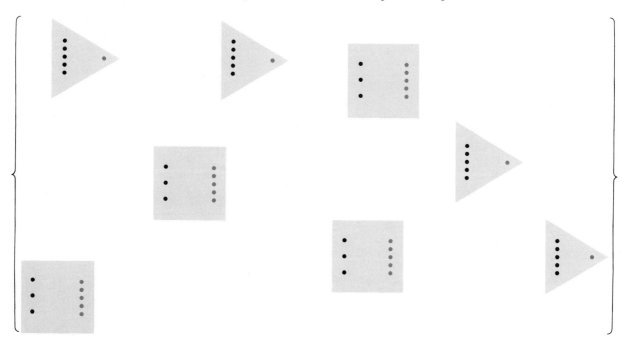

Figure 11

The number of grams of carbohydrates, protein, and fat in each of six food exchanges of the American Diabetic Associa-tion is listed in Table 7. For Problems 32–36, you are asked to plan meals for a diabetic patient. Each meal will contain

TABLE 7
Nutrition Information for Problems 32–40

	Food Exchange					
	Milk (*m*)	Vegetable (*v*)	Fruit (*f*)	Bread (*b*)	Meat (*t*)	Fat (*a*)
Carbohydrates	12	7	10	15	0	0
Protein	8	2	0	2	7	0
Fat	10	0	0	0	5	5

two of the six food exchanges. Write the system of two inequalities that must be satisfied. The number of food exchanges can be denoted by the variables m, v, f, b, t, and a. Solve the system of inequalities to determine all possible combinations of the two food exchanges that will meet the minimum requirements described by each problem.

			Minimum Requirements		
	Exchanges in Meal		Carbohydrates	Protein	Fat
32.	Milk	Bread	25	12	5
33.	Bread	Meat	25	15	10
34.	Vegetable	Fat	10	5	8
35.	Fruit	Meat	15	7	10
36.	Vegetable	Fruit	10	10	0

For Problems 37–40, you are asked to plan meals for a diabetic patient. Each meal will contain three of the six food exchanges. Write the system of three inequalities that must be satisfied. Do not solve the system of inequalities.

				Minimum Requirements		
	Exchanges in Meal			Carbohydrates	Protein	Fat
37.	Milk	Vegetable	Meat	30	15	20
38.	Milk	Fruit	Fat	25	15	20
39.	Fruit	Meat	Fat	20	15	10
40.	Vegetable	Bread	Fat	30	5	10

2.2   Linear Programming in Two Dimensions

PURPOSE

We introduce linear programming problems with two variables and show how these problems can be solved by both the graphical method (method of isoprofit/isocost lines) and the corner-point method. Some of the major ideas introduced are

- the objective function (what we maximize or minimize),
- the decision variables (what we have control over),
- the constraints (physical limitations),
- the feasible set (solutions of the constraints),
- the isoprofit/isocost lines (lines where the objective function is a constant),
- the corner-points of the feasible set (candidates for the optimal feasible point), and
- the optimal feasible point (the corner-point that maximizes or minimizes the objective function).

Introduction

Linear programming is probably the most used and best known mathematical technique in the general area of management science and operations research. Some researchers in these areas have even gone so far as to say that linear programming is one of the most important mathematical discoveries of the twentieth century. Whether that is true or not, the variety of problems in business, science, and industry that can be solved by using the linear programming model is impressive, to say the least. Problems in transportation scheduling, capital budgeting, production planning, inventory control, advertising, and many more areas can be solved by the linear programming approach. In recent years, people have discovered applications of this technique in areas that were completely unimagined in the early days following its discovery. For example, AT&T uses linear programming to route long-distance telephone calls through hundreds or thousands of cities with maximum efficiency.

HISTORICAL NOTE

Linear programming is truly a contemporary subject. Much of the original work dates back only to 1947 and the work of the U.S. mathematician George B. Dantzig (1914–   ), who was working on problems in military logistics for the U.S. Air Force. The original name given to the subject was Programming of Independent Activities in a Linear Structure. Later the name was changed to linear programming. The word programming refers to "planning" and has nothing to do with computer programming, although the technique often used for solving large linear programming problems (the simplex method) involves a sufficient number of arithmetic operations to require the services of a computer. New theories and applications of linear programming are being discovered every year. Over the past ten years, two new methods for solving the linear programming problem have been discovered:

1. the Khachian algorithm (1979) and

2. the Karmarkar algorithm (1984).

Whether these new methods will replace the simplex method remains to be seen.

We introduce linear programming by means of the diabetic snack problem.

Linear Programming and the Snack Problem

In the previous section we saw that the number of bread and meat exchanges in a diabetic snack must satisfy a system of three linear inequalities so that the snack meets minimum requirements for carbohydrates, protein, and fat. The number of grams of carbohydrates, protein, and fat (and hence calories) in each of these two exchanges is given in Table 8.

TABLE 8
Number of grams of
carbohydrates, protein, and
fat in bread and meat
exchanges, plus the
minimum requirements for
carbohydrates, protein, and
fat

| | Exchange | | Minimum |
	Bread	*Meat*	**Requirements**
Carbohydrates	15	0	20 gm
Protein	2	7	14 gm
Fat	0	5	5 gm
Calories	60	75	

In particular, we showed that if

$$x = \text{number of exchanges of bread in the snack}$$

$$y = \text{number of exchanges of meat in the snack}$$

then for a snack consisting of x bread exchanges and y meat exchanges to contain at least 20 grams of carbohydrates, 14 grams of protein, and 5 grams of fat, the values of x and y must satisfy the following system of linear inequalities:

$$15x \qquad \geq 20 \qquad \text{(carbohydrates constraint)}$$
$$2x + 7y \geq 14 \qquad \text{(protein constraint)}$$
$$5y \geq \ 5 \qquad \text{(fat constraint)}$$

Since the values of x and y are numbers of exchanges, it is obvious that they satisfy the nonnegative constraints

$$x \geq 0$$
$$y \geq 0$$

The feasible set for this linear system of five inequalities is shown in Figure 12.

The feasible set of points in Figure 12 represents all possible combinations of bread and meat exchanges that meet the carbohydrate, protein, and fat requirements.

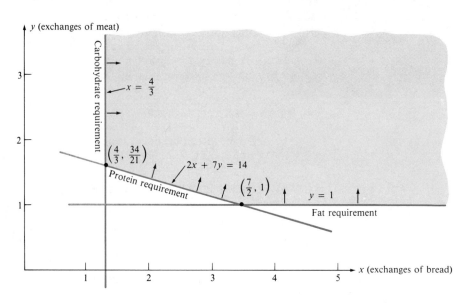

Figure 12
*Feasible bread and meat
exchanges that meet carbohydrate,
protein, and fat requirements*

Suppose now that we go one step further. Consider finding the point (x, y) in the feasible set that has the smallest number of calories. In other words, find the snack consisting of bread and meat exchanges that meet the minimum requirements for carbohydrates, protein, and fat and at the same time contains the least number of calories. To find this minimum calorie snack, we first write the expression for the number of calories in a snack that contains x exchanges of bread and y exchanges of meat. We get the following linear function:

$$\text{Number of calories} = 60x + 75y$$

We are now ready to examine the linear programming problem.

Definition of the Major Concepts of Linear Programming

- *Decision variables.* The variables x and y in the snack problem are called **decision variables**. Decision variables generally represent quantities that we have some control over (such as how many bread and meat exchanges should be included in a snack). The number of decision variables in a linear programming problem is always two or more. Some problems have as many as several hundred decision variables.

- *Objective function.* The expression being **maximized** or **minimized** (minimized in the previous problem) is called the **objective function**. Any expression of the form

$$C = ax + by$$

  where a and b are real numbers is called a **linear function** of x and y. One reason for the name "linear programming" is that the objective function is a linear function. We often call the objective function P (for profit) when it is being maximized and C (for cost) when it is being minimized.

- *Constraints.* The system of five inequalities in the snack problem represents restrictions on the decision variables x and y. In this problem they represent restrictions on the number of exchanges of bread and meat. Points (x, y) that satisfy these **constraints** are called **feasible points**, and the collection of all these points is called the **feasible set**. The constraints are subdivided into two types, the *functional constraints* and the *nonnegative constraints.*

We can now state the general linear programming problem.

The Linear Programming Problem

A linear programming problem consists of maximizing or minimizing a linear objective function of two or more decision variables that are subject to constraints in the form of a system of linear inequalities. All of the decision variables must be nonnegative.

We now determine the number of bread and meat exchanges that should be included in a snack to meet the minimum requirements for carbohydrates, protein, and fat, while at the same time having a minimum number of calories. In other words, we will solve the following linear programming problem.

Linear Programming Problem for the Snack Problem

Find the values of x and y that will minimize

$$C = 60x + 75y \qquad \text{(objective function)}$$

subject to

$$15x \qquad \geq 20 \qquad \text{(functional constraints)}$$
$$2x + 7y \geq 14$$
$$5y \geq 5$$

and

$$x \geq 0 \qquad \text{(nonnegative constraints)}$$
$$y \geq 0$$

Solution by the Graphical Method

We begin by graphing the feasible set (solution of the system of five inequalities), which will give us the values of x and y that meet the carbohydrate-protein-fat requirements. (See Figure 13.)

What we know so far is that all of the points in the shaded region of Figure 13 will meet the minimum requirements for carbohydrates, protein, and fat. We would like to know which feasible point minimizes the objective function. To find this minimizing point, draw the straight lines

$$60x + 75y = c$$

called **isocost lines** for different values of c. Since the objective function has the same value at all points on a single line, all of the points on a line represent

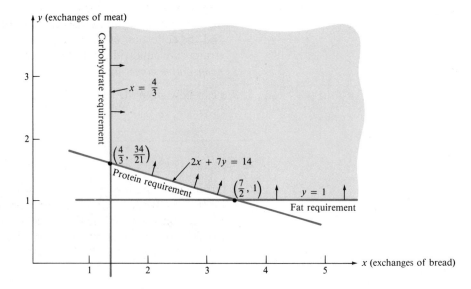

Figure 13
Feasible bread and meat exchanges that meet carbohydrate, protein, and fat requirements

snacks with the same number of calories. For example, all the points (x, y) on the line

$$60x + 75y = 500$$

describe snacks that have 500 calories. By drawing these lines for smaller and smaller values of c, we can determine the point in the feasible set at which the objective function (or number of calories) is a minimum. It is clear from Figure 14 that the smallest value of c for which an isocost line

$$60x + 75y = c$$

intersects the feasible set is the c for which the isocost line passes through the intersection of the two boundary lines

$$15x \qquad = 20$$
$$2x + 7y = 14$$

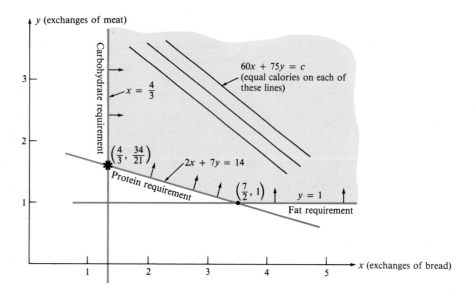

Figure 14
Lines where the objective function is a minimum

Solving these two equations to find the point of intersection, we get

$$x^* = 4/3 \cong 1.33$$
$$y^* = 34/21 \cong 1.62$$

Hence the optimal feasible point (or solution) of the linear programming problem is (1.33, 1.62). The value of the objective function at this point is

$$C^* = 60x^* + 75y^*$$
$$= 60(1.33) + 75(1.62)$$
$$= 201.4$$

Interpretation for Snacks

In other words, a snack consisting of 1.33 exchanges of bread and 1.62 exchanges of meat will meet the carbohydrates-protein-fat requirements and at the same time have a smaller number of calories (201.4 calories) than any other snack that meets these requirements.

The above problem illustrates the **graphical** (or isoprofit/isocost) **method** for solving linear programming problems in two variables. We summarize this method here.

The Graphical Solution of Linear Programming Problems

- **Step 1.**   Shade the feasible set.
- **Step 2.**   For maximizing problems, graph the straight lines

$$\text{Objective function} = c$$

for larger and larger values of c and determine the point in the feasible set that is the one last touched by one of these lines. This point maximizes the objective function and is called the **optimal feasible point**.

The lines are called **isoprofit lines** when the objective function is being maximized, and the method is often called the **isoprofit method**.

Minimizing problems are solved in the analogous way by graphing the **isocost lines**

$$\text{Objective function} = c$$

for smaller and smaller values of c. The point in the feasible set that is the one last touched by an isocost line is the point that minimizes the objective function and is called the optimal feasible point. For minimizing problems, the method is called the **isocost method**.

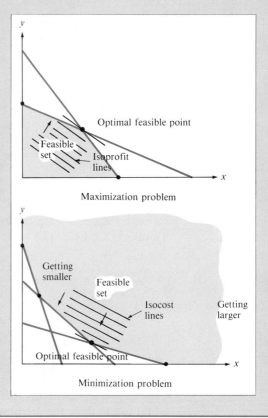

Maximization problem

Minimization problem

Example 1

Graphical Method   Solve the following maximizing linear programming problem.

$$\text{Maximize } P = 2x + y \qquad \text{(objective function)}$$

subject to

$$4x + 3y \le 12$$
$$5x + 2y \le 10$$

(functional constraints)

and

$$x \ge 0$$
$$y \ge 0$$

(nonnegative constraints)

Solution

· **Step 1** (shade the feasible set).   The feasible set is shown in Figure 15.

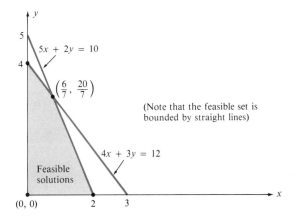

Figure 15
Feasible set with four corner-points

· **Step 2** (draw isoprofit lines).   Any point inside the shaded region is a feasible point and a candidate for the optimal feasible point to the problem. To determine which point will maximize the objective function, we graph the isoprofit lines

$$2x + y = c$$

for different values of c. (See Figure 16.)

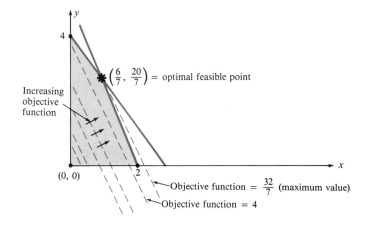

Figure 16
Last corner-point touched by increasing isoprofit lines is the optimal feasible point

It is clear from the graphs of the isoprofit lines that the maximizing point occurs at the intersection of the boundary lines defined by the equations

$$4x + 3y = 12$$
$$5x + 2y = 10$$

The intersection of any two boundary lines is called a **corner-point** of the feasible set and in this case is found to be

$$x^* = \ \ 6/7$$
$$y^* = 20/7$$

Hence the optimal feasible point is (6/7, 20/7). Substituting these values into the objective function, we get the maximum value of the objective function

$$P^* = 2x^* + y^*$$
$$= 2(6/7) + (20/7)$$
$$= 32/7$$
$$\cong 4.57 \qquad\qquad \square$$

We now present a second method for solving linear programming problems in two variables.

The Corner-Point Method

The feasible set of a linear programming problem always has a boundary consisting of straight lines. The points of intersection of pairs of these straight lines are called corner-points of the boundary. Figure 17 illustrates some boundary lines and corner-points of a feasible set.

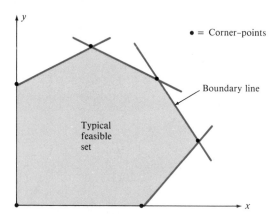

Figure 17
Typical feasible set with six boundary lines and six corner-points

The following result, known as the **Fundamental Theorem of Linear Programming**, is one of the most important results in the theory of linear programming. It tells exactly why there is such an interest in corner-points of the feasible set.

Fundamental Theorem of Linear Programming

If the feasible set of a linear programming problem is bounded (see diagram below), then the objective function will have both a maximum and a minimum value, and these values will occur at corner-points of the feasible set.

If the feasible set is not bounded, the objective function may not have a maximum or minimum value; but if it does, it will occur at a corner-point.

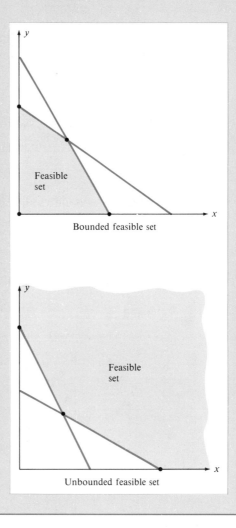

Bounded feasible set

Unbounded feasible set

The fact that the objective function attains its maximum and minimum values at corner-points of the feasible set leads to a useful method for solving linear programming problems in two variables.

Corner-Point Method for Solving Linear Programming Problems

The following steps can be used to solve a standardized (maximization) linear programming problem in two variables.

- Graph the feasible set.
- Looking at the feasible set, make a list of all 2×2 systems of equations whose solutions define the corner-points. Solve these systems to find the coordinates of these corner-points.
- Evaluate the objective function at each of the corner-points.
- Find the largest value of the objective function found in the previous step. This largest value is the maximum of the objective function. The corner-point where this largest value occurs is the optimal feasible point.

When solving minimization linear programming problems, use the same steps, except now find the minimum value of the objective function at the corner-points.

Example 2 _____

Corner-Point Method   Solve the following linear programming problem by the corner-point method.

$$\text{Maximize } P = x + 3y$$

subject to
$$7x + 4y \leq 28$$
$$5x + 9y \leq 45$$
$$5x - 3y \leq 15$$

and
$$x \geq 0$$
$$y \geq 0$$

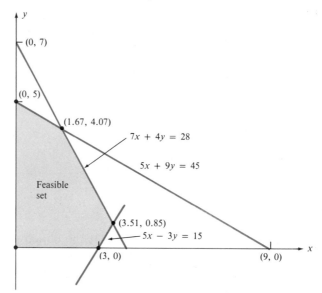

Figure 18
Shaded feasible set

Solution

- **Step 1** (shade the feasible set).   The feasible set is shown in Figure 18.
- **Steps 2–4** (find corner-points by solving 2×2 systems).   Using Figure 18, we determine all pairs of lines that determine the corner-points of the feasible set. Table 9 lists the straight lines that intersect on the boundary, the point of intersection of these lines, and the value of the objective function at these points.

TABLE 9
Corner-Points and Values
of Objective Function at
the Corner-Points

Intersecting Boundary Lines	Corner-Point	Value of Objective Function
$x = 0$ $y = 0$	$(0, 0)$	0
$x = 0$ $5x + 9y = 45$	$(0, 5)$	15 (maximum)
$5x + 9y = 45$ $7x + 4y = 28$	$(1.67, 4.07)$	13.88
$7x + 4y = 28$ $5x - 3y = 15$	$(3.51, 0.85)$	6.06
$5x - 3y = 15$ $y = 0$	$(3, 0)$	3

It is clear from Table 9 that the maximum of the objective function occurs at the point $(0, 5)$ and has a value of 15.   □

Exceptional Types of Problems

Certain exceptional situations arise in solving linear programming problems. They are not common in practical work, but computer programs must consider them as distinct possibilities. They often come up when problems are not formulated correctly. Hence when these situations arise, one should recheck the statement of the problem.

No Points in the Feasible Set

Consider the following linear programming problem:

$$\max \quad P = x + 2y$$

subject to
$$x + y \geq 3$$
$$x + 2y \leq 2$$
$$2x + v \leq 2$$

and
$$x \geq 0$$
$$y \geq 0$$

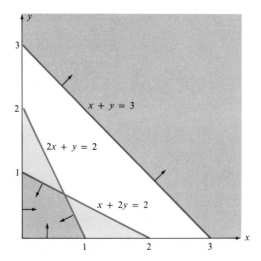

Figure 19
Three inequalities with no feasible solution

If we try to solve the system of five inequalities, we will find that there are *no points* that simultaneously satisfy all five inequalities. (See Figure 19.)

More Than One Optimal Feasible Point

Consider the linear programming problem

$$\max \quad P = 2x + y$$

subject to
$$2x + \ y \le 2$$
$$x + 2y \le 2$$

and
$$x \ge 0$$
$$y \ge 0$$

Note that all the isoprofit lines

$$2x + y = c$$

in Figure 20 are parallel to the boundary line

$$2x + y = 2$$

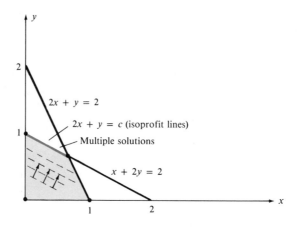

Figure 20
Multiple solutions occur when the isoprofit lines are parallel to one of the boundary lines

Hence as the value of c is increased, it is clear that *any* feasible point on that boundary line is an optimal feasible point. The objective function will have the maximum value of 2 at each point on that line.

Unbounded Feasible Set (No Maximum for Objective Function)

Consider the problem

$$\max \quad P = x + y$$

subject to

$$-x + y \leq 1$$
$$y \leq 2$$

and

$$x \geq 0$$
$$y \geq 0$$

The feasible set is shown in Figure 21. It is clear that the feasible set is unbounded (it cannot be enclosed in a circle of finite radius). The objective function can be made as large as we please by selecting feasible points farther and farther to the right. Hence it does not have a maximum value.

Often, when situations like these occur, the problem has been incorrectly formulated and should be restudied.

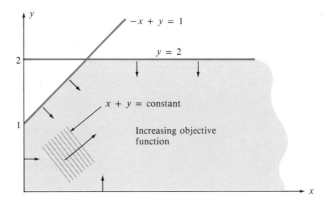

Figure 21
Unbounded feasible set

Where We Are Headed

We have seen in this section that linear programming problems with two decision variables can be solved by either the graphing method or the corner-point method. These methods, however, will generally fail when the number of decision variables is greater than two. Later in this chapter we will study a more powerful method, which will allow us to solve problems with more than two decision variables. This is the *simplex method*.

Problems

For Problems 1–5, draw isoprofit lines for the following objective functions for $c = -1, 0, 1, 2$, and 3.

1. $P = x + y$ **2.** $P = 2x + y$

3. $P = 6x + 4y$ **4.** $P = 10x + y$

5. $P = x + 10y$

For Problems 6–15, solve the following linear programming problems by the graphing method.

6. Maximize $P = x + y$

subject to
$$2x + y \leq 2$$
$$x + 2y \leq 2$$

and
$$x \geq 0$$
$$y \geq 0$$

7. Maximize $P = 3x + y$

subject to
$$x + y \leq 1$$

and
$$x \geq 0$$
$$y \geq 0$$

8. Maximize $P = 2x + y$

subject to $2x + y \leq 5$ (many solutions)
$$x + y \leq 3$$

and
$$x \geq 0$$
$$y \geq 0$$

9. Maximize $P = 3x + 5y$

subject to $x + y \leq 1$ (mystery feasible set)
$$-x + y \leq -1$$

and
$$x \geq 0$$
$$y \geq 0$$

10. Maximize $P = 3x + 9y$

subject to $-x + y \leq 1$ (tricky problem)
$$-x + y \leq -1$$

and
$$x \geq 0$$
$$y \geq 0$$

11. Maximize $P = 2x + y$

subject to $x + 10y \leq 50$
$$x + y \leq 5 \quad \text{(interesting feasible set)}$$
$$10x + y \leq 50$$

and
$$x \geq 0$$
$$y \geq 0$$

12. Minimize $C = x + y$

subject to $2x + y \geq 2$ (unbounded feasible set)
$$x + 2y \geq 2$$

and
$$x \geq 0$$
$$y \geq 0$$

13. Minimize $C = 2x + 2y$

subject to
$$2x + y \geq 1$$
$$x + 2y \geq 1$$

and
$$x \geq 0$$
$$y \geq 0$$

14. Minimize $C = 2x + 4y$

subject to
$$2x + y \geq 8$$
$$4x + 3y \geq 22$$
$$2x + 5y \geq 18$$

and
$$x \geq 0$$
$$y \geq 0$$

15. Minimize $C = 7x + 9y$

subject to
$$x + y \geq 5$$

and
$$x \geq 0$$
$$y \geq 0$$

For Problems 16–25, solve the linear programming problems by the corner-point method.

16. Solve Problem 6.   **17.** Solve Problem 7.

18. Solve Problem 8.   **19.** Solve Problem 9.

20. Solve Problem 10.   **21.** Solve Problem 11.

22. Solve Problem 12.   **23.** Solve Problem 13.

24. Solve Problem 14.   **25.** Solve Problem 15.

Empty Feasible Set

For Problems 26–28, determine whether the feasible set has any points.

26. Maximize $P = x + y$

subject to
$$2x + y \leq 2$$
$$x + 2y \leq 1$$

and
$$x \geq 0$$
$$y \geq 0$$

27. Maximize $P = 2x + y$

subject to
$$2x + y \geq 2$$
$$x + 2y \leq 4$$

and
$$x \geq 0$$
$$y \geq 0$$

28. Maximize $P = 2x + 4y$

subject to
$$2x + y \geq 4$$
$$x + 2y \leq 4$$

and
$$x \geq 0$$
$$y \geq 0$$

Unbounded Objective Function

For Problems 29–32, draw the feasible set of the given linear programming problem and determine whether the problem has a solution.

29. Maximize $P = x + y$

subject to
$$-x + y \leq 2$$
$$-x + 2y \leq 5$$

and
$$x \geq 0$$
$$y \geq 0$$

30. Maximize $P = 2x + 5y$

subject to
$$x - y \geq 2$$
$$-x + y \leq 2$$

and
$$x \geq 0$$
$$y \geq 0$$

31. Minimize $C = x + y$

subject to
$$2x + y \geq 2$$
$$x + 2y \geq 2$$

and
$$x \geq 0$$
$$y \geq 0$$

32. Maximize $P = 4x + y$

subject to
$$x - y \geq 2$$

and
$$x \geq 0$$
$$y \geq 0$$

Linear Programming in Business and Science

33. Finance A financial analyst is advising a client on the investment of $500,000 in class AAA and AA bonds. AAA bonds yield 8% annually, while AA bonds earn 10%. The only constraint is that the client requires twice as much money be invested in AAA bonds as in AA bonds. How much should be invested in each type of bond to maximize the overall return?

34. Marketing A cheese company is offering two new kinds of cheese mixes for sale, Old Wisconsin and Old Minnesota. The cheese packages contain the following mixes:

Old Wisconsin	Old Minnesota
20 oz Cheddar	10 oz Cheddar
10 oz Swiss	10 oz Swiss
5 oz Brie	10 oz Brie

The company has the following amounts of cheeses available every week:

- 5000 oz Cheddar
- 2000 oz Swiss
- 1000 oz Brie

The company determines that it can make a profit of $1.00 and $1.50 on packages of Old Minnesota and Old Wisconsin, respectively.

(a) How much of each product should the company produce to maximize its profits?

(b) What are the company's weekly profits if it uses this optimal strategy?

35. Nutrition One of the authors feeds his cat, Alexander, a combination of two cat foods, Brands A and B. A listing of the relevant ingredients in each of the two brands (8-ounce cans) is listed below.

Ingredients	Brand A	Brand B
Carbohydrates	1 unit	2 units
Protein	7 units	3 units
Fat	5 units	8 units

The cat requires a daily amount of the following nutrients:

- 6 units carbohydrates
- 15 units protein
- 25 units fat

Cat food costs 50 cents per can for Brand A and 40 cents per can for Brand B.

(a) How many cans of each brand should the author feed his cat per day so that the cat gets his daily requirements of protein, carbohydrates, and fat at minimum cost?

(b) How much does it cost the author per day for cat food if he uses this optimal strategy?

36. Blending—Oil Refinery An oil company has two refineries, one in Galveston and one in Pensacola. Each of the two refineries can refine the amounts of gasoline and kerosene in a 24-hour period shown in Table 10. It costs

TABLE 10
Refinery Production Information for Problem 36

Maximum Production	Galveston	Pensacola
High-octane gasoline	500 barrels	600 barrels
Low-octane gasoline	1000 barrels	1000 barrels
Kerosene	1000 barrels	800 barrels

the Galveston refinery $25,000 per day to operate, whereas it costs $20,000 per day for the operation of the Pensacola plant. The production manager of the company receives an order for the following items:

- 100,000 barrels of high-octane gas
- 80,000 barrels of low-octane gas
- 150,000 barrels of kerosene

(a) How many days should each facility operate to fill this order at a minimum cost?

(b) How much will it cost the company to refine these amounts?

37. Assembly Line A manufacturing company produces two products, A and B. Both products require three machines

for their completion, and the time required on each machine is given in Table 11. The last column lists the total time the machines are available each week. Suppose

TABLE 11
Assembly Line Information for Problem 37

Machine	Product A	Product B	Maximum
1	6 min	6 min	60 hr
2	12 min	3 min	120 hr
3	3 min	12 min	120 hr

the products can be sold at a profit of $10 and $6, respectively.

(a) How many units each of products A and B should the company produce each week to maximize profits?

(b) What will be the company's weekly profits?

Table 12 lists the number of grams of carbohydrates, protein, and fat in each of the six food exchanges as measured by the American Diabetic Association. A snack is to be prepared using two of the six food exchanges. For Problems 38–42, determine the number of the two exchanges required so the snack meets the minimum requirements of carbohydrates, protein, and fat and at the same time has a minimum number of calories.

	Exchanges in Snack		Minimum Requirements (grams)		
			Carbohydrates	Protein	Fat
38.	Milk	Bread	25	15	8
39.	Bread	Meat	25	15	10
40.	Vegetable	Fat	10	5	8
41.	Fruit	Meat	15	7	10
42.	Vegetable	Fruit	10	10	0

TABLE 12
Nutrition Information for Problems 38–42

	Food Exchange					
	Milk (m)	Vegetable (v)	Fruit (f)	Bread (b)	Meat (t)	Fat (a)
Carbohydrates	12	7	10	15	0	0
Protein	8	2	0	2	7	0
Fat	10	0	0	0	5	5
Calories	60	75	40	70	75	45

| **2.3** |

Applications of Linear Programming

PURPOSE

We present two general problems that illustrate how linear programming problems arise in the business and scientific world. The problems discussed are

- the dietician's problem (minimization problem) and
- the allocation of resources problem (maximization problem).

Introduction

If there is one general principle that permeates modern business and finance, it is the idea of **optimization**. The allocation of a company's resources to maximize profits is certainly a major axiom of business. The minimization of costs is another major problem facing modern business.

We now show how the optimization technique of linear programming can be used to solve some typical problems in business management. We start with a well-known problem, called **the dietician's problem**. In spite of the name, it actually refers to a class of problems, most of which are not at all related to nutrition. For example, a mining company mines two grades of ore, each contains a different amount of copper, zinc, and molybdenum. Suppose it is more costly to mine one of the ores. If the company receives an order to ship given amounts of copper, zinc, and molybdenum, how much of each ore should the company mine to meet its order at a minimum cost? This problem is essentially the same as the dietician's problem. We state the dietician's problem here, not in the context of mining, but as the Campbell Soup Problem.

The Dietician's Problem
(the Campbell Soup Problem*)

We would like to determine the most economical monthly menu plan that consists entirely of two kinds of soups, Campbell's Cheddar Cheese and Campbell's Tomato Garden. Of course, monthly menus planned by dietary staffs would certainly contain many other foods as well, but we simplify the problem by including only these two soups. The reader can gain an understanding of the power of this general problem by generalizing from this simple problem to more complex problems.

We now seek to find the number of servings (8 oz) of Campbell's Cheddar Cheese and Tomato Garden soups that have a minimum total cost but at the same time satisfy the following nutritional requirements:

- 200 grams carbohydrates
- 40 grams protein
- 2000 calories

According to a recent supermarket survey, a serving of Cheddar Cheese soup costs 21 cents, and a serving of Tomato Garden soup costs 19 cents. The amounts of carbohydrates, protein, and calories per serving are listed on the sides of the cans and are shown on the next page.

---

* The authors would like to thank the Campbell Soup Company for its cooperation in the development of this example.

To solve this problem, we first formulate it as a linear programming problem.

Formulating the Campbell Soup Problem

This problem has been organized into a series of steps. You can use these steps as a guide to solve similar problems.

- **Step 1** (organize the information of the problem). We begin by organizing the relevant information in Table 13.

TABLE 13

Number of Grams of Carbohydrates, Protein, and Calories per serving of Cheddar Cheese and Tomato Garden Soups

	Cheddar Cheese	Tomato Garden	Minimum Requirements
Carbohydrates	10 g	18 g	200 g
Protein	4 g	1 g	40 g
Calories	130 cal	80 cal	2000 cal
Cost/serving	21 cents	19 cents	—

The numbers in Table 13 under "Cheddar Cheese" and "Tomato Garden" give the number of grams of carbohydrates and protein in one serving of soup as well as the number of its calories and its cost. The last column gives the requirements for carbohydrates, protein, and calories.

- **Step 2** (identify the decision variables). To identify the decision variables, remember that these variables represent things that can be controlled. In this problem the decision variables are

$$C = \text{number of servings of Cheddar Cheese}$$
$$T = \text{number of servings of Tomato Garden}$$

- **Step 3** (determine the objective function).   The objective function is found by determining what is being maximized or minimized. In this problem the total cost of the soup is being minimized. Since a serving of Cheddar Cheese costs $0.21 and a serving of Tomato Garden costs $0.19, the total cost for C servings of Cheddar Cheese soup and T servings of Tomato Garden soup is given by

$$\text{Total Cost} = 0.21C + 0.19T \qquad \text{(objective function)}$$

- **Step 4** (find the constraints).   The constraints represent limitations on the decision variables and are described mathematically by inequalities. In this problem the decision variables C and T are constrained by the fact that we require minimum amounts of carbohydrates, protein, and calories in the two servings. To find these constraints, we first find the mathematical expressions for the number of grams of carbohydrates and protein and the number of calories contained in C servings of Cheddar Cheese soup and T servings of Tomato Garden soup. By a simple analysis we find the following equations:

$$\text{Grams of carbohydrates} = (10 \text{ g/serving})(C \text{ servings}) + (18 \text{ g/serving})(T \text{ servings})$$
$$= 10C + 18T$$
$$\text{Grams of protein} = (4 \text{ g/serving})(C \text{ servings}) + (1 \text{ g/serving})(T \text{ servings})$$
$$= 4C + T$$
$$\text{Calories} = (130 \text{ cal/serving})(C \text{ servings}) + (80 \text{ cal/serving})(T \text{ servings})$$
$$= 130C + 80T$$

Hence the constraints are

$$10C + 18T \geq 200 \qquad \text{(carbohydrates)}$$
$$4C + T \geq 40 \qquad \text{(protein)}$$
$$130C + 80T \geq 2000 \qquad \text{(calories)}$$

Since C and T represent servings of soup, they must also satisfy

$$C \geq 0$$
$$T \geq 0$$

The above equations can be summarized as the following linear programming problem.

Linear Programming Model for the Soup Problem

Find C and T that minimize

$$\text{COST} = 0.21C + 0.19T$$

subject to
$$10C + 18T \geq 200$$
$$4C + T \geq 40$$
$$130C + 80T \geq 2000$$

and
$$C \geq 0$$
$$T \geq 0$$

Solution of the Campbell Soup Problem

We use the corner-point method to solve the above problem.

· **Step 1** (draw the feasible set).   The feasible set is shown in Figure 22.
· **Step 2** (find the corner-points).   The coordinates of the corner-points have been found and are listed in Table 14.

TABLE 14
Corner-Points and Value of
Objective Function

Corner-Point (C, T)	Cost $= 0.21C + 0.19T$
(0, 40)	$7.60
(6.3, 14.7)	$4.12
(13, 3.9)	$3.47 (minimum)
(20, 0)	$4.20

Selecting the minimum of the values above, we conclude that the optimal feasible point is

$$C^* = 13.0 \text{ servings of Cheddar Cheese}$$
$$T^* = 3.9 \text{ servings of Tomato Garden}$$

The minimum cost is

$$\text{Minimum cost} = 0.21C^* + 0.19T^*$$
$$= 0.21(13) + 0.19(3.9)$$
$$= \$3.47$$

We conclude that a menu plan consisting of 13 servings of Cheddar Cheese and 3.9 servings of Tomato Garden soups meets the nutritional requirements for carbohydrates, protein, and calories at minimum cost.

Further Analysis of the Solution

Figure 22 shows that the optimal solution of

$$C^* = 13.0 \text{ servings of Cheddar Cheese}$$
$$T^* = 3.9 \text{ servings of Tomato Garden}$$

lies at the intersection of the constraint lines for carbohydrates and calories. In other words, the point (C^*, T^*) exactly meets the minimum requirements for these nutrients. On the other hand, it more than meets the minimum requirements for protein. Substituting C^* and T^* into the expression developed earlier for the number of grams of protein consumed, we have

$$\text{Grams protein} = 4C^* + T^* = 55.9$$

This means that the optimal menu has an excess of protein of

$$\text{Excess of protein} = \text{Protein in menu} - \text{Protein required}$$
$$= 55.9 - 40$$
$$= 15.9 \text{ grams}$$

This completes the analysis of the dietician's problem.

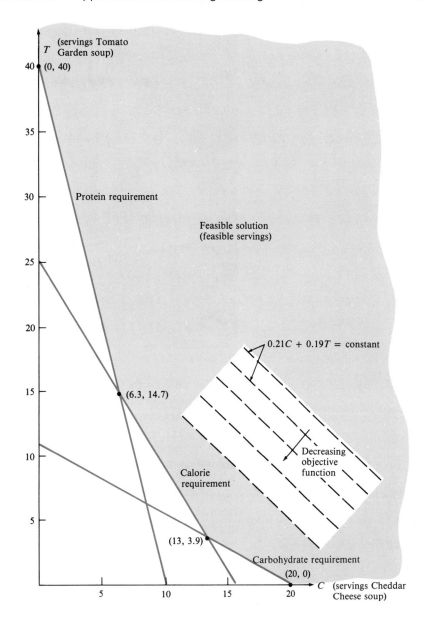

Figure 22
*Feasible solution and
corner-points for the
Campbell Soup problem*

The Allocation of Resources Problem
(Old Town Canoe Problem*)

Another important problem that can be solved by the linear programming
model is the **allocation of resources problem**. It tells how a company should use
resources, such as labor, raw materials, and time in an optimal manner in order
to maximize profits.

---

* This example was developed from material supplied to the authors by the Old Town
Canoe Company. The authors would like to thank the management of The Old Town
Canoe Company for their cooperation. Much of the material has been fabricated to
simplify calculations.

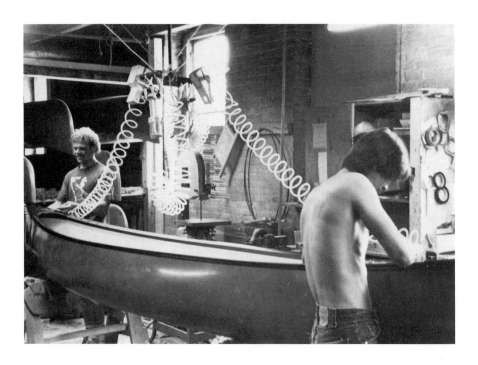

The Old Town Canoe Company in Old Town, Maine, is one of the few canoe makers in the United States. The company produces several synthetic-based canoes. Two major synthetic models are the Columbia and Katahdin canoes.

A problem facing the company is the determination of the number of Columbias and Katahdins to produce each week with a given amount of available resources. Some of the resources that go into the production of a canoe are fiberglass mat, fiberglass cloth, resin, and labor. The difference between a good production schedule and a poor one may result in a difference in profits of tens of thousands of dollars per year. A poor production schedule for a company may result in many of the raw materials going unused. Such scheduling results in increased inventory costs and a shortage of other raw materials, resulting in excessive labor costs. The following problem shows how the production manager at the Old Town Canoe Company might go about planning the schedule for the production of these two types of canoes.

Whatever schedule the production manager decides upon must ultimately depend on the overall availability of resources and the relative profit gained from each of the products.

Formulation of the Old Town Canoe Problem

First, collect and organize the information. The information about the production requirements for these two models of canoes is collected and displayed in Table 15. The columns of the table list the amount of each resource that is used in the production of each kind of canoe. The last column lists the total amount of each resource available. Finally, the bottom row of the table gives the profit the company earns on each unit of each product.

The information in Table 15 can be formulated as the following linear programming problem.

TABLE 15
Resource Requirements for
Columbia and Katahdin
Canoes

Resources	Products		Maximum Resources Available
	Columbia	*Katahdin*	
Labor	20 hr	15 hr	4000 hr
Fiberglass mat	2 sq ft	5 sq ft	900 sq ft
Resin	5 gal	5 gal	1500 gal
Fiberglass cloth	11 sq ft	12 sq ft	2640 sq ft
Profits	$350	$300	—

Linear Programming Formulation of the Old Town Canoe Problem

We begin by defining the two decision variables:

$$C = \text{number of Columbias to produce each week}$$
$$K = \text{number of Katahdins to produce each week}$$

The problem is to schedule the production of a given number of Columbia and Katahdin canoes so that the profit is maximized while not using more resources than are available. Stated mathematically, we wish to find C and K so that we maximize

$$P = 350C + 300K$$

subject to

$$20C + 15K \leq 4000 \quad \text{(labor)}$$
$$2C + 5K \leq 900 \quad \text{(fiberglass mat)}$$
$$5C + 5K \leq 1500 \quad \text{(resin)}$$
$$11C + 12K \leq 2640 \quad \text{(fiberglass cloth)}$$

and

$$C \geq 0$$
$$K \geq 0$$

We solve this problem by the corner-point method. The feasible set is shown in Figure 23.

A listing of the corner-points and the value of the objective function at these points is given in Table 16.

TABLE 16
Corner-Point Solution of the
Old Town Canoe Problem

Corner-Point (C, K)	Value of Objective Function $P = 350C + 300K$
(0, 0)	0
(0, 180)	$54,000
(78,149)	$72,000
(112, 117)	$74,300 (maximum)
(200, 0)	$70,000

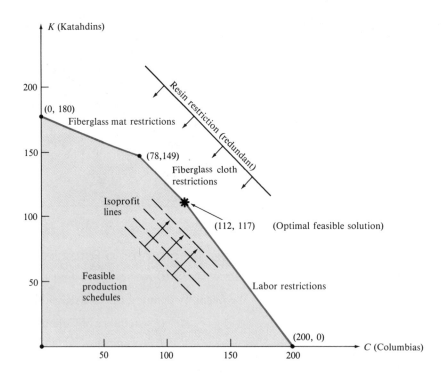

Figure 23
Illustration of the solution for the Old Town Canoe problem

From Table 16 we see that the profit is a maximum when the production level is

$$C^* = 112 \text{ Columbias}$$
$$K^* = 117 \text{ Katahdins}$$

The value of the objective function at this production level is $74,300.

Comments on the Old Town Canoe Problem

We conclude that the company should produce 112 Columbias and 117 Katahdins each week with the resources available in order to maximize its profits. The weekly profits from this production will be $74,300.

The reader may ask why the company does not order more resources or hire more employees in order to increase profits. To answer this question, remember that we are making the assumption that the company can actually sell all of the product it produces. If the company can in fact sell all of its product, then it might be a good idea to expand its facilities and enlarge its stock of resources. The question that we are asking in the above *linear programming* model is: Under the *given amount* of availability of resources, what is the number of each product that should be produced?

Critical Resources

The optimal production of Columbias and Katahdins uses all the inventory of fiberglass cloth and labor, whereas some resin and fiberglass mat are left over.

The amount of resin used in the optimal production schedule was

$$\text{Resin used} = 5C^* + 5K^*$$
$$= 5(112) + 5(117)$$
$$= 1145 \text{ gallons}$$

This is much less than the 1500 gallons available. Also, the amount of fiberglass mat used was

$$\text{Fiberglass mat used} = 2C^* + 5K^*$$
$$= 2(112) + 5(117)$$
$$= 809 \text{ square feet}$$

This is less than the 900 square feet available. We say that fiberglass cloth and labor are the critical resources, since they are completely used by the optimal production schedule, whereas resin and fiberglass mat are not. If the company were able to sell more canoes and wanted to increase production, then the amount of the critical resources would have to be increased.

Problems

1. **War Against the Gnolgs** In defending the Empire, we must destroy the attacking, villainous Gnolgs with our powerful weapons, the Neutron Vaporizer and the Laser Disintegrator. The Neutron Vaporizer is more effective, killing seven Gnolgs per firing in contrast to the Laser Disintegrator, which destroys only two. However, firing the vaporizer requires more power in the form of Neorons and Orgons than firing the laser disintegrator. In particular, the Neutron Vaporizer takes three Neorons per firing and four Orgons per firing. On the other hand the Laser Disintegrator takes only one Neoron and one Orgon per firing. The detailed specifications for these weapon systems are given in Table 17.

TABLE 17
Neorons and Orgons Needed to Fire the NV and LD

| | Weapons | | |
| | Neutron | Laser | Available |
Ingredients	Vaporizer (NV)	Disintegrator (LD)	Energy
Neorons	3 Neorons/firing	1 Neoron/firing	270
Orgons	4 Orgons/firing	1 Orgon/firing	320
Gnolgs Destroyed	7	2	—

Let

NV = the number of firings of the Neutron Vaporizer

LD = the number of firings of the Laser Disintegrator

(a) Draw the feasible set for NV and LD.
(b) How should NV and LD be chosen in order to destroy as many Gnolgs as possible with the limitations on the number of Neorons and Orgons?
(c) What is the maximum number of Gnolgs that can be destroyed?
(d) Is it possible to destroy an entire Ergle (80 Gnolgs)?
(e) How many more Gnolgs can be destroyed if the energy level for Neorons is increased from 270 million Neorons to 300 million Neorons?

2. **Gnolgs Revisited** Solve Problem 1, but change the number of Gnolgs destroyed to seven for both the Neutron Vaporizer and the Laser Disintegrator.

3. **Nut Problem** A company sells two mixes of nuts, Plain and Fancy. Each mix is a blend of the three basic kinds: peanuts, cashews, and Brazil nuts. Table 18 shows

   - the exact blend of the two mixes,
   - the company's profits from each of the mixes, and
   - the daily inventory of the three kinds of nuts.

   (a) What is the linear programming problem that describes this problem? Let

   x = the number of bags of Plain nuts to fill per day

   y = the number of bags of Fancy nuts to fill per day

   (b) Solve the LP problem in part (a) to find the values of x and y that maximize profits under the above restrictions.
   (c) What quantities of each ingredient are left over at the end of the day?
   (d) Will the optimal feasible point and total profit change if the price charged for Plain nuts is raised from $0.40 to $0.50?

4. **Nut Problem Revisited** Answer the questions in Problem 3, but assume now that the daily amount of nuts available is 1000 pounds of each kind of nuts.

TABLE 18
Information on Nuts for Problems 3 and 4

Ingredients	Mix		Daily Amount Available
	Plain	*Fancy*	
Peanuts	0.35 lb/bag	0.20 lb/bag	5000 lb
Cashews	0.10 lb/bag	0.20 lb/bag	2000 lb
Brazil Nuts	0.05 lb/bag	0.10 lb/bag	1000 lb
Profit	$0.40 per bag	$0.55 per bag	—

5. **Nutrition Problem** A nutritionist is responsible for planning lunches. The nutritionist plans a meal consisting of two major foods. Each ounce of Food 1 contains 9 grams of protein, 2 grams of carbohydrates, and 15 milligrams of vitamin B. Each ounce of Food 2 contains 3 grams of protein, 3 grams of carbohydrates, and 15 milligrams of vitamin B. The minimum requirements for protein, carbohydrates, and vitamin B along with the cost of the two foods are given in Table 19.

   (a) What number of servings of each food (x, y) will meet the minimum daily requirements of protein, carbohydrates, and vitamin B at minimum cost?
   (b) What is the minimum cost of a serving?
   (c) Given the optimal serving of each food, which of the ingredients of protein, carbohydrates, and vitamin B exactly meet the minimum daily requirements?

TABLE 19
Nutrition Information for Problems 5 and 6

Ingredients	Meal		Minimum Requirements
	Food 1	*Food 2*	
Protein	9 g/oz	3 g/oz	36 g
Carbohydrates	2 g/oz	3 g/oz	80 g
Vitamin B	15 mg/oz	15 mg/oz	240 mg
Cost of Foods	$0.15/oz	$0.20/oz	—

6. **Nutrition Problem Revisited** Answer the questions in Problem 5, but assume now that the cost of the foods has changed to $0.20 for Food 1 and $0.25 for Food 2.

7. **Transportation** In the transportation sector a major problem is the determination of shipping schedules. Here we restrict the problem to a corporation with two fac-

tories and three warehouses. A manufacturer of automobile engines wishes to ship engines from factories in Cleveland and Birmingham to assembly plants in Los Angeles, Omaha, and Baltimore. Table 20 and Figure 24 give the

- output of each factory (last column in Table 20),
- demand of each assembly plant (the bottom row in Table 20), and

- cost in dollars of transporting an engine from a factory to an assembly plant (the numbers in the table).

Let

x_{ij} = the number of engines shipped from factory i to assembly plant j

as illustrated in Table 21.

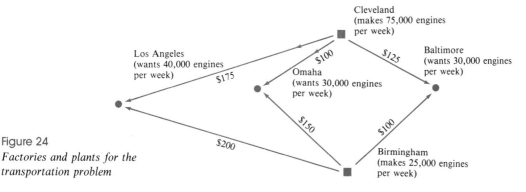

Figure 24
Factories and plants for the transportation problem

TABLE 20
Outputs, Demands, and Costs to Deliver Goods from Factories to Assembly Plants

| Source | Assembly Plant (destination) | | | Output |
	Los Angeles	Omaha	Baltimore	
Cleveland	$175	$100	$125	75,000
Birmingham	$200	$150	$100	25,000
Demand	40,000	30,000	30,000	100,000

TABLE 21
Number of Engines to be Shipped from Factories to Assembly Plants

| Source | Destination | | |
	Los Angeles	Omaha	Baltimore
Cleveland	x_{11}	x_{12}	x_{13}
Birmingham	x_{21}	x_{22}	x_{23}

The goal of this problem is to find $x_{11}, x_{12}, x_{13}, x_{21}, x_{22}, x_{23}$ that will satisfy the output and demand requirements (every engine produced at the factories gets shipped, and every assembly plant gets its demands) at minimum cost. Write the linear programming problem that describes this problem.

(a) How many variables does this problem have?

(b) How many constraints does this problem have?
(c) What is unusual about the constraints of this problem compared to the constraints in the linear programming problems we have seen thus far?
8. Transportation Problem Revisited Answer the questions in Problem 7 with the change that Cleveland and Birmingham each supply 50,000 engines.

9. **Allocation of Resources** A furniture company specializes in maple tables and chairs. Two resources used in these products are labor and bird's-eye maple. The amount of maple and labor used in the production of the tables and chairs, the maximum amount of maple and labor available, and the profit per table and chair are summarized in Table 22.

   (a) How many tables and chairs should the production manager schedule to maximize profits with the resources available?

   (b) What are the company's profits using the optimal schedule?

   (c) Does the company use all of its resources using the optimal schedule?

10. **Furniture Company Revisited** Answer the questions in Problem 9 if the profit per table is $100.

TABLE 22
Furniture Company Information for Problems 9 and 10

	Product		Ingredients Available
Ingredients	*Table*	*Chair*	
Maple	7 board feet	14 board feet	1000 board feet
Labor	5 hours labor	15 hours labor	800 hours
Profit	$70	$30	—

11. **Mining Company** A gold-mining company in Idaho operates two mines that are in production seven days a week. In addition to gold, the mines yield varying amounts of silver and lead. Table 23 lists the daily output from each mine.

TABLE 23
Daily Mine Output for Problem 11

Metal	Mine 1	Mine 2
Gold	5 lb/day	15 lb/day
Silver	50 lb/day	75 lb/day
Lead	500 lb/day	500 lb/day

It costs the company $10,000 per day to operate Mine 1 and $20,000 per day to operate Mine 2. The company receives an order for 100 pounds of gold, 400 pounds of silver, and 1000 pounds of lead. How many days should the company operate each mine in order to meet this demand and at the same time minimize its costs?

   (a) Write the linear programming problem that describes this problem.

   (b) Solve the linear programming problem by the corner-point method.

12. **Mining Company Revisited** Answer the questions in Problem 11 with the change that Mine 1 extracts 25 pounds of gold per day from the mine.

13. **Linear Programming Form of the ST Game** A player starts with a large supply of squares and triangles as shown in Figure 25. On each square is drawn ten color dots and five black dots. On each triangle is drawn five color dots and ten black dots. The game consists of putting squares and triangles into a pot of some kind. A reward of $4 is given for each square placed in the pot, and $5 for each triangle. The only restriction is that the pot cannot contain more than 145 color dots and 125 black dots. What is the number of squares and triangles that should be placed in the pot in order to maximize the total reward? What is the maximum reward a player can win?

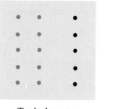
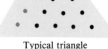

Typical square   Typical triangle

Figure 25
Square and triangle for the ST game

14. **ST Game Revisited** Play the ST game described in Problem 13 with the change that the pot can contain 600 color dots and 675 black dots.

2.4

Slack Variables and the Simplex Tableau

PURPOSE

We introduce the concept of slack variables and show how they help in the description of a feasible set. We will learn

- how to find the slack variable of an inequality,
- how to construct a simplex tableau, and
- how to find corner-points of a feasible set by manipulation of the simplex tableau.

Introduction

We have seen that the optimal feasible point for a linear programming problem occurs at a corner-point of the feasible set. We can solve linear programming problems with two variables by drawing the feasible set, examining the corner-points one at a time, and concluding from this examination the location of the optimal feasible point. In higher dimensions, however, this strategy fails, since we cannot easily graph the feasible set.

Fortunately, there is a way of "seeing" these higher dimensional corner-points through a "mathematical lens," known as the simplex tableau. To understand this tableau, however, we must first understand slack variables.

Slack Variables

To understand slack variables, consider the following system of four inequalities whose solution is shown in Figure 26:

$$2x_1 + x_2 \leq 2$$
$$x_1 + 2x_2 \leq 2$$
$$x_1 \geq 0$$
$$x_2 \geq 0$$

In the feasible set shown in Figure 26, the cartesian coordinates (x_1, x_2) of a point tell where the point is located in relation to the two boundary lines $x_1 = 0$ and $x_2 = 0$. The cartesian coordinates do not, however, tell where the point is located in relation to the other two boundary lines. We now introduce a new augmented coordinate system that gives the position of a point in the plane in relationship to all the boundary lines of the feasible set. The slack variables allow us to introduce this augmented coordinate system.

To define this new coordinate system for the system of inequalities above, we introduce two new coordinates s_1 and s_2, called **slack variables**. These two variables are defined as the difference (slack) between the right-hand and left-hand sides of the two inequalities

$$2x_1 + x_2 \leq 2$$
$$x_1 + 2x_2 \leq 2$$

That is,
$$s_1 = 2 - (2x_1 + x_2)$$
$$s_2 = 2 - (x_1 + 2x_2)$$

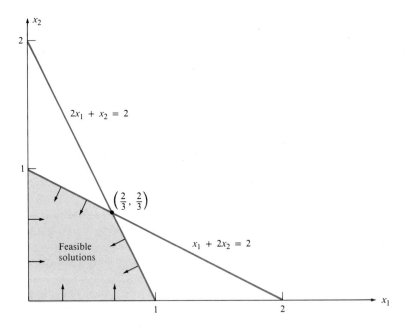

Figure 26
Feasible solutions to four inequalities

Rewriting these two equations in the form

$$2x_1 + \; x_2 + s_1 \qquad = 2$$
$$x_1 + 2x_2 \qquad + s_2 = 2$$

we see that the two original inequalities have been rewritten as equalities with the introduction of the two slack variables. Now that the slack variables s_1 and s_2 have been defined, we assign four coordinates (x_1, x_2, s_1, s_2) to each point in the plane. To find these four coordinates for any point, remember that x_1 and x_2 are the usual cartesian coordinates and s_1 and s_2 are the slack variables determined by the equations

$$s_1 = 2 - (2x_1 + x_2)$$
$$s_2 = 2 - (x_1 + 2x_2)$$

Augmented Coordinates (Decision-Slack Coordinate System)

The **augmented coordinates** of a point in the plane consist of the cartesian coordinates followed by the slack variables. For example, a point (x_1, x_2) in the $x_1 x_2$-plane that has two slack variables s_1 and s_2 would have the augmented coordinates (x_1, x_2, s_1, s_2).

Figure 27 shows the augmented coordinates of feasible and infeasible points for the given system of inequalities.

The question we now ask is: Why is the augmented coordinate system so valuable? It would be much simpler to label points with their cartesian coordinates rather than with their augmented coordinates. To answer this question,

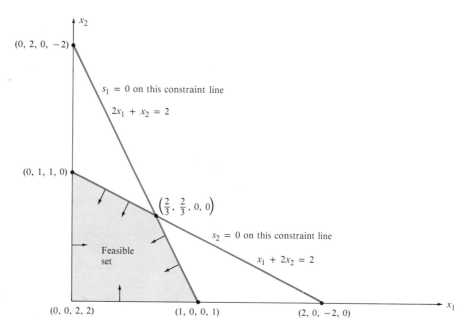

Figure 27
Sample points with (x_1, x_2, s_1, s_2) coordinates

note that in Figure 27 the feasible points are the only points with all four augmented coordinates (x_1, x_2, s_1, s_2) greater than or equal to zero. Note too that s_1 and s_2 give a measure of the distance (not the actual distance, but proportional to the distance) from the point (x_1, x_2, s_1, s_2) to the boundary lines:

$$2x_1 + x_2 = 2$$
$$x_1 + 2x_2 = 2$$

This is similar to the way x_1 and x_2 measure the distance from the x_1 and x_2 axes. Hence the four coordinates (x_1, x_2, s_1, s_2) tell exactly where the point is in relation to all four boundary lines. The arithmetic signs of the slack variables tell on what side of the boundary lines the point lies. We summarize the reasons for introducing slack variables and augmented coordinates.

Reasons for Augmented Coordinates

Additional information. We can always determine the location of a point in the x_1x_2-plane by specifying the cartesian coordinates (x_1, x_2). Adding s_1 and s_2 to the cartesian coordinates allows us to tell where the point is with respect to each boundary line of the feasible set.

Easy identification of feasible set. Points in the feasible set in Figure 27 can be determined by simply observing that all four coordinates (x_1, x_2, s_1, s_2) of such points are nonnegative. Points with at least one negative coordinate are not feasible.

Easy identification of boundary and corner-points. The corner-points of the feasible set in Figure 27 are characterized by having two of the four coordinates x_1, x_2, s_1, s_2 equal to zero and the other two positive. The boundary lines of the feasible set are characterized by having one of the four coordinates x_1, x_2, s_1, s_2 equal to zero and the other three positive. For example, if we set one of the four coordinates x_1, x_2, s_1, s_2 to zero, then the point (x_1, x_2, s_1, s_2) lies on one of the four boundary lines. In particular,

- if $x_1 = 0$, the point (x_1, x_2, s_1, s_2) lies on $x_1 = 0$.
- if $x_2 = 0$, the point (x_1, x_2, s_1, s_2) lies on $x_2 = 0$.
- if $s_1 = 0$, the point (x_1, x_2, s_1, s_2) lies on $2x_1 + x_2 = 2$.
- if $s_2 = 0$, the point (x_1, x_2, s_1, s_2) lies on $x_1 + 2x_2 = 2$.

Table 24 shows the augmented coordinates (x_1, x_2, s_1, s_2) of the corner-points of the feasible set defined by the system of inequalities

$$2x_1 + x_2 \leq 2$$
$$x_1 + 2x_2 \leq 2$$
$$x_1 \geq 0$$
$$x_2 \geq 0$$

TABLE 24
Cartesian Coordinates and
Augmented Coordinates of
Corner-Points

Cartesian Coordinates (x_1, x_2)	Augmented Coordinates (x_1, x_2, s_1, s_2)
$(0, 0)$	$(0, 0, 2, 2)$
$(0, 1)$	$(0, 1, 1, 0)$
$(2/3, 2/3)$	$(2/3, 2/3, 0, 0)$
$(1, 0)$	$(1, 0, 0, 1)$

These points are shown in Figure 27.

Example 1

Augmented Coordinates   Given the two inequalities

$$x_1 + x_2 \leq 5$$
$$3x_1 + x_2 \leq 6$$

find the augmented coordinates (x_1, x_2, s_1, s_2) of the points whose cartesian coordinates are the following values:

(a) $(x_1, x_2) = (0, 0)$ (b) $(x_1, x_2) = (0, 5)$
(c) $(x_1, x_2) = (5, 0)$ (d) $(x_1, x_2) = (1, 1)$

Solution   The two slack variables corresponding to the two inequalities above are

$$s_1 = 5 - (x_1 + x_2)$$
$$s_2 = 6 - (3x_1 + x_2)$$

Substituting the values of x_1 and x_2 into these equations, we have

(a) $(0, 0)$ has augmented coordinates $(0, 0, 5, 6)$
(b) $(0, 5)$ has augmented coordinates $(0, 5, 0, 1)$
(c) $(5, 0)$ has augmented coordinates $(5, 0, 0, -9)$
(d) $(1, 1)$ has augmented coordinates $(1, 1, 3, 2)$

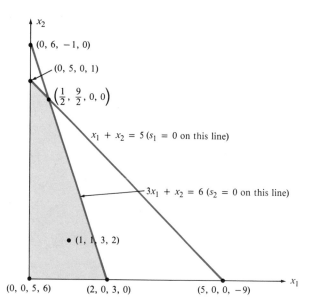

Figure 28
Sample points labeled with augmented coordinates

These points are illustrated in Figure 28. □

Simplex Tableau (Searching for Corner-Points)

The remainder of this section will be devoted to showing how slack variables are used to find the corner-points of the feasible set. To do this, we need to introduce the simplex tableau.

Consider finding the corner-points of the feasible set for the following linear programming problem:

$$\text{maximize } P = 3x_1 + x_2$$

subject to

$$2x_1 + 3x_2 \le 6$$
$$4x_1 + 2x_2 \le 8$$
$$1.5x_1 - 2x_2 \le 1$$

and

$$x_1 \ge 0$$
$$x_2 \ge 0$$

The feasible set for this problem is shown in Figure 29.

Although we can find the corner-points by finding the points of intersection of boundary lines, we will use a completely different approach. The approach will work for linear programming problems with any number of variables.

We introduce the slack variables s_1, s_2, and s_3 (one for each constraint), defined by

$$s_1 = 6 - (2x_1 + 3x_2)$$
$$s_2 = 8 - (4x_1 + 2x_2)$$
$$s_3 = 1 - (1.5x_1 - 2x_2)$$

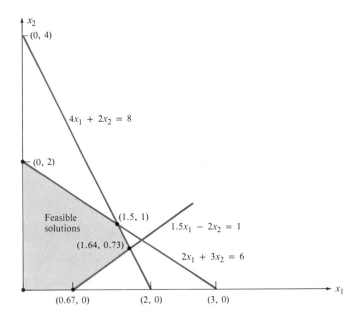

Figure 29
Solution of five simultaneous inequalities

Rewriting these equations, we can restate the linear programming problem above using the slack variables.

Linear Programming Problem with Equality Constraints

$$\text{maximize } P = 3x_1 + x_2$$

subject to
$$2x_1 + 3x_2 + s_1 \qquad\qquad = 6$$
$$4x_1 + 2x_2 \qquad + s_2 \qquad = 8$$
$$1.5x_1 - 2x_2 \qquad\qquad + s_3 = 1$$

and
$$x_1 \geq 0$$
$$x_2 \geq 0$$
$$s_1 \geq 0$$
$$s_2 \geq 0$$
$$s_3 \geq 0$$

We now show how to find the corner-points of the feasible set of this linear programming problem. To find these points, observe that two of the augmented coordinates $(x_1, x_2, s_1, s_2, s_3)$ are zero at each corner-point. Hence to find a corner-point, set any two of the coordinates x_1, x_2, s_1, s_2, s_3 equal to zero and solve for the other three using the following constraint equations:

$$2x_1 + 3x_2 + s_1 \qquad\qquad = 6$$
$$4x_1 + 2x_2 \qquad + s_2 \qquad = 8$$
$$1.5x_1 - 2x_2 \qquad\qquad + s_3 = 1$$

If the three variables for which we solve are nonnegative, we know that the point $(x_1, x_2, s_1, s_2, s_3)$ is a corner-point of the feasible set.

It does not make any difference from a mathematical point of view which two variables are set to zero, but it makes a lot of difference from a practical point of view. To determine which two variables should be set to zero, observe that the coefficients of the above slack variables s_1, s_2, and s_3 are 1's and 0's, while the coefficients of x_1 and x_2 are not quite as simple. Hence the easiest way to find a corner-point is to set x_1 and x_2 to zero and solve for s_1, s_2, and s_3. The variables that are set to zero are called the **nonbasic variables**, and the variables for which we solve are called **basic variables**. Generally we are interested in the basic variables when they are all positive. In this example, x_1 and x_2 are the nonbasic variables, and s_1, s_2, and s_3 are basic variables. By setting x_1 and x_2 to zero in these equations and solving for s_1, s_2, and s_3 we find

$$x_1 = 0 \qquad \text{(nonbasic variable)}$$
$$x_2 = 0 \qquad \text{(nonbasic variable)}$$
$$s_1 = 6 \qquad \text{(basic variable)}$$
$$s_2 = 8 \qquad \text{(basic variable)}$$
$$s_3 = 1 \qquad \text{(basic variable)}$$

In so doing, we have found the corner-point:

$$(x_1, x_2, s_1, s_2, s_3) = (0, 0, 6, 8, 1)$$

which is the origin $(0, 0)$ in the $x_1 x_2$-plane with slack variables 6, 8, and 1. (See Figure 30.)

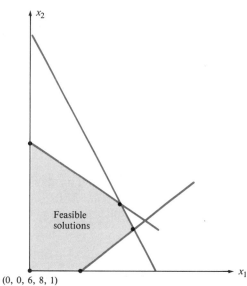

Figure 30
Augmented coordinates of the origin

To carry out the steps of the simplex method, it is important from a computational point of view that we rewrite the constraint equations as an augmented matrix, called the **simplex tableau** as shown in Table 25.

In this tableau the variables x_1 and x_2 are the nonbasic variables $(=0)$, and the slack variables s_1, s_2, and s_3 are the basic variables (>0). Observe that the BV column at the left of the tableau lists the names of the basic variables.

TABLE 25

Simplex Tableau Identifying
Basic and Nonbasic
Variables

BV	Nonbasic Variables (= 0)		Basic Variables (≠ 0)			Right-Hand Side (RHS)
	x_1	x_2	s_1	s_2	s_3	
s_1	2	3	1	0	0	6
s_2	4	2	0	1	0	8
s_3	1.5	−2	0	0	1	1

Identity matrix

Comments on Basic and Nonbasic Variables

The variables selected to be zero in the simplex tableau are the nonbasic variables. The remaining variables (for which we solve) are the basic variables. The assigning of basic and nonbasic variables is always made in such a way that the values of the basic variables are located in the RHS column. In other words, the variables that have the identity matrix as coefficients will be the basic variables (nonzero), and the variables that have other coefficients will be the nonbasic (zero) variables.

Note that in the simplex tableau in Table 25 we set

$$x_1 = 0 \qquad \text{(nonbasic variables)}$$
$$x_2 = 0$$

and solved for the remaining variables (whose values are in the RHS column):

$$s_1 = 6$$
$$s_2 = 8 \qquad \text{(basic variables)}$$
$$s_3 = 1$$

In summary, the simplex tableau in Table 25 would be identified as the corner-point (0, 0, 6, 8, 1), which is the corner-point located at the origin in the $x_1 x_2$-plane. (See Figure 30.)

Moving from Corner-Point to Corner-Point

We have just seen how a corner-point corresponds in a natural way to a simplex tableau by setting two of the variables to zero and solving for the others. We now show how to move from one corner-point (represented by a simplex tableau) to a new corner-point (represented by another simplex tableau). In this way a systematic search of all the corner-points of the feasible set can be done.

To find a new simplex tableau, we use the elementary row operations introduced in Chapter 1 to solve systems of equations by the Gauss-Jordan method. These operations allow us to rewrite the simplex tableau in an alternate form by placing the identity matrix in front of different variables. In that way, different corner-points will be identified with different tableaux. For example, if we use elementary row operations, the initial simplex tableau shown in Table 26 can be transformed to a new tableau in which the identity matrix is located in "front of" the variables x_2, s_2, s_3 (instead of s_1, s_2, s_3 as in the initial tableau). The identity matrix that identifies the basic variables is printed in color for convenience.

TABLE 26
Initial Simplex Tableau

BV	$\overbrace{x_1 \qquad x_2}^{(=0)}$		$\overbrace{s_1 \qquad s_2 \qquad s_3}^{(\neq 0)}$			RHS	Initial Algebraic Form of Constraints
s_1	2	3	1	0	0	6	$\begin{bmatrix} 2x_1 + 3x_2 + s_1 & & = 6 \\ 4x_1 + 2x_2 & + s_2 & = 8 \\ 0.5x_1 - 2x_2 & + s_3 & = 1 \end{bmatrix}$
s_2	4	2	0	1	0	8	
s_3	1.5	−2	0	0	1	1	

$$\text{want} \begin{bmatrix} 1 \\ 0 \\ 0 \end{bmatrix}$$

By using consecutively the three elementary row operations

$$(1/3)R_1 \longrightarrow R_1$$
$$R_2 + (-2)R_1 \longrightarrow R_2$$
$$R_3 + 2R_1 \longrightarrow R_3$$

the above equations will be transformed into the new system shown in Table 27.

TABLE 27
New Simplex Tableau

BV	x_1	x_2	s_1	s_2	s_3	RHS	New Algebraic Form of Constraints
x_2	2/3	1	1/3	0	0	2	$\begin{bmatrix} 2/3x_1 + x_2 + 1/3s_1 & & = 2 \\ 8/3x_1 & - 2/3s_1 + s_2 & = 4 \\ 17/6x_1 & + 2/3s_1 & + s_3 = 5 \end{bmatrix}$
s_2	8/3	0	−2/3	1	0	4	
s_3	17/6	0	2/3	0	1	5	

Conclusion

The new simplex tableau has nonbasic variables x_1 and s_1 and basic variables x_2, s_2, and s_3. Note that the identity matrix is sitting in front of the basic variables. By setting the nonbasic variables x_1 and s_1 to zero, we can find the values of the basic variables x_2, s_2, and s_3 by simply reading off the values in the RHS

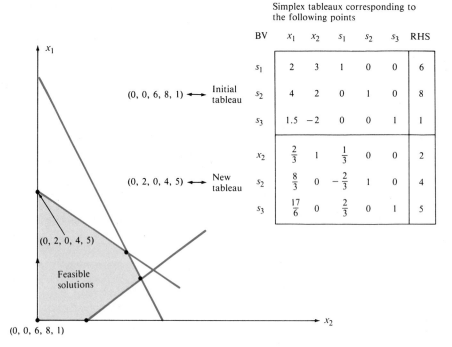

Simplex tableaux corresponding to the following points

BV	x_1	x_2	s_1	s_2	s_3	RHS
s_1	2	3	1	0	0	6
s_2	4	2	0	1	0	8
s_3	1.5	-2	0	0	1	1
x_2	$\frac{2}{3}$	1	$\frac{1}{3}$	0	0	2
s_2	$\frac{8}{3}$	0	$-\frac{2}{3}$	1	0	4
s_3	$\frac{17}{6}$	0	$\frac{2}{3}$	0	1	5

$(0, 0, 6, 8, 1) \longleftrightarrow$ Initial tableau

$(0, 2, 0, 4, 5) \longleftrightarrow$ New tableau

$(0, 2, 0, 4, 5)$

Feasible solutions

$(0, 0, 6, 8, 1)$

Figure 31
Moving from one corner-point to another corner-point

column. Hence the new tableau is identified as the corner-point $(0, 2, 0, 4, 5)$. This is the point in the $x_1 x_2$-plane with cartesian coordinates $(0, 2)$ and slack variables 0, 4, and 5. Geometrically, we have moved from the initial corner-point $(0, 0, 6, 8, 1)$ to the new corner-point $(0, 2, 0, 4, 5)$ by carrying out elementary row operations on the simplex tableau. (See Figure 31.)

Where We Are Headed

In the next section we will learn a very powerful method for solving linear programming problems. The method starts with an initial simplex tableau and steps from tableau to tableau, using elementary row operations. Geometrically, the method starts at the origin and moves successively from one corner-point to an adjacent corner-point in such a way that the objective function is always increased (or decreased for minimizing problems). Finally, the method will stop at the corner-point for which the objective function is a maximum (or minimum). This method is Dantzig's simplex method.

Problems

For Problems 1–5, draw the feasible sets. Then introduce the slack variables for each constraint and label the corner-points of the feasible set using augmented coordinates.

1. $x_1 + x_2 \leq 1$
$x_1 \geq 0$
$x_2 \geq 0$

2. $2x_1 + x_2 \leq 2$
$x_1 + 2x_2 \leq 2$
$x_1 \geq 0$
$x_2 \geq 0$

3. $x_1 + x_2 + x_3 \leq 1$
$x_1 \geq 0$
$x_2 \geq 0$
$x_3 \geq 0$

4. $x_1 + x_2 \leq 2$
$x_2 \leq 1$
$x_1 \geq 0$
$x_2 \geq 0$

5. $x_1 + x_2 \leq 5$
$-x_1 + x_2 \leq 1$ (interesting
$x_1 - x_2 \leq 1$ feasible set)
$x_1 \geq 0$
$x_2 \geq 0$

Problems 6–12 use the following inequalities:

$$x_1 + x_2 \le 2$$
$$x_2 \le 1$$
$$x_1 \ge 0$$
$$x_2 \ge 0$$

Draw the point or region of the $x_1 x_2$-plane described by the following points or equations.

6. $(0, 0, 2, 1)$ **7.** $(0, 1, 1, 0)$ **8.** $(1/2, 1/2, 1, 1/2)$
9. $(2, 2, -2, -1)$ **10.** $s_1 = 0$ **11.** $s_1 = 0, s_2 \ge 0$
12. $x_1 = 0, s_2 \ge 0$

For Problems 13–16, label the points A, B, C, and D using the augmented coordinates (x_1, x_2, s_1, s_2).

13.

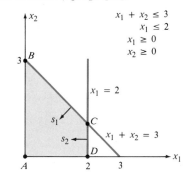

$$x_1 + x_2 \le 3$$
$$x_1 \le 2$$
$$x_1 \ge 0$$
$$x_2 \ge 0$$

14.

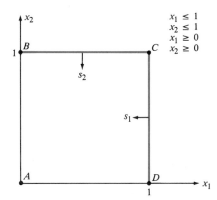

$$x_1 \le 1$$
$$x_2 \le 1$$
$$x_1 \ge 0$$
$$x_2 \ge 0$$

15.

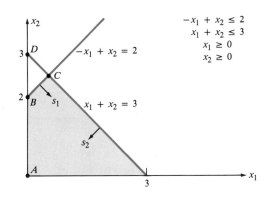

$$-x_1 + x_2 \le 2$$
$$x_1 + x_2 \le 3$$
$$x_1 \ge 0$$
$$x_2 \ge 0$$

16.

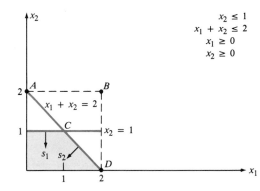

$$x_2 \le 1$$
$$x_1 + x_2 \le 2$$
$$x_1 \ge 0$$
$$x_2 \ge 0$$

For Problems 17–20, carry out the following steps.

(a) Write the inequalities as equalities by introducing slack variables.
(b) Using the equalities from part (a), construct a simplex tableau. Be sure to label the basic variables.
(c) Using elementary row operations, transform the simplex tableau found in part (b) to a new tableau in which the basic variables are x_1 and s_2. *Hint:* The new simplex tableau should have the 2×2 identity matrix in the columns under x_1 and s_2.
(d) What are the augmented coordinates (x_1, x_2, s_1, s_2) associated with the new tableau?
(e) Graph the feasible set and plot the two points associated with the two tableaux found in parts (b) and (c).

17. $x_1 + 2x_2 \le 2$ **18.** $x_1 + 3x_2 \le 6$
$2x_1 + x_2 \le 2$ $x_1 + x_2 \le 3$
$x_1 \ge 0$ $x_1 \ge 0$
$x_2 \ge 0$ $x_2 \ge 0$
19. $x_1 + x_2 \le 2$ **20.** $x_1 + x_2 \le 10$
$x_2 \le 1$ $5x_1 + x_2 \le 5$
$x_1 \ge 0$ $x_1 \ge 0$
$x_2 \ge 0$ $x_2 \ge 0$

21. In Figure 32, what are the signs (P = positive or N = negative) of the coordinates (x_1, x_2, s_1, s_2) of the points A, B, C, D, E, F, and G? For example, the signs of the coordinates of A are (P, P, P, P).

For Problems 22–25, use the simplex tableau described to carry out the following steps.

(a) Identify the basic and nonbasic variables.
(b) Carry out elementary row operations in such a way that the new basic variables are x_2 and s_2 (2×2 identity matrix in the columns labeled x_2 and s_2).
(c) What are the coordinates (x_1, x_2, s_1, s_2) of the points that correspond to the new tableau?

22.

BV	x_1	x_2	s_1	s_2	RHS
s_1	1	2	1	0	2
s_2	2	1	0	1	2

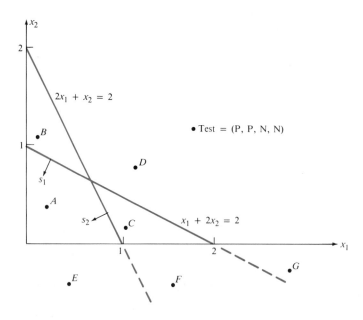

Figure 32
Labeling points with slack variables

23.

BV	x_1	x_2	s_1	s_2	RHS
s_1	1	1	1	0	2
s_2	1	3	0	1	1

24.

BV	x_1	x_2	s_1	s_2	RHS
s_1	1	0	1	0	2
s_2	1	1	0	1	4

25.

BV	x_1	x_2	s_1	s_2	RHS
s_1	0	1	1	0	2
s_2	1	1	0	1	4

For Problems 26–29, write out in algebraic form the equations described by the following simplex tableaux.

26. Use the tableau in Problem 22.
27. Use the tableau in Problem 23.
28. Use the tableau in Problem 24.
29. Use the tableau in Problem 25.

Problems 30–32 describe linear programming problems. Carry out the following steps for each problem.
(a) Draw the feasible sets.
(b) Change the problem to a linear programming problem with equality constraints by adding slack variables.

(c) Label all corner-points of the feasible set in the augmented coordinate system.

30. Maximize $P = x_1 + x_2$

subject to
$$2x_1 + x_2 \le 2$$
$$x_1 + 2x_2 \le 2$$

and
$$x_1 \ge 0$$
$$x_2 \ge 0$$

31. Maximize $P = 5x_1 + 4x_2$

subject to
$$x_1 + x_2 \le 2$$
$$x_1 \qquad \le 1$$

and
$$x_1 \ge 0$$
$$x_2 \ge 0$$

32. Maximize $P = 6x_1 + 6x_2$

subject to
$$6x_1 + 4x_2 \le 9$$

and
$$x_1 \ge 0$$
$$x_2 \ge 0$$

For the points $(x_1, x_2, s_1, s_2, s_3)$ in Problems 33–37, answer the following questions.
(a) Which of the variables are basic and which are nonbasic?
(b) Where would the points be plotted in the $x_1 x_2$-plane?
(c) How many constraints does the corresponding linear programming problem have?

33. $(0, 0, 2, 1, 3)$ **34.** $(0, 6, 9, 0, 4)$
35. $(2, 1, 0, 0, -1)$ **36.** $(1, 1, -3, 0, 0)$
37. $(1, 0, 2, 1, 0)$

Consider the feasible set described by the following four inequalities:

$$2x_1 + x_2 \le 2$$
$$x_1 + 2x_2 \le 2$$
$$x_1 \ge 0$$
$$x_2 \ge 0$$

For Problems 38–42, define the slack variables s_1 and s_2 for the inequalities above and describe the points in the x_1x_2-plane that are described by the following equations.

38. $s_1 = 0$ **39.** $s_2 = 0$

40. $x_1 = 0$ **41.** $x_2 = 0$

42. $s_1 = 0$ and $s_2 = 0$

2.5   The Simplex Method

PURPOSE

We will show how to solve maximizing linear programming problems by the simplex method. This method starts with an initial simplex tableau, which represents a feasible point at the origin, and then moves from one corner-point to an adjacent corner-point of the feasible set (by repeated transformations of the simplex tableau) in such a way that we eventually arrive at the optimal feasible point.

Introduction

The corner-point method for solving linear programming problems gives a geometric feeling for the subject and an understanding of some of the basic ideas. However, from a practical point of view the method breaks down for problems with more than two variables. The **simplex method** provides a means of moving from corner-point to corner-point in higher dimensions by a strictly algebraic procedure. We explain the simplex method using the following example.

Step-by-Step Description of the Simplex Method

Consider the linear programming problem that was discussed in the previous section.

$$\text{maximize } P = 3x_1 + x_2$$

subject to

$$2x_1 + 3x_2 \le 6$$
$$4x_1 + 2x_2 \le 8$$
$$1.5x_1 - 2x_2 \le 1$$

and

$$x_1 \ge 0$$
$$x_2 \ge 0$$

Although the simplex method does not require that we draw the feasible set, we present it in Figure 33 to show how the algebra of the method corresponds to the geometry of the corner-points.

- **Step 1** (write the initial simplex tableau). We begin by introducing a slack variable for each inequality and rewriting the inequality constraints as equality constraints. Hence the original linear programming problem is restated in the following form:

$$\text{maximize } P = 3x_1 + x_2$$

subject to

$$2x_1 + 3x_2 + s_1 \qquad\qquad = 6$$
$$4x_1 + 2x_2 \qquad + s_2 \qquad = 8$$
$$1.5x_1 - 2x_2 \qquad\qquad + s_3 = 1$$

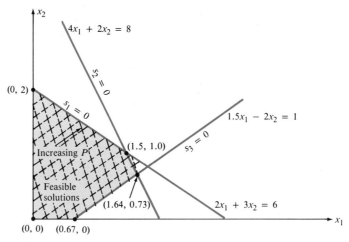

Figure 33
Physical illustration of a linear programming problem

and

$$x_1 \geq 0$$
$$x_2 \geq 0$$
$$s_1 \geq 0$$
$$s_2 \geq 0$$
$$s_3 \geq 0$$

It is important to understand that the three equality constraints and the five nonnegative constraints in the above problem describe the same feasible set as was shown in Figure 33.

We now write the above equations in the compact form of the initial simplex tableau in Table 28. The basic coefficients are in color.

TABLE 28
Initial Simplex Tableau

Zero variables

BV	ROW	P	x_1	x_2	s_1	s_2	s_3	RHS	Algebraic Form of Tableau
*	0	1	-3	-1	0	0	0	0	$P - 3x_1 - x_2 \qquad\qquad = 0$
s_1	1	0	2	3	1	0	0	6	$2x_1 + 3x_2 + s_1 \qquad = 6$
s_2	2	0	4	2	0	1	0	8	$4x_1 + 2x_2 \qquad + s_2 \quad = 8$
s_3	3	0	1.5	-2	0	0	1	1	$1.5x_1 - 2x_2 \qquad\quad + s_3 = 1$

Comments on the Initial Simplex Tableau

ROW 0 of the initial simplex tableau contains the objective function

$$P = 3x_1 + x_2$$

written in the form

$$P - 3x_1 - x_2 - 0s_1 - 0s_2 - 0s_3 = 0$$

It is useful to include this equation in the tableau, since the objective function is needed to find the optimal solution. You should note that the simplex tableau described here contains this extra row, unlike the simplex tableau described in the previous section. They are both called simplex tableaux, but the tableau here contains more information. In the previous section the idea of a tableau was introduced without any reference to an optimization process. This section adds the optimization process to the tableau in the form of ROW 0 to complete the description of the linear programming problem.

The basic variable column (BV) at the left of the tableau tells which three of the variables x_1, x_2, s_1, s_2, s_3 are basic (the ones for which we solve after setting the others to zero). The values of the basic variables can always be found in the right-hand side column (RHS). In this tableau they are $s_1 = 6$, $s_2 = 8$, and $s_3 = 1$.

Since s_1, s_2, s_3 are the basic variables in the above tableau, x_1, x_2 are the nonbasic variables (the ones we set to zero). The nonbasic variables in the simplex tableau are the variables listed at the very top of the tableau that are always circled.

The value of the objective function P at the corner-point $(0, 0, 6, 8, 1)$ is located in ROW 0 in the RHS column. The initial simplex tableau always corresponds to the corner-point at the origin.

- **Step 2** (find a better corner-point).
- Substep 2a: What direction to go? The initial simplex tableau corresponds geometrically to starting at the corner-point $(0, 0, 6, 8, 1)$. This means that we are starting at the origin $(0, 0)$ in the $x_1 x_2$-plane. The strategy of the simplex method is to move from the origin to one of the adjacent corner-points. In Figure 34 we see that there are two adjacent corner-points at $(0.67, 0)$ and $(0, 2)$.

The question we ask is: In what direction do we go (up or to the right)? Another way of stating this is: Should we increase x_1 or x_2? Here is the rule.

Find the Pivot Column (Determining the Variable to Increase)

The simplex method always increases that variable that gives rise to the largest increase in the objective function. To find this variable, select the column in the simplex tableau that has the most negative number in ROW 0. The variable above this **pivot column** is the variable for which the objective function increases the most.

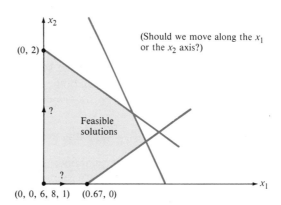

Figure 34
*Finding the next corner-point
after the origin*

To understand why this rule works, note that the objective function

$$P = 3x_1 + x_2$$

increases faster if we increase x_1 than if we increase x_2. This is due to the fact that the coefficient of x_1 is larger than the coefficient of x_2. For this reason we move from the origin in the x_1x_2-plane to the right along the positive x_1-axis as seen in Figure 34. In terms of the simplex tableau, we increase the variable that is written over that column with the most negative number in ROW 0. In Table 29 the most negative number in ROW 0 is -3. Hence the simplex method moves along the positive x_1 axis.

TABLE 29

Finding the Pivot Column

BV	ROW	P	x_1	x_2	s_1	s_2	s_3	RHS
*	0	1	-3	-1	0	0	0	0
s_1	1	0	2	3	1	0	0	6
s_2	2	0	4	2	0	1	0	8
s_3	3	0	1.5	-2	0	0	1	1

$\uparrow$
*Pivot
column*

- Substep 2b: When do we stop?   The second question is: How large should we let x_1 become before the point $(x_1, 0)$ lies outside the feasible set? Looking at the feasible set in Figure 34, we see that the point becomes infeasible when x_1 is greater than 0.67. How do we know that we have reached a corner-point of the feasible set? The following rule gives the answer.

Find the Pivot Row (When to Stop Increasing the Variable)

To find the pivot row, carry out the following steps:

- Divide each positive value in the pivot column into the number in the RHS column in the same row.
- Pick the smallest of these ratios. The row with the smallest ratio is called the pivot row.

In Table 30 we see that ROW 3 is the pivot row. The number that lies at the intersection of the pivot row and the pivot column is called the **pivot element**. Here, 1.5 is the pivot element.

Comments on Step 2

The purpose of finding the pivot row and the pivot column is to determine new basic and nonbasic variables. In this example the initial basic variables were s_1, s_2, and s_3. After finding the pivot row and pivot column, we replace the

TABLE 30
Finding the Pivot Row and Element

BV	ROW	P	x_1	x_2	s_1	s_2	s_3	RHS	
*	0	1	−3	−1	0	0	0	0	
s_1	1	0	2	3	1	0	0	6	$6/2 = 3$
s_2	2	0	4	2	0	1	0	8	$8/4 = 2$
Pivot → s_3	3	0	(1.5)	−2	0	0	1	1	$1/1.5 = 0.67$ (minimum)

↑ Pivot column

Pivot element

basic variable s_3 (the variable in the pivot row) with x_1 (the variable above the pivot column). Hence the new basic and nonbasic variables are the following:

	Initially		After Finding Pivot Row and Column	
	Basic	*Nonbasic* (=0)	*Basic*	*Nonbasic* (=0)
	s_1, s_2, s_3	x_1, x_2	s_1, s_2, x_1	s_3, x_2

In other words, the new nonbasic (zero) variables are x_2 and s_3, which means that we should set x_2 and s_3 to zero in the constraint equations and solve for the other variables. By so doing, we will find a new corner-point. The only

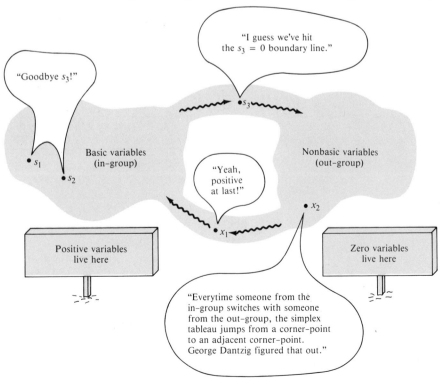

problem is that the equations are not written in a convenient form. The last step of the simplex method uses elementary row operations to rewrite the constraint equations in such a manner that when we set x_2 and s_3 to zero, the values of the remaining variables $x_1, s_1,$ and s_2 will be located in the RHS column.

- **Step 3** (rewriting the simplex tableau). The goal now is to use the elementary row operations to make the pivot element 1 and place 0's above and below it in the pivot column. These operations will not change the solution of the equations but will allow us to find more easily the new corner-point. We start with the tableau in Table 31.

TABLE 31
Initial Simplex Tableau

BV	ROW	P	x_1	x_2	s_1	s_2	s_3	RHS
*	0	1	-3	-1	0	0	0	0
s_1	1	0	2	3	1	0	0	6
s_2	2	0	4	2	0	1	0	8
s_3	3	0	(1.5)	-2	0	0	1	1

want 1   Pivot element

We begin by transforming the pivot element (the element that lies at the intersection of the pivot row and column) to 1. We do this with the elementary operation

$$(2/3)R_3 \longrightarrow R_3$$

This gives the new tableau in Table 32.

TABLE 32
Intermediate Tableau

BV	ROW	P	x_1	x_2	s_1	s_2	s_3	RHS
*	0	1	-3	-1	0	0	0	0
s_1	1	0	2	3	1	0	0	6
s_2	2	0	4	2	0	1	0	8
s_3	3	0	(1)	$-4/3$	0	0	2/3	2/3

Pivot element

Now transform ROWS 0, 1, and 2 so that the pivot column contains 0's above the pivot element. This can be done by using the following three elementary row operations:

$$R_0 + 3R_3 \longrightarrow R_0$$
$$R_1 + (-2)R_3 \longrightarrow R_1$$
$$R_2 + (-4)R_3 \longrightarrow R_2$$

The calculations for these operations are carried out in Table 33.

The new simplex tableau is given in Table 34. The new basic variables are $s_1, s_2,$ and $x_1,$ and hence the new nonbasic variables (zero variables) are x_2 and

TABLE 33
Intermediate Tableau

		P	x_1	x_2	s_1	s_2	s_3	RHS	
		$[1$	-3	-1	0	0	0	$0]$	R_0
	$(+3)$	$[0$	1	$-4/3$	0	0	$2/3$	$2/3]$	R_3
New ROW 0		$[1$	0	-5	0	0	2	$2]$	R_0
		$[0$	2	3	1	0	0	$6]$	R_1
	(-2)	$[0$	1	$-4/3$	0	0	$2/3$	$2/3]$	R_3
New ROW 1		$[0$	0	$17/3$	1	0	$-4/3$	$14/3]$	R_1
		$[0$	4	2	0	1	0	$8]$	R_2
	(-4)	$[0$	1	$-4/3$	0	0	$2/3$	$2/3]$	R_3
New ROW 2		$[0$	0	$22/3$	0	1	$-8/3$	$16/3]$	R_2

Zero variables

TABLE 34
New Simplex Tableau
at Corner-Point
$(2/3, 0, 14/3, 16/3, 0)$

BV	ROW	P	x_1	(x_2)	s_1	s_2	(s_3)	RHS
*	0	1	0	-5	0	0	2	2
s_1	1	0	0	$17/3$	1	0	$-4/3$	$14/3$
s_2	2	0	0	$22/3$	0	1	$-8/3$	$16/3$
x_1	3	0	1	$-4/3$	0	0	$2/3$	$2/3$

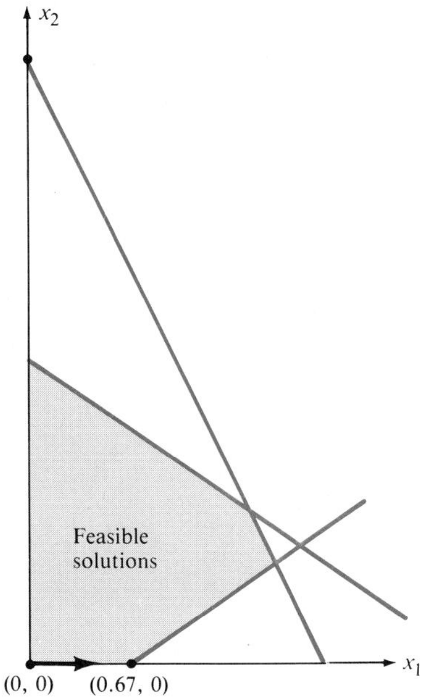

Figure 35
First two points found by the simplex method labeled only with cartesian coordinates

s_3. The zero variables are circled. Note that the ones and the zeros appear in front of the new basic variables in the new tableau.

This new simplex tableau is identified with the corner-point

$$(x_1, x_2, s_1, s_2, s_3) = (2/3, 0, 14/3, 16/3, 0).$$

Geometrically, this is located at the point $(0.67, 0)$ in the $x_1 x_2$-plane. The values of the slack variables are $s_1 = 14/3$, $s_2 = 16/3$; and $s_3 = 0$. This means that the corner-point lies on the boundary line determined by $s_3 = 0$ and is a positive distance from the boundary lines determined by the other two slack variables. This corner-point is illustrated in Figure 35.

Note too that at this point the objective function P has the value of 2. This value is given in the RHS column of ROW 0.

When the Simplex Method Stops

Stopping Rule

If all the numbers in ROW 0 are greater than or equal to zero, the simplex method stops, and the corner-point corresponding to the current simplex tableau is the optimal feasible point. If at least one of the numbers in ROW 0 is less than 0, repeat Steps 2 and 3.

In the present example, ROW 0 consists of

$$\begin{array}{ccccccc} P & x_1 & x_2 & s_1 & s_2 & s_3 & \text{RHS} \\ [1 & 0 & -5 & 0 & 0 & 2 & 2] \end{array}$$

Since this row contains a negative number (namely, -5), we should repeat Steps 2 and 3 at least one more time.

Table 35 shows the new pivot column (column under x_2) and the new pivot row (ROW 2). Note that when finding the pivot row, we divided only the positive entries in the pivot column into the right-hand-side entries.

TABLE 35
Finding the New Pivot
Column and Pivot Row

	BV	ROW	P	x_1	(x_2)	s_1	s_2	(s_3)	RHS	
	*	0	1	0	-5	0	0	2	2	
Pivot	s_1	1	0	0	$17/3$	1	0	$-4/3$	$14/3$	$(14/3)(3/17) = 0.82$
row ⟶	s_2	2	0	0	$(22/3)$	0	1	$-8/3$	$16/3$	$(16/3)(3/22) = 0.73$
	x_1	3	0	1	$-4/3$	0	0	$2/3$	$2/3$	

↑
Pivot   *Pivot*
column   *element*

The next step is to transform the above pivot element of 22/3 into a 1 and then transform the entries above and below the new 1 into 0's. Doing this, we get the tableau in Table 36 (note that x_2 has become a basic variable and s_2 has become a nonbasic variable):

TABLE 36
Final Simplex Tableau
(1.64, 0.73, 0.53, 0, 0)

Zero variables

BV	ROW	P	x_1	x_2	s_1	(s_2)	(s_3)	RHS
*	0	1	0	0	0	0.68	0.18	5.64
s_1	1	0	0	0	1	-0.77	0.73	0.53
x_2	2	0	0	1	0	0.14	-0.36	0.73
x_1	3	0	1	0	0	0.18	-0.19	1.64

Observe that all the entries in ROW 0 are positive or zero. Hence the simplex method comes to a halt.

Observations About the Solution

- The final corner-point is

$$(x_1, x_2, s_1, s_2, s_3) = (1.64, 0.73, 0.53, 0, 0)$$

- The optimal feasible point is

$$(x_1{}^*, x_2{}^*) = (1.64, 0.73)$$

- The maximum value of the objective function is $P = 5.64$.

- The final values of the slack variables are

$$(s_1, s_2, s_3) = (0.53, 0, 0)$$

The above statements say that the optimal feasible point lies at the intersection of the second and third constraint lines. This is illustrated in Figure 36.
We summarize the steps of the simplex method that we have used for solving a maximization problem.

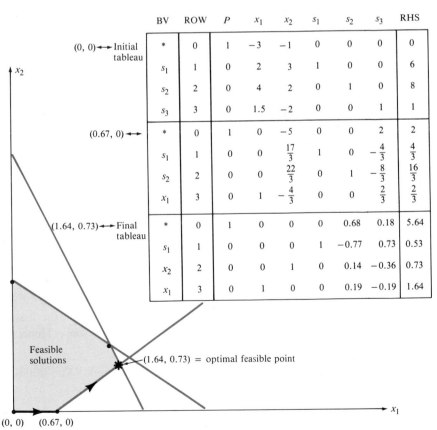

Simplex tableaux corresponding to corner-points of the feasible set

BV	ROW	P	x_1	x_2	s_1	s_2	s_3	RHS
*	0	1	−3	−1	0	0	0	0
s_1	1	0	2	3	1	0	0	6
s_2	2	0	4	2	0	1	0	8
s_3	3	0	1.5	−2	0	0	1	1
*	0	1	0	−5	0	0	2	2
s_1	1	0	0	$\frac{17}{3}$	1	0	$-\frac{4}{3}$	$\frac{4}{3}$
s_2	2	0	0	$\frac{22}{3}$	0	1	$-\frac{8}{3}$	$\frac{16}{3}$
x_1	3	0	1	$-\frac{4}{3}$	0	0	$\frac{2}{3}$	$\frac{2}{3}$
*	0	1	0	0	0	0.68	0.18	5.64
s_1	1	0	0	0	1	−0.77	0.73	0.53
x_2	2	0	0	1	0	0.14	−0.36	0.73
x_1	3	0	1	0	0	0.19	−0.19	1.64

$(0, 0) \longleftrightarrow$ Initial tableau

$(0.67, 0) \longleftrightarrow$

$(1.64, 0.73) \longleftrightarrow$ Final tableau

Feasible solutions

$(1.64, 0.73) =$ optimal feasible point

$(0, 0)$ $(0.67, 0)$

Figure 36
Three tableaux to get to the optimal feasible solution

Simplex Method for Maximization Problems

- **Step 1** (initialization). Write the initial simplex tableau for the problem.
- **Step 2** (find pivot column, row, and element).

  (a) Find the pivot column by selecting the column with the most negative numbers in ROW 0. Break ties arbitrarily.

  (b) Find the pivot row by dividing the positive entries in the pivot column into the corresponding numbers in the RHS column and selecting the smallest of these ratios. The row where the smallest ratio occurs is the pivot row.

  (c) The number in the tableau that lies in both the pivot column and the pivot row is the pivot element.

- **Step 3** (transform the simplex tableau). Use elementary row operations to transform the simplex tableau into a new tableau in which the pivot element is 1 and all other entries in the pivot column are 0.

- **Step 4** (test for completion). Repeat Steps 2 and 3 until all entries in ROW 0 are positive or zero. When this occurs, stop the procedure. The final tableau identifies both the optimal feasible point and the maximum value of the objective function. The nonbasic variables (the ones circled at the top of the tableau) will be 0, and the basic variables (listed in the BV column) will have values listed in the RHS column. The value of the objective function is located in the RHS column in ROW 0.

Example 1

Simplex Method   Solve the following linear programming problem with three decision variables by the simplex method:

$$\text{maximize } P = x_1 + 6x_2 + 3x_3$$

subject to

$$x_1 + x_2 + 2x_3 \leq 4$$
$$x_1 + 2x_2 + x_3 \leq 4$$

and

$$x_1 \geq 0$$
$$x_2 \geq 0$$
$$x_3 \geq 0$$

Solution   It is difficult to graph the feasible set for this problem, since it is a subset of three-dimensional space. Hence we use the algebraic approach of the simplex method.

- **Step 1** (initialization). The initial simplex tableau (nonbasic variables circled) is given by Table 37.
- **Step 2** (find pivot element).
- **Substep 2a:** Find the pivot column. Since the most negative number in ROW 0 is -6, the column under x_2 is the pivot column. (See Table 38.)

TABLE 37
Initial Simplex Tableau

BV	ROW	P	x_1	x_2	x_3	s_1	s_2	RHS
*	0	1	-1	-6	-3	0	0	0
s_1	1	0	1	1	2	1	0	4
s_2	2	0	1	2	1	0	1	4

- **Substep 2b:** Find the pivot row. Divide the positive numbers in the pivot column into the corresponding values of the RHS column and select the smaller of these ratios. The smaller ratio, as shown in Table 37, occurs in ROW 2; hence this is the pivot row. The intersection of the pivot row and pivot column gives the pivot element.

TABLE 38
Finding the Pivot Row and Column

BV	ROW	P	x_1	x_2	x_3	s_1	s_2	RHS	
*	0	1	-1	-6	-3	0	0	0	
s_1	1	0	1	1	2	1	0	4	$4/1 = 4$
s_2	2	0	1	②	1	0	1	4	$4/2 = 2$ ⟵ Pivot row

Pivot column   *Pivot element*

- **Step 3** (transforming the simplex tableau). Using elementary row operations, transform the tableau in Table 38 so that the pivot element is 1 and all other entries in the pivot column are 0. We can do this with the following three operations:

$$(1/2)R_2 \longrightarrow R_2$$
$$R_0 + \quad 6R_2 \longrightarrow R_0$$
$$R_1 + (-1)R_2 \longrightarrow R_1$$

Carrying out these three row operations, we arrive at a new tableau (Table 39). The new basic variables are s_1 and x_2, which means that the new zero variables are x_1, x_3, and s_2.

TABLE 39
Final Simplex Tableau

BV	ROW	P	x_1	x_2	x_3	s_1	s_2	RHS
*	0	1	2	0	0	0	3	12
s_1	1	0	1/2	0	3/2	1	$-1/2$	2
x_2	2	0	1/2	1	1/2	0	1/2	2

Now all the entries in ROW 0 are greater than or equal to zero. Hence the simplex method stops with nonbasic (zero) variables x_1, x_3, and s_2 and basic (nonzero) variables $s_1 = 2$ and $x_2 = 2$.

Conclusion: The interpretation of these values is that the optimal feasible point of the linear programming problem is $(x_1{}^*, x_2{}^*, x_3{}^*) = (0, 2, 0)$. The maximum value of the objective function P is 12 and is found in column RHS of ROW 0.

$\square$

Problems

For Problems 1–7, solve the linear programming problems by (a) the corner-point method and (b) the simplex method. Identify each simplex tableau found in the simplex method with a corner-point in the feasible set.

1. Maximize $P = x_1 + x_2$

subject to $\qquad 2x_1 + x_2 \leq 2$

and $\qquad\qquad x_1 \geq 0$

$\qquad\qquad\qquad x_2 \geq 0$

2. Maximize $P = x_1 + x_2$

subject to $\qquad 2x_1 + x_2 \leq 2$

$\qquad\qquad x_1 + 2x_2 \leq 2$

and $\qquad\qquad x_1 \geq 0$

$\qquad\qquad\qquad x_2 \geq 0$

3. Maximize $P = 2x_1 + x_2$

subject to $\qquad 3x_1 + x_2 \leq 9$

$\qquad\qquad x_1 + x_2 \leq 4$

$\qquad\qquad x_1 + 4x_2 \leq 12$

and $\qquad\qquad x_1 \geq 0$

$\qquad\qquad\qquad x_2 \geq 0$

4. Maximize $P = 2x_1 + x_2 + x_3$

subject to $\qquad x_1 + x_2 + x_3 \leq 1$

and $\qquad\qquad x_1 \geq 0$

$\qquad\qquad\qquad x_2 \geq 0$

$\qquad\qquad\qquad x_3 \geq 0$

Hint: This problem has three variables, but the corner-points are found in the same way as in two dimensions.

5. Maximize $P = 2x_1 + x_2 + x_3$

subject to $\qquad x_1 + x_2 + x_3 \leq 2$

and $\qquad\qquad\qquad x_1 \geq 0$

$\qquad\qquad\qquad x_2 \geq 0$

$\qquad\qquad\qquad x_3 \geq 0$

See the hint for Problem 4.

6. Maximize $P = 4x_1 + 3x_2$

subject to $\qquad x_1 + x_2 \leq 4$

$\qquad\qquad 2x_1 + x_2 \leq 6$

and $\qquad\qquad x_1 \geq 0$

$\qquad\qquad\qquad x_2 \geq 0$

7. Maximize $P = 4x_1 + 3x_2$

subject to $\qquad x_1 + x_2 \leq 2$

$\qquad\qquad -x_1 + x_2 \leq 1$

and $\qquad\qquad x_1 \geq 0$

$\qquad\qquad\qquad x_2 \geq 0$

For Problems 8–11, solve the linear programming problems by the simplex method.

8. Maximize $P = x_1 + x_2 + 2x_3$

subject to $\qquad x_1 + 4x_2 + 3x_3 \leq 1$

and $\qquad\qquad x_1 \geq 0$

$\qquad\qquad\qquad x_2 \geq 0$

$\qquad\qquad\qquad x_3 \geq 0$

9. Maximize $P = x_1 + 2x_2 + 3x_3$

subject to $\qquad 2x_1 + 2x_2 + x_3 \leq 5$

$\qquad\qquad x_1 + 2x_2 + x_3 \leq 5$

$\qquad\qquad x_1 + x_2 + 2x_3 \leq 5$

and $\qquad\qquad x_1 \geq 0$

$\qquad\qquad\qquad x_2 \geq 0$

$\qquad\qquad\qquad x_3 \geq 0$

10. Maximize $P = x_1$

subject to $\qquad x_1 + 4x_2 + x_3 \le 5$

$\qquad\qquad 3x_1 + x_2 + x_3 \le 5$

and $\qquad\qquad\qquad x_1 \ge 0$

$\qquad\qquad\qquad x_2 \ge 0$

$\qquad\qquad\qquad x_3 \ge 0$

11. Maximize $P = x_1 + x_3$

subject to $\qquad x_1 + 5x_2 \qquad\quad \le 5$

$\qquad\qquad x_1 \qquad + x_3 \le 1$

$\qquad\qquad\quad x_2 + 2x_3 \le 2$

and $\qquad\qquad\qquad x_1 \ge 0$

$\qquad\qquad\qquad x_2 \ge 0$

$\qquad\qquad\qquad x_3 \ge 0$

12. Answer the following questions concerning the simplex tableau in Table 40.
(a) What are the basic variables of this tableau?
(b) What are the nonbasic variables of this tableau?
(c) What variables should be set to zero?
(d) What is the value of the objective function in this tableau?
(e) What is the point $(x_1, x_2, s_1, s_2, s_3)$ associated with this tableau?
(f) Is the point $(x_1, x_2, s_1, s_2, s_3)$ associated with this tableau optimal?
(g) What is the optimal value of P?
(h) Write out in algebraic form the linear programming problem that corresponds to this tableau.
(i) What are the values of the basic variables in this problem?

TABLE 40
Simplex Table for Problem 12

BV	ROW	P	x_1	x_2	s_1	s_2	s_3	RHS
*	0	1	0	0	3	0	4	19
x_1	1	0	1	0	3	0	4	2
s_2	2	0	0	0	5	1	6	4
x_2	3	0	0	1	9	0	9	6

13. Answer the following questions concerning the simplex tableau in Table 41.
(a) What is the linear programming problem that corresponds to this tableau?
(b) Is the given tableau the initial simplex tableau for the simplex method? Why?
(c) What are the basic variables in this tableau?
(d) What are the nonbasic variables in this tableau?
(e) How many constraints does the given linear programming problem have?
(f) How many decision variables does the given linear programming problem have?
(g) What point $(x_1, x_2, s_1, s_2, s_3)$ corresponds to this tableau? Is this point a feasible point? Why or why not?
(h) What is the pivot column in this tableau?

TABLE 41
Simplex Tableau for Problem 13

BV	ROW	P	x_1	x_2	s_1	s_2	s_3	RHS
*	0	1	-3	-2	0	0	0	0
s_1	1	0	1	2	1	0	0	2
s_2	2	0	2	1	0	1	0	2
s_3	3	0	2	2	0	0	1	3

(i) What is the pivot row in this tableau?

(j) What is the pivot element in this tableau?

(k) Would the simplex method stop at this tableau? Why or why not?

(l) If the simplex method does not stop at this tableau, find the next tableau. Is the next tableau an optimal tableau?

14. Answer the following questions concerning the simplex tableau in Table 42.

(a) What is the point $(x_1, x_2, x_3, s_1, s_2)$ that corresponds to this tableau?

(b) What are the basic variables in this tableau?

(c) What are the nonbasic variables in this tableau?

(d) What are the variables that should be set to zero in this tableau?

(e) What are the variables that we should solve after setting some of them equal to zero? Are the nonzero variables positive? What does this mean?

(f) What is the pivot column of this tableau?

(g) What is the pivot row of this tableau?

(h) What is the pivot element of this tableau?

(i) Find the next tableau. Will the simplex method stop after finding this tableau?

(j) What is the corner-point that corresponds to the new tableau?

(k) What is the value of the objective function in the new tableau?

TABLE 42
Simplex Table for Problem 14

BV	ROW	P	x_1	x_2	x_3	s_1	s_2	RHS
*	0	1	-3	-4	-3	0	0	0
s_1	1	0	1	1	1	1	0	2
s_2	2	0	1	0	0	0	1	1

15. Answer the following questions concerning the simplex tableau in Table 43.

(a) What is the linear programming problem that corresponds to this tableau?

(b) What are the basic variables in this tableau?

(c) What are the nonbasic variables in this tableau?

(d) Will the simplex method stop after finding this tableau?

(e) What is the pivot column of this tableau?

(f) What is the pivot row of this tableau?

(g) What is the pivot element of this tableau?

(h) Find the next tableau.

(i) What is the corner-point that corresponds to the next tableau?

(j) What is the value of the objective function in the next tableau? Has it been increased from the present tableau?

TABLE 43
Simplex Tableau for Problem 15

BV	ROW	P	x_1	x_2	s_1	s_2	s_3	RHS
*	0	1	0	-5	0	0	2	2
s_1	1	0	0	5.67	1	0	-1.33	4.67
s_2	2	0	0	7.33	0	1	-2.67	5.33
x_1	3	0	1	-1.33	0	0	0.67	0.67

2.6

Duality and Postoptimal Analysis

PURPOSE

We begin by showing how to find the dual linear programming problem to a given linear programming problem. We then show how to use the dual problem to solve a minimizing linear programming problem.

We also show how to analyze the optimal solution of a linear programming problem once it is found. This type of analysis is called *postoptimal analysis* or *what-if analysis.*

Duality and the Dual Problem

Until now we have discussed the maximizing and minimizing linear programming problems as two separate types of problems. But in fact, corresponding to every maximizing linear programming problem, there always exists a counterpart minimizing linear programming problem that has special properties in relation to the original maximizing problem. In other words, maximizing and minimizing linear programming problems come in pairs. Each maximizing linear programming problem has a "twin" minimizing problem and vice versa.

The subject of duality and the dual linear programming problem is a vast subject. The subject includes concepts such as shadow prices, marginal values, postoptimal analysis, and a host of other topics. Entire books have been written about the implications of duality, and we will not do it justice by devoting only a few pages to it now. Nevertheless, we give the reader a glimpse into this elegant theory and show how it can be used.

Duality refers to the fact that every linear programming problem has two forms—the primal form and the dual form. As we will see, it is irrelevant which form we call primal and which form we call dual, since each is the dual of the other.

The Concept of Duality

To begin the study of duality, consider the following maximizing linear programming problem with $\leq$ constraints. Since we are starting with this problem, it is the convention to call it the **primal problem**.

Primal Problem

$$\text{maximize } P = 350x_1 + 300x_2$$

subject to
$$20x_1 + 15x_2 \leq 4000$$
$$2x_1 + 5x_2 \leq 1000$$
$$5x_1 + 5x_2 \leq 1500$$
$$11x_1 + 12x_2 \leq 2640$$

and
$$x_1 \geq 0$$
$$x_2 \geq 0$$

Associated with this primal problem is the following dual form, or **dual problem**.

Dual Problem

$$\text{minimize } C = 4000y_1 + 1000y_2 + 1500y_3 + 2640y_4$$

subject to
$$20y_1 + 2y_2 + 5y_3 + 11y_4 \geq 350$$
$$15y_1 + 5y_2 + 5y_3 + 12y_4 \geq 300$$

and
$$y_1 \geq 0$$
$$y_2 \geq 0$$
$$y_3 \geq 0$$
$$y_4 \geq 0$$

We now state the rules for finding the dual problem of a linear programming problem.

Rules for Finding the Dual Problem

Given either a maximizing linear programming problem with $\leq$ constraints or a minimizing problem with $\geq$ constraints, the following rules tell how to find the dual problem.

- The "max" in the primal problem should be changed to a "min" in the dual problem. Likewise, if the primal problem is a min problem, the dual will be a max problem.
- The number of variables in the dual problem is the same as the number of functional constraints in the primal problem.
- The number of functional constraints in the dual problem is the same as the number of variables in the primal problem.
- The numbers on the right-hand side of the inequalities in the primal problem become the coefficients in the objective function in the dual problem.
- The coefficients in the objective function in the primal problem become the numbers on the right-hand side of the inequalities in the dual problem.
- The direction of the inequalities changes in the dual problem. "Less than or equal to" inequalities become "greater than or equal to" inequalities.
- The coefficients of the decision variables in the functional constraints are interchanged in the primal and dual problems in the sense that "rows" become "columns" and vice versa. For example, in Figure 37 the coefficients 20 and 15 in the first functional constraint (think of them as being in the first row) become the coefficients of y_1 (think of them as being in the leftmost column) in the functional constraints of the dual problem. Also, the numbers 20, 2, 5, and 11 in the leftmost column of the primal problem lie in the first row of the dual problem.

The rules for finding the dual problem are illustrated in Figure 37.

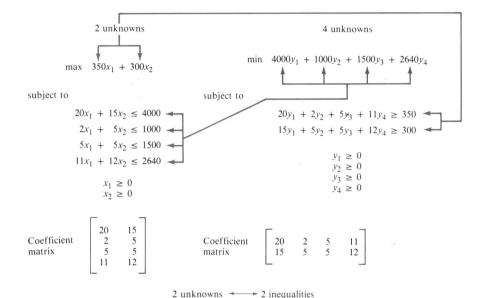

Figure 37
The primal-dual relationship

Example 1

Finding the Dual Problem   Find the dual problem for the following maximizing linear programming problem:

$$\text{maximize } P = 4x_1 + 7x_2 + 9x_3$$

subject to
$$x_1 + x_2 + x_3 \leq 5$$
$$x_1 + 7x_2 + 5x_3 \leq 4$$

and
$$x_1 \geq 0$$
$$x_2 \geq 0$$
$$x_3 \geq 0$$

Solution   Using the rules for finding the dual problem, we have

$$\text{minimize } C = 5y_1 + 4y_2$$

subject to
$$y_1 + y_2 \geq 4$$
$$y_1 + 7y_2 \geq 7$$
$$y_1 + 5y_2 \geq 9$$

and
$$y_1 \geq 0$$
$$y_2 \geq 0$$

Note that the number of variables in the primal is the same as the number of constraints in the dual (and vice versa). The reader should study the relationship between these two forms.   ☐

Example 2

Dual of a Minimizing Problem   Find the dual problem of the following minimizing linear programming problem:

$$\text{minimize } C = 5x_1 + 4x_2$$

subject to
$$x_1 + 7x_2 \geq 7$$
$$x_1 + 3x_2 \geq 9$$

and
$$x_1 \geq 0$$
$$x_2 \geq 0$$

Solution We call the above minimizing problem the primal problem, and in this case the dual problem will be a maximizing problem. Using the rules for finding the dual problem, we have the following problem:

$$\text{maximize } P = 7y_1 + 9y_2$$

subject to
$$y_1 + \ y_2 \leq 5$$
$$7y_1 + 3y_2 \leq 4$$

and
$$y_1 \geq 0$$
$$y_2 \geq 0 \qquad \qquad \square$$

The Dual Problem and Shadow Prices

We now give an economic interpretation of the dual form of a linear programming problem. For purposes of illustration we consider a specific type of company. Suppose we wish to determine the number of tables and chairs that should be made with certain resources of labor and lumber so that the profit from the sale of these items is maximized. We define x_1 and x_2 to be

$$x_1 = \text{number of tables made}$$
$$x_2 = \text{number of chairs made}$$

A typical linear programming problem would be to find x_1 and x_2 so that

$$\text{maximize } P = x_1 + x_2$$

subject to
$$2x_1 + \ x_2 \leq 2 \qquad \text{(labor resource)}$$
$$x_1 + 3x_2 \leq 3 \qquad \text{(lumber resource)}$$

and
$$x_1 \geq 0$$
$$x_2 \geq 0$$

The feasible set is drawn in Figure 38. The dual form of the above problem is the minimizing problem

$$\text{minimize } C = 2y_1 + 3y_2$$

subject to
$$2y_1 + \ y_2 \geq 1$$
$$y_1 + 3y_2 \geq 1$$

and
$$y_1 \geq 0$$
$$y_2 \geq 0$$

The reason for the interest in the dual problem lies in the interpretation of the optimal solution $(y_1{}^*, y_2{}^*)$ of the dual problem. The values $y_1{}^*$ and $y_2{}^*$ are called **shadow prices**. In the context of the allocation of resources problem they have the following interpretation:

$$y_1{}^* = \text{value of one unit of labor} \qquad \text{(resource 1)}$$
$$y_2{}^* = \text{value of one unit of lumber} \qquad \text{(resource 2)}$$

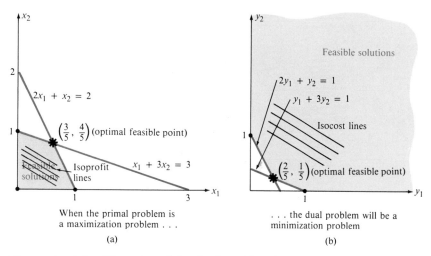

Figure 38
Graphs of primal and dual problems

When the primal problem is
a maximization problem . . .

(a)

. . . the dual problem will be a
minimization problem

(b)

The optimal feasible point of the dual problem is

$$y_1{}^* = 2/5$$
$$y_2{}^* = 1/5$$

as illustrated in Figure 38(b).

The optimal solution $(x_1{}^*, x_2{}^*)$ of the primal problem would represent the number of tables and chairs that the company should produce to maximize profits with the given resources. On the other hand, the optimal solution $(y_1{}^*, y_2{}^*)$ of the dual problem, or the shadow prices, represents the increase in profit brought about by increasing the amounts of labor and lumber by one unit.

More precisely, $y_1{}^*$ represents the increase in profit brought about by increasing the availability of labor by one hour. Likewise, $y_2{}^*$ is the increase in profit brought about by increasing the amount of the second resource, lumber, by one unit. From these interpretations we can understand why $y_1{}^*$ and $y_2{}^*$ represent the value of a single unit of each of the two resources. In this problem, for each additional hour of labor that the company has available, it will earn an additional profit of 2/5, or \$0.40. Also, for each additional square foot of lumber that the company has available, it will earn an additional profit of 1/5, or \$0.20, on the sale of tables and chairs.

For the reasons described here, the resources of labor and lumber have values of \$0.40 per hour for labor and \$0.20 per square feet for lumber. These are the two shadow prices, which are found by solving the dual problem of the allocation of resources problem.

A second reason for our interest in duality theory and the dual problem can be seen in a statement of the Fundamental Theorem of Duality. We state it here without proof.

Fundamental Theorem of Duality

Suppose two linear programming problems, one a maximizing and the other a minimizing linear programming problem, are dual problems to each other. The maximum value of the objective function of the maximizing problem is equal to the minimum value of the objective function of the minimizing problem.

The significance of this theorem lies in the fact that it provides a method for solving minimization linear programming problems with $\geq$ constraints. The following steps show how minimization problems can be solved by using the Fundamental Theorem of Duality.

Solving Minimization Problems by Solving the Dual Problem

Consider the following minimizing linear programming problem with two variables and two constraints:

$$\text{minimize } C = 2x_1 + 3x_2$$

subject to

$$2x_1 + x_2 \geq 1$$
$$x_1 + 3x_2 \geq 1$$

(primal problem)

and

$$x_1 \geq 0$$
$$x_2 \geq 0$$

Since we start with this problem, it is called the primal problem. We cannot solve this problem as stated, since the simplex method is designed to solve only maximizing problems with $\leq$ constraints. However, the dual problem of the above minimizing problem is the following maximizing problem:

$$\text{maximize } P = y_1 + y_2$$

subject to

$$2y_1 + y_2 \leq 2$$
$$y_1 + 3y_2 \leq 3$$

(dual problem)

and

$$y_1 \geq 0$$
$$y_2 \geq 0$$

We now solve this dual problem by the simplex method. Without listing all of the intermediate tableaux, the final tableau of the simplex method is shown in Table 44.

TABLE 44
Final Simplex Tableau of
Maximizing Problem

BV	ROW	P	y_1	y_2	s_1	s_2	RHS
*	0	1	0	0	2/5	1/5	7/5
y_1	1	0	1	0	3/5	−1/5	3/5
y_2	2	0	0	1	−1/5	2/5	4/5

From Table 44 we see that the maximum of the objective function is 7/5. However, by the Fundamental Theorem of Duality, this is also the minimum value of the objective function in the original minimizing problem.

To find the optimal feasible point $(x_1{}^*, x_2{}^*)$ of the minimizing problem, we make use of another result from duality theory at the top of the facing page.

In the tableau in Table 44 the numbers 2/5 and 1/5 are located beneath s_1 and s_2 in ROW 0; hence (2/5, 1/5) is the optimal feasible point in the dual of this maximizing problem. But the dual of the maximizing problem is the orig-

---

Primal Solution Located in ROW 0 of the Dual Tableau

The optimal feasible point of a minimizing linear programming problem with $\geq$ constraints can be found by looking in ROW 0 under the slack variables of the final simplex tableau of the dual maximizing problem. In particular, the value of the first variable $x_1{}^*$ will be located under the first slack variable s_1, the value of the second variable under the second slack variable, and so on.

BV	ROW	P	y_1	y_2	s_1	s_2	s_3	RHS
	0	1	*	*	$x_1{}^*$	$x_2{}^*$	$x_3{}^*$	*

Optimal solution
of primal problem

---

inal minimizing problem. Hence the optimal feasible point of the original minimizing problem is

$$x_1{}^* = 2/5$$
$$x_2{}^* = 1/5$$

We now summarize the process of solving the minimizing objective function with $\geq$ constraints by duality theory.

Solving Minimizing Problems with $\geq$ Constraints

To solve minimizing linear programming problems with $\geq$ constraints, carry out the following steps.

- **Step 1.** Write the dual problem, which will be a maximizing problem with $\leq$ constraints.
- **Step 2.** Solve the maximizing problem by the simplex method and be sure to record the final simplex tableau.
- **Step 3.** (location of dual variables in final simplex tableau). The numbers in ROW 0 under the slack variables of the final simplex tableau of the maximizing problem (dual problem) will be the coordinates of the optimal feasible point of the original minimizing problem.
- **Step 4** (Fundamental Theorem of Duality). The minimum value of the objective function of the minimizing problem is the same as the maximum value of the objective function of the maximizing problem and hence is located in ROW 0 in the RHS column of the final simplex tableau found in Step 2.

Example 3

Solving Minimizing Problems   Solve the minimizing linear programming problem

$$\text{minimize } C = 20x_1 + 35x_2 + 12x_3$$

subject to

$$x_1 + 2x_2 - 3x_3 \geq 5$$
$$x_1 + x_2 + x_3 \geq 4$$

and

$$x_1 \geq 0$$
$$x_2 \geq 0$$
$$x_3 \geq 0$$

Solution

- **Step 1** (write the dual problem).   The dual problem for the above minimizing problem is the maximizing problem

$$\text{maximize } P = 5y_1 + 4y_2$$

subject to

$$y_1 + y_2 \leq 20$$
$$2y_1 + y_2 \leq 35$$
$$-3y_1 + y_2 \leq 12$$

and

$$y_1 \geq 0$$
$$y_2 \geq 0$$

- **Step 2** (solve the dual problem).   Without listing all the intermediate tableaux, the final simplex tableau is given in Table 45.

TABLE 45
Final Simplex Table of Maximizing Problem

BV	ROW	P	y_1	y_2	s_1	s_2	s_3	RHS
*	0	1	0	0	3	1	0	95
y_2	1	0	0	1	2	-1	0	5
y_1	2	0	1	0	-1	1	0	15
s_3	3	0	0	0	-5	4	1	52

- **Step 3** (find the optimal feasible point).   The optimal feasible point $(x_1{}^*, x_2{}^*, x_3{}^*)$ of the primal problem is located in ROW 0 under the slack variables of the final simplex tableau of the dual problem. From Table 45, we see that

$$x_1{}^* = 3$$
$$x_2{}^* = 1$$
$$x_3{}^* = 0$$

- **Step 4** (find the minimum value of objective function).   The primal and dual problems have the same optimal value for their respective objective functions. From the tableau above, they both have the value 95.   □

This concludes the brief introduction to duality theory and the dual problem. We close this section, and in fact the chapter, on linear programming, appropriately with a short discussion of postoptimal analysis.

Postoptimal Analysis

The work of a scientist or business manager is generally not complete even after the simplex method has found the optimal solution. The parameters of a linear programming problem, such as the coefficients in the objective function and the numbers on the right-hand side of the inequalities, are generally only estimates of their true values. These parameters may even be overestimated or underestimated to give upper or lower bounds on the real solution. It is often important that the business manager try different combinations of these parameters to determine the effects of changes in these variables on the optimal solution found by the simplex method.

In other words, the original solution of a linear programming problem is often meant only to be a starting point for further analysis. This further analysis is called **postoptimal analysis**.

Consider the general linear programming problem with two decision variables and two functional constraints:

$$\text{maximize } P = c_1 x_1 + c_2 x_2$$

subject to
$$a_{11} x_1 + a_{12} x_2 \leq b_1$$
$$a_{21} x_1 + a_{22} x_2 \leq b_2$$

and
$$x_1 \geq 0$$
$$x_2 \geq 0$$

The goal of postoptimal analysis is to determine the effects that various changes in the parameters c_1, c_2, a_{11}, a_{12}, a_{21}, a_{22}, b_1, and b_2 have on the solution of the problem. To illustrate these ideas, consider the Old Town Canoe problem, in which we determined the optimal weekly production of canoes under certain restrictions on the availability of resources. In that problem the goal was to find the number of Columbia and Katahdin canoes the company should produce each week in order to maximize its profit. The problem was described mathematically by the linear programming problem

$$\text{maximize } P = 350C + 300K$$

subject to
$$20C + 15K \leq 4000 \quad \text{(labor)}$$
$$2C + 5K \leq 900 \quad \text{(fiberglass mat)}$$
$$5C + 5K \leq 1500 \quad \text{(resin)}$$
$$11C + 12K \leq 2640 \quad \text{(fiberglass cloth)}$$

and
$$C \geq 0$$
$$K \geq 0$$

The optimal solution, shown in Figure 39, was found to be

$$C^* = 112 \text{ Columbia canoes per week}$$
$$K^* = 117 \text{ Katahdin canoes per week}$$
$$\text{Profit} = \$74{,}300 \text{ per week}$$

Now that the optimal solution has been found, what should be done next? The following questions could be asked concerning the above optimal solution.

How will the optimal solution (C^*, K^*) and objective function change if

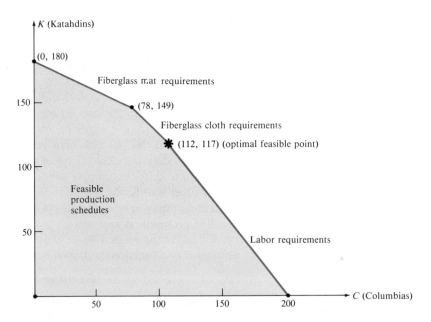

Figure 39
Solution of the Old Town Canoe problem

we modify the above linear programming problem? For example, what if we change

- the numbers b_i on the right-hand sides of the functional constraints (amounts available)?
- the numbers c_i in the objective function (the price of each item)?
- the numbers a_{ij} in the functional constraints (the quantity of resource i in each unit of product j)?
- the problem by adding a new decision variable x_i (introduce a new product)?
- the problem by adding another functional constraint (an additional restriction of some kind)?

Often, an analysis of questions of this kind provides more information than the optimal solution itself. We simulate a typical postoptimal analysis session that involves the business manager for the Old Town Canoe Company.

Postoptimal Analysis of the Old Town Canoe Problem

Suppose we have just examined a computer printout that gives the optimal feasible solution for the weekly production of Columbia and Katahdin canoes as illustrated in Figure 39. The question is: Do we accept these numbers at face value and make management decisions on the basis of this information, or should we carry out some further analysis before making any decisions?

The optimal solution says that the optimal weekly production of canoes should be

$$C^* = 112 \text{ Columbias}$$
$$K^* = 117 \text{ Katahdins}$$

In Figure 39 we see that the optimal feasible point

$$(C^*, K^*) = (112, 117)$$

lies at the intersection of the constraint lines for fiberglass cloth and labor. In other words, the company completely uses its weekly inventory of these critical resources in the suggested production schedule of Columbias and Katahdins. Figure 39 also shows that the optimal point $(C^*, K^*) = (112, 117)$ lies below the constraint lines for fiberglass mat and resin. This means there is a surplus of these resources. From these observations an important question to ask would be: How much will the current profit of $74,300 increase if the company purchases more of the critical resources of fiberglass cloth and labor? To answer this question, we analyze the effect of increasing the company's labor supply (hiring more employees).

What If More People Are Employed?

Figure 40 shows the effect of raising the number of hours of labor per week from 4000 to 4800 (from 100 to 120 employees). The labor constraint moves upward and to the right. The net effect of this action will be to change the labor constraint from

$$20x_1 + 15x_2 \leq 4000$$

to

$$20x_1 + 15x_2 \leq 4800$$

If we solve the new linear programming problem using the new functional constraint, we will find the new solution

$$C^* = 240 \text{ Columbias}$$
$$K^* = 0 \text{ Katahdins}$$
$$\text{Profit} = \$84,000$$

In other words, hiring 20 more employees allows the company to increase the weekly profit increases by

$$\text{Change in } P = \$84,000 - \$74,300$$
$$= \$9700$$

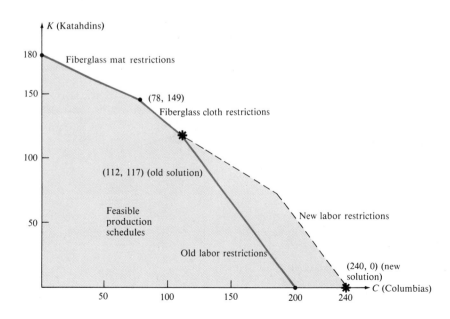

Figure 40

Effect on the optimal feasible point (C*, K*) by adding more labor

If we divide this weekly increase in profit by the number of hours per week added, we see that the weekly profit will increase by $9700/800 = $12.12 for each extra hour of labor added to the work force. If wages and benefits cost less per hour than $12, then additional hiring might be a possibility. Note that the new solution suggests a weekly production of 240 Columbias and no Katahdins. This may be an unreasonable solution because of demand requirements. Then, too, we are assuming that the company can actually sell all the canoes it produces. In other words, you should keep in mind that several aspects of the overall problem have not been built into the linear programming model.

Other Postoptimal Analysis

The above analysis showing the effects of a change in available labor on the weekly production schedule and profit is only one type of postoptimal analysis that could be performed. In addition to determining the effect of increasing the work force, we could have measured the effect of increasing the availability of fiberglass cloth.

On the other hand, we could measure the effect of a price increase in one of the two canoes on the production schedule. Stated in terms of the linear programming model, we want to see how the optimal feasible point (C^*, K^*) will change if we make changes in the coefficients of the objective function

$$P = 350C + 300K$$

Figure 41 illustrates how the optimal feasible point will change if, for example, we change the objective function to

$$P = 350C + 600K$$

This new objective function corresponds to raising the price of the Katahdins so that the new profit is $600. Assuming that the company can sell all the Katahdins that it produces at this price, the new production schedule should be 180 Katahdins per week (and no Columbias).

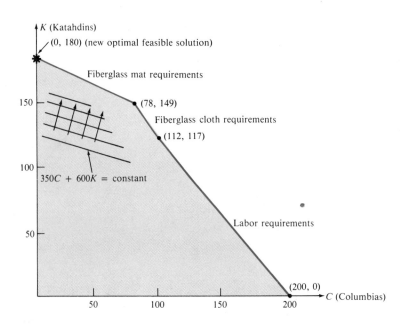

Figure 41

Change in solution as a result of charging $600 per Katahdin (and selling them all)

Problems

For Problems 1–8, write the dual form of the following linear programming problems.

1. Maximize $P = x_1 + x_2$

subject to
$$2x_1 + x_2 \leq 2$$
$$x_1 + 2x_2 \leq 2$$

and
$$x_1 \geq 0$$
$$x_2 \geq 0$$

2. Maximize $P = x_1 + x_2 + x_3$

subject to
$$x_1 + 2x_2 + x_3 \leq 4$$
$$x_1 \quad\quad + x_3 \leq 5$$

and
$$x_1 \geq 0$$
$$x_2 \geq 0$$
$$x_3 \geq 0$$

3. Maximize $P = 4x_1 + 5x_2$

subject to
$$2x_1 + x_2 \leq 2$$
$$x_1 - x_2 \leq 4$$
$$-3x_1 - x_2 \leq 1$$

and
$$x_1 \geq 0$$
$$x_2 \geq 0$$

4. Maximize $P = x_1 + 2x_2$

subject to
$$x_1 + x_2 \leq 1$$

and
$$x_1 \geq 0$$
$$x_2 \geq 0$$

5. Maximize $P = x_1 + 5x_2 + 9x_3$

subject to
$$6x_1 + 7x_2 - 3x_3 \leq 5$$

and
$$x_1 \geq 0$$
$$x_2 \geq 0$$
$$x_3 \geq 0$$

6. Minimize $C = x_1 + x_2 + x_3$

subject to
$$5x_1 \quad\quad + x_3 \geq 3$$
$$x_1 + 2x_2 \quad\quad \geq 3$$

and
$$x_1 \geq 0$$
$$x_2 \geq 0$$
$$x_3 \geq 0$$

7. Minimize $C = x_1 + 4x_2$

subject to
$$x_1 + 4x_2 + 3x_3 \geq 1$$

and
$$x_1 \geq 0$$
$$x_2 \geq 0$$
$$x_3 \geq 0$$

8. Minimize $C = x_1 + 3x_2$

subject to
$$4x_1 - x_2 \geq 9$$
$$-x_1 - 4x_2 \geq 10$$

and
$$x_1 \geq 0$$
$$x_2 \geq 0$$

Problems 9–12 refer to the following linear programming problem, which we call the primal problem:

$$\text{maximize } P = 7x_1 + 3x_2$$

subject to
$$-x_1 + x_2 \leq 5$$
$$x_1 + x_2 \leq 5$$

and
$$x_1 \geq 0$$
$$x_2 \geq 0$$

9. Find the dual problem and solve both it and the primal problem by the corner-point method. What is true about the solutions of these two problems?
10. If the above linear programming problem describes an allocation of resources problem, what are the shadow prices of the two resources represented by the two constraints? Interpret the meaning of the shadow prices.
11. Which of the two resources is more valuable per unit insofar as having an effect on the profit?
12. Evaluate the objective functions of both the maximizing (primal) and minimizing (dual) problems at a few points in their respective feasible sets and observe that the objective function of the maximizing problem is always less than or equal to the objective function of the minimizing problem.
13. Answer the following questions concerning the simplex tableau in Table 46.

TABLE 46
Simplex Tableau for Problem 13

BV	ROW	P	x_1	x_2	s_1	s_2	RHS
*	0	1	0	0	2/5	3/5	6
x_1	1	0	1	0	2/5	1/2	2
x_2	2	0	0	1	1/3	1/4	2

(a) What is the optimal point $(x_1{}^*, x_2{}^*)$?
(b) What are the final basic variables?
(c) What are the final nonbasic variables?

(d) What are the shadow prices $(y_1{}^*, y_2{}^*)$?

(e) If resource 1 is increased by one unit, how much will the objective function increase?

(f) If resource 2 is increased by one unit, how much will the objective function increase?

14. Answer the following questions concerning the final simplex tableau in Table 47 of a maximizing linear programming problem (primal problem).

(a) What is the optimal solution $(x_1{}^*, x_2{}^*)$ of the primal problem?

(b) What are the final basic variables?

(c) What are the final nonbasic variables?

(d) What is the optimal feasible point of the dual problem?

(e) What are the shadow prices?

(f) What is the minimum value of objective function of the dual problem?

(g) If resource 3 is increased by one unit, how much is the objective function increased?

(h) Which is the most valuable of the three resources described by the problem above? Why?

TABLE 47

Final Simplex Tableau for Problem 14

BV	ROW	P	x_1	x_2	$\textcircled{s_1}$	$\textcircled{s_2}$	s_3	RHS
*	0	1	0	0	1/2	3/4	0	9
x_1	1	0	1	0	1/3	1/2	0	4
x_2	2	0	0	1	2/5	1/4	0	2
s_3	3	0	0	0	1/2	2/5	1	1

15. **Information in the Final Tableau** A furniture company produces chairs (x_1), tables (x_2), and cabinets (x_3) from walnut and maple lumber. The optimal production schedule for these three products is found by solving the linear programming problem

$$\max \quad P = 5x_1 + 4x_2 + 5x_3 \quad \text{(profit in dollars)}$$

subject to

$$x_1 + 2x_2 + x_3 \le 100 \quad \text{(walnut in square feet)}$$
$$2x_1 + 8x_2 + 2x_3 \le 150 \quad \text{(maple in square feet)}$$
$$x_1 + x_2 + x_3 \le 200 \quad \text{(labor in hours)}$$

and

$$x_1 \ge 0$$
$$x_2 \ge 0$$
$$x_3 \ge 0$$

The above linear programming problem has been solved and the final simplex tableau is listed in Table 48.

(a) How many tables, chairs, and cabinets should the company make?

(b) What is the company's profit?

(c) How much should the company be willing to pay for an additional square foot of walnut lumber?

(d) How much should the company be willing to pay for an additional hour of labor?

(e) How much should the company be willing to pay for an additional square foot of maple lumber?

(f) If the company purchased another 10 square feet of maple lumber, how would it affect the profit of the company?

(g) If the company had a shortage of 10 square feet of walnut, how would it affect the profit of the company?

TABLE 48

Final Simplex Tableau for Problem 15

BV	ROW	P	x_1	$\textcircled{x_2}$	$\textcircled{x_3}$	s_1	$\textcircled{s_2}$	s_3	RHS
*	0	1	0	16	0	0	5/2	0	375
s_1	1	0	0	−2	0	1	−1/2	0	25
x_1	2	0	1	4	1	0	1/2	0	75
s_3	3	0	0	−3	0	0	−1/2	1	125

For Problems 16–19, solve the minimizing linear programming problems by using duality theory.

16. Minimize $C = 2x_1 + 2x_2$

  subject to $\qquad 2x_1 + x_2 \geq 1$

  $\qquad\qquad\qquad x_1 + 2x_2 \geq 1$

  and $\qquad\qquad\qquad x_1 \geq 0$

  $\qquad\qquad\qquad\qquad x_2 \geq 0$

17. Minimize $C = 9x_1 + 4x_2 + 12x_3$

  subject to $\qquad 3x_1 + x_2 + x_3 \geq 2$

  $\qquad\qquad\qquad x_1 + x_2 + 4x_3 \geq 1$

  and $\qquad\qquad\qquad x_1 \geq 0$

  $\qquad\qquad\qquad\qquad x_2 \geq 0$

  $\qquad\qquad\qquad\qquad x_3 \geq 0$

18. Minimize $C = x_1 + x_2$

  subject to $\qquad x_1 + 3x_2 \geq 1$

  $\qquad\qquad\qquad 3x_1 + x_2 \geq 1$

  and $\qquad\qquad\qquad x_1 \geq 0$

  $\qquad\qquad\qquad\qquad x_2 \geq 0$

19. Minimize $C = x_1 + x_2 + x_3 + x_4$

  subject to $\quad x_1 + 2x_2 + 3x_3 + 4x_4 \geq 1$

  and $\qquad\qquad\qquad x_1 \geq 0$

  $\qquad\qquad\qquad\qquad x_2 \geq 0$

  $\qquad\qquad\qquad\qquad x_3 \geq 0$

  $\qquad\qquad\qquad\qquad x_4 \geq 0$

Postoptimal Analysis

Consider the following linear programming problem:

$$\max \quad P = x_1 + x_2$$

subject to $\qquad 2x_1 + x_2 \leq 10$

$\qquad\qquad\qquad x_1 + 2x_2 \leq 10$

and $\qquad\qquad\qquad x_1 \geq 0$

$\qquad\qquad\qquad\qquad x_2 \geq 0$

For Problems 20–24, first solve the above problem by the corner-point method, and then determine the effect on the optimal solution (x_1^*, x_2^*) and corresponding objective function by the indicated changes.

20. Change the objective function to

$$P = 1.5x_1 + x_2$$

This means that we change the price of product 1 from $1 to $1.50.

21. Change the first constraint to

$$2x_1 + x_2 \leq 10.2$$

This means that we change the amount available of resource 1 from 10 to 10.2 units.

22. Change the first constraint to

$$2.5x_1 + x_2 \leq 10$$

This means that we change the amount of resource 1 that goes into product 1 from 2 to 2.5 units.

23. Change the first constraint to

$$2x_1 + x_2 \leq 100$$

This means that we essentially add an unlimited amount of resource 1.

24. Add the new constraint

$$x_1 + x_2 \leq 6$$

This means that we add a new constraint to the existing ones. Maybe there is a shortage of a new resource.

25. What are possible interpretations of the changes in Problems 20–24? Why would a researcher or business manager be interested in knowing how (x_1^*, x_2^*) and P^* change as a result of these changes in the parameters?

Epilogue: New Trends in Linear Programming

Linear programming is a branch of mathematics that is part of a more comprehensive area of mathematics called *operations research*. In addition to linear programming, operations research also includes the areas of inventory control, game theory, parts of probability theory, network theory, and many more. The Opera-

tions Research Society of America (ORSA) meets twice a year to provide a forum for the presentation of new results and ideas. The following article, taken from the *Boston Globe* (April 14, 1985), tells of some important results in the area of linear programming.

Methods of the master problem solvers*
While operations research is complex, it remains apart from economics

ECONOMIC PRINCIPALS
DAVID WARSH

Narendra Karmarkar's name is not exactly a household word, nor will it ever be. Within certain precincts, however, his immortality is already assured. He just might be rich and successful, too, but that depends.

Karmarkar is the Bell Laboratories fellow who came up with a new algorithm designed to replace the simplex method that's been used to solve a wide variety of scheduling problems since its discovery in 1947. Mathematicians can assess the novelty of this sort of thing very rapidly, and Karmarkar's breakthrough was advertised, among other places, on the front page of the New York Times last autumn.

And why not? Supplanting the simplex method is like beating an aging but powerful champ. Anyone who has made a phone call, bought a gallon of gas, taken an airplane trip or called a big city cop is beneficiary of it—and the method has hung onto its central place in the scheduler's tool kit for an unusually long time. "If you look at the way that progress is proceeding in all fields of applied math, the simplex method, which is 40 years old, should be dead. Hardly anything lasts that long," says Robert Dorfman of Harvard University.

It now turns out that Karmarkar's problem-solving recipe may not be more practical than simplex after all. The great joy of simplex is its applicability—it works, planners say, better than it should. But it is limited in the kinds of problems it can handle. "The simplex method is fine for problems with a few thousand variables, but above 16,000 or 20,000 variables, it runs out of steam," says Karmarkar—and of course that range is just where the 28-year-old Indian's method is thought to come in handy.

But some mathematicians and scientists fear that the new solution could turn out to

* Reprinted courtesy of *The Boston Globe.*

be like a Russian discovery of a few years back, called the ellipsoid method, which was beautiful but not very useful in solving the workaday problems at which simplex excels. Indeed, applied mathematicians are somewhat polarized in their reaction to news of the discovery. Specialists will be flocking to a major session in Boston tonight on the matter, but it won't be settled there. "Mind you, any breakthrough in science gets attacked," says Richard C. Larson of MIT.

The occasion for a Karmarkar session—and for nearly 300 other meetings of varying degrees of accessibility—is a convention of the Operations Research Society of America, of which Larson is the president. Operations research was formed in the cauldron of World War II. The task of linking newly developed radar and Hurricane fighter planes was its first big problem; the results of the Battle of Britain its first big success. Suddenly there were dozens of wartime problems of unprecedented complexity to be solved: bombing patterns, merchant fleet operations, antisubmarine warfare searches all had to be optimized, to say nothing of the logistical trains behind the front.

While economists, especially the bunch under William J. (Wild Bill) Donovan in the Office for Strategic Services, had as much to offer as any other set of keen professional intelligences, it was a different bunch that made the beginning that, after the war, became operations research. Philip M. Morse, the MIT physicist widely regarded as the American father of the field, noted in his presidential address in 1953 that what had begun with borrowing gave way quickly to original research.

Central figure

The central figure in this process was George B. Dantzig, the discoverer of the simplex algorithm. An Air Force planner charged with figuring out how to build the best air corps at minimal cost, Dantzig wrestled with postwar problems of coordinating base construction, troop recruitment and training, equipment manufacture and so on. In 1947, he hit on an algebraic method he described as "climbing the beanpole," after the pattern it made in intuitive geometry. It turned out to be the solution to any number of problems, from calculating the cheapest combination of nine nutrients that would comprise a satisfactory diet to allocating resources in the most complicated ways.

So fundamental was his contribution that many felt that Dantzig should have shared the 1975 Nobel Award in Economics with Tjalling Koopmans and Leonid V. Kantrovitch—but was snubbed instead because he worked for the most part outside of economics. Other economists say the line between economics and all else has to be drawn someplace and that the leading edge of Dantzig's wingtips is as good a place as any to start.

Key Ideas for Review

allocation of resources problem, 163
basic variables, 177, 185
corner-points of feasible set, 150
dual problem, 197, 199
feasible set, 145
functional constraints, 145
graphical solution, 148
history of linear programming, 142

initial simplex tableau, 184
isocost lines, 148
isoprofit lines, 148
linear inequalities, 130
nonbasic variables, 177, 185
optimal feasible point, 148
postoptimal analysis, 205
primal problem, 197
shadow prices, 200
simplex method, 177

simplex tableau, 177, 184
slack variables, 171
solution of an inequality, 132
stopping rule for simplex method, 186

Review Exercises

Problems 1–8 refer to the following linear programming problem:

$$\max \quad P = x_1 + x_2$$

subject to
$$x_1 + 10x_2 \leq 10$$
$$3x_1 + x_2 \leq 3$$

and
$$x_1 \geq 0$$
$$x_2 \geq 0$$

1. Draw the feasible set for the above problem.
2. What are the corner-points of the feasible set?
3. Solve the problem graphically by drawing isoprofit lines.
4. Solve the problem by the corner-point method.
5. Find the equivalent problem with equality constraints by introducing slack variables.
6. Label the corner-points of the feasible set, using the augmented coordinates (x_1, x_2, s_1, s_2).
7. What is the initial simplex tableau for this problem?
8. Solve the problem above by the simplex method.

Problems 9–14 refer to the following linear programming problem:

$$\min \quad C = x_1 + x_2$$

subject to
$$x_1 + 10x_2 \geq 10$$
$$3x_1 + x_2 \geq 3$$

and
$$x_1 \geq 0$$
$$x_2 \geq 0$$

9. Draw the feasible set for the above problem.
10. What are the corner-points of the feasible set?
11. Solve the above problem by drawing isocost lines.
12. Solve the above problem by the corner-point method.
13. Find the dual problem to the above problem.
14. Solve the dual problem to the above problem.

Problems 15–22 refer to the following linear programming problem:

$$\max \quad P = 3x_1 + 4x_2$$

subject to
$$2x_1 + x_2 \leq 2$$
$$x_1 + 2x_2 \leq 2$$

and
$$x_1 \geq 0$$
$$x_2 \geq 0$$

15. Find the dual problem for the above primal problem and solve both the primal and dual problems by the corner-point method.
16. Solve the primal problem by the simplex method.
17. Solve the dual problem for the above problem by looking in ROW 0 of the final simplex tableau of the primal problem.
18. What are the two shadow prices for the above linear programming problem? What is the interpretation of these values?
19. If the amount of resource 1 available were increased by one unit, how would this affect the objective function?
20. If the coefficient of x_1 in the objective function were increased from 3 to 4, how would this affect the optimal feasible point (x_1^*, x_2^*) and the value of the objective function?
21. If the coefficient of x_1 in the objective function is changed from 3 to 10, how will this affect the optimal feasible point (x_1^*, x_2^*) and the value of the objective function? Illustrate this graphically.
22. If the coefficient of x_1 in the first constraint is changed from 2 to 1, how does this affect the optimal feasible point? What does this mean graphically?

Jeweler Problem

Problems 23–30 refer to the following linear programming problem. A jeweler adds small amounts of gold and silver to rings and bracelets, where

$$x_1 = \text{number of rings made}$$
$$x_2 = \text{number of bracelets made}$$

The following linear programming model will determine the optimal number of rings and bracelets the jeweler should make to maximize profit.

$$\max \quad P = 50x_1 + 70x_2 \quad \text{(profit in dollars)}$$

subject to

$$50x_1 + 20x_2 \leq 200 \quad \text{(gold in grams)}$$
$$10x_1 + 20x_2 \leq 300 \quad \text{(silver in grams)}$$
$$x_2 \leq 5 \quad \text{(bracelets)}$$

and
$$x_1 \geq 0$$
$$x_2 \geq 0$$

23. How much gold (grams) is used in each ring?
24. How much silver (grams) is used in each ring?

25. How much gold (grams) is used in each bracelet?
26. How much silver (grams) is used in each bracelet?
27. What is the interpretation of the constraint $x_2 \leq 5$?
28. What is the profit from each ring?
29. What is the profit from each bracelet?
30. What is the maximum profit?

Fertilizer Problem

Problems 31–40 refer to the following allocation of resources problem. A company markets FAST GRO, a lawn fertilizer. FAST GRO is a mixture of two chemicals, called chemicals I and II. These two chemicals are both mixtures of the nutrients nitrogen, phosphate, and potash. The problem the company would like to answer is how many pounds, x_1 and x_2, of the two chemicals should be included in each bag of FAST GRO so that given requirements of nitrogen, phosphate, and potash are met at a minimum cost. The company has determined that the following linear programming problem describes this problem:

$$\min \quad C = 2x_1 + 3x_2 \quad \text{(cost in dollars)}$$

subject to

$$7x_1 + 3x_2 \geq 25 \quad \text{(nitrogen in ounces)}$$
$$x_1 + 2x_2 \geq 15 \quad \text{(phosphate in ounces)}$$
$$2x_1 \qquad \geq 5 \quad \text{(potash in ounces)}$$

and
$$x_1 \geq 0$$
$$x_2 \geq 0$$

31. On the basis of the above linear programming model, how much does the company pay for each pound of the two chemicals?
32. How many ounces of nitrogen are there in each pound of chemical I?
33. How many ounces of nitrogen are there in each pound of chemical II?
34. How many ounces of phosphate are there in each pound of chemical I?
35. How many ounces of phosphate are there in each pound of chemical II?
36. How many ounces of potash are there in each pound of chemical I?
37. How many ounces of potash are there in each pound of chemical II?
38. What are the minimum requirements of nitrogen in a bag of FAST GRO?
39. What are the minimum requirements of phosphate in a bag of FAST GRO?
40. What are the minimum requirements of potash in a bag of FAST GRO?

Chapter Test

1. Solve the system of linear inequalities

$$-x_1 + x_2 < 1$$
$$x_1 - x_2 < 1$$
$$x_1 + x_2 \leq 1$$
$$x_1 + x_2 \geq -1$$

2. Write the initial simplex tableau for

$$\max \quad P = 3x_1 + 5x_2 + 7x_3 + 2x_4$$

subject to $\quad x_1 + x_2 - x_3 + 2x_4 \leq 4$

and
$$x_1 \geq 0$$
$$x_2 \geq 0$$
$$x_3 \geq 0$$
$$x_4 \geq 0$$

3. Solve both the following primal problem and its dual by the corner-point method:

$$\max \quad P = x_1 + x_2$$

subject to
$$2x_1 + x_2 \leq 1$$
$$x_1 + 2x_2 \leq 1$$

and
$$x_1 \geq 0$$
$$x_2 \geq 0$$

Given the simplex tableau in Table 49.

TABLE 49
Simplex Tableau for Problems 4–11

BV	ROW	P	x_1	x_2	s_1	s_2	RHS
*	0	1	0	-1	2/3	0	6
x_1	1	0	1	1/2	1/3	0	2
s_2	2	0	0	2/3	1/4	1	6

4. Is this tableau an optimal tableau?
5. What is the current corner-point (x_1, x_2, s_1, s_2)?
6. What are the current basic and nonbasic variables?
7. What is the value of the objective function at the current corner-point?
8. What is the pivot column?
9. What is the pivot row?
10. What is the pivot element?
11. Find the next tableau. Is it optimal?

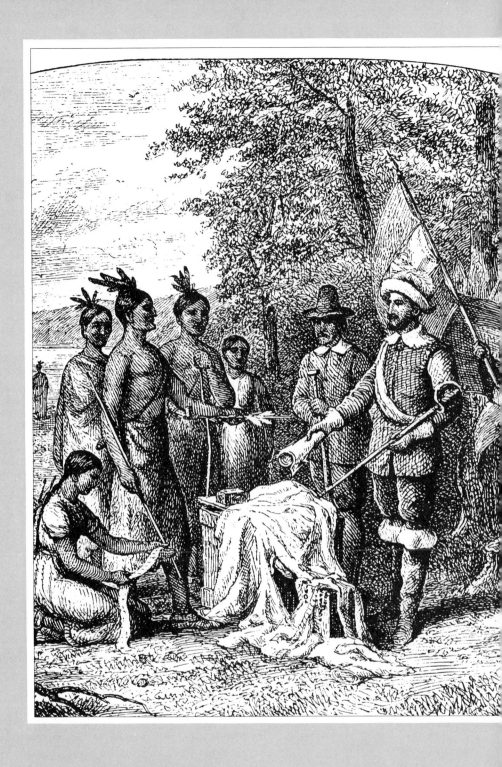

3

The Mathematics
of Finance

With the continuing deregulation of the banking industry, financial institutions have introduced new options for lenders and borrowers. Innovations as varied as money-market funds, NOW accounts, jumbo certificates of deposit, and zero-coupon bonds are common marketing devices for the financial industry. The focus of the money manager, regardless of how large or small the account, must be to understand and use the appropriate analytic tools to compare various financial alternatives.

This chapter explains the basic mathematical concepts that are used in evaluating short- and long-term costs of money. The basic formulas will be explained and applied to the evaluation of typical financial problems. Only by under-standing these formulas and the role of various parameters can business people make informed decisions and develop appropriate financial strategies.

Most of the calculations performed in this chapter can be facilitated with the use of a simple four-key $(+, -, \times, /)$ calculator. Good business calculators containing present and future value keys, annuity keys, along with other financial keys can be bought today for under $25.

Simple Interest

PURPOSE

We introduce the concept of simple interest and show

- how to calculate the simple interest owed on a loan,
- how to calculate the discount and proceeds on a discount note, and
- how to calculate the effective rate of interest for discount notes.

Interest is the seed of money.
—Jean Jacques Rousseau

The Concept of Interest

In financial transactions one must pay for the use of borrowing money. This fee paid is called the **interest** on the borrowed amount. In short, interest is the fee, sum, or rent paid for the use of money over a period of time. Interest depends on the amount borrowed, the length of time the money is borrowed, and the **interest rate** of the loan, which is a fraction or percent of the amount borrowed. The interest rate is either quoted as a fraction or a percent, and is generally stated as an annual interest rate. At one time banks were limited to paying an **annual percentage rate** (APR) of 5% on the money deposited in savings accounts. When the banking industry was partially deregulated, however, the rate paid became responsive to the marketplace. Money market accounts have paid more than twice that rate. The current rate paid by money market accounts fluctuates in response to market conditions and government actions.

To illustrate these ideas suppose Dorothy Gravlund is opening a new travel agency. To pay for office furniture and computer equipment, she requests a loan of $10,000 from the First Bank of Ayrshire. The bank, knowing of her good reputation for honesty and business sense, accepts Dorothy's written promise to repay the loan plus an interest fee in one year's time. This written form is called a **promissory note**. Although there is no standard form for promissory notes, a typical one is shown in Figure 1.

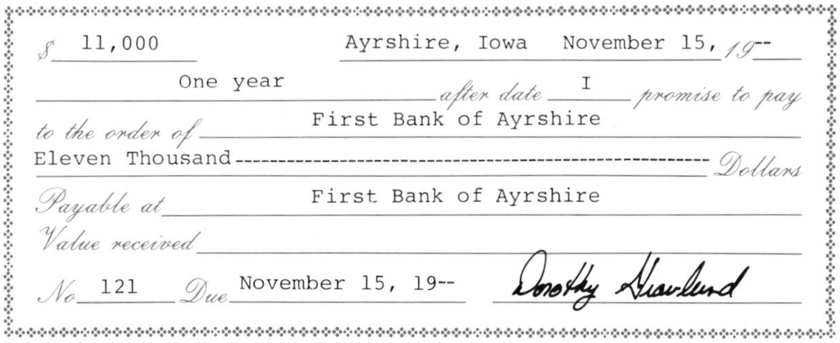

Figure 1
A promissory note

In this promissory note, Dorothy agrees to repay the First Bank of Ayrshire $10,000 plus a 10% interest fee of $1000 at the end of one year. The amount of money that is promised by the note, $11,000 in this case, is known as the **face value** of the note. The $1000 charged by the bank for the use of the money is called the **interest charged**. The date when the note is due, November 15, 19— is called the **maturity date**. The length of time over which the money is borrowed, one year in the case of Dorothy's loan, is called the **period**, or **term** of the loan. Dorothy is called the **maker** of the note, and the First Bank of Ayrshire is called the **payee**.

Note that for this loan the First Bank of Ayrshire collects the $1000 interest charge at the **time of maturity**. Sometimes, however, banks collect the interest due at the beginning of the term. If this were the policy of the First Bank of Ayrshire, the bank would in fact deduct $1000 from Dorothy's $10,000 loan and give her only $9000. This type of loan is sometimes referred to as a **discount note**, and the $1000 interest charged at the beginning of the term is called the **discount** of the note. The $9000 that Dorothy would actually receive is called the **proceeds** of the note. At the end of one year's time, Dorothy would repay the bank the amount of $10,000. Naturally, Dorothy would rather pay the $1000 interest at the end of the term rather than at the beginning of the term. In this way she would have the use of this $1000 during the term of the loan rather than the bank. We will analyze both of these two types of loans in this section.

For loans of a year or less, such as Dorothy's loan from the First Bank of Ayrshire, interest is often taken to be **simple interest** or interest charged only on the amount borrowed. The interest owed by the borrower to the lender on a simple interest loan is computed from the following formula.

> ### Interest Owed on a Simple Interest Loan
>
> The simple interest owed for borrowing P dollars for t years at an annual interest rate of r is given by
>
> $$I = Prt$$
>
> where
>
> $$I = \text{simple interest owed (dollars)}$$
> $$P = \text{principal (amount of money borrowed)}$$
> $$r = \text{annual interest rate}$$
> $$t = \text{length of time money is borrowed (years)}$$

Example 1

Simple Interest Owed   A local bank pays 7% per year for money deposited in savings accounts. How much simple interest is earned in one year by a depositor who has $100 deposited in a savings account?

Solution   In this problem the individual who deposits money in the bank is the lender and the bank is the borrower. Hence the bank pays the depositor interest for the privilege of using the depositor's money. The variables in the interest formula are

$$P = \$100$$
$$r = 0.07$$
$$t = 1$$

Hence the interest earned by the depositor and paid by the bank is

$$I = Prt$$
$$= (\$100)(0.07)(1)$$
$$= \$7$$

The annual interest rate r is the fraction of the amount borrowed that the borrower must pay the lender every year for the use of the principal. In this example the annual interest rate of $r = 0.07$ (often stated 7%) means the bank must pay the lender 7 cents each year for the use of each dollar borrowed.   □

Example 2

Brazil's Annual Interest Payments   The World Bank has loaned the country of Brazil $73,000,000,000. If the annual interest on this loan is set at 9% simple interest, how much interest should Brazil pay each year for the use of this money?

Solution   The variables in the interest formula are

$$P = \$73{,}000{,}000{,}000$$
$$r = 0.09$$
$$t = 1$$

Hence the annual interest owed at the end of one year to the World Bank is

$$I = Prt$$
$$= (\$73{,}000{,}000{,}000)(0.09)(1)$$
$$= \$6{,}570{,}000{,}000$$

After paying this interest for using the principal for one year, Brazil will still owe the World Bank 73 billion dollars. The $6.57 billion is paid only for the privilege of using the 73 billion dollars. Stated another way, just to pay the interest on this loan, Brazil would have to pay the World Bank $750,000 every hour of the day, every day of the year.   ☐

HISTORICAL NOTE

Shakespeare's advice, "Neither a borrower nor a lender be," may be sound advice in personal relationships, but it does not apply to modern financial planning.

No one knows when credit was first used as a lubricant for business transactions, but banks were in existence in ancient Mesopotamia,

Greece, and Rome. Long before Christians denounced the charging of interest as a sin, Greek philosophy called charging interest immoral, since "coins do not wear out."

Today, of course, charging interest on the use of money is the norm.

Example 3

Loan Sharking A loan shark offers a financially strapped businessman a friendly loan of $100,000 at an annual interest rate of 35%. How much must the businessman pay the loan shark each month just to pay off the interest on the loan?

Solution We have

$$P = \$100{,}000$$
$$r = 0.35$$
$$t = 1/12$$

Hence the interest owed each month is

$$I = Prt$$
$$= (\$100{,}000)(0.35)(1/12)$$
$$= \$2916.67$$

The businessman will still owe the loan shark the principal of $100,000; the monthly payment of $2916.67 is strictly for the privilege of using the amount borrowed during that month.   ☐

Future Value of Money

A deposit of P dollars today in a bank paying an annual interest rate r earns $I = Prt$ over the course of t years. This earned value, when added to the original deposit or **principal** P, gives the **future value** F of the principal P. The amount of a future value is given by the following future value formula.

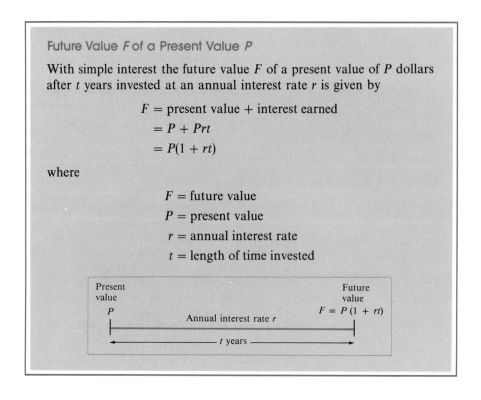

Future Value F of a Present Value P

With simple interest the future value F of a present value of P dollars after t years invested at an annual interest rate r is given by

$$F = \text{present value} + \text{interest earned}$$
$$= P + Prt$$
$$= P(1 + rt)$$

where

$$F = \text{future value}$$
$$P = \text{present value}$$
$$r = \text{annual interest rate}$$
$$t = \text{length of time invested}$$

Present value
P

Annual interest rate r

Future value
$F = P(1 + rt)$

t years

Example 4

Future Value   A company puts $100,000 in a bank that pays 9% annual interest. What is the value of this $100,000 after two months?

Solution   Here we have

$$P = \$100,000$$
$$r = 0.09$$
$$t = 2/12$$

Hence the future value is

$$F = P(1 + rt)$$
$$= \$100,000[1 + (0.09)(2/12)]$$
$$= \$101,500$$

See Figure 2.

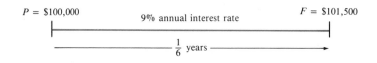

$P = \$100,000$ 9% annual interest rate $F = \$101,500$

$\frac{1}{6}$ years

Figure 2
Future value

Present Value

The amount of money deposited today that yields some larger value in the future is called the **present value** of that future amount. One dollar today, for example, might be two dollars in five years at the current rate of simple interest.

To find the present value in terms of its future value, we solve the equation

$$F = P(1 + rt)$$

for the present value P. We do this by dividing each side of the equation by the factor $(1 + rt)$, which gives the following result.

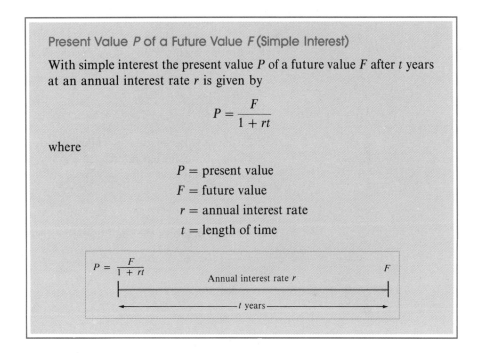

Present Value P of a Future Value F (Simple Interest)

With simple interest the present value P of a future value F after t years at an annual interest rate r is given by

$$P = \frac{F}{1 + rt}$$

where

$$P = \text{present value}$$
$$F = \text{future value}$$
$$r = \text{annual interest rate}$$
$$t = \text{length of time}$$

Example 5

Present Value   How much should be invested in a bank that pays 9% simple interest so that after six months the future value of the account is $10,000?

Solution   The variables in the present value formula are

$$F = \$10{,}000$$
$$r = 0.09$$
$$t = 6/12 = 0.5$$

Hence the present value is

$$P = \frac{F}{1 + rt}$$
$$= \frac{\$10{,}000}{1 + (0.09)(0.5)}$$
$$= \$9569.38$$

See Figure 3.

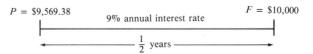

Figure 3
Present value

$P = \$9,569.38$ 9% annual interest rate $F = \$10,000$

$\frac{1}{2}$ years

Doubling Time for Simple Interest

The length of time it takes a given amount of money P to double in value if the money is deposited in a bank that pays an annual interest rate r can be found by solving for t in the equation

$$2P = P(1 + rt)$$

We do this by first dividing each side of the equation by P, getting

$$2 = 1 + rt$$

Hence we find

$$t = \frac{1}{r}$$

The above value of time is called the doubling time for simple interest. For example, if the annual interest rate is $r = 0.10$, then the time it takes a sum of money to double in value is

$$t = \frac{1}{0.10} = 10 \text{ years}$$

Figure 4 shows the doubling time for various annual interest rates.

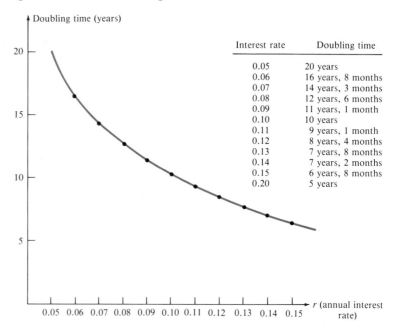

Interest rate	Doubling time
0.05	20 years
0.06	16 years, 8 months
0.07	14 years, 3 months
0.08	12 years, 6 months
0.09	11 years, 1 month
0.10	10 years
0.11	9 years, 1 month
0.12	8 years, 4 months
0.13	7 years, 8 months
0.14	7 years, 2 months
0.15	6 years, 8 months
0.20	5 years

Figure 4
Doubling times for simple interest

Simple Discount Notes

Normally, when you borrow money from someone for a given length of time, you are expected to repay the amount of the loan (principal) at the end of the term, plus interest. Often, however, banks collect the interest due at the beginning of the term. As we mentioned earlier, when discussing Dorothy's loan from the First Bank of Ayrshire, these are called **discount notes**. The interest that is deducted at the beginning of the time period, is called the **discount** of the note, and the money actually given to the borrower is called the **proceeds**. At the end of the term the borrower will repay the bank the **face value** of the loan (amount before the discount was deducted).

To compute the discount (or interest charged at the beginning of the term) of a simple discount note, we use the simple interest formula.

Discount and Proceeds

The discount D on a discount note of M dollars borrowed at an annual interest rate r for t years charging simple interest is

$$D = Mrt$$

where

$$D = \text{discount (interest paid up front)}$$
$$M = \text{maturity value (amount borrowed)}$$
$$r = \text{annual interest rate (called discount rate here)}$$
$$t = \text{length of loan}$$

The proceeds P of the loan is the amount the borrower actually receives at the beginning of the time period and is calculated by the formula

$$P = M - D$$

Example 6

Discount Note   A businessman signs a discount note for $50,000 at 14% discount for 8 months. What is the amount the bank deducts from the face value, and what are the proceeds that the businessman actually receives?

Solution   The variables for the needed formulas are

$$M = \$50,000$$
$$r = 0.14$$
$$t = 8/12 = 2/3$$

Hence the bank will deduct the interest from the $50,000 principal at the beginning of the period in the amount

$$D = Mrt$$
$$= (\$50,000)(0.14)(2/3)$$
$$= \$4666.67 \qquad \text{(cost of using \$50,000 for 8 months)}$$

This will leave the businessman with the proceeds

$$P = M - D$$
$$= \$50,000 - \$4666.67$$
$$= \$45,333.33$$

After signing this $50,000 discount note and taking the $45,333.33 the businessman must repay the banker the maturity value of $50,000 after 8 months. Note again that the banker takes the interest payment out of the principal at the beginning of the time period (which is beneficial to the banker).   □

Example 7 _____

Discount Note with Given Proceeds   Suppose the businessman in the previous example objects that the bank takes the discount off the top of the loan, since he needs exactly $50,000 to pay off a debt. How can he negotiate a new note with the same terms so that the proceeds are exactly $50,000?

Solution   Solve for the proceeds M in the equation

$$P = M - D$$
$$= M - Mrt$$

where

$$P = \$50,000 \qquad \text{(amount given to the businessman)}$$
$$r = 0.14$$
$$t = 2/3$$

Substituting these values into the equation above, we have

$$\$50,000 = M - M(0.14)(2/3)$$

Solving for M, we get

$$M = \frac{\$50,000}{1 - (0.14)(2/3)}$$
$$= \$55,147.06$$

In other words, the businessman must borrow $55,147.06 so that the proceeds will be $50,000.   □

Nominal and Effective Rates of Interest

In any loan the government requires that the annual percentage rate (APR) be stated. In the case of a simple discount note the "actual interest rate" can be quite different from the amount quoted by the bank. The rate of interest actually

paid is called the **effective rate** (true interest rate), while the advertised rate is called the **nominal rate**. The next example shows how to find the effective rate of interest, given the nominal rate. The effective rate of interest is important, since it is one of the major ways by which loans with different terms can be compared. For simple interest loans the effective and nominal rates are the same. For simple discount notes, however, the effective rate can be quite a bit larger than the nominal rate.

Example 8

Effective Rate of Interest   Sue Anderson agrees to pay $6000 to her banker in six months. The banker subtracts a discount of 8% and gives the balance to Sue. Find the discount, the proceeds, and the effective rate of interest of this discount note.

Solution

- **Step 1** (find the discount).   The discount on a $6000 loan at a discount rate of 8% over six months is

$$D = Mrt$$
$$= \$6000(0.08)(0.5)$$
$$= \$240$$

- **Step 2** (find the proceeds).   The proceeds that Sue receives is

$$P = M - D$$
$$= \$6000 - \$240$$
$$= \$5760$$

- **Step 3** (find the effective rate of interest).   Since Sue had the use of $5760 (proceeds) for six months, and since she was charged an interest of $240 (discount), the effective rate of interest on this loan can be found by solving the general simple interest formula

$$I = Prt$$

for r, where

$$I = \$240 \qquad \text{(discount of loan)}$$
$$P = \$5760 \qquad \text{(proceeds of loan)}$$
$$r = \text{unknown} \qquad \text{(simple interest)}$$
$$t = 0.5 \qquad \text{(period of loan)}$$

Solving for r and calling this value r_{eff} to denote the effective rate of interest, we find

$$r_{\text{eff}} = \frac{\$240}{(\$5760)(0.5)}$$
$$= 0.0833$$

In other words, the effective rate of interest of this discount loan (which is normally advertised 8% discount) is 8.33%.

If Example 8 had been solved in general, we would have found the following formula for computing the effective rate of interest r_{eff} of a discount note.

Effective Rate of Interest of a Discount Note

The effective rate of interest r_{eff} of a discount note is computed by

$$r_{\text{eff}} = \frac{D}{Pt} = \frac{Mrt}{(M - Mrt)t} = \frac{r}{1 - rt}$$

where

$$D = \text{discount of note}$$
$$P = \text{proceeds of note}$$
$$r = \text{advertised discount rate}$$
$$t = \text{length of note}$$

Example 9

Effective Rate of Interest   A bank advertises discount notes at an annual rate of 10%. If Ed takes out a discount loan for six months, what is the effective rate of this note?

Solution   From the above formulas for r_{eff}, note that it is possible to find the effective rate of interest by knowing only the discount rate and the length of the note. Since.

$$r = 0.10 \qquad \text{(advertised annual discount rate)}$$
$$t = 1/2 \qquad \text{(term of discount note)}$$

we have the effective rate of interest of

$$r_{\text{eff}} = \frac{r}{1 - rt}$$

$$= \frac{0.10}{1 - (0.10)(0.5)}$$

$$= 0.1053$$

In other words, a six-month bank note advertised at an annual rate of 10% has an effective annual rate of interest of 10.53%   □

Problems

For Problems 1–6, the amounts of money P are borrowed for a length of time t at the annual interest rate r. What is the simple interest owed for the use of this money?

1. $P = \$5000$, $r = 7.5\%$, $t = 3$ months
2. $P = \$2500$, $r = 9\%$, $t = 2$ years
3. $P = \$4000$, $r = 11\%$, $t = 30$ months
4. $P = \$10,000$, $r = 6.75\%$, $t = 2$ years

5. $P = \$25,000$, $r = 10\%$, $t = 5$ years
6. $P = \$10,000$, $r = 5\%$, $t = 1$ year

Exact Simple Interest

In most simple interest problems, time is measured in days or parts of a year. If a loan is made on March 1 and repaid on

March 31, then the time of the loan in days d is 30, and the exact simple interest owed for this loan is given by

$$I = Pr\frac{d}{365} = Pr\frac{30}{365}$$

where P is the amount of the loan and r is the annual interest rate. For Problems 7–10, compute the exact simple interest on the described loans.

7. $P = \$5000$, $r = 5\%$, January 3–March 10
8. $P = \$10,000$, $r = 6\%$, January 1–April 15
9. $P = \$7500$, $r = 9\%$, June 6–September 6
10. $P = \$9000$, $r = 10\%$, August 3–December 24

Finding Interest Rates

For Problems 11–14, the simple interest earned I, the amount loaned P, and the length of time t of the loan are given. Find the simple interest r of the loan.

11. $I = \$100$, $P = \$500$, $t = 1$ year
12. $I = \$500$, $P = \$3000$, $t = 2$ years
13. $I = \$1000$, $P = \$5000$, $t = 1$ year, 6 months
14. $I = \$5000$, $P = \$5000$, $t = 4$ years, 6 months

15. **Future Value**   A principal of $2500 is borrowed at 9.5% per year simple interest. Find the interest earned and the future value of the amount after 2 years, 4 years, 5 years, and 10 years. Estimate at what point in time the future value will be $5000.

16. **Interest Earned and Doubling Time**   An investor lends $7500 at 15% per year simple interest. Determine how much the investor earns after 2 years, 3 years, 5 years, and 7 years. Estimate how long it takes for the original investment to double.

17. **Doubling and Tripling Time**   How long will it take a simple interest loan to double in value if interest charged on the loan is 8% per annum? How long will it take the money to triple in value?

18. **Present Value**   How much should be invested at 11% simple interest so that the future value of this amount is $5000 after one year?

19. **Interest Owed**   Sue borrowed $1000 from her employer and promised to repay the full amount at the end of nine months, plus interest at the rate of 10% per year. How much interest must Sue pay?

20. **Interest Rates**   At the end of 1985 the national debt of the United States was 2 trillion dollars, and the annual interest charge was 180 billion dollars. What was the annual interest rate?

Simple Discount Note

For Problems 21–24, calculate the simple discount, the proceeds, and the effective annual rate of interest for the simple discount loans.

21. $M = \$500$, $r = 8\%$, $t = 9$ months
22. $M = \$2000$, $r = 9\%$, $t = 2$ years
23. $M = \$2500$, $r = 12\%$, $t = 30$ months
24. $M = \$5000$, $r = 15\%$, $t = 5$ years

25. **Simple Discount Note**   Dorothy Gravlund obtains a $500 loan from the Bank of Ayrshire. She is charged 6% discount on a six-month note.
    (a) What is the discount of this loan?
    (b) How much will Dorothy receive from the bank?
    (c) How much will Dorothy owe the bank after six months?
    (d) What is the effective rate of interest of this loan?

26. **Discount Note**   Mr. Smith asks for a loan of $500 from the First National Bank of Portland. He is charged 9% discount on a 90-day bank note. What is the discount of the note? How much must Mr. Smith repay the bank after 90 days?

27. **Discount Note**   The bank discount note or promissory note in Figure 5 is offered by Ms. Rouse to the First National Bank of Oklahoma in exchange for a loan of $2000. She is charged 10% discount on the loan. What is the proceeds of the loan? What is the maturity value of the loan? What is the effective rate of interest of this loan?

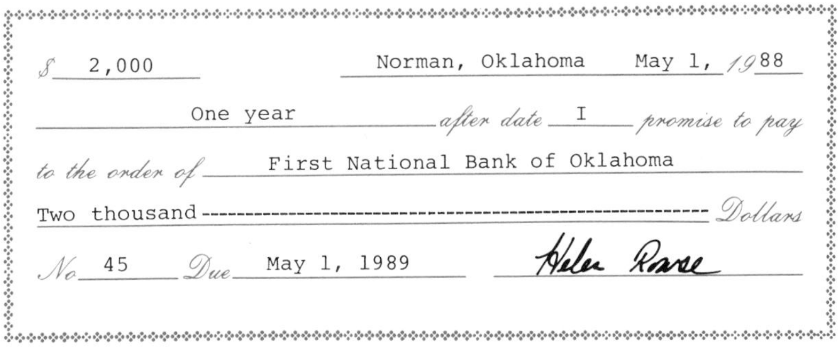

$ ___2,000___   Norman, Oklahoma   May 1, *19*88

_____One year_____*after date* ___I___ *promise to pay*

to the order of _____First National Bank of Oklahoma_____

Two thousand -- *Dollars*

No. ___45___ *Due* __May 1, 1989__   *Helen Rouse*

Figure 5
A bank discount note

3.2

Remember that Time is Money.
    —Benjamin Franklin

Compound Interest

PURPOSE

> We introduce the concept of compound interest and show
>
> - how to find the future value or compound amount of a bank account earning compound interest,
> - how to find the present value of a future value when compound interest is used, and
> - how to find the future value of an amount that is compounded continuously.

Introduction to Compound Interest

Simple interest is generally used for short-term loans of a year or less. Some corporate and municipal bonds pay simple interest every six months. For longer periods, compound interest is generally used. To introduce this idea, suppose a bank pays a depositor $100 interest for the use of $10,000 during a one-month period. As payment, the bank puts this $100 interest into the depositor's account at the end of the month. Then, at the end of the next month, the bank again pays monthly interest, but this time on the $10,100 in the account. In other words, the bank pays interest on both the original principal and the previously paid interest.

By making periodic interest payments to the depositor's account, based on the current amount of the account, the bank is paying **compound interest**. The time period of one month in this example is called the **compounding period** or simply the **period**. The number of times a year interest is compounded will be denoted by k. For example, $k = 1$ means that interest is compounded annually, $k = 2$ means that interest is compounded semiannually, and so on. If interest is compounded k times per year, the **interest per compounding period** is computed by $i = r/k$, where r is the **annual interest rate**.

Interest per Compounding Period

The interest rate i per compounding period is given by

$$i = r/k$$

where

$$r = \text{annual interest rate}$$
$$k = \text{number of compounding periods per year}$$

Example 1

Interest per Period   A bank pays an annual interest rate of $r = 0.10$. Determine the rate of interest i for each compounding period when the number of compoundings k per year is given below.

(a) $k = 1$ (compounded annually).

(b) $k = 2$ (compounded semiannually).

(c) $k = 4$ (compounded quarterly).

(d) $k = 12$ (compounded monthly).

(e) $k = 365$ (compounded daily).

Solution

(a) Annual interest rate $= 0.10$ (10%).

(b) Semiannual interest rate $= 0.10/2 = 0.050$ (5%).

(c) Quarterly interest rate $= 0.10/4 = 0.025$ (2.5%).

(d) Monthly interest rate $= 0.10/12 = 0.0083$ (0.83%).

(e) Daily interest rate $= 0.10/365 = 0.000274$ (0.0274%). □

We now show how an initial deposit of money grows when interest is compounded.

Compound Interest Earned and Future Value of an Account

Suppose P dollars are deposited in a bank that pays compound interest at a rate of i percent per compounding period. We now compute the value of the account at the end of each compounding period.

- **Period 1.** By definition of interest per period i, the interest earned during the first time period is

$$\text{First period earnings} = Pi$$

At the end of the first period the value of the account will be

$$\text{Principal after 1 period} = P + Pi$$
$$= P(1 + i)$$

- **Period 2.** Interest earned during period 2 is found in the same way. We multiply the interest per period i times the current principal:

$$\text{Second period earnings} = P(1 + i)(i)$$

Adding these earnings to the principal from the first period, we have

$$\text{Principal after 2 periods} = P(1 + i) + P(1 + i)(i)$$
$$= P(1 + i)(1 + i)$$
$$= P(1 + i)^2$$

- **Period 3.** By a similar analysis the total amount in the account after the third period will be

$$\text{Principal after 3 periods} = P(1 + i)^3$$

Generalizing, after n periods the value of the deposit will be given by the following formula.

Interest Earned and Future Value of an Account

For an initial principal P earning compound interest at the rate of i percent per compounding period the **future value** or **compound amount** of this principal after n periods will be

$$F = P(1 + i)^n$$
$$= P(1 + r/k)^n$$

where
$$F = \text{future value or compound amount}$$
$$P = \text{principal (amount deposited)}$$
$$r = \text{annual interest rate}$$
$$k = \text{number of periods per year}$$
$$i = \text{interest rate per period } (i = r/k)$$
$$n = \text{number of compounding periods}$$

The interest earned will be the future value minus the initial principal. That is,

$$\text{Interest earned} = F - P$$

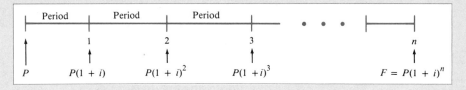

The determination of $(1 + i)^n$ for different values of i and n can be done in a number of ways. A hand calculator with an x^y key or a computer could be used. The appendix in the back of the book as well as Table 1 lists different values of $(1 + i)^n$.

Example 2

Compound Interest   Find the future value or compound amount of $5000 invested for 3 years at 6% compounded every two months. Also find the interest earned.

Solution   The variables in the future value formula are

$$P = \$5000 \qquad \text{(amount invested)}$$
$$i = r/k = 0.06/6 = 0.01 \qquad \text{(interest per period)}$$
$$n = 3 \cdot 6 = 18 \qquad \text{(number of periods)}$$

Hence the compound amount is

$$F = P(1 + i)^n$$
$$= \$5000(1 + 0.01)^{18}$$
$$= \$5000(1.01)^{18}$$
$$= \$5000(1.19614748)$$
$$= \$5980.74$$

TABLE 1

Future Value $(1 + i)^n$ of $1 Invested for n Interest Periods at Interest Rate of i per Period

	Interest per Period i				
n	*0.02%*	*0.5%*	*1%*	*2%*	*6%*
1	1.000200000	1.00500000	1.01000000	1.02000000	1.06000000
2	1.000400040	1.01002500	1.02010000	1.04040000	1.12360000
3	1.000600120	1.01507513	1.03030100	1.06120800	1.19101600
4	1.000800240	1.02015050	1.04060401	1.08243216	1.26247696
5	1.001000400	1.02525125	1.05101005	1.10408080	1.33822558
6	1.001200600	1.03037751	1.06152015	1.12612640	1.41851911
7	1.001400840	1.03552940	1.07213534	1.14868567	1.50363026
8	1.001601120	1.04070704	1.08285671	1.17165938	1.59384807
9	1.001801447	1.04591058	1.09368527	1.19509257	1.68947896
10	1.002001806	1.05114013	1.10462213	1.21809442	1.79084770
11	1.002202201	1.05639583	1.11566835	1.24337431	1.89329856
12	1.002402642	1.06167781	1.12682503	1.26824179	2.01219647
13	1.002603122	1.06698620	1.13809328	1.29360663	2.13292826
14	1.002803643	1.07232113	1.14947421	1.31947876	2.26090396
15	1.003004204	1.07768274	1.16096896	1.34586834	2.39655819
16	1.003204804	1.08307115	1.17257864	1.37278571	2.54035168
17	1.003405445	1.08848651	1.18430443	1.40024142	2.69277279
18	1.003606145	1.09392894	1.19614748	1.42824625	2.85433915
19	1.003806848	1.09939858	1.20810895	1.45681117	3.02559950
20	1.004007609	1.10489558	1.22019004	1.48594740	3.20713547

See Figure 6.

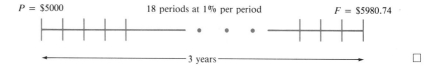

$P = \$5000$ 18 periods at 1% per period $F = \$5980.74$

← ———— 3 years ———— →

Figure 6
Future value

Time Is Money (the Dutch Explorers and the Canarsie Indians)

The story of the Dutch explorers paying the Canarsie Indians $24 in 1626 for the island of Manhattan is one every school child knows. What is not so well known is that if the Canarsie Indians had invested the $24 in a long-term bond yielding 10% compound interest, compounded annually, then the 1990 value of that $24 would be roughly $27,999,394,470,000,000 (almost 28 quadrillion dollars).

Table 2 shows the future value of that $24 had the Canarsies invested this money at just 5% compounded annually.

TABLE 2
Future Value of $24
When Invested at 5%
Compounded Annually

Year	Future Value of $24	
1626	$24	= $24.00
1627	$24(1 + 0.05)	= $25.20
1628	$24(1 + 0.05)^2$	= $26.46
1629	$24(1 + 0.05)^3$	= $27.78
1630	$24(1 + 0.05)^4$	= $29.17
1650	$24(1 + 0.05)^{24}$	= $77.40
1700	$24(1 + 0.05)^{74}$	= $887.60
1750	$24(1 + 0.05)^{124}$	= $10,178.51
1800	$24(1 + 0.05)^{174}$	= $116,721.08
1850	$24(1 + 0.05)^{224}$	= $1,338,487.30
1900	$24(1 + 0.05)^{274}$	= $15,348,968.21
1950	$24(1 + 0.05)^{324}$	= $176,012,754.70
1990	$24(1 + 0.05)^{364}$	= $1,239,127,806.00

Example 3

Increasing the Frequency of Compoundings   Find the future value or compound amount of $1000 invested for 10 years at 8% compounded

(a) annually   (b) semiannually   (c) quarterly
(d) monthly   (e) weekly   (f) daily

Solution   Using the future value formula

$$F = P(1 + i)^n$$

we calculate Table 3.   □

TABLE 3
Future Value of $1000
Invested at 8% after 10 Years

Compoundings	Number of Periods	Future Value after 10 Years	
Simple interest	*		= $1800
Annually	10	$1000(1 + 0.08)^{10}$	= $2158.92
Semiannually	20	$1000(1 + 0.08/2)^{20}$	= $2191.12
Quarterly	40	$1000(1 + 0.08/4)^{40}$	= $2208.04
Monthly	120	$1000(1 + 0.08/12)^{120}$	= $2219.64
Weekly	520	$1000(1 + 0.08/52)^{520}$	= $2224.17
Daily	3650	$1000(1 + 0.08/365)^{3650}$	= $2225.35

How Long Does It Take Money to Grow?

How many compounding periods n does it take for an initial principal of P to grow to a future value of F if the principal earns interest of i percent per compounding period? To answer this question, we begin with the formula for future value

$$F = P(1 + i)^n$$

The goal is to solve this equation for the number of periods n in terms of F, P, and i. We can do this by taking the common logarithm of each side of the above equation. Doing this and using properties of logarithms developed in the Finite Mathematics Preliminaries, we get

$$\log F = \log P + n \log (1 + i)$$

Solving for n gives the formula below.

Number of Periods It Takes Money to Grow from P to F

The number of compounding periods n it takes for a present value P to reach the future value F with an interest rate of i percent per period is given by

$$n = \frac{\log (F/P)}{\log (1 + i)}$$

Example 4

Doubling Time   Find the time it takes money invested at 10% compounded quarterly to double in value.

Solution   Money will double in value when $F = 2P$. Since the interest rate of i percent per compounding period is $i = 0.10/4 = 0.025$, we have

$$n = \frac{\log 2}{\log (1 + i)}$$

$$= \frac{0.3010}{\log (1.025)}$$

$$= \frac{0.3010}{0.010724}$$

$$= 28 \text{ compounding periods}$$

Hence the initial amount will double in value after 28 compounding periods. Since there are four compounding periods per year, it will take $28/4 = 7$ years. Figure 7 shows the doubling times for different interest rates when interest is compounded annually.   □

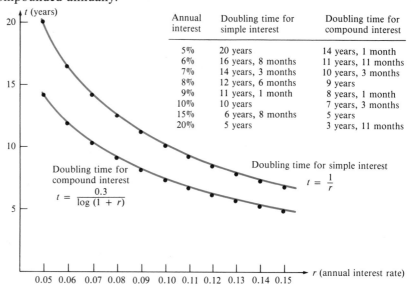

Annual interest	Doubling time for simple interest	Doubling time for compound interest
5%	20 years	14 years, 1 month
6%	16 years, 8 months	11 years, 11 months
7%	14 years, 3 months	10 years, 3 months
8%	12 years, 6 months	9 years
9%	11 years, 1 month	8 years, 1 month
10%	10 years	7 years, 3 months
15%	6 years, 8 months	5 years
20%	5 years	3 years, 11 months

Figure 7

Doubling times for simple interest versus compound interest

Present Value

The **present value** of a sum of money refers to the amount of money that, by earning interest, will be equal to a specified amount at some specified future date. The initial principal P is often called the present value of the future value F. The present value P can be found in terms of the future value F by solving for P in the formula

$$F = P(1 + i)^n$$

Hence we have the following formula.

Present Value of a Future Amount

The present value P that must be invested at an interest rate of i percent per compounding period to generate a future value of F after n compoundings is given by

$$P = \frac{F}{(1 + i)^n} = F(1 + i)^{-n}$$

where

$$F = \text{future value after } n \text{ periods}$$
$$P = \text{present value of } F$$
$$i = \text{interest rate per period}$$
$$n = \text{number of compounding periods}$$

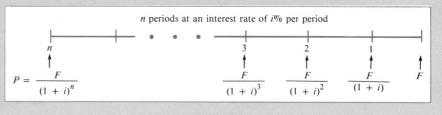

Example 5

Computation of Present Value   How much money must be deposited today in a bank paying 9% compounded semiannually so that after seven years the future value will be $10,000?

Solution   The values for the variables in the present value formula are

$$F = \$10{,}000 \qquad \text{(future value)}$$
$$i = 0.09/2 = 0.045 \qquad \text{(interest rate per period)}$$
$$n = 7 \cdot 2 = 14 \qquad \text{(number of compounding periods)}$$

Hence we have

$$P = F(1 + i)^{-n}$$
$$= \$10{,}000(1 + 0.045)^{-14}$$
$$= \$10{,}000(0.539973) \qquad \text{(use calculator)}$$
$$= \$5399.73$$

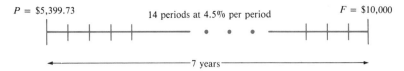

Figure 8
Present value

In other words, an investment of $5399.73 today will be worth $10,000 in seven years if invested at 9%, compounded semiannually. (See Figure 8.)   □

Effective Rate of Interest

A recurring problem in financial planning is to find the rate of return on an investment. In Section 3.1 we saw that the effective rate of interest of a simple discount note was higher than the advertised rate. We now examine the effective rate of interest of a deposit that pays compound interest.

Suppose, for example, that $1000 is deposited in a bank that pays 8% annual interest, compounded monthly. After one year the future value of this deposit will be

$$F = P(1 + i)^n$$
$$= \$1000(1 + 0.08/12)^{12}$$
$$= \$1083.00$$

This says that the original principal of $1000 has grown in value to $1083 over a one-year period. In other words, this investment earns the same interest as an investment that pays 8.3% simple interest. The 8% rate of interest is called the **nominal** or **advertised rate** of interest, and the 8.3% is the **effective rate** of interest. The relationship between the nominal and effective rates of interest can be found by setting the interest earned (in one year) from simple interest equal to the interest earned from compound interest. Setting these equal gives

$$P(1 + r_{eff}) = P(1 + i)^n$$

If we solve this equation for r_{eff}, we will get the following result.

Effective Rate of Interest

The effective annual rate of interest r_{eff} corresponding to the nominal (advertised) rate of interest i, compounded n times per year, is

$$r_{eff} = (1 + i)^n - 1$$

Example 6

Effective Rate of Interest   The nominal (4.85%) and effective (4.97%) rates of interest were advertised in a 1964 copy of the *New York Times*. Is it true that the advertised rate of 4.85% has an effective rate of 4.97%?

Solution   Using the formula for the effective rate of interest, we have

$$r_{eff} = (1 + i)^n - 1$$
$$= (1 + 0.0485/365)^{365} - 1$$
$$= (1.000132877)^{365} - 1$$
$$= 1.0497 - 1$$
$$= 0.0497 \quad (4.97\%)$$

In other words, the effective interest rate is 4.97%. which is what the advertisement claims. □

You can check the financial section of almost every daily newspaper to find the current advertised and effective rates of interest.

Example 7

Which Bank Pays the Higher Rates? Bank A offers depositors 9% compounded annually, while bank B offers investors $8\frac{3}{4}\%$ compounded monthly. Bank B claims that its effective rate is higher than bank A's effective rate, owing to its increased number of compoundings per year. Bank A claims that its higher advertised or nominal rate of 9% is the more important factor. Which bank is correct?

Solution The effective rates for both banks are computed below.
Bank A (9% compounded annually):

$$r_{eff} = (1 + i)^n - 1$$
$$= (1 + 0.09) - 1$$
$$= 0.09 \quad (9\% \text{ effective rate})$$

Bank B (8.75% compounded monthly):

$$r_{eff} = (1 + i)^n - 1$$
$$= (1 + 0.0875/12)^{12} - 1$$
$$= 0.0911 \quad (9.11\% \text{ effective rate})$$

In other words, $8\frac{3}{4}\%$ compounded monthly has a higher effective rate of interest than does 9% compounded annually. □

Continuous Compounding

The future value of a bank account is increased if the bank increases the frequency of compoundings per year. Table 4 gives the future value of $1000 after one year if the rate of interest is 12% per annum and interest is compounded n times a year for various values of n.

TABLE 4
Changing the Future Value by Changing the Frequency of Compounding

Number of Periods per Year, n	Interest per Period, i	Compound Amount, $P(1 + i)^n$	Effective Rate, $(1 + i)^n - 1$
1	0.12	$1120.00	12%
2	0.06	$1123.60	12.36%
3	0.04	$1124.86	12.49%
4	0.03	$1125.51	12.55%
6	0.02	$1126.16	12.62%
8	0.015	$1126.49	12.65%
12	0.01	$1126.83	12.68%
24	0.005	$1127.16	12.71%
48	0.0025	$1127.33	12.73%

Will the effective rate of interest increase without bound by increasing the frequency of the number of compoundings? The answer is no, and an upper bound on the effective interest rate can be approached by compounding interest with increasingly smaller periods such as every day, hour, minute, second, . . . , until one thinks that interest is being compounded continuously.

To understand **continuous compounding**, assume that an initial principal P is deposited in a bank that pays an annual interest rate of r percent, compounded k times per year. After n compounding periods this principal will have a future value of

$$F = P(1 + r/k)^n$$

Since the total number of compounding periods n can be written

$$n \text{ periods} = (k \text{ periods per year})(t \text{ years})$$

the future value formula can be rewritten as

$$F = P\left(1 + \frac{r}{k}\right)^n$$

$$= P\left[1 + \frac{1}{(k/r)}\right]^{kt}$$

$$= P\left\{\left[1 + \frac{1}{(k/r)}\right]^{(k/r)}\right\}^{rt}$$

It can be shown by using methods in calculus that as the number of compounding periods per year k increases without bound, the number

$$\left[1 + \frac{1}{(k/r)}\right]^{(k/r)}$$

gets closer and closer to the number $2.718\ldots$, which we denote by the letter e and which is sometimes called the **banker's constant**. Hence the future value above will approach the value

$$F = P(2.718\ldots)^{rt} = Pe^{rt}$$

We summarize this result now.

Future Value Using Continuous Compounding

For an initial principal P invested at an annual rate of interest of r percent compounded continuously the future value F of this account after t years is given by

$$F = Pe^{rt}$$

where

F = future value after t years

P = principal invested (present value)

$e = 2.718281828\ldots$ (banker's constant)

r = annual rate of interest

t = length of time in years principal earns interest

> The effective rate of interest r_{eff} for continuous compounding is found from the formula
>
> $$r_{\text{eff}} = e^r - 1$$
>
> The present value P can be also be solved in terms of the future value:
>
> $$P = Fe^{-rt}$$
>
>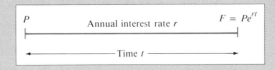
>
> P —————— Annual interest rate r —————— $F = Pe^{rt}$
>
> ←————————— Time t —————————→

Example 8 ——————

Continuous Compounding   The advertisement in Figure 9 offers an investor an 11% annual percentage rate (APR) for making a deposit in a five-year Certificate of Deposit. What is the future value of $1000 invested in this five-year Certificate of Deposit after five years? Also, is the bank telling the truth when it says that the effective rate of interest is 11.63%?

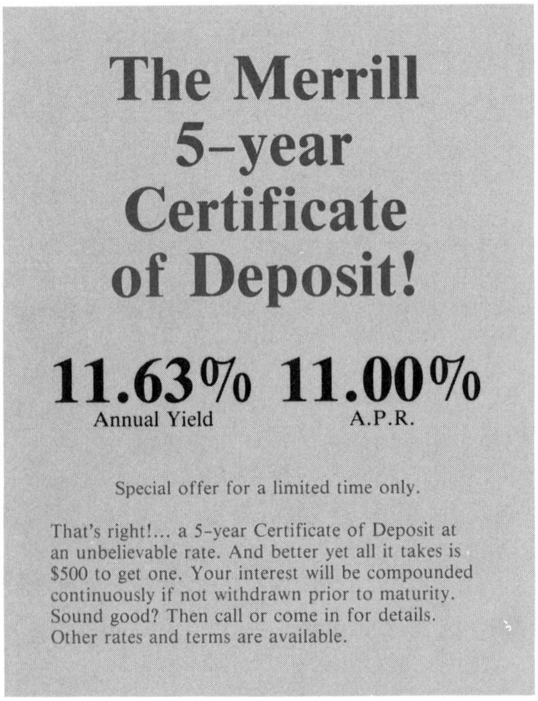

The Merrill 5–year Certificate of Deposit!

11.63% 11.00%
Annual Yield   A.P.R.

Special offer for a limited time only.

That's right!... a 5-year Certificate of Deposit at an unbelievable rate. And better yet all it takes is $500 to get one. Your interest will be compounded continuously if not withdrawn prior to maturity. Sound good? Then call or come in for details. Other rates and terms are available.

Figure 9
Advertisement for a certificate of deposit. (Courtesy of The Merrill Trust Company.)

Solution   The values of the variables in the future value formula for continuous compounding are

$$P = \$1000 \qquad \text{(principal invested)}$$
$$r = 0.11 \qquad \text{(annual interest)}$$
$$t = 5 \qquad \text{(length of time invested)}$$

Hence the future value after five years is

$$F = Pe^{rt}$$
$$= \$1000e^{(0.11)(5)}$$
$$= \$1000e^{(0.55)}$$
$$= \$1000(1.73325)$$
$$= \$1733.25$$

The quantity $e^{0.55} = 1.73325$ can be computed on a calculator by using the e^x key with $x = 0.55$.

To find the effective rate of interest, we find

$$r_{\text{eff}} = e^r - 1$$
$$= e^{(0.11)} - 1$$
$$= 0.1163$$

In other words, the effective rate of interest is 11.63%, which is exactly what the bank claims.

The interest earned from this account over the five-year period is the difference between the final and initial values of the account. That is,

$$\text{Interest earned} = F - P$$
$$= \$1733.25 - \$1000.00$$
$$= \$733.25 \qquad \square$$

Problems

Future Value and Interest Earned

For Problems 1–6, calculate the future value and interest earned for the described investments at the end of two years and five years.

	Present Value, P	Annual Interest Rate, r	Compounded
1.	$10,000	8%	semiannually
2.	$15,000	10%	quarterly
3.	$5000	9%	annually
4.	$20,000	12%	monthly
5.	$10,000	5%	continuously
6.	$5000	8%	continuously

the given future value for given rates of interest. Convert the number of compounding periods to time in years.

	Present Value, P	Future Value, F	Interest Rate per Period, i	Length of Period
7.	$1000	$2000	0.01	one month
8.	$1	$500	0.08	one year
9.	$10,000	$12,000	0.05	six months
10.	$10,000	$100,000	0.02	three months
11.	$10,000	$500,000	0.12	one year
12.	$10,000	$1,000,000	0.18	one year

Number of Periods for Money to Grow

For Problems 7–12, calculate the number of compounding periods it takes the given present value to grow in value to

Present Value of Money

For Problems 13–18, calculate the present value of the following future values.

	Future Value, F	Annual Interest Rate, r	Length of Loan, t	Compounded
13.	$3500	8%	5 years	monthly
14.	$10,000	9%	8 years	annually
15.	$50,000	5%	10 years	quarterly
16.	$75,000	10%	15 years	monthly
17.	$5000	15%	25 years	continuously
18.	$10,000	10%	5 years	continuously

Effective Rate of Interest

For Problems 19–22, determine the better investment.

19. 8% compounded monthly or 9% compounded semiannually.

20. 10% compounded annually or 10% compounded continuously.

21. 12% compounded annually or 11.5% compounded daily.

22. 11% compounded annually or 10.85% compounded monthly.

The Real Value of Money

For Problems 23–26, a banker offers you one of two options. Which one will you take? Why?

23. $10,000 now or $17,000 in ten years if the bank pays compound interest of 7% compounded annually?

24. $10,000 now or $19,000 in nine years if the bank pays compound interest of 7.5% compounded annually?

25. $50,000 now or $55,000 in two years if the bank pays compound interest of 5% compounded annually?

26. $25 now or $100 in 25 years if the bank pays an annual interest rate of 10% compounded semiannually?

27. Future Value In 1860, when the Bank of Texas was chartered, an account was opened at the birth of Robert L. Thomas with an initial deposit of $25. No further deposits or withdrawals were made. One hundred and twenty years later, in 1980, a descendant of Mr. Thomas unexpectedly came into possession of this account in a will. Assuming that the bank paid 2% compounded continuously during the first 50 years and 5% compounded annually during the next 70 years, how much was the account worth to this descendant?

28. Man Repays Old Loan In 1924 in Hackensack, N.J., a man dropped into the police station and repaid a 20-cent loan plus interest that a motorcycle policeman had lent him 20 years earlier when the man found himself stranded in Hackensack with no funds to get home. The man left $2 at the station to repay the loan with interest. How much interest did the man pay, assuming that the man paid interest compounded annually? *Hint:* Set $F = P(1 + i)^n$ where $F = 2$, $P = 0.20$, and $n = 20$ and solve for the interest i.

29. Finding Compound Interest Rate A portfolio of ten common stocks sold on the New York Stock Exchange has increased in value from $29,245 to $156,945 over a ten-year period. What rate of interest compounded annually would have accounted for this capital increase? *Hint:* Set $F = P(1 + i)^n$ where F, P, and n are given and solve this equation for i.

30. Finding Compound Interest Rates A house purchased for $60,000 six years ago sold for $95,000 today. What rate of interest compounded annually would have given the same return? *Hint:* Set $F = P(1 + i)^n$ and solve for i.

31. Effective Rate of Interest What is the simple rate of interest that is equivalent to compound interest of 7.5% compounded (a) quarterly, (b) monthly, (c) weekly, (d) daily, (e) continuously?

32. Effective Rate of Interest What is the simple rate of interest that is equivalent to compound interest of 9% compounded (a) quarterly, (b) monthly, (c) weekly, (d) daily, (e) continuously?

33. Present Value A trust fund is being established to underwrite a scholarship fund for the grandchildren of the Joneses. The aim of the fund is to have $50,000 available when disbursements begin in 15 years. What sum should be given to the trust if the investment pays an annual interest of 8% compounded semiannually?

34. Future Value How much would a seller be willing to accept for an apartment purchased six years ago at $99,000 if an acceptable rate of return is 10% compounded annually?

Problems 35–38 refer to Figure 10 showing the growth of P dollars over one year compounded at an annual interest rate r compounded k times per year. The bottom curve shows the growth for continuous compounding.

35. Use a calculator to compute the future value F at each quarter if the initial principal is $1 and the annual interest rate is 8%.

36. Use a calculator to compute the future value F at each month if the initial principal is $1 and the annual interest rate is 12%.

37. Use a calculator to compute the future value F after 1, 2, 3, 4, and 5 years if the initial principal is $1 and the annual interest rate is 8%.

38. Use a calculator to compute the future value F after one year if the initial principal is $1 and the interest rate is 12% compounded
(a) annually   (b) quarterly
(c) monthly   (d) continuously

The following two stories involving compound interest were paraphrased from the interesting book, *Neifeld's Guide to Installment Computations* (Mack Publishing Co., 1953).

39. Hollywood Stunt Thwarted by Compound Interest In 1944, Hollywood publicity agents were going to dramatize the opening of the motion picture *Knickerbocker Holiday*

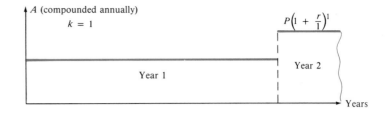

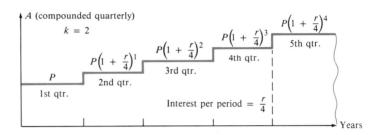

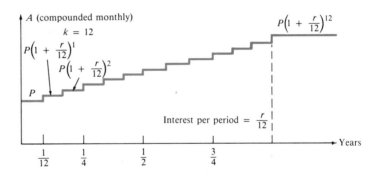

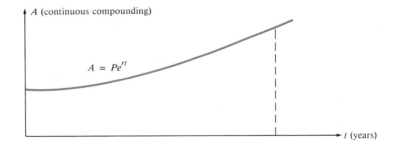

Figure 10

The growth of money as a function of the number of compoundings

by arranging a publicity stunt in which three bottles of whiskey, originally thought to have been given to the Canarsie Indians for the island of Manhattan, were to have been returned to the mayor of New York City, plus interest at 6% compounded annually. To their horror, just before the gala event the agents discovered that the compound interest on three bottles of whiskey over a period of 300 years would be more than 100 million bottles. As one agent put it, "The stunt just ain't worth it." Exactly how many bottles of whiskey should be repaid to the Canarsie Indians?

40. Money Fails to Grow   In Binghampton, England, in 1874 a General Edward F. Jones deposited the British equivalent of 5 cents in the Binghampton Savings Bank at the time the bank opened its doors. Although the bank pays 4% interest compounded annually, 70 years later the bank returned the 5 cents to the executor of General Jones's estate saying that "there was no interest because the deposit was less than one dollar." Assuming that the bank had a more generous policy, what would be the value of the general's account?

Annuities

3.3

PURPOSE

We introduce the financial concepts of

* an annuity,
* the future value of an annuity,
* the rent payments of an annuity,
* the present value of an annuity, and
* a sinking fund.

Introduction

The cost of a college education increases every year. Besides increases in tuition and room-and-board costs, a student finds that the costs of books, supplies, and transportation are also rising. Government loans, grants, and work-study funds can help, but changing eligibility requirements for these programs make it prudent to have other alternatives available. Ideally, this financial problem would be faced as a long-range problem many years before the student matriculates. Businesses too are often faced with making large expenditures. For example, new equipment needed to comply with health and safety requirements or to modernize production methods often involves large capital expenditures. Companies usually find it difficult to fund large expenditures out of current funds alone.

This section explains a financial tool that can be used by both individuals and companies to prepare for large expenditures at a future time. The tool is called an annuity.

An oil company may make payments into an annuity in order to finance a $100 million offshore drilling platform.

The Concept of an Annuity

An **annuity** is a series of equal payments, paid at equal intervals of time. Typical annuities might be rent payments, interest payments on bonds or mortgages, preferred dividends on stocks, pension premiums on life insurance, and installment payments for a debt.

In many cases the interval between payments, called the **payment period**, is a year, but other common periods are half-years, quarters, months, and sometimes periods even greater than a year. Payments are generally made at the end of each period. This type of annuity is called an **ordinary annuity**. The total time period between the beginning of the first period and the last payment is called the **term** of the annuity. Some annuities, called **perpetuities**, go on forever, but we will study **fixed-term annuities** here. (See Figure 11.)

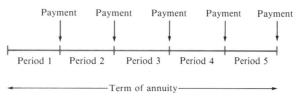

Figure 11
An ordinary annuity with five periods

In financial programs, annuities are generally used in one of two ways.

1. **Future Value Problem** (sinking funds). When a sum of money will be needed at some future date, a good practice is to make periodic payments to a bank account, for which the payments and the interest earnings will equal the desired amount at the time it is needed. Such an account is called a **sinking fund**. The problem is to determine the future value of the series of payments.

2. **Present Value Problem** (generating payments). An individual or company deposits a lump sum (present value) of money in a bank for the purpose of generating a series of equal payments at later dates. The problem is to determine the lump sum needed so that together with the interest earned on the account, a series of payments of a given size can be made at given required times.

Future Value of an Annuity

Dorothy plans on making deposits of $100 to a bank account at the end of each year for the next five years. The bank pays interest at a rate of 9% compounded annually. Under these conditions, what will be the future value of Dorothy's account at the end of the five years? Figure 12 illustrates this payment schedule.

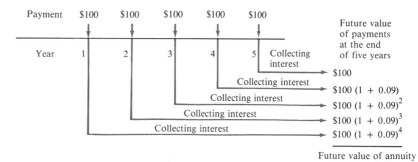

Figure 12
Future value of an annuity after five years

To determine the future value of Dorothy's account, we first observe that at the end of the first year after she deposits the first $100, the value of her account is simply

$$\text{Value after one year} = \$100$$

At the end of the second year, Dorothy's initial deposit of $100 has grown in value to $100(1 + 0.09)$, and she then deposits another $100. Hence the value of her account at the end of two years is

$$\text{Value after two years} = \$100(1 + 0.09) + \$100$$

At the end of the third year, Dorothy's initial deposit has now grown in value to $\$100(1 + 0.09)^2$, and her second deposit has grown to $\$100(1 + 0.09)$. She then deposits a third $100. Hence her account will have a total value of

$$\text{Value after three years} = \$100(1 + 0.09)^2 + \$100(1 + 0.09) + \$100$$

By the same type of reasoning, the future value F of Dorothy's account after five years will be

$$
\begin{aligned}
F &= \$100(1.09)^4 + \$100(1.09)^3 + \$100(1.09)^2 + \$100(1.09) + \$100 \\
&= \$100\left[(1.09)^4 + (1.09)^3 + (1.09)^2 + (1.09) + 1\right] \\
&= \$100\left[\frac{(1.09)^5 - 1}{0.09}\right] \\
&= \$598.47 \qquad \text{(use a calculator)}
\end{aligned}
$$

The third step is a result of the formula for geometric series

$$p^4 + p^3 + p^2 + p + 1 = \frac{p^5 - 1}{p - 1}$$

with $p = 1.09$. To verify it, simply multiply each side of the equation by $p - 1$ and show that both sides of the equation are the same.

In other words, if Dorothy makes five yearly payments of $100 to a bank that pays 9% annual interest compounded annually, then at the end of this time the value of her account will be $598.47 (note that she has earned $98.47 in interest). This series of payments is called an annuity, and the future value of the annuity is $598.47.

This example generalizes to give the formula for the future value of an annuity.

Future Value of an Annuity

The future value F of an annuity having n payments of R dollars each, with payments made at the end of each payment period and compounded at a rate of i percent per period, is given by

$$F = R\frac{(1 + i)^n - 1}{i}$$

> where
>
> $$F = \text{future value of the annuity}$$
> $$R = \text{periodic payments of the annuity}$$
> $$i = \text{interest rate per period}$$
> $$n = \text{number of payment periods}$$
>
> As a convenience in using tables, we define
>
> $$s_{\overline{n}|i} = \frac{(1 + i)^n - 1}{i}$$
>
> which is called the **future value of dollar deposits** and is read "s angle n sub i." This gives an alternate form for the future value of an annuity:
>
> $$F = Rs_{\overline{n}|i}$$

When periodic payments R are one dollar, the future value F is equal to $s_{\overline{n}|i}$. Hence $s_{\overline{n}|i}$ can be interpreted as the future value of an annuity after n payments of $1 have been made, and compound interest earned is at a rate of i percent per period. Values for $s_{\overline{n}|i}$ can be found in the table entitled "Future Value of an Annuity" in the Appendix.

Example 1

Future Value of an Annuity   To pay for college, Sue's parents decide to invest $100 each quarter in a bank from the time Sue is 6 years old until she is 18. If the bank pays an annual interest of 10% compounded quarterly, what will be the value of this annuity when Sue is 18?

Solution   Here we have

$$R = \$100 \qquad \text{(payments)}$$
$$i = 0.10/4 = 0.025 \qquad \text{(interest per payment period)}$$
$$n = 12 \cdot 4 = 48 \qquad \text{(number of payment periods)}$$

Hence

$$F = Rs_{\overline{n}|i}$$
$$= \$100 s_{\overline{48}|0.025}$$
$$= \$100(90.8595)$$
$$= \$9085.96$$

In other words, if Sue's parents make quarterly payments of $100 each quarter for 12 years, at the end of that time they will have available a lump sum of $9085.96 to pay for Sue's college education.   □

Finding Payments of an Annuity

In practical problems an individual or corporation often makes periodic payments into an annuity for the purpose of generating a given future amount.

The question arises: What should the payments be to generate the given amount? To find the payments R of an annuity, we solve for R in the formula

$$F = Rs_{\overline{n}|i}$$

in terms of the future value F and $s_{\overline{n}|i}$. We get the following result.

Payments of an Annuity

The payment R per period for n periods that results in a future value F, when money is compounded at a rate of i percent per period, is given by

$$R = \frac{F}{s_{\overline{n}|i}}$$

where

$$R = \text{payments of the annuity}$$
$$F = \text{future value of the annuity}$$
$$n = \text{number of payment periods}$$
$$i = \text{interest rate per payment period}$$

Example 2

Payments   Coles Express needs \$150,000 in five years to replace delivery vehicles. The company intends to make semiannual payments to a bank that pays 8% compounded semiannually to finance the expenditures. What payments must the company make every 6 months to generate this amount?

Solution   We have

$$F = \$150,000 \qquad \text{(future value of annuity)}$$
$$i = 0.08/2 = 0.04 \qquad \text{(interest rate per period)}$$
$$n = 5 \cdot 2 = 10 \qquad \text{(number of payment periods)}$$

Hence the payments should be

$$R = \frac{F}{s_{\overline{n}|i}}$$

$$= \frac{\$150,000}{s_{\overline{10}|0.04}}$$

$$= \frac{\$150,000}{12.006}$$

$$= \$12,493.75$$

In other words, if the company makes ten semiannual payments of \$12,493.75 each, then at the end of five years the annuity will have a value of \$150,000.

□

Sinking Funds

In the two previous examples, Sue's parents and Coles Express paid money into funds for the purpose of meeting some future financial obligation. Both of these funds are special kinds of annuities called sinking funds. In general, a **sinking fund** is any account that is established for the purpose of generating funds to meet future obligations. Often a firm will anticipate a capital expenditure for new equipment at some time in the future. The company then deposits monies at regular intervals into a sinking fund to finance this future expenditure.

Example 3

Sinking Fund   A major shipbuilder refits one of its shipyards every 15 years. To ensure that money is available, it sets up a sinking fund by making semiannual payments into an account paying an annual interest of 8% compounded semiannually. The company determines that it will need six million dollars in 15 years. What are the payments for this sinking fund?

Solution   The values of the variables in the payment formula are

$$F = \$6,000,000 \qquad \text{(future value of annuity)}$$
$$i = 0.08/2 = 0.04 \qquad \text{(interest rate per period)}$$
$$n = 15 \cdot 2 = 30 \qquad \text{(number of payment periods)}$$

Hence the payments are

$$R = \frac{F}{s_{\overline{n}|i}}$$
$$= \frac{\$6,000,000}{s_{\overline{30}|0.04}}$$
$$= \frac{\$6,000,000}{56.084938}$$
$$= \$106,980.59$$

In other words, if the company makes semiannual payments of $106,980.59 for the next 15 years into the sinking fund, the company will have exactly $6,000,000 when it is needed for its refitting program. The total amount of money paid into the sinking fund (multiply the payment by the number of payments) is $3,209,417.70. The difference between the six million dollars future value and this amount is $2,790,582.30, which is the interest collected over the 15 years on the amount in the sinking fund.   □

Sinking funds are often used by companies to provide future funds for major expenditures.

Present Value of an Annuity

Until now, we have made periodic payments into an annuity for the purpose of obtaining a final lump sum of money. We now consider the reverse problem. We start with a lump sum of money and make periodic withdrawals over a period of time. Here, payments are made to an individual or company from a bank, which is the reverse from the previous type of annuity, in which periodic payments were made to a bank from an individual or corporation.

This new type of annuity occurs in pension plans. A person may have $100,000 on deposit in a savings institution at the age of 65. The money is then disbursed for retirement purposes in equal payments over a period of time. It should be pointed out that several types of annuities are popular in retirement programs. Two common types of annuities are **varying annuities**, in which payments vary over time, and **whole life annuities**, in which the payments last for the life of the individual. We restrict ourselves here to annuities with fixed payments and a given number of payment periods.

We will now determine the initial lump sum that must be deposited in an account paying 7% compounded annually to allow a withdrawal of $10,000 every year for the next five years. This type of annuity is illustrated in Figure 13.

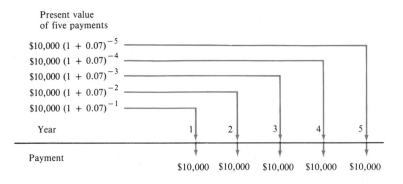

Figure 13

Flow of money in an annuity with a given present value

We can interpret various amounts or portions of the initial deposit as funding the various payments. In particular, a present value of

$$\frac{\$10,000}{(1 + 0.07)} \quad \text{will pay for the } \textit{first } \$10,000 \text{ payment}$$

$$\frac{\$10,000}{(1 + 0.07)^2} \quad \text{will pay for the } \textit{second } \$10,000 \text{ payment}$$

$$\frac{\$10,000}{(1 + 0.07)^3} \quad \text{will pay for the } \textit{third } \$10,000 \text{ payment}$$

$$\frac{\$10,000}{(1 + 0.07)^4} \quad \text{will pay for the } \textit{fourth } \$10,000 \text{ payment}$$

$$\frac{\$10,000}{(1 + 0.07)^5} \quad \text{will pay for the } \textit{fifth } \$10,000 \text{ payment}$$

Hence the total amount that will "pay" for all five payments is

$$P = \$10,000(1.07)^{-5} + \$10,000(1.07)^{-4} + \cdots + \$10,000(1.07)^{-1}$$

$$= \$10,000(1.07)^{-5}[1 + (1.07) + (1.07)^2 + \cdots + (1.07)^4]$$

$$= \$10,000(1.07)^{-5}\left[\frac{1 - (1.07)^5}{-0.07}\right]$$

$$= \$10,000\left[\frac{1 - (1.07)^{-5}}{0.07}\right]$$

$$= \$41,001.97$$

The third step uses the formula for geometric series

$$1 + p + p^2 + p^3 + p^4 = \frac{1 - p^5}{1 - p}$$

with $p = 1.07$.

Conclusion

An initial deposit of $41,001.97 will fund five equal payments of $10,000 each for the next five years if the bank pays interest at a rate of 7% compounded annually. At the end of five years the account will be empty.

We can generalize this example to give the formula for the present value of an annuity.

Present Value of an Annuity

The present value P necessary to provide for n withdrawals of R dollars each, when interest is paid at the rate of i percent per withdrawal period, is given by

$$P = R \frac{1 - (1 + i)^{-n}}{i}$$

where

$$P = \text{present value of the annuity}$$
$$R = \text{amount of withdrawals}$$
$$i = \text{rate of interest per withdrawal period}$$
$$n = \text{number of withdrawal periods}$$

For convenience we denote

$$a_{\overline{n}|i} = \frac{1 - (1 + i)^{-n}}{i}$$

which is called the present value of $1 payments and is read "$a$ angle n sub i." This gives an alternative form for the present value of an annuity:

$$P = R a_{\overline{n}|i}$$

In the description of the present value of an annuity, annuity payments are called **withdrawals**, and the payment periods are called **withdrawal periods**. The quantity $a_{\overline{n}|i}$ can be interpreted as the value of an account necessary to provide for n withdrawals of $1 each when interest is paid at the rate i percent per period. Values of $a_{\overline{n}|i}$ are listed in the Financial Tables in the Appendix.

Example 4

Present Value of an Annuity   A retirement policy offers 20 yearly payments of \$15,000 beginning at the age of 65. What would be the cash equivalent of these payments if money can earn interest at an annual rate of 8% compounded annually?

Solution   Here the values of the variables in the present value formula for an annuity are

$$R = \$15,000 \qquad \text{(periodic withdrawals)}$$
$$i = 0.08 \qquad \text{(rate of interest per withdrawal period)}$$
$$n = 20 \qquad \text{(number of withdrawals)}$$

Hence the present value is

$$P = Ra_{\overline{n}|i}$$
$$= \$15,000 a_{\overline{20}|0.08}$$
$$= \$15,000(9.81815)$$
$$= \$147,272.25$$

In other words an initial deposit of \$147,272.25 will fund 20 payments of \$15,000 each if interest is paid at 8% compounded annually. Note that the first withdrawal will be made one year after the initial deposit is made.   □

Example 5

Figure 14
A lottery ticket

Amount of Payments   Joe has won the Mega-State Lottery (see Figure 14), which consists of a tax-free jackpot of \$1,000,000. He deposits this amount in a bank that pays interest at an annual rate of 10% compounded quarterly. He requests that his winnings be paid to him in equal payments every three months for the next 25 years. What will be the amount of each payment?

Solution   We find the size of the withdrawals R from the present value equation

$$P = Ra_{\overline{n}|i}$$

Solving for R, we have

$$R = \frac{P}{a_{\overline{n}|i}}$$

The known values of the variables are

$$P = \$1,000,000 \qquad \text{(initial deposit or present value)}$$
$$i = 0.10/4 = 0.025 \qquad \text{(interest rate per withdrawal period)}$$
$$n = 25 \cdot 4 = 100 \qquad \text{(number of withdrawals)}$$

Hence we have

$$R = \frac{P}{a_{\overline{n}|i}}$$
$$= \frac{\$1,000,000}{a_{\overline{100}|0.025}} \qquad \text{(use calculator)}$$
$$= \frac{\$1,000,000}{36.614105}$$
$$= \$27,311.88$$

Therefore Joe will receive a check for $27,311.88 every three months for the next 25 years. After 25 years the account will be empty. □

Problems

For Problems 1–6, find the future value of the given annuity with the following values.

	Value of Payments	Interest per Period	Number of Periods
1.	$1000	2%	30
2.	$5000	1%	48
3.	$10,000	8%	10
4.	$100	4%	10
5.	$500	10%	5
6.	$10	1%	50

7. Future Value A farmer rents his farm for $25,000 per year, payable at the end of each year. He can earn 8% interest compounded annually from a bank. What amount will he have to his credit after 10 years?

8. Future Value At the end of each year, Jane Johnson deposits $1000 in the Bank of Houston. The bank pays an annual rate of interest of 10% compounded annually. What is the value of Jane's account after 15 years? What is the total amount of interest that she will collect during these 15 years?

9. Future Value Find the future value of an annuity after 25 years if payments of $250 are made every quarter when interest is earned at an annual rate of 8% compounded quarterly?

For Problems 10–13, find the future value of the annuities described at the end of t years.

	Payment, R	Annual Interest Rate, r	Number of Periods per Year, k	Number of Years, t
10.	$500	8%	4	5
11.	$1000	12%	12	3
12.	$250	8%	2	10
13.	$400	6%	1	20

In Problems 14–17, determine the payments R of an annuity with future value F, annual interest rate r, number of payment periods per year k, and length of term t as given.

	Future Value, F	Annual Interest Rate, r	Number of Payments per Year, k	Number of Years, t
14.	$10,000	12%	1	10
15.	$7500	10%	2	5
16.	$25,000	8%	4	10
17.	$100,000	7%	1	25

18. Sinking Fund Payments Barbara agrees to pay Jerry $5000 five years from now. What annual payments should she make into a sinking fund that pays an annual interest rate of 8% compounded annually in order that she will have $5000 after five years?

19. Sinking Fund Payments Richard starts a sinking fund so that he will have $25,000 after 10.5 years. He decides to make payments into this fund every three months. What must his payments be if interest is being paid at an annual rate of 12% compounded quarterly?

20. Sinking Fund The Phoenix Printing Co. will need a new computer system in three years. The company would like to generate $250,000 in preparation for this expansion. What monthly payments should they make into a sinking fund that pays an annual rate of interest of 7% compounded monthly?

21. Sinking Fund Phyllis plans on taking a round-the-world cruise in five years and will need $25,000 available at that time. What quarterly payments should she make into an annuity paying an annual rate of interest of 9% compounded quarterly to achieve this goal?

22. Sinking Fund Parks' Hardware determines that it will need a second floor in five years. The cost of this construction project is estimated at $175,000. Mr. Parks sets up a sinking fund to pay for the project. What quarterly payments are necessary if the sinking fund pays an annual rate of interest of 8% compounded quarterly?

23. Sinking Fund A couple just graduating from college have decided to buy a new house. To make a down payment on a house, they set up a sinking fund in which they deposit $250 per month in a bank for three years. If the bank pays an annual rate of interest of 9% compounded monthly, and if the down payment is 20% of the cost of the house, how much can they pay for a house in three years?

INTERVIEW WITH AETNA LIFE & CASUALTY

The Aetna Life Insurance Company and Aetna Casualty and Surety Company of Hartford, Connecticut, is one of the world's largest life and casualty insurance companies.

Random House: When one thinks of annuities, one naturally thinks of life insurance companies. We have just studied the ordinary annuity, in which payments are made at the end of each period. I understand that the insurance industry uses a whole host of annuities.

Aetna: That is correct. The insurance industry offers a variety of products designed to meet the needs of individuals from all walks of life. For instance, the deferred annuity with payments commencing after a certain period of time is designed to meet the needs of retirees. The survivor annuity

TOPIC: Mathematics in the Insurance Industry

In Problems 24–28, determine the present value P of an annuity with payments R, interest rate per period i percent, and number of periods n as given.

	Payments, R	Interest Rate per Period, i	Number of Periods, n
24.	$500	3%	40
25	$250	8%	10
26.	$1000	1%	48
27.	$5000	2%	24
28.	$50,000	10%	5

29. **Present Value**   A life insurance company offers an optional settlement of cash or $1500 a month for ten years, paid in equal installments at the end of each month. What would be the equivalent cash settlement by the company if money earns interest at an annual rate of 12% compounded monthly?

30. **Present Value**   What lump sum would have to be invested at an annual interest rate of 9% compounded semiannually to provide an annuity that pays $200 every six months for the next ten years?

31. **Present Value**   Calculate the present value of an annuity of $500 per quarter for five years at an annual interest rate of 8% compounded quarterly.

32. **Present Value**   To pay the total amount for books, fees, and tuition to a private school, the school offers a year-long payment plan of $250 per month. How much would be needed at the start of the school year to make these payments if interest is paid at the annual rate of 10% compounded monthly?

33. **Present Value**   Find the present value of an annuity of $100 paid at the end of each month for 18 years if the annual interest rate is 9% compounded monthly.

34. **Present Value**   What lump sum would have to be invested at an annual interest rate of 9% compounded semiannually to provide an annuity that pays $200 every six months for the next ten years?

35. **Present Value**   A refrigerator can be purchased for $50 down and $30 a month for 12 months. What is the equivalent cash price of the refrigerator if the annual interest rate is 10% compounded monthly?

Comparing Investments

For Problems 36–38, compare the investment alternatives.

36. **Sinking Fund**   The RST Company must retire (pay off) $200,000 worth of bonds in 20 years. They have decided to use a sinking fund. They have found one that pays an

meets the needs of the spouse and children upon the untimely death of an income-earner. Some annuities provide monthly payments, while others offer annual payments. The variety of annuities is almost endless.

Random House: If a student is interested in the mathematics of finance, what are the professional possibilities in the insurance industry?

Aetna: The insurance industry provides all the opportunities available in most industries plus the unusual opportunity of becoming an actuary. For instance, a financier could work in the investment division dealing with stocks, bonds, and real estate. Further opportunities exist in the planning and economic areas of the company, where the financial expert deals with the effects of short- and long-range planning. Perhaps the most unique opportunity, however, lies in the actuarial field. Actuaries are insurance experts who apply their knowledge of

insurance, law, accounting, and investments as well as mathematical skills to the insurance industry. To become an actuarial scientist, one must take a rigorous series of exams, the first four of which focus on such mathematical skills as calculus, linear algebra, and operations research.

Random House: Briefly, could you describe a typical project that a young actuary might become involved with at Aetna?

Aetna: A typical project would be designing innovative and competitive insurance products. The products must be designed to meet the needs of a group of individuals at a fair price, while providing a sufficient return to the company's shareholders. The actuary might be asked to investigate legal considerations and estimate the product's future financial cash flow. The actuary would then determine the appropriate price for the product, keeping in mind the competitive constraints.

---

annual rate of interest of 6.5% compounded monthly and another that pays an annual rate of interest of 7% compounded quarterly. What would the payments per year be in each case? Which is the better investment?

37. **Present Value** A lot on Trout Lake is for sale for $10,000 down and 18 quarterly payments of $250. The owner will accept $13,000 cash for the lot. If money earns interest at the annual interest rate of 12% compounded quarterly, which is the better deal?

38. **Present Value** Is it cheaper to pay $8500 cash for a car or to pay $900 down and $275 per month until the loan

is paid off in 30 months? Assume that money can earn interest at an annual rate of 9% compounded monthly.

Something To Think About

39. **Giving up Smoking** Eddie decided when he entered college at the age of 18 to quit smoking and take the money spent on cigarettes and put it in a bank. If he puts $10 each week in a bank that pays annual interest of 8% compounded weekly, then how much will Eddie's bank account be worth by the time he turns 40 years of age? Make a guess.

3.4

The Amortization Method for Repaying Loans

PURPOSE

We show how to repay loans by the amortization method by finding

- the payments of an amortized loan,
- the payments applied to the principal,
- the payments applied to the interest,
- the amortization schedule, and
- the number of periods of an amortized loan.

We will compare the sinking fund method for repaying loans with the amortization method.

Introduction

The purchase of a new car or a home represents a different kind of financial transaction than we have seen thus far. In the purchase of a new car a bank loans the full amount to pay for the car, and then the borrower repays the loan in equal periodic payments. Loans that permit repayment in equal periodic installments are said to be **amortized**. Stated another way, the process of paying a debt by making a series of equal payments, from which a portion of each payment is applied to the interest due and the remaining portion of the payment is applied to the principal, is called **amortizing a loan**.

Amortized Loans

Suppose $10,000 is borrowed from a bank for the purpose of buying a new car. If the terms of the loan require that the borrower repay the loan in 60 equal monthly payments, then this would be an amortized loan. A major problem involving amortized loans, such as car loans or home mortgages, is the determination of the size of the periodic payments. To understand how these payments are determined, suppose you amortize a loan for the purpose of buying a car. What actually happens is that the bank gives you a lump sum or present value to purchase the car. You then repay the bank with a fixed number of equal payments. Suppose, however, that you take the lump sum, say P, from the bank and invest it in a second bank. This amount invested in the second bank would act as a present value P that would pay n annuity payments of value

$$R = \frac{P}{a_{\overline{n}|i}}$$

This is exactly the value of the payments that you must make to the first bank to make payments on the car. This discussion leads to the following result.

Loans for the purchase of a new car are usually amortized.

Payments on an Amortized Loan

If an amortized loan of P dollars is repaid in n equal payments to a lender who charges interest at a rate of i percent per payment period, then the amount R of each payment is given by

$$R = \frac{P}{a_{n|i}}$$

where

$R = $ value of payment

$P = $ amount borrowed

$n = $ number of payment periods

$i = $ interest rate per period

$$a_{n|i} = \frac{1 - (1+i)^{-n}}{i}$$

The total interest paid on the loan is given by

$$\text{Total interest paid} = nR - P$$

Example 1

Payments Monthly payments are to be made on an $8000 amortized car loan over a period of three years. For an annual interest rate of 9% compounded monthly, what are the payments for this loan?

Solution The values of the variables in the periodic payment formula are

$P = \$8000$ (amount borrowed)

$i = 0.09/12 = 0.0075$ (interest per payment period)

$n = 3 \cdot 12 = 36$ (number of payments)

Hence the monthly payments will be

$$R = \frac{P}{a_{n|i}}$$

$$= \frac{\$8000}{a_{36|0.0075}} \quad \text{(use calculator)}$$

$$= \$254.40$$

In other words, 36 monthly payments of $254.40 will retire the loan.

It is always interesting to compute the total interest that is paid on an amortized loan. It is the amount paid nR minus the amount borrowed P. In the above example it is

$$\text{Total interest paid} = nR - P$$

$$= 36(\$254.40) - \$8000$$

$$= \$1158.40 \qquad \square$$

When a family borrows money from a bank to buy a home, they often **mortgage** their home. What this means is that the family gives the bank claim on the home as security for payment of the loan.

Example 2

Buying a Home   When the Joneses bought a home, they took out a 25-year mortgage of $50,000 at 12% to be repaid with monthly payments. What are the payments for this mortgage?

Solution   The values of the variables in the periodic payment formula are

$$P = \$50,000 \qquad \text{(amount borrowed)}$$
$$i = 0.12/12 = 0.01 \qquad \text{(interest per payment period)}$$
$$n = 25 \cdot 12 = 300 \qquad \text{(number of payments)}$$

Hence the monthly payments are

$$R = \frac{P}{a_{\overline{n}|i}}$$

$$= \frac{\$50,000}{a_{\overline{300}|0.01}}$$

$$= \frac{\$50,000}{94.94655} \qquad \text{(use calculator)}$$

$$= \$526.61$$

The mortgage payments to the bank are $526.61 every month.   □

Using the periodic payment formula, we construct a table of monthly payments for a $50,000 amortized loan as a function of interest rates. We include in the table the total interest paid during the life of the loan and the recommended salary for persons taking out such a loan. As a rule of thumb, a home loan is affordable if no more than 28% of a person's gross income (before taxes) is needed to pay for loan payments, property taxes, and insurance premiums. In Table 5 we assume that taxes and insurance will be about 8% of the gross salary.

TABLE 5
Payments on a $50,000
Amortized Loan over 25 Years

Annual Interest Rate, r	Monthly Payments, R	Interest Paid, $nR - P$	Recommended Salary
8%	$386	$65,800	$23,160
9%	$420	$76,000	$25,200
10%	$454	$86,200	$27,240
11%	$490	$97,000	$29,403
12%	$527	$108,100	$31,620
13%	$564	$119,200	$33,840
14%	$602	$130,600	$36,120

Outstanding Principal on an Amortized Loan

During the time an amortized loan is being repaid, it is important to know after every payment exactly how much of the loan is still outstanding. A family making monthly house payments on a $30,000 loan would often like to know how much of the debt still must be repaid. It is necessary for the family to know this if they sell the home during the term of the loan.

The unpaid principal on a loan is simply the present value of the loan. It can be found by using the present value formula for an annuity with the number of payments being the number of remaining payments. This gives the following result.

Amount Still Owed on an Amortized Loan

The amount P of an amortized loan still owed when m of the n payments of R dollars each have been made is given by

$$P = Ra_{\overline{(n-m)}|i}$$

where

$$P = \text{amount of loan still owed}$$
$$R = \text{amount of each payment}$$
$$n = \text{total number of payments}$$
$$m = \text{number of payments made}$$

The amount of the loan that has been repaid after m payments is the amount of the loan minus the amount still owed.

Example 3

Outstanding Principal   Janet has obtained a $10,000 loan from a bank for the purpose of buying a new car. The loan is to be repaid in 48 monthly payments at an interest rate of 12% per year compounded monthly. Monthly payments are $263.34. How much money does Janet still owe the bank after 2.5 years?

Solution   The values of the variables in the amount owed formula are

$$R = \$263.34 \qquad \text{(amount of each payment)}$$
$$n = 48 \qquad \text{(total number of payments)}$$
$$m = (2.5)(12) = 30 \qquad \text{(number of payments made)}$$
$$n - m = 18 \qquad \text{(remaining payments)}$$
$$i = 0.12/12 = 0.01 \qquad \text{(interest per payment period)}$$

Hence we have

$$P = Ra_{\overline{(n-m)}|0.01}$$
$$= \$263.34 a_{\overline{18}|0.01}$$
$$= \$263.34(16.39827)$$
$$= \$4318.32$$

That is, Janet still owes the bank a balance of \$4318.32. The amount of the loan that Janet has repaid is

$$\begin{aligned} \text{Repaid amount} &= \$10{,}000 - \$4318.32 \\ &= \$5681.68 \end{aligned}$$ □

By using the present value formula for an annuity it is possible to determine the amount from each payment of an amortized loan that is applied to the principal. We begin by writing the expressions for the amount owed after $m-1$ and m payments. They are

$$\text{Amount owed after } m-1 \text{ payments} = Ra_{\overline{(n-m+1)}|i}$$
$$\text{Amount owed after } m \text{ payments} = Ra_{\overline{(n-m)}|i}$$

The difference between these two values is the amount of the mth payment that is applied to the principal.

Payment Against Principal

For an amortized loan being retired with n payments of R dollars each, and interest being charged at a rate of i percent per payment period, the amount of the mth payment that is applied to the principal is

$$\text{Payment against principal} = R(a_{\overline{n-m+1)}|i} - a_{\overline{(n-m)}|i})$$

where

$$\begin{aligned} R &= \text{periodic payments} \\ n &= \text{total number of payments} \\ m &= \text{the given payment (from 1 to } n) \\ i &= \text{interest rate per period} \end{aligned}$$

To find the portion of the payment that goes to the interest, subtract the payment against principal from the rent R.

Example 4

Amount Paid to Principal To buy a new home, the Petersons borrowed \$75,000 at 9% interest compounded monthly, which is amortized over 30 years in equal monthly payments. The monthly payments have been determined to be \$603.47. How much of the first payment goes to the principal and how much goes to the interest?

Solution We are given

$$\begin{aligned} R &= \$603.47 && \text{(periodic payments)} \\ n &= 30 \cdot 12 = 360 && \text{(total number of payments)} \\ m &= 1 && \text{(the given payment)} \\ i &= 0.09/12 = 0.0075 && \text{(interest per payment period)} \end{aligned}$$

Hence we have

$$\text{Payment against principal} = R[a_{\overline{(n-m+1)}|i} - a_{\overline{(n-m)}|i}]$$
$$= \$603.47(a_{\overline{360}|0.0075} - a_{\overline{359}|0.0075})$$
$$= \$603.47(124.28187 - 124.21398)$$
$$= \$40.97$$

In other words, $\$40.97$ of the first payment goes to the principal, and the remaining $\$562.50$ of the payment of $\$603.47$ goes to paying the interest.   □

Amortization Schedules

It is convenient when amortizing a loan to make a table listing the amount of each payment that goes to principal and interest. This table is called an **amortization schedule**. Most amortization schedules also list several other useful items, such as the accumulation of the interest and the principal payments as well as the remaining balance of the loan.

To illustrate the amortization schedule, suppose a loan for $\$1000$ is amortized by making 12 equal monthly payments. The bank charges an annual interest rate of 12% compounded monthly. The amortization schedule for this loan is shown in Table 6.

TABLE 6

Amortization Schedule for a $1000 Loan, Repayable (12 × $88.85) Monthly at 1% Per Month

Month Number	Monthly Payment	Payment to Principal	Cumulative Principal	Payment to Interest	Cumulative Interest	Unpaid Balance
1	$88.85	$78.85	$78.85	$10.00	$10.00	$921.15
2	$88.85	$79.64	$158.49	$9.21	$19.21	$841.51
3	$88.85	$80.43	$238.92	$8.42	$27.63	$761.08
4	$88.85	$81.24	$320.16	$7.61	$35.24	$679.84
5	$88.85	$82.05	$402.21	$6.80	$42.04	$597.79
6	$88.85	$82.87	$485.08	$5.98	$48.01	$514.92
7	$88.85	$83.70	$568.78	$5.15	$53.16	$431.22
8	$88.85	$84.54	$653.32	$4.31	$57.48	$346.68
9	$88.85	$85.38	$738.70	$3.47	$60.94	$261.30
10	$88.85	$86.24	$824.93	$2.61	$63.56	$175.07
11	$88.85	$87.10	$912.03	$1.75	$65.31	$87.97
12	$88.85	$87.97	$1000.00	$.88	$66.19	000

The flow diagram in Figure 15 can be used to compute an amortization schedule.

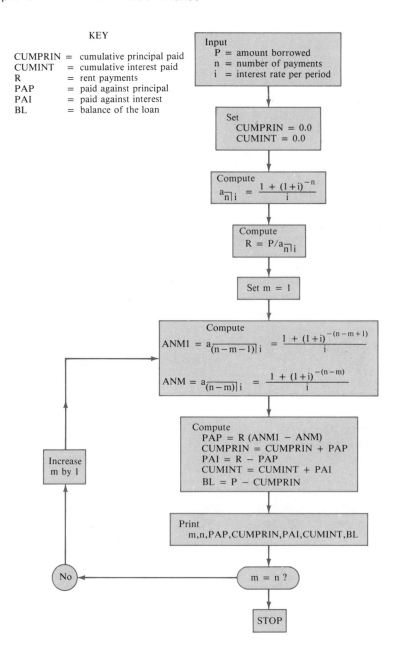

KEY

CUMPRIN = cumulative principal paid
CUMINT = cumulative interest paid
R   = rent payments
PAP   = paid against principal
PAI   = paid against interest
BL   = balance of the loan

Figure 15
Flow diagram of an amortization schedule

This flow diagram was programmed in BASIC, as follows:

```
REM    PROGRAM TO COMPUTE AMORTIZATION SCHEDULE
REM    INPUT P = AMOUNT BORROWED
REM          N = NUMBER OF PAYMENTS
REM          I = INTEREST RATE PER PERIOD
PRINT 'INPUT THE AMOUNT BORROWED, THE NUMBER OF'
PRINT 'PAYMENTS AND THE INTEREST RATE PER PERIOD'
INPUT P,N,I
REM    INITIALIZE CUMULATIVE PRINCIPAL AND INTEREST
CUMPRIN = 0.0
```

```
CUMINT = 0.0
ANI = (1 - (1 + I)^( -N) )/I
REM    COMPUTE RENT PAYMENTS
R = P/ANI
FOR M = 1 TO N
  ANM1 = (1 - (1 + I)^( -N + M + 1) )/I
  ANM  = (1 - (1 + I)^( -N + M) )/I
  REM      COMPUTE PAYMENT AGAINST PRINCIPAL
  PAP = R*(ANM1 - ANM)
  CUMPRIN = CUMPRIN + PAP
  REM      COMPUTE PAYMENT AGAINST INTEREST
  PAI = R - PAP
  CUMINT = CUMINT + PAI
  REM      COMPUTE BALANCE OF LOAN
  BL = P - CUMPRIN
  PRINT M, R, PAP, CUMPRIN, PAI, CUMINT, BL
NEXT M
END
```

Number of Payments for an Amortized Loan

A final calculation using the present value formula

$$P = Ra_{\overline{n}|i}$$

$$= R \frac{1 - (1 + i)^{-n}}{i}$$

arises when the payments R for an amortized loan are specified in advance. For example, it may be determined that only \$650 per month is available for making payments. If a \$60,000 home is to be purchased, we would like to know the number of payments required to pay off this loan. To find the number of payments required, solve the present value equation for n. Doing this, we get the following result.

Finding the Number of Payments

To retire a loan of P dollars with payments of R dollars each and an interest rate of i percent per period will require n periods, where

$$n = \frac{\log\left[\dfrac{R}{R - Pi}\right]}{\log(1 + i)}$$

where

n = number of payments

R = amount of each payment

P = amount borrowed

i = interest rate per payment period

The logarithm in the formula above is taken as the common logarithm.

Example 5

Length of a Loan   To buy a new car, a graduating senior borrows $10,000 at an annual interest rate of 13% compounded monthly. The senior determines that payments of $225 per month can be afforded. How long will it take her to retire this loan?

Solution   The values of the variables in the number of payments formula are:

$$R = \$225 \qquad \text{(monthly payments)}$$
$$P = \$10,000 \qquad \text{(amount borrowed)}$$
$$i = 0.13/12 = 0.010833 \qquad \text{(interest rate per period)}$$

Hence we have

$$n = \frac{\log\left[\dfrac{R}{R - Pi}\right]}{\log\left(1 + i\right)}$$

$$= 60.95$$

In other words, it will take 61 months (five years and one month) to retire this loan with payments of $225 per month. It is interesting to note that the total interest paid on this $10,000 loan is

$$\text{Total interest paid on loan} = nR - P$$
$$= (60.95)(\$225) - \$10,000$$
$$= \$13,713.75 - \$10,000$$
$$= \$3713.75$$

It's like Ben Franklin said, "Time is money."   □

Problems

For Problems 1–6, determine the periodic payment for the amortized loans described.

	Principal	Number of Periods, n	Rate of Interest per Period, i
1.	$10,000	10	12%
2.	$5000	24	9%
3.	$7500	24	2%
4.	$15,000	12	1%
5.	$20,000	5	8%
6.	$50,000	48	2%

For Problems 7–12, given the amortized loans of R dollars with n payment periods and rate of interest per period i, determine for the mth payment the following quantities: (a) amount still owed, (b) amount paid to the principal, (c) amount paid to interest.

7. $R = \$175,$ $i = 10\%,$ $n = 20,$ $m = 1$

8. $R = \$200,$ $i = 8\%,$ $n = 10,$ $m = 5$

9. $R = \$250,$ $i = 2\%,$ $n = 24,$ $m = 24$

10. $R = \$685,$ $i = 1\%,$ $n = 48,$ $m = 10$

11. $R = \$735,$ $i = 2\%,$ $n = 36,$ $m = 5$

12. $R = \$500,$ $i = 3\%,$ $n = 48,$ $m = 1$

For Problems 13–18, determine the number of payments needed to retire the amortized loans described. A hand calculator will be useful.

	Value of Payments, R	Present Value, P	Rate of Interest per Period, i
13.	$500	$3000	15%
14.	$1750	$15,000	9%
15.	$3000	$20,000	12%
16.	$400	$4000	8%
17.	$800	$7500	9%
18.	$600	$10,500	2%

19. **Payments**   Phil Locke borrows $3500 for two years at an annual rate of interest of 12% compounded monthly and promises to repay the amount, principal, and interest by making equal monthly payments. What is the amount of the monthly payments?

20. **Payments**   Joe Jones borrows $10,000 to buy a new car. The bank charges 12% annual interest compounded monthly and requires that Joe repay the loan by making 30 equal monthly payments. How much is each payment?

21. **Payments**   A debt of $10,000 bearing interest at 2% per payment period is to be paid, principal and interest, in ten equal installments. What is the payment per period of this loan?

22. **Payments**   Ocean View Properties sells lots with an ocean view for $37,000. The financial package offered by the company is for the buyers to make quarterly payments for five years at an annual rate of interest of 16% compounded quarterly. What are the quarterly payments?

23. **Comparing Rents**   An appliance company offers two financing plans on purchases of more than $600. The first is 30 monthly payments at an annual interest rate of 12%. The second is 48 monthly payments at an annual interest rate of 10%. For a $750 purchase, what would the payments be in each case?

24. **Payments**   The Smiths are retiring to Ft. Myers, Florida, and have financed a new $75,000 condominium with a 30-year mortgage, which they agree to repay in equal monthly payments. The bank charges 9% annual interest compounded monthly. What are the monthly payments?

25. **Principal Still Owed**   A new car loan for $7500 from the Central Credit Union is amortized over 40 months at 16% annual interest compounded quarterly. What are the quarterly payments? How much of the loan is still owed after 20 months? How much has the borrower paid?

26. **Principal Still Owed**   The Smiths have retired and moved to Ft. Myers, Florida. They have borrowed $75,000 to buy a condominium and have promised to repay the loan by making equal monthly payments for the next 30 years. If the bank charges 9% annual interest compounded monthly, how much of the loan is still unpaid after five years? How much of the loan have they paid by this time?

27. **Principal Still Owed**   A family purchases a $100,000 home and makes a down payment of $10,000. They then get a mortgage for the balance at an annual interest rate of 9% compounded monthly to be repaid in monthly installments over a 20-year period. What are the monthly payments on this mortgage and how much of the principal is still owed after five years (60 payments)? How much of the loan have they paid by this time?

28. **Principal Still Owed**   A home loan for $75,000 is amortized over 25 years after a 20% down payment. Interest is charged at an annual rate of 10.5% compounded monthly. What are the monthly payments and how much of the principal remains to be repaid after 10 years? How much of the loan has been repaid by this time?

29. **Payment Going to Principal**   The Youngs buy a $100,000 home with a $75,000 mortgage for 25 years at an annual interest rate of 10% compounded monthly. They agree to repay the loan by making equal monthly payments. What are the monthly payments? How much of the first payment goes to pay off the loan, and how much goes to pay for the interest?

30. **Payment Going to Principal**   Ed amortizes a $15,000 loan to buy a new car and plans on repaying the bank by making 36 equal payments. The bank charges an annual interest rate of 10% compounded monthly. How much of the last payment goes to repay the loan and how much pays for the interest

31. **Number of Payments**   Jane has borrowed $8000 to buy a new car and would like to make monthly payments of $200. If the bank charges an annual interest rate of 12% compounded monthly, how long will it take to pay off this loan?

32. **Number of Payments**   How many monthly payments of $300 are required to amortize a $5000 loan if the annual interest rate is 10% compounded monthly?

33. **Amortization Schedule**   The amortized schedule in Table 7 shows the process of a $1000 amortized loan over five payment periods if interest charged is 6% per period.
    (a) What was the total amount paid for interest?
    (b) What is the sum of the numbers in the "Payment to Principal" column?
    (c) Why are the numbers in the "Payment to Interest" column getting smaller?

34. **Amortization Schedule**   Make an amortization schedule for a $100 amortized loan if the loan is to be repaid in four equal payments. Interest charged per payment period is 2%.

TABLE 7
Amortization Schedule for Problem 33

Payment Number	Total Payment	Payment to Principal	Payment to Interest	Balance of Loan
1	$237.40	$177.40	$60.00	$822.60
2	$237.40	$188.04	$49.36	$634.56
3	$237.40	$199.33	$38.07	$435.23
4	$237.40	$211.29	$26.11	$223.94
5	$237.40	$223.94	$13.46	$000

Computer-Generated Amortization Schedules

For Problems 35–40, use the BASIC program on page 262 or write a new computer program to construct an amortization schedule for the following amortized loans.

	Amount Loaned	Annual Interest Rate	Payments	Length of Loan
35.	$100	25%	monthly	1 year
36.	$10,000	18%	monthly	5 years
37.	$50,000	12%	annually	30 years
38.	$100,000	9%	monthly	20 years
39.	$200,000	10%	semiannually	30 years
40.	$500,000	8%	annually	25 years

Epilogue: The PinRF Variables and the Seven Equations

The first time the mathematics of finance is studied, the student may get the impression that the subject is an endless parade of equations relating dozens of variables. The fact of the matter is, the mathematics of finance is essentially the interplay of only five variables and seven equations. The five variables are the *PinRF variables.* They are

1. present value (P)
2. interest rate per period (i)
3. number of periods (n)
4. value of payments (R)
5. future value (F)

We write "PinRF" because money begins with a present value P and, with the help of i, n, and R, arrives at a future value F.

An interesting conclusion of the mathematics of finance is that most of the results are derived from *seven basic equations.* The seven equations are as follows.

1. $F = P(1 + i)^n$ (future value of compounded amount)
2. $P = \dfrac{F}{(1 + i)^n}$ (present value of compounded amount)
3. $R = P\dfrac{i}{1 - (1 + i)^{-n}}$ (withdrawals from an annuity)
4. $P = R\dfrac{1 - (1 + i)^{-n}}{i}$ (present value of an annuity)
5. $R = F\dfrac{i}{(1 + i)^n - 1}$ (payments in an annuity)
6. $F = R\dfrac{(1 + i)^n - 1}{i}$ (future value of an annuity)
7. $n = \dfrac{\log [R/(R - Pi)]}{\log (1 + i)}$ (number of payments in an amortized loan)

Key Ideas for Review

$a_{\overline{n}|\,i}$, 251
amortization schedule, 261
amortizing a loan, 256
annuity, 245
bank discount, 225
compound amount, 232
compound interest, 232
compounding period, 232
continuous compounding, 238
discount note, 224
doubling time
  commpound interest, 235
  simple interest, 224

effective rate of interest, 226
future value
  annuity, 246
  compound interest, 232
  continuous compounding, 238
  simple interest, 222
geometric series, 246
interest
  compound, 230
  effective rates, 226
  nominal rates, 226
  simple, 220
maturity value, 225

payment against interest, 261
payment against principal, 260
periodic payments
  amortization, 257
  annuity, 248, 252
present value
  annuity, 251
  compound interest, 236
  simple interest, 223
proceeds, 225
$s_{\overline{n}|}$, 247
sinking fund, 249
term of a loan, 219

Review Exercises

Simple Interest

1. **Interest Owed** Find the interest owed on $2500 borrowed for four months at 10% simple interest.
2. **Time of Loan** If $960 interest was paid on a loan of $21,000 at 8% simple interest, for what length of time was the money borrowed?
3. **Rate of Interest** If $325 interest was paid on a loan of $6000 for nine months, what was the rate of interest?
4. **Simple Discount** Find the discount on $15,000 due in ten months if simple interest is charged at 9%.
5. **Proceeds** Find the proceeds for a simple discount (promissory) note for $14,000 due in 30 months at 10% simple discount.
6. **Proceeds** What are the proceeds of a $10,000 loan over six months when the interest rate is 12% simple discount?
7. **Proceeds** What are the proceeds of a $5000 loan over four months when the interest rate is 9% simple discount?
8. **Discount** Find the discount for a bank note of $1500 charging 12% simple interest written on Octorber 15 and paid in full on the next July 15.
9. **Proceeds** What are the proceeds of a $10,000 loan over six months when the interest rate is 10% simple discount?
10. **Present Value of Simple Interest** What amount of money would be invested now at 8% simple interest to yield a future value of $10,000 in nine months?
11. **Present Value of Simple Interest** Sally needs $750 in seven months. Joe offers her 12% simple interest on her money. How much should she loan him in order to reach her financial objective?

Compound Interest

12. **Present Value** What sum of money should be invested for five years at an annual interest rate of 9% compounded quarterly to provide a compound amount of $10,000?
13. **Present Value and Interest Earned** What sum of money should be invested for ten years at an annual interest rate of 8% compounded monthly to provide a future value of $100,000? How much interest will be earned?
14. **Effective Annual Rate of Interest** Is an interest rate of 8.25% per year compounded annually a better rate of return than an interest rate of 8% per year compounded semiannually?
15. **Interest Rates** A sum of money triples in 20 years. What rate of interest compounded annually would cause this to happen? What would be the annual rate of interest if it takes 11 years to triple the original amount?
16. **Time for Growth** How long will it take for $2500 to grow to $6500 if the original amount is invested at an annual interest rate of 9% per year compounded monthly?
17. **Present Value** A child will inherit $25,000 on her eighteenth birthday. If money grows at the rate 10% per year compounded monthly, what is the present value of the inheritance when the child turns 12?
18. **Future Value** Ed makes payments of $2000 per year into a retirement annuity. How much will this retirement fund be worth in 25 years if interest is paid at the annual rate of 9% per year compounded annually?
19. **Effective Rate of Interest** What is the effective rate of interest for an annual interest rate of 18% compounded monthly? This is the rate charged by many credit card companies.

20. Future Value An account of $7500 is established when a child is five years old. If the account earns an annual interest rate of 8% compounded monthly, how much will the account be worth on the child's eighteenth birthday?

21. Present Value How much must be deposited in an account on a child's sixth birthday if $20,000 is needed when the child is 18? Assume that money earns interest at the annual rate of 9% compounded semiannually.

Annuities

22. Future Value When her son was one year old, a mother started a savings account by depositing $100 to the son's account. She kept this up each year until the twentieth payment was made. The bank paid an annual interest rate of 7% compounded annually. What was the value of the account after the twentieth payment was made?

23. Sinking Fund A retirement bonus of $475,000 for 30 employees is needed in ten years. What annual payments should be made into a sinking fund that earns an annual interest rate of 7% compounded annually to ensure that the amount will be available?

24. Present Value A trust fund will pay five heirs $500 per month for 20 years. If money earns an annual interest rate of 6% per year compounded monthly, what amount is needed to support this trust?

25. Length of Term A down payment of $15,000 is needed to purchase a house. If payments of $250 are made each month into an annuity earning an annual interest rate of 8% compounded monthly, how long will it take to generate the down payment?

26. Future Value Denise deposits $200 every six months in an account paying an annual interest rate of 9% compounded semiannually. What is the value of the account after ten years? How much will Denise have earned in interest in that time?

Amortization

27. Amortization Schedule Construct an amortization schedule for a loan of $500 for one year at an annual interest rate of 10% compounded quarterly.

28. Payments What are the monthly payments to amortize a $60,000 loan over 25 years at an annual interest rate of 10% compounded monthly?

29. Payments and Balance Due A loan of $7500 is amortized over five years at an annual interest rate of 9% compounded monthly. What are the monthly payments? How much is still owed after three years?

30. Payments A company amortizes a $20,000 loan over ten years at an annual interest rate of 12% compounded monthly. The company repays the loan by making monthly payments for ten years. What are the montly payments for this loan?

31. Balance Due To purchase a $100,000 home, the Snows amortize a loan over 30 years by making 360 monthly payments of $800. How much of the $100,000 loan is still unpaid after the 180th payment has been made if the annual rate of interest is 9% compounded monthly?

32. Payment Against Principal In Problem 31, how much of the 180th payment will go toward the principal of the loan? How much will go toward the payment of the interest?

33. Number of Payments To buy a new boat, Jim amortizes a $15,000 loan in which the bank charges an annual rate of interest of 12% compounded monthly. So that the monthly payments are kept to only $100, how many payments must Jim make to retire the loan? Does the answer seem reasonable?

Chapter Test

1. Find the interest on $2500 borrowed for four months at 10% per year simple interest.

2. If $960 was paid on a loan of $21,000 at 8% simple interest, for what length of time was the money borrowed?

3. Calculate the amount earned on $5000, if invested at an annual interest rate of 15% compounded monthly for three years?

4. What is the effective rate of interest if the nominal rate is 9% compounded monthly?

5. An alumnus is giving City University a gift sufficient to provide for an annual award of $1000 for the next 20 years. The money can be invested in a bank paying an annual interest rate of 8% compounded annually. How large must the gift be?

6. Find the future value of an annuity with payments of $150 per month for ten years if money earns an annual interest rate of 8% compounded monthly.

7. Determine the monthly payments to repay an amortized loan of $50,000 over 25 years if the annual rate of interest is 12% compounded monthly.

4

Sets and
Counting

Although sets were not formally introduced into mathematics until the late nineteenth century, they now form the foundation on which most of mathematics is built. In this chapter we introduce the necessary background in sets and counting that is needed for a rigorous study of probability. In particular, we will learn how to count the number of elements in complicated sets by the use of the multiplication principle. We will then use the multiplication principle to count permutations and combinations.

4.1 Introduction to the Algebra of Sets

PURPOSE

We present a brief introduction to the algebra of sets. The basic topics in the section are

- the meaning of a set,
- membership in a set $a \in A$,
- subsets of sets $A \subseteq B$,
- the universal set U,
- Venn diagrams,
- union of sets $A \cup B$,
- intersection of sets $A \cap B$,
- complement of a set A', and
- De Morgan's laws: $(A \cup B)' = A' \cap B'$ and $(A \cap B)' = A' \cup B'$.

Sets and Their Origins

The definition of a set used here is an intuitive one.

> **Meaning of a Set**
>
> A **set** is any collection of objects. The objects that form the set are called **elements** or **members** of the set.

Logicians would object to such an intuitive definition as this. For one thing the definition is circular. If a set is a collection of objects, then just what is a

Georg Cantor

HISTORICAL NOTE

The first person to realize the importance of sets as a mathematical subject for study was the German mathematician Georg Cantor (1845–1918). Cantor's first paper on the use of sets, published in 1874, was very controversial because it was innovative and differed with the thinking of the time. To understand the trouble surrounding Cantor's set theory, one must understand the social climate of the time. Most scientists were conservative and proud of the successes of science. Many scientists thought that most scientific facts had already been discovered. One scientist went so far as to say, "Most physical principles have already been uncovered with the exception of a few refinements here and there."

Leopold Kronecker attacked Cantor's set theory as more theology than mathematics. Although Kronecker's animosity toward Cantor was based on scientific fact, more often than not the attack became petty and degenerated into a personal vendetta. The hypersensitive Cantor finally could take no more. At the age of 40, he suffered a mental breakdown and briefly entered a mental institution at Halle. Although he made many of his major contributions to set theory after that time, he suffered nervous breakdowns for the last 33 years of his life. Cantor died despondent in a mental institution in 1918 at the age of 74.

collection of objects? A collection of objects is merely a group of things. But a group of things is an aggregate, which might be called a heap, which could be called a bunch, which could be called a set.

Notation

We denote sets by capital letters A, B, C, and so on; members of sets are denoted by lowercase letters a, b, c, and so on.

Membership in a Set

If an element a is a member of a set A, we write $a \in A$. If an object a is *not* a member of a set A, we write $a \notin A$. We read $a \in A$ as "a is an element of A." (See Figure 1.)

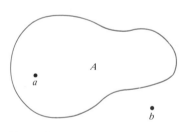

Figure 1
$a \in A$ but $b \notin A$

Representation of Sets

There are two standard ways for representing sets.

1. **Roster Method.** The most common way, the roster method, lists the elements of a set in curly braces.

$$A = \{a, b, c\} \qquad (A \text{ contains the three elements } a, b, \text{ and } c)$$

$$B = \{1, 2, \ldots, 99, 100\} \qquad (B \text{ is understood to be the integers from 1 to 100})$$

$$C = \{1, 2, 3, \ldots\} \qquad (C \text{ is understood to contain all the integers greater than or equal to 1})$$

An element is not repeated in a set. For example, the sets $\{a, a\}$ and $\{a\}$ are considered to be the same set. Similarly, we have the same sets $\{a, b, a, c, a\} = \{a, b, c\}$.

2. **Defining Property Method.** It is often convenient to describe a set by using the notation

$$A = \{x \mid x \text{ satisfies certain properties}\}$$

the set of all x　　*such that*　　*x satisfies these properties*

The elements of the set are not listed, but well-defined membership properties are given against which elements can be tested to determine whether they belong to the set.

$A = \{x \mid -1 \le x \le 1\}$　　(*A* is the set of real numbers in the closed interval $[-1, 1]$)

$B = \{x \mid x \text{ is a positive integer}\}$　　(*B* is the set of integers 1, 2, 3, . . .)

$C = \{(x, y) \mid x^2 + y^2 \le 1\}$　　(*C* is the set of points on and inside the unit circle)

Note that it would be impossible to use the roster method to list the points of the unit circle.

Universal Set

The set of all things under discussion at any given time is called the **universal set**, denoted by U. Everything outside of U is excluded from consideration. To a political scientist who is studying voting patterns of U.S. senators, the universal set is the set of U.S. senators. Anyone not in this set would not be considered in the study.

Subsets

If every element of a set A also belongs to a set B, we say that A is a **subset** of B and denote this by $A \subseteq B$. (See Figure 2.) If there is at least one element of C that does not belong to D, then C is not a subset of D; we denote this by $C \nsubseteq D$.

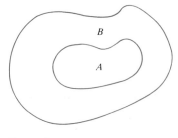

Figure 2
A is a subset of B

Example 1 ──────────── Subsets and Nonsubsets

(a) $\{a, b, c\} \subseteq \{a, b, c, d\}$

(b) $\{5, 2, 0, -1\} \subseteq \{2, 6, 9, 5, -1, \text{Chicago}, 0\}$

(c) $\{x \mid 0 < x < 1\} \subseteq \{x \mid 0 < x < 1\}$

(d) $\{x \mid x \text{ is a solution of } x^2 - 1 = 0\} \subseteq \{0, -1, 1\}$

(e) $\{\text{Chicago, Denver, Richmond}\} \nsubseteq \{x \mid x \text{ is a state capital}\}$

(f) $\{\text{Notre Dame}\} \nsubseteq \{x \mid x \text{ is a past NCAA wrestling champion}\}$　　□

It's not much, but you'd be surprised how useful it is!

ϕ

Empty Set

It is convenient to talk about a set $\varnothing$ that contains no elements. The set that contains no elements is called the **empty set**. The empty set is a subset of every other set (including itself).

Example 2

Subsets of a Set of Size 3   Find the subsets of $A = \{a, b, c\}$.

Solution   The strategy is to enumerate systematically the subsets of A that have 0, 1, 2, and 3 elements. This gives the eight subsets in Table 1.

TABLE 1
Subsets of $A = \{a, b, c\}$

Number of Elements in Subset	Subsets of $\{a, b, c, d\}$
0	$\varnothing$
1	$\{a\}, \{b\}, \{c\}$
2	$\{a, b\}, \{a, c\}, \{b, c\}$
3	$\{a, b, c\}$

Equality of Sets

Two sets A and B are equal, denoted $A = B$, when A and B contain exactly the same elements.

Example 3

Equality of Sets

(a) $\{x \mid x^2 - 1 = 0\} = \{1, -1\}$
(b) $\{x \mid x \text{ is the smallest state capital}\} = \{\text{Carson City}\}$
(c) $\{x \mid x \text{ is the largest state capital}\} = \{\text{Boston}\}$
(d) $\{a, c, b, d\} = \{a, b, c, d\}$

We now study a number of useful operations that can be performed on sets.

Union of Sets

The union of two sets A and B, denoted $A \cup B$, is the set of all elements that belong to either A or B (or both).

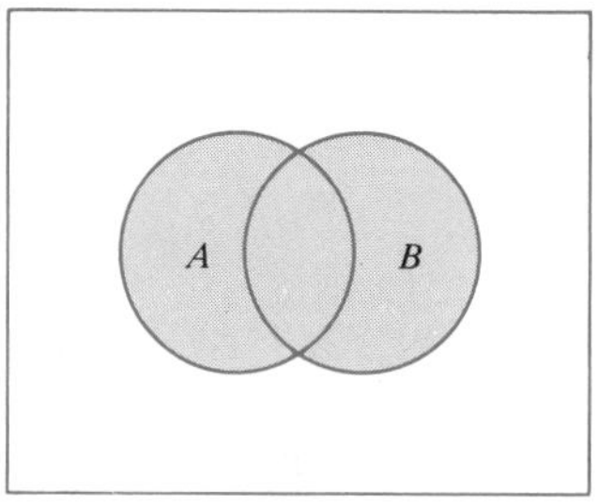

> **Union of Sets**
> $$A \cup B = \{x \mid x \in A \text{ or } x \in B\}$$

Figure 3
Union of two sets

This can be visualized by means of the **Venn diagram** in Figure 3.

> **Venn Diagrams**
>
> Venn diagrams are drawings that illustrate abstract ideas. The universal set U is represented by a rectangle, and subsets of U by circles inside U. Generally, the circles are drawn so that they overlap to illustrate set theory concepts. Venn diagrams can be drawn for more than two subsets of U, although they may become quite complicated for more than three subsets.

Example 4

Unions

(a) $\{x \mid 0 < x < 1\} \cup \{x \mid 0.5 < x < 5\} = \{x \mid 0 < x < 5\}$
(b) $\{\text{Cleveland}, 6, 2, 0\} \cup \{0, 2, c\} = \{\text{Cleveland}, 6, 2, 0, c\}$
(c) $\{a, b\} \cup \{a\} = \{a, b\}$ □

The Intersection of Two Sets

The intersection of two sets A and B, denoted $A \cap B$, consists of the set of all elements belonging to both A and B.

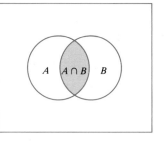

Figure 4
Intersection of two sets

> **Intersection of Sets**
>
> $$A \cap B = \{x \mid x \in A \text{ and } x \in B\}$$

We can illustrate the idea of the intersection of two sets by means of the Venn diagram in Figure 4.

Example 5

$[0, 1] \cap [0.5, 2] = [0.5, 1]$

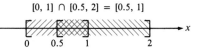

Figure 5
Intersection of two intervals

Intersection of Sets

(a) $\{x \mid 0 \leq x \leq 1\} \cap \{x \mid 0.5 \leq x \leq 2\} = \{x \mid 0.5 \leq x \leq 1\}$ (See Figure 5.)
(b) $\{a, b, c, d\} \cap \{e, f, g, h\} = \varnothing$
(c) $\{a, b, c, d\} \cap \{a, b, c\} = \{a, b, c\}$ □

Disjoint Sets

If A and B have no elements in common, they are called **disjoint**. (See Figure 6.) Disjoint sets A and B satisfy the property that $A \cap B = \varnothing$.

Example 6

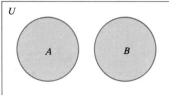

Figure 6
Two disjoint sets

Disjoint Sets

(a) $\{(x, y) \mid x + y = 1\} \cap \{(x, y) \mid x + y = 2\} = \varnothing$
(b) $\{(x, y) \mid x + y < 1\} \cap \{(x, y) \mid x + y > 1)\} = \varnothing$
(c) $\{a, b, c\} \cap \{d, e, f\} = \varnothing$
(d) $\{1, 3, 5, 7\} \cap \{2, 4, 6, 8\} = \varnothing$ □

Complement of a Set

The **complement of a set** A, denoted A', is the set of elements in the universal set U that do *not* belong to A. (See Figure 7.)

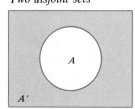

Figure 7
Complement of a set

> **Complement of a Set**
>
> $$A' = \{x \in U \mid x \notin A\}$$

The rules relating how the union, intersection, and complement of two sets interact are **De Morgan's Laws**.

> **De Morgan's Laws**
>
> If A and B are any two sets, then
>
> - $(A \cup B)' = A' \cap B'$ (the complement of the union is the intersection of the complements).
> - $(A \cap B)' = A' \cup B'$ (the complement of the intersection is the union of the complements).

Figure 8
Collection of Venn diagrams

Some common sets are illustrated in Figure 8 by means of Venn diagrams.

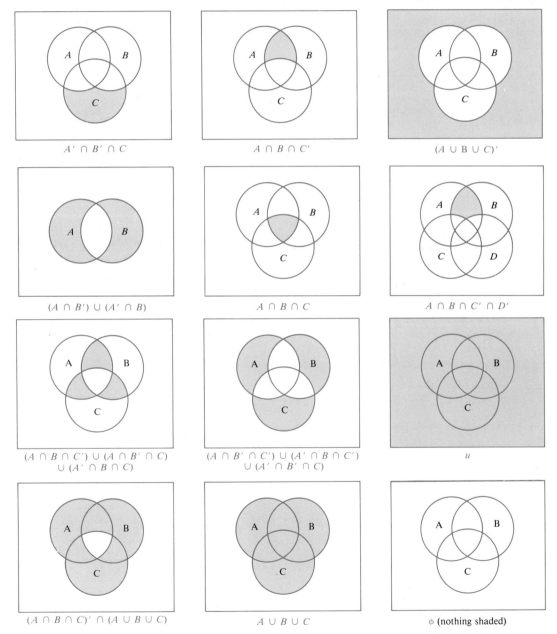

Augustus De Morgan

HISTORICAL NOTE

Both Augustus De Morgan (1806–1871) and John Venn (1834–1923) were English logicians who made major contributions to logic and probability. John Venn was an ordained minister but resigned his ministry in 1883 to concentrate on logic, which he taught at Cambridge. The diagrams for which he is remembered were actually used earlier by the Swiss mathematician Leonhard Euler (and are sometimes called Euler-Venn diagrams), but they were perfected by Venn.

De Morgan was a brilliant mathematician who introduced the slash notation for representing fractions, such as 1/2 and 3/4. Once asked when he was born, De Morgan replied, "I was once x years old in the year x^2." Can you determine the year he was born?

Problems

For Problems 1–8, use the roster method or defining property to represent the following collection of objects.

1. All even integers from 2 to 20.
2. All odd integers from 1 to 19.
3. All subsets of the set $\{1, 2, 3, 4\}$.
4. All points in the xy-plane in the first quadrant.
5. All points in the xy-plane with integer coordinates.
6. All 3×3 matrices with real numbers as entries.
7. All 3×3 matrices with 1's on the main diagonal and integer values in other locations.
8. All feasible points in the xy-plane satisfying

$$x + 2y \leq 2$$
$$2x + y \leq 2$$
$$x \geq 0$$
$$y \geq 0$$

For Problems 9–18, shade the specified set in Figure 9.

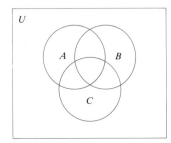

Figure 9
Venn diagram for Problems 9–18

9. $A \cup B$
10. $A \cup C$
11. $A \cap B$
12. $A \cap (B \cup C)'$
13. $A' \cap B' \cap C'$
14. $A' \cap B \cap C'$
15. $A \cup (A \cap C)'$
16. $(A \cup B) \cap (A \cup C)$
17. $A \cap (B \cup C)$
18. $(A \cap B) \cup (A \cap C)$
19. Draw the Venn diagrams for $A \cup (B \cap C)$, $A \cup B$, and $A \cup C$. What is true about $A \cup (B \cap C)$ and $(A \cup B) \cap (A \cup C)$?
20. Draw the Venn diagram for $A \cap (B \cup C)$. What is true about $A \cap (B \cup C)$ and $(A \cap B) \cup (A \cap C)$?

True or False

For Problems 21–40, place a check in the column labeled T if the statement is true; otherwise, place a check in the column labeled F.

	T	F
21. $0 \in \{x \mid x^2 = 1\}$	___	___
22. $\{1, 2, 3\} = \{3, 2, 1\}$	___	___
23. $2 \in \{1, 2, 3\}$	___	___
24. $\{\text{Notre Dame}\} = \varnothing$	___	___
25. $2 \in \{1, \text{Joan Benoit}, 2\}$	___	___
26. $\{0, 1, 2\} = \{1, 0, 2\}$	___	___
27. $\varnothing \subseteq \{\varnothing\}$	___	___
28. $\varnothing \in \{\varnothing\}$	___	___
29. $A \subseteq A \cup B$	___	___
30. $A \cup B \subseteq B$	___	___
31. $A \cap B = A' \cup B'$	___	___

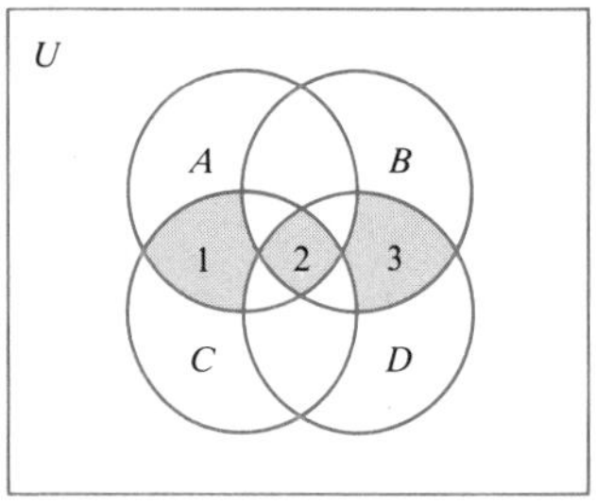

Figure 10
Venn diagram for Problems 41–45

32. $(A \cup B)' = A' \cup B'$ T   F

33. $\{2, 2, 2, 2, 2, 2, 2, 2, 2\} = \{2\}$

34. If $A \subseteq B$, then $B' \subseteq A'$

35. $A \cup B \subseteq A \cap B$

36. All subsets of $\{a, b\}$ are $\varnothing, \{a\}, \{b\}, \{a, b\}$

37. $A \cap B \subseteq A \cup B$

38. $\{1, 1, 2\} = \{1, 2\}$

39. $\{1, 2, 3\} \cap \{1, 3\} = \{1, 3\}$

40. $\{x \mid x^2 = 1\} \cap \{0, 1, 2, 3, 4\} = \varnothing$

Problems 41–45 refer to the Venn diagram in Figure 10.

     T   F

41. Region 1 is the set $(A \cup C) \cap (B' \cap D')$.

42. Region 2 is $A \cap B \cap C \cap D$.

     T   F

43. Region 3 is a subset of $A \cap C$.

44. Region 3 is a subset of $C \cap D$.

45. Region 3 is $A' \cap B' \cap C \cap D$.

4.2

Four Counting Techniques

PURPOSE

We introduce some techniques for counting the number of elements in a set. The four major techniques studied are the following:

- inclusion/exclusion principle,
- Venn diagrams,
- tree diagrams, and
- the multiplication principle.

Introduction

Counting is so commonplace that we do not give it a second thought. If you were going to count the number of students in your math class, you would probably mentally point to the students one at a time, saying to yourself: one, two, three, . . . until you were done. Counting in this way is not too different from the way the Babylonians counted 4000 years ago. The concepts of number and counting were developed so long ago that their origins are mostly conjectural. Early counting probably consisted of making a one-to-one correspondence between one's goats and a pile of stones. Every time a goat was born, one stone was added to the pile. When a goat was butchered or died, a stone was removed from the pile. This is essentially how you counted your classmates, except that you assigned to each student a number instead of a stone.

As we will see, there are other sets that cannot as easily be counted in a one-by-one manner. Consider the following problem.

The Coach's Problem

The University of Maine women's basketball team has twelve players on the traveling squad. A basketball team consists of a center, two forwards, and two guards; these are called the three different positions of the team. The women playing each position are given in Table 2.

TABLE 2
Positions of the Players on
the Basketball Team

Centers	Forwards	Guards
Ellen	Peggy	Ann
Susan	Barbara	Anita
	Jane	Nancy
	Karen	Wendy
	Betty	Jean

Can you determine how many different teams the coach can put on the court? (Each player plays only her position.) It may be surprising to learn that the coach can play 200 different teams. We will learn how to determine this number in this chapter using basic counting principles.

We introduce first the **inclusion/exclusion principle**. This counting principle relates the size of the four sets A, B, $A \cup B$, and $A \cap B$.

Inclusion/Exclusion Principle

Let A and B be two arbitrary finite sets, and let $n(\cdot)$ represent the number of elements in a set. Then

$$n(A \cup B) = n(A) + n(B) - n(A \cap B)$$

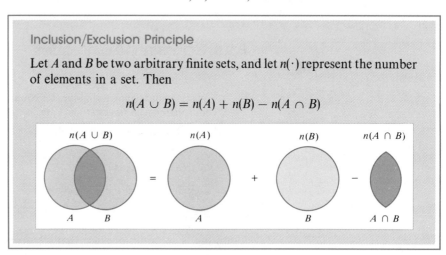

Verification of the Inclusion/Exclusion Principle

The reason is simple enough. To find the size of $A \cup B$, start by adding the size of A and the size of B. The sum, however, has added the elements of $A \cap B$ twice. Hence we must subtract this number from the sum. This is the inclusion/exclusion principle.

Example 1

Inclusion/Exclusion Principle   A set A contains eight elements, and a set B contains six elements. Is it possible to tell from this information the number of elements in the union $A \cup B$? If we are told that four elements belong to both A and B, then how many elements belong to $A \cup B$?

Solution   Knowing only $n(A)$ and $n(B)$ does not allow us to determine $n(A \cup B)$. However, if we are given the three values

$$n(A) = 8$$
$$n(B) = 6$$
$$n(A \cap B) = 4$$

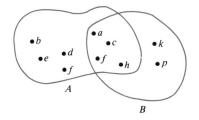

Figure 11
Illustration of the inclusion/exclusion principle

then there is only one unknown term in the inclusion/exclusion principle, and we solve for that term:

$$n(A \cup B) = n(A) + n(B) - n(A \cap B)$$
$$= 8 + 6 - 4$$
$$= 10$$

See Figure 11. ☐

Example 2 ———

Inclusion/Exclusion Principle   The number of football players in the National Football League in 1986 who could run a forty-yard sprint in 4.5 seconds was 62. The number of players who could bench press 550 pounds was 114. There were 19 players who could *both* run a 4.5 forty-yard sprint *and* bench press 550 pounds. How many players in the National Football League could run a 4.5 forty-yard sprint *or* bench press 550 pounds?

Solution   Calling

F = the set of players who can run a 4.5 forty-yard dash

S = the set of players who can bench press 550 pounds

we have

$$n(F \cup S) = n(F) + n(S) - n(F \cap S)$$
$$= 62 + 114 - 19$$
$$= 157 \text{ players}$$ ☐

We saw in the previous section that Venn diagrams allow us to visualize relationships between sets. We will see now that Venn diagrams can also be used as an aid in counting.

Use of Venn Diagrams in Counting Problems

Venn diagrams for two sets and three sets are shown in Figure 12. There are four disjoint subsets in the Venn diagram for two sets and eight disjoint subsets in the Venn diagram for three sets, as listed in Table 3.

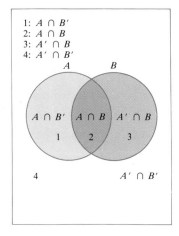

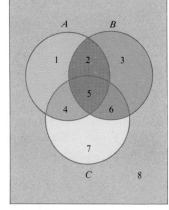

Figure 12
Venn diagrams of two sets and three sets

(a) Four disjoint sets

(b) Eight disjoint sets

Table 3
Disjoint Subsets in Figure 12

Disjoint Subsets of A and B	Disjoint Subsets of A, B, and C
$A \cap B$	$A \cap B \cap C$
$A \cap B'$	$A \cap B \cap C'$
$A' \cap B$	$A \cap B' \cap C$
$A' \cap B'$	$A \cap B' \cap C'$
	$A' \cap B \cap C$
	$A' \cap B \cap C'$
	$A' \cap B' \cap C$
	$A' \cap B' \cap C'$

Venn Diagrams and Blood Typing

Human blood can be tested for the absence or presence of one or more of the antigens A, B, and Rh. The antigens A and B belong to the ABO blood group, and Rh is the antigen in the Rhesus blood group. Blood is categorized or typed according to which of the three antigens A, B, and Rh are present. Table 4 lists the eight different blood types (the asterisks indicate that the antigen is not present).

Table 4
Classification of Blood Types

Blood Type	Antigens Present		
AB+	A	B	Rh
AB−	A	B	*
A+	A	*	Rh
A−	A	*	*
B+	*	B	Rh
B−	*	B	*
O+	*	*	Rh
O−	*	*	*

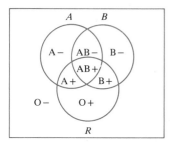

Set	Blood type
$A \cap B \cap R$	AB+
$A \cap B \cap R'$	AB−
$A \cap B' \cap R$	A+
$A \cap B' \cap R'$	A−
$A' \cap B \cap R$	B+
$A' \cap B \cap R'$	B−
$A' \cap B' \cap R$	O+
$A' \cap B' \cap R'$	O−

Figure 13
Venn diagram of blood types

Designate the universal set as

$$U = \{u \,|\, u \text{ is a person tested for blood type}\}$$

and three subsets of the universal set as

$$A = \{u \,|\, u \text{ is a tested person having antigen A}\}$$
$$B = \{u \,|\, u \text{ is a tested person having antigen B}\}$$
$$R = \{u \,|\, u \text{ is a tested person having antigen Rh}\}$$

There are eight different blood types, and they are illustrated in the Venn diagram in Figure 13.

Example 3

Counting Blood Types with Venn Diagrams   A group of individuals was tested to determine their blood types. Unfortunately, a fire destroyed some of the records. The following records were recovered:

- 181 persons had antigen A;
- 146 persons had antigen B;
- 196 persons had antigen Rh;
- 36 persons had antigens A and B;
- 46 persons had antigens A and Rh;
- 56 persons had antigens B and Rh;
- 10 persons had antigens A, B, and Rh (AB+);
- 40 persons did not have any antigens (O−).

Can you determine from this partial information the number of persons with each blood type?

Solution We begin by labeling the sizes of the two sets AB+ and O− shown in the Venn diagram in Figure 14. By subtracting the size of $A \cap B \cap R$ (the individuals having blood type AB+) from the three intersections $A \cap B$, $A \cap R$, and $B \cap R$, we can find the number of individuals in blood groups AB−, B+, and A+. Doing this, we get

$$n(A \cap B \cap R') = n(A \cap B) - n(A \cap B \cap R) = 36 - 10 = 26 \quad \text{(AB−)}$$
$$n(A' \cap B \cap R) = n(B \cap R) - n(A \cap B \cap R) = 56 - 10 = 46 \quad \text{(B+)}$$
$$n(A \cap B' \cap R) = n(A \cap R) - n(A \cap B \cap R) = 46 - 10 = 36 \quad \text{(A+)}$$

We now label the size of these sets as shown in Figure 15. By looking at the Venn diagram we can find the number of individuals having blood types A−, B−, and O+ by making the following subtractions:

$$n(A \cap B' \cap R') = 181 - 26 - 10 - 36 = 109 \quad \text{(A−)}$$
$$n(A' \cap B \cap R') = 146 - 26 - 10 - 46 = 64 \quad \text{(B−)}$$
$$n(A' \cap B' \cap R) = 196 - 36 - 10 - 46 = 104 \quad \text{(O+)}$$

We have now found the number of individuals having all eight blood types. These are illustrated in Figure 16. □

This brings us to the third counting technique, the tree diagram.

Counting with Trees

In 1971 the Pittsburg Pirates beat the Baltimore Orioles in a seven-game World Series, winning games 3, 4, 5, and 7. An interesting counting problem is to find the total number of different series that can be played in a best-of-seven-game series like the World Series. This type of counting problem, along with many others, can be solved easily by using a **tree diagram**. Before taking on the World Series, however, we consider a best-of-five-game series.

Example 4

Boston Versus Philadelphia   The Celtics and 76ers are playing a best-of-five-game series for the Eastern Championship of the NBA. How many different ways can this series be played? For example, PBBPP represents a possible series outcome in which Philadelphia wins games 1, 4, and 5 and hence wins the series.

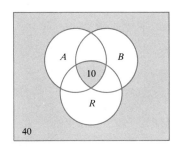

Figure 14
Shade the two known regions

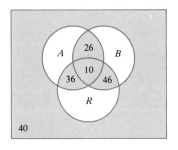

Figure 15
Shade the five known regions

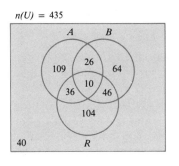

Figure 16
All regions are now known

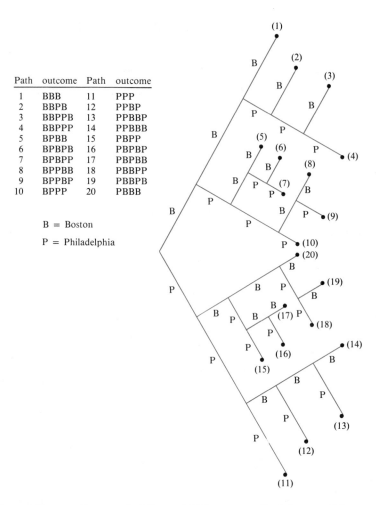

Path	outcome	Path	outcome
1	BBB	11	PPP
2	BBPB	12	PPBP
3	BBPPB	13	PPBBP
4	BBPPP	14	PPBBB
5	BPBB	15	PBPP
6	BPBPB	16	PBPBP
7	BPBPP	17	PBPBB
8	BPPBB	18	PBBPP
9	BPPBP	19	PBBPB
10	BPPP	20	PBBB

B = Boston

P = Philadelphia

Figure 17

Tree diagram of all possible best-of-five-game series

Solution   The tree diagram in Figure 17 illustrates all of the possible outcomes for this series. Each path of the tree (going from left to right) represents a specific five-game series. By simply counting all of these paths we find a total of 20 five-game series.   □

The following example shows how trees can be used to count the total number of ways in which one can order a word processor and a database program.

Example 5 _____

How Many Choices Are There?   Suppose, for example, that you are going to buy a word processor and a database program. You have decided upon either WordStar or EasyWriter as your word processor and DataStar, dBASE, or FirstBase as your database program. To count the different purchases you could make, a diagram called a tree diagram can conveniently display all the different purchase combinations of word processors and database programs.

Solution   To count these combinations using a tree diagram, it is useful to think of selecting these products in sequence. For example, think of selecting your word processor first. You clearly have two choices. For each choice of word processor, you can select three different database programs. This process of

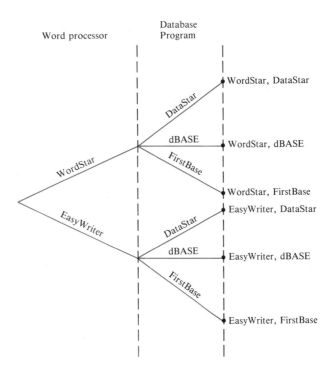

Word processor

Database Program

WordStar, DataStar

WordStar, dBASE

WordStar, FirstBase

EasyWriter, DataStar

EasyWriter, dBASE

EasyWriter, FirstBase

WordStar

EasyWriter

DataStar

dBASE

FirstBase

DataStar

dBASE

FirstBase

Figure 18
Tree diagram showing six different choices of word processors and database programs

enumeration is illustrated in the tree diagram in Figure 18. Note that there are six possible selections of word processors and database programs, as illustrated by the six **paths** in the tree diagram. □

Example 6

Cards and Coins   Select a card at random from a deck of 52 playing cards, and record whether the card is a heart, a diamond, a club, or a spade. After

the card has been selected, flip a coin and observe whether it turns up a head or a tail. How many different outcomes are there to this experiment?

Solution   The tree diagram in Figure 19 illustrates the different possibilities of the experiment. From this tree diagram we can easily count the paths of the tree and determine that there are eight possible outcomes to the experiment. They can be denoted as the elements of the set:

Outcomes = {(Ht, H), (Ht, T), (D, H), (D, T), (C, H), (C, T), (S, H), (S, T)}   □

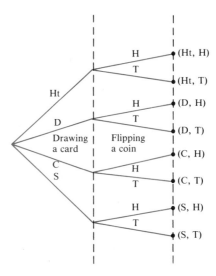

Figure 19
Tree diagram of a two-step process

Example 7

World Series Time   In how many different ways can a team win the World Series?

Solution   The World Series is a best-of-seven-game series (the first team that wins four games wins the series). We denote each series by a sequence of W's and L's that indicate wins and losses. For example, WWWW denotes a series in which the winner wins four straight games, and the sequence WLLWWLW represents a seven-game series in which the winner is victorious in games 1, 4, 5, and 7.

The tree diagram shown in Figure 20 illustrates all possible outcomes of such a series. It shows 35 different paths starting at the left (game one) and moving to the right. Each path can be thought of as a sequence of W's and L's (wins go upward in the tree, losses go to the right) always ending with a W (the winner always wins the last game). For example, the end point numbered 20 in the tree represents the series WLLLWWW, in which the winner wins games 1, 5, 6, and 7. The number 0 in parentheses after the end point 20 indicates that there has never been a World Series among the 77 World Series played from 1903 to 1986 in which the winner has won games 1, 5, 6, and 7. It is interesting that of the 35 possible World Series scenarios, 11 of them have never occurred. They are the ones shown in Table 5. The most common series have been those shown in Table 6.   □

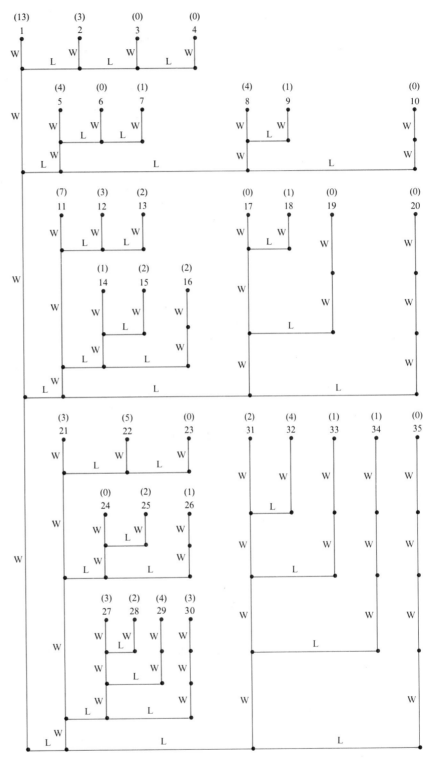

Figure 20
World series tree diagram (numbers in parentheses show the number of series that finished in each sequence)

TABLE 5

The 11 World Series That
Have Never Occurred

Number	Type of Series	Tree Location
1	WWWLLW	(path 3 in tree)
2	WWWLLLW	(path 4 in tree)
3	WWLWLW	(path 6 in tree)
4	WWLLLWW	(path 10 in tree)
5	WLLWWW	(path 17 in tree)
6	WLLWLWW	(path 19 in tree)
7	WLLLWWW	(path 20 in tree)
8	LWWWLLW	(path 23 in tree)
9	LWWLWW	(path 24 in tree)
10	LLWWLWW	(path 33 in tree)
11	LLLWWWW	(path 35 in tree)

TABLE 6

The Most Common Types of
World Series

Type of Series	Times Occurred	Years
WWWW	13	1907, 1914, 1922, 1927, 1928, 1932, 1938, 1939, 1950, 1954, 1963, 1966, 1976
WLWWW	7	1905, 1913, 1941, 1943, 1949, 1961, 1974
LWWWLW	5	1911, 1935, 1936, 1948, 1959
WWLWW	4	1908, 1916, 1929, 1933
WWLLWW	4	1917, 1930, 1953, 1980
LWLWLWW	4	1924, 1940, 1946, 1952
LLWWWLW	4	1955, 1956, 1965, 1971

The Multiplication Principle

How many dots are in the following diagram?

We suspect that you counted the 11 columns and then multiplied this number times 3 to get 33. If this is true, then you used the **multiplication principle** of counting. As another example, suppose that you can travel from Newport to Corvallis in four different ways and from Corvallis to Eugene in three ways.

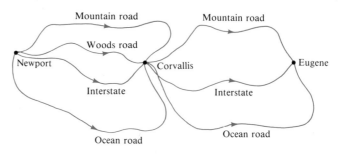

Figure 21

*How many paths from Newport
to Eugene?*

(See Figure 21.) In how many ways can you travel from Newport to Eugene by way of Corvallis? You can find the $4 \cdot 3 = 12$ ways by using the following general multiplication principle.

Multiplication Principle for Counting

Suppose k **operations** are performed in succession, where

- Operation 1 can be performed in n_1 ways,
- Operation 2 can be performed in n_2 ways,

$$\vdots \quad \vdots \quad \vdots \quad \vdots$$

- Operation k can be performed in n_k ways.

The total number of ways the k operations can be performed in succession is

$$n = n_1 \cdot n_2 \cdot n_3 \cdots n_k$$

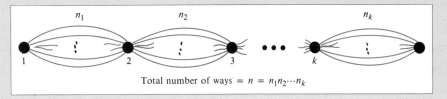

Total number of ways $= n = n_1 n_2 \cdots n_k$

You should note that the number of paths in the tree diagrams in Figures 17 and 20 could be counted by using the multiplication principle. However, the number of paths in the tree diagrams in Figures 18 and 19 could not be counted by using the multiplication principle.

Example 8

True-False Exam   Most people who are not well prepared for examinations tend to favor true-or-false tests, hoping they might get lucky. Suppose a student takes a true-false examination that has 25 questions. In how many different ways can the student answer the examination? Make a guess: 500, 1000, 5000?

Solution   The answer can be found by using the multiplication principle. A 25-question true-false examination can be interpreted as 25 successive operations, where each operation can be performed in two ways. Applying the multiplication principle, the total number of ways to answer the examination is

$$n = n_1 \cdot n_2 \cdot n_3 \cdots n_{25}$$
$$= 2^{25}$$
$$= 33{,}554{,}432 \qquad \qquad \square$$

Example 9

Electronic Numerals   Electronic numerals such as the one shown in Figure 22 are used in digital watches and electronic calculators. Each of the seven bars marked a through g is a molded light-emitting diode (LED) that can be turned

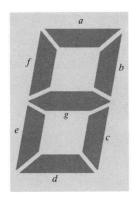

Figure 22
Electronic numeral

on or off. By turning on specific LED's it is possible to produce the numbers 0 through 9. For example, the number 9 is produced when g, f, a, b, c, and d are turned on. What is the total number of symbols that can be produced by using these seven LED's?

Solution We can interpret the turning on and off of each of the seven LED's as a sequence of seven operations in which each operation has two possibilities. Hence by the multiplication rule the total number of possibilities will be

$$\text{Number of symbols} = 2^7$$
$$= 128$$

It is surprising how fast the number of possibilities grows as additional LED's are added. For instance, along the same lines a black-and-white television set with 525 rows and 525 columns of pixels (dots) will have a total of $525 \times 525 = 275{,}625$ dots on the screen. If each dot is either black or white, then the total number of "pictures" that can be displayed on this television screen will be $2^{275,625}$. This number is so large that it can almost be described as unfathomable. For instance, if television sets completely covered the earth (one television set for every square foot), and if each television set could display 100 trillion different pictures every second, then the total length of time it would take these television sets to display all of the possible pictures would be longer than the age of the universe! In fact, it would take longer than

$$100{,}000{,}000{,}000{,}000{,}000{,}000{,}000{,}000{,}000{,}000{,}000{,}000{,}000{,}000{,}000$$

times the age of the universe. As a matter of fact, after this time has passed, the total fraction of pictures displayed will be less than

$$.00000000000000000000000000000000000000000000000000000000000001$$

of the total number of possible pictures.

The next time you are watching television, realize that in your lifetime you will never see all the possible "pictures" that can be displayed.   □

Example 10 —————— Multiplication Rule A pair of dice are rolled. In how many ways can they turn up?

Solution We can think of rolling a pair of dice as rolling one of the die first (Die I) and the other die second (Die II). Since each die can turn up in six ways, the multiplication principle says that there are $n = 6 \cdot 6 = 36$ ways to roll the dice. We can enumerate the 36 possibilities by drawing the tree diagram in Figure 23.   □

Example 11 —————— The License Plate Problem The Iowa Department of Transportation has suggested that the state change the Iowa license plate so that it consists of three letters of the alphabet followed by three single-digit numbers, such that no letter or digit occurs more than once. (In this way the police can read them more easily.) The suggestion was contested by a state legislator who believed that there would not be enough plates to go around. The state licenses 1,750,000 vehicles. Who is correct? Will the state run out of license plates?

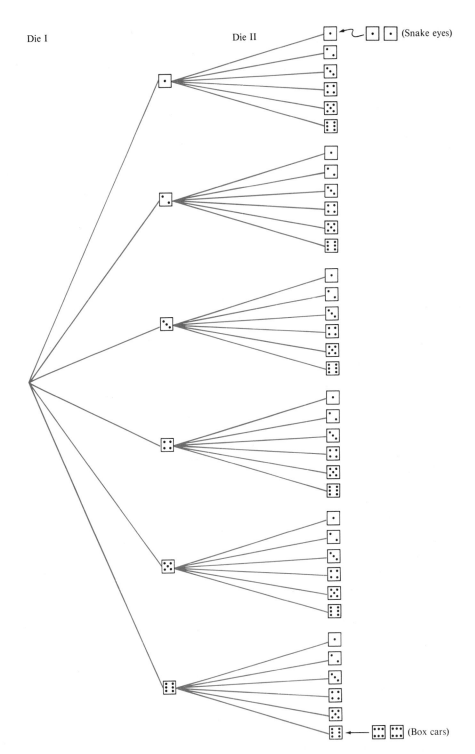

Figure 23
*Tree diagram illustrating the
multiplication rule*

Solution The suggested design for the Iowa license plates is shown in Figure 24. We can think of assigning the three letters and the three numbers as a sequence of six operations as we work from left to right. The number of ways

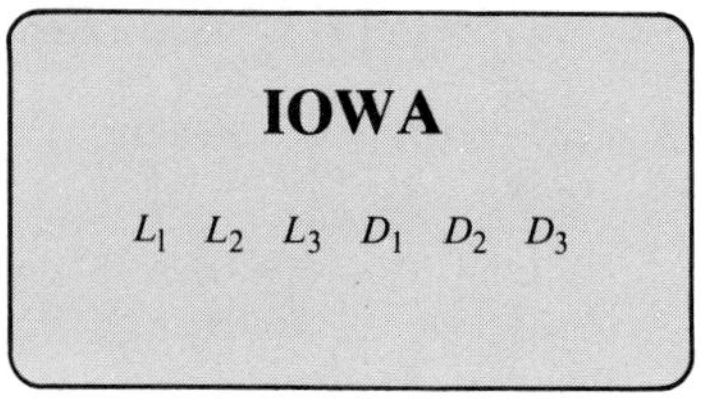

IOWA

L_1 L_2 L_3 D_1 D_2 D_3

Figure 24
New Iowa license plate?

that we can assign each successive letter and digit is given by the following values:

$n_1 = 26$ (number of ways for the first letter)
$n_2 = 25$ (number of ways for the second letter)
$n_3 = 24$ (number of ways for the third letter)
$n_4 = 10$ (number of ways for the first number)
$n_5 = \ 9$ (number of ways for the second number)
$n_6 = \ 8$ (number of ways for the third number)

Hence by the multiplication principle the total number of different license plates that can be made is

$$n = n_1 \cdot n_2 \cdot n_3 \cdot n_4 \cdot n_5 \cdot n_6$$
$$= 26 \cdot 25 \cdot 24 \cdot 10 \cdot 9 \cdot 8$$
$$= 11,232,000$$

In other words, there are more than enough possible license plates so that each vehicle can have its own plate. □

Problems

Counting with Venn Diagrams

1. **Counting Students at Alabama** The University of Alabama currently has an enrollment of 35,486 students. Suppose that of these students 4500 know WordStar, 3000 know EasyWriter, and 2000 know both WordStar and EasyWriter. For parts (a)–(e), how many students know the following word processors?
   (a) How many students know WordStar but not Easy-Writer?
   (b) How many students know EasyWriter but not Word-Star?
   (c) How many students know either EasyWriter or WordStar?
   (d) How many students know neither WordStar nor EasyWriter?
   (e) How many students know exactly one of WordStar or EasyWriter?
2. **Counting Students at Delaware** The University of Delaware currently has an enrollment of 13,241 students. Of these students, 4500 know Lotus 1-2-3, 4000 know SuperCalc, and 1500 know Symphony. The following facts are also known:

   - 2000 students know Lotus 1-2-3 and SuperCalc,
   - 300 students know Lotus 1-2-3 and Symphony,
   - 1000 students know SuperCalc and Symphony,
   - 75 students know all three packages.

   On the basis of these facts, answer the following questions.

(a) How many students know Lotus 1-2-3 but not Super-Calc or Symphony?
(b) How many students know SuperCalc but not Lotus 1-2-3 or Symphony?
(c) How many students know Symphony but not Lotus 1-2-3 or SuperCalc?
(d) How many students know at least one of these systems?
(e) How many students do not know any of these systems?
(f) How many students know SuperCalc and Lotus 1-2-3 but not Symphony?
(g) How many students know Symphony and Lotus 1-2-3 but not SuperCalc?
(h) How many students know Symphony and SuperCalc but not Lotus 1-2-3?

3. **Counting with Venn Diagrams** There are eight disjoint subsets labeled 1, 2, 3, 4, 5, 6, 7, and 8 in the Venn diagram in Figure 25. Denoting the number of elements in a set by $n(\cdot)$, we see

$$n(U) = 100$$
$$n(A) = \ 20$$
$$n(B) = \ 20$$
$$n(C) = \ 15$$
$$n(A \cap B) = \ 9$$
$$n(A \cap C) = 12$$
$$n(B \cap C) = \ 7$$
$$n(A \cap B \cap C) = \ 5$$

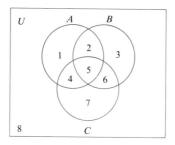

Figure 25
Venn diagram for Problem 3

For parts (a)–(h), how many members are in each of the eight disjoint sets 1–8?

(a) $n(1) =$ _____ (b) $n(2) =$ _____

(c) $n(3) =$ _____ (d) $n(4) =$ _____

(e) $n(5) =$ _____ (f) $n(6) =$ _____

(g) $n(7) =$ _____ (h) $n(8) =$ _____

4. **The Dragon Problem** A thousand dragons have been terrorizing the village. Suppose that 60 of them breathe fire and have bad breath, while 300 have at least one of these bad habits. Suppose too that 20 dragons only breathe fire. Find the following quantities.
   (a) How many dragons only have bad breath?
   (b) How many dragons have neither bad breath nor breathe fire?
   (c) How many dragons breathe fire or have bad breath but not both?

Some Interesting Counting Problems

5. **Number of TV Stations** A congressman claims that if the FCC keeps licensing new radio and TV stations, there will be no names left to assign to the stations. Radio and TV station names, like KICD and WKRP, always begin with a K or W followed by three arbitrary letters (although the last letter can be a blank, as in the case of WHO). How many names of stations are possible?
6. **The Key Problem** The key in Figure 26 is subdivided into seven parts, each part having three patterns. How many different keys of this variety can be made?

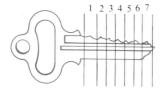

Figure 26
Key for Problem 6

Instant Insanity

In the puzzle Instant Insanity a player is given four blocks. The various sides of the blocks are colored red, green, blue, and white. The goal is to place the blocks adjacent to each other in such a way that the four colors are not duplicated along any of the four fronts. (See Figure 27.) For Problems 7–8, answer the following questions about Instant Insanity.

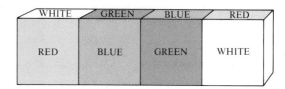

Figure 27
Instant Insanity puzzle for Problems 7–8

7. In how many ways can one block be placed on a table?
8. In how many ways can all four blocks be placed in a row (keeping the blocks in a given order but rotating the blocks)?
9. **Coloring a Venn Diagram** The Venn diagram shown in Figure 28 has the sets A and B and the four disjoint sets 1, 2, 3, and 4. Since there are 16 subsets of a set with four elements, can you find the 16 different ways to shade this Venn diagram?

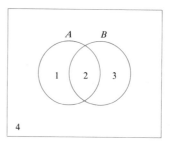

Figure 28
Venn diagram for Problem 9

10. **Five-Game Series** In how many ways can the New Jersey Nets and the Philadelphia 76ers play a best-of-three-game series? Draw a tree diagram showing all these ways.
11. **Number of Batteries** A baseball team has five pitchers and three catchers. How many different batteries (catcher/pitcher combinations) are possible?
12. **Telephone Numbers** Long-distance phone numbers consist of ten digits (xxx) xxx-xxxx. Will the population of the United States ever get so large that more numbers will be needed?
13. **Zip Codes** The U.S. Post Office Zip Code consists of five digits. How many different Zip Codes are possible with this system?

14. Extended Zip Codes   The Post Office is adding four more digits to the five-digit Zip Code. How many different Zip Codes will be possible with this new system?

15. Alphabetical Zip Codes   Some people have suggested a more elaborate Zip Code to replace the present five-digit postal system. The new system would use the letters of the alphabet in addition to numbers that are currently used. Would it be possible for each person in the United States to have his or her own Zip Code under this system? Assume that the population of the United States is 240 million.

Trees or Multiplication Rule?

For Problems 16–21, solve by using a tree diagram, the multiplication rule, or both.

16. Find the number of Canadian Zip Codes that can be formed. Each code has the general form ANANAN where A stands for any letter of the alphabet and N any digit 0–9.

17. Find the number of ways in which the Lakers and the Nuggets can play a five-game series (first team to win three games wins).

18. In how many ways can four coins be tossed? A typical toss of four coins would be (HTTH), which means that the first coin turns up a head, the second and third coins turn up tails, and the fourth coin turns up heads.

19. In how many ways can one roll a die (singular of dice) and then flip a coin?

20. In how many ways can five horses Win, Place, and Show (finish a race in first, second, and third places)?

21. How many different license plates can be made if each plate consists of one letter followed by three digits?

4.3   Permutations and Combinations

PURPOSE

We introduce two important concepts of counting theory, permutations and combinations. We will learn that for any set of n elements, any ordered arrangement of r of these n elements is called a permutation, and the number of these permutations is

$$P(n, r) = \frac{n!}{(n - r)!}$$

We then learn that a combination is any subset of r elements, taken from a set of size n. The number of these combinations is

$$C(n, r) = \frac{P(n, r)}{r!} = \frac{n!}{r!(n - r)!}$$

Introduction

The family photographer has lined up in a row the entire 25-member clan for the annual family portrait. Suppose someone does not like the way everyone is lined up. After rearranging the family to that person's liking, a second photograph is taken. After the second picture is taken, someone else complains and asks that the order be rearranged again. Someone then says that if the photographer takes a picture of every possible arrangement, they will be there all day. How long will it take the photographer to take a picture of every possible arrangement of the group if it takes five seconds to rearrange the group and photograph the new arrangement?

In this problem, each arrangement of people is called a permutation of the people. The problem here is to find the number of such permutations. Before solving this problem we introduce formally the concept of a permutation.

Permutations

A **permutation** is simply an arrangement of a collection of objects. More generally, we have the following definition.

> ### Permutation of r Elements from a Set of Size n
>
> A permutation of r elements, selected from a set of n elements, is an arrangement or ordering of the r elements.

Example 1

Permutations in a Set of Size 3   List all the possible arrangements or permutations of the three letters a, b, and c when the permutations contain one letter, two letters, and three letters.

Solution   A permutation of r letters (for $r = 1, 2, 3$) of the set $\{a, b, c\}$ is an arrangement of r letters chosen from a, b, and c. The permutations of sizes 1, 2, and 3 are listed in Table 7.

TABLE 7
Permutations of Size r
from $\{a, b, c\}$

$r = 1$	$r = 2$	$r = 3$
a	ab	abc
b	ac	acb
c	ba	bac
	bc	bca
	ca	cab
	cb	cba

□

Example 2

Solution of the Photographer's Problem   In how many ways can the 25 people mentioned earlier be arranged? If it takes five seconds to change from one arrangement to another arrangement, how long will it take for all arrangements to be achieved?

Solution   The solution of this problem lies at the heart of permutations. The total number of permutations of these 25 people can be found by the multiplication rule. Think of assigning a person to each of the 25 positions (from left to right), one after the other. In this way, assigning 25 people can be interpreted as a sequence of operations where

$$n_1 = \text{number of ways of assigning the first person} = 25$$
$$n_2 = \text{number of ways of assigning the second person} = 24$$
$$n_3 = \text{number of ways of assigning the third person} = 23$$
$$\vdots \qquad\qquad \vdots \quad \vdots \quad \vdots$$
$$n_{25} = \text{number of ways of assigning the 25th person} = 1$$

Hence the number of ordered arrangements of 25 people is

$$n = n_1 \cdot n_2 \cdots \cdots n_{25}$$
$$= 25 \cdot 24 \cdot 23 \cdots \cdots 3 \cdot 2 \cdot 1$$
$$= 15{,}511{,}210{,}042{,}908{,}563{,}210{,}000{,}000$$

If it takes five seconds to shoot each picture, then it would take approximately

$$1,000,000,000,000,000,000 \text{ years}$$

to take pictures of all possible arrangements of the family. In other words, it would take more than 100,000,000 times the estimated age of the universe!   □

---

n Factorial

In counting theory, one encounters the product of consecutive integers from 1 to n. This product is called **n factorial** and is denoted by

$$n! = n \cdot (n - 1) \cdot (n - 2) \cdots 2 \cdot 1$$

It is convenient to define **zero factorial** as 1.

Example 3

Growth of Factorials   Compute the first ten factorials.

Solution

$0! = 1$	$=$	1
$1! = 1$	$=$	1
$2! = 2 \cdot 1$	$=$	2
$3! = 3 \cdot 2 \cdot 1$	$=$	6
$4! = 4 \cdot 3 \cdot 2 \cdot 1$	$=$	24
$5! = 5 \cdot 4 \cdot 3 \cdot 2 \cdot 1$	$=$	120
$6! = 6 \cdot 5 \cdot 4 \cdot 3 \cdot 2 \cdot 1$	$=$	720
$7! = 7 \cdot 6 \cdot 5 \cdot 4 \cdot 3 \cdot 2 \cdot 1$	$=$	5040
$8! = 8 \cdot 7 \cdot 6 \cdot 5 \cdot 4 \cdot 3 \cdot 2 \cdot 1$	$=$	$40,320$
$9! = 9 \cdot 8 \cdot 7 \cdot 6 \cdot 5 \cdot 4 \cdot 3 \cdot 2 \cdot 1$	$=$	$362,880$
$10! = 10 \cdot 9 \cdot 8 \cdot 7 \cdot 6 \cdot 5 \cdot 4 \cdot 3 \cdot 2 \cdot 1$	$=$	$3,628,800$

Note the fast rate of growth of this function.   □

HISTORICAL NOTE

The use of the exclamation point $n!$ to represent n factorial was introduced in 1808 by Christian Kramp (1760–1826) of Strasbourg, France. He used this symbol to avoid the printing difficulties that were incurred by a previously used symbol.

Now that we know the meaning of a permutation, the next step is to determine the number of permutations of a given size.

Calculation of the Number of Permutations

The formula for the number of permutations of r elements from a set of n elements, denoted by $P(n, r)$, is determined by using the multiplication principle. To determine the number of ways in which we can arrange r elements of a set of size n, consider how many ways we can select the first element, then the second element, and the third, and so on until r elements have been selected. We see that

- the first element can be selected in n ways,
- the second element can be selected in $n - 1$ ways,

$$\vdots \quad \vdots \quad \vdots$$

- the rth element can be selected in $n - r + 1$ ways.

Hence by use of the multiplication principle the number of arrangements or permutations of r of the n elements is

$$P(n, r) = n \cdot (n - 1) \cdot (n - 2) \cdots (n - r + 1)$$

This leads us to the general equation for the number of permutations.

Number of Permutations $P(n, r)$

The **number of permutations** of r elements taken from a set of size n is denoted by $P(n, r)$ and given by

$$P(n, r) = \frac{n!}{(n - r)!} = n \cdot (n - 1) \cdot (n - 2) \cdots (n - r + 1)$$

It is useful to note that

$$P(n, n) = n \cdot (n - 1) \cdot (n - 2) \cdots 2 \cdot 1 = n!$$

Example 4

Number of Permutations   Evaluate the quantities $P(4, 2)$, $P(7, 3)$, $P(4, 1)$, $P(10, 3)$, and $P(4, 4)$.

Solution   The easy way to remember how to evaluate $P(n, r)$ is to start at n and multiply r factors, each factor decreasing by 1 from the previous factor. We have

$$
\begin{aligned}
P(4, 2) &= 4 \cdot 3 & &= 12 \\
P(7, 3) &= 7 \cdot 6 \cdot 5 & &= 210 \\
P(4, 1) &= 4 & &= 4 \\
P(10, 3) &= 10 \cdot 9 \cdot 8 & &= 720 \\
P(4, 4) &= 4 \cdot 3 \cdot 2 \cdot 1 & &= 24 \quad \square
\end{aligned}
$$

Example 5

Permutations of a Set of Four Elements   List the permutations of a, b, c, and d of different sizes.

Solution The lists in Table 8 give the permutations of different sizes. We have listed the permutations in alphabetical order.

$r = 1$	$r = 2$	$r = 3$		$r = 4$	
$P(4, 1) = 4$	$P(4, 2) = 12$	$P(4, 3) = 24$		$P(4, 4) = 24$	
a	ab	abc	cab	abcd	cabd
b	ac	abd	cad	abdc	cadb
c	ad	acb	cba	acbd	cbad
d	ba	acd	cbd	acdb	cbda
	bc	adb	cda	adbc	cdab
	bd	adc	cdb	adcb	cdba
	ca	bac	dab	bacd	dabc
	cb	bad	dac	badc	dacb
	cd	bca	dba	bcad	dbac
	da	bcd	dbc	bcda	dbca
	db	bda	dca	bdac	dcab
	dc	bdc	dcb	bdca	dcba

Example 6

Ways to Finish a Race   Four runners are running a ten-kilometer race.

(a) In how many ways can first place be determined?

(b) In how many ways can the race end if only the first two finishers are recorded?

(c) In how many ways can the race end if only the first three finishers are recorded?

(d) In how many ways can the race end if all four finishers are recorded?

Solution The different number of ways the race can end is the same as the number of arrangements of 1, 2, 3, and 4 elements of a set of size four. That is,

(a) The number of ways in which first place can be determined is

$$P(4, 1) = 4$$

(b) The number of ways in which the race can end when the first two finishers are recorded is

$$P(4, 2) = 4 \cdot 3 = 12$$

(c) The number of ways in which the race can end when the first three finishers are recorded is

$$P(4, 3) = 4 \cdot 3 \cdot 2 = 24$$

(d) The number of ways in which the race can end when all four runners are recorded is

$$P(4, 4) = 4 \cdot 3 \cdot 2 \cdot 1 = 24$$

It is interesting to note that the number of ways in which the runners can finish 1, 2, 3 is the same as the number of ways in which they can finish 1, 2, 3, 4. The reason, of course, is that once the first three places have been determined, the fourth place finisher is automatically determined.   □

Example 7

Distinguishable Permutations   How many permutations are there of the following letters?

(a) EYE

(b) DISTRESS

(c) TOOTS

Solution

(a) Permutations of EYE.   If we tried to solve this problem by using the same argument that we used in Example 6, we would get $P(3, 3) = 3 \cdot 2 \cdot 1 = 6$ permutations. However, if we actually listed the possibilities, we would find only three different permutations. The reason for this difference is that two of the letters in EYE are indistinguishable. If we labeled the two E's as in E_1YE_2, then of course we would consider the two arrangements E_1YE_2 and E_2YE_1 to be separate permutations, and we would indeed get all six permutations:

$$E_1YE_2 \quad YE_1E_2 \quad E_1E_2Y$$
$$E_2YE_1 \quad YE_2E_1 \quad E_2E_1Y$$

However, since we cannot distinguish between the two E's in EYE, we must divide by the number of ways that we can permute these indistinguishable letters. In other words,

$$\text{Number of permutations of EYE} = \frac{3!}{2!} = \frac{3 \cdot 2}{2} = 3$$

These three permutations are

$$EYE \quad YEE \quad EEY$$

(b) Permutations of DISTRESS.   There are eight letters in DISTRESS. If we consider them as distinguishable, as in

$$DIS_1TRES_2S_3$$

there are $P(8, 8) = 40,320$ permutations. However, since we cannot distinguish between the three different S's, the permutations

$$DIS_1TRES_2S_3 \quad DIS_2TRES_1S_3 \quad DIS_3TRES_1S_2$$
$$DIS_1TRES_3S_2 \quad DIS_2TRES_3S_1 \quad DIS_3TRES_2S_1$$

would be considered the same, and hence we divide by $3! = 6$ (number of permutations of $S_1S_2S_3$). Hence we have

$$\text{Number of permutations of DISTRESS} = \frac{8!}{3!} = \frac{8 \cdot 7 \cdot 6 \cdot 5 \cdot 4 \cdot 3 \cdot 2}{3 \cdot 2} = 6720$$

(c) Permutations of TOOTS.   Since TOOTS has five letters, of which two are T's, two are O's, and one is an S, we divide 5! by the number of ways in which two T's, two O's, and 1 S can be permuted. Hence we have

$$\text{Number of permutations of TOOTS} = \frac{5!}{2!2!1!} = \frac{5 \cdot 4 \cdot 3 \cdot 2}{2 \cdot 2 \cdot 1} = 30$$

Letter T occurs two times   *Letter O occurs two times*   *Letter S occurs one time*

The six permutations with S at the beginning are

SOOTT	STOOT
SOTOT	STOTO
SOTTO	STTOO

Can you find the other 24 arrangements? Try systematically putting S in the second, third, fourth, and last positions but leaving all other letters in the same order. □

Number of Distinguishable Permutations

The number of distinguishable permutations of n objects in which n_1 are of one type, n_2 are of another type, and n_k are of yet another type so that

$$n_1 + n_2 + \cdots + n_k = n$$

is given by

$$\text{Number of distinguishable permutations} = \frac{n!}{n_1!n_2! \cdots n_k!}$$

Combinations

With permutations we are concerned about the number of different ways in which a set of items can be arranged. In many situations, however, we are concerned only with finding the number of ways in which a set of items can be selected without any concern about their order. For instance, we may be interested in knowing the number of committees of three people that can be formed from a group of five people. If we denote a group of five people by

$$\text{Group} = \{A, B, C, D, E\}$$

we would like to find the number of subsets of size three from this group. One subset or combination is $\{A, B, C\}$, while another is $\{A, B, D\}$. A combination is essentially a subset, and hence we use the curly brace notation used for denoting sets to denote a combination. Realize that $\{A, B, C\}$ and $\{A, C, B\}$ are the same combination. The question we ask is: How many distinct combinations are there?

If the sets are small, we can simply enumerate all of the possibilities. It is important, however, to have a general formula that applies no matter how many elements are in the sets. We state without proof the following general result, giving the number of subsets of size r in a set of size n.

> **Number of Combinations of r Elements Selected from a Set of Size n**
>
> A subset of r elements, selected from a set of size n, is a combination of r elements taken from the set of size n. The number of these combinations is denoted by either $C(n, r)$ or $\binom{n}{r}$ and is given by
>
> $$C(n, r) = \binom{n}{r} = \frac{n!}{r!(n-r)!}$$

From this we can see that the number of committees of size three that can be selected from a group of five people is

$$C(5, 3) = \frac{5!}{3!(5-3)!}$$

$$= \frac{5 \cdot 4 \cdot 3 \cdot 2 \cdot 1}{3 \cdot 2 \cdot 1 \cdot 2 \cdot 1}$$

$$= 10$$

Calling the people in the group $\{A, B, C, D, E\}$, you should enumerate these ten committees. The numbers $C(n, r)$ are often denoted by using the alternative notation $\binom{n}{r}$ and in either case are called **binomial coefficients** (read n choose r, or n over r). They are called binomial coefficients because they are the coefficients of terms in the expansion of the binomial expression $(a + b)^n$.

Example 8

Binomial Coefficients   Given the first three letters of the alphabet a, b, and c, find all the combinations of these letters of size 1, size 2, and size 3.

Solution   By direct enumeration, we find the combinations in Table 9. A direct enumeration of the combinations yields the same number as the formulas:

$$\binom{3}{1} = \frac{3!}{1!2!} = \frac{3 \cdot 2 \cdot 1}{1 \cdot 2 \cdot 1} = 3 \qquad \text{(3 choose 1)}$$

$$\binom{3}{2} = \frac{3!}{2!1!} = \frac{3 \cdot 2 \cdot 1}{2 \cdot 1} = 3 \qquad \text{(3 choose 2)}$$

$$\binom{3}{3} = \frac{3!}{0!3!} = \frac{3 \cdot 2 \cdot 1}{3 \cdot 2 \cdot 1} = 1 \qquad \text{(3 choose 3)} \qquad \square$$

TABLE 9
Combinations of Size r
Taken from $\{a, b, c\}$

$r = 1$	$r = 2$	$r = 3$
$\{a\}$	$\{a, b\}$	$\{a, b, c\}$
$\{b\}$	$\{a, c\}$	
$\{c\}$	$\{b, c\}$	

> **Important Distinction**
>
> For permutations, order is important. For example, ab is *not* the same permutation as ba. For combinations, order does not matter. For instance, $\{a, b\}$ *is* the same combination as $\{b, a\}$.

The following example makes clear this distinction.

Example 9

Difference Between Permutations and Combinations   Find the permutations and combinations of two elements, where the elements are selected from a, b, c, and d.

Solution   By direct enumeration we find the values in Table 10.

TABLE 10
Permutations and
Combinations of
Two Elements

Permutations of Two Elements	Combinations of Two Elements
ab	$\{a, b\}$
ac	$\{a, c\}$
ba	$\{b, c\}$ $\binom{3}{2} = 3$
bc $P(3, 2) = 6$	
ca	
cb	

Example 10

Binomial Coefficients   Evaluate the binomial coefficients

(a) $C(10, 5)$ (b) $C(6, 4)$ (c) $C(6, 6)$ (d) $C(7, 0)$ (e) $C(4, 2)$

Solution   Using the general formula

$$C(n, r) = \binom{n}{r} = \frac{n!}{r!(n - r)!}$$

we have the following:

(a) $C(10, 5) = \dfrac{10!}{5!5!} = \dfrac{10 \cdot 9 \cdot 8 \cdot 7 \cdot 6 \cdot 5 \cdot 4 \cdot 3 \cdot 2}{5 \cdot 4 \cdot 3 \cdot 2 \cdot 5 \cdot 4 \cdot 3 \cdot 2} = 252$

(b) $C(6, 4) = \dfrac{6!}{4!2!} = \dfrac{6 \cdot 5 \cdot 4 \cdot 3 \cdot 2}{4 \cdot 3 \cdot 2 \cdot 2} = 15$

(c) $C(6, 6) = \dfrac{6!}{6!0!} = \dfrac{6 \cdot 5 \cdot 4 \cdot 3 \cdot 2}{6 \cdot 5 \cdot 4 \cdot 3 \cdot 2} = 1$

(d) $\dbinom{7}{0} = \dfrac{7!}{0!7!} = \dfrac{7 \cdot 6 \cdot 5 \cdot 4 \cdot 3 \cdot 2}{7 \cdot 6 \cdot 5 \cdot 4 \cdot 3 \cdot 2} = 1$

(e) $\dbinom{4}{2} = \dfrac{4!}{2!2!} = \dfrac{4 \cdot 3 \cdot 2}{2 \cdot 2} = 6$

Example 11

Record of the Month Club   The Record of the Month Club has a total of 50 different selections it can offer to its members. As an introductory offer, new members may choose five records for free. In how many different ways can a new member choose five records?

Solution Each selection of five records is a combination of five elements chosen from a set of size 50. Hence the number of different selections that a new member can choose is

$$\binom{50}{5} = \frac{50!}{5!45!}$$

$$= \frac{50 \cdot 49 \cdot 48 \cdot 47 \cdot 46}{5 \cdot 4 \cdot 3 \cdot 2}$$

$$= 2,118,760 \qquad \square$$

Example 12
_____

Number of Baseball Teams The New York Yankees are taking 23 players on a road trip. Assume that the traveling squad consists of

- 3 catchers,
- 6 pitchers,
- 8 infielders, and
- 6 outfielders.

There are one catcher, one pitcher, four infielders, and three outfielders on a team. Each player can play only his own position. How many different teams can the manager put on the field?

Solution This problem uses two rules of counting: counting combinations and the multiplication principle. We first find the following combinations:

n_1 = number of ways of assigning the 3 catchers

= number of combinations of 1 element taken from 3 things

= $C(3, 1) = 3$ (or use common sense to find this number)

n_2 = number of ways of assigning the 6 pitchers

= number of combinations of 1 element taken from 6 things

= $C(6, 1) = 6$ (or use common sense to find this number)

n_3 = number of ways of assigning the 8 infielders

= number of combinations of 4 elements taken from 8 things

= $C(8, 4) = 70$

n_4 = number of ways of assigning the 6 outfielders

= number of combinations of 3 elements taken from 6 things

= $C(6, 3) = 20$

Using these values, along with the multiplication principle, we see that the total number of teams that can be put on the field is

$$n = n_1 \cdot n_2 \cdot n_3 \cdot n_4$$

$$= C(3, 1) \cdot C(6, 1) \cdot C(8, 4) \cdot C(6, 3)$$

$$= 3 \cdot 6 \cdot 70 \cdot 20$$

$$= 25,200$$

In other words, if the team played a game a day, it would take roughly 69 years to play all the different lineups. $\qquad \square$

Similarly, we can determine the number of women's basketball teams that the University of Maine can put on the court (this problem was stated on page 277 in Section 4.2). The total number of teams will be

$$\text{Number of women's basketball teams} = C(2, 1) \cdot C(5, 2) \cdot C(5, 2)$$
$$= 2 \cdot 10 \cdot 10$$
$$= 200$$

Example 13 ————

The Committee Problem   Recently, the Society for the Protection of the Snail Darter, which consists of 25 members, decided to elect a four-person executive committee and a three-person entertainment committee. It was decided that no person should serve on both committees. In how many different ways can the Society elect these two committees?

Solution   For this problem it is best to think of the two elections being carried out one after the other. Which election comes first makes no difference. Assuming that the election for the executive committee is held first, we can determine the number of executive committees that can be elected. Since we are choosing four members from a set of 25 members, the number of combinations is $C(25, 4)$, or 25 choose 4.

After this election is completed, there are 21 people left as candidates for the entertainment committee. Since we are electing three members to this committee, the possible number of combinations of the entertainment committee is 21 choose 3, or $C(21, 3)$.

By the multiplication principle the total number of combinations of both committees is the product of the above numbers. Evaluating this product gives the total number of ways of forming the two committees:

$$C(25, 4) \cdot C(21, 3) = \frac{25!}{4!21!} \frac{21!}{3!18!}$$
$$= \frac{25 \cdot 24 \cdot 23 \cdot 22}{4 \cdot 3 \cdot 2} \frac{21 \cdot 20 \cdot 19}{3 \cdot 2}$$
$$= 16{,}824{,}500 \text{ committees}$$

You should rework this problem assuming that the election for the entertainment committee was held first. The total number of committees is found to be $C(25, 3) \cdot C(22, 4)$. This turns out to be the same value as $C(25, 4) \cdot C(21, 3)$.

It is interesting to note that if the members of the Snail Darter Society can serve on both the executive and entertainment committees, then the total number of committees that can be elected increases to the following value:

$$C(25, 4) \cdot C(25, 3) = \frac{25!}{4!21!} \frac{25!}{3!22!}$$
$$= \frac{25 \cdot 24 \cdot 23 \cdot 22}{4 \cdot 3 \cdot 2} \frac{25 \cdot 24 \cdot 23}{3 \cdot 2}$$
$$= 29{,}095{,}000 \text{ committees} \qquad \square$$

Problems

For Problems 1–10, evaluate the following.

1. $P(5, 3)$ **2.** $P(4, 1)$ **3.** $P(3, 3)$
4. $P(30, 2)$ **5.** $C(4, 1)$ **6.** $C(4, 4)$

7. $C(10, 8)$ **8.** $\binom{7}{2}$ **9.** $\binom{9}{2}$

10. $\binom{10}{0}$

11. Evaluation of Factorials Compute the following quantities:

(a) $\dfrac{(10,000)!}{(9999)!}$ (b) $\dfrac{n!(n + 1)!}{(n - 1)!}$ (c) $\dfrac{n!}{(n - 1)!}$

12. The Growth of n! A leading computer software company introduces a new computer program for solving the linear programming problem. Suppose the company advertises that the number of steps it takes to solve a linear programming problem is

$$m!/3 + n!/5$$

where

m = number of constraints in the problem

n = number of decision variables in the problem

Suppose each step of this program takes one nanosecond (one billionth of a second) on a modern computer. How long will it take the program to solve a linear programming problem with 25 constraints and 20 variables? You should use a calculator to evaluate the necessary factorials.

The purpose of this problem is to impress upon you the rate at which the factorial function grows. You can see that factorial growth methods are not very good.

Counting Permutations

13. Kentucky Derby Ten horses are entered in the Kentucky Derby. In how many different ways can the horses Win, Place, and Show (finish in first, second, and third places)?

14. Permutations How many permutations are there of the following letters?
(a) *ab*
(b) *abcd*
(c) *JKLMNOP*

15. Permutations and Trees Draw a tree diagram that represents all the permutations of size two of the letters *abcd*. How many permutations are there?

16. Permutations of Size Three How many permutations of size three can be formed from the letters *abcdef*?

17. Permutations in Botany Four pollen grains (one pine, one fir, one spruce, and one oak) are on a microscope slide preparation. In how many orders can these slides be presented?

18. Permutations of Numbers How many permutations are there of the digits of the following numbers?
(a) 146
(b) 2934
(c) 145096

For Problems 19–32, compute the number of distinguishable permutations of the following words.

19. TO	**20.** TWO
21. TOO	**22.** TOOT
23. SNOOT	**24.** BOSTON
25. DALLAS	**26.** MIAMI
27. SEATTLE	**28.** OHIO
29. ILLINOIS	**30.** ALABAMA
31. TENNESSEE	**32.** MISSISSIPPI

Counting Combinations

33. Combinations The people in the picture below are the members of the Protection of the Maine Coon Cat Society. In how many different ways can this society choose its president, vice-president, and secretary?

34. Combinations   One of the authors of this book makes a daily pilgrimage to the faculty lounge for coffee. Every day he takes with him three students from his business math class. There are 90 students in the class. How long will it take him to have coffee with every combination of three students (two combinations are considered different if only one student is different)?

35. Hands in Poker   When playing five-card draw poker, each player is dealt five cards from an ordinary playing deck of 52 cards. How many different hands can a player be dealt?

Counting Poker Hands

When playing five-card draw poker, each player is dealt five cards from an ordinary deck of 52 playing cards. How many different hands can a player be dealt that contain the cards in Problems 36–44?

36. Royal Flush   In how many ways can a player be dealt a *royal flush* (A, K, Q, J, 10 of the same suit)?

37. Straight Flush   In how many ways can a player be dealt a *straight flush* (five consecutive cards of the same suit—but not a royal flush)?

38. Four of a Kind   In how many ways can a player be dealt *four of a kind* (four cards alike)?

39. Full House   In how many ways can a player be dealt a *full house* (three cards alike and a pair)?

40. Flush   In how many ways can a player be dealt a *flush* (all five cards of the same suit: hearts, diamonds, clubs, or spades, but not a straight or royal flush)?

41. Straight   In how many ways can a player be dealt a *straight* (any run of five consecutive cards from 2, 3, . . . , 10, J, Q, K, A, except a straight or royal flush)?

42. Three of a Kind   In how many ways can a player be dealt *three of a kind* (three cards alike, the other two cards different)?

43. Two Pairs   In how many ways can a player be dealt *two pairs* (two different pairs, the fifth card different)?

44. Pair   In how many ways can a player be dealt *one pair* (two cards matching, the other three not matching)?

45. Coaching Problem   Pat Riley, coach of the Los Angeles Lakers, once said that he would try every combination of players on his squad of 12 players in order to find a winning team. How many teams can he play?

In Problems 46–49, suppose we toss a penny ten times and observe the sequence of heads and tails.

46. How many different outcomes or sequences of heads and tails are possible?

47. How many different outcomes have exactly one head (or nine tails)? Can you enumerate them?

48. How many different outcomes have exactly four heads? Can you list two of these outcomes?

49. How many outcomes have at least eight heads?

50. Tours of the Mystery City   The picture in Figure 29 shows square blocks of a mystery city. One of the authors once walked from the upper left corner to the lower right corner (always walking either south or east). He wondered just how many different paths there actually were. At that time he did not realize that he was trying to count combinations.

With a little thought the author discovered that each path can be thought of as a list of 18 E's and S's, since each path has a total length of 18 blocks. The path in the diagram above is

Typical path
= (E, E, S, S, E, S, S, E, E, S, S, S, E, E, E, S, E, E)

The letter E means that the author travels east on a given block, and S means that the author travels south. Also, note that each path contains 8 E's and 10 S's to travel from the northwest (NW) to the southeast corner (SW) of the map.

Given this information, how many different walks could the author take from the northwest to the southeast corner, always traveling either eastward or southward? Would the author have time to make all these walks in one afternoon?

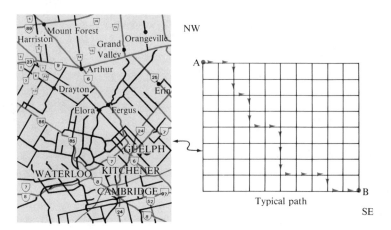

Figure 29
How many paths from A to B?

51. **Combinations** In how many ways can four players choose sides with two players on each side? Enumerate the different teams, calling the four players A, B, C, and D.
52. **Combinations** In how many ways can six players choose sides with three players on each side?
53. **Combinations** In how many ways can eight players choose sides with four players on each side?
54. **Combinations** In how many ways can $2n$ players choose sides with n players on each side?

The Committee Problem

The Society for the Protection of the Snail Darter has a membership of 25 members. In Problems 55–58, determine the number of different committees that the Society can elect.

55. **One-Committee Problem** In how many ways can the Snail Darter Society elect an executive committee of two members?
56. **Two-Committee Problem** In how many ways can the Snail Darter Society elect an executive committee of two members and an entertainment committee of four members if no member of the society can serve on both committees?
57. **Three-Committee Problem** In how many ways can the Snail Darter Society elect an executive committee of two members, an entertainment committee of three members, and a welcoming committee of two members if no member of the society can serve on more than one committee?
58. **Four-Committee Problem** In how many ways can the Snail Darter Society elect four different committees of

two members each if no member of the society can belong to more than one committee?

The Committee Problem (Serving on More Than One Committee)

Problems 59–61 refer to the problems above.

59. **Society Members Serving on Two Committees** Answer Problem 56 if members of the Society can serve on both committees.
60. **Society Members Serving on Three Committees** Answer Problem 57 if members of the Society can serve on all three committees.
61. **Society Members Serving on Four Committees** Answer Problem 58 if members of the Society can serve on all four committees.
62. **Enumerating Committees.** A very small select organization consisting of five members A, B, C, D, and E are going to elect two committees of two members each. If no member of the organization can belong to both committees, list the possible committees that can be formed. One possibility for the two committees is

Committee I $= \{A, C\}$ Committee II $= \{B, E\}$

63. **Enumerating Committees** Solve Problem 62 if it is possible for two members of the organization to belong to both committees. One possibility for the two committees is

Committee I $= \{A, C\}$ Committee II $= \{A, B\}$

Epilogue: The Ultimate Counting Problem

A brush salesman, Billie Loman, was required to visit each of four towns in his territory. He left from his home town, visited each of the other four towns exactly once, and then returned home. There were a sufficient number of roads that he could travel from one town to any other town if necessary. To minimize the cost for gasoline, Billie found what he claimed was the route that had the shortest total distance. His "minimal distance tour" of the four towns was

Home $\longrightarrow$ Curlew $\longrightarrow$ Rolfe $\longrightarrow$
        Plover $\longrightarrow$ Ayrshire $\longrightarrow$ Home

The total length of this route was 500 miles. It is shown in Figure 30.

Billie's boss was furious. He thought that there was a shorter route and that Billie was wasting company money. To appease his boss, Billie spent an entire weekend trying to list all of the possible routes. By Monday

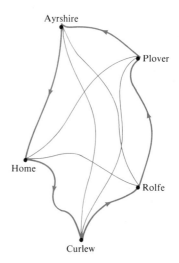

Figure 30
Simple traveling salesman problem

morning he had found 24 different routes. Fortunately, he had a mathematician friend who told him that he had found all of them. Billie's boss was impressed and promoted Billie to District Manager in charge of 48 cities.

Billie was so excited in his new position that he decided to list all possible routes that a salesman could travel among the 48 cities of his new district. What Billie did not realize was the nature of combinatorics. His mathematician friend told him that if there are n cities on a route (counting the home city), then the number of routes that a salesman can traverse, starting from the home city, visiting each of the other cities exactly once, and returning to the home city is

$$\text{Number of routes} = (n - 1)!$$

In the case of Billie's new district with 48 cities, the possible number of routes a salesman could take is

$$\text{Number of routes} = 47!$$

or

258,623,244,593,423,434,545,480,823,543,245,678,608,210,000,000,000,000,000,000

Billie decided not to list all of the possibilities.

To put this number in perspective, the great Greek mathematician Archimedes once estimated that there are less than 10^{51} grains of sand in the universe. The number of routes is roughly one hundred million times larger than 10^{51}.

The determination of the shortest route for making a "complete tour" of n cities and returning to the original city is called the Traveling Salesman Problem. Many strategies have been proposed to find this minimum-distance path (other than by enumeration of all the routes, of course), but no quick method has ever been found. One recent strategy was proposed by Shen Lin of AT&T. He has applied his method to finding the shortest path between the 48 capitals in the lower 48 states. He cannot prove that his method gives the shortest possible distance, but he is willing to pay anyone $100 if they can find a route shorter than 10,628 miles that visits each of the state capitals exactly once and returns to the starting capital (start at your own state capital if you like).

Key Ideas for Review

binomial coefficient, 299
combination, 298
counting principles
   inclusion/exclusion principle, 278
   multiplication principle, 286
   tree diagrams, 281
   Venn diagrams, 273

De Morgan's laws, 275
factorial, 294
permutations, 293
sets
   complement, 274
   disjoint, 274
   empty set, 272

intersection, 274
membership in a set, 271
subsets, 272
union, 273
universal set, 272
trees, 281
Venn diagrams, 273

Review Exercises

For Problems 1–7, describe the set by either the roster method or the defining property method.
1. Mary, John, and Sally are lined up for a photograph. List the set of all possible arrangements.
2. List the set of integers strictly between 1 and 13.
3. List the set of real numbers between 1 and 13.
4. Describe the feasible set

$$2x_1 + x_2 \le 2$$
$$x_1 \ge 0$$
$$x_2 \ge 0$$

5. Describe the solution set of the equation $x + 3 = 2$.

6. Describe the set of real numbers that satisfy $x^2 + 1 = 0$.
7. Describe the solution set of $x^2 - 1 = 0$.

Blood Types and Venn Diagrams

In testing blood types, the three important types of antigens are A, B, and Rh. Figure 31 shows the results of a test given to a group of individuals. The numbers in each of the eight disjoint sets are the numbers of persons having the given antigens.

For Problems 8–15, how many people have the given blood types?

8. A—   9. AB—   10. B—

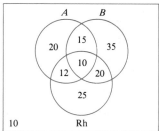

Figure 31
Eight blood types

11. A+ **12.** AB+ **13.** B+
14. O+ **15.** O−

Public Relations and Venn Diagrams

A public relations company for an ocean cruise company surveys passengers to determine the acceptability of a newly christened luxury cruise ship. The results are given in Table 11.

TABLE 11
Results of Cruise Ship Survey

	Disliked Greatly	Disliked	Liked	Liked Much
Men	15	25	150	200
Women	5	5	250	150
U.S. Citizens	0	10	200	300
Non-Citzens	20	20	200	50

We define the following sets:

M = set of all men in the survey

C = set of all U.S. citizens in the survey

D = set of all people in the survey who disliked the ship

L = set of all people in the survey who liked the ship

V = set of all people in the survey who liked the ship very much

In Problems 16–28, describe the meaning of the given set, and tell how many people belong to the set.

16. M **17.** M' **18.** V
19. V' **20.** L **21.** L'
22. $M \cap L$ **23.** $M \cap L'$ **24.** $M' \cap L'$
25. $C \cap V'$ **26.** $C \cap D$ **27.** $M' \cap (L \cup V)'$
28. $M \cap (L \cup V)'$

Medical Study and Venn Diagrams

Problems 29–32 are concerned with medical studies on lung cancer. A survey was made of 1000 persons to determine the relationship between smoking and lung cancer. The set C is the set of persons in the survey that have lung cancer, and S is the set of smokers in the survey. The Venn diagram in Figure

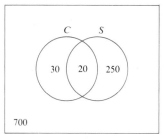

Figure 32
Survey of smokers and cancer patients
C = cancer patients
S = smokers

32 describes the results of the survey. For Problems 29–32, determine the following quantities.

29. The fraction of people with lung cancer who smoke.
30. The fraction of people without lung cancer who smoke.
31. The fraction of people who smoke or have lung cancer.
32. The fraction of people who do not smoke and have lung cancer.

Campus Survey and Venn Diagrams

The percentages of students at a major university who like different kinds of music are as follows:

- 60% of the students like rock music;
- 40% of the students like classical music;
- 50% of the students like country music;
- 30% of the students like rock and classical music;
- 15% of the students like classical and country music;
- 40% of the students like rock and country music;
- 10% of the students like all three kinds of music.

For Problems 33–40, determine the following quantities.

33. The percentage of students who like rock and classical but not country music.
34. The percentage of students who do not like any of the three kinds of music.
35. The percentage of students who like exactly one kind of music.
36. The percentage of students who like exactly two kinds of music.
37. The percentage of students who like at least one kind of music.
38. The percentage of students who like rock but not country music.
39. The percentage of students who like country but not rock music.
40. The percentage of students who like classical but neither rock nor country music.

41. The diagram in Figure 33 shows a set A with three elements and its eight subsets. Add a rectangle to the square, circle, and triangle of A, and list the subsets of this new set of size four.

$$A = \left\{ \blacksquare \; \blacktriangle \; \bullet \right\}$$

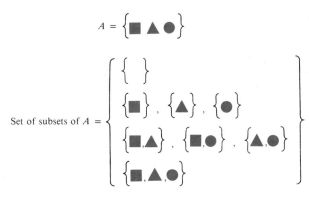

Figure 33
A set of size three and its subsets

46. $\binom{12}{4}$

47. $\binom{9}{1}$

48. $\dfrac{m!}{(m-1)!}$

49. $\dfrac{(m+3)!}{(m-1)!}$

Interesting Counting Problems

50. Combinatorics in Baseball A baseball squad has three catchers, three pitchers, five infielders, and four outfielders on its roster. How many different lineups can this team field if each player can play any position within his own area (the areas are catcher, pitcher, infielder, and outfielder)? Remember that there are one catcher, one pitcher, four infielders, and three outfielders in a lineup.

51. Another Picture Problem There are five people in a family. Are there more ways to take pictures of family members with two people in the picture or with three people in the picture? Do you see why your answer is as it is?

Computation Problems (Try Using a Calculator)

For Problems 42–49, compute the following quantities.

42. $C(1000, 999)$

43. $P(1000, 999)$

44. $P(10, 4)$

45. $C(20, 4)$

Chapter Test

*1. The principal of the Willow Junction High School says that the senior class at Willow Junction has 45 students. The principal also makes the claim that

- 35 of the students are girls,
- 28 students play sports,
- 18 girls play sports,
- 17 girls are B students,
- 15 girls are B students and play sports,
- 14 students who play sports are B students.

Why is this impossible? Explain.

2. A modern language department at a university is staffed with 80 people. Of these people, 50 speak Spanish, 40 speak German, and 19 speak both Spanish and German. How many speak neither Spanish nor German?

3. Enumerate all subsets of the set $\{a, b, c, d, e\}$ of size two.

4. Enumerate all permutations of the set $\{a, b, c, d, e\}$ of size two.

*5. NASA is selecting a three-person crew to go to Mars: one commander, one engineer, and one doctor. The candidates for this voyage are the following:

- four commanders, whom we denote by c_1, c_2, c_3, and c_4;
- three engineers, whom we denote by e_1, e_2, and e_3;
- three doctors, whom we denote by d_1, d_2, and d_3.

Because NASA would like the crew of this voyage to be as compatible as possible, a psychologist has drawn up the following compatibility table for the voyage.

- Commander 1 is compatible with Engineers 1 and 3 and Doctors 2 and 3;
- Commander 2 is compatible with Engineers 1 and 2 and all three doctors;
- Commander 3 is compatible with Engineers 1 and 2 and Doctors 1 and 3;
- Commander 4 is compatible with all three engineers and Doctor 2;
- Engineer 1 is incompatible with Doctor 3;
- Engineer 2 is incompatible with Doctor 1;
- Engineer 3 is incompatible with Doctor 2.

Draw a tree diagram to illustrate all possible crews.

---

* Problems 1 and 5 were paraphrased from problems taken from the fascinating book *Combinatorics* by N. A. Vilenkin (Academic Press, 1971).

5

Probability

The word "chance" is used so often today that we generally do not give it a second thought. We talk about the chance that the Dow Jones average will reach 3000 by the end of the year or the chance that the Chicago Cubs will win the National League Pennant. The word "chance" is a synonym for "probability," and the subject of probability attempts to give a precise meaning to this word. It is hard to believe that a subject such as probability, which had its origins in French gambling parlors, would become one of the most important intellectual developments of modern times.

Today the subject of probability has outgrown its disreputable origins. It is extremely important in the natural and social sciences, business management, physics, and engineering. For example, it has been said that the single most important result in population genetics is the Hardy-Weinberg Law, which is basically a theorem in probability.

This chapter provides the answers to such questions as:

- What is probability?

- How do we find probabilities?

- What is a conditional probability?

- What is a random variable?

If you understand these concepts, you will have laid the foundations for the study of probabilistic models in business and science.

5.1 The Probability Experiment and Sample Spaces

Probability is the very guide of life.
—*Cicero*

PURPOSE

We present the fundamental concepts of probability. In particular, we introduce

- the probability experiment (an experiment that has two or more outcomes),
- the sample space of an experiment (the collection of all possible outcomes), and
- an event (a collection of outcomes).

The Probability Experiment

Central to the subject of probability is the probability experiment. Any experiment that has two or more outcomes is called a **probability experiment**. Examples are tossing a coin, rolling a die, testing products on an assembly line, measuring some physical characteristic of a spruce budworm, and even asking a family its annual income. The set of all possible outcomes of a probability experiment is called the **sample space** of the experiment. Any collection of outcomes, or subset of the sample space, is called an **event** in the sample space. For instance, when we toss a coin and observe whether the coin is a head (H) or a tail (T), the sample space is $S = \{H, T\}$. The subset $\{H\}$ is an event in S and would be called "the event of tossing a head."

We summarize these basic definitions.

Probability Experiment, Sample Space, and Event

- **Probability Experiment.**   A probability experiment is an experiment that has two or more outcomes.

- **Sample Space.**   The set of all outcomes of a probability experiment is called the sample space of the experiment. It is denoted by S.

- **Event.**   A subset of a sample space is called an event.

Example 1

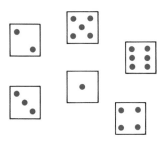

Figure 1
The sample space for tossing one die and observing the top face

Rolling a Die   Consider the experiment of rolling a die (singular of dice) and recording the number that turns up on the top face. What is the sample space of this experiment?

Solution   This experiment is a probability experiment because it has more than one outcome. The set of all possible outcomes or the sample space of the experiment would be denoted by $S = \{1, 2, 3, 4, 5, 6\}$ (See Figure 1.) Typical events in S would be

$$A = \{1\} \qquad \text{Event of rolling a one}$$
$$B = \{4, 5, 6\} \qquad \text{Event of rolling a four, five, or six}$$
$$C = S \qquad \text{Certain event (rolling a 1, 2, 3, 4, 5, or 6)}$$
$$D = \varnothing \qquad \text{Impossible event} \qquad \square$$

Example 2

Pair of Dice   Roll a pair of dice (one red and one green), and observe the numbers that turn up on the top faces. What is the sample space of this experiment?

Solution   This experiment has 36 possible outcomes. Each individual outcome is represented by an ordered pair (r, g), where

$$r = \text{Number showing on the red die} \qquad r \in \{1, 2, 3, 4, 5, 6\}$$
$$g = \text{Number showing on the green die} \qquad g \in \{1, 2, 3, 4, 5, 6\}$$

The 36 elements of the sample space are shown in Figure 2. In this sample space, three typical events are

- Event of rolling a match:

$$MD = \{(r, g) \,|\, r = g\} = \{(1, 1), (2, 2), (3, 3), (4, 4), (5, 5), (6, 6)\}$$

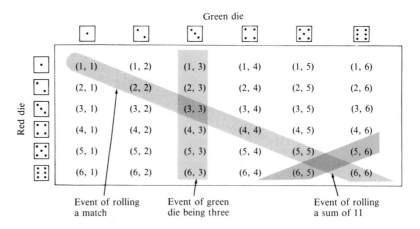

Green die

Red die

	(1, 1)	(1, 2)	(1, 3)	(1, 4)	(1, 5)	(1, 6)
(2, 1)	(2, 2)	(2, 3)	(2, 4)	(2, 5)	(2, 6)
(3, 1)	(3, 2)	(3, 3)	(3, 4)	(3, 5)	(3, 6)
(4, 1)	(4, 2)	(4, 3)	(4, 4)	(4, 5)	(4, 6)
(5, 1)	(5, 2)	(5, 3)	(5, 4)	(5, 5)	(5, 6)
(6, 1)	(6, 2)	(6, 3)	(6, 4)	(6, 5)	(6, 6)

Event of rolling a match

Event of green die being three

Event of rolling a sum of 11

Figure 2

Typical events when rolling a pair of dice

• Event that the sum of the top faces is 11:

$$S11 = \{(r, g)\,|\,r + g = 11\} = \{(5, 6), (6, 5)\}$$

• Event of rolling a pair of dice where the green die is a 3:

$$G3 = \{(r, g)\,|\,g = 3\} = \{(1, 3), (2, 3), (3, 3), (4, 3), (5, 3), (6, 3)\} \qquad \square$$

Example 3

Events   Flip a coin and observe whether the coin turns up heads or tails. The sample space of this experiment is

$$S = \{H, T\}$$

where H denotes heads and T denotes tails. Find all the events in this sample space.

Solution   There are four different subsets or events of S. They are

Event	Name
$\varnothing$	Impossible event
$\{H\}$	Event of flipping a head
$\{T\}$	Event of flipping a tail
$S = \{H, T\}$	Certain event

$\square$

Example 4

Typical Events   The sample space of the experiment of flipping three different coins is illustrated by means of a tree diagram in Figure 3. The sample spaces can be written mathematically as

$$S = \{TTT, TTH, THT, THH, HTT, HTH, HHT, HHH\}$$

Find the following three different events:

(a) Event of getting one head,

(b) Event of getting two heads,

(c) Event of getting three heads.

HISTORICAL NOTE

Probability had its origin during the fifteenth and sixteenth centuries when Italian mathematicians attempted to evaluate the odds in various games of chance. The great Italian mathematician Geronimo Cardano (1501–1576) wrote a gambler's guidebook in which he discussed how to cheat and how to detect those who do. He later used probability theory to predict the exact date of his death. When the date of his prediction arrived and he seemed to be in good health, Cardano committed suicide to uphold his theory. Although Cardano's work was important, it is generally agreed that the rigorous origins of probability sprang from the minds of the two great French mathematicians Pierre de Fermat (1601–1665) and Blaise Pascal (1623–1662) around 1654. By studying games of chance, they developed much of the elementary theory of probability.

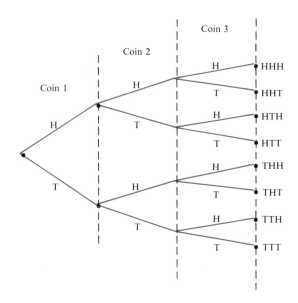

Figure 3

The sample space for tossing three coins

Solution

(a) Event of getting one head = {HTT, THT, TTH}

(b) Event of getting two heads = {HHT, HTH, THH}

(c) Event of getting three heads = {HHH}   □

Confusion Between the Physical Experiment and the Sample Space

Flip a pair of coins and observe whether they turn up heads or tails. Here the sample space is $S = \{HH, HT, TH, TT\}$. However, we can associate other sample spaces with the tossing of the same pair of coins. The tossing of the pair of coins merely represents the physical activity that we perform. The outcome of the probability experiment is what we observe or measure. For instance, if we toss two coins, we can measure several different things.

· We can measure the total number of heads. Here the sample space is

$$S = \{0, 1, 2\}$$

· We can measure whether or not the coins match or do not match. Here the sample space is

$$S = \{\text{Match, Not match}\}$$

The point that we are illustrating is that the sample space represents the thing we are actually observing or measuring and not the physical experiment.

Example 5

Sample Spaces in Business   A quality control engineer for a high-technology electronics company tests a critical computer component and records the length of time it takes before the component fails. What is the sample space of this experiment?

Solution   The time in hours it takes a component to fail can be any positive number. Hence the sample space of this experiment is written as

$$S = \{t \mid t > 0\} \qquad (t \text{ is time it takes to fail})$$

An important subset of S would be the event

$$D = \{t \mid 0 < t < 500\} \qquad (\text{failure within 500 hours})$$

which might be called "the defective event." Any experiment the engineer conducts whose outcome falls within this event would represent a component that failed within the first 500 hours. On the other hand, the event

$$ND = \{t \mid 500 \leq t < \infty\} \qquad (\text{lasting at least 500 hours})$$

might be given the name "nondefective event." Any experiment the engineer conducts whose outcome falls within this event would represent a component that lasted at least 500 hours. (See Figure 4.) The company might like to know for a given type of component the chances that the outcome will lie in each of these events.

Figure 4
Possible lifetimes for a computer component

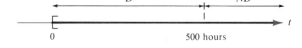

Example 6

Infinite Sample Space   A coin is flipped until a head appears. What is the sample space of this experiment?

Solution   If a head appears on the first toss, the experiment is stopped, and the outcome of the experiment is H. If a tail appears on the first toss but a head appears on the second toss, then the experiment is stopped, and the outcome is TH. Continuing in this way, we can see that the sample space is

$$S = \{\text{H, TH, TTH, TTTH, TTTTH}, \ldots\}$$

Since there are an infinite number of outcomes in this sample space, the sample space is called an **infinite sample space**.

In Section 4.1 we showed how Venn diagrams can be used to illustrate some of the basic concepts of sets. We now show how Venn diagrams can also illustrate some basic concepts of probability.

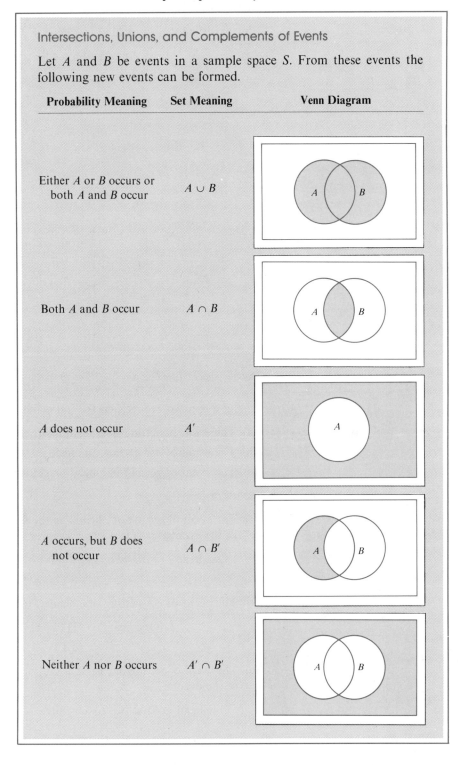

Intersections, Unions, and Complements of Events

Let A and B be events in a sample space S. From these events the following new events can be formed.

Probability Meaning	Set Meaning	Venn Diagram
Either A or B occurs or both A and B occur	$A \cup B$	
Both A and B occur	$A \cap B$	
A does not occur	A'	
A occurs, but B does not occur	$A \cap B'$	
Neither A nor B occurs	$A' \cap B'$	

We now illustrate how unions and intersections of events come about in typical problems.

Example 7 ——————————————

Combining Events A penny is tossed three times. The sample space S of this experiment is illustrated in Figure 5. Consider the two events in S:

$$\text{One head tossed} = \{HTT, THT, TTH\}$$

$$\text{First two tosses tails} = \{TTT, TTH\}$$

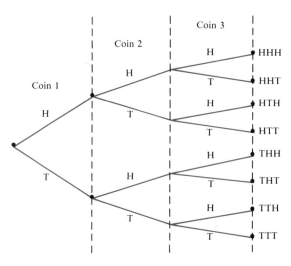

Figure 5
Sample space of tossing a penny three times

Using these two events, find the following new events:

(a) The event of getting one head *and* the first two tosses tails.

(b) The event of *not* getting one head.

(c) The event of getting one head *or* getting the first two tosses tails.

Solution

(a) The event of getting one head *and* the first two tosses tails is the intersection of the events:

$$\text{One head} \cap \text{First two tosses tails} = \{HTT, THT, TTH\} \cap \{TTT, TTH\}$$
$$= \{TTH\}$$

(b) The event of *not* getting one head is the complement

$$(\text{One head})' = \{HTT, THT, TTH\}'$$
$$= \{TTT, THH, HTH, HHT, HHH\}$$

(c) The event of getting one head *or* getting the first two tosses tails is the union of the events

$$\text{One head} \cup \text{First two tosses tails} = \{HTT, THT, TTH\} \cup \{TTT, TTH\}$$
$$= \{HTT, THT, TTH, TTT\} \qquad \square$$

Problems

For Problems 1–6, find the sample space when three coins are tossed if the following quantities are observed.

1. If we observe whether each coin is a head or a tail.
2. If we observe the number of heads tossed.
3. If we compute the sum when we let Head = 2 and Tail = 5.
4. If we observe the difference

    Number of tails − Number of heads

5. If we observe the difference

    (Number of heads)^2 − (Number of tails)^2

6. If we observe the number of switches from head to tail or tail to head (the first toss does not count as a switch) if the three coins are tossed sequentially.

For Problems 7–11, illustrate by means of a diagram the sample space of rolling a red die and a green die. Shade the following events.

7. Shade the event that the sum of the two dice is greater than or equal to 10.
8. Shade the event that the number shown on the red die is 2 or more larger than the number shown on the green die.
9. Shade the event that the sum of squares of the numbers on the two dice is less than 9.
10. Shade the event that the sum of the two dice is 7 or 11 (winning roll in craps).
11. Shade the event that the difference (red die − green die) is 2.

Events in Poker

For Problems 12–15, you are dealt five cards from an ordinary playing deck of 52 cards.

12. How many possible hands are there in the sample space of all possible hands?
13. How many elements are there in the event of getting a royal flush (A, K, Q, J, 10 of the same suit)?
14. How many elements are there in the event of getting a three of a kind (three cards the same, the other two cards neither matching each other nor the same as the three matching cards)?
15. How many elements are there in the event of getting a straight flush (all cards in a run of the same suit—except 10, J, Q, K, A, which is a royal flush)?
16. **Finding the Sample Space** A physician measures a person's weight and systolic blood pressure (which is generally between 90 mg and 210 mg of mercury). What

would seem to be a reasonable sample space for this experiment?

For Problems 17–18, a couple has three children.

17. Describe the sample space if you are interested in knowing the sex of each child (oldest, middle, youngest).
18. Describe the sample space if you are interested only in knowing the number of girls.

Finding the Sample Space

For Problems 19–27, describe the sample space of the given experiment.

19. An economist is studying the incomes of families in a given city.
20. A sociologist is studying the size of families.
21. A political scientist is computing the percentage of voters who favor a political candidate.
22. A computer scientist is tabulating how many users of a computer are logged onto a computer system at various times of the day.
23. A sports reporter is recording the number of points Bernard King gets in different games.
24. A zoologist is sampling the sizes of lizards in Sumatra.
25. An interviewer is questioning consumers to find their preference in brands of ice cream.
26. A biologist is recording the diameters of tumors in mice five days after the mice have been injected with tumor cells.
27. A zoologist is measuring the weight change in pigs being fed a given diet.

For Problems 28–32, a coin and a die are tossed simultaneously.

28. Enumerate the elements of the sample space of this experiment.
29. Enumerate the elements of the event that the coin turns up heads.
30. Enumerate the elements of the event that the die turns up a 3.
31. Enumerate the elements of the event that the die turns up a 3 or a 6.
32. Enumerate the elements of the event that the coin turns up a head and the die turns up a 2.

For Problems 33–37, a red die and a green die are rolled.

33. Enumerate the elements of the event that the sum is a 6 or a 9.
34. Enumerate the elements of the event that the sum is even.
35. Enumerate the elements of the event that the red die is a 2 or 3.

36. Enumerate the elements of the event that the sum is a 7 or the red die is a 5.

37. Enumerate the elements of the event that the sum is a 2.

Unions, Intersections, and Complements of Events

For Problems 38–50, a coin is tossed three times. The sample space of this experiment is

$$S = \{TTT, TTH, THT, THH, HTT, HTH, HHT, HHH\}$$

We define the following events:

$$H1 = \text{Event that the first toss is a head}$$

$$T = \text{Event that two of the coins are heads}$$

$$A = \text{Event that all three coins are heads}$$

Enumerate the elements in the indicated events.

38. $H1 \cup T$ **39.** $H1 \cap T$
40. $H1 \cup A$ **41.** $H1 \cap A$
42. $T \cup A$ **43.** $T \cap A$
44. $H1 \cup T \cup A$ **45.** $H1 \cap T \cap A$
46. $H1'$ **47.** T'
48. A' **49.** $(H1 \cup T)'$
50. $(T \cap A)'$

5.2

The Concept of Probability

PURPOSE

We introduce the concept of probability and show

- how to find the probability $P(A)$ of an event,
- how to find the probability of the complement of an event, and
- how to find the probability of the union of events.

The most important questions of life are, for the most part, really only problems of probability.

—Laplace

Meaning of Probability

Probability has to do with our confidence about knowing whether something will happen. One often hears something like "There is a 40% chance of rain tomorrow" or "The chances that the Yankees will win the pennant are 1 in 100." The subject of mathematical probability provides a framework for making these ideas precise.

Consider a probability experiment that can have several possible outcomes. Suppose that none of the outcomes can be predicted with 100% accuracy but that a given "degree of confidence" P can be assigned to each outcome. The value of P measures the likelihood that a given outcome will occur. For instance, if we assign the value $P = 0.50$ to an outcome, this means that the likelihood that an outcome occurs on any trial of the experiment is 50%. The number P is called the **probability** that the outcome occurs. The concept of probability is made precise in the following definition.

Definition of Probability

Let S be the finite sample space

$$S = \{s_1, s_2, s_3, \ldots, s_n\}$$

Assign to each outcome s_i of S any number $P(s_i)$ such that the value of $P(s_i)$ satisfies the two **fundamental laws of probability**:

1. $0 \leq P(s_i) \leq 1 \qquad (i = 1, 2, \ldots, n)$

2. $P(s_1) + P(s_2) + \cdots + P(s_n) = 1$

The number $P(s_i)$ is called the **probability** that the outcome s_i occurs.

Example 1

Coin Toss   Toss a coin and observe whether it turns up heads or tails. The sample space of this probability experiment is

$$S = \{H, T\}$$

Determine both the probability that the coin turns up heads and the probability that it turns up tails.

Solution   Since both heads and tails have the same chance of occurring, we would assign the probabilities

$$P(H) = 1/2$$
$$P(T) = 1/2$$

This assignment of probabilities satisfies the two fundamental laws of probability.   □

When every outcome of a probability experiment has the same chance of occurring (as in the above example), the outcomes are called **equally likely**.

Example 2

Coin Toss Three Times   Suppose a fair coin is tossed three times. The sample space of this experiment consists of the eight equally likely outcomes:

$$S = \{HHH, HHT, HTH, HTT, THH, THT, TTH, TTT\}$$

Assign probabilities to each outcome of the experiment.

Solution   Since each of the eight outcomes has the same chance of occurring, we assign

$$P(HHH) = 1/8 = 0.125$$
$$P(HHT) = 1/8 = 0.125$$
$$P(HTH) = 1/8 = 0.125$$
$$P(HTT) = 1/8 = 0.125$$
$$P(THH) = 1/8 = 0.125$$
$$P(THT) = 1/8 = 0.125$$
$$P(TTH) = 1/8 = 0.125$$
$$P(TTT) = 1/8 = 0.125$$ □

Example 3

Unequal Probabilities   Suppose New York Mets pitcher Dwight Gooden has already thrown 2793 pitches this season. Of these pitches, he has thrown 546 on a 3–2 count (three balls, two strikes), which are further classified in Table 1.

TABLE 1
Dwight Gooden's 3–2 Pitches

Type of Pitch	Number of Pitches Thrown
Fastball	340
Slider	113
Curve	81
Knuckleball	12
Total	546

A batter facing Gooden on this count would conclude that the sample space of Gooden's pitches is

$$S = \{\text{Fastball, Slider, Curve, Knuckleball}\}$$

Assign probabilities to the outcomes in the above sample space.

Solution   From Table 1 we would assign the following probabilities:

$$
\begin{aligned}
P(\text{Fastball}) &= 340/546 = 0.62 \\
P(\text{Slider}) &= 113/546 = 0.21 \\
P(\text{Curve}) &= 81/546 = 0.15 \\
P(\text{Knuckleball}) &= 12/546 = 0.02 \\
\hline
&1.00
\end{aligned}
$$

□

Probability of an Event

We have defined the probability P that an outcome of a probability experiment occurs. We can also define the probability that an event occurs. We have said before that an event occurs when the outcome of the experiment is one of the outcomes in the event. We can then define the **probability of an event** to mean the probability that the outcome of the experiment is one of the outcomes in the event. The following rules show how to find the probability of an event, whether all the outcomes of the event are equally likely or not.

Computation of Probabilities of Events

Let S be a sample space with n equally likely outcomes, and let A be an event in S that contains k of these outcomes. Then the probability that the event A occurs is

$$P(A) = \frac{k}{n}$$

In a sample space that is not equally likely, the probability of an event A can be found by adding the probabilities of each of the outcomes in A. For the two special events $\varnothing$ and S, called the "impossible event" and the "certain event," respectively, we assign the probabilities

$$P(\varnothing) = 0$$
$$P(S) = 1$$

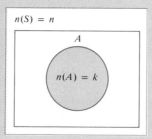

Example 4

Baseball Probabilities   In Example 3 we listed the probabilities that Dwight Gooden would throw four different pitches. Using these probabilities, what is the probability that Dwight Gooden will throw

(a) a fastball *or* a curve?

(b) a fastball *or* a curve *or* a knuckleball?

Solution   Recalling the probabilities of the individual outcomes, we have

$$P(\text{Fastball}) = 0.62$$
$$P(\text{Slider}) = 0.21$$
$$P(\text{Curve}) = 0.15$$
$$P(\text{Knuckleball}) = 0.02$$

Note that the sample space

$$S = \{\text{Fastball, Slider, Curve, Knuckleball}\}$$

is not an equally likely sample space.

(a) Let A be the desired event of throwing a fastball *or* a curve. That is,

$$A = \{\text{Fastball, Curve}\}$$

To find $P(A)$, simply add the probabilities of the outcomes in A, getting

$$P(A) = P(\text{Fastball}) + P(\text{Curve})$$
$$= 0.62 + 0.15$$
$$= 0.77$$

(b) Let A be the desired event of throwing a fastball *or* a curve *or* a knuckleball. That is,
$$A = \{\text{Fastball, Curve, Knuckleball}\}$$

Adding the individual probabilities for the outcomes in this event gives

$$P(A) = 0.62 + 0.15 + 0.02$$
$$= 0.79 \qquad \qquad \Box$$

Example 5

Probabilities in Poker   The table in Figure 6 lists the ten different poker hands and the probability that a player will be dealt each hand. Using this table, find the probability of being dealt the following hands:

(a) Three of a kind or better.

(b) Two pairs or better.

(c) Full house or better.

Solution   Each of the ten hands can be interpreted as an outcome in the sample space:

$$S = \{\text{Royal flush, Straight flush}, \ldots, \text{Pair, Other}\}$$

with given probabilities. Hence we have

(a) $P(\text{Three of a kind or better}) = 0.0211285 + \cdots + 0.0000015$
$$= 0.0287145$$

Poker hand	Hand	Number of favorable events	Probability
Royal flush		4	.00000153908
Other straight flush		36	.00001385169
Four of a kind		624	.00024009604
Full house		3,744	.00144057623
Flush		5,108	.00196540155
Straight	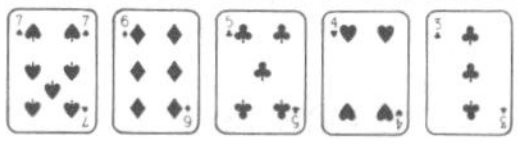	10,200	.00392464682
Three of a kind		54,912	.02112845138
Two pairs	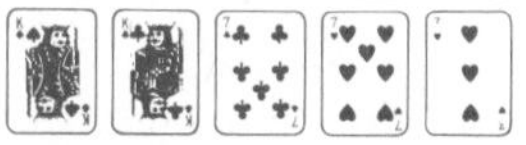	123,552	.04753901561
One pair	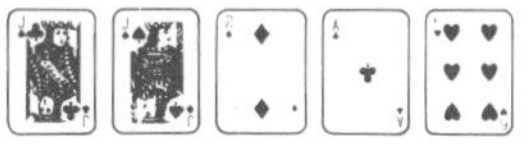	1,098,240	.42256902761
Other hands		1,302,540	.50117739403
Totals		2,598,960	1.00000000000

Figure 6
Poker hands

(b) $P(\text{Two pairs or better}) = 0.0475390 + \cdots + 0.0000015$
$$= 0.0762535$$

(c) $P(\text{Full house or better}) = 0.0014406 + \cdots + 0.0000015$
$$= 0.0016961$$

Example 6

Equally Likely Outcomes   Roll a pair of dice. The sample space of this experiment is shown below. What is the probability that the sum turned up is a 7 or an 11?

Solution   There are $n = 36$ outcomes in the sample space and $k = 8$ outcomes in the desired event:

Rolling 7 or 11 = {(6, 1), (5, 2), (4, 3), (3, 4), (2, 5), (1, 6), (5, 6), (6, 5)}

(See Figure 7.)
Since the outcomes of the sample space are equally likely, we have

$$P(\text{Rolling 7 or 11}) = \frac{k}{n}$$

$$= \frac{8}{36}$$

$$= 0.22$$

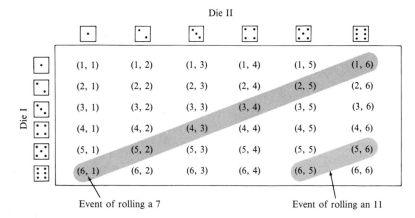

Figure 7
Event of rolling a 7 or 11

In other words, there is a 22% chance of rolling a 7 or an 11. □

Probability of a Complement

The **complement law of probability** makes it possible to find the probability of an event in terms of the probability of its complement.

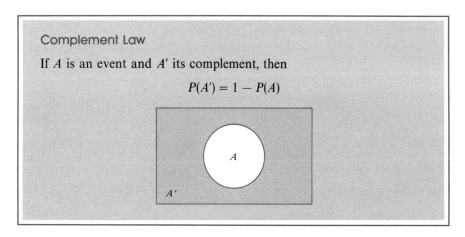

Complement Law

If A is an event and A' its complement, then

$$P(A') = 1 - P(A)$$

The importance of this rule lies in the fact that finding the probability of an event is sometimes difficult but finding the probability of its complement is easy.

We present a famous problem, the *birthday problem*, which makes use of the complement rule.

Example 7

The Birthday Problem Suppose there are five people in a room. What is the probability that at least two of these poeple's birthdays fall on the same day of the year?

Solution The process of selecting five people at random and recording their birthdays is a probability experiment whose sample space has

$$356^5 = 6,478,348,728,585$$

outcomes. One of these outcomes is

(May 4, June 8, August 21, January 13, November 28)

We now define the event

$$B = \text{Event that } at \ least \text{ two people have the same birthday}$$

A typical outcome in this event is

$$(\text{May 1, June 4, September 13, June 4, December 20})$$

Note that the second and fourth people have the same birthday. The difficulty in computing $P(B)$ lies in the fact that it is very hard to count the number of elements in B. The rules of counting that we learned in the previous chapter (trees, multiplication rule, permutations and combinations) cannot be used to determine the size of B. There is, however, a way to solve this problem by using the complement rule. Note that the complement of B is

$$B' = \text{Event that } no \text{ two people have the same birthday}$$

We can count the elements in this set by using the multiplication principle. The first person's birthday can be selected in 365 ways. After the first person's birthday has been assigned, the second person's birthday can be assigned in 364 ways (and not have the same birthday as the first person). The third person's birthday can then be assigned in 363 ways (and not have the same birthday as either of the first two people). Likewise, the fourth and fifth people can have their birthdays assigned in 362 and 361 ways, respectively. Hence by the multiplication rule the total number of ways that five people can have their birthdays with no matching birthdays is

$$\text{Size of } B' = P(365, 5) = 365 \cdot 364 \cdot 363 \cdot 362 \cdot 361$$

Hence we can find $P(B')$ from the formula

$$P(B') = P(\text{No two people have the same birthday})$$

$$= \frac{\text{Number of ways 5 people can have birthdays without a match}}{\text{Total ways 5 people can have birthdays}}$$

$$= \frac{365 \cdot 364 \cdot 363 \cdot 362 \cdot 361}{(365)^5}$$

$$= 0.973$$

This gives us the probability of B'. By the complement rule we can find the probability of the desired event:

$$P(\text{At least two people have the same birthday}) = 1 - 0.973$$
$$= 0.027$$

In other words, in any group of five people there is a 2.7% chance that two or more people will have the same birthday.   □

The interesting fact about the birthday problem is the manner in which the probability changes as a function of the number of people in the room. If there are 25 people in the room, the probability that at least two of these people will have the same birthday is

$$P(\text{At least two people have the same birthday}) = 1 - \frac{P(365, 25)}{(365)^{25}}$$

$$= 0.569$$

It turns out that in a room of 100 people, the probability that at least two people have the same birthday is 0.9999996. Stated another way, if you walked into a room of 100 people every day for the next 9000 years, you would expect to find people with matching birthdays every day but one!

Table 2 shows some results of the birthday problem with different numbers of people.

TABLE 2
Summary of the Birthday
Problem

	Number of People	Probability of Matching Birthdays
	5	0.03
	10	0.12
	15	0.25
	20	0.41
Over 50% chance	21	0.44
of matching	22	0.48
birthdays ⟶	**23**	**0.51**
	24	0.54
	25	0.57
	30	0.71
	40	0.89
	50	0.97
	60	0.994
	70	0.9991673
	80	0.9999166
	90	0.9999937
	100	0.9999996

Probabilities can always be restated in the language of odds. The language of odds is not often used in scientific work, but it is often used in gaming, sports, and other areas.

Odds and Probabilities

The **odds** in favor of an event are $r:s$ (r to s) if

$$\frac{r}{s} = \frac{\text{Probability the event happens}}{\text{Probability the event does not happen}}$$

Odds $r:s$ are generally quoted in lowest terms. Thus 10:5 odds would be the same as 2:1.

If the odds in favor of an event are $r:s$, then the probability p the event occurs is

$$p = \frac{r}{r+s}$$

Example 8

Changing Probability to Odds   Table 3 gives the probability that a college football player drafted on a given round will make a team in the National Football League.* What are the odds that a player drafted in a given round will make a team in the NFL?

TABLE 3
Probability of Making
the NFL

Round	Probability of Making NFL
1st	0.99
2nd	0.91
3rd–4th	0.83
5th–7th	0.55
8th–10th	0.57
11th–14th	0.50
15th–lower	0.40

Solution   The odds in favor of making the NFL are $r:s$, where

$$\frac{r}{s} = \frac{\text{Probability of making NFL}}{\text{Probability of not making NFL}}$$

Hence we can compute Table 4.

TABLE 4
Odds of Making the NFL

Round	Odds of Making NFL		
1st	99:1	or roughly	100:1
2nd	91:9	or roughly	10:1
3rd–4th	83:17	or roughly	5:1
5th–7th	55:45	or	11:9
8th–10th	57:43	or roughly	4:3
11th–14th	50:50	or	1:1
15th–lower	40:60	or	2:3

☐

Example 9

Changing Odds to Probability   Racetracks generally quote odds for horses before each race. The odds are always stated for the horse's losing. Suppose the odds in Table 5 are posted. What horse is the favorite? What horse is the longshot?

TABLE 5
Odds at the Racetrack

Horse	Odds to Lose
Instant Fader	5:3
Crazy Glue	3:1
Momma's Boy	4:1
Falling Star	17:3
Don't Bet on Me	39:1

Solution   We change the odds $r:s$ to a probability p of losing by use of the formula

$$p = \frac{r}{r+s}$$

Hence we have Table 6. In other words, Instant Fader is the favorite, and Don't Bet on Me is the longshot.   ☐

---

* Table 3 was taken from *The World Book of Odds* by Neft, Cohen, and Deutsch (Grosset and Dunlop, 1984).

TABLE 6
Probability at the Racetrack

Horse	Probability of Losing	Probability of Winning
Instant Fader	$\dfrac{5}{8} = 0.625$	$\dfrac{3}{8} = 0.375$
Crazy Glue	$\dfrac{3}{4} = 0.750$	$\dfrac{1}{4} = 0.250$
Momma's Boy	$\dfrac{4}{5} = 0.800$	$\dfrac{1}{5} = 0.200$
Falling Star	$\dfrac{17}{20} = 0.850$	$\dfrac{3}{20} = 0.150$
Don't Bet on Me	$\dfrac{39}{40} = 0.975$	$\dfrac{1}{40} = 0.025$

Probability of a Union of Two Events

Suppose you randomly select a card from a deck of 52 playing cards. How would you find the probability that the card is a king *or* a red card? In this problem the subsets of "all kings" and "all red cards" are two events. We are trying to find the probability that one event *or* another event will occur. This brings us to an important rule, the rule that finds the probability of a union of two events.

Probability Rule for Unions

If A and B are events, then

$$P(A \cup B) = P(A) + P(B) - P(A \cap B)$$

If A and B are disjoint events, then

$$P(A \cup B) = P(A) + P(B)$$

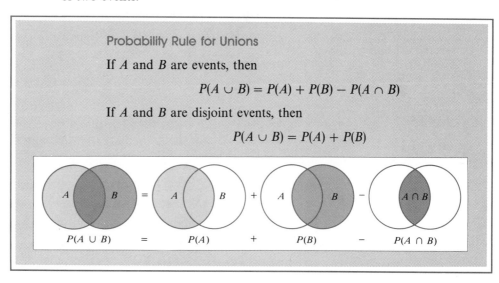

$$P(A \cup B) \quad = \quad P(A) \quad + \quad P(B) \quad - \quad P(A \cap B)$$

Verification of $P(A \cup B) = P(A) + P(B) - P(A \cap B)$

Consider the Venn diagram in Figure 8 illustrating two events A and B. It is convenient to denote

- the elements that are in A but not B as $a_1, a_2, \ldots, a_m$;
- the elements that are in B but not A as $b_1, b_2, \ldots, b_n$;
- the elements that are in A and B as $s_1, s_2, \ldots, s_k$.

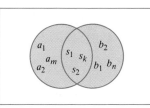

Figure 8
Proof that
$(A \cup B) = P(A) + P(B) - P(A \cap B)$

Hence we have

$$A = \{s_1, s_2, \ldots, s_k, a_1, a_2, \ldots, a_m\}$$
$$B = \{s_1, s_2, \ldots, s_k, b_1, b_2, \ldots, b_n\}$$
$$A \cup B = \{s_1, s_2, \ldots, s_k, a_1, a_2, \ldots, a_m, b_1, b_2, \ldots, b_n\}$$

We now compute the probabilities of the following events by summing the probabilities of the outcomes:

$$P(A \cap B) = P(s_1) + \cdots + P(s_k)$$
$$P(A \cup B) = P(s_1) + \cdots + P(s_k) + P(a_1) + \cdots + P(a_m) + P(b_1) + \cdots + P(b_n)$$
$$P(A) = P(s_1) + \cdots + P(s_k) + P(a_1) + \cdots + P(a_m)$$
$$P(B) = P(s_1) + \cdots + P(s_k) + P(b_1) + \cdots + P(b_n)$$

By taking a close look at the above probabilities and performing the necessary arithmetic we get the desired result:

$$P(A \cup B) = P(A) + P(B) - P(A \cap B)$$

Example 10

Probability of Disjoint Events   Roll a pair of dice. What is the probability that the sum is a 7 or an 11?

Solution   Since the events

- Event of rolling a 7
- Event of rolling an 11

are disjoint events, as the reader can verify, we can use the rule

$$P(\text{Rolling a 7 or 11}) = P(\text{Rolling a 7}) + P(\text{Rolling an 11})$$
$$= 6/36 + 2/36$$
$$= 2/9$$

$\square$

Example 11

Probability of Overlapping Events   Roll a pair of dice (call them Die I and Die II). What is the probability that the sum of the rolled numbers is a 7 or that Die II is a 3?

Solution   The two events

$$S7 = \text{Event that the sum is a 7}$$
$$= \{(1, 6), (2, 5), (3, 4), (4, 3), (5, 2), (6, 1)\}$$
$$D3 = \text{Event that Die II is a 3}$$
$$= \{(1, 3), (2, 3), (3, 3), (4, 3), (5, 3), (6, 3)\}$$

are not disjoint, since the outcome (4, 3) is common to both events. Hence, using the probability law for unions, we have

$$P(S7 \cup D3) = P(S7) + P(D3) - P(S7 \cap D3)$$
$$= 6/36 + 6/36 - 1/36$$
$$= 11/36$$

See Figure 9.

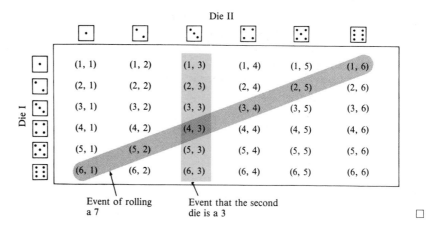

Figure 9
Two intersecting events

Die II

Event of rolling a 7

Event that the second die is a 3

Problems

Flipping Coins

For Problems 1–8, the probability tree in Figure 10 illustrates the sample space of tossing a coin four times. Find the indicated probabilities.

1. What is the probability of tossing no heads (four tails)?
2. What is the probability of tossing one head (three tails)?
3. What is the probability of tossing two heads (two tails)?
4. What is the probability of tossing three heads (one tail)?

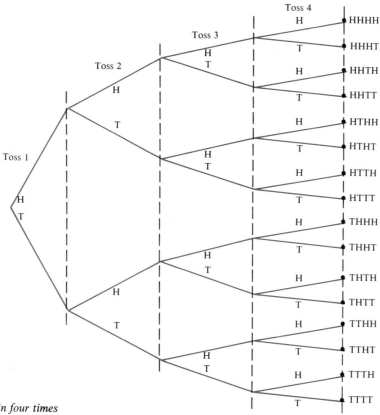

Figure 10
Outcomes of tossing a coin four times

5. What is the probability of tossing four heads (no tails)?
6. What is the probability of tossing two or four heads?
7. What is the probability of tossing two more heads than tails or three tails?
8. What is the probability of tossing three more heads than tails?

For Problems 9–14, select one card from a deck of 52 playing cards. What is the probability, and what are the odds in favor of the indicated events?

9. The card is an ace.
10. The card is a heart.
11. The card is an ace or a heart.
12. The card is a king or a queen.
13. The card is black or a four.
14. The card is a face card (A, K, Q, J) or black.

Rolling Dice

For Problems 15–20, roll three dice. What is the probability that the sum of the three numbers is the following values?

15. 3	**16.** 5	**17.** 18
18. 20 or 5	**19.** 3 or 18	**20.** 3 or 4 or 18

Venn Diagram

Problems 21–32 refer to the Venn diagram for events A and B in an equally likely sample space as shown in Figure 11. Find each of the following probabilities.

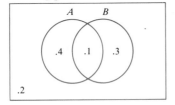

Figure 11
Probabilities of disjoint events

21. $P(A)$	**22.** $P(A')$	**23.** $P(B)$
24. $P(B')$	**25.** $P(A \cup B)$	**26.** $P(A \cap B)$
27. $P(A' \cup B)$	**28.** $P(A \cup B')$	**29.** $P(A \cap B')$
30. $P(A' \cap B)$	**31.** $P(A' \cup B')$	**32.** $P(A' \cap B')$

Pascal's Triangle

Someone once said that there are 1000 ideas hidden in *Pascal's triangle*. Pascal himself wrote a book on the triangle and confessed that there was much that he did not know. The numbers in Figure 12 are the binomial coefficients studied in Section 4.3. Each row of the triangle gives useful information. For example, the third row says that if three coins are tossed, then:

Number of heads	0	1	2	3
Number of outcomes	1	3	3	1
Probability of outcomes	1/8	3/8	3/8	1/8

Use Pascal's triangle to answer the questions in Problems 33–37.

33. What are the probabilities of getting 0, 1, 2, 3, 4, or 5 heads, when a coin is tossed five times?
34. What is the probability that one coin is a head and the other a tail when two coins are tossed?
35. What is the probability of a couple having at least four girls if they have six children (assuming equally likely outcomes for both sexes)?
36. What row shows the number of ways in which five coins can turn up?
37. The probability of getting one head is the same as the probability of getting another number of heads when tossing five coins. What is this number?

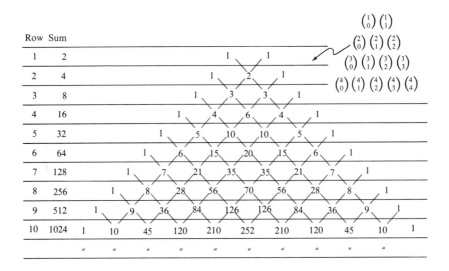

Figure 12
Pascal's triangle

38. Complement Rule Suppose you toss ten pennies. What is the probability that at least one penny will come up heads? What are the odds that this happens?

39. Complement Rule Pick three cards at random from an ordinary playing deck. What is the probability that at least one of these cards is a king? What are the odds against this happening?

40. Complement Rule The probability that a CAT-Scan will detect a given type of brain tumor is 0.9. What is the probability that at least one of two scans will detect the tumor?

41. Complement Rule Larry Bird shoots a free throw percentage of 95%. What is the probability that he will make at least one of two free throws? Note that this problem is essentially the same as Problem 40.

42. Presidents' Birthdays What is the probability that at least two past presidents had the same birthday? It turns out that Presidents Polk and Harding had birthdays on November 2. At the time of this writing, President Reagan is the 40th president.

43. Presidents' Deaths What would one think the probability should be that at least two presidents died on the same day of the year (Jefferson, Adams, and Monroe all died on July 4).

Gaming Odds

For Problems 44–47, compute the odds in favor of the following events.

44. Drawing a heart from a deck of 52 playing cards.
45. Tossing exactly two heads on three flips of a coin.
46. Drawing a 7 of hearts from a deck of 52 playing cards.
47. Drawing a 7 or a heart from a deck of 52 playing cards.

48. Sports Odds The probability that North Texas State will beat Notre Dame in football this Saturday is 0.65. What are the odds in favor of this happening?

49. Racing Odds A race track is giving 5:1 odds on Crazy Glue to win in the seventh race. This means that any one who bets on this horse will win $5 if the horse wins and lose $1 if the horse loses. Another way of saying this is

the gambler will win $5 on every $1 bet if the horse wins. What is the probability that this horse will win?

50. Racing Odds A race track is giving 4:1 odds on Crazy Glue in the seventh race. What is the probability that Crazy Glue will win?

51. Basketball Odds The following odds have been taken from *The World Book of Odds*. The odds against NCAA and NBA basketball games having various point spreads (the difference in the final score) are given in Table 7. Change the odds in Table 7 to the probability that the event will not happen.

TABLE 7

Basketball Odds for Problem 51

	Odds Against Happening	
Point Spread	*NCAA*	*NBA*
1 or 2 points	5:1	7:1
Less than 5 points	2:1	2:1
More than 10 points	1:1	5:4

Biology Problem

For Problems 52–56, 100 fish of species A and 150 fish of species B are released into a lake containing 500 fish of species C. Some fish are recaptured at a later date, one at a time, for inspection. After each fish is examined, it is released. If each fish has an equally likely chance of being captured, find the following probabilities.

52. What is the probability of capturing a fish of species A?
53. What is the probability of capturing a fish of species B?
54. What is the probability of capturing a fish of species C?
55. What is the probability of capturing a fish of species A or B?
56. What is the probability of capturing a fish of species A or C?

5.3

Conditional Probability and Independence of Events

PURPOSE

We introduce the important ideas of

- conditional probability,
- joint probability of two arbitrary events,
- joint probability of two independent events, and
- probability trees.

Conditional Probability

The probability that an individual will buy a tube of SuperWhite toothpaste, given that the individual has seen an advertisement for the product, is called a **conditional probability**. Often, we would like to compare different conditional probabilities. For example, what is the relationship between the probability of getting lung cancer given that an individual smokes and given that an individual does not smoke? The following game illustrates the concept of conditional probability.

The Marble Game (Game of Conditional Probability)

As a simple geometric illustration of conditional probability, suppose we play the following game. A marble is dropped from a considerable height into a box. On the bottom of the box is drawn two overlapping circles. (See Figure 13.) This game constitutes a probability experiment with the points on the bottom of the box acting as the sample space. The numbers shown in Figure 13 are the probabilities that the marble will roll to rest in each of the four disjoint regions (events).

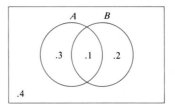

Figure 13

Probabilities that a marble will stop in each disjoint set

It is clear that there is a 40% chance that the marble will roll to rest in Region A, since we can add the probabilities of the two disjoint events that are labeled 0.3 and 0.1. That is,

$$P(A) = 0.3 + 0.1 = 0.4$$

Suppose now that we drop the marble but immediately close our eyes so that we do not see where the marble stops. Someone then gives us a clue by saying the marble has stopped within Region B. The question now is: What is the probability that the marble has stopped within Region A, given that we know that it lies within Region B? This probability is the **conditional probability** of A given B and is written as $P(A|B)$. To find it, we simply find the following ratio:

$$P(A|B) = \frac{P(A \cap B)}{P(B)} = \frac{0.1}{0.3} = \frac{1}{3}$$

This conditional probability is illustrated in Figure 14.

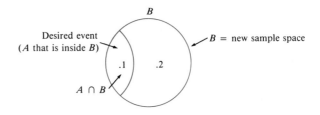

Figure 14

Relevant events and sample space for $P(A|B)$

We now state the formal definition of conditional probability.

Definition of Conditional Probability

Let A and B be events with $P(B) > 0$. The conditional probability of A given B, written $P(A|B)$, is defined by

$$P(A|B) = \frac{P(A \cap B)}{P(B)}$$

If the sample space S has a finite number of equally likely outcomes, then $P(A|B)$ can be written

$$P(A|B) = \frac{n(A \cap B)}{n(B)}$$

where $n(\cdot)$ is the number of outcomes in an event.

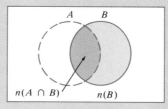

Example 1

Conditional Probability   Two coins are tossed. Assume that it is known that at least one of the coins has turned up a head. What is the probability that both coins turned up heads?

Solution   The sample space of tossing a pair of coins is

$$S = \{HH, HT, TH, TT\}$$

The two events of interest are

$$BH = \text{Both coins turning up heads} = \{HH\}$$
$$AL = \text{At least one coin turning up heads} = \{HT, TH, HH\}$$

The conditional probability that both coins turn up heads, given that at least one coin is a head, is

$$P(BH|AL) = \frac{n(BH \cap AL)}{n(AL)} = \frac{1}{3}$$

In other words, if nothing is known about the outcome of the two coins, the probability that both coins turned up heads is 1/4; but if we know that one of the coins is a head, then the probability they both are heads rises to 1/3. □

Social Science Data (Sample Surveys)

A questionnaire was distributed to 500 university students asking the following questions.

- Are you a regular cigarette smoker?
- Is your GPA below 2.00?

Consider the events

$$C = \text{Event of being a cigarette smoker}$$
$$L = \text{Event of being student with GPA below 2.00}$$
$$S = \text{Sample space of all 500 students}$$

The results of the survey are summarized in Table 8. These values are also illustrated in Figure 15.

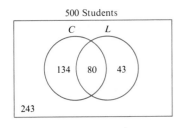

500 Students

Figure 15
Profile of student questionnaire

TABLE 8
Results of Survey on
Smoking and GPA

Type of Student	Number of Students
Low grades	$n(L) = 123$
Cigarette smoker	$n(C) = 214$
Low grades and cigarette smoker	$n(L \cap C) = 80$
Low grades or cigarette smoker	$n(L \cup C) = 257$
High grades and nonsmoker	$n(L' \cap C') = 243$
Smoker and high grades	$n(C \cap L') = 134$
Nonsmoker and low grades	$n(C' \cap L) = 43$

The following example illustrates a typical analysis of this type of data.

Example 2

Smoking Survey   Assume that the data illustrated in Figure 15 are representative of university students.

(a) What is the probability that any university student selected at random will be a smoker?

(b) What is the probability that any university student with a GPA below 2.00 selected at random will be a smoker?

Solution

(a) From the values in Figure 15 the probability that a student will be a cigarette smoker is

$$P(C) = \frac{n(C)}{n(S)}$$

$$= \frac{214}{500}$$

$$= 0.43$$

(b) The conditional probability of being a smoker given that the student has a GPA below 2.00 is

$$P(C|L) = \frac{n(C \cap L)}{n(L)}$$

$$= \frac{80}{123}$$

$$= 0.65$$

In other words, the study indicates that 65% of students with GPAs below 2.00 smoke, while 43% of students in general smoke. This indicates a 22% increase in smokers among students with low GPAs. □

Example 3

Accident Victims   The set S represents all people in a western state who were drivers responsible for an automobile accident in 1985. The following events are of interest:

$$D = \text{Drivers who were drinking}$$
$$Y = \text{Drivers who were young (under 25 years of age)}$$

Past data indicates that

·   65% of the drivers responsible for accidents were drinking,
·   40% of the drivers responsible for accidents were young,
·   30% of the drivers responsible for accidents were young and drinking.

Is it more likely that an accident is drinking-related when a young person is involved?

Solution   We would like to compare the two probabilities

$$P(D) = \text{Probability that an accident is related to drinking}$$
$$P(D|Y) = \text{Probability that an accident is related to drinking given that the driver was young}$$

We know that

$$P(D) = 0.65$$
$$P(Y) = 0.40$$
$$P(D \cap Y) = 0.30$$

Hence we can find the conditional probability

$$P(D|Y) = \frac{P(D \cap Y)}{P(Y)}$$

$$= \frac{0.30}{0.40}$$

$$= 0.75$$

On the basis of the given data, for drivers under 25 years of age the chance that an accident is alcohol-related is 0.75 in contrast to a probability of 0.65 for the general driving population. Figure 16 illustrates these probabilities. □

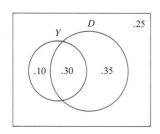

Figure 16
Probabilities of four disjoint events

TABLE 9
Results of Test for
Liver Cancer

	Disease Present, D	Disease Not Present, D'	Totals
Disease Detected, T_+	19	10	29
Disease Not Detected, T_-	5	26	31
Totals	24	36	60

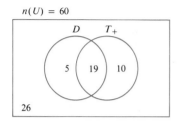

$n(U) = 60$

Figure 17
Venn diagram showing D and T_+

Conditional Probabilities in Medical Tests

A preliminary screening test for detecting cancer of the liver is being developed. Each individual is tested and then classified as either having the disease or not having the disease. After this, each individual is tested exhaustively to determine whether the disease is or is not present. The results of the experiment are shown in Table 9. The four events illustrated in Table 9 are shown in Figure 17.

The following example illustrates four conditional probabilities that are of interest in the above study.

Example 4

Conditional Probabilities in Medicine*   Find and interpret the four conditional probabilities $P(T_+|D')$, $P(T_-|D)$, $P(D|T_+)$, and $P(D'|T_-)$.

Solution   The first two probabilities above are called **error probabilities**, and the second two probabilities are called **detection probabilities**. More specifically, we have

- Error probabilities (want small):

$$P(T_+|D') = P\,[\text{test says disease present when not present}]$$
$$P(T_-|D) = P\,[\text{test says disease not present when it is present}]$$

- Detection probabilities (want big):

$$B(D|T_+) = P\,[\text{disease present when test says present}]$$
$$P(D'|T_-) = P\,[\text{disease not present when test says not present}]$$

To evaluate these probabilities, we compute

- Error probabilities:

$$P(T_+|D') = \frac{P(T_+ \cap D')}{P(D')} = \frac{(10/60)}{(36/60)} = \frac{10}{36} = 0.28$$

$$P(T_-|D) = \frac{P(T_- \cap D)}{P(D)} = \frac{(5/60)}{(24/60)} = \frac{5}{24} = 0.21$$

- Detection probabilities:

$$P(D|T_+) = \frac{P(D \cap T_+)}{P(T_+)} = \frac{(19/60)}{(29/60)} = \frac{19}{29} = 0.66$$

$$P(D'|T_-) = \frac{P(D' \cap T_-)}{P(T_-)} = \frac{(26/60)}{(31/60)} = \frac{26}{31} = 0.84 \qquad \square$$

---

* This example was based in part on the article, "Conditional Probability and Medical Tests," by J. S. Milton and J. J. Corbet, *UMAP Journal* Vol. 3, No. 2, 1982.

Conclusion of Test

Note that the two error probabilities are lower than the two detection probabilities. If the disease is life-threatening, then it is critical that the error probability $P(T_-|D)$ of a **false negative** be very small. Here the error probability of 0.21 means that (on the average) 21% of individuals who have the disease will be judged as not having the disease by the test (failure to detect). The error probability $P(T_+|D')$ of a **false positive** should also be small, but it is not as critical. The value of $P(T_+|D') = 0.28$ means that 28% of individuals who do not have the disease will be judged by the test as having the disease (false alarm error). For people who are judged by this test as having the disease, a second more extensive test could be performed to make a more precise determination (if possible).

Most medical tests have lower error probabilities than the test illustrated here. The test that should be used depends on several things: the error probabilities, the medical side-effects of the test, and the cost of performing the test as well as other factors.

The Joint Probability $P(A \cap B)$

The formula

$$P(A|B) = \frac{P(A \cap B)}{P(B)}$$

gives the conditional probability $P(A|B)$ in terms of $P(A \cap B)$ and $P(B)$. It is also possible to turn this formula around and solve for the **joint probability** $P(A \cap B)$.

Product Rule for Joint Probabilities

Let A and B be events. The joint probability that both events A and B occur is given by

$$P(A \cap B) = P(B)P(A|B) \qquad \text{(both right-hand sides are the same)}$$
$$= P(A)P(B|A)$$

Since we can interchange A and B on the left-hand side of the above equation, we can also interchange A and B on the right-hand side. Hence we have the two alternative expressions for $P(A \cap B)$.

Example 5 _____

Joint Probability   Select two cards without replacement from a deck of 52 playing cards. What is the probability both cards will be hearts?

Solution   The two events of interest are

$$H1 = \text{Event of getting a heart on draw 1}$$
$$H2 = \text{Event of getting a heart on draw 2}$$

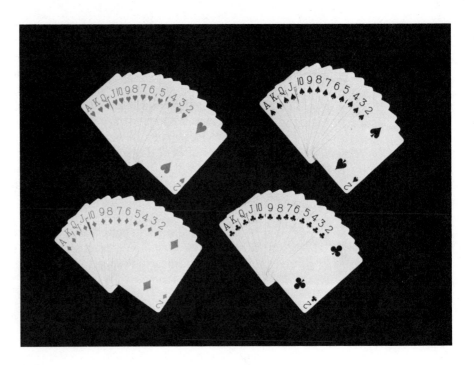

The product rule gives the joint probability

$$P(H1 \cap H2) = P(H1)P(H2|H1)$$
$$= (13/52)(12/51)$$
$$= 0.06 \qquad \square$$

Example 6 ————

Probability of Hitting a Home Run   Eddie hopes that Billy will throw him a fastball on the next pitch since he can only hit a fastball out of the park. Suppose that the probability that Billy will throw a fastball on the next pitch is 0.3 and the probability Eddie can hit a home run off a fastball is 0.4. What is the probability that Eddie will hit a home run on the next pitch?

Solution   This is a typical problem in conditional probability. If we call

$$F = \text{Event that Billy will throw a fastball}$$
$$H = \text{Event that Eddie will hit a home run}$$

then the joint probability that a fastball will be pitched *and* a home run will be hit is

$$P(F \cap H) = P(F)P(H|F)$$
$$= (0.3)(0.4)$$
$$= 0.12 \qquad \square$$

Use of Tree Diagrams in Assigning Probabilities

A tree diagram is useful for describing a sequence of experiments. If we call any direct line between two points of the tree a branch, then a path through the tree diagram consists of a sequence of branches connecting the beginning

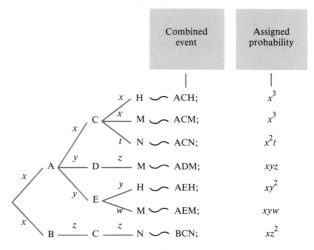

Combined event	Assigned probability
x — H — ACH;	x^3
x — M — ACM;	x^3
t — N — ACN;	$x^2 t$
z — M — ADM;	xyz
y — H — AEH;	xy^2
w — M — AEM;	xyw
z — C — z — N — BCN;	xz^2

Figure 18
The probability of following any path of a tree is equal to the product of the probabilities along that path (from James E. Mosimann, Elementary Probability for the Biological Sciences, *© 1968, pp. 201–202. Adapted by permission of Prentice-Hall, Inc., Englewood Cliffs, New Jersey)*

point of the tree with an end point of the tree. On each branch we assign the probability that the given experiment will "follow" along that specific branch. We can find the probability that the sequence of experiments takes a given path through the tree by computing the product of the probabilities along that path. The tree diagram in Figure 18 illustrates how probabilities are assigned to each path of a tree.

Tree Diagrams in Ecology*

The diagram in Figure 19 illustrates how phosphorus circulates in a simple

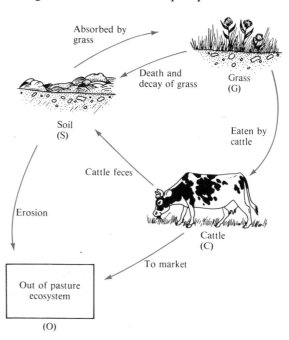

Figure 19
Diagram of phosphorus movement

---

* For a more thorough understanding of how probability is used in the biological sciences, you should consult the fascinating book *Elementary Probability for the Biological Sciences* by James Mosimann (Appleton-Century-Crofts, 1968).

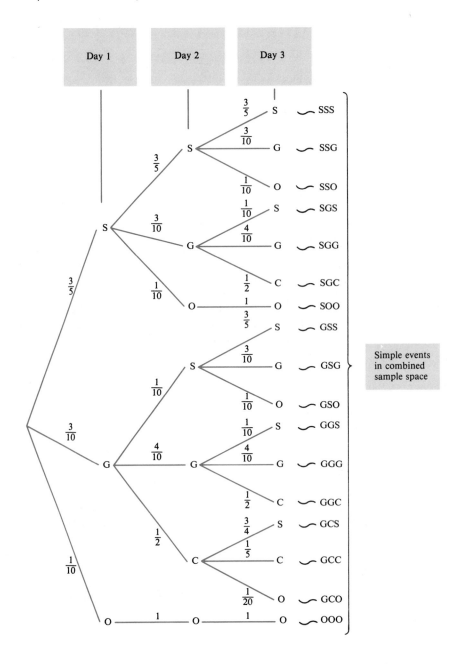

Figure 20
Tree diagram of phosphorus movement in an ecological system

ecosystem. The numbers on the branches in Figure 20 represent the probabilities that an atom of phosphorus will move among the states of soil (S), grass (G), cattle (C), or out (O). For example, the value of 1/2 on the branch connecting G with C means that there is a 50% chance that an atom of phosphorus located in the grass at the beginning of a day will be eaten by cattle by the end of the day. Also, note the high probability (3/4) that an atom of phosphorus ingested by cattle will be returned to the soil via feces by the end of the day. Finally, if an atom of phosphorus is outside the ecosystem, it stays outside.

An ecologist who is interested in knowing how fast phosphorus leaves the system would like to know the probability that an atom of phosphorus currently in the soil will be outside the system in three days time. We can find this probability with the help of the tree diagram in Figure 20. Each path of this tree denotes a three-day history of an atom of phosphorus. For example, the path GSO means that an atom of phosphorus was in the ground on day 1, in the soil on day 2, and out of the ecosystem on day 3.

From Figure 20 we can see that the event that a phosphorus atom currently in the soil will be outside the ecosystem in three days time is

$$OUT = \{SOO, SSO, GSO, GCO, OOO\}$$

By adding the probabilities that each of these simple outcomes occur, we have

$$P(OUT) = P(SOO) + P(SSO) + P(GSO) + P(GCO) + P(OOO)$$

$$= \frac{3}{50} + \frac{9}{250} + \frac{3}{1000} + \frac{3}{400} + \frac{1}{10}$$

$$= \frac{413}{2000}$$

$$= 0.2065$$

Conclusion

The probability that an atom of phosphorus that is currently in the soil will be removed from the ecosystem in three days time is 0.2065. Another way of saying this is that we can expect roughly 21% of the phosphorus atoms present in the soil today to be gone from the ecosystem in three days time.

Trees Used in Answering Sensitive Questions

One of the more interesting applications of probability trees was discovered only about 20 years ago.* Suppose we ask a sensitive question that might subject someone to embarrassment, social stigma, or even arrest. In such a situation it is unreasonable to expect truthful answers. Questions such as "Do you smoke marijuana at least once a week?" and "Have you ever been arrested for gross sexual conduct?" and "Have you ever cheated on your spouse?" are questions that many people would not answer truthfully. To find the answers to questions of this type, we use what are called **randomized responses**.

When using randomized responses, a pollster does not ask the sensitive question directly. Instead, the pollster asks the individual to carry out (in private) the following steps.

---

* Interested readers should consult the article "Randomized Response: A Survey Technique for Eliminating Evasive Answer Bias" by S. L. Warner, *Journal of the American Statistical Association* Vol. 60, pp. 63–69, 1965.

Steps for Answering a Randomized Response Questionnaire

- **Step 1.**   Flip a coin.
- **Step 2.**   If the coin turns up heads, answer the sensitive question (yes or no). If the coin turns up tails, answer the question, "Is the last digit of your Social Security number a 0, 2, 4, 6, or 8?" This innocuous question can be replaced by any other innocuous question as long as the probability of a yes-response is known. The pollster, of course, does not know the individual's Social Security number, only that the probability that the last digit is a 0, 2, 4, 6, or 8 is 1/2.

Confidential Marijuana Survey

Instructions

1. Toss the enclosed 25¢.
   —If the coin turns up a head, answer Question A.
   —If the coin turns up a tail, answer Question B.

A Do you smoke marijuana at least once a week?

B Is the last digit of your Social Security a 0, 2, 4, 6, or 8?

Indicate your answer by making an X in the appropriate box below. Do not indicate which question is being answered.

- -

Tear on dotted line and return answer in enclosed envelope.

YES   NO
☐   ☐

   Note that if the pollster receives a yes-response from an individual, the individual's privacy is still protected, since the pollster does not know what question is being answered. What might be hard to believe, however, is that by collecting randomized responses from a given number of individuals, the pollster can still estimate the fraction of individuals who answered the sensitive question with a yes-response. In other words, although the pollster does not know how each individual has answered the sensitive question, it is possible to estimate how the individuals as a group answered the sensitive question. The following example shows how probability trees can be used to answer this question.

Example 7 _____   The Marijuana Question   A pollster would like to determine the percentage of students at a midwestern university who smoke marijuana at least once a week. One hundred students were asked to follow the above rules of the randomized response questionnaire. The sensitive question was "Do you smoke marijuana at least once a week?" The innocuous question was "Is the last digit of your Social Security number a 0, 2, 4, 6, or 8?" Suppose that of these 100 students, 32 responded with a yes answer. What is the pollster's estimate of the percentage of students at this university who smoke marijuana at least once a week?

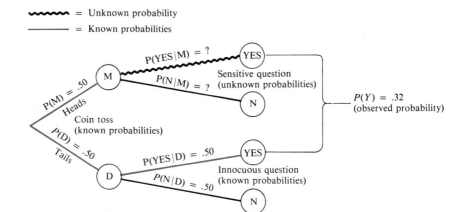

Figure 21
Probability tree illustrating the randomized response method

Solution   We draw the probability tree shown in Figure 21. In this tree we know the probabilities of the events

$$P(M) = P \text{ [student answers the marijuana question]}$$
$$= 1/2 \quad \text{(probability of tossing a head)}$$
$$P(D) = P \text{ [student answers the Social Security question]}$$
$$= 1/2 \quad \text{(probability of tossing a tail)}$$
$$P(YES|D) = P \text{ [last digit of the student's Social Security number is a 0, 2, 4, 6,}$$
$$\text{or 8]}$$
$$= 1/2$$

Since 32 students out of 100 answered yes, we have

$$P(YES) = P \text{ [student answers yes]}$$
$$= 0.32$$

From the tree diagram in Figure 21 we can see that the probability that a student will answer yes is the sum of the probabilities along the two "yes" paths. That is,

$$P(YES) = P(YES|M)P(M) + P(YES|D)P(D)$$

Since the only unknown in this equation is $P(YES|M)$, which is the probability of answering yes to smoking marijuana, we can write

$$0.32 = (0.50)P(YES|M) + (0.50)(0.50)$$

or, solving for $P(YES|M)$, we have

$$P(YES|M) = \frac{0.32 - 0.25}{0.50}$$

$$= 0.14 \qquad \square$$

Conclusion

On the basis of this survey of 100 students the pollster estimates that 14% of the students at this university smoke marijuana at least once a week. Studies indicate that the randomized response method does encourage cooperation.

Nowadays this method is being used by many social scientists to find the answers to sensitive questions.

Independence of Events

There is a story about a doctor who comforts his patient by saying, "You have a very serious disease. Of every ten patients who get this disease, only one survives. But do not worry. It is lucky that you came to me, for I recently had nine patients with this disease, and they all died."

The above story can be reformulated in terms of a coin-tossing experiment. Suppose you toss a fair coin nine times, and suppose the coin turns up heads on every toss. Based on the occurrence of the nine heads, what is the probability that the coin will turn up heads on the tenth toss? The answer, of course, is 0.50. The outcome of the first nine tosses will not affect the outcome of the tenth toss. We say that the event of getting heads on the first nine tosses and the event of getting a head on the tenth toss are **independent events**.

On the other hand, suppose that you select two cards from a deck of playing cards (without replacing the first card in the deck). In this experiment the event of selecting a red card on the first draw clearly affects the occurrence of the event of selecting a red card on the second draw. We would say that these two events are not independent events but are **dependent events**. The following definition makes these intuitive ideas precise.

Definition of Independent Events

Two events A and B with nonzero probabilities are independent if *any one* of the following equations holds.

- $P(A \cap B) = P(A)P(B)$
- $P(A|B) = P(A)$
- $P(B|A) = P(B)$

If one of the three equations is satisfied, the other two equations are also satisfied. If two events are not independent, they are called dependent.

Quite often, it is intuitively clear that two events are independent. Other times you can be fooled. If it is not intuitively clear, you should check one of the three conditions.

Example 8

Independent Events   Flip a coin, and then roll a die. The sample space of this experiment consists of the 12 elements

$$S = \{(c, d) | c \in \{\text{H, T}\}, \quad d \in \{1, 2, 3, 4, 5, 6\}\}$$

Define the two events

$$H = \text{Event of tossing a head on the coin toss}$$
$$D3 = \text{Event of rolling a three on the roll of the die}$$

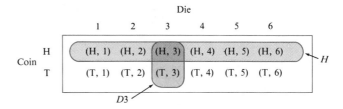

Figure 22
Two independent events

Both of these events are illustrated in Figure 22. Are these two events independent?

Solution It is intuitively clear that knowing whether one of these events has occurred provides no information about whether the other event has occurred. Hence we are inclined to say that the two events are independent.

We could also use one of the three mentioned conditions for independence in order to check our intuition. For example, if we evaluate

$$P(H) = 1/2$$
$$P(D3) = 1/6$$
$$P(H \cap D3) = 1/12$$

we see that these values satisfy the independence condition

$$P(H \cap D3) = P(H)P(D3)$$

Hence H and $D3$ are independent.　　　　　□

Example 9

Dependent Events Roll a pair of dice. The sample space consists of the 36 equally likely outcomes shown in Figure 23. Two typical events in S are the following.

$$M = \text{Matching dice} = \{(1, 1), (2, 2), (3, 3), (4, 4), (5, 5), (6, 6)\}$$
$$S4 = \text{Sum of 4} = \{(3, 1), (2, 2), (1, 3)\}$$

Are these events independent? (See Figure 23.)

Solution We check the independence condition

$$P(M \cap S4) \stackrel{?}{=} P(M)P(S4)$$

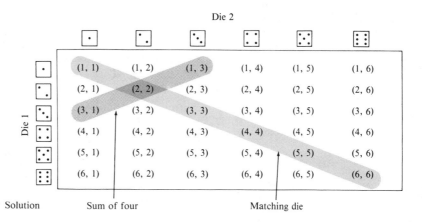

Figure 23
Dependent events

Since $M \cap S4 = \{(2, 2)\}$, we have $P(M \cap S4) = 1/36$. However,

$$P(M) = \frac{6}{36} = \frac{1}{6}$$

$$P(S4) = \frac{3}{36} = \frac{1}{12}$$

and so

$$P(M \cap S4) = \frac{1}{36} \neq \frac{1}{6}\frac{1}{12} = P(M)P(S4)$$

Hence the two events are dependent.   □

Example 10

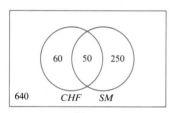

Figure 24
Venn diagram showing CHF and SM

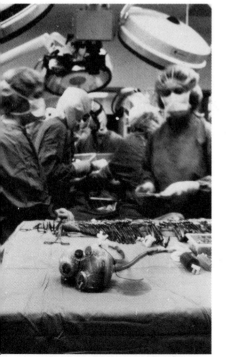

Congestive Heart Failure In a study of congestive heart failure, 1000 males over the age of 60 are examined. From this population it is discovered that 300 smoke cigarettes and that among these smokers, 50 show evidence of congestive heart failure. Among the nonsmokers, 60 show evidence of congestive heart failure. Are men who smoke more likely to have congestive heart failure?

Solution We begin by defining the two events:

$$CHF = \text{Men with evidence of congestive heart failure}$$
$$SM = \text{Men who smoke}$$

Using the information given, we can draw the Venn diagram in Figure 24. From the Venn diagram we can compute

$$P(CHF) = 110/1000 = 0.11$$
$$P(CHF|SM) = 50/300 = 0.17$$

Hence we see that

$$P(CHF|SM) \neq P(CHF)$$

 In other words, the events *CHF* and *SM* are dependent events, and we would say that the occurrence of congestive heart failure and smoking are related in some way. Note that smokers have a higher chance of getting congestive heart failure than individuals in the general population.   □

Confusion Between Independent and Disjoint Events

Students often think that independent events are disjoint. This is not true. Independence simply means that knowledge of one event does not provide knowledge of the other event.

 In fact, if A and B are two independent events (with nonzero probabilities), then they *cannot* be disjoint. Note that the two independent events in Figure 22 are independent and have the element $(H, 3)$ in common.

 On the other hand, disjoint events (with nonzero probabilities) are *automatically* dependent, since the occurrence of one event tells us something about the occurrence of the other event (since if one event occurs, then obviously the other disjoint event will not occur).

If two events are independent then there is a convenient formula for computing the joint probability $P(A \cap B)$.

Product Rule for Independent Events

If A and B are independent events, then the joint probability of A and B is
$$P(A \cap B) = P(A)P(B)$$

The formula also can be extended to find the joint probability of several independent events $A_1, A_2, \ldots, A_n$:

$$P(A_1 \cap A_2 \cap \cdots \cap A_n) = P(A_1)P(A_2) \cdots P(A_n)$$

Example 11

Several Independent Events   Toss a coin twenty times. What is the probability that the coin will appear heads every time?

Solution   The events
$$H1 \ = \text{Head on toss 1}$$
$$H2 \ = \text{Head on toss 2}$$
$$\vdots \qquad \vdots \quad \vdots$$
$$H20 = \text{Head on toss 20}$$

are clearly independent of each other. Hence we can use the general product formula

$$P(H1 \cap H2 \cap \cdots \cap H20) = P(H1)P(H2) \cdots P(H20)$$
$$= (1/2)^{20}$$
$$= \frac{1}{1,048,576}$$

In other words, the odds are $1:1,048,575$ against tossing 20 consecutive heads.
□

We close this section by presenting a summary of the laws of probability in Table 10.

TABLE 10
Summary of Probability of Two Events A, B

Events A, B	Addition Rules, $P(A \cup B)$	Multiplication Rules, $P(A \cap B)$
Neither independent nor disjoint	Three equivalent formulas: $P(A \cup B) = P(A) + P(B) - P(A \cap B)$ $= P(A) + P(B) - P(A)P(B\|A)$ $= P(A) + P(B) - P(B)P(A\|B)$	Two equivalent formulas: $P(A \cap B) = P(A)P(B\|A)$ $= P(B)P(A\|B)$
Independent events	$P(A \cup B) = P(A) + P(B) - P(A)P(B)$	$P(A \cap B) = P(A)P(B)$
Disjoint events	$P(A \cup B) = P(A) + P(B)$	$P(A \cap B) = 0$

Problems

Conditional Probability

For Problems 1–4, two playing cards are drawn from a deck of 52. Find the following probabilities.

1. The second card is black given that the first card is red.
2. The second card is black given that the first card is black.
3. The second card is a king given that the first card is a four.
4. The second card is a king given that the first card is a king.

Conditional Probability

For Problems 5–6, the Venn diagram in Figure 25 illustrates grades given by a professor to the students in a mathematics class.

5. What is the probability that a student gets an A given that the student is a business major?
6. Are A and B independent events? Does your intuition agree with the test for independence?

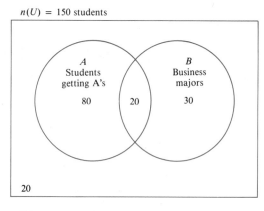

$n(U) = 150$ students

A — Students getting A's — 80

20

B — Business majors — 30

20

Figure 25
Classification of students with Venn diagram

Conditional Probability in Advertising

A survey of 100 individuals asked the two questions: "Have you seen an advertisement for SuperWhite toothpaste in the last month?" and "Did you buy SuperWhite toothpaste in the last month?" The result of this survey is shown in Table 11.

TABLE 11
Results of SuperWhite Toothpaste Survey

	Buy (B)	Not Buy (B')	Totals
Seen Ad (SA)	15	25	40
Not Seen Ad (SA')	10	50	60
Totals	25	75	100

Define the events

$$SA = \text{Event of seeing the ad}$$
$$B = \text{Event of buying the toothpaste}$$

Find the probabilities in Problems 7–11.

7. $P(B)$ 8. $P(SA)$ 9. $P(SA \cap B)$
10. $P(B|SA)$ 11. $P(B'|SA)$

12. Are the events of B and SA independent? What does this mean for the advertising director for SuperWhite toothpaste?

Product Rule for Probabilities

For Problems 13–16, two playing cards are selected without replacement from a deck of 52. Draw the probability tree that represents each experiment and find the following probabilities.

13. The first card is red and the second is black.
14. The first card is red and the second is red.
15. The first card is a king and the second is a king.
16. The first card is a king and the second is a queen.

Sampling with Replacement

For Problems 17–20, two playing cards are selected with replacement from a deck of 52. Draw the probability tree that represents this experiment and find the following probabilities.

17. The first card is red and the second is black.
18. The first card is red and the second is red.
19. The first card is a king and the second is a king.
20. The first card is a king and the second is a queen.

Conditional Probability

For Problems 21–24, let A and B be two events with

$$P(A) = 0.4$$
$$P(B) = 0.3$$
$$P(A \cap B) = 0.2$$

Draw a Venn diagram illustrating these probabilities and find

21. $P(A|B)$ 22. $P(B|A)$
23. $P(A|B')$ 24. $P(A'|B')$

25. In Problems 21–24, are the events A and B independent?

Finding Conditional Probabilities

For Problems 26–29, let A and B be events with

$$P(A) = 0.4$$
$$P(B) = 0.5$$
$$P(A \cup B) = 0.6$$

Draw a Venn diagram illustrating these probabilities and find

26. $P(A \cap B)$　**27.** $P(A|B)$　**28.** $P(B|A)$　**29.** $P(A|B')$

30. Actual Problem　An actuary knows that a 70-year-old woman has a 70% chance of living at least another 15 years and that a 70-year-old man has a 55% chance of living at least another 15 years. What are the chances that a married couple, each 70 years of age, will both reach their 85th birthdays?

31. Free Throws　Larry Bird is fouled in the act of shooting and is given two free throw attempts. His free throw average is 90%. What is the probability that he will make at least one free throw? What is the probability that he will make both free throws?

32. Brain Tumors　A new screening test for detecting brain tumors will detect tumors 90% of the time. To be on the safe side, a physician orders that two applications of the test be made on a patient. Assuming that the patient has a brain tumor, what is the probability that at least one of the tests will detect the tumor? Note the similarity between this problem and the previous free throw problem.

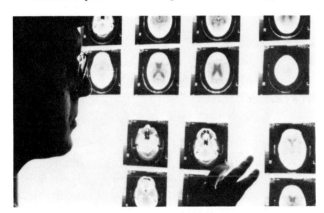

An Urn Problem (Joint Probabilities)

Although most probabilists have probably never seen an urn, they always like to pose problems in terms of urns full of red marbles and black marbles. It is a way of illustrating difficult ideas in a clear and precise fashion. Here is a typical urn problem that will test your knowledge of combinations and the multiplication rule. Suppose an urn contains

- four red marbles,
- three white marbles,
- two blue marbles.

Select two marbles, without replacement, at random from this urn. Find the probabilities in Problems 33–41.

33. Both marbles are red.
34. Both marbles are white.
35. Both marbles are blue.
36. One marble is red, and one is white.
37. One marble is white, and one is blue.
38. One marble is blue, and one is red.
39. Neither is blue.
40. Neither is red.
41. Neither is white.

Randomized Responses

Problems 42–43 are concerned with the randomized response method for finding the answer to sensitive questions.

42. Do You Cheat on Your Taxes?　To determine the percentage of individuals who cheat on their taxes, a pollster gives a randomized response questionnaire to 1000 people. The instructions are to begin by rolling a die (not in the presence of the pollster). If the die turns up a 1, 2, or 3, the individual is instructed to answer the question, "Have you cheated on your taxes within the past five years?" If the die turns up 4, 5, or 6, the individual is instructed to flip a coin and answer the question, "Did the coin turn up heads?" The individual's answer (yes or no) is then told to the pollster. The pollster, of course, does not know which of the two questions is being answered, only the answer. Suppose that of the 1000 individuals, 400 reply with a yes-answer. What can the pollster conclude that about the fraction of people who cheat on their taxes?

43. Do You Cheat on Your Taxes?—Part II.　The randomized response method in Problem 42 has been modified. The pollster conducts a new test and instructs the participants that if the die turns up 1, 2, 3, or 4, they should answer the question, "Have you cheated on your taxes within the past five years?" If the die turns up a 5 or 6, they should toss a coin and answer the question, "Is the outcome of the coin toss a head?" Suppose that of 1000 individuals who were given this test, 200 replied with a yes-answer. What can the pollster conclude about the fraction of people who cheat on their taxes? Are there any obvious advantages or disadvantages in this test compared to the test in Problem 42?

Probability of Independent Events

For Problems 44–47, find the probabilities of the following events, assuming that all events are independent.

44. Tossing ten heads in a row.
45. All eight children of a husband and wife being girls.
46. Drawing two aces from a deck of playing cards when the sampling is done with replacement.
47. Three batters getting three hits in a row when

- the first batter hits 0.350,
- the second batter hits 0.275,
- the third batter hits 0.300.

To Catch a Thief

Problems 48–50 are concerned with an interesting situation described in *Mathematics, A Human Endeavor* by Harold

Jacobs. It is based on a story from *Time* magazine (January 8, 1965). Roughly, it goes something like this. According to eye-witnesses, an elderly man is mugged while walking in the park. Witnesses describe a young *blonde-haired woman* running off and getting into a *dark Chevrolet* driven by a man wearing a *broad-brimmed* hat. Other than that, about the only other details were that she wore a *leather jacket* and was wearing *several bracelets* on her left wrist. A week later, police picked up a couple matching this description. The prosecutor made the argument that since

P(Woman has blonde hair)	$= 1/4$
P(Woman wears bracelets)	$= 1/10$
P(Woman wears a leather jacket)	$= 1/5$
P(Man wears a broad-brimmed hat)	$= 1/25$
P(Driving a dark Chevrolet)	$= 1/15$

the chance of all five occurring is

$$P(\text{All five occur}) = (1/4)(1/10)(1/5)(1/25)(1/15)$$
$$= 1/75{,}000$$

In other words, the chance that a couple chosen at random would have all five of these attributes, is 1 in 75,000. Partly on the basis of these observations the couple was convicted and sent to jail. Later, however, a higher court reversed the decision, saying that the above probabilities were suspect and in fact are not independent.

48. Do you agree with the original jail sentence or with the higher court?

49. Which of the above events would you think might not be independent of one another?

50. How might you estimate the above probabilities?

5.4

Random Variables and Their Probability Distributions

PURPOSE

We introduce three important concepts:

- the random variable,
- the probability distribution of a random variable, and
- the expectation of a random variable.

The Random Variable

When one performs probability experiments, such as tossing a coin, rolling a pair of dice, measuring vital signs of a patient, or measuring daily inventory, numbers are generally recorded. More often than not, when a probability experiment is performed, some real number is computed for each outcome of the experiment. This brings us to one of the most important ideas in probability, the random variable.

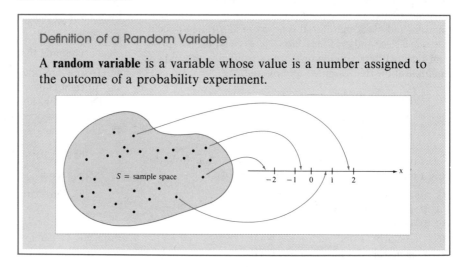

Definition of a Random Variable

A **random variable** is a variable whose value is a number assigned to the outcome of a probability experiment.

Random variables are generally denoted by capital letters X, Y, or Z.

Example 1

Typical Random Variable   Roll a pair of dice. List three typical random variables that could be associated with this experiment.

Solution   There are many ways to assign numbers to the 36 outcomes of this experiment. We list three here.

(a) One random variable is

$$X = \text{Sum of the two numbers on the top faces}$$

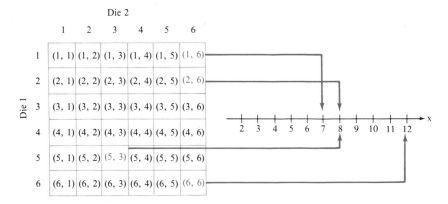

Figure 26
Random variable assigning the sum of the dots to each outcome

Figure 26 illustrates how the random variable X assigns a number to each outcome of the sample space of the experiment.

(b) A second random variable is

$$Y = \text{Difference between larger and smaller numbers}$$

Here the random variable Y would be 2 when the roll of the dice turned up a 5 and a 3.

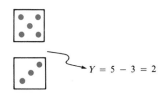

Figure 27
Dice showing a difference of 2

(c) A third random variable is

$$Z = \text{Sum of squares of the numbers on the top faces}$$

The value of Z would be 37 when the roll of the dice turns up a 6 and 1.   □

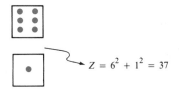

Figure 28
Dice showing a sum of squares equal to 37

A random variable is called a **finite random variable** if it can take on only a finite number of different values. The following example illustrates this idea.

Example 2

Finite Random Variable   Roll a pair of dice and define the random variable

$$X = \text{Sum of the two numbers on the top faces}$$

What are the possible values of X?

Solution   The **range of values** of X, or the values that X can take on, are clearly the integers $2, 3, \ldots, 12$. Since these are the only 11 possible values for X, we call X a finite random variable. (See Figure 29.)

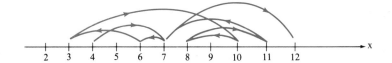

We say that a random variable is an **infinite random variable** if it can take on infinitely many different values. The following example illustrates such a random variable.

Example 3

Infinite Random Variable   Consider the experiment of tossing a coin repeatedly until a head occurs. Define the random variable

$$X = \text{Number of tosses until a head occurs}$$

Show X is an infinite random variable.

Solution   The possible outcomes of this experiment and values of X are given in Table 12. Since there is a possibility that X can take on any positive integer, it is an infinite random variable.

TABLE 12

Outcomes of Experiment in Example 3

Outcome of Experiment	Random Variable X (number of tosses)
H	1
TH	2
TTH	3
TTTH	4
TTTTH	5
TTTTTH	6
⋮	⋮

A random variable is called **continuous** if it can take on values on an entire interval. A marine biologist measuring the length of lobster claws would be measuring a continuous random variable, the length of the claw. We will not study continuous random variables now; to do so would require a knowledge of integral calculus.

Probability Distribution of a Random Variable

Intuitively, we often think of a random variable as a variable "jumping" from number to number while the probability experiment is being performed over and over. For example, for repeated rolls of a pair of dice, the sum of the numbers on the top faces "jumps" randomly among the values 2 through 12. The question we would like to answer is whether there is a pattern to the way in which this sum jumps from number to number. In other words, is it possible to find the probability that the sum takes on a given value? This brings us to the concept of the probability distribution of a random variable.

Probability Distribution of a Finite Random Variable X

Let X be a finite random variable that can take on values $x_1, x_2, \ldots, x_n$. The **probability distribution** of X is the function p defined for $1 \leq i \leq n$ by

$$p(x_i) = P(X = x_i)$$

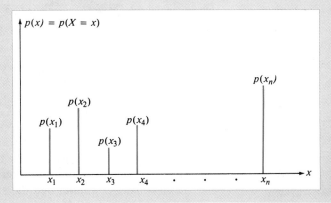

In other words, the probability distribution $p(x)$ at x is the probability that the random variable X takes on the value x. From this definition of the probability distribution, the following two properties hold.

Two Properties of Probability Distributions

A probability distribution p of a random variable satisfies

- $0 \leq p(x_i) \leq 1 \qquad (i = 1, 2, \ldots, n)$
- $p(x_1) + p(x_2) + \cdots + p(x_n) = 1$

where $x_1, x_2, \ldots, x_n$ are all the possible values taken on by the random variable.

Example 4 ──────────   Probability Distribution   Roll a pair of dice, and define the random variable

$$X = \text{Sum of the numbers on the top faces}$$

The probability distribution for this random variable is graphed in Figure 30 and listed in Table 13.

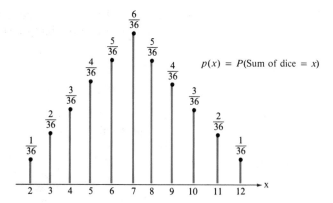

$p(x) = P(\text{Sum of dice} = x)$

Figure 30
Probability distribution of the sum of the rolled numbers of two dice

TABLE 13
Probability Distribution for Example 4

x	$p(x)$
2	1/36
3	2/36
4	3/36
5	4/36
6	5/36
7	6/36
8	5/36
9	4/36
10	3/36
11	2/36
12	1/36

Expectation of a Random Variable

Suppose we perform a probability experiment several times, each time recording the value of a random variable X. When finished, we compute the arithmetic average of the values of X that have occurred (just add them up and divide by the number of terms). This average value will approximate what is called the **expectation** or **expected value** of X. Sometimes people say intuitively that the expected value of X is the average value of an infinite number of values of X. Of course, we cannot find the average value of an infinite number of terms, but we can compute the average value of a large number of terms. A more precise definition of the expectation of X is now given.

Expectation of a Random Variable X

Let the probability distribution p of a finite random variable X be defined by the following table.

Possible Values of X	Probability of X Taking on This Value
x	$p(x)$
x_1	$p(x_1)$
x_2	$p(x_2)$
$\vdots$	$\vdots$
x_n	$p(x_n)$

The expectation of X (or expected value of X), denoted by $E[X]$, is given by

$$E[X] = x_1 p(x_1) + x_2 p(x_2) + \cdots + x_n p(x_n)$$

$$= \sum_{i=1}^{n} x_i p(x_i) \quad \text{(summation notation)}$$

Steps for Finding $E[X]$

To find the expectation of a random variable X, perform the following steps.

- **Step 1.** Find the possible values $x_1, x_2, \ldots, x_n$ that X can take on.
- **Step 2.** Find the probabilities

$$p(x_1) = P(X = x_1)$$
$$p(x_2) = P(X = x_2)$$
$$\vdots \quad \vdots \quad \vdots$$
$$p(x_n) = P(X = x_n)$$

- **Step 3.** Compute

$$E[X] = x_1 p(x_1) + x_2 p(x_2) + \cdots + x_n p(x_n)$$

The words expectation or expected value are often used when discussing games of chance. The following game illustrates this idea.

Example 5

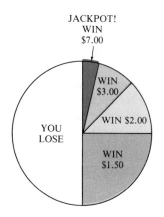

JACKPOT!
WIN
$7.00

Figure 31
What are the average winnings?

Spin the Wheel   Let's play Spin the Wheel. The player spins the wheel and waits for it to stop. The wheel is subdivided into five sectors, and the winnings for each sector are shown in Figure 31. Would you be willing to pay $1 for a spin of this wheel?

Solution   Define the random variable

$$W = \text{Winnings from one spin of the wheel}$$

To find $E[W]$, we carry out the following steps.

- **Step 1** (find the values that W takes on).   The possible winnings are 0, $1.50, $2, $3, and $7. These are listed in the first column of Table 14.
- **Step 2** (find the probability that W takes on its values).   The probability that the winnings will take on any of these values is determined from Figure 31. These are listed in the second column of Table 14.
- **Step 3** (compute the expected value).   Compute the product of the values in the first and second columns found in Steps 2 and 3 and put these products in the third column. Summing the values in the third column, we have the expected value.

TABLE 14
Steps in Finding an Expected Value

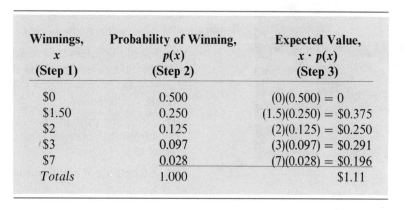

Winnings, x (Step 1)	Probability of Winning, $p(x)$ (Step 2)	Expected Value, $x \cdot p(x)$ (Step 3)
$0	0.500	$(0)(0.500) = 0$
$1.50	0.250	$(1.5)(0.250) = \$0.375$
$2	0.125	$(2)(0.125) = \$0.250$
$3	0.097	$(3)(0.097) = \$0.291$
$7	0.028	$(7)(0.028) = \$0.196$
Totals	1.000	$1.11

In other words, on the average you will collect $1.11 per spin while playing this game. Of course, if it costs $1 to play the game, then your expected net winnings per game will be reduced to only $0.11. This does not mean that you will win 11 cents on every game (in fact you never win 11 cents on a single game), but after, say, an entire weekend of playing this game, your average winnings per game will approximate this value. If you play 1000 games over the course of a weekend, your expected take will be $110. The graph of the probability distribution of the winnings W is shown in Figure 32.

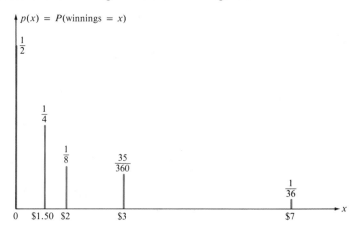

Figure 32
Distribution of winnings in Spin the Wheel

Actuaries at the Aetna Life and Casualty Company are keenly interested in probability theory.

Example 6

Actuarial Problem   The insurance industry maintains tables, called *actuarial tables*, that give the probability that a person of any current age will live to a specified age. Suppose an actuarial table states the probability that a 20-year-old person will survive the following year as 0.995. An insurance company sells a $2000 one-year life policy for $25 (if the insured person dies during the year, the beneficiary of the policy collects $2000). What is the company's expected gain for this type of policy?

Solution   The company can interpret the selling of a policy as a probability experiment. The sample space of this experiment is

$$S = \{\text{Insured person lives, Insured person dies}\}$$

For each of these two outcomes we define the random variable

$$R = \text{Company's annual revenue from one policy}$$

The company is interested in the expected revenue from a policy. To find it, we carry out the following steps.

· **Step 1** (find the values that R can take on).   The revenue R can be either −$1975 if the person dies and the company must pay $2000 to the beneficiary of the policy, or $25 if the person lives and the company earns the cost of the policy.

· **Step 2** (find the probability that R takes on different values). Using the results from the actuarial table, we can construct the table and graph in Figure 33.

· **Step 3** (compute the expectation of R).   The expectation of the company's revenue R is

$$E[R] = (-\$1975)(0.005) + (\$0.25)(0.995) = \$15.00$$

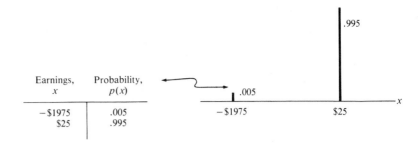

Figure 33
Probability distribution of R

Earnings, x	Probability, p(x)
−$1975	.005
$25	.995

On the average, the company will collect a revenue of $15.00 for each policy sold. □

Example 7

Magazine Sweepstakes   In 1973 a popular magazine sponsored a Summertime Sweepstakes. The official list of prizes is shown in Table 15.

TABLE 15
Summertime Sweepstakes
Portfolio

Prize	Winnings	Odds of Winning	Probability of Winning
Grand Prize	$24,000	1:6,046,614	(0.000000165)
Second Prize	$10,000	1:5,060,228	(0.000000198)
Third Prize	$7500	1:2,530,114	(0.000000395)
Fourth Prize	$1000	1:404,818	(0.00000247)
Fifth Prize	$500	1:202,409	(0.00000494)
Sixth Prize	$25	1:5060	(0.000198)

Using the information in Table 15, calculate the expected payoff to a person entering this sweepstakes.

Solution   Table 16 lists the possible amounts x that a player can win (including zero), the probability $p(x)$ of winning each amount, and the expected value $x \cdot p(x)$ of winning the amount x. The probability of not winning any prize is 1 minus the sum of the probabilities of winning *some* prize.

TABLE 16
Analysis of the
Summertime Sweepstakes

Winnings, x	Probability of Winning, p(x)	Expected Value, x · p(x)
$24,000	0.000000165	$0.00397
$10,000	0.000000198	$0.00198
$7500	0.000000395	$0.00296
$1000	0.000002470	$0.00247
$500	0.000004940	$0.00247
$25	0.000198000	$0.00495
0	0.999793832	0
Totals	1.000000000	$0.01880

Summing the values in the Expected Value column in Table 16, we find that the expected winnings for playing this Sweepstakes is $0.0188, or roughly 1.9 cents. □

Conclusion

This example illustrates why many companies use lotteries, sweepstakes, and other contests as sales and promotion tools. In most contests of this type, although the expected payoff to the player is very low, many people still play these games and take their chances. Most companies do not charge a fee *per se* (except for proof of purchase of their product) to play the game but gain from exposure and added sales.

Problems

For Problems 1–3, an urn contains two black marbles and two white marbles, as in Figure 34. Select two marbles at random from this urn and define the random variable

$$W = \text{Number of white marbles selected}$$

Figure 34

1. What are the possible values that W can take on?
2. What is the probability distribution of W?
3. What is the probability that W will be 1 or 2? *Hint:* Is the event of getting $W = 1$ disjoint from the event of getting $W = 2$?

For Problems 4–8, which of the following random variables are finite, infinite, or continuous?

4. The number of delinquent accounts a utility company contends with every month.
5. The amount of sulfur (in milligrams per cubic foot of air) emitted from a smokestack.
6. The number of shoes a worker can stitch in a day.
7. The results (favorable, no opinion, unfavorable) of a personal preference poll for a new product.
8. The length of time observed in a quality control study until a battery dies.

9. **Quality Control** Suppose you are the plant manager at a large shoe factory. A study has been conducted to determine the number of machines that fail each day. It has been determined that the random variable

$$X = \text{Number of machines that fail each day}$$

has the probability distribution given in Table 17. What is the expected number of machines that will fail each day?

TABLE 17
Probability Distribution for Problem 9

Failures, x	Probability of Failure, $P(x)$
0	0.40
1	0.35
2	0.20
3	0.05

Coin-Flipping Game

Problems 10–11 concern themselves with the game described by the probability tree in Figure 35. The possible outcomes of this game are

$$S = \{T, HT, HHT, HHH\}$$

Define the random variable

$$W = \text{Winnings per play}$$

10. What is the probability distribution of W?
11. What is the expected value of W?

Coin Toss

For Problems 12–13, toss two coins and define the random variable

$$H = \text{Total number of heads}$$

12. Graph the probability distribution of H.
13. What is the expected value of H?

Rolling Dice

Roll a pair of dice and define the random variables:

$$D = \text{Difference between the high and low dice}$$
$$P = \text{Product of the numbers on the dice}$$

14. What is the probability distribution of D?
15. What is the probability distribution of P?

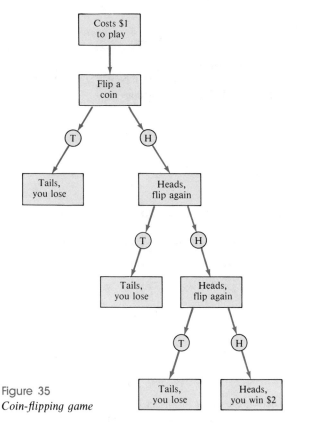

Figure 35
Coin-flipping game

16. What is the expected value of *D*?

17. What is the expected value of *P*?

Expected Values of Distributions

For Problems 18–23, what are the expectations of the indicated probability distributions?

18.

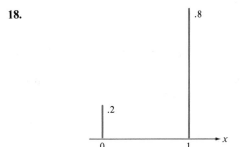

19.

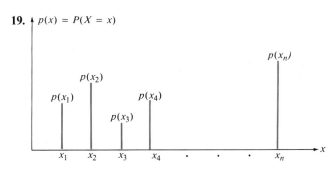

20.

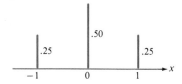

21.

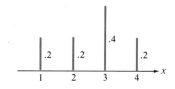

22.

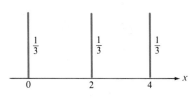

23.

Odds and Ends

24. Baby Needs New Shoes Cut out ten cards from a deck of 52 playing cards and mark them as follows:

- $1 on four of them,
- $2 on three of them,
- $3 on two of them,
- $5 on one of them.

Shuffle the ten cards and pick one at random. You win the value shown on the card. How much should you expect to win?

Roulette—Monte Carlo Style

The French word *roulette* refers to a ball that rolls around on a wheel. The wheel is subdivided into 37 divisions (38 in Las Vegas), 36 of which are alternately colored red and black and numbered 1–36 (*not* in numerical order). The 37th, which is numbered 0, is colored differently. To play the game, the croupier spins the wheel and tosses the ball onto the wheel. Eventually, the ball settles into one of the numbered indentations. Possible bets, the probabilities of winning, and their payoffs are shown in Table 18 for bets of $1.

25. Suppose a player plays *passé* (the player bets that the ball will land on a given number 19–36). What is the probability that a player will win this bet ten times in a row?

TABLE 18
Probabilities and Payoffs in Roulette

Player's Bet	Probability of Winning	Payoff on $1	Expected Payoff (dollars)	Expected Net Winnings
A specific number	1/37	$36	$36/37	−$1/37
Two consecutive numbers (for example, a 25 and 26)	2/37	$18	$36/37	−$1/37
Three consecutive numbers	3/37	$12	$36/37	−$1/37
Square (four numbers selected by the player)	4/37	$9	$36/37	−$1/37
Any red number	18/37	$2	$36/37	−$1/37
Any black number	18/37	$2	$36/37	−$1/37
Any number 1–18 (*manque*)	18/37	$2	$36/37	−$1/37
Any number 19–36 (*passe*)	18/37	$2	$36/37	−$1/37

In Monte Carlo in 1952 an unidentified man walked up to a table and made 28 consecutive *passés*.

26. What are a player's expected winnings per game when betting $1?

27. In Monte Carlo approximately 500 games are played at each table per day, each table hosting an average of five players. Each player bets an average of $10 per game. If the casino has ten tables, what is the casino's expected take every day?

28. Spin the Wheel   The wheel in Figure 36 costs $1 to play. Is it a bargain?

Figure 36
Wheel for Problem 28

Probability in the Insurance Industry

Problems 29–32 are concerned with the insurance industry. Suppose an insurance company sells an accident insurance policy for $5000 per year. The policy pays liable policyholders $250,000 in the event of an accident. The probability that a policyholder will have a liable accident is $p = 0.01$.

29. What is the probability distribution of the following random variable?

$$P = \text{Company's profit per policy each year}$$

30. What is the probability distribution of the following random variable?

$$M = \text{Money paid to a policyholder}$$

31. What is the company's expected profit per policy each year?

32. What must the company charge for a policy in order for its expected profit to be $2,000?

Some Interesting Expected Values

33. Expected Value   Roll a pair of dice and define the random variable

$$SS = \text{Sum of squares of the numbers on the top faces}$$

What is the expected value of SS?

34. Expected Value   Suppose you have five nickels, four dimes, two quarters, two fifty-cent pieces, and a Susan B. Anthony silver dollar in your pocket. Suppose you draw one coin at random from your pocket and tip a waitress. What is the expected tip?

35. Numbers Game   People still play "the numbers" in many parts of the country. It consists of guessing a three-digit number, like 391, that will occur in tomorrow's newspaper. Suppose the local bookie pays $250 on a $1 bet. What is a player's expected winnings?

36. **Punch-Out cards** Back in the 1940's, a game that was popular was punch-out cards. A person could go into a drug store and pay the druggist a dollar for the privilege of punching out one hole in a gigantic board. Each hole contained a rolled-up wad of paper, which told the player whether he or she had won a prize. Unfortunately, most wads were blank. Suppose a board contains 1000 holes with one $100 prize, five $50 prizes, ten $25 prizes, and twenty-five $10 prizes. What is a player's expected winnings for this game?

37. **Carnival Game** A carnival game offers $100 to anyone who can knock off three milk bottles on a stand. A player thinks she has a 25% chance of success. What is her expected payoff?

38. **Coin Game** A game consists of tossing two coins. If both coins turn up heads, a player wins $10. If both coins turn up tails, the player wins $5. Finally, if one coin turns up a head and the other a tail, the player wins $2. What is the expected value of this game?

Maine Lottery

Problems 39–44 are concerned with the Maine Lottery. The lottery ticket in Figure 37 is sold in the state of Maine for $0.50. To win a Straight bet, which pays $2500, the player must select the correct four-digit number. On another type of bet, a Box 24-Way bet, the player again selects a four-digit number but this time wins by simply choosing the correct four digits, not necessarily in any order. In this case the player wins $104 on a $0.50 bet.

39. How much can a player expect to win on a Straight bet?

HOW TO PLAY THE NUMBERS GAME
4 DIGIT

WAGER	BET TYPE	PRIZE
50¢	Straight	$2500

HOW WON EXAMPLE: Pick; 1234. Win With;
1, 2, 3, 4.

WAGER	BET TYPE	PRIZE
50¢	Box 24–Way	$104

HOW WON EXAMPLE: Pick; 1234. Win With;

1324, 2134, 3124, 4123, 1342, 2143,
3142, 4132, 1234, 2314, 3214, 4213,
1243, 2341, 3241, 4231, 1423, 2413,
3412, 4312, 1432, 2431, 3421, 4321.

Figure 37
Maine lottery ticket

40. How much can the state of Maine expect to earn when a player makes a Straight bet?
41. How much can the state of Maine expect to earn if 50,000 people make the Straight bet?
42. How much can a player expect to win on a Box 24-Way bet?
43. How much can the state of Maine expect to earn when a player makes a Box 24-Way bet?
44. How much can the state of Maine expect to earn if 50,000 people make the Box 24-Way bet?

Epilogue: The Buffon Needle Problem

In 1654 the Chevalier de Mere, gambler and gentleman, asked the French mathematician Blaise Pascal why it was unprofitable to bet on the outcome that at least one pair of sixes would come up in 24 rolls of a pair of dice. With this question, probability was born. Over three hundred years later, the end is still not in sight.

Probability has outgrown its disreputable origins and currently is one of the most active areas of mathematical research. Today, probability is more likely to be used in a medical research center for the design of a complex experiment or a computerized expert system.

Probability can be deceptively complicated. Problems are often easy to state but hard to solve. For example, toss a toothpick onto a floor whose narrow boards are just as wide as the toothpick is long (see Figure 38). What is the probability that the toothpick

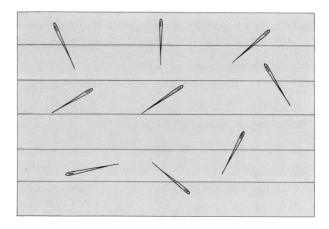

Figure 38
The Buffon needle problem

will land in such a way that it crosses one of the cracks in the floor?

This problem was posed and solved by the French naturalist and iconoclast Comte de Buffon (1707–1788). The answer is $2/\pi \cong 0.636619\ldots$. (Buffon is better known for estimating the age of the earth as 75,000 years. At that time, the commonly accepted estimate was around 6000 years.)

Buffon was one of the first people to bring probability out of the gaming parlors and into the real world. Today, medical research, biology, physics, and many other areas of science use probabilistic models. No doubt in the years ahead, probability models, linked to sophisticated computerized expert systems, will help society make important decisions. Students today should be prepared to use these important analytical tools in a challenging and exciting future.

It is remarkable that as long ago as 50 B.C., when probability was not much more than a synonym for uncertainty, Roman orator and statesman Marcus Cicero said, "Probability is the very guide of life."

Key Ideas for Review

birthday problem, 323
equally likely outcomes, 319
events
  dependent, 344
  independent, 344
expectation of a random variable, 354
odds, 325

probability
  complement rule, 323
  conditional, 333
  definition of, 318
  distribution, 352
  of equally likely outcomes, 320
  experiment, 311
  of an intersection, 337, 344

joint probability, 337
tree, 338
of a union, 327
random variable
  continuous, 352
  finite, 351
  infinite, 352
sample space, 311

Review Exercises

For Problems 1–5, roll a pair of dice. Draw the sample space for this experiment and find the following probabilities.

1. Find the probability that the sum of the two numbers on the top faces is 2.
2. Find the probability that the two numbers match.
3. Find the probability that the larger number is 2 greater than the smaller number.
4. Find the probability that the larger number is 2 or more greater than the smaller number.
5. Find the probability that the larger number is 2 larger than the smaller and that the two numbers match.

For Problems 6–10, consider the sample space

$$S = \{1, 2, 3, 4, 5, 6\}$$

with equally likely outcomes to find the indicated probabilities.

6. Find the probability that the outcome of the experiment is 2, 5, or 6.
7. Find the probability that the outcome of the experiment is odd or divisible by 3 (no remainder when divided by 3).
8. Find the probability that the outcome of the experiment is divisible by 3 given that the outcome is odd.
9. Find the probability that the outcome of the experiment is divisible by 3 given that the outcome is even.
10. Find the probability that the outcome of the experiment is not even and not divisible by 3.

11. **Empirical Probabilities** A pair of dice are rolled 1000 times with the frequencies of outcomes given in Table 19. Use these frequencies to find the empirical probabilities by dividing each frequency by 1000. How do these compare with the theoretical probabilities?

Drawing Cards

Select five cards from a deck of 52 playing cards.

12. What is the probability of getting at least one heart?
13. What is the probability of getting no hearts?
14. What is the probability of getting at least one ace?

TABLE 19
Frequency Distribution for Problem 11

Sum of Dice	2	3	4	5	6	7	8	9	10	11	12
Frequency	20	35	45	120	150	200	140	125	80	60	25

Rolling Dice

For Problems 15–22, roll three dice (red, green, and blue) and observe the numbers on the top faces.

15. What is the sample space of this experiment?
16. What is the probability of getting a sum of 3?
17. What is the probability of getting a sum of 4?
18. What is the probability of getting a sum of 5?
19. What is the probability of getting a sum of 6?
20. What is the probability of getting a sum of 7?
21. What is the probability that all three dice match?
22. What is the probability that exactly two dice match?

Amazing Coincidences

On June 11, 1950, the *New York Herald Tribune* reported that an unidentified man walked up to a dice table in Las Vegas and made an amazing 28 consecutive passes with the dice (he beat the house 28 times in a row). He left the casino as a hero, especially to the side bettors, who cleaned out the casino to the tune of $150,000. The man himself was cautious and won only $750.

23. What is the probability of someone winning 28 consecutive times against the house rolling dice? In Las Vegas the probability of making a single passé of the dice is 1/2.
24. What are the odds against winning 28 consecutive times against the house rolling dice?

The Envelope Problem (Simple Version)

In Problems 25–26 a student writes letters to three friends but absentmindedly does not pay attention when the letters are enclosed in the envelopes.

25. What is the sample space of the writer's actions, and what is the probability that all three letters are put in the correct envelopes?
26. What is the sample space of the writer's actions, and what is the probability that none of the three letters are put in the correct envelopes?

27. The Man Who Broke the Bank at Monte Carlo It has been reported that a man betting on even numbers in roulette won 28 consecutive *coups* (games) at Monte Carlo. The roulette table at Monte Carlo has 37 indentations in the wheels, of which 18 are marked even; hence the probability of the ball settling in an even number is

18/37. What is the probability of a player winning 28 consecutive times?

28. Two Hands in a Row A player is dealt 13 cards from a deck of 52 playing cards, but another player calls for a misdeal, since the dealer has accidentally turned over a card. What is the probability that the player will be dealt exactly the same cards on the new deal?

Family Probabilities

A couple is planning on having four children. Assuming that the probability of a boy being born is the same as the probability of a girl being born, answer the questions in Problems 29–32.

29. What is the sample space of all possible outcomes?
30. What is the probability of having three boys and one girl?
31. What is the probability of having three children of one sex?
32. What is the probability of having two boys and two girls?

Probabilities of Various Events

Let A and B be events in an equally likely sample space, with $P(A) = 0.5$, $P(B) = 0.5$, and $P(A \cap B) = .2$. Find the probabilities in Problems 33–39.

33. $P(A \mid B)$ **34.** $P(B \mid A)$
35. $P(A')$ **36.** $P(A' \cap B')$
37. $P(A \cap B')$ **38.** $P(A' \cap B)$
39. $P(A \cup B)$

Card Probabilities

For Problems 40–42, select two cards from a deck of 52 playing cards, and observe whether the cards are kings. Define

$K1$ = Event of selecting a king on the first draw

$K2$ = Event of selecting a king on the second draw

The probability tree in Figure 39 illustrates the selection of the two cards.

40. What is the probability of getting two kings?
41. What is the probability of getting a king on the first draw but not a king on the second draw?
42. What is the probability of getting a king on draw 2?

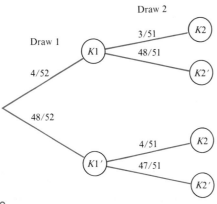

Figure 39
Probability tree for Problems 40–42

College Survey

For Problems 43–49, a total of 1000 college students with varying grade point averages were asked survey questions that tested their political attitudes. The survey classified each individual as Conservative (C), Moderate (M), or Liberal (L). Table 20 shows the results.

43. Change the above frequency table to a probability table by dividing all the numbers by 1000.
44. What is the probability that a student with a GPA below 2.5 is a conservative?
45. What is the probability that a student with a GPA below 2.5 is a moderate?
46. What is the probability that a student with a GPA below 2.5 is a liberal?

47. Are the events "GPA below 2.5" and "C" independent?
48. Are the events "GPA below 2.5" and "M" independent?
49. Are the events "GPA below 2.5" and "L" independent?

Expectation

50. Expected Gain   A player rolls a pair of dice and receives in dollars the sum of the two numbers on the top faces. What is the player's expected gain?
51. Expected Gain   A player rolls a pair of dice and receives in dollars the sum of the numbers on the top faces, except when the result is snake-eyes (both ones), in which case the player pays $100. What is the player's expected gain per play of the game?
52. Original Problem in Probability   What is the probability of at least one pair of sixes occurring in 24 rolls of a pair of dice? As was previously noted, this problem was posed by the Chevalier de Mere to Blaise Pascal in 1654. *Hint:* Use the complement rule and the concept of independent events.
53. AND or OR   Why is it true that it is at least likely that "something *or* something" occurs more often than "something *and* something" occurs? In other words why is it always true that

$$P(A \cap B) \leq P(A \cup B)?$$

Flip a coin and define the random variable

$$X = \begin{cases} 0 & \text{if the coin turns up tails} \\ 1 & \text{if the coin turns up heads} \end{cases}$$

54. What is the probability distribution of X?
55. What is the expected value of X?

TABLE 20
Survey Results for Problems 43–49

	Conservative (C)	Moderate (M)	Liberal (L)	Totals
GPA > 2.5	75	175	300	550
GPA ≤ 2.5	125	150	175	450
Totals	200	325	475	1000

Chapter Test

1. Toss two coins and define the random variable

$$X = \begin{cases} 0 & \text{if both coins are tails} \\ 1 & \text{if the coins are unmatched} \\ 2 & \text{if both coins are heads} \end{cases}$$

Draw the probability distribution of X. What is the expected value of X?
2. Roll a pair of dice, one red and one green, and define a random variable as the difference of the values on the top faces

$$D = \text{Red die} - \text{Green die}$$

Find the probability distribution of D and plot its graph.

3. Let A and B be two events in a sample space with equally likely outcomes. Assume that

$$P(A) = 0.9$$
$$P(B) = 0.2$$
$$P(A \cap B) = 0.2$$

Draw a Venn diagram of these events and determine the following probabilities.

(a) $P(A \cup B)$
(b) $P(A \cap B')$
(c) $P(A|B)$
(d) Are A and B independent events?

4. Draw two cards without replacement from a deck of 52 playing cards. What is the probability that

(a) the second card is an ace given that the first card is a king?
(b) the second card is an ace given that the first card is an ace?

5. The probability that it will rain today is 0.4. The probability that it will rain tomorrow given that it rains today is 0.6. What is the probability that it will rain on both days? Is it possible to determine from this information the probability that it will rain tomorrow?

6. An urn contains three red marbles and two black marbles. Select two marbles from the urn. Define the random variable

$$BM = \text{Number of black marbles selected}$$

(a) What are the possible values of BM?
(b) What is the probability distribution of BM?
(c) What is the expected value $E[BM]$?

6

Applications
of Probability

A recent trend in business management and many other fields of science has been the merging of these subjects with uncertainty or probability. As a typical example, in management science the nonprobabilistic or deterministic approach would be to estimate a company's profit. The alternate probabilistic approach would be to find the probability that a company's profit is a given amount. Since we do not live in a perfectly predictable world, the probabilistic approach is often more in tune with reality than a definite yes-or-no answer provided by a deterministic model.

In this chapter we study the important probabilistic models of Markov chains, Bayes' formula, binomial experiments, and simulation.

6.1 Markov Chains

PURPOSE

We introduce the concept of a Markov chain and illustrate a number of applications. The important ideas are

- the definition of a Markov chain,
- the states and trials of a Markov chain,
- the transition matrix of a Markov chain,
- the state vector of a Markov chain,
- the steady state vector of a Markov chain, and
- the average payoff of a Markov chain.

Introduction

A problem that faces university administrators across the country is predicting student enrollments in coming years. The ability to predict the number of current students that will still be enrolled in future years is a critical factor in the determination of future enrollments. For that reason an administrator would like to know the probability that a currently enrolled freshman will be enrolled at the same university as a senior three years hence.

In a business environment, suppose General Foods has 23% of the instant cake mix market today. What will its market share be in five years? Still another problem: What will Wendy's share of the fast-food market be in three years if it undertakes a million-dollar advertising campaign?

What is a common theme in the above problems? It may not be clear at this time, but you will soon discover that all of the previous problems can be modeled by what is referred to in probability as a *Markov chain*.

In many problems it is convenient to classify people or things into one of a finite number of different states and observe how they move from state to state. Problems like this can be described by a probabilistic model known as a Markov chain. We introduce the subject of Markov chains by means of a simple game.

Three-Wheeler (Simple Markov Chain)

A player begins by spinning any one of the three wheels shown in Figure 1. After the spinner stops, the player then moves to the indicated wheel and spins again. This process is continued as long as the player likes.

		Probability of going to		
		Wheel 1	Wheel 2	Wheel 3
If player is spinning	Wheel 1	.60	.30	.10
	Wheel 2	.20	.70	.10
	Wheel 3	.25	.25	.50

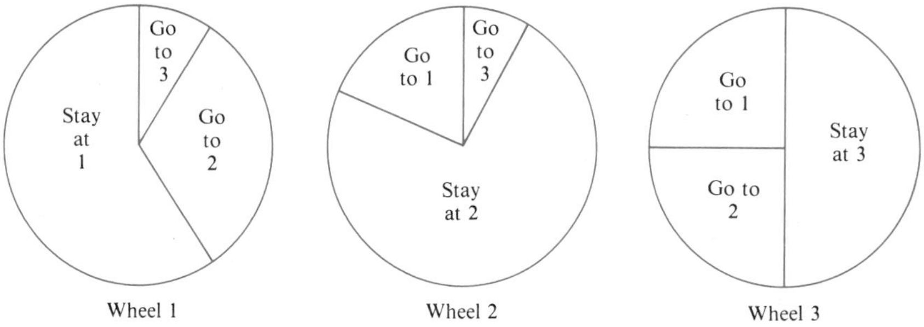

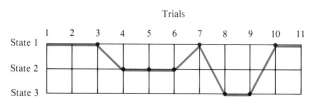

Figure 1
Game of Three-Wheeler

Wheel 1   Wheel 2   Wheel 3

A typical "tour," starting at Wheel 1 (or state 1), for the first ten spins or trials is shown in Figure 2.

This game may not seem very complex, but there are many phenomena in the natural and social sciences whose random movement from state to state is analogous to the movement of the player from wheel to wheel. This game is a simple example of a Markov chain.

Figure 2
Sample run of the first ten trials of a Markov chain

The Markov Chain

A **Markov chain** is a probabilistic model that describes the random movement over time of some activity. At each period of time the activity is in one of a finite number of different states. These states may be the weather conditions (rainy or sunny), the earnings of a company, the size of a biological population, or even the state of mind of an individual. We define this concept precisely, using the language of probability.

Definition of a Markov Chain

A Markov chain is a sequence of probability experiments, called **trials**, that satisfy the following conditions:

- The outcome of every trial is one of the elements, called **states**, of a finite set.
- The outcome of a trial depends only on the present state and not on any previous state.

Three-Wheeler, described previously, is a simple example of a Markov chain with three states, the states being the three wheels. We would say that the Markov chain "is in state i" if a player is currently spinning Wheel i for $i = 1, 2, 3$. A player starts in a given state (starts by spinning a given wheel) and then moves randomly from one state to another according to the probabilities listed on the three wheels. The way the game is played, the movement to a new state (wheel) depends only on the present state and not on any previous state that the Markov chain has visited. This is the second property described in the definition of a Markov chain, called the **memoryless property**.

Transition Matrix

As was noted, the outcome of any trial in a Markov chain depends only on the current state of the Markov chain. This gives rise to the most fundamental concept of a Markov chain, the transition matrix.

Definition of the Transition Matrix

The **transition matrix** P of a Markov chain having m states is the $m \times m$ matrix given by

Present State	New State			
	State 1	*State 2*		*State m*
State 1	p_{11}	p_{12}	$\cdots$	p_{1m}
State 2	p_{21}	p_{22}	$\cdots$	p_{2m}
$\vdots$				
State m	p_{m1}	p_{m2}	$\cdots$	p_{mm}

The entries p_{ij} for $1 \le i, j \le m$ are the **transition probabilities**:

$$p_{ij} = P(\text{Moving to state} = j \,|\, \text{Present state } i)$$

In other words, p_{ij} is the *conditional probability* of moving to state j, given that the Markov chain is currently in state i.

Note that the m elements in the first row of the transition matrix represent the probabilities that the Markov chain will move from state 1 to each of the m other states. Since the Markov chain must naturally move to one of these states, we can say that the sum of the elements in the first row is 1. In fact, the sum of the elements in any row of a transition matrix will always be 1.

Example 1

Transition Matrix of Three-Wheeler Find the transition matrix for the Markov chain that describes Three-Wheeler.

TABLE 1
Transition Matrix for
Three-Wheeler

Present Wheel	New Wheel		
	Wheel 1	*Wheel 2*	*Wheel 3*
Wheel 1	0.60	0.30	0.10
Wheel 2	0.20	0.70	0.10
Wheel 3	0.25	0.25	0.50

Figure 3
*Transition diagram for
Three-Wheeler*

Solution The transition matrix for Three-Wheeler is determined by the probabilities in Figure 1. By the definition of P the three numbers in the ith row for $i = 1, 2, 3$ are the three probabilities that the player will move to Wheel 1, 2, or 3, respectively, given that the player is currently spinning wheel i. The transition matrix P is shown in Table 1.

For example, $p_{12} = 0.30$ means that the probability of moving to Wheel 2, given that a player is spinning Wheel 1, is 0.30. This would be written

$$p_{12} = P(\text{Moving to wheel } 2 \,|\, \text{Player is spinning wheel } 1) = 0.30$$

The transition diagram in Figure 3 provides a visual representation of the transition matrix. □

HISTORICAL NOTE

Markov chains are named after Andrei Andreevich Markov (1856–1922), the man who first discovered them and realized their importance. Markov was born in Ryazan, Russia. He received a gold medal at the age of 22 from St. Petersburg University and remained there as a professor for 25 years.

As a mathematician, Markov was considered extremely exacting, toward both himself and others. As a lecturer, he was not so critical. He often never bothered to write the equations on the board in any order, and he paid even less attention to his personal appearance.

Markov introduced the study of Markov chains in 1906. Markov was not overly concerned with the applications of Markov chains, but more with the theoretical aspects of the subject. He did, however, apply his theories to the distribution of vowels and consonants in Pushkin's long poem *Eugene Onegin*. This work of Markov's is often considered the seminal work in the field of mathematical linguistics.

Example 2

Five & Ten Game   Imagine having an *unfair* nickel and dime that have the following probabilities of turning up heads and tails when tossed in the air.

$$\text{Nickel:} \quad P(\text{Head}) = 0.7 \quad P(\text{Tail}) = 0.3$$
$$\text{Dime:} \quad P(\text{Head}) = 0.8 \quad P(\text{Tail}) = 0.2$$

Begin by tossing either one of these coins. If the coin turns up heads, toss the same coin again. Otherwise, toss the other coin. Repeat this process several times. What are the states and transition matrix for the Markov chain that describe this process?

Solution   This process can be interpreted as a Markov chain with two states. The states are the two coins. We could call state 1 the nickel and state 2 the dime. We would say that we are in state 1 if we are currently tossing the nickel and in state 2 if we are tossing the dime. The transition matrix describing the dynamics of this process is the following 2×2 matrix:

Current State	New State Nickel	Dime
Nickel	0.7	0.3
Dime	0.2	0.8

State Vector P_k

Imagine all of your classmates who are also reading this book playing the Five & Ten game at the same time. For the sake of argument, assume that your professor initially assigns 50% of the class to start by tossing a nickel (state 1) and the other 50% to start by tossing a dime (state 2). In the language of Markov chains we would say that the initial distribution of students in each state or the **initial state vector** is

$$P_0 = (0.5, 0.5)$$

After the first toss, however, the distribution of students who are in each state will change. Depending on the outcomes of the tosses, some students will change coins, and some will keep tossing the same coin. We would like to know the expected distribution of students in each state after the first toss (trial). The answer depends on the transition probabilities in the transition matrix. It turns out that after the first toss we will expect that 45% of the students will move to the nickel and 55% will change to the dime. We would say that the new distribution of students in each state, or the **new state vector**, is

$$P_1 = (0.45, 0.55)$$

The two elements of P_1 (0.45 and 0.55) can be interpreted either as the probabilities that a student will be in each of the two states after the first coin toss (trial) or as the expected fraction of students that will be in each state after the first toss.

We would now like to determine the expected fraction of students in each state after the next trial, and the next, and so on. This brings us to the following definition.

Definition of the State Vector P_k

The **state vector P_k** of a Markov chain with m states is a vector of the form

$$P_k = (p_1, p_2, \ldots, p_m)$$

where the elements p_i are the probabilities of being in state i for $1 \leq i \leq m$ after the kth trial, where $k = 1, 2, 3, \ldots$.

Since the state vector gives the probabilities of being in a given state, we also have the following probability condition:

$$p_1 + p_2 + \cdots + p_m = 1$$

The state vector can also be interpreted, in the context of the game Three Wheeler, as the expected fraction of the individuals playing the Markov chain who will be in each state after the kth trial. For that reason the state probability vector is sometimes called the *state distribution vector*.

Finding the State Vector

We now show how to find the state vectors $P_1, P_2, \ldots$ of Five & Ten. Using this game as a guide, we can then find the state vectors for any Markov chain. We begin by drawing the probability tree as shown in Figure 4. This tree says that initially, 50% of the students are in state 1 (tossing the nickel), and 50% of the students are in state 2 (tossing the dime). By the rules of the game, 70% of those students who are in state 1 (nickel) stay in state 1 (on the average), and 30% move to state 2. On the other hand, for those students who are in state 2 (dime), 20% change to state 1 (on the average), and 80% stay in state 2. These dynamics are illustrated in the probability tree in Figure 4.

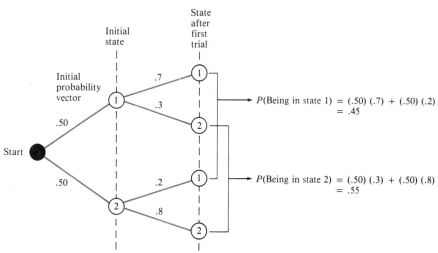

Figure 4
Finding state vector
$P_1 = (0.45, 0.55)$ from
$P_0 = (0.50, 0.50)$

To determine the probabilities that a student will be in each of the two states after the first trial (after the first toss of the coins), we add the probabilities along the different paths of the probability tree, getting

$$P(\text{Player will be in state 1}) = (0.50)(0.7) + (0.50)(0.2) = 0.45$$
$$P(\text{Player will be in state 2}) = (0.50)(0.3) + (0.50)(0.8) = 0.55$$

In other words, the new state vector P_1 is

$$P_1 = (0.45, 0.55)$$

The above relationship between the initial state vector P_0 and the new state vector P_1 can be written simply as the following matrix product:

$$(0.50, 0.50)\begin{bmatrix} 0.7 & 0.3 \\ 0.2 & 0.8 \end{bmatrix} = (0.45, 0.55)$$

Initial State   Transition   New State
Vector   Matrix   Vector

The above matrix equation gives a hint as to the general formula for finding a new state vector from the previous state vector.

Formula for Finding the New State Vector

Let P be the transition matrix of a Markov chain with present state vector

$$p = (p_1, p_2, \ldots, p_m)$$

The new state vector

$$p' = (p_1', p_2', \ldots, p_m')$$

is found by computing the matrix product

$$(p_1, p_2, \ldots, p_m)\begin{bmatrix} p_{11} & p_{12} & \cdots & p_{1m} \\ p_{21} & p_{22} & \cdots & p_{2m} \\ & & \vdots & \\ p_{m1} & p_{m2} & \cdots & p_{mm} \end{bmatrix} = (p_1', p_2', \ldots, p_m')$$

Present State Vector   Transition Matrix   New State Vector

In matrix language we can write the above relationship as

$$pP = p'$$

where P is the transition matrix.

Example 3

Finding the New State Find the first eight state vectors for the Five & Ten game if the initial state vector is

$$P_0 = (0.5, 0.5)$$

Solution Starting with P_0, we compute the following matrix products:

$$P_0 = (0.500, 0.500) \qquad \textit{Initial State Vector}$$

$$(0.500, 0.500) \begin{bmatrix} 0.7 & 0.3 \\ 0.2 & 0.8 \end{bmatrix} = (0.450, 0.550) \qquad \textit{First State Vector}$$

$$(0.450, 0.550) \begin{bmatrix} 0.7 & 0.3 \\ 0.2 & 0.8 \end{bmatrix} = (0.425, 0.575) \qquad \textit{Second State Vector}$$

$$(0.425, 0.575) \begin{bmatrix} 0.7 & 0.3 \\ 0.2 & 0.8 \end{bmatrix} = (0.412, 0.587) \qquad \textit{Third State Vector}$$

$$(0.412, 0.587) \begin{bmatrix} 0.7 & 0.3 \\ 0.2 & 0.8 \end{bmatrix} = (0.406, 0.594) \qquad \textit{Fourth State Vector}$$

$$(0.406, 0.594) \begin{bmatrix} 0.7 & 0.3 \\ 0.2 & 0.8 \end{bmatrix} = (0.403, 0.597) \qquad \textit{Fifth State Vector}$$

$$(0.403, 0.597) \begin{bmatrix} 0.7 & 0.3 \\ 0.2 & 0.8 \end{bmatrix} = (0.401, 0.599) \qquad \textit{Sixth State Vector}$$

$$(0.401, 0.599) \begin{bmatrix} 0.7 & 0.3 \\ 0.2 & 0.8 \end{bmatrix} = (0.400, 0.600) \qquad \textit{Seventh State Vector}$$

$$(0.400, 0.600) \begin{bmatrix} 0.7 & 0.3 \\ 0.2 & 0.8 \end{bmatrix} = (0.400, 0.600) \qquad \textit{Eighth State Vector}$$

Note that the state vectors are quickly approaching or converging to a **limiting vector**. After only eight trials the state vector has stabilized (to three places) to the vector

$$P_{ss} = (0.400, 0.600)$$

The limiting vector of the above sequence of state vectors is called the **steady state vector** of the Markov chain. The steady state vector will give the proportion or fraction of players that will ultimately be in each of the states. Theoretically, the steady state vector will never be reached exactly, but for most practical purposes we can say that a Markov chain reaches a "steady state" when the first two or three digits stabilize. In terms of the Five & Ten game, the above steady state vector P_{ss} tells us that ultimately, 40% of the students should expect to be in state 1 (tossing a nickel), and 60% of the students to be in state 2 (tossing a dime). An interesting fact about the steady state vector is that, no matter what the initial distribution P_0 of students in each state, eventually 40% will be expected to be in state 1 and 60% in state 2. □

Warning

Not all Markov chains reach a steady state vector by repeated multiplication of the current state vector by the transition matrix. However, if the transition matrix P has the property (called a **regular transition matrix**) that some power P^n of P has all positive elements, then it can be shown that for any initial state vector the Markov chain will approach some steady state vector.

Market Share Problems in Markov Chains

Suppose that Brite-Cola and Splash are competing for the soft drink market and that the Brite-Cola Company introduces a new formula. The company hires a marketing firm to test consumer acceptance of the new product. The marketing company has determined that among consumers who last purchased Brite-Cola, 60% remained loyal on the next purchase, 30% switched to Splash, and 10% changed to some other brand.

The movement of consumers among the three brands (Brite-Cola, Splash, and Other) can be described by the transition matrix in Table 2. The numbers in the transition matrix are the probabilities that a consumer's next purchase will be the column brand, given that the last brand purchased was the row brand. The survey also determined the current market share for each brand, as shown in Table 3.

The Brite-Cola Company would like to know what market share it should expect if the company introduces the new formula on a national scale.

This market share problem can be described by a Markov chain with three states: Brite-Cola, Splash, and Other. A consumer will move from state to state with the probabilities described in the above transition matrix. The initial state of the Markov chain will be the current fraction of consumers in each state. That is,

$$P_0 = (0.40, 0.25, 0.35)$$

The next three state vectors P_1, P_2, and P_3 and the steady state vector P_{ss} are found by performing the following matrix multiplications:

$$
(0.40, 0.25, 0.35)
\begin{bmatrix}
0.60 & 0.30 & 0.10 \\
0.20 & 0.70 & 0.10 \\
0.25 & 0.25 & 0.50
\end{bmatrix}
= (0.38, 0.38, 0.24) \qquad \text{First State } P_1
$$

$$
(0.38, 0.38, 0.24)
\begin{bmatrix}
0.60 & 0.30 & 0.10 \\
0.20 & 0.70 & 0.10 \\
0.25 & 0.25 & 0.50
\end{bmatrix}
= (0.36, 0.44, 0.20) \qquad \text{Second State } P_2
$$

$$
(0.36, 0.44, 0.20)
\begin{bmatrix}
0.60 & 0.30 & 0.10 \\
0.20 & 0.70 & 0.10 \\
0.25 & 0.25 & 0.50
\end{bmatrix}
= (0.35, 0.46, 0.18) \qquad \text{Third State } P_3
$$

$$
(0.35, 0.49, 0.16)
\begin{bmatrix}
0.60 & 0.30 & 0.10 \\
0.20 & 0.70 & 0.10 \\
0.25 & 0.25 & 0.50
\end{bmatrix}
= (0.35, 0.49, 0.16) \qquad \text{Steady State } P_{ss}
$$

Initial State   *Brite Splash Others*

The state (or distribution) vectors for the first six trials are plotted in Figure 6.

Analysis

If the data from the market survey are accurate and preferences remain constant, then by introducing the new flavor, Brite-Cola will expect to lose 5% of its market share, decreasing its market share from 40% to 35%. At the same time,

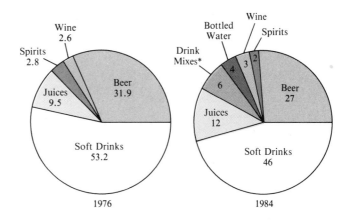

Figure 5
*Soft drink share of the beverage
market, in percent*

1976   1984

TABLE 2
Transition Matrix for Drinks

Last Preference	Next Preference		
	Brite-Cola	*Splash*	*Other*
brite-Cola	0.60	0.30	0.10
Splash	0.20	0.70	0.10
Other	0.25	0.25	0.50

TABLE 3
Current Market Share of
Soft Drink

	Type of Drink	**Current Market Share**
Brite-Cola	1	40%
Pepsi	2	25%
Other	3	35%

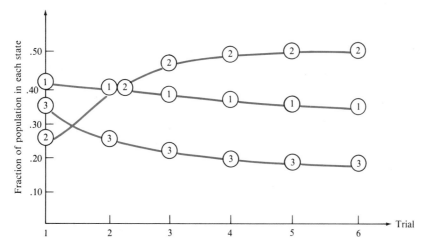

	Fraction of population in		
Trial	State 1	State 2	State 3
1	.40	.25	.35
2	.38	.38	.24
3	.36	.44	.19
4	.35	.47	.18
5	.348	.482	.171
10	.34729	.48599	.16671
15	.34722	.48611	.16666
20	.34722	.48611	.16666

Figure 6
Fractions of population in state 1, state 2, and state 3

Splash will increase its market share from 25% to 49%. The Other brands are the big losers in this readjustment, moving all the way down from a market share of 35% to a market share of 16%.

Finding the Steady State Vector by Solving a System of Equations

It is possible to find the steady state vector

$$P_{ss} = (p_1, p_2, p_3, \ldots, p_m)$$

of a Markov chain with m states by solving a certain system of m equations. To find the m equations, we select any $m - 1$ of the m equations

$$P_{ss}P = P_{ss}$$

along with equation

$$p_1 + p_2 + \cdots + p_m = 1$$

For instance, suppose the transition matrix P is the 2×2 matrix

$$P = \begin{bmatrix} 1/2 & 1/2 \\ 1/3 & 2/3 \end{bmatrix}$$

Calling the steady state vector

$$P_{ss} = (p_1, p_2)$$

we write

$$(p_1, p_2) \begin{bmatrix} 1/2 & 1/2 \\ 1/3 & 2/3 \end{bmatrix} = (p_1, p_2)$$

Multiplying the above matrices and setting the elements of the matrices equal to one another, we get the two equations

$$\frac{1}{2} p_1 + \frac{1}{3} p_2 = p_1$$

$$\frac{1}{2} p_1 + \frac{2}{3} p_2 = p_2$$

The reader should simplify the above two equations and see that they are in fact the same equation. To find a second equation, we use the condition

$$p_1 + p_2 = 1$$

Hence we find the two equations by selecting

$$\frac{1}{2} p_1 + \frac{1}{3} p_2 = p_1$$

$$p_1 + p_2 = 1$$

Solving this linear system, we get the steady state probabilities

$$P_{ss} = (p_1, p_2)$$
$$= (2/5, 3/5)$$

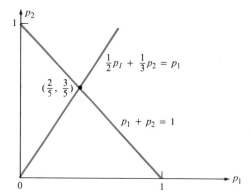

Figure 7
Solving for a steady state vector

This solution can be seen in Figure 7. The process can be extended to systems of any order.

Expected Payoff of a Markov Chain

Suppose you are given a reward of some kind every time you pass through a given state of the Markov chain. For instance, if you are playing Five & Ten, you might receive $1 every time you flip a nickel (state 1) and $2 every time you flip a dime (state 2). After several flips (trials), what will be your expected reward per flip? These ideas lead us to the concept of the **expected payoff** of a Markov chain.

Expected Payoff of a Markov Chain

Let payoffs of $c_1, c_2, \ldots, c_m$ be given every time or trial a Markov chain is in the state $1, 2, \ldots, m$, respectively. If a steady state vector of the Markov chain exists and is given by

$$P_{ss} = (s_1, s_2, \ldots, s_m)$$

then the expected payoff per trial is

$$\text{Expected payoff per trial} = c_1 s_1 + c_2 s_2 + \cdots + c_m s_m$$

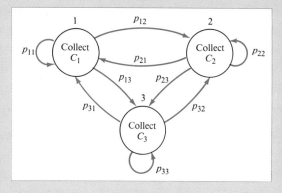

Example 4

Expected Payoff of Three-Wheeler   Bob plays Three-Wheeler as described earlier, but now Bob receives the following rewards each time he spins one of the three wheels:

Wheel	Reward Bob Receives
1	$1.50
2	$0.75
3	$1.00

Bob can start at any wheel he pleases (which means that he will probably pick Wheel 1). What will be Bob's expected reward per spin if he plays the game 100 times (total of 100 spins)?

Solution   We begin by reprinting the transition matrix in Table 4, which gives the probabilities that Bob will move from any one of the three wheels (states) to any other wheel (possibly the same wheel). This transition matrix is the same as the transition matrix in the market study example, which had the steady state vector

$$P_{ss} = [0.35, 0.49, 0.16]$$

This means that if Bob plays the game a large number of times, the percentage of times he will spend at each wheel will be

Wheel	% Time at Wheel
1	35%
2	49%
3	16%

Hence the expected payoff per spin can be calculated as

$$\text{Expected payoff per spin} = 0.35(\$1.50) + 0.49(\$0.75) + 0.16(\$1.)$$
$$= \$1.05$$

In the long run, Bob should expect to profit $1.05 for each spin of the wheel.

   For the above formula for the expected reward to be accurate, we are making the tacit assumption that the Markov chain has come to a steady state. Since Bob plays the game 100 times and since the Markov chain reaches its approximate steady state in only eight trials, we can feel confident about the answer.

TABLE 4
Transition Matrix for
Three-Wheeler

Old Wheel	New Wheel		
	Wheel 1	*Wheel 2*	*Wheel 3*
Wheel 1	0.60	0.30	0.10
Wheel 2	0.20	0.70	0.10
Wheel 3	0.25	0.25	0.50

□

Problems

For Problems 1–10, determine whether the given matrix is a transition matrix for a Markov chain.

1. $\begin{bmatrix} 1 & 0 \\ 0 & 1 \end{bmatrix}$

2. $\begin{bmatrix} 1/2 & 1/2 \\ 1/2 & 1/2 \end{bmatrix}$

3. $\begin{bmatrix} 1/4 & 3/4 \\ 3/4 & 1/4 \end{bmatrix}$

4. $\begin{bmatrix} 1 & 0 \\ 3/2 & 1/2 \end{bmatrix}$

5. $\begin{bmatrix} 0 & 1 \\ 1 & 0 \end{bmatrix}$

6. $\begin{bmatrix} 1/3 & 1/3 & 1/3 \\ 1 & 0 & 1 \\ 0 & 0 & 1 \end{bmatrix}$

7. $\begin{bmatrix} 1 & 0 & 0 \\ 0 & 1/2 & 0 \end{bmatrix}$

8. $\begin{bmatrix} 1 & 0 & 0 \\ 0 & 0 & 0 \\ 0 & 1 & 0 \end{bmatrix}$

9. $\begin{bmatrix} 1/2 & 3/2 \\ 1/2 & 1/2 \end{bmatrix}$

10. $\begin{bmatrix} 0 & 1/3 & 2/3 \\ 1/2 & 0 & 1/2 \\ 0 & 1/2 & 1/2 \end{bmatrix}$

For Problems 11–20, determine whether the given vector is a valid state vector for a Markov chain.

11. $(1, 0)$

12. $(1, 0, 0, 0)$

13. $(1, 1)$

14. $(-1, 1, 1)$

15. $(1/2, 1/2, 1/2)$

16. (1)

17. $(1, 0, 1)$

18. $(0, 0)$

19. $(1/4, 1/4, 1/4, 1/4)$

20. $(2, -1)$

For Problems 21–30, find the state vectors P_1, P_2, and P_3, given the transition matrix P and the initial state vector P_0.

21. $P = \begin{bmatrix} 1/2 & 1/2 \\ 1/2 & 1/2 \end{bmatrix}$ and $P_0 = (1, 0)$

22. $P = \begin{bmatrix} 3/4 & 1/4 \\ 1/4 & 3/4 \end{bmatrix}$ and $P_0 = (0, 1)$

23. $P = \begin{bmatrix} 1 & 0 \\ 1/2 & 1/2 \end{bmatrix}$ and $P_0 = (1/2, 1/2)$

24. $P = \begin{bmatrix} 0.9 & 0.1 \\ 0.5 & 0.5 \end{bmatrix}$ and $P_0 = (1, 0)$

25. $P = \begin{bmatrix} 1/8 & 7/8 \\ 1/2 & 1/2 \end{bmatrix}$ and $P_0 = (0, 1)$

26. $P = \begin{bmatrix} 1/3 & 1/3 & 1/3 \\ 1/3 & 1/3 & 1/3 \\ 1/3 & 1/3 & 1/3 \end{bmatrix}$ and $P_0 = (1, 0, 0)$

27. $P = \begin{bmatrix} 0 & 1 & 0 \\ 1 & 0 & 0 \\ 0 & 0 & 1 \end{bmatrix}$ and $P_0 = (1, 0, 0)$

28. $P = \begin{bmatrix} 1 & 0 & 0 \\ 0 & 1/2 & 1/2 \\ 1/2 & 1/2 & 0 \end{bmatrix}$ and $P_0 = (1/3, 1/3, 1/3)$

29. $P = \begin{bmatrix} 1 & 0 & 0 \\ 0 & 1 & 0 \\ 0 & 0 & 1 \end{bmatrix}$ and $P_0 = (1, 0, 0)$

30. $P = \begin{bmatrix} 0.7 & 0.2 & 0.1 \\ 0 & 0.1 & 0.9 \\ 0.1 & 0.2 & 0.7 \end{bmatrix}$ and $P_0 = (0.5, 0.5, 0)$

For Problems 31–34, draw the transition diagram analogous to the one drawn in Figure 3 for the given transition matrix.

31. $P = \begin{bmatrix} 0.5 & 0.5 \\ 0.6 & 0.4 \end{bmatrix}$

32. $P = \begin{bmatrix} 1/2 & 1/2 \\ 1/3 & 2/3 \end{bmatrix}$

33. $P = \begin{bmatrix} 0.1 & 0.1 & 0.8 \\ 0.2 & 0.7 & 0.1 \\ 0.6 & 0.2 & 0.2 \end{bmatrix}$

34. $P = \begin{bmatrix} 0.5 & 0.5 \\ 0.5 & 0.5 \end{bmatrix}$

For Problems 35–36, find the transition matrix for the given transition diagram.

35.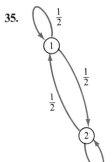

36.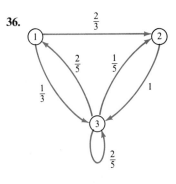

Typical Markov Chains

37. Markov Game Put ten slips of paper in each of two boxes, Box I and Box II. Of the ten slips of paper in Box I, six should be marked "Go to Box I," and four should be marked "Go to Box II." Of the ten slips of paper in Box II, three should be marked "Go to Box I," and seven should be marked "Go to Box II." Start the game by randomly selecting a slip of paper from Box I. Replace the slip of paper in the box and go to the box indicated by the paper. Carry out this process 20 times and record the sequence of boxes from which slips of paper were drawn. When you have done this, you will have just carried out the first 20 trials of the Markov chain described by the transition matrix

$$P = \begin{bmatrix} 0.6 & 0.4 \\ 0.3 & 0.7 \end{bmatrix} \quad \text{and} \quad P_0 = (0, 1)$$

38. Assembly Line Problem Items coming off an assembly line are either defective or nondefective. If an item is defective, then the chances that the next item is defective is 0.05. If an item is nondefective, then the probability that the next item is defective is 0.005. What are the states of this Markov chain, and what is the transition matrix?

Finding the Steady State

For Problems 39–44, find the steady state probabilities of the given transition matrices.

39. $\begin{bmatrix} 1/2 & 1/2 \\ 1/3 & 2/3 \end{bmatrix}$ **40.** $\begin{bmatrix} 0 & 1 \\ 1 & 0 \end{bmatrix}$ **41.** $\begin{bmatrix} 0.8 & 0.2 \\ 0.4 & 0.6 \end{bmatrix}$

42. $\begin{bmatrix} 1/2 & 1/2 \\ 1/2 & 1/2 \end{bmatrix}$ **43.** $\begin{bmatrix} 1 & 0 & 0 \\ .8 & .2 & 0 \\ .2 & .8 & 0 \end{bmatrix}$ **44.** $\begin{bmatrix} 1/2 & \cdot 1/2 & 0 \\ 0 & 1 & 0 \\ 0 & 1/2 & 1/2 \end{bmatrix}$

45. Rain, Rain, Go Away If it rains today, the probability of rain tomorrow is 0.4. If it is sunny today, then the probability of sun tomorrow is 0.90. What are the states of this Markov chain, and what is the transition matrix? What is the steady state vector, and what is the interpretation of the steady state probabilities?

46. More Weather Some people think that the severity of one winter can be determined by the severity of the previous winter. Historical records indicate that if last year's winter was severe, then the probability of a severe winter this year is 0.45. On the other hand, if last year the winter was mild, then the chances for a mild winter this year are 0.35. What are the states of this Markov process, and what is the transition matrix? What is the steady state vector?

47. The Success of Brand A Suppose a company introduces a new product, Brand A, and launches a major advertising campaign. The company currently has 25% of the overall market. A market research company discovers that among the users of the new product, 75% will still be using it next month. For other brands it is found that 50% of the users will buy Brand A next month. How much of the market should the company expect Brand A to have next month? In two months? In three months? Will the market share of Brand A reach a steady state?

48. Five & Ten Game A hundred people are flipping nickels and dimes. The coins are unfair and have the following probabilities:

Nickel: $P[\text{Head}] = 0.8$

Dime: $P[\text{Head}] = 0.6$

The rules of the game are as follows: If a player flips a head, the player should stay with the same coin; otherwise,

switch coins. Initially, 60 people are flipping nickels, and 40 people are flipping dimes. How many people will be expected to be flipping each kind of coin after one toss? After two tosses? Estimate the steady state vector on the basis of seeing P_0, P_1, P_2, and P_3.

49. Experiment versus Theory Place four slips of paper in each of two boxes, Box I and Box II. In Box I, mark one of the slips with an X. In Box II, mark two of the slips with an X. Starting at Box I, select at random one of the slips of paper. If the slip of paper is marked with an X, return the slip to the box and go to the other box and repeat the same process. If an X does not appear on a slip of paper, return the paper to the box and draw again. Carry out this process, drawing a total of 25 slips of paper. How many times did you draw a slip of paper from Box I? From Box II? Does this agree with the theory of Markov chains?

Expected Gain in a Markov Chain

50. Expected Gain per Trial Suppose that in Problem 49 you are rewarded $1 every time you draw a slip of paper from Box I and $2 every time you draw from Box II. What should your expected reward per draw be after several trials?

For Problems 51–54, find the average expected gain per trial using the given transition matrices and rewards for being in a given state.

51. $\begin{bmatrix} 1/2 & 1/2 \\ 3/4 & 1/4 \end{bmatrix}$ Reward for being in state 1 = $0.50·
Reward for being in state 2 = $2

52. $\begin{bmatrix} 0 & 1 \\ 1 & 0 \end{bmatrix}$ Reward for being in state 1 = $5
Reward for being in state 2 = $10

53. $\begin{bmatrix} 1/2 & 1/2 \\ 9/10 & 1/10 \end{bmatrix}$ Reward for being in state·1 = $0.05
Reward for being in state 2 = $10

54. $\begin{bmatrix} 1 & 0 & 0 \\ 0 & 1/2 & 1/2 \\ 1/2 & 1/2 & 0 \end{bmatrix}$ Reward for being in state 1 = $1
Reward for being in state 2 = $5
Reward for being in state 3 = $5

55. Genetics Problem A certain variety of rose has red, pink, or white flowers depending on the genotype RR, RW, and WW, respectively. If flowers of these three kinds

TABLE 5
Transition Matrix for Problem 55

Present Generation	Next Generation		
	Red	Pink	White
Red	1/2	1/2	0
Pink	1/4	1/2	1/4
White	0	1/2	1/2

are crossed with a RW rose (pink-flowering), one can expect that the offspring as indicated in Table 5 will result.

If this variety of rose is repeatedly crossed with RW roses (pink-flowering) and if the initial mixture of red, pink, and white roses is 1/2:1/2:0, then what will be the expected mixture of roses in the next generation? In two generations? In three generations? Does the mixture of red, pink, and white roses seem to be reaching a steady state?

56. Steady State Probabilities Find the steady state probabilities of the following general 2 × 2 transition matrix:

$$P = \begin{bmatrix} a & b \\ c & d \end{bmatrix}$$

57. All-Star Game For the 30 years from 1956 to 1985 the Major League All-Star game between the American League (A.L.) and the National League (N.L.) has been won as shown in Table 6. Can we view these annual games as satisfying the assumptions of a Markov chain? If so, what is the transition matrix of this Markov chain? What are the steady state probabilities of this Markov chain?

TABLE 6
Winners of All-Star Games, 1956–1985

Year	Winner	Year	Winner
1956	N.L.	1971	A.L.
1957	A.L.	1972	N.L.
1958	A.L.	1973	N.L.
1959	N.L.	1974	N.L.
1960	N.L.	1975	N.L.
1961	N.L.	1976	N.L.
1962	N.L.	1977	N.L.
1963	N.L.	1978	N.L.
1964	N.L.	1979	N.L.
1965	N.L.	1980	N.L.
1966	N.L.	1981	N.L.
1967	N.L.	1982	N.L.
1968	N.L.	1983	A.L.
1969	N.L.	1984	N.L.
1970	N.L.	1985	N.L.

6.2 Bayes' Formula

PURPOSE

We introduce the concept of Bayes' formula. The important concepts introduced in this section are

- the prior probability,
- the posterior probability, and
- Bayes' formula for finding the posterior probability from the prior probability.

Introduction

In the previous chapter we learned that if we selected two cards in succession from a deck of playing cards, the probability that the second card is a heart (called the event *H2*) given that the first card was a heart (called the event *H1*) is $P(H2|H1) = 12/51$. We now determine the **inverse probability**. That is, given that the second card selected is a heart, what is the probability that the first card was a heart? Believe it or not, we can find this probability, and it is Bayes' formula that we use to find it. This remarkable formula relating the two conditional probabilities $P(A|B)$ and $P(B|A)$ can be used to solve a variety of problems in medicine, expert systems, genetics, business management, and many more. We introduce Bayes' formula by means of the following urn problem.

The Urn Problem (Karnap Reveals the Past)

Karnap the Magician places two urns on a table. Urn 1 contains four red marbles and two black marbles, while Urn 2 contains one red marble and three black marbles. (See Figure 8.)

Karnap is then blindfolded while his assistant rolls a die. Karnap has instructed his assistant to draw a marble from Urn 1 if the die turns up 1, 2, 3, or 4

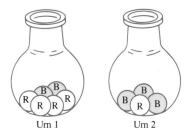

Figure 8
Karnap's urns

and to draw a marble from Urn 2 if the die turns up 5 or 6. After performing the experiment, the assistant removes the blindfold from Karnap's eyes and presents him with the chosen marble, which we assume is red. After a drumroll, Karnap announces that although he cannot be absolutely sure from what urn the marble was selected, there is a 4/7 chance it came from Urn 1. How did Karnap perform this magic?

Karnap's Solution

We first define the relevant events. They are

$$U_1 = \text{Event that the marble was selected from Urn 1}$$
$$U_2 = \text{Event that the marble was selected from Urn 2}$$
$$R = \text{Event that a red marble was selected}$$
$$B = \text{Event that a black marble was selected}$$

We next draw the probability tree that illustrates the selection of the urn followed by the selection of a marble. (See Figure 9.)

Karnap's announcement of the probability that the marble was drawn from Urn 1 says that

$$P(U_1|R) = 4/7$$

This conditional probability is different from any that we have found thus far in the sense that we are finding the probability that something happened in the *past* (which urn was chosen) given that we know something in the *present* (the color of the marble selected). To find these conditional probabilities, called **posterior probabilities**, we recall the conditional probability formula

$$P(U_1|R) = \frac{P(U_1 \cap R)}{P(R)}$$

However, since $U_1 \cap R = R \cap U_1$, the probability $P(U_1 \cap R)$ on the right-hand side can be turned around and written in terms of the new conditional probability

$$P(U_1 \cap R) = P(R|U_1)P(U_1)$$

Hence we can write the conditional probability $P(U_1|R)$ in terms of the inverted conditional probability $P(R|U_1)$. We have

$$P(U_1|R) = \frac{P(R|U_1)P(U_1)}{P(R)}$$

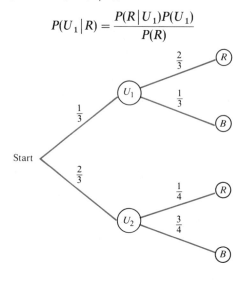

Figure 9
Sample space of Karnap's experiment

Now, since the event of drawing a red marble is located at the end of two disjoint paths of the probability tree in Figure 9, we have

$$P(R) = P(U_1 \cap R) + P(U_2 \cap R) = P(R|U_1)P(U_1) + P(R|U_2)P(U_2)$$

Substituting this expression into the previous equation, we get **Bayes' formula** (and the results of Karnap's claim). We have

$$P(U_1|R) = \frac{P(U_1 \cap R)}{P(U_1 \cap R) + P(U_2 \cap R)} = \frac{P(R|U_1)P(U_1)}{P(R|U_1)P(U_1) + P(R|U_2)P(U_2)}$$

$$= \frac{(2/3)(1/3)}{(2/3)(1/3) + (1/4)(2/3)}$$

$$= .571$$

In other words, Karnap performed his magic by using Bayes' formula. He determined that selection of a red marble had a 57.1% chance of being a selection from Urn 1.

The following gives the general Bayes' formula and a visual way to interpret it.

Bayes' Formula

Consider a two-stage probability tree. Suppose that at the first stage, three mutually exclusive events can occur, which we denote A_1, A_2, and A_3 (we could consider more). At the second stage we examine the path that leads to one specific event, say, R. Suppose we know that the event R has occurred and would like to know the probability that the event A_1 (or A_2 or A_3) has already occurred (or that the path to R passes through A_1, A_2, or A_3).

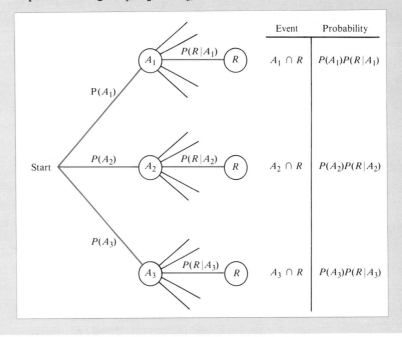

> Given that the event R has occurred, the probabilities that the previous event A_1, A_2, or A_3 have occurred are given by the three variations of Bayes' formula:
>
> $$P(A_1|R) = \frac{P(R|A_1)P(A_1)}{P(R|A_1)P(A_1) + P(R|A_2)P(A_2) + P(R|A_3)P(A_3)}$$
>
> $$P(A_2|R) = \frac{P(R|A_2)P(A_2)}{P(R|A_1)P(A_1) + P(R|A_2)P(A_2) + P(R|A_3)P(A_3)}$$
>
> $$P(A_3|R) = \frac{P(R|A_3)P(A_3)}{P(R|A_1)P(A_1) + P(R|A_2)P(A_2) + P(R|A_3)P(A_3)}$$
>
> The three probabilities $P(A_1)$, $P(A_2)$, and $P(A_3)$ are called **prior probabilities**, whereas the three probabilities $P(A_1|R)$, $P(A_2|R)$, and $P(A_3|R)$ are called **posterior probabilities** (note that they are in bold print in the above equations).

How Bayes' Formula Is Used

Bayes' formula is useful in the following way. Suppose we know the (prior) probabilities of three events A_1, A_2, and A_3. With the additional knowledge that a second event R has occurred, we can update these (prior) probabilities to obtain more accurate probabilities of A_1, A_2, and A_3—namely, the posterior probabilities $P(A_1|R)$, $P(A_2|R)$, and $P(A_3|R)$. Since the events A_1, A_2, and A_3 would have taken place before the event R has occurred, we interpret these posterior probabilities as probabilities that something happened in the past, given knowledge of something happening in the present. The above formulas extend naturally to more general trees. However, the simple tree above illustrates the basic ideas. We now show how Bayes' formula can be used to solve problems in medicine and the social sciences.

Expert Systems in Medicine

There has been a lot of talk recently about expert systems, knowledge-based systems, and artificial intelligence. An **expert system** is an "intelligent" computer program that uses factual knowledge and logical inference to solve problems that are difficult enough to require significant human expertise for their solution. The first large expert system was MYCIN, a medical expert system, developed at Stanford University in the mid-1970's to help physicians in the diagnosis and treatment of meningitis and bacteremia infections.

Many expert systems use Bayes' formula. When a medical expert (the computer program) is first written, it might tell a physician who lists the symptoms of a patient, "There is a probability of 0.56 that the patient has spinal meningitis." Later, when new information is supplied to the expert, the expert will incorporate this information, using Bayes' formula to update the original probability. The computer might say, "The probability is now 0.45 that the same patient has spinal meningitis." The original probability is the prior probability, and the new updated probability is the posterior probability.

The following example shows how an expert system (or a physician) can use Bayes' formula.

Example 1

Medical Expert Systems   A physician has ordered a diabetes test for a patient. Suppose it is known that if a person has diabetes, the test shows positive 90% of the time. On the other hand, when a person does not have diabetes, the test shows positive 6% of the time. It is also known that 3% of the general population has undetected diabetes. Find the probability that a person who is tested positive will actually have diabetes.

Solution   The probability tree in Figure 10 illustrates the possible outcomes of a person being tested for diabetes.

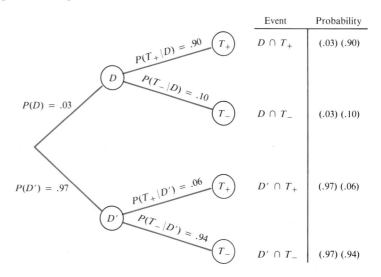

Figure 10
Sample space of a person being tested for diabetes

We define the following events.

$$D = \text{Event that the patient has diabetes}$$
$$T_+ = \text{Event that the test is positive (indicates diabetes)}$$

The physician knows the probabilities

$$P(T_+|D) = 0.90$$
$$P(T_+|D') = 0.06$$
$$P(D) = 0.03 \qquad \text{(prior probability)}$$
$$P(D') = 0.97 \qquad \text{(prior probability)}$$

To find the desired posterior probability

$$P(D|T_+) = P[\text{patient has diabetes given that the test is positive}]$$

we compute

$$P(D|T_+) = \frac{P(T_+|D)P(D)}{P(T_+|D)P(D) + P(T_+|D')P(D')}$$

$$= \frac{(0.90)(0.03)}{(0.90)(0.03) + (0.06)(0.97)}$$

$$= 0.32$$

In other words, the prior probability of 0.03 that a patient from the general population will have diabetes will be raised to the revised posterior probability of 0.32 if the test turns out to be positive. On the other hand, if the test turns out negative, then the posterior probability that the patient has diabetes is

$$P(D|T_-) = \frac{P(T_-|D)P(D)}{P(T_-|D)P(D) + P(T_-|D')P(D')}$$

$$= \frac{(0.10)(0.03)}{(0.10)(0.03) + (0.94)(0.97)}$$

$$= 0.00328$$

In this case the prior probability of 0.03 that a patient from the general population will have diabetes will be lowered to the revised or posterior probability of 0.00328 if a test turns out negative. ☐

HISTORICAL NOTE

Bayes' formula is named for an English clergyman, Thomas Bayes (1702–1761), who gave a mechanism for finding $P(A|B)$ in terms of $P(B|A)$ and some related information. He called these probabilities "inverse probabilities."

Social Science Problems

Bayes' formula has been applied to a variety of problems in the social sciences. The following example shows how Bayes' formula could be used in a court case.

Example 2

Determining the Father   In a recent paternity suit a man denied that he was the father of a child. The blood type of the child was different from that of the mother and hence was inherited from the father. The man was required by the court to take a blood test. It is common medical knowledge that if a man's blood type is different from that of a child, then the man is not the father. On the other hand, if a man has the same blood type as a child, then the probability that the man is the father is increased. In this example the blood type of the child occurred in only 10% of the general public. The court estimated prior to the testing that there was a 50% chance that the man was the father (based on court testimony, subjective feelings, and so on). The test determined that the man had the same blood type as the child. Using this new information, what is the probability that the man is the father?

Solution   We can illustrate this problem with the probability tree in Figure 11. The relevant events are

$$F = \text{Event that the man is the father}$$
$$F' = \text{Event that the man is not the father}$$
$$T_+ = \text{Event that the test is positive (man has the same blood type as the child)}$$
$$T_- = \text{Event that the test is negative}$$

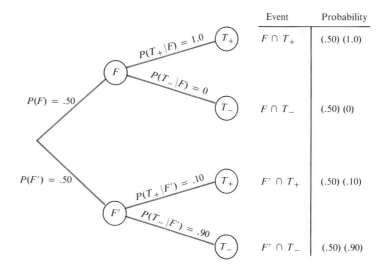

Event	Probability
$F \cap T_+$	$(.50)(1.0)$
$F \cap T_-$	$(.50)(0)$
$F' \cap T_+$	$(.50)(.10)$
$F' \cap T_-$	$(.50)(.90)$

Figure 11
Probability tree for determining the father

Here we have

$$P(F) = 0.50 \qquad \text{(prior probability)}$$
$$P(F') = 0.50 \qquad \text{(prior probability)}$$
$$P(T_+ \,|\, F) = 1.0$$
$$P(T_+ \,|\, F') = 0.10$$

Hence by Bayes' formula we can find the posterior probability that the man is the father given that the test was positive. It is

$$P(F \,|\, T_+) = \frac{P(T_+ \,|\, F)P(F)}{P(T_+ \,|\, F)P(F) + P(T_+ \,|\, F')P(F')}$$

$$= \frac{(1)(0.50)}{(1)(0.50) + (0.10)(0.50)}$$

$$= 0.91$$

In other words, the prior probability of the man's being the father has been modified from 0.50 to the posterior probability of 0.91 as a result of the positive test. □

Problems

Problems 1–6 refer to the following procedure. We start by flipping a coin. If the coin turns up a head, select a marble from Urn 1 in Figure 12. If the coin turns up a tail, select a marble from Urn 2. Call the events

H = Event of getting a head on the coin toss

T = Event of getting a tail on the coin toss

R = Event of drawing a red marble from an urn

B = Event of drawing a black marble from an urn

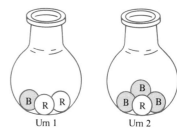

Figure 12

Find the following probabilities.

1. $P(R\,|\,H)$
2. $P(R\,|\,T)$
3. $P(H\,|\,R)$
4. $P(T\,|\,R)$
5. $P(H\,|\,B)$
6. $P(T\,|\,B)$

For Problems 7–14, two cards are drawn in succession from a deck of 52 playing cards (without replacement). Let

H_1 = Event that a heart is drawn on the first draw

H_2 = Event that a heart is drawn on the second draw

Find the following probabilities.

7. $P(H_1)$
8. $P(H_1')$
9. $P(H_2\,|\,H_1)$
10. $P(H_2\,|\,H_1')$
11. $P(H_2)$
12. $P(H_2')$
13. $P(H_1\,|\,H_2)$
14. $P(H_1\,|\,H_2')$

Fraudulent Tax Forms

The probability tree in Figure 13 illustrates the possible things that can happen when a tax form is submitted to the Internal Revenue Service. First, the tax form may or may not be audited. Second, the tax form may or may not be fraudulent. Call the events

A = Event that the tax form is audited

F = Event that the tax form is fraudulent

and
$$P(A) = 0.01$$
$$P(F\,|\,A) = 0.80$$
$$P(F\,|\,A') = 0.10$$

Find and interpret the probabilities in Problems 15–18.

15. $P(A\,|\,F)$
16. $P(A'\,|\,F)$
17. $P(A\,|\,F')$
18. $P(A'\,|\,F')$

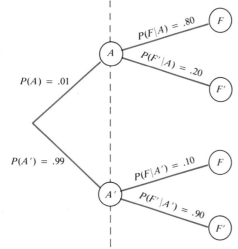

Figure 13
Probability tree for
Problems 15–18

Puzzling Problems

19. **The Convict Problem**   Let C be the event that a person is an ex-convict and R the event that he is a rapist. The following probabilities are given:

$$P(C) = 0.01$$
$$P(R) = 0.001$$
$$P(C\,|\,R) = 0.95$$
$$P(C\,|\,R') = 0.01$$

Find $P(R\,|\,C)$ and interpret its meaning.

20. **College Exams**   An exam is given to high school students to determine admission into college. Suppose the probability that a qualified student will pass the exam is 95% and the probability that an unqualified student will pass the exam is 5%. Assume that 70% of the students from a given high school are qualified. What is the probability that a student who passes the exam is in fact qualified? *Hint:* It is always useful to draw a probability tree and label all the prior and posterior probabilities.

The Capture-Recapture Method for Counting Bears

The capture-recapture method is a common way to count the size of various wildlife populations (deer, fish, moose, wolves, etc.). Here we estimate the number of bears in a wildlife preserve. It is known that the preserve contains one, two, three, or four bears. Since bears normally do not like to be counted, biologists have developed the capture-recapture method. The method consists of catching, tagging, and releasing a bear back into the preserve. Suppose that in this preserve a biologist has captured, tagged, and released a bear into the preserve. Later another bear (maybe the same bear) is captured, and the biologists observe whether the bear is tagged.

21. Draw a probability tree that illustrates the possible number of bears in the preserve (and label these branches of the tree with the equal probabilities 1/4), followed by the observation of whether the recaught bear is tagged or not tagged (second stage). *Hint:* The tree should have eight possible paths.

22. Given that the captured bear was not tagged, what is the probability that there is one bear in the preserve?

23. Given that the captured bear was not tagged, what is the probability that there are two bears in the preserve?

24. Given that the captured bear was not tagged, what is the probability that there are three bears in the preserve?

25. Given that the captured bear was not tagged, what is the probability that there are four bears in the preserve?

26. Given that the captured bear was tagged, what is the probability that there is one bear in the preserve?

27. Given that the captured bear was tagged, what is the probability that there are two bears in the preserve?

28. Given that the captured bear was tagged, what is the probability that there are three bears in the preserve?

29. Given that the captured bear was tagged, what is the probability that there are four bears in the preserve?

Bayes' Formula in Medicine

30. Cancer Diagnosis A test for cancer will detect 95% of the cancers it is designed to detect. If a patient does not have cancer, the test will be positive (say that the patient has cancer) 1% of the time. If 2% of the general population has this type of cancer, what is the probability that a person who tests positive actually has the cancer? *Hint:* Draw a probability tree and label the prior and posterior probabilities.

31. Medical Diagnosis Mary goes to the doctor, complaining of tiredness. The physician considers three different alternatives: an underactive thyroid, mononucleosis, and allergies. The physician assigns the prior probabilities of 0.5, 0.3, and 0.2, respectively, to the three ailments. The physician then orders a blood test, which is found to be normal. Normal blood tests occur in 20% of patients having underactive thyroids, 10% of patients having mononucleosis, and 40% of patients having allergies. On the basis of this new information, find the three posterior probabilities. That is, find

(a) P(Mary has a thyroid condition|Test was negative)

(b) P(Mary has mononucleosis|Test was negative)

(c) P(Mary has allergies|Test was negative)

Bayes' Formula in Expert Systems

32. PROSPECTOR The expert system PROSPECTOR is a real computer-based system that helps geologists and mineral prospectors hunt for commercially exploitable minerals. Suppose a team of gold prospectors hypothesize that an exploitable gold deposit exists in Sutter's Canyon. They define this hypothesis in terms of the event

$$G = \text{A rich gold vein exists in Sutter's Canyon}$$

Before exploration, the prospectors assign to the expert (the computer) an educated guess, say, $P(G) = 0.01$. Suppose the prospectors then find a two-ounce nugget of gold in Sutter's Canyon. The expert knows that the probability of finding such a gold nugget in a neighborhood of a rich vein is, say, 0.75, while the probability of finding such a nugget randomly is 0.05. What probability will the expert system now assign to the event that a rich gold vein will be found in Sutter's Canyon's *Hint:* Letting

$$N = \text{Event of finding a gold nugget}$$

we must find $P(G|N)$.

6.3

Binomial Experiments

PURPOSE

We introduce the binomial distribution. The major topics studied are

- the binomial experiment,
- the repeated binomial experiment,
- the binomial random variable, and
- the binomial distribution.

Repeated Binomial Experiments

In many problems, experiments have only two possible outcomes. For instance, in marketing surveys a consumer may be asked whether he prefers Brand A or Brand B. In political polling a potential voter may be asked whether she prefers Candidate A or Candidate B. In the natural sciences a biologist may test for the presence or absence of a fungus in Dutch elm trees. Experiments such as these can all be characterized by one thing: They each have two possible outcomes. These experiments are called **binomial experiments** (or **Bernoulli trials**).

Binomial Experiments

To gain a better appreciation of the binomial experiment, we list a few examples.

- Flipping a coin and observing whether the coin turns up heads or tails.
- Testing a floppy disk and determining whether it is defective or non-defective.

- Examining a patient to see whether or not the patient has a particular disease.
- Examining a flower to see whether it has a blight or is healthy.
- Rolling a die and observing whether the die turns up 1, 2, 3 (which we call success) or 4, 5, 6 (which we call failure).
- Shooting a free throw in a basketball game and making the basket or not.

In this section we are not so much interested in performing a binomial experiment once, but several times. This introduces the idea of the **repeated binomial experiment**.

Repeated Binomial Experiment (Tossing a Coin n Times)

A repeated binomial experiment is a sequence of binomial experiments that satisfy the following four properties:

1. The number of experiments is fixed (called n).
2. Each experiment has two possible outcomes (called success and failure).
3. The probability of success p and failure $q = 1 - p$ does not change from experiment to experiment.
4. All experiments are independent of each other. That is, the success or failure of one experiment does not affect the success or failure of later experiments.

The following are examples of repeated binomial experiments.

Example 1 ——————— Repeated Binomial Experiment

(a) Coin Flipping.   Flip a coin nine times. This experiment clearly satisfies the above conditions. Here $n = 9$ and $p = 0.5$.

(b) Shooting Free Throws.   Shooting ten free throws of a basketball is a repeated binomial experiment, provided that the shooter's accuracy remains constant and is not affected by previous shots. If these conditions hold, we would say that the repeated binomial experiment model is appropriate.

Nonrepeated Binomial Experiments

The following experiments illustrate situations that might appear at first glance to be repeated binomial experiments but fail one of the above four conditions.

- Property 1 fails (not fixed number of trials).   Toss a coin until a head occurs. This is not a repeated binomial experiment, since the number of tosses is not determined in advance.
- Property 2 fails (all trials do not have two outcomes).   A medical researcher examines skin tissue from subjects and classifies the tissue as one of five possible types. This experiment is not a repeated binomial experiment because each experiment has more than two possible outcomes.
- Property 3 fails (probabilities change).   In succession:

Flip a coin (success = tossing a head)

Roll a die (success = rolling a 1 or 2)

Select a card from a deck of playing cards (success = selecting a heart)

Here we have three successive binomial experiments, but they do not qualify as a repeated binomial experiment, since the probability of success changes from 1/2 to 1/3 and finally to 1/4.

· **Property 4 fails (independence fails).** Using a deck of 52 playing cards, turn them over one by one, recording whether the card is red or black. This experiment has a fixed number of $n = 52$ trials and has two outcomes per trial, but since the chances of getting a card of a given color will depend on previously drawn cards, the individual binomial experiments are not independent. Hence the experiment is not a repeated binomial experiment.

HISTORICAL NOTE

The binomial experiment is also known as the Bernoulli experiment, named after the brilliant Swiss mathematician, Jakob Bernoulli (1654–1705). He was born in Basel, Switzerland, and belonged to a family that produced no fewer than ten eminent mathematicians and scientists.

Jakob's brother, Johann Bernoulli, was an even more prolific contributor to mathematics than Jakob. The two brothers were extremely jealous of one another and often tried to steal each other's mathematical secrets. On one occasion, Johann attempted to replace his own incorrect solution with his brother's correct solution.

On another occasion, Johann expelled his son Daniel from his house for obtaining a prize from the French Academy that he himself had expected to win.

The Binomial Random Variable

In a repeated binomial experiment we are primarily interested in the number of successes X that occur. The number of successes is a random variable and is called the **binomial random variable**. The following example illustrates this random variable.

Example 2

Binomial Random Variable   Three typical binomial random variables are given below.

1. Flip a coin $n = 20$ times. The number

$$X = \text{Number of heads that turn up}$$

is a binomial random variable.

2. Examine $n = 100$ trees for budworm infestation. The number

$$X = \text{Number of infested trees}$$

is a binomial random variable.

3. A basketball player shoots $n = 100$ free throws. The number

$$X = \text{Number of shots made}$$

is a binomial random variable.   □

Every random variable has a probability distribution, and the binomial random variable is no exception.

The Binomial Distribution

The probability distribution of a binomial random variable is called the **binomial distribution**. We introduce this distribution by means of the following example. Suppose a basketball player makes an average of two free throws out of every three attempted. In addition, the success or failure of any single shot does not affect the outcome of other shots. Suppose now the player shoots ten free throws. What is the probability that the player will make exactly seven shots out of the ten attempted?

We answer this question in two steps.

- **Step 1** (find the probability of making a specific combination). First, consider the probability that the player makes the following specific seven shots:

$$\text{SSFFSSSFSS}$$

where

$$S = \text{Shot made}$$
$$F = \text{Shot failed}$$

Since each attempt is independent of every other attempt, the probability that the player will make these specific seven shots is the product of the individual probabilities of making the shots. That is,

$$P(\text{SSFFSSSFSS}) = (2/3)(2/3)(1/3)(1/3)(2/3)(2/3)(2/3)(1/3)(2/3)(2/3)$$
$$= (2/3)^7(1/3)^3$$
$$= 0.002168 \quad \text{(using a calculator)}$$

- **Step 2** (find the number of combinations). We have found the probability that the player will make the above seven specific shots out of ten. However, there are other ways in which the player can make seven shots. We must find this number and multiply it by

$$(2/3)^7(1/3)^3$$

The number of ways to make seven shots in ten attempts can be found by determining the number of ways in which we can place seven S's in ten squares. One possibility is shown in Figure 14.

Finding the number of ways to place seven S's in ten boxes is the same as finding the number of combinations of seven things taken from a set of size 10.

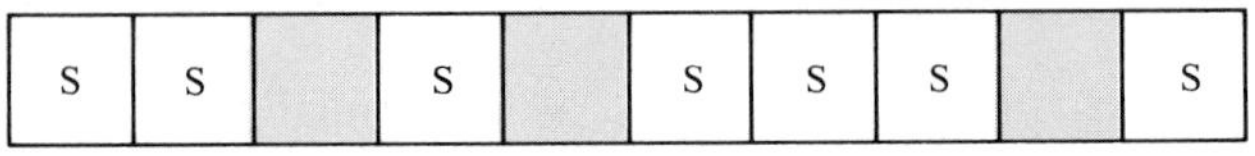

Figure 14
One of $\binom{10}{7} = 120$ ways to place seven S's in ten squares

In other words, the number of possibilities is the binomial coefficient:

$$C(10, 7) = \binom{10}{7} = \frac{10!}{3!7!} = 120$$

Hence multiplying this number times the results from Step 1, we have

$$P(X = 7) = \binom{10}{7}(2/3)^7(1/3)^3 = (120)(0.002168) = 0.2602$$

In other words, the basketball player has a 26% chance of making exactly seven shots out of ten. You can carry out a similar analysis to find the probability that the player will make a different number of shots. The results of such an analysis are summarized by the probability distribution shown in Figure 15.

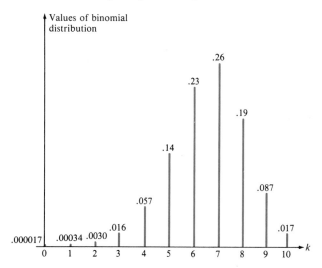

Figure 15
Binomial distribution with n = 10 and p = 2/3

The above discussion leads us to the general definition of the **binomial probability distribution**.

The Binomial Distribution

In a repeated binomial experiment of n trials, with each experiment having a probability of success p, the probability that exactly k successes occur, denoted by $P(X = k)$ or $b(k; n, p)$, is given by the binomial distribution

$$b(k; n, p) = P(X = k) = \binom{n}{k}p^k(1 - p)^{n-k} \qquad (k = 0, 1, 2, \ldots, n)$$

For simplicity we often write $b(k) = b(k; n, k)$.

Example 3

Binomial Distribution   The binomial distributions graphed in Figure 16 illustrate the shape of the binomial distribution for different values of n and p. The height of the graphs at k give the probability of getting exactly k successes when n repeated binomial experiments are performed, provided that each binomial experiment has probability p of success.

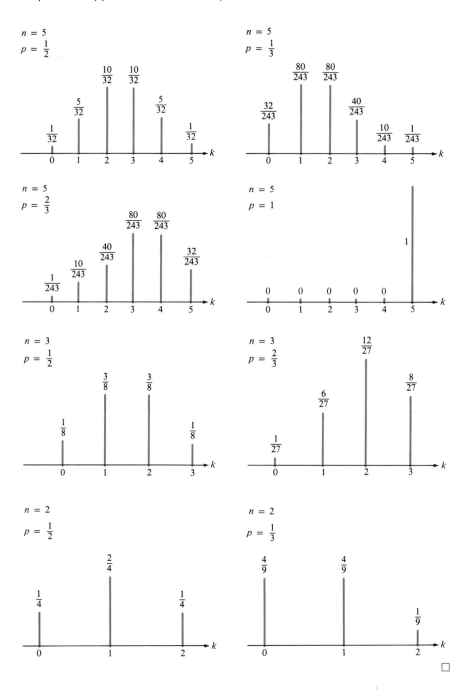

Typical binomial distributions

Quality Control and Repeated Binomial Experiments

A computer company purchases microchips, knowing that the percentage of nondefective or "good" ones is 95%. The company installs ten of these chips in one of its computers.

- What is the probability that *all* ten chips are good?
- What is the probability that *exactly* nine chips are good?
- What is the probability that *at least* eight chips are good?

Good quality control using concepts from probability and statistics is essential for many companies.

To answer these questions, we realize that testing ten microchips to determine which of the chips is good is a repeated binomial experiment with $n = 10$ and $p = 0.95$. Hence the expression

$$P(X = k) = b(k; n, p) = \binom{n}{p} p^k (1 - p)^{n - k}$$

is the probability of getting exactly k good chips (k successes). The exact numerical value can be computed by using a calculator or computer or by means of the tables in Table 8 of the Appendix. Hence we have the following probabilities:

· To find the probability that all ten chips are good, we compute

$$P(X = 10) = b(10; 10, 0.95) = \binom{10}{10} (0.95)^{10} (0.05)^0$$

$$= (0.95)^{10}$$

$$= 0.5987$$

In other words, there is a 59.87% chance that all ten chips are good.

· To find the probability that exactly nine chips are good, we compute

$$P(X = 9) = b(9; 10, 0.95) = \binom{10}{9} (0.95)^9 (0.05)^1$$

$$= 0.3151$$

In other words, there is a 31.51% chance that exactly nine chips are good.

· To find the probability of at least eight chips being good, we compute

$$P(X \geq 8) = P(X = 8) + P(X = 9) + P(X = 10)$$

$$= b(8; 10, 0.95) + b(9; 10, 0.95) + b(10; 10, 0.95)$$

$$= 0.0746 + 0.3151 + 0.5987$$

$$= 0.9884$$

In other words, if the circuitry of the computer operates normally when eight or more of the chips are operational, then the computer has a 98.84% chance of running properly.

The graph of the binomial distribution when $n = 10$ and $p = 0.95$ is drawn in Figure 17.

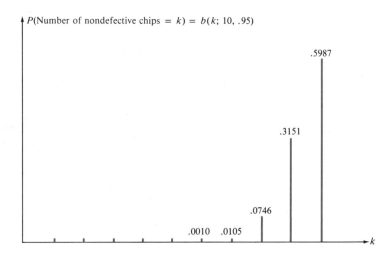

Figure 17
Binomial distribution with n = 10 and p = 0.95

Psychic Powers

The Great Karnap claims to have psychic powers. A skeptical psychologist makes the following proposal to the Great Karnap: "I will test your powers by flipping a coin ten times. You must tell me whether the outcome was a head or a tail after each flip."

Suppose that Karnap predicts the occurrence of the coin

- five times
- seven times
- nine times

What would the psychologist conclude from these results?

For the sake of argument, let us suppose that Karnap has no ability and has been faking it all along. If this is true, then the experiment proposed by the psychologist is simply a repeated binomial experiment with values $n = 10$ and $p = 0.50$. Hence the probability that Karnap picks exactly k correct coin tosses is

$$b(k) = b(k; 10, 0.50) = \binom{10}{k}(1/2)^k(1/2)^{10-k}$$

$$= \binom{10}{k}(1/2)^{10}$$

$$= \frac{10!}{(10-k)!k!}\left(\frac{1}{2}\right)^k$$

The graph of this binomial distribution is shown in Figure 18.

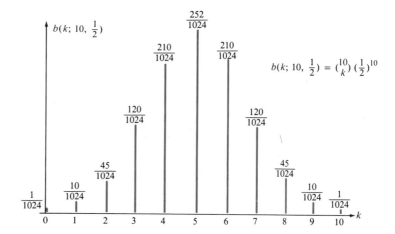

Figure 18
Binomial distribution for n = 10,
p = 1/2

Karnap Gets Five Correct

If Karnap were guessing, the probability that he would guess five or more correct is

$$P(X \geq 5) = b(5) + b(6) + b(7) + b(8) + b(9) + b(10)$$

$$= \frac{252 + 210 + 120 + 45 + 10 + 1}{1024}$$

$$= 0.6230$$

In other words, any person guessing has a 62% chance of getting five or more correct. Hence the psychologist would not be overly impressed with Karnap.

Karnap Gets Seven Correct

If Karnap is guessing, the probability that he gets seven or more correct is

$$P(X \geq 7) = b(7) + b(8) + b(9) + b(10)$$

$$= \frac{120 + 45 + 10 + 1}{1024}$$

$$= 0.1719$$

In other words, any person guessing has a 17% chance of getting seven or more correct. The psychologist is more impressed, but the chance factor is still fairly large.

Karnap Gets Nine or Ten Correct

If Karnap is guessing, the probability of getting nine or ten correct is

$$P(X \geq 9) = b(9) + b(10)$$

$$= \frac{10 + 1}{1024}$$

$$= 0.0107$$

In other words, on the average, only 1% of the general public will get this many guesses correct. Whether or not the psychologist decides to call Karnap a legitimate psychic depends on the psychologist's criteria for determining legitimacy. Possibly, more tests would be conducted.

Example 4

Medical Testing   The Center for Infectious Disease in Atlanta has determined that 10% of the general population of adults in the United States has a potentially contagious virus. What is the probability that exactly three people out of a randomly selected group of eight will have the virus?

Solution   Testing eight adults selected at random constitutes a repeated binomial experiment with values $n = 8$ and $p = 0.10$. Hence the probability that exactly $k = 3$ people in this group will have the virus (ironically, we call this outcome the "success") is

$$P(\text{Number of successes} = 3) = \binom{8}{3}(0.1)^3(0.9)^5$$

$$= 0.033$$

In other words, there is a 3.3% chance that exactly three people out of eight will have the virus.   □

Problems

Repeated Binomial Experiment Fails

For Problems 1–6, determine which experiments are repeated binomial experiments and which are not. If an experiment is not a repeated binomial experiment, indicate the properties that do not hold.

1. Roll a die 25 times, observing whether the die turns up even or odd on each roll.
2. Roll a die until a 6 appears, observing whether the die is even or odd on each roll.
3. Roll a die 25 times, observing the number that turns up on each roll.
4. Select a card 25 times from a deck of playing cards and observe whether the card is red or black. Return the card to the deck after the card has been observed.
5. Select a card 25 times from a deck of playing cards and observe whether the card is red or black. Do not replace the card in the deck after the card has been observed.
6. Select a card from a deck of playing cards and observe whether the card is a heart, diamond, club, or spade. Replace the card back in the deck after the card has been observed.

For Problems 7–13, a repeated binomial experiment is performed in which the probability of success is p. What is the probability of getting the exact sequence of successes (S) and failures (F) for the given values of p?

7. SSFF   where $p = 0.4$
8. SSSSSSSF   where $p = 0.2$
9. SFFSSSFFFS where $p = 0.5$
10. SFFFFF   where $p = 0.1$
11. FFFF   where $p = 0.1$
12. SSSS   where $p = 0.9$
13. SFSFSFSF where $p = 0.5$

For Problems 14–19, perform a repeated binomial experiment five times with probability of success $p = 2/3$. Find the following probabilities.

14. $P(\text{Number of successes} \geq 3)$
15. $P(\text{Number of successes} \geq 0)$
16. $P(\text{Number of failures} = 2)$
17. $P(\text{Number of failures} \geq 2)$
18. $P(\text{Number of failures} < 2)$
19. $P(\text{Number of failures} = 0)$

For Problems 20–23, toss a coin ten times. Find the probability of getting the following outcomes.

20. Seven heads.
21. Eight or more heads.
22. At least one head.
23. At least one tail.

Finding the Binomial Distribution

For Problems 24–25, a box contains one red marble and two black marbles. Three marbles are selected randomly one at a time, with replacement after each selection.

24. What is the binomial distribution of the number of red marbles that are selected? In other words, what is the probability that the number of red marbles selected will be zero, one, two, and three?

25. What is the binomial distribution of the number of black marbles selected?

Batter's Problem

Dave Winfield is hitting 0.250. Assuming that his hitting can be modeled by a repeated binomial experiment, find the probability that he will get the outcomes in Problems 26–30.

26. At least one hit in the next two times at bat.
27. At least one hit in the next three times at bat.
28. At least one hit in the next four times at bat.
29. At least two hits in the next four times at bat.
30. At least three hits in the next four times at bat.
31. What is the probability distribution of the number of hits Dave Winfield will get in the next four times at bat?

For Problems 32–35, Magic Johnson is shooting two free throws for the Lakers at the end of regulation play with the following score: Boston Celtics 105, Los Angeles Lakers 104. Magic Johnson makes 85% of his free throws. Assuming that his free throw shooting follows a repeated binomial experiment, find the following probabilities. For those readers that are unfamiliar with the game of basketball, a free throw counts one point.

32. The Celtics win.
33. The game goes into overtime.
34. The Lakers win.

35. The Lakers win or the game goes into overtime.
36. Famous Game One of the authors is writing this problem approximately one hour after Dennis Johnson made a 15-foot jump shot at the buzzer at the end of the fourth quarter in game 4 of the 1985 NBA Championship Series between the Celtics and the Lakers. The Celtics won that game 104–102 and tied the series 2–2. "DJ" had previously made 8 out of 12 shots during that game. Assuming that his shooting represented a repeated binomial process, what was the probability that he would make that last shot?
37. Rat Problem A rat is trained to touch one of two levers upon command. The probability that the rat will touch the correct lever on hearing the command is 0.85. If the rat's responses to the commands are independent, what is the probability that out of ten tries, the rat will touch the correct lever at least seven times?

Color Blindness

For Problems 38–40, in a given population the fraction of individuals who are color blind is 0.03. Find the following probabilities.

38. At least two individuals out of every ten selected will be color blind.
39. At least one individual out of every six selected will be color blind.
40. At least one individual out of every five selected will be color blind.

For Problems 41–53, graph the binomial distribution for the given values of n and p. If necessary, use a calculator. One can also use the fact that the nth row of the Pascal triangle shown below contains the $n + 1$ binomial coefficients $C(n, k)$, $k = 0, 2, \ldots, n$.

Row 1 $\longrightarrow$						1		1			
Row 2 $\longrightarrow$					1		2		1		
Row 3 $\longrightarrow$				1		3		3		1	
			1		4		6		4		1
		1		5		10		10		5		1
$C(6, k) \longrightarrow$		1		6		15		20		15		6		1
$C(7, k) \longrightarrow$	1		7		21		35		35		21		7		1
	1		8		28		56		70		56		28		8		1

41. $n = 2, p = 1/2$
42. $n = 2, p = 1/4$
43. $n = 2, p = 3/4$
44. $n = 3, p = 1/2$
45. $n = 3, p = 1/4$
46. $n = 3, p = 3/4$
47. $n = 4, p = 1/2$
48. $n = 4, p = 1/4$
49. $n = 5, p = 1/2$
50. $n = 5, p = 1/3$
51. $n = 8, p = 1/2$
52. $n = 8, p = 4/5$
53. $n = 11, p = 1/2$

6.4 Simulation (If All Else Fails)

PURPOSE

We introduce the basic ideas of simulation and show how this powerful technique can be used to solve problems that cannot be solved by other means. Simulation refers to a process that "acts out" or imitates real-world situations. In particular, we will study

- why simulation is used,
- the concept of Monte Carlo sampling,
- how to simulate baseball, and
- how to simulate the scheduling of surgical operations in a hospital.

Introduction

Mathematical models involving systems of equations, linear programming, or probability are used to *represent* reality. Simulation is different. Simulation *imitates* reality.

In this section we study computer simulation, in contrast to physical simulation. In physical simulation, one models or studies complex systems by constructing a scaled-down version of the real thing. Flight simulators used in the airline industry are excellent examples of scaled-down versions of the real thing.

In recent years, however, computer simulation has replaced many of the physical simulators of the past. The Navy now simulates on a computer many of the physical characteristics of a ship instead of building scaled-down models. The Bureau of Reclamation in Denver, Colorado, simulates many of the characteristics of dams on computers, whereas in the past it relied on miniature dams.

An engineer in the automobile industry can simulate a new suspension system 1000 times before an automobile is ever built.

The computer can simulate a hundred flash floods in a few hours and see whether a given dam design will meet engineering specifications. A business management simulation system can simulate the results of a management decision.

Reasons for Computer Simulation

With the advent of high-speed computers in the 1950's the tool of simulation left the physics laboratory and became useful in the areas of social science, biology, and business management. Today, business simulation games offer excellent ways to get a feeling for the impact of management decisions. Simulation languages such as GPSS, DYNAMO, GASP, and SIMSCRIPT have been developed to help the modeler simulate complicated problems. It has been said that simulation is the most commonly used mathematical technique in business today, being even more popular than linear programming.

In computer simulation the researcher supplies all the relevant information about the physical system to the computer, and the computer acts out the real-world situation. On the basis of the results of the computer's computation, the researcher draws conclusions about the real world. There are a number of different reasons why one would employ computer simulation.

- Speed and Economy.   The computer can act out a real-world situation cheaply and quickly.
- Flexibility.   It is easy to make modifications in a problem by simply changing the values of certain numbers in the computer. For instance, an engineer testing dam structures could change the entire design of a dam by simply changing some numbers that are fed to the computer.
- Training Device.   Computer simulation is often an excellent training device. The current abundance of business games supplies a valuable tool for learning how to act and respond to new situations. One game may give an executive experience with price strategies, while another may help a manager schedule the production of a product.

There are new simulators today that are a hybrid of physical and computer simulators, such as those that train physicians in the diagnosis of physical ailments. A lifelike mannequin, complete with many physical characteristics of the human body, is attached to a computer, which in turn is programmed to cause the mannequin to act and respond in realistic ways. The physician can examine the mannequin, take vital signs involving up to 25 different variables, and then make a diagnosis. After the physician has made a diagnosis the computer can evaluate the physician's performance. The computer is then reset for a new ailment, and the physician makes a new diagnosis.

Generating Random Numbers

When simulating most real-world situations, the computer is ultimately forced to generate a sequence of random numbers. In this way the computer can introduce an element of reality, or randomness, into the problem. The most common method for generating random numbers is the **congruential method**.

> **Random Number Generator (Congruential Method)**
>
> Start with three numbers a, b, P and a seed integer r_0 (first random integer).
>
> ▪ **Step 1.** Multiply the most recent random integer by a.
> ▪ **Step 2.** Add b to the result of Step 1.
> ▪ **Step 3.** Divide the result of Step 2 by P, and select the remainder as the next random integer.
>
> Steps 1, 2, and 3 are repeated to find more random integers.

In most computer programs the values of a, b, and P have already been selected, and the user of the program need only select the initial seed number r_0.

Example 1 ——————

Generating Random Numbers   Generate four random numbers from the values

$$r_0 = 4$$
$$a = 6$$
$$b = 3$$
$$P = 10$$

Solution   The first four random numbers are illustrated in Table 7. Note that after a new random integer has been found, it is divided by P (in this case 10) to find a random number between 0 and 1.

Table 8 gives a list of numbers that is a sample of random numbers generated by the congruential method using the following values:

$$a = 456,787$$
$$b = 546,786$$
$$P = 454,345$$
$$r_0 = 54,587$$

Given a sequence of random numbers, we can simulate other kinds of random phenomena. We present a few possibilities now.

TABLE 7 ——————
Generation of
Random Numbers

n	r_n	$6r_n + 3$	$\dfrac{6r + 3}{10}$	r_{n+1}	$\dfrac{r_{n+1}}{10}$
0	4	27	$2 + 7/10$	7	0.7
1	7	45	$4 + 5/10$	5	0.5
2	5	33	$3 + 3/10$	3	0.3
3	3	21	$2 + 1/10$	1	0.1
4	1	9	$0 + 9/10$	9	0.9

TABLE 8
50 Six-Digit Random
Numbers

0.593749	0.340923	0.043213	0.505457	0.045347
0.468750	0.560432	0.931232	0.054364	0.654568
0.109373	0.450321	0.543021	0.296501	0.120934
0.934656	0.344971	0.109324	0.219654	0.453565
0.345439	0.540034	0.234530	0.609121	0.309345
0.564530	0.550633	0.109345	0.850921	0.543455
0.548990	0.530535	0.294532	0.120543	0.093435
0.300345	0.540934	0.504592	0.434023	0.987565
0.540345	0.430912	0.400344	0.450544	0.092344
0.013435	0.892323	0.204567	0.103264	0.550566

☐

Example 2

Flipping Coins on a Computer   Show how a computer can simulate the flipping of a coin and obtain a sequence of heads and tails.

Solution   We start with a sequence of random numbers between 0 and 1. If a random number lies between 0 and 0.5, call the outcome a Head. If the random number lies between 0.5 and 1, call the outcome a Tail. If the random number is exactly 0.5, assign either a Head or a Tail to the outcome (we will select a Tail).

A flow diagram that describes a computerized coin tossing experiment is shown in Figure 19.

The list in Table 9 of random numbers shows how a computer can simulate the flipping of a coin.

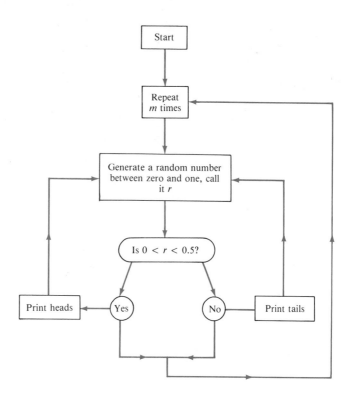

Figure 19
Flow diagram for generating a sequence of heads and tails

TABLE 9
Random Number Simulation
of Coin Tosses

Random Number		Result of Coin Toss
0.4554	$\longrightarrow$	Head
0.9302	$\longrightarrow$	Tail
0.5349	$\longrightarrow$	Tail
0.3436	$\longrightarrow$	Head
0.5649	$\longrightarrow$	Tail
0.1547	$\longrightarrow$	Head
0.0358	$\longrightarrow$	Head
$\vdots$		$\vdots$

The Monte Carlo Method

John Von Neumann once called the Monte Carlo method any probabilistic method for solving nonprobabilistic problems. In other words, if a problem is not related to probability, but it is solved by a method that is based on probability, then this method of solution is called a **Monte Carlo method**. We present such an example here.

Stanislaw Ulam

John Von Neumann

HISTORICAL NOTE

Simulation is a contemporary subject. It began with the work of John Von Neumann (1903–1957) and Stanislaw Ulam (1909–1984) when they were working on the Manhattan Project during the Second World War. The name *Monte Carlo method* referred to simulation methods that were used in solving nuclear-shielding problems too complicated to study with analytical methods. The name *Monte Carlo* was picked because the method ultimately depended on a random process not unlike the spinning of a roulette wheel. Today, the phrase Monte Carlo methods refers to methods for solving nonprobabilistic problems by probabilistic means.

The Area of a Circle Using a Monte Carlo Method

To illustrate how problems unrelated to probability can be solved by using ideas of probability, consider finding the area of a circle. To find this area,

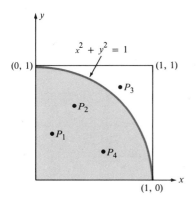

Figure 20
Finding the area of a circle by the Monte Carlo method

it is convenient to find the shaded area shown in Figure 20, and multiply it by 4.

We first use a random number generator to find a sequence of random numbers $r_1, r_2, r_3, \ldots$, all between 0 and 1. From these numbers we can obtain a sequence of random points:

$$P_1 = (r_1, r_2)$$
$$P_2 = (r_3, r_4)$$
$$P_3 = (r_5, r_6)$$
$$\vdots \qquad \vdots$$

These points can be interpreted as random points spread uniformly in the square shown in Figure 20. We can think of these points as the locations struck by darts being thrown at the square by a dart thrower. For each of these points we determine whether it lies inside the circle (shaded area of Figure 20). After several points have been generated, we compute the ratio of points inside the circle to the total number N of points. From this ratio we can approximate the area of the quarter circle from the obvious relation:

$$\frac{\text{Number of points under quarter circle}}{\text{Total number points tossed}} \cong \frac{\text{Area of quarter circle}}{\text{Area of square}}$$

Since the area of the square is 1, an estimate of the area of the circle is given by

$$\text{Area of the circle} \cong 4\,\frac{\text{Number of points under quarter circle}}{N}$$

Table 10 gives an approximation to the exact area of the unit circle (which is $\pi = 3.1415927\ldots$) by means of the Monte Carlo method.

TABLE 10

Approximate Area of a Circle by the Monte Carlo Method

Number of Tosses, N	Monte Carlo Approximation to π	Error
50	3.21043	0.068840
100	3.17034	0.028750
200	3.10433	−0.037260
300	3.14205	0.000460
400	3.15311	0.011520
500	3.14145	−0.000145
600	3.14210	0.000507
700	3.14146	−0.000133
800	3.14167	0.000077
900	3.14154	−0.000053
1000	3.14149	−0.000103

A flow diagram that describes these computations and the results from a computer program are shown in Figure 21.

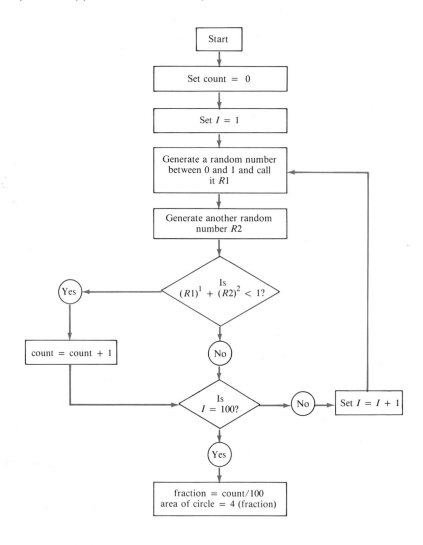

Figure 21
*Finding the area of a circle
by Monte Carlo sampling*

Example 3 ──────────────   Rolling a Die   Show how a computer can simulate the rolling of a die.

Solution   We carry out the following steps:

    1. Start with a random number r between 0 and 1,

    2. Compute the value $6r + 1$,

    3. Find the largest integer less than or equal to $6r + 1$.

The integer found in Step 3 will be the rolled value of the die. For example, the random number $r = 0.2340$ will give rise to a rolled die of 2 as can be seen from

$$r = 0.2340 \longrightarrow 6(0.2340) + 1 = 2.4040 \longrightarrow 2$$

This process essentially subdivides the interval (0, 1) into six equal portions and assigns each of the values 1, 2, 3, 4, 5, and 6 to one of the subintervals. The process is illustrated in Figure 22.

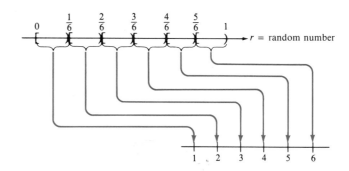

Figure 22
Rolling a die using a computer

Using a sequence of ten random numbers, the rolls of a die in Table 11 are found.

TABLE 11
Random Number Simulation of Die Tosses

Random Number, r	$6r + 1$	Die
0.234	2.404	2
0.593	4.558	4
0.731	5.386	5
0.103	1.618	1
0.924	6.544	6
0.364	3.184	3
0.401	3.406	3
0.530	4.180	4
0.206	2.236	2
0.821	5.926	5

Two Sample Simulations

TABLE 12
Player's Batting Statistics

Result	Probability
Fly out	0.110
Ground out	0.210
Strike out	0.220
Walk	0.140
Single	0.220
Double	0.050
Triple	0.025
Home run	0.025
	1.000

We will now show how physical processes can be simulated by means of Monte Carlo sampling.

Baseball

Baseball is an example of a process that can be successfully simulated. The game is complex enough that a theoretical analysis is difficult, but it is an orderly and well-documented process. Past data can be fed into a computer, and the computer can then "act out" the game.

On the basis of two teams' statistics (batting and fielding averages, slugging percentages, steals, and so on), it is possible to simulate different situations (hit, strike out, fly out, bunt, double steal, etc).

For example, let us assume that a player's batting statistics are shown in Table 12. We can simulate this batter at the plate by generating a random number $0 \leq r < 1$ and selecting the appropriate outcome. Figure 23 shows how this can be done.

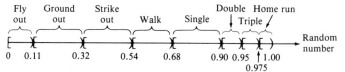

Figure 23
*The value of a random number
determines the fate of the batter*

Other batters are treated in an analogous manner. Base runners also have statistics, such as the number of bases that they will advance in different situations. In this way an entire game can be simulated.

The model would be ideal as a learning tool for a new manager. He could place the relevant statistics of two teams into the computer and call for various plays. The manager could call for a hit-and-run or a steal or could send in a pinch hitter. The manager could play a couple of hundred games before the season actually began.

Simulation of Operating Room Scheduling

This example describes how simulation can help determine the resources needed by a hospital. Deaconess Hospital in St. Louis contemplated the need for more operating room facilities.* The management of the hospital decided that by adding 144 more beds to its existing facilities, they could increase the number of operations each year by the amounts in Table 13.

TABLE 13

Additional Operations per Year as a Result of New Beds

Type of Surgery	Increased Case Load per Year
Ophthalmology	132
Gynecology	282
Urology	264
Orthopedic	202
ENT	1098
Dental	715
Other	683
	3376

The increase of 3376 in the expected number of surgical procedures was added to the previous year's case load of 6293 operations per year. This gave a new total of 9669 operations per year, or $9669/365 \cong 27$ operations per day that could be performed.

The question the management of Deaconess Hospital had was how many operating rooms would be required to handle the new total of 27 operations per day. The hospital had three operating rooms in use at the time and the management suspected that an increase in the number of operations from 6293 to 9669 per year (roughly a 50% increase) would mean that the hospital would require five operating rooms (60% increase). However, they felt that more information could be obtained by performing a computer simulation. To simulate the sched-

---

* H. Schmitz and N. Kwak, "Monte Carlo Simulation of Operating Rooms and Recovery Room Usage," *Operations Research* Vol. 20, 1972, pp. 1171–1180. (Table reprinted with permission from *Operations Research*, Volume 20, issue 6, 1972, Operations Research Society of America. No further reproduction permitted without the consent of the copyright owner.)

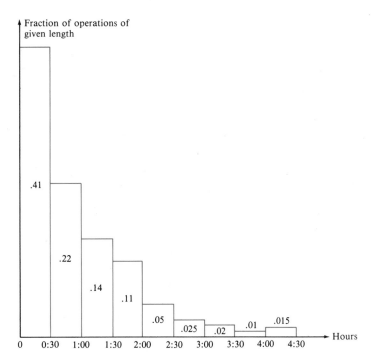

Figure 24
Frequency distributions of lengths of operations

uling of operating rooms, the management collected the necessary information to construct the histogram in Figure 24. This histogram shows the fraction of operations that require a given length of time.

From the histogram, along with a sequence of random numbers, it was possible to generate a series of surgical procedures that had the same distribution of lengths as operations that were normally performed in the hospital. The relationship between a generated random number and the corresponding "generated operation" can be seen in Table 14.

Twenty-seven random numbers were then generated, and hence 27 operations were generated having various lengths. These 27 operations represented a typical daily operating schedule in the hospital. To determine whether five operating rooms could handle 27 operations in one day, the hospital simulated this exact situation. The results of this simulation are shown in Table 15.

TABLE 14
Relationship Between
Random Numbers and
Length of Operations

Random Number	Length of Operation (hr:min)
[0, 0.41)	0:30
[0.41, 0.63)	1:00
[0.63, 0.77)	1:30
[0.77, 0.88)	2:00
[0.88, 0.93)	2:30
[0.93, 0.955)	3:00
[0.955, 0.975)	3:30
[0.975, 0.985)	4:00
[0.985, 1.]	4:30

TABLE 15
Simulation of 27 Operations
in Five Operating Rooms

Operation Number	Random Number	Length of Operation (hr:min)	Time Operation Begins	Time Operation Ends	Operating Room Number
1	0.889	2:15	7:30 A.M.	9:45 A.M.	1
2	0.396	0:45	7:30 A.M.	8:15 A.M.	2
3	0.358	0:30	7:30 A.M.	8:00 A.M.	3
4	0.715	1:15	7:30 A.M.	8:45 A.M.	4
5	0.502	0:45	7:30 A.M.	8:15 A.M.	5
6	0.068	0:30	8:15 A.M.	8:45 A.M.	3
7	0.604	0:45	8:30 A.M.	9:15 A.M.	2
8	0.270	0:30	8:30 A.M.	9:00 A.M.	5
9	0.228	0:30	9:00 A.M.	9:30 A.M.	4
10	0.782	1:45	9:00 A.M.	10:45 A.M.	3
11	0.379	0:30	9:15 A.M.	9:45 A.M.	5
12	0.093	0:30	9:30 A.M.	10:00 A.M.	2
13	0.011	0:30	9:45 A.M.	10:15 A.M.	4
14	0.648	1:15	10:00 A.M.	11:15 A.M.	1
15	0.527	0:45	10:00 A.M.	10:45 A.M.	5
16	0.987	4:30	10:15 A.M.	14:45 P.M.	2
17	0.214	0:30	10:30 A.M.	11:00 A.M.	4
18	0.474	0:45	11:00 A.M.	11:45 A.M.	3
19	0.238	0:30	11:00 A.M.	11:30 A.M.	5
20	0.045	0:30	11:15 A.M.	11:45 A.M.	4
21	0.408	0:45	11:30 A.M.	12:15 P.M.	1
22	0.116	0:30	11:45 A.M.	12:15 P.M.	5
23	0.209	0:30	12:00 A.M.	12:30 P.M.	3
24	0.048	0:30	12:00 A.M.	12:30 P.M.	4
25	0.393	0:45	12:30 P.M.	13:15 P.M.	1
26	0.550	0:45	12:30 P.M.	13:15 P.M.	5
27	0.306	0:45	12:45 P.M.	13:30 P.M.	3

One can follow along and see how the simulation was performed. The first random number was 0.889, which means that Operation 1 will take 2 hours and 15 minutes. If we add 15 minutes clean-up time for the operating room, this means that Operating Room 1 will be ready for the second operation at 7:30 + 2:15 + 0:15 = 10:00 A.M. Since we are assuming five operating rooms, the next four rows of the table are found in the same way as row 1.

Note that Operating Room 3 is the first room that becomes empty (8:00 A.M.) and after cleanup is ready for Operation 6 at 8:15 A.M.

The last operation, Operation 27, begins at 12:45 P.M. and is completed at 1:30 P.M. The reason for the early quitting time is that most patients spend upwards of three hours in the recovery room after surgery.

Summary of Simulation

The above simulation was only a small part of the simulation carried out by Schmitz and Kwak. They also simulated the stay of patients in the recovery room. It was discovered that the last patient left the recovery room at 4:34 P.M. and that there was a maximum of 12 patients in the recovery room at one time.

Surgical procedures were simulated for four days. On three of the days, 11 recovery room beds were needed; on one of the days, 12 beds were needed. The latest departure from the recovery room during these four days was 8:36 P.M.

Surgical procedures were simulated using three, four, five, and six operating rooms. The hospital management decided that three or four rooms were not enough (patients left the recovery room too late) but that six operating rooms were more than enough. Hence they decided that five were optimum.

The above simulation was verified by simulating the hospital's past data. In other words, during the previous year, when the hospital had three operating rooms and 17 operations per day, the hospital kept records similar to Table 14. To test the accuracy of the simulation process, the hospital management assumed the same conditions (three operating rooms and 17 operations per day) and determined whether the computer simulation produced a table similar to the observed table. The checking of a simulation model against past data is called verification of the model. It is important that it be carried out.

Problems

For Problems 1–9, use the given sequence of random numbers to simulate the following random phenomena. Start the sequence at any location in the list, and proceed in any orderly fashion, selecting the numbers you want to use. Be aware that answers will vary, since they depend on the order in which the random numbers are used.

0.734932	0.304532	0.045343	0.038653	0.504532
0.409453	0.230654	0.045610	0.874370	0.395610
0.451043	0.548771	0.330454	0.919402	0.545011
0.540934	0.543877	0.205438	0.104534	0.120944
0.670923	0.720945	0.698324	0.823093	0.910454

1. Simulate the flipping of a coin ten times.
2. Simulate the rolling of a die ten times.
3. Simulate the drawing of ten cards, with replacement, from a deck of playing cards.
4. Simulate the rolling of a pair of dice ten times.
5. Simulate a softball player batting ten times if the player's statistics are those in Table 16.
6. **Weather Problem** Simulate the weather for the next ten days, assuming that it is rainy today and the weather

TABLE 16
Baseball Player's Statistics for Problem 5

Result	Probability
Walk	0.10
Out	0.65
Single	0.15
Double	0.05
Triple	0.03
Home run	0.02
	1.00

pattern from day to day can be modeled by the following transition matrix of a Markov chain:

Weather Today	Weather Tomorrow	
	Fair	*Rainy*
Fair	0.75	0.25
Rainy	0.50	0.50

7. Spin the Wheel   Simulate the spinning of the wheel in Figure 25, where

$$P(\text{I}) = 0.50$$
$$P(\text{II}) = 0.25$$
$$P(\text{III}) = 0.25$$

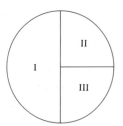

Figure 25
Spin the Wheel

8. Sampling from the Binomial Distribution   Simulate the sampling of ten numbers from the binomial distribution in Figure 26. Does your sample of ten values agree roughly with what they should be?

Figure 26
Simple binomial distribution

9. Sampling from the Binomial Distribution   Simulate the sampling of ten numbers from the binomial distribution with $n = 4$ and $p = 0.25$.

10. Random Number Generator   Generate a sequence of five random integers using the congruential method with values $a = 2$, $b = 1$, $P = 5$, and the seed integer $r_0 = 2$. Does this sequence repeat itself? What can you say about this choice of values for a, b, and P?

11. Random Number Generator   Generate a sequence of ten random integers using the congruential method with values $a = 17$, $b = 41$, $P = 100$, and seed integer $r_0 = 30$. Divide the random integers by 100 to get a sequence of two-digit random numbers between 0 and 1. Do they appear to be random?

12. Approximate Area of a Circle   Use the random number list at the beginning of the problem set to approximate the area of the unit circle

$$x^2 + y^2 = 1$$

Use ten random points.

13. Approximate Area of Circle   Write a flow diagram to find the approximate area of the circle

$$x^2 + y^2 = 2$$

using the Monte Carlo method.

14. Approximate Area of Ellipse   Write a flow diagram to find the approximate area of the ellipse as shown in Figure 27.

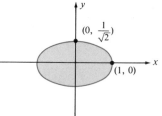

Figure 27
Ellipse for Problem 14

15. Approximate Area Under a Curve   Write a flow diagram to find the approximate area under the parabola as shown in Figure 28:

$$y = x^2 \qquad 0 \le x \le 1$$

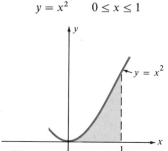

Figure 28
Parabola for Problem 15

16. Hospital Simulation   Use the random number list from the beginning of the problem set to simulate ten surgical procedures and one operating room, where the lengths of the operations are distributed the same as in Figure 24 in the text. At what time is the last operation completed?

17. Buffon's Needle Problem   The Buffon needle problem is a classic example of the Monte Carlo method. It allows one to approximate the value of π by tossing toothpicks. On a smooth table, draw a series of parallel lines. The lines should be drawn as far apart as the length of the toothpick. It can be shown that if one tosses a toothpick on the table several times, then the value of π can be approximated by the following ratio:

$$\pi \cong \frac{\text{Number of times the toothpick is tossed}}{\text{Number of times the toothpick crosses a line}}$$

Approximate the value of π by tossing a toothpick on a surface where lines are drawn (see Figure 29) and computing the above ratio.

For Problems 18–23, how would you use a sequence of random numbers between 0 and 1 to simulate the following processes?

18. Selecting a card from a deck of 52 playing cards.
19. Rolling a pair of dice.

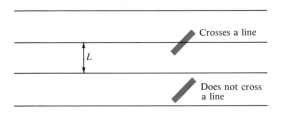

Figure 29
Buffon needle problem

20. Generating random successes and failures if the probability of success is 0.10.
21. Generating random successes and failures if the probability of success is 0.25.
22. Generating a binomial random variable with $n = 5$ and $p = 0.50$.
23. Generating outcomes for a batter in baseball, given the probabilities in Table 17.
24. **Baseball Game** Play a three-inning game using Figure 30 and the following rules. Player I places a button on the starting position and then tosses three pennies. The button is moved on the chart as follows:

   - one position down if one head appears;

   - two positions down if two heads appear;

   - three positions down if three heads appear.

TABLE 17
Baseball Player's Batting Statistics
for Problem 23

Outcome	Probability
Walk	0.100
Single	0.200
Double	0.050
Triple	0.025
Home run	0.075
Strike out	0.200
Fly out	0.200
Ground out	0.150
	1.000

The player then acts accordingly, depending on where the button has landed. Extra buttons can be used for base runners, and all base runners advance the same number of bases as the batter. After three outs, Player II comes to bat, with the button starting in the position where the last out was made. If the toss of the coins results in the button's running over the bottom of the list, the player should return the button to the top of the list.

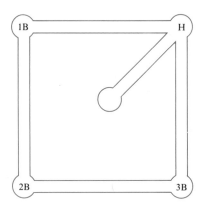

Start

Strike out
Ground out
Single
Double
Fly out
Single
Walk
Walk
Strike out
Double
Home run
Strike out
Single
Ground out
Single
Double
Fly out
Triple
Triple
Strike out
Single
Home run

Figure 30
Simulated baseball

INTERVIEW WITH MR. DON PETERSEN
Chairman of the Board
Ford Motor Company

The world's second largest producer of automobiles, the Ford Motor Company, started with one man, a small garage, and a quadricycle. Products now include replacement parts for cars, trucks, tractors, and farm implements as well as glass, steel, radios, television sets, special military vehicles, control systems, satellite tracking systems, space communications, and many more.

TOPIC: Probability and Quality Control at the Ford Motor Company

Random House: It is understandable why many engineers and scientists at a company like the Ford Motor Company would need to study mathematics and probability, but is it really useful for your people in management to learn mathematics?

Mr. Petersen: Yes. Mathematics is an important part of both the culture and the language of management. Now, more than ever before, managers need to understand all kinds of business decision making and analytical factors as well as "big picture" issues. An increasing number of decisions, involving such areas as product design, process quality control, economics studies, and financial analysis, require a grounding in probability. A manager without a mathematics background is at a disadvantage.

Random House: Many readers of this book probably haven't the slightest idea of what they will be doing in ten years, and some probably even have doubts about making a serious effort to learn mathematics. What would you say to them?

Mr. Petersen: I consider mathematics a fundamental tool. Knowing how to use it automatically broadens an individual's understanding and analytical abilities. This helps greatly to extend the range of vocational and personal options.

Random House: Granted, it is impossible to give a precise answer to this question, but could you tell us how the Ford Motor Company uses probability? Maybe in the general area of quality control. We understand there are some revolutionary new techniques in statistical quality control that will make the next generation of cars far superior to any we have seen thus far.

Mr. Petersen: Probability is the basis for statistical process control (SPC), an extremely important tool in our manufacturing processes. We employ SPC to ensure the quality of our products. Many of our supplier companies are making SPC an integral factor in their operations as well.

Random House: This question is really off the beaten track, but we have just spent a few days studying simulation, and some students have asked exactly how useful a tool it really is.

Mr. Petersen: You may be very surprised to discover the many ways in which simulation techniques are used in business. For example, we have used simulation quite successfully in numerous training applications and in operations research. In developing new manufacturing facilities, we use simulation for planning the numbers and kinds of discrete machines needed for a given production volume. We also use simulation in product design, where CAD/CAM (computer-aided design/computer-aided manufacture) is replacing the traditional drawing boards. Simulation helps us to compare and evaluate real-world alternatives in an experimental setting.

Epilogue: Probability Crossword Puzzle

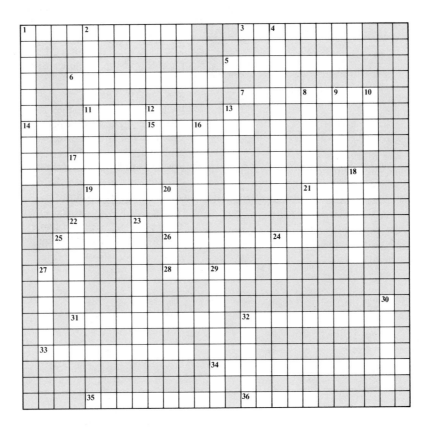

Across

1. The probability of an event given another event is a _____ probability.

3. A special kind of sequence of binomial experiments is called a _____ binomial experiment.

5. The _____ distribution tells how the number of successes of a coin-flipping experiment are distributed.

6. The property of a Markov chain that says that the next state depends only on the current state is called the _____ property.

7. Free Question A state of a Markov chain is called transient if there is a positive probability of leaving the state once you arrive.

11. A collection of outcomes of a probability experiment is called an _____.

14. Disjoint events are critical in computing the probability of the _____ of two events.

15. Free Question A Markov chain is called regular when the Markov chain has a steady state.

17. Free Question A game played in casinos that has very poor odds is Keno.

19. Individual experiments of a Markov chain are called _____.

21. A _____ of a Markov chain is one of the experiments.

25. A _____ random variable can take on only a finite number of values.

26. The _____ matrix allows one to find the new state vector from the previous vector.

28. Free Question If a Markov chain is regular, then it will always have a steady state vector.

31. A method discovered by John Von Neumann and Stanislaw Ulam for solving problems that cannot be approached by other ways is called the _____ method.

32. When all else fails, try _____.

33. Mathematical models describe the real world; _____ imitates it.

34. Independent events are critical in computing the probability of _____.

35. Free Question A state of a Markov chain is called <u>absorbing</u> if once you arrive there you always remain there.

36. A Markov chain is always in a given _____.

Down

1. A random variable is called _____ if it can take on all values in an interval.

2. If $P(A|B) = P(A)$, then A and B are called _____.

4. The _____ vector gives the probability of being in the different states of a Markov chain.

8. One of the most popular ways of solving problems today is _____.

9. The probability of the _____ set is always 0.

10. A probability _____ allows one to find the probability of a sequence of experiments.

12. A useful device that allows us to find the probability of intersections is the probability _____.

13. A _____ variable assigns a number to each outcome of a probability experiment.

16. Disjoint events are critical in finding the probability of the _____ of two events.

18. The French mathematician responsible for a useful device for computing binomial coefficients is _____.

20. The _____ of a Markov chain are listed along the side and top of the transition matrix.

22. The transition _____ is a visual way of looking at a Markov chain.

23. If $P(A|B) = P(A)P(B)$, then A and B are _____.

24. There are 36 outcomes when rolling a pair of _____.

27. The probability of intersections is never greater than the probability of _____.

29. Free Question If one of the diagonal entries of the transition matrix is 1, then the row where the 1 appears is called an <u>absorbing</u> state.

30. Disjoint events are critical in computing the probability of the _____ of two events.

32. The outcomes of a Markov chain are the _____ of the Markov chain.

C	O	N	D	I	T	I	O	N	A	L			R	E	P	E	A	T	E	D			
O			N												R								
N			D					B	I	N	O	M	I	A	L								
T		M	E	M	O	R	Y	L	E	S	S				B								
I			P							T	R	A	N	S	I	E	N	T					
N			E	V	E	N	T			R		B		I		M		R					
U	N	I	O	N			R	E	G	U	L	A	R		I		M	P		E			
O			D				E		N		N		I		L		U	T		E			
U		K	E	N	O		E		I		D		I		L		Y						
S			N						O		O		T		A			P					
			T	R	I	A	L	S		N		M		Y		T	R	I	A	L			
							T									I			S				
		D			D		A					D		O		C							
	F	I	N	I	T	E		T	R	A	N	S	I	T	I	O	N		A				
		A			P		E						C					L					
	U	G			E		E	S	T	E	A	D	Y		E								
	N	R			N		B																
	I	A			D		S										U						
	O	M	O	N	T	E	C	A	R	L	O		S	I	M	U	L	A	T	I	O	N	
	N			N			O				R	T					I						
	S	I	M	U	L	A	T	I	O	N	B			A			O						
									I	N	T	E	R	S	E	C	T	I	O	N	S		
							N		E								S						
		A	B	S	O	R	B	I	N	G		S	T	A	T	E							

Key Ideas for Review

Bayes' formula, 386
Bernoulli experiment, 391
binomial distribution, 394
binomial experiment, 391
binomial random variable, 393
Markov chain
   average payoff, 379

memoryless property, 370
state of, 370
state vector, 372
steady state vector, 378
transition matrix, 370
Monte Carlo method, 406
posterior probabilities, 384, 386

prior probabilities, 386
random numbers, 403
repeated binomial experiment, 392
simulation, 402
simulation of operating rooms, 410

Review Exercises

For Problems 1–3, draw transition diagrams for the following transition matrices.

1. $\begin{bmatrix} 1/2 & 1/2 \\ 1/3 & 2/3 \end{bmatrix}$

2. $\begin{bmatrix} 1/3 & 0 & 2/3 \\ 0 & 1 & 0 \\ 1/2 & 1/2 & 0 \end{bmatrix}$

3. $\begin{bmatrix} 1/2 & 1/2 & 0 \\ 0 & 1/2 & 1/2 \\ 0 & 0 & 1 \end{bmatrix}$

For Problems 4–5, write the transition matrices for the following transition diagrams.

4.

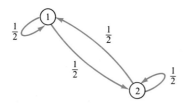

5.

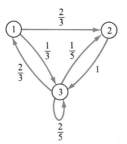

6. The Urn Markov Chain Initially, two urns each contain three marbles as shown in Figure 31. Roll a die and then move the marble showing the rolled number to the opposite urn. Repeat this process over and over. Call the states of this Markov chain the number of marbles in Urn 1.

Figure 31
Markov chain problem

What are the different states of this Markov chain, and what is the transition matrix?

7. Sociological Markov Chain A person's profession is classified as blue collar or white collar. Assume that the child of a blue collar worker has a 50% chance of being a blue collar worker and the child of a white collar worker has a 40% chance of being a white collar worker. What are the states of this Markov chain, and what is the transition matrix? What assumptions are present that allow one to say that this sociological process can be modeled by a Markov chain?

For Problems 8–13, which of the following could be state vectors?

8. (1/3, 2/3)

9. (2/3, 1/2, 1/3)

10. (1, 1)

11. (0, −1, 2)

12. (1/10, 4/10, 1/2)

13. (1, 0, 0, 0)

For Problems 14–17, which of the following could be transition matrices?

14. $\begin{bmatrix} 1 & 0 \\ 0 & 1 \end{bmatrix}$

15. $\begin{bmatrix} 2/3 & 2/3 \\ 0 & 1 \end{bmatrix}$

16. $\begin{bmatrix} 1/3 & 1/3 & 1/3 \\ 0 & 1 & 0 \\ 0 & 0 & 1 \end{bmatrix}$

17. $\begin{bmatrix} 1/2 & 1 \\ 1 & 0 \end{bmatrix}$

For Problems 18–20, use coins, dice, or marked slips of paper in boxes to run or simulate the following Markov chains for ten trials.

18. $\begin{bmatrix} 1/2 & 1/2 \\ 1/2 & 1/2 \end{bmatrix}$ **19.** $\begin{bmatrix} 1/3 & 2/3 & 0 \\ 0 & 1/2 & 1/2 \\ 0 & 1/5 & 4/5 \end{bmatrix}$

20. $\begin{bmatrix} 1/2 & 1/2 & 0 \\ 0 & 1/2 & 1/2 \\ 0 & 0 & 1 \end{bmatrix}$

Use the transition matrix from Problem 18 and the initial state vectors in Problems 21–25 to find the state vectors P_1, P_2, P_3, and P_4. What can one conclude about the steady state vector? Do these results agree with your intuition?

21. $(1, 0)$ **22.** $(0, 1)$
23. $(1/2, 1/2)$ **24.** $(1/4, 3/4)$
25. $(0.1, 0.9)$

For Problems 26–29, use the transition matrices found in Problem 19 and the following initial state vectors to find the state vectors P_1, P_2, P_3, and P_4. What can one conclude about the steady state vector? Do these results agree with your intuition?

26. $(1, 0, 0)$ **27.** $(0, 1, 0)$
28. $(0, 0, 1)$ **29.** $(1/3, 1/3, 1/3)$

Powers of the Transition Matrix

To find the state vector P_2, one generally multiples the state vector P_1 times the transition matrix P, or

$$P_2 = P_1P$$

This computation could also be carried out by

$$P_2 = P_1P = (P_0P)P = P_0P^2$$

In general, one has the following formula:

$$P_n = P_0P^n$$

Use the above formula to compute P_2, P_3, and P_4 if

30. $P = \begin{bmatrix} 1/2 & 1/2 \\ 1/3 & 2/3 \end{bmatrix}$ and $P_0 = (1, 0)$

31. $P = \begin{bmatrix} 1 & 0 \\ 1/2 & 1/2 \end{bmatrix}$ and $P_0 = (1/2, 1/2)$

32. $P = \begin{bmatrix} 1/2 & 1/2 & 0 \\ 0 & 0 & 1 \\ 0 & 1/2 & 1/2 \end{bmatrix}$ and $P_0 = (1/3, 1/3, 1/3)$

33. Genetic Mutation A gene in a chromosome is either Type 1 or Type 2. The probability of a Type 1 gene mutating to Type 2 over any generation is 0.001, while the probability of a Type 2 gene mutating to a Type 1 gene is

0.0001. What is the transition matrix for this Markov chain? If the initial state vector is $(1, 0)$, what fraction of genes will be in each of the two types after one generation?

34. All-Star Game In the 20 years from 1966–1985, the All-Star game between the American League (A.L.) and the National League (N.L.) has been won by the teams listed in Table 18. Does this process satisfy the assumption of a Markov chain? If so, how would one go about finding the transition matrix from the information?

TABLE 18
All-Star Game Winners for Problem 34

Year	Winner	Year	Winner
1966	N.L.	1976	N.L.
1967	N.L.	1977	N.L.
1968	N.L.	1978	N.L.
1969	N.L.	1979	N.L.
1970	N.L.	1980	N.L.
1971	A.L.	1981	N.L.
1972	N.L.	1982	N.L.
1973	N.L.	1983	A.L.
1974	N.L.	1984	N.L.
1975	N.L.	1985	N.L.

Posterior Probabilities in Medical Testing

Problems 35–37 illustrate how Bayes' theorem can be used in medical testing. Computerized expert systems can be designed on the basis of these general ideas.

Harry goes to the doctor complaining of stomach pains. The doctor considers three alternatives: stomach gas, an ulcer, and gallstones. The doctor assigns the prior probabilities of 0.6, 0.3, and 0.1, respectively, to the three ailments based on prior experience. The doctor then orders a test that turns out negative. It is known that negative tests occur in 75% of people with stomach gas, 20% of people with ulcers, and 5% of people with gallstones.

35. Find $P(\text{Harry has stomach gas} \mid \text{Test was negative})$.
36. Find $P(\text{Harry has an ulcer} \mid \text{Test was negative})$.
37. Find $P(\text{Harry has a gallstone} \mid \text{Test was negative})$.

Medical Testing

Problems 38–42 are concerned with the following problem in medical testing. Suppose among people having high blood pressure, three-fourths are unaware of their condition. Out of every ten people who have high blood pressure, find the probability that the following number of people are ignorant of their condition.

38. Six people.   **39.** Seven people.
40. Eight people.   **41.** Nine people.
42. All ten people.

43. Recurrent States A Markov chain may trap a process in a set of states. What states described by the following transition matrix will trap the described process? Draw the Markov diagram for this process.

$$P = \begin{bmatrix} 1/2 & 1/2 & 0 & 0 \\ 1/4 & 3/4 & 0 & 0 \\ 0 & 1/3 & 1/3 & 1/3 \\ 0 & 0 & 1/2 & 1/2 \end{bmatrix}$$

Graphing the Binomial Distribution

For Problems 44–46, draw the graphs for the binomial distributions for $n = 3$ and p as given in the problem.

44. $p = 1/4$ **45.** $p = 1/2$ **46.** $p = 3/4$

Simulation with a Penny

How would you use a coin to simulate the processes in Problems 47–52?

47. Rolling a die.
48. Rolling a pair of dice.
49. Spinning the wheel in Figure 32.

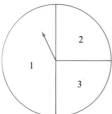

Figure 32
Spin the Wheel for Problem 49

50. Selecting an outcome from a binomial distribution with $n = 3$ and $p = 1/2$.
51. Selecting an outcome from a binomial distribution with $n = 4$ and $p = 1/2$.
52. Selecting an outcome from a binomial distribution with $n = 4$ and $p = 1/4$.
53. Baseball Simulation Use the set of random numbers in Table 19 to simulate two innings of a baseball game between the two teams in Table 20. Assume that base runners always advance the same number of bases as does the batter. Choose your own lineup.

TABLE 19
Random Numbers

0.45	0.45	0.76	0.94	0.58	0.76	0.56
0.65	0.04	0.34	0.95	0.34	0.90	0.98
0.34	0.67	0.44	0.09	0.34	0.12	0.34
0.01	0.76	0.34	0.78	0.15	0.09	0.78
0.95	0.67	0.38	0.87	0.23	0.65	0.56
0.35	0.67	0.46	0.45	0.45	0.87	0.43
0.87	0.23	0.09	0.24	0.87	0.81	0.98
0.45	0.04	0.01	0.98	0.76	0.24	0.23
0.98	0.35	0.69	0.57	0.34	0.77	0.12
0.23	0.85	0.93	0.78	0.09	0.09	0.67
0.45	0.88	0.26	0.46	0.45	0.45	0.98
0.87	0.23	0.47	0.24	0.54	0.67	0.34
0.13	0.74	0.29	0.78	0.87	0.56	0.22
0.46	0.57	0.13	0.84	0.34	0.45	0.56
0.78	0.35	0.37	0.54	0.65	0.34	0.67
0.32	0.58	0.50	0.04	0.96	0.12	0.79

TABLE 20
Listing of Baseball Teams
(continued on page 422)

Atlanta Braves								
Batting	*BA*	*G*	*AB*	*R*	*H*	*2B*	*3B*	*HR*
Thompson	0.370	29	54	9	20	2	1	0
Horner	0.288	91	340	47	98	22	2	20
Murphy	0.287	110	422	90	121	22	2	30
Oberkfell	0.283	93	279	23	79	13	4	2
Ramirez	0.283	98	421	46	119	20	3	4
Washington	0.278	86	309	47	86	11	5	11
Harper	0.250	96	352	37	88	9	2	14
Hubbard	0.226	96	292	30	66	14	0	4
Komminsk	0.223	79	233	42	52	9	3	1
Chambliss	0.219	73	146	13	32	7	0	1
Cerone	0.213	62	207	12	44	4	0	2
Benedict	0.211	40	123	6	26	2	0	0
Zuvella	0.192	47	99	7	19	2	0	0
Perry	0.191	75	162	12	31	3		1
Runge	0.190	26	42	9	8	0	0	1

TABLE 20 (cont.)

Batting	BA	G	AB	R	H	2B	3B	HR
			Chicago Clubs					
Bosley	0.302	67	106	16	32	5	2	4
Sandberg	0.293	103	417	74	122	21	3	16
Moreland	0.286	109	392	43	112	21	1	9
Lopes	0.285	78	221	41	63	7	0	9
Durham	0.279	104	366	36	102	21	2	11
Hatcher	0.277	24	94	16	26	10	0	1
Hebner	0.247	55	93	7	23	1	0	1
Dernier	0.245	73	277	32	68	12	2	0
Speier	0.244	78	176	13	43	10	0	3
Davis	0.236	93	326	33	77	18	0	12
Dayett	0.231	22	26	1	6	0	0	1
Matthews	0.229	57	175	24	40	7	0	8
Cey	0.224	100	348	43	78	11	1	13
Woods	0.203	60	64	9	13	2	0	0
Dunston	0.187	24	75	9	14	4	1	1
Lake	0.136	41	81	4	11	2	0	1

Chapter Test

1. A basketball player who normally makes 75% of her free throws shoots four times. What is the probability that she makes
   (a) The specific shots HHMM?
   (b) Two shots out of the four?
2. A multiple-choice exam has ten questions, each question having five possibilities. If a student guesses at all ten questions, what is the probability that the student gets exactly nine questions correct?
3. Draw the transition diagram for the following transition matrix:

$$\begin{bmatrix} 1/2 & 1/2 & 0 \\ 0 & 1 & 0 \\ 1/3 & 1/3 & 1/3 \end{bmatrix}$$

4. Show that the steady state probability vector for the transition matrix

$$P = \begin{bmatrix} a & b \\ c & d \end{bmatrix}$$

is

$$\begin{bmatrix} \dfrac{c}{b+c}, & \dfrac{b}{b+c} \end{bmatrix}$$

5. A study of brand loyalties in a market survey shows that 40% of the people who used Brand A toothpaste plan on buying it next time. For Brand B, only 30% plan on remaining loyal on the next purchase. Using the steady state probability given in Problem 4, what are the steady state probabilities for this Markov chain? Why might this process not be a Markov chain?
6. Describe how one would use a pair of coins to simulate the following processes.
   (a) Rolling a pair of dice.
   (b) Selecting a value from the binomial distribution with $n = 2$ and $p = 0.25$.
7. Let B be the event that a student is a business major and A the event that the student will get an A in a business math course. The following probabilities are given:

$$P(B) = 0.3$$
$$P(A) = 0.2$$
$$P(B|A) = 0.15$$

Find $P(A|B)$ and interpret its meaning.

7

Introduction to
Statistics

In a nutshell, statistics is a subject in which we learn facts about the real world through observations. In Sections 7.1 and 7.2 we introduce the "what's going on here" side of statistics, known as *descriptive statistics*. In Section 7.3 we introduce the more mainstream and deeper side of statistics, known as *statistical inference*. In statistical inference we learn facts about a large group of people or things (a population) from a subset of these people or things (a sample).

7.1 Frequency Distributions and Polygons

Purpose

We introduce some of the basic techniques used for classifying, summarizing, and graphically displaying a set of observed measurements. The major concepts are

- frequency and relative frequency distributions,
- cumulative and cumulative relative frequency distributions, and
- histograms and polygons.

Introduction

To many people, statistics is associated with endless columns of numbers, tables of data, and reams of computer output. To others, like the Canadian economist/humorist Stephen Leacock, statistics arouses images of subversive activities. To mathematicians, scientists, and business managers, however, statistics constitutes a body of knowledge that deals with drawing conclusions from collected data.

English writer H. G. Wells might be right; statistical thinking has become so ingrained into modern thought that a day does not pass without statistical ideas being found in the news. Newspaper articles often contain statistics showing such things as the relationship between interest rates and the stock market, smoking and lung cancer, acid rain and the environment, and on and on. Often, an article reports on experiments conducted by scientists and the conclusions drawn from the experiments.

Statistics is subdivided into two major areas: *descriptive statistics* and *statistical inference*. This section and the following one cover many of the important ideas of descriptive statistics. The final section of this chapter gives you a glimpse into the far more sophisticated area of statistical inference. The diagram in Figure 1 illustrates the difference between descriptive statistics and statistical inference.

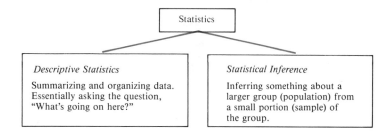

Figure 1

The two major areas of statistics

Descriptive Statistics (What's Going on Here)

Descriptive statistics or *exploratory data analysis* is that area of statistics that involves the classifying, summarizing, and presentation of data. Someone once said that descriptive statistics answers the question, "What's going on here?" This was the first area of statistics to be developed during the nineteenth century and the first area that is studied in a beginning statistics course. With the introduction of computer graphics systems during recent years, many new ways for illustrating and presenting information have been developed. The graphical display on page 426 was generated by CHART-MASTER, a computer graphics system for personal computers.

We illustrate some of the basic ideas of descriptive statistics by analyzing data taken from a summary of the New York City Marathon.

Descriptive Statistics and the NYC Marathon

Table 1, on page 427, lists the finishing times and ages for the first 100 women runners in the 1980 New York City Marathon.* The 100 observations list the runners' times and ages.

To analyze the data, we begin by putting the runners into different age groups. This process will give us what is called a **frequency distribution** of the ages of these runners. Frequency distributions often reveal patterns that otherwise go undetected. Here, we subdivide the runners into the following age groups or **categories**: 15 to 17, 18 to 20, 21 to 23, . . . , 45 to 47. These categories are selected so that each observation can be placed in one of the categories. In this example we have chosen as the first category those runners whose ages are 15, 16, or 17. In this way we include the youngest runner, who is 15 years old. We have chosen the **width** of each category as three years. In so doing, we have made sure that the number of categories is neither too large nor too small. Many people feel that the number of categories should be between six and twenty. The width of each category should also be the same.

The three-year categories in this example are convenient, too, since each category has a **midpoint** year (the 15–17 category has a midpoint of 16). Later we will use these midpoints in formulas involving grouped data.

---

* Table 1 was taken from J. B. Pomeranz, "Exploring Data from the 1980 New York City Marathon" *The Journal of Undergraduate Mathematics and Its Applications.*

Government can show spending.

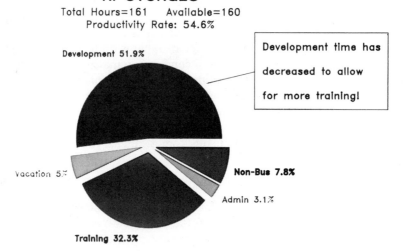

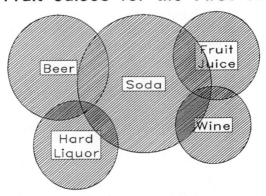

Some computer-generated graphs, showing different ways of presenting data.

TABLE 1
Top 100 Women Runners in
1980 New York City Marathon

Rank	Runner	Time	Rank	Runner	Time
1.	Waitz, Grete, 27-Norway	2:25:41	51.	Sork, Victoria L, 29-Gateway Athl, MO	3:08:43
2.	Lyons-Catalano, Patti, 27-MA	2:29:33	52.	Shasteen, Elaine, 30-FL	3:08:43
3.	Christensen, Ingrid, 24-Norway	2:34:24	53.	Mudway, Frances M, 32-Great Britain	3:08:47
4.	Gould, Carol T, 36-Great Britain	2:35:04	54.	Lane, Lucy H, 26-NY	3:08:58
5.	Adams, Gillian P, 25-Great Britain	2:37:55	55.	Gilchrist, Anne, 31-WSSAC-NY, NY	3:09:03
6.	Binder, Laurie, 33-Adidas, CA	2:38:09	56.	Kelly, Barbara J, 31-VT	3:09:19
7.	Sweigart, Kiki, 29-WSSAC-NJ, CT	2:40:34	57.	Lerner, Mimi, 43-NY	3:09:20
8.	Mosling, Oddrun, 27-Norway	2:41:00	58.	Stachecki, Catherine G, 29-Motor City Str	3:09:50
9.	Olinek, Gayle, 27-Adidas, CA	2:41:32	59.	Maury, Marie-Jeanne M, 30-France	3:09:52
10.	Chodnicki, Jean C, 21-WSSAC-NJ, NJ	2:43:33	60.	Williams, Brigitte E, 40-TX	3:10:12
11.	Laxton, Sonja, 32-Zimbabwe	2:43:48	61.	Blakeslee, Ruth A, 35-CPTC, NY	3:10:14
12.	Arenz, Janice M, 30-Northern Lights, NY	2:44:15	62.	Rosen, Yvonne J, 29-CPTC, NY	3:10:15
13.	Burge, Donna J, 27-Gr Houston TC, TX	2:44:47	63.	Cave, Elyane, 29-France	3:10:28
14.	Watson, Leslie, 32-Great Britain	2:45:39	64.	Phillips, Leticia G, 34-Canada	3:10:30
15.	Cook, Carol L, 26-MO	2:46:09	65.	Murphy, Claire, 25-PPTC, NY	3:10:40
16.	Erickson, Karlene, 15-NB	2:47:23	66.	Carre, Brigitte, 29-France	3:11:30
17.	Horton, Kathleen A, 32-NY	2:48:08	67.	Hulak, Marilyn A, 22-NY	3:11:51
18.	Grottenberg, Sissel, 24-Norway	2:48:28	68.	Deckert, Margarete L, 47-NY	3:11:56
19.	Martinez, Iliar, 33-Spain	2:49:01	69.	Spencer, Alicia M, 30-NY	3:12:08
20.	Israel, Diane J, 20-NY	2:50:35	70.	Murray, Patricia M, 20-IL	3:12:11
21.	Kerr, Jean, 21-Westchester RR, NY	2:50:45	71.	Engelby, Vlla, 26-Sweden	3:12:39
22.	Forster, Vreni, 24-Switzerland	2:51:29	72.	Thomas, Wendy M, 28-IN	3:12:43
23.	Heiskanen, Ulla, 30-Finland	2:52:14	73.	Barnard, Bonnie G, 21-NY	3:12:45
24.	Mcdonald, Katie, 29-WSSAC-NY, NY	2:53:39	74.	Anderson, Donna M, 21-Lehigh ValAA	3:12:51
25.	King, Barbara A, 22-Great Britain	2:54:00	75.	Duplichan, Donna L, 24-Houston Har	3:13:30
26.	Reinhardt, Marilyn J, 29-IN	2:56:31	76.	Downey, Julie, 22-NJ	3:14:06
27.	Sappl, Edith, 34-Switzerland	2:57:24	77.	Spiering, Adriana W, 27-NY	3:14:17
28.	Koncelik, Burke M, 26-WSSAC-NYJ, NY	2:57:32	78.	Tessier, Christine P, 28-MA	3:14:28
29.	Ammermuller, Carol L, 23-Mercer-Bucks	2:57:33	79.	Heffernan, Beverly K, 34-CT	3:14:33
30.	Billington, Carolyn, 35-Great Britain	2:58:03	80.	Schonfeld, Polly E, 41-Millrose, NY	3:14:45
31.	Wuss, Cindy A, 23-NY	2:58:38	81.	Shipp, Stephanie S, 27-VA	3:14:46
32.	Lowell, Linda L, 32-Millrose, NY	2:59:10	82.	Sher, Patricia R, 33-Jacksonville TC	3:14:55
33.	Harmeling, Maddy A, 35-OVTC, NY	3:00:14	83.	Ripley, Laurene S, 31-Mercer-Bucks	3:14:56
34.	Bartee, Hermine M, 41-CPTC, NY	3:00:34	84.	Heaton, Phyllis G, 47-So Vermont RR	3:14:57
35.	Bateman, Maureen, 37-WSSAC-NY, NY	3:02:26	85.	Terry, Jane E, 28-NY	3:15:32
36.	Millspaugh, Jane A, 32-FL	3:03:16	86.	Heilbronn, Mary K, 27-Cape Cod AC, MA	3:15:39
37.	Bevans, Marilyn T, 31-Baltimore Suns	3:03:54	87.	Ellis, Patricia A, 37-CPTC, NY	3:15:53
38.	Thornhill, Anna, 40-Millrose, NY	3:03:55	88.	Goldin, Beth A, 28-NY	3:16:20
39.	Carmichael, Isabelle, 30-WSSAC-NY, NY	3:04:25	89.	Weber, Toni L, 25-OH	3:16:31
40.	Meade, Patricia A, 26-Waltham TC, MA	3:04:30	90	Hoffmann, Garima, 32-Sri Chinmoy M, CA	3:16:34
41.	Thurston, Linda J, 38-NJ	3:04:47	91.	Sommerville, Christine A, 30-Bermuda	3:16:40
42.	Foster, Susan A, 24-Front Rnrs, NY	3:05:23	92.	Wisniewski, Meredith A, 31-Van Cort TC	3:16:40
43.	Higgins, Monica M, 27-NY	3:05:45	93.	Obrien, Robin, 40-No Jersey Mst, NJ	3:16:44
44.	Barnett, Posie H, 24-MA	3:05:49	94.	Hoiska, Elaine K, 36-VT	3:16:54
45.	Hill, Marian L, 39-NJ	3:05:52	95.	Connell, Anette L, 34-FL	3:16:55
46.	Griffith, Maureen, 25-Great Britain	3:05:59	96.	Ponser, Judith M, 24-Gateway Athl	3:16:56
47.	Kuscsik, Nina L, 41-GNYAA, NY	3:06:25	97.	Cahill, Regina F, 26-PPTC, NY	3:16:58
48.	Kremer, Elyse, 25-NY	3:07:30	98.	Loebl, Judith H, 26-West Side Y, NY	3:17:32
49.	Campbell, Marvie, 33-VT	3:07:45	99.	Sheehy, Nancy R, 32-NY	3:17:40
50.	Given, Sharon L, 30-CA	3:08:34	100.	Hoogerhoud, Mariam M, 32-Holland	3:17:46

The Summary Table

We now use the above ideas to summarize much of the information in Table 1 in a **summary table**. To construct this table, it is convenient to carry out the calculations in a series of steps.

- **Step 1** (categories and midpoints). Determine the categories and midpoints of the categories.
- **Step 2** (tally observations). Place a tally mark in the appropriate category for each observation to count the observations in each category.
- **Step 3** (count the observations in each category). Count the tally marks in each category and record this number in the frequency distribution column. The frequency distribution is the number of observations in each category.
- **Step 4** (find the cumulative tallies). Find the cumulative distribution of the observations by adding the number of observations in each category to the number of observations in the preceding categories.
- **Step 5** (find the relative frequencies). Determine the relative frequencies by dividing the number of observations in a category by the total number of observations. The relative frequencies are the fraction of the total number of observations in each category.
- **Step 6** (find the cumulative relative frequency). To find the cumulative relative frequencies, add the cumulative frequency of each category to the cumulative frequencies of the preceding categories.

Using the steps just explained, we construct the summary table for the ages of the top 100 runners in the 1980 NYC marathon, as shown in Table 2.

TABLE 2

Summary Table for Ages of the Top 100 Women Runners in the 1980 New York City Marathon

Age Category	Tally	Midpoint	Frequency	Cumulative Frequency	Relative Frequency	Cumulative Relative Frequency
15–17	I	16	1	1	0.01 (1%)	0.01 (1%)
18–20	II	19	2	3	0.02 (2%)	0.03 (3%)
21–23	|||| IIII	22	9	12	0.09 (9%)	0.12 (12%)
24–26	|||| |||| |||| IIII	25	19	31	0.19 (19%)	0.31 (31%)
27–29	|||| |||| |||| |||| I	28	21	52	0.21 (21%)	0.52 (52%)
30–32	|||| |||| |||| |||| II	31	22	74	0.22 (22%)	0.74 (74%)
33–35	|||| |||| I	34	11	85	0.11 (11%)	0.85 (85%)
36–38	||||	37	5	90	0.05 (5%)	0.90 (90%)
39–41	|||| II	40	7	97	0.07 (7%)	0.97 (97%)
42–44	I	43	1	98	0.01 (1%)	0.98 (98%)
45–47	II	46	2	100	0.02 (2%)	1.00 (100%)
Totals			100		1.00 (100%)	

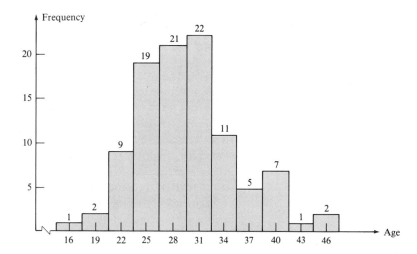

Figure 2
Histogram of ages of the top 100 women finishers in the 1980 New York City Marathon

Graphical Representation of Data

Two common devices for graphically displaying frequency distributions are histograms and frequency polygons.

Histograms

A **histogram** is simply a bar graph of a frequency distribution of the type shown in Figure 2. In the construction of the histogram the categories are plotted on the horizontal axis, and the frequencies are plotted on the vertical axis. A horizontal line is drawn above each category at a height representing the frequency.

Frequency Polygons

A **frequency polygon** is a series of straight lines connecting the midpoints of the horizontal lines that form the top of a histogram. It is customary to close the ends of the polygon by extending the ends of the lines to the horizontal axis as shown in Figure 3.

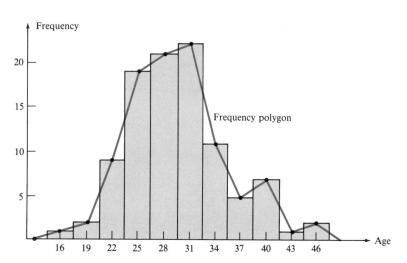

Figure 3
Frequency polygon of ages of the top 100 women finishers in the 1980 New York City Marathon

The Mark Twain Problem (Use of Frequency Distributions)

During the year 1861, in a series of letters written to the *New Orleans Daily Crescent*, a Quintus Curtius Snodgrass (QCS) told of his experiences with the Louisiana militia during the Civil War. Some historians believe that the letters were in fact written by the American humorist, Mark Twain.*

From 1934, when the letters were first brought to light, to the 1950's, a bitter debate was waged by both the believers of this claim and the nonbelievers. In an attempt to determine whether the letters were in fact authentic, statisticians have made a comparative study between the distribution of word lengths in a sample of Twain's works and the distribution of word lengths in the ten letters written by Snodgrass.

Linguists have known for some time that frequency distributions of word lengths can act like "fingerprints" in the determination of authorship. Strange as it may seem, the frequency with which an author uses words of a given length often remains constant throughout the author's career. For example, compare carefully the two samples of word length distributions, taken from the collected works of Mark Twain, in Table 3.

TABLE 3
Relative Frequency
Distributions of Word Lengths
in Two Works of Mark Twain

Word Length	Collection of Letters, 1858 to 1861		Collection of Letters, 1863	
	Frequency	*Relative Frequency*	*Frequency*	*Relative Frequency*
1	74	0.039 (3.9%)	312	0.051 (5.1%)
2	349	0.185 (18.5%)	1146	0.188 (18.8%)
3	456	0.242 (24.2%)	1394	0.228 (22.8%)
4	374	0.198 (19.8%)	1177	0.193 (19.3%)
5	212	0.113 (11.3%)	661	0.108 (10.8%)
6	127	0.067 (6.7%)	442	0.072 (7.2%)
7	107	0.057 (5.7%)	367	0.060 (6.0%)
8	84	0.045 (4.5%)	231	0.038 (3.8%)
9	45	0.024 (2.4%)	181	0.030 (3.0%)
10	27	0.014 (1.4%)	109	0.018 (1.8%)
11	13	0.007 (0.7%)	50	0.008 (0.8%)
12	8	0.004 (0.4%)	24	0.004 (0.4%)
13+	9	0.005 (0.5%)	12	0.002 (0.2%)
Totals	1885	1.000 (100%)	6106	1.000 (100%)

We now draw the relative frequency polygons for the frequency distributions in Table 3. These are shown in Figure 4. The reader can observe just how close these polygons are to each other.

---

* Brinegar, Claude S., "Mark Twain and the Quintus Curtius Snodgrass Letters: A Statistical Test of Authorship," *Journal of the American Statistical Association*, Vol. 58, 1963, pp. 85–96.

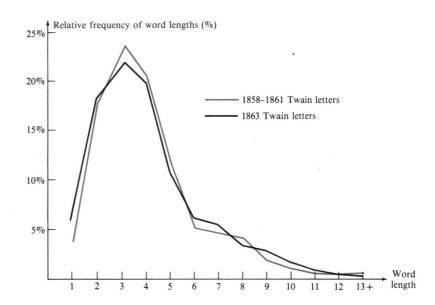

Figure 4
Relative frequency polygons of word lengths (graph from "Mark Twain and the Quintus Curtius Snodgrass. Letters," by Claude Brinegar, Volume 58, 1963, pp. 85–96, *Journal of the American Statistical Association*)

The next step in the analysis was to make a word length count of the ten Snodgrass letters. The frequency and relative frequency distributions in Table 4 were found.

The relative frequency polygon of word lengths of the Snodgrass letters is now drawn on the same scale as two relative frequency polygons of word lengths of Twain letters. These three graphs are shown in Figure 5. Note that we compare the relative frequency polygons (and not the frequency polygons), since the number of words used to construct the frequency distributions is not the same in each case.

By comparing these polygons we begin to suspect that Mark Twain was not the same person as Quintus Curtius Snodgrass. This conclusion can be

TABLE 4
Word Length Distribution for Ten Snodgrass Letters

Word Length	Frequency	Relative Frequency
1	424	0.032 (3.2%)
2	2685	0.204 (20.4%)
3	2752	0.209 (20.9%)
4	2302	0.175 (17.5%)
5	1431	0.109 (10.9%)
6	992	0.075 (7.5%)
7	896	0.068 (6.8%)
8	638	0.048 (4.8%)
9	465	0.035 (3.5%)
10	276	0.021 (2.1%)
11	152	0.011 (1.1%)
12	101	0.008 (0.8%)
13 +	61	0.005 (0.5%)
Total	13,175	1.000 (100%)

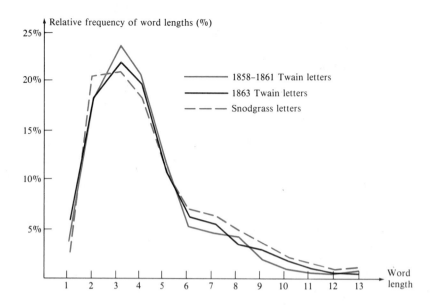

Figure 5
Comparison of relative frequencies of word lengths

reinforced by further statistical analysis and by examining other works of the two men. It is also the general belief, held by Mark Twain biographers and the literary editors of the Mark Twain papers, that the two authors were, in fact, not the same person.

HISTORICAL NOTE

Statistics as a discipline traces its origins to 1662 and an Englishman by the name of John Graunt of London. Graunt was a haberdasher who wrote a small treatise, *Natural and Political Observations Made Upon the Bills of Mortality*. It was the first attempt to interpret biological and social phenomena strictly on the basis of numerical observations.

Another pioneer of the discipline of statistics was the eminent astronomer Sir Edmund Halley, for whom Halley's comet is named. In 1675, Halley constructed the first mortality tables and founded the subject of actuarial science, on which the solvency of life insurance companies depends.

Problems

1. **Summarizing Room and Board Fees** The numbers in Table 5 represent the average amounts that students paid for room and board at various four-year universities in the United States during the academic year 1986–1987. Construct a summary table as shown in the text, using six categories, the first category being [$1600, $1900). Follow the steps in parts (a)–(h).
   (a) Determine the categories and the midpoints of each category.
   (b) Make a tally mark for each observation in the appropriate category.
   (c) Count the tally marks in each category and record this number in a frequency distribution column.
   (d) Compute the cumulative frequency column.
   (e) Compute the relative frequency column.
   (f) Compute the cumulative relative frequency column.
   (g) Draw histograms of the frequency distribution and the cumulative frequency distribution.

TABLE 5
Room-and-Board Fees for Problem 1

$2410	$1870	$2085	$2675	$2310
$1885	$1975	$1955	$2355	$2850
$2155	$1635	$2010	$3105	$2845
$1850	$2460	$2465	$2155	$2255
$1955	$2055	$2360	$1855	$1995

(h) Draw polygons of the frequency distribution and the cumulative frequency distribution.

2. **Summarizing Business Data**   A company is interested in analyzing the mileage traveled by its salesmen. The total mileages (rounded to the nearest hundred miles) in Table 6 for a given month were reported by each of the 50 salesmen. Construct a summary table, for the data with ten categories, the first category being [2200, 2500). Follow the steps in parts (a)–(h).

(a) Determine the categories and midpoints of each category.

(b) Make a tally mark for each observation within the appropriate category.

(c) Compute the frequency distribution by adding the number of tallies within each category.

(d) Compute the cumulative frequency distribution column.

TABLE 6
Mileages for Problem 2

3300	3700	3500	3900	3400
4200	2800	4000	4400	3200
4500	4100	3800	3600	3000
4000	3300	2900	3700	5100
3800	5000	2500	4600	3400
3200	3600	4000	4300	3500
3100	4400	3300	3900	2700
3800	4200	3400	3100	4800
4100	3000	2200	3700	3800
4200	3400	3100	4800	4100

(e) Compute the relative frequency distribution column.

(f) Compute the relative cumulative frequency distribution column.

(g) Draw histograms representing the frequency distribution and the cumulative frequency distribution.

(h) Draw polygons representing the frequency distribution and the cumulative frequency distribution.

3. **Speeding Tickets**   Table 7 gives the cost of a ticket a driver must pay if caught speeding in each state.

(a) Determine the categories and midpoints of each category.

TABLE 7
What a Ticket Will Cost You: Penalties for going 10 mph over the speed limit vary widely across the United States

More Than $45		$31–$45		$30 or Less	
Alabama	$60	Delaware	$31.50	Alaska	$20
Arkansas	$47	Florida	$44	Arizona*	$15
California*	$50	Georgia*	$45	Idaho*	$12.50
Colorado	$64	Hawaii*	$35	Illinois	$20
Connecticut	$60	Iowa	$33	Kansas*	$28
Indiana	$50	Louisiana	$45	Montana*	$ 5
Kentucky*	$65	Maine	$40	Nebraska*	$10
Massachusetts*	$50	Maryland	$40	Nevada*	$ 5
Missouri*	$50	Michigan	$40	New Mexico*	$20
New Jersey	$50	Minnesota	$44	New York	$28
North Carolina	$47.50	Mississippi	$39.50	North Dakota*	$10
Ohio	$100	New Hampshire*	$33	Oregon*	$26
Oklahoma	$56	South Carolina	$45	Rhode Island*	$30
Pennsylvania	$72.50	Tennessee	$40	South Dakota*	$10
Texas	$48	Vermont	$35	Utah	$28
West Virginia	$60	Virginia	$40	Wyoming*	$15
		Washington*	$38		
*No License Points Or Penalties Are Assessed		Wisconsin	$44.50	Some Fines Are Average Minimums Or Estimates	

(b) Make a tally mark for each observation within the appropriate category.

(c) Compute the frequency distribution by adding the number of tallies within each category.

(d) Compute the relative frequency distribution.

4. **Summarizing Medical Data**   Table 8 shows the lengths of 30 surgical operations performed in a hospital during a 24-hour period. Construct a summary table for the data using six categories, the first category being [0:30, 1:15). Follow the steps in parts (a)–(h).

(a) Determine the categories and midpoints of each category.

(b) Make a tally mark for each observation within the appropriate category.

(c) Compute the frequency distribution by adding the number of tallies within each category.

(d) Compute the cumulative frequency distribution column.

(e) Compute the relative frequency distribution column.

(f) Compute the relative cumulative frequency column.

(g) Draw a histogram representing the frequency distribution and the cumulative frequency distribution.

(h) Draw polygons representing the frequency distribution and the cumulative frequency distribution.

TABLE 8

Data on Operations for Problem 4

Operation	Length (hr:min)	Operation	Length (hr:min)
1	2:25	16	1:40
2	0:45	17	1:30
3	1:30	18	2:50
4	0:30	19	3:50
5	2:00	20	1:45
6	1:45	21	4:25
7	1:30	22	3:15
8	0:35	23	0:55
9	0:50	24	0:45
10	1:45	25	1:50
11	2:45	26	1:45
12	3:50	27	2:45
13	1:25	28	0:30
14	0:40	29	1:25
15	1:30	30	0:50

5. **Baseball Statistics**   Table 9, on the following page, shows the American League pennant winners from 1901 to 1985 and their won-loss record. Summarize Table 9 by carrying out the steps in parts (a)–(c).

(a) Make a frequency distribution of the games won during this period. Use five games as the width of the category interval.

(b) Make a frequency distribution of the percentage of games won. Use 0.25 as the width of the category interval.

(c) Draw a histogram and a polygon that represent the frequency distribution of the games won.

6. **Summarizing Market Research**   A soft drink company is testing a new flavor of soda to determine public acceptability. Controlled conditions were obtained by bringing individuals into a central testing location and having trained personnel prepare, serve, and interview the tasters. Table 10 gives the results obtained from testing 25 women and 25 men. Summarize the table by carrying out the steps in parts (a) and (b).

TABLE 10

Frequency Distribution of Tasting Preferences

Score	Men	Women
+3	1	0
+2	2	3
+1	7	8
0	8	9
−1	5	3
−2	2	1
−3	0	1
Totals	25	25

(a) Draw histograms representing the frequency distributions of the men's and women's preferences.

(b) Draw polygons representing the frequency distributions of the men's and women's preferences.

7. **Summarizing Rat Data**   A psychologist tested 61 rats to determine the number of trials it would take for them to escape from a maze. The frequency distribution in Table 11 was found.

TABLE 11

Trials for Rats to Learn to Escape from Maze

Number of Trials	Number of Rats
1–3	4
4–6	12
7–9	20
10–12	21
13–15	3
16–18	1

TABLE 9
American League Pennant Winners 1901–1985

Year	Club	Manager	Won	Lost	Pct
1901	Chicago	Clark C. Griffith	83	53	0.610
1902	Philadelphia	Connie Mack	83	53	0.610
1903	Boston	Jimmy Collins	91	47	0.659
1904	Boston	Jimmy Collins	95	59	0.617
1905	Philadelphia	Connie Mack	92	56	0.622
1906	Chicago	Fielder A. Jones	93	58	0.616
1907	Detroit	Hugh A. Jennings	92	58	0.613
1908	Detroit	Hugh A. Jennings	90	63	0.588
1909	Detroit	Hugh A. Jennings	98	54	0.645
1910	Philadelphia	Connie Mack	102	48	0.680
1911	Philadelphia	Connie Mack	101	50	0.669
1912	Boston	J. Garland Stahl	105	47	0.691
1913	Philadelphia	Connie Mack	96	57	0.627
1914	Philadelphia	Connie Mack	99	53	0.651
1915	Boston	William F. Carrigan	101	50	0.669
1916	Boston	William F. Carrigan	91	63	0.591
1917	Chicago	Clarence H. Rowland	100	54	0.649
1918	Boston	Ed Barrow	75	51	0.595
1919	Chicago	William Gleason	88	52	0.629
1920	Cleveland	Tris Speaker	98	56	0.636
1921	New York	Miller J. Huggins	98	55	0.641
1922	New York	Miller J. Huggins	94	60	0.610
1923	New York	Miller J. Huggins	98	54	0.645
1924	Washington	Stanley R. Harris	92	62	0.597
1925	Washington	Stanley R. Harris	96	55	0.636
1926	New York	Miller J. Huggins	91	63	0.591
1927	New York	Miller J. Huggins	110	44	0.714
1928	New York	Miller J. Huggins	101	53	0.656
1929	Philadelphia	Connie Mack	104	46	0.693
1930	Philadelphia	Connie Mack	102	52	0.662
1931	Philadelphia	Connie Mack	107	45	0.704
1932	New York	Joseph V. McCarthy	107	47	0.695
1933	Washington	Joseph E. Cronin	99	53	0.651
1934	Detroit	Gordon Cochrane	101	53	0.656
1935	Detroit	Gordon Cochrane	93	58	0.616
1936	New York	Joseph V. McCarthy	102	51	0.667
1937	New York	Joseph V. McCarthy	102	52	0.662
1938	New York	Joseph V. McCarthy	99	53	0.651
1939	New York	Joseph V. McCarthy	106	45	0.702
1940	Detroit	Delmar D. Baker	90	64	0.584
1941	New York	Joseph V. McCarthy	101	53	0.656
1942	New York	Joseph V. McCarthy	103	51	0.669
1943	New York	Joseph V. McCarthy	98	56	0.636
1944	St. Louis	Luke Sewell	89	65	0.578
1945	Detroit	Steve O'Neill	88	65	0.575
1946	Boston	Joseph E. Cronin	104	50	0.675
1947	New York	Stanley R. Harris	97	57	0.630
1948	Cleveland	Lou Boudreau	97	58	0.626
1949	New York	Casey Stengel	97	57	0.630
1950	New York	Casey Stengel	98	56	0.636
1951	New York	Casey Stengel	98	56	0.636
1952	New York	Casey Stengel	95	59	0.617
1953	New York	Casey Stengel	99	52	0.656
1954	Cleveland	Al Lopez	111	43	0.721
1955	New York	Casey Stengel	96	58	0.623
1956	New York	Casey Stengel	97	57	0.630
1957	New York	Casey Stengel	98	56	0.636
1958	New York	Casey Stengel	92	62	0.597
1959	Chicago	Al Lopez	94	60	0.610
1960	New York	Casey Stengel	97	57	0.630
1961	New York	Ralph Houk	109	53	0.673
1962	New York	Ralph Houk	96	66	0.593
1963	New York	Ralph Houk	104	57	0.646
1964	New York	Yogi Berra	99	63	0.611
1965	Minnesota	Sam Mele	102	60	0.630
1966	Baltimore	Hank Bauer	97	53	0.606
1967	Boston	Dick Williams	92	70	0.568
1968	Detroit	Mayo Smith	103	59	0.636
1969	Baltimore	Earl Weaver	109	53	0.673
1970	Baltimore	Earl Weaver	108	54	0.667
1971	Baltimore	Earl Weaver	101	57	0.639
1972	Oakland	Dick Williams	93	62	0.600
1973	Oakland	Dick Williams	94	68	0.580
1974	Oakland	Alvin Dark	90	72	0.556
1975	Boston	Darrell Johnson	95	65	0.594
1976	New York	Billy Martin	97	62	0.610
1977	New York	Billy Martin	100	62	0.617
1978	New York	Billy Martin and Bob Lemon	100	63	0.613
1979	Baltimore	Earl Weaver	102	57	0.642
1980	Kansas City	Jim Frey	97	65	0.599
1981	New York	Gene Michael-Bob Lemon	59	48	0.551
1982	Milwaukee	Harvey Kuenn	95	67	0.586
1983	Baltimore	Joe Altobelli	98	64	0.605
1984	Detroit	Sparky Anderson	104	58	0.642
1985	Kansas City	Dick Howser	91	71	0.562

Summarize the above data by performing the calculations in parts (a) and (b).

(a) Draw a histogram and a polygon of the frequency distribution of the number of trials it takes for a rat to escape from the maze.

(b) Draw a histogram and a polygon of the cumulative frequency distribution of the number of trials it takes for a rat to escape from the maze.

8. **Summarizing Categorical Data**   Sometimes the outcome of an experiment is not a collection of numbers, but a classification of the observations into a number of different categories. For example, suppose 100 people were sampled and asked their favorite flavor of Brite-Cola. The results are shown in Table 12.

Represent the above information graphically by using both a histogram and a frequency polygon.

9. **Linguistic Study**   Pick two paragraphs from the same book, and construct a frequency distribution of the word

TABLE 12
Preference of Brite-Cola Flavor

Kind	Number
Brite Classic	53
New Brite	32
Cherry Brite	15
Totals	100

lengths. Are the styles of these two paragraphs consistent as indicated from the distribution of word lengths?

10. **Linguistic Study**   Pick two pages from different books, and construct a frequency distribution of the word lengths. Are the styles of these two books similar as indicated by the distribution of word lengths?

7.2 Measures of Central Tendency and Dispersion

PURPOSE

We introduce several measures that are useful for summarizing observations. They are measures of central tendency:

- mean (average value)
- median (middle value)
- mode (most common value)

and measures of dispersion:

- range (largest value minus the smallest value)
- standard deviation

Introduction

In the previous section we started with a collection of numerical observations. We then summarized the content of these observations by classifying them into categories and making frequency distributions and histograms. We now continue the process of summarizing the observations by introducing measures of central tendency and dispersion.

Measures of Central Tendency (the Three M's)

It is useful to find a single number that can act as a representative of a whole set of numbers. Three different values are commonly used as a representative value. These values are called the *mean*, the *median*, and the *mode*. Each of these quantities is, in its own way, a measure of the central value of the observations.

For instance, psychologists say that 50% of the general adult population in the United States has an IQ of 90 and above on the Stanford-Binet intelligence test, and 50% has an IQ under 90. The score of 90 is the median IQ. The modal IQ would be the IQ with the largest frequency of occurrence. The *arithmetic average* or **mean**, denoted by $\bar{x}$, is simply the sum of the values divided by the number of terms.

Mean of Ungrouped Data

The mean $\bar{x}$ of a collection $x_1, x_2, \ldots, x_n$ of n numerical observations is given by

$$\bar{x} = \frac{1}{n}(x_1 + x_2 + \cdots + x_n)$$

Example 1

Computing the Mean   Find the mean of the quiz grades 80, 65, 90, 74, 84, and 37.

Solution

$$\bar{x} = \frac{80 + 65 + 90 + 74 + 84 + 37}{6} = \frac{430}{6} = 71.7$$

We may wish to round off the computed value of the mean to 72.   □

Computing the Mean from Grouped Observations

When the observations have been grouped into categories and a frequency distribution has been made, we can approximate the mean from the frequency distribution.

Mean from the Frequency Distribution

Let a collection of n observations $x_1, x_2, \ldots, x_n$ be grouped into k categories and the frequency distribution be found. The mean $\bar{x}$ can be approximated by the grouped mean formula:

$$\bar{x} \cong \frac{m_1 f_1 + m_2 f_2 + \cdots + m_k f_k}{n}$$

where

$$k = \text{number of categories}$$
$$f_i = \text{number of observations in the } i\text{th category}$$
$$m_i = \text{midpoint of the } i\text{th category}$$

You should be aware that the above formula requires adding only k terms in contrast to n terms in the ungrouped formula for the mean. This would be a great savings of time if $n = 500$ observations had already been classified into $k = 10$ categories. Instead of adding 500 terms, we would have to add only ten terms.

Example 2

Mean Computed from Frequency Distribution   Fifty representative families were selected in a given midwestern city, and the number of children in

Number of Children in the Family	Number of Families
0	9
1	12
2	14
3	11
4	3
5	1
6	0
Total	50

each family was recorded. The result of the survey is given in Table 13. From this frequency distribution, approximate the mean family size.

Solution   The families have been classified according to the number of children. Here the observations are placed in one of seven categories. The midpoints of the categories would be chosen as $m_1 = 0, m_2 = 1, \ldots, m_7 = 6$. Using the formula for the mean of categorized observations, we find

$$\bar{x} \cong \frac{f_1 m_1 + f_2 m_2 + \cdots + f_7 m_7}{n}$$

$$= \frac{9 \cdot 0 + 12 \cdot 1 + 14 \cdot 2 + 11 \cdot 3 + 3 \cdot 4 + 1 \cdot 5 + 0 \cdot 6}{50}$$

$$= 1.80 \text{ children} \qquad \qquad \square$$

Conclusion

In other words, the average number of children in the above 50 families is approximately 1.8. Note that when we computed the above mean from the categorized observations, it was necessary to sum only seven terms. If we had computed the mean directly from the 50 original observations, without classifying them, it would have required summing 50 terms. This is the major advantage of approximating the mean from the grouped mean formula.

Mean as the Center of Gravity

To better understand the mean of a set of numbers, consider the observations in Figure 6, whose values are plotted on the real axis. Think now of the

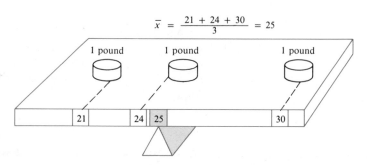

$$\bar{x} = \frac{21 + 24 + 30}{3} = 25$$

Figure 6
The mean is located where the beam balances

real axis as a weightless beam with one-pound weights sitting at the points specified by the observations. If we place a fulcrum under the beam at the point where the beam balances, then this point will be the mean of the observations.

The Median and the Mode

Two other measures of central tendency are the median and the mode. If there is an odd number of observations, then the **median** is simply the middle number when the numbers are arranged in order of size. For example, the median of the five observations 3, 9, 11, 2, 8 is 8, since there are two numbers smaller and two numbers larger than 8. When the number of observations is even, the median is taken as the mean of the two middle values. For example, the median of the six numbers 6, 2, 8, 3, 5, 7 is 5.5 (the mean of the two middle values 5 and 6). The median income is often quoted in discussing a representative salary for a group of wage earners.

On the other hand, the **mode** of a set of observations is the observation that occurs most frequently. If no value occurs more than once, we say that the set of observations has no mode. For example, the mode of the observations 3, 4, 2, 4, 9 is 4, since 4 occurs more often than the other observations. For the numbers 3, 4, 5, 4, 3, 7, 6 we would say that both values 3 and 4 are modes, since these values both occur twice, which is more often than any other values occur. In other words, a collection of observations can have more than one mode. The mode is useful when we are interested in the most likely value, such as a shoe size or a coat size.

Example 3

Mean, Median, and Mode   Jerry and Susan play nine holes of golf. The scores for the nine holes are as shown in Table 14. Jerry was leading until he hooked his tee shot over the Interstate on the ninth hole and ended up with a 13. Susan claimed victory, but Jerry, being a statistician, says he still won. Compute the mean, median, and mode of both Susan's and Jerry's scores and tell why Jerry claimed victory.

Solution   According to the rules of golf, Susan won because she had the lower score. However, if we compute the mean, median, and mode for both of their scores, we will find the values in Table 15.

TABLE 14
Nine Holes of Golf

Hole	Jerry	Susan
1	4	5
2	5	5
3	4	4
4	8	5
5	4	6
6	6	7
7	3	4
8	5	7
9	13	5
Totals	52	48

TABLE 15
Mean, Median, and Mode for Golf Scores

	Jerry	Susan
Mean (average score)	5.78	5.33
Median (middle score)	5	5
Mode (most common score)	4	5

The reason that Jerry claimed victory was that he had the lower modal value. Would you agree with Jerry's argument? □

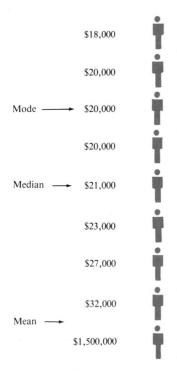

$18,000

$20,000

Mode ——→ $20,000

$20,000

Median ——→ $21,000

$23,000

$27,000

$32,000

Mean ——→

$1,500,000

Figure 7
Comparison of Mean, Medium, and Mode

Which Is the Best: Mean, Median, or Mode?

The diagram in Figure 7 illustrates how the mean can be misleading. The mean salary of $186,777.78 is heavily influenced by one very large salary ($1,500,000). Hence the mean is not as representative of this set of incomes as the median (the salary in the middle) and the mode (salary that occurs most often).

The measure of central tendency that should be used to represent a set of observations depends on the specific type of observations and on the way that will most convincingly represent the data.

Two Measures of Dispersion

The mean, median, and mode are useful tools that help summarize the information in a collection or sample of observations. However, there is information related to the variability of the observations that is not reflected by these measures of central tendency.

Measures of dispersion summarize the extent to which observations are scattered. For example, we would intuitively expect to say that the numbers 30.0, 30.1, 29.9, 30.0, 30.1, and 29.9 are scattered less than the numbers 2, 29, 1, 77, 35, and 36. It is even possible for two sets of numbers, one scattered more than the other, to have the same mean. (Would you believe that the above two sets of numbers each have a mean of 30?) In other words, the mean of a set of observations tells only part of the story.

For example, if Jerry is playing golf, he might hook his first nine tee shots off to the right and then slice the next nine shots off to the left. If someone asked him how he did, he could say that on the average he did pretty well.

Range

One important reason for measuring the dispersion of a collection of observations is to form a judgment about the reliability of the observations. Sets of observations whose values are scattered greatly tend to be less reliable measurements of a phenomenon.

One useful measure of dispersion for a set of observations is the range. The **range**, the simplest measure of dispersion, is simply the difference between the largest and smallest observations. The range is often used in reporting the variability of temperature over a given period of time. The weakness of the range is that many observations are completely ignored. Only the largest and smallest values are used to calculate the range. The range of a set of values can also be misleading. For instance, in 1972 the daily temperature measurements in Caribou, Maine, had the same range as the temperature measurements in Charleston, South Carolina. The temperature in Caribou ranged from a high of 85°F to a low of −20°F (range of 105°); in Charleston the temperature ranged from a high of 109°F to a low of 4°F (also a range of 105°). The climate certainly was *not* the same.

Standard Deviation

The second measure of dispersion is the standard deviation. Unlike the range, the standard deviation depends on every observation. The **standard deviation** measures the amount of variation of the observations about the mean of the observations. The more the observations are spread out, the larger the standard deviation will be. It can be computed by the following rules.

> **Computing the Standard Deviation**
>
> The standard deviation s of a collection of observations $x_1, x_2,$ $x_3, \ldots, x_n$ is found by completing Steps 1–5.
>
> - **Step 1.** Find the mean $\bar{x}$ of the observations.
> - **Step 2.** Compute the differences $d_i = x_i - \bar{x}$ between the observations and the mean.
> - **Step 3.** Find the squares $d_i^2 = (x_i - \bar{x})^2$ of the differences found in Step 2.
> - **Step 4.** Compute the sum of the squares d_i^2 and divide the sum by $n - 1$. This will be
>
> $$s^2 = \frac{(x_1 - \bar{x})^2 + (x_2 - \bar{x})^2 + \cdots + (x_n - \bar{x})^2}{n - 1}$$
>
> The value s^2 is called the **variance** of the observations.
>
> - **Step 5.** Compute the square root of the variance. This gives the standard deviation
>
> $$s = \sqrt{\frac{(x_1 - \bar{x})^2 + (x_2 - \bar{x})^2 + \cdots + (x_n - \bar{x})^2}{n - 1}}$$

Example 4

Standard Deviation   Determine from visual inspection which of the following data sets has the larger standard deviation. Check your educated guess by computing the standard deviation of each set of observations.

Data Set A	Data Set B
5	5
3	3
4	4
4	6
4	2

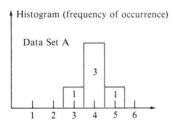

Solution   From observation we suspect that Data Set B has the larger standard deviation. This suspicion is strengthened by drawing histograms of the data sets as shown in Figure 8.

We begin by computing the mean of each data set. We find

$$\bar{x}_A = \frac{5 + 3 + 4 + 4 + 4}{5} = 4$$

$$\bar{x}_B = \frac{5 + 3 + 4 + 6 + 2}{5} = 4$$

Figure 8
Two different sets of observations

Table 16 shows the calculations that are needed to compute the two standard deviations s_A and s_B.

TABLE 16
Calculations for Standard
Deviations s_A and s_B

	Data Set A			Data Set B	
x_i	$x_i - \bar{x}_A$	$(x_i - \bar{x}_A)^2$	x_i	$x_i - \bar{x}_B$	$(x_i - \bar{x})^2$
5	1	1	5	1	1
3	-1	1	3	-1	1
4	0	0	4	0	0
4	0	0	6	2	4
4	0	0	2	-2	4
Totals		$\overline{2}$			$\overline{10}$

Hence we have the following two standard deviations.

$$s_A = \sqrt{\frac{2}{4}} = 0.707 \qquad s_B = \sqrt{\frac{10}{4}} = 1.58$$

Both sets of observations have the same mean; but as suspected, the observations in Data Set B have more dispersion. ☐

The standard deviation can also be approximated from the frequency distribution of a set of observations.

Robotics are used to maintain quality control in industry.

Computing the Standard Deviation from the Frequency Distribution

When n observations $x_1, x_2, \ldots, x_n$ have been classified into k categories, the standard deviation s can be approximated from the grouped standard deviation formula:

$$s = \sqrt{\frac{(m_1 - \bar{x})^2 f_1 + (m_2 - \bar{x})^2 f_2 + \cdots + (m_k - \bar{x})^2 f_k}{n - 1}}$$

where

$$k = \text{Number of categories}$$
$$n = \text{Number of observations}$$
$$\bar{x} = \text{Mean of the observations}$$
$$m_i = \text{Midpoint of the } i\text{th category}$$
$$f_i = \text{Number of observations in the } i\text{th category}$$

Statistical Quality Control

Quality control in manufacturing processes consists of verifying and maintaining a desired level of quality in a product by careful inspection, followed by corrective action when required. Measurements by machines and human observers provide critical data for statisticians to analyze.

Example 5

Quality Control and Standard Deviations   An automobile manufacturer wishes to determine which of two robots should be installed on an as-

Width of Cut (inches) Categories	Midpoint m_i	Robot A (frequency) f_i	Robot B (frequency) f_i
0.2400–0.2433	0.2417	10	11
0.2434–0.2466	0.2451	70	13
0.2467–0.2499	0.2484	15	25
0.2500–0.2533	0.2517	2	27
0.2534–0.2566	0.2551	2	15
0.2567–0.2599	0.2584	1	9
Totals		100	100

sembly line to perform certain metal-cutting operations. An experiment was performed in which each robot was programmed to cut pieces of metal 0.25 inch wide. The observations in Table 17 give the results of the experiment.

On the basis of these observations, which robot should be chosen?

Solution The mean of each set of observations gives a measure of the accuracy of each robot. The standard deviations give a measure of their consistency. Computing the means first, we have

Robot A:
$$\bar{x}_A = \frac{m_1 f_1 + m_2 f_2 + \cdots + m_6 f_6}{n}$$

$$= \frac{(0.2417)(10) + (0.2451)(70) + \cdots + (0.2584)(1)}{100}$$

$$= 0.2460 \text{ inch}$$

Robot B:
$$\bar{x}_B = \frac{m_1 f_1 + m_2 f_2 + \cdots + m_6 f_6}{n}$$

$$= \frac{(0.2417)(11) + (0.2451)(13) + \cdots + (0.2584)(9)}{100}$$

$$= 0.2500 \text{ inch}$$

The standard deviations are

$$s_A = \sqrt{\frac{(m_1 - \bar{x}_A)^2 f_1 + (m_2 - \bar{x}_A)^2 f_2 + \cdots + (m_6 - \bar{x}_A)^2 f_6}{n - 1}}$$

$$= \sqrt{\frac{(0.2417 - 0.2460)^2(10) + \cdots + (0.2584 - 0.2460)^2(1)}{99}}$$

$$\cong 0.0026 \text{ inch}$$

$$s_B = \sqrt{\frac{(m_1 - \bar{x}_B)^2 f_1 + (m_1 - \bar{x}_B)^2 f_2 + \cdots + (m_6 - \bar{x}_B)^2 f_6}{n - 1}}$$

$$= \sqrt{\frac{(0.2417 - 0.2500)^2(11) + \cdots + (0.2584 - 0.2500)^2(9)}{99}}$$

$$\cong 0.0052 \text{ inch} \qquad \square$$

Conclusion

On the basis of the above calculations we could say that either robot should be chosen. On one hand, Robot B is better because the average width of the metal strips it produces is exactly equal to the desired 0.25 inch (to four decimal places). On the other hand, an argument could be made for Robot A because, although the average width (0.2460) differs from the desired 0.25 inch, the standard deviation of the observed widths is only half that of the standard deviation of the observed widths found from Robot B. If $\bar{x}_A = 0.2460$ is within the tolerance limits, then Robot A might be chosen over Robot B.

In other words, both the mean and the standard deviation are useful in their own right.

Problems

Mean, Median, and Mode

For Problems 1–10, find the mean, median, and mode of the given numbers.

1. 1, 2, 3, 4, 5, 6, 7
2. 4, 5, 7, 4, 8, 6, 5
3. 2, 1, 1, 4, 7, 7, 3
4. 3, 5, 9, 5, 1, 8, 8
5. 3, 5, 9, 5, 6, 8, 3, 5, 500
6. 17, 29, 20, 30, 25, 70, 30
7. 1.2, 1.5, 1.3, 1.9, 2.0, 1.8, 1.7
8. 50, 0, 50, 0, 50, 0, 50, 0
9. 50, 0, 50, 0, 50, 0, 50, 0, 50
10. −50, 0, 50, 0, −50, 0, 50
11. From the numbers in Problems 8 and 9, what can you say about relying solely on the median?
12. From the numbers in Problem 5, what can you say about relying solely on the mean?
13. From the numbers in Problem 9, what can you say about relying solely on the mode?

For Problems 14–19, find five numbers whereby the mean, median, and mode satisfy the indicated conditions.

	Smallest ≤	Middle ≤	Largest
14.	mean	median	mode
15.	mean	mode	median
16.	median	mean	mode
17.	median	mode	mean
18.	mode	median	mean
19.	mode	mean	median

For Problems 20–22, fifteen students in a statistics class receive the following quiz grades:

Quiz 1:   85, 50, 59, 84, 74, 70, 59, 86, 40, 100, 95, 88, 55, 40, 85

Quiz 2:   95, 80, 50, 100, 85, 25, 95, 85, 75, 80, 83, 88, 90, 95, 90

20. Find the mean, median, and mode for Quiz 1. Which measure best represents the grades?
21. Find the mean, median, and mode for Quiz 2. Which measure best represents the grades?
22. Find the mean, median, and mode for both quizzes together (30 scores). Which measure best represents the grades?

For Problems 23–26, compute the mean, median, and mode of the given data.

23. Cigarettes Consumed per Capita (1930)

Country	Consumption
Iceland	220
Norway	250
Sweden	310
Canada	510
Denmark	380
Australia	455
United States	1280
Holland	460
Switzerland	530
Finland	1115
Great Britain	1145

24. Lung Cancer Deaths per Million (1950)

Country	Deaths
Iceland	58
Norway	90
Sweden	115
Canada	150
Denmark	165
Australia	170
United States	190
Holland	245
Switzerland	250
Finland	350
Great Britain	465

25. Weight in Grams of Both Kidneys of 15 Normal Men

Subject	Weight (gm)
1	350
2	340
3	310
4	250
5	380
6	300
7	290
8	340
9	280
10	240
11	350
12	360
13	320
14	330
15	260

26. Mean Distance from the Sun

Planet	Distance (miles)
Mercury	36,000,000
Venus	67,200,000
Earth	93,000,000
Mars	141,710,000
Jupiter	483,880,000
Saturn	887,140,000
Uranus	1,783,980,000
Neptune	2,795,460,000
Pluto	3,675,270,000

Range and Standard Deviation

For Problems 27–36, find the range, and the standard deviation of the following numbers.

27. 3, 3, 3, 3, 3, 3, 3, 3, 3, 3
28. 4, 4, 4, 4, 4, 4, 4, 4, 4, 4
29. 1, 0, 1, 0, 1, 0, 1, 0, 1, 0
30. 2, 0, 2, 0, 2, 0, 2, 0, 2, 0
31. 1, 2, 3, 4, 5, 6, 7, 8, 9, 10
32. 7, 9, 6, 3, 8, 1, 0, 3, 5, 7
33. -1, 2, 3, 2, 0, 0, -2, 3, 3, 4
34. 20, 25, 15, 22, 18
35. 40, 45, 35, 42, 38
36. 40, 50, 30, 44, 36

37. What do Problems 27 and 28 tell you about the standard deviation?

38. What do Problems 34 and 35 tell you about the standard deviation?

39. World Series Time   From 1947 to 1960 it was observed that the number of runs scored per half inning in the World Series was given by the numbers in Table 18.
   (a) Compute the range in the number of runs scored per half inning.
   (b) Compute the mean of the number of runs scored per half inning, using the formula for grouped observations.
   (c) Compute the standard deviation of the number of runs scored per half inning, using the formula for grouped observations.

TABLE 18
World Series, 1947–1960

Number of Runs	Number of Times Occurred
0	1023
1	222
2	87
3	32
4	18
5	11
6	6
7	3

40. Football Data   In 1967 the statistician Frederick Mosteller collected the observations in Table 19 on page 446. The numbers show the number of times during the 1966 season that football teams won or lost when they scored a given number of points. For example, 34 teams won that year by scoring seven points, and 186 teams lost when scoring only seven points. Ties were not included in the observations.
   (a) Find the mean winning score during the 1966 season.
   (b) Find the mean losing score during the 1966 season.
   (c) Find the standard deviation of the winning scores.
   (d) Find the standard deviation of the losing scores.

41. Population Problem   It has been predicted that the

TABLE 19
Distribution of Team Scores up to 29, 1966

Score	Winning Team	Losing Team	Score	Winning Team	Losing Team
0	0	219	15	15	23
1	0	0	16	27	29
2	0	6	17	52	23
3	7	36	18	12	14
4	0	1	19	22	14
5	0	0	20	55	29
6	7	114	21	82	36
7	34	186	22	20	14
8	3	32	23	35	8
9	13	19	24	53	13
10	24	48	25	8	5
11	1	4	26	27	2
12	13	42	27	42	10
13	39	72	28	61	11
14	74	128	29	18	4
			Totals	1158	1158

TABLE 20
Predicted Number (in Millions) of Girls and Women under 45 Years of Age in the United States

Age	1990	2035
0–14	37.7	65.0
15–29	28.9	53.8
30–44	26.7	43.0
45–older	38.7	51.3
Totals	132.0	213.1

TABLE 21
Distribution of the Number of Utterances in Bout of Song Type D

Number of Utterances per Bout	Number of Times Occurred
1	132
2	52
3	34
4	9
5	7
6	5
7	5
8	6
Total	250

number of girls and women under the age of 45 in the United States over the next 50 years is given by Table 20.

(a) Find the average age of females in the year 1990.
(b) Find the standard deviation of the ages of females in the year 1990.
(c) Find the predicted average age of females in the year 2035.
(d) Find the predicted standard deviation of the ages of females in the year 2035.

42. **Bird Data** In the book *Statistics in the Real World* by R. J. Larson and D. F. Stroup the authors analyze song types of the male cardinal. The list of observations in Table 21 gives the distribution of the number of times male cardinals uttered a "Song Type D" during a single bout (a series of utterances).

(a) What is the range of the number of utterances?
(b) What was the mean number of utterances that the 250 observed male cardinals made per bout?
(c) What was that standard deviation in the number of utterances?

7.3

The Basics of Statistical Inference

PURPOSE

We continue the introduction of statistics by presenting a few of the basic ideas of statistical inference. We will learn

- the meaning of a population,
- the meaning of a population parameter,
- the meaning of a random sample, and
- how a statistician can estimate a population parameter by constructing a 95% confidence interval.

As an example of statistical inference, we show how pollsters estimate the outcome of presidential elections.

Introduction

There are two major forms of logical thinking: deduction and induction. For deductive thinking, we are indebted to the Greeks, who saw the great power of making general axioms or assumptions and deducing from these assumptions useful conclusions.

Inductive thinking, which has been called the second stage of intellectual development, did not become a scientific tool until the latter part of the eighteenth century. Induction proceeds in the opposite direction from deduction. Whereas deduction tries to determine specific things from general principles, induction tries to infer general conclusions from facts of experience.

The area of statistics known as statistical inference is inductive in nature.

Statistical Inference (Estimating the Whole from a Part)

To understand statistical inference, consider the problem a pollster faces when predicting the outcome of a presidential election. Owing to the enormous size of the electorate, the pollster cannot ask the preference of each eligible voter. The strategy for predicting the outcome of an election before the election occurs is to select a subset or *sample* of voters from the *population* of all voters and infer, from the response of this sample, the response of the entire population. This process of determining information about a collection from a portion of the collection is called **statistical inference**. For instance, suppose you are given a box that contains 100 marbles, each of which is colored black or white. Blindfolded, you reach into the box and select ten marbles. Suppose that seven of the marbles selected are white and three are black. What can you say about the number of black and white marbles in the box? This is a typical problem in statistical inference. Someone once said that statistical inference is the opposite of probability. Figure 9 illustrates the difference between the two subjects.

Figure 9
In probability, an experimenter knows the proportion of red and black marbles in the box and finds the probability of drawing a given proportion of red and black marbles. Statistics solve the opposite problem. The experimenter observes the proportion of red and black marbles drawn from the box and infers the proportion of red and black marbles in the box.

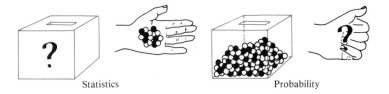

Statistics   Probability

Any collection of objects is called a **population**. Generally, the size of the population is very large. Ordinarily, it is impossible or impractical to examine all the members of a population to draw conclusions about its behavior. For that reason we examine only a subset of this population, called a **sample**. In the case of the pollster problem the population is the set of all eligible voters, and a sample from this population is simply any subset of these voters. If each individual voter in the population has the same chance of being chosen for the sample, the sample is called a **random sample**. The purpose of choosing a random sample is to ensure that the sample is representative of the population.

The Literary Digest Fiasco

There have been pre-election polls that have had embarrassing results because they were based on nonrandom samples. In 1936 the *Literary Digest*, a popular magazine (similar to *Time* and *Newsweek* these days) with an established reputation for prognostication of political elections, predicted that Senator Alfred Landon would defeat Franklin D. Roosevelt in the 1936 presidential election. Ten million prospective voters were sent questionnaires to determine their preferences. Of these ten million, approximately two million questionnaires were returned. On the basis of the responses so obtained, the *Literary Digest* predicted that Landon would win by a 3:2 ratio. The prediction turned out to be a disaster. Roosevelt defeated Landon in a landslide victory, winning 62% of the popular vote and carrying 46 of the 48 states. Because of this mistake, the *Literary Digest* became a laughingstock throughout the country. From that point on, the reputation of the *Literary Digest* went downhill, and the magazine eventually disappeared from the national scene. Several statisticians have suggested reasons for the inaccuracy in the *Literary Digest*'s poll.

A statistical myth has been perpetuated that the *Literary Digest* sampled only people who owned telephones or automobiles, thus making the sample biased in favor of upper-middle income voters. This is not true. The *Literary Digest* sampled all registered voters. As a matter of fact, since voter participation tends to be highest among the well-to-do, telephone owners should not have been that bad a sample. Interested readers should consult Maurice Bryson, "The *Literary Digest*: Making of a Statistical Myth," *The American Statistician* Vol. 30, 1976, pp. 184–85.

The accepted reason for the failure of this poll is that the recipients who responded to the questionnaire felt quite differently than those who did not respond. Often, individuals who respond to questionnaires are people who have strong beliefs and do not represent the general population. Voluntary response to mailed polls is perhaps the most common method of collecting social science data, and perhaps the worst. U.S. representatives and senators use mailed questionnaires to see how their constituents feel about various issues, and they justify subsequent votes on the results. People should be aware that the results of such voluntary response polls are, for the most part, meaningless. Professional pollsters today are very careful in selecting random samples.

Estimation of Parameters (Confidence Intervals)

In the pollster problem we would like to estimate the fraction of voters in a population who prefer a given candidate, based on the fraction of voters in a random sample who prefer the candidate. We assume that there are only two candidates in the election and that all of the voters have decided for whom they will vote. The fraction of voters in the population who prefer a given candidate is called a **population parameter**. Unless every voter can be polled by taking a **census** of all voters, this population parameter can only be approximated by the fraction of voters in a sample who prefer the candidate. This fraction of voters in a sample who prefer a candidate is called a **statistic**. The error, or difference between the population parameter and the statistic, is called the **sampling error** in the prediction. It is important to know the relationship between the accuracy of a statistic and the size of a random sample upon which the statistic is based.

Consider now the problem of estimating the fraction p of voters in a given population who prefer a given candidate. Hence if the fraction of voters who prefer one candidate is p, then the fraction of voters who prefer the other candidates is $1 - p$. The only way to be absolutely sure of the exact value of p is to take a census of all the voters in the population. Usually, this is too expensive or impossible. We also assume that the size of the random sample is small in comparison to the size of the population. In presidential elections the size of the electorate (population) is close to 100 million, whereas the size chosen for a random sample may be only 1500.

To estimate p, we select a random sample of a given size n and compute the fraction of voters in the sample who prefer one of the candidates. This fraction, denoted by $\bar{p}$, is the statistic used to estimate p. The difference $p - \bar{p}$ is the sampling error, or simply error, in the statistic. The goal here is to determine the accuracy of the statistic as an estimator of p.

To determine the accuracy of the statistic $\bar{p}$, imagine the pollster selecting a random sample of n voters and computing the value of $\bar{p}$. Suppose now that the pollster selects another random sample of the same size n from the same population and computes the new value of $\bar{p}$. The new value of $\bar{p}$ will probably not be exactly the same as the first value, but it probably will be fairly close to the first value. The pollster then performs this process over and over, each time selecting a new random sample (but of the same size n) and computing a new value of $\bar{p}$. The interesting thing about these repeated evaluations is that the frequency distribution of $\bar{p}$ "piles up" around the population parameter p. This frequency distribution of $\bar{p}$ is shown in Figure 10. That is, the values of $\bar{p}$, although they are not exactly equal to the unknown p, do the next best thing. It can also be proved, using the theory of statistical inference, that on the average at least 95% of the values of $\bar{p}$ lie between the values of

$$p - \frac{1}{\sqrt{n}} \quad \text{and} \quad p + \frac{1}{\sqrt{n}}$$

In the language of probability we can write

$$P\left(p - \frac{1}{\sqrt{n}} \leq \bar{p} \leq p + \frac{1}{\sqrt{n}}\right) \geq 0.95$$

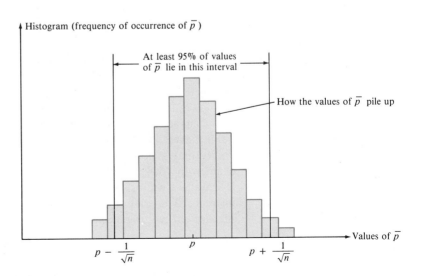

Figure 10
Histogram of the values of the statistic $\bar{p}$

Using elementary properties of inequalities, we can rewrite the above probability statement in the following form:

$$P\left(-\frac{1}{\sqrt{n}} \le \bar{p} - p \le \frac{1}{\sqrt{n}}\right) \ge 0.95$$

By multiplying the inequality inside the argument of P by -1, we change its direction. Hence we have

$$P\left(\frac{1}{\sqrt{n}} \ge p - \bar{p} \ge -\frac{1}{\sqrt{n}}\right) \ge 0.95$$

or

$$P\left(\bar{p} - \frac{1}{\sqrt{n}} \le p \le \bar{p} + \frac{1}{\sqrt{n}}\right) \ge 0.95$$

These results are summarized below.

Confidence Interval for p

If a random sample of size n is selected from a population, where $\bar{p}$ is the fraction of voters in the sample who prefer a given candidate, then there is at least a 95% chance that the fraction of voters p in the entire population who prefer the given candidate will lie in the interval

$$\left[\bar{p} - \frac{1}{\sqrt{n}}, \bar{p} + \frac{1}{\sqrt{n}}\right]$$

This interval is called the 95% **confidence interval** for the population parameter p.

We can be at least 95% sure that p is somewhere in this interval

---

The subject of confidence intervals falls within the area of statistical inference. Note that in the above confidence interval the value of the unknown parameter p has not been determined as a specific number in the same way as unknowns in an algebraic equation. What we are saying is that there is at least a 95% chance or probability that p lies in the given interval. Statistical inference never makes claims with 100% certainty. It says only that there is a given probability that a population parameter lies within a given interval. To make a claim with 100% certainty, a census must be taken of the entire population. By taking a census, however, we are no longer within the realm of statistical inference. The goal of statistical inference is to learn about populations from *samples* of these populations.

Example 1 ─────────── Confidence Intervals and Presidential Elections In 1984 a Gallup poll, based on a random sample of 1600 registered voters, predicted that President

Ronald Reagan would receive 60% of the popular vote and candidate Walter Mondale would receive 40%. On the basis of these statistics, determine the 95% confidence interval for p.

Solution The two important quantities in finding the confidence interval for p are the size of the sample size n and the computed statistic $\bar{p}$. Here we have

$$n = 1600$$
$$\bar{p} = 0.60$$

Hence we can state with at least 95% certainty that the fraction of voters who planned on voting for President Reagan was some fraction in the range

$$95\% \text{ confidence interval} = \left[\bar{p} - \frac{1}{\sqrt{n}}, \bar{p} + \frac{1}{\sqrt{n}}\right]$$

$$= \left[0.60 - \frac{1}{\sqrt{1600}}, 0.60 + \frac{1}{\sqrt{1600}}\right]$$

$$= \left[0.60 - \frac{1}{40}, 0.60 + \frac{1}{40}\right]$$

$$= [0.60 - 0.025, 0.60 + 0.025]$$

$$= [0.575, 0.625] \qquad \square$$

Conclusion

HISTORICAL NOTE

Confidence interval were introduced into statistics in 1937 by the famous Polish-American statistician Jerzy Neyman (1907–1981).

On the basis of the results of the poll, a pollster can claim that there is at least a 95% chance that the percentage of voters in the electorate who plan on voting for President Reagan is between 57.5% and 62.5%. A pollster might state that President Reagan will earn 60% of the popular vote, with a 95% assurance that the true value is within 2.5% of this estimate.

Random samples of size 1600 are popular because they offer a good compromise between accuracy and economy. (It is also easy to compute the square root of 1600.)

Example 2

Confidence Interval with a Smaller Sample Suppose a second pollster asks 100 individuals from the same population as that in Example 1 their presidential preference. In this sample, 60% plan on voting for President Reagan. What can this pollster claim?

Solution On the basis of the random sample of size $n = 100$ and the computed statistic of $\bar{p} = 0.60$, we conclude that there is at least a 95% chance that the population parameter p lies in the interval

$$95\% \text{ confidence interval} = \left[\bar{p} - \frac{1}{\sqrt{n}}, \bar{p} + \frac{1}{\sqrt{n}}\right]$$

$$= \left[0.60 - \frac{1}{\sqrt{100}}, 0.60 + \frac{1}{\sqrt{100}}\right]$$

$$= [0.50, 0.70]$$

In other words, there is a 95% chance that the percentage of voters who prefer President Reagan is between 50% and 70%. Of course, if we had to use one number to estimate p, we would use the statistic $\bar{p} = 0.60$ (60%). The confidence interval is important because it provides more information about p than the single estimate $\bar{p}$.

It should be noted that the estimate $\bar{p} = 0.60$ in this example is not as reliable as the estimate in Example 1. Although the pollster in each example would make the prediction that 60% of the population prefers President Reagan, the pollster with the larger random sample can make it with more assurance. □

Problems

Census or Samples?

Statisticians often prefer to take random samples from a population instead of a census because they are cheaper, practical, and often more accurate. Sometimes the results of a census can be inaccurate because of failure to observe every individual in the sample (people cannot be found and so on). For Problems 1–10,

- Would a sample or a census be preferable? Why?
- What is the population?
- What is the population parameter being estimated?
- What is the statistic that estimates the population parameter in the cases when a sample is taken?

1. An economist working for the U.S. government wishes to determine information about the unemployment rate in the United States.
2. A quality control engineer at Hershey's plant in Hershey, Pennsylvania, wishes to determine whether candy bars pass a taste test.
3. A quality control engineer at General Motors wishes to determine whether the engines in automobiles start after coming off the assembly line.
4. A week before a presidential election a pollster wishes to determine who will win the election.
5. An agricultural inspector wishes to determine the amount of weed seed in a shipment of wheat to the Soviet Union.
6. A medical researcher wishes to determine the average weight of infants in the United States.
7. An ecologist wishes to determine the amount of insects that a certain species of bird eats per day.
8. A business manager wishes to determine the number of days per year employees for a certain company are absent on sick leave.
9. A government inspector wishes to determine information about the average life of batteries imported from Korea.
10. A building inspector wishes to determine information about whether houses in New York City pass a fire code.

Confidence Interval for p

11. **Fruit Flies**   A biologist examines 1000 fruit flies for the presence or absence of a given characteristic. It is discovered that among the 1000 fruit flies, 770 have the trait. What can the biologist say about the fraction p of fruit flies in the entire population of fruit flies that have the trait?
12. **Fruit Flies**   Suppose the biologist in Problem 11 counted only 100 fruit flies and found the same fraction of fruit flies that had the characteristic (77 fruit flies). What could be concluded from this result?
13. **Increasing the Accuracy**   Suppose the biologist in Problem 12 would like the sampling error to be cut in half. In other words, the confidence interval for p is to be halved. How many fruit flies must be tested?
14. **Consumer Survey**   A random sample of 1600 men and women in the United States was asked whether they preferred frozen vegetables to canned vegetables. Among these 1600 individuals, 550 preferred frozen vegetables. What is the value of the statistic? What is the 95% confidence interval for p?
15. **Consumer Survey**   Suppose that in a new consumer survey of 1600 men and women, similar to the one in Problem 14, a total of 800 men and women preferred frozen vegetables. What is the 95% confidence interval for p?
16. **Wade Boggs Problem**   Wade Boggs is batting .400. This statement is both a historical fact and a statistic. It is a historical fact in the sense that he has already collected hits 40% of the time so far this season. On the other hand, it is a statistic in the sense that it can be used to estimate the probability of getting a hit on the next time at bat. Suppose Boggs has batted 100 times (40 hits). What is the 95% confidence interval for the probability that he will get a hit on his next time at bat.
17. **Wade Boggs Again**   Resolve Problem 16 with the change that Boggs has batted 400 times (160 hits).
18. **Financial Analysis**   A financial analyst has studied corporate bond issues sold during the past ten years. The analyst has discovered that among the 100 issues studied,

38 were sold at a premium (the issue price exceeded the face value of the bond). On the basis of this random sample, what could the analyst conclude about the fraction p of bond issues for all corporate bonds that are sold at a premium?

19. **Financial Analysis** Suppose the financial analyst in Problem 18 studies 400 corporate bonds and discovers that 175 of them are sold at a premium. What can the analyst conclude from this information?

20. **Customer Survey** A garden supply company wishes to estimate the proportion p of its customers who make use of its mail-order services. A telephone survey is made of 100 customers, and 35 of these customers say that they use this service. What can the company conclude from these results? What is the statistic in this survey? What is the population parameter being estimated?

21. **Political Science** A political scientist has conducted in-depth interviews with a random sample of 25 lawyers and has learned that 14 of these lawyers are in favor of judicial reform. Find the 95% confidence interval for the population proportion p that favor judicial reform.

22. **Political Science** The political scientist in Problem 21 has interviewed 100 lawyers and has discovered that 65 favor judicial reform. Construct a 95% confidence interval for the population proportion p that favor judicial reform.

Nielsen Ratings

The A. C. Nielsen Company, a Chicago-based marketing research firm, provides ratings for the TV industry. The company monitors 1200 television sets across the country to get an accurate cross section of the 72.5 million U.S. households that have TV sets. The Nielsen rating is the estimated percentage of households that are watching a particular show. In 1984–1985, *Dallas* had the highest season average rating with 25.6. This means that on the average, 25.6% of the TV households in the United States watched *Dallas* every week (not 25.6% of the TV sets that were turned on, but 25.6% of the total TV sets in the United States). Table 22 lists the top 15 regular prime time TV programs for 1984–1985. For Problems 23–32, find a 95% confidence interval for the fraction of households that watched the indicated program.

23. *Dallas* (CBS)   24. *Dynasty* (ABC)
25. *60 Minutes* (CBS)
26. *NBC Monday Night Movies*
27. *Simon & Simon* (CBS)
28. *A Team* (NBC)   29. *Falcon Crest* (CBS)
30. *ABC Sunday Night Movie*
31. *Bill Cosby Show* (NBC)   32. *Hotel* (ABC)

33. Is CBS justified in saying with 95% reliability that *Dallas* is watched by at least 25% of all TV households?

34. Is ABC justified in saying with 95% reliability that *Dynasty* is watched by at least 20% of all TV households?

35. Is CBS justified in saying with 95% reliability that *Simon & Simon* is watched by at least 20% of all TV households?

How Large the Sample?

A pollster would like to select the size n of a random sample in advance so that the length of the 95% confidence interval will be a given value. For Problems 36–39, determine the sample size that will make the 95% confidence intervals have

---

TABLE 22

Top 15 Regular Prime Time TV Programs of 1984–1985

Rank	Program (network)	Total Percent of TV Households
1	*Dallas* (CBS)	25.6
2	*Dynasty* (ABC)	24.6
3	*60 Minutes* (CBS)	23.9
4	*NBC Monday Night Movies*	23.0
5	*Simon & Simon* (CBS)	20.9
6	*A Team* (NBC)	20.6
7	*Falcon Crest* (CBS)	20.5
8	*ABC Sunday Night Movie*	20.1
9	*Bill Cosby Show* (NBC)	20.1
10	*Hotel* (ABC)	19.4
11	*Magnum, P. I.* (CBS)	19.1
12	*Knots Landing* (CBS)	18.9
13	*NBC Sunday Night Movies*	18.8
14	*Murder, She Wrote* (CBS)	18.6
15	*Family Ties* (NBC)	18.4

the given lengths. *Hint:* Set the length of the confidence interval

$$\text{Length of confidence interval} = \frac{2}{\sqrt{n}}$$

equal to the given length and solve for *n*.

36. The length of the confidence interval is 0.05 (in other words, 5% or 2.5% on either side of the statistic).

37. The length of the confidence interval is 0.01 (0.5% on either side of the statistic).

38. The length of the confidence interval is 0.02 (1% on either side of the statistic).

39. The length of the confidence interval is 0.10 (5% on either side of the statistic).

Epilogue: Chernoff Faces—Innovation or Fad?

We have seen that the histogram and frequency polygon are useful to visually represent observations consisting of a single column of numbers. Many times, however, observations consist of several columns of numbers.

In 1971, Stanford statistician Herman Chernoff proposed an unorthodox method for visually displaying observations consisting of several columns of numbers. The idea was to represent the observations as cartoon faces (called "Chernoff faces" nowadays). By matching properties of the observations with different facial characteristics—nose shape, eye slant, mouth shape, ear shape, and so on—it was hoped that complex interrelationships in the data could be detected in the Chernoff faces. A subtle expression in a Chernoff face could possibly reveal information that could not be seen in the rows of numbers.

Chernoff faces were used recently by Howard Wainer to show several characteristics of the 50 states. "We found that the use of Chernoff faces for this task yielded an evocative and easy understanding of the display," Wainer said. By letting different features of the Chernoff face represent the different characteristics of the states, he was able to represent each state with a given face. Table 23 shows the equivalences between the facial features and the variables that Wainer observed.

The following Chernoff faces give visual information about the above seven variables.

Take a look at the Chernoff face for Texas. Note the peanut-shaped head, which means that it is hot in Texas. Also, note that Texas has four Chernoff faces, which means that the population is about the same size as Florida. The face is not smiling very much, which indicates that the average salary is not too large (in

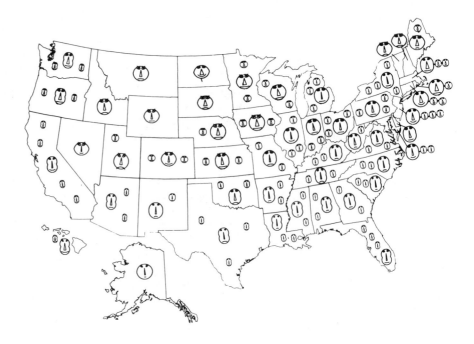

Chernoff faces illustrating several variables of each state.

TABLE 23
Equivalence Between
Chernoff Face Features
and Variables

Variable	Facial Feature
State population	Number of faces per state
Literacy rate	Eye size (higher = bigger)
Percent of high school graduates	Eye slant (higher = more slant)
Life expectancy	Length of mouth (longer = longer)
Homicide rate	Width of nose (lower = wider)
Income	Curvature of mouth (larger = smile)
Temperature	Shape of face (warm = peanut-shaped)

contrast, look at the smile on the face of California). A trained reader, knowing more about the way the faces were constructed, might be able to determine the average income for Texas from the smile. You can try to analyze the faces of your own state. Whether Chernoff faces are only a passing fad or will be a major research tool in the future remains to be seen.

Key Ideas for Review

confidence interval, 448
cumulative frequency, 428
frequency distribution, 425
frequency polygon, 429
histogram, 429
mean
    grouped data, 437

ungrouped data, 437
median, 439
mode, 439
population, 448
population parameter, 448
random sample, 447
range, 440

relative frequency, 425
standard deviation
    grouped data, 442
    ungrouped data, 441
statistic, 448
statistical inference, 446

Review Exercises

1. **Summarizing the Sizes of the National Parks** Table 24 on page 456 gives the sizes of all the national parks in acres, according to the National Park Service. Construct a summary table using the following steps, with categories of length 200,000. The first category should be [0, 200,000).
   (a) Determine the categories and their midpoints.
   (b) Make a tally mark for each observation in the appropriate category.
   (c) Count the tally marks in each category and record this number in the frequency distribution column.
   (d) Compute the cumulative frequency column.
   (e) Compute the relative frequency column.

   (f) Compute the cumulative relative frequency column.
   (g) Draw a histogram of the frequency distribution.
   (h) Draw a frequency polygon of the frequency distribution.
2. **Finding the Mean from Grouped Data** Use the summary table constructed in Problem 1 to compute the average size of the National Parks.
3. **Finding the Standard Deviation from Grouped Data** Use the summary table constructed in Problem 1 to compute the standard deviation of the sizes of the National Parks.
4. What is the range of the sizes of the National Parks?
5. What is the median of the sizes of the National Parks?

TABLE 24
Sizes of National Parks

Park	Location	Size
Acadia	Maine	41,642
Arches	Eastern Utah	82,953
Big Bend	Western Texas	708,221
Bryce Canyon	SW Utah	36,010
Canyonlands	SE Utah	257,640
Capitol Reef	Southern Utah	254,242
Carlsbad Caverns	SE New Mexico	46,753
Crater Lake	SW Oregon	160,290
Everglades	Southern Florida	1,400,533
Glacier	NW Montana	1,013,101
Grand Canyon	NW Arizona	673,575
Grand Teton	NW Wyoming	310,443
Great Smoky Mountains	N.C., Tennessee	516,626
Guadalupe Mountains	Western Texas	81,077
Haleakala	Hawaii, on Maui	27,283
Hawaii Volcanoes	Hawaii, on Hawaii	229,616
Hot Springs	Central Arkansas	3,535
Isle Royale	NW Michigan	539,341
Kings Canyon	Eastern California	460,331
Lassen Volcanic	Northern California	106,934
Mammoth Cave	Central Kentucky	51,354
Mesa Verde	SW Colorado	52,074
Mount McKinley	Central Alaska	1,939,493
Mount Rainier	SW Washington	241,992
North Cascades	Northern Washington	505,000
Olympic	NW Washington	896,599
Petrified Forest	Eastern Arizona	94,189
Platt	Southern Oklahoma	912
Redwood	NW California	56,201
Rocky Mountain	Central Colorado	262,191
Sequoia	Eastern California	386,863
Shenandoah	Northern Virginia	193,537
Virgin Islands	Virgin Islands	14,419
Voyageurs	Northern Minnesota	219,431
Wind Cave	SW South Dakota	28,059
Yellowstone	Wyo., Mont., Idaho	2,221,773
Yosemite	Central California	761,320
Zion	SW Utah	147,035

TABLE 25
Percent of Population Ever Married

Age Group (years)	1980		1984	
	Males	*Females*	*Males*	*Females*
15–19	2.7%	8.8%	1.5%	6.6%
20–24	31.2%	49.8%	25.2%	43.1%
25–29	67.0%	79.1%	62.2%	74.1%
30–34	84.1%	90.5%	79.1%	86.7%
35–44	92.5%	94.5%	90.8%	93.6%
45–54	93.9%	95.1%	93.8%	95.4%

(a) For 1980, draw a histogram and frequency polygon of the percent of males ever married.

(b) For 1980, draw a histogram and frequency polygon of the percent of females ever married.

(c) For 1984, draw a histogram and frequency polygon of the percent of males ever married.

(d) For 1984, draw a histogram and frequency polygon of the percent of females ever married.

8. **Summarizing NBA Data**   Table 26 lists the National Basketball League scoring champions from 1953 to 1985. For parts (a)–(k), construct a summary table for the data in Table 25. Use ten categories, the width of each category being three points. The first category should be [22, 24).

(a) Group the winning averages into categories with a width of three points. What are the midpoints of the categories?

(b) Make a tally mark for each observation in the appropriate category.

(c) Count the tally marks in each category and record this number in a frequency distribution column.

(d) Compute the cumulative frequency column.

(e) Compute the relative frequency column.

(f) Draw a histogram of the frequency distribution.

(g) Draw the frequency polygon for the frequency distribution.

(h) Use the formula for the mean of grouped observations to find the mean of the winning averages.

(i) What is the median of the winning averages?

(j) What is the mode of the winning averages?

(k) What is the range of the winning averages?

Measures of Central Tendency

For Problems 9–15, find the mean, median, and mode of the given numbers.

9. $-1, 1, -1, 1$

10. $0, 1$

11. $0, 0, 0, 0, 1$

12. $0, 1, -1, 2, -2, 3, -3$

13. $7, 8, 9, 10, 9, 8, 7$

14. $23, 23, 23, 23, 23$

15. $20, 15, 20$

6. What is the mode of the sizes of the National Parks?

7. **Summarizing Sociological Data**   Table 25 gives the percent of the United States population that has ever been married.

TABLE 26
Individual NBA Scoring Champions

Season	Player	Average
1953–54	Neil Johnston, Philadelphia Warriors	24
1954–55	Neil Johnston, Philadelphia Warriors	23
1955–56	Bob Pettit, St. Louis Hawks	26
1956–57	Paul Arizin, Philadelphia Warriors	26
1957–58	George Yardley, Detroit Pistons	28
1958–59	Bob Pettit, St. Louis Hawks	29
1959–60	Wilt Chamberlain, Philadelphia Warriors	38
1960–61	Wilt Chamberlain, Philadelphia Warriors	38
1961–62	Wilt Chamberlain, Philadelphia Warriors	50
1962–63	Wilt Chamberlain, San Francisco Warriors	45
1963–64	Wilt Chamberlain, San Francisco Warriors	37
1964–65	Wilt Chamberlain, San Francisco Warriors–Philadelphia 76ers	35
1965–66	Wilt Chamberlain, Philadelphia 76ers	34
1966–67	Rick Barry, San Francisco Warriors	36
1967–68	Dave Bing, Detroit Pistons	27
1968–69	Elvin Hayes, San Diego Rockets	28
1969–70	Jerry West, Los Angeles Lakers	31
1970–71	Lew Alcindor, Milwaukee Bucks	32
1971–72	Kareem Abdul-Jabbar, Milwaukee Bucks	35
1972–73	Nate Archibald, Kansas City–Omaha Kings	34
1973–74	Bob McAdoo, Buffalo Braves	31
1974–75	Bob McAdoo, Buffalo Braves	35
1975–76	Bob McAdoo, Buffalo Braves	31
1976–77	Pete Maravich, New Orleans Jazz	31
1977–78	George Gervin, San Antonio Spurs	27
1978–79	George Gervin, San Antonio Spurs	30
1979–80	George Gervin, San Antonio Spurs	33
1980–81	Adrian Dantley, Utah Jazz	31
1981–82	George Gervin, San Antonio Spurs	32
1982–83	Alex English, Denver Nuggets	28
1983–84	Adrian Dantley, Utah Jazz	31
1984–85	Bernard King, New York Knicks	33

For Problems 16–21, find the mean, range, and standard deviation of the numbers. Plot the numbers on the real number line and indicate the point where the numbers would balance.

16. 1, 1, 2, 2, 3
17. 2, 2, 4, 4, 6
18. 0, 0, 0, 0, 0, 0, 0, 0, 0, 1
19. 10, 0, 10, 0, 10
20. -1, 0, 1, -2, 2
21. 3, 3, 3, 3, 3

Statistical Inference

In Problems 22–26, determine the population and the random sample.

22. An agricultural inspector is testing a shipment of coffee from Brazil for mold by taking periodic samples.
23. An economist is trying to determine the average income of U.S. households by interviewing 100 families.
24. A zoologist is trying to determine the proportion of a certain species of beetle that is resistant to a new insecticide. The zoologist tests 100 beetles and then makes a determination.
25. A geologist is measuring core samples from a mining exploration experiment in Indonesia.
26. A quality control engineer at General Electric would like to know the expected lifetime of a new design of lightbulb. A test measures the lifetime of 1000 bulbs.
27. Flu Vaccine A group of 400 high school students were given a vaccine. Of these students, 120 experienced side effects of discomfort. Based on these results, what is the 95% confidence interval for the proportion of all students who will experience some discomfort? What is the assumption that one makes when constructing the confidence interval?
28. Flu Vaccine To ensure more reliability, the public health officers in Problem 27 have decided to enlarge the sample. In a new study, 435 out of 1600 students have discomfort after being given the vaccine. What can the health officers say about the proportion of individuals in the entire population that will suffer some discomfort?
29. Accurate Polling In an attempt to get a greater degree of reliability on a very important question, a pollster samples 10,000 individuals. Of the 10,000 sampled, 6945 claim to prefer Brite Classic over New Brite. What can the pollster claim from this experiment? What is the statistic and what is the population parameter in this problem?
30. More Accurate Polling The pollster in Problem 29 would like to get even greater accuracy. How large must the sample be so that the length of the confidence interval found in Problem 29 is cut in half?
31. Double Accuracy What is the relationship between the lengths of the confidence intervals with the sample sizes 100, 400, 1600? What would one suspect in general from this result?

Chapter Test

1. Find five numbers (and verify your results) where

   mean > median > mode

2. Find five numbers (and verify your results) where

   median = mode > mean

3. Find five numbers (and verify your results) where

   range > standard deviation

4. Find five numbers (and verify your results) where

   range < standard deviation

5. What is the mean and standard deviation of the observations in Table 27?

6. In a sample survey, 214 out of 1000 people interviewed in a large city said that they opposed public transportation. Construct the 95% confidence interval for the true proportion of people in the city who oppose public trans-

TABLE 27
Observations for Problem 5

Values	Frequency
1–3	20
4–6	35
7–9	10

portation. What is the statistic and what is the population parameter in this problem?

7. In Problem 6, how many people would have to be sampled so that the length of the confidence interval is 2%? Suppose the same fraction of people oppose public transportation as in Problem 6. What would be the new confidence interval with this new sample?

PART II
CALCULUS

Calculus Preliminaries

Scientific Notation

If we always dealt with numbers of the size 3, 5, 4.2, −3.2, and so on, there would be no need to introduce the concept of scientific notation. The word "scientific" refers to the fact that scientists often work with very large or very small numbers. For example, astronomers can actually determine the weight of the sun. It weighs approximately

$$3,600,000,000,000,000,000,000,000,000,000 \text{ pounds}$$

Writing this number in **scientific notation** is much more convenient. In this form it weighs

$$3.6 \times 10^{30} \text{ pounds}$$

One can see the advantage in scientific notation. Small numbers also can become awkward. The weight of an electron is approximately

$$0.0000000000000000000000000009 \text{ gram}$$

In scientific notation this number is represented as

$$9.0 \times 10^{-27} \text{ gram}$$

This discussion leads us to the formal definition of scientific notation.

Definition of Scientific Notation

The scientific form of a number is that number written as a power of ten or as a decimal number between 1 and 10 times a power of ten.

Example

The following numbers are written in scientific notation.

Number	Scientific Notation
(a) 1,000,000,000	10^9
(b) 1,000,000,000,000	10^{12}
(c) 5,870,000,000,000	5.87×10^{12}
(d) 0.000000354	3.54×10^{-7}
(e) 0.002	2.0×10^{-3}

Problems

For Problems 1–9, write the numbers in scientific notation.

1. One light year = 5,880,000,000,000 miles.
2. The distance to the nearest star = 6,000,000,000,000 miles.
3. The weight of water in all the oceans = 1,580,000,000,000,000,000 tons.
4. The thickness of the Milky Way = 300,000,000,000,000,000 miles.
5. The approximate number of different bridge hands = 600,000,000,000.
6. The probability of winning the Irish Sweepstakes = 0.0000001.
7. The width of a human hair = 0.005 inch.
8. The number of hemoglobin molecules in one red blood cell = 270,000,000.
9. The weight of a fly = 0.01 gram.

For Problems 10–17, write the numbers out in longhand.

10. The number of candles whose light is equivalent to the light emitted by the sun = 3.0×10^{27}.
11. The number of ways 20 boxcars on a train can be arranged = 2.4×10^{18}.
12. The distance to the center of Milky Way = 6.0×10^{12} miles.
13. The number of grains of sand on Coney Island = 1.0×10^{20}.
14. The number of poker hands = 2.6×10^{6}.
15. A googol = 1.0×10^{100}.
16. The number of words ever spoken by humans = 1.0×10^{16}.
17. One angstrom = 1.0×10^{-8} centimeter.

"IF IT'S TRUE THAT THE WORLD ANT POPULATION IS 10^{15} THEN IT'S NO WONDER WE NEVER RUN INTO ANYONE WE KNOW."

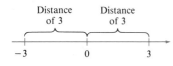

Figure 1
Absolute value

Absolute Value

You are studying this book on about the 500th anniversary of the introduction of negative numbers into mathematics. In 1489 the first appearance in print of the plus and minus signs was made in a German arithmetic book written by Johann Widman of Bohemia. Until that time, people were interested only in the magnitude of the number. Today, we refer to the magnitude of a number as the absolute value of a number. We define the **absolute value** of a number a, denoted as $|a|$, as

$$|a| = \begin{cases} a & a \geq 0 \\ -a & a < 0 \end{cases}$$

The geometric interpretation of the absolute value $|a|$ is that it is the distance from the number a to 0 on the real number line. For example, as shown in Figure 1,

$$|3| = 3$$
$$|-3| = 3$$

Problems

1. What is $|0|$?
2. What is $|-3|$?
3. What is $|3 - 4 + 8|$?
4. Which is larger, $|3 - 4|$ or $|4 - 3|$?
5. Is it true that $|x| = |-x|$ for any real number x?
6. Is it true that $|x^2| = |x|^2$ for any real number x?
7. Is it true that $|xy| = |x| \cdot |y|$ for all numbers x and y?
8. What are the numbers x that satisfy $|x| \geq 1$?
9. **Hmmmmm** Which is larger, $|-x - y|$ or $|x + y|$?

Roots and Radicals

In the Finite Mathematics Preliminaries chapter we defined a^n for any nonzero real number a and any integer n. We now define $a^{1/n}$ for any real number a and any positive integer n. To begin we say that r is a square root of a if $r^2 = a$. In general, we say that r is an **nth root** of a if $r^n = a$. We are interested in finding the real roots (in contrast to complex roots), and so when we say "roots," we mean **real roots**. Table 1 illustrates the nature of nth roots of a. For the number zero, the only nth root is zero.

TABLE 1
Properties of nth Roots
of a Real Number a

	a Positive	**a Negative**
n Even	Two Roots (9 has two square roots, namely, 3 and -3)	No Roots (-2 has no square root)
n Odd	One Root (8 has one cube root, namely, 2)	One Root (-8 has one cube root, namely, -2)

The notation $\sqrt[n]{a}$ or $a^{1/n}$ is called the **nth root radical**, and it represents the nth root if there is only one root. If there are two real roots, it represents the positive root. We also define $a^{-1/n} = 1/a^{1/n}$.

Example

The following are examples of nth root radicals.

(a) $\sqrt{4} = +2$ positive root
(b) $\sqrt[3]{8} = +2$ only root
(c) $\sqrt[3]{-8} = -2$ only root
(d) $\sqrt[4]{16} = +2$ positive root
(e) $\sqrt{-2} =$ no root

Problems

Evaluate the roots in Problems 1–7. Use a calculator if necessary.

1. $8^{1/3}$

2. $\sqrt[3]{27}$

3. $\sqrt[3]{64}$

4. $\sqrt[3]{-125}$

5. $\sqrt{2}$

6. $\sqrt[3]{512}$

7. $\sqrt[8]{256}$

8. Hmmmmm The next time you owe someone half a dollar, just say, "Since

$$25 \text{ cents} = \tfrac{1}{4} \text{ dollar}$$

I'll just take the square root of these equal quantities and obtain the new equal quantities

$$5 \text{ cents} = \tfrac{1}{2} \text{ dollar}$$

So I'll give you a nickel." What is wrong with your argument?

Fractional Exponents

We have introduced powers of numbers a^m, and roots of numbers $a^{1/n}$ where m and n are integers. We can now put these two ideas together and construct a new type of exponent, the fractional exponent. You may think that fractional exponents represent pure mathematics and have no meaning in the real world. This is not the case. Many natural phenomena are described by functions having fractional exponents, and many answers to real world problems are written in terms of fractional exponents.

Definition of $a^{m/n}$

For any real number a for which the nth root $a^{1/n}$ exists, the **fractional exponent** $a^{m/n}$, where m and n are integers, is defined by

$$a^{m/n} = (a^{1/n})^m$$

In other words, we first take the nth root radical, then take the mth power.

Example

Sample Fractional Exponents

(a) $8^{4/3} = (8^{1/3})^4 = 2^4 = 16$

(b) $8^{-4/3} = \dfrac{1}{8^{4/3}} = \dfrac{1}{2^4} = \dfrac{1}{16}$

(c) $(-8)^{4/3} = (-8^{1/3})^4 = (-2)^4 = 16$

(d) $(-8)^{5/2}$ not defined

(e) $4^{5/2} = (4^{1/2})^5 = 2^5 = 32$

(f) $2^{5/2} = (2^{1/2})^5 = (1.4142\ldots)^5 = 5.6568\ldots$

HISTORICAL NOTE

The Middle Ages were barren when it came to mathematical discovery. However, the greatest mathematician of the period,

Nicole Oresme (c.1323–1382) of Normandy, introduced the notation for fractional exponents.

The following properties are useful when working with radicals.

> ## Properties of Radicals
>
> If a and b are any real numbers and m and n are any integers such that $a^{1/n}$ and $b^{1/n}$ are defined, then
>
> - $(\sqrt[n]{a})^n = a;$
> - $\sqrt[n]{a^n} = \begin{cases} |a| & \text{if } n \text{ is even,} \\ a & \text{if } n \text{ is odd;} \end{cases}$
> - $\sqrt[n]{a} \cdot \sqrt[n]{b} = \sqrt[n]{ab};$
> - $\dfrac{\sqrt[n]{a}}{\sqrt[n]{b}} = \sqrt[n]{\dfrac{a}{b}} \quad (b \neq 0);$
> - $\sqrt[m]{\sqrt[n]{a}} = \sqrt[mn]{a}.$

Example

Sample Radical Manipulations

(a) $\sqrt{45} = \sqrt{9 \cdot 5} = \sqrt{9} \cdot \sqrt{5} = 3\sqrt{5}$

(b) $\sqrt{50} = \sqrt{25 \cdot 2} = \sqrt{25}\sqrt{2} = 5\sqrt{2}$

(c) $\sqrt[3]{54} = \sqrt[3]{27 \cdot 2} = 3\sqrt[3]{2}$

(d) $3\sqrt{18} + 4\sqrt{50} = 3\sqrt{9 \cdot 2} + 4\sqrt{25 \cdot 2}$
$$= 3\sqrt{9}\sqrt{2} + 4\sqrt{25}\sqrt{2}$$
$$= 3(3)\sqrt{2} + 4(5)\sqrt{2}$$
$$= 9\sqrt{2} + 20\sqrt{2}$$
$$= 29\sqrt{2}$$

(e) $\sqrt{9x^2} = 3x$

Problems

Simplify the expressions in Problems 1–15.

1. $\sqrt{32}$

2. $\sqrt{1000}$

3. $2\sqrt{5} + 5\sqrt{45}$

4. $\sqrt{1000x^4y^6}$

5. $\sqrt{32x^5}$

6. $\sqrt{a^5b^5} + \sqrt{a^4b^7}$

7. $\sqrt{90x^5y^4}$

8. $3\sqrt{3} + \sqrt{12}$

9. $\sqrt[3]{3x^4y^6}$

10. $\sqrt[4]{32x^4y^5}$

11. $\sqrt{2x^2y^3z^4}$

12. $\sqrt[5]{64x^5}$

13. $\sqrt{80x^5w^9}$

14. $\sqrt{50} + \sqrt{60}$

15. $\sqrt{x^2} + \sqrt{x^4}$

Polynomials

A polynomial of degree n is an expression of the form

$$P_n(x) = a_n x^n + a_{n-1} x^{n-1} + \cdots + a_1 x + a_0$$

where $a_0, a_1, \ldots, a_n$ are any real numbers, a_n is different from zero, and n is any nonnegative integer. Polynomials of degree two are called **quadratic polynomials**, while polynomials of degree three are called **cubic polynomials**.

Polynomials are extremely useful in business management and mathematical economics. They are easy to work with and allow us to describe complex relationships between economic variables. Later, when we study the economic theory called *theory of the firm*, we will see that a company's profit, revenue, and cost can often be described by means of polynomials.

HISTORICAL NOTE

The letter x in the above polynomial is called the **variable**, while the letters $a_n, a_{n-1}, \ldots, a_0$ are called **coefficients**. The person responsible for the convention of using the last letters of the alphabet ($u, v, w, x, y,$ and z) to represent variables, or unknown quantities, and the early letters of the alphabet ($a, b, c, \ldots$) to represent constants, or known quantities, was the French mathematician René Descartes (1596–1650).

Adding, Subtracting, Multiplying, and Dividing Polynomials

Polynomials can be added, subtracted, multiplied, and divided in much the same way as numbers. We will need to perform these operations on polynomials throughout the book. Later, when we study polynomials representing the profit of a firm, we will see that the profit can be expressed as the difference between two other polynomials; the revenue polynomial and the cost polynomial.

Adding Polynomials

Two or more polynomials are added by adding the coefficients of terms with similar powers.

Example

Addition of Polynomials

(a) $(x^3 + 3x - 4) + (2x^2 + 5x + 5) = x^3 + 2x^2 + 8x + 1$

(b) $(3p^2 + 4p - 5) + (2p + 1) = 3p^2 + 6p - 4$

Subtracting Polynomials

One polynomial is subtracted from another polynomial by subtracting terms with similar powers.

Example

Subtraction of Polynomials

(a) $(x^4 + 3x^3 - x^2 + 3) - (x^3 - 1) = x^4 + 2x^3 - x^2 + 4$

(b) $(x^2 - x + 1) - (x^5 + x^3 + 2) = -x^5 - x^3 + x^2 - x - 1$

Multiplying Polynomials

To multiply one polynomial by another polynomial, multiply each term of one polynomial by each term of the other polynomial, and take the sum of all the several products.

Multiplication of Polynomials

(a) $2x(x + 3) = 2x^2 + 6$

(b) $(x + 1)(x + 2) = (x + 1)(x) + (x + 1)(2)$
$$= (x^2 + x) + (2x + 2)$$
$$= x^2 + 3x + 2$$

(c) $(x^3 + 4)(3x + 5) = (x^3 + 4)(3x) + (x^3 + 4)(5)$
$$= 3x^4 + 12x + 5x^3 + 20$$
$$= 3x^4 + 5x^3 + 12x + 20$$

Dividing Polynomials

When one polynomial is divided into another polynomial, the net result is a quotient polynomial such that

$$\frac{\text{Numerator polynomial}}{\text{Denominator polynomial}} = \text{Quotient} + \frac{\text{Remainder}}{\text{Denominator polynomial}}$$

The results can be checked by multiplication.

Division of Polynomials　If $x^2 - 3x + 5$ is divided by $x - 2$, the quotient is $x - 1$, and the remainder is 3. That is,

$$\frac{x^2 - 3x + 5}{x - 2} = x - 1 + \frac{3}{x - 2}$$

This result may be checked by showing that

$$x^2 - 3x + 5 = (x - 2)(x - 1) + 3$$

The steps for dividing polynomials are as follows.

- **Step 1** (arrange the terms).　Arrange the terms of the polynomials in descending powers.
- **Step 2** (divide the first term).　Divide the first term of the denominator into the first term of the numerator; this gives the first term of the quotient.
- **Step 3** (multiply the denominator).　Multiply the denominator by the first term of the quotient and subtract the product from the numerator; this gives the first remainder.
- **Step 4** (repeat Steps 1–3).　Repeat Steps 1–3, using each successive remainder as a new numerator; continue the process until the remainder is zero or a polynomial of degree lower than the denominator.

Example

$$
\begin{array}{r}
2x^2 - x - 4 \leftarrow quotient \\
\text{denominator } \quad 3x^2 - 2x + 5 \overline{)\; 6x^4 - 7x^3 \qquad + 10x - 23} \leftarrow numerator \\
\underline{6x^4 - 4x^3 + 10x^2} \\
-3x^3 - 10x^2 + 10x - 23 \\
\underline{-3x^3 + 2x^2 - 5x} \\
-12x^2 + 15x - 23 \\
\underline{-12x^2 + 8x - 20} \\
7x - 3 \leftarrow remainder
\end{array}
$$

Problems

For Problems 1–7, carry out the four arithmetic operations of addition, subtraction, multiplication, and division on the polynomials. In the case of subtraction, subtract the second polynomial from the first polynomial. In the case of division, divide the second polynomial into the first polynomial.

1. $x^2 + 2$ and $3x$
2. $s^2 + 1$ and $s^2 - 1$
3. $q^2 + 3q - 1$ and $q^2 - 2q + 3$
4. $q^3 + 2q^2 + 1$ and $q^3 + q^2 + 1$

5. $x^2 - 1$ and $x^4 + 1$
6. $t^4 + 1$ and $t^2 - 1$
7. $p^4 + p^3 + p^2 + p + 1$ and $p + 1$

George Brett just hit another
$y = -x^2 + 400x$!

Uh-huh.

It is also possible to add, subtract, and multiply algebraic expressions other than polynomials. The rules are essentially the same.

Example

(a) $(x^{1/2} + 2) + (x^{3/2} + 2x^{1/2} - 3) = x^{3/2} + 3x^{1/2} - 1$
(b) $(x^{1/2} + y^{1/2})(x^{1/2} - y^{1/2}) = x - y$
(c) $(2\sqrt{x} - 3)(\sqrt{x} + 1) = 2x - \sqrt{x} - 3$

Problems

Perform the indicated operations in Problems 1–4.

1. $(x^{3/2} + x^{1/2}) + (x^{1/2} - 1)$
2. $(s^{5/2} - s^{1/2} + 1) - (2s^{1/2} + 4)$
3. $(x^{1/2} + 2)(x^{3/2} + 1)$
4. $(s^{7/2} - t^{1/2})(s^{3/2} + t^{1/2})$

5. **Hmmmmmm** Algebra can be used to impress the unknowing. Karnap the Magician asks you to think of a number. He then asks you to
   (a) add 3,
   (b) multiply by 2,
   (c) subtract 4,
   (d) divide by 2,
   (e) subtract the original number.

Karnap then says, "Your final number is your I.Q." (He's a real kidder.) What is the final number, and how does Karnap perform this magic?

6. **Another Karnap Puzzle** Karnap has a new puzzle. He asks the audience to pick a number. He then asks them to
   (a) add 2,
   (b) square this number,
   (c) subtract 4,
   (d) divide by the original number,
   (e) subtract the original number.

Karnap then says, "You have the number 4." How did Karnap perform this new puzzle?

7. **Make Your Own Magic** Solve Problem 6 and design your own magic puzzle.

Factoring

Multiplying expressions together is a much easier task than the opposite problem of "taking them apart," that is, factoring them into two or more factors. For example,

$$x^2 - 4 = (x - 2)(x + 2)$$
$$x^3 - 2x^2 - x + 2 = (x - 1)(x + 1)(x - 2)$$
$$x^4 - 1 = (x - 1)(x + 1)(x^2 + 1)$$

This leads us to the following definition.

Definition of Prime Polynomials

A polynomial is called a **prime polynomial** if it cannot be factored as the product of two other polynomials.

For instance,

$x^2 - 4$ is not prime because $x^2 - 4 = (x - 2)(x + 2)$

$x^2 + 1$ is prime

$x^2 - 2$ is not prime because $x^2 - 2 = (x - \sqrt{2})(x + \sqrt{2})$

A polynomial is completely factored when it is written as the product of prime polynomials.

Example

$x^2 - 4 = (x - 2)(x + 2)$ is completely factored.

$x^3 - 2x^2 - 4x + 8 = (x - 2)(x^2 - 4)$ is not completely factored.

It is not always easy to factor a polynomial, and there is no general method that will produce all the factors of an expression. The factors depend largely on the form of the expression to be factored, and they normally must be determined by inspection. However, the results may be checked quite easily by showing that the product of the factors is equal to the original expression. We shall identify some special types of polynomials that we can factor. If a polynomial can be identified as belonging to one of these special types of polynomials, the factors can be found by inspection.

Common Factor

If all terms of a polynomial have a common factor, this factor can be taken out of the polynomial. We illustrate this idea by the general expression

$$ax + ay + az = a(x + y + z)$$

Example

(a) $3x^3 + 6x - 9 = 3(x^3 + 2x - 3)$

(b) $abx^4 + a^2bx^2 - ab^2 = ab(x^4 + ax^2 - b)$

Difference of Two Squares

The difference of two squares may be factored into the product of the sum and differences of the quantities that are squared. Thus

$$x^2 - a^2 = (x + a)(x - a)$$

Example

(a) $x^2 - 9 = (x + 3)(x - 3)$

(b) $16x^4 - 25 = (4x^2)^2 - (5)^2 = (4x^2 + 5)(4x^2 - 5)$

Quadratic Polynomial

The quadratic polynomials

$$x^2 + bx + c$$
$$_, ax^2 + bx + c$$

can often be factored into a product of factors of the form

$$x^2 + bx + c = (x + r)(x + s) \qquad (rs = c, r + s = b)$$
$$ax^2 + bx + c = (ux + r)(vx + s) \qquad (uv = a, rs = c, us + rv = b)$$

Example

(a) $x^2 + 5x + 6 = (x + 3)(x + 2)$

(b) $x^2 - x - 12 = (x - 4)(x + 3)$

(c) $6x^2 + 7x - 20 = (2x + 5)(3x - 4)$

(d) $9x^2 + 9x - 4 = (3x - 1)(3x + 4)$

Sum and Difference of Cubes

The sum and difference of cubes can be factored in the following way:

$$x^3 - a^3 = (x - a)(x^2 + ax + a^2)$$
$$x^3 + a^3 = (x + a)(x^2 - ax + a^2)$$

Example

(a) $x^3 + 8 = (x + 2)(x^2 - 2x + 4)$

(b) $x^3 - 27 = (x - 3)(x^2 + 3x + 9)$

(c) $8x^3 - 27a^6 = (2x - 3a^2)(4x^2 + 6a^2x + 9a^4)$

(d) $27x^3 + 1 = (3x + 1)(9x^2 - 3x + 1)$

Factoring by Special Grouping

When the previous four methods cannot be used, it is sometimes possible to group the terms in such a manner that the factors may be found by inspection.

Example

(a) $2x^3 + 8x^2 - 3x - 12 = (2x^3 + 8x^2) - (3x + 12)$
$$= 2x^2(x + 4) - 3(x + 4)$$
$$= (x + 4)(2x^2 - 3)$$

(b) $9x^4 + 15x^2 + 16 = (9x^4 + 24x^2 + 16) - 9x^2$
$$= (3x^2 + 4)^2 - (3x)^2$$
$$= [(3x^2 + 4) + 3x][(3x^2 + 4) - 3x]$$
$$= (3x^2 + 3x + 4)(3x^2 - 3x + 4)$$

Problems

For Problems 1–20, write the factors of the expressions.

1. $x^2 - 10x + 16$
2. $a^2 + 4ab + 3b^2$
3. $2x^2 + 7xy - 15y^2$
4. $4y^4 - 12y^2 + 9$
5. $8x^3 - 125$
6. $a^2 - 9$
7. $x^4 - a^4$
8. $x^3 - 27$
9. $a^6 - b^6$
10. $x^4 + 4y^4$
11. $8x^3 + 125$
12. $mx - 3my + 4m$
13. $9 - x^2$
14. $t^3 - t^2 - t + 1$
15. $xy - x^2 + y^2 - xy$
16. $x^2 - 6x + 9 - 16y^2$
17. $x - x^2 + (1 - x)^2$
18. $10 - a - 3a^2$
19. $m^2n^2 - 10mn + 24$
20. $20x^3 - 5x^2 + 4x - 1$

HISTORICAL NOTE

The equal sign ($=$) was introduced by the English mathematician Robert Recorde (1510–1558) "bicause noe 2 thynges can be moare equalle." Recorde's mathematics books, which were very influential, were written as dialogues between master and student.

Quadratic Equations

A quadratic equation is an equation of the form

$$ax^2 + bx + c = 0$$

where a, b, and c are any real numbers with $a \neq 0$. We show how to find the solutions of this equation by factoring and by using the quadratic formula.

Example

Factoring Quadratic Equations   Solve the following equation for x:

$$x^2 - x - 12 = 0$$

Solution   Since the quadratic on the left-hand side of the equation can be factored:

$$x^2 - x - 12 = (x - 4)(x + 3)$$

we have

$$(x - 4)(x + 3) = 0$$

From this we can see that the two values of x that satisfy this equation are

$$x = \quad 4$$
$$x = -3$$

□

Example

An Investor and the Quadratic Equation   Suppose you deposit $100 in a bank that pays interest compounded semiannually. A year later (after two compoundings) the value of your deposit has increased to $115. What is the annual interest rate paid by the bank?

Solution   In general, if an amount P is deposited in a bank that pays an annual interest rate of r compounded semiannually, then after two compoundings (one year) the value F of the account will be

$$F = P(1 + r/2)^2$$

Hence we set

$$115 = 100(1 + r/2)^2$$

and solve for r. In this equation it is easiest first to take the square root of each side of the equation, getting

$$10(1 + r/2) = \sqrt{115}$$

Hence we have

$$1 + r/2 = \frac{\sqrt{115}}{10} = 1.072$$

$$\frac{r}{2} = 0.072$$

$$r = 0.144$$

In other words, the annual interest rate is 14.4%.   □

Example ——————————   Factoring a Cubic   Solve the cubic polynomial equation

$$x^3 - 2x^2 - 11x + 12 = 0$$

Solution   The cubic polynomial on the left-hand side of the equation can be written as

$$x^3 - 2x^2 - 11x + 12 = (x - 1)(x + 3)(x - 4)$$

In other words, any of the three numbers $x = 1, -3, 4$ will satisfy the cubic equation.   □

Some polynomials of degree two cannot be easily factored and hence are difficult to solve. When a quadratic equation cannot easily be factored, we generally use the quadratic formula to find the solution.

Quadratic Formula

The two solutions of the quadratic equation

$$ax^2 + bx + c = 0 \qquad (a \neq 0)$$

are given by the **quadratic formula**:

$$x = \frac{-b \pm \sqrt{b^2 - 4ac}}{2a}$$

If $b^2 - 4ac < 0$, the solutions are complex numbers. If $b^2 - 4ac = 0$, the quadratic equation is said to have a double root of $-b/2a$.

Example

Quadratic Formula   Solve the quadratic equation

$$x^2 + 5x - 2 = 0$$

Solution   Since the quadratic polynomial on the left-hand side of this equation is difficult to factor, we use the quadratic formula. Here, we have

$$x = \frac{-5 \pm \sqrt{5^2 - 4(1)(-2)}}{2(1)}$$

$$= \frac{-5 \pm \sqrt{25 + 8}}{2}$$

$$= \frac{-5 \pm \sqrt{33}}{2}$$

Evaluating these two numbers, we find

$$x \cong 0.372$$
$$x \cong 5.372 \qquad \square$$

HISTORICAL NOTE

The quadratic equation was solved over 2000 years ago by Babylonian mathematicians (although not in the form we know it today). The cubic equation was solved by the Italian mathematicians Scipione dal Ferro (1465?–1526) and Niccolò Tartaglia (1500?–1557) during the 1500s. The quartic equation (fourth degree) was solved in 1540 by the Italian mathematician Ludovico Ferrari (1522–1565). In 1832 the French mathematician Evariste Galois (1811–1832) proved that there exist fifth-order polynomial equations and higher-order polynomial equations, which cannot be solved in terms of the sums, differences, products, and quotients of the roots of their coefficients.

Problems

For Problems 1–12, solve by factoring.

1. $x^2 + 5x - 14 = 0$
2. $3x^2 + x - 10 = 0$
3. $3x^2 - 27 = 0$
4. $6x^2 - 4x - 10 = 0$
5. $16x^2 - 9 = 0$
6. $3x^2 + 13x = -14$
7. $10x^2 + 4x - 6 = 0$
8. $4x^2 - 12x + 9 = 0$
9. $\dfrac{x}{4} - \dfrac{x}{x+4} = \dfrac{x-1}{6}$
10. $\dfrac{3x-6}{x-1} = \dfrac{2-x}{3+x}$
11. $x^3 - 2x^2 - x + 2 = 0$
12. $x^4 - 6x^3 + 9x^2 = 0$

For Problems 13–22, find real solutions by using the quadratic formula.

13. $3x^2 + 2x - 10 = 0$
14. $5x^2 - 2x - 7 = 0$
15. $12y^2 - 17y + 6 = 0$
16. $4x^2 - 2x - 7 = 0$
17. $x^2 + 8x + 2 = 0$
18. $3x^2 + 2px + q = 0$
19. $x^2 - 4mx + m = 0$
20. $x^2 + x + 1 = 0$
21. $\dfrac{3}{x+1} + \dfrac{2}{x} - \dfrac{1}{1-x} = 0$
22. $\dfrac{4y^2}{y-1} - 5y + 6 = 0$

Algebraic Fractions

Algebraic fractions are quotients of algebraic expressions. Some examples are

$$\frac{1}{\sqrt{x}} \qquad \frac{2x + 1}{x} \qquad \frac{x^2 + 2x + 3}{x + 2}$$

The basic rules for manipulating algebraic fractions are the same as those for manipulating real numbers.

Basic Rules for Manipulating Fractions

If a, b, c, and d represent algebraic expressions, then (provided that the denominators are nonzero) the following rules hold:

1. $\dfrac{a}{b} + \dfrac{c}{d} = \dfrac{ad + bc}{bd}$ finding a common denominator

2. $\dfrac{a}{b} - \dfrac{c}{d} = \dfrac{ad - bc}{bd}$ finding a common denominator

3. $\dfrac{a}{b} \cdot \dfrac{c}{d} = \dfrac{ac}{bd}$ product of fractions

4. $\dfrac{\dfrac{a}{b}}{\dfrac{c}{d}} = \dfrac{ad}{bc}$ quotient of fractions (invert and multiply rule)

Example

The following equations illustrate the above rules:

$$\frac{1}{x} + \frac{2}{x + 1} = \frac{(x + 1) + 2x}{x(x + 1)} = \frac{3x + 1}{x(x + 1)}$$

$$\frac{x}{2x - 3} \cdot \frac{x + 1}{x + 5} = \frac{x(x + 1)}{(2x - 3)(x + 5)}$$

$$\frac{\dfrac{1}{x}}{\dfrac{2x + 5}{4x + 3}} = \frac{4x + 3}{x(2x + 5)}$$

The cancellation property is another important property used with algebraic fractions.

> **Cancellation of Common Factors**
>
> If a, b, and c are algebraic expressions, with $c \neq 0$, then
>
> $$\frac{ac}{bc} = \frac{a}{b}$$

Example

$$\frac{(2x + 1)(x^2 + 1)}{x(x^2 + 1)} = \frac{2x + 1}{x}$$

Problems

Rewrite the expressions in Problems 1–7 as a single fraction, and simplify by cancelling all common factors.

1. $\dfrac{1}{x} + \dfrac{x}{2}$

2. $\dfrac{1}{x - 1} + \dfrac{x + 1}{3}$

3. $\dfrac{1}{x - 1} - \dfrac{x}{x + 2}$

4. $\dfrac{x - 1}{x + 1} \cdot \dfrac{x}{x + 3}$

5. $\dfrac{(x - 1)(x + 2)}{(x^2 - 4)}$

6. $\dfrac{1}{x - 1} + \dfrac{1}{x + 1}$

7. $\dfrac{\sqrt{x} + 1}{1 - \sqrt{x}} + \dfrac{1}{x}$

8. Hmmmmm   Starting with the equation

$$\frac{x - 10}{7 - x} = \frac{x - 10}{13 - x}$$

we divide each side by $x - 10$ to get

$$\frac{1}{7 - x} = \frac{1}{13 - x}$$

Since the numerators are the same, we must have

$$7 - x = 13 - x$$

Adding x to each side then gives $7 = 13$. Can you discover what is wrong with this argument?

Rationalizing the Numerator and Denominator

It is often possible to eliminate square roots from either the numerator or denominator of a fraction by a process called **rationalization.**

Example

Rationalizing the Denominator   Eliminate the square roots from the denominator of

$$\frac{1}{\sqrt{x} - \sqrt{y}}$$

Solution   Multiply the numerator and the denominator of the fraction by the conjugate of the denominator (change the sign between the square roots) of the original expression:

$$\sqrt{x} + \sqrt{y} \quad \text{(conjugate expression)}$$

This gives

$$\frac{1}{\sqrt{x} - \sqrt{y}} \cdot \frac{\sqrt{x} + \sqrt{y}}{\sqrt{x} + \sqrt{y}} = \frac{\sqrt{x} + \sqrt{y}}{x - y} \qquad \square$$

Example

Rationalizing the Numerator Eliminate the square roots from the numerator of

$$\frac{\sqrt{x + h} - \sqrt{x}}{h}$$

Solution Multiplying both numerator and denominator by $\sqrt{x + h} + \sqrt{x}$, we have

$$\frac{\sqrt{x + h} - \sqrt{x}}{h} \cdot \frac{\sqrt{x + h} + \sqrt{x}}{\sqrt{x + h} + \sqrt{x}} = \frac{1}{\sqrt{x + h} + \sqrt{x}} \qquad \square$$

Problems

Rationalize the expressions in Problems 1–6.

1. $\dfrac{1}{\sqrt{x} + \sqrt{y}}$

2. $\dfrac{5}{\sqrt{z} - 2\sqrt{x}}$

3. $\dfrac{1}{\sqrt{x + a} - \sqrt{x}}$

4. $\dfrac{\sqrt{x - 3} + \sqrt{x + 3}}{\sqrt{x - 3} - \sqrt{x + 3}}$

5. $\dfrac{1}{\sqrt{x + 2} + \sqrt{x - 2}}$

6. $\dfrac{1}{x - \sqrt{y}}$

Nonlinear Inequalities

In the Finite Mathematics Preliminaries chapter we solved linear inequalities of the form

$$2x + 3 < 4$$

We now solve nonlinear inequalities, such as

$$x^2 + 2 < 3x$$

To solve this inequality, we first write the inequality with zero on the right-hand side

$$x^2 - 3x + 2 < 0$$

We then factor the left-hand side to get

$$(x - 2)(x - 1) < 0$$

For the expression on the left to be negative, we find those values of x for which the factors $x - 2$ and $x - 1$ have opposite sign. To do this, we draw the sign diagram shown in Table 2.

TABLE 2

Sign Analysis of $(x - 2)(x - 1)$

Sign of $x - 1$	$- - - - - - - \; 0 + + + + + + + + + + + + + +$
Sign of $x - 2$	$- - - - - - - - - - - - - - - - \; 0 + + + + + +$
Value of x	$\qquad\qquad\qquad 1 \qquad\qquad\quad 2$
	$\underline{- - - - - - - - +\!- - - - - - - - - +\!- - - -}$
Sign of $(x - 1)(x - 2)$	$+ + + + + + + \; 0 - - - - - - - - \; 0 + + + + + +$
	$- - - - - - - \qquad\quad - - - - - - - \qquad - - -$

Note that the sign of $(x - 1)(x - 2)$ is negative when $1 < x < 2$.

Problems

Solve the inequalities in Problems 1–7 by using a sign diagram.

1. $x^2 \leq 9$

2. $x^2 + 3x + 2 < 0$

3. $x^2 - 4 < 3x$

4. $x^2 < x$

5. $x^2 + 1 < 0$

6. $x^2 \leq 0$

7. $x^2 + 3x + 2 > 0$

Rational Inequalities

The sign diagram previously described can also be used to solve inequalities involving rational expressions (ratios of polynomials). We illustrate this below.

Example

Solve the rational inequality

$$\frac{x^2 - x - 4}{x - 1} \leq 1$$

Solution We might be tempted to multiply each side of the inequality by $x - 1$, as we do for equations. However, since $x - 1$ can be either positive or negative, depending on the value of x, it is best to use another method. First, however, we rewrite the inequality as a single term. A simple approach is to draw a sign diagram in which we illustrate the relevant factors in the above expression.

$$\frac{x^2 - x - 4}{x - 1} - 1 \leq 0 \qquad \text{(move all terms to left-hand side)}$$

$$\frac{x^2 - x - 4 - (x - 1)}{x - 1} \leq 0 \qquad \text{(write as one fraction)}$$

$$\frac{x^2 - 2x - 3}{x - 1} \leq 0 \qquad \text{(simplify)}$$

$$\frac{(x + 1)(x - 3)}{x - 1} \leq 0 \qquad \text{(factor the numerator)}$$

There are three factors in the above rational expression: $x + 1$, $x - 3$, and $x - 1$. The inequality holds when either

- one factor is negative (and two are positive) or
- all three factors are negative.

Do you understand this observation? Since the entire expression changes sign each time one of the three factors changes sign, we plot the points where the three factors are zero (these are the points where the factors change sign). Calling these points the **zero points**, we have

Zero points: $-1, 1, 3$

We now draw the sign diagram shown in Table 3 and observe where either one or three of the above three factors are negative. □

TABLE 3
Sign Analysis

Sign of $(x + 1)$	− − − − 0 + + + + + + + + + + + + + + + + + +
Sign of $(x - 3)$	− − − − − − − − − − − − − − − − − 0 + + + + + +
Sign of $(x - 1)$	− − − − − − − − − − − 0 + + + + + + + + + + + +
Value of x	−1 1 3
Sign of Fraction	− − − − 0 + + + + + + * − − − − − − 0 + + + + + +

Hence the values of x that satisfy the inequality are

Inequality notation: $x \leq -1$ or $1 < x \leq 3$

Interval notation: $(-\infty, -1] \cup (1, 3]$

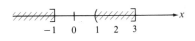

Figure 2
The union of two intervals

Note that the inequality holds at the two points -1 and 3 but not at the point $+1$, where the rational expression is not defined. The solution of the inequality is shown in Figure 2.

HISTORICAL NOTE

The symbols for inequalities (< and >) were introduced around 1600 by the English algebraist Thomas Harriot (1560–1621). It has been rumored that he got his inspiration by viewing some Indian decorations while he was on a surveying expedition to Virginia. More likely, however, they were the natural progeny of the notation for equality, =.

Problems

Solve the inequalities in Problems 1–5 for x.

1. $\dfrac{x^2 - 2x - 8}{x} < 0$

2. $\dfrac{(x + 2)(x - 3)}{x - 1} > 0$

3. $\dfrac{1}{x + 3} \leq \dfrac{1}{x - 2}$

4. $\dfrac{2}{x} + \dfrac{x}{2} < 2$

5. $\dfrac{x + 4}{x} < \dfrac{x}{2}$

The Circle

The circle is the most famous of all mathematical curves. It is the most symmetrical. It encloses the most area for its circumference. It was even thought at one time to be the shape of the path on which the planets moved about the sun.

Mathematically, the circle consists of all points in a plane that are at the same distance from a fixed point in the plane.

To find the equation that is satisfied by all points on a circle, remember that the distance between two arbitrary points (a, b) and (c, d) in the plane is

$$\sqrt{(a - c)^2 + (b - d)^2} \qquad \text{(Pythagorean theorem)}$$

Hence any point (x, y) on a circle of radius r, with center at (a, b), will satisfy the equation

$$r = \sqrt{(x - a)^2 + (y - b)^2}$$

Squaring each side of this equation gives the following result.

Equation of Circle with Center (a, b) and Radius r

All points (x, y) on a circle of radius r, centered at the point (a, b), will satisfy the equation

$$(x - a)^2 + (y - b)^2 = r^2$$

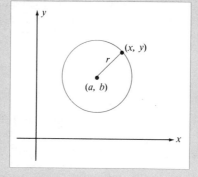

Example _____

Equation of a Circle The circle with center $(-1, -1)$ and radius 3 (see Figure 3) is described by the equation

$$(x + 1)^2 + (y + 1)^2 = 9$$

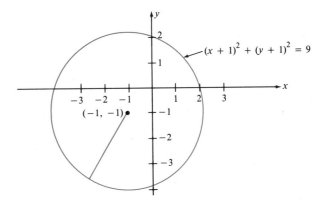

Figure 3
A typical circle

Completing the Square

A useful technique in graphing circles is an algebraic technique called **completing the square**. Although we could spend a page or two describing the method in general, it is best to simply illustrate how it works by means of an example.

Example _____

Completing the Square Write the quadratic expression

$$2x^2 + 2x + 5$$

in the form

$$2x^2 + 2x + 5 = a(x - h)^2 + k$$

Solution

- **Step 1** (factor out the coefficient of x^2). Factoring out the coefficient of x^2 from the terms involving x, we have

$$2x^2 + 2x + 5 = 2(x^2 + x) + 5$$

- **Step 2** (add and subtract). Add and subtract the square of one-half the coefficient of x to the right-hand side. Since the coefficient of x is 1, the square of one-half this number is $(1/2)^2 = 1/4$. Hence we add and subtract 1/4 (inside the parentheses); and carrying out a few algebraic steps, we get the desired result. We have

$$2x^2 + 2x + 5 = 2\left(x^2 + x + \frac{1}{4} - \frac{1}{4}\right) + 5$$

$$= 2\left(x^2 + x + \frac{1}{4}\right) - \frac{2}{4} + 5$$

$$= 2\left(x + \frac{1}{2}\right)^2 + \frac{9}{2} \qquad \square$$

We now show why this new form of a quadratic is useful.

Example _____

Completing the Square   Graph the equation

$$x^2 + 2x + y^2 - 4y + 1 = 0$$

Solution   Completing the square separately for the quadratic expressions

$$x^2 + 2x$$
$$y^2 - 4y$$

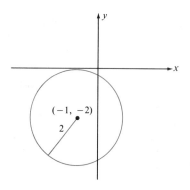

in x and y, we get

$$x^2 + 2x = x^2 + 2x + 1 - 1 = (x + 1)^2 - 1$$
$$y^2 - 4y = y^2 - 4y + 4 - 4 = (y + 2)^2 - 4$$

Hence the above equation can be written as

$$[(x + 1)^2 - 1] + [(y + 2)^2 - 4] + 1 = 0$$

or

$$(x + 1)^2 + (y + 2)^2 = 4$$

This is the equation of the circle centered at $(-1, -2)$ of radius 2, shown in Figure 4. $\square$

Figure 4
Circle whose equation is
$(x + 1)^2 + (y + 2)^2 = 4$

HISTORICAL NOTE

Many of the early discoveries in algebra were made by Arab mathematicians during the Arabian empire (600–1200 A.D.). The word "algebra" comes from the Arab word *al-jabr*. The word "zero" comes from the Latin word *zephirum*, which in turn comes from the Arab word *sifr*, which means "void" or "empty."

Figure 5
*All points on a parabola are
equidistant from a fixed point
and a fixed line*

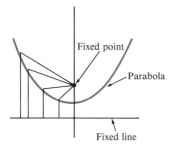

"I TEND TO AGREE WITH YOU — ESPECIALLY
SINCE $6 \cdot 10^{-4} \sqrt{c}$ IS MY LUCKY NUMBER."

The Parabola

When a baseball is hit into the air, it moves along a path called a **parabola**. In fact, Galileo Galilei first proved around 1632 that the path of an object thrown in space follows such a curve. Geometrically, a parabola is defined to be all points in a plane such that the distance of each point on the curve to a fixed point is the same as its distance to a fixed line. This is shown in Figure 5.

The equation of a parabola is given by

$$y = ax^2 + bx + c$$

where a, b, and c are constants such that $a \neq 0$.

Graph of $y = ax^2 + bx + c$

The graph of a parabola, defined by the quadratic function

$$y = ax^2 + bx + c \qquad (a \neq 0)$$

has its axis (line of symmetry) parallel to the y-axis. It opens upward if $a > 0$ and downward if $a < 0$. The point of intersection of the line of symmetry and the parabola is called the **vertex** of the parabola. The vertex will be the high point of the parabola if the parabola turns downward; it will be the low point of the parabola if the parabola turns upward. The x-coordinate of the vertex point is always given by $x = -b/(2a)$.

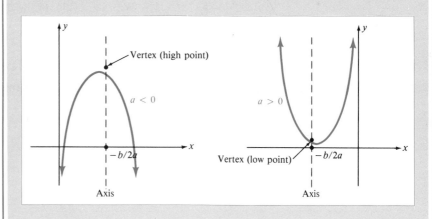

Example

Graph the parabola

$$y = x^2 - x - 2$$

and find the vertex point.

Solution   To sketch the graph of a parabola in the xy-plane, we evaluate y for different values of x. It is always convenient to first set x to zero and find the point where the graph crosses the y-axis. In this example the y-intercept is $(0, -2)$. It is also sometimes convenient to set y to zero and solve for x. If the resulting quadratic equation has two real solutions, these solutions will give the points where the parabola crosses the x-axis. If the quadratic equation has complex roots, the parabola will not cross the x-axis.

Table 4 gives the values of y for different values of x. Figure 6 shows the results of "filling in" between the points.

TABLE 4

Values for $y = x^2 - x - 2$

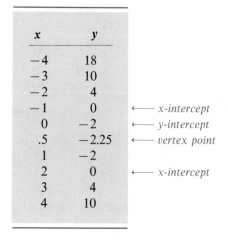

x	y	
-4	18	
-3	10	
-2	4	
-1	0	← *x-intercept*
0	-2	← *y-intercept*
.5	-2.25	← *vertex point*
1	-2	
2	0	← *x-intercept*
3	4	
4	10	

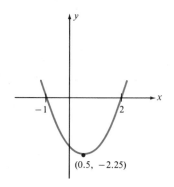

(0.5, −2.25)

Figure 6
Parabola $y = x^2 - x - 2$

Since $a = 1$ and $b = -1$, we find the x-coordinate of the vertex point to be $x = -b/(2a) = 1/2$. We then evaluate the quadratic function

$$y = x^2 - x - 2$$

at this point to find the vertex point $(0.5, -2.25)$. Later, when we study the derivative, we will be able to prove that this point is the "lowest" point of the parabola.   □

Problems

For Problems 1–19, graph the circles and parabolas described by the equations.

Circles

1. $x^2 + y^2 = 9$
2. $x^2 + (y - 2)^2 = 9$
3. $(x + 3)^2 + y^2 = 16$
4. $(x - 1)^2 + (y + 1)^2 = 25$
5. $(x + 5)^2 + (y + 5)^2 = 1$

6. $x^2 + y^2 - 8x + 2y + 8 = 0$ (complete the square)
7. $x^2 + y^2 - 2x - 2y - 7 = 0$ (complete the square)
8. $4x^2 + 4y^2 = 1$

Parabolas

9. $y = x^2$
10. $y = 2x^2$
11. $y = x^2 + 1$
12. $y = 2x^2 - 1$

13. $y = x^2 + x$

14. $y = 3x^2 + 2x - 3$ (try completing the square)

15. $y = 2x^2 + 4x + 2$ (try completing the square)

16. $y = -x^2$

17. $y = -2x^2 + 1$

18. $y = -x^2 - 2x + 3$ (try completing the square)

19. $y = -2x^2 + x$

Manipulation of Graphs

It is surprising how many different ways graphs of functions can be moved around simply by changing the function in a minor manner. By learning a number of little "tricks," you will be able to invert, shrink, expand, shift, and manipulate graphs very easily. Computer graphics systems often use rules such as these to manipulate pictures on a computer screen.

Rules for Manipulating Graphs

The following table lists some common manipulations.

Movement of the Graph	How to Change $f(x)$
Graph moves up or down	$f(x) \longrightarrow f(x) \pm h$
Graph moves to the left or right	$f(x) \longrightarrow f(x + h)$
Graph stretches or contracts in the horizontal direction	$f(x) \longrightarrow f(hx)$
Graph stretches or contracts in the vertical direction	$f(x) \longrightarrow hf(x)$
Graph is reflected through the y-axis	$f(x) \longrightarrow f(-x)$
Graph is reflected through the x-axis	$f(x) \longrightarrow -f(x)$
Graph is reflected through the origin, $(0, 0)$	$f(x) \longrightarrow -f(-x)$

We now illustrate the above movements by means of examples.

Moving Graph Up or Down

To move the graph $y = f(x)$ up or down, add or subtract a positive number to or from $f(x)$. (See Figure 7.)

Moving Graph Left or Right

To move the graph $y = f(x)$ to the left or right, replace x by

- $x - h$ for a shift h units to the right ($h > 0$),
- $x + h$ for a shift h units to the left ($h > 0$).

See Figure 8.

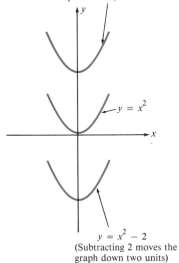

$y = x^2 + 2$
(Adding 2 moves the graph up two units)

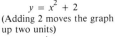

$y = x^2$

$y = x^2 - 2$
(Subtracting 2 moves the graph down two units)

Figure 7

The result of adding or subtracting a constant

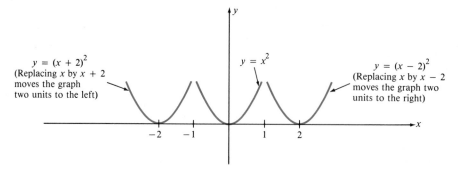

Figure 8
*The result of replacing x by $x + c$
or $x - c$*

Horizontal Stretching and Contracting

To stretch or contract the graph $y = f(x)$ in the horizontal direction, replace x in $f(x)$ by hx, where h is a positive number. If $h > 1$, the graph will be contracted toward the y-axis. If $h < 1$, the graph will stretch away from the y-axis. (See Figure 9.)

Figure 9
*Replacing x by hx in a function
stretches or contracts in the x
direction, depending on the size of
the constant*

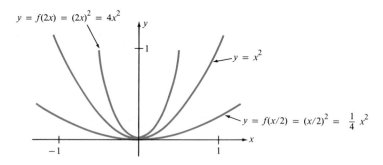

Vertical Stretching and Contracting

To stretch or contract the graph $y = f(x)$ in the vertical direction, replace $f(x)$ by $hf(x)$, where $h > 0$. (See Figure 10.)

Figure 10
*Multiplying a function by a
constant stretches or contracts the
graph in the y direction, depending
on the size of the constant*

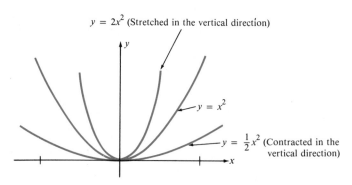

Reflecting Through the *y*-Axis

To reflect the graph $y = f(x)$ through the *y*-axis, replace x by $-x$. (See Figure 11.)

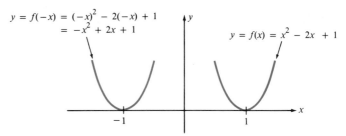

$$y = f(-x) = (-x)^2 - 2(-x) + 1$$
$$= -x^2 + 2x + 1$$

$$y = f(x) = x^2 - 2x + 1$$

Figure 11
Changing x to −x reflects the graph through the y-axis

Reflecting Through the *x*-Axis

To reflect the graph $y = f(x)$ through the *x*-axis, replace $f(x)$ by $-f(x)$. (See Figure 12.)

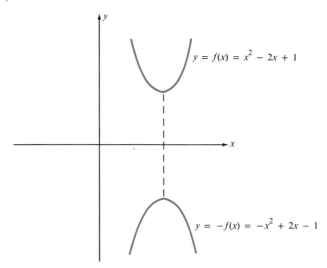

$$y = f(x) = x^2 - 2x + 1$$

$$y = -f(x) = -x^2 + 2x - 1$$

Figure 12
Changing the sign of a function reflects the graph through the x-axis

Reflecting Through the Origin

To reflect the graph $y = f(x)$ through the origin $(0, 0)$, replace $f(x)$ by $-f(-x)$. (See Figure 13.)

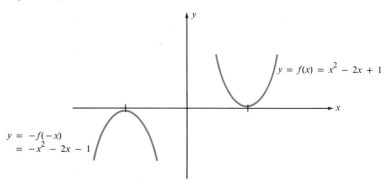

$$y = f(x) = x^2 - 2x + 1$$

$$y = -f(-x)$$
$$= -x^2 - 2x - 1$$

Figure 13
Changing x to −x and then changing the sign of the function reflects the graph through the origin

We now apply the above manipulations to the square root function.

Function Manipulations at Work

The graphs in Figure 14 represent manipulations of the square root function

$$f(x) = \sqrt{x}$$

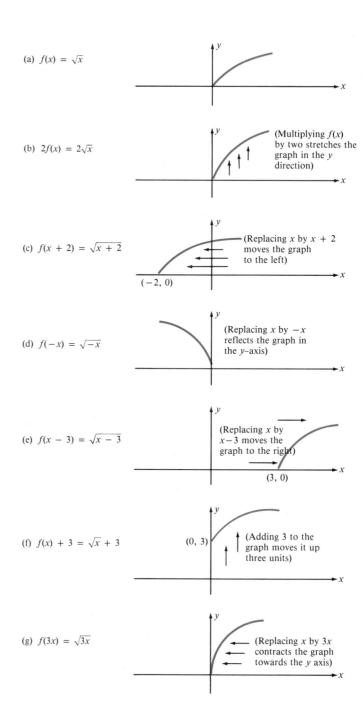

(a) $f(x) = \sqrt{x}$

(b) $2f(x) = 2\sqrt{x}$ — (Multiplying $f(x)$ by two stretches the graph in the y direction)

(c) $f(x + 2) = \sqrt{x + 2}$ — (Replacing x by $x + 2$ moves the graph to the left) — $(-2, 0)$

(d) $f(-x) = \sqrt{-x}$ — (Replacing x by $-x$ reflects the graph in the y-axis)

(e) $f(x - 3) = \sqrt{x - 3}$ — (Replacing x by $x - 3$ moves the graph to the right) — $(3, 0)$

(f) $f(x) + 3 = \sqrt{x} + 3$ — $(0, 3)$ — (Adding 3 to the graph moves it up three units)

(g) $f(3x) = \sqrt{3x}$ — (Replacing x by $3x$ contracts the graph towards the y axis)

Figure 14
Some manipulations of the square root function

Problems

For each of the operations in Problems 1–11, determine how to change the function $f(x) = x^2$ to bring about the indicated change. Sketch the graphs of both $f(x) = x^2$ and the manipulated function.

1. Move the parabola up three units.
2. Move the parabola down five units.
3. Move the parabola to the left two units.
4. Move the parabola to the right seven units.
5. Stretch the parabola in the horizontal direction by a factor of 2.
6. Contract the parabola in the horizontal direction by a factor of 3.
7. Stretch the parabola in the vertical direction by a factor of 5.
8. Shrink the parabola in the vertical direction by a factor of 4.
9. Reflect the parabola through the y-axis.
10. Reflect the parabola through the x-axis.
11. Reflect the parabola through the origin.

Composition of Functions

We often think of complicated things as being made up of several simpler things. In mathematics a "complicated" function can be interpreted as a composition of "simpler" functions. To make this idea more precise, consider the function defined by

$$h(x) = \sqrt{x^2 + 1}$$

where

$$f(x) = \sqrt{x}$$
$$g(x) = x^2 + 1$$

The function h can be interpreted as the **composition** of the function g and f. That is, the function h does the same thing as g and f together. The first function g takes a real number x and assigns to it the value $x^2 + 1$; the second function f takes this result $x^2 + 1$ and assigns the square root $\sqrt{x^2 + 1}$. But this team effort by g and f is exactly what the function h does. (See Figure 15.)

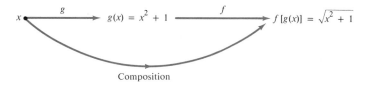

Figure 15
The composition of two functions

Another way of understanding compositions is in terms of the black box idea of a function, as illustrated in Figure 16. Here we consider two machines, working in sequence. We imagine a real number x being fed into the first machine (the "g machine"), which acts on this number and turns out a new number, called $g(x)$. This new number $g(x)$ is then fed into the second machine (the "f machine"), which in turn acts on this number and turns out another number,

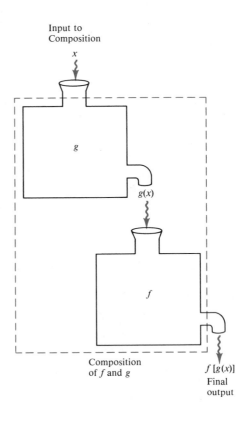

Input to
Composition

x

g

$g(x)$

f

Composition
of f and g

$f[g(x)]$
Final
output

Figure 16
Composition of functions as
"machines"

which we call $f[g(x)]$. These two machines acting in succession would be called
the composition of g followed by f.

Example _____   Composition of Functions   If the functions f, g, and h are defined by

$$f(x) = x^2 + 1 \qquad g(x) = \sqrt{x} \qquad h(x) = 1 + 2x$$

find the following compositions:

(a) $f[g(x)]$
(b) $h[g(x)]$
(c) $g[f(x)]$

Solution

(a) $f[g(x)] = [g(x)]^2 + 1$
$\qquad\qquad = [\sqrt{x}]^2 + 1$
$\qquad\qquad = x + 1$

(b) $h[g(x)] = 1 + 2g(x)$
$\qquad\qquad = 1 + 2\sqrt{x}$

(c) $g[f(x)] = \sqrt{f(x)}$
$\qquad\qquad = \sqrt{x^2 + 1}$ □

Problems

Find the composition of functions in Problems 1–6 if

$$f(x) = 3x \quad g(x) = x^2 + 2x + 1 \quad h(x) = 1 + \frac{1}{x}$$

1. $f[g(x)]$
2. $f[h(x)]$
3. $g[f(x)]$
4. $g[h(x)]$
5. $h[f(x)]$
6. $h[g(x)]$

Calculus Practice Test

The following test can be used by students to determine whether they need additional work in the Calculus Preliminaries. Before studying calculus, it is advisable for all students to spend some time and take this test. There are 11 well-chosen questions. Experience has shown that students should use the guidelines in Table 5.

TABLE 5
Guidelines for Calculus Practice Test

Correct Answers	Review Strategy
11	Excellent, no review necessary
10–9	Good, just a little touch-up needed
8–7	Average, review the weak sections
6–5	Advisable to review, quick review
4–0	Necessary to review, detailed review

You should also talk with your professor about your weaknesses.

1. Evaluate the roots (if they are defined) of the following expressions.
   (a) $16^{1/4}$
   (b) $(1/64)^{1/5}$
   (c) $(-8)^{2/3}$
   (d) $(-16)^{1/4}$
   (e) $(1/27)^{-2/3}$
2. Simplify the following expressions.
   (a) $\sqrt{16x^2}$
   (b) $\sqrt{a^2b^2} + \sqrt{c^2}$
   (c) $\sqrt[3]{8x^3y^6}$

3. Perform the following addition and multiplication of polynomials.
   (a) $(x^2 + 2x - 1) + (x^3 - 3x^2 + 4x - 3)$
   (b) $(x^2 + 3x)(2x - 1)$
4. Factor the following polynomials as the product of first-order polynomials.
   (a) $x^2 - 9$
   (b) $x^2 - x - 12$
5. Find the roots of the equation
   $$x^2 + 5x - 2 = 0$$
6. Rewrite the expression
   $$\frac{x-1}{x+1} - \frac{x}{x+3}$$
   as a single fraction and simplify as much as possible.
7. Eliminate the square roots from the denominator of
   $$\frac{1}{\sqrt{x} - \sqrt{y}}$$
   by rationalization.
8. Solve the inequality $2x + 3 < 4$.
9. Determine the center and radius of the circle defined by
   $$(x + 3)^2 + y^2 = 16$$
10. Find the center and radius of the circle
    $$x^2 + 2x + y^2 - 4y + 1 = 0$$
    by completing the square.
11. If $f(x) = x^2$ and $g(x) = 2x + 3$, find
    (a) $f[g(x)]$
    (b) $g[f(x)]$

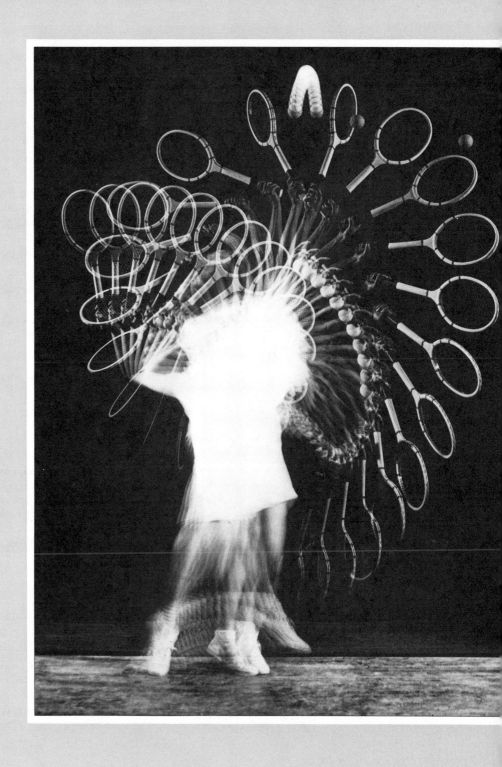

8

The Derivative

Many physical phenomena involve changing quantities. Some examples are the change in economic variables such as interest rates and profit curves, the change in biological populations such as virus and bacteria populations, and the change in physical variables such as the speed of an airplane or automobile. In this chapter we introduce the mathematical tools that are necessary to understand precisely the concept of the rate of change of a quantity.

8.1

An Historical Look at Calculus

PURPOSE

We present an historical overview of the calculus and give a glimpse into the lives of the individuals who were most responsible for making this monumental discovery. We will also introduce the two problems that were partly responsible for motivating the development of the calculus, the tangent problem and the problem of areas. It was the problem of tangents that motivated the study of the derivative (one of the two major areas of the calculus) and the area problem that motivated the study of the integral (the other major area of the calculus).

Introduction

Calculus is many things to many people. Economists often think of calculus in terms of mathematical models for forecasting discount rates, inflation, or the national debt. To an electrical engineer, calculus might represent a body of knowledge that will help to understand the circuitry of a computer. Today, in fact, practically every area of the natural and social sciences uses calculus in some form or another. We will show how calculus is used to describe phenomena in the modern world. Our first goal, however, is to outline its historical development.

Two Major Areas of Calculus

The most basic classification in calculus is the one shown in Figure 1. One of the two major subareas of the calculus is the study of *differential calculus* and the *derivative*. The other major subarea is the study of the *integral calculus* and the *integral*.

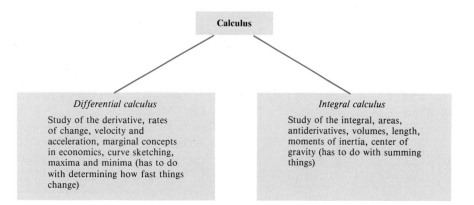

Figure 1
The two major areas of the calculus

The **derivative** is one of the most basic concepts in all of mathematics. It is a mathematical quantity that describes the rate of change of one quantity with respect to another. It is used today to study the motion of objects in space, marginal revenue in economics, the growth of biological populations, and a host of other phenomena.

On the other hand, the **integral** can be interpreted as a process of summing. One use of the integral is to represent the area enclosed by a curve.

Two Historic Problems of Calculus

During the 1600's, two major problems occupied the attention of many leading mathematicians. They were called the **tangent problem** and the **area problem**. We restate them here.

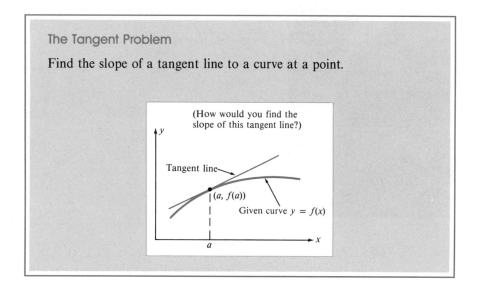

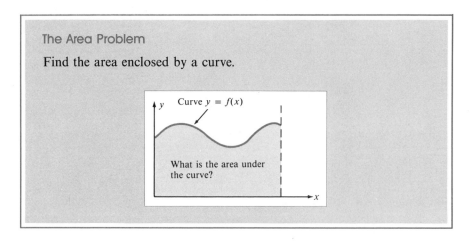

The Area Problem

Find the area enclosed by a curve.

The problems above were solved in the late 1600's by the English mathematician Sir Isaac Newton and the German scholar Gottfried Wilhelm von Leibniz. In addition to solving these problems they developed an extensive body of knowledge that we now call the **calculus**. We now give a glimpse into the lives of these brilliant individuals.

HISTORICAL NOTE

Sir Isaac Newton (1642–1727) was born on Christmas day in 1642 in Woolsthorp, England. As a youth, he showed no exceptional mathematical ability. In 1661 he entered Trinity College at Cambridge University. At Cambridge, Newton was stimulated by a very creative professor, Isaac Barrow (who in fact developed much of the calculus himself). After leaving Cambridge in 1665, Newton returned home and initiated his great works on mechanics, mathematics, and optics. It was in the years 1665–1666, during the great plague of London, that Newton developed his ideas for the calculus. Isaac Barrow, knowing of Newton's towering brilliance, resigned his prestigious position as Lucasian Professor of Mathematics in favor of the young Newton.

Newton was shy about publishing his discoveries, afraid of the criticism that his novel ideas might cause. In fact, when Newton did publish his results on optics in 1672, his theories were criticized by many scientists, including Robert Hooke and Christiaan Huygens. After that experience, Newton decided never to publish again. It is now known that the integral and derivative were developed by Newton during the years 1665–1670 but were not published until the years 1710–1735.

The other mathematician who was responsible for the development of the calculus was Gottfried Wilhelm von Leibniz.

HISTORICAL NOTE

Gottfried Wilhelm von Leibniz (1646–1716) was four years younger than Newton and was born in Leipzig, Germany. Unlike Newton, however, Leibniz was a child prodigy, having mastered Greek and Latin as a young boy. By the time he was twenty, he had mastered mathematics, philosophy, theology, and law. In contrast to Newton, Leibniz never held an academic chair at a university. His life was spent in the diplomatic service, which gave him plenty of time to devote to scholarly work. Most of his creative work in coinventing the calculus was carried out during the years 1682–1692 and appeared in the mathematical journal *Acta Eruditorum*, of which he was the editor. This journal had wide circulation among the leading scholars of Europe, and it was here that the outside world first learned of the calculus.

There is, of course, much more to the story of these two men. It seemed that Newton's followers in England claimed that Leibniz had plagiarized Newton's earlier works. Leibniz too had his own admirers in Europe (Huygens, L'Hôpital, and the Bernoulli brothers), and the fight was on. The controversy grew and grew until many scholars in England and Europe refused to talk to one another.

An interesting point in the development of the calculus was that Newton's approach turned out to be superior to the manner in which Leibniz developed it. Newton introduced a concept known as the *limit*, while Leibniz used another device called the *infinitesimal*. The limit concept won out, and limits are what are used in this book.

It appeared that the infinitesimal concept was dead, but in 1961 the mathematician Abraham Robinson showed how the infinitesimal approach could be used in a much better manner than Leibniz originally imagined. Hence in the last 30 years the infinitesimal has made something of a comeback. In fact there are a few new calculus books that use this "infinitesimals over limits" approach. One is *Infinitesimal Calculus* by James M. Henle and Eugene M. Klienberg (MIT Press, 1979).

8.2 Limits and Continuity

PURPOSE

We introduce the fundamental concept of the calculus, the limit. It is by means of limits that the derivative and integral are defined. Using the limit, we introduce the concept of continuity and continuous functions.

Introduction

If there is one basic concept that might be called the essence of calculus, it is the limit. In later chapters the limit will be used to define the two basic tools of calculus, the derivative and the integral. Indeed, the basic idea of infinity, of which mathematicians are the academic custodians, can be described in terms of limits. Of course, other disciplines are also aware of the infinite. Theologians have been known to say, "Infinity is the hand of God," while one poet has written:

> *Infinity, it ebbs and flows with the sea*
> *It takes and gives from you and me.*
> *But in the end it's all for naught*
> *The ebbing and flowing of the sea.*

When mathematicians speak of infinity and limits, the prose is generally not so lush. Mathematicians do not speak in such superlatives as do poets and theologians. Mathematicians only claim is that they speak of infinity in a precise manner. This is where the limit comes in.

Limits of Functions

One dictionary defines a limit as something that can be "approached but never reached." One wonders exactly what type of thing this would be. It would seem that if something could be approached, it could be reached. In mathematics

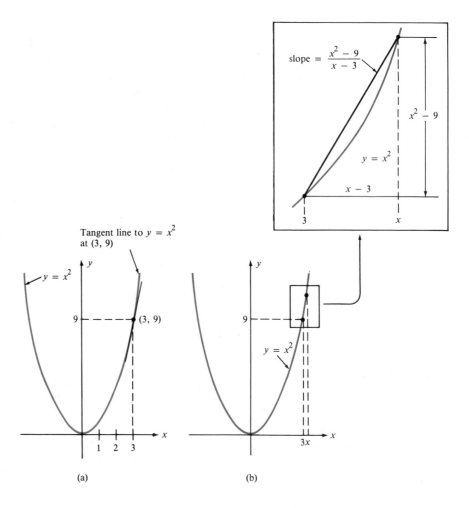

Figure 2
Finding the slope of the secant line

there are two major reasons why things cannot be reached. The reason is that these things are located either at infinity or at a point where division by zero is attempted.

To illustrate this idea, consider the important problem in calculus of finding the slope of a tangent line to a curve. Intuitively, a tangent line to a curve is a straight line that just grazes a curve at a point. The tangent line to the graph of $f(x) = x^2$ at the point $(3, 9)$ is drawn in Figure 2a.

The calculus approach for finding the slope of this tangent line is first to find the slope of the straight line that passes through the two points $(3, 9)$ and some nearby point (x, x^2) on the graph. This straight line is called a **secant line**. Since the slope of a straight line that passes through two points (x_1, y_1) and (x_2, y_2) is

$$m = \frac{y_2 - y_1}{x_2 - x_1} \qquad x_1 \neq x_2$$

the slope of the secant line passing through $(3, 9)$ and (x, x^2) is

$$m = \frac{x^2 - 9}{x - 3}$$

See Figure 2b.

The second step in finding the slope of the tangent line is more difficult. If we let $x = 3$ in the above equation for m, we discover that the expression for m has the form $0/0$, which is undefined in mathematics. In other words, the number 3 is the "something" we referred to earlier that can be approached but not reached. Table 1 shows the slopes of the secant lines for values of x close to 3.

TABLE 1

Slopes of Secant Lines Near
$x = 3$

x	Slope of Secant Lines
2.5	5.5
2.7	5.7
2.9	5.9
2.95	5.95
2.99	5.99
2.999	5.999
Can only approach → **3.000**	**limit**
3.001	6.001
3.01	6.01
3.05	6.05
3.1	6.1
3.3	6.3
3.5	6.5

Note that although we cannot let $x = 3$, the slopes of the secant lines seem to approach 6 as x approaches 3. We say that 6 is the limit of the slope of the secant lines passing through $(3, f(3))$ and $(x, f(x))$ as x approaches 3. We write this as

$$\lim_{x \to 3} \frac{x^2 - 9}{x - 3} = 6$$

In this example the number 6 will represent the slope of the tangent line to the graph $f(x) = x^2$ at the point $(3, 9)$.

The above discussion motivates the following definition of the limit of a function.

Definition of the Limit of a Function

If the value of $f(x)$ approaches a number L as x approaches a number a from the left side, and if $f(x)$ also approaches L as x approaches a from the right side, we say that L is the **limit** of $f(x)$ at $x = a$. We denote the limit as

$$\lim_{x \to a} f(x) = L$$

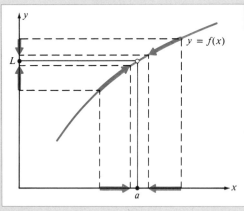

Note that the *limit* of $f(x)$ at $x = a$ is not determined by the *value* of $f(x)$ at $x = a$. In fact, $f(x)$ may not even be defined when $x = a$.

Example 1

Limit of a Function   Find the limit

$$\lim_{x \to 2} \frac{x^3 - x^2 + x - 6}{x - 2}$$

Solution   First note that this rational function is not defined when $x = 2$, since the denominator is 0. However, we can use a calculator and evaluate the function for values of x approaching 2, as shown in Table 2. From this table we conclude intuitively that the limit is 9. We write this as

$$\lim_{x \to 2} \frac{x^3 - x^2 + x - 6}{x - 2} = 9 \qquad \square$$

x	$f(x)$
1.95	8.75250
1.99	8.95010
1.999	8.99464
1.9999	8.99761
2.0000	**limit**
2.0001	9.00239
2.001	9.00477
2.01	9.05019
2.05	9.25251

Can only approach ⟶ (pointing to **2.0000**)

Example 2

Limit of a Function **Find the limit**

$$\lim_{x \to 0} \frac{x}{\sqrt{x+9}-3}$$

Solution   The function has no meaning at $x = 0$, since the denominator is 0 when $x = 0$ (again, so is the numerator). Using a calculator, we evaluate the function at points closer and closer to zero. (See Table 3.)

x	$f(x)$
-0.5	5.915476
-0.1	5.983287
-0.01	5.998333
-0.001	5.999833
-0.0001	5.999952
0	**limit**
0.0001	6.000024
0.001	6.000168
0.01	6.001667
0.1	6.016621
0.5	6.082207

Can only approach ⟶ (pointing to **0**)

We can see that although the function is not defined at 0, the value of $f(x)$ gets closer and closer to 6 as x gets closer and closer to 0. Hence the limit is 6, which we write as

$$\lim_{x \to 0} \frac{x}{\sqrt{x+9}-3} = 6$$

The graph of this function in the neighborhood of 0 is shown in Figure 3.   □

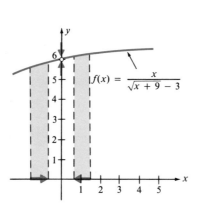

Figure 3

Illustration of $\lim\limits_{x \to 0} \dfrac{x}{\sqrt{x + 9} - 3}$

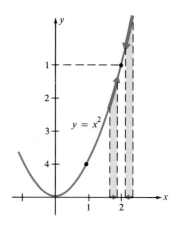

Figure 4

Illustration of $\lim\limits_{x \to 2} x^2$

Example 3

Limit of a Function   Find the limit

$$\lim_{x \to 2} x^2$$

Solution   If it is easy to graph the function, this should always be done. In this case the graph of $y = x^2$ is shown in Figure 4. We can see that x^2 approaches 4 as x approaches 2. Hence we have

$$\lim_{x \to 2} x^2 = 4$$

In this example the value of the function x^2 at 2 is the same as the limit of x^2 as x approaches 2. This is not always true, but it is true in some cases. Later, we will see that this is true when a function is continuous at a point.   □

Example 4

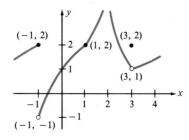

Figure 5
Finding limits

Limits of Functions   Find the following limits for the function f graphed in Figure 5.

(a) $\lim\limits_{x \to -1} f(x)$

(b) $\lim\limits_{x \to 1} f(x)$

(c) $\lim\limits_{x \to 2} f(x)$

(d) $\lim\limits_{x \to 3} f(x)$

Solution

(a)   The limit

$$\lim_{x \to -1} f(x)$$

does not exist, since $f(x)$ does not approach a unique value as x approaches -1 from different sides. In this case, $f(x)$ approaches 2 as x approaches -1 from the left side and -1 as x approaches -1 from the right side. You should also be aware that $f(-1) = 2$ has nothing to do with the limit at -1.

(b) We have

$$\lim_{x \to 1} f(x) = 2$$

since $f(x)$ approaches 2 as x approaches 1 from both the left and right side of 1.

(c) The limit

$$\lim_{x \to 2} f(x)$$

does not exist, since $f(x)$ does not approach any finite value as x approaches 2 from either side.

(d) As x approaches 3, we have the limit

$$\lim_{x \to 3} f(x) = 1 \qquad \square$$

Example 5 ────────────   Limits of Rational Functions   Find the limit

$$\lim_{x \to 3} \frac{1}{x - 3}$$

Solution   We evaluate the function for x close to 3. (See Table 4.)

TABLE 4 ────────────
Values of $f(x)$ Close to 3

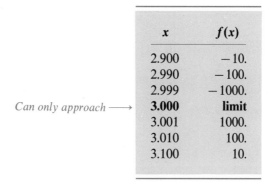

x	$f(x)$
2.900	$-10.$
2.990	$-100.$
2.999	$-1000.$
3.000	**limit**
3.001	1000.
3.010	100.
3.100	10.

Can only approach ⟶ (at row 3.000)

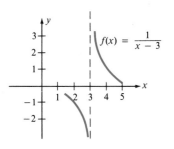

Figure 6
No limit at $x = 3$

It is clear that the values of $f(x)$ do not approach any limiting value as x approaches 3. This can also be seen by looking at the graph of $f(x)$ in Figure 6. $\qquad \square$

The following rules for limits are often helpful in the evaluation of complicated limits.

Rules for Limits

If f and g are functions such that the limits

$$\lim_{x \to a} f(x) \qquad \lim_{x \to a} g(x)$$

exist then the following rules for limits hold.

- The limit of a sum is the sum of the limits:

$$\lim_{x \to a} [f(x) + g(x)] = \lim_{x \to a} f(x) + \lim_{x \to a} g(x)$$

- The limit of a difference is the difference of the limits:

$$\lim_{x \to a} [f(x) - g(x)] = \lim_{x \to a} f(x) - \lim_{x \to a} g(x)$$

- The limit of a product is the product of the limits:

$$\lim_{x \to a} [f(x)g(x)] = \lim_{x \to a} f(x) \cdot \lim_{x \to a} g(x)$$

- The limit of a quotient is the quotient of the limits:

$$\lim_{x \to a} \frac{f(x)}{g(x)} = \frac{\lim_{x \to a} f(x)}{\lim_{x \to a} g(x)} \qquad \text{provided that} \quad \lim_{x \to a} g(x) \neq 0$$

- The limit of a root is the root of the limit:

$$\lim_{x \to a} \sqrt[n]{f(x)} = \sqrt[n]{\lim_{x \to a} f(x)} \qquad \text{provided} \ \sqrt[n]{\lim_{x \to a} f(x)} \ \text{exists}$$

- The limit of a power is the power of the limit:

$$\lim_{x \to a} [f(x)]^n = \left[\lim_{x \to a} f(x) \right]^n$$

Two other useful limits are as follows:
- The limit of a constant is the constant

$$\lim_{x \to a} c = c \qquad \text{for any constant } c$$

- The limit of x at a is a

$$\lim_{x \to a} x = a$$

Example 6 _____ Limit of a Polynomial   Find the limit

$$\lim_{x \to 2} (3x^2 - 4x + 1)$$

Solution First, using the rules that the limit of a sum or difference is the sum or difference of the limits, we write

$$\lim_{x \to 2} (3x^2 - 4x + 1) = \lim_{x \to 2} (3x^2) - \lim_{x \to 2} (4x) + \lim_{x \to 2} (1)$$

$$= 3 \lim_{x \to 2} x^2 - 4 \lim_{x \to 2} (x) + \lim_{x \to 2} (1)$$

$$= 3(4) - 4(2) + (1)$$

$$= 5 \qquad \square$$

Example 7

Limit of a Root Find the limit

$$\lim_{x \to 2} \sqrt{x^2 - 1}$$

Solution Using the rule that the limit of a root is the root of a limit, we have

$$\lim_{x \to 2} \sqrt{x^2 - 1} = \sqrt{\lim_{x \to 2} (x^2 - 1)}$$

$$= \sqrt{\lim_{x \to 2} (x^2) - \lim_{x \to 2} (1)}$$

$$= \sqrt{4 - 1}$$

$$= \sqrt{3} \qquad \square$$

Comment on the Above Limit

Note that in finding the above limit, we used the fact that

$$\lim_{x \to 2} (x^2 - 1) > 0$$

Example 8

Limit of a Product Find the limit of the product

$$\lim_{x \to 2} (x\sqrt{x^2 - 1})$$

Solution Using the rule that the limit of a product is the product of the limits, we have

$$\lim_{x \to 2} (x\sqrt{x^2 - 1}) = \lim_{x \to 2} (x) \cdot \lim_{x \to 2} \sqrt{x^2 - 1}$$

$$= 2\sqrt{3} \qquad \square$$

Example 9

Limit of a Quotient Find the limit of the quotient

$$\lim_{x \to 4} \frac{x^2 + 3x + 1}{\sqrt{x} - 1}$$

Solution   Using the rule that the limit of a quotient is the quotient of the limits, we can write

$$\lim_{x \to 4} \frac{x^2 + 3x + 1}{\sqrt{x} - 1} = \frac{\lim_{x \to 4} (x^2 + 3x + 1)}{\lim_{x \to 4} (\sqrt{x} - 1)}$$

$$= \frac{29}{1} \qquad \square$$

Comment on the Above Limit

Note that in finding the above limit we used the fact that

$$\lim_{x \to 4} (\sqrt{x} - 1) \neq 0$$

Example 10 _____

Limit of a Power   Find the limit

$$\lim_{x \to -1} (x + 3)^5$$

Solution   Using the fact that the limit of a power is the power of a limit, we have

$$\lim_{x \to -1} (x + 3)^5 = \left[\lim_{x \to -1} (x + 3) \right]^5$$

$$= [2]^5$$

$$= 32 \qquad \square$$

The Concept of Continuity

Most of the functions graphed thus far could be drawn without lifting the pencil from the paper. Intuitively, a function f is continuous at a point $x = a$ if its graph passes through the point $(a, f(a))$ without a break. Figure 7 shows the graphs of continuous functions from business management and science.

We now define the concept of continuity in terms of the limit.

Definition of Continuity

A function f is **continuous** at a point $x = a$ if

$$\lim_{x \to a} f(x) = f(a)$$

If f is not continuous at $x = a$, we say that it is **discontinuous** or has a **discontinuity** at $x = a$.

    If a function f is continuous at all points on an open interval (a, b), we simply say the function is continuous on that interval.

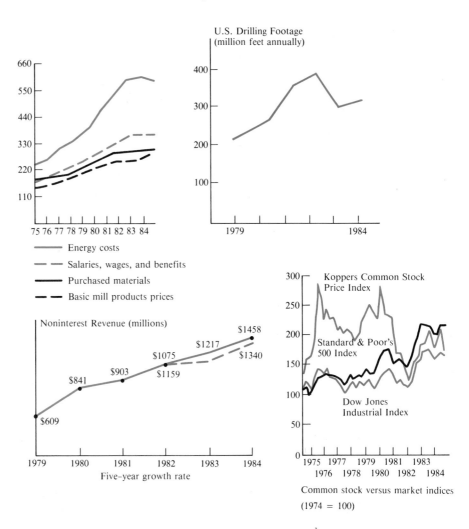

Figure 7
Some continuous functions in the business world

The above definition of continuity at $x = a$ implies that if a function is continuous at $x = a$, then the function f is defined at $x = a$ and that

$$\lim_{x \to a} f(x)$$

exists.

Figure 8 shows four different ways that a function can fail to be continuous at a point. All four of these functions are discontinuous at $x = 0$. The functions are, however, continuous at all other points.

Example 11 ――――――

Continuity on an Interval   The graph in Figure 9 of $N(t)$ shows the number of cats that lived in one of the author's home over the past ten years. Determine whether $N(t)$ is continuous on the following intervals.

(a) (0, 2)

(b) (4, 7)

(c) (7, 9)

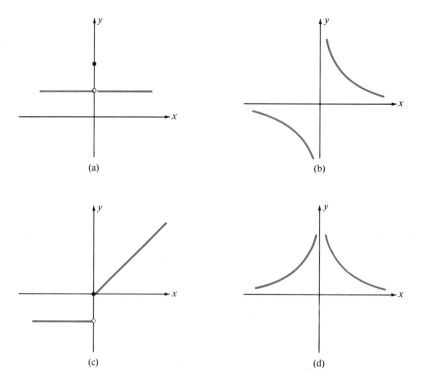

Figure 8
All functions discontinuous at
$x = 0$

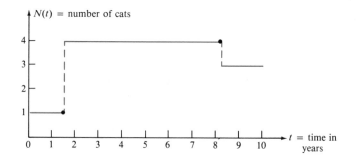

Figure 9
The cat function

Solution

(a) Interval $(0, 2)$: The "cat function" $N(t)$ is not continuous on $(0, 2)$, since the function had a jump discontinuity at $t = 1.5$ years (Daisy had kittens).

(b) Interval $(4, 7)$: The function $N(t)$ is constant and hence continuous on the open interval $(4, 7)$.

(c) Interval $(7, 9)$: The jump discontinuity at $t = 8.2$ means that $N(t)$ is not continuous on the interval $(7, 9)$.   ☐

Constructing New Continuous Functions

Earlier, we said that the limit of a sum of functions was the sum of the limits of the functions. This property can be used to show that the sum of two continuous functions is also continuous. In the same way, the other limit properties can be used to show that the difference, product, and quotient of continuous functions are continuous. We now present an important result stating that polynomials and rational functions (when defined) are continuous functions.

Continuity of Polynomials and Rational Functions

Polynomials of the form

$$p(x) = a_0 x^{n+1} + a_1 x^n + \cdots + a_n x + a_{n+1}$$

are continuous for all values of x.

Rational functions of the form

$$r(x) = \frac{p(x)}{q(x)}$$

where $p(x)$ and $q(x)$ are polynomials are continuous for all values of x except for those x where $q(x) = 0$.

Example 12

Continuity of Polynomials and Rational Functions Where is each of the following two functions continuous?

(a) $p(x) = x^4 - 3x^3 + 2x^2 + 6x + 7$

(b) $q(x) = \dfrac{x^3 + 3x^2 + 1}{x^2 - 1}$

Solution

(a) The polynomial $p(x)$ is continuous for all values of x.

(b) The rational function $q(x)$ is continuous for all values of x except when x is equal to 1 or -1 where the denominator is equal to 0.　□

Continuity in the Real World

The reason for our interest in continuity and continuous functions is that many phenomena in the real world change or move continuously. The height of a ball thrown into the air is a continuous function of time. Temperature, air pressure, electric current, light intensity, and many other physical phenomena change continuously with time. In fact, you might find it difficult to think of a physical phenomenon that does not change continuously with time.

In the business world, although most phenomena such as interest rates and a company's profit change by jumps, it is possible to approximate these changes by continuous models. For instance, although interest rates change by tenths (9.5% and 10.3% are typical interest rates), it is convenient to study continuous interest rate curves. Other curves such as the number of unemployed persons, profit curves, and cost curves also change in jumps, but for simplicity they are often described by such continuous functions as polynomials.

In biology, too, population growth changes in discrete jumps but is generally described by continuous models when populations are large.

Discontinuity in the Real World

Although many phenomena are modeled by continuous functions, several types of problems must be described by discontinuous functions. Figure 10 shows some discontinuous phenomena.

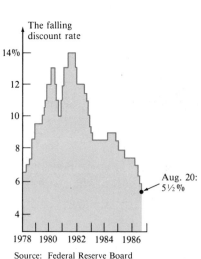

Source: Federal Reserve Board

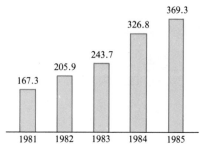

Net Interest Income

Net interest income, the corporation's principal source of earnings, represents the difference between interest and fees received on earning assets and the interest expense associated with the funding sources utilized to support them.

Net interest income (fully taxable equivalent basis – millions of dollars)

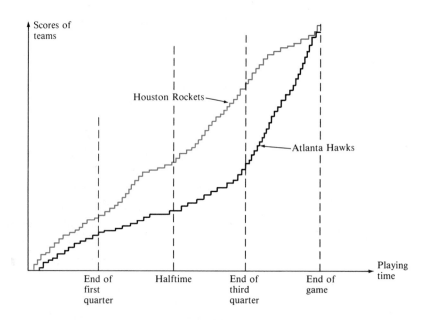

Figure 10
Discontinuous phenomena

Problems

For Problems 1–11, find the limits. If possible, graph the function.

1. $\lim\limits_{x \to 2} 3x$

2. $\lim\limits_{x \to 1} (3x^2 + 1)$

3. $\lim\limits_{x \to 4} \dfrac{x - 2}{x + 2}$

4. $\lim\limits_{x \to -3} \dfrac{x^2 - 9}{x^2 + 9}$

5. $\lim\limits_{x \to 2} \sqrt{9 - x^2}$

6. $\lim\limits_{x \to 0} (x^2 + 2x + 1)$

7. $\lim_{x \to 2} \dfrac{x - 2}{x^2 - 4}$

8. $\lim_{x \to 3} \dfrac{x^3 - 27}{x^2 - 9}$

9. $\lim_{x \to 3} \dfrac{2}{x + 1}$

10. $\lim_{x \to 3} (x^2 \sqrt{x^2 + 7})$

11. $\lim_{x \to 3} \sqrt[3]{x^2 - 1}$

For Problems 12–17, find the indicated quantities if they exist. Use Figure 11.

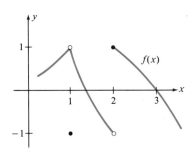

Figure 11
Graph for Problems 12–17

12. $\lim_{x \to 1} f(x)$

13. $\lim_{x \to 2} f(x)$

14. $\lim_{x \to 3} f(x)$

15. $f(1)$

16. $f(2)$

17. $f(3)$

For Problems 18–21, find the limits by evaluating (on a calculator) the function at values of x close to the limiting value of x. For problems in which you can rationalize the denominator, find the limit by the rationalizing technique to check the calculator solution. You can review the process of rationalizing the denominator in the Calculus Preliminaries.

18. $\lim_{x \to 0} \dfrac{x}{\sqrt{x + 4} - 2}$

19. $\lim_{x \to 0} \dfrac{x}{\sqrt{x + 9} - 3}$

20. $\lim_{x \to 0} \dfrac{x}{\sqrt{x + 16} - 4}$

21. $\lim_{x \to 0} \dfrac{x}{\sqrt{x + 25} - 5}$

For Problems 22–24, find the following limits. Indicate the limit rules used at each step.

22. $\lim_{x \to 3} \left[\dfrac{1}{x + 7} - \dfrac{2x}{3x - 2} \right]$

23. $\lim_{t \to 4} \dfrac{2t + 4}{t - 3}$

24. $\lim_{x \to 1} (2x^3 + 4x + 3)$

For Problems 25–31, tell whether the given phenomenon represents a continuous or discontinuous function of time. If the function is discontinuous, tell where the function fails to be continuous.

25. The total score of a basketball game.

26. The number of people infected with a given type of bacteria.

27. The temperature during the day.

28. The amount of inventory of a given product.

29. The daily revenue of a company.

30. Your weight throughout your life.

31. A company's annual profits.

For Problems 32–42, determine the points for which the functions are continuous. Also find any points where the functions are discontinuous.

32. $f(x) = x + 3$

33. $f(x) = x^3 + 3x^2 - 2x + 1$

34. $f(x) = x^5 + x^4 + 3x - 1$

35. $f(x) = \dfrac{x + 1}{x^2 + 1}$

36. $f(x) = \dfrac{2x^2 + 5x + 1}{x^2 - 1}$

37. $f(x) = \dfrac{x^2 + 1}{x + 1}$

38. $f(x) = (x^2 + x + 1)(x^3 - x^2 + 4)$

39. $f(x) = (x^2 + 2)(x^5 - x^3 + x + 1)$

40. $f(x) = (x^3 + 1)\left(\dfrac{x^2 + x + 6}{x^2 + 1} \right)$

41. $f(x) = \dfrac{1}{x + 1} \cdot \dfrac{2x^5 + 1}{x - 3}$

42. $f(x) = \dfrac{1}{x + 5} \cdot \dfrac{x^3 + 3}{x^2 + 5} \cdot \dfrac{5}{2x - 3}$

For Problems 43–45, answer the questions concerning the graph in Figure 12.

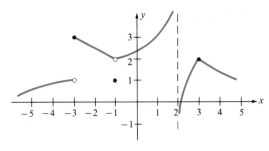

Figure 12
Graph for Problems 43–45

43. At what points (if any) is the function discontinuous? Tell why in terms of limits.

44. Where is the function discontinuous as a result of the limit not existing at a point?

45. Where is the function discontinuous as a result of the limit existing but not being equal to the value of the function at the point?

46. Population Problem   Let $S(t)$ be the number of people in a store at time t, where t ranges from 9:00 A.M. to 5:00 P.M. When is such a function continuous? When does this function have a discontinuity?

47. Postage Problem   Postage rates in 1986 were 22 cents for a first-class letter weighing no more than one ounce and 17 cents for each additional ounce or fraction thereof. Determine the function that describes the amount of postage required for a first-class letter weighing no more than 12 ounces. At what values is this function continuous? At what value is this function discontinuous?

8.3   An Analysis of Change

PURPOSE

We introduce the three important types of change of a function:

- total change,
- average rate of change, and
- instantaneous rate of change.

We will see that the computation of the total change and the average rate of change of a function involve only simply algebra. However, to introduce the idea of the instantaneous rate of change, we will need to use the concept of the limit of a function. After we introduce the instantaneous rate of change, we will introduce the concept of instantaneous velocity.

Introduction

Throughout the ages, mathematics has been called upon to solve problems that had previously gone unsolved. Four thousand years before Christ, the Sumerians used mathematics in carrying out financial dealings such as computing simple and compound interest, mortgages, and deeds of sale. Later, Egyptian builders used mathematics and geometry to construct the great pyramids of the Pharaohs.

Centuries later, Portuguese and Italian sailors needed mathematics to understand navigational charts and complex navigational equipment. The Renaissance period introduced complex mechanical devices that could be understood and explained by using mathematical principles.

However, the mathematics inherited from previous times was ill-equipped for studying motion. The discoveries by the pioneer Italian astronomer Galileo Galilei, who might be called the "father of dynamics," introduced the study of motion and change into science. It was in this setting that Newton and Leibniz developed the calculus, a mathematical discipline for understanding problems involving changing quantities. It is interesting to note that Newton was born in 1642, the same year that Galileo died.

It has often been said that, with the beginning of the calculus, the modern age of mathematics began. We now show how Newton and Leibniz analyzed change. Since a function is the mathematical device that shows how two quantities are related, we naturally analyze the change in functions.

Total and Average Change of a Function

We illustrate the concept of change by studying the growth of a biological organism. Suppose a medical researcher is monitoring the diameter $D(t)$ of a malignant tumor at day t. Table 5 represents observations recorded at the same time of day over a six-day period (day 20 to day 25).

TABLE 5
Daily Tumor Readings

Day	Diameter of Tumor (mm)	Change in Diameter (mm)
20	150.0	*
21	157.3	7.3
22	166.5	9.2
23	176.4	9.9
24	188.1	11.7
25	200.0	11.9

The total change D in the diameter of the tumor during the period from day 20 to day 25 is given by

$$\Delta D = D(25) - D(20)$$
$$= 200 \text{ mm} - 150 \text{ mm}$$
$$= 50 \text{ mm}$$

This change is shown in Figure 13.

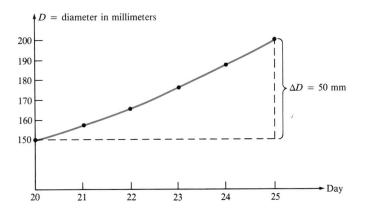

Figure 13
Diameter of tumor

A second type of change, the average daily change in the diameter, is found by dividing the total change by the number of days elapsed (five in this case). This gives

$$\text{Average daily change} = \frac{D(25) - D(20)}{5}$$
$$= \frac{200 - 150}{5}$$
$$= 10.0 \quad \text{mm per day}$$

In other words, the average daily growth of the tumor during this time period is 10 millimeters per day. Note in Table 5 that during the first two days, growth is less than 10 millimeters per day, while during the last two days the growth is greater than 10 millimeters per day. The average growth, however, is exactly 10 millimeters per day.

The above discussion leads us to the following general definition for the average change in a function.

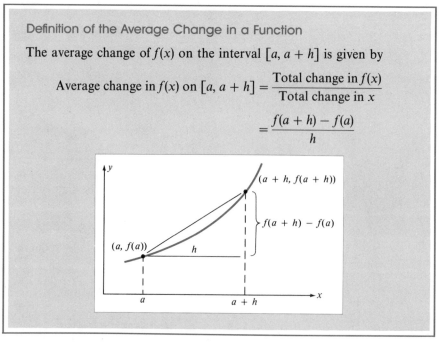

Definition of the Average Change in a Function

The average change of $f(x)$ on the interval $[a, a + h]$ is given by

$$\text{Average change in } f(x) \text{ on } [a, a + h] = \frac{\text{Total change in } f(x)}{\text{Total change in } x}$$

$$= \frac{f(a + h) - f(a)}{h}$$

The diagram above shows that the average change in a function of an interval is simply the total change of the value of the function over the interval divided by the length of the interval. Geometrically the average change of a function f over an interval $[a, a + h]$ is the slope of the secant line connecting $(a, f(a))$ and $(a + h, f(a + h))$.

Example 1 —————————

Suppose the diameter D of a cancerous tumor grows over a given period of time according to the "power" law

$$D(t) = 0.015t^2 \text{ millimeters}$$

where t is time measured in days and $t = 0$ denotes the time when the tumor begins growth. Find the total change in the diameter during the time period from day 10 to day 60. Use the total change to determine the average daily growth during this time period.

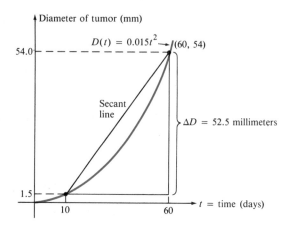

Figure 14
Total and average rate of change in diameter of tumor

Solution The graph of the diameter $D(t)$ is shown in Figure 14. The total change in diameter ΔD during this time period is given by

$$\begin{aligned}
\Delta D &= D(60) - D(10) \\
&= 0.015[(60)^2 - (10)^2] \\
&= 52.5 \text{ millimeters}
\end{aligned}$$

The average daily change during these days is

$$\text{Average daily change} = \frac{52.5}{50}$$

$$= 1.05 \text{ millimeters per day} \qquad \square$$

Instantaneous Rate of Change

Until now, we have used algebra only to define the total and average rate of change in a function. To introduce the concept of the instantaneous rate of change, however, we need to use the idea of the limit.

The **instantaneous rate of change** (or simply the *rate of change*) refers to the rate of change of a function at a point. We have seen that the average change in a function f over an interval $[a, a + h]$ is the slope of the line connecting $(a, f(a))$ and $(a + h, f(a + h))$. This line is called the secant line connecting these two points. If we now let the length h of the interval $[a, a + h]$ approach 0, the secant lines connecting $(a, f(a))$ and $(a + h, f(a + h))$ will approach the tangent line to the graph at $(a, f(a))$. This tangent line is the line that just grazes the graph of $y = f(x)$ at the point $(a, f(a))$. The instantaneous rate of change of a function f at the point $(a, f(a))$ is the slope of the tangent line at this point. We can state this more precisely in terms of limits.

Definition of the Instantaneous Rate of Change

The instantaneous rate of change of $f(x)$ at $x = a$ is given by

$$\text{Instantaneous rate of change} = \lim_{h \to 0} \frac{f(a + h) - f(a)}{h}$$

provided that the limit exists.

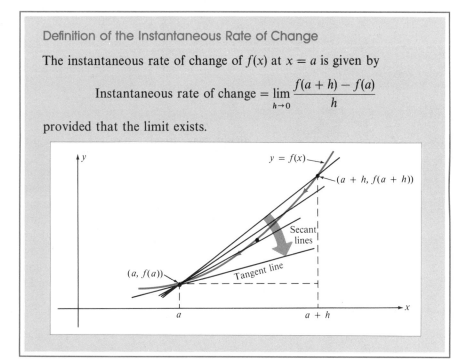

Example 2 ———————   *Instantaneous Rate of Change*   Using the same growth function

$$D(t) = 0.015t^2$$

as discussed in Example 1, find the instantaneous rate of change in the diameter D of the tumor on the 20th day.

Solution   The instantaneous rate of change in the diameter D when $t = 20$ days is found by computing the limit:

$$\text{Instantaneous rate of change} = \lim_{h \to 0} \frac{D(20 + h) - D(20)}{h}$$

$$= \lim_{h \to 0} \frac{0.015(20 + h)^2 - 0.015(20)^2}{h}$$

$$= 0.015 \lim_{h \to 0} \frac{(400 + 40h + h^2 - 400)}{h}$$

$$= 0.015 \lim_{h \to 0} (40 + h)$$

$$= 0.6 \quad \text{millimeter per day}$$

That is, on day 20 the size of the tumor is increasing at the rate of 0.6 millimeter per day. This instantaneous rate of change is the slope of the tangent line at (20, 6) as shown in Figure 15.   □

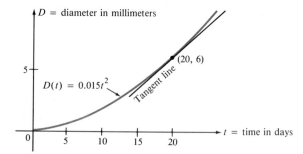

Figure 15
The slope of the tangent line gives the instantaneous growth rate

Instantaneous Velocity

Calculus was introduced originally to study the movement of the earth and the other planets around the sun. This brings us in a natural way to the concept of instantaneous velocity. This book is mostly devoted to the applications of the calculus to business and the life sciences; but since the study of motion is so basic to the calculus, we discuss the basics of motion here.

To illustrate the concept of velocity, imagine driving from Denver to Omaha on Interstate 80. The distance traveled from Denver after t hours is denoted by the position versus time curve $s(t)$ shown in Figure 16.

From Figure 16 we can make the following observations concerning the journey. The total distance traveled from Denver to Omaha during the 10 hours was

$$\text{Distance traveled} = s(10) - s(0)$$

$$= 540 - 0$$

$$= 540 \quad \text{miles}$$

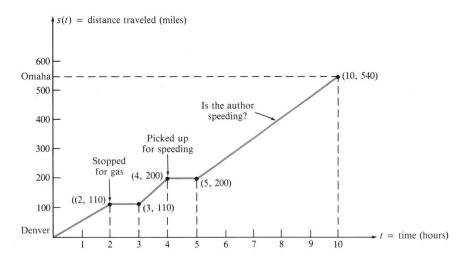

Figure 16
Road trip from Denver to Omaha

Since the car traveled 540 miles during a 10-hour period, we say that the average velocity of the car during the course of the trip was $540/10 = 54$ miles per hour. We write this as

$$\text{Average velocity} = \frac{s(10) - s(0)}{10}$$

$$= 54 \quad \text{miles per hour}$$

However, just because the average velocity of the car was 54 miles per hour during the 10-hour trip, it does not necessarily follow that the car traveled at exactly that rate of speed during the entire trip. Sometimes the car may have gone faster, and sometimes it may have gone slower.

While the average velocity may be useful for some purposes, it is not always important. If we were to run head-on into a tree, we would no doubt be more concerned with the instantaneous velocity at the exact instant of impact than with the average velocity over the entire trip.

The concept of instantaneous velocity has baffled mathematicians from the very earliest times. It was not until the advent of the calculus that mathematicians gained a clear understanding of the concept. The difficulty of understanding instantaneous velocity lies in the fact that when we use the word "instantaneous," we are referring to something that happens at a certain *instant* of time. This is in contrast to something that happens over an interval—one hour, one minute, one second, and so on. Instantaneous means that something happens so fast that no time elapses. If this is true, then in the case of instantaneous velocity the distance traveled is 0, the elapsed time is 0, and so the average velocity must be 0/0, which of course is meaningless.

The way we understand instantaneous velocity today, having the power of the calculus at our disposal, is to think in terms of limits. For instance, suppose we wish to find the instantaneous velocity at exactly 25 seconds after leaving Denver for Omaha. Although the velocity of the car may be varying at this instant of time, it is clear that over a short period of time—say, from 25 seconds to 25.1 seconds—the change in velocity will be very small. We can then approximate the instantaneous velocity at 25 seconds by computing the average velocity over the time interval from 25 seconds to 25.1 seconds. This average velocity can be computed by measuring the distance traveled during the time period from 25 seconds to 25.1 seconds and then dividing by the time elapsed, which is 0.1 second. The error that results from approximating the instantaneous velocity by the average velocity will decrease as the length of the time interval over which the average velocity was computed decreases. We would expect that the average velocity between 25 seconds and 25.01 seconds would be a better approximation to the instantaneous velocity at 25 seconds than the average velocity between 25 seconds and 25.1 seconds. Thus if the average velocity is computed for smaller and smaller time intervals, the average velocity should approach the instantaneous velocity.

This discussion leads us to the formal definition of instantaneous velocity.

Definition of Instantaneous Velocity

Let $s(t)$ be a function that gives the location of an object (such as a car) moving along a line (such as a road) as a function of time t. The **instantaneous velocity** of the object at time t_0 is the limiting value of the average velocity of the object during the time from t_0 to $t_0 + h$ as h approaches zero. Since the average velocity of the object during the time from t_0 to $t_0 + h$ is

$$\text{Average velocity on } [t_0, t_0 + h] = \frac{s(t_0 + h) - s(t_0)}{h}$$

the instantaneous velocity at t_0 is the limit

$$v(t_0) = \lim_{h \to 0} \frac{s(t_0 + h) - s(t_0)}{h}$$

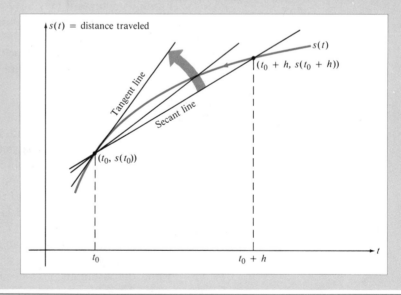

Observe that velocity is the instantaneous rate of change of the position $s(t)$. If $s(t)$ is measured in miles and t in hours, then the instantaneous velocity will have units of miles per hour.

Instantaneous Velocity of a Falling Ball

Suppose you drop a ball from the top of a tall building. Have you ever wondered how far it falls after one second? After two seconds? After three seconds? Galileo discovered that (in the absence of air friction) the exact distance $s(t)$ in feet that the ball will fall in t seconds is given by the quadratic function of time

$$s(t) = 16t^2$$

as illustrated in Figure 17. Table 6 gives the distance the ball falls during given periods of time.

TABLE 6
Object Falling Under the
Influence of Gravity

Elapsed Time (sec)	Distance Fallen (feet)
0	0
1	16
2	64
3	144
4	256
5	400
6	576
7	784
8	1024

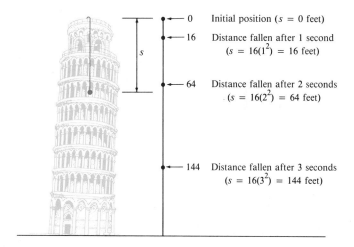

Figure 17
*Distance that an object falls in
t seconds*

To determine the velocity (instantaneous) of the ball after t seconds, we compute the limit of the average velocity during the time interval $[t, t + h]$ as h goes to zero. That is,

$$v(t) = \lim_{h \to 0} \frac{s(t + h) - s(t)}{h}$$

$$= \lim_{h \to 0} \frac{16(t + h)^2 - 16t^2}{h}$$

$$= \lim_{h \to 0} \frac{16(t^2 + 2ht + h^2) - 16t^2}{h}$$

$$= \lim_{h \to 0} 16(2t + h)$$

$$= 32t \quad \text{feet per second}$$

In other words, after t seconds we have

$$\text{Position of the ball} = 16t^2 \text{ feet}$$
$$\text{Velocity of the ball} = 32t \text{ feet per second}$$

Table 7 gives the distance fallen and the velocity of the ball after different periods of elapsed time.

TABLE 7
State of Ball Dropped
from Building

Time Elapsed (seconds)	Distance Fallen (feet)	Velocity (feet/seconds)
0	0	0
1	16	32
2	64	64
3	144	96
4	256	128
5	400	160
6	576	192
7	784	224
8	1024	256
9	1296	288
10	1600	320

Problems

For Problems 1–8, graph the function

$$f(x) = 3x + 2$$

and find the following quantities. Interpret the meaning of each quantity.

1. $f(2)$

2. $f(2 + 0.5) - f(2)$

3. $\dfrac{f(2 + 0.5) - f(2)}{0.5}$

4. $f(2 + h) - f(2)$

5. $\dfrac{f(2 + h) - f(2)}{h}$

6. $\lim\limits_{h \to 0} \dfrac{f(2 + h) - f(2)}{h}$

7. $\lim\limits_{h \to 0} \dfrac{f(3 + h) - f(3)}{h}$

8. $\lim\limits_{h \to 0} \dfrac{f(4 + h) - f(4)}{h}$

Problems 9–22 concern themselves with the function $f(x) = x^2$ graphed in Figure 18. For each problem, find the indicated change.

9. Total change in the function over the interval $[0, 1]$.
10. Total change in the function over the interval $[0, 2]$.
11. Total change in the function over the interval $[1, 3]$.
12. Total change in the function over the interval $[-1, 1]$.
13. Total change in the function over the interval $[-2, 0]$.

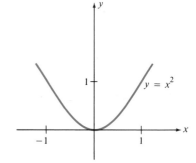

Figure 18
Graph for Problems 9–22

14. Average change in the function over the interval $[0, 1]$.
15. Average change in the function over the interval $[0, 2]$.
16. Average change in the function over the interval $[1, 3]$.
17. Average change in the function over the interval $[-1, 1]$.
18. Average change in the function over the interval $[-2, 0]$.
19. Instantaneous rate of change at $x = 1$.
20. Instantaneous rate of change at $x = 2$.
21. Instantaneous rate of change at $x = -1$.
22. Instantaneous rate of change at $x = -2$.

Falling Ball

For Problems 23–27, the function

$$s(t) = 16t^2$$

represents the distance in feet a ball falls (in the absence of air friction) in t seconds when dropped from a tall building. Find the indicated quantities.

23. Find the distance fallen during the first 3 seconds.
24. Find the distance fallen during the first 10 seconds.
25. Find the velocity of the ball at time $t = 3$ seconds.
26. Find the velocity of the ball at time $t = 10$ seconds.
27. At what time will the instantaneous velocity be the same as the average velocity over the first 5 seconds?

Instantaneous Velocity

For Problems 28–32, a particle moves along a straight line where $s(t)$ is the distance traveled in feet at time t seconds. Find the velocity of the particle at time t for the given functions.

28. $s(t) = 5$
29. $s(t) = t$
30. $s(t) = 5t + 7$
31. $s(t) = t^2 + 4$
32. $s(t) = 4t^2 + 3t$

Other Rates of Change

Find the instantaneous rate of change of the quantities in Problems 33–34 with respect to the radius r.

33. The circumference C of a circle, where $C = 2\pi r$.
34. The area A of a circle, where $A = \pi r^2$.

Changing Demand for Gasoline

Suppose the daily demand $D(p)$ for automobile gasoline in the United States as a function of price p has been determined to be

$$D(p) = \frac{(p - 5)^4}{10} \qquad (0.5 \le p \le 5)$$

where D is measured in millions of gallons and p is measured in dollars. The graph of $D(p)$ is shown in Figure 19.

35. Find the total decrease in demand when the price of gasoline is raised from \$1 to \$5.
36. Find the average decrease in demand in millions of gallons per dollar when the price is raised from \$1 to \$5.
37. Find the total decrease in demand when the price is raised from \$2 to \$5.
38. Find the average decrease in demand when the price is increased from \$2 to \$5.

Seborrheic Lesion

A seborrheic lesion or sore has been treated with a new form of hydrocortisone, and its size is measured over the course of ten days. The observations in Table 8 were taken. Find the quantities in Problems 39–44.

TABLE 8
Observations of Seborrheic Lesion

Day	Lesion Size (cm)
0	5.1
1	4.6
2	3.9
3	3.3
4	2.7
5	2.2
6	1.7
7	1.3
8	0.7
9	0.3
10	0.0

39. The total decrease in lesion size during the ten-day period.
40. The average decrease in lesion size during the ten-day period.
41. The total decrease in lesion size during the first five days.
42. The average decrease in lesion size during the first five days.

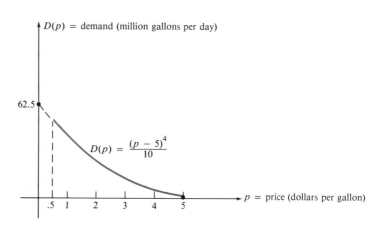

Figure 19
Demand curve for gasoline

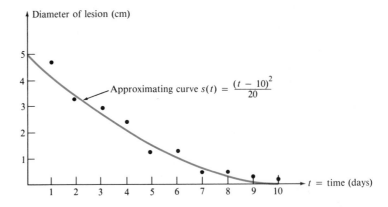

Figure 20
Shrinking of a seborrheic lesion

43. The total decrease in lesion size during the last five days.
44. The average decrease in lesion size during the last five days.

Medical Problem

Suppose a medical researcher constructs a mathematical model or equation describing the size of a lesion. The equation that describes the diameter of a lesion (cm) as a function of time (days) is given by

$$D(t) = \frac{(t - 10)^2}{20} \qquad 0 \le t \le 10$$

as illustrated in Figure 20.

45. What is the instantaneous rate of change of the lesion size on day 5?
46. What is the instantaneous rate of change of the lesion size on day 7?
47. What is the instantaneous rate of change of the lesion size on day 1?

Olympic Swimming Times (Are women catching up?)

Table 9 shows the winning Olympic swimming times for the men's and women's 100-meter freestyle event during the time period 1912–1984. The last column shows the percentage dif-

• **TABLE 9**
Winning 100-Meter Freestyle Times (1912–1984 Olympic Games)

Year	Men's 100-Meter Freestyle (min:sec)		Women's 100-Meter Freestyle (min:sec)		Percentage Difference in Times
1912	1:03.4	(U.S.)	1:22.2	(Australia)	29.7%
1920	1:01.4	(U.S.)	1:13.6	(U.S.)	19.9%
1924	0:59.0	(U.S.)	1:12.4	(U.S.)	22.7%
1928	0:58.6	(U.S.)	1:11.0	(U.S.)	21.2%
1932	0:58.2	(Japan)	1:06.8	(U.S.)	14.8%
1936	0:57.6	(Hungary)	1:05.9	(Netherlands)	14.4%
1948	0:57.3	(U.S.)	1:06.3	(Denmark)	15.7%
1952	0:57.4	(U.S.)	1:06.8	(Hungary)	16.0%
1956	0:55.4	(Australia)	1:02.0	(Australia)	11.9%
1960	0:55.2	(Australia)	1:01.2	(Australia)	10.5%
1964	0:53.4	(U.S.)	0:59.5	(Australia)	11.4%
1968	0:52.2	(Australia)	1:00.0	(U.S.)	14.5%
1972	0:51.22	(U.S.)	0:58.59	(U.S.)	14.4%
1976	0:49.99	(U.S.)	0:55.65	(GDR)	11.3%
1980	0:50.40	(GDR)	0:54.79	(GDR)	8.7%
1984	0:49.80	(U.S.)	0:55.92	(U.S.)	12.8%

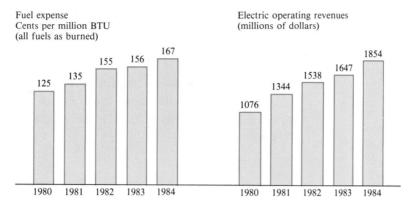

Figure 21
Fuel expense and revenues of a power and light company.

ference between the men's and women's times. That is,

Percentage difference

$$= 100 \frac{\text{Women's times} - \text{Men's times}}{\text{Men's times}}$$

48. What was the total change in the percentage difference during the years from 1912 to 1984?
49. What was the average change per year in the percentage difference from 1912 to 1984?
50. What was the average change per Olympic Games in the percentage difference during the years from 1912 to 1984? *Hint:* The Olympic Games occurred 28 times in that span of years.
51. What was the average change in the men's times during the period from 1912 to 1984?

52. What was the average change in the women's times during the period from 1912 to 1984?

Annual Report

Figure 21 shows the annual fuel expenses and revenues (what they take in) during the period 1980–1984 for a given power and light company.

53. Find the average yearly increase in cost of fuel during this period.
54. Find the average yearly increase in revenues during this period.

Annual Report

Figure 22, which shows circulation during the period 1976–1985, was taken from the annual report of the Gannett Co., Inc. Gannett has become one of the most innovative media companies in the United States in recent years. For Problems 55–56, find the indicated quantities.

Figure 22
Gannett circulation revenues (in millions of dollars)

55. Find the total change in circulation revenue during the time period from 1976 to 1985.

56. Find the average change in circulation during the time period 1976–1985.

Introduction to the Derivative

PURPOSE

We define the derivative of a function f at a point x and give

- a geometric interpretation of the derivative and
- a precise definition of the tangent line to a graph.

We will also show how the derivative is used in economics to study the rate of change of supply and demand curves.

Introduction

In mathematics a function relates the "size" of one quantity to the "size" of another quantity. When we write $y = f(x)$, we are describing a functional relationship that gives the value of one quantity y in terms of another quantity x. The derivative of a function, however, does not give the size of y but the instantaneous rate of change of y as a function of x. We now make this idea precise.

The Concept of the Derivative

We begin by selecting an arbitrary point $(a, f(a))$ on the graph $y = f(x)$ and drawing a straight line from this point to the neighboring point on the graph, $p_1 = (a + h, f(a + h))$. This line is drawn in Figure 23 and is called a **secant line**. It clearly has the slope

$$\text{Slope of secant line} = \frac{f(a + h) - f(a)}{h}$$

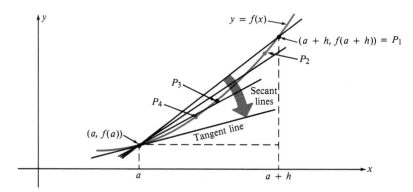

Figure 23
Derivative as the slope of the tangent line

If we now allow h to approach 0, the point p_1 will "slide along" the graph, reaching the successive points p_2, p_3, and p_4. We can see from Figure 23 that the secant lines will approach a line called the **tangent line** to the graph of $f(x)$ at $(a, f(a))$. It is clear that the **limiting value** of the slopes of the secant lines will approach the slope of this tangent line. This slope is called the **derivative** of the function f at the point a and is denoted by $f'(a)$. We now summarize these ideas.

Definition of the Derivative

The derivative of a function f at a point a, denoted by $f'(a)$, is defined by

$$f'(a) = \lim_{h \to 0} \frac{f(a + h) - f(a)}{h}$$

provided that the limit exists.

Note in Figure 23 that the secant lines approach the tangent line that just grazes the graph at $(a, f(a))$. In addition to this intuitive definition of the tangent line, we can define the tangent line more precisely in terms of the derivative.

Calculus Definition of the Tangent Line

The tangent line to the graph of f at the point $(a, f(a))$ is the straight line passing through $(a, f(a))$ with slope

$$f'(a) = \lim_{h \to 0} \frac{f(a + h) - f(a)}{h}$$

Using the point-slope formula, the equation of this tangent line is

$$y - f(a) = f'(a)(x - a)$$

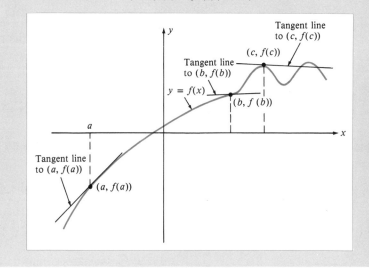

Example 1

Secant and Tangent Line   Find the secant line to the graph of $f(x) = x^2$ that passes through the points $(1, 1)$ and $(2, 4)$. Then find the tangent line to the graph at $(1, 1)$.

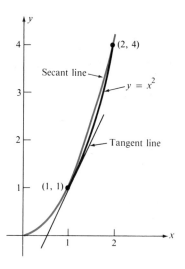

Figure 24
The secant line for $y = x^2$ through (1, 1) and (2, 4)

Solution   The graph of $f(x) = x^2$ along with the secant and tangent lines are drawn in Figure 24.

To find the secant line that passes through the points (1, 1) and (2, 4), we first compute its slope:

$$\text{Slope} = \frac{4 - 1}{2 - 1} = 3$$

Hence from the point-slope formula the equation of the secant line is

$$\frac{y - 1}{x - 1} = 3$$

or, written in slope-intercept form

$$y = 3x - 2$$

To find the tangent line to the curve at (1, 1), we first find the derivative of $f(x) = x^2$ when $x = 1$. That is,

$$f'(1) = \lim_{h \to 0} \frac{f(1 + h) - f(1)}{h}$$

$$= \lim_{h \to 0} \frac{(1 + h)^2 - 1^2}{h}$$

$$= \lim_{h \to 0} \frac{(1 + 2h + h^2) - 1}{h}$$

$$= \lim_{h \to 0} (2 + h)$$

$$= 2$$

In other words, when $x = 1$, the value of x^2 is changing by two units for each unit change in x.

Using this derivative, we find the tangent line to the graph of $f(x) = x^2$ at (1, 1) to be

$$y - f(1) = f'(1)(x - 1)$$
$$y - 1 = 2(x - 1)$$

or $\qquad\qquad\qquad\qquad y = 2x - 1$ $\qquad\qquad\qquad\qquad$ □

Example 2 _____   Computing the Derivative   Find the derivative of $f(x) = x^2$ at an arbitrary point $x = a$.

Solution   Using the definition of the derivative, we have

$$f'(a) = \lim_{h \to 0} \frac{f(a + h) - f(a)}{h}$$

$$= \lim_{h \to 0} \frac{(a + h)^2 - a^2}{h}$$

$$= \lim_{h \to 0} \frac{(a^2 + 2ha + h^2) - a^2}{h}$$

$$= \lim_{h \to 0} (2a + h)$$

$$= 2a$$

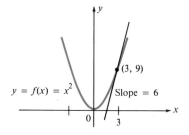

Figure 25
The derivative represents the slope of the tangent line

Since $x = a$ is an arbitrary real number, it is often customary to replace a by x and refer to the derivative at an arbitrary real number x. In this example the derivative of $f(x) = x^2$ at an arbitrary x is

$$f'(x) = 2x$$

This derivative gives the slope of the tangent line to the curve $f(x) = x^2$ at any point $(x, f(x))$. Keep in mind that $f'(x)$ gives the rate of change of $f(x)$ at x, whereas $f(x)$ gives the size of $f(x)$ at x. Figure 25 illustrates this idea.

HISTORICAL NOTE

Originally, the subject of the calculus was a subject for debate. The concept of "infinitely small" quantities, which were the forerunners of the limit, were awfully hard to take by many people. Even the brilliant Johann Bernoulli once said, "A quantity which is increased or decreased by an infinitely small quantity is neither increased nor decreased."

One of the more able critics of the early calculus was the eminent metaphysician Bishop George Berkeley (1685–1753). He once referred to the derivative as "ghosts of departed quantities."

Derivative Notation

The process of finding a derivative is called **differentiation**. If the independent variable is x, then differentiation of some quantity $[*]$ is often denoted by

$$\frac{d}{dx}[*]$$

which is read "the derivative of $*$ with respect to x."

For example, we would write

$$\frac{d}{dx}[x^2] = 2x$$

The d/dx notation can be used to find derivatives of functions involving variables other than x. For example, if the independent variable is t, then we would replace x by t and write

$$\frac{d}{dt}[t^2 + 2t + 1] = 2t + 2$$

The major failing of the d/dx notation is that it is difficult to express the derivative at a specific point. Here, the **prime notation** is better suited. For example, to represent the value of the derivative of f at the specific point $x = a$,

the two notations are

Prime notation: $f'(a)$

d/dx notation: $\left.\dfrac{df(x)}{dx}\right|_{x=a}$

Clearly, the prime notation is simpler.

HISTORICAL NOTE

Baron Gottfried Wilhelm von Leibniz was very careful to develop proper mathematical notation. His work in logic had made him aware that symbols should be chosen with great care and should appeal to one's intuition. It was Leibniz who devised the *d/dx* notation for the derivative.

The followers of Newton in England did not use the Leibniz *d/dx* notation; they used Newton's inferior "fluxions." As a result, Newton's followers were unable to solve many problems that Leibniz and his followers found almost trivial, and Newton's followers became helplessly bogged down in a notational quagmire. As a result of these notational difficulties, English mathematics lagged behind the mathematics of Europe during the 1700's. It was not until the 1800's and the arrival of the English school of algebraists, led by Arthur Cayley and James Joseph Sylvester, that English mathematics fully recovered.

Example 3 ————— Finding the Derivative   **Find the derivative of**

$$f(x) = \frac{1}{x}$$

at an arbitrary point $x \neq 0$.

Solution   Using the limit definition of the derivative, we can write

$$f'(x) = \lim_{h \to 0} \frac{f(x+h) - f(x)}{h} \qquad \text{(definition of the derivative)}$$

$$= \lim_{h \to 0} \frac{\dfrac{1}{x+h} - \dfrac{1}{x}}{h} \qquad \text{(direct substitution)}$$

$$= \lim_{h \to 0} \frac{x - (x+h)}{x(x+h)h} \qquad \text{(simple algebra)}$$

$$= \lim_{h \to 0} \frac{-1}{x(x+h)} \qquad \text{(more algebra)}$$

$$= \frac{-1}{\lim_{h \to 0} (x^2 + hx)} \qquad \text{(the limit of a quotient is the quotient of the limits)}$$

$$= -\frac{1}{x^2} \qquad \text{(evaluating the above limit)} \qquad \square$$

Example 4

Finding Derivatives   Find the derivative of $f(x) = \sqrt{x}$.

Solution   Using the definition of the derivative, we write

$$f'(x) = \lim_{h \to 0} \frac{\sqrt{x+h} - \sqrt{x}}{h}$$ (definition of the derivative)

$$= \lim_{h \to 0} \frac{\sqrt{x+h} - \sqrt{x}}{h} \cdot \frac{\sqrt{x+h} + \sqrt{x}}{\sqrt{x+h} + \sqrt{x}}$$ (multiply the numerator and denominator by the conjugate)

$$= \lim_{h \to 0} \frac{(x+h) - x}{h(\sqrt{x+h} + \sqrt{x})}$$ (algebra)

$$= \lim_{h \to 0} \frac{1}{\sqrt{x+h} + \sqrt{x}}$$ (algebra)

$$= \frac{1}{\lim_{h \to 0} (\sqrt{x+h}) + \sqrt{x}}$$ (the limit of a quotient is the quotient of the limits)

$$= \frac{1}{2\sqrt{x}}$$ (perform the above limit)

Figure 26 interprets this derivative by showing the tangent lines to the function $f(x) = \sqrt{x}$.   □

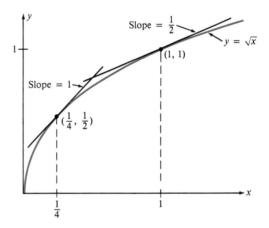

Figure 26
Slopes of tangent lines to $y = x$ at $x = 1/4$ and $x = 1$

Example 5

Derivative at a Point   If $y = x^2$, find

$$\frac{dy}{dx}\bigg|_{x=0} \quad \text{and} \quad \frac{dy}{dx}\bigg|_{x=3}$$

Solution   The derivative of $y = x^2$ at x is

$$\frac{dy}{dx} = 2x$$

Hence simply evaluating this derivative at the points 0 and 3, we have

$$\frac{dy}{dx}\bigg|_{x=0} = 0 \qquad \frac{dy}{dx}\bigg|_{x=3} = 6$$ □

> ## Steps for Finding the Derivative
>
> Given $y = f(x)$:
>
> - **Step 1.** Compute the value $f(x + h)$.
> - **Step 2.** Compute the difference $f(x + h) - f(x)$.
> - **Step 3.** Compute the quotient
>
> $$\frac{f(x + h) - f(x)}{h}$$
>
> - **Step 4.** Compute the limit
>
> $$\frac{df(x)}{dx} = \lim_{h \to 0} \frac{f(x + h) - f(x)}{h}$$

Derivatives of the Supply and Demand Curves

In recent years, economists have put calculus to good use. Of the past 20 recipients of Nobel prizes in Economics, many have used concepts from calculus. For example, the 1978 winner, Herbert Simon, used ideas of calculus to analyze decision-making processes in economic organizations. Two years later, Lawrence Klein from the University of Pennsylvania won the Nobel prize in economics for his studies in economic forecasting. His ideas, too, use concepts of calculus. We now show how differential calculus can be used to analyze the supply and demand curves in economics.

Let us suppose that corn is being sold on the Chicago Board of Trade. The amount of corn offered every day by farmers and other sellers depends on the price offered. Some farmers will be willing to sell their corn at a fairly low price, while others will be willing to sell only at a higher price. The total quantity q of corn that farmers and other sellers will offer for sale at a given price p defines a **supply curve**:

$$q = S(p)$$

On the other hand, from the buyers' point of view the amount q of corn purchased by exporters, traders, and others also depends on the price p of the corn. The higher the price of corn, the smaller the quantity of corn that will be purchased. The lower the price of corn, the greater the quantity of corn that will be purchased by the buyers. The exact relationship between the quantity q that buyers are willing to buy (that they demand) and the price of the corn p defines the **demand curve**:

$$q = D(p)$$

Typical supply and demand curves are graphed in Figure 27.

In Figure 27, we see that the supply curve is negative when the price p is less than some value p_L. The number p_L is called the **lower price limit**, and it represents the lowest price at which someone will be willing to sell some corn. If the price is less than p_L, no one will be willing to sell any corn.

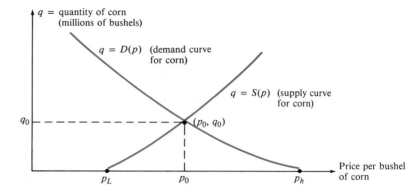

Figure 27

Supply and demand curves for corn

Farther up the price scale we have the number p_h, which is called the **upper price limit**. This number represents the highest price at which any buyer will buy corn.

Observe that the supply curve is always increasing with price, while the demand curve is always decreasing with price. (Do you understand why this should be?) The curves will intersect at a point (p_0, q_0). This point of intersection is called the **market equilibrium point**. In a competitive market the price p_0 of the commodity and the amount q_0 of commodity demanded and sold will tend to stabilize at this point. In other words, the price of corn will stabilize at p_0, and the amount of corn demanded and sold will stabilize at q_0.

To see how calculus can be used to analyze the supply and demand curves, first consider the derivative of the supply curve:

$$\frac{dq}{dp} = \frac{dS(p)}{dp}$$

In this problem the derivative represents the rate of change of the corn supplied by sellers as a function of the price of the corn. It can be interpreted as the reaction by the sellers to a given change in price. For example, if

$$\left.\frac{dS}{dp}\right|_{p=\$3.50} = 2 \quad \text{million bushels per dollar}$$

then for every dollar increase (or fraction thereof) in the price of corn up from \$3.50, there will be an increase in the quantity of corn offered of two million bushels. (See Figure 28.)

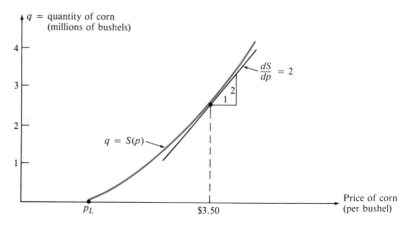

Figure 28

Slope of the supply curve for corn when the price is \$3.50

On the other hand, the derivative of the demand curve

$$\frac{dq}{dp} = \frac{dD(p)}{dq}$$

represents the reaction by the suppliers to a change in the price. Suppose, for example, that corn is selling at $3.50 per bushel and that the derivative of the demand curve is

$$\left.\frac{dD}{dp}\right|_{p=\$3.50} = -3$$

This means that when the price of corn is $3.50 per bushel, a one-dollar increase (or some fraction thereof) in price will cause a decrease of three million bushels in the quantity demanded (or some fraction thereof). For example, an increase of $0.10 in the price for a bushel of corn from $3.50 to $3.60 (one-tenth unit) will result in a decrease in demand of 0.3 million bushels (three-tenths unit) of corn by the consumers. (See Figure 29.)

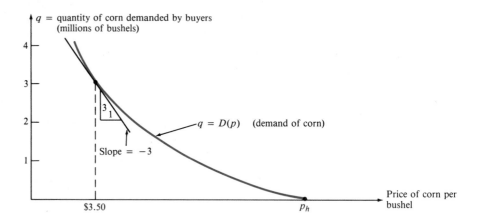

Figure 29
Slope of the demand curve for corn when the price is $3.50

Example 6

Supply and Demand   Suppose the supply and demand curves for corn on the Chicago Board of Trade have been determined by economists to be

$$q = S(p) = -1 + 0.10p^2 \qquad \text{(supply curve)}$$
$$q = D(p) = 4 - 0.04p^2 \qquad \text{(demand curve)}$$

Here q is the quantity of corn measured in millions of bushels and p is the price for a bushel of corn measured in dollars.

Suppose corn is currently selling for $4.00 a bushel. Find and interpret the derivatives of the supply and demand curves at this price.

Solution   The graphs of the supply and demand curves are shown in Figure 30. The derivative of the supply curve is given by

$$S'(p) = 0.20p$$

When $p = 4.00$, we have

$$S'(4.00) = 0.2(4.00)$$
$$= 0.80 \quad \text{million bushels per dollar}$$

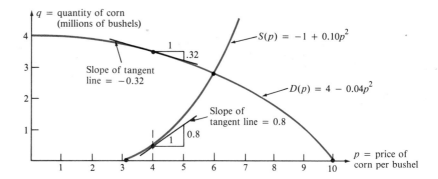

Figure 30

Derivative of supply and demand curves at p = $4.00

On the other hand, the derivative of the demand curve is

$$D'(p) = -0.08p$$

When $p = \$4.00$, we have

$$D'(4.00) = -0.08(4.00)$$
$$= -0.32 \quad \text{million bushels per dollar} \qquad \square$$

Interpretation

Supply derivative: $S'(4.00) = 0.8$

When the price of corn is $4.00 per bushel, a one-dollar increase in price will result in an increase in supply of 800,000 bushels. If the price of corn increases from $4.00 to $4.10 per bushel (one-tenth unit of price), we would expect that sellers (collectively) would increase the supply by 80,000 bushels. On the other hand, if the price falls from $4.00 to $3.90, then the supply will decrease by 80,000 bushels.

Demand derivative: $D'(4.00) = -0.32$

When the price of corn is $4.00 per bushel, a one-dollar increase in price will result in a decrease in demand of 320,000 bushels. If the price of corn increases from $4.00 to $4.10 per bushel (one-tenth unit of price), then the demand for corn by the buyers will decrease by 32,000 bushels. On the other hand, if the price falls from $4.00 to $3.90, then the demand will increase by 32,000 bushels.

When the Derivative Fails to Exist

There are three common ways in which a function can fail to have a derivative at a point. Roughly, they are

1. The graph of the function has a break.
2. The graph of the function has a corner-point.
3. The graph of the function has a vertical tangent line.

In all of these cases it is impossible to draw a unique, nonvertical tangent line to the graph at the given point. We illustrate in Figure 31 some functions that fail to have derivatives at different points.

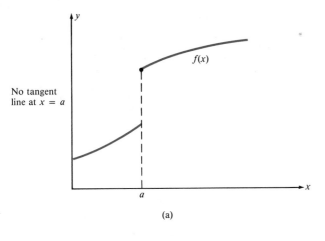

(a)

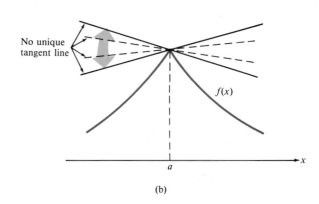

(b)

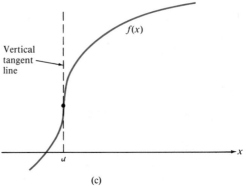

(c)

Figure 31
*Three ways in which a function can
fail to have a derivative*

Problems

For Problems 1–4, graph the function

$$f(x) = x^2 + 1$$

and answer the indicated questions. Interpret graphically the
quantities computed.

1. What is the equation of the secant line that passes
   through the points $(1, 2)$ and $(2, 5)$ on the graph of $f(x)$?

2. $f(1 + h) - f(1)$

3. $\dfrac{f(1 + h) - f(1)}{h}$

4. $\lim\limits_{h \to 0} \dfrac{f(1 + h) - f(1)}{h}$

For Problems 5–19, use the definition of the derivative to
compute $f'(x)$ for each function. After finding the derivative,
compute $f'(0)$ and $f'(1)$.

5. $f(x) = x$
6. $f(x) = 2x$
7. $f(x) = 2x + 3$
8. $f(x) = 5x - 4$

9. $f(x) = x^2$
10. $f(x) = 5x^2$
11. $f(x) = x^2 + x$
12. $f(x) = 3x^2 + 2x$
13. $f(x) = 7x^2 + 5x + 3$
14. $f(x) = 10x^2 - 10x + 20$

15. $f(x) = \dfrac{1}{x}$
16. $f(x) = \dfrac{1}{x + 1}$

17. $f(x) = \dfrac{1}{x^2}$
18. $f(x) = \sqrt{x}$

19. $f(x) = \sqrt{x} + x$

For Problems 20–24, find the slope of the tangent line to the
graph at the indicated points. Find the equation of the tangent
line.

20. $f(x) = 2x + 3$; $(1, 5)$
21. $f(x) = x^2$; $(2, 4)$

22. $f(x) = \dfrac{1}{x + 1}$; $(0, 1)$
23. $f(x) = \dfrac{1}{x^2}$; $(1, 1)$

24. $f(x) = \dfrac{1}{x^2}$; $(-1, 1)$

For Problems 25–30, find the indicated quantities if

$$y = f(x) = x^2$$

25. $f'(0)$ **26.** $f'(1)$

27. $f'(-1)$ **28.** $\dfrac{dy}{dx}\Big|_{x=3}$

29. $\dfrac{dy}{dx}\Big|_{x=4}$ **30.** $\dfrac{dy}{dx}\Big|_{x=-1}$

31. If $y = \sqrt{x} + 1$, find $\dfrac{dy}{dx}\Big|_{x=16}$

32. If $y = \dfrac{1}{x+1}$, find $\dfrac{dy}{dx}\Big|_{x=0}$

Tangent Lines

For Problems 33–36, find the indicated tangent lines.

33. Find the equation of the tangent line to the graph of

$$f(x) = 3x + 5$$

at the point $(0, 5)$.

34. Find the equation of the tangent line to the graph of

$$f(x) = x^2 + 3$$

at the point $(-1, 4)$.

35. Find the equation of the tangent line to the graph of

$$f(x) = x^2 + 2x$$

at the point $(2, 8)$.

36. Find the equation of the tangent line to the graph of

$$f(x) = \sqrt{x}$$

at the point $(1. 1)$.

Supply and Demand Derivatives

For Problems 37–42, carry out the following steps for the indicated supply and demand curves.
(a) Plot the supply and demand curves.
(b) Find the lower and upper price limits.
(c) Find the equilibrium point.
(d) Find and interpret the derivatives of the supply and demand curves at the indicated prices.

37. $S(p) = 4p - 1$ (linear supply)
   $D(p) = 4 - p$ (linear demand)
   Price = $0.50

38. $S(p) = 3p - 5$ (linear supply)
   $D(p) = 10 - 2p$ (linear demand)
   Price = $2.50

39. $S(p) = 4p - 1$ (linear supply)
   $D(p) = 4 - p^2$ (quadratic demand)
   Price = $1.00

40. $S(p) = p^2 - 16$ (quadratic supply)
   $D(p) = -p^2 + 36$ (quadratic supply)
   Price = $4.50

41. $S(p) = p^2 - 2p - 15$ (quadratic supply)
   $D(p) = 100 - p^2$ (quadratic supply)
   Price = $7.50

42. $S(p) = p^2 - p - 5$ (quadratic supply)
   $D(p) = 50 - p - p^2$ (quadratic supply)
   Price = $4.00

43. Price Elasticity of Demand   Economists refer to the *price elasticity of a demand function* as a measure of the responsiveness between the proportional change in the quantity demanded to the proportional change in the price. That is,

$$E(p) = -\frac{dD/dp}{D/p} \qquad \text{(price elasticity of demand)}$$

(the minus sign is simply to make the sign of the elasticity positive). For the demand function

$$D(p) = 16 - p^2$$

(a) find the price elasticity $E(p)$ for any price p,
(b) find and interpret the price elasticity when the price is $2.00.

44. Price Elasticity of Supply   Economists refer to the *price elasticity of a supply function* as the responsiveness that measures the relationship between the proportional change in the quantity supplied to the proportional change in the price. That is,

$$E(p) = \frac{dS/dp}{S/p} \qquad \text{(price elasticity of supply)}$$

For the supply function

$$S(p) = p^2 - 9$$

(a) find the price elasticity $E(p)$ for any price p,
(b) find and interpret the price elasticity when the price is $5.00.

Points Where Derivatives Do Not Exist

For Problems 45–48, find the points (if any) where the functions do not have derivatives.

45.

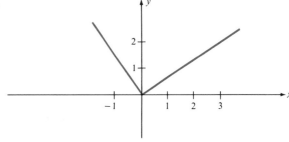

46.

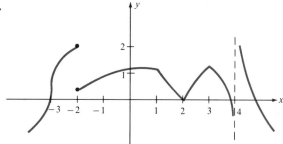

47.

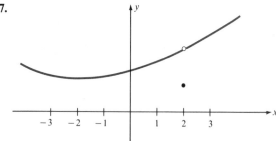

48.

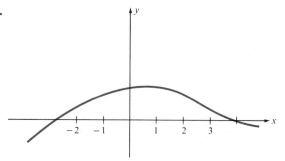

8.5

Derivatives of Polynomials and Sums

PURPOSE

We introduce the general power rule for derivatives and prove some important rules for derivatives. We also continue our analysis of how the derivative is used in economics by introducing the concept of marginal analysis.

Introduction

Until now, we have found the derivative of simple functions at any point x by using the definition of the derivative:

$$f'(x) = \lim_{h \to 0} \frac{f(x + h) - f(x)}{h}$$

We now develop rules that will help simplify the computation of the derivative We start by learning how to differentiate any constant power of x.

Derivative of x^a (a Any Real Number)

In the previous section we differentiated $f(x) = x^2$, using the definition of the derivative. We now show how to find the derivative of any power of x. To do this we begin by drawing **Pascal's triangle** as shown in Table 10. This triangle shows how to expand the binomial expression

$$(a + b)^n$$

in terms of powers of a and b. The numbers in the Pascal triangle are the coefficients of the individual terms in the expansion.

TABLE 10
Expansions of $(a + b)^n$ by
Means of Pascal's Triangle

Binomial Expression	Binomial Expansion	
$(a + b)^0 \ =$	1	
$(a + b)^1 \ =$	$1a + 1b$	
$(a + b)^2 \ =$	$1a^2 + 2ab + 1b^2$	*Pascal's*
$(a + b)^3 \ =$	$1a^3 + 3a^2b + 3ab^2 + 1b^3$	*Triangle*
$\vdots$	$\vdots \qquad\qquad \vdots$	
$(a + b)^n \ = 1a^n + na^{n-1}b +$	$\cdots \qquad + nab^{n-1} + 1b^n$	

Using the results of the above table, we can write

$$\frac{d}{dx}[x^n] = \lim_{h \to 0} \frac{(x + h)^n - x^n}{h}$$

$$= \lim_{h \to 0} \frac{[x^n + nx^{n-1}h + \cdots + h^n] - x^n}{h} \qquad \text{(first and last terms cancel)}$$

$$= \lim_{h \to 0} [nx^{n-1} + n(n-1)x^{n-2}h + \cdots + h^{n-1}] \qquad \text{(cancel } h\text{)}$$

$$= nx^{n-1}$$

Although we have proven the above power rule only for integer powers of x, it also holds for all real powers. We state the more general result without proof.

> **Power Rule for Derivatives**
>
> If a is any real number, then the power function $f(x) = x^a$ has the derivative
>
> $$\frac{d}{dx} x^a = ax^{a-1}$$

Example 1

Derivatives of Powers of x Table 11 demonstrates the power rule for particular values of a.

TABLE 11
The Power Rule for Some
Values of a

a	$f(x) = x^a$	$f'(x) = ax^{a-1}$
0	1	0
1	x	1
2	x^2	$2x$
5	x^5	$5x^4$
-1	x^{-1}	$-x^{-2}$
-2	x^{-2}	$-2x^{-3}$
1.5	$x^{1.5}$	$1.5x^{1/2}$
7.3	$x^{7.3}$	$7.3x^{6.3}$

Factoring Constants Out of the Derivative

We now introduce some useful properties of the derivative. The first property says that we can factor constants outside the derivative. This rule is useful because, although we know the derivative of x^2, we do not know the derivatives of $3x^2$, $2x^2$, and other such constant multiples.

Derivative of a Constant Times a Function

If c is a real number, then for any function $f(x)$, we have

$$\frac{d}{dx}[cf(x)] = c\frac{d}{dx}f(x)$$

Verification of the Above Rule

Since the derivative is a limit, the verification must rely on properties of limits. For a function $f(x)$ and a constant c the function $cf(x)$ has the derivative

$$\frac{d}{dx}[cf(x)] = \lim_{h \to 0}\frac{cf(x+h) - cf(x)}{h}$$

$$= \lim_{h \to 0} c\frac{f(x+h) - f(x)}{h}$$

$$= c\lim_{h \to 0}\frac{f(x+h) - f(x)}{h} \qquad \text{(by the limit property}$$
$$\lim_{x \to a} cf(x) = c\lim_{x \to a} f(x))$$

$$= c\frac{d}{dx}[f(x)] \qquad \text{(by the definition of the derivative)}$$

Example 2

Factoring Out Constants The following derivatives illustrate the use of the constant rule:

$$\frac{d}{dx}[3x^2] = 3\frac{d}{dx}[x^2] = 6x$$

$$\frac{d}{dx}[5x^{-1}] = 5\frac{d}{dx}[x^{-1}] = -5x^{-2}$$

$$\frac{d}{dx}[8x^{1/2}] = 8\frac{d}{dx}[x^{1/2}] = 4x^{-1/2}$$

$$\frac{d}{dx}[-4x] = -4\frac{d}{dx}[x] = -4$$

Example 3

Derivative of Constants Find the derivative of the constant function

$$f(x) = c \qquad (c \text{ any real number})$$

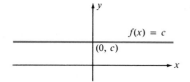

Figure 32
The derivative of a constant function is zero

Solution   Using the definition of the derivative, we can write

$$f'(x) = \lim_{h \to 0} \frac{c - c}{h}$$

$$= 0$$

The interpretation of this result is geometrically obvious, since the slope of any tangent line to a horizontal graph is zero. (See Figure 32.) □

Derivatives of Sums and Differences of Functions

The next rules allow us to find the derivatives of sums and differences of functions. For example, we would like to find the derivative of $3x^2 + 2x^4 - 7x^5$ in terms of the derivatives of the individual terms.

Derivative of Sums and Differences

Let f and g be differentiable functions. Then

- The derivative of a sum is the sum of the derivatives:

$$\frac{d}{dx}[f(x) + g(x)] = \frac{d}{dx}f(x) + \frac{d}{dx}g(x)$$

- The derivative of a difference is the difference of the derivatives:

$$\frac{d}{dx}[f(x) - g(x)] = \frac{d}{dx}f(x) - \frac{d}{dx}g(x)$$

Verification of the Sum Rule

The statement that the derivative of a sum is the sum of the derivatives is verified by using the limit rule: The limit of a sum is the sum of the limits. Starting with the definition of the derivative of a sum, we write

$$\frac{d}{dx}[f(x) + g(x)] = \lim_{h \to 0} \frac{[f(x + h) + g(x + h)] - [f(x) + g(x)]}{h}$$

$$= \lim_{h \to 0} \frac{[f(x + h) - f(x)] + [g(x + h) - g(x)]}{h}$$

$$= \lim_{h \to 0} \frac{f(x + h) - f(x)}{h} + \lim_{h \to 0} \frac{g(x + h) - g(x)}{h}$$

$$= \frac{df(x)}{dx} + \frac{dg(x)}{dx}$$

This completes the verification of the sum rule.

The verification of the difference rule is essentially the same.

Example 4

Derivatives of Sums and Differences   Find the derivative of

$$f(x) = 3x^2 + x^{1/2}$$

Solution   Using the needed derivative rules, we have

$$\frac{d}{dx}(3x^2 + x^{1/2}) = \frac{d}{dx}(3x^2) + \frac{d}{dx}(x^{1/2})$$

$$= 6x + \frac{1}{2}x^{-1/2}$$

Some other functions and their derivatives are shown here:

$$\frac{d}{dx}[\sqrt{x} + 5x^{5/2}] = \frac{1}{2}x^{-1/2} + \frac{25}{2}x^{3/2}$$

$$\frac{d}{dx}[3x^2 + 6x - 5] = 6x + 6$$

$$\frac{d}{dx}[-x^{-1} - x] = x^{-2} - 1$$

$$\frac{d}{dx}[8x^3 - 1/x] = 24x^2 + x^{-2} \qquad \square$$

Using the previous rules for differentiating powers x^n, sums, and differences, along with the rule for factoring out constants, we can now state the general rule for differentiating polynomials.

Derivative of Polynomials

The derivative of a polynomial is given by

$$\frac{d}{dx}[a_0x^n + a_1x^{n-1} + a_2x^{n-2} + \cdots + a_{n-1}x + a_n]$$

$$= a_0nx^{n-1} + a_1(n-1)x^{n-2} + \cdots + a_{n-1}$$

Example 5

Derivative of a Polynomial   Differentiate the polynomial

$$\frac{d}{dx}[x^2 - 3x + 5]$$

Solution   Using the rule for differentiating polynomials, we have

$$\frac{d}{dx}[x^2 - 3x + 5] = 2x - 3 \qquad \square$$

Marginal Analysis in Economics

The adjective *marginal* is used in economics to describe the rate of change or derivative of an economic quantity. For instance, suppose a firm produces q

units of a product (computers, candy bars, airplanes, bars of soap, textbooks, or whatever) and we assume that the firm sells all that it produces. Not only is the firm interested in the basic quantities

$$C(q) = \textbf{Cost} \text{ to produce } q \text{ units of the product}$$
$$R(q) = \text{Firm's } \textbf{revenue} \text{ when } q \text{ units are produced}$$
$$P(q) = \text{Firm's } \textbf{profit} \text{ when } q \text{ units are produced}$$

but also the marginal quantities

$$MC = C'(q) \qquad \textbf{(marginal cost)}$$
$$MR = R'(q) \qquad \textbf{(marginal revenue)}$$
$$MP = P'(q) \qquad \textbf{(marginal profit)}$$

Whereas the *total cost* $C(q)$ represents the cost to produce q units of the product, the *marginal cost* $C'(q)$ represents the rate of change in the cost. By the definition of the derivative the marginal cost is approximately equal to

$$C'(q) \cong \frac{C(q + 1) - C(q)}{1} = C(q + 1) - C(q)$$

which is simply the increase in cost of producing an additional unit of the product when q units are currently being produced. For instance, if the firm produces computers and if the marginal cost is $C'(10{,}000) = \$250$, then when the production level is 10,000 computers (per week, month, or whatever), it costs the company roughly \$250 to produce each additional computer.

The *total revenue function* $R(q)$ represents the proceeds to the firm when the company sells q units of the product (in other words, the price charged per unit times the number of units sold). The *marginal revenue* $R'(q)$ represents rate of change in revenue. That is, it is the approximate increase in revenue as a result of an additional sale of one more unit of the product when the sales level is q units.

Finally, the *profit* $P(q)$ represents the profit to the firm when it produces (and hence sells) q units of the product. It is simply the revenue minus the cost. That is

$$\text{Profit} = \text{Revenue} - \text{Cost}$$

The *marginal profit* $P'(q)$ is the rate of change in the profit, or approximately the increase in profit as a result of producing one more unit of the product when the current production level is q.

Throughout the remainder of this book we will see how these three important marginal quantities are useful in economics.

Example 6

Revenue, Cost, and Profit   For a company that manufactures pocket calculators suppose the revenue and profit from the daily production (and hence sale) of q calculators are given by

$$R(q) = 15q \qquad\qquad \text{(revenue function)}$$
$$C(q) = 1000 + 5q + 0.01q^2 \quad \text{(cost function)}$$

(a) Find the company's daily profit if it produces q calculators.

(b) Find the marginal revenue, marginal cost, and marginal profit.

(c) Find the number q of items produced that makes the marginal cost equal to the marginal revenue.

Solution

(a) The product $P(q)$ resulting from the sale of q items is given by

$$P(q) = R(q) - C(q)$$
$$= 15q - (1000 + 5q + 0.01q^2)$$
$$= -0.01q^2 + 10q - 1000$$

(b) The marginal quantities are

$$MR = R'(q) = 15 \qquad \text{(marginal revenue)}$$
$$MC = C'(q) = 0.02q + 5 \qquad \text{(marginal cost)}$$
$$MP = P'(q) = -0.02q + 10 \qquad \text{(marginal profit)}$$

(c) The marginal revenue is equal to the marginal cost when

$$R'(q) = C'(q)$$

or

$$15 = 0.02q + 5$$
$$0.02q = 10$$
$$q = 500 \qquad \qquad \square$$

Interpretation of Results

The fact that the marginal revenue is always equal to 15 for all q means that the company will always take in an additional \$15 for each additional calculator produced. In other words, \$15 is the selling price of the calculator. Also note that the marginal cost is larger for higher production levels. This could be due to decreased efficiency in the factory at higher production levels. For example, when 100 calculators per day are produced, it costs the company

$$C'(100) = 0.02(100) + 5$$
$$= \$7$$

to produce an additional calculator. However, when 500 calculators per day are produced, the cost to produce an additional calculator has risen to

$$C'(500) = 0.02(500) + 10$$
$$= \$15$$

Note that when 500 calculators are produced, the cost to produce an additional calculator is the same as the selling price of the calculator. Hence there is no reason to produce any more calculators than 500.

Example 7

Marginal Cost   A company that produces microcomputers has determined that the cost $C(q)$ in dollars to produce q computers per week can be approximated by the quadratic polynomial

$$C(q) = 100,000 + 400q + 0.06q^2$$

(a) Find the marginal cost when the production level is 1000 computers.

(b) Find the marginal cost when the production level is 2000 computers.

Solution

(a) Since the total cost is

$$C(q) = 100{,}000 + 400q + 0.06q^2$$

the marginal cost, which we denote by MC, is

$$MC = \frac{d}{dq} C(q)$$

$$= \frac{d}{dq} (100{,}000 + 400q + 0.06q^2)$$

$$= 400 + 0.12q \qquad \text{(dollars per computer)}$$

When the production level is 1000 computers per week, the marginal cost is

$$\left. \frac{dC}{dq} \right|_{q=1000} = 400 + 0.12(1000)$$

$$= \$520 \quad \text{per computer}$$

(b) When the level of production has risen to 2000 computers per week, the marginal cost is

$$\left. \frac{dC}{dq} \right|_{q=2000} = 400 + 0.12(2000)$$

$$= \$640 \quad \text{per computer} \qquad \square$$

Interpretation of Results

When production increases from 1000 to 2000 computers, the marginal cost changes from \$520 to \$640. This means that the production cost per computer has risen by \$120. The reason for this increase might be decreased efficiency caused by insufficient plant facilities, overtime wages, or a combination of these and other factors. It is important that a firm keep track of unit operating expenses for different levels of production. This is where marginal cost analysis plays a key role.

Problems

For Problems 1–10, differentiate the powers of x using the power rule for derivatives.

1. $f(x) = 1$

2. $f(x) = x^5$

3. $f(x) = x^{2.5}$

4. $f(x) = \sqrt{x}$

5. $f(x) = \dfrac{1}{x^2}$

6. $f(x) = \dfrac{1}{\sqrt{x}}$

7. $f(x) = \sqrt{x^3}$

8. $f(x) = x^{7/3}$

9. $f(x) = x^{-3}$

10. $f(x) = x^{-3.5}$

For Problems 11–25, use the rules learned in this section to differentiate the functions. Do not use the definition of the derivative to find the derivatives.

11. $f(x) = 4$

12. $f(x) = c$ (c a constant)

13. $f(x) = 2^{1/2}$ **14.** $f(x) = 1 - x$
15. $f(x) = 2x + 1$ **16.** $f(x) = 2x^2$
17. $f(x) = (x + 2)^2$ **18.** $f(x) = (x - 1)^2$
19. $f(x) = ax^2 + bx + c$ (a, b, and c are constants)
20. $f(x) = x^4 + x^3 + x^2 + x + 1$
21. $f(x) = x^{3/2}$ **22.** $f(x) = 2x^{-1} + 3x^2 + 3$
23. $f(x) = x^2 + 1/x$ **24.** $f(x) = 1/x^3 - 7x + 8$
25. $f(x) = 10x^{-4} + 3x^2 + 2x + 7$

Falling Object

A ball dropped from a tall building has fallen a distance of $s(t) = 16t^2$ feet after t seconds.

26. Find the velocity of the ball as a function of time t. What is the velocity after 1 second? After 2 seconds?

27. Find the acceleration (change in velocity) of the ball as a function of time t. What is the acceleration after 1 second? After 2 seconds?

28. Fish Population A new species of fish is introduced into a lake ecosystem. Initially, 100,000 fish are stocked. A marine biologist predicts that the population $P(t)$ of the fish after t years will be

$$P(t) = 100t^2 + 5000t + 100,000$$

What is the rate of change in fish per year after year 1? After year 2? After year 5?

29. Flea Jumping A flea leaping vertically reaches a height of $h(t)$ feet after t seconds where $h(t)$ is given by

$$h(t) = 12t - 16t^2$$

What is the velocity in feet per second of the flea after 1 second? At what time will the velocity of the flea be zero? What is your interpretation of this result?

30. Botany Problem The proportion of seeds of a certain species of tree that scatter farther than the distance r (measured in feet) is given by

$$p(r) = \frac{2}{r} + \frac{3}{r^{1/2}}$$

Find the rate of change in this proportion as a function of r. What is the value of this rate of change when $r = 100$ feet? What is your interpretation of this result?

31. Protein Synthesis After t days, protein synthesizing in a cell was found to have a mass (in grams) given by

$$M(t) = 3t^2 + 1/2t$$

What is the rate of change (grams per day) at day 1? At day 2?

Bacteria Populations

For Problems 32–35, two types of bacteria grow according to the laws

$$p_1(t) = 50t^2 - 20t + 1000$$
$$p_2(t) = 30t^2 - 80t + 2000$$

where $p_1(t)$ and $p_2(t)$ represent the number of each type of bacteria present at time t (measured in hours).

32. Find the instantaneous rate of growth of Bacteria 1.
33. What is the rate of growth of Bacteria 1 after 4 hours? After 6 hours?
34. Find the instantaneous rate of growth of Bacteria 2.
35. Find the instantaneous rate of growth of the total population.

Finding Tangent Lines to Graphs

36. Tangent Line Find the points on the graph of

$$y = 2x^3 + 3x^2 - 6x + 1$$

where the curve has a horizontal tangent line.

37. Tangent Line Find the equation of the tangent line to the graph of

$$y = x^2 + 3x + 3$$

at the point (0, 3).

38. Tangent Line Find the points on the graph of

$$y = x^2$$

where the slope is equal to its height. Draw a picture to illustrate your conclusions.

39. Airplane Problem An airplane takes off, starting from rest. The distance in feet that it travels during the first few seconds is given by the function

$$s(t) = 2t^2 + 4t + 1$$

How fast is the plane traveling after 10 seconds? After 20 seconds?

40. Velocity of a Ball A ball thrown straight up with an initial velocity of v_0 feet per second will have a height $h(t)$ after t seconds where $h(t)$ is given by the function

$$h(t) = -16t^2 + v_0 t$$

What will be the velocity of the ball after 5 seconds if the initial velocity is $v_0 = 12$?

Marginal Analysis in Economics

41. Revenue, Cost, and Profit If the revenue and cost functions are given by

$$R(q) = 10q - 0.01q^2$$
$$C(q) = 5000 + 5q$$

(a) Find the profit function.
(b) Find the marginal revenue, marginal cost, and marginal profit.
(c) For what value of q is the marginal profit equal to zero?

42. Marginal Profit Why is it true that the marginal profit is equal to the marginal revenue minus the marginal cost? What rule for derivatives do you need to verify this fact?

43. Revenue Found from Demand Curve The revenue function can always be found by the formula

$$R(q) = \text{(Number of items sold)(Price per item)}$$

$$= (q)\text{(Price per item)}$$

If the demand function is given by

$$q = D(q) = 4p - 10$$

find the revenue function as a function of q. *Hint:* Solve for the price p in the demand function as a function of q, and substitute this value for the price per item in $R(q)$. This is one way that economists can find the equation for the revenue function.

44. Finding Profit from Demand Curve and Cost The demand function is given by

$$q = D(p) = 2p - 20 \quad \text{(demand curve)}$$

and the cost function is

$$C(q) = 10{,}000 + 4q \quad \text{(cost function)}$$

Find the profit function. *Hint:* First find the revenue function by

$$R(q) = \text{(Number of items sold)(Price per item)}$$

$$= (q)\text{(Price solved from demand curve)}$$

Then compute $P(q) = R(q) - C(q)$.

45. Marginal Cost The cost to produce q units of a product is given by

$$C(q) = 3q^2 - 3q + 12$$

Find the marginal cost of the above cost function. What is the marginal cost when $q = 100$? What is the interpretation of this marginal cost?

46. Marginal Cost The daily cost in dollars to produce q automobiles is found to be given by

$$C(q) = 400{,}000 + 2000q - 10q^2$$

What is the marginal cost when 100 cars are produced in a day? Is the marginal cost increasing or decreasing when more than 100 cars are produced?

8.6 Derivatives of Products and Quotients

PURPOSE

We continue the development of the derivative by presenting the product and the quotient rules. These rules allow us to find the derivative of the product or quotient of two functions in terms of the derivatives of the individual functions. We also introduce more marginal concepts of economics.

Introduction

The product and quotient rule introduced in this section will add substantially to the repertoire of functions that we can easily differentiate. For instance, after mastering this section you will be able to differentiate such formidable-looking expressions as

$$f(x) = \frac{\sqrt{x} + x^2}{x^3 + 4x - x^{-2}}$$

We start with the product rule.

Differentiation of Products

In the previous section we learned how to differentiate sums and differences of differentiable functions. We now make the natural progression to differentiating *products* and *quotients*. Although the natural temptation is to think that the derivative of a product is the product of the derivatives, we will see that this is *not* the case. It is also not true that the derivative of a quotient is the quotient of the derivatives. These rules are more involved.

We begin by presenting a geometric interpretation of the product rule that will help understand how the derivative should be found.

Geometric Interpretation of the Product Rule

Imagine a rectangle as shown in Figure 33 in which both the width $W(t)$ and the height $H(t)$ of the rectangle change with time. Close your eyes and imagine the rectangle changing in height $H(t)$ and width $W(t)$, maybe getting tall and skinny and then later becoming short and fat. Since the area $A(t)$ of the rectangle is the product $A(t) = W(t)H(t)$, the rate of change of the area is the derivative of the product $W(t)H(t)$. That is,

$$\frac{d}{dt} A(t) = \frac{d}{dt} \left[W(t)H(t) \right]$$

To find this rate of change, we allow time to increase by a small amount h from an arbitrary time t to $t + h$. The total change in area of the rectangle will then be

$$\text{Total change in area} = A(t + h) - A(t)$$

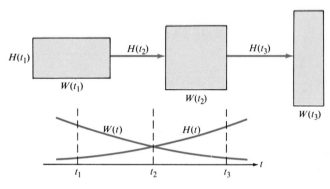

Figure 33
The changing dimensions of a rectangle

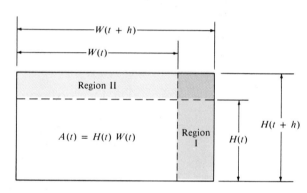

Figure 34
The change in area is essentially the change in Region I and Region II

This change is represented in Figure 34 by the shaded region. If we now ignore the tiny rectangle in the upper right-hand corner of the shaded region, which becomes insignificant when the increase in time h is small, then from Figure 34 we see that the total change in the area is essentially

$$\text{Total change in area} = \text{Region I} + \text{Region II}$$
$$= H(t)\left[W(t + h) - W(t)\right] + W(t)\left[H(t + h) - H(t)\right]$$

To find the instantaneous rate of change of this area, and hence the derivative of $H(t)W(t)$, we divide this total change by h to get the average change and then take the limit as h approaches zero, getting

$$\frac{d}{dt} \left[W(t)H(t) \right] = H(t) \lim_{h \to 0} \frac{W(t + h) - W(t)}{h} + W(t) \lim_{h \to 0} \frac{H(t + h) - H(t)}{h}$$
$$= H(t)W'(t) + W(t)H'(t)$$

The above discussion motivates the general rule for finding the derivative of a product of two functions.

> **Product Rule for Derivatives**
>
> Let f and g be differentiable functions at x. Then
>
> $$\frac{d}{dx}\left[f(x)g(x)\right] = f(x)g'(x) + g(x)f'(x)$$
>
> This rule can be stated by saying that the derivative of a product of two factors is the first factor times the derivative of the second factor plus the second factor times the derivative of the first factor.

We now show how the above derivative rule for products can be used to differentiate various functions.

Example 1

Derivative of a Product   Find $h'(x)$ where

$$h(x) = 3x^3(3x + 4)$$

Solution   Calling the two factors

$$f(x) = 3x^3$$
$$g(x) = 3x + 4$$

we first differentiate each factor separately, getting

$$f'(x) = 9x^2$$
$$g'(x) = 3$$

Now, applying the product rule, we have

$$\frac{d}{dx}\left[f(x)g(x)\right] = f(x)g'(x) + g(x)f'(x)$$

$$= 3x^3(3) + (3x + 4)(9x^2)$$
$$= 36x^2(x + 1) \qquad \qquad \square$$

Example 2

Derivative of a Product   Differentiate

$$h(x) = (\sqrt{x} + x)(x^2 + 3)$$

Solution   We identify the two factors

$$f(x) = \sqrt{x} + x$$
$$g(x) = x^2 + 3$$

We first differentiate each factor separately, getting

$$f'(x) = \frac{1}{2\sqrt{x}} + 1$$

$$g'(x) = 2x$$

Applying the product rule, we have

$$\frac{d}{dx}[f(x)g(x)] = f(x)g'(x) + g(x)f'(x)$$

$$= (\sqrt{x} + x)(2x) + (x^2 + 3)\left(\frac{1}{2\sqrt{x}} + 1\right) \qquad \square$$

Example 3 _____

Differentiation of a Product   Find $h'(x)$ where

$$h(x) = (x^3 + 3x)(x + 3)$$

Solution   Calling

$$f(x) = x^3 + 3x$$
$$g(x) = x + 3$$

we first differentiate each factor separately, getting

$$f'(x) = 3x^2 + 3$$
$$g'(x) = 1$$

Now, applying the product rule, we have

$$\frac{d}{dx}[f(x)g(x)] = f(x)g'(x) + g(x)f'(x)$$

$$= (x^3 + 3x)(1) + (x + 3)(3x^2 + 3)$$
$$= 4x^3 + 9x^2 + 6x + 9 \qquad \square$$

The Derivative of a Quotient

Until now, we have found rules for differentiating sums, differences, and products of functions. Another important rule is the **quotient rule**. We state it here without proof.

Quotient Rule for Derivatives

Let f and g be differentiable functions. Then

$$\frac{d}{dx}\frac{f(x)}{g(x)} = \frac{g(x)f'(x) - f(x)g'(x)}{\{g(x)\}^2}$$

for all points where $g(x) \neq 0$.

This rule can be stated by saying that the derivative of the quotient of two functions is the denominator times the derivative of the numerator minus the numerator times the derivative of the denominator all divided by the denominator squared. (Whew!)

Example 4 _____

Derivative of a Quotient   Find dy/dx if

$$y = \frac{1}{x^2 + 1}$$

Solution   Calling the numerator and the denominator

$$f(x) = 1$$
$$g(x) = x^2 + 1$$

respectively, we have

$$f'(x) = 0$$
$$g'(x) = 2x$$

Hence using the quotient rule, we get

$$\frac{dy}{dx} = \frac{g(x)f'(x) - f(x)g'(x)}{\{g(x)\}^2}$$

$$= \frac{(x^2 + 1)(0) - (1)(2x)}{(x^2 + 1)^2}$$

$$= \frac{-2x}{(x^2 + 1)^2} \qquad \qquad \square$$

Example 5 ——————

Derivative of a Quotient   Find $h'(x)$ where

$$h(x) = \frac{2x + 3}{x^2 + 2x - 3}$$

Solution   Calling the numerator and denominator

$$f(x) = 2x + 3$$
$$g(x) = x^2 + 2x - 3$$

respectively, we have

$$f'(x) = 2$$
$$g'(x) = 2x + 2$$

Using the quotient rule, we get

$$\frac{d}{dx}\frac{f(x)}{g(x)} = \frac{g(x)f'(x) - f(x)g'(x)}{\{g(x)\}^2}$$

$$= \frac{(x^2 + 2x - 3)(2) - (2x + 3)(2x + 2)}{(x^2 + 2x - 3)^2}$$

$$= \frac{-2(x^2 + 3x + 6)}{(x^2 + 2x - 3)^2} \qquad \qquad \square$$

Average Cost and Marginal Average Cost

In the previous section we introduced the idea of marginal analysis in economics and business. Essentially, for every economic quantity that changes over time, there is a corresponding marginal economic quantity that gives the rate of change of this quantity. Suppose, for example, that $C(q)$ is the cost to a firm to manufacture q items. It is clear then that the average cost to produce each item is

$$\text{Average cost per item} = \frac{C(q)}{q}$$

The marginal average cost is the rate of change of this average cost and is found by

$$\text{Marginal average cost} = \frac{d}{dq} \frac{C(q)}{q}$$

To illustrate how the marginal average cost is used in the business world, suppose it has been determined that the monthly cost to a company to manufacture q computers is

$$C(q) = 200\sqrt{q} + 500{,}000$$

From the above cost function it is an easy matter to find the average cost per computer when q computers are produced. It is simply

$$\text{Average cost} = \frac{C(q)}{q} = \frac{200\sqrt{q} + 500{,}000}{q}$$

Knowing the average cost, we can now find the marginal average cost, or the derivative of the average cost. Differentiating the above quotient, we get

$$\frac{d}{dq} \frac{C(q)}{q} = \frac{q(100q^{-1/2}) - (200\sqrt{q} + 500{,}000)(1)}{q^2}$$

$$= \frac{-100[\sqrt{q} + 5000]}{q^2}$$

Graphs of the cost, average cost, and marginal average cost are shown in Figure 35.

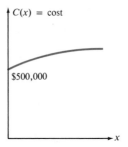
$C(x) = \text{cost}$

$500,000

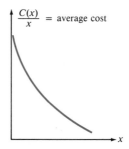
$\frac{C(x)}{x} = \text{average cost}$

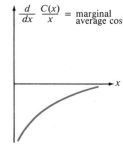
$\frac{d}{dx} \frac{C(x)}{x} = \text{marginal average cost}$

Figure 35
Cost, average cost, and marginal average cost

Interpretation

To interpret the above results, suppose the firm has orders for 10,000 computers for the coming month. The total cost or the cash flow needed to produce these computers will be

$$C(10{,}000) = 200(100) + 500{,}000$$

$$= \$520{,}000$$

As a measure of the efficiency of the firm's production facilities, it is valuable to know the average cost per computer. For this firm it is

$$\frac{C(10{,}000)}{10{,}000} = \frac{520{,}000}{10{,}000}$$

$$= \$52 \quad \text{per computer}$$

This average cost gives a measure of the efficiency to produce all 10,000 computers. On the other hand, the marginal average cost gives a measure of how efficient the firm is at the given production level. For this firm the marginal average cost is

$$\text{Marginal average cost} = \frac{(10,000)(100/100) - (200 \cdot 100 + 500,000)}{(10,000)^2}$$

$$= \frac{10,000 - 520,000}{100,000,000}$$

$$= -\$0.0051 \quad \text{per computer}$$

In other words, the average cost to produce a computer (which is \$52 per computer) is coming down at a half cent for each additional computer produced. The reason for this is that at a production rate of 10,000 computers per month it costs much less to make each new computer than it did when the production level was lower. If we compute the marginal cost (rate of change in the cost), we will find

$$C'(x) = \frac{100}{\sqrt{x}}$$

Hence we get the surprising result that

$$C'(10,000) = \frac{100}{\sqrt{10,000}}$$

$$= \$1$$

which means that it costs the company only \$1 to make a computer when the production level is 10,000 computers. This explains why the average cost per computer is coming down by one-half cent per new computer produced when the production level is 10,000 computers.

Problems

For Problems 1–20, use the product rule or the quotient rule to find the indicated derivatives.

1. $f(x) = x^2(x^2 + x + 3)$

2. $f(x) = (x + 1)(x - 1)$

3. $f(x) = (x^2 + 1)(x^2 - 1)$

4. $f(x) = (x^2 + x + 1)(x^2 - 3x - 4)$

5. $y = (t^2 + 1)(t^2 - 1)$

6. $f(x) = \sqrt{x} \cdot (x^2 - x^{-1})$

7. $g(u) = (u^2 + 1)(u^2 + u - 1)$

8. $f(x) = x^{3/2}(1 - x)$

9. $f(x) = \frac{1}{x^2}$

10. $f(x) = \frac{1}{x^2 + x + 1}$

11. $f(x) = \frac{5}{x^3 + 5}$

12. $g(x) = \frac{x}{x - 1}$

13. $h(x) = \frac{x^2 - 1}{x + 2}$

14. $f(x) = \frac{x(x^2 + 1)}{2x + 3}$

15. $f(x) = x^2 + \frac{2}{x + 1}$

16. $f(x) = \frac{2x^2 + x + 3}{x^2(x^2 + 1)}$

17. $y = \frac{x^2 + 1}{(2x + 1)(x^2 + 3)}$

18. $y = x + \frac{1}{x^2 + 1}$

19. $y = \frac{1}{\sqrt{x}}$

20. $y = \frac{\sqrt{x}}{\sqrt{x} + 1}$

21. **Extended Product Rule**   Show that for functions f, g, and h differentiable at x that the derivative rule

$$\frac{d}{dx}\left[f(x)g(x)h(x)\right] = f'(x)g(x)h(x) + f(x)g'(x)h(x)$$

$$+ f(x)g(x)h'(x)$$

holds.

22. **Extended Product Rule**   Use the extended product rule developed in Problem 21 to differentiate

$$\frac{d}{dx}\left[(x^2 + 1)(x^2 + x + 3)(1 + 2x)\right]$$

23. **Marginal Revenue**   If $p(x)$ is the price per unit at which x units of a commodity can be sold, then $R = x \cdot p(x)$ is the revenue of the commodity. Show that the marginal revenue is given by

$$\frac{d}{dx}R(x) = p(x) + x\frac{d}{dx}p(x)$$

24. **Marginal Revenue**   Assume that the price per unit is given by

$$p(x) = \frac{50}{x + 5}$$

Find the marginal revenue.

25. **Capital Depreciation**   A firm has determined that a capital investment depreciates according to the rule

$$C(t) = \frac{100,000}{t + 2}$$

where $C(t)$ is the value of the investment measured in dollars and t is the age of the investment in years. Find the rate of depreciation of the investment (in dollars per year) after 1 year. After 2 years. After 3 years.

26. **Adrenalin Injection**   Medical data predict that the electrical response y (in millivolts) of a muscle to the amount of adrenalin injected x (in cubic centimeters) is given by the function

$$y(x) = \frac{x}{a + bx}$$

What is the rate of change of y with respect to x? What is your interpretation of this result?

27. **Velocity of a Particle**   The displacement $s(t)$ in feet of a particle from a given reference point is given by

$$s(t) = \frac{100}{t + 1} - 100$$

where t is time measured in seconds. What is the velocity of the particle after 1 second? After 2 seconds? After 3 seconds?

28. **Velocity of a Particle**   The position $s(t)$ of an object on the x-axis is given by

$$s(t) = 10 - \frac{20}{t^2 + 1}$$

where t is measured in seconds. What is the velocity of the object at any time t?

29. **Marginal Revenue**   Find the marginal revenue when x units of a product are sold if the price p charged for each unit is

$$p(x) = 60 - 3x$$

Note that the price charged for each item depends on the number sold. *Hint:* The revenue is given by $R(x) = x \cdot p(x)$.

An Analysis of Costs

A firm has determined that it costs

$$C(x) = 100\sqrt{x} + 100,000$$

to produce x units of a given product. For Problems 30–38, determine the types of costs.

30. Find the marginal cost when the production rate is x units.

31. Find the marginal cost when the production rate is 100 units.

32. Find the marginal cost when the production rate is 10,000 units.

33. Find the average cost per unit when the production rate is x units.

34. Find the average cost per unit when the production rate is 100 units.

35. Find the average cost per unit when the production rate is 10,000 units.

36. Find the marginal average cost when the production rate is x units.

37. Find the marginal average cost when the production rate is 100 units.

38. Find the marginal average cost when the production rate is 10,000 units.

39. Blood Flow According to *Poiseuille's law*, the total resistance R to blood flow in a blood vessel of constant length L and radius r is given by

$$R = \frac{kL}{r^4}$$

where k is a constant of proportionality determined by the viscosity of the blood. How fast is the resistance decreasing when $r = 0.2$ mm?

40. Rate of Change Assume that the width W and height H of a rectangle depend on time t according to the functions

$$H(t) = 3t^2 + 2t + 1$$
$$W(t) = t - 3$$

What is the area of the rectangle and what is the rate of change of the area after 2 seconds? After 4 seconds?

41. General Derivative Formula Find y' if

$$y = \frac{xf(x)}{g(x)}$$

Use the result to find the derivative of

$$y = \frac{x(x^2 + 1)}{(x^2 + 3)^2}$$

42. Misconception A common error for beginners is to differentiate a product of two functions by use of the formula

$$\frac{d}{dx}\left[f(x)g(x)\right] = \frac{d}{dx}f(x)\frac{d}{dx}g(x)$$

Are there any pairs of functions from the five functions listed below for which this identity actually holds? Show that this relation does not hold for the other pairs.
(a) $a(x) = 13$
(b) $b(x) = x + 1$
(c) $c(x) = x$

(d) $d(x) = \dfrac{1}{x - 1}$

(e) $e(x) = \dfrac{1}{x}$

43. Fish Population The total weight W in pounds of a fish population in a certain lake is given by the product

$$W = nw$$

where

$$n = \text{number of fish in the lake}$$
$$w = \text{average weight of the fish}$$

Suppose a biologist estimates that n and w will change according to the following functions:

$$n(t) = 0.5t^2 + t + 1000$$
$$w(t) = 0.3t^2 + 2t + 1$$

where t is time measured in years. What is the total weight W of the fish population, and what is the rate of change in the total weight after one year? After two years?

44. Algae Study The density D of algae in a container is given by the quotient

$$D = \frac{n}{V} \quad \text{(algae per cubic centimeter)}$$

where

$$n = \text{number of algae}$$
$$V = \text{volume of the container} \quad \text{(cubic centimeters)}$$

Suppose that both the number of algae and the volume of the container change with time according to the functions

$$n(t) = \sqrt{t} + 100 \quad \text{(number of algae)}$$

$$V(t) = \frac{1000}{\sqrt{t}} \quad \text{(cubic centimeters)}$$

where t is time measured in hours. What is the density D and the rate of change D' of the density after 100 hours?

45. Challenge Problem The following function was mentioned as a complicated function that could be differentiated after learning the material of this section. Can you differentiate this function?

$$f(x) = \frac{\sqrt{x} + x^2}{x^3 + 4x - x^{-2}}$$

8.7

The Chain Rule

PURPOSE

We introduce a rule that shows how to find the derivative of the composition of two differentiable functions

$$y = f[g(x)]$$

in terms of the derivative of $f(x)$ and the derivative of $g(x)$. This rule, called the chain rule, allows us to differentiate a wide variety of functions.

Introduction

We are coming to the end of the line of the rules for differentiation. Although it does not really make sense to say that one rule is more important than any other rule, most people would probably say that the chain rule is the most important. You should understand how this rule works, since it will be used over and over throughout the remainder of this book. In the next chapter, when we study related rates, it will be critical that you understand this rule. To master its use, we must recognize complicated functions as compositions of simpler functions. When the individual components of the composition have been recognized, differentiation by the chain rule is relatively simple. You should review how compositions of functions are formed, as explained in the Calculus Preliminaries.

The Chain Rule

We begin by presenting a geometric interpretation of the chain rule, as shown in Figure 36. This will give you an intuitive understanding of this important rule. Let us assume that y depends on u by means of some relationship

$$y = f(u)$$

and that u depends on x by means of a second relationship

$$u = g(x)$$

To find the derivative of y with respect to x, we must determine the relationship between the change in y and the change in x. To find this, suppose a small change in x of $\Delta x = 0.1$ gives rise to a change in u of $\Delta u = 0.4$. Suppose, too, that a small change in u of $\Delta u = 0.1$ gives rise to a change in y of $\Delta y = 0.2$. By knowing how x affects u and how u affects y, we can determine how x affects y. We simply have the ratio of the change in y divided by the change in x as

$$\frac{\Delta y}{\Delta x} = \frac{\Delta y}{\Delta u}\frac{\Delta u}{\Delta x} = \frac{0.2}{0.1}\frac{0.4}{0.1} = (2)(4) = 8$$

In other words, y is changing 8 times faster than x. If we now let Δx tend to zero, we arrive at the following formula:

$$\frac{dy}{dx} = \frac{dy}{du}\frac{du}{dx}$$

This is the **chain rule**. Intuitively, the chain rule says that for small changes in x,

$$\frac{\text{Change in } y}{\text{Change in } x} = \frac{\text{Change in } y}{\text{Change in } u} \cdot \frac{\text{Change in } u}{\text{Change in } x}$$

It is useful to note that in the formula for the chain rule, dy/du is the derivative of $y = f(u)$ with respect to u and du/dx is the derivative of $u = g(x)$ with respect to x.

Figure 36

Here y depends on u, and u depends on x

x $u = g(x)$ $y = f(u) = f[g(x)]$

We now give a formal statement of the chain rule.

Chain Rule for Differentiating Compositions

Let

$$y = f(u)$$

be a differentiable function of u, where

$$u = g(x)$$

is a differentiable function of x. Then the composition of these functions

$$y = f(u) = f[g(x)]$$

is also a differentiable function of x whose derivative is

$$\frac{dy}{dx} = \frac{dy}{du}\frac{du}{dx} \qquad \text{(chain rule)}$$

An alternative way of writing the above chain rule is

$$\frac{dy}{dx} = f'[g(x)] \cdot g'(x) \qquad \text{(alternative form)}$$

Example 1

Chain Rule   Find dy/dx if

$$y = (x^2 + 2x + 1)^5$$

Solution   The above function is a composition, or is "built up" from the two simpler functions

$$y = u^5$$

and

$$u = x^2 + 2x + 1$$

To use the chain rule, first compute the following two derivatives with respect to u and x, respectively:

$$\frac{dy}{du} = 5u^4$$

$$\frac{du}{dx} = 2x + 2$$

The chain rule then gives

$$\frac{dy}{dx} = \frac{dy}{du}\frac{du}{dx}$$

$$= 5u^4(2x + 2)$$
$$= 5(x^2 + 2x + 1)^4(2x + 2)$$

The last step above consisted of replacing the value of u by its expression in terms of x so that the final derivative dy/dx is given completely in terms of the variable x.   ☐

Example 2

Find dy/dx if y is given by

$$y = \left(\frac{x-1}{x+1}\right)^3$$

Solution   The above function can be interpreted as the composition of

$$y = u^3$$

where u is

$$u = \frac{x-1}{x+1}$$

If we now differentiate y with respect to u and u with respect to x, we have

$$\frac{dy}{du} = 3u^2 \qquad \text{(power rule)}$$

$$\frac{du}{dx} = \frac{(x+1)-(x-1)}{(x+1)^2} \qquad \text{(quotient rule)}$$

$$= \frac{2}{(x+1)^2}$$

Using the chain rule, we get

$$\frac{dy}{dx} = \frac{dy}{du}\frac{du}{dx}$$

$$= 3u^2 \frac{2}{(x+1)^2}$$

$$= 3\frac{(x-1)^2}{(x+1)^2}\frac{2}{(x+1)^2}$$

$$= 6\frac{(x-1)^2}{(x+1)^4} \qquad\qquad ☐$$

Example 3

Chain Rule   Find the derivative dy/dx if

$$y = \frac{1}{x^2+1}$$

Solution   The above expression is a composition of the two functions

$$y = \frac{1}{u}$$

and

$$u = x^2 + 1$$

If we now differentiate y with respect to u and u with respect to x, we get

$$\frac{dy}{du} = -\frac{1}{u^2} \qquad \text{(power rule)}$$

$$\frac{du}{dx} = 2x \qquad \text{(power rule)}$$

Applying the chain rule, we have

$$\frac{dy}{dx} = \frac{dy}{du}\frac{du}{dx}$$

$$= \frac{-1}{u^2}(2x)$$

$$= \frac{-2x}{(x^2 + 1)^2} \qquad \text{(substituting for } u\text{)} \qquad \square$$

Remembering the Chain Rule

The chain rule

$$\frac{dy}{dx} = \frac{dy}{du}\frac{du}{dx}$$

is easy to remember by saying "dee y dee x is equal to dee y dee u times dee u dee x."

Example 4 ────────

Alternative Form of the Chain Rule Find the derivative $h'(x)$ of the expression

$$h(x) = \frac{1}{(x^2 + 1)^2}$$

Solution We use a slightly different notation in this problem so that we can use the alternative form of the chain rule. The above function can be interpreted as the composition

$$h(x) = f[g(x)]$$

of the two simpler functions

$$f(u) = \frac{1}{u^2}$$

where u is

$$g(x) = x^2 + 1$$

First, we differentiate $f(u)$ with respect to u and $g(x)$ with respect to x, getting

$$f'(u) = -\frac{2}{u^3}$$

$$g'(x) = 2x$$

The alternative form of the chain rule gives the derivative of $h'(x)$ as

$$h'(x) = f'[g(x)] \cdot g'(x)$$

$$= -\frac{2}{[g(x)]^3}(2x)$$

$$= -\frac{4x}{(x^2 + 1)^3} \qquad \square$$

The chain rule allows us to state the following useful rule.

> **Power Rule for Functions**
>
> Let g be a differentiable function and let a be a real constant. The power rule for differentiating powers of functions states that
>
> $$\frac{d}{dx}[g(x)]^a = a[g(x)]^{a-1}g'(x)$$

Example 5

Power Rule for Functions   Find dy/dx if

$$y = (x^3 + 3x + 4)^6$$

Solution   Letting

$$g(x) = x^3 + 3x + 4$$

we have

$$g'(x) = 3x^2 + 3$$

Hence the power rule with $a = 6$ states that

$$\frac{d}{dx}(x^3 + 3x + 4)^6 = 6(x^3 + 3x + 4)^5(3x^2 + 3) \qquad \square$$

Example 6

Find dy/dx if

$$y = \frac{1}{\sqrt{(x^2 + 1)^3}}$$

Solution   Calling

$$g(x) = x^2 + 1$$

we have

$$g'(x) = 2x$$

We now write the above expression as

$$y = \frac{1}{\sqrt{g(x)^3}}$$

$$= [g(x)]^{-3/2}$$

Using the power rule for functions with $a = -3/2$, we get

$$\frac{dy}{dx} = -\frac{3}{2}[g(x)]^{-5/2}g'(x)$$

$$= -\frac{3}{2}(x^2 + 1)^{-5/2}(2x)$$

$$= \frac{-3x}{\sqrt{(x^2 + 1)^5}} \qquad \square$$

Example 7 ——————————

Product and Chain Rule   Find the derivative dy/dx if
$$y = 3x^2 \sqrt{x^3 + 4}$$

Solution   We can interpret y as the product of the two functions
$$f(x) = 3x^2$$
$$g(x) = \sqrt{x^3 + 1}$$

Applying the product and power rule, we have

$$\frac{dy}{dx} = \frac{d}{dx}[f(x)g(x)]$$

$$= f(x)\frac{d}{dx}g(x) + g(x)\frac{d}{dx}f(x) \qquad \text{(product rule)}$$

$$= 3x^2 \frac{d}{dx}\left[\sqrt{x^3 + 4}\right] + \sqrt{x^3 + 4}\,\frac{d}{dx}\left[3x^2\right]$$

$$= 3x^2 \frac{1}{2}(x^3 + 4)^{-1/2}(3x^2) + \sqrt{x^3 + 4}(6x) \qquad \text{(power rule)}$$

$$= \frac{9}{2}x^4(x^3 + 4)^{-1/2} + 6x\sqrt{x^3 + 4} \qquad\qquad \square$$

Example 8 ——————————

Quotient and Chain Rule   Find the derivative of

$$y = \frac{(7x - 4)^5}{(x + 1)}$$

Solution   Calling the numerator and denominator
$$f(x) = (7x - 4)^5$$
$$g(x) = x + 1$$

respectively, we use the quotient rule

$$\frac{dy}{dx} = \frac{d}{dx}\frac{f(x)}{g(x)}$$

$$= \frac{g(x)f'(x) - f(x)g'(x)}{[g(x)]^2} \qquad \text{(quotient rule)}$$

$$= \frac{(x + 1)\frac{d}{dx}(7x - 4)^5 - (7x - 4)^5\frac{d}{dx}(x + 1)}{(x + 1)^2}$$

$$= \frac{(x + 1)(5)(7x - 4)^4(7) - (7x - 4)^5(1)}{(x + 1)^2} \qquad \text{(chain rule)}$$

$$= \frac{(7x - 4)^4(28x + 39)}{(x + 1)^2} \qquad\qquad \square$$

All of the rules that we have learned in this chapter are summarized in Table 12.

Rule	Function	Derivative
Constant Function	$y = c$	$y' = 0$
Constant Times a Function	$y = cf(x)$	$y' = cf'(x)$
Power Rule	$y = x^a$	$y' = ax^{a-1}$
Derivative of a Sum	$y = f(x) + g(x)$	$y' = f'(x) + g'(x)$
Derivative of a Difference	$y = f(x) - g(x)$	$y' = f'(x) - g'(x)$
Derivative of a Product	$y = f(x)g(x)$	$y' = f(x)g'(x) + g(x)f'(x)$
Derivative of a Quotient	$y = \dfrac{f(x)}{g(x)}$	$y' = \dfrac{g(x)f'(x) - f(x)g'(x)}{[g(x)]^2}$
Chain Rule	$y = f(u)$ $u = g(x)$	$\dfrac{dy}{dx} = \dfrac{dy}{du}\dfrac{du}{dx}$
Power Rule for Functions	$y = [g(x)]^a$	$y' = a[g(x)]^{a-1}g'(x)$

More Theory of the Firm (Profits and Marginal Profits)

The **total profit** $P(q)$ of a firm depends on the number q of items the firm produces if we always assume that the firm sells all that it produces. The derivative $P'(q)$ of the profit is called the **marginal profit** and represents the rate of change of the profit. In practice, it is often interpreted as the additional profit to the firm for making one additional item when the current production level is q.

Example 9

Marginal Profit   A firm's weekly profit P (in thousands of dollars) for producing q units of a given product has been determined to be

$$P(q) = \sqrt{500q - q^2} \qquad (0 \le q \le 500)$$

A graph of this profit function is shown in Figure 37. Find the marginal profit when 100 units are produced and when 250 units are produced.

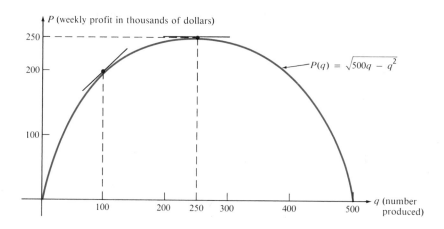

Figure 37
Typical profit function

Solution   To find the marginal profit, we differentiate $P(q)$ with respect to q. Using the power rule for functions, we have

$$\frac{d}{dq}\sqrt{500q - q^2} = \frac{d}{dq}(500q - q^2)^{1/2}$$

$$= \frac{1}{2}(500q - q^2)^{-1/2}(500 - 2q)$$

$$= \frac{250 - q}{\sqrt{500q - q^2}} \qquad (0 \le q \le 500)$$

Hence the marginal profit when production levels are 100 and 250 is given by

$$P'(100) = \frac{250 - 100}{\sqrt{50{,}000 - 10{,}000}} = 0.75$$

$$P'(250) = \frac{250 - 250}{\sqrt{125{,}000 - 62{,}500}} = 0 \qquad \square$$

Interpretation of Results

The interpretation of the above marginal profit calculations is that the firm's profit will increase by an additional 0.75 unit (or $750) when production is raised from 100 units to 101 units per week. On the other hand, when production is at the higher level of 250 items per week, the marginal profit is zero. This means that essentially no additional profit is earned by producing item 251. From Figure 37 we can see that when production is less than 250 items per week, it makes sense to produce more items, since the profit will be increased (provided that the firm can sell the items). When production reaches 250 items per week, however, the marginal profit becomes zero, and no additional profit is earned by increased production. In fact, when production becomes larger than 250 items per week, the marginal profit becomes negative, which means that the firm will lose money on each additional item produced. The reason for this negative return might be decreased efficiency as a result of crowded working conditions, insufficient plant facilities, overtime wages, or a wide variety of other reasons.

Problems

For Problems 1–13, use the chain rule to find the indicated derivatives.

1. $f(x) = (x^2 + 3x + 1)^4$
2. $f(x) = (x^3 + x^2 + 3)^2$
3. $f(x) = \sqrt{x^2 + 3x + 1}$

4. $f(x) = \dfrac{1}{x^2 + 1}$ (Also do this problem using the quotient rule.)

5. $f(x) = \dfrac{1}{\sqrt{(x^2 + x + 1)^3}}$

6. $f(x) = \dfrac{(x^2 + 1)^3}{(x - 2)^2}$

7. $f(x) = \left(x^2 + \dfrac{1}{x}\right)^3$

8. $f(x) = (x^{-1} + x^{-2})^{-3}$

9. $f(x) = [(x + 1)^2 + x]^3$

10. $f(x) = (2x + 1)^{50}$

11. $f(x) = \dfrac{1}{(x^2 + 1)^{10}}$

12. $f(x) = (3x^2 - x)^5(2x + 4)^5$

13. $f(x) = \dfrac{(x + 5)^5}{(x + 4)^3}$

Chain Rule Exercise

For Problems 14–19, let

$$f(x) = (x^2 + 1)^3$$
$$g(x) = 3x - 4$$

Find the indicated quantities.

14. $f[g(x)]$ **15.** $g[f(x)]$

16. $\dfrac{d}{dx} f[g(x)]$ **17.** $\dfrac{d}{dx} g[f(x)]$

18. $\dfrac{d}{dx} f[g(x)]$ when $x = 1$ **19.** $\dfrac{d}{dx} g[f(x)]$ when $x = 1$

20. Find $f'(2)$ if

$$f(z) = \dfrac{1}{z^2 - 1}$$

For Problems 21–23, find the equation of the tangent line to the graphs of the following functions at the indicated points.

21. $y = (x^2 + 1)^3;\ (1, 8)$ **22.** $y = \dfrac{(x + 1)}{(x - 1)^{1/2}};\ (2, 3)$

23. $y = (x^2 + x + 1)^3;\ (1, 27)$

24. General Derivative Rule   Use the derivative rule

$$\frac{d}{dx} \frac{1}{u(x)} = -\frac{u'(x)}{[u(x)]^2}$$

to find the derivative of

$$y = \frac{1}{(x^2 + 1)^2}$$

Alternative Form of Chain Rule

For Problems 25–28, use the alternative form of the chain rule,

$$y' = f'[g(x)] \cdot g'(x)$$

to differentiate the functions. Identify each of the functions $f(x)$, $g(x)$, $f'(x)$, $g'(x)$, $f'[g(x)]$, and dy/dx.

25. $y = (x^2 + 1)^5$ **26.** $y = (x^2 + x + 3)^{-5}$
27. $y = \sqrt{2x + 1}$ **28.** $y = (3x^2 + 1)^3$

Derivatives in the Sciences

For Problems 29–35, the functions have been taken from textbooks in economics, zoology, ecology, sociology, and botany describing phenomena in these areas. Find the rates of change of these functions with respect to the relevant independent variable.

29. $R(q) = 1000\left(2 - \dfrac{q}{500}\right)^2$ (sales revenue)

30. $c(t) = \dfrac{50}{\sqrt{t + 2}}$ (concentration of a hormone)

31. $N(t) = 50\sqrt{t + 4} + 100$ (animal population)
32. $y(t) = 30\sqrt{6t}$ (substrate produced by a hormone)
33. $y(x) = k(x - a)^{1/3}$ (response to visual brightness)
34. $y(x) = k(x - a)^{8/5}$ (response to warmth)
35. $y(x) = k(x - a)^{7/2}$ (response to an electrical stimulus)

36. Tangent Line   Show that there are exactly two tangent lines to the graph of the function

$$y = (x + 1)^3$$

that have x-intercept zero. Find the equation of each of these lines.

37. Related Rates   If Bill is running twice as fast as Mary, and Mary is running three times as fast as Joe, how many times as fast as Joe is Bill running? Set this problem up as a composition and solve it by the chain rule.

38. Related Rates   The radius r of a spherical balloon is expanding according to the function

$$r(t) = 2t^2 + 2t + 3$$

where r is measured in centimeters and t in seconds. What is the rate of change in the radius of the balloon when $t = 2$ seconds? What are the units of this rate of change?

39. Chain Rule   Find dy/dx if

$$y = 2u^2 + 5$$
$$u = 3x^2 - 7$$

40. Chain Rule   Let

$$y = [1 - (2x - 1)^2]^2$$

Identify $f(x)$ and $g(x)$ so that

$$y = f[g(x)]$$

and find dy/dx.

41. Marginal Profit   The daily profit P a company earns from producing q items is given by

$$P(q) = \sqrt{1200q - q^3} \qquad (0 \le q \le 35)$$

What is the marginal profit when the production level is 10 units per day? When the production level is 20 units per day? What are your conclusions?

42. **Marginal Revenue**   The daily revenue R received from selling q units of a certain product is given by

$$R(q) = \sqrt{400q - q^2} \qquad (0 \le q \le 400)$$

What is the marginal revenue when sales are 50 units a day? When sales are 100 units a day? What can you conclude?

Epilogue: Marginality and the Derivative

Economics is very much concerned with change. For this reason the derivative is a very important tool. In economics the word **marginality** refers to the change in a quantity. We list some of the major marginal quantities in economics.

Economic Concept	Mathematical Representation	Meaning
Cost Function	$C(q)$	Cost to a company to make q items
Marginal Cost	$C'(q)$	Approximate change brought about by the production of one extra item when q units are produced
Demand Function	$D(p)$	Consumer demand as a function of price p
Marginal Demand	$D'(p)$	Approximate change in demand brought about by a unit change in cost when the price is p
Revenue Function	$R(q)$	Amount of money taken in as a result of selling q items
Marginal Revenue	$R'(q)$	Approximate change in revenue brought about by the sale of an additional item when q units are sold
Supply Function	$S(p)$	Amount industry is willing to supply at a price p
Marginal Supply	$S'(p)$	Approximate change in the amount supplied brought about by a unit change in price when the price is p
Consumption Function	$C(I)$	Amount a country spends as function of its income I
Marginal Propensity to Consume	$C'(I)$	Approximate change in spending brought about by a unit change in income

Key Ideas for Review

Review Exercises

For Problems 1–10, find the limits.

1. $\lim\limits_{x \to 1} (x^2 + 2x + 1)$

2. $\lim\limits_{x \to -1} \dfrac{x^2 - 1}{x + 1}$

3. $\lim\limits_{x \to 1} (\sqrt{x + 1} - \sqrt{x})$

4. $\lim\limits_{x \to 0} (\sqrt{x^2 + x} - \sqrt{x})$

5. $\lim\limits_{x \to 2} \dfrac{2x^2 + 3}{5x^2 + 4}$

6. $\lim\limits_{x \to -1} \dfrac{2x^2 + 3}{5x^2 - 6}$

7. $\lim\limits_{x \to 4} \dfrac{\sqrt{x} - 2}{x^2 - 4}$

8. $\lim\limits_{x \to 2} \sqrt{x^2 + 2x}$

9. $\lim\limits_{x \to 0} \dfrac{\sqrt{x + 9} - 3}{x}$

10. $\lim\limits_{x \to 1} \dfrac{x - 1}{|x| - 1}$

For Problems 11–20, differentiate each of the functions.

11. $f(x) = 2x + 1/x$

12. $f(x) = (x + 1)^2$

13. $f(x) = (3x^2 + 4)^3$

14. $f(x) = \sqrt{x} \cdot (2x + 3x^2)$

15. $f(t) = 3\sqrt{t} + t^{-1/2}$

16. $f(z) = \dfrac{z}{1 - z}$

17. $f(z) = (z + \sqrt{z})^{1/2}$

18. $f(z) = (1 + z^2)^{1/3}$

19. $f(r) = 4r^2$

20. $f(x) = (x^2 + 1)(3x^2 + x - 4)$

For Problems 21–31, find the indicated quantities.

21. $\dfrac{d}{dx} (x^2 + 3x + 1) \Big|_{x = 1}$

22. $\dfrac{d}{dx} (x^2 + \sqrt{x}) \Big|_{x = 1}$

23. $\dfrac{d}{dx} \left(\dfrac{x + 1}{x - 1} \right) \Big|_{x = -2}$

24. $\dfrac{d}{dz} \sqrt{z^2 + 1} \Big|_{z = 0}$

25. Find ds/dt if $s(t) = 16t^2$

26. Find du/dv if $u = v^4 + v$

27. Find $\dfrac{d}{dt} (3u^2)$ if $u = 2t + 1$

28. Find $\dfrac{d}{dx} u(x)$ if $u = 3x^2 + x$

29. Find $f'(4)$ if $f(x) = \sqrt{x}$

30. Find $f'(1)$ if $f(z) = 3z^2 + z - 1$

31. Find $\dfrac{d}{dx} [(x^2 + 1)(x^2 - 1)(x^2 + 2)]$

32. Moving Water Wave   A circular water wave is moving outward at the rate of 1 foot per second and hence the area of the circle inside this wave is

$$A = \pi t^2$$

How fast is the enclosed area changing after 10 seconds?

33. Related Rates   A right circular cylinder has radius r and height h that are both changing with respect to time according to the functions

$$r(t) = 3t + \frac{1}{t}$$

$$h(t) = 2t^2 + 3t - 1$$

What is the rate of change of the surface area of the cylinder at time $t = 1$ second? *Hint:* The surface area is given by $S = r^2 \pi h$.

Marginality

The cost C and revenue R when x units of a product are produced are given by

$$C(x) = x^2 + x + 5$$

$$R(x) = 9x$$

For Problems 34–37, determine the indicated quantities.

34. What is the marginal cost?

35. What is the marginal revenue?

36. What production level makes the marginal revenue equal to the marginal cost? What is your interpretation of this result?

37. What is the profit function $P(x)$? What is the value of the profit when the marginal profit is zero?

38. Marginal Cost   The cost of producing x widgets is given by

$$C(x) = 0.36x^2 - 0.14x + 2$$

What is the marginal cost when the production is 100 widgets?

39. Marginal Profit   If the revenue function for the widgets in Problem 38 is

$$R(x) = 7.06x$$

what is the marginal profit $P(x)$?

Chapter Test

1. Find the limit

$$\lim_{x \to 36} \frac{6 - \sqrt{x}}{x + 36}$$

2. Find the derivative of

$$f(x) = \frac{1}{x - 3}$$

by using the definition.

3. Find the derivative of

$$f(x) = [\sqrt{x} + 1]^{1/2}$$

4. Find the derivative of

$$f(x) = (x^2 + 1)(5x - 1)$$

using the product rule.

5. Find the derivative of

$$f(x) = x^{4/3} + x^{3/4}$$

6. An object moves along a line so that after t seconds the distance $s(t)$ in feet from a given reference point is

$$s(t) = 50 - 4t - 5t^2$$

Find $s'(1)$ and $s'(2)$. Interpret the results.

Problems 7–10 refer to the graph in Figure 38.

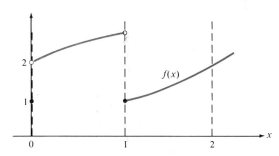

Figure 38
Graph for Problems 7–10

Problems 7–10 refer to the graph in Figure 38.

7. Find the following limit:

$$\lim_{x \to 1} f(x)$$

8. Is f continuous on the interval $(0, 1)$? Why?

9. Is f continuous on the interval $(1, 2)$? Why?

10. Is f continuous on the interval $(0, 2)$? Why?

9

Applications of
the Derivative

The derivative is one of the most useful concepts in all of mathematics. In this chapter we show how it can be used to solve a variety of problems in business and science. In Sections 9.1 and 9.2 we begin by showing how the derivative can be used to analyze in detail the graph of a function. In Sections 9.3 and 9.4 we show how the derivative can be used to find both the maximum and the minimum of a function and to solve an important class of problems known as optimization problems. Finally, in Section 9.5 we introduce what are known as related rates problems, which allow us to find the rate of change of one variable in terms of the rate of change of another variable.

<table>
<tr><td>9.1</td><td></td></tr>
</table>

Use of the Derivative in Graphing

PURPOSE

We introduce several tools used in graphing functions. The major tools are

- the determination of increasing and decreasing regions of a function and
- the determination of critical and stationary points of a function.

Introduction

Beginning students of mathematics usually graph functions by plotting a series of points and then drawing curves of some type connecting the plotted points. Although this point-plotting approach may give a rough shape of the curve, the method does not provide enough detailed information about the function to solve many advanced graphing problems. No matter how many points are plotted, the values of the function between the points can only be surmised. For example, between two successive plotted points, we must guess whether the graph exhibits a downward curvature (Figure 1a), an upward curvature (Figure 1b), or both (Figure 1c).

In this section and the next we will show how the derivative can aid in graphing functions. We begin by determining when a function is increasing and when it is decreasing.

Increasing and Decreasing Functions

The adjectives *increasing* and *decreasing*, when applied to a function, refer to the behavior of the graph of a function as we move from left to right along the real axis. If the graph is rising at a point as we move from left to right, we say

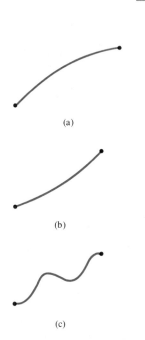

(a)

(b)

(c)

Figure 1
Different shapes of a graph

that the function is **increasing** at this point. On the other hand, if the graph is falling at a point as we move from left to right, we say that the function is **decreasing** at this point. The function graphed in Figure 2 is rising or increasing at points in the interval $(-\infty, -2)$, falling or decreasing at points in the interval $(-2, 3)$, and increasing again in the interval $(3, \infty)$.

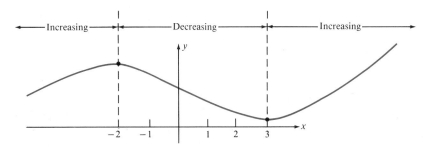

Figure 2
Intervals where a function is increasing and decreasing

It is clear from Figure 2 that on the intervals where the graph is increasing, the derivative $f'(x)$ is positive, and on the intervals where the curve is decreasing, the derivative $f'(x)$ is negative. This leads us to a useful test for determining where the graph of a function is increasing and where it is decreasing.

Finding Increasing and Decreasing Intervals of a Function

Assume that a function f has a derivative at $x = c$. The sign of the derivative $f'(c)$ will determine whether the graph $y = f(x)$ is increasing or decreasing at c. Specifically,

- if $f'(c) > 0$, then f is increasing at c,
- if $f'(c) < 0$, then f is decreasing at c, and
- if $f'(c) = 0$, then f is neither increasing nor decreasing at c.

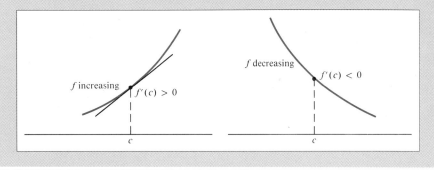

Example 1

Increasing and Decreasing Functions   Given the function

$$f(x) = \frac{1}{3}x^3 - \frac{3}{2}x^2 + 2x + 1$$

find the values of x where the graph $y = f(x)$ is increasing and where it is decreasing. Use this information to graph the function.

Solution   We first compute the derivative

$$f'(x) = x^2 - 3x + 2$$
$$= (x - 1)(x - 2)$$

To determine where this derivative is positive and where it is negative, it is useful to draw the following diagram showing the intervals where each factor is positive or negative.

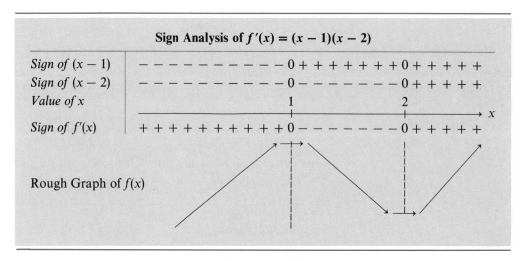

Sign Analysis of $f'(x) = (x - 1)(x - 2)$

Sign of $(x - 1)$ — — — — — — — — — — $0 + + + + + + + 0 + + + + +$
Sign of $(x - 2)$ — — — — — — — — — — $0 — — — — — — — 0 + + + + +$
Value of x 1   2

Sign of $f'(x)$ $+ + + + + + + + + + 0 — — — — — — — 0 + + + + +$

Rough Graph of $f(x)$

Examining the numerical sign of the derivative (whether it is positive or negative), we see that the graph of f is increasing for x less than 1 or greater than 2 and decreasing for x satisfying $1 < x < 2$. Using this information and plotting a few relevant points (Figure 3), we can draw an accurate graph of the function.   □

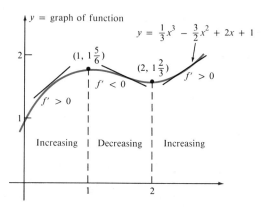

y = graph of function

$y = \frac{1}{3}x^3 - \frac{3}{2}x^2 + 2x + 1$

$(1, 1\frac{5}{6})$

$(2, 1\frac{2}{3})$

$f' < 0$

$f' > 0$

$f' > 0$

Increasing | Decreasing | Increasing

Figure 3
First derivative test

"A Few Relevant Points"

When we drew the graph in Figure 3, we used the fact that we knew where the function was increasing and where it was decreasing. We then selected a few relevant points to draw the graph. By selecting certain points we are using the old point-plotting method, except that now we know much more about what happens between the points. It is a rule of thumb that one of the relevant points is usually chosen by letting $x = 0$ so that we can find the point where the graph crosses the y-axis (the y-intercept). Also, by setting $y = 0$ and solving for x we can find the places where the graph crosses the x-axis. The number of relevant points needed to plot the graph depends on the behavior of the function and how certain we are of the form of the graph.

In the graph in Figure 3 the points $x = 1, 2$ where $f'(x) = 0$ play an important role in the graph of the function. These points are called stationary points. This leads us to the next important tool for graphing.

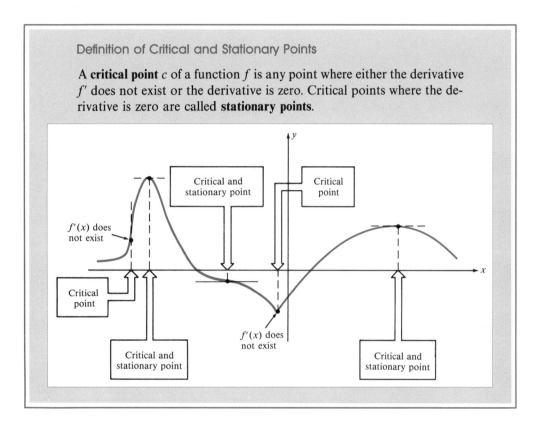

Definition of Critical and Stationary Points

A **critical point** c of a function f is any point where either the derivative f' does not exist or the derivative is zero. Critical points where the derivative is zero are called **stationary points**.

Comments on Critical and Stationary Points

The diagram above illustrates the different ways in which the graph of a function can behave at a critical point. For example, when a function fails to have a derivative at a point, the graph can have a corner-point, it can go straight up, or it can have a discontinuity.

Stationary points are different. They correspond to points on the graph of the function where the function has momentarily stopped increasing or decreasing. The fundamental property that distinguishes a stationary point is the fact that the tangent line to the graph is horizontal at a stationary point. Often the graph of a function reaches its maximum or its minimum values at stationary points. At other times the graph of a function momentarily stops increasing or decreasing at a stationary point. Knowledge about the behavior of a function at its critical points is extremely valuable in graphing functions.

Example 2

Critical and Stationary Points   Find both the critical and the stationary points, if any, of the function

$$f(x) = x^2 - 4x + 3$$

Solution   Computing the first derivative, we have

$$f'(x) = 2x - 4$$

This derivative exists for all values of x and is zero when x is 2. Hence the function has a critical point (which is also a stationary point) at 2. The graph of the function is shown in Figure 4.   □

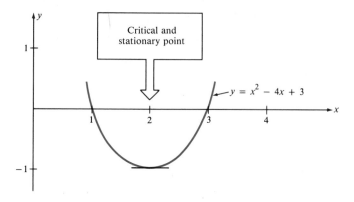

Figure 4
Critical and stationary point at 2

Example 3

Critical and Stationary Points   Find both the critical and the stationary points of the absolute value function

$$f(x) = |x|$$

Solution   The graph of this function can easily be drawn without resorting to ideas of calculus. It is shown in Figure 5.

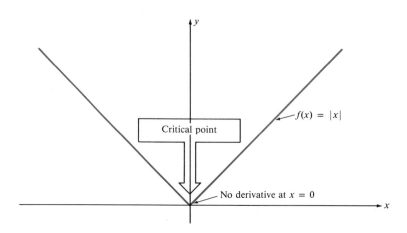

Figure 5
Critical point at 0

The slopes of the tangent lines to the graph of $f(x) = |x|$ are given in Table 1 for different values of x.

TABLE 1
Slope of Tangent
Lines for Values of x

Value of x	Slope of Tangent Line to $f(x) = |x|$
$x < 0$	-1
$x = 0$	does not exist
$x > 0$	$+1$

Hence the derivative of $f(x) = |x|$ is

$$\frac{d}{dx}|x| = \begin{cases} -1 & (x < 0) \\ \text{does not exist} & (x = 0) \\ +1 & (x > 0) \end{cases}$$

Thus the absolute value function has a critical point at zero. Since the derivative is never zero, the function does not have a stationary point. ☐

Example 4

Critical and Stationary Points Find both the critical and the stationary points of

$$f(x) = x^{2/3}$$

Also find the regions where the function is increasing and where it is decreasing.

Solution Using the power rule to compute the derivative, we have

$$f'(x) = \frac{2}{3}x^{-1/3}$$

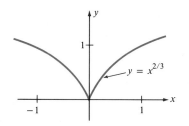

Figure 6
Graph of $y = x^{2/3}$

Since $f'(x)$ does not exist at $x = 0$, we conclude that f has a critical point at zero (observe no stationary points since the derivative is never zero). To find the regions where f is increasing and where it is decreasing, we carry out the following sign analysis.

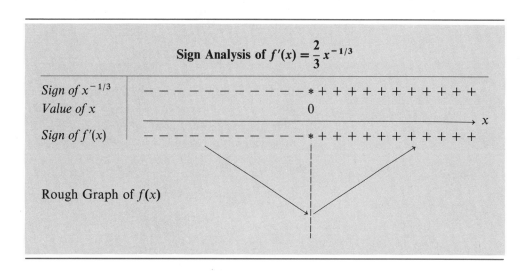

By plotting a few relevant points and using the above information, the graph of $f(x) = x^{2/3}$ can be plotted as shown in Figure 6. ☐

Finding the Revenue Function from the Demand Curve

The study of a company's revenue, cost, and profit is known as the **theory of the firm** in mathematical economics. We continue our ongoing study of this important topic by showing why the **demand function** $q = D(p)$ is so important in this theory.

Until now we have expressed the revenue $R(q)$ in terms of the number of items q produced. However, revenue can always be computed by the formula

$$R = pq$$

where

$$p = \text{Price of each item}$$
$$q = \text{Number of items produced}$$

If we now substitute the value of q in the demand curve $q = D(p)$ into the above revenue equation, we have

$$R(p) = pD(p) \qquad \text{(revenue in terms of price)}$$

This equation is important because it gives the revenue in terms of the price of a product and not the number of items produced. In other words, it gives an alternative way of finding the revenue.

Also, since the profit P from the production of a given product is the revenue R minus the cost C, we can also write the profit in terms of the price. Doing this, we have

$$P(p) = R(p) - C(p)$$
$$= pD(p) - C(p) \qquad \text{(profit in terms of price)}$$

Example 5

Theory of the Firm   The number q of tons of molybdenum that the Colorado Molybdenum Co. can sell in one day depends on the market price p in dollars for a ton of molybdenum. Suppose the relationship between the daily demand for molybdenum and the market price is

$$q = D(p) = 100 - \frac{p}{10} \qquad (\$100 \le p \le \$1000)$$

(a) Find the company's daily revenue in terms of the price p of molybdenum and evaluate the revenue when the price is $400 per ton.

(b) Find the company's daily revenue in terms of the number of tons q of molybdenum produced.

Solution

(a) Using the general revenue formula $R = pq$, we can write

$$R(p) = pD(p)$$

$$= p\left(100 - \frac{p}{10}\right)$$

$$= -\frac{p^2}{10} + 100p \qquad (\$100 \le p \le \$1000)$$

When the price of molybdenum is \$400 per ton, the company's weekly revenue will be

$$R(400) = 400\left(100 - \frac{400}{10}\right)$$

$$= \$24,000$$

(b) To find the revenue $R = pq$ in terms of q, we simply solve for p in the demand function

$$q = D(p) = 100 - \frac{p}{10}$$

in terms of q. This gives

$$p = 1000 - 10q$$

Hence we have

$$R(q) = pq$$
$$= (1000 - 10q)q$$
$$= -10q^2 + 1000q \qquad\qquad \square$$

Final Comment

The previous two examples illustrate one of the reasons why the demand function $q = D(p)$ is useful in the theory of the firm. It allows the revenue, cost, and profit of a firm to be rewritten completely in terms of p.

Problems

Problems 1–6 refer to the graph of $y = f(x)$ in Figure 7.

1. Identify the points where $f(x)$ is increasing.
2. Identify the points where $f(x)$ is decreasing.
3. Identify the critical points.
4. Identify the stationary points.
5. Identify the points where the derivative does not exist.
6. Identify the points where the derivative is zero.

For Problems 7–8, draw a graph $y = f(x)$ that has the given properties.

7. (a) $f'(-1)$ does not exist
    (b) $f(0) = 1$

(c) $f'(1) = 0$
(d) $f'(2) = 1$
(e) $f'(3)$ is negative
(f) f has a stationary point at $x = 4$
(g) $f(6) = 0$

8. (a) f has a critical point at $x = 0$
    (b) f has a stationary point at $x = 1$
    (c) $f(1) = 2$
    (d) $f'(2) = 1$
    (e) $f'(3) = -1$
    (f) $f'(4) = 0$

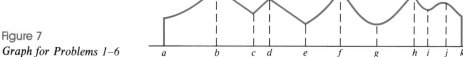

Figure 7
Graph for Problems 1–6

a   b   c   d   e   f   g   h i   j   k

For Problems 9–10, use the information given to make a rough sketch of a plausible graph $y = f(x)$.

9.

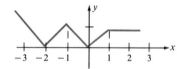

x	$f(x)$
0	1
1	4

10.

x	$f(x)$
1	0
2	1
3	0

11. Sign of the Derivative In the graph in Figure 8, give the sign of the derivative in each of the intervals:
(a) $(-2, -1)$
(b) $(-1, 0)$
(c) $(0, 1)$
(d) $(1, 2)$

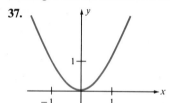

Figure 8
Graph for Problem 11

For Problems 12–36, find the critical and stationary points, and determine the intervals where the functions are increasing and where they are decreasing. Use this information to graph the functions.

12. $f(x) = 5$ 　　　　 **13.** $f(x) = 3x + 1$

14. $f(x) = x^2$ 　　　　 **15.** $f(x) = x^2 - 1$

16. $f(x) = x^2 - 2x - 3$ 　 **17.** $f(x) = (x - 4)^2$

18. $f(x) = x^3$ 　　　　 **19.** $f(x) = x^3 + 5$

20. $f(x) = x^3 - 4x$ 　　 **21.** $f(x) = (x - 3)^3$

22. $f(x) = 3x^3 - 3x^2 + 1$ 　 **23.** $f(x) = x + \dfrac{1}{x^2}$

24. $f(x) = x + \dfrac{1}{x}$ 　　 **25.** $f(x) = x^4$

26. $f(x) = (x - 1)^4$ 　　 **27.** $f(x) = (x + 2)^4$

28. $f(x) = \dfrac{x - 1}{x + 1}$ 　　 **29.** $f(x) = x^{1/2}$

30. $f(x) = x^{1/3}$ 　　　 **31.** $f(x) = x^{1/5}$

32. $f(x) = x^{4/3}$ 　　　 **33.** $f(x) = (x - 4)^{2/3}$

34. $f(x) = 2x - 3x^{2/3}$ 　 **35.** $f(x) = 1 - x^{2/3}$

36. $f(x) = (x - 1)^{5/3}$

For Problems 37–42, determine the intervals where the functions are increasing and those where the functions are decreasing. Find the critical and stationary points.

37.

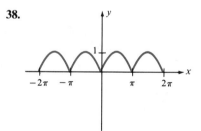

38.

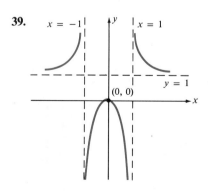

39.

40.

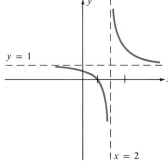

41.

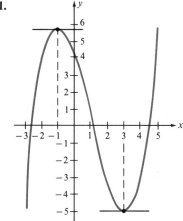

42.

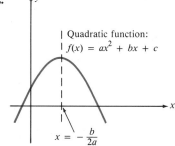

For Problems 43–45, state in common language how the slope of the functions change as one moves from left to right on the real axis.

43.

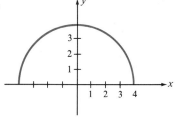

44.

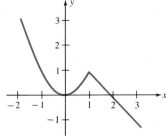

45.

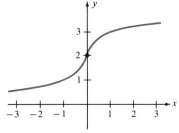

More Theory of the Firm

46. Finding Revenue from the Demand Curve   Find the revenue function $R(p)$ if the demand curve is given by

$$q = D(p) = 20 - 5p$$

47. Finding Revenue from the Demand Curve   Find the revenue function $R(p)$ if the demand curve is given by

$$q = D(p) = 9 - p^2$$

48. Finding Profit from Demand and Cost   Find the profit function $P(q)$ if the demand curve is

$$q = D(p) = 25 - p^2$$

and the cost function is

$$C(q) = 10 + 3q$$

49. Finding Profit from Demand and Cost   Find the profit function $P(q)$ if the demand curve is

$$q = D(p) = 36 - p^2$$

and the cost function is

$$C(q) = 5 + 0.5q$$

50. Finding Marginal Revenue and Marginal Profit   Find the marginal revenue $R'(q)$ and the marginal profit $P'(q)$ if the demand curve is

$$q = D(p) = 25 - p$$

and the cost function is

$$C(q) = 50 + 0.5q$$

51. Finding Marginal Revenue and Marginal Profit   Find the marginal revenue $R'(q)$ and the marginal profit $P'(q)$ if the demand curve is

$$q = D(p) = 64 - p^2$$

and the cost function is

$$C(q) = 100 + 0.5q + 0.001q^2$$

9.2

Use of the Second Derivative in Graphing

PURPOSE

We introduce the concept of higher derivatives and show how the second derivative also gives useful information about the graph of a function. Important topics include

- higher derivatives,
- the concavity and inflection points of a graph, and
- the use of second derivatives for determining concavity.

Introduction

In the previous section we learned how the first derivative can be used to determine where the graph of a function is increasing and where it is decreasing. However, several characteristics of the graph of a function cannot be determined by the first derivative. For instance the two graphs drawn in Figure 9 are both increasing, yet they have a basic qualitative difference in their behavior. Although they both are headed in an "upward" direction, one graph is curved "downward," and the other curved "upward."

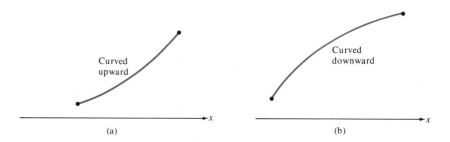

Figure 9
Two increasing functions

    To understand some of these additional properties of graphs, we introduce higher derivatives.

Higher Derivatives

The derivative f' is called the **first derivative** or simply the derivative of f. Since this derivative is a new function, we can differentiate it and obtain the derivative of the derivative. This is called the **second derivative** of f. In fact, if we continue this process we can find a third, fourth, and even **higher derivatives**. The general notation for derivatives of all orders are given in Table 2.

TABLE 2
General Notation for
Higher Derivatives

Notation	Meaning
f	Function
f'	First derivative
$f'' = (f')'$	Second derivative
$f''' = (f'')'$	Third derivative
$f^{(4)} = (f''')'$	Fourth derivative
$f^{(5)} = (f^{(4)})'$	Fifth derivative
$\vdots \qquad \vdots$	$\vdots$
$f^{(n)} = (f^{(n-1)})'$	nth derivative

Example 1

Higher Derivatives   Find the first six derivatives of the polynomial

$$f(x) = x^4 + 5x^3 + 3x^2 + 3x + 7$$

Solution   Starting with

$$f(x) = x^4 + 5x^3 + 3x^2 + 3x + 7$$

we simply differentiate to get

$$f'(x) = 4x^3 + 15x^2 + 6x + 3$$
$$f''(x) = 12x^2 + 30x + 6$$
$$f'''(x) = 24x + 30$$
$$f^{(4)} = 24$$
$$f^{(5)} = 0$$
$$f^{(6)} = 0$$

Note that all derivatives of order five or higher are zero.   □

Leibniz "dee" Notation

It is also possible to represent higher-order derivatives by means of the Leibniz
"dee" notation. If f is a function of x, then we write

$$f'(x) = \frac{d}{dx}\left[f(x)\right]$$

$$f''(x) = \frac{d}{dx}\left[\frac{d}{dx}\left[f(x)\right]\right] = \frac{d^2}{dx^2}\left[f(x)\right]$$

$$f'''(x) = \frac{d}{dx}\left[\frac{d}{dx}\left[\frac{d}{dx}\left[f(x)\right]\right]\right] = \frac{d^3}{dx^3}\left[f(x)\right]$$

$$\vdots \qquad \vdots$$

$$f^{(n)}(x) = \frac{d^n}{dx^n}\left[f(x)\right]$$

In general, we have two alternative notations for the nth derivative, which are

$$f^{(n)}(x) \quad \text{and} \quad \frac{d^n}{dx^n}\left[f(x)\right]$$

Both are read "the nth derivative of f with respect to x."

Example 2

Higher Derivatives   Find the first three derivatives of the function

$$y = x^2 + \frac{1}{x} \qquad (x \neq 0)$$

Solution   Using the rules of differentiation, we repeatedly differentiate, getting

$$\frac{dy}{dx} = 2x - \frac{1}{x^2}$$

$$\frac{d^2y}{dx^2} = 2 + \frac{2}{x^3}$$

$$\frac{d^3y}{dx^3} = -\frac{6}{x^4} \qquad \qquad \qquad \square$$

Concavity and Inflection Points

Earlier in this chapter we saw that the first derivative provided us with information about whether a function is increasing or decreasing. It is natural to wonder whether the second and higher derivatives provide any information about the graph of a function. Although derivatives higher than second order do not provide much information about the graph of a function, the second derivative is very important to the understanding of the graph of a function.

To understand how the second derivative is useful in graphing a function, it is important to realize that the second derivative is the derivative of the first derivative. In other words, at those points where the first derivative is increasing, the second derivative will be positive. On the other hand, at those points where the first derivative is decreasing, the second derivative will be negative. These ideas lead to the following concept of concavity.

Definition of Concavity and Inflection Points

A function f is said to be **concave up** on an interval (a, b) if the derivative $f'(x)$ increases as we move from left to right on the interval. A function is **concave down** on an interval (a, b) if the derivative $f'(x)$ decreases as we move from left to right on the interval.

If a function changes concavity at a point c from up to down or vice versa, the function is said to have an **inflection** point at c.

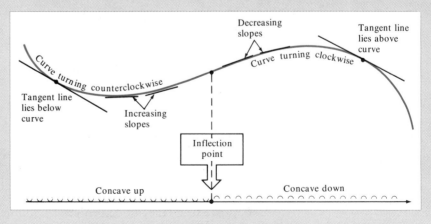

Geometrically, a function f is concave up on an interval if for each point x in the interval the graph of the function lies above the tangent line at the point $(x, f(x))$. Likewise, a function is concave down on an interval if for each point x in the interval the graph of the function lies below the tangent line at the point $(x, f(x))$.

Use of the Second Derivative

The role of the second derivative in graphing lies in the fact that it can be used to determine where a function is concave up or concave down. We make this clear here.

Second Derivative in Determining Concavity

Let f define a function that has a second derivative $f''(c)$ at $x = c$. The importance of this second derivative is that

- if $f''(c) > 0$, then f is concave up at c,
- if $f''(c) < 0$, then f is concave down at c, and
- if $f''(c) = 0$, then no conclusion can be drawn.

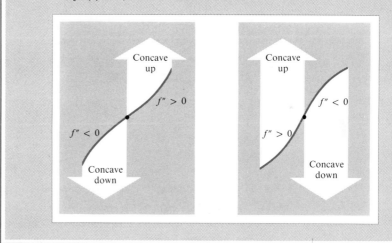

Although no conclusion can be drawn in general when $f''(c) = 0$, it often happens that f has an inflection point at c when $f''(c) = 0$.

Example 3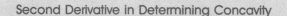

Determining Concavity   Let $f(x)$ be given by

$$f(x) = \frac{1}{3}x^3 - \frac{3}{2}x^2 + 2x + 1$$

Find

(a) the intervals where f is increasing and where it is decreasing,

(b) the critical and the stationary points,

(c) the intervals on which f is concave up or concave down, and

(d) the inflection points.

Use the above information to graph the function.

Solution

(a) *Increasing and decreasing intervals:* We begin by computing the first derivative. We have

$$f'(x) = x^2 - 3x + 2$$
$$= (x - 1)(x - 2)$$

To determine where the graph $y = f(x)$ is increasing and where it is decreasing, we determine where the derivative $f'(x)$ is positive and where it is negative.

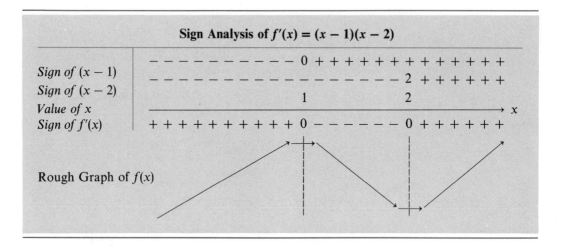

Sign Analysis of $f'(x) = (x - 1)(x - 2)$

The above sign analysis says that the graph $y = f(x)$ is increasing on the intervals $(\infty, 1)$ and $(2, \infty)$ and decreasing on $(1, 2)$.

(b) *Critical and stationary points:* To find the critical and the stationary points, we observe that the derivative $f'(x)$ always exists and is zero when $x = 1, 2$. Hence these two points are critical points (which are also stationary points).

(c) *Concave up and down:* To determine where the graph is concave up and where it is concave down, we compute the second derivative:

$$f''(x) = 2x - 3 \qquad \begin{cases} \text{positive for } x > 3/2 \\ \text{zero for } x = 3/2 \\ \text{negative for } x < 3/2 \end{cases}$$

The second derivative says that the graph is concave up on the interval $(3/2, \infty)$, since $f''(x) > 0$ on that interval. Also the graph is concave down on the interval $(-\infty, 3/2)$, since $f''(x) < 0$ on that interval.

(d) *Inflection point (changing concavity):* The function has an inflection point at $x = 3/2$, since the second derivative $f''(x)$ changes sign at this point. It is often the case that the inflection point occurs when the second derivative is zero.

To graph the function, we first plot the points in Table 3. The graph of f is shown in Figure 10. $\square$

TABLE 3

Relevant Points of the Graph

x	$f(x)$	Why Relevant?
0	1.00	Easy point to plot
1.0	1.83	Stationary point
1.5	1.75	Inflection point
2.0	1.67	Stationary point
4.0	6.33	Typical point

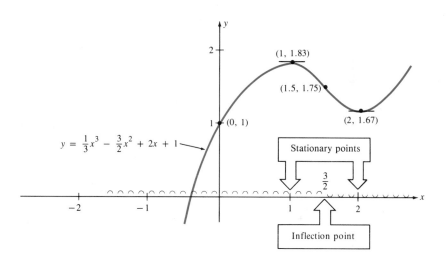

Figure 10
Graph of the cubic equation

Using the first and second derivatives, the following steps can be followed when graphing functions.

Steps in Graphing $y = f(x)$

- **Step 1.** Calculate $f'(x)$ and $f''(x)$.
- **Step 2.** Use $f'(x)$ to determine the critical and the stationary points and find the intervals where f is increasing and the intervals where f is decreasing.
- **Step 3.** Use $f''(x)$ to determine inflection points and find the intervals where f is concave up and the intervals where f is down.
- **Step 4.** Plot a few strategic points and use the above information to sketch the graph.

Example 4 _____ Advanced Graphing   Graph the function

$$f(x) = \frac{1}{3}x^3 - x^2 + 4$$

Solution   Using the above four-step process, we have

- **Step 1** (calculate $f'(x)$ and $f''(x)$). Finding the first and second derivatives, we have

$$f'(x) = x^2 - 2x \quad \begin{cases} \text{Positive} & (x < 0 \text{ or } x > 2) \\ \text{Zero} & (x = 0, 2) \\ \text{Negative} & (0 < x < 2) \end{cases}$$

$$f''(x) = 2x - 2 \quad \begin{cases} \text{Positive} & (x > 1) \\ \text{Zero} & (x = 1) \\ \text{Negative} & (x < 1) \end{cases}$$

- **Step 2** (find intervals where f is increasing and decreasing).   The first derivative tells us

   (i) f is increasing on $(-\infty, 0)$ and $(2, \infty)$,

   (ii) f has stationary points at 0 and 2, and

   (iii) f is decreasing on $(0, 2)$.

- **Step 3** (find the intervals of concavity).   The second derivative tells us

   (i) f is concave up on $(1, \infty)$,

   (ii) f has an inflection point at 1, and

   (iii) f is concave down on $(-\infty, 1)$.

- **Step 4** (plot some relevant points).   We now plot the points in Table 4. The graph of f is shown in Figure 11.   □

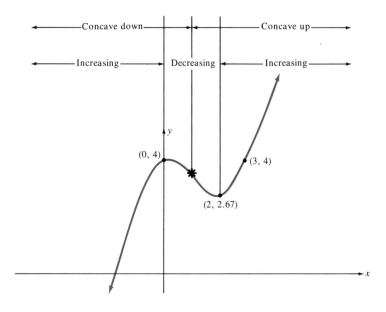

Figure 11
Graph of a cubic polynomial

TABLE 4
Points Plotted for Example 4

Point Plotted	Why Plotted?
(0, 4)	The graph crosses the *y*-axis at this point. The graph also has a horizontal tangent line at this point.
(2, 8/3)	The graph of the function has a horizontal tangent line at this point (also an inflection point of the curve).
(3, 4)	Selected at random.
(−3, −14)	Selected at random.

Asymptotes and Limits at Infinity

We are often interested in how the graph of a function behaves as its argument becomes large positively or large negatively. As an example, consider the rational function defined by

$$f(x) = \frac{5x + 4}{2x} \qquad (x \neq 0)$$

It is interesting to evaluate this function for larger and larger values of x as shown in Table 5.

TABLE 5
Illustration of the Limit
at Infinity

x	*f*(*x*)
1	4.5
10	2.7
100	2.52
1,000	2.502
10,000	2.5002
100,000	2.50002
1,000,000	2.500002
10,000,000	2.5000002

The graph of this function is shown in Figure 12. Note that as x increases, the values of $f(x)$ get closer and closer to the horizontal line $y = 2.5$. This line

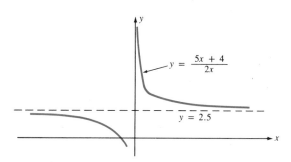

Figure 12
Horizontal asymptote y = 2.5

is called the **horizontal asymptote** of the graph $y = f(x)$ at infinity. We use the notation $x \to \infty$ (read "x approaches infinity") to indicate that x increases without bound. This discussion leads us to the formal definition of horizontal asymptotes and limits at infinity.

Definitions of Horizontal Asymptotes and Limits at Infinity

A function f has a limit or approaches a **limiting value** of L as $x \to \infty$ if the value of $f(x)$ can be made as close to L as we like simply by making x a sufficiently large positive number. We write this limit at infinity as

$$\lim_{x \to \infty} f(x) = L$$

On the other hand, we write the limit of f at minus infinity

$$\lim_{x \to -\infty} f(x) = L$$

to mean that the value of $f(x)$ can be made as close to L as we like simply by making x a sufficiently large negative number.

If either of these limits exists, we say that the horizontal line $y = L$ approached by the graph $y = f(x)$ is a horizontal asymptote of the graph. It is possible for any graph $y = f(x)$ to have either 0, 1, or 2 horizontal asymptotes.

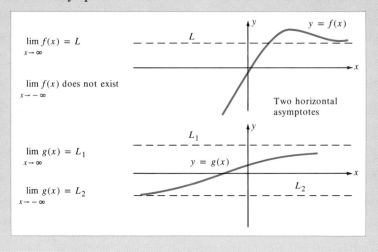

$\lim\limits_{x \to \infty} f(x) = L$

$\lim\limits_{x \to -\infty} f(x)$ does not exist

Two horizontal asymptotes

$\lim\limits_{x \to \infty} g(x) = L_1$

$\lim\limits_{x \to -\infty} g(x) = L_2$

Horizontal Asymptotes of Rational Functions

We have already studied one class of functions that always have limits at infinity. These are the **rational functions** (ratios of polynomials), where the degree of the numerator polynomial is less than or equal to the degree of the denominator polynomial. Some examples are shown in Table 6.

TABLE 6
Examples of
Rational Functions

Function	Limit at Infinity	Graph of Function
$f(x) = \dfrac{1}{x-1}$	$y = 0$	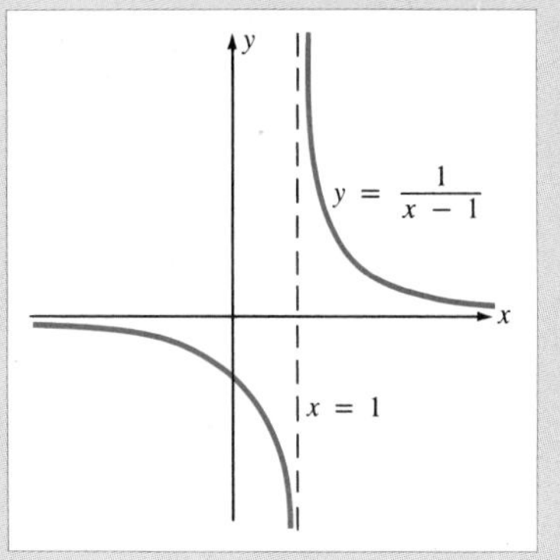
$f(x) = \dfrac{x}{x-2}$	$y = 1$	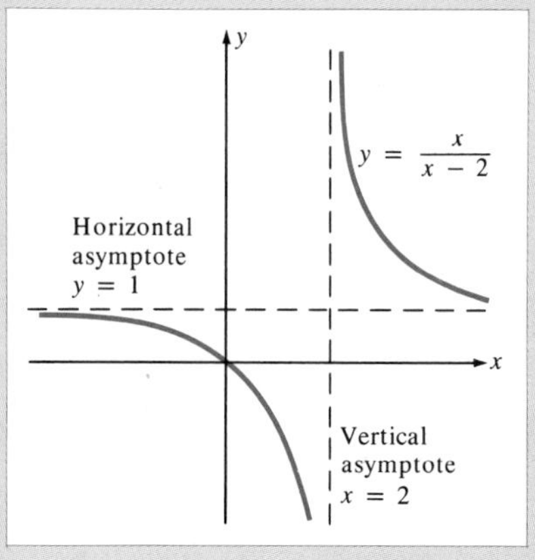
$f(x) = \dfrac{2 + x - x^2}{(x-1)^2}$	$y = -1$	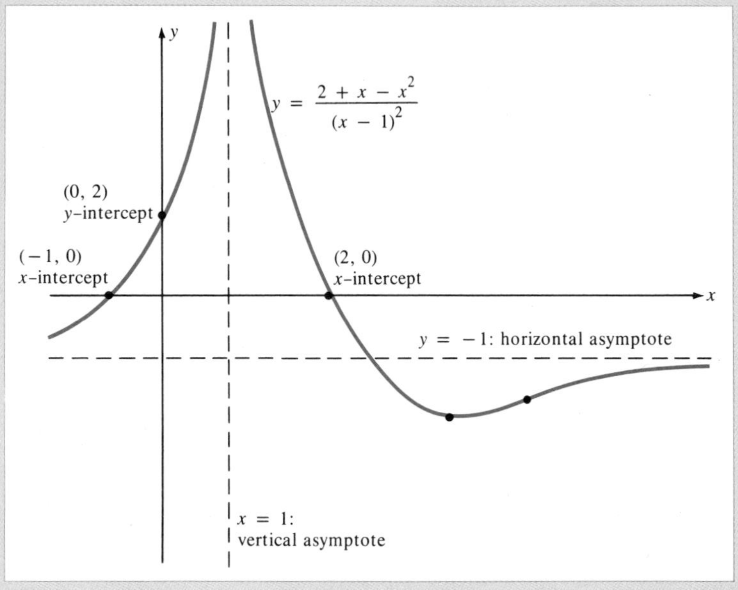

Graph 1: $y = \dfrac{1}{x-1}$, with $x = 1$

Graph 2: $y = \dfrac{x}{x-2}$

Horizontal asymptote $y = 1$

Vertical asymptote $x = 2$

Graph 3: $y = \dfrac{2 + x - x^2}{(x-1)^2}$

$(0, 2)$ y-intercept

$(-1, 0)$ x-intercept

$(2, 0)$ x-intercept

$y = -1$: horizontal asymptote

$x = 1$: vertical asymptote

Example 5

Horizontal Asymptotes   Find the limit at plus infinity (if it exists) of the function

$$f(x) = \frac{3x + 5}{4x + 7}$$

and find the horizontal asymptote there.

Solution   To calculate the limiting value at infinity of a rational function, a general rule is to divide both the numerator and denominator by the highest power of x in the denominator. This gives

$$f(x) = \frac{3x + 5}{4x + 7} = \frac{3 + \dfrac{5}{x}}{4 + \dfrac{7}{x}}$$

As $x \to \infty$, the two terms with x in the denominator will go to zero. This gives the limiting value at infinity of

$$\lim_{x \to \infty} f(x) = \frac{3 + 0}{4 + 0} = \frac{3}{4}$$

In this case, both the numerator and the denominator polynomials had the same degree. Hence the limit at plus infinity is the ratio of the coefficients of the highest powers of x. Note, too, that the limit of $f(x)$ at minus infinity is also 3/4. The graph of $f(x)$ is shown in Figure 13.   □

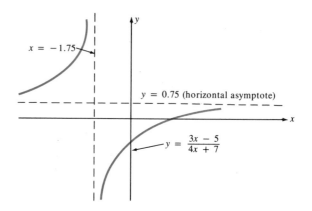

Figure 13
A rational function with a horizontal asymptote

Example 6

Horizontal Asymptotes   Find the limit at infinity of the rational function

$$f(x) = \frac{5x^2 + 2x + 3}{2x^2 - x + 9}$$

and find the horizontal asymptote.

Solution   Again, the degrees of the numerator and the denominator are the same. Hence we rewrite $f(x)$ as

$$f(x) = \frac{5x^2 + 2x + 3}{2x^2 - x + 9} = \frac{5 + \dfrac{2}{x} + \dfrac{3}{x^2}}{2 - \dfrac{1}{x} + \dfrac{9}{x^2}}$$

Clearly, as $x \to \infty$, we get the limit

$$\lim_{x \to \infty} f(x) = \frac{5 + 0 + 0}{2 - 0 + 0} = \frac{5}{2}$$

Hence the line $y = 5/2$ is the horizontal asymptote of $f(x)$ at plus infinity. □

Vertical Asymptotes

It is often the case that the value of a function $f(x)$ "approaches" plus or minus infinity as x approaches some finite number $x = c$. The graph of the function

$$f(x) = \frac{2x}{x - 1}$$

is shown in Figure 14. In this case the vertical line $x = 1$ is called a **vertical asymptote** of the graph of $y = f(x)$. The following is a more formal definition of this concept.

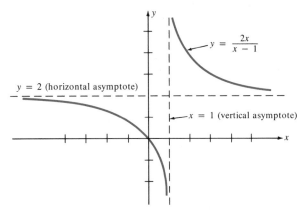

Figure 14
*A rational function with both
vertical and horizontal asymptotes*

Definition of a Vertical Asymptote

If the value of a function $f(x)$ gets large without bound or becomes an arbitrarily large negative number as x approaches a value c from either the left or the right, we say that the vertical line $x = c$ is a vertical asymptote of the graph $y = f(x)$.

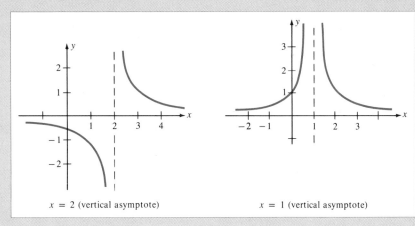

The graph of a function may have either no vertical asymptotes or any number of vertical asymptotes. In fact, there are functions that have an infinite number of vertical asymptotes.

Example 7

Vertical Asymptote   Find the vertical asymptotes (if any) of the function

$$f(x) = \frac{x+1}{x-2}$$

Solution   Since the denominator of $f(x)$ is zero at $x = 2$ while the numerator is not zero at $x = 2$, it is clear that the function will approach plus or minus infinity as x approaches 2. To draw the graph of $y = f(x)$ in a small region around $x = 2$, we perform the following sign analysis.

Sign Analysis of $f(x) = \dfrac{x+1}{x-2}$ near the Vertical Asymptote $x = 2$

Sign of $x + 1$	$- - - - - - - - - 0 + + + + + + + + + + + +$
Sign of $x - 2$	$- - - - - - - - - - - - - 0 + + + + + + +$
Value of x	$\qquad\qquad\qquad -1 \qquad\quad 2$
	$\qquad\qquad\qquad\qquad\qquad\qquad\qquad\qquad\longrightarrow x$
Sign of $f(x)$	$+ + + + + + + 0 - - - - - * + + + + + +$

We now use the additional fact that the sign of $f(x)$ is negative in the interval $(-1, 2)$ and positive elsewhere to draw its graph as shown in Figure 15. Note that $f(x)$ has the horizontal asymptote $y = 1$ and the vertical asymptote $x = 2$.   □

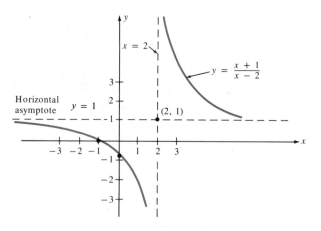

Figure 15

Graph with both a horizontal and a vertical asymptote

In general, if the denominator of a rational function is zero at $x = c$ and the numerator is not zero at c, then $x = c$ will be a vertical asymptote. Hence the rational function

$$f(x) = \frac{x^2 + 1}{(x-1)(x+2)}$$

has two vertical asymptotes, $x = 1$ and $x = -2$. It also has one horizontal asymptote, $y = 1$.

Problems

For Problems 1–12, find $f''(x)$, $f''(0)$, $f''(1)$, and $f''(2)$.

1. $f(x) = x^3 + x^2 + x + 1$ **2.** $f(x) = 4x^2 + 1$

3. $f(x) = ax^2 + bx + c$ **4.** $f(x) = \dfrac{1}{x}$

5. $f(x) = \sqrt{x}$ **6.** $f(x) = \sqrt{x^2 + 1}$

7. $f(x) = \dfrac{x+1}{x-1}$ **8.** $f(x) = \dfrac{1}{x^2 + 1}$

9. $f(x) = (x+1)^2$ **10.** $f(x) = \dfrac{1}{\sqrt{x}}$

11. $f(x) = (x^2 + 1)(x + 2)$ **12.** $f(x) = (x+1)(x-1)$

For Problems 13–17, find $f''(x)$, $f'''(x)$, and $f^{(4)}(x)$.

13. $f(x) = 3x^3 + 4x^2 + x - 5$

14. $f(x) = \dfrac{1}{x}$ **15.** $f(x) = x^n$ (where $n \geq 5$)

16. $f(x) = \dfrac{x}{x+1}$ **17.** $f(x) = x^{-3} + x^{1/2}$

For Problems 18–33, find the first and second derivatives of the given functions. Use this information to find
(a) the critical points,
(b) the stationary points,
(c) the intervals where the functions are increasing,
(d) the intervals where the functions are decreasing,
(e) the inflection points,
(f) intervals where the functions are concave up, and
(g) intervals where the functions are concave down.
Use this information to graph the functions.

18. $f(x) = 2x + 1$ **19.** $f(x) = x^2$
20. $f(x) = x^2 - 1$ **21.** $f(x) = x^2 - 4x + 3$
22. $f(x) = (x - 1)^2$ **23.** $f(x) = x^3$
24. $f(x) = x^3 + 1$ **25.** $f(x) = x^3 - x$

26. $f(x) = 3x^3 + 3x^2 - 2$ **27.** $f(x) = x + \dfrac{1}{x^2}$

28. $f(x) = x + \dfrac{1}{x}$ **29.** $f(x) = (x - 2)^3$

30. $f(x) = (x + 5)^3$ **31.** $f(x) = (x + 2)^4$
32. $f(x) = x^{1/3}$ **33.** $f(x) = x^{1/5}$

For Problems 34–39, determine the intervals where the graphs are concave up and intervals where the graphs are concave down. Find the inflection points.

34.

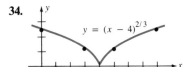
$y = (x - 4)^{2/3}$

35.

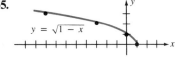
$y = \sqrt{1 - x}$

36.

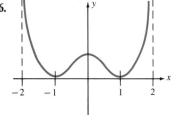

37.

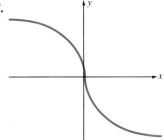

38.

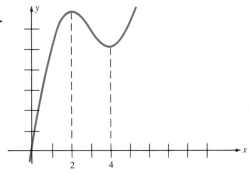

39.

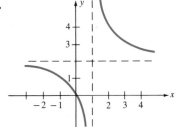

40. First and Second Derivatives   In Figure 16, give the sign of the first and second derivatives in each of the intervals:
(a) $(-2, -1)$
(b) $(-1, 0)$
(c) $(0, 1)$
(d) $(1, 2)$

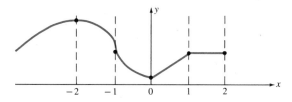

Figure 16
Graph for Problem 40

Limits at Infinity and Horizontal Asymptotes

For Problems 41–50, find the limits both at plus infinity and at minus infinity for the given functions. Find the horizontal asymptotes (if any).

41. $f(x) = 3$

42. $f(x) = \dfrac{1}{x}$

43. $f(x) = 3 + \dfrac{2}{x}$

44. $f(x) = \dfrac{1}{x - 1}$

45. $f(x) = \dfrac{1}{x^2}$

46. $f(x) = \dfrac{2x + 5}{3x - 8}$

47. $f(x) = \dfrac{9x + 10}{x - 4}$

48. $f(x) = \dfrac{x + 1}{x^2 + 1}$

49. $f(x) = \dfrac{3x^2 + 1}{x + 1}$

50. $f(x) = \dfrac{3x^3 + x^2 - 1}{4x^3 - x^2 + x + 5}$

Using Asymptotes to Assist in Graphing

For Problems 51–59, find all vertical and horizontal asymptotes of the graphs determined by the following functions. Use these asymptotes, along with the x- and y-intercepts of the graphs, to plot the graphs of the functions. If necessary, use the first and second derivatives to assist in graphing.

51. $f(x) = \dfrac{1}{x + 1}$

52. $f(x) = \dfrac{x}{x - 1}$

53. $f(x) = \dfrac{x - 1}{x + 1}$

54. $f(x) = \dfrac{x}{x^2 + 1}$

55. $f(x) = \dfrac{1}{x^2 - 1}$

56. $f(x) = \dfrac{x}{x^2 - 1}$

57. $f(x) = \dfrac{x^2 - 1}{x^2 - 3x + 2}$

58. $f(x) = \dfrac{x^2 + 1}{x}$

59. $f(x) = \dfrac{1}{(x - 1)(x - 2)(x - 3)}$

9.3 Maximum and Minimum Values of Functions

PURPOSE

We show how to find both the relative maximum and the relative minimum values of a function and then use these values to find both the absolute maximum and the absolute minimum values. To find the relative maximum and minimum values, we use

- the first derivative test and
- the second derivative test.

   We continue our ongoing development of the theory of the firm by determining the optimal pricing strategy of a product.

Introduction

Many problems ultimately reduce to finding the largest and smallest value of a function. In a business environment, profit is always maximized and cost is always minimized. In fact there are very few disciplines in which the concept of optimizing something does not play some role. If the "something" can be described mathematically by a differentiable function, then differential calculus can possibly play an important role in solving such problems. Problems that involve the finding of maximum or minimum points on the graph of a function are called **optimization problems**. To find these points, we first introduce the concept of relative maximum and relative minimum points.

Relative Maximum and Minimum Points of Graphs

The graphs of many functions form hills and valleys. The tops of the hills are called **relative maximum points**, and the low points of the valleys are called **relative minimum points**. (See Figure 17.) The top of any hill need not be the highest point on the entire graph, but it is a high point relative to its neighbors. Likewise, the bottom of a valley need not be the lowest point on the graph, but it is a low point relative to its neighbors. These ideas are formalized in the following definition.

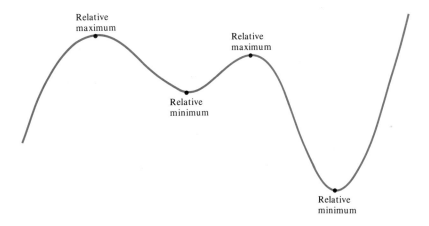

Figure 17
Hills and valleys showing relative maximum and minimum points

Definition of Relative Maximum and Minimum Values

The graph $y = f(x)$ of a function f has a relative maximum value at the point c if there is an open interval (a, b) containing c such that

$$f(c) \geq f(x)$$

for all x in the interval (a, b).

Similarly, the graph $y = f(x)$ has a relative minimum value at the point c if there is an open interval (a, b) containing c such that

$$f(c) \leq f(x)$$

for all x in the interval (a, b).

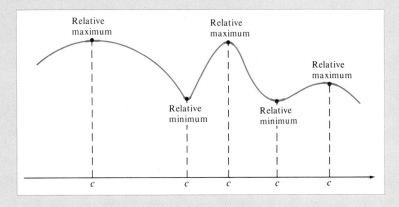

Example 1

Relative Maximum and Minimum Points   Identify the points where the graph in Figure 18 has relative maximum or relative minimum values.

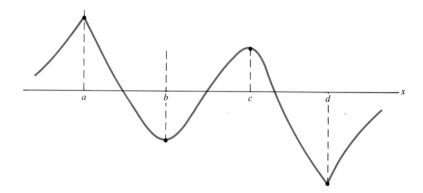

Figure 18
Where are the relative maximum and minimum points?

Solution   The graph has relative maximum values at $x = a$ and $x = c$ and relative minimum values at $x = b$ and $x = d$.   □

In the previous example we were fortunate to be given a drawing of the graph of a function. Often, however, we may have only an algebraic expression for the function. How is it possible in these situations to find the relative maximum and relative minimum points? This question is answered by the first and second derivative tests. We first study the first derivative test.

The First Derivative Test

The first derivative is used to determine relative maximum and relative minimum points in the following way. If a continuous function is increasing to the left of a point c and decreasing to the right of c, then clearly the graph of f has a relative maximum point at c, as shown in Figure 19. On the other hand, if a function is decreasing to the left of c and increasing to the right of c, then the graph of the function has a relative minimum point at c. These observations are stated more formally below.

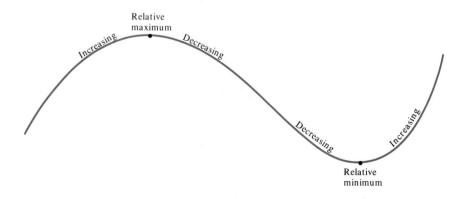

Figure 19
Relative maximum and minimum points

First Derivative Test (Relative Maximum and Minimum Points)

Let f be continuous at a real number c where either $f'(c) = 0$ or $f'(c)$ does not exist. The following two conditions can be used to determine whether $f(c)$ is a relative maximum or relative minimum value.

Relative maximum test: If

- $f'(x) > 0$ for x close to c on the left and
- $f'(x) < 0$ for x close to c on the right,

then

- $f(x)$ has a relative maximum value at c.

Relative minimum test: If

- $f'(x) < 0$ for x close to c on the left and
- $f'(x) > 0$ for x close to c on the right,

then

- $f(x)$ has a relative minimum value at c.

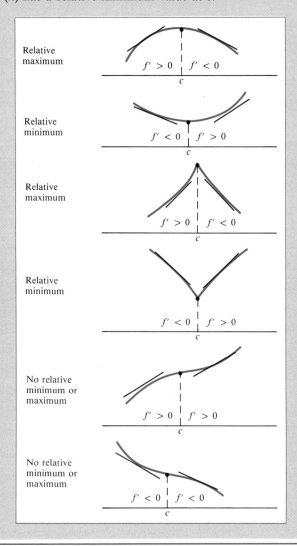

Example 2

Relative Maximum and Minimum Values   Find the points on the graph of the function

$$f(x) = \frac{x}{1 + x^2}$$

where the relative maximum and the relative minimum occur.

Solution   Computing the first derivative by means of the quotient rule, we have

$$\frac{df(x)}{dx} = \frac{(1 + x^2)\dfrac{dx}{dx} - x\dfrac{d}{dx}(1 + x^2)}{(1 + x^2)^2}$$

$$= \frac{1 - x^2}{(1 + x^2)^2}$$ □

The following sign analysis tells where this derivative is positive and where it is negative.

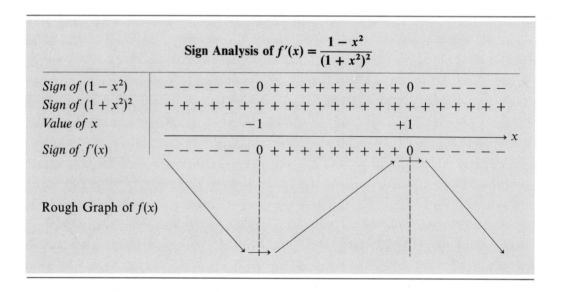

Sign Analysis of $f'(x) = \dfrac{1 - x^2}{(1 + x^2)^2}$

Sign of $(1 - x^2)$	$- - - - - - \; 0 \; + + + + + + + + + \; 0 \; - - - - - -$
Sign of $(1 + x^2)^2$	$+ +$
Value of x	$\qquad\qquad -1 \qquad\qquad\qquad +1$
Sign of $f'(x)$	$- - - - - - \; 0 \; + + + + + + + + + \; 0 \; - - - - - -$

Rough Graph of $f(x)$

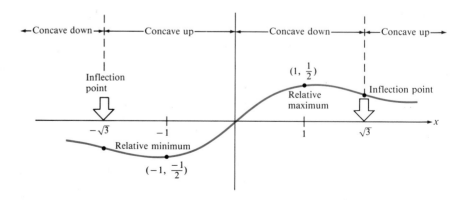

Figure 20
Graph showing relative minimum and maximum and inflection points

Using the first derivative test, we conclude that f has a relative minimum value at -1 (the sign of $f'(x)$ changes from minus to plus there) and a relative maximum value at $+1$ (the sign of $f'(x)$ changes from plus to minus there). The graph of $y = f(x)$ is shown in Figure 20.

We have seen that the second derivative of a function contains information about the graph of a function that is not found by the first derivative. Hence it is not surprising that the second derivative can also be used to provide information about both relative maximum and relative minimum points. We now show the role that the second derivative plays in this important problem.

The Second Derivative Test

Since the second derivative of a function determines whether the graph of the function turns up or down, it seems reasonable that it might be used to find either relative maximum or relative minimum points. We see now how the second derivative is used in finding both relative maximum and relative minimum points.

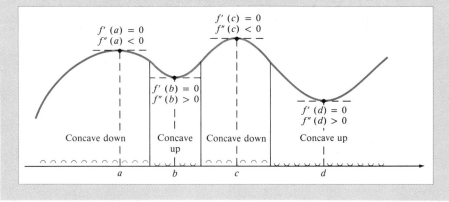

Second Derivative Test (Relative Maximum and Minimum Values)

Assume that a function f has both first and second derivatives at a point c and that $f'(c) = 0$.

Relative maximum test: If $f''(c) < 0$, then the graph of $y = f(x)$ has a relative maximum value at $x = c$.

Relative minimum test: If $f''(c) > 0$, then the graph of $y = f(x)$ has a relative minimum value at $x = c$.

When $f''(c) = 0$ no conclusion can be drawn.

Although the second derivative test requires that we compute the first two derivatives, it is often easier to apply than the first derivative test. The second derivative test is based on the geometrical fact that a relative maximum value occurs at those points where the tangent line to the graph is horizontal and the graph is concave down. Likewise, a graph has a relative minimum value at those points where the tangent line to the graph is horizontal and the graph is concave up.

Example 3

Second Derivative Test   Find the relative maximum and minimum points of the function

$$f(x) = x^2 + \frac{1}{x^2} \qquad (x \ne 0)$$

Solution   Differentiating $f(x)$ twice, we get

$$f'(x) = 2x - \frac{2}{x^3}$$

$$f''(x) = 2 + \frac{6}{x^4}$$

[handwritten:]
$$2x - 2x^{-3} = 0$$
$$-2x^{-3} = -2x$$
$$x^{-2} = 1$$
$$x = \pm\sqrt{1} = \pm 1$$

We find the stationary points by setting $f'(x)$ to zero and solving for x. This gives

$$f'(x) = 2x - \frac{2}{x^3} = 0$$

Solving for the real roots of this equation, we get the stationary points $x = -1$ and $+1$. Computing the second derivative, we get

$$f''(x) = 2 + \frac{6}{x^4}$$

Evaluating this second derivative at the two stationary points, we have

$$f''(-1) = 8 > 0$$
$$f''(+1) = 8 > 0$$

Since these second derivatives are both positive, we conclude that $y = f(x)$ has relative minimum values at $x = -1$ and $x = +1$. The actual relative minimum values can be found by substituting $x = -1$ and $+1$ into $f(x)$, getting

$$f(-1) = 2$$
$$f(1) = 2$$

To graph the function, we make the following observations:

- The graph has relative minimum points at $(-1, 2)$ and $(1, 2)$.
- The second derivative $f''(x)$ is always positive, and hence the graph $y = f(x)$ is always concave up.
- The graph is not defined at zero.
- The graph is symmetric about the y-axis, since $f(x) = f(-x)$ for all x.

Using these observations, we graph the function as shown in Figure 21.   □

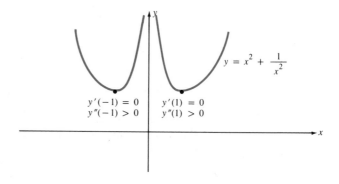

Figure 21
Second derivative test

Absolute Maximum and Minimum Values

Many problems in business and science require that we perform a task in an "optimal manner." For example, every firm that sells a product, whether the firm is a manufacturing or a service company, is faced with the problem of determining the best price for its product. If the firm sets the price too low, the volume of sales might be great, but the revenue per item might be so low that the overall revenue will not be maximized. If the price is too high, the revenue per item will be high, but the number of items sold might be so low that again the overall revenue will not be maximized. This is a typical problem in **revenue maximization**. It leads to the mathematical problem of finding the **absolute maximum value** of the revenue function, which in turn leads us to the following definition.

Definition of Absolute Maximum and Minimum Values

A function f has an absolute maximum value at c if

$$f(c) \geq f(x)$$

for all x in the domain of f.

A function f has an absolute minimum value at c if

$$f(c) \leq f(x)$$

for all x in the domain of f.

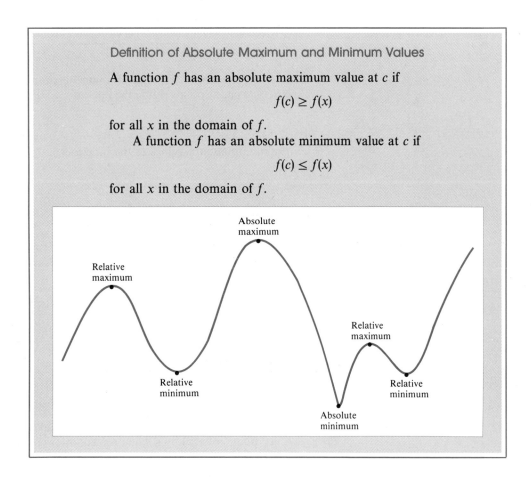

We now give a set of rules that can be used to find the absolute maximum and minimum values of a function. We restrict ourselves to continuous functions defined on closed intervals. It will not be shown here, but it can be proven that for any continuous function defined on a closed interval $[a, b]$ there is always a point in the interval where the function attains its absolute maximum value and a point where it attains its absolute minimum value.

> ### Rules for Finding Absolute Maximum and Minimum Values
>
> To find the points where a continuous function f attains its absolute maximum and minimum values on an interval $[a, b]$ carry out the following steps.
>
> - **Step 1.** Evaluate the function at the endpoints a and b to obtain the values $f(a)$ and $f(b)$.
> - **Step 2.** Evaluate $f(x)$ at each critical point (where $f'(x)$ does not exist or where $f'(x) = 0$).
> - **Step 3.** Find the largest of all the above computed values of $f(x)$. The largest of these values is the absolute maximum of the function on $[a, b]$. The smallest of these values is the absolute minimum of the function on $[a, b]$.

It is useful to note that the absolute maximum and absolute minimum values may occur at points inside the interval $[a, b]$ or at the endpoints a and b.

Example 4

Absolute Maximum and Minimum Values   Find the absolute maximum and absolute minimum values of the function

$$f(x) = \frac{1}{3} x^3 - 2x^2 + 3x + 2$$

on the interval $[-1, 4]$.

Solution   Using the above three-step process, we have the following.

- **Step 1** (evaluate $f(x)$ at the endpoints).   Evaluating $f(x)$ at the endpoints -1 and 4, we have

$$f(-1) = -10/3$$
$$f(4) = 10/3$$

- **Step 2** (evaluate $f(x)$ at the critical points).   Computing the derivative of $f(x)$, we get

$$f'(x) = x^2 - 4x + 3$$
$$= (x - 1)(x - 3)$$

Since $f'(x)$ exists at all points in the interval $(-1, 4)$, we find the stationary points of $f(x)$ by solving the equation

$$f'(x) = 0$$

or

$$(x - 1)(x - 3) = 0$$

This gives the two stationary points 1, 3. Evaluating $f(x)$ at these points gives

$$f(1) = 10/3$$
$$f(3) = 2$$

- **Step 3** (find the largest of the above values).   To find the absolute maximum and the absolute minimum values, we find the largest and smallest values among

$$
\begin{aligned}
f(-1) &= -10/3 && \text{(left endpoint)} \\
f(4) &= 10/3 && \text{(right endpoint)} \\
f(1) &= 10/3 && \text{(stationary point)} \\
f(3) &= 2 && \text{(stationary point)}
\end{aligned}
$$

Hence the absolute maximum value of $f(x)$ is 10/3, which occurs at the two points 1 and 4. The absolute minimum value of $f(x)$ is $-10/3$, which occurs when $x = -1$. The graph of $y = f(x)$ is shown in Figure 22.   □

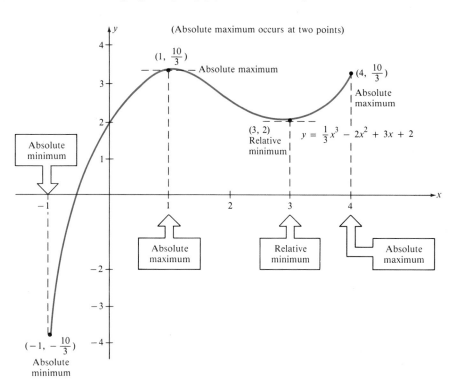

Figure 22
Absolute maximum and minimum values

Determining Price to Maximize Profit

The pricing of a product to maximize profit is a major problem facing companies today. The incorrect pricing of a product can be the difference between success and failure. The overpricing of oil by the OPEC cartel in 1973 resulted in a combination of increased oil exploration and a consumer backlash. The net result was that by the year 1985 there was a worldwide glut of oil, and the price of oil tumbled from $35 a barrel to $8 on the spot market. Most economists have felt that if the OPEC cartel had increased the price of oil in a more gradual manner, there would not have been this consumer resistance, and the long-range revenue to OPEC would have been greater.

Although the following example has been simplified from the pricing problems that most companies face, many of the ideas are present in this example.

Example 5

Optimal Pricing in the Restaurant Industry Imagine yourself as the operational director for an entire chain of family restaurants (typical chains are Big Boy, Friendly's, and Hot Shoppes). Upon your hiring, you discover that the current price the company charges for their popular Family Platter is $8.00. Your first job is to determine whether the company is charging too little, too much, or just the right amount to maximize profit. To determine the optimal price, you carry out a market survey and discover that on the average each increase in price of $0.25 per meal will decrease the company's demand by 20,000 meals per week and that each decrease of $0.25 per meal will increase demand by 20,000 meals per week. You also know that at the current price of $8.00 per meal a total of 250,000 meals are sold per week.

It has also been determined that the weekly cost to the company in dollars to serve q Family Platter meals is

$$C(q) = 100{,}000 + 3q$$

What price should be charged for the Family Platter in order that the total revenue be maximized?

Solution The company's weekly profit $P(q)$ when q meals per week are sold is given by

$$P(q) = R(q) - C(q)$$

where

$$R(q) = \text{Revenue from the sale of } q \text{ meals}$$
$$C(q) = \text{Cost to produce } q \text{ meals}$$

With the help of the demand curve $q = D(p)$, the revenue and cost functions can be written as

$$R(q) = pq$$
$$= pD(p)$$
$$C(q) = 100{,}000 + 3q$$
$$= 100{,}000 + 3D(p)$$

where

$$p = \text{Price charged per meal}$$
$$D(p) = \text{Number of meals sold when the price per meal is } p$$

Hence the profit can be written in terms of the price charged per meal as

$$P(p) = R(p) - C(p)$$
$$= pD(p) - [100{,}000 + 3D(p)]$$

To maximize the profit $P(p)$, we must first find the equation for the demand $D(p)$. The information given in this problem essentially says that the demand function $D(p)$ is a linear function of p, with slope minus 80,000 (the demand goes down by 80,000 for each additional dollar charged). Since the demand function also satisfies $D(8) = 250{,}000$, we can use the point-slope formula for straight lines to write

$$D(p) = 250{,}000 - 80{,}000(p - 8)$$
$$= -80{,}000p + 890{,}000$$

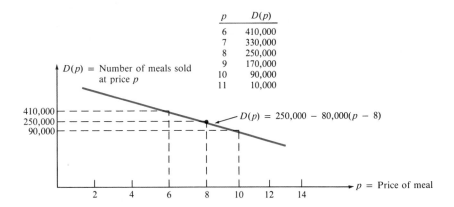

p	$D(p)$
6	410,000
7	330,000
8	250,000
9	170,000
10	90,000
11	10,000

Figure 23
Number-of-meals function

The graph of $D(p)$ is shown in Figure 23. Substituting the expression for $D(p)$ into the equation for $P(p)$ and carrying out a few algebraic steps, we have

$$P(p) = pD(p) - [100,000 + 3D(p)]$$
$$= -80,000p^2 + 1,130,000p - 2,770,000$$

The graph of $P(p)$ is a parabola that, if we set $P(p) = 0$, can be seen to be zero when the price p is (to the nearest penny) \$3.16 or \$10.97. A rough graph is shown in Figure 24.

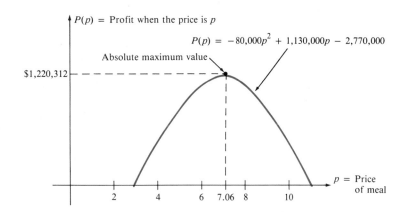

Figure 24
Profit as a function of price p

By a quick examination of the first and second derivatives,

$$P'(p) = -160,000p + 1,130,000 = \begin{cases} \text{positive for } p < 7.06 \\ \text{negative for } p > 7.06 \end{cases}$$

$$P''(p) = -160,000 < 0 \qquad \text{for all } p$$

we conclude that $P(p)$ will be positive only when p is between \$3.16 and \$10.97. Hence we need only find the absolute maximum value of $P(p)$ on the interval [3.16, 10.97]. To do this we carry out the following three-step process.

- **Step 1** (evaluate the function at the endpoints). Evaluating $P(p)$ at the endpoints, we have

$$P(3.16) = 0$$
$$P(10.97) = 0$$

• **Step 2** (evaluate the function at the critical points).   Setting the derivative $P'(p)$ equal to zero and solving for p, we have

$$P'(p) = -160{,}000p + 1{,}130{,}000 = 0$$

This gives

$$p = \$7.06$$

Evaluating $P(p)$ at this stationary point, we have

$$P(7.06) = \$1{,}220{,}312$$

• **Step 3** (find the largest of the above values).   The values found in Steps 1 and 2 are

$$P(3.16) = 0 \quad \text{(not exactly zero but very small)}$$
$$P(7.06) = 1{,}220{,}312$$
$$P(10.97) = 0 \quad \text{(not exactly zero but very small)}$$

It is clear that the absolute maximum profit is $1,220,312, which is attained when $p = 7.06$.

Summary of the Problem

The original price of $8.00 per meal was too high by $0.94 per meal. Table 7 gives the weekly profit to the company as a function of the price of a Family Platter.

TABLE 7
Weekly Results for Various Pricing Strategies

	Price per Meal	Meals Sold per Week	Weekly Revenue	Weekly Cost	Weekly Profit
	$3.16	637,200	$2,013,552	$2,011,600	$1,952
	$4.00	570,000	$2,280,000	$1,810,000	$470,000
	$5.00	490,000	$2,450,000	$1,570,000	$880,000
	$6.00	410,000	$2,460,000	$1,330,000	$1,130,000
	$7.00	330,000	$2,310,000	$1,090,000	$1,220,000
Optimal ⟶	$7.06	325,200	$2,295,912	$1,075,600	$1,220,312
Original ⟶	$8.00	250,000	$2,000,000	$850,000	$1,150,000
	$9.00	170,000	$1,530,000	$610,000	$920,000
	$10.00	90,000	$900,000	$370,000	$530,000
	$10.97	12,400	$136,028	$137,200	-$1,172

Note from Table 7 that the company's weekly profit will increase from $1,150,000 to $1,220,312 (an increase of $70,312) when lowering the price from $8.00 to $7.06. This example shows the importance of careful mathematical analysis in the pricing of goods and services.

Problems

For Problems 1–9, find the intervals or points where the graph in Figure 25 has the indicated properties.

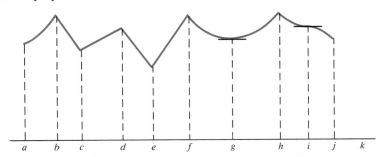

Figure 25
Graph for Problems 1–9   *a   b   c   d   e   f   g   h   i   j   k*

1. $f(x)$ is increasing.
2. $f(x)$ is decreasing.
3. The derivative is zero.
4. The derivative does not exist.
5. $f(x)$ has a relative minimum.
6. $f(x)$ has a relative maximum.
7. $f(x)$ has an absolute minimum.
8. $f(x)$ has an absolute maximum.
9. The second derivative is zero.

For Problems 10–13, use the values of x and $f(x)$ along with the sign of $f'(x)$ on the indicated intervals to give a rough sketch of the graph of f.

10.

Function Information		Derivative Information	
x	$f(x)$	Region	Sign
-1	4	$(-1, 1)$	Negative
1	0	1	Zero
3	4	$(1, 3)$	Positive

11.

Function Information		Derivative Information	
x	$f(x)$	Region	Sign
-1	-2	$(-1, 1/2)$	Positive
1/2	1/4	1/2	Zero
2	-2	$(1/2, 2)$	Negative

12.

Function Information		Derivative Information	
x	$f(x)$	Region	Sign
1/2	5/2	$(1/2, 1)$	Negative
1	2	1	Zero
3/2	13/6	$(1, 3)$	Positive
2	5/2		
5/2	29/10		
3	10/3		

13.

Function Information		Derivative Information	
x	$f(x)$	Region	Sign
-2	-8	$(-2, -1)$	Positive
-1	1	-1	Zero
$-1/2$	7/16	$(-1, 0)$	Negative
0	0	0	Zero
1/2	7/16	$(0, 1)$	Positive
1	1	1	Zero
2	-8	$(1, 2)$	Negative

For Problems 14–30, when possible use the first derivative test to determine the points where $f(x)$ has relative maximum or relative minimum values. If the derivative test does not apply, indicate the reason.

14. $f(x) = 3$
15. $f(x) = 3x - 5$
16. $f(x) = x^2 + 1$
17. $f(x) = 5x^2 - 5x + 3$
18. $f(x) = x^2 + 12x - 10$
19. $f(x) = x^3 - x + 1$
20. $f(x) = x^3 - 18x$
21. $f(x) = 3x^2 - 72x + 15$

22. $f(x) = x^3 + 3x^2 - 24x + 5$

23. $f(x) = x^3 + \dfrac{9}{4}x^2 - 3x - 6$

24. $f(x) = \dfrac{x^4}{x^2 - 1}$ **25.** $f(x) = \sqrt{x}$

26. $f(x) = \sqrt{x^2 - 1}$ **27.** $f(x) = x^{1/3}$
28. $f(x) = x^{2/3}$ **29.** $f(x) = x^{4/5}$
30. $f(x) = 6x^{2/3} - 4x$

For Problems 31–43, use the second derivative test to determine whether any of the following functions have relative maximum or relative minimum values at the critical points. If the second derivative test does not apply, explain why.

31. $f(x) = 3x$ **32.** $f(x) = 3x - 6$
33. $f(x) = x^2$ **34.** $f(x) = x^2 - 1$
35. $f(x) = x^2 - 2x - 3$ **36.** $f(x) = x^2 + x + 1$
37. $f(x) = 3x^4 + 8x^3 - 6x^2 - 24x + 10$

38. $f(x) = x^3 - 15x + 4$ **39.** $f(x) = \dfrac{x}{x^2 + 2}$

40. $f(x) = x + \dfrac{1}{x}$ **41.** $f(x) = 2x^3 - x^6$

42. $f(x) = \dfrac{x}{\sqrt{x^2 + 4}}$ **43.** $f(x) = \dfrac{x^2}{9} + \dfrac{6}{x}$

For Problems 44–58, find both the absolute maximum and the absolute minimum values of each of the continuous functions on the indicated intervals.

44. $f(x) = 4$ on $[0, 1]$
45. $f(x) = x$ on $[-1, 2]$
46. $f(x) = 3x + 2$ on $[1, 5]$
47. $f(x) = -x + 3$ on $[2, 5]$
48. $f(x) = x^2 - 4x - 3$ on $[-1, 4]$
49. $f(x) = 2x^3 - 15x^2 + 12$ on $[-1, 3]$
50. $f(x) = (x - 1)^3$ on $[-1, 2]$
51. $f(x) = 2x^2 + 2x + 3$ on $[-1, 1]$
52. $f(x) = x^4 - 8x^3 + 22x^2 - 24x + 7$ on $[2, 4]$
53. $f(x) = 2x^3 - 9x^2 + 12x - 5$ on $[2, 4]$

54. $f(x) = x^4 - \dfrac{4}{3}x^3 - 2$ on $[-1, 1]$

55. $f(x) = \sqrt{x - 1} - \dfrac{x}{2}$ on $[3/2, 3]$

56. $f(x) = x^{1/3}$ on $[-1, 1]$
57. $f(x) = x^{4/3}$ on $[-1, 1]$
58. $f(x) = 6 - 3x - (x - 1)^{3/2}$ on $[2, 6]$

Theory of the Firm

59. Average Cost   The average production cost per item, when q items are produced, is found by dividing the total cost $C(q)$ by the number of items q. That is,

$$\text{Average cost:}\quad \bar{C}(q) = \dfrac{C(q)}{q}$$

Assume that the cost to produce q items is given by

$$C(q) = 625 + \dfrac{q^2}{10}$$

(a) Find the number of items that should be produced to minimize the average cost. Assume that the feasible number of items lies somewhere in the interval $[10, 100]$.
(b) If the value of q that minimizes the average cost is not an integer, how would you find an integer value to use in trying to minimize the average cost?

60. Maximum Revenue   The total revenue $R(p)$ that a company receives depends on the price p the company charges for its product and is given by

$$R(p) = -p^2 + 7p \qquad (0 \le p \le 8)$$

Find the price that will maximize the revenue.

61. Maximum Revenue   The total revenue $R(p)$ that a company receives depends on the price p the company charges for its product and is given by

$$R(p) = -25p^2 + 300p \qquad (0 \le p \le 8)$$

Find the price that will maximize the total revenue.

Maximum Profit

The total profit $P(q)$ that a company earns for producing q units of a product is the total revenue $R(q)$ from the sale of q units of the product minus the total cost $C(q)$ of producing q units of the product. That is,

$$P(q) = R(q) - C(q)$$

Often a firm can determine expressions for $R(q)$ and $C(q)$ as functions of q. For Problems 62–65, find the indicated quantities.

62. Maximum Profit   Suppose $R(q)$ and $P(q)$ can be described by the polynomials

$$R(q) = 100q \qquad\qquad (0 \le q \le 10)$$
$$C(q) = q^3 - 3q^2 + 2q \qquad (0 \le q \le 10)$$

Find the number of items that maximizes the total profit. What is the maximum profit for this number of items produced?

63. Maximum Profit   For total revenue and cost functions given by

$$R(q) = 500q - q^2 \qquad\qquad\qquad (0 \le q \le 35)$$
$$C(q) = q^3 - 50q^2 + 500q + 250 \qquad (0 \le q \le 35)$$

find the number of items that maximizes the total profit. What is the maximum profit for this number of items produced?

64. Maximum Profit Suppose $R(q)$ and $P(q)$ are described by the functions

$$R(q) = 50q \qquad (0 \le q \le 100)$$
$$C(q) = 0.5q^2 + 50 \qquad (0 \le q \le 100)$$

Graph these cost and revenue functions. Find the number of items that will maximize the profit. What is the maximum profit?

65. Maximum Profit Suppose $R(q)$ and $C(q)$ are described by the functions

$$R(q) = 5q \qquad (0 \le q \le 5)$$
$$C(q) = 2q^2 \qquad (0 \le q \le 5)$$

Draw graphs of these functions and find the number of items q that will maximize the profit. What is the maximum profit?

66. Gasoline Mileage The Department of Transportation has determined that the number of miles per gallon $M(s)$ that a certain model of automobile will get when it travels s miles per hour is

$$M(s) = -0.025s^2 + 1.40s + 5.5 \qquad (20 \le s \le 55)$$

(a) At what speed will the automobile get the largest number of miles per gallon?
(b) What speed will minimize the miles per gallon?
(c) How many miles per gallon will this model get at these speeds?

Max–Min Puzzlers

67. Can You Find the Maximum? What is the absolute maximum value of the function

$$f(x) = x$$

for $0 < x < 1$?

68. Can You Find the Maximum? What is the absolute maximum value of the function

$$f(x) = \frac{1}{x}$$

for $-1 < x < 1$?

69. Mystery Function Can you think of a function $f(x)$ defined on some interval that has an absolute maximum value but not an absolute minimum value?

9.4 Optimization Problems (Max–Min Problems)

PURPOSE

We present examples to show how absolute maximum and absolute minimum points of functions are useful in business management and science. These types of problems are called optimization or max–min problems. Some of the optimization problems in this section differ from problems that we have seen so far because they have constraints.

Introduction

One often asks, "What is the best way of doing something?" A business manager might wish to determine how to price a given product in order to maximize the company's profit. A fisheries biologist may wish to determine an optimal feeding program for fish growing in a controlled environment. In many cases the best way of doing something can be reduced to finding the maximum or minimum value of a function. It is here that calculus can come to the rescue. We present some examples of optimization problems that can be solved by using differential calculus.

The Apple Orchard Problem

An agronomist has been asked to determine the density of apple trees in an orchard that will maximize the total yield of apples. From experiments it is known that if the density of trees is 30 trees or fewer per acre, then each tree will produce 150 bushels of apples on the average. On the other hand, if the density of trees exceeds 30 trees per acre, the yield of each tree will decrease by three bushels per tree for each additional tree planted above 30. How many trees should be planted per acre to maximize the total yield of apples?

From the above information we can plot the graph of

$$N(t) = \text{Number of bushels per tree}$$

as a function of

$$t = \text{Number of trees per acre}$$

This graph is shown in Figure 26.

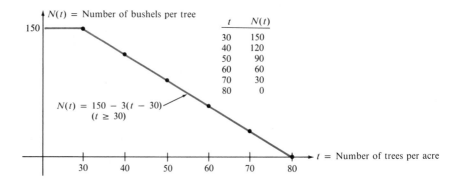

Figure 26

Bushels of apples per tree as a function of the number of trees per acre

Since the yield per tree is always 150 bushels per tree for a tree density less than 30 trees per acre, it makes sense that the number t of trees per acre should be at least 30. However, when the density is greater than 30, the number $N(t)$ of bushels per tree falls by three for each additional tree planted. This information says that $N(t)$ is a linear function with slope -3 that passes through the point $N(30) = 150$. Hence it can be written in point-slope form as

$$N(t) = 150 - 3(t - 30)$$
$$= -3t + 240 \qquad (t \geq 30)$$

If we now call

$$Y(t) = \text{Number of bushels per acre}$$

then we can write

$$\begin{array}{c}\text{Number of bushels} \\ \text{per acre}\end{array} = \left(\begin{array}{c}\text{Number of bushels} \\ \text{per tree}\end{array}\right) \cdot \left(\begin{array}{c}\text{Number of trees} \\ \text{per acre}\end{array}\right)$$

or

$$Y(t) = N(t)t$$
$$= (-3t + 240)t$$
$$= -3t(t - 80) \qquad (t \geq 30)$$

The goal now is to find the value of t where $Y(t)$ has its absolute maximum value. To do this, we first observe that the total yield of apples $Y(t)$ becomes negative when the density t becomes greater than 80 trees per acre. Hence we seek the absolute maximum yield $Y(t)$ of apples when the density t is between 30 and 80 trees per acre (inclusive). Stated mathematically, we wish to find the absolute maximum value of

$$Y(t) = -3t^2 + 240t \qquad (30 \le t \le 80)$$

To find this value, we first find the stationary points by solving the equation

$$Y'(t) = -6t + 240 = 0$$

We see that $Y(t)$ has one stationary point when $t = 40$. Evaluating $Y(t)$ at this point, we get

$$Y(40) = -3(40)(40 - 80)$$
$$= 4800$$

To find the absolute maximum value of $Y(t)$, we pick the largest of the three values

$$Y(30) = 4500 \text{ bushels} \qquad \text{(left endpoint)}$$
$$Y(40) = 4800 \text{ bushels} \qquad \text{(stationary point)}$$
$$Y(80) = 0 \text{ bushels} \qquad \text{(right endpoint)}$$

Conclusion

The maximum yield of apples occurs when the tree density is 40 trees per acre. With this density, 4800 bushels will be harvested per acre. Any larger or smaller density of trees per acre will result in a smaller yield of apples per acre. The graph of $Y(t)$ is shown in Figure 27.

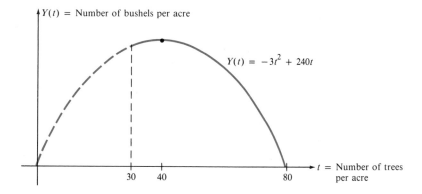

Figure 27

Bushels of apples per acre as a function of the number of trees per acre

Another well-known optimization problem is the maximum-volume box problem. We organize the solution of this problem by subdividing it into small steps.

The Maximum-Volume Box Problem

We start with a square sheet of cardboard for which the length of each side is 12 inches. Next we cut the same size square from each corner of the cardboard and fold the resulting piece of cardboard along the dotted lines as shown in Figure 28 to form a topless box.

The question we ask is, "How many inches should we cut from each corner of the original cardboard square to maximize the volume of the topless box?" How many inches would you guess: 1, 2, 3, 4, or 6 inches?

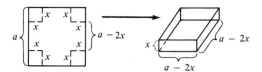

Figure 28
Illustration of the box problem

To find the absolute maximum of the volume of the resulting box, we carry out the following steps.

- **Step 1** (draw a picture illustrating the relevant variables). In Figure 28 we have denoted x (in inches) as the length of the sides of the small square cut from each corner.

- **Step 2** (find the equation that relates the relevant variables). The volume $V(x)$ of the topless box is given by

$$V(x) = \text{(length)(width)(height)}$$
$$= (12 - 2x)(12 - 2x)(x)$$
$$= x(12 - 2x)^2 \qquad (0 \leq x \leq 6)$$

Note that we cannot physically cut out squares with sides larger than 6. Hence the restriction on the variable x.

- **Step 3** (find the absolute maximum value of $V(x)$). To find the absolute maximum value of $V(x)$ on the interval $[0, 6]$, we first evaluate it at the endpoints, getting

$$V(0) = 0 \quad \text{cubic inches}$$
$$V(6) = 0 \quad \text{cubic inches}$$

We now find the stationary points by computing the derivative $V'(x)$. We find

$$V'(x) = 12x^2 - 96x + 144$$
$$= 12(x - 6)(x - 2)$$

Setting this derivative to zero and solving for x, we get the two stationary points $x = 2$ and $x = 6$.

Hence the absolute maximum of $V(x)$ can be found by taking the largest of the quantities

$$V(0) = 0 \qquad \text{(left endpoint)}$$
$$V(2) = (2)(8)^2 = 128 \qquad \text{(stationary point)}$$
$$V(6) = 0 \qquad \text{(right endpoint and stationary point)}$$

Conclusion

To obtain a topless box with a maximum volume, we should cut a square that has 2-inch sides from each corner of the sheet of cardboard. This will result in a topless box with a maximum volume of

$$V(2) = \text{(length)(width)(height)}$$
$$= (8)(8)(2)$$
$$= 128 \quad \text{cubic units}$$

The graph of the volume $V(x)$ of the box as a function of x is drawn in Figure 29.

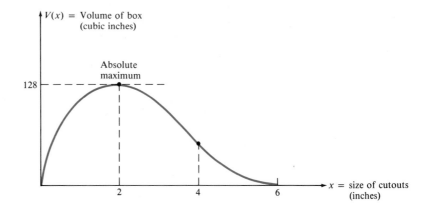

Figure 29
Volume of the box as a function of the cutouts

Max–Min Problems with Constraints

In both the apple orchard problem and the maximum-volume box problem we maximized a function of a single variable. In the remaining problems we will maximize a function of f that depends on two variables x and y which we denote by $f(x, y)$. The method of solution consists of finding a relationship or constraint between the two variables. By substituting this constraint, which we write as $y = g(x)$, into the function $f(x, y)$ we are trying to maximize, we will get an expression $f(x, g(x))$ in one variable. We then find the absolute maximum or minimum value of this function, using the methods previously learned. We illustrate these ideas with two fascinating problems.

The Cat Food Can Problem (Minimum Surface Area)

The Carnation Co. of Los Angeles produces and distributes Friskies Buffet cat foods. All flavors come in cylindrical containers, and a 6-ounce can of Friskies Buffet Mixed Grill has a volume of roughly 14.5 cubic inches and dimensions

$$\text{Height} = 1.75 \text{ inches}$$
$$\text{Radius} = 1.62 \text{ inches}$$

The object of our study is to find the radius and the height of the can having a volume of 14.5 cubic inches that has the smallest surface area. In other words, how would we make a can for Friskies Mixed Grill that contains the least amount of metal in its construction? And finally, does the can for Friskies Mixed Grill that the Carnation Co. makes have this smallest surface area?

We begin by drawing typical cans with various diameters and heights as shown in Figure 30. Note that when the can is very tall or very short, the surface area will be large. For instance, when the can is short and fat, although the surface area around the sides is small, the surface area of the top and bottom is large. Also, if the can is tall and skinny, the top and bottom have small surface areas, but the surface area around the sides is large. The can with the smallest surface area lies somewhere between two such extremes.

To find the dimensions of the can having volume 14.5 cubic inches with minimum surface area, we begin by writing the surface area S of a cylinder in terms of its radius r and height h. We have

$$S(r, h) = 2\pi r^2 + 2\pi rh \qquad \text{(surface area of a cylinder)}$$

Area of top and bottom   Area of side

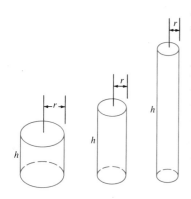

Figure 30
Three cylinders with the same volume but different surface areas

The goal is to find the values of r and h that minimize $S(r, h)$. Since the surface area S depends on two variables r and h, this maximization problem is different from any we have seen thus far. However, the two variables are not independent of each other, since they are related by the volume constraint:

$$\pi r^2 h = 14.5 \qquad \text{(volume of a cylinder)}$$

Hence we can solve for one of these variables in terms of the other and substitute the solved variable into the equation for S. This will give us $S(r, h)$ as a function of a single variable. Since it is easier in this case to solve for h, we have

$$h = \frac{14.5}{\pi r^2}$$

Substituting this value into the formula for the surface area S, we have

$$S(r) = 2\pi r^2 + \frac{29\pi r}{\pi r^2}$$

$$= 2\pi r^2 + \frac{29}{r} \qquad (0 < r < \infty)$$

Although this function is not defined on a closed interval, it is still possible to find the absolute minimum value of $S(r)$. We begin by computing its first two derivatives:

$$S'(r) = 4\pi r - \frac{29}{r^2}$$

$$S''(r) = 4\pi + \frac{58}{r^3} \qquad \text{(always positive)}$$

Solving the equation $S'(r) = 0$, we find the stationary point

$$r = \sqrt[3]{29/4\pi}$$
$$\cong 1.32 \quad \text{inches}$$

Since the second derivative is always positive, the graph of $S(r)$ is always concave up. It is clear that the surface area $S(r)$ obtains its absolute minimum value at the above stationary point. The graph of the surface area $S(r)$ as a function of the radius r is shown in Figure 31.

To calculate the corresponding height of the can, we substitute the value $r = 1.32$ inches into the volume constraint, getting

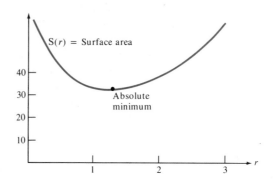

Figure 31
Surface area of a can as a function of its radius

$$h = \frac{14.5}{\pi r^2} = \frac{14.5}{\pi (1.32)^2}$$

$$\cong 2.64 \quad \text{inches} \qquad \text{(twice the radius)}$$

Conclusion

If the volume of a cylindrical can is 14.5 cubic inches, then the dimensions of the can with the minimum surface area will be one for which both the diameter and the height are 2.64 inches. The surface area for such a minimal surface area can was found to be 32.9 square inches.

We conclude that a can of Friskies Buffet Mixed Grill with a diameter of 3.24 inches and a height of 1.75 inches is too short. It is easy to show that the Carnation Co. can save 6.7 square inches of metal for every can of cat food manufactured if they design the cans with height equal to the diameter.

Interested in knowing why the Carnation Co. packaged cat food in such short cans (like many other products such as tuna and canned chicken), we sent a letter to the California office of the Carnation Co. asking the reason why. Here is their reply:

(arnation

Corporate Offices

5045 Wilshire Boulevard
Los Angeles, California 90036
Telephone: (213) 932-6000

April 13, 1987

Jerry Farlow
Professor of Mathematics
Dept. of Mathematics
University of Maine
Orono, ME 04473

Dear Professor Farlow:

We appreciate the interest you expressed in examining the height-to-diameter relationship of containers used in our food products. A 1:1 ratio of height versus diameter is the most efficient use of material, if only the surface area of material is considered. However, there are many other factors which must be considered when designing a can for a particular product. Listed below are some of these other factors:

1) Thermal Processing — There is an inverse relationship between the most efficient design for cans relative to surface area and the amount of processing time required to sterilize the product contained within. In other words, a tall thin can or short wide can will require considerably less processing time and energy to achieve commercial sterility than a can which is nearly equal in height and diameter.

2) Strength Requirements — During thermal processing, considerable internal pressure develops. This pressure can cause the ends of the can to become permanently distorted. Because of this, ends on most cans are made of metal which is substantially thicker than that used in the can cylinder. Therefore, there is not a simple cost-to-surface area relationship relative to metal. As this can becomes taller and the end becomes smaller, thinner metal can be used in both the cylinder and the ends.

3) Can Manufacturing Line Changeover Time — Virtually all can lines run a variety of can sizes. The time required to change over from one can size to another is considerably less if only can height is changed, rather than height and diameter. In addition, since the same ends can be used if only the height is changed, the machinery used to manufacture ends does not have to be changed over to a different diameter. Reduced changeover time translates into reduced downtime and increased line efficiency.

4) Scrap Loss — Generally, more metal scrap is generated as the diameter is increased.

5) Warehouse and Shipping Efficiency — Smaller diameter cans make more efficient use of packaging and shipping space.

As you can see, cost and efficiency of a container are related to factors other than just the amount of material used. These are just a few of the factors which must be taken into consideration when designing a can. We hope that you now better understand that container design is not quite as simple as minimizing surface area.

Once again, thank you for your genuine interest.

Sincerely,

Vince Daukas

Vince Daukas
Assistant Product Manager
Friskies Buffet

The following set of rules can be used for solving optimization problems with constraints. The rules were used in solving the cat food can problem.

Steps in Solving Optimization Problems with Constraints

- **Step 1** (draw a picture).   If possible, draw a picture illustrating the problem and label the relevant variables.
- **Step 2** (find the objective function).   Determine the function to be maximized or minimized, and write it in terms of its variables. This function is called the objective function.
- **Step 3** (find the constraint).   If the objective function depends on two variables, find a relationship or constraint between these variables, and solve for one in terms of the other.
- **Step 4** (write the objective function in terms of one variable). Substitute the solved variable from Step 3 into the objective function so that the objective function is a function of a single variable.
- **Step 5** (find the absolute maximum and minimum values).   Find the absolute maximum or minimum values of the objective function.

The Salmon Problem (How Fast Should a Salmon Swim?)

The amount of energy $E(v)$ that a salmon expends in swimming upstream with velocity v (relative to the stream) over a period of time T has experimentally been shown to be

$$E(v, T) = cv^3 T$$

where c is a constant. Suppose the velocity of the stream is 4 miles per hour and the salmon swims upstream 200 miles. How fast should a fish swim in order to minimize its use of energy?

- **Step 1** (draw a picture).   We start by drawing an illustration of the relevant variables, as shown in Figure 32.
- **Step 2** (find the objective function).   The quantity or objective function that should be minimized is the energy

$$E(v, T) = cv^3 T$$

which depends on the two variables v (velocity) and T (time).

- **Step 3** (find the constraint).   The variables v and T are not independent of one another but are related by an equation. To find this equation, observe that the velocity of the fish relative to the ground is 4 miles per hour less than it is relative to the stream. Hence by using the general

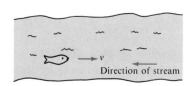

Figure 32
The fish moves with velocity v relative to the velocity of the stream

formula for velocity,

$$\text{Velocity} = \frac{\text{Distance}}{\text{Time}}$$

we can write the constraint equation, which relates v and T:

$$v - 4 = \frac{200}{T} \qquad \text{(constraint equation)}$$

Solving for T in terms of v, we have

$$T = \frac{200}{v - 4}$$

- **Step 4** (write the objective function in terms of one variable). Substituting the above value for T into the objective function $E(v, T)$, we get a function of v alone:

$$E(v) = cv^3 T$$

$$= 200c\, \frac{v^3}{v - 4} \qquad (4 < v < \infty)$$

- **Step 5** (minimize the objective function). To find the absolute minimum value of $E(v)$, we first find its derivative:

$$\frac{dE}{dv} = 400cv^2\, \frac{v - 6}{(v - 4)^2}$$

Setting dE/dv to zero, we get a single stationary point at $v = 6$. It is also clear that the derivative dE/dv is negative for v less than 6 and positive for v greater than 6. Hence we can conclude that the absolute minimum value of $E(v)$ occurs at $v = 6$.

Conclusion for the Fish Velocity

We conclude that for the salmon to minimize its use of energy, it should swim at a rate of $v = 6$ miles per hour or 50% faster than the velocity of the stream (which is 4 miles per hour). Note that the velocity of the fish is 2 miles per hour relative to the bank of the stream. If the fish swims faster than this velocity, it will end up spending more energy. It will, of course, arrive at its destination sooner, but its increased velocity will result in a larger total expended energy. On the other hand, if the fish swims slower than 6 miles per hour, it ends up taking so long that again the total energy expended is not a minimum. It is interesting to note that most salmon do swim at a velocity roughly 50% greater than the velocity of the stream.

The Fence Problem (Minimum Cost)

A nursery wants to add a 1000-square-foot rectangular area to its greenhouse to sell seedlings. For aesthetic reasons they have decided to border the area on three sides by cedar siding at a cost of $10 per running foot. The remaining side of the enclosure is to be a wall with a brick mosaic that costs $25 per running foot. What should the dimensions of the sides be so that the cost of the project will be minimized?

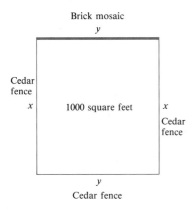

Brick mosaic

y

Cedar
fence

x   1000 square feet   *x*

Cedar
fence

y

Cedar fence

Figure 33
Diagram for the fence problem

- **Step 1** (draw a picture).   Calling

$$x = \text{Length of the two cedar fences adjacent to the brick wall}$$
$$y = \text{Length of the brick wall (and the opposite cedar fence)}$$

we draw a diagram and label these variables, as shown in Figure 33.

- **Step 2**   (find the objective function).   The function that we wish to maximize is the total cost of the project. Since the general formula is given by

$$\text{Total cost of fence} = (\text{Cost per foot})(\text{Length in feet})$$

we have

$$\text{Total cost for cedar fence} = (\$10)(2x + y)$$
$$\text{Total cost for brick mosaic} = (\$25)(y)$$

Hence the total cost in dollars of the project will be

$$C(x, y) = 10(2x + y) + 25y$$
$$= 20x + 35y \quad \text{(objective function)}$$

- **Step 3** (find the constraint).   The two variables x and y are not independent but are related by the area constraint

$$xy = 1000 \quad \text{(constraint equation)}$$

Solving for y in terms of x gives

$$y = \frac{1000}{x}$$

- **Step 4** (write the objective function in terms of one variable).   Substituting the above value of y into the objective function $C(x, y)$, we have

$$C(x) = 20x + 35\frac{1000}{x} \quad (0 < x < \infty)$$

- **Step 5** (minimize the objective function).   Computing the derivative of $C(x)$, we have

$$C'(x) = 20 - \frac{35,000}{x^2}$$

Setting this derivative to zero, we get

$$20 - \frac{35,000}{x^2} = 0$$

or

$$x^2 = \frac{35,000}{20}$$

Solving for x, we obtain the stationary point

$$x = 41.83$$

It is clear from the formula for $C'(x)$ that $C'(x)$ is always negative to the left of 41.83 and positive to the right of 41.83. From these observations it is clear that $C(x)$ has an absolute minimum value when $x = 41.83$. Using this value of x, we can find the second dimension y from the constraint equation $xy = 1000$. This gives $y = 23.91$.

Conclusion

The most economical fence that will enclose an area of 1000 square feet is one 23.91 × 41.83 feet. The total cost of the fence is

$$\text{Total cost} = 20x + 35y$$
$$= 20(41.83) + 35(23.91)$$
$$= \$1673.45$$

Any other rectangular design will cost more. A scale drawing of this optimal design is shown in Figure 34.

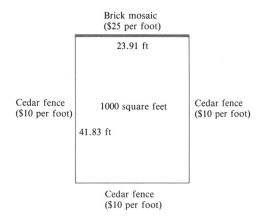

Figure 34
Minimum-cost rectangle

Problems

1. Maximum Product Find two numbers whose sum is 15 and whose product is a maximum.

2. Minimum Sum of Squares Find two numbers whose sum is 10 for which the sum of their squares is a minimum.

3. Constrained Maximum Find nonnegative numbers x and y whose sum is 75 and for which the value of xy^2 is as large as possible.

4. Constrained Maximum Find the maximum value of the function of two variables

$$f(x, y) = x^2 y$$

when

$$x + y = 150$$

Theory of the Firm

5. Minimum Average Cost A manufacturer has determined that the total cost C to produce q units of a given product is

$$C(q) = -0.05q^2 + 10q + 250$$

Find the level of production for which the **average cost** per unit is a minimum.

6. Minimum Operating Costs A trucking company has determined that the cost per hour to operate a single truck is given by

$$C(s) = -0.001s^2 + 0.10s + 0.12$$

where s is the speed that the truck travels. At what speed is the total cost per hour a minimum? What is the hourly cost to operate the truck?

7. Maximum Revenue A company has determined that the weekly demand $D(p)$ for its product is a function of the price p that the company charges. The relationship has been determined to be

$$D(p) = -3p + 300$$

How much should the company charge to maximize the revenue? *Hint:* The revenue R is given by $R(p) = pD(p)$.

8. Marginal Revenue A company's total revenue is

$$R(q) = 10q - 0.01q^2$$

where q is the number of items produced. Find the number of items produced that will make the marginal revenue zero. What is your interpretation of this value of q?

9. **Maximum Profit**   A manufacturing firm can sell all the items it can produce at a price of $5 each. The cost in dollars to produce q items per week is given by

$$C(q) = 1000 + 10q - 0.002q^2$$

What value of q should be selected to maximize the total profit? *Hint:* Profit $= 5q - C(q)$.

10. **Maximum Profit**   Suppose the unit sales volume V for a given product depends on the offering price p, according to the equation

$$V(p) = 40,000 - 8000p$$

Suppose the cost C to manufacture each item depends on the number V of items produced (same as the number sold) and is given by

$$C(V) = 100,000 + 25V$$

Find the price that should be charged for the product that will maximize the profit.
Hint: Profit $=$ (Price)(Volume) $-$ (Cost).

11. **Optimal Pricing**   A real estate company owns 100 apartments in New York City. At $1000 per month, each apartment can be rented. However, for each $50 increase, there will be two additional vacancies. How much should the real estate company charge for rent to maximize its revenues?

12. **Optimal Pricing**   A national chain of service stations charges $28 to replace a muffler. At this rate the company replaces 75,000 mufflers every week. For each additional dollar that the company charges, it tends to lose 1000 customers every week. For each dollar the company subtracts from the $28, the company gains 1000 customers every week. How much should the company charge to change a muffler to maximize the revenue?

13. **Optimal Density of Pear Trees**   Normally a pear tree will produce 30 bushels of pears per tree when 20 or fewer pear trees are planted per acre. However, for each additional pear tree planted above 20 trees per acre, the yield per tree will fall by one bushel per tree. How many trees should be planted per acre to maximize the total yield?

Minimum-Cost Container Problems

14. **Minimum-Cost Box**   A closed box with a square base is to have a volume of 1600 cubic inches. The material for the top and bottom of the box costs $3 per square inch, while the material for the sides costs $1 per square inch. Find the dimensions of the box that will lead to the minimum total cost. What is the minimum total cost?

15. **Minimum Surface Area**   Oversea Containers Inc. builds large wooden containers for secure overseas shipping of goods. Find the dimensions of such a container with a square top, minimal surface area, and a volume of 1000 cubic feet.

16. **Minimum Surface Area**   Find the dimensions of a box with a minimal surface area, a square base, a volume of 100 cubic feet, and no top.

17. **Minimum Surface Area**   Find the dimensions of a carport that is attached to a house, encloses 2000 cubic feet, and has minimal surface area and a height equal to its length. The carport has a flat roof and a side parallel to the house but no front or back wall.

18. **Maximum Volume**   The U.S. Postal Service will accept a package for parcel post only if its length plus girth (shortest distance around the package) does not exceed 108 inches. Mail-It-Secure Co. plans to market a box that will satisfy this condition and have maximum volume. If the box is to have square ends and rectangular sides, what should be the dimensions of the box?

19. **Maximum Volume**   A rectangular sheet of cardboard $3' \times 5'$ is to be cut as shown in Figure 35 to form a topless box. What dimensions of the box will enclose the maximum volume? What is the maximum volume?

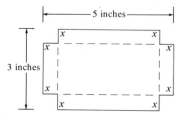

Figure 35
Illustration for Problem 19

20. **Minimum-Cost Box**   A box with a top is three times as long as it is wide and holds a fixed volume V. The material for the sides and bottom costs 20 cents per square foot, while the material for the top costs 30 cents per square foot. What dimensions of the box are most economical?

21. **Wire Problem**   A wire 25 inches long is cut into two pieces. One piece is to be shaped into a square, and the other piece into a circle, as shown in Figure 36. How should the wire be cut to maximize the total area enclosed by the square and the circle?

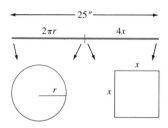

Figure 36
Illustration for Problem 21

22. Rectangle in a Circle What is the rectangle of largest area that can be cut from a circle of radius 20 inches? (See Figure 37.)

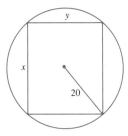

Figure 37
Illustration for Problem 22

23. Optimal Printing Ready Words needs to set up several templates for the printing requirements of its major customers. One type of page should contain 80 square inches of printed matter with two-inch margins on each side and a three-inch margin at the top and bottom. Find the dimensions of this type of page that has the minimum total page area.

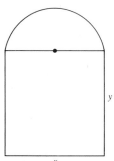

Figure 38
Norman window

24. Norman Window Problem A designer of custom windows wishes to build a Norman window with an outside perimeter of 40 feet as shown in Figure 38. How should one design the window to maximize the area of the window? A Norman window consists of a rectangular region bordered above by a semicircle.

25. Maximum Area A strip of metal 12 inches wide is to be formed into an open gutter with a rectangular cross section, as shown in Figure 39. What is the maximum possible area of the cross section?

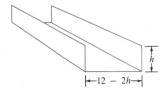

Figure 39
Gutter for Problem 25

26. Maximum Surface Area A tent with a volume of 432 cubic feet is formed with sides shaped as rectangles and ends as equilateral triangles. What should be the dimensions of the tent to minimize its surface area? Does adding a floor to the tent change the answer?

27. Telephone Lines A telephone wire is to be laid to an island seven miles off shore at a cost of $2000 per mile along the shore and $3000 per mile under the sea. How should the project be planned to minimize the cost if the distance from A to B is 12 miles? (See Figure 40.)

28. Maximum Height of a Ball A ball is thrown straight up in the air. Its height after t seconds is given by

$$s(t) = -16t^2 + 50t$$

When does the ball reach its maximum height? What is its maximum height?

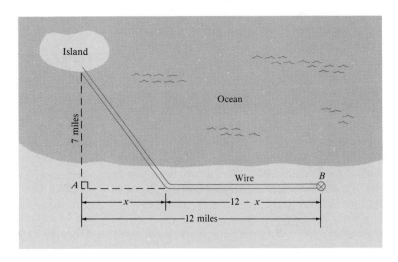

Figure 40

Telephone line to an island

INTERVIEW WITH GARY C. MCDONALD
Department Head, Mathematics Department
General Motors Research Laboratories

TOPIC: Calculus at General Motors

Random House: Of course, it's obvious that all General Motors engineers have a strong mathematics background, but is it necessary for your business managers to study mathematics?

Mr. McDonald: The answer is yes. To some extent, the business manager today can be confronted with accounting systems, statistical analysis, and various scenarios involving decision making. Being able to understand the assumptions and limitations of these concepts often requires a foundation in mathematics.

Random House: When one thinks of applying the derivative to business management and economics, one often thinks of the revenue, costs, and profit curves and their derivatives, the marginal revenue, cost, and profit curves. Most books in business mathematics always describe these functions by polynomials. Can polynomials actually describe these economic phenomena adequately?

Mr. McDonald: Within certain limitations, polynomials can do a fine job. More complicated functions can often be approximated by polynomials. Over an extended domain, it may be necessary to piece together several polynomials (so-called spline functions).

Random House: Many readers of this book probably haven't the slightest idea what they will be doing ten years from now and probably even have doubts about making a serious effort to learn mathematics. What kind of mathematical background does General Motors look for when hiring business majors?

Mr. McDonald: This, of course, depends on the area for which the individual is being considered. I would recommend business majors at the B.A. level to study mathematics through calculus. I would also encourage those students to take a course in finite mathematics and linear programming.

Random House: I understand that microcomputers are being used in the latest design of automobiles to optimize engine performance. Do any of these optimization techniques require calculus? In other words, do these microcomputers solve optimization problems similar to the ones studied in this book in order to get the best fuel mixture?

Mr. McDonald: The answer is yes and no. Computers attached to the engine collect engine information about 125 times every second. This information is then used to control certain key engine parameters, such as the air–fuel ratio, to ensure good fuel economy. Thus you might say we solve 125 optimization problems every second that you are driving your car. However, they may not all be of the formal max–min variety encountered in freshman calculus classes.

29. Maximum Receipts   A concert promoter knows that 5000 people will attend an event with tickets set at $10. For each dollar less in ticket price an additional 1000 tickets will be sold. What should the ticket price be to maximize total receipts?

30. Group Travel   A travel agent is offering charter holidays to Europe for college students. For groups of size up to 100, the fare is $1000 per person. For larger groups the fare per person decreases by $5 for each additional person in excess of 100. Find the size of the group that maximizes the travel agent's revenues.

31. Maximum Area of a Field   A farmer has 2000 feet of fencing to enclose a pasture area. The field should have the shape of a rectangle with one side bordered by a currently existing rock fence, as shown in Figure 41. What should be the dimensions of the field to enclose the maximum area?

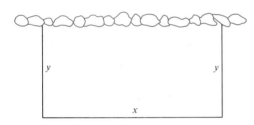

Figure 41
Field bordered by a brick wall

32. Fish-Stocking Problem   A fisheries biologist is stocking fish in a lake. She knows that when there are n fish per unit area of water, the average weight $W(n)$ of each fish after one season will be

$$W(n) = 500 - 20n \quad \text{grams}$$

What is the value of n that will maximize the total fish weight after one season? *Hint*: Total weight = $nW(n)$.

33. Bacteria Growth   The size $N(t)$ of a population of bacteria introduced to a nutrient grows according to the formula

$$N(t) = \frac{5000t}{50 + t^2}$$

where t is measured in weeks. Determine when the bacteria will reach maximum size. What is the maximum size of the population?

34. Medical Testing   Blood pressure in a patient will drop by an amount $D(x)$, where

$$D(x) = 0.025x^2(30 - x)$$

and x is the amount of drug injected. Find the dosage that provides the greatest drop in blood pressure. What is the drop in blood pressure?

9.5

Implicit Differentiation and Related Rates

PURPOSE

This section introduces the idea of implicit differentiation and uses it to study the topic of related rates of change. The major ideas introduced are

- implicit differentiation,
- tangent lines to general curves, and
- related rates of change.

Introduction

Until now, we have found the derivative dy/dx when y was related to x by means of an explicit functional relationship $y = f(x)$. We now show how to find the derivative dy/dx when x and y are related *implicitly*, as in a formula such as $f(x, y) = 0$. The process used to find dy/dx when x and y are related implicitly is called *implicit differentiation*.

The second portion of this section, which makes use of implicit differentiation, introduces the idea of *related rates*. Here, we find the rate of change of one quantity in terms of the rate of change of other quantities.

Implicit Differentiation

Until now, to find the derivative dy/dx, it was necessary to start with the functional relationship $y = f(x)$. Now, however, we show how to find the derivative dy/dx given an implicit relationship between x and y, such as

$$xy = 1$$

without first solving for y in terms of x. The conventional way to find the derivative dy/dx from the above equation is to first solve for y in terms of x (which we can easily do in this case), getting

$$y = \frac{1}{x}$$

Straightforward differentiation then gives

$$\frac{dy}{dx} = -\frac{1}{x^2}$$

It is possible, however, to find this derivative by another procedure known as implicit differentiation. In implicit differentiation we differentiate both sides of the equation

$$xy = 1$$

before solving for x, still treating y as a function of x. Using this technique, we get

$$\frac{d}{dx}[xy] = \frac{d}{dx}[1]$$

$$x\frac{d}{dx}[y] + y\frac{d}{dx}[x] = 0 \qquad \text{(product rule)}$$

$$x\frac{dy}{dx} + y = 0$$

$$\frac{dy}{dx} = -\frac{y}{x} \qquad \text{(solving for } dy/dx)$$

We now have the derivative except that it is in terms of both x and y. If we now solve for y in terms of x from the original equation, we get

$$y = \frac{1}{x}$$

Substituting this expression for y into the above equation for dy/dx, we have

$$\frac{dy}{dx} = -\frac{1}{x^2}$$

Note that this derivative agrees with the derivative found earlier when we solved for y first and differentiated second. This new method of finding derivatives by first differentiating and then solving for dy/dx is called **implicit differentiation**. In the above problem it made little difference whether we used explicit or implicit differentiation. However, there are problems in which the derivative dy/dx can best be found by implicit differentiation, since it is difficult and even impossible to solve for y explicitly in terms of x. Imagine solving for

y in terms of x when they are related by the equation

$$xy^7 + xy^3 - xy^4 = 5$$

In cases such as this we use implicit differentiation.

Example 1

Comparison of Differentiation Techniques   Find dy/dx using both implicit and explicit differentiation when x and y are related by the equation

$$x^2y - 1 = 0$$

Solution   For the above equation it is possible to find dy/dx by both techniques. Table 8 compares the two types of differentiation.   □

TABLE 8

Comparison of Explicit and Implicit Differentiation

Explicit Differentiation	**Implicit Differentiation**
Step 1. Solve for y: $$y = \frac{1}{x^2}$$ **Step 2.** Differentiate $$\frac{dy}{dx} = -\frac{2}{x^3}$$ *We Are Done*	**Step 1.** Differentiate each side of the equation with respect to x, treating y as a function of x: $$\frac{d}{dx}[x^2y - 1] = \frac{d}{dx}[0]$$ $$\frac{d}{dx}[x^2y] - \frac{d}{dx}[1] = 0$$ $$x^2\frac{d}{dx}[y] + y\frac{d}{dx}[x^2] = 0$$ $$x^2\frac{dy}{dx} + 2xy = 0$$ Solving for dy/dx, we have $$\frac{dy}{dx} = -\frac{2y}{x}$$ (gives the derivative in terms of x and y). **Step 2.** If possible, solve for y in terms of x in the original equation: $$x^2y - 1 = 0$$ Here we have $$y = \frac{1}{x^2}$$ **Step 3.** Substituting the above value of y into the expression for dy/dx, we get dy/dx strictly in terms of x. Here we have $$\frac{dy}{dx} = -\frac{2}{x^3}$$ *We Are Done*

Note that both implicit and explicit differentiation give the same result. In this example explicit differentiation was the easier technique. However, in the remainder of this section and later sections, we will see that implicit differentiation is a valuable and necessary tool.

Example 2

Implicit Differentiation   Find dy/dx given

$$x^3 + y^3 - xy = 0$$

Solution   Since it is difficult to solve for y in terms of x, implicit differentiation is the more convenient way to find dy/dx. Differentiating each side of the above equation with respect to x and remembering that y is a function of x, we get

$$\frac{d}{dx}[x^3 + y^3 - xy] = \frac{d}{dx}[0]$$

$$\frac{d}{dx}[x^3] + \frac{d}{dx}[y^3] - \frac{d}{dx}[xy] = 0$$

$$3x^2 + 3y^2\frac{d}{dx}[y] - x\frac{d}{dx}[y] - y\frac{d}{dx}[x] = 0$$

$$3x^2 + 3y^2\frac{dy}{dx} - x\frac{dy}{dx} - y = 0$$

TABLE 9
Curves Written in
Implicit Form

$f(x, y) = 0$	Type of Curve	
$x^2 + y^2 - 1 = 0$	Circle	
$x^2 - y^2 - 1 = 0$	Hyperbola	
$x^2 + 4y^2 - 16 = 0$	Ellipse	

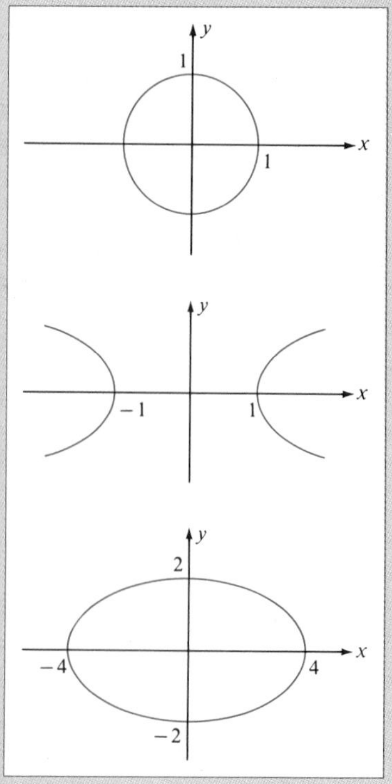

Solving for dy/dx, we have

$$\frac{dy}{dx} = \frac{y - 3x^2}{3y^2 - x}$$

We have found the derivative dy/dx in terms of both x and y.   □

In the preceding example, since it was difficult to solve algebraically for y in terms of x in the original equation, we left the result in terms of x and y. Often the derivative is desired at some point (x, y). In those cases there is no disadvantage in having both x and y in the formula for dy/dx.

Tangent Lines to Curves

One of the major applications of implicit differentiation lies in finding tangent lines to curves that are not defined explicitly as $y = f(x)$ but are defined implicitly as $f(x, y) = 0$. Some examples are given in Table 9. If the derivative dy/dx of an implicit function $f(x, y) = 0$ is evaluated at a point (x, y) on the curve $f(x, y) = 0$, then the derivative is the slope of the tangent line to the curve at that point. We illustrate this idea with the following example.

Example 3   _____   *Slope of a Tangent Line*   Let x and y be related by the equation

$$x^2 - y^2 = 1$$

which describes the hyperbola shown in Figure 42. Find the slope of the tangent line to this hyperbola at the point $(2, \sqrt{3})$.

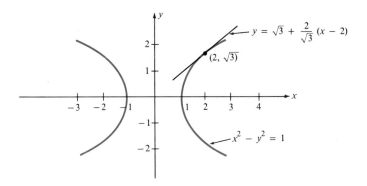

Figure 42
Finding tangent line by implicit differentiation

Solution   To find dy/dx at the desired point, we first differentiate implicitly each side of the equation with respect to x. This gives

$$\frac{d}{dx}[x^2 - y^2] = \frac{d}{dx}[1]$$

$$\frac{d}{dx}[x^2] - \frac{d}{dx}[y^2] = 0$$

$$2x - 2y\frac{dy}{dx} = 0$$

Solving for dy/dx gives

$$\frac{dy}{dx} = \frac{x}{y}$$

This derivative says that the slope of the tangent line at any point (x, y) on the hyperbola is given by the ratio x/y. By a simple check we can verify that the point $(2, \sqrt{3})$ lies on the hyperbola $x^2 - y^2 = 1$, and so the slope of the tangent line at this point is $2/\sqrt{3}$. Using the point-slope form of the straight line, we find that the equation of the tangent line at this point is

$$y - \sqrt{3} = \frac{2}{\sqrt{3}}(x - 2)$$

This line is shown in Figure 42.   □

We now summarize the technique of implicit differentiation by stating the following rules.

Rules for Implicit Differentiation

To find dy/dx when x and y are related implicitly by $f(x, y) = 0$, use the following three-step procedure.

- **Step 1.** Differentiate $f(x, y) = 0$ with respect to x. Keep in mind that y is a function of x.
- **Step 2.** Solve for dy/dx in terms of x and y.
- **Step 3.** If possible, solve for y in terms of x in the original equation and substitute this value of y into the equation for dy/dx found in Step 2.

Example 4

Tangent Lines to Circles   Find the tangent line to the circle

$$x^2 + y^2 = 1$$

at any point (x_0, y_0) on the circle.

Solution   Using the above three-step procedure to find dy/dx, we do the following:

- **Step 1** (differentiate the equation with respect to x).   Differentiating the given equation with respect to x gives

$$\frac{d}{dx}[x^2 + y^2] = \frac{d}{dx}[1]$$

$$2x + 2y\frac{dy}{dx} = 0$$

- **Step 2** (solve for dy/dx).   Solving this equation for dy/dx gives

$$\frac{dy}{dx} = -\frac{x}{y}$$

- **Step 3** (if possible, solve for y in the original equation).   Since we cannot solve for a unique value of y in the original equation (we get two values), we leave the derivative dy/dx in terms of both x and y.

Hence the tangent line to any point (x_0, y_0) on the circle

$$x^2 + y^2 = 1$$

will have the slope

$$\frac{dy}{dx} = -\frac{x_0}{y_0}$$

Using the point-slope formula for a straight line, this tangent line can be written as

$$y - y_0 = -\frac{x_0}{y_0}(x - x_0)$$

Some of the tangent lines to the circle are shown in Figure 43.   □

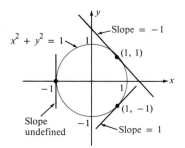

Figure 43
Tangent lines to a circle

The remainder of this section will be devoted to the concept of related rates. Implicit differentiation is often required to evaluate many of the derivatives encountered.

Related Rates

If y depends on x, then y will change in some manner when x changes. Suppose now that x changes with time t. Inasmuch as y changes when x changes and x changes when t changes, it follows that y changes when t changes. More specifically, if x and y are related by $y = f(x)$ and x depends on time by means of some function $x = x(t)$, then x and y will be related by the equation $y(t) = f[x(t)]$. From this equation we can find the rate of change in y (dy/dt) in terms of the rate of change in x (dx/dt) by simply using the chain rule and differentiating this equation with respect to t. This gives

$$\frac{dy}{dt} = \frac{d}{dx}f(x)\frac{dx}{dt} \qquad \text{(related rates equation)}$$

Rate of change of y with   Rate of change of x with
respect to time   respect to time

This **related rates equation** gives the relationship between the rate of change in y with respect to time (dy/dt) and the rate of change in x with respect to time (dx/dt). We now present examples to show how this related rates equation can be used to find some interesting and useful relationships.

Example 5

The Soft Drink Problem   Eddie is drinking a soda from a cylindrical can through a straw. The level of the soda in the can is decreasing at a rate of one-half cubic inch per second. If the radius across the top of the can is one inch, how fast is the level of soda in the can going down?

Solution   We wish to find the rate of change dh/dt of the height h of the soda and are given the rate of change dV/dt of the volume V of the soda. We start by finding the algebraic relationship between V and h. We first write the equation $V = \pi r^2 h$, which expresses the volume V of a cylinder in terms of its radius r and height h. These variables are illustrated in Figure 44.

As Eddie drinks the soda, the volume V and height h of the soda change with time t. On the other hand, the radius r remains the same. If we now differentiate the equation

$$V(t) = \pi r^2 h(t)$$

with respect to time (note that we have denoted the volume and height as $V(t)$ and $h(t)$, since they both change with time), we will get the related rates equation relating dV/dt and dh/dt. Doing this, we get

$$\frac{dV}{dt} = \pi r^2 \frac{dh}{dt} \qquad \text{(related rates equation)}$$

We can now solve for the desired quantity dh/dt in terms of the known quantities r and dV/dt. Since we were given

$$r = 1 \text{ inch}$$

$$\frac{dV}{dt} = -0.5 \quad \text{cubic inch per second}$$

we have

$$\frac{dh}{dt} = \frac{1}{\pi r^2} \frac{dV}{dt}$$

$$= \frac{1}{\pi (1)^2} (-0.5)$$

$$\cong -0.16$$

Note that we let dV/dt be a negative number, since V is decreasing with respect to time as Eddie drinks the soda.

If Eddie drinks the soda at a rate of 0.5 cubic inch per second and if the radius of the can is 1 inch, then the height of the soda will go down (since dh/dt is negative) at a rate of 0.16 inch per second.   □

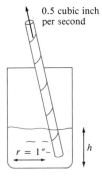

0.5 cubic inch
per second

$r = 1''$ h

Figure 44
How fast is the soda going down?

Example 6

The Balloon Problem   Imagine yourself blowing up a balloon as in Figure 45, blowing air into a balloon at a rate of 15 cubic inches per second. (That is pretty close to normal for most people.) At the exact moment when the radius of the balloon is 1 inch, how fast will the radius of the balloon be increasing? Make a guess: 1 inch per second? 2 inches per second?

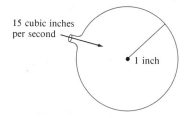

Figure 45
Blowing air into a balloon

Solution The volume and radius of a balloon (sphere) are related by the equation

$$V(t) = \frac{4}{3}\pi r^3(t)$$

To find the relationship between dV/dt and dr/dt, we differentiate with respect to time t, getting the related rates equation

$$\frac{d}{dt}\,V(t) = \frac{4}{3}\pi\frac{d}{dt}\left[r^3(t)\right]$$

$$= 4\pi r^2\frac{dr}{dt}$$

In this problem we are given the two quantities

$$\frac{dV}{dt} = 15 \text{ cubic inches per second}$$

$$r = 1 \text{ inch}$$

Solving for the unknown dr/dt in the related rates equation, we get   □

$$\frac{dr}{dt} = \frac{1}{4\pi r^2}\frac{dV}{dt}$$

$$= \frac{1}{4\pi(1)^2}(15)$$

$$\cong 1.2 \quad \text{inches per second}$$

Conclusion

If you blow air into a balloon at a constant rate of 15 cubic inches per second, then when the radius of the balloon is 1 inch (about the size of a tennis ball), the radius will be increasing at a rate of 1.2 inches per second. As the balloon gets larger, however, the rate at which the radius increases will not be as large. In fact, the graph in Figure 46 shows the relationship between the rate of change dr/dt in the radius and the radius r.

Figure 46
The rate of change in the radius of a balloon as a function of the radius when the air is being blown into it at a rate of 15 cubic inches per second

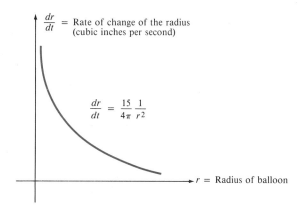

> Steps for Solving Related Rates Problems
>
> · **Step 1.** Draw a picture that describes the problem (if possible) and label all relevant variables (especially the ones that change with time).
>
> · **Step 2.** Find the algebraic equation that relates the variables of the problem.
>
> · **Step 3.** Differentiate the equation from Step 2 with respect to time.
>
> · **Step 4.** Solve for the unknown rate of change (one of the time derivatives).

Example 7

The Ladder Problem   Suppose a young man and woman are eloping. The young man has a 20-foot ladder leaning up against the house. However, at the exact moment he is standing at the top of the ladder, the young woman's father starts pulling the ladder away from the house at the rate of 5 feet per second. How fast is the young man coming down the side of the house when the bottom of the ladder is 10 feet from the base of the house?

Solution

· **Step 1** (draw a picture and label the variables).   Figure 47 shows the relevant variables of the problem. The two important variables are

$$x(t) = \text{Distance from the bottom of the ladder to the house}$$
$$y(t) = \text{Height of the top of the ladder from the ground}$$

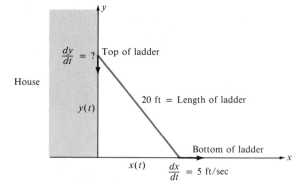

Figure 47
Relevant variables for the ladder problem

· **Step 2** (find a relationship between the variables).   The variables $x(t)$ and $y(t)$ are always related by the Pythagorean theorem:

$$x^2(t) + y^2(t) = 400$$

provided that the ladder is leaning against the house.

- **Step 3** (differentiate with respect to time). We now differentiate this equation implicitly with respect to t to get the related rates equation

$$2x\frac{dx}{dt} + 2y\frac{dy}{dt} = 0$$

- **Step 4** (solve for dy/dt). Solving for dy/dt, which represents the rate at which the young man is coming down the side of the house, we find

$$\frac{dy}{dt} = -\frac{x}{y}\frac{dx}{dt}$$

This equation gives dy/dt in terms of x, y, and dx/dt.

Unfortunately, we are given only dx/dt and x. Hence we must solve for y in terms of x, using the original equation

$$x^2 + y^2 = 400$$

This gives

$$y = \sqrt{400 - x^2}$$

We use the positive root because it represents the height of the ladder above the ground. Substituting this value into the related rates equation, we obtain the unknown rate

$$\frac{dy}{dt} = -\frac{x}{\sqrt{400 - x^2}}\frac{dx}{dt}$$

We now substitute into this equation the given values

$$\frac{dx}{dt} = 5 \text{ ft/sec} \qquad \text{(rate that the ladder is pulled from house)}$$

$$x = 10 \text{ ft} \qquad \text{(distance of ladder from the base of house)}$$

Hence we have

$$\frac{dy}{dt} = -\frac{10}{\sqrt{300}}(5)$$

$$\cong -2.89 \quad \text{feet per second} \qquad \square$$

Interesting Conclusion

The value $dy/dt = -2.89$ feet per second means that the function $y(t)$ is decreasing at this rate. This means that when the bottom of the ladder is 10 feet from the house, the top of the ladder is coming down the side of the house at the rate of 2.89 feet per second. We can see in Figure 48 that as the father starts pulling the ladder away from the house, the young man starts falling immediately. Even though the father pulls the ladder at a constant rate of 10 feet per second, the young man falls faster and faster. In fact, the interesting thing is that the speed at which the young man falls approaches infinity immediately before he hits the ground. Is this really true?

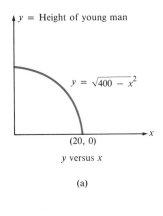

y = Height of young man

$y = \sqrt{400 - x^2}$

(20, 0)

x

y versus x

(a)

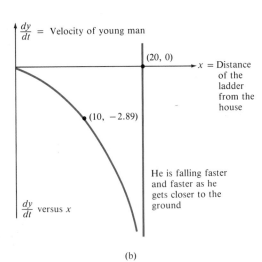

$\frac{dy}{dt}$ = Velocity of young man

(20, 0)

x = Distance of the ladder from the house

(10, −2.89)

$\frac{dy}{dt}$ versus x

He is falling faster and faster as he gets closer to the ground

(b)

Figure 48
Graphs of y versus x and dy/dx versus x

Problems

In Problems 1–17, find dy/dx using implicit differentiation.

1. $y - x^2 = 0$ **2.** $y - x^3 = 0$
3. $x^2 y = 1$ **4.** $x^3 y = 1$
5. $xy + x = 1$ **6.** $x^2 y + y = 4$
7. $x^3 + xy = x$ **8.** $xy - 3 + 7x = 0$
9. $xy + y^2 = x$ **10.** $y^2 + y = x^2(x + 1)$
11. $9x^2 + 16y^2 = 36$ **12.** $y^5 - 7xy - 18x^3 = 0$

13. $y^2 + xy = x(x - y)$ **14.** $\dfrac{y + x}{y - x} = x^2$

15. $y^2 = \dfrac{x^2}{x^2 - 1}$

16. $\sqrt{x} + \sqrt{y} = \sqrt{a}$ where a is a positive constant
17. $xy^2 + x^2 y = x^3 + 2x^2$

In Problems 18–26, find dy/dx using implicit differentiation and then evaluate dy/dx at the given point.

18. $xy = 4$ at $(1, 4)$ **19.** $x^2 y + y = 1$ at $(0, 1)$
20. $x^3 y + x = 0$ at $(1, -1)$ **21.** $x^2 y = 1$ at $(1, 1)$
22. $x^2 - y^2 = 3$ at $(\sqrt{3}, \sqrt{2})$
23. $x^2 - 3xy = 10$ at $(2, -1)$
24. $x^2 y - xy^2 + y - 8 = 0$ at $(-1, 2)$
25. $(\sqrt{x} + 1)(\sqrt{y} + 1) = 8$ at $(1, 4)$
26. $(x - y^2)(x + y) = 15$ at $(4, 1)$

In Problems 27–36, find the equation of the tangent line to the curve at the given point. If possible, draw the curve and the tangent line.

27. $x^2 y = 1$ at $(1, 1)$ **28.** $x^2 y + y = 1$ at $(0, 1)$
29. $xy = -1$ at $(1, -1)$ **30.** $x^3 y + xy = 1$ at $(1, 0.5)$

31. $x^2 + y^2 = 9$ at $(0, 3)$ **32.** $x^2 - y^2 = 9$ at $(3, 1)$
33. $x^2 y + xy = 5$ at $(1, 2)$ **34.** $y^2 - 3x^2 = 1$ at $(-1, 2)$
35. $x^2 - y^2 = 2$ at $(2, \sqrt{3})$
36. $(\sqrt{x} + 1)(\sqrt{y} + 1) = 8$ at $(1, 4)$

Related Rates

37. Particle Moving on a Curve   A particle p is moving along the curve $x^2 = 4y$. Remember that

$$\frac{dx}{dt} = \text{Velocity of the particle in the } x\text{-direction}$$

$$\frac{dy}{dt} = \text{Velocity of the particle in the } y\text{-direction}$$

What is the velocity of the particle in the y-direction when the particle is located at the point $(4, 4)$ and the velocity in the x-direction is 2 units per second?

38. Growing Square   The sides of a square are growing at the rate of 2 ft/min. How fast is the area of the square growing when $x = 3$?

39. Soda Problem   Emma is drinking a soda through a straw. She is drinking at a constant rate of 1 cubic inch per second. If the radius of the can is 1 inch, how fast is the level of soda going down?

40. Price Fluctuations   A certain commodity is priced (in dollars) at the opening of the market each day at

$$p(x) = 5 + \frac{50}{x}$$

where x is the amount (measured in hundreds of pounds) of the commodity available on a given day. At what rate will the price be changing when there are 75 hundred pounds available and the amount of the commodity is decreasing at a rate of 500 pounds per day?

41. Oil Spill Hurricane winds have damaged an oil rig and caused a circular oil slick two inches thick. Suppose the radius of the slick is currently 100 feet and growing at the rate of 0.5 ft/min. What is the rate at which the oil is spilling? *Hint:* The volume of the oil slick is $V = \pi r^2 h$, where r is the radius and h is the height (or thickness) of the oil slick.

42. Water Waves Sally throws a rock onto the surface of a lake. A circular pattern of ripples is formed. If the radius of the outermost circle increases at the rate of 10 inches/second, how fast is the area within the outer circle changing after nine seconds?

43. Related Velocities A highway that runs north and south is crossed at right angles by railroad tracks running east and west. At 8:00 A.M. a car is 20 miles north of the intersection, heading south at 60 mph. At the same time a train is 10 miles east of the intersection heading west at 80 mph. (See the figure.) When are the car and the train closest to each other? How far apart are the car and the train at the time they are closest together? *Hint:* The distance between the vehicles at time t is given by

$$s(t) = \sqrt{(60t - 20)^2 + (80t - 10)^2}$$

44. Related Velocities Two bicyclists start off from the intersection of two roads, one on a road running north–south and the other on a road running east–west. The speed of the cyclist traveling north is 5 mph, while the speed of the cyclist traveling east is 10 mph. At what rate are the cyclists separating after 30 minutes?

45. Sanitation Analysis A waste product that decomposes in water is dumped into holding ponds. The amount of waste dumped after t weeks is given by

$$w(t) = (3t + 5)^{3/2}$$

If the waste decomposes at a constant rate of 45 gallons per week, how long will it take before the waste is dumped faster than it can be decomposed? Assume that the size of the holding ponds is adequate.

46. The Street Lamp Problem A street light 21 feet off the ground casts a shadow of a pedestrian 6 feet tall who is walking away from the light at a rate of 3 ft/sec. (See the figure.) How fast is the shadow lengthening when the pedestrian is 25 feet from the light? How fast is the tip of the shadow moving?

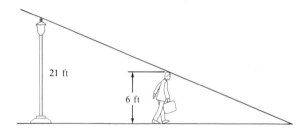

Figure 50
Diagram for Problem 46

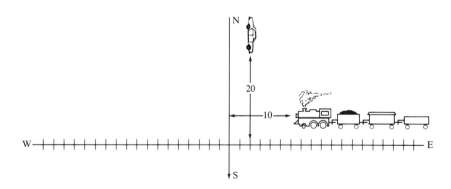

Figure 49
Diagram for Problem 43

9.6

Differentials

PURPOSE

We introduce the concept of the differentials dy and dx. We will show that while a functional relationship relates the size of x to the size of y, the differential relationship shows how a small change in x will give rise to a small change in y.

Introduction

The symbol dy/dx is a bit confusing, since it represents a derivative and not the quotient of two quantities dy and dx. The reason for the dy/dx notation is mostly historical and due to Leibniz. He imagined the derivative dy/dx as the ratio of two infinitely small quantities dy and dx. Nowadays, we interpret dy/dx not as a quotient, but as the limiting value of the quotient

$$\frac{dy}{dx} = \lim_{h \to 0} \frac{f(x+h) - f(x)}{h}$$

It is possible, however, to assign meanings to the quantities dy and dx. We now show how dy and dx can be interpreted in their own right.

Differential Approximation

Consider the function defined by the graph $y = f(x)$ as shown in Figure 51. Pick a point x_0. Let x_0 change by a small amount dx from x_0 to a new value, $x_0 + dx$. The value of $f(x_0)$ will change from the initial value to the new value $f(x_0 + dx)$. The difference between these values is the change f in the function. That is

$$\Delta f = f(x_0 + dx) - f(x_0)$$

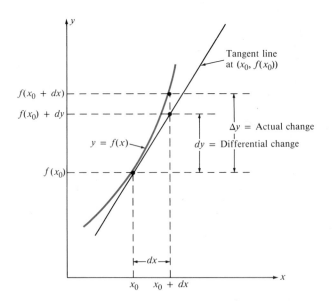

Figure 51

Illustration of the differential

Now consider the tangent line to the curve $y = f(x)$ at the point $(x_0, f(x_0))$. As we have seen before, the equation of the tangent line is

$$y = f(x_0) + f'(x_0)(x - x_0)$$

Since the tangent line remains close to the graph $y = f(x)$ for values of x close to x_0, the change in y along the tangent line will be approximately the same as the change in y along the graph $y = f(x)$. We see in Figure 51 that as x changes from x_0 to $x_0 + dx$, the change in y along the tangent line will be

$$dy = f'(x_0)\, dx$$

This change is called the **differential change** (or *approximate change*) in $y = f(x)$. When the change dx in x is small, dy approximates the real change in $f(x)$.

Definition of the Differentials *dx* and *dy*

Let $y = f(x)$ have a derivative at the point x_0. The differential change in x, denoted by dx, is simply any change in x (generally taken to be small).

The differential change in y, denoted by dy, which results from a differential change dx from x_0 to $x_0 + dx$, is given by

$$dy = f'(x_0)\, dx$$

Example 1 ——————— Differential Change   If

$$y = f(x) = x^2$$

find the differential change in y when x changes from 1 to 1.1.

Solution   Here we have

$$x_0 = 1 \qquad \text{(starting point)}$$
$$dx = 0.1 \qquad \text{(change in } x\text{)}$$

We first find the derivative

$$f'(x) = 2x$$

The differential dy is found by multiplying the derivative evaluated at $x_0 = 1$ times the differential dx. In other words,

$$
\begin{aligned}
dy &= f'(x_0)\, dx \\
&= 2x_0\, dx \\
&= 2(1)(0.1) \\
&= 0.2
\end{aligned}
$$

The geometric interpretation of the differential dy can be seen in Figure 52.   □

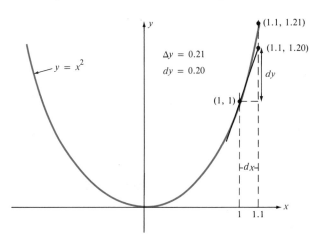

Figure 52
Differential when $x_0 = 1$ and $dx = 0.1$

Reason for an Interest in Differentials

Differentials approximate the exact change in the dependent variable y as a result of a change in the independent variable x. There are a number of reasons why we are satisfied with this approximate change

$$dy = f(x_0)\,dx$$

in contrast to the real change

$$\Delta y = f(x_0 + dx) - f(x_0)$$

The major reason is that it is often impossible to compute the real change, owing to the fact that we cannot evaluate the function at the point $x_0 + dx$. The differential approximation dy uses information only at the point x_0. Also, the differential change dy is generally easier to compute than the real change y, and it is often a good enough approximation for the purpose at hand.

Extrapolation by Differentials

Often, the value of a function and its derivative are known at a point x_0, and from this information we would like to approximate the value of the function at some nearby point $x_0 + dx$. If we draw the tangent line to the graph at $(x_0, f(x_0))$, it is possible to approximate the value $f(x_0 + dx)$ by following along the tangent line, getting the differential approximation

$$f(x_0 + dx) = f(x_0) + dy$$
$$= f(x_0) + f'(x_0)\,dx$$

See Figure 53.

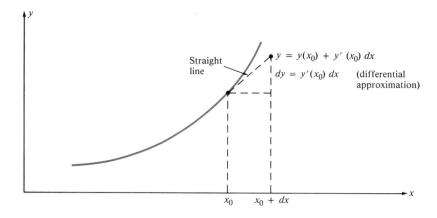

Figure 53
Differential extrapolation is linear extrapolation

Differential Approximation to Extrapolate the World Mile Record

Table 10 lists the previous eight world records in the men's mile run (prior to 1987). These points are plotted in the Figure 54.

TABLE 10
World Records in
Men's Mile
Run, 1975–1985

Runner	Country	Time	Year
Filbert Bayi	Tanzania	3:51.0	1975
John Walker	New Zealand	3:49.4	1975
Sebastian Coe	England	3:49.0	1979
Steve Ovett	England	3:48.8	1980
Sebastian Coe	England	3:48.53	1981
Steve Ovett	England	3:48.40	1981
Sebastian Coe	England	3:47.33	1981
Steve Cram	England	3:46.31	1985

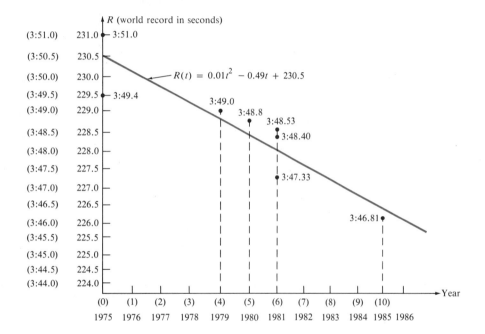

Figure 54
*World record times in men's
mile run, 1975–1985*

Statisticians have ways of finding curves that best approximate data points. A quadratic polynomial was found that approximates the data points shown in Table 10. It is given by

$$R(t) = 0.01t^2 - 0.49t + 230.5$$

In this equation, $R(t)$ is the mile record (in seconds), and t is the year the record was made ($t = 0$ stands for 1975, $t = 1$ stands for 1976, ..., $t = 10$ stands for 1985). This approximating curve is also drawn in Figure 54. We now differentiate the fitted curve $R(t)$, getting

$$R'(t) = 0.02t - 0.49$$

The value of this derivative at $t = 10$ (which corresponds to 1985) is

$$R'(10) = (0.02)(10) - 0.49$$
$$= -0.29$$

To extrapolate the graph $R(t)$ from the year 1985 to the years 1986, 1987, 1988, 1989, and 1990, we compute the different values

$$R(10) + R'(10)\,dt$$

for $dt = 1, 2, 3, 4$, and 5. These values are given in Table 11. You can check the accuracy of these differential approximations by comparing new world records that have occurred since Steve Cram's 1985 record run.

TABLE 11
Differential Extrapolation
of World Mile Records

dt	Year	Differential dR $dR = R'(10)\,dt$	Estimated World Mile Record $R(t) = R(10) + dR$
1	1986	$(-0.29)(1) = -0.29$	$226.60 - 0.29 = 226.31\ (3{:}46.31)$
2	1987	$(-0.29)(2) = -0.58$	$226.60 - 0.58 = 226.02\ (3{:}46.02)$
3	1988	$(-0.29)(3) = -0.87$	$226.60 - 0.87 = 225.73\ (3{:}45.73)$
4	1989	$(-0.29)(4) = -1.16$	$226.60 - 1.16 = 225.44\ (3{:}45.44)$
5	1990	$(-0.29)(5) = -1.45$	$226.60 - 1.45 = 225.15\ (3{:}45.15)$

Differentials Used in Error Analysis

Differentials are often used in studying the propagation of errors. For example, suppose that x is a variable that we measure and that y is a variable that is computed by means of a formula

$$y = f(x)$$

If there is an error in the measurement of x, this error will give rise to an error in y. It is important to know how errors in the measured value of x will give rise to errors in y.

To determine these relationships, we introduce some of the language of error analysis. If the **true value** of some quantity is x, but the value determined by some experiment is $x + dx$, then the **error** in the measurement is dx, and the **relative error** in the measurement is

$$\text{Relative error} = \frac{dx}{x}$$

If we multiply the relative error times 100, we get the **percentage error**:

$$\text{Percentage error} = 100\,\frac{dx}{x}\,\%$$

Example 2

Relative Error   Suppose you approximate the length of the line in Figure 55 using a ruler, and suppose your estimate is 6.06 inches. Suppose that the real length of the line is exactly six inches. What are the absolute error, relative error, and percentage error in your measurement?

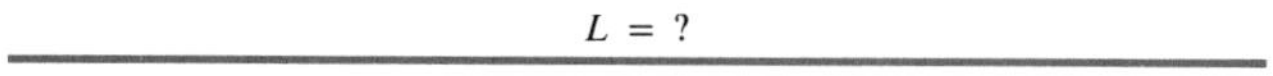

Figure 55
A straight line of unknown length

$$L = ?$$

Solution   The error (or absolute error) is simply the difference between the measurement and the real value. Although you can never really determine this value in practice (since you never know the exact real value), in this case it is

$$\text{Absolute error} = 0.06 \text{ inch}$$

The absolute error is always taken to be a positive number. The relative error is the absolute error divided by the real value. In this case it is

$$\text{Relative error} = \frac{0.06}{6} = 0.01$$

If we multiply this quantity times 100, we get the percentage error:

$$\text{Percentage error} = 100(0.01)\% = 1\% \qquad \qquad \square$$

Comment on Relative Error

The relative error is a much more useful measure of the accuracy of a measurement than the absolute error. Someone may tell you that the absolute error in a measurement is 5 pounds but unless you know the size of the measurement, this value is almost meaningless. After all, an error of 5 pounds could be the error in weighing a child or an elephant. If an elephant were being weighed, an error of 5 pounds would not be as startling as an error of 5 pounds in weighing a child.

Example 3

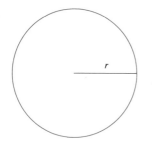

Figure 56
Can you find the area of this circle by measuring the radius?

Propagation of Errors   Suppose you approximate the area A of the circle in Figure 56 by estimating the radius r using a ruler and then using the formula $A = \pi r^2$. Suppose you know that your measurement of the radius has an error of less than 0.02 inches. What is the relative error in your estimate in the area?

Solution

- **Step 1** (find the algebraic relationship between the variables).   The algebraic relationship between the area of a circle A and its radius r is given by

$$A = \pi r^2$$

- **Step 2** (find the differential relationship between the variables).   Finding the derivative of A with respect to r, we compute the differential dA to be

$$dA = \frac{dA}{dr} \, dr$$

$$= (2\pi r) \, dr$$

- **Step 3** (find dA/A in terms of dr/r).   Since the goal is to find the relative error dA/A in the area, we divide each side of the above equation by A, getting

$$\frac{dA}{A} = \frac{2\pi r}{A} \, dr$$

Substituting the expression $A = \pi r^2$ for the value of A in the right-hand side of the above equation and rewriting the equation algebraically so that the relative error dr/r appears on the right, we finally have

$$\frac{dA}{A} = \frac{2\pi r}{\pi r^2}\, dr$$

$$= 2\frac{dr}{r}$$

$$= 2(0.02)$$

$$= 0.04 \qquad\qquad \square$$

Conclusion

If the radius of the circle can be measured with a relative error dr/r less than 0.02, then the area of the circle will have a relative error dA/A less than 0.04. In fact, from the above formula for the relative error dA/A we have the general result that the relative error in the area of the circle is always exactly twice the relative error in the radius.

Differentials in Business

Differentials are also useful in determining the effects of errors in business. Suppose you are the sales manager for a business firm and have estimated sales for the coming month to be 2500 items. Suppose you are sure that this value is accurate with a maximum percentage error of 2%. If the profit $P(q)$ in dollars from selling q items is given by

$$P(q) = 20q - 0.0003q^2$$

what is the maximum percentage error in your estimation of the monthly profit of

$$P(2500) = 20(2500) - 0.0003(2500)^2$$
$$= \$48,125$$

To answer this question, we must find the equation that relates the unknown percentage error of P (which is $100\,dP/P$) to the known percentage error in q (which is $100\,dq/q$). We begin by finding the derivative:

$$\frac{d}{dq}P(q) = 20 - 0.0006q$$

Hence the differential of P is

$$dP = P'(q)\,dq$$
$$= (20 - 0.0006q)\,dq$$

To get $100\,dP/P$ on the left-hand side of the above equation and $100\,dq/q$ on the right-hand side of the equation, we carry out the following steps.

- **Step 1.** Start with
$$dP = (20 - 0.0006q)\,dq$$

- **Step 2.** Divide by P:
$$\frac{dP}{P} = \frac{(20 - 0.0006q)}{P}\,dq$$

- **Step 3.** Multiply and divide the right-hand side by q:

$$\frac{dP}{P} = \frac{(20 - 0.0006q)q}{P} \frac{dq}{q}$$

- **Step 4.** Multiply each side by 100:

$$100 \frac{dP}{P} = \frac{(20 - 0.0006q)q}{P} \left(100 \frac{dq}{q}\right)$$

We now substitute the known values

$$q = 2500$$
$$P(2500) = 48{,}125$$

$$100 \frac{dq}{q} = 2\%$$

into the right-hand side of the above equation. This gives the percentage error in the cost

$$100 \frac{dP}{P} = \frac{[20 - 0.0006(2500)](2500)}{48{,}125} \tag{2}$$

$$= 1.9\% \qquad \qquad \square$$

Conclusion

The maximum percentage error in the estimated profit is 1.9%. This means that the real profit will be within 1.9% of the predicted value of $48,125 or between the values of

$$\$48{,}125 - (0.019)(\$48{,}125) = \$47{,}210.62$$

and

$$\$48{,}125 + (0.019)(\$48{,}125) = \$49{,}039.37$$

Problems

For Problems 1–10, determine the value of dy at the indicated point x_0 and for the differential in x, dx.

1. $f(x) = 2$, $x_0 = 0, dx = 0.1$
2. $f(x) = 2x + 1$, $x_0 = 2, dx = 0.1$
3. $f(x) = x^2$, $x_0 = 1, dx = 0.01$
4. $f(x) = x^2 + x$, $x_0 = 3, dx = 0.1$
5. $f(x) = 1/x$, $x_0 = 1, dx = 0.1$
6. $f(x) = 1/x^2$, $x_0 = 1, dx = 0.1$
7. $f(x) = 3x^3 - 9x^2 + 5$, $x_0 = 1, dx = 0.01$
8. $f(x) = (2x^2 + x)(x^3 - 2)$, $x_0 = 2, dx = -0.01$

9. $f(x) = \dfrac{x - 1}{x + 2}$, $x_0 = 0, dx = 0.05$

10. $f(x) = \dfrac{1}{x^2 - 2}$, $x_0 = 1, dx = 0.001$

For Problems 11–13, a company had estimated that the next year's net income would be 15 million dollars. Suppose the real net income turned out to be 15.3 million dollars. Determine the following quantities.

11. The absolute error.
12. The relative error.
13. The percentage error.

14. **Comparison of Errors** A biologist determines the weight of a rat to be 1 pound. The biologist then determines the weight of an elephant to be 7000 pounds. Suppose the rat

actually weighs 1.1 pounds, and the elephant actually weighs 7500 pounds. Determine the following.
(a) Which measurement had the smaller absolute error?
(b) Which measurement had the smaller relative error?
(c) Which measurement had the smaller percentage error?

For Problems 15–19, determine the following errors.

15. Error in the Area of a Square   Find the relative error in the area of a square if the relative error in measuring the side is 2%.

16. Error in the Volume of a Sphere   Find the relative error in the volume of a sphere if the relative error in measuring the radius is 3%.

17. Error in the Volume of a Cube   Find the relative error in the volume of a cube if the relative error in measuring a side is 1%.

18. Error in the Volume of a Cylinder   Find the relative error in the volume of a cylinder with radius equal to one-half the height if the relative error in measuring the radius is 4%.

19. Error in the Surface Area of a Cube   Find the relative error in the surface area of a cube if the relative error in measuring the side is 3%.

Using Differentials to Extrapolate

For Problems 20–26, the value of a function and its derivative are given at a point. Use this information and the differential to extrapolate the value of the function at the indicated point.

20. Given $f(3) = 2$ $f'(3) = 1$,   approximate $f(3.1)$.
21. Given $f(2) = 1.3$ $f'(2) = -0.2$, approximate $f(2.05)$.
22. Given $f(10) = 2.4$ $f'(10) = 0.5$,   approximate $f(10.3)$.
23. Given $f(-1) = 0$ $f'(-1) = 0$,   approximate $f(-0.8)$.
24. Given $f(0) = 35.4$ $f'(0) = 0.25$,   approximate $f(0.01)$.
25. Given $f(50) = 2500$ $f'(50) = 0.35$, approximate $f(52)$.
26. Given $f(50) = 2500$ $f'(50) = 0.35$, approximate $f(54)$.

For Problems 27–30, extrapolate the functions at the indicated points, using the differential.

27.

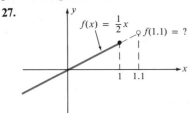

28.

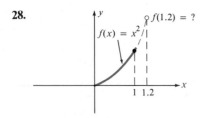

29.

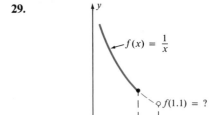

30.

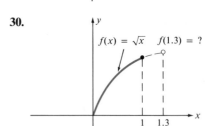

Differentials in Business

31. Relative Error in the Profit   The monthly profit of a firm when it produces q items per month is given by

$$P(q) = 3q - 0.001q^2$$

Sales estimates are 1000 items for the coming month with a maximum relative error of 0.02. What is the estimated profit for the next month? What is the relative error in this estimate?

32. Percentage Error in the Profit   The monthly profit of a firm when it produces q items per month is given by

$$P(q) = 20q - 0.0005q^3$$

Sales estimates for the coming month are 3000 with a percentage error of 5%. What is the percentage error in the profits?

33. Relative Error in Revenue   The monthly revenue $R(q)$ of a firm is given by the formula

$$R(q) = 25q - 0.001q^2$$

where q is the number of items sold during the month. The firm estimates that sales during the coming month will be 5000 items and that this estimate has a maximum relative error of 0.01. What is the estimated revenue for the coming month? What is the relative error in this estimate?

34. Percentage Error in Revenue   The monthly revenue $R(q)$ of a firm is given by the formula

$$R(q) = 50q - 0.001q^2$$

where q is the number of items sold during the month. The firm estimates that sales during the coming month will be 3000 items and that this estimate will have a maximum error of 250 items. What is the estimated revenue for the coming month? What is the percentage error in this estimate?

Differentials in Science

35. Light Intensity The intensity of light entering the iris of the eye is given by the function

$$I = cr^2$$

where r is the radius of the pupil and c is a constant depending on the eye. How much does a small change in the radius of the opening of the eye affect the intensity of light entering the iris?

36. Determination of g The acceleration g of an object due to gravity is sometimes determined by measuring the period of the swing of a pendulum. If the length of the pendulum is L and the measured period is T, then g is

given by the formula

$$g = \frac{4\pi^2 L}{T^2}$$

Find the relative error in g if
(a) L is measured accurately, but T has an error of 5%.
(b) T is measured accurately, but L has an error of 1%.

37. Error in Cardiac Output A common way to measure cardiac output F (the number of cubic centimeters per minute of blood pumped by the heart) is by using *Fick's formula*:

$$F = \frac{k_1}{x - k_2}$$

The values k_1 and k_2 are constants, and x is the concentration of carbon dioxide in the blood entering the lung. Suppose the measurement of x has a maximum error of 2%. What is the percentage error in the cardiac output?

Epilogue: A Derivative Puzzler

Now that you have mastered differential calculus, we would like to prove that $2 = 1$ by using the derivative. You can look for the error in the proof.

Claim $2 = 1$

Proof We start with the rather obvious algebraic identity

$$x^2 = x \cdot x = \underbrace{x + x + x + \cdots + x}_{x \ x\text{'s}}$$

If we now differentiate each side of the above equation with respect to x, we get the obvious result

$$\frac{d}{dx} x^2 = \frac{d}{dx} [x + x + x + \cdots + x]$$

or

$$2x = \underbrace{1 + 1 + 1 + \cdots + 1}_{x \ 1\text{'s}}$$

Hence we have

$$2x = x$$

If we now divide by x, we get the desired result

$$2 = 1 \qquad\qquad \square$$

Error in Proof

The very first equation

$$x^2 = \underbrace{x + x + x + \cdots + x}_{x \ x\text{'s}}$$

is true when $x = 1, 2, 3, \ldots$ but has no meaning when x is not a positive integer.

Key Ideas for Review

Review Exercises

For Problems 1–6, locate the critical points of the given function. Determine which critical points are relative maximum or relative minimum points using the first derivative test.

1. $f(x) = x^3 - 2x^2 - 4x + 3$　　**2.** $f(x) = x^{2/3}$

3. $f(x) = \dfrac{x}{x^2 + 2}$　　　　**4.** $f(x) = \dfrac{5x - 6}{3x + 11}$

5. $f(x) = x^2 - x$　　　　　**6.** $f(x) = (x + 1)^3$

For Problems 7–12, use the second derivative test to determine whether any of the critical points of the functions are relative minimum or relative maximum points. If the second derivative test does not apply, indicate that nothing is learned.

7. $f(x) = x^3 - 3x^2 - 9x + 5$　　**8.** $f(x) = 3x^4 - 4x^3 - 15$
9. $f(x) = (x - 1)^3$　　　　　**10.** $f(x) = (x - 1)^4$
11. $f(x) = x^{12/5}$　　　　　　**12.** $f(x) = 3x^4 - 4x^3$

For Problems 13–21, determine both the absolute maximum and the absolute minimum of each function on the indicated interval.

13. $f(x) = 2x^3 - 15x^2 + 12$ on $[-1, 3]$
14. $f(x) = x^4 - 8x^3 + 22x^2 - 24x + 7$ on $[-4, 4]$
15. $f(x) = (x - 2)^4$ on $[0, 4]$
16. $f(x) = (x - 2)^3$ on $[2, 3]$
17. $f(x) = 2x^3 - 15x^2 + 36x$ on $[0, 1]$
18. $f(x) = x$ on $[0, 1]$
19. $f(x) = \sqrt{x - 1}$ on $[1, 4]$
20. $f(x) = \sqrt[3]{x - 1}$ on $[1, 4]$
21. $f(x) = \sqrt{x}$ on $[0, 1]$

22. Marginal Cost The Express Delivery Co. is planning to add new routes. The added cost of the next expansion phase is given by

$$C(x) = 2x^3 + x^2 - 4x$$

where x is the number of deliveries on the routes added. When is the marginal cost increasing on the interval $[1, 50]$?

23. Inflection Points of a Cubic Equation Find the inflection point(s) of a function of the form

$$f(x) = ax^3 + bx^2 + cx + d$$

where a, b, c, and d are constants with a different from zero.

For Problems 24–31, find dy/dx using implicit differentiation.

24. $x^2 + y^2 + 1 = 0$　　**25.** $x^2 + y^2 = 16$
26. $x^3 - y^3 = 1$　　　　**27.** $x^2 y = 1$
28. $x + y = 1$　　　　　**29.** $y - x = 0$
30. $\sqrt{x} + \sqrt{y} = 1$　　**31.** $\sqrt{xy} = x + 1$

32. Related Rates Let A be the area of a square whose sides have length x. Assume that x changes with time t. How are dA/dt and dx/dt related?

33. Related Rates Let A be the area of a circle with radius r. Assume that r changes with time t. How are dA/dt and dr/dt related?

34. Ladder Problem Revisited A 16-foot ladder is leaning against a house. If the bottom of the ladder is being pulled away from the house at a rate of 10 ft/sec, how fast is the top of the ladder coming down when the bottom of the ladder is 5 feet from the house?

35. Street Lamp Problem Revisited A man who is 6 feet tall is walking at the rate of 4 ft/sec toward a street light that is 20 feet tall. At what rate is the size of his shadow changing?

36. Balloon Tracking Problem A weather balloon is rising vertically at a rate of 5 ft/sec. An observer is situated 100 feet from a point on the ground directly beneath the balloon. At what rate is the distance between the balloon and the observer changing when the height of the balloon is 300 feet?

For Problems 37–43, use differentials to approximate the given values.

37. $\sqrt{99}$　　　　　　　**38.** $\sqrt{101}$
39. $\sqrt[3]{101}$　　　　　　**40.** $(0.99)^5$

41. $(1.001)^9$　　　　　　**42.** $\dfrac{1}{100.4}$

43. $\dfrac{1}{99.9}$

44. Error in the Area of a Square Use differentials to find the approximate relative error in the area of a square. The sides of the square have been measured and found to be 10 inches, with a maximum relative error of 0.05 inch.

45. Error in the Volume of a Cube Use differentials to find the approximate relative error in the volume of a cube. The sides of the cube have been measured and found to be 10 inches, with a maximum relative error of 0.10 inch.

46. Box Problem Revisited An open box with a rectangular base is to be constructed from a rectangular piece of cardboard that is 16 inches wide and 25 inches long by cutting out a square from each corner and turning up the sides. Determine the box that will have the largest volume.

47. Number Theory Find two integers whose difference is 10 and whose product is a minimum.

48. Number Theory Find two integers whose sum is 10 and whose product is a maximum.

Chapter Test

1. Answer the following question for the function

$$f(x) = 5x^2 - 5x + 7$$

   (a) Where is the function increasing and where is it decreasing?
   (b) What are the critical points?
   (c) What are the stationary points?
   (d) What are the relative maximum and the relative minimum points?
   (e) What are the inflection points?
   (f) Where is the function concave up and where is it concave down?
   (g) What is the absolute maximum of this function in the interval $[0, 5]$?

2. Use implicit differentiation to find dy/dx if

$$xy + \sqrt{y} = 1$$

3. If

$$y = x^3$$

   and x changes with time t, what is the relationship between dy/dt and dx/dt?

4. What are the two positive real numbers whose sum is 100 that have the largest product?

5. If the sides of a square are measured and found to be 10 inches with a maximum error of 0.05 inch, approximate the relative error in the area of the square.

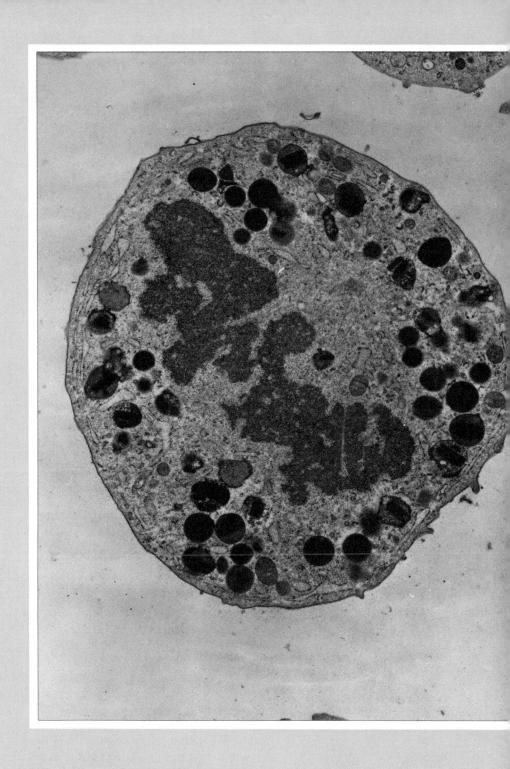

10

Exponential and Logarithmic Functions

In the preceding chapters we saw how polynomials can be used to describe many physical phenomena. However, many natural processes, such as the growth and decay of biological populations, are best described by exponential functions. Also, there are natural phenomena such as earthquake measurements that are best described by logarithmic types of functions. In this chapter we examine these two important types of functions and show how their derivatives can be used to solve a variety of important problems.

10.1 Exponential Growth and Decay

PURPOSE

We introduce the idea of a geometric or exponential sequence and show how it gives rise to the exponential function $f(x) = b^x$.

We then introduce the special exponential growth and decay curves and show how these curves are used to describe phenomena such as biological growth and decay and the growth of money due to continuous compounding of interest.

Introduction

Suppose that there are ten fungi spores in a culture and that each spore subdivides into two spores on the average of once every hour. At the end of one hour there will be 20 spores, at the end of two hours 40 spores, at the end of three hours 80 spores, and so on. A related situation occurs when a radioactive substance decays. In the case of the carbon isotope ^{14}C the number of atoms decreases by one-half every 5550 years, which length of time is the **half-life** of the substance. These types of **growth and decay** phenomena are known as **geometric** or **exponential** growth and decay phenomena. They are characterized by the fact that if their population were measured at equally spaced time periods (of any length), then the population at the end of any of these periods would always be some constant b times the population at the end of the previous time period. For example, the population of fungi spores can be found at any time by simply multiplying the size of the population of the spore colony an hour earlier by $b = 2$. Likewise, the number of ^{14}C atoms in a sample can be found at any time by multiplying the number of atoms present 5760 years earlier by $b = 0.5$.

Since the types of physical phenomena in which exponential growth and decay occur are almost endless, it is important that an entire chapter be spent studying the mathematical properties of the functions that describe these phenomena.

Geometric Sequences

A sequence of numbers

$$(1, b, b^2, b^3, b^4, \ldots)$$

where each number in the sequence is found by multiplying the preceding number times a fixed constant is called a **geometric sequence**. The fixed constant used to define the geometric sequence is called the **multiplier** and is denoted by b. A few examples of geometric sequences are

$$(1, 2, 4, 8, 16, \ldots) \qquad\qquad (b = 2)$$
$$(1, 1/2, 1/4, 1/8, \ldots) \qquad\qquad (b = 1/2)$$
$$(1, 0.8, 0.64, 0.512, 0.4096, \ldots) \qquad (b = 0.8)$$
$$(1, -1, 1, -1, \ldots) \qquad\qquad (b = -1)$$

It is not necessary that a geometric sequence begin with the number 1. If each number in the sequence is multiplied by a constant A, the new numbers

$$(A, Ab, Ab^2, Ab^3, Ab^4, \ldots)$$

still constitute a geometric sequence. More examples of geometric sequences are

$$(2, 4, 8, 16, 32, \ldots) \qquad\qquad (A = 2, b = 2)$$
$$(8, 4, 2, 1, 0.5, \ldots) \qquad\qquad (A = 8, b = 1/2)$$
$$(3, -0.6, 0.12, -0.024, \ldots) \qquad (A = 3, b = -0.2)$$

Example 1

Population Growth In 1988 the population of the world climbed to 4.9 billion people, and it is currently increasing at an annual rate of 2% per year (every year the population is 1.02 times the population of the year before). If the world's population continues to grow at this rate, what will be its population by the year 2000?

Solution An annual population growth of 2% means that the population at any time can be found by multiplying the population exactly one year earlier times 1.02. This is the same as saying that the population is described by the geometric sequence

$$(P, Pb, Pb^2, Pb^3, \ldots)$$

with $P = 4.9$ and $b = 1.02$. The first few numbers of this sequence, giving the world's population from 1988–2000, are shown in Table 1.

TABLE 1
Geometric Population
Growth, 1988–2000

Year	Population (billions)		
1988	P		$= 4.900$
1989	$P(1.02)$		$= 4.998$
1990	$[P(1.02)](1.02)$	$= P(1.02)^2$	$= 5.098$
1991	$[P(1.02)^2](1.02)$	$= P(1.02)^3$	$= 5.200$
1992	$[P(1.02)^3](1.02)$	$= P(1.02)^4$	$= 5.304$
$\vdots$	$\vdots$		$\vdots$
2000	$[P(1.02)^{13}](1.02)$	$= P(1.02)^{14}$	$= 6.465$

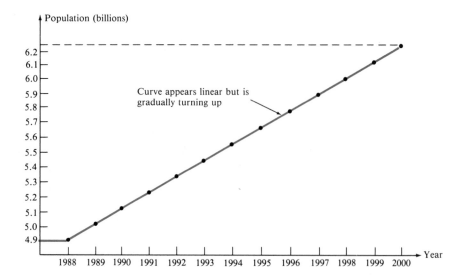

Figure 1
U.S. population, 1988–2000

In other words, if the world's population grows exponentially at the rate of 2% per year until the year 2000, the population of the world will then be 6.465 billion people. A graph of this population curve is shown in Figure 1.

□

The Exponential Function b^x

We have seen earlier in the Finite Mathematics Preliminaries that the exponential b^n for a positive integer n denotes the product

$$b^n = b \cdot b \cdot b \cdot b \cdots b \qquad (n \text{ factors})$$

We have also defined the fractional exponent $b^{m/n}$ by

$$b^{m/n} = (b^{1/n})^m$$

We now find the meaning of b^x when x is not an integer or a fraction. For example, what would be the meaning of 2^π with the irrational number

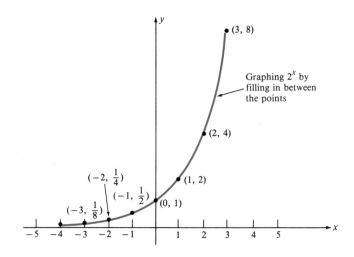

Figure 2
Graph of $y = 2^x$

$\pi = 3.1415927\ldots$ as an exponent? We certainly cannot multiply 2 times itself π times. For the present time we merely interpret the exponential function $y = b^x$ as the smooth curve (called an exponential curve) "filled in" between the points of $y = b^n$ as shown in Figure 2.

Difference Between Exponential and Power Functions

Students often confuse the **power function**

$$y = x^b$$

and the **exponential function**

$$y = b^x$$

To distinguish the two functions, just remember that for the exponential function b^x the variable x occurs in the exponent, while for the power function x^n the variable x is the *base* of the exponent. Table 2 compares the power function $y = x^2$ with the exponential function $y = 2^x$. Note how fast the exponential function grows in Table 2. The following example illustrates just how fast.

TABLE 2
Growth of Power and
Exponential Functions

x	Power Function x^2	Exponential Function 2^x	
0	$0 \cdot 0 \;=\; 0$	$2^0 = 1$	$= 1$
1	$1 \cdot 1 \;=\; 1$	$2^1 = 2$	$= 2$
2	$2 \cdot 2 \;=\; 4$	$2^2 = 2 \cdot 2$	$= 4$
3	$3 \cdot 3 \;=\; 9$	$2^3 = 2 \cdot 2 \cdot 2$	$= 8$
4	$4 \cdot 4 \;=\; 16$	$2^4 = 2 \cdot 2 \cdot 2 \cdot 2$	$= 16$
5	$5 \cdot 5 \;=\; 25$	$2^5 = 2 \cdot 2 \cdot 2 \cdot 2 \cdot 2$	$= 32$
6	$6 \cdot 6 \;=\; 36$	$2^6 = 2 \cdot 2 \cdot 2 \cdot 2 \cdot 2 \cdot 2$	$= 64$
7	$7 \cdot 7 \;=\; 49$	$2^7 = 2 \cdot 2 \cdot 2 \cdot 2 \cdot 2 \cdot 2 \cdot 2$	$= 128$
8	$8 \cdot 8 \;=\; 64$	$2^8 = 2 \cdot 2 \cdot 2 \cdot 2 \cdot 2 \cdot 2 \cdot 2 \cdot 2$	$= 256$
9	$9 \cdot 9 \;=\; 81$	$2^9 = 2 \cdot 2 \cdot 2 \cdot 2 \cdot 2 \cdot 2 \cdot 2 \cdot 2 \cdot 2$	$= 512$
10	$10 \cdot 10 = 100$	$2^{10} = 2 \cdot 2 \cdot 2 \cdot 2 \cdot 2 \cdot 2 \cdot 2 \cdot 2 \cdot 2 \cdot 2$	$= 1024$

Amazing Exponential Growth

Suppose you tear a page from a newspaper $1/250 = 0.004$ inch thick and fold it in half 50 times. How thick will the folder paper be? What is your guess: $2''$, $4''$, $6''$, or even two feet?

To find the thickness is simply an exercise in arithmetic. Each time the paper is folded, the thickness of the folded paper is doubled. Hence the thickness of the paper will grow exponentially according to the geometric sequence

$$(0.004, 0.008, 0.016, 0.032, 0.064, \ldots)$$

Table 3 shows the thickness of the paper as a function of the number of foldings.

TABLE 3
Growth of the Exponential
Function $(0.004)2^n$

Number of Foldings	Thickness (inches)	
0	$0.004 \cdot 2^0 = 0.004$	⟵ *Initial thickness*
1	$0.004 \cdot 2^1 = 0.008$	
2	$0.004 \cdot 2^2 = 0.016$	
3	$0.004 \cdot 2^3 = 0.032$	
4	$0.004 \cdot 2^4 = 0.064$	
⋮	⋮	
50	$0.004 \cdot 2^{50} = 4.503 \times 10^{12}$	⟵ *Final thickness*

By using a calculator the thickness of the paper after 50 foldings was found to be approximately 4.503×10^{12} inches. Converting inches to feet and feet to miles, we have

$$
\begin{aligned}
\text{Thickness} &= 4.503 \times 10^{12} \quad \text{inches} \\
&= (4.503 \times 10^{12})(1/12) \quad \text{feet} \\
&= (4.503 \times 10^{12})(1/12)(1/5280) \quad \text{miles} \\
&= 71{,}079{,}539 \quad \text{miles}
\end{aligned}
$$

In other words, the height of the folded paper is roughly 300 times more than the distance from the earth to the moon. Believe it or not! If you think that this is a rather tall stack of paper, imagine how *wide* the stack of paper will be. After all, the width of the paper will be the width of the original paper *divided* by 2^{50}. It will have a width that is smaller than the diameter of an electron.

We now illustrate how certain populations also satisfy exponential growth over given periods of time.

Example 2

Bacteria Growth   A microbiologist has determined that every 12 hours, the size of a colony of *Salmonella bacteria* increases by 15%. Stated another way, as long as the size of the colony obeys this exponential growth, the size of the colony will always be 1.15 times the size of the colony 12 hours earlier. If the initial population of this colony is 200, what will be the population of this colony at any future time t?

Solution   Calling P_0 the initial population of the bacteria, the population size $P(t)$ at any time t (and not just at 12 one-hour intervals) can be described by means of the exponential growth function

$$
\begin{aligned}
P(t) &= P_0 b^t \\
&= 200(1.15)^t
\end{aligned}
$$

where t is time measured in 12-hour periods. In other words, $t = 1$ means 12 hours, $t = 2$ means 24 hours, and so on. Using a calculator, we can find the population at any future value of time t, and not just at integer values of t. (See Table 4.) This growth curve is shown in Figure 3.   □

TABLE 4
Exponential Growth
of Bacteria

Time (hours)	t	Population	
0 ←— *Present*	0	200	= 200
6	0.5	$200(1.15)^{0.5}$	= 214
12	1	$200(1.15)$	= 230
24	2	$200(1.15)^2$	= 264
36	3	$200(1.15)^3$	= 304
48	4	$200(1.15)^4$	= 350
54	4.5	$200(1.15)^{4.5}$	= 375

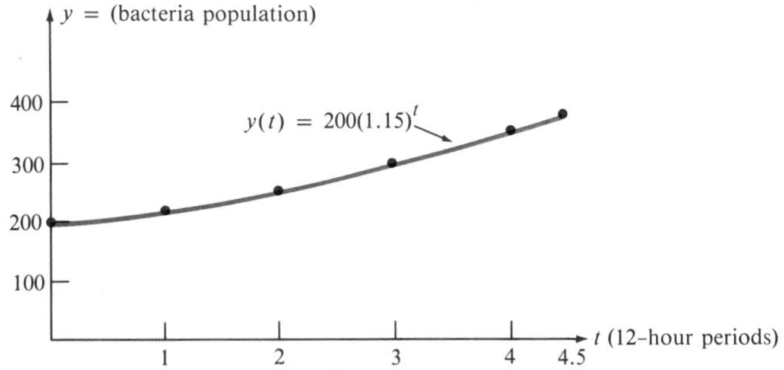

Figure 3
*Exponential growth of
bacteria*

We should point out here that the previous example could just as well have been stated in terms of the growth of a company's revenue. For example, if a company's revenue is 200 million dollars this year and it predicts a 15% annual increase for the next several years, then Table 4 could just as well represent the company's revenue (in millions of dollars) over the next t years.

The Growth and Decay Curves

Although there are many exponential functions such as 2^t, 3^t, $(1/2)^t$ that describe exponential growth and decay, there is one class of exponential functions that are special. They are the exponential functions with base e, defined by

$$f(t) = e^{kt}$$

The constant k is an arbitrary constant that describes the rate of growth or decay. In growth problems, k is positive; in decay problems, k is negative. The constant e is a number, sometimes called the **banker's constant*** in financial circles, that has the value $e = 2.71828\ldots$. We will show in the next section (when we can use logarithms) that any exponential function b^x can be rewritten in terms of e^{kt}. Hence all exponential growth and decay phenomena can be studied by using this basic function.

---

* Although we have called e the banker's constant, mathematicians would probably call it *Euler's constant*. The constant was introduced into mathematics in about 1731 by the Swiss mathematician, Leonhard Euler. The notation was probably chosen as an abbreviation for the word "exponential" and not, as some people think, an abbreviation for Euler's name.

To numerically evaluate $f(t) = e^{kt}$, it is possible to use either the table of values listed in the Appendix of this book or (probably more conveniently) to use a calculator with an e^t or equivalent key. Graphs of $f(t) = e^{kt}$ for various values of k are shown in Figure 4.

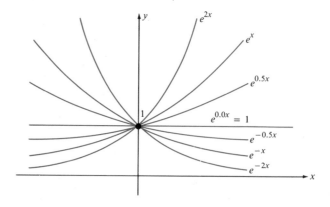

Figure 4
Exponentials e^{kx}

To gain a better understanding of the constant e, use your calculator to verify the computations made to form the second column in Table 5. We see that as k gets larger and larger, the value of

$$\left[1 + \frac{1}{k}\right]^{k}$$

approaches a limiting value. This limiting value is the constant

$$e = 2.718281828\ldots$$

TABLE 5
Sequence of Numbers
Converging to e

k	$\left[1 + \dfrac{1}{k}\right]^{k}$
1	2.00000
2	2.25000
5	2.48832
25	2.66584
50	2.69159
100	2.70481
1,000	2.71692
10,000	2.71815
100,000	2.71825

Banker's Constant e

$$e = \lim_{k \to \infty}\left[1 + \frac{1}{k}\right]^{k} = 2.718281828\ldots$$

The Banker's Constant e and the Four Banks

To understand how the number e is used in the calculations of financial matters, consider the following four banks. Each bank pays the same compound interest of 8% per year. However;

- Bank 1 makes interest payments every year (annually);
- Bank 2 makes interest payments every six months (semiannually);
- Bank 3 makes interest payments every month (monthly);
- Bank 4 makes interest payments "continuously" (continuous compounding).

Suppose you deposit $P = \$1000$ in each bank. What will be the value of each of these accounts after five years?

To determine these final amounts, we use the formula for computing the future value F of an initial deposit P. This is given by

$$F = P(1 + i)^n$$

where

$$i = \text{interest rate per compounding period}$$

$$n = \text{number of compoundings}$$

We can now find the amount F in each account from the following formulas:

Bank 1: $F_1 = \$1000(1 + 0.08)^5 = \$1469.33$

Bank 2: $F_2 = \$1000\left[1 + \dfrac{0.08}{2}\right]^{(5)(2)} = \$1480.24$

Bank 3: $F_3 = \$1000\left[1 + \dfrac{0.08}{12}\right]^{(5)(12)} = \$1489.85$

Bank 4: $F_4 = ?$

To determine the future value of the account in Bank 4, suppose for a moment that Bank 4 makes k interest payments per year (the interest rate will then be $0.08/k$ per period). After five years the initial \$1000 deposit will be worth

$$\text{Future value for } k \text{ compoundings} = \$1000\left[1 + \dfrac{0.08}{k}\right]^{5k}$$

For values $k = 1, 2,$ and 12 this formula gives the amounts in Banks 1, 2, and 3 at the end of the five years because they compound interest annually, semi-annually, and monthly, respectively. If we now let k get larger and larger (interest payments made more often), the above future value will approach the future value F_4 when interest is "compounded continuously." That is, we can perform a little fancy algebra and get

$$F_4 = \lim_{k \to \infty}\left[1000\left(1 + \dfrac{0.08}{k}\right)^{5k}\right]$$

$$= 1000 \lim_{k \to \infty}\left[1 + \dfrac{1}{(k/0.08)}\right]^{(k/0.08)(0.08)(5)}$$

However, by looking at the definition of the banker's constant e we see that the expression

$$\left[1 + \frac{1}{(k/0.08)}\right]^{(k/0.08)}$$

will have a limiting value of e as $k \to \infty$. (The fact that the expression contains $k/0.08$ instead of k only makes the expression approach e more slowly.) Hence we can write the above future value F_4 as

$$F_4 = \$1000e^{(0.08)(5)}$$
$$= \$1491.82$$

Conclusion of the Four Banks Problem

Although each of the four banks pays the same 8% annual rate of interest, in reality the bank that compounds interest "continuously" pays the most (compare $1491.82 to the other future values). Of course, no bank can really compound interest continuously. Continuous compounding is simply a mathematical idealization of compounding many, many times with a small time interval between compoundings. Still, the concept is very useful. In reality there is very little difference between the future value of an account when interest is "compounded continuously" and when interest is compounded daily as is done in many banks today. One reason that financial people like to work with continuous compounding is that the exponential function is easy to manipulate mathematically. In general, if interest is compounded continuously at an annual rate of interest r, then after t years an initial deposit of P dollars will have a future value of

$$F = Pe^{rt}$$

Example 3

Compound Interest   Sally deposits $100 in a bank that pays 8% annual interest, compounded continuously. Graph the amount of money in her account after each year for the next ten years.

HISTORICAL NOTE

Three of the most important constants in mathematics are π, i, and e, where π is the ratio of the circumference of a circle to its diameter, i is the unit complex number $i = \sqrt{-1}$, and e is the constant **e**. The symbols π, **i** and **e** that are used to denote these constants were all introduced into mathematics by the great Swiss mathematician Leonhard Euler (1707–1783). It is fascinating that these three famous constants are related by the celebrated equation $e^{2\pi i} = 1$. This is the reason that students of mathematics often wear tee-shirts with the slogan, "Mathematicians, We're Number $e^{2\pi i}$."

Solution   To find this graph, we can use a calculator with an e^x key and construct a table of values for

$$F = Pe^{rt}$$
$$= 100e^{(0.08)t}$$

for $t = 1, 2, \ldots, 10$, as shown in Table 6. Knowing these points, we then connect the points with a smooth curve, as shown in Figure 5.

TABLE 6
Values for Example 3

t	$100e^{(0.08)t}$	
0	$100.00	⟵ *Initial value*
1	$108.33	
2	$117.35	
3	$127.12	
4	$137.71	
5	$149.18	
6	$161.61	
7	$175.07	
8	$189.65	
9	$205.44	
10	$222.55	⟵ *Final value*

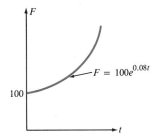

Figure 5
Graph for Example 3

The Banker's Constant *e* (an Interpretation)

An interesting interpretation of the banker's constant e is the following. If you were to make an initial deposit of $P = \$1$ in a bank that pays an annual interest r of 10% compounded continuously, then the future value F of the account after ten years would be

$$F = Pe^{rt}$$
$$= \$1e^{(0.10)(10)}$$
$$= e \quad \text{(the banker's constant } e)$$

In other words, the future value would be $2.71828\ldots$.

Growth and Decay Phenomena

One class of phenomena that can be studied with the help of exponential functions is growth and decay phenomena. We now define this important class of curves.

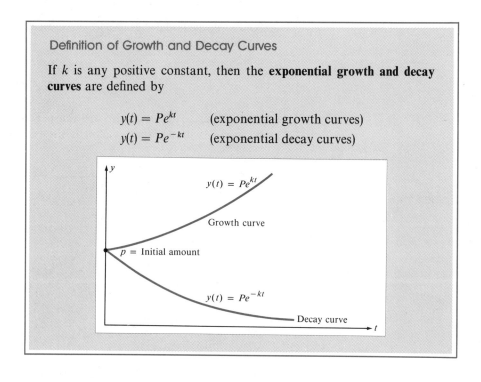

Definition of Growth and Decay Curves

If k is any positive constant, then the **exponential growth and decay curves** are defined by

$$y(t) = Pe^{kt} \qquad \text{(exponential growth curves)}$$
$$y(t) = Pe^{-kt} \qquad \text{(exponential decay curves)}$$

Light Intensity as Exponential Decay

A typical decay phenomenon in science is the decay of sunlight filtering down through water. Anyone who has dived to a depth of 100 feet in most ocean waters will tell you that there is very little light at that depth. Most divers could probably not tell you, however, that the light intensity decays exponentially as a function of depth.

The **Bougour-Lambert law** states that the intensity $I(x)$ of sunlight filtering down through water at a depth x decreases according to the exponential decay function

$$I(x) = I_0 e^{-kx} \qquad (k > 0)$$

Here, I_0 is the intensity of light at the surface, and $k > 0$ is an absorption constant that depends on the murkiness of the water. The murkier the water, the

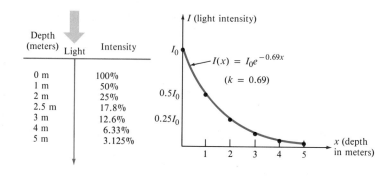

Figure 6
Light intensity decaying exponentially

Depth (meters)	Intensity
0 m	100%
1 m	50%
2 m	25%
2.5 m	17.8%
3 m	12.6%
4 m	6.33%
5 m	3.125%

larger the value of k. Some examples of light intensity curves for different values of k are shown in Figure 6.

Example 4

Light Intensity in Water   Suppose that the light intensity at the surface of a lake is some given value I_0 and that the absorption constant k has been determined experimentally to be $k = 0.69$. How much light has been absorbed by the water at a depth of 2.5 meters?

Solution   Using the basic absorption equation

$$I(x) = I_0 e^{-kx}$$

we simply substitute in the given values, getting

$$I(2.5) = I_0 e^{-(0.69)(2.5)}$$
$$= 0.178 I_0$$

This means that at a depth of 2.5 meters, the light intensity is only 17.8% as intense as on the surface of the water. (See Figure 7.)   □

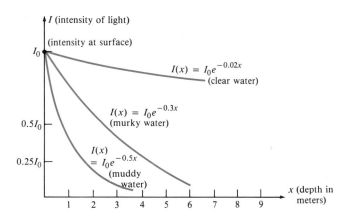

Figure 7
Three typical lakes

Example 5

Future Value as a Growth Curve   Dorothy has deposited $1000 in a bank that pays an annual interest rate of 9% compounded continuously. What is the exponential growth function that describes the value of her account after t years?

Solution   Here, we are given

$$r = 0.09 \qquad \text{(annual rate of interest)}$$
$$P = \$1000 \qquad \text{(initial amount deposited)}$$

The future value F of her account after t years is given by the exponential growth curve

$$F = Pe^{rt}$$
$$= \$1000 e^{0.09t}$$

This growth curve is plotted in Figure 8.   □

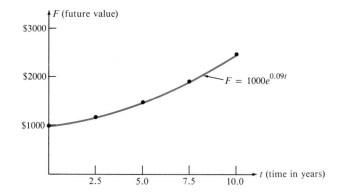

Figure 8
*Future value of an account
that pays 9% compounded
continuously*

Problems

For Problems 1–10, find the first six terms of the geometric sequences with given values of A and b.

1. $A = 1$, $b = 3$
2. $A = 5$, $b = 2$
3. $A = 1/2$, $b = 4$
4. $A = 1$, $b = 1/3$
5. $A = 2$, $b = 3$
6. $A = 8$, $b = 1/2$
7. $A = 1$, $b = 0.1$
8. $A = 4$, $b = -1/2$
9. $A = 3$, $b = -3$
10. $A = 6$, $b = -2$

World Growth

Problems 11–14 refer to Table 7, which gives the world's population (in millions) for different regions of the world for the years 1960–1980.

11. Does the population in North America appear to be increasing exponentially? If so, what would you say is the fractional increase every ten years? In other words, what is the multiplier b in the geometric sequence?
12. Does the population in South America appear to be increasing exponentially? If so, what would you say is the percentage increase every ten years?
13. Assuming the population in North America is increasing exponentially, what would you predict the population to be in 1990? In the year 2000?
14. Which regions of the world have the lowest and highest rate of increase?

TABLE 7
Population Figures for Problems 11–14

Year	North America	South America	Europe	U.S.S.R.	Asia	Africa
1960	199	215	425	214	1683	275
1970	226	283	460	244	2091	354
1980	252	365	484	266	2613	472

For Problems 15–20, graph the given exponential function by plotting the values of the functions at the integers -2, -1, 0, 1, 2 and filling in between these points with a smooth curve. Computations can be made by hand or by using a calculator.

15. $y = 2^{-x}$
16. $y = 4^x$
17. $y = 4 \cdot 2^{-x}$
18. $y = 4 \cdot 3^{-2x} + 5$
19. $y = 3 \cdot 2^{-x} + 3$
20. $y = 2^x + 2^{-x}$

For Problems 21–26, solve the exponential equations for x.

21. $(1/8)^x = 8$
22. $7^{-x} = 49^{(x+3)}$
23. $5^{-2x} = 1/25$
24. $2^{x^2 - 4x} = 32$
25. $8^{x^2} = 2^{5x}$
26. $3^{(2x+1)} = 3^x$

27. **Solving Exponential Equations** Find the points of intersection of the graphs of the functions

$$f(x) = 3^{x^2} \quad \text{and} \quad g(x) = 3^{x+2}$$

28. **Solving Exponential Equations** Find the points of intersection of the graphs of the functions

$$f(x) = 5^{x(x^2 - 3)} \quad \text{and} \quad g(x) = 5^{-2x^2}$$

Annual Growth of Leading Industrial Countries

The Gross National Product (GNP) of a country for a single year is the total value of all goods and services produced by that country for that year. Table 8 lists the 1986 GNP and the average annual growth rate (over the previous ten years) of the GNP for leading industrial countries. For Problems 29–32, assume that the GNPs of the above countries are growing exponentially according to the growth curve

$$\text{GNP}(n) = P(1 + i)^n$$

where

$$i = \text{average annual growth rate}$$

$$n = \text{number of years starting from 1986}$$

$$P = \text{GNP in 1986}$$

TABLE 8
GNP Figures for Problems 29–32

Country	Gross National Product, 1986 (in billions)	Average Annual Growth Rate of GNP, 1977–1986
Canada	420	2.50%
France	569	3.90%
United Kingdom	460	2.75%
Italy	369	2.20%
Japan	1204	4.85%
United States	3662	3.50%
U.S.S.R.	1843	2.20%
West Germany	655	3.10%

29. **United States GNP** From the above information, what is the exponential growth function that describes the GNP for the United States? Using this function, what would you expect the GNP of the United States to be in the year 1990?

30. **Equation of GNP** From the above information, what would the equation of the GNP be for each of the countries?

31. **Leading Economic Country** If economic growth continues as it has during the period 1977–1986, will Japan's GNP surpass the GNP of the United States by the year 2000?

32. **Number Two Country** If economic growth continues as it has during the period 1977–1986, will Japan's GNP surpass the Soviet Union's by the year 2000? In the next section, when we study logarithms, we will learn to determine the exact year when these two growth curves intersect.

33. **Semiannual Compounding of Money** Sally deposits $100 in a bank account that pays 10% annual interest, compounded every six months (that is, compounded semiannually). What will be the value of her account after six months, after one year, after 18 months, after two years, and after n compoundings? *Hint:* 10% annual interest compounded semiannually means that every six months the bank pays $10/2 = 5\%$ interest on the current value of the account.

34. **Quarterly Compounding of Money** Richard deposits $100 in a bank account that pays 12% annual interest, compounded every three months (quarterly). What will be the value of his account after three months, after six months, after nine months, and after n compoundings? *Hint:* 12% annual interest compounded quarterly means that every three months the bank pays $12/4 = 3\%$ interest on the current value of the account.

35. **World's Population** Suppose the world's population is falling at the rate of 1% per year. Suppose the current population is 5 billion people. Estimate the world's population for each of the next five years.

36. **Bacteria Growth** Find a function that represents the growth of a bacteria population that triples every day if the initial amount of the bacteria is ten cells. Sketch a graph of this function.

37. **Bacteria Decay** A bacterial infection is being treated with a new form of hydrocortizone. Suppose that every week the bacteria population falls to 80% of the population of the preceding week. If the initial population is 500, what is the exponential function that describes the population at any time $t > 0$? Sketch a graph of this function.

38. **Temperature of Coffee** Suppose that every ten minutes you measure the temperature of your coffee and determine that the temperature is 90% of the temperature ten minutes earlier. If the initial temperature was 180 degrees, what is the exponential function that describes the temperature at any time t? What is the temperature of the coffee after one hour? (Cooling of temperatures is one of many natural phenomena that satisfies an exponential decay law.)

39. **Stopping an Oil Tanker** It has been shown that if an oil tanker stops its engines, then its velocity will decrease exponentially. Suppose that the captain shuts off the

engines when the speed of the tanker is 20 miles per hour and that every minute the speed of the tanker is only 90% of the speed at the previous minute. What is the exponential decay function that describes the speed of the tanker for any positive time t? What is the speed of the tanker after one hour? Note that the exponential function is never zero, but clearly the tanker will eventually stop. What is the meaning of this discrepancy?

40. Virus Growth A population of a certain virus grows so that its size $S(t)$ at the end of t days is given by

$$S(t) = S_0 2^{kt}$$

where S_0 is the size of the initial population. For an initial population of 1000 that doubles in 20 days, what is the value of k? What will be the size of the population at the end of 15 and 30 days?

41. Bacteria Population The size of the population E of a bacteria colony is described by the function

$$E(t) = E_0 2^{t/30}$$

where E_0 is the size of the initial population and t is measured in months. For the initial population size of 500, calculate the population size at times $t = 1, 2, 3, 4$, and 10 months.

42. The Clever Rat Psychologists have long held that much about human nature can be learned from rats. Suppose an experimenter places a rat in a maze and that it takes the rat one minute to escape. After that, each time the experimenter places the rat in the maze, it takes only 80% as long for the rat to escape. What is the geometric sequence that describes how long it takes the rat to escape on every trial?

43. Animal Population The size of an animal population in a given habitat is described by the function

$$P(t) = P_0 2^t$$

where P_0 is the initial population size and t is time measured in years. For an original population size of 400, calculate the population size after 1, 2, 4, 10, and 20 years.

44. Exponential Growth Would you rather have $1,000,000 or the value calculated by doubling the amount you have every day for 100 days starting with one cent on the first day? What is the value of each amount at the end of 100 days?

45. Continuous Compounding Andy deposits $500 in a bank that pays 9% annual interest compounded continuously. What is the function that describes the value of his account at any time t? What will be the value of the account after one year? After 18 months?

46. Continuous Compounding Karen deposits $1000 in a bank that pays 8% annual interest compounded contin-

uously. What is the function that describes the value of her account at any time t? What will be the value of her account after ten years? After 15.5 years?

47. Long-Term Account Mary inherited the amount contained in a bank account taken out by her grandfather 75 years earlier. The initial deposit in the account was $10, and the bank paid a constant annual interest rate of 5% compounded continuously. How much did Mary inherit?

48. Which Account Has More Money? Twenty years ago, Harry deposited $100 in a bank that pays an annual interest rate of 9% compounded continuously. John deposited $250 in the same bank ten years ago. Which person's account contains more money?

49. Business Learning Curves Businesses often measure training programs by how fast new employees can learn new skills. Using a given training program, suppose the average number of units L produced by a new employee on the tth week at work can be described by the exponential learning curve

$$L(t) = 100(1 - e^{-0.1t})$$

Use tables or a calculator to find the number of units L that an average employee can produce on week $t = 1, 2, 3, 4$, and 5. Draw a graph of this learning curve.

50. Sales Decay Sales of a new product tend to decrease over time. For sales levels described by the function

$$S(t) = 1500 + 750e^{-0.2t}$$

where t is measured in years evaluate the sales level for $t = 0, 1, 2, 3, 4$, and 10 years.

51. Radioactive Dating of Fossils Radioactive carbon isotope ^{14}C dating of fossils of plants and animals depends on knowing the amount $C(t)$ of ^{14}C left in the fossil after a given period of time t. The function

$$C(t) = A_0 e^{-t/5500} \qquad (t \geq 0)$$

gives the amount of ^{14}C after t years from the time the plant or animal has died where A_0 is the amount of ^{14}C that was present at the time of death. Find the amount of ^{14}C present $t = 0, 1000, 2000, 4000$, and 8000 years after the plant or animal has died if $A_0 = 1$ gram.

52. Long-Distance Phone Calls Conversations in long-distance telephone calls usually fade out owing to a damping effect on the lines. If I_0 is the initial strength of the signal, then the signal can be measured by

$$I(x) = I_0 e^{-kx}$$

where x is the distance the signal is sent measured in miles and $k > 0$ is the damping constant that depends on the type of wire used in the communication (along with other factors). If $k = 0.002$, what fraction of the signal is lost at 10 miles? After 100 miles? After 500 miles?

53. Suspension Bridge and the Banker The middle span of any suspension bridge is always shaped in the form of a curve called a *catenary*. The equation of this curve is

$$y = \frac{e^x + e^{-x}}{2} \qquad (-1 \le x \le 1)$$

It is interesting that the banker's constant e is the main ingredient of the equation that describes the shape of a suspension bridge. Use a calculator to evaluate this catenary for $x = -1, -0.5, 0, 0.5, 1$. Then fill in between the points with a smooth curve.

Amazing but True Stories (Exponential Growth)

54. Wise Young Man A story is told about a king who offered a young peasant a gold piece for saving the life of the king's daughter. Being humble, however, the young peasant declined and agreed instead to accept one kernel of wheat on the first day, two kernels of wheat on the second day, four kernels of wheat on the third day, eight

kernels of wheat on the fourth day, and so on for the next 100 days. How many pieces of grain would the king have to give the young peasant on the 100th day? Assuming there are 1,000,000 kernels of wheat in a bushel, how many bushels of wheat would the king have to give the peasant on the last day?

55. The Rat's Progeny The progeny of a rat numbers about 500 per year. Assume that half of these are females and half are males and that all of them survive. How many progeny will the rat have in 25 years?

56. A Fast-Growing Function We have seen the amazing growth of the exponential function 2^n (remember the paper-folding problem). There is, however, another function that grows much, much faster. That function is the *factorial function*

$$n! = n \cdot (n - 1) \cdot (n - 2) \cdots 3 \cdot 2 \cdot 1$$

Use a calculator to convince yourself of this fact by computing 2^{25} and 25!.

10.2 Logarithms and Logarithmic Scales

PURPOSE

We introduce the general logarithmic function with base b and show how it can be used to

- solve algebraic equations and
- describe physical phenomena.

We also introduce the two specific logarithmic functions, the common logarithm and the natural logarithm.

Introduction

Roughly 400 years before you are reading these pages, in around 1590, the Scottish laird John Napier looked at two rows of numbers, similar to the rows of numbers in Table 9, and made a discovery that changed the world.

TABLE 9
Napier's Mystery Numbers

0	1	2	3	4	5	6	7	8	9	10
1	2	4	8	16	32	64	128	256	512	1024

This was a time before computers, calculators, and even slide rules. It was a time when all computations were done by hand with pencil and paper. What

[8]

Napier saw in the above rows was an ingenious way to multiply numbers by adding. Since addition is far easier and quicker to perform, he discovered what might be called a sixteenth-century computer.

To understand Napier's discovery, let us suppose we wish to multiply the numbers 16 and 64. If we carry out this multiplication, we will see that the product is 1024. However, there is a far easier way to find this product using the above numbers. (Do you see what it is?) What Napier saw in the above rows of numbers was: if the two numbers directly above 16 and 64 (namely, 4 and 6) are added, then the sum (10) is the number directly above the product of 16 and 64. In fact, this scheme will work for the product of any two numbers in the bottom row (4 times 128 is the number below the sum 2 + 7 = 9, or 512). Of course, Napier still had a few bugs to work out. Namely, how could he multiply numbers that were not in the bottom row? These problems were all worked out, however, and after 20 years of refinement he published his results in 1614 in a paper entitled "A Description of the Marvelous Rule of Logarithms." We will now learn about these refinements.

The Logarithmic Function

The columns of numbers in Table 10 illustrate the relationship between the exponential function $y = 10^x$ and what is called its **logarithmic inverse**. To understand the logarithmic function, first consider the exponential 10^x. To evaluate 10^x, we start with a value of x—say, $x = 3$—and raise 10 to this power:

$$10^3 = 1000$$

To evaluate the logarithm, we ask the reverse question. That is, given the answer (say 1000), to what power x should we raise 10 in order to get that answer? In other words, find x such that

$$10^x = 1000$$

Clearly, the exponent is 3, and this value is called the logarithm of 1000 to the base 10 and is denoted by

$$\log_{10} 1000 = 3$$

TABLE 10
Relationship Between
$y = 10^x$ and $y = \log_{10} x$

Exponential Function		Logarithmic Function	
x	$y = 10^x$	x	$y = \log_{10} x$
-3	$10^{-3} = 0.001$	0.001	$10^y = 0.001 \longrightarrow y = -3$
-2	$10^{-2} = 0.01$	0.01	$10^y = 0.01 \longrightarrow y = -2$
-1	$10^{-1} = 0.1$	0.1	$10^y = 0.1 \longrightarrow y = -1$
0	$10^0 = 1$	1	$10^y = 1 \longrightarrow y = 0$
1	$10^1 = 10$	10	$10^y = 10 \longrightarrow y = 1$
2	$10^2 = 100$	100	$10^y = 100 \longrightarrow y = 2$
3	$10^3 = 1000$	1000	$10^y = 1000 \longrightarrow y = 3$

Table 10 illustrates the relationship between the logarithmic function $\log_{10} x$ and the exponential function 10^x. The relationship between the graphs of exponential and logarithmic functions can be seen in Figure 9.

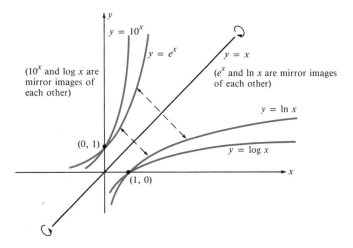

Figure 9
Comparison of log x and ln x

The above discussion leads to the general definition of the logarithm to any base b.

Definition of the Logarithm

The **logarithm** of a positive number x to a base b, denoted $\log_b x$, where b is any positive number different from 1, is the number to which the base b must be raised to equal the number x. If the base is 10, then the logarithm is called the **common logarithm** and is denoted as

$$y = \log x \qquad (x > 0)$$

If the base is the constant $e = 2.718\ldots$, then the logarithm is called the **natural logarithm** and is denoted as

$$y = \ln x \qquad (x > 0)$$

Table 11 illustrates the logarithms with different bases. We can see from Table 11 the general pattern

$$y = \log_b x \quad \text{is the same as} \quad b^y = x$$

TABLE 11

Relationship Between
Exponentials and Logarithms

Exponential Form	Logarithmic Form
$10^2 = 100$	$\log 100 = 2$
$4^3 = 64$	$\log_4 64 = 3$
$10^3 = 1000$	$\log 1000 = 3$
$2^{-4} = 1/16$	$\log_2(1/16) = -4$
$10^{1.2} = 15.8$	$\log 15.8 = 1.2$
$(1/2)^6 = 1/64$	$\log_{1/2}(1/64) = 6$
$20^2 = 400$	$\log_{20} 400 = 2$

HISTORICAL NOTE

The invention of the logarithm was to the sixteenth century what the computer is to the twentieth century—an evolutionary jump in the speeding up of arithmetic operations. While the logarithm decreased the time that sixteenth century astronomers had to spend on arithmetic computations by more than tenfold, computers have increased the speed of arithmetic operations at least another millionfold.

The inventor of the logarithm, John Napier (1550–1617), was born in Scotland and spent most of his life at the family estate of Merchiston Castle. He was violently anti-Catholic and published a bitter and widely read tirade against the Church of Rome. Napier was very creative and novel in all of his endeavors and thought of mathematics as a diversion from his activities in religion and politics.

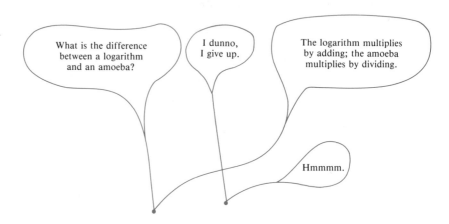

Useful Properties of Logarithms

Logarithms are a valuable tool that are used in many areas of science. They can aid in the solution of algebraic equations when the unknown is in the exponent. They are also useful in the description of physical phenomena. Many applications come about as a result of the following properties of the logarithm.

Six Properties of Logarithms

The following properties of logarithms hold provided all the arguments are positive. Some of the examples use the logarithms $\log 2 = 0.301$ and $\log 3 = 0.477$.

- **Property 1:** $\log_b uv = \log_b u + \log_b v$
  Examples:

  $$\log 200 = \log 2 + \log 100 = 0.301 + 2 = 2.301$$
  $$\log 6000 = \log 6 + \log 1000 = 0.778 + 3 = 3.778$$

- **Property 2:** $\log_b \dfrac{u}{v} = \log_b u - \log_b v$
  Examples:

  $$\log \frac{80}{3} = \log 80 - \log 3 = 1.903 - 0.477 = 1.426$$

  $$\log \frac{1}{10} = \log 1 - \log 10 = -\log 10 = -1$$

- **Property 3:** $\log_b u^a = a \log_b u$
  Examples:

  $$\log 3^2 = 2 \log 3 = 0.954$$
  $$\log 2^{14} = 14 \log 2 = 4.214$$

- **Property 4:** $\log_b b = 1$
  Examples:

  $$\log 10 = 1$$
  $$\ln e = 1$$
  $$\log_2 2 = 1$$

- **Property 5:** $\log_b 1 = 0$
  Examples:

  $$\log 1 = 0$$
  $$\ln 1 = 0$$

- **Property 6:** $b^{\log_b x} = x$
  Examples:

  $$10^{\log_{10} 5} = 5$$
  $$2^{\log_2 3} = 3$$

- **Property 7:** $\log_b b^x = x$
  Examples:

  $$\log 10^4 = 4$$
  $$\ln e^3 = 3$$

Verification of Logarithm Rules

We illustrate the general idea of why the above rules hold by proving Property 1:

$$\log_b uv = \log_b u + \log_b u$$

The proofs of the other rules follow similar lines.

To simplify matters, we let

$$x = \log_b u$$
$$y = \log_b v$$

We rewrite these equations in equivalent exponential form so that we can use properties of exponentials. This gives

$$u = b^x \quad \text{(exponential form)}$$
$$v = b^y \quad \text{(exponential form)}$$

We now multiply these quantities, getting

$$uv = b^x b^y$$

However, the product on the right can be written as

$$b^x b^y = b^{x+y} \quad \text{(basic step in proof)}$$

and hence

$$uv = b^{x+y}$$

Finally, rewriting this equation back in logarithmic form, we have

$$\log_b uv = x + y$$

which is nothing more than

$$\log_b uv = \log_b u + \log_b v$$

Example 1

Uses of Logarithm Properties   Provided that the following expressions in x represent positive numbers, we can write the following relationships:

(a) $\log_b x + \log_b(x^2 + 1) = \log_b x(x^2 + 1)$

(b) $\log_b \dfrac{x^2 + 4}{x} = \log_b(x^2 + 4) - \log_b x$

(c) $\log_b(5x^4) = \log_b 5 + \log_b x^4$
$$= \log_b 5 + 4\log_b x$$

Solving Equations Involving Logarithms and Exponentials

Using the properties of logarithms, we can solve many types of equations that would otherwise be extremely difficult to handle. The following examples illustrate how logarithms are used to solve a variety of problems.

Example 2 _____

Equation Involving Logarithms   Solve for x:

$$\log x^2 + \log x = 3$$

Solution   Using Property 1, we write the left-hand side of this equation as

$$\log x^2 + \log x = \log x^2 \cdot x$$
$$= \log x^3$$

Hence the original equation becomes

$$\log x^3 = 3$$

By the definition of the common logarithm, we get

$$x^3 = 10^3$$

or

$$x = 10 \qquad\qquad \square$$

Example 3 _____

Equation Involving Logarithms   Solve for x:

$$\ln (x + 1) - \ln (x - 1) = 1$$

Solution   Using Property 2, we can write

$$\ln \left[\frac{x + 1}{x - 1} \right] = 1$$

By the definition of the natural logarithm we have

$$\frac{x + 1}{x - 1} = e$$

Multiplying by $x - 1$ and solving for x gives

$$x + 1 = e(x - 1)$$
$$x(1 - e) = -1 - e$$
$$x = \frac{e + 1}{e - 1}$$
$$\cong 2.16 \qquad\qquad \square$$

When the unknown variable x is located in an exponent, the general strategy is to take the logarithm of each side of the equation. While it is possible to use a logarithm with any base, the specific problem will generally dictate which logarithm is the most convenient.

Example 4 _____

Unknown in Exponent   Solve for x:

$$10^{x^2} = 73$$

Solution   Taking the common logarithm of each side of the equation, we have

$$\log 10^{x^2} = \log 73$$

Property 7 says that the left-hand side of this equation is x^2. Hence we have

$$x^2 = \log 73$$

Solving for x, we get

$$x = \sqrt{\log 73}$$
$$\cong 1.365 \qquad \qquad \Box$$

Example 5

Unknown in Exponent   Solve for x:

$$2^{2x-1} = 10^x$$

Solution   Taking the common logarithm of each side of the equation, we have

$$\log 2^{2x-1} = \log 10^x$$

Using Property 3, we can bring down the exponents, getting

$$(2x - 1) \log 2 = x \log 10$$
$$= x$$

Finally, solving for x gives

$$x = \frac{\log 2}{2 \log 2 - 1}$$
$$\cong -0.756 \qquad \qquad \Box$$

In the above problem we could have used a logarithm with a different base. However, the final value of x would have been written in terms of this new logarithm. Since common and natural logarithm tables are generally the only ones published, it is best to use these logarithms. Also, most calculators have common and natural logarithm keys.

Logarithm Functions in Growth and Decay Phenomena

In the previous section we introduced exponential growth and decay curves. The main emphasis of the study was to find the exponential functions that described various growth and decay phenomena and to evaluate these functions at different times. We now study what could be called the inverse problem. That is, given an exponential function, such as

$$y = Ae^{kt}$$

what is the value of time t that will make the value of the function y a given amount? In other words, y is the known quantity, and t is the unknown quantity.

For example, suppose Eddie has deposited $4000 in a bank account that pays 8% annual interest compounded continuously. He hopes that in four years, when he graduates from college, the value of the account will be $10,000 and he can buy a new car. The question is: When will Eddie's account be worth $10,000?

To solve this problem, we set the future value of Eddie's account

$$F(t) = \$4000e^{0.08t}$$

where t is time in years equal to $10,000 and solve for t. In other words, we solve for t in the equation

$$4000e^{0.08t} = 10,000$$

Hence

$$e^{0.08t} = 2.5$$

Take the natural logarithm of each side of this equation (remember $\ln e^x = x$) and then solve for t. We get

$$0.08t = \ln (2.5)$$

$$t = \frac{\ln (2.5)}{0.08}$$

$$\cong \frac{0.916}{0.08}$$

$$= 11.45 \quad \text{years} \qquad\qquad \square$$

In other words, it will take about 11.45 years for a $4000 deposit to grow to $10,000 if interest is paid at an annual rate of 8% compounded continuously. Hopefully, Eddie will not be in college that long.

Example 6

Doubling Time of Money   How long will it take money to double in value if invested in a bank that pays 12% compounded continuously?

Solution   An initial amount P_0, deposited at 12% annual interest compounded continuously will have a future value F of

$$F(t) = P_0 e^{0.12t} \qquad \text{(future value)}$$

after t years. The rate that this future value increases can be seen in Figure 10.

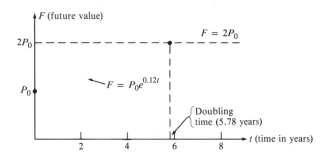

Figure 10
Computation of doubling times

To find the time t where this graph intersects the doubling line

$$F(t) = 2P_0 \qquad \text{(twice the initial amount)}$$

we set the future value equal to this amount and solve for t. That is, we solve for t in the equation

$$P_0 e^{0.12t} = 2P_0$$

Dividing each side of this equation by P_0 gives

$$e^{0.12t} = 2$$

Taking the natural logarithm of each side of the equation, we get

$$\ln e^{0.12t} = \ln 2$$

Finally, using the property $\ln e^x = x$ and solving for x, we can write

$$0.12t = \ln 2$$
$$t \cong 5.78 \quad \text{years}$$

In other words, every 5.78 years, money will double in value if invested at 12% compounded continuously. This means that if a student graduates from high school at the age of 18 and deposits a $1000 graduation gift in a bank that pays 12% annual interest compounded continuously, then by the time the student reaches the age of 65 (roughly eight doubling times), the value of the account will be

$$\text{Future value at age 65} = \$1000 \cdot 2^8$$
$$= \$1000(256)$$
$$= \$256,000$$

This gives real meaning to Ben Franklin's adage that "Time is Money." □

Example 7

Japan Number Two?   The 1986 Gross National Product (GNP) of Japan was 1.204 trillion dollars, with an average annual rate of increase over the previous ten years of 4.85%. On the other hand, the GNP of the Soviet Union in 1986 was 1.843 trillion dollars, with an average annual rate of increase over the previous 10 years of 2.2%. Assuming that both GNPs will continue to grow at these rates of growth, when will Japan's GNP surpass the GNP of the Soviet Union?

Solution   The GNPs of the two countries are described by the following exponential functions:

Japan: $J(n) = 1.204(1.0485)^n$ (trillions of dollars)
Soviet Union: $S(n) = 1.843(1.0220)^n$ (trillions of dollars)

where n is the number of years measured from 1986 ($n = 0$ means 1986, $n = 1$ means 1987, and so on). Although n is generally thought to be 0, 1, 2, . . . , we will allow it to be any nonnegative real number. In this way we will be able to interpret a GNP after a fractional number of years (0.5 year, 4.7 years, and so on). These curves are shown in Figure 11.

It is clear that if the present growth rates continue, Japan's GNP will eventually surpass the GNP of the Soviet Union. By examining the curves in Figure 11 it appears that this will happen after roughly 16 or 17 years. To find a better estimate of when the two GNPs intersect, we can solve for n in the equation $J(n) = S(n)$. Doing this, we have

$$1.204(1.0485)^n = 1.843(1.022)^n$$

Moving the quantities involving n to the left-hand side of the equation, we have

$$\frac{(1.0485)^n}{(1.022)^n} = \frac{1.843}{1.204} = 1.531$$

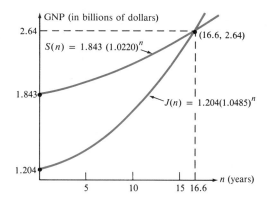

Figure 11
Forecasts of GNPs of Japan and the Soviet Union

Using the property $a^n/b^n = (a/b)^n$ and simplifying, we see

$$(1.026)^n = 1.531$$

Finally, taking the common logarithm of each side of this equation, we have

$$n \log (1.026) = \ln (1.531)$$

or

$$n = \frac{\log (1.531)}{\log (1.026)}$$

$$\cong 16.6 \text{ years} \qquad\qquad \square$$

Interpretation of the GNPs of Japan and the Soviet Union

In the GNPs of Japan and the Soviet Union continue to grow exponentially at the current rate, then sometime between January 1, 2002, and January 1, 2003, Japan will have the second largest GNP, behind the United States.

Logarithmic Scales (the Richter Scale)

One of the important uses of the logarithm is in describing phenomena whose measurements are very large, are very small, or range over large intervals. Since logarithms are essentially exponents, their values do not become as large or as small as quickly as other quantities. Observe how the logarithms in the column back in Table 11 do not vary as much as their arguments.

To understand how logarithms are applied in science, we show how the logarithm is used to define the **Richter scale** in seismology. We will see that adding one unit on the Richter (logarithm) scale corresponds to multiplying the seismic activity by ten units. Hence a change of 3 on the Richter scale corresponds to a factor of 1000 in seismic activity.

The energy released by an earthquake at the epicenter is generally measured in units called ergs. In a good-sized quake the energy released might be somewhere around

$$100,000,000,000,000,000,000 \text{ ergs}$$

Since it is difficult to work with numbers of this tremendous size, seismologists use the Richter scale. Imagine the headline

Earthquake Measuring 34,468,346,012,438,547,983 Ergs

Hits Los Angeles!

For this reason, earthquakes are always reported in units R "on the Richter scale" with R defined as

$$R = 0.67 \log E - 7.9 \qquad \text{(Richter scale)}$$

where E is the energy in ergs released by the quake. We illustrate the Richter scale with a few examples.

Example 8

Richter Scale   The devastating Mexico City earthquake of 1985 released roughly 4×10^{24} ergs of energy. What was the magnitude of this quake on the Richter scale?

Solution

$$
\begin{aligned}
R &= 0.67 \log (4 \times 10^{24}) - 7.9 \\
&= 0.67[\log 4 + 24 \log 10] - 7.9 \\
&\cong 0.67[0.60 + 24] - 7.9 \\
&= 8.6 \qquad\qquad \square
\end{aligned}
$$

Table 12 shows the effects of various size quakes. Since the Richter scale is a logarithmic scale with base 10, an increase of 1 on the Richter scale represents a tenfold increase in energy released by the quake. Hence an earthquake that measures 8.6 on the Richter scale is 1000 times more powerful than one that measures 5.6 on the Richter scale.

TABLE 12

Types of Earthquakes on the Richter Scale

Magnitude (Richter Scale)	Result at Epicenter	Number per Year
1.0–1.9	Detectable only on a seismograph	Many
2.0–2.9	Noticeable by a few people	800,000
3.0–3.9	Felt by most people, no cause for alarm	20,000
4.0–4.9	Felt by all, some glass broken	2800
5.0–5.9	Some furniture turned over	1000
6.0–6.9	Ground cracks, some buildings fall	185
7.0–7.9	Bridges and dams fall	14
8.0+	Disaster on a massive scale, the worst	0.2

Example 9

Energy Released by an Earthquake   In 1976 a devastating earthquake measuring 8.9 on the Richter scale struck Guatemala, killing 23,000 people. How much energy was released by that earthquake?

Solution   Setting

$$0.67 \log E - 7.9 = 8.9$$

we solve for E. We get

$$\log E = 25.07$$

or

$$E = 10^{25.07}$$
$$\cong 1.175 \times 10^{25} \quad \text{ergs} \qquad \Box$$

Problems

For Problems 1–14, use $\log 5 = 0.70$ and $\log 7 = 0.85$ to determine the numerical value of the expressions.

1. $\log 35$

2. $\log (5/7)$

3. $\log \sqrt[8]{35}$

4. $\log \sqrt[3]{5}$

5. $\log (0.2)$

6. $\log (125/49)$

7. $\log (5^4 \cdot 7^5)$

8. $\log \sqrt[5]{7 \cdot 25}$

9. $\log 125$

10. $\log (625/49)^{10}$

11. $\log \sqrt{5^8 \cdot 7^3}$

12. $\log (1/\sqrt{5^8 \cdot 7^3})$

13. $\log (49/5)$

14. $\log \sqrt[5]{7^{10}/5^6}$

For Problems 15–20, simplify and/or expand the expressions.

15. $e^{\ln 3 - \ln 4}$

16. $e^{4 \ln 2}$

17. $4^{\log_2 3}$

18. $\log (rt/5)$

19. $\log (e^3/a)$

20. $\log \left[\dfrac{x^2(x + 3)}{(x + 1)} \right]$

For Problems 21–36, solve for x.

21. $\log x = 3 \log 2 + 2 \log 3$

22. $\log x = \log 2 - \log 3 + \log 5 - \log 7$

23. $\log \sqrt{x} = 4 \log 2 - 3 \log (1/2)$

24. $x = \log_2 4$

25. $x = \log_6 36 \log_{25}(1/5)$

26. $\log x + \log x^2 = 3$

27. $\ln x^2 - \ln x = \ln 18 - \ln 6$

28. $\ln x^2 = 2$

29. $\log_x \dfrac{9}{4} = 2$

30. $\log_3 x^2 - \log_3 x = \log_3 18 - \log_3 6$

31. $\log_4(3x + 5) = \log_4 72 - 3 \log_4 2$

32. $10^{x^2} = 150$

33. $2^{x+1} = 10^{2x}$

34. $3^{4x} = 5^{x+1}$

35. $5^{3x} = 6^{x-1}$

36. $10^x - 2 \cdot 10^{-x} = 1$

37. Doubling Time for Annual Compounding Henry's father and mother have deposited \$1000 in a bank account in his name. The bank pays interest at a rate of 10.5% compounded annually. How long will it take for this account to double in value?

38. General Doubling Time for Annual Compounding How long will it take for a bank account to double in value if the bank pays an annual interest rate of r, compounded annually? How long will it take for this account to triple in value?

39. Doubling Time for Continuous Compounding Sally's father and mother have deposited \$1000 in a bank account in her name. The bank pays an annual interest rate of 10.5% compounded continuously. How long will it take for this account to double in value? Note the difference between this answer and the answer to Problem 37.

40. General Doubling Time for Continuous Compounding How long will it take for a bank account to double in value if the bank pays an annual interest rate of r compounded continuously? How long will it take for this account to triple in value?

41. Business Growth A franchised car repair outlet has $N(t)$ outlets and is expanding this number at the rate of 15% per year. The function that describes this number is

$$N(t) = N_0 e^{0.15t}$$

where t is time measured in years and N_0 is the initial number of franchises. How long will it take for the number of franchises to double? To triple?

42. How Much Interest Does the Bank Pay? Suppose you deposited \$100 in a bank that compounds interest continuously, but you forgot to ask the annual interest rate. All you know is that your deposit of \$100 today will be worth \$150 in five years. What is the annual rate of interest that this bank pays?

43. How Much Interest Does the Bank Pay? Another bank down the street from the one in Problem 42 also pays compound interest, but this bank compounds interest only every year. Suppose this banker tells you that if you deposit \$100 in the bank today, the deposit will be worth \$175 in six years. How much annual interest does this bank pay?

44. Population Growth In 1988 the population of a certain city was one million people and was increasing at the rate of 4% per year. When will the population of this city reach 1.5 million people?

45. **Drinking and Driving**   After a person drinks an alcoholic beverage, the alcohol level $L(t)$ in the person's blood rises to a level of 0.3 mg of alcohol per milliliter of blood. After that time, the amount of alcohol decreases exponentially according to the law

$$L(t) = 0.3(0.5)^t$$

where t is time measured in hours from the time when the peak level was reached. Sketch a graph of the function $L(t)$. If the legal driving limit of alcohol is 0.1 mg of alcohol per milliliter of blood, how long will it be until a person will be able to legally drive?

46. **Surface Area of the Body**   Biologists have found an equation that approximates the surface area S in square feet of the human body in terms of the height h and weight w of the body. The equation is

$$S = 0.1w^{0.425}h^{0.725}$$

where w is a person's weight in pounds and h is a person's height in inches. Approximate the surface area of your own body by substituting your height and weight in the above equation. Rewrite this equation in terms of log S.

47. **Exponential Learning Curve**   Rita is in charge of employee development at a shoe factory. Using her new training methods, she has discovered that the number of shoes that an average new employee can stitch in a day is given by the learning curve

$$L(t) = 25(1 - e^{-0.125t})$$

where t is number of days the employee has been on the job. How many days will it take before the average new employee can stitch 20 shoes per day? Graph the learning curve.

48. **Biological Growth**   The mass in grams of a given bacteria culture grows according to the logistic growth model of the type

$$y(t) = \frac{8}{1 + 7(2^{-t})}$$

where t is time measured in days. (See Figure 12.) When will the mass of the culture reach 7 grams?

49. **Logarithmic Production Curve**   The price $P(x)$ of an item depends on the number of units x produced. For the price function

$$P(x) = \log\left(10 + \frac{x}{4}\right)$$

how many units will be produced at a price of $3?

50. **Hydrogen Ion Concentration**   The acidity of a substance is measured by the concentration (measured in moles per liter) of hydrogen ions (H^+) in the substance. The standard way to describe this concentration is to define the pH of a substance as the negative logarithm of the hydrogen ion concentration. That is,

$$pH = -\log[H^+]$$

What is the pH of distilled water if its hydrogen ion concentration is 10^{-7} moles/liter?

51. **Hydrogen Ion Concentration**   A substance with pH smaller than 7 is termed an acid, and a substance with pH greater than 7 is called a base. Using the formula in Problem 50, determine the pH for the substances whose hydrogen ion concentrations are given below, and indicate whether they are acids or bases.
    (a) 4.2×10^{-6}
    (b) 8×10^{-6}
    (c) 0.6×10^{-7}
    (d) 6.3×10^{-4} (grapefruit juice)

Decibels

A decibel is a unit of relative loudness. One decibel is the smallest unit of sound detected by the human ear. If x denotes the intensity of a sound wave (measured in watts per square centimeter), then the noise level $L(x)$ in decibels of the wave is given by

$$L(x) = 10 \log\left(\frac{x}{I_0}\right)$$

where I_0 is the intensity of the sound wave at the threshold of audibility (approximately 10^{-16} watt of energy per square centimeters). Use this formula to solve Problems 52–55.

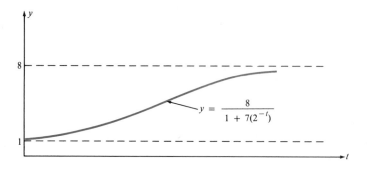

Figure 12
Graph for Problem 48

52. What are the noise levels $L(x)$ in decibels for the following intensities, x:
   (a) 1,000 watts per square centimeter
   (b) 5,000 watts per square centimeter
   (c) 10,000 watts per square centimeter
53. The noise level of a whisper is about 20 decibels, and that of ordinary conversation about 65 decibels. Determine the ratio of the intensity of a conversation to the intensity of a whisper.

54. How many decibels are the following phenomena, if they have the following energy levels.
   (a) a motor with an intensity of 10^{-11} watt/square cm
   (b) a typewriter with an intensity of 10^{-4} watt/square cm
   (c) a jet engine with an intensity of 10^{-2} watt/square cm
   (d) a jet with afterburner with an intensity of 10^{-1} watt/ square cm
55. What is the difference in the noise levels of two sounds, one of which is 1000 times the intensity of the other?

Differentiation of Exponential Functions

PURPOSE

We show the role that exponential functions play in the calculus. The major topics introduced are

- how to write b^x in terms of e^{kx},
- how to differentiate e^{kx},
- how to differentiate $e^{g(x)}$, and
- how to differentiate $b^{g(x)}$.

We then introduce the economic concept of discounting and determine the optimal time for a timber company to harvest trees.

Introduction

We have seen that the exponential growth curve

$$G(t) = Ae^{kt} \qquad (k > 0)$$

describes numerous growth phenomena in economics, biology, botany, physics, and other areas. In this section we will learn a remarkable fact about this curve, namely, that the rate of change dG/dt is proportional to the value G. That is,

$$\frac{dG(t)}{dt} = kG(t)$$

The reason that many phenomena obey such a law is the way populations grow. In many insect populations the size increases as a result of random mating between insects. That is, the larger the population, the larger the growth becomes (more insects, more mating). Hence we have the phenomenon that the rate of growth is proportional to the size of the population.

Since we will learn how to differentiate the exponential function e^{kx}, we first learn how to rewrite other exponential functions b^x in terms of e^{kx}. In that way we will be able to differentiate them as well.

Rewriting All Exponential Functions b^x in Terms of e^{kx}

Earlier, when we studied exponential growth and decay phenomena, we introduced various exponential functions such as 2^x, $(1.03)^n$, and $(1/2)^x$. The interesting thing about these functions is that they can be rewritten in terms of e^{kx}. In this way, all exponential growth and decay phenomena can be restated in terms of the one basic function e^{kx}. Since the independent variable in the following discussion does not necessarily stand for time, we use x instead of t to represent the independent variable.

Rewriting b^x in Terms of e^{kx}

If b is a positive real number and x any real number, then

$$b^x = e^{kx}$$

where $k = \ln b$.

Verification: Starting with an arbitrary exponential function b^x, we set

$$e^{kx} = b^x$$

and solve for k. Taking the natural logarithm of each side of the equation gives

$$\ln b^x = \ln e^{kx}$$

Using properties of the logarithm (bringing down the exponents), we have

$$x \ln b = kx \ln e \qquad (\ln e = 1)$$
$$= kx$$

Hence we have

$$k = \ln b$$

Example 1 _____

Rewriting Exponentials   Rewrite the exponential functions 2^x, 3^x, $(0.5)^x$, $(0.8)^x$ in terms of e^{kx}.

Solution   Using the basic equation $b^x = e^{(\ln b)x}$, we have

$$2^x = e^{(\ln 2)x} \cong e^{0.69x} \qquad (b = 2, k = 0.69)$$
$$3^x = e^{(\ln 3)x} \cong e^{1.1x} \qquad (b = 3, k = 1.1)$$
$$(0.5)^x = e^{x \ln (0.5)} \cong e^{-0.69x} \qquad (b = 0.5, k = -0.69)$$
$$(0.8)^x = e^{x \ln (0.8)} \cong e^{-0.22x} \qquad (b = 0.8, k = -0.22)$$

The graphs of these exponential functions are shown in Figure 13.   □

The Derivative of e^{kx}

Since we have seen how the exponential function b^x can be expressed in terms of e^{kx}, it is important we know how to differentiate e^{kx}. In this way we will be able to differentiate the exponential function b^x as well. We begin by stating the derivative of e^x.

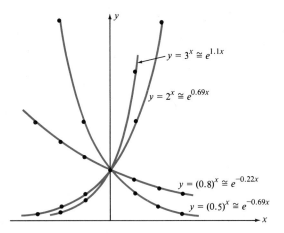

Figure 13
*Rewriting exponential
functions*

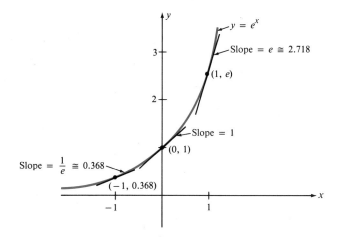

Figure 14
*The slope of the tangent line is
equal to the height of the graph
for $y = e^x$*

A Function Whose Derivative Is the Same as the Function

The exponential function e^x has a property that no other function possesses (except for constant multiples Ce^x): The derivative is the same as the function. That is,

$$\frac{d}{dx} e^x = e^x$$

Although we will not prove this fact here, the graph shown in Figure 14 illustrates the basic idea. Note that the slope of the tangent line to the graph of $y = e^x$ is always the same as the height of the graph of $y = e^x$.

We now state a number of useful derivatives involving exponential functions.

Summary of Useful Exponential Derivatives

By use of the basic derivative formula $de^x/dx = e^x$ along with the chain rule, we can find the derivative of several other exponential functions. We have the following rules:

$$\frac{d}{dx} e^{g(x)} = e^{g(x)}g'(x) \qquad \text{(general exponential)}$$

$$\frac{d}{dt} e^{kt} = ke^{kt} \qquad \text{(derivative of growth curve)}$$

$$\frac{d}{dt} e^{-kt} = -ke^{-kt} \qquad \text{(derivative of decay curve)}$$

The first rule above for differentiating the general exponential function can easily be remembered by simply realizing that it says

$$\frac{d}{dx} e^{(\text{exponent})} = e^{(\text{exponent})} \frac{d}{dx} (\text{exponent})$$

Example 2 ———————   Differentiating Exponentials   Find y' if

$$y = e^{(x^2 + 2x)}$$

Solution   Here

$$g(x) = x^2 + 2x$$
$$g'(x) = 2x + 2$$

Using the rule for differentiating general exponents, we have

$$y' = e^{g(x)}g'(x)$$
$$= (2x + 2)e^{x^2 + 2x}$$ □

Example 3 ———————   Differentiating Exponentials   Find y' if

$$y = e^{1/x}$$

Solution   Here

$$g(x) = \frac{1}{x}$$

$$g'(x) = -\frac{1}{x^2}$$

Using the rule for differentiating general exponents, we have

$$y' = e^{g(x)}g'(x)$$

$$= -\frac{1}{x^2} e^{1/x}$$ □

Example 4 ———————   Differentiating a Growth Curve   Find the derivative of the growth curve

$$y(t) = 30e^{0.05t}$$

Solution   Using the rule for exponential derivatives, we have

$$\frac{dy}{dt} = 30 \frac{d}{dt} e^{0.05t}$$

$$= (30)(0.05)e^{0.05t}$$
$$= 1.5e^{0.05t}$$

See Figure 15.   □

Example 5 ———————   Differentiating a Decay Curve   Find the derivative of the decay curve

$$y(t) = 100e^{-0.25t}$$

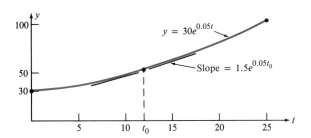

Figure 15
Derivative of an exponential curve

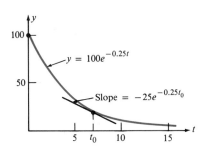

Figure 16
*Derivative of a negative
exponential curve*

Solution Using the rule for exponential derivatives, we have

$$\frac{dy}{dt} = 100 \frac{d}{dt} e^{-0.25t}$$

$$= (100)(-0.25)e^{-0.25t}$$

$$= -25e^{-0.25t}$$

See Figure 16.   □

The Derivative of b^x

Since we now know how to write the exponential b^x for any base b in terms of e^{kx} by means of the formula

$$b^x = e^{(\ln b)x}$$

we can also differentiate b^x.

The Derivative of b^x

If b is any positive number, then the derivative of b^x is

$$\frac{d}{dx} b^x = (\ln b)b^x$$

Verification

Since b^x can be written

$$b^x = e^{(\ln b)x}$$

we have that

$$\frac{d}{dx} b^x = \frac{d}{dx} e^{(\ln b)x}$$

$$= (\ln b)e^{(\ln b)x}$$

$$= (\ln b)b^x$$

This completes the verification.

The derivative we have just derived leads naturally to the following more general formula.

General Derivative for Exponential Function with Base b

Using the above derivative and the chain rule, we can state the more general rule

$$\frac{d}{dx} b^{g(x)} = b^{g(x)} \cdot g'(x) \cdot \ln b$$

Example 6

Differentiating Exponentials with Base 2   Find y' if

$$y = 2^x$$

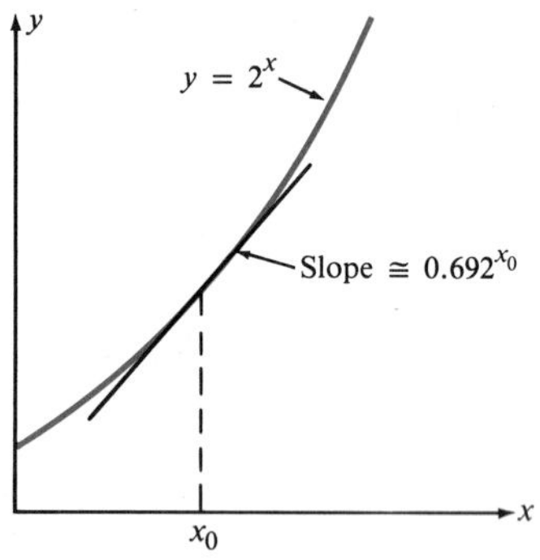

Solution We have

$$b = 2$$
$$g(x) = x$$
$$g'(x) = 1$$

Using the rule for finding the derivative of $b^{g(x)}$ gives

$$\frac{d}{dx} 2^x = b^{g(x)} \cdot g'(x) \cdot \ln b$$

$$= 2^x \cdot \ln 2$$
$$\cong 0.69(2^x)$$

See Figure 17.

Figure 17
Derivative of 2^x

In the figure: $y = 2^x$, Slope $\cong 0.692^{x_0}$, x_0

Example 7

Differentiating Exponentials   Find y' if

$$y = 2^{(x^2 + 1)}$$

Solution We have

$$b = 2$$
$$g(x) = x^2 + 1$$
$$g'(x) = 2x$$

Using the rule for finding the derivative of $b^{g(x)}$ gives

$$\frac{d}{dx} 2^{(x^2 + 1)} = 2^{(x^2 + 1)} \cdot 2x \cdot \ln 2$$

Example 8

Derivative of Exponentials   Find y' if

$$y = 3^{4x}$$

Solution   Here

$$b = 3$$
$$g(x) = 4x$$
$$g'(x) = 4$$

Using the rule for finding the derivative of $b^{g(x)}$ gives

$$\frac{d}{dx} 3^{4x} = 3^{4x} \cdot 4 \cdot \ln 3$$

$$\cong 4.39(3^{4x}) \qquad \square$$

Example 9

Tangent Line to the Exponential Curve   Find the tangent line to the curve

$$y = e^x$$

at the point $(1, e)$.

Solution   The tangent line to a curve $y = f(x)$ at a point $(a, f(a))$ is given by

$$y - f(a) = f'(a)(x - a)$$

Since

$$\frac{d}{dx} e^x = e^x$$

we have $f(1) = e$, and $f'(1) = e$. Hence the tangent line will be

$$y - e = e(x - 1)$$

or

$$y = ex$$
$$\cong 2.718x$$

See Figure 18. $\square$

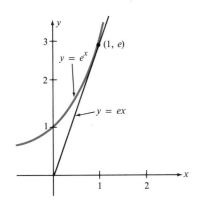

Figure 18
Tangent line to e^x at $(1, e)$

Discounting in Economics

In the problem of continuous compounding of interest the goal is to find the future value F from a given present value P. The problem of discounting is the opposite one. In such cases we seek to find the present value P of a given amount F that will be available t years from now. In the case of continuous compounding, an initial principal P will grow to a future value F of

$$F = Pe^{rt}$$

where r is the annual rate of interest. We can also find the corresponding **discount formula** simply by solving for the present value P in terms of the future value F

$$P = Fe^{-rt}$$

The above value of P is called the **discounted value** (or present value) of F; and in this new context, r is often called the **discount rate**. This negative exponential illustrates the fact that if someone were to give you F dollars t years from now, then the value of this money today is only Fe^{-rt} (assuming that money can earn interest at a rate r, compounded continuously. Figure 19 gives the discounted value (or present value) of $1000 over a 100-year period for various interest rates. The moral of this table is that the value of $1000 today is not the same as the value of $1000 in 100 years.

	Discount Rate			Time (years)
	0.02	0.05	0.15	
	$1000.00	$1000.00	$1000.00	0
	$ 670.32	$ 367.88	$135.34	20
	$ 449.33	$ 135.34	$ 18.32	40
	$ 301.94	$ 49.79	$ 2.48	60
	$ 201.90	$ 2.48	$ .12	80
	$ 135.34	$ .12	$ .05	100

P (discounted value)

$1000

$750 $P = 1000e^{-0.02t}$

$500 $P = 1000e^{-0.05t}$

$250

20   40   60   80   100 t (time in years)

$P = 1000e^{-0.15t}$

Figure 19
Discounted value of $1000 over 100 years

Discounting in the Wood Products Industry

To understand the importance of discounting in business management, consider the problem facing a large wood products company in the state of Washington. The company has just planted hybrid tree seedlings on several tracts of land and has determined that the value $V(t)$ of this timber (in millions of dollars) is increasing over time according to the exponential function

$$V(t) = e^{\sqrt{t}}$$

where t is time in years measured from the date the trees were planted. Assuming the world's financial situation to be such that money is discounted at a rate of 8% per year ($1 today will be worth $1e^{0.08t}$ in t years), when should the company cut the timber for maximum revenue? We assume the cost to maintain the trees is negligible in comparison to the value of the trees, and so to maximize the revenue is actually the same as to maximize the profit.

To solve this problem, we proceed as follows. At each future point in time the value of the timber is given by

$$V(t) = e^{\sqrt{t}}$$

However, we must keep in mind that the value $V(t)$ is measured in tomorrow's dollars (the company is paid when the timber is harvested). Hence the present or discounted value P of the timber in today's dollars if the timber is harvested after t years is

$$P(t) = V(t)e^{-0.08t} \qquad \text{(present value)}$$
$$= e^{\sqrt{t}}e^{-0.08t}$$
$$= e^{\sqrt{t}-0.08t}$$

The problem now is to find the maximum value of $P(t)$. To do this, we begin by finding the derivative $P'(t)$. Calling

$$g(t) = \sqrt{t} - 0.08t$$

$$g'(t) = \frac{1}{2\sqrt{t}} - 0.08$$

we have

$$\frac{d}{dt} P(t) = \frac{d}{dt} e^{g(t)}$$

$$= e^{g(t)} g'(t)$$

$$= e^{(\sqrt{t} - 0.08t)} \left[\frac{1}{2\sqrt{t}} - 0.08 \right]$$

Setting the derivative to zero to find the critical points, we have

$$e^{(\sqrt{t} - 0.08t)} \left[\frac{1}{2\sqrt{t}} - 0.08 \right] = 0$$

Since the exponential function on the left is never zero, we have that

$$\frac{1}{2\sqrt{t}} - 0.08 = 0$$

Solving this equation for t gives

$$\sqrt{t} = \frac{1}{(2)(0.08)}$$

$$t = \frac{1}{(4)(0.08)^2} \cong 39 \quad \text{years}$$

Although we have not formally shown that $P(t)$ attains a relative maximum or even its absolute maximum value at this critical point, it is not hard to do so.

Interpretation of the Tree-Harvesting Problem

The company should harvest the trees 39 years after planting to maximize the profit. If this strategy is followed, then the value of the trees in today's dollars, immediately after planting, will be

$$P(39) = e^{\sqrt{39} - (0.08)(39)}$$

$$\cong 22.76 \quad \text{million dollars}$$

Of course, when the trees are harvested in 39 years, the value of the timber will be much more. In fact the value received at that time will be

$$V(39) = e^{\sqrt{39}}$$

$$\cong 515.43 \quad \text{million dollars}$$

This value is the same amount of money the company would have after 39 years if it sold the newly planted timber today for 22.76 million and invested the money in a bank that paid 8% annual interest, compounded continuously.

It is also interesting to note from the above equation for $P(t)$ that if the annual discount rate were r instead of 0.08, then the time the company should harvest their trees would be given by

$$t = \frac{1}{4r^2} \quad \text{years}$$

Table 13 gives the time when the company should harvest their trees as a function of the discount rate r. The discount rate is often reasonably close to the annual inflation rate.

TABLE 13
When to Harvest Timber
Given the Discount Rate

Annual Discount Rate, r	Time to Harvest (years)
0.06	69
0.07	51
0.08	39
0.09	31
0.10	25
0.11	21
0.12	17

Problems

For Problems 1–10, write the exponential functions in the form e^{kx}.

1. 2^x

2. 3^x

3. 4^x

4. 8^x

5. $(1.5)^x$

6. $(0.7)^x$

7. $(0.9)^x$

8. 10^x

9. 100^x

10. 500^x

For Problems 11–26, find the derivative of the expressions.

11. $y = 2e^{3x}$

12. $y = e^{-x}$

13. $y = e^{5x}$

14. $y = e^{-9x}$

15. $y = e^{x^2}$

16. $y = e^{x+2}$

17. $y = e^{\sqrt{x}}$

18. $y = xe^{2x}$

19. $y = \dfrac{e^x}{x}$

20. $y = x^2 e^{3x}$

21. $y = e^{x^2 + x}$

22. $y = x^2 e^{-x^2}$

23. $y = e^x + e^{-x}$

24. $y = \sqrt{e^x + 1}$

25. $y = 2x - e^{4x} + 3e^{x^2}$

26. $y = \dfrac{1 + e^x}{x}$

For Problems 27–34, find the derivative of the functions.

27. $y = 5^x$

28. $y = 9^x$

29. $y = 2^x$

30. $y = 10^x$

31. $y = 2^{x^2}$

32. $y = 3^{x^2 + 1}$

33. $y = 9^{(2x + 5)}$

34. $y = 10^{-x^2}$

Graphing Problems

35. Shape of e^x Show that the graph of $y = e^x$ is concave up for all x.

36. Shape of e^{-x} Is e^{-x} an increasing or decreasing function? Is the graph of e^{-x} concave up or concave down? Sketch the graph of e^{-x}.

37. Shape of 2^x Show that the graph of $y = 2^x$ is always increasing. What is the concavity of the graph of $y = 2^x$?

38. Tangent Line Find the equation of the tangent line to

$$y = 2e^{-3x}$$

at $x = 0$. Sketch the graph of the exponential curve and the tangent line at $x = 0$.

39. Tangent Line Find the equation of the tangent line to

$$y = xe^x$$

at $x = 0$.

Business Management Problems

40. Marginal Revenue The revenue R in dollars a manufacturer receives when q units of a product are sold is given by

$$R(q) = 0.75qe^{-0.002q}$$

What is the marginal revenue when $q = 100$?

41. When to Sell Coins A coin and stamp dealer calculates that the value $V(t)$ in dollars that a collection will appreciate after t years is given by the formula

$$V(t) = \$1000e^{\sqrt{t/4}}$$

If the annual discount rate is 8%, when should the collection be sold to maximize the return?

42. Sell No Wine Before Its Time A vinter can either sell wine at the present time or age it further for future sales. Suppose the value $V(t)$ of the wine in thousands of dollars is increasing according to the function

$$V(t) = e^{0.2\sqrt{t}}$$

where t is time in years. If the discount rate is 9% per year, when should the vinter sell the wine to maximize profit? We assume that it does not cost anything to store the wine and so profit and revenue are the same thing. What is the value of the wine in today's dollars if this strategy is used?

43. Wine Problem Solve Problem 42 if the value of the wine grows according to the function

$$V(t) = 2^{\sqrt{t}}$$

44. When to Sell Land A landowner owns a piece of property that is increasing in value according to the function

$$V(t) = 100,000e^{0.08t}$$

where V is measured in dollars and t is time measured in years. If the discount rate of money is 7% per year, when should the owner sell the land?

45. Land Speculating Another landowner who lives next to the owner mentioned in Problem 44 knows the value of his land is increasing according to the function

$$V(t) = 50,000e^{0.06t}$$

where V is measured in dollars and t in years. Assuming the same discount rate of 7% per year, when should this owner sell the land?

46. Learning Model The number of units $P(t)$ produced per day after t days of training is given by

$$P(t) = 150(1 - e^{-kt})$$

where k varies with the individual. Estimate k for an individual who produces 75 units per day after one day of training. Find the derivative $dP(t)/dt$ using this value for k. Find the rate at which $P(t)$ is increasing after five days of training for this individual.

Other Growth Models in Biology

47. Logistic Growth Curve A biologist observes that the weight $w(t)$ of a colony of bacteria is growing according to the logistic growth curve

$$w(t) = \frac{500}{1 + 49e^{-kt}}$$

where t is time measured in days. Here the population starts to grow rapidly; but after some time, certain influences begin to restrict the rate of growth. Figure 20 shows the above logistic curve. Find the rate of change dw/dt of the size of this population of bacteria.

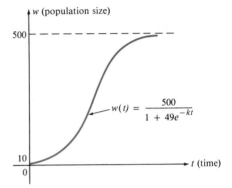

Figure 20
Graph for Problem 47

48. Gompertz Growth Curve A biologist observes that the size of a population of yeast cells is growing according to the Gompertz growth curve

$$N(t) = 500e^{e^{-0.05t}}$$

where t is time measured in days. (See Figure 21.) Find the rate of change dN/dt of the size of this population.

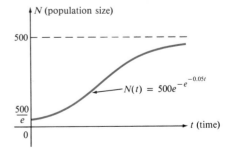

Figure 21
Graph for Problem 48

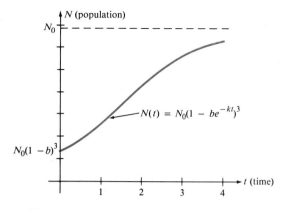

Figure 22
Graph for Problem 49

49. Von Bertalanffy Growth Curve   A biologist observes that the size of a certain virus culture is growing according to the Von Bertalanffy growth curve

$$N(t) = N_0(1 - be^{-kt})^3$$

where t is time measured in hours. (See Figure 22.) Find the rate of change dN/dt of this population.

10.4 Differentiation of Logarithmic Functions

PURPOSE

We begin by learning how to differentiate the natural logarithm $\ln x$. After differentiating this basic logarithmic function, we then use the chain rule to differentiate other logarithmic functions, such as $\ln [g(x)]$. We also show how differential calculus is used to introduce the elasticity of the demand function.

Introduction

We have seen that the exponential function 2^x increases very rapidly (remember the paper-folding problem). In fact, all of the exponential functions e^x, 2^x, 3^x, ... are extremely fast-growing functions. On the other hand, the logarithm functions $\log x$, $\ln x$, ... are extremely slow-growing functions. They do approach infinity as the argument x gets large, but they do it very slowly. For example, the value of the common logarithm $y = \log x$ has reached only $y = 10$ by the time x has reached 10,000,000,000. We now investigate more than the size of the logarithm functions. We study the rate of growth and the concavity of these functions.

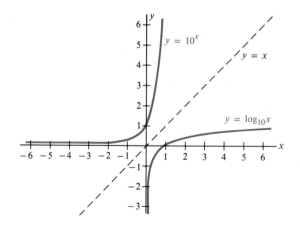

Figure 23
The common logarithm and its inverse function 10^x

Differentiation of the Natural Logarithm

One property of the logarithm functions that can be seen in Figure 23 is that the graphs of the functions are always increasing and are always concave down. These observations lead us to believe that the first derivative is always positive and that the second derivative is always negative.

We begin by stating and verifying the derivative formula for the natural logarithm.

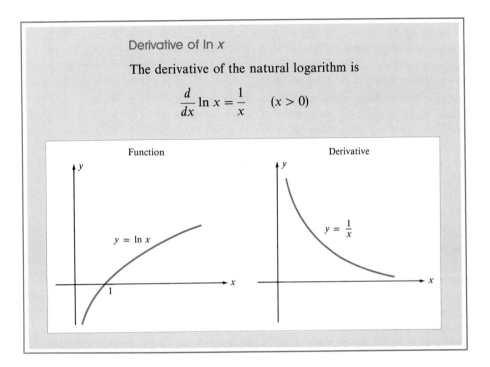

Derivative of ln x

The derivative of the natural logarithm is

$$\frac{d}{dx} \ln x = \frac{1}{x} \qquad (x > 0)$$

Function — $y = \ln x$

Derivative — $y = \frac{1}{x}$

Verification of the Derivative of ln x

We start with the property of logarithms that states

$$e^{\ln x} = x \qquad (x > 0)$$

We now differentiate implicitly each side of this equation with respect to x. Doing this, we get

$$\frac{d}{dx} e^{\ln x} = \frac{dx}{dx}$$

By using the chain rule to find the derivative on the left, the above equation becomes

$$e^{\ln x} \frac{d}{dx} \ln x = 1$$

$$x \frac{d}{dx} \ln x = 1$$

Solving for the derivative

$$\frac{d}{dx} \ln x$$

gives

$$\frac{d}{dx} \ln x = \frac{1}{x}$$

This completes the verification.

Example 1

Differentiating Logarithms   If

$$y = \ln x$$

(a) find y' and y'',

(b) determine the regions where $\ln x$ is increasing and where it is decreasing,

(c) determine the regions where $\ln x$ is concave up and where it is concave down.

Solution

(a) Using the formula for the derivative of the natural logarithm, we have

$$y' = \frac{d}{dx} \ln x = \frac{1}{x} \qquad (x > 0)$$

$$y'' = \frac{d^2}{dx^2} \ln x = -\frac{1}{x^2} \qquad (x > 0)$$

(b) Since y' is always positive, we conclude that $\ln x$ is always increasing.

(c) Since y'' is always negative, we conclude that $\ln x$ is always concave down.

Figure 24 shows the derivative of the logarithm at various points.   □

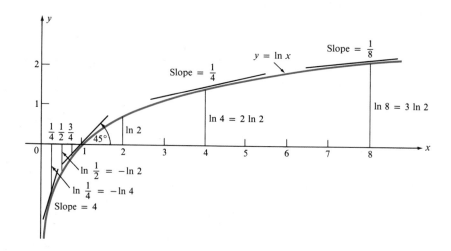

Figure 24
Slopes of the logarithm

Example 2

Differentiating Logarithms   Find y' if

$$y = x \ln x \qquad (x > 0)$$

Solution   Using the product rule, we have

$$y' = x\frac{d}{dx}\ln x + \ln x$$

$$= 1 + \ln x \qquad \square$$

Example 3

Differentiating Logarithms   Find y' if

$$y = (\ln x)^3 \qquad (x > 0)$$

Solution   Let

$$g(x) = \ln x$$

and apply the generalized power rule

$$\frac{d}{dx}g(x)^n = ng(x)^{n-1}g'(x)$$

In this case we have

$$\frac{d}{dx}(\ln x)^3 = 3(\ln x)^2\frac{d}{dx}\ln x$$

$$= \frac{3(\ln x)^2}{x} \qquad \square$$

Example 4

Differentiating Logarithms   **Differentiate**

$$y = \ln(x^2 + 3x + 5)$$

Solution   Letting

$$u = x^2 + 3x + 5$$

we have

$$y = \ln u$$

Using the chain rule

$$\frac{dy}{dx} = \frac{dy}{du}\frac{du}{dx}$$

we can write

$$\frac{dy}{dx} = \frac{1}{u} \cdot (2x + 3)$$

Finally, substituting for u gives

$$\frac{d}{dx}\ln(x^2 + 3x + 5) = \frac{2x + 3}{x^2 + 3x + 5} \qquad \square$$

The Derivative of $y = \ln[g(x)]$

Example 4 suggests that we can differentiate the logarithm function with a more general argument such as $g(x)$. Inasmuch as $\ln[g(x)]$ is the composition of the two functions $g(x)$ and $\ln x$, we can apply the chain rule and obtain the following derivative formula.

Formula for the Derivative of $y = \ln [g(x)]$

If $g(x)$ is differentiable and positive at the value x, then

$$\frac{d}{dx} \ln [g(x)] = \frac{g'(x)}{g(x)}$$

The above derivative formula can be stated verbally as

$$\frac{d}{dx} \ln [\text{argument}] = \frac{1}{(\text{argument})} \frac{d}{dx} (\text{argument})$$

If we select $g(x)$ as the absolute value function $g(x) = |x|$, then we get the formula

$$\frac{d}{dx} \ln |x| = \frac{\dfrac{d}{dx} |x|}{|x|} = \frac{1}{x} \qquad (x \neq 0)$$

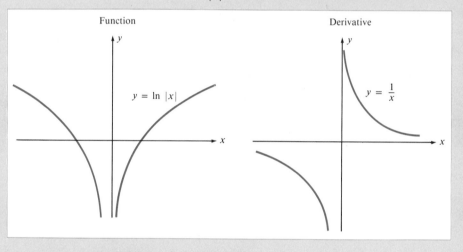

Function

$y = \ln |x|$

Derivative

$y = \dfrac{1}{x}$

Example 5

Differentiation of Logarithms Find y' if

$$y = \ln (x^2)$$

Solution Here

$$g(x) = x^2$$
$$g'(x) = 2x$$

By use of the formula for the derivative of $\ln [g(x)]$ we get

$$\frac{dy}{dx} = \frac{g'(x)}{g(x)}$$

$$= \frac{2x}{x^2}$$

$$= \frac{2}{x}$$

□

Example 6

Differentiation of Logarithms   Find y' if

$$y = \ln(e^x + x^2)$$

Solution   Here

$$g(x) = e^x + x^2$$
$$g'(x) = e^x + 2x$$

Using the formula for the derivative of $\ln[g(x)]$, we have

$$\frac{dy}{dx} = \frac{g'(x)}{g(x)}$$

$$= \frac{e^x + 2x}{e^x + x^2} \qquad\qquad \square$$

We now present a useful formula to show how other logarithms, in particular the common logarithm $\log x$, can be written in terms of the natural logarithm $\ln x$.

Change of Base Formula for Logarithms

The logarithms $\log_a x$ and $\log_b x$ with bases a and b are related by the **change of base formula**:

$$\log_b x = \frac{\log_a x}{\log_a b}$$

When the base a is the constant e and $b = 10$, then the change of base formula acts as a conversion formula from the natural to common logarithm

$$\log x = \frac{\ln x}{\ln 10} \cong 0.434 \ln x$$

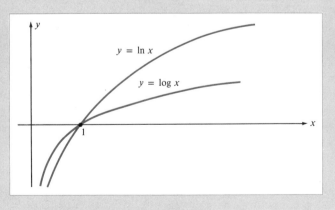

Example 7

Differentiation of the Common Logarithm   Find y' if

$$y = \log(x^2 + 1)$$

Solution   Rewriting the above common logarithm in terms of a natural logarithm, we have

$$\log(x^2 + 1) \cong 0.434 \ln(x^2 + 1)$$

Hence

$$\frac{d}{dx}\log(x^2 + 1) \cong 0.434 \frac{d}{dx}\ln(x^2 + 1)$$

$$= 0.434 \frac{2x}{x^2 + 1} \qquad \square$$

Elasticity of the Demand Function

Earlier, we studied the economic function known as the demand curve:

$$q = D(p) \qquad \text{(demand curve)}$$

This function gives the number of items q that consumers will buy at a given price p. As everyone knows, q will normally decrease as p increases. A typical demand function is shown in Figure 25.

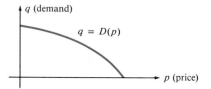

Figure 25
A typical demand function

We also learned that the derivative dD/dp of the demand curve represents the rate of change in the quantity demanded as a function of the price of the goods. For example, if $dD/dp = -2$ when $p = \$5$, then an increase in price of one dollar from \$5 to \$6 would result in a decrease in demand of two units.

The concept of the elasticity of the demand function is closely related to the derivative of the demand function. The difference is that whereas the derivative represents the total change in the demand, the elasticity represents the relative change in the demand. The elasticity is often a more useful measure of change. For example, a change in price of \$2 for a movie ticket would be large, but a change of \$2 would be insignificant in talking about the price of an automobile. For this reason the total change in itself is not too useful, and it is often better to find the relative change in demand. This introduces us to the economic concept of elasticity.

We first define the **relative rate of change** in the demand function. We say that

$$\text{Relative rate of change in demand} = \frac{D'(p)}{D(p)}$$

We can then write the ratio of the relative rate of change in demand to the relative rate of change in the price as

$$\frac{[\text{Relative rate of change in quantity}]}{[\text{Relative rate of change in price}]} = \frac{D'(p)/D(p)}{1/p}$$

$$= \frac{pD'(p)}{D(p)}$$

Students are often bothered by the above expression of $1/p$ representing the relative rate of change in the price. Just remember that the derivative of the price p (with respect to p) is 1. Hence the relative rate of change in the price (the derivative of the price divided by the price) is $1/p$. Note, too, that since the demand function $D(p)$ is always decreasing, the above ratio will always be negative. Since economists dislike working with negative numbers, the elasticity of demand is taken to be the above ratio multiplied by -1. This leads us to the definition of the elasticity of the demand function.

Definition of Elasticity of Demand

Let $q = D(p)$ be a demand function. The **elasticity** $E(p)$ of the demand function at a given price p is defined to be

$$E(p) = -\frac{pD'(p)}{D(p)}$$

The elasticity $E(p)$ will always be a nonnegative real number.

Example 8

Elasticity of Demand   Suppose the weekly demand for a certain high grade steel is

$$q = 250 - 25p$$

where p is the price per pound and q is the quantity of steel demanded in millions of pounds.

(a) How much steel will be sold per week at $3 a pound?

(b) Evaluate the derivative $dD(p)/dp$.

(c) Find the elasticity of the demand function.

(d) Evaluate the elasticity of the demand function when $p = \$3$.

Solution   For convenience we have graphed the demand function in Figure 26.

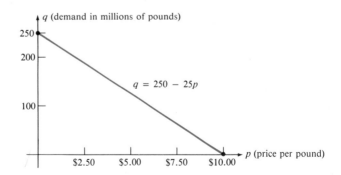

Figure 26
Demand for steel in millions of pounds

(a) *Demand at $3:*   When the price is $3 per pound, the weekly demand will be

$$D(3) = 250 - (25)(3)$$
$$= 175 \quad \text{million tons}$$

(b) *Derivative of the demand function:* Given the demand function

$$D(p) = 250 - 25p$$

we have

$$D'(p) = -25$$

(c) *Elasticity of the demand function:* The elasticity of the demand function is given by

$$E(p) = -\frac{pD'(p)}{D(p)}$$

$$= \frac{-p(-25)}{250 - 25p}$$

$$= \frac{p}{10 - p}$$

The graph of the elasticity is shown in Figure 27.

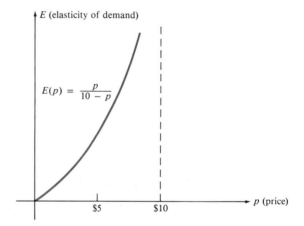

Figure 27
Elasticity of the demand function

(d) *Evaluate elasticity:* When the price of steel is $3 per pound, the elasticity $E(p)$ is

$$E(3) = \frac{3}{10 - 3}$$

$$= \frac{3}{7}$$

$$= 0.43 \qquad\qquad \square$$

Conclusion

When the price of steel is $3 per pound a small percentage increase in price will result in a percentage decrease in the demand of 0.43 times the percentage increase in the price. For example, if the price is increased by 5% (from $3 to $3.15 per pound), then the weekly demand for steel will fall by $0.43 \cdot (5\%) = 2.15\%$ (or from 175 million tons to 171.24 million tons).

Also, when the price of steel is $8 then the elasticity is

$$E(8) = \frac{8}{10 - 8}$$

$$= 4$$

This means that when steel is priced at $8 per pound, a 1% increase (or 8 cents) in the price of steel will result in a corresponding drop in the demand of $(4)(1\%) = 4\%$. Of course, it also means that a 1% drop in price would result in a 4% increase in the demand.

Elasticity and Inelasticity

The elasticity of a demand function is a valuable tool for studying the dynamics of consumer behavior. Are there certain prices at which consumers will resist price increases more than others? Why are many consumer goods often priced at $0.99 or $1.99? For a company to price its goods effectively, it is useful to study the concepts of elasticity and inelasticity.

Definition of Elasticity and Inelasticity

At a given price p_0 the demand function

$$q = D(p)$$

is called

- **inelastic** if $E(p_0) < 1$,
- of **unit elasticity** if $E(p_0) = 1$,
- **elastic** if $E(p_0) > 1$.

Example 9

Elastic and Inelastic   Suppose the daily demand for airline tickets from New York to Los Angeles is given by

$$q = 5000\sqrt{900 - p} \qquad (0 \le p \le 900)$$

where p is the number of dollars charged for a ticket. This demand function is shown in Figure 28.

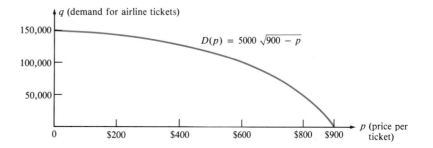

Figure 28
Demand for airline tickets

(a) How many tickets could be sold at $500 a ticket?

(b) What is the elasticity of the demand function?

(c) At what prices is the demand function elastic and at what prices is it inelastic?

Solution

(a) *Demand:* When the price of a ticket is $500, the daily demand is

$$D(500) = 5000\sqrt{900 - 500}$$
$$= 5000(20)$$
$$= 100{,}000 \quad \text{tickets}$$

(b) *Elasticity:* The elasticity of the demand function can be found to be

$$E(p) = -\frac{pD'(p)}{D(p)}$$

$$= \frac{p}{2(900 - p)} \qquad (0 \le p < 900)$$

The graph of this function is shown in Figure 29.

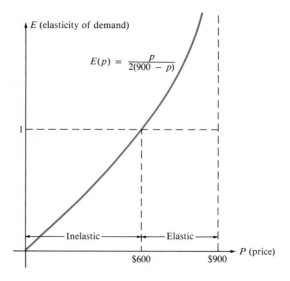

Figure 29
Elasticity of the demand function

(c) *Elastic and inelastic demand:* We begin by finding the prices for which the demand is elastic. That is, find the prices for which

$$E(p) = \frac{p}{2(900 - p)} > 1 \qquad (0 \le p < 900)$$

Subtracting 1 from each side of the inequality, we get

$$\frac{p}{2(900 - p)} - 1 > 0$$

Hence

$$\frac{p - 2(900 - p)}{2(900 - p)} > 0$$

or

$$\frac{3p - 1800}{2(900 - p)} > 0$$

Since the denominator is always positive for p in the interval $[0, 900)$, we seek those prices p in this region for which the numerator is also positive. Hence we write

$$3p - 1800 > 0$$

or

$$p > \$600$$

It is also a simple matter to determine those prices for which the demand function is inelastic ($E < 1$) and those prices for which it has unit elasticity ($E = 1$). We arrive at the following conclusion:

· Demand is inelastic for $0 \le p \le \$600$ ($E < 1$).
· Demand has unit elasticity for $p = \$600$ ($E = 1$).
· Demand is elastic for $\$600 < p < \$900$ ($E > 1$).   □

Interpretation

If the price of an airline ticket is under $600, the demand is inelastic, and a percentage increase in price is met with a percentage decrease in demand that is not as great as the percentage increase in price. Hence the airlines will take in more money for each increase in price. However, when the price of a ticket is greater than $600, the demand is elastic, and a percentage increase in price is met with a larger percentage decrease in demand. Hence an airline company will lose revenue by increasing the price higher than $600. Therefore, the ideal price is $600.

Relationship Between Revenue and Elasticity

The elasticity of demand also has a useful relationship to the marginal revenue $R'(p)$. Consider the previous example of selling airline tickets from New York to Los Angeles. If we write the revenue function as

$$R(p) = (\text{Price per ticket})(\text{Tickets sold})$$
$$= p \cdot D(p)$$

we can differentiate to get the marginal revenue

$$R'(p) = \frac{d}{dp}\big[D(p) \cdot p\big]$$

$$= D(p) \cdot 1 + D'(p) \cdot p$$

$$= D(p)\left[1 + \frac{pD'(p)}{D(p)}\right]$$

$$= D(p)[1 - E(p)]$$

Hence we have the interesting economic law that revenue increases ($R'(p) > 0$) with increasing prices when the demand for a product is inelastic ($E(p) < 1$) and revenue decreases ($R'(p) < 0$) with increasing prices when the demand for a product is elastic ($E(p) > 1$).

Problems

For Problems 1–15, find dy/dx.

1. $y = \ln (3x + 5)$ **2.** $y = \ln (x^2 + 4x)$

3. $y = \ln \sqrt{4 - x^2}$ **4.** $y = \ln (1/x)$

5. $y = \ln (x + 1)$ **6.** $y = \ln \left(\dfrac{x + 1}{x^2 + 1}\right)$

7. $y = \ln (e^x + e^{-x})$ **8.** $y = \ln (1 + e^x)$

9. $y = \ln (1/x^2)$ **10.** $y = \dfrac{\ln x}{x^3}$

11. $y = \dfrac{\ln x^2}{x^2}$ **12.** $y = \dfrac{\ln (x^2 + 1)}{x^2}$

13. $y = x \ln x - x$ (interesting derivative)

14. $y = \ln x + \ln 2x$ **15.** $y = e^x \ln (2x)$

Rewriting Logarithms in Terms of Natural Logarithms

For Problems 16–25, write the given logarithms in terms of the natural logarithm.

16. $\log_2 x$ **17.** $\log_5 x$
18. $\log_{20} x$ **19.** $\log 10^x$
20. $\log (x + 1)$ **21.** $\log (x^2 + 3x + 5)$
22. $\log (e^x + 1)$ **23.** $\log_2(e^x + 1)$
24. $\log_2(2^x)$ **25.** $\log_2(\log x + 1)$

Differentiation Logarithms with Any Base

For Problems 26–35, find the derivatives of the indicated function. *Hint:* First rewrite the logarithm in terms of the natural logarithm.

26. $\log (9x)$ **27.** $\log (x + 1)$
28. $\log (3x + 5)$ **29.** $\log (x^2 + 1)$
30. $\log e^x$ **31.** $\log (1/x)$
32. $\log x^2$ **33.** $\log (x^2 + 3x + 1)$
34. $\log_2(x + 1)$ **35.** $\log_2(1/x)$

Related Mathematical Problems

36. Implicit Differentiation Involving Logarithms Find dy/dx by implicit differentiation if

$$y + \ln xy = 1$$

37. Implicit Differentiation Involving Logarithms Find dy/dx by implicit differentiation if

$$\ln y = x$$

Does this agree with the derivative found by solving for y first and then finding dy/dx?

38. Slope of Tangent Line Find the slope of the tangent line to the graph

$$y = e^x \ln (x + 1)$$

at the point $(0, 0)$.

39. Maximum Value Find the maximum value of the function

$$f(x) = \dfrac{\ln x}{\sqrt{x}} \quad (x > 0)$$

40. Maximum Value Find the maximum value of the function

$$f(x) = \dfrac{\ln (x + 1)}{x} \quad (x > 0)$$

41. Logarithm Derivative on Entire Real Line Find the derivative $f'(x)$ where

$$f(x) = \ln |x| \quad (x \neq 0)$$

Sketch the graph of $f(x)$. *Hint:* Find the derivative for $x < 0$ and then find it for $x > 0$.

Elasticity of the Demand Function

For Problems 42–47, find the elasticity $E(p)$ of the demand functions. Find the prices where the demand functions are elastic and when they are inelastic.

42. $q = 100 - 2p$ **43.** $q = 10{,}000 - 50p$

44. $q = \dfrac{100}{p^2}$ **45.** $q = \dfrac{1000}{p^3}$

46. $q = 10{,}000\sqrt{100 - p}$ **47.** $q = 50{,}000\sqrt{1600 - 2p}$

48. Movie Theater Tickets A movie theater has a seating capacity of 1000 people. The number of tickets q that can be sold at p dollars is given by

$$q = D(p) = \dfrac{5000}{p} - 250 \quad (p > 0)$$

(a) How many tickets could be sold at \$3?
(b) Find $D'(p)$.

(c) Find $E(p)$.
(d) Is the demand elastic or inelastic when the price of a ticket is $5?
(e) What is the marginal revenue $R'(p)$ when the price is $5? From the value of the marginal revenue, would you increase or decrease the price per ticket to increase revenues?

49. **Balance of Trade**   To improve its balance of trade position with Korea, the United States has decided to lower the price for some heavy machinery equipment. The demand function is

$$q = \frac{5000}{p^2}$$

(a) Find $E(p)$.
(b) Find $R'(p)$.
(c) Will the United States succeed in raising its revenue by decreasing prices? What is the reason?

50. **Regions of Elasticity and Inelasticity**   Consider the demand function

$$q = \frac{A}{p^m}$$

where A and m are constants with m a positive integer.
(a) Find $E(p)$.
(b) Find those prices for which the demand function is elastic.
(c) Find the marginal revenue $R'(p)$.

51. **Mathematics in the Cruise Ship Industry**   The Stella Solaris cruise ship is sailing from Miami on an Amazon cruise. The number of cabins that can be filled is determined by the demand function

$$q = 20\sqrt{1600 - 0.2p} \qquad (0 \le p \le 8000)$$

(a) Find $E(p)$.
(b) Is the demand elastic or inelastic when the price of a cabin is $2000?
(c) Is the demand elastic or inelastic when the price of a cabin is $5000?
(d) When the price of a cabin is $2000, should the price be raised or lowered to increase the total revenue?
(e) When the price of a cabin is $5000, should the price be raised or lowered to increase the total revenue?

Relative Rates of Change

52. **Relative Rate of Change**   Suppose the value of a business investment is given by the function

$$I(t) = 100{,}000e^{0.5\sqrt{t}} \qquad (t > 0)$$

Determine the relative rate of increase in this investment after 10 years. *Hint:* It is easiest simply to take the natural logarithm of each side of the equation and then differen-

tiate. This will give the desired value

$$\frac{d}{dt}\ln I(t) = \frac{I'(t)}{I(t)}$$

53. **Constant Relative Rate of Change**   Show that if the value of an investment increases according to the function

$$y(t) = Pe^{rt}$$

then the relative rate of increase is a constant. What is this relative rate of increase?

54. **Changing Interest Rates**   Suppose the Federal Reserve Board forecasts that interest rates will increase over the next several months according to the function

$$I(t) = 0.05 + 0.05t + 0.10e^{-0.5t}$$

where t is time measured in months. What is the relative change in interest rates over the next six months?

55. **Bacteria Growth**   The size of a population of bacteria growing in a liquid medium is given by the function

$$p(t) = 100e^{0.04t}$$

where t is time measured in months. Draw a graph of the relative rate of increase $p'(t)/p(t)$ in the population over the next ten months.

Interesting Derivatives of Logarithms

56. **Rate of Change in Doubling Time**   We have seen before that if money is deposited in a bank that pays interest at an annual rate of r compounded continuously, then the time T it takes a deposit to double in value is

$$T = \frac{\ln 2}{\ln(1 + r)}$$

Find dT/dr.

57. **The Logarithmic Derivative**   Find the derivative of the function

$$y = \frac{x^3}{(x^2 + 1)(3x - 4)}$$

Hint: One of the virtues of the logarithm is its ability to turn products into sums and quotients into differences. Take the natural logarithm of each side of this equation and write

$$\ln y = \ln \frac{x^3}{(x^2 + 1)(3x - 4)}$$
$$= \ln(x^3) - \ln(x^2 + 1) - \ln(3x - 4)$$

Now differentiate implicitly with respect to x, and then solve for dy/dx.

Epilogue: History of Mathematical Economics

Only in the past 40 years has mathematics gained widespread acceptance among economic theorists. At one time many people thought that mathematics was used in economics only to confuse outsiders or to add some dignity to an otherwise vacuous subject.

Originally, only the values of graphs and symbols were acknowledged as being important tools to speak the language of economics. Today, the entire subject of economics has been embodied into an axiomatic framework. To gain a perspective on the importance of mathematics in modern economic theory, you should be aware that the work of most of the recent Nobel Prize winning economists has been formulated in the language of mathematics.

Some of the important early developments in mathematical economics, such as marginal revenue and cost, supply and demand curves, and discounting, which are the basis for much of economic development in this book, are listed below. All of the references that follow, with the exception of the first two, rely heavily on differential calculus.

- *1711 (Early Rumblings)*. First book dealing with the applications of mathematics to economics was written by an Italian engineer, Giovanni Ceva.

- *1760 (Early Rumblings)*. An Italian, Cesare Beccaria, used algebra to show the hazards of "profits and smuggling" in the book *Tentativo Analytico Sui Contrabbandi*.

- *1776 (The Wealth of Nations)*. In 1776 the Scottish economist Adam Smith wrote the first genuine analytical work in economics with the two-volume work, *The Wealth of Nations*. His work formed the foundations for the work of later mathematical economists such as David Ricardo.

- *1817 (Mathematical Rigor Raised to a New Level)*. The British economist David Ricardo published the book *Principles*, in which he established a theory of value. Ricardo raised the mathematical level of economics from intuition and common sense to a higher mathematical level.

- *1838 (Supply and Demand Introduced)*. Antoine Augustin Cournot wrote the first major economic treatise, *Researches into the Mathematical Principles of the Theory of Wealth*. Many people believe that this book initiated the study of mathematical economics. Cournot (1801–1877) was a French mathematician who first defined the concepts of supply and demand as mathematical functions. Unfortunately, the book was too advanced and was not appreciated by economists of the time.

- *1871 (Marginality Introduced)*. William Stanley Jevons published the treatise *Theory of Political Economy*. In this book, Jevons introduced the idea of marginality, which is familiar to every modern-day student of economic theory. The book was a major influence on researchers of the time. Jevons once stated, "If economics is to be a science at all, it must be a *mathematical science*."

- *1874 (Utility Theory and Equilibrium Theory Introduced)*. *Elements of Pure Economics* was published by Leon Walras and introduced the concept of marginal utility theory and general equilibrium theory. Walras, Austrian Carl Menger, and Englishman William Stanley Jevons are often given credit for bringing the "marginal revolution" into economics.

- *1884 (Theory of Interest)*. The Austrian economist Eugan Bohm-Bawerk introduced many of the ideas that ultimately led to the theory of interest.

- *1907 (Present and Future Value of Money Introduced)*. *The Rate of Interest* was published by Irving Fisher. In this book, Fisher introduces the ideas of present value, future value, and discounting.

- *1909 (Theory of Equilibrium)*. The most famous untranslated book in the history of economics, *Manuel d'économie Politique*, was published by the Swiss economist Vilfredo Pareto. It covers much of the theory of equilibrium theory in economics. The book uses calculus extensively.

- *1920 (Price Elasticity Introduced)*. One of the epic contributions to mathematical economics was the book *Principles of Economics* by English economist Alfred Marshall. Marshall made major contributions to utility theory and demand curves. Marshall was the first economist to define mathematically the concept of price elasticity.

- *1928 (Theory of Savings Introduced by Ramsey).* A research paper, "A Mathematical Theory of Savings," was published by F. P. Ramsey. In this paper a mathematical theory based on calculus is developed to determine how much capital to place in savings.

- *1928 (Cobb-Douglas Production Function Introduced).* The now famous Cobb-Douglas production function was introduced by Charles Cobb and Paul Douglas in a research paper, "A Theory of Production," in the *American Economic Review*. This function shows the relationship between the amount of a product produced as a function of the amounts of labor and capital used.

- *1936 (Keynesian Economics).* John Maynard Keynes has dominated twentieth century econo-

mics more than any other economist. Although it is hard to pinpoint any single accomplishment, his 1936 book *The General Theory of Employment, Interest, and Money* is a well-known classic. This is the work on which all the Keynesian models are based. Although this specific work contains little mathematics, much of Keynes's work is highly mathematical.

We have presented here a listing of some of the major contributions to the history of mathematical economics. However, the list is in no way complete. Other important names come to mind in such a review. Names like Edgeworth, Wicksell, Wicksteed, Pigou, and Carey are important.

In more recent years, areas of mathematics other than calculus have played an important role in economics. We mention the introduction of game theory into economics by Oskar Morgenstern and John von Neumann in 1931.

Key Ideas for Review

decay curves, 658
discounting, 683
exponential functions
  decay curves, 658
  derivative of, 679

growth curves, 658
geometric sequence, 649
logarithmic functions
  change of base, 693
  common, 665

derivative of, 688
natural, 665
properties, 667

Review Exercises

For Problems 1–5, solve the equations for x.

1. $3^{-x} = 1/9$
2. $8^x = 2^{x^2}$
3. $7^x = 49^{(x+1)}$
4. $2^{x+1} = 1/16$
5. $7^x = 49$

For Problems 6–11, solve for x.

6. $\log x^2 = 3 \log 2 - 4 \log 5$
7. $\log_2 x = 2$
8. $\ln x^2 + \ln x = \ln 8$
9. $\ln x - \ln x^2 = 1$
10. $\ln (x + 1) = 1$
11. $\ln (x + 1) = 0$

For Problems 12–26, find the derivative of the given function.

12. $f(x) = xe^x$
13. $f(x) = e^{(2x+1)}$
14. $f(x) = e^x \ln x$
15. $f(x) = e^x \log x$
16. $f(x) = [\log x]^2$
17. $f(x) = \ln \sqrt{x}$
18. $f(x) = x \ln x$
19. $f(x) = \ln [e^x]$
20. $f(x) = e^{\ln x}$
21. $f(x) = \dfrac{e^x + e^{-x}}{2}$
22. $f(x) = e^{x^2} \ln x^2$
23. $f(x) = \log_2 x$
24. $f(x) = (\ln x^2)^3$
25. $f(x) = \ln \left(\dfrac{x^2 + 1}{x^2 - 1} \right)$
26. $f(x) = \ln [(x^2 + 1)(x^2 - 1)]$

27. **Normal Distribution in Statistics** The exponential function

$$y = e^{-x^2}$$

plays an important role in probability and statistics. Find y' and y''.

Economics and Financial Problems

28. Tripling Time   An amount P_0 is deposited in a bank that pays interest at an annual interest rate of r, compounded continuously. How long will it take this deposit to triple in value?

29. Marginal Revenue   The revenue R in dollars that a manufacturer receives when q units of a product are sold is given by

$$R(q) = 0.5qe^{-0.01q}$$

What is the marginal revenue when $q = 100$?

30. Discounted Investments   A speculator in precious stones has purchased a ruby that is increasing in value according to the function

$$V(t) = V_0 e^{\sqrt{t}}$$

where t is time measured in years. If the discount rate is 8% per year, when should the speculator sell the ruby to maximize profits?

31. Emerald Speculator   The same speculator as in Problem 30 has now invested in an emerald that will increase in value according to the function

$$V(t) = V_0 e^{0.2\sqrt{t}}$$

When should the emerald be sold to maximize profits?

32. Which Is Increasing Faster?   The price for a new car is $12,000. This price is increasing at the rate of $800 per year. The cost of silver is $6.00 per ounce and is increasing at the rate of 25 cents per year. What is the relative change in each item per year? Which relative change is larger?

33. When To Sell   An investment is valued approximately by the function

$$f(t) = 50{,}000 e^{0.2\sqrt{t}}$$

What will be the relative change in the value of this investment in ten years?

34. Art Investment   A painting has its value given approximately by

$$v(t) = 5000 e^{\sqrt[3]{t/2}}$$

What is the relative rate of change in the investment after five years? After 15 years? What is the percentage rate of change after this length of time?

35. Changing Annual Compounding to Continuous Compounding   If interest is paid by a bank at an annual interest rate of i, compounded annually, then an initial amount P will have a future value of

$$F = P(1 + i)^n$$

after n years. Rewrite the above equation in the form

$$F = Pe^{rn}$$

to determine the effective rate r that this bank pays if it compounded interest continuously.

36. Elasticity of the Demand Function   Given the demand function

$$q = 5000 - 5p^2$$

(a) Find $E(p)$.
(b) Determine the values of p for which the demand is inelastic and those for which it is elastic.
(c) Find $R(p)$.
(d) Find $R'(p)$.

37. Elasticity of the Demand Function   Suppose that the demand q is 100 when the price p is 5 and that the demand is 90 when the price is 10. Assuming the demand function is a straight line, find the demand function

$$q = a + bp.$$

Use this demand function to find:
(a) $E(p)$;
(b) the regions where the demand function is inelastic and where the demand function is elastic;
(c) $R(p)$;
(d) $R'(p)$.

Biology and Scientific Problems

38. Finding the Growth Equation   Every day the size of a bacteria population grows by 10%. If the initial size of the population is 100, write the size of this population in the form

$$P(t) = P_0 e^{kt}$$

39. Finding the Decay Equation   Every day the size of a bacteria population declines by 15%. If the initial size of the population is 65, write the size of this population in the form

$$P(t) = P_0 e^{-kt}$$

40. Finding the Proper Dosage   A virus has been exposed to X-rays. The number of surviving virus cells N depends on the number of roentgens r applied and is given by

$$N(r) = N_0 e^{-0.2r}$$

How many roentgens must be applied for 95% of the virus cells to be destroyed?

41. Radioactive Decay   The radioactive carbon isotope ^{14}C decays according to the exponential decay

$$C(t) = C_0 e^{-t/5500}$$

where C_0 is the initial amount and t is time measured in years. How long will it take for the substance to decrease to 1/10 of the original amount?

42. Richter Scale The Richter scale is given by

$$R = 0.67 \log E - 7.9$$

where E is the energy in ergs released by an earthquake.

If the Richter scale measures 4.5, how many ergs of energy are released?

43. Richter Scale Two earthquakes are reported. One earthquake is in Colombia and measures 4.5 on the Richter scale; the other is in Indonesia and measures 5.3. How many times more energy is released by the Indonesia earthquake?

Chapter Test

1. Find y' if

$$y = x^2 \ln (x^2 + 1)$$

2. Use properties of logarithms to find y' if

$$y = \log \frac{\sqrt{x^2 + 2x + 1}}{\sqrt{x}}$$

3. If the population of a bacteria is increasing exponentially at the rate of 5% every 12 hours, what is the population at any time t? Write the growth function in the form

$$P(t) = P_0 e^{kt}$$

4. Suppose the population of bacteria is increasing accord-

ing to the function

$$P(t) = P_0 e^{0.05t}$$

where t is time measured in months. How long will it take the bacteria population to quadruple in size?

5. Find the maximum value of the function

$$f(x) = \frac{\ln x^2}{x} \qquad (x \neq 0)$$

6. Suppose you deposit $500 in a bank that compounds interest annually. Suppose that in ten years your account will be worth $1500. What is the annual rate of interest paid by this bank?

11

Integration

In this chapter we introduce the second of the two major processes of the calculus, integration. We begin in Section 11.1 by introducing the concept of the antiderivative, which may be thought of as the reverse or inverse of the derivative. In Section 11.2 we then define the definite integral of a continuous function and show how it can be used to find areas of regions in the plane. In Section 11.3 we link Sections 11.1 and 11.2 together by means of the Fundamental Theorem of Calculus, which shows how an antiderivative of a function can be used to evaluate the definite integral of a function. Finally, in Section 11.4 we introduce the method of substitution, which is an aid in finding antiderivatives, and hence the definite integral of a function.

11.1 The Antiderivative (Indefinite Integral)

PURPOSE

We begin the study of integral calculus by introducing the concept of the antiderivative. In particular, we introduce

- the meaning of an antiderivative and
- the way to find antiderivatives.

We finish by showing how the antiderivative can be used in mathematical economics.

Introduction

You can open a door, and then you can close it. You can square a positive number, and then you can take the positive square root of the answer. These pairs of operations have the property that the second operation undoes the first. The door is returned to its original position, and the number that we squared was refound by taking the positive square root.

Operations like these are **inverse operations**. In mathematics, operations like the operations of squaring and taking the positive square root are inverses, as are addition and subtraction. It is now time to study the inverse operation of the derivative, the antiderivative.

Doing the "Inverse" of Differentiation (the Antiderivative)

For demonstration purposes, consider the function defined by $f(x) = x^2$. So far, we have learned that its derivative is $f'(x) = 2x$. However, we can perform another operation on $f(x) = x^2$ called **antidifferentiation**. Instead of "going forward" and finding the derivative

$$x^2 \longrightarrow 2x \qquad \text{(differentiation)}$$

we can "go backward" and find the function whose derivative is x^2.

$$\frac{1}{3}x^3 \longleftarrow x^2 \qquad \text{(antidifferentiation)}$$

That is, we find the antiderivative of x^2.

We are now ready for the formal definition of the antiderivative.

Definition of an Antiderivative of f

A function F is an **antiderivative** of f on the interval (a, b) if

$$\frac{d}{dx}F(x) = f(x)$$

for all x in (a, b).

The word "an" in this definition indicates that there is more than one antiderivative of a function. We now see from an example why we always expect a function to have more than one antiderivative.

Example 1

Finding Antiderivatives   Find the antiderivative(s) of

$$f(x) = x^3$$

Solution   Using the power rule for the derivative, we know that

$$\frac{d}{dx}\frac{1}{4}x^4 = x^3$$

Hence an antiderivative of $f(x) = x^3$ is

$$F(x) = \frac{1}{4}x^4$$

But the two functions

$$F_1(x) = \frac{1}{4}x^4 + 2 \qquad \text{and} \qquad F_2(x) = \frac{1}{4}x^4 - 100$$

are also antiderivatives. In fact, any function of the form

$$F(x) = \frac{1}{4}x^4 + C$$

where C is any constant is an antiderivative of f. (See Figure 1.)

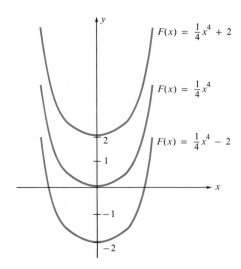

Figure 1
A few antiderivatives of $f(x) = x^3$

In other words, if $F(x)$ is an antiderivative of $f(x)$, then so is each of the functions

$$F(x) + C$$

where C is any real number. □

Notation for Antiderivatives

In differential calculus we denote the derivative of $f(x)$ as $f'(x)$ or $df(x)/dx$. In integral calculus the antiderivative or **indefinite integral** of $f(x)$ is denoted by

$$\int f(x)\,dx = F(x) + C$$

The elongated S, or $\int$, is called the **integral sign**, and $f(x)$ is called the **integrand** of the indefinite integral. The symbol dx indicates the variable in which the anti-derivative is taken and is called the **variable of integration**. Later in this chapter, when we study the integration technique of the method of substitution, we will learn why it is convenient to use this differential notation.

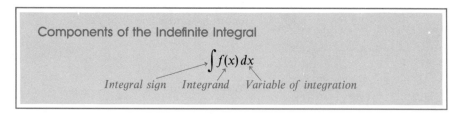

Antiderivatives of Powers of x

In differential calculus we learned the power rule

$$\frac{d}{dx}\frac{x^{a+1}}{a+1} = x^a \qquad \text{(power rule for differentiation)}$$

This derivative formula leads to the following antiderivative formula.

> **Power Rule for Antiderivatives**
>
> Let a be any real number such that $a \neq -1$. The **power rule for anti-derivatives** says that
>
> $$\int x^a \, dx = \frac{x^{a+1}}{a+1} + C$$

Verification of the Power Rule for Antiderivatives

It is easy to see that this antiderivative formula is correct. All we need to do is observe that the derivative of the antiderivative is the integrand. That is,

$$\frac{d}{dx}\left[\frac{x^{a+1}}{a+1} + C\right] = x^a$$

Example 2

Power Rule for Antiderivatives   Find the antiderivatives of

$$f(x) = \sqrt{x}$$

Solution   Using the power rule for antiderivatives with $a = 1/2$, we have

$$F(x) = \int x^{1/2} \, dx = \frac{x^{3/2}}{3/2} + C$$

$$= \frac{2}{3} x^{3/2} + C$$

Check:

$$\frac{d}{dx}\left[\frac{2}{3} x^{3/2} + C\right] = x^{1/2} \qquad \square$$

Just as there are many general rules for derivatives, there are general rules for antiderivatives. These rules will allow us to find the antiderivatives of more complex expressions.

> **Rules for Antiderivatives**
>
> Let f and g both have antiderivatives, and let c be a constant. Then
>
> Constant rule: $\int cf(x) \, dx = c \int f(x) \, dx$
>
> Sum rule: $\int [f(x) + g(x)] \, dx = \int f(x) \, dx + \int g(x) \, dx$
>
> Difference rule: $\int [f(x) - g(x)] \, dx = \int f(x) \, dx - \int g(x) \, dx$

Verification of the Rules for Antiderivatives

To convince yourself of the correctness of the above rules, try restating the meaning of each rule in words. For example, to convince yourself that the sum rule is reasonable, ask yourself whether it is true that a function whose derivative is $f(x) + g(x)$ is equal to the sum of two functions, one whose derivative is $f(x)$ and another whose derivative is $g(x)$. From your knowledge of the properties of the derivative, you know that this is correct. You should also restate in words the other two antiderivatives.

Example 3

Using Rules for Antiderivatives   Find

$$\int (3x^2 + 2x + 1)\,dx$$

Solution   Using the sum followed by the constant rule, we can write

$$\int (3x^2 + 2x + 1)\,dx = \int 3x^2\,dx + \int 2x\,dx + \int dx$$

$$= 3\int x^2\,dx + 2\int x\,dx + \int dx$$

$$= 3\left[\frac{x^3}{3} + C_1\right] + 2\left[\frac{x^2}{2} + C_2\right] + [x + C_3]$$

$$= x^3 + x^2 + x + (3C_1 + 2C_2 + C_3)$$

$$= x^3 + x^2 + x + C$$

Since the three constants C_1, C_2, and C_3 represent arbitrary real numbers, the sum $3C_1 + 2C_2 + C_3$ is also an arbitrary real number. Hence we denote it simply by C and usually call it the **constant of integration**.

Check:

$$\frac{d}{dx}[x^3 + x^2 + x + C] = 3x^2 + 2x + 1 \qquad \square$$

How do we pick out a specific one of these antiderivatives? Is there one antiderivative that is the "desirable" one? For example, the entire "family" of antiderivatives

$$F(x) = \frac{1}{3}x^3 + C$$

of the function $f(x) = x^2$ is shown in Figure 2.

Many problems dictate that we should find the specific antiderivative that satisfies an additional side or initial condition of the form $F(x_0) = y_0$. Figure 2 illustrates that in the above example we seek the specific antiderivative that passes through the point $(0, 1)$. In other words, we seek the antiderivative that satisfies the side condition $F(0) = 1$. Since this condition says that

$$F(0) = \frac{1}{3}0^3 + C = 1$$

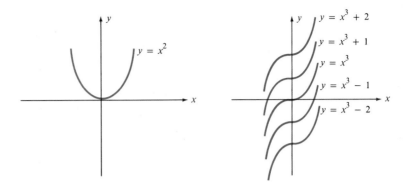

Figure 2
A function and a few of its derivatives

it implies that the constant of integration C is 1 and hence the desired antiderivative is

$$F(x) = \frac{1}{3}x^3 + 1$$

Example 4

Finding a Particular Antiderivative   Find the specific function $F(x)$ whose derivative is

$$f(x) = 3x^2 + 4x + 5$$

that also satisfies the condition $F(1) = 2$.

Solution   Using the rules for antiderivatives, we obtain

$$F(x) = x^3 + 2x^2 + 5x + C$$

Substituting this expression into the condition $F(1) = 2$, we get the equation

$$F(1) = 1^3 + 2(1^2) + 5(1) + C = 2$$

$$1 + 2 + 5 + c = 2$$
$$c = 2 - 8$$

Solving for C, we find

$$C = -6 \qquad = -6$$

Hence the desired function is

$$F(x) = x^3 + 2x^2 + 5x - 6$$

Check:

$$\frac{d}{dx}(x^3 + 2x^2 + 5x - 6) = 3x^2 + 4x + 5$$

$$F(1) = 1^3 + 2(1^2) + 5(1) - 6 = 2 \qquad \square$$

General Power Rule for Antiderivatives

In Section 8.7, when we studied the chain rule, we learned the general power rule for derivatives

$$\frac{d}{dx}[g(x)]^{a+1} = (a+1)[g(x)]^a g'(x)$$

This equation also says that the antiderivative of the expression on the right-hand side is equal to $[g(x)]^{a+1}$. This fact leads to the following rule (stated slightly differently).

Rule for Finding the Antiderivative of General Powers

Let g be a differentiable function and a be any real number such that $a \neq -1$. Then

$$\int [g(x)]^a g'(x)\,dx = \frac{[g(x)]^{a+1}}{a+1} + C$$

Verification of the General Power Rule

Antiderivative formulas are generally easy to verify. To verify such a rule, all we have to do is differentiate the antiderivative to see that we get the integrand of the indefinite integral. In the above rule we need to verify

$$\frac{d}{dx}\left[\frac{1}{a+1}[g(x)]^{a+1} + C \right] = [g(x)]^a g'(x)$$

By using the chain rule and the properties of the derivative we see that this statement is true.

Example 5　Indefinite Integral　Find the indefinite integral

$$\int (x^2+1)^3 2x\,dx$$

Solution　For this integral we interpret the integrand as $[g(x)]^a g'(x)$ where

$$g(x) = x^2 + 1$$
$$g'(x) = 2x$$
$$a = 3$$

Using the generalized power rule, we get

$$\int (x^2+1)^3 2x\,dx = \int g(x)^3 g'(x)\,dx$$
$$= \frac{1}{4}[g(x)^4] + C$$
$$= \frac{1}{4}(x^2+1)^4 + C$$

Check:

$$\frac{d}{dx}\left[\frac{1}{4}(x^2+1)^4 + C \right] = (x^2+1)^3 2x \qquad \square$$

A New Look at Marginal Analysis

In the theory of the firm we have seen that the revenue, cost, and profit functions (called **total functions** by economists) have derivatives which are called the marginal revenue, marginal cost, and marginal profit functions, respectively. We can now "go backward" and find the antiderivatives of the marginal functions to get the total functions. Since it is often easy for economists to empirically find the marginal functions from economic data, it is possible to find the total functions by finding antiderivatives. For instance, the marginal cost $C'(q)$ represents approximately the added cost of producing one more item of a product when the production level is q units. A company may be able to determine this information much more readily than finding the total cost $C(q)$ of producing q items. Often the total cost $C(q)$ is difficult to determine, since it depends on the fixed cost (the cost when no items are produced), whereas the marginal cost does not depend on the fixed cost.

Example 6

Finding Total Cost from Marginal Cost Suppose the marginal cost of producing q units of a given product is

$$C'(q) = 3q^2 - 2q$$

with a fixed cost of $10,000. Find the total cost function $C(q)$. What is the cost of producing 25 items?

Solution The problem is to find $C(q)$, given

$$C'(q) = 3q^2 - 2q$$
$$C(0) = 10,000$$

We begin by finding the indefinite integral or antiderivative of $C'(q)$. This is given by

$$C(q) = \int C'(q)\, dq$$
$$= \int (3q^2 - 2q)\, dq$$
$$= q^3 - q^2 + K$$

We use K instead of C to denote the constant of integration so as not to confuse it with the cost $C(q)$. To find K, we evaluate $C(q)$ at the point $q = 0$, knowing that at this point the cost function satisfies the condition $C(0) = 10,000$. This gives

$$C(0) = (0)^3 - (0)^2 + K = 10,000$$

Solving for K gives

$$K = 10,000$$

Hence the total cost is

$$C(q) = q^3 - q^2 + 10,000$$

The cost to the firm to produce 25 items will be

$$C(25) = (25)^3 - (25)^2 + 10,000$$
$$= \$25,000$$

Example 7

Finding Total Revenue from Marginal Revenue   A wholesaler charges its retailers $250 for a television set. To encourage retailers to buy in volume, the wholesaler deducts $2 from the price of each additional set sold up to 50 sets. In other words, the price of the second set is $248, that of the third set is $246, and so on, until the 50th set and thereafter, each of which costs $150. For this wholesaling scheme:

(a) What is the marginal revenue?

(b) What is the total revenue?

(c) What is the total revenue when 30 sets are sold?

Solution

(a) *Marginal revenue.* Since the marginal revenue $R'(q)$ is approximately the amount of money taken in on the sale of the next television set when q sets have been sold, we can write $R'(q)$ mathematically as

$$R'(q) = 250 - 2q \qquad (0 \le q \le 50)$$

(b) *Total revenue.* The total revenue R is the antiderivative of the marginal revenue function. Hence

$$R(q) = \int R'(q)\,dq$$

$$= \int (250 - 2q)\,dq$$

$$= 250q - q^2 + K$$

To find the constant K, we use the fact that the revenue is zero when the number of sets sold is zero. That is,

$$R(0) = 0$$

Using the condition that $R(0) = 0$, we get

$$R(0) = 250(0) - 0^2 + K = 0$$

Finally, solving for K, we find

$$K = 0$$

and substituting this value into $R(q)$, we have the total revenue function

$$R(q) = 250q - q^2$$

(c) *Finding $R(30)$.* When 30 television sets are sold, the revenue will be

$$R(30) = (250)(30) - (30)^2$$
$$= \$6600 \qquad\qquad \square$$

Comparison Between the Calculus and the Exact Solution

You may have observed that in the theory of the firm the number of items q produced or sold is actually an integer $(0, 1, 2, \ldots)$, whereas we have been treating it as a continuous variable. When q is large, say 100 or larger, this approximation is quite accurate. If q were smaller, say in the range of 5–10, the calculus approach of writing the revenue, cost, and profit as continuous functions of a real variable would be subject to significant errors. Using calculus, however, makes the analysis much easier for the normal situation in which large numbers of units are involved. For that reason this model is appropriate and closely reflects the actual situation being modeled.

The revenue problem that we have just solved is more or less on the boundary region of when we can and when we cannot use the calculus effectively. It is useful to compare the calculus solution with the exact solution. For example, if the previous wholesaler were to sell three television sets, the revenue function $R(q)$ found by computing the antiderivative of $R'(q)$ would say that the revenue (amount taken in) is

$$R(3) = 250(3) - (3)^2$$
$$= \$741$$

On the other hand, without using calculus the company accountant could simply add up the receipts from the sale of the three sets, getting

$$\text{Revenue from the sale of three sets} = \$250 + \$248 + \$246$$
$$= \$744$$

In other words, there is an error of $3 (or 0.4% percentage error) in the calculus solution.

Table 1 shows the difference between the exact revenue from the sale of q television sets and the value of the revenue function $R(q)$ found from the marginal revenue function.

TABLE 1
Error in Treating the Number Sold q as a Continuous Variable

Number of Sets Sold, q	Exact Revenue from the Sale of q Sets	Value of $R(q)$	Absolute and Percentage Error
0	0	0	0
1	$250	$249	$1 (0.4%)
2	$498	$496	$2 (0.4%)
3	$744	$741	$3 (0.4%)
4	$988	$984	$4 (0.4%)
5	$1,230	$1,225	$5 (0.4%)
10	$2,410	$2,400	$10 (0.4%)
20	$4,620	$4,600	$20 (0.4%)
30	$6,630	$6,600	$30 (0.5%)
40	$8,440	$8,400	$40 (0.5%)
50	$10,050	$10,000	$50 (0.5%)

Indefinite Integral of the Exponential Function

In the previous chapter we learned the derivatives

$$\frac{d}{dx} e^{kx} = k e^{kx}$$

$$\frac{d}{dx} \ln x = \frac{1}{x} \qquad (x > 0)$$

It is possible to restate these two equations in terms of antiderivative equations by saying that the antiderivatives of the expressions on the right-hand side are equal to the expressions being differentiated on the left-hand side. These antiderivative forms of the above equations are stated now.

Antiderivatives of Exponential and Logarithm Functions

Exponential: $\int e^{kx}\, dx = \dfrac{1}{k} e^{kx} + C$ $(k \neq 0)$

Special Exponential: $\int e^{x}\, dx = e^{x} + C$

Logarithm: $\int \dfrac{dx}{x} = \ln |x| + C$ $(x \neq 0)$

Special Logarithm: $\int \dfrac{dx}{x} = \ln x + C$ $(x > 0)$

Example 8

Antiderivatives of Exponentials   Find the antiderivative of

$$f(x) = e^{0.08x}$$

Solution   Using the rule for antiderivatives of exponential function with $k = 0.08$, we have

$$F(x) = \int e^{0.08x}\, dx$$

$$= \frac{1}{0.08} e^{0.08x} + C$$

Check:

$$\frac{d}{dx}\left[\frac{1}{0.08} e^{0.08x} + C \right] = e^{0.08x} \qquad \square$$

Example 9

Antiderivatives of Exponentials   Find the antiderivative of

$$f(t) = e^{-t}$$

Solution   Using the rule for the antiderivative of exponentials with $k = -1$, we have

$$F(t) = \int e^{-t}\, dt$$

$$= -e^{-t} + C$$

Check:

$$\frac{d}{dt}[-e^{-t} + C] = e^{-t} \qquad \square$$

More Uses for Antiderivatives (Finding Totals from Rates)

Indefinite integrals have many applications. In many areas of science we are often able to measure by an experiment the rate of change at which something occurs. For instance, we can often determine the rate of change in the size of a population of bacteria. The problem then is to find the size of the population after a given period of time.

A typical experience that is familiar to most people is driving a car. Suppose you drive your car for a given period of time at varying rates of speed but your odometer is broken, so you cannot determine how far you have traveled. However, you have kept a record of your velocity $v(t)$ during the trip. Knowing the velocity, how can you determine the distance you have traveled? The answer to this question is obtained by finding the antiderivative of the velocity function. If $v(t)$ represents the velocity (miles per hour) at time t (hours), then the antiderivative

$$s(t) = \int v(t)\, dt$$

represents the distance traveled. We illustrate this idea with an example.

Example 10

Finding Distance from Velocity   You are driving along an interstate and your velocity is initially 40 miles per hour. Over the next hour you gradually increase your velocity $v(t)$ to 55 miles per hour according to the function

$$v(t) = 40 + 15t^2 \qquad (0 \le t \le 1)$$

How far did you travel during this hour?

Solution   By finding the antiderivative of a velocity function $v(t)$ we can find the distance $s(t)$ that an object has traveled. In the case of the car we have

$$s(t) = \int v(t)\, dt$$

$$= \int (40 + 15t^2)\, dt$$

$$= 40t + 5t^3 + C$$

To find the constant C, we simply use the fact that $s(0) = 0$. This gives

$$s(0) = 40(0) + 5(0)^3 + C = 0$$

or

$$C = 0$$

Hence the position of your car at any time t (between 0 and 1) is given by

$$s(t) = 40t + 5t^3$$

After one hour you would have traveled

$$s(1) = 40(1) + 5(1)^2$$

$$= 45 \quad \text{miles}$$

Table 2 shows the velocity and distance traveled during this one-hour period.

TABLE 2
Relationship Between
Velocity and Distance

Time, t (hours)	Velocity, $v(t) = 40 + 15t^2$ (miles/hr)	Distance Traveled, $s(t) = 40t + 5t^3$ (miles)
0 (0 min)	40.0	0
0.2 (12 min)	40.6	8.04
0.4 (24 min)	42.4	16.32
0.6 (36 min)	45.4	25.08
0.8 (48 min)	49.6	34.56
1.0 (60 min)	55.0	45.00

Problems

For Problems 1–16, find the antiderivatives for each derivative.

1. $\dfrac{dy}{dx} = x + 3$

2. $\dfrac{ds}{dt} = t^2 - 2t + 3$

3. $\dfrac{dy}{dx} = \dfrac{1}{2}x^3 - \dfrac{2}{3}x^2 + \dfrac{1}{5}$

4. $\dfrac{dy}{dt} = t^3 + \dfrac{2}{t}$

5. $\dfrac{dy}{dx} = 4x^3 - 3x^{-2} + x - 3$

6. $\dfrac{dx}{dt} = t^{5/2} + 3t^{3/2} - t^{1/2} + 3t^{-1/2}$

7. $\dfrac{dx}{dt} = e^{2t} + 3e^{-3t}$

8. $\dfrac{dx}{dt} = \dfrac{2}{3}t + \dfrac{3}{t^3} + e^{4t} - \dfrac{6}{t}$

9. $\dfrac{dy}{dx} = 7\sqrt{x} + \dfrac{2}{e^{2x}}$

10. $\dfrac{dy}{dx} = e^x + \dfrac{5}{x^2} + x^3$

11. $\dfrac{dy}{dt} = e^{-3t}$

12. $\dfrac{dy}{dx} = e^x + e^{-x}$

13. $\dfrac{dy}{dt} = e^{-0.5t}$

14. $\dfrac{dy}{dx} = (x^2 + 1)^3(2x)$

15. $\dfrac{dy}{dx} = \sqrt{x^3 + 2x + 3}\,(3x^2 + 2)$

16. $\dfrac{dy}{dx} = (5x^3 - 1)^{50}\,(15x^2)$

For Problems 17–30, find the general antiderivative.

17. $\int (2x^2 + 5x^3 + x^4)\,dx$

18. $\int \left[\dfrac{y-1}{y^2} + y^2\right] dy$ (*Hint:* First perform division.)

19. $\int (2t + 3)^2\,dt$ (*Hint:* Expand the integrand.)

20. $\int (2x + 1)(x - 3)\,dx$ (*Hint:* First perform multiplication.)

21. $\int \dfrac{6x^5}{(x^6 - 3)^3}\,dx$

22. $\int 3x^2(x^3 + 1)^{1/2}\,dx$

23. $\int e^{5x}\,dx$

24. $\int (e^{2x} + e^{-2x})\,dx$

25. $\int e^{-0.04t}\,dt$

26. $\int \left(x^2 + \dfrac{1}{x^2}\right)dx$

27. $\int 3\sqrt{3x + 2}\,dx$

28. $\int 4(1 + 4x)\sqrt{1 + 2x + 4x^2}\,dx$

29. $\int (x^2 + 1)^3\,(2x)\,dx$

30. $\int (x^3 + 2x + 10)^5\,(3x^2 + 2)\,dx$

For Problems 31–50, find the particular antiderivative of the given derivative determined by the value of the function.

31. $f'(x) = 0,\ f(0) = 1$

32. $f'(t) = 1,\ f(0) = 0$

33. $f'(s) = -1,\ f(0) = 1$

34. $f'(t) = t + 3,\ f(2) = 9$

35. $f'(x) = 3x + 4,\ f(6) = 70$

36. $f'(x) = \dfrac{x^3}{2} + 2x,\ f(0) = 100$

37. $f'(u) = (u - 2)^2,\ f(4) = 2$

38. $f'(x) = x^3 - \dfrac{2}{3}x^2 - x - 3,\ f(6) = 200$

39. $f'(u) = 2u + u^{-2},\ f(1) = -1$

40. $f'(t) = \dfrac{(t + 2)(t - 5)}{t^3},\ f(1) = 4$

41. $f'(x) = x^{-2/3}(x + 16)$, $f(8) = 11.4$

42. $f'(u) = \dfrac{1 - u}{u^2}$, $f(e) = e$ **43.** $f'(t) = t + \sqrt{t}$, $f(4) = 2$

44. $f'(x) = e^{2x}$, $f(0) = 0$ **45.** $f'(t) = e^{-0.01t}$, $f(0) = 1$
46. $f'(t) = e^t + e^{-t}$, $f(0) = 1$
47. $f'(x) = 2x\sqrt{x^2 - 9}$, $f(5) = 5$

48. $f'(t) = \dfrac{1}{t}$ $(t > 0)$, $f(1) = 2$

49. $f'(x) = (x^2 + 1)^3(2x)$, $f(0) = 0$
50. $f'(x) = (x^3 + 2x + 5)^2(3x^2 + 2)$, $f(0) = -1$

Marginal Analysis

For Problems 51–59, find the total cost, revenue, and profit given the marginal cost, revenue, and profit.

51. $C'(q) = 40q - 0.01q^2 + 100$ $C(0) = 100$
52. $C'(q) = 3 + 5q - 0.05q^2$ $C(0) = 250$
53. $C'(q) = 2q - 0.02q^2$ $C(0) = 500$
54. $C'(q) = 0.5q^3 - 2q^2 + 6q + 3$ $C(0) = 1000$
55. $C'(q) = 7.5e^{0.15q}$ $C(0) = 80$
56. $P'(q) = 5 - 0.2q$ $P(0) = 0$
57. $P'(q) = 10 - 0.5q$ $P(0) = 0$
58. $R'(q) = 15 - 0.05q - 0.05q^2$ $R(0) = 0$
59. $R'(q) = 10 - 0.08q - 0.10q^2$ $R(0) = 0$

60. Finding Total Cost from Marginal Cost   The marginal cost MC of producing the qth unit is given by

$$MC(q) = 8 + 0.03q^2$$

Find the total cost of producing q units if the fixed cost is $20.

61. Finding Total Cost from Marginal Cost   The Acme Hardware Co. has determined its marginal cost is

$$MC(q) = \frac{q^2}{50} - q + 75$$

where q is the number of items produced per day. Find the total cost of producing 100, 200, and 500 units per day. Assume that the fixed cost is zero.

62. Finding Total Revenue from Marginal Revenue   A marginal revenue function MR is given by

$$MR(q) = 75 + 0.4q + 0.2q^2$$

where q is the number of units sold. Find the total revenue from selling q units.

Finding Total Size from Growth Rates

63. Spread of Contagious Diseases   There have been many studies about the spread of diseases. Assume that the rate at which a certain disease spreads has been determined

to be

$$p'(t) = e^{-0.02t}$$

where $p(t)$ is the percentage of the general population that has contracted the disease in year t. If no one has the disease initially (negligible number), how many people will have the disease after t years?

64. Fruit Flies   A biologist has determined that the rate of growth in the size of a colony of fruit flies on day t is given by

$$P'(t) = 100\left(2 + \frac{1}{t^2}\right)   (t \geq 1)$$

where the population on day one is $P(1) = 100$. Find the population for any day $t \geq 1$.

65. How Far Have You Traveled?   Suppose you are initially driving your car at 40 miles per hour, and during a one-hour period you gradually increase your velocity $v(t)$ to 60 miles per hour according to the function

$$v(t) = 40 + 20t$$

where t is time in hours. How far have you traveled after one hour?

66. How Far Have You Traveled?   Suppose you drive your car for two hours at the constant velocity of

$$v(t) = 50$$

miles per hour. It is clear that you have traveled 100 miles. Find this answer using calculus.

67. How Far Has the Ball Fallen?   If a ball is dropped from the top of a tall building, the velocity at which the ball falls is given by

$$v(t) = 32t$$

feet per second, where t is time measured from the time that the ball is dropped. How far has the ball fallen after five seconds?

Reverse Problem to Finding the Tangent Line

68. Finding the Curve from the Tangent Line   Suppose the slope of the tangent line to a curve is given by

$$f'(x) = 2x$$

If the curve passes through the point $(0, 0)$, what is the curve?

69. Finding the Curve from the Tangent Line   Suppose the slope of the tangent line to a curve is given by

$$f'(x) = e^x$$

If the curve passes through the point $(0, 2)$, what is the curve?

11.2

Area and the Definite Integral

PURPOSE

> We introduce the definite integral and show how this integral solves one of the two original problems of calculus, the problem of areas. Major topics are
>
> · the calculation of an area using the concept of the limit and
> · the definite integral.

Introduction

Mathematicians have struggled with the problem of finding the area of a region in the plane for over 3000 years. Until the invention of the integral calculus, however, the regions considered were mostly those regions bounded by straight lines, called polygons, with a few exceptions such as the circle and the ellipse. For example, it is not necessary to use integral calculus to find the area of a polygon. Figure 3 illustrates how the Greek mathematicians found the area of a polygon by

1. first finding the area of a rectangle;

2. using the area of a rectangle to find the area of a parallelogram;

3. using the area of a parallelogram to find the area of a triangle;

4. using the area of a triangle to find the area of a polygon.

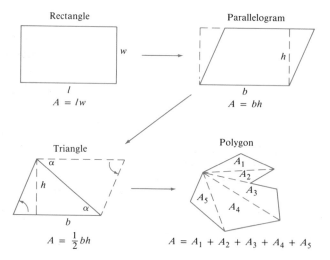

Figure 3
Steps in finding the area of a polygon

Continuing Where the Greeks Left Off (General Areas)

If we know how to find the area of a polygon, we can use this information to approximate, and sometimes even find exactly, the area of a region bordered by curved boundaries. Consider finding the area of a circle. Over 2000 years ago the brilliant Greek mathematician Archimedes (287(?)–212 B.C.) found the area of a circle by drawing a sequence of **inscribed polygons** $P_3, P_4, P_5, \ldots$ that

Figure 4
Three typical polygons whose areas are getting closer to the area of a circle

approximate the circle more and more accurately. Figure 4 shows a few representative polygons.

Although mathematicians have studied the concept of area throughout the ages, it was the discovery of the integral calculus by Newton and Leibniz that allowed for the more general determination of areas. For that reason it would certainly be reasonable to give much of the credit for solving the problem of areas (discussed earlier) to Newton and Leibniz. However, the mathematical determination of areas understood by Newton and Leibniz is quite primitive in comparison to the ideas that we will learn in this book. The common mathematical tool for determining area that is learned by students today, the definite integral, was discovered much later, in the mid-1800's, by the French mathematician Augustin Louis Cauchy (1789–1857) and the German mathematician Georg Friedrich Bernhard Riemann (1826–1866).

The way in which we find areas today using integral calculus is to find a formula for the area $A(P_n)$ of an inscribed polygon P_n with n sides (in other words treat n as a variable). Then by taking the limit of $A(P_n)$ as n grows without bound ("goes to infinity") this expression will approach the exact area of the circle. That is,

$$\text{Area of circle} = \lim_{n \to \infty} A(P_n)$$

We illustrate how this process works by finding the exact area under the curve $y = x^2$. It is not critical that the approximating polygons be inscribed polygons but merely approximate the area of the region in question more closely as the number of polygons increases.

Example 1

Approximating an Area by Rectangles   Approximate the area under the graph of the function

$$f(x) = x^2 \qquad (0 \le x \le 1)$$

as shown in Figure 5, using four rectangles as the approximating polygons.

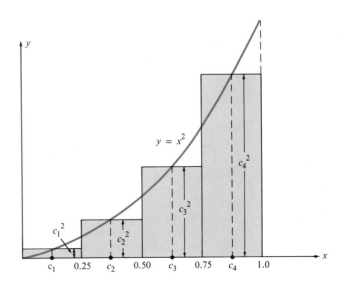

Figure 5
Approximation of the area under a curve by four rectangles

Solution

- **Step 1** (find the length of the subintervals).   Subdivide the interval $[0, 1]$ into four subintervals of equal length and compute the length h of each subinterval from the formula

$$h = \frac{\text{Length of interval}}{\text{Number of subintervals}} = \frac{1 - 0}{4} = \frac{1}{4}$$

- **Step 2** (find midpoints and heights of rectangles).   Find the midpoint of each subinterval:

$$c_1 = \frac{h}{2} = \frac{0.25}{2} = 0.125$$

$$c_2 = c_1 + h = 0.125 + 0.250 = 0.375$$
$$c_3 = c_2 + h = 0.375 + 0.250 = 0.625$$
$$c_4 = c_3 + h = 0.625 + 0.250 = 0.875$$

Next draw the four rectangles; the width of each is $h = 0.25$, and the heights are given by

$$f(c_1) = c_1{}^2 = (0.125)^2 = 0.0156$$
$$f(c_2) = c_2{}^2 = (0.375)^2 = 0.1406$$
$$f(c_3) = c_3{}^2 = (0.625)^2 = 0.3906$$
$$f(c_4) = c_4{}^2 = (0.875)^2 = 0.7656$$

Note that the resulting rectangles (polygons) are not entirely beneath the graph of $y = x^2$. It is generally convenient to select the midpoint of each subinterval as the location where we determine the height of each rectangle. In other problems we may decide to choose other points in the subintervals as dictated by computational convenience.

- **Step 3.**   Compute the total area of the four rectangles:

$$\begin{aligned}
\text{Area of rectangles} &= f(c_1) \cdot h + f(c_2) \cdot h + f(c_3) \cdot h + f(c_4) \cdot h \\
&= 0.25[\, f(0.125) + f(0.375) + f(0.625) + f(0.875)] \\
&= 0.25[(0.125)^2 + (0.375)^2 + (0.625)^2 + (0.875)^2] \\
&= 0.25[0.0156 + 0.1406 + 0.3906 + 0.7656] \\
&= 0.3281 \qquad \text{(the exact area is } 1/3) \qquad \square
\end{aligned}$$

Example 2 ——————   Better Approximation   Approximate the area under the graph of the function

$$f(x) = x^2 \qquad (0 \le x \le 1)$$

using $n = 8$ rectangles.

Solution   Using the same steps as in Example 1 but now using eight subintervals, we find the midpoints, heights, and areas of the eight rectangles. Table 3 shows the results of these computations.

TABLE 3
Eight Rectangles Used to
Approximate the Area Under
$f(x) = x^2$

Subintervals	Midpoints, c_i	Heights of Rectangles, $f(c_i)$	Areas of Rectangles, $nf(c_i)$
$[0.000, 0.125]$	0.0625	0.0039	0.0005
$[0.125, 0.250]$	0.1875	0.0352	0.0044
$[0.250, 0.375]$	0.3125	0.0977	0.0122
$[0.375, 0.500]$	0.4375	0.1903	0.0238
$[0.500, 0.625]$	0.5625	0.3164	0.0396
$[0.625, 0.750]$	0.6875	0.4727	0.0591
$[0.750, 0.875]$	0.8125	0.6602	0.0825
$[0.875, 1.00]$	0.9375	0.8789	0.1099
		Total:	0.3320

Note in Table 3 that the length of each subinterval, which is the width of each rectangle, is $1/8 = 0.125$. Note too that the areas of the eight rectangles are computed by multiplying the heights of the rectangles by their common width of 0.125. The rectangles are shown in Figure 6.

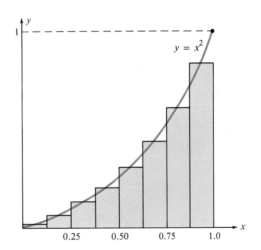

Figure 6
Approximation of area by eight rectangles

When the number of rectangles is doubled from $n = 4$ to $n = 8$, the total area of the rectangles changes from 0.3281 to 0.3320, which is closer to the exact area of $1/3$.   □

Table 4 gives the approximation of the area under the curve $y = x^2$ using n rectangles for different values of n. It seems clear that the area of n rectangles is approaching $1/3$ as n gets larger and larger.

TABLE 4
Approximating Area Under
$y = x^2$ $(0 \leq x \leq 1)$ with n
Rectangles

Number of Rectangles, n	Area of Rectangles
1	0.2500
2	0.3125
3	0.3241
4	0.3281
5	0.3300
6	0.3310
7	0.3316
8	0.3320
9	0.3323
10	0.3325
15	0.3330
20	0.3331
30	0.3332
40	0.3333
50	0.3333

The Definite Integral

We have seen that it is easy to find the area of squares, triangles, and other polygons by simple methods. However, finding the areas of regions with curved boundaries, such as the area under a portion of the curve $y = x^2$, requires the use of the limiting process. For the most part we will concentrate on finding areas between the graphs of functions and the x-axis. (See Figure 7.)

Geometrically, the definite integral can be interpreted as finding the area between the graph of a function and the x-axis by finding an algebraic expression for the area of n "approximating rectangles" and then letting n approach infinity. If this limit exists, it can often be shown to approach the exact area under the graph and above the x-axis. (See Figure 8.)

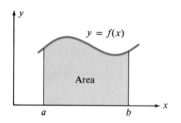

Figure 7
Finding the area under the graph of a function by the definite integral

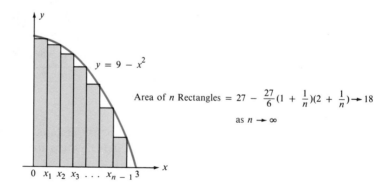

Area of n Rectangles $= 27 - \dfrac{27}{6}(1 + \dfrac{1}{n})(2 + \dfrac{1}{n}) \rightarrow 18$

as $n \rightarrow \infty$

Figure 8
Area under n rectangles approaches 18 as n gets big

Intuitive Comments About the Definite Integral

When the graph of a function $y = f(x)$ lies above the x-axis, the limiting value of the area of n rectangles as n approaches infinity represents the area under the curve. When all of these rectangles have the same width h, height $f(c_i)$, and area $hf(c_i)$, the general expression for these n approximating rectangles is

$$A(R_n) = hf(c_1) + hf(c_2) + \cdots + hf(c_n)$$

As n gets larger and larger, the "shape" of these n rectangles will look more and more like the region between the graph of $y = f(x)$ and the x-axis. This limit

$$\lim_{n \to \infty} A(R_n)$$

regardless of whether the graph of $y = f(x)$ is always above the x-axis is called the definite integral.

If the graph of the function does not lie completely above the x-axis, the definite integral does not represent the area between the graph and the x-axis but can be modified so that this area between the graph and the x-axis can be found.

We now give the formal definition of the definite integral.

Definition of the Definite Integral

Let f be a continuous function on the interval $[a, b]$. The **definite integral** of f is defined as the limiting value of the sum

$$\int_a^b f(x)\,dx = \lim_{n \to \infty} \left[f(c_1)h + f(c_2)h + \cdots + f(c_n)h \right]$$

provided that the limit exists (note that n represents the number of terms in the series). The constant h is given by

$$h = \frac{b - a}{n}$$

and the constant c_i for $i = 1, \ldots, n$ is an arbitary number in the ith subinterval

$$x_{i-1} \le c_i \le x_i$$

The function f is called the **integrand** of the definite integral, and the numbers b and a are the **upper limit** and **lower limit** of integration, respectively. The following diagram illustrates the relevant variables.

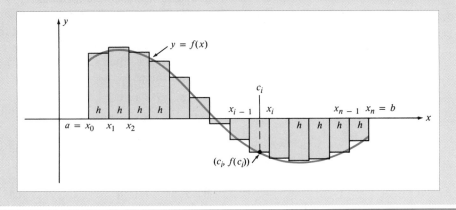

Note that a major difference between the definite integral of this section and the indefinite integral of the previous section is that the definite integral is a number, whereas the indefinite integral is a function.

Example 3 ——————————

Definite Integral   Evaluate the definite integral

$$\int_0^2 3x\,dx$$

Since the graph of $y = 3x$ is always above the x-axis over the interval $[0, 2]$, the value of its definite integral will be equal to the area between its graph and the x-axis. (See Figure 9.)

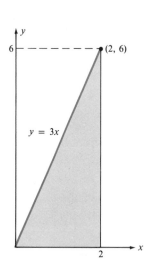

Figure 9
Find the shaded area

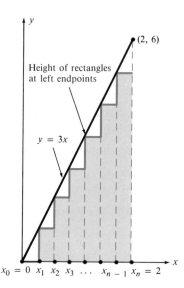

Figure 10
An approximation by n rectangles

Solution   Note that the upper and lower limits of integration are

$$b = 2 \qquad \text{(upper limit of integration)}$$
$$a = 0 \qquad \text{(lower limit of integration)}$$

We begin by finding an algebraic formula A_n for the area under n rectangles. We then compute the limit of A_n as n increases without bound. (See Figure 10.) For convenience we have chosen the heights of the rectangles to be the values of $y = 3x$ at the left endpoints of the subintervals (instead of at the midpoints). If we choose n rectangles, then the widths and heights of these rectangles will be given by the formulas:

$$h = \frac{b - a}{n} = \frac{2 - 0}{n} = \frac{2}{n} \qquad \text{(width of the rectangles)}$$

$$c_i = 0 + (i - 1)\left(\frac{2}{n}\right) = (i - 1)\left(\frac{2}{n}\right) \qquad \text{(left endpoint of \textit{i}th subinterval)}$$

$$f(c_i) = 3c_i = 3(i - 1)\left(\frac{2}{n}\right) \qquad \text{(height of \textit{i}th rectangle)}$$

for $i = 1, 2, \ldots, n$.

Hence the formula for the total area of these n rectangles is simply the sum of the areas of each individual rectangle. That is,

$$A_n = f(c_1)h + f(c_2)h + \cdots + f(c_n)h$$

$$= (3c_1)\frac{2}{n} + (3c_2)\frac{2}{n} + \cdots + (3c_n)\frac{2}{n}$$

$$= \frac{6}{n}[c_1 + c_2 + \cdots + c_n]$$

$$= \frac{6}{n}\left[0 + \frac{2}{n} + \frac{4}{n} + \frac{6}{n} + \cdots + \frac{2(n-1)}{n}\right]$$

$$= \frac{12}{n^2}[1 + 2 + 3 + \cdots + (n-1)]$$

$$= \frac{12}{n^2}\frac{n(n-1)}{2} \qquad *$$

$$= \frac{6n^2 - 6n}{n^2}$$

$$= 6 - \frac{6}{n} \qquad \text{(area under } n \text{ rectangles)}$$

This expression gives the area of n inscribed rectangles under the curve $y = 3x$. For example, if the number of rectangles is 10,000,000, the area would be

$$6 - \frac{6}{10,000,000} = 5.9999994$$

The beauty of this formula lies in the fact that it gives the area for any number of rectangles. It is easy to see that as more and more rectangles are included, the total area gets closer and closer to the value 6. In other words, the limiting value and hence the definite integral are given by

$$\int_0^2 3x\,dx = \lim_{n \to \infty} [f(c_1)h + f(c_2)h + \cdots + f(c_n)h]$$

$$= \lim_{n \to \infty}\left[6 - \frac{6}{n}\right] = 6 \qquad \qquad \square$$

Comment on the Choice of c_i

In the above example we chose the left endpoints of the subintervals as the locations c_i at which to determine the heights of the rectangles. This decision was for convenience more than anything else. We can choose these locations c_i at any

---

* This step is the result of the following formula which gives the sum of the sequence of n consecutive integers beginning with 1:

$$1 + 2 + 3 + \cdots + (n-1) = \frac{n(n-1)}{2}$$

You can check this result by substituting different values of n.

place in the ith subinterval and the limiting value of the total area of the rectangles will approach the same limit. Of course, for a given finite number of rectangles n the areas of the rectangles will be different from the value of the definite integral, but for larger and larger n this difference will go to zero.

Problems

For Problems 1–9, approximate the areas under the curves, using rectangles with the indicated number of subintervals. For each problem, calculate the approximate area by using the following three ways to determine the height of the rectangles:

- Left endpoint: Pick the left endpoint of each subinterval.

- Right endpoint: Pick the right endpoint of each subinterval.

- Midpoint: Pick the midpoint of each subinterval.

1. One rectangle.

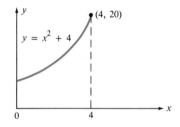

2. Two rectangles.

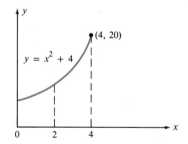

3. Four rectangles.

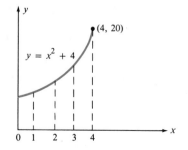

4. One subinterval.

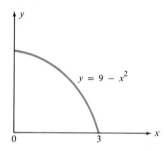

5. Two subintervals.

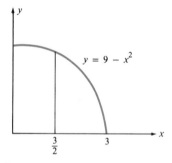

6. Three subintervals.

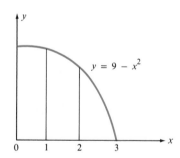

7. One subinterval.

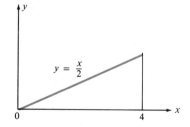

8. Two rectangles.

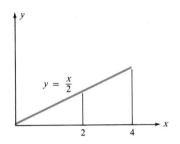

9. Four rectangles.

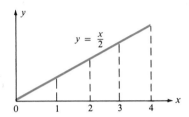

For Problems 10–18, approximate the area under the curves over the indicated intervals. Find approximations for both two and four subintervals, and let the height of the rectangles be the value of the function at the midpoint of each subinterval.

10. $f(x) = 6$ $[0, 1]$ **11.** $f(x) = x$ $[0, 2]$
12. $f(x) = 3x - 2$ $[1, 5]$ **13.** $f(x) = x + 1$ $[1, 5]$
14. $f(x) = x^2$ $[-1, 1]$ **15.** $f(x) = x^2 + 3$ $[4, 6]$

16. $f(x) = x^3$ $[1, 3]$ **17.** $f(x) = \dfrac{1}{x}$ $[1, 2]$

18. $f(x) = \dfrac{1}{x^2}$ $[1, 2]$

For Problems 19–22, find the exact area under the curves over the indicated intervals using the definition of the definite integral. Use the value of the function at the right endpoint of each subinterval to calculate the height of a rectangle.

19. $\displaystyle\int_2^5 5\,dx$

20. $\displaystyle\int_1^3 (x + 4)\,dx$ $\left(\text{Hint: } 1 + 2 + \cdots + n = \dfrac{n(n + 1)}{2}\right)$

21. $\displaystyle\int_0^1 x^2\,dx$

$\left(\text{Hint: } 1^2 + 2^2 + 3^2 + \cdots + n^2 = \dfrac{n(n + 1)(2n + 1)}{6}\right)$

22. $\displaystyle\int_0^1 x^3\,dx$ $\left(\text{Hint: } 1^3 + 2^3 + 3^3 + \cdots + n^3 = \dfrac{n^2(n + 1)^2}{4}\right)$

Areas in Business

23. Total Sales Find the total sales for the last four years by finding the area under the curve described in Figure 11. The units for the y-axis are in \$100,000 of sales. Only formulas from geometry are needed.

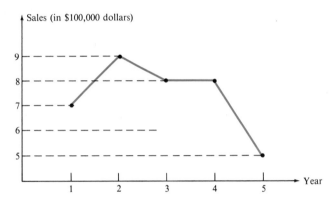

Figure 11
Graph for Problem 23

24. Total Cost Estimate the total advertising cost for the past six months by finding the area under the curve in Figure 12. The graph below gives the advertising costs by day. The vertical axis has units of \$100. Use any convenient point to determine the heights of the rectangles over each of the six subintervals. The location of the point need not be the same for each subinterval.

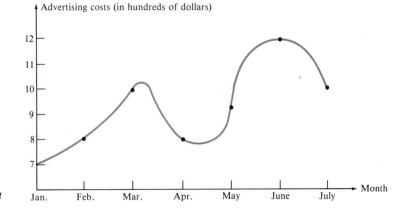

Figure 12
Graph for Problem 24

The Fundamental Theorem of Calculus

PURPOSE

We introduce the fundamental theorem of calculus. This theorem tells us how to evaluate the definite integral of a function using the antiderivative of the integrand without having to evaluate the limit found in the definition. Using the fundamental theorem, we then evaluate definite integrals used in economics.

Fundamental Theorem

In the previous section we evaluated the definite integral of a function f defined on a given interval $[a, b]$ by evaluating the limit

$$\int_a^b f(x)\,dx = \lim_{n \to \infty} \left[f(c_1)h + f(c_2)h + \cdots + f(c_n)h \right]$$

This definition is hardly an operational one. It does not provide a practical method for integrating functions and giving answers. Trying to find the definite integral of a function such as

$$f(x) = x^5 - e^x + \ln x$$

using the definition would be almost impossible. There is an important theorem in calculus that allows us to bypass the complicated limit in the definition of the definite integral and evaluate definite integrals by using the antiderivative. This theorem is the **fundamental theorem of calculus**.

Statement of the Fundamental Theorem of Calculus

Let f be a continuous function defined on the interval $[a, b]$ and let F be any antiderivative of f. The definite integral of f can be evaluated from the formula

$$\int_a^b f(x)\,dx = F(b) - F(a)$$

The difference $F(b) - F(a)$ is often denoted in shorthand by the expression

$$F(b) - F(a) = F(x)\Big|_a^b$$

Comments on the Fundamental Theorem

Many students wonder what is so significant about this theorem that it warrants the special name, the fundamental theorem of calculus. What is there about it that gives it this central place in the theory of calculus? Of course, from a practical point of view it is obviously an extremely useful tool for evaluating definite integrals. By applying this theorem, all we have to do to evaluate the definite integral

$$\int_a^b f(x)\,dx$$

is carry out the following steps.

Evaluating Definite Integrals Using the Fundamental Theorem

- **Step 1.** Find any antiderivative $F(x)$ of the integrand $f(x)$.
- **Step 2.** Evaluate the antiderivative found in Step 1 at the upper and lower limits b and a, getting $F(b)$ and $F(a)$.
- **Step 3.** Subtract $F(b) - F(a)$ to obtain the value of the definite integral.

Although the above steps for evaluating definite integrals are extremely useful, they are still not the reason for the name "fundamental theorem." The reason can be seen if we rewrite the theorem in a slightly different manner. Since an integrand $f(x)$ and its antiderivative $F(x)$ are related by

$$F'(x) = f(x)$$

we can rewrite the fundamental theorem in the alternative form

$$\int_a^x F'(x)\,dx = F(x) - F(a)$$

where we have decided to replace the upper limit b by the variable x. (We can do this because b was arbitrary.) If the lower limit of integration a is chosen so that $F(a) = 0$, then we have

$$\int_a^x F'(x)\,dx = F(x)$$

In other words, the fundamental theorem relates the two basic operations of the calculus, differentiation and integration. It says that differential calculus is not a completely unrelated subject from integral calculus. But it even says more. It says that the operation of integration "undoes" the operation of differentiation. In other words, the two basic operations of the calculus are inverse operations.

We now apply the fundamental theorem of calculus to evaluate some definite integrals.

Example 1

Evaluating Definite Integrals   Evaluate the definite integral

$$\int_0^1 x^3\,dx$$

Solution

- **Step 1** (find an antiderivative).   Finding an antiderivative of $f(x) = x^3$, we get

$$F(x) = \frac{1}{4}x^4 + C$$

- **Step 2** (evaluate $F(b)$ and $F(a)$). Evaluating the antiderivative at the upper and lower limits, we have

$$F(1) = \frac{(1)^4}{4} + C = \frac{1}{4} + C$$

$$F(0) = \frac{(0)^4}{4} + C = C$$

- **Step 3** (calculate $F(b) - F(a)$). Finding the difference between the antiderivative at the upper and lower limits, we have

$$\int_0^1 x^3\,dx = F(1) - F(0)$$

$$= \left(\frac{1}{4} + C\right) - C$$

$$= \frac{1}{4}$$

Note that the constant C in the antiderivative has canceled and hence does not affect the value of the definite integral. For this reason we will always pick an antiderivative $F(x)$ with $C = 0$ when evaluating definite integrals.

The definite integral represents the area under the curve $y = x^3$ over the interval $[0, 1]$ as shown in Figure 13. ☐

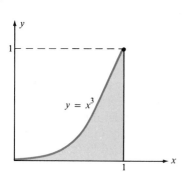

Figure 13
Area under $y = x^3$

Example 2

Finding Areas   Find the area under the curve

$$y = 1 + e^x$$

over the interval $[0, 2]$ as shown in Figure 14.

Solution   Since the given function is positive, the area is the value of the definite integral

$$\int_0^2 (1 + e^x)\,dx$$

Now that we have seen the three-step process carried out in Example 1, we will streamline our operation and simply write

$$\int_0^2 (1 + e^x)\,dx = F(x)\Big|_0^2$$

$$= (x + e^x)\Big|_0^2$$

$$= (2 + e^2) - (0 + e^0)$$

$$= e^2 + 1 \qquad ☐$$

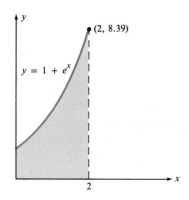

Figure 14
Area under an exponential curve

Example 3

Finding a Definite Integral   Evaluate the definite integral

$$\int_{-1}^1 (t^2 + 1)\,dt$$

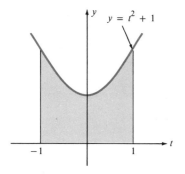

Figure 15
Finding the area under a parabola

Solution   Using the fundamental theorem of calculus, we have

$$\int_{-1}^{1}(t^2+1)\,dt = \left[\frac{1}{3}t^3+t\right]\Bigg|_{-1}^{1}$$

$$= \left[\frac{1}{3}(1)^3+1\right] - \left[\frac{1}{3}(-1)^3+(-1)\right]$$

$$= \frac{4}{3}+\frac{4}{3}$$

$$= \frac{8}{3}$$

This value represents the value of the area shown in Figure 15.   □

Rules of Integration

Just as we have rules for evaluating derivatives of sums, differences, products, quotients, and compositions of functions, we can also develop rules for the integration of such combinations of functions. These rules, along with the fundamental theorem of calculus, will allow us to evaluate a wide assortment of definite integrals. If we interpret the integral of a positive function as an area, then for such functions these rules can be intuitively verified.

Rules for Evaluating Definite Integrals

Let f and g functions such that the following integrals exist and let a, b, and c be constants. Then

· Constants can be factored outside the integral.

$$\int_{a}^{b} cf(x)\,dx = c\int_{a}^{b} f(x)\,dx$$

· The integral of a sum is the sum of the integrals.

$$\int_{a}^{b}\left[f(x)+g(x)\right]dx = \int_{a}^{b} f(x)\,dx + \int_{a}^{b} g(x)\,dx$$

· The integral of a difference is the difference of the integrals.

$$\int_{a}^{b}\left[f(x)-g(x)\right]dx = \int_{a}^{b} f(x)\,dx - \int_{a}^{b} g(x)\,dx$$

Verification of the Integration Rules

The previous rules can be verified intuitively for positive functions by interpreting the definite integral as an area. For instance, the second rule stating that the integral of a sum is the sum of the integrals has a visual interpretation as shown in Figure 16. Essentially, the rule says that the area under the graph of the sum of two functions is the sum of the areas under each individual graph.

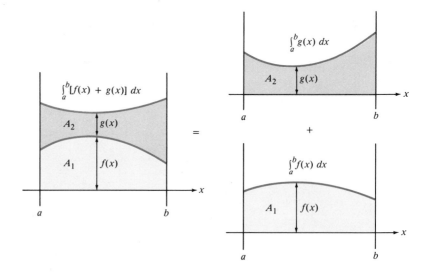

Figure 16
Geometric interpretation of the sum rule for integrals

Example 4

Using Rules of Integration   Evaluate the definite integral

$$\int_0^1 5x \, dx$$

Solution   Bringing the constant 5 outside the integral, we have

$$\int_0^1 5x \, dx = 5 \int_0^1 x \, dx$$

$$= 5 \left[\frac{1}{2} x^2 \Big|_0^1 \right]$$

$$= 5 \left[\frac{1}{2} (1)^2 - \frac{1}{2} (0)^2 \right]$$

$$= \frac{5}{2}$$
☐

Example 5

Using the Rules of Integration   Evaluate the definite integral

$$\int_0^1 (3x^2 + 2x + 4) \, dx$$

Solution   Using the rules of integration, we can write

$$\int_0^1 (3x^2 + 2x + 4) \, dx = \int_0^1 3x^2 \, dx + \int_0^1 2x \, dx + \int_0^1 4 \, dx \quad \text{(sum rule)}$$

$$= 3 \int_0^1 x^2 \, dx + 2 \int_0^1 x \, dx + 4 \int_0^1 dx \quad \text{(constant rule)}$$

$$= 3 \left[\frac{1}{3} x^3 \Big|_0^1 \right] + 2 \left[\frac{1}{2} x^2 \Big|_0^1 \right] + 4 \left[x \Big|_0^1 \right]$$

$$= 3 \left(\frac{1}{3} \right) + 2 \left(\frac{1}{2} \right) + 4(1)$$

$$= 6$$
☐

The Use of the Definite Integral in Economics

If $C(q)$ denotes the total cost of producing q units of a given commodity, then $C'(q)$ represents the marginal cost. Applying the fundamental theorem of calculus, we can write

$$\int_a^b C'(q)\, dq = C(b) - C(a)$$

The value of this formula lies in the fact that the quantity $C(b) - C(a)$ simply represents the change in cost when the production level is raised from a items to b items. Hence we have the useful rules.

Integral Rules in Economics

Let MR, MP, and MC denote marginal revenue, marginal profit, and marginal cost, respectively. The changes in the total revenue R, total profit P, and total cost C, when the level of production q is changed from a units to b units are given by the following definite integrals.

$$\text{Change in revenue} = \int_a^b MR(q)\, dq$$

$$\text{Change in profit} = \int_a^b MP(q)\, dq$$

$$\text{Change in cost} = \int_a^b MC(q)\, dq$$

Example 6

Finding Increased Costs   Suppose you have just been hired as production manager for a large firm that makes stuffed bears. When the production of bears is q bears per hour, your analysis of various costs has determined that the cost (in dollars) of making one more bear is

$$MC(q) = 22 - 0.02q$$

In other words, the first bear costs \$22 to produce, while the cost of making each succeeding bear goes down by two cents. Find the increase in the company's cost when the hourly production level of bears is raised from 100 to 200 bears.

Solution   The increase in total cost C, which is $C(200) - C(100)$, can be found by using the fundamental theorem of calculus:

$$\begin{aligned}
\text{Increase in cost} &= \int_{100}^{200} MC(q)\, dq \\[2mm]
&= \int_{100}^{200} (22 - 0.02q)\, dq \\[2mm]
&= (22q - 0.01q^2)\Big|_{100}^{200} \\[2mm]
&= (4400 - 400) - (2200 - 100) \\[2mm]
&= 1900
\end{aligned}$$

In other words, the hourly increase in cost necessary to increase production from 100 units to 200 units is \$1900. It is important to note that we did not need to know the value of the fixed costs to find this answer. This is important, since it is often difficult to find this value.   □

Problems

For Problems 1–25, evaluate the definite integrals using the fundamental theorem of calculus.

1. $\int_{-3}^{2} 2\,dx$ **2.** $\int_{0}^{4} 5x\,dx$

3. $\int_{2}^{5} (x+3)\,dx$ **4.** $\int_{1}^{5} (7t-3)\,dt$

5. $\int_{1}^{8} (y^3-1)\,dy$ **6.** $\int_{1}^{4} (x^2-2x+2)\,dx$

7. $\int_{0}^{2} (x^3+x+3)\,dx$ **8.** $\int_{-1}^{2} (9x^2+2)\,dx$

9. $\int_{-2}^{4} (5-3x+2x^2)\,dx$

10. $\int_{1}^{3} (x-1)(x+2)\,dx$ (*Hint:* Do the multiplication first.)

11. $\int_{1}^{4} (9-x)\sqrt{x}\,dx$ **12.** $\int_{3}^{4} (x+1)^2\,dx$

13. $\int_{1}^{5} 2x(x^2+3)\,dx$ **14.** $\int_{0}^{1} (8x^2+4)(x^2+2)\,dx$

15. $\int_{1}^{e} \left(x+\dfrac{1}{x^2}\right)dx$ **16.** $\int_{0}^{1} \left(\dfrac{3}{2}e^x - \dfrac{1}{2}e^{-x}\right)dx$

17. $\int_{1}^{4} \left(\sqrt{x}+\dfrac{1}{x^2}\right)dx$

18. $\int_{0}^{1} (e^{-x}+x^2)^3(2x-e^{-x})\,dx$

19. $\int_{2}^{5} \dfrac{x^2+1}{x^2}\,dx$ **20.** $\int_{1}^{3} 20x(x^2+7)^3\,dx$

21. $\int_{2}^{3} \dfrac{6x^2+4}{\sqrt{x^3+2x}}\,dx$ **22.** $\int_{1}^{4} \dfrac{x^2+3x+4}{\sqrt{x}}\,dx$

23. $\int_{0}^{1} \dfrac{2x}{(3+x^2)^2}\,dx$ **24.** $\int_{5}^{10} \dfrac{8}{\sqrt{4u+3}}\,du$

25. $\int_{0}^{1} (ax^2+bx+c)(2ax+b)\,dx$
(a, b, c are positive constants)

For Problems 26–45, find the area between the nonnegative functions and the x-axis over the indicated interval.

26. $f(x) = 2x - 5$, $[3, 6]$ **27.** $f(t) = 9 - t^2$, $[-2, 1]$
28. $f(x) = x^3$, $[0, 2]$ **29.** $f(u) = 1 + 2u$, $[0, 3]$
30. $f(x) = x^2 + x^3$, $[0, 2]$
31. $f(x) = x^3 - 3x^2 + 5$, $[0, 2]$
32. $f(t) = 4t + t^3$, $[1, 2]$ **33.** $f(x) = 2x + x^2$, $[0, 5]$

34. $f(x) = \sqrt[3]{x}$, $[1, 8]$ **35.** $f(w) = \sqrt{w} + \dfrac{1}{w^2}$, $[2, 4]$

36. $f(x) = 5 + 2x\sqrt{x^2+1}$, $[1, 2]$
37. $f(x) = e^x$, $[0, 1]$
38. $f(x) = 4e^{-4x}$, $[1, 3]$ **39.** $f(t) = 3e^{-t/5}$, $[0, \ln 2]$

40. $f(u) = 2e^{4u}$, $[-1, 1]$ **41.** $f(t) = e^{-t} + 1$, $[0, 2]$
42. $f(u) = xe^{-x^2}$, $[0, 1]$ **43.** $f(w) = xe^{x^2}$, $[0, 1]$
44. $f(t) = e^{-t}$, $[0, 5]$ **45.** $f(t) = 3e^t + 1$, $[1, 2]$

Finding the Change in Functions

46. Increase in Revenue The marginal revenue for a firm when sales are q is given by

$$MR(q) = 10 - 0.05q$$

Find the increase in revenue when the sales level increases from 100 to 200 units.

47. Increase in Revenue The marginal revenue when sales are q items is given by

$$MR(q) = 2 - 0.05q + 0.006q^2$$

What is the revenue increase resulting from a sales increase from 50 to 100 units?

48. Increase in Cost The marginal cost for Gadgets Inc. is given by

$$MC(q) = 2 - 0.03q + 0.005q^2$$

where q is the number of units produced. What is the total cost to raise production from 100 to 200 units?

49. Increase in Profits The marginal cost of a certain firm is given by

$$MC(q) = 15 - 0.002q$$

whereas the marginal revenue is

$$MR(q) = 20 - 0.003q$$

where q is the number of units produced. Find the increase in profits when sales are increased from 500 to 700 units.

50. Increased Population A population of bacteria adds $2000/\sqrt{t}$ new bacteria to its colony each week. How many bacteria are added to the population from week 3 to week 5?

51. Total Revenue Find the total revenue for 540 units given the marginal revenue

$$MR(q) = 3000 - 4q$$

Hint: Use $R(0) = 0$ to evaluate the constant of integration.

52. Total Cost Find the total cost of producing 100 items if the marginal cost is

$$MC(q) = 3q^2 - 4q + 5$$

and the fixed cost is zero.

53. Distance Traveled If a ball is dropped from a tall building, the velocity of the ball in feet per second will be

$$v(t) = 32t$$

where t is time in seconds measured from the time that the ball is dropped. How far will the ball fall between the third second and the fifth second?

54. **Lifting Weights**   A weight lifter is exercising by repeatedly stretching a spring to its full extension of two feet. The amount of work W done in foot-pounds to stretch this spring L feet is given by

$$W = \int_0^L 50x\,dx$$

How much work does the weight lifter perform on each stretch of the spring?

11.4 Evaluating Integrals by the Method of Substitution

PURPOSE

We introduce the method of substitution, which is a method that is helpful in finding an antiderivative $F(x)$ of a function $f(x)$. Knowing an antiderivative, we can then use the fundamental theorem of calculus to evaluate the definite integral. We also present a new economic application of the definite integral, the coefficient of inequality for income distributions.

Introduction

The evaluation of the definite integral

$$\int_a^b f(x)\,dx$$

by the use of the fundamental theorem of calculus is dependent on finding an antiderivative $F(x)$ of the integrand $f(x)$. However, finding an antiderivative of a function is generally much more difficult than finding the derivative. When we find the derivative, we can always use one of the useful rules such as the sum, difference, product, quotient, or chain rule. However, when we seek to find an antiderivative, there are no clear guidelines to follow, and we can often try several different approaches.

The diagram in Figure 17 illustrates our predicament. When we go to the right to find the derivative, there is always a clear path to follow. However, when we go to the left, choices must be made. Here is where the method of substitution is useful. It helps us to make these choices.

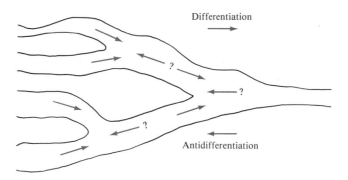

Figure 17
There are specific rules for differentiation but not for antidifferentiation

The Method of Substitution

To understand the method of substitution, it is useful that we review the concept of the differential. For example, if

$$y = (x^2 + 1)^3$$

then the differential is

$$dy = f'(x)\,dx$$
$$= 3(x^2 + 1)^2(2x)\,dx$$

Suppose now that we wish to find the antiderivative

$$\int (x^2 + 1)^2(2x)\,dx$$

Here is where the method of substitution comes into play. To find the above antiderivative, we make a substitution by letting

$$u = x^2 + 1$$

which has the differential

$$du = 2x\,dx$$

Substituting these values into the above indefinite integral, we get

$$\int (x^2 + 1)^2(2x)\,dx = \int u^2\,du$$

We can now find the antiderivative on the right-hand side of this equation by using the power rule. It is simply

$$\int u^2\,du = \frac{1}{3}u^3 + C$$

Finally, substituting $u = x^2 + 1$ for u in the above antiderivative, we get the desired antiderivative in terms of x:

$$\int (x^2 + 1)^2(2x)\,dx = \frac{1}{3}(x^2 + 1)^3 + C$$

The hardest part of the method of substitution is knowing how to define the substitution variable (what should u be). Expertise comes with practice, and one often ends up making a few trial substitutions to find one that actually works.

Example 1 _____

Method of Substitution   Find the antiderivative

$$\int \sqrt{2x + 3}\,dx$$

Solution   It is often unclear how to make the proper substitution; in fact, sometimes no substitution will work. A good rule of thumb when solving problems involving square roots is to set u equal to the expression within the radical. We carry out the following steps.

- **Step 1.**   Introduce the new variable $u = 2x + 3$ and find the differential $du = 2\,dx$.

- **Step 2.**   Solve for dx and $2x + 3$ in terms of u and du so that it is possible to transform the original indefinite integral in x to an indefinite integral in u. Doing this, we have

$$2x + 3 = u$$

$$dx = \frac{1}{2}\,du$$

and hence

$$\int \sqrt{2x + 3} \, dx = \int \frac{\sqrt{u}}{2} \, du$$

- **Step 3.** Find the antiderivative of the new integral in u. Doing this, we get

$$\int \sqrt{u} \, du = \frac{1}{3} u^{3/2} + C$$

- **Step 4.** Substitute back in terms of the original variable. Doing this, we find

$$\int \sqrt{2x + 3} \, dx = \frac{1}{3}(2x + 3)^{3/2} + C$$

Check:

$$\frac{d}{dx}\left[\frac{1}{3}(2x + 3)^{3/2} + C\right] = \sqrt{2x + 3} \qquad \Box$$

In the previous problem we found the antiderivative in four steps. We list these steps now.

Steps in the Method of Substitution

To find the definite integral

$$\int_a^b f(x) \, dx$$

carry out the following steps.

- **Step 1** (find u and du).   Introduce a new variable u that will be some function of the original variable x and compute the differential du.

- **Step 2** (find the new integral).   Substitute u and du found in Step 1 into the original integral. This gives the new indefinite integral in u:

$$\int f(x) \, dx = \int g(u) \, du$$

where $g(u)$ is the new integrand.

- **Step 3** (find the new indefinite integral).   Find the antiderivative $G(u)$ of $g(u)$:

$$\int g(u) \, du = G(u) + C$$

Hopefully, this step will be easy.

- **Step 4** (substitute back).   Substitute the expression that defined u in terms of x in Step 1 into the antiderivative $G(u)$ to get the final answer in terms of x.

Example 2 —————— Method of Substitution   Evaluate the indefinite integral

$$\int \frac{3x^2 + 4}{(x^3 + 4x)^2}\, dx$$

Solution   Here we recognize that the numerator is the derivative of the expression being squared in the denominator. This suggests the following substitution.

- **Step 1** (find u and du).   We let

$$u = x^3 + 4x$$
$$du = (3x^2 + 4)\, dx$$

- **Step 2** (find the new integral).   Substituting these expressions into the above integral gives the new integral in u:

$$\int \frac{3x^2 + 4}{(x^3 + 4x)^2}\, dx = \int \frac{1}{u^2}\, du$$

- **Step 3** (integrate the new integral).   Finding the antiderivative of the new integral, we have

$$\int \frac{1}{u^2}\, du = -\frac{1}{u} + C$$

- **Step 4** (substitute back).   Finally, the last step is to substitute back $u = x^3 + 4x$ to get the desired antiderivative in terms of x. This gives

$$\int \frac{3x^2 + 4}{(x^3 + 4x)^2}\, dx = \frac{-1}{x^3 + 4x} + C \qquad \square$$

We illustrate the method once again with another example.

Example 3 —————— Method of Substitution   Find the indefinite integral

$$\int \frac{x}{\sqrt{x^2 + 2}}\, dx$$

Solution   Here we note that the derivative of the quantity inside the radical sign is $2x$, or twice the numerator. This suggests the following substitution.

- **Step 1** (find u and du).   We let

$$u = x^2 + 2$$
$$du = 2x\, dx$$

- **Step 2** (find the new integral).   We now manipulate these quantities so that we can rewrite the original integral in terms of u and du. We write

$$\sqrt{x^2 + 2} = \sqrt{u}$$

$$x\, dx = \frac{1}{2}\, du$$

This will result in the new integral in u and du:

$$\int \frac{x}{\sqrt{x^2 + 2}}\, dx = \frac{1}{2}\int \frac{du}{\sqrt{u}}$$

- **Step 3** (integrate the new integral).   Finding the above indefinite integral in u, we get

$$\frac{1}{2}\int \frac{du}{\sqrt{u}} = \sqrt{u} + C$$

- **Step 4** (substitute back).   Replacing the u in the above equation by $u = x^2 + 2$, we get the desired result:

$$\int \frac{x}{\sqrt{x^2 + 2}}\, dx = \sqrt{x^2 + 2} + C$$

Check:

$$\frac{d}{dx}\left[\sqrt{x^2 + 2} + C\right] = \frac{x}{\sqrt{x^2 + 2}} \qquad \square$$

Comment on the Method of Substitution

At this point you should note that it would have been possible (although maybe difficult) to find the previous antiderivatives by a purely trial-and-error approach. In other words, simply asking yourself what is the function whose derivative is the integrand in the original integral. If one's differentiation skills are sharp, one can sometimes use this approach. In fact, students with a great deal of experience may try to find antiderivatives directly and, if that fails, *then* use the method of substitution.

Method of Substitution and Definite Integrals

Until now we have seen how the method of substitution can be used to find antiderivatives. We now use this method to find definite integrals. Here, the strategy is to find the antiderivative of the integrand by the method of substitution and then use the fundamental theorem of calculus. The following example illustrates this idea.

Example 4

Method of Substitution   Evaluate the definite integral

$$\int_0^1 x^2 e^{x^3}\, dx$$

Solution

- **Step 1** (find u and du).   Observe that the integral is of the form

$$\int f'(x)e^{f(x)}\, dx$$

where

$$f(x) = e^{x^3}$$

except for a constant factor of 3. This suggests the substitution

$$u = x^3$$
$$du = 3x^2\, dx$$

· **Step 2** (find the new integral).   The expressions found in Step 1 can be rewritten as

$$x^3 = u$$
$$x^2\, dx = \frac{1}{3}\, du$$

Hence we have the new integral

$$\int x^2 e^{x^3}\, dx = \int \frac{1}{3} e^u\, du$$

· **Step 3** (integrate the new integral).

$$\int \frac{1}{3} e^u\, du = \frac{1}{3} e^u + C$$

· **Step 4** (substitute back).   Letting $u = x^3$, we have

$$\int x^2 e^{x^3}\, dx = \frac{1}{3} e^{x^3}$$

where we have let the constant of integration be zero, since it always cancels in applying the fundamental theorem. Now that we know the antiderivative, we can apply the fundamental theorem of calculus:

$$\int_0^1 x^2 e^{x^3}\, dx = \left. \frac{1}{3} e^{x^3} \right|_0^1$$

$$= \frac{1}{3}(e - 1) \qquad \square$$

Example 5   Method of Substitution   Find the area under the curve

$$y = x\sqrt{1 - x}$$

for x between 0 and 1. (See Figure 18.)

Solution   The above area is represented by the definite integral

$$\int_0^1 x\sqrt{1 - x}\, dx$$

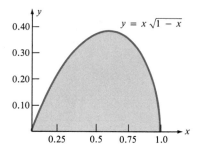

Figure 18
Area under $y = x\sqrt{1 - x}$

We first must find the indefinite integral

$$\int x\sqrt{1-x}\,dx$$

Here we let

$$u = 1 - x$$
$$du = -dx$$

These expressions can be rewritten as

$$x = 1 - u$$
$$dx = -du$$

Hence we have the new integral in u:

$$\int x\sqrt{1-x}\,dx = \int (1-u)\sqrt{u}(-du)$$

$$= \int (u-1)u^{1/2}\,du$$

$$= \int (u^{3/2} - u^{1/2})\,du$$

$$= \frac{2}{5}u^{5/2} - \frac{2}{3}u^{3/2}$$

Substituting back in terms of x gives

$$\int x\sqrt{1-x}\,dx = \frac{2}{5}(1-x)^{5/2} - \frac{2}{3}(1-x)^{3/2}$$

Calling the above antiderivative $F(x)$ and applying the fundamental theorem of calculus, we have

$$\int_0^1 x\sqrt{1-x}\,dx = F(1) - F(0)$$

$$= \left[\frac{2}{5}(1-1)^{5/2} - \frac{2}{3}(1-1)^{3/2}\right] - \left[\frac{2}{5}(1-0)^{5/2} - \frac{2}{3}(1-0)^{3/2}\right]$$

$$= \frac{4}{15} \qquad\qquad \square$$

The following rules should be kept in mind when using the method of substitution.

Common Forms for Using the Method of Substitution

Power of function: $\quad\displaystyle\int [u(x)]^n u'(x)\,dx = \frac{[u(x)]^{n+1}}{n+1} + C$

Function in exponent: $\quad\displaystyle\int e^{u(x)}u'(x)\,dx = e^{u(x)} + C$

Numerator is derivative of denominator: $\quad\displaystyle\int \frac{u'(x)}{u(x)}\,dx = \ln|u(x)| + C$

Example 6

Recognizing Special Forms   Find the antiderivative

$$\int \frac{x}{x^2+1}\,dx$$

Solution   Except for a constant factor, this integral has the form

$$\int \frac{f'(x)}{f(x)}\,dx$$

We let

$$u = x^2 + 1$$
$$du = 2x\,dx$$

Transforming the original integral to the new integral, we get

$$\int \frac{x}{x^2+1}\,dx = \frac{1}{2}\int \frac{1}{u}\,du$$

$$= \frac{1}{2}\ln|u| + C$$

$$= \frac{1}{2}\ln(x^2+1) + C$$

Note that we do not need the absolute value sign, since $x^2 + 1$ is always positive.

Check:

$$\frac{d}{dx}\left[\frac{1}{2}\ln(x^2+1) + C\right] = \frac{x}{x^2+1}\qquad\square$$

Example 7

Algebraic Manipulation Before Integration   Find the antiderivative

$$\int \frac{t}{t+1}\,dt \qquad (t > 0)$$

Solution   This integrand does not fall within any of the standard forms that can be solved by the method of substitution. However, whenever integrating a rational function in which the degree of the numerator is greater than or equal to the degree of the denominator, the first step is to algebraically divide the numerator by the denominator. In this instance we get

$$\frac{t}{t+1} = 1 - \frac{1}{t+1}$$

Hence we can write

$$\int \frac{t}{t+1}\,dt = \int \left[1 - \frac{1}{t+1}\right]dt$$

$$= \int 1\,dt - \int \frac{1}{t+1}\,dt$$

The first indefinite integral is clearly t. The second indefinite integral

$$\int \frac{1}{t+1}\,dt$$

can be found by letting

$$u = t + 1$$
$$du = dt$$

Thus we have

$$\int \frac{1}{t+1} \, dt = \int \frac{du}{u}$$
$$= \ln |u| + C$$
$$= \ln (t + 1) + C$$

Note that we do not need the absolute value sign in the above expression, since we have restricted t to be positive. Finally, putting the above two antiderivatives together, we have

$$\int \frac{t}{t+1} \, dt = t - \ln (t + 1) + C$$

Note too that the algebraic sign in front of the constant C is immaterial, since C represents an arbitrary constant. Hence the convention is to always place a plus before the C.

Check:

$$\frac{d}{dt} [t - \ln (t + 1) + C] = 1 - \frac{1}{t+1}$$

$$= \frac{t}{t+1} \qquad \square$$

The Lorentz Curve and the Coefficient of Inequality

One of the more interesting applications of the integral calculus in economics is the concept of the Lorentz curve and the coefficient of inequality (*CI*). Both the Lorentz curve and the coefficient of inequality are useful tools in the study of income distribution in a society. For instance, are wages of workers more uniformly distributed in the United States than they are in England? Are wages more uniformly distributed today in the United States than they were in 1950? Such questions can be analyzed with the help of the Lorentz curve and the coefficient of inequality.

To introduce the Lorentz curve, consider a study of wage earners in the United States in 1985. According to studies carried out by the U.S. Department of Labor, the lowest-paid 25% of all wage earners earned only about 10% of the total wages paid in the country during that year. The bottom 50% of all wage earners earned 26% of the total wages paid. The lower 75% of all wage earners earned roughly 50% of the total wages. Table 5 gives in more detail the fraction of the total wages earned by wage earners in various income categories.

The information in Table 5 can also be displayed by drawing the Lorentz curve as shown in Figure 19. The **Lorentz curve** $y = L(x)$ is defined as

$$L(x) = \text{fraction of total income earned by lowest } 100x\% \text{ of wage earners}$$

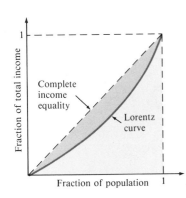

Figure 19
A typical Lorentz curve

TABLE 5

Points on a Lorentz Curve

Fraction of Lowest-Paid Wage Earners, x	Fraction of Total Wages Earned, $y = L(x)$
0	0
0.10 (10%)	0.02 (2%)
0.20 (20%)	0.06 (6%)
0.30 (30%)	0.12 (12%)
0.40 (40%)	0.17 (17%)
0.50 (50%)	0.26 (26%)
0.60 (60%)	0.32 (32%)
0.70 (70%)	0.46 (46%)
0.80 (80%)	0.58 (58%)
0.90 (90%)	0.72 (72%)
1.00 (100%)	1.00 (100%)

For example, the fact that $L(0.40) = 0.17$ in Figure 19 means that 17% of the total income is earned by the lowest 40% of wage earners.

Every economic society has its own Lorentz curve, which will change over time. The usefulness of the Lorentz curve lies in its ability to be an effective tool for comparing different economic societies and for comparing a single economic environment over time. All Lorentz curves $L(x)$ are defined for $0 \leq x \leq 1$, and all lie below the 45 degree line $y = x$. They also all satisfy the two conditions

$$L(0) = 0 \qquad \text{(``nobody earns nothing'')}$$
$$L(1) = 1 \qquad \text{(``everybody earns everything'')}$$

We say that the line $y = x$ represents complete equality of income because when $L(x) = x$, the lowest $100x\%$ of all wage earners earn $100x\%$ of the total income. In other words, the bottom 25% of the workers earn 25% of the wages, the bottom 50% of the workers earn 50% of the total wages, and so on. The farther the Lorentz curve lies below the line $y = x$, the more inequity in the wages of the workers. We can measure this inequity by computing the area under the curve $x - L(x)$, which is the same as the area between the Lorentz curve

TABLE 6

Coefficient of Inequality for Different Countries

Country	Coefficient of Inequality
Brazil	0.34
United Kingdom	0.31
Italy	0.28
United States	0.26
France	0.24
Canada	0.22
U.S.S.R.	0.19
Sweden	0.18

$L(x)$ and the line $y = x$. This area multiplied times 2 is called the **coefficient of inequality** (CI) and is defined by the definite integral

$$CI = 2 \int_0^1 [x - L(x)] \, dx$$

The factor of 2 is included simply for convenience to make this quantity vary from 0 to 1 (instead of from 0 to 0.5). When CI is 0, the Lorentz curve is $L(x) = x$, and there is complete uniform distribution of income. As CI gets closer and closer to 1, the inequity in the distribution of income is greater. Table 6 lists some coefficients of inequalities for countries in 1980.

Example 8

Coefficient of Inequality The following Lorentz curves describe the distribution of wages in Squareland and Cubeland:

$$\text{Squareland:} \quad L_s(x) = x^2$$
$$\text{Cubeland:} \quad L_c(x) = x^3$$

Graphs of these curves are shown in Figure 20. Which country has the fairer income distribution among its workers?

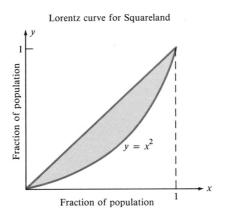

Lorentz curve for Squareland

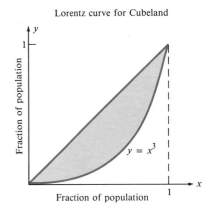

Lorentz curve for Cubeland

Figure 20
Comparing Lorentz curves

Solution The two coefficients of inequalities are found in Table 7. Since Squareland has a smaller coefficient of inequality, there is more equity in its wage distribution. □

TABLE 7
Coefficients of Inequality for Squareland and Cubeland

Squareland	Cubeland
$CE_s = 2 \int_0^1 [x - L_s(x)] \, dx$	$CE_c = 2 \int_0^1 [x - L_c(x)] \, dx$
$= 2 \int_0^1 [x - x^2] \, dx$	$= 2 \int_0^1 [x - x^3] \, dx$
$= 2 \left[\left(\dfrac{x^2}{2} - \dfrac{x^3}{3} \right) \Big|_0^1 \right]$	$= 2 \left[\left(\dfrac{x^2}{2} - \dfrac{x^4}{4} \right) \Big|_0^1 \right]$
$= \dfrac{1}{3}$	$= \dfrac{1}{2}$

Problems

For Problems 1–25, use the method of substitution to find the antiderivatives.

1. $\int x(3x^2 - 5)^3 \, dx$

2. $\int (t^3 - 2)(t^4 - 8t + 6)^3 \, dt$

3. $\int 2w(w^2 + 3)^2 \, dw$

4. $\int \frac{x + 1}{x^2 + 2x + 1} \, dx$

5. $\int 6(2x - 5)^{3/2} \, dx$

6. $\int (7t - 11)^{5/6} \, dt$

7. $\int \frac{1}{t + 3} \, dt$

8. $\int \frac{3}{1 + 2y} \, dy$

9. $\int \frac{2}{x - 3} \, dx$

10. $\int \frac{5x^2}{(x^3 - 7)^2} \, dx$

11. $\int \frac{(\sqrt{x} + 2)^5}{\sqrt{x}} \, dx$

12. $\int 6u^4 \sqrt{1 - u^5} \, du$

13. $\int \sqrt{x^3 + 12x^2 + 3}(3x^2 + 24x) \, dx$

14. $\int \frac{e^x}{e^x + 1} \, dx$

15. $\int e^x \sqrt{e^x + 1} \, dx$

16. $\int 7x^3 e^{x^4} \, dx$

17. $\int \frac{e^{2u} - e^{-2u}}{e^{2u} + e^{-2u}} \, du$

18. $\int \frac{\ln 5x}{3x} \, dx$

19. $\int \frac{2x \ln (x^2 + 4)}{x^2 + 4} \, dx$

20. $\int (1 + e^{-2t})(2t - e^{-2t})^3 \, dt$

21. $\int xe^{-x^2} \, dx$

22. $\int x\sqrt{x - 3} \, dx$

23. $\int \frac{x}{x + 2} \, dx$ (*Hint:* Do the division first.)

24. $\int \frac{t}{t - 3} \, dt$ (*Hint:* Do the division first.)

25. $\int (u - 2)\sqrt{u - 1} \, du$

For Problems 26–45, use the method of substitution to evaluate the definite integrals.

26. $\int_{-3}^{1} x^2(1 + x^3)^2 \, dx$

27. $\int_{2}^{3} (3w - 2)^3 \, dw$

28. $\int_{2}^{4} \frac{2x - 1}{(x^2 - x + 3)^2} \, dx$

29. $\int_{1}^{3} \frac{t + 3/2}{t^2 + 3t - 2} \, dt$

30. $\int_{1}^{2} \frac{3u}{u^2 + 1} \, du$

31. $\int_{1}^{4} \frac{1}{2u + 3} \, du$

32. $\int_{0}^{7} \sqrt{t + 2} \, dt$

33. $\int_{0}^{1} \frac{1}{\sqrt{3w + 1}} \, dw$

34. $\int_{-2}^{1} \frac{x}{\sqrt{x^2 + 16}} \, dx$

35. $\int_{0}^{3} s\sqrt{9 - s^2} \, ds$

36. $\int_{0}^{3} \frac{r}{\sqrt{4 - r}} \, dr$

37. $\int_{2}^{6} \sqrt{10 + 3x} \, dx$

38. $\int_{0}^{1} (s^5 + 3)\sqrt{s^6 + 18s} \, ds$

39. $\int_{1}^{4} 2x^3 \sqrt{1 + 5x^4} \, dx$

40. $\int_{-1/3}^{2/3} \frac{1}{\sqrt{2 + 3t}} \, dt$

41. $\int_{2}^{4} \frac{2 + 2x^3}{\sqrt[4]{4x + x^4}} \, dx$

42. $\int_{1}^{3} \sqrt{4s + 5} \, ds$

43. $\int_{0}^{1} \frac{3u}{\sqrt{4 - u^2}} \, du$

44. $\int_{0}^{2} x^2 \sqrt{x^3 + 1} \, dx$

45. $\int_{0}^{\sqrt{3}} \frac{t}{(4 - t^2)^{3/2}} \, dt$

Antiderivative in Business

46. Finding Total Cost   Marginal cost is given by

$$C'(q) = \frac{6}{\sqrt{q + 5}}$$

where q is the number of units produced. What is the cost in increasing production from 50 to 100 units? From 100 to 200 units?

Coefficient of Inequality

47. Coefficient of Inequality   Let the Lorentz curve be

$$L(x) = \frac{7}{8}x^2 + \frac{1}{8}x$$

Find the coefficient of inequality.

48. Coefficient of Inequality   The Lorentz curves for two countries are given below:

$$\text{Country A:} \quad L_A(x) = \frac{15}{16}x^2 + \frac{1}{16}x$$

$$\text{Country B:} \quad L_B(x) = \frac{7}{8}x^2 + \frac{1}{8}x$$

What are the coefficients of inequalities for each of the countries? Can you say anything about the economic environment in these two countries?

A common mathematical model that is often used to describe the Lorentz curve is

$$L(x) = \frac{k - 1}{k}x^2 + \frac{1}{k}x$$

for $k = 1, 2, \ldots$ and $0 \leq x \leq 1$. In other words, a given country may be described by some constant k.

49. Show that $L(x)$ satisfies the two boundary conditions $L(0) = 0$ and $L(1) = 1$.

50. Find the coefficient of inequality as a function of k.

51. Sketch the graphs of $L(x)$ for $k = 1, 2, 3,$ and 10.

52. Verify that for any k the above Lorentz curve lies below the line $y = x$.

Epilogue: Typical Lorentz Curves

The following Lorentz curves and their coefficients of inequality illustrate how economists compare incomes between different groups and classes of people.*

Coefficient of inequality as a function of age. See Figure 21 and Table 8.

Coefficient of inequality as a function of education. See Figure 22 and Table 9.

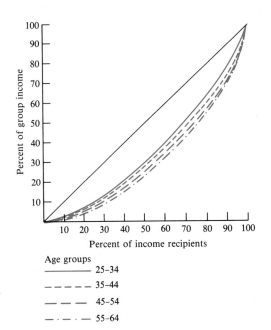

Age groups

— 25–34

– – – – 35–44

– – – 45–54

– · – · 55–64

Figure 21

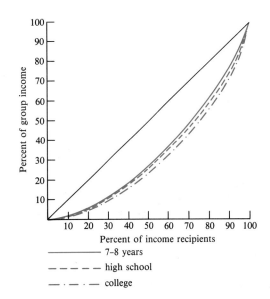

——— 7–8 years

– – – – high school

– · – · – college

Figure 22

TABLE 8

Age Group	Coefficient of Inequality
25–34	0.25
35–44	0.27
45–54	0.29
55–64	0.31

TABLE 9

Education	Coefficient of Inequality
7–8 years	0.31
high school	0.32
college	0.34

* These Lorentz curves were taken from the Ph.D thesis of Jacob Mincer, "A Study of Personal Income Distribution," Columbia University, 1957.

Coefficient of inequality as a function of occupation. See Figure 23 and Table 10.

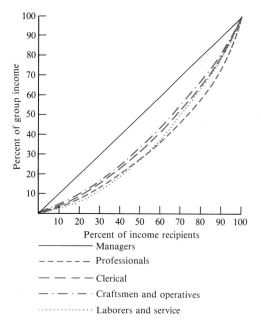

Figure 23

Coefficient of inequality as a function of sex. See Figure 24 and Table 11.

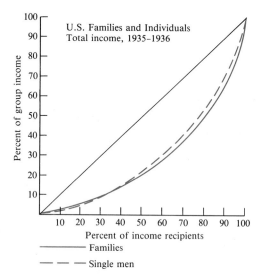

Figure 24

TABLE 10

Occupation	Coefficient of Inequality
Managers	0.25
Professionals	0.27
Clerical	0.29
Craftsmen	0.31
Laborers and Service	0.34

TABLE 11

Sex	Coefficient of Inequality
Female	0.23
Male	0.26

Key Ideas for Review

Review Exercises

For Problems 1–6, find the indefinite integrals.

1. $\int 7\, dx$

2. $\int (x^4 + 3x^2 + 1)\, dx$

3. $\int \sqrt{u}(u^2 + 1)\, du$

4. $\int \sqrt{x}(x + 1)(x - 1)\, dx$

5. $\int (e^x + e^{-x})\, dx$

6. $\int \frac{t + t^2}{t}\, dt$

For Problems 7–12, find the particular antiderivative of the given derivative determined by the value of the function.

7. $f'(t) = 2t + 5$ $f(0) = 0$ **8.** $f'(t) = t^2$ $f(1) = 0$

9. $f'(x) = e^x$ $f(0) = 1$ **10.** $g'(x) = 2x + e^x$ $g(0) = 5$

11. $h'(u) = u\sqrt{u^2 - 9}$ $h(5) = 0$

12. $f'(x) = 0$ $f(0) = 0$

For Problems 13–15, find the total cost given the marginal cost.

13. $C'(q) = 45q - 2q^2$, $C(0) = 10{,}000$

14. $C'(q) = 100q - q^2$, $C(0) = 1000$

15. $C'(q) = 10 - 2e^{-0.5q}$, $C(0) = 1000$

For Problems 16–22, evaluate the definite integrals.

16. $\int_0^1 (x + 1)\, dx$

17. $\int_{-1}^1 (1 - x^2)\, dx$

18. $\int_5^{10} 5\, dx$

19. $\int_0^1 (x + e^{-x})\, dx$

20. $\int_1^3 2x(x^2 + 3)\, dx$

21. $\int_1^e \left(1 + \frac{1}{x}\right) dx$

22. $\int_0^1 \frac{1}{\sqrt{4u + 3}}\, du$

For Problems 23–30, sketch the graph of the function over the indicated interval. Evaluate the integral that gives the area between the curve and the x-axis over the indicated interval.

23. $f(x) = x + 1$ $[0, 5]$

24. $f(x) = 4 - x$ $[0, 4]$

25. $f(x) = 4 - x^2$ $[0, 2]$

26. $f(x) = 1 + e^x$ $[0, 1]$

27. $f(x) = e^x + e^{-x}$ $[-1, 1]$

28. $f(x) = x^n$ $[0, 1]$ for $n \geq 0$

29. $f(x) = \frac{1}{x^n}$ over $[1, 2]$ for $n \geq 0$

30. $f(x) = \sqrt{x} + 1$ over $[0, 1]$

For Problems 31–39, use the method of substitution to find the indefinite integral.

31. $\int x(x^2 - 1)^3\, dx$

32. $\int (x + 1)(x - 1)^4\, dx$

33. $\int u(u + 1)^5\, du$

34. $\int u(u^2 + 4)^2\, du$

35. $\int 5(x + 1)^{10}\, dx$

36. $\int \frac{1}{1 + 2x}\, dx$

37. $\int \frac{5}{x - 1}\, dx$

38. $\int \frac{x^2}{(x - 1)^2}\, dx$

39. $\int \frac{1}{t + 1}\, dt$

For Problems 40–44, use the method of substitution to evaluate the definite integrals.

40. $\int_0^1 x^2(1 + x^2)^3\, dx$

41. $\int_0^4 s\sqrt{s^2 + 1}\, ds$

42. $\int_0^1 \frac{r}{\sqrt{1 + r}}\, dr$

43. $\int_1^6 \sqrt{10 + x}\, dx$

44. $\int_0^2 \frac{s}{(4 - s)^{5/2}}\, ds$

Business Management Problems

45. Total Revenue   The marginal revenue of a firm is given by
$$MR(q) = 100 - 5q$$
Find the total revenue.

46. Change in Revenue   If the marginal revenue is given by
$$MR(q) = 250 - 10q$$
determine the increase in revenue if the number of items produced q is increased from 100 to 150.

47. Change in Cost   If the marginal cost is given by
$$MC(q) = 10$$
find the increase in cost if the number of items produced q is increased from 1000 to 1500.

48. Change in Profit   If the marginal profit is
$$MP(q) = 10 - q$$
find the increase in profit if the number of items produced q is increased from 50 to 75.

49. Lorentz Curve   The wages in a given country are described by the Lorentz curve
$$L(x) = 0.9x^2 + 0.1x$$
(a) What percent of the total wages of the country are earned by the lowest-paid 10% of the work force?
(b) What percent of the total wages of the country are earned by the lowest-paid 50% of the work force?

(c) Show that $L(x)$ is an increasing function.

(d) Show that $L(x)$ is concave up.

(e) What is the coefficient of inequality?

50. Puzzle You have inherited 50 annual payments from a rich uncle. You can receive the payments in one of two ways: in increasing order or decreasing order as shown in Table 12. Which order should you choose if banks pay an interest of 8% annually compounded continuously? What is the present value of each continuous cash flow? *Hint:* The decreasing payments can be approximated by the continuous revenue function

$$R_1(t) = \$1000(0.9)^t = \$1000e^{-0.105t}$$

and the increasing payments can be approximated by the continuous revenue function

$$R_D(t) = \$1000(0.9)^{49-t} = \$5.73e^{0.105t}$$

TABLE 12
Payments for Problem 51

Decreasing Payments		Increasing Payments	
$1000	= $1000.00	$1000(0.9)^{49} =	$5.73
$1000(0.9)	= $900.00	$1000(0.9)^{48} =	$6.36
$1000(0.9)^2	= $810.00		
$1000(0.9)^3	= $729.00	$\vdots$	
$\vdots$		$1000(0.9)^3 =	$729.00
		$1000(0.9)^2 =	$810.00
$1000(0.9)^{48} =	$6.36	$1000(0.9) =	$900.00
$1000(0.9)^{49} =	$5.73	$1000 =	$1000.00

Chapter Test

1. Find the following antiderivative:

$$\int \frac{(t^2 + t + 1)}{t}\, dt$$

2. Evaluate the definite integral

$$\int_0^1 \frac{x}{(x^2 + 1)^3}\, dx$$

3. Use the method of substitution to find the antiderivative of

$$\int u^2 \sqrt{1 - u^3}\, du$$

4. The marginal cost of a product is given by

$$C'(q) = 10 + 2e^{-0.10q}$$

Find the increase in cost if the number of items produced q is increased from 10 to 20.

5. If the marginal profit is

$$MP(q) = 5 - 2q$$

find the increase in profit if the number of items produced q is increased from 100 to 200.

12

Integration Techniques
and Applications

In this chapter we introduce a number of important ideas that are related to the integral. We begin in Section 12.1 by showing how to find the area of an unbounded region in the plane using the concept of the improper integral. Next, in Section 12.2 we introduce a technique known as integration by parts, which helps to find the antiderivative of a function and thus evaluate its definite integral. Then in Section 12.3 we show how the definite integral can be used to find the area between two graphs as well as to find the volume of certain solids in three dimensions. In Section 12.4 we show how to approximate the integral of functions that cannot be integrated by the usual methods using numerical methods. Finally, in Section 12.5 we introduce differential equations and show they can be used to describe growth and decay phenomena.

12.1 Improper Integrals

PURPOSE

We show how it is possible to find the area of an unbounded region of the plane by taking the limit of a sequence of integrals, each defined over a finite interval. This introduces the idea of the improper integral.

The concept of the improper integral is then used to find the present value of a continuous cash flow.

Introduction

Consider for a moment a region in the plane that is unbounded. That is, it goes off in some direction to infinity as illustrated by the regions in Figure 1. One is tempted to say that the area of such an unbounded region is infinite. It is often true that the area of an unbounded region is infinite. Certainly, we would say that the area of the entire plane is infinite. However, there are unbounded regions to which it is perfectly reasonable to assign finite areas. A region of the plane may be unbounded, but if the portion of the region that goes off to infinity becomes narrower and narrower, it might be the case that the total area is finite. (See Figure 2.)

In fact, there are unbounded regions in the plane that may reach to the moon from where you are now sitting but have an area equal to the area of the page you are now reading. The problem is to find these areas, and the tool for doing this is the improper integral.

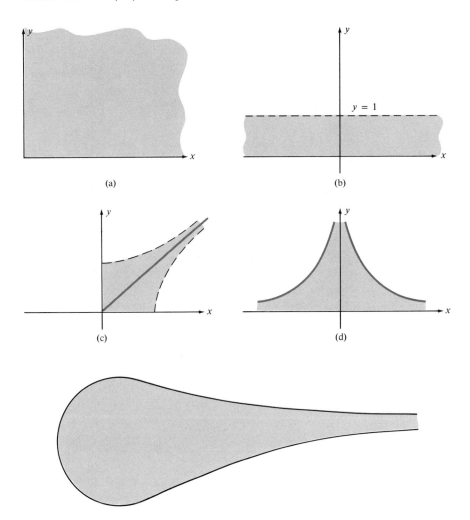

Figure 1
Four unbounded regions in the plane

(a)

(b)

$y = 1$

(c)

(d)

Figure 2
Is this area finite?

Improper Integrals

We begin with a simple unbounded region. We consider the region that lies between the graph of the function

$$f(x) = \frac{1}{x^2} \qquad (1 \le x < \infty)$$

and the x-axis for x greater than or equal to 1. (See Figure 3.) We ask if it is possible to assign a finite area to this region. The problem is fascinating because, although the region is unbounded, its height becomes shorter and shorter as we move farther and farther to the right. Clearly, as we enclose more and more of the region by moving to the right, the area of the enclosed region will become larger and larger. The important question is, "Does the enclosed area grow without bound, or does it approach some finite limit?" If it approaches a finite limit, we call this limit the **area** of the region. If the limit does not exist, we say that the area of the region is infinite or that the area is **diverging**.

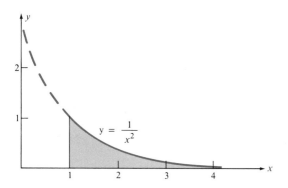

Figure 3
Finding the area of
an unbounded region

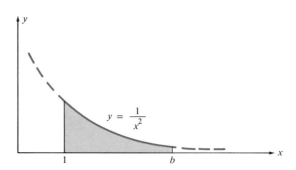

Figure 4
Finding the area over
a finite interval

To determine whether the area exists, we begin by selecting an arbitrary number $b > 1$ and evaluate the integral

$$\int_1^b \frac{dx}{x^2}$$

(See Figure 4.) Evaluating this integral, we have

$$\int_1^b \frac{dx}{x^2} = \int_1^b x^{-2}\, dx$$

$$= -x^{-1}\Big|_1^b$$

$$= -\left(\frac{1}{b} - 1\right)$$

$$= 1 - \frac{1}{b}$$

This expression in b gives the area of the shaded region between 1 and b as shown in Figure 4. For example, when $b = 1000$, we see that the area under the graph between 1 and 1000 is

$$\text{Area between 1 and 1000} = 1 - \frac{1}{1000} = 0.999$$

To find the area under the entire curve for $x \geq 1$, we take the limit as $b \rightarrow \infty$. This gives

$$\text{Area of unbounded region} = \lim_{b \rightarrow \infty} \left(1 - \frac{1}{b}\right) = 1$$

In other words, although the region is unbounded, the area is only one square unit. This limiting value (1 in this case) is called the value of the improper integral and is denoted by

$$\int_1^\infty \frac{dx}{x^2}$$

The above discussion suggests the following definition of improper integrals.

Definition of Improper Integrals

Let f be a continuous function defined over the indicated intervals and assume the following limits exist. We define the **improper integrals** as follows:

- Right infinite integral:

$$\int_a^\infty f(x)\,dx = \lim_{b \to \infty} \int_a^b f(x)\,dx$$

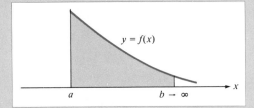

- Left infinite integral:

$$\int_{-\infty}^b f(x)\,dx = \lim_{a \to -\infty} \int_a^b f(x)\,dx$$

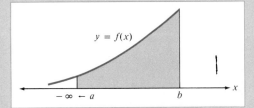

- Doubly infinite integral:

$$\int_{-\infty}^\infty f(x)\,dx = \int_{-\infty}^c f(x)\,dx + \int_c^\infty f(x)\,dx$$

$$= \lim_{a \to -\infty} \int_a^c f(x)\,dx + \lim_{b \to \infty} \int_c^b f(x)\,dx$$

The value of c can be chosen in any way one likes. It will not affect the result. It is often convenient to pick zero for the value of c.

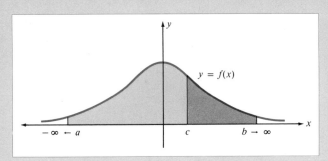

Example 1

Improper Integral   Determine whether the area between the graph of $y = 1/x$ and the x-axis for $x \geq 1$ is finite or infinite. (See Figure 5.)

Solution   The area is given by the improper integral

$$\int_1^\infty \frac{dx}{x}$$

Using the definition of the improper integral, we have

$$\int_1^\infty \frac{dx}{x} = \lim_{b \to \infty} \int_1^b \frac{dx}{x}$$

$$= \lim_{b \to \infty} \ln x \Big|_1^b$$

$$= \lim_{b \to \infty} (\ln b - \ln 1)$$

$$= \lim_{b \to \infty} \ln b$$

Since $\ln b \to \infty$ as $b \to \infty$, we say that the improper integral diverges or that the area is infinite.   □

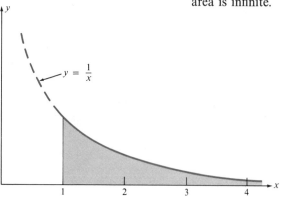

Figure 5

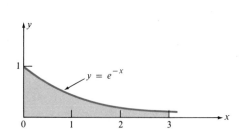

Figure 6
Improper integral

Example 2

Improper Integral   Find the area between graph of $y = e^{-x}$ and the x-axis for $x \geq 0$. (See Figure 6.)

Solution   The area is described by the improper integral

$$\int_0^\infty e^{-x}\, dx$$

Using the definition of the improper integral, we have

$$\int_0^\infty e^{-x}\, dx = \lim_{b \to \infty} \int_0^b e^{-x}\, dx$$

$$= \lim_{b \to \infty} -e^{-x} \Big|_0^b$$

$$= \lim_{b \to \infty} (-e^{-b} + 1)$$

$$= \lim_{b \to \infty} \left(1 - \frac{1}{e^b}\right)$$

$$= 1$$

In other words, although the region has a "width" that is wider than the length of the United States (much, much wider), the area shown in Figure 6 is only one square unit (less than one square inch).   □

Example 3 ──────────── Improper Integral   Evaluate the improper integral

$$\int_{-\infty}^{\infty} xe^{-x^2}\, dx$$

See Figure 7.

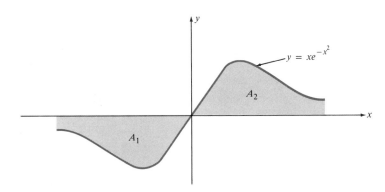

Figure 7

Solution   The area of this region is given by the value of the improper integral

$$\int_{-\infty}^{\infty} xe^{-x^2}\, dx$$

By definition we can write

$$\int_{-\infty}^{\infty} xe^{-x^2}\, dx = \int_{-\infty}^{0} xe^{-x^2}\, dx + \int_{0}^{\infty} xe^{-x^2}\, dx$$

$$= \lim_{a \to -\infty} \int_{a}^{0} xe^{-x^2}\, dx + \lim_{b \to \infty} \int_{0}^{b} xe^{-x^2}\, dx$$

$$= \lim_{a \to -\infty} \left[-\frac{1}{2}e^{-x^2} \right]\Big|_{a}^{0} + \lim_{b \to \infty} \left[-\frac{1}{2}e^{-x^2} \right]\Big|_{0}^{b}$$

$$= -\frac{1}{2} \lim_{a \to -\infty} \left[e^{-0^2} - e^{-a^2} \right] - \frac{1}{2} \lim_{b \to \infty} \left[e^{-b^2} - e^{-0^2} \right]$$

$$= -\frac{1}{2}(1 - 0) - \frac{1}{2}(0 - 1)$$

$$= -\frac{1}{2} + \frac{1}{2}$$

$$= 0$$

The improper integral has a value of 0. Since the graph of the function is not always positive, the integral will not represent any area. Notice, however, in Figure 7 by the symmetric nature of the graph that the area A_1 that lies below the x-axis is the same as the area A_2 that lies above the x-axis. When this is true, we expect the value of the definite integral to be 0.   □

Present Value of a Continuous Cash Flow

It often happens in financial transactions that an individual or firm will receive a continuous cash flow or revenue stream over a period of time. For example, an oil well might yield a company continuous revenue over a period of time. A pension fund might be paid to a retiree in varying amounts for a number of years.

Suppose a baseball player has just signed a five-year contract that calls for annual payments to the player (or to his estate in event of death) over the next 50 years according to Table 1. The first payment consists of $75,000, and each succeeding payment is 2% less than the previous payment.

TABLE 1

Revenue Stream of Fifty Annual Payments

Year	Annual Payment	
1988	$75,000	= $75,000.00
1989	$75,000(0.98)	= $73,500.00
1990	$75,000(0.98)^2	= $72,030.00
1991	$75,000(0.98)^3	= $70,598.40
1992	$75,000(0.98)^4	= $69,177.61
⋮	⋮	⋮
2035	$75,000(0.98)^{47}	= $29,019.29
2036	$75,000(0.98)^{48}	= $28,438.91
2037	$75,000(0.98)^{49}	= $27,870.13

Although the revenue flow described in Table 1 is not paid to the player continuously (it is paid in 50 discrete payments), it can fairly accurately be described as a continuous cash flow and represented by the continuous revenue function

$$R(t) = \$75{,}000e^{-0.02t}$$

where t is time in years measured from when the payments are begun. This function is shown in Figure 8.

The value of the continuous revenue function for $t = 0, 1, \ldots, 50$ gives the same payments as those listed in Table 1. Also, the area between the revenue function and the t-axis for t between 0 and 50 is very close to the sum of the 50 payments made to the player. Since it is easier to analyze $R(t)$ than an entire

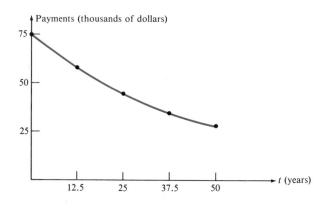

Figure 8

table of numbers, it is easy to see the usefulness of approximating the table of payments by a continuous function $R(t)$.

In the mathematics of finance, one learns that if a present amount P is deposited in a bank that pays an annual interest of r percent compounded continuously, then the future value F of that account after t years will be given by

$$F = Pe^{rt}$$

Turning this equation around and solving for P, we see that if an account will be worth F dollars t years from now, then it has a present value P today of

$$P = Fe^{-rt}$$

However, if a steady cash flow $R(t)$ is deposited into this account during these t years, then the present value of the account is given by the following **present value integral**.

Present Value of a Continuous Cash Flow

If $R(t)$ is the rate at which money flows into an account (dollars per year) over the next T years, and if the account pays an annual interest of r percent compounded continuously, then the present value of this account in today's dollars is

$$P = \int_0^T R(t)e^{-rt}\,dt$$

In the special case of perpetual cash flow (goes on forever), the present value of the account is given by the improper integral

$$P = \int_0^\infty R(t)e^{-rt}\,dt$$

Example 4

Perpetual Bond An opera singer signs a contract that calls for her and her heirs to receive constant payments of $100,000 per year forever. What is the value of this contract in today's dollars? Assume that money will always earn an annual interest of $r = 10\%$, compounded continuously.

Solution If the annual interest rate paid for borrowing money is 10%, then the current value of a constant perpetual cash flow of $R = \$100,000$ per year is

$$P = \int_0^\infty Re^{-rt}\,dt$$

$$= \$100,000 \lim_{b \to \infty} \int_0^b e^{-0.10t}\,dt$$

$$= \$100,000 \lim_{b \to \infty} \left[-\frac{1}{0.10} e^{-0.10t} \right]\Big|_0^b$$

$$= \frac{\$100,000}{0.10} \lim_{b \to \infty} \left[1 - e^{-0.10b} \right]$$

$$= \frac{\$100,000}{0.10}$$

$$= \$1,000,000$$

In other words, the contract is worth $1,000,000 in today's value. This is the same as saying that the owner of the opera company can make the annual $100,000 payments out of a special $1,000,000 bank account that pays 10% annual interest compounded continuously. These payments would draw on the principle of the account but would represent interest paid by the bank. Of course, the owner will pay a much larger sum than $1,000,000 over the infinite period of time (in fact the total sum of annual $100,000 payments in perpetuity is infinite). Fortunately for the owner, this number is meaningless. A payment of $100,000 has no meaning unless one says *when* it is made. One should realize that a payment of $100,000 today is not the same as this same payment in 100 years. If one assumes that inflation increases at the same 10% rate that the bank pays for borrowed money, then $100,000 in 100 years will be worth (in today's dollars) roughly the price of the movie ticket ($4.54).   □

Example 5 ———————   Present Value of a Perpetual Fund   You have inherited an old gold mine that generates a cash flow given by

$$R(t) = \$50,000e^{-0.2t}$$

where t is time measured in years. In other words, the mine is dwindling out at the rate given by the values in Table 2. If the annual interest rate paid by banks is 8% compounded continuously, what is the present value of this mine?

TABLE 2 ———————
Cash Flow of a Mine

Year, t	Yearly Earnings
0 (present)	$50,000
1 (next year)	$50,000e^{-0.2} = \$40,936.54$
2	$50,000e^{-0.4} = \$33,516.00$
3	$50,000e^{-0.6} = \$27,440.58$
4	$50,000e^{-0.8} = \$22,466.45$
5	$50,000e^{-1} = \$18,393.97$
⋮	⋮

Solution The current value of this mine is given by the present value integral

$$P = \int_0^\infty R(t)e^{-0.08t}\,dt$$

$$= \$50,000 \int_0^\infty e^{-0.2t}e^{-0.08t}\,dt$$

$$= \$50,000 \int_0^\infty e^{-0.28t}\,dt$$

$$= \$178,571.43$$

In other words, your inheritance is worth $178,571.43, which is the present value of the mine.   □

Problems

For Problems 1–25, evaluate (if the integrals exist) the following integrals.

1. $\int_0^\infty dx$

2. $\int_1^\infty \frac{1}{x^2}\, dx$

3. $\int_1^\infty \frac{1}{x^{3/4}}\, dx$

4. $\int_1^\infty \frac{1}{(x+1)^3}\, dx$

5. $\int_0^\infty t\, dt$

6. $\int_1^\infty \frac{t}{1+t^2}\, dt$

7. $\int_0^\infty \sqrt{t}\, dt$

8. $\int_0^\infty \frac{1}{x^{2/3}}\, dx$

9. $\int_{-\infty}^{-1} x^{-3}\, dx$

10. $\int_0^\infty \frac{x}{\sqrt{x^2+5}}\, dx$

11. $\int_{-\infty}^0 \frac{1}{1-x}\, dx$

12. $\int_{-\infty}^0 \frac{1}{(x-3)^2}\, dx$

13. $\int_{-\infty}^0 \frac{dx}{x-3}$

14. $\int_{-\infty}^{-4} \frac{2-3x}{x^2}\, dx$

15. $\int_{-\infty}^0 \frac{x+3}{x-3}\, dx$

16. $\int_{-\infty}^\infty \frac{3t^2}{(1+t^3)^3}\, dt$

17. $\int_{-\infty}^\infty \frac{x^3}{(x^4+1)^3}\, dx$

18. $\int_{-\infty}^\infty \frac{x}{\sqrt{x^2+2}}\, dx$

19. $\int_0^\infty e^{3x}\, dx$

20. $\int_0^\infty e^{-x}\, dx$

21. $\int_0^\infty e^{-kt}\, dt\ (k>0)$

22. $\int_0^\infty xe^{-x^2}\, dx$

23. $\int_{-\infty}^0 xe^{-x^2}\, dx$

24. $\int_{-\infty}^\infty \frac{e^x - e^{-x}}{e^x + e^{-x}}\, dx$

25. $\int_{-\infty}^\infty f(x)\, dx$ where $f(x) = \begin{cases} e^x & \text{for } -\infty < x \le 0 \\ 1 & \text{for } \quad 0 < x < \infty \end{cases}$

Continuous Cash Flows (over Finite Time)

26. **Value of Graduation Present** You have been given a graduation present consisting of a continuous cash flow of $500 per year for the next ten years. What is the present value of the gift if banks pay an annual rate of interest of 10% compounded continuously?

27. **Insurance Policy** Upon retirement an annuity will pay a retiree $25,000 for the first year with payments increasing by 3% per year thereafter for 25 years. If banks pay interest at an annual rate of 10%, what is the value of this policy? *Hint:* Approximate this revenue stream by the formula

$$R(t) = \$25{,}000e^{0.03t}$$

where R is the annual payment and t is time measured from the time payments begin.

28. **Basketball Contract** According to the newspapers, a basketball player has just signed a ten-million-dollar contract that calls for the player to be paid $250,000 this year, with an increase of 2% per year for the next 25 years. If banks are paying an annual interest of 12%, compounded continuously, how much is this contract really costing the owner?

29. **Company Buyout** Grandma's Cookies is being bought out by General Cookie Corporation. As payment, General Cookie Corp. has agreed to make annual payments according to the revenue function

$$R(t) = \$100{,}000e^{0.03t}$$

for the next 50 years. If banks pay annual interest at an annual rate of 9%, how much is General Cookie Corp. actually paying for the new company?

Continuous Cash Flow in Perpetuity

30. **Perpetuity Trust** What is the value of a trust that pays $2000 per year in perpetuity (forever) if interest is compounded continuously at a rate of 8% per year?

31. **Perpetuity Trust** Find the present value of a perpetual income trust that pays a yearly amount given by

$$R(t) = \$10{,}000e^{-0.05t}$$

Assume that interest rates are 8% per year, compounded continuously. What would interest rates have to be for the present value to be infinite?

32. **Perpetuity Trust** What is the present value of a perpetual income trust that pays a yearly amount given by

$$R(t) = \$10{,}000e^{-0.05t}$$

if interest rates are 6% compounded continuously?

33. **Value of Constant Payments** Someone offers to set up a trust fund in which you (and your estate) will be given $10 per year in perpetuity (forever). Assuming that banks pay annual interest of 8% compounded continuously, how much is this person giving you?

34. **Value of Constant Payments** Someone offers to set up a trust fund in which you (and your estate) will be given R dollars per year in perpetuity. Assuming that banks pay annual interest of r percent compounded continuously, how much is this person actually giving you? This is the general formula for the present value of constant payment in perpetuity funds.

35. **Puzzler** Would you rather collect $150 per year for 50 years or $100 per year in perpetuity? Assume that banks pay an annual interest of 10% compounded continuously.

36. Super Baseball Contract   In their attempt to finally win the Central League pennant, the Toledo Cubs have signed the Puerto Rican sensation, Rico Hernandez. They have agreed to give him almost everything except the clubhouse. The final 20-year contract calls for a yearly salary $R(t)$ of

$$R(t) = \$500{,}000e^{-0.04t}$$

where t is time measured in years. If annual interest rates being paid by the banks are 8% compounded continuously, how much is this contract worth?

12.2   Integration by Parts

PURPOSE

We introduce the powerful integration by parts formula, which will allow us to evaluate many integrals that cannot be evaluated by other methods. We will see that the integration by parts formula is "analogous" to the product formula for differentiation.

Introduction

There are many types of integrals that we cannot yet evaluate. For example, the integral

$$\int xe^x \, dx$$

cannot be evaluated by the method of substitution or any other method that we have seen thus far. It would take a clever reader indeed to find this antiderivative by picking and choosing functions by a trial-and-error method. We have already seen how the chain rule for finding derivatives motivated the substitution method for finding antiderivatives. We will see now that the product rule for finding derivatives gives rise to another method for finding antiderivatives, the integration by parts method.

The Integration by Parts Formula

Earlier, we learned the following rule for differentiating products:

$$\frac{d}{dx} f(x)g(x) = f'(x)g(x) + f(x)g'(x)$$

If we set the antiderivative of the left-hand side of this equation (with $C = 0$) equal to the antiderivative of the right-hand side, we get

$$f(x)g(x) = \int f'(x)g(x) \, dx + \int f(x)g'(x) \, dx$$

By simply rewriting this equation as

$$\int f(x)g'(x) \, dx = f(x)g(x) - \int f'(x)g(x) \, dx$$

we get a powerful formula for finding the antiderivative on the left, provided that we can find the antiderivative of the right-hand side of the equation. This equation is the integration by parts equation. To use this equation, it is convenient to let

$$u = f(x)$$
$$v = g(x)$$

and hence

$$du = f'(x)\,dx$$
$$dv = g'(x)\,dx$$

The integration by parts formula can now be rewritten in the following more concise form.

Integration by Parts Formula for Indefinite Integrals

The integration by parts formula for indefinite integrals states that

$$\int u\,dv = uv - \int v\,du$$

The integration by parts method consists of using this formula.

To remember this formula just say, "*u* dee *v* equals *uv* minus *v* dee *u*."

The strategy to follow when using the integration by parts formula is to identify part of the integral in question as *u* and the other part as *dv*. If we identify *u* and *dv* correctly, the integration by parts formula will reduce the problem to evaluating an integral simpler than the original integral. We illustrate the process with a few examples.

Example 1

Integration by Parts   Find the antiderivative of

$$\int xe^x\,dx$$

Solution   There are two possibilities for *u*, either $u = x$ or $u = e^x$. Let us try

$$u = x$$
$$dv = e^x\,dx$$

Hence we have

$$du = dx$$
$$v = e^x$$

The integration by parts formula then gives

$$\underset{u\,dv}{\int xe^x\,dx} = \underset{uv}{xe^x} - \underset{v\,du}{\int e^x\,dx}$$

The new integral on the right-hand side of the above equation is easy to integrate. Integrating it, we get the original antiderivative as

$$\int xe^x\,dx = xe^x - e^x + C$$

Check:

$$\frac{d}{dx}\left[xe^x - e^x + C\right] = (xe^x + e^x) - e^x = xe^x \qquad \square$$

Note that if we made the alternative substitution

$$u = e^x$$
$$dv = x\,dx$$

the integration by parts formula would give us

$$\int xe^x\,dx = \frac{x^2}{2}e^x - \int x^2 e^x\,dx$$

Although this formula is mathematically correct, the integral on the right-hand side of the equation is more difficult to integrate than is the original integral.

Example 2

Integration by Parts   Find the antiderivative

$$\int \ln x\,dx$$

Solution   It is convenient to construct a table in which the substitutions for u and dv along with v and du are displayed. We call this table the Integration by Parts (IBP) Table. (See Table 3.)

TABLE 3
IBP Table for Example 2

Original Integral: $\int \ln x\,dx$	
Let	*Then*
$u = \ln x$	$du = \dfrac{dx}{x}$
$dv = dx$	$v = x$
IBP formula:	$\int u\,dv = uv - \int v\,du$
	$\int \ln x\,dx = x \ln x - \int dx$

Rewriting the above IBP formula and evaluating the simple integral on the right-hand side with the integrand 1, we get the desired antiderivative

$$\int \ln x\,dx = x \ln x - \int dx$$
$$= x \ln x - x + C$$

Check:

$$\frac{d}{dx}\left[x \ln x - x + C\right] = \ln x + \frac{x}{x} - 1 = \ln x \qquad \square$$

It is possible to find antiderivatives by repeated application of the integration by parts formula. In the following example we must use the integration by parts formula two times.

Example 3

Repeated Integration by Parts   Find the antiderivative of

$$\int x^2 e^x \, dx$$

Solution   Looking at the IBP formula in Table 4, we see that we have reduced the problem to finding a new antiderivative

$$2\int xe^x \, dx$$

TABLE 4
IBP Table for Example 3

Original Integral: $\int x^2 e^x \, dx$	
Let	*Then*
$u = x^2$	$du = 2x \, dx$
$dv = e^x \, dx$	$v = e^x$

IBP formula: $\int u \, dv = uv - \int v \, du$

$$\int x^2 e^x \, dx = x^2 e^x - 2\int xe^x \, dx$$

We have already found this antiderivative in Example 1 to be

$$\int xe^x \, dx = xe^x - \int e^x \, dx$$
$$= xe^x - e^x + C$$

Substituting this antiderivative into the IBP formula, we get

$$\int x^2 e^x \, dx = x^2 e^x - 2\int xe^x \, dx$$
$$= x^2 e^x - 2[xe^x - e^x + C]$$
$$= (x^2 - 2x + 2)e^x + C_1$$

where $C_1 = -2C$. (Do not be bothered by the notation; C_1 is still just an arbitrary constant.)

Check:

$$\frac{d}{dx}\left[(x^2 - 2x + 2)e^x + C_1\right] = (2x - 2)e^x + (x^2 - 2x + 2)e^x$$

$$= x^2 e^x \qquad \square$$

We now recap the properties that we look for in an integrand when deciding whether the integration by parts method is an appropriate method for integration.

Tips for Choosing u and dv

- You must be able to integrate dv.
- The IBP formula should produce an integral that is easier to integrate.
- For antiderivatives of the form $\int x^p e^{kx}\,dx$, let $u = x^p$ and $dv = e^{kx}\,dx$.
- For antiderivatives of the form $\int x^p(\ln x)^q\,dx$, let $u = (\ln x)^q$ and $dv = x^p\,dx$.

Evaluating Definite Integrals by Integration by Parts

We can evaluate definite integrals as well as indefinite integrals using the integration by parts formula. By using the fundamental theorem of calculus it is a simple matter to verify the following formula.

Integration by Parts Formula for Definite Integrals

$$\int_a^b u\,dv = uv\Big|_a^b - \int_a^b v\,du$$

Example 4

Integration by Parts   Evaluate

$$\int_0^1 x\sqrt{x+1}\,dx$$

Solution   We make the substitutions shown in Table 5.

TABLE 5
IBP Table for Example 4

Original Integral: $\int_0^1 x\sqrt{x+1}\,dx$

Let	Then
$u = x$	$du = dx$
$dv = \sqrt{x+1}\,dx$	$v = \dfrac{2}{3}(x+1)^{3/2}$

IBP formula: $\int_a^b u\,dv = uv\Big|_a^b - \int_a^b v\,du$

$$\int_0^1 x\sqrt{x+1}\,dx = \frac{2x}{3}(x+1)^{3/2}\Big|_0^1 - \frac{2}{3}\int_0^1 (x+1)^{3/2}\,dx$$

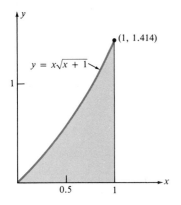

Figure 9
The area under $y = x\sqrt{x+1}$

By evaluating the integral on the right-hand side of the above IBP formula we get

$$\int_0^1 x\sqrt{x+1}\,dx = \frac{2x}{3}(x+1)^{3/2}\Big|_0^1 - \frac{(2)(2)}{(3)(5)}(x+1)^{5/2}\Big|_0^1$$

$$= \frac{2}{3}(2^{3/2}-0) - \frac{4}{15}(2^{5/2}-1)$$

$$\cong 0.6438$$

The numerical value of the definite integral represents the area under the curve shown in Figure 9.   □

Phenomena That Start Big and Die Out

Many phenomena initially increase in size or population with a burst of activity but decrease in size or population after a period of time. For instance, a population of bacteria might grow initially but then die out owing to natural factors. Also, the flow of oil from a newly drilled well might increase initially for a number of years but then start to decrease. Phenomena like this are common and can often be described by an equation of the form

$$f(t) = Ate^{-kt}$$

where A and k are positive constants. The general shape of the graph of this function is shown in Figure 10. The exact values of A and k depend on the phenomenon being described.

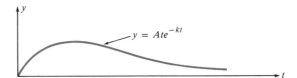

Figure 10
The general curve of $y = Ate^{-kt}$

Example 5

Gas Production   Oil engineers have started pumping gas from a new well in the Gulf of Mexico. On the basis of preliminary tests and past experience they predict that the monthly production of gas t months after pumping begins will be given by the function

$$p(t) = 3te^{-0.02t}$$

where $p(t)$ is measured in millions of cubic feet of gas. Estimate the total production in the first 12 months of operation.

Solution   The graph of the production curve $p(t)$ is shown in Figure 11. The total production during the first 12 months is found by evaluating the integral

$$\int_0^{12} 3te^{-0.02t}\,dt$$

To evaluate this definite integral, we construct the IBP table in Table 6.

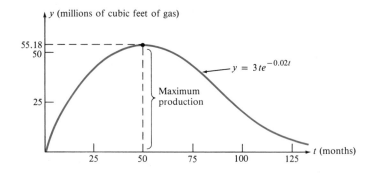

Figure 11
Monthly production of gas

TABLE 6
IBP Table for Example 5

Original Integral: $\int_0^{12} 3te^{-0.02t}\,dt$	
Let	*Then*
$u = 3t$ $dv = e^{-0.02t}\,dt$	$du = 3\,dt$ $v = -50e^{-0.02t}$

IBP formula: $\int_a^b u\,dv = uv\Big|_a^b - \int_a^b v\,du$

$$\int_0^{12} 3te^{-0.02t}\,dt = -150te^{-0.02t}\Big|_0^{12} + 150\int_0^{12} e^{-0.02t}\,dt$$

We can now evaluate the new integral on the right-hand side of the IBP formula, getting

$$150\int_0^{12} e^{-0.02t}\,dt = -(150)(50)e^{-0.02t}\Big|_0^{12}$$
$$= -7500[e^{-0.24} - 1]$$
$$= 7500[1 - e^{-0.24}]$$
$$= 1600$$

Substituting this back into the IBP formula, we have

$$\int_0^{12} 3te^{-0.02t}\,dt = -150te^{-0.02t}\Big|_0^{12} + 1600$$
$$= -150[(12)e^{-0.24} - 0] + 1600$$
$$\cong 184 \quad \text{million cubic feet}$$

Interpretation: The engineers can expect to pump 184 million cubic feet of gas during the first 12 months of production. We could also find the total production for any time period by integrating $p(t)$ over the appropriate interval. Table 7 gives the total production of this well for a number of time periods.

It would be interesting to evaluate the improper integral

$$\int_0^{\infty} 3te^{-0.02t}\,dt$$

to determine the total amount of gas in the well as predicted by the model. Without carrying out all the details of the computations we find that the improper

TABLE 7

Gas Pumped Over the First
t Years

First *t* Years	Production (million cubic feet)
1	184
2	1221
3	1871
4	2530
5	3164
6	3754
7	4289
8	4767
9	5186
10	5566
50	7499
100	7500

integral has the value

$$\int_0^\infty 3te^{-0.02t}\,dt = \lim_{b\to\infty} \int_0^b 3te^{-0.02t}\,dt$$

$$= \lim_{b\to\infty}\left[-150te^{-0.02t}\Big|_0^b - 7500e^{-0.02t}\Big|_0^b \right]$$

$$= 7500 \quad \text{million cubic feet of gas}$$

In other words, the engineers estimate that the well contains 7.5 billion cubic feet of gas. ☐

Problems

For Problems 1–10, use integration by parts to evaluate the indefinite integrals.

1. $\int xe^{3x}\,dx;$ $\qquad u = x,$ $\qquad dv = e^{3x}\,dx$

2. $\int x \ln 2x\,dx$ $\qquad u = \ln 2x,$ $\quad dv = x\,dx$

3. $\int (\ln x)^2\,dx$ $\qquad u = (\ln x)^2,$ $\quad dv = dx$

4. $\int \sqrt{x}\ln x\,dx$ $\qquad u = \ln x,$ $\quad dv = \sqrt{x}\,dx$

5. $\int x^3(x^2 + 2)^{1/2}\,dx$ $\quad u = x^2,$ $\quad dv = x(x^2 + 2)^{1/2}\,dx$

6. $\int 3xe^{-(2x+3)}\,dx$ $\qquad u = 3x,$ $\quad dv = e^{-(2x+3)}\,dx$

7. $\int x(\ln x)^2\,dx$ $\qquad u = (\ln x)^2,$ $\quad dv = x\,dx$

8. $\int \dfrac{x^3}{\sqrt{x^2 - 1}}\,dx$ $\quad u = x^2,$ $\quad dv = \dfrac{x}{\sqrt{x^2 - 1}}\,dx$

9. $\int \dfrac{x^3}{\sqrt{1 - x^2}}\,dx$ $\quad u = x^2,$ $\quad dv = \dfrac{x}{\sqrt{1 - x^2}}\,dx$

10. $\int x(x + 2)^4\,dx$ $\qquad u = x,$ $\quad dv = (x + 2)^4\,dx$

For Problems 11–14, find a formula for the indefinite integrals using integration by parts.

11. $\int axe^{-bx}\,dx,\ a, b > 0$ $\qquad$ **12.** $\int axe^{bx}\,dx,\ a, b > 0$

13. $\int ax \ln (bx)\,dx,\ a, b > 0$

14. $\int ax\sqrt{bx + c}\, dx$, $a, b, c > 0$

For Problems 15–16, find a formula for the indefinite integrals using integration by parts. The answers should be given in terms of the integrals in Problems 11–12. Let a and b be positive constants.

15. $\int ax^2 e^{-bx}\, dx$ **16.** $\int ax^2 e^{bx}\, dx$

For Problems 17–20, evaluate the definite integrals using integration by parts.

17. $\int_1^e x \ln x\, dx$ **18.** $\int_0^4 xe^{-2x}\, dx$

19. $\int_1^5 x^2 e^{2x}\, dx$ **20.** $\int_0^8 x^2\sqrt{1 + x}\, dx$

21. Area Under the Curve Find the area bounded by the curve $y = xe^x$, the x-axis, and the lines $x = 1$ and $x = e$.

22. Area Under the Curve Find the area under the curve $y = x \ln x$ for x between 1 and e.

23. Total Revenue Find the total revenue for a product for the next ten years if the demand is given by

$$D(t) = 100(1 - e^{-t})$$

and the price is given by

$$P(t) = 2t + 3$$

Hint: Revenue at time t is $R(t) = D(t)P(t)$.

24. Total Production On the basis of preliminary estimates the number of barrels of oil per day (in millions) that a new oil field in Alaska is estimated to produce is given by

$$p(t) = 2te^{-0.10t}$$

where t is time measured in years. How much oil will be pumped in the first ten years?

25. Finding the Robot's Batting Average On the basis of artificial intelligence principles a computer scientist is building a robot that can learn to hit a baseball. Initially, the robot cannot hit anything, but it learns quickly. A pitching machine is set up to throw 250 pitches every hour to the robot. On the basis of the design of the robot, the number of hits per hour the robot is expected to get satisfies the learning curve

$$H(t) = 250(1 - e^{-0.02t})$$

Sketch a rough graph of the learning curve. What is the total number of hits the robot is expected to get during the first 50 hours? With the total number of hits the robot gets, what will be its batting average after the first 50 hours?

26. Average Life of Carbon Isotope The amount of radioactive material remaining in a substance at time t is given by the decay curve

$$f(t) = Ae^{-kt} \qquad (k > 0)$$

where A is the amount of radioactive substance present in the material at time zero. The average life M of an atom of the radioactive substance is given by

$$M = \frac{1}{A}\int_0^\infty tf(t)\, dt$$

Find the average life of a carbon 14 atom if $k = 1.24 \times 10^{-4}$. *Hint:*

$$\lim_{n \to \infty} \frac{n}{e^n} = 0$$

Present Value Problems

27. Present Value of an Oil Well Suppose a newly discovered oil well is expected to earn income of

$$f(t) = 5t$$

where the income is given in millions of dollars and t is time measured in years. If money is discounted at the rate of 10% per year, then the present value of this oil well in today's dollars is given by

$$V = \int_0^\infty 5te^{-0.10t}\, dt$$

Find the present value of this new oil well. Would you be willing to sell the drilling rights for this well for 25 million dollars?

28. Present Value of an Oil Well A second oil well, much larger than the one in Problem 27, is found, and geologists predict a steady stream of income from this new well of

$$f(t) = 10t$$

How much is this well worth in today's dollars with money discounted at the rate of 10% per year? *Hint:*

$$\lim_{n \to \infty} \frac{n}{e^{kn}} = 0 \qquad (k > 0)$$

INTERVIEW WITH AETNA LIFE AND CASUALTY
151 Farmington Ave.
Hartford, Connecticut 06156

The *Aetna Life Insurance Company* and *Aetna Casualty and Surety Company* of Hartford, Connecticut, is one of the world's largest life and casualty insurance companies.

TOPIC: Calculus in the Insurance Industry

Random House: I suspect most people would be hard pressed to think of a reason why calculus and in particular integration would be useful in the insurance industry. Could you give an instance of where the integral is used in your industry?

Spokesperson: The theory behind the frequency and severity of automobile accidents uses a probability distribution formula to descibe the number and size of payments we expect to make from various kinds of accidents. When we want to know how much we will pay on all policies under $50,000, we integrate this formula from 0 to $50,000.

Random House: When I think of the insurance industry, I think of exponential functions describing mortality tables. Is it true that many of the functions that needed to be integrated in your industry are exponential functions?

Spokesperson: Yes. In fact the probability function that I just mentioned is an exponential function.

Random House: I know that to become a full-fledged actuarial scientist, you need to take a rigorous examination in calculus. Could a good student

in calculus take this examination immediately after taking a college course in calculus?

Spokesperson: Yes, in fact that is the best time to do so. It would be wise, however, to take a practice test provided by the Society of Actuaries to ensure you've covered all the right materials.

Random House: I hate to ask, being a college professor, but what exactly would a young college graduate who had passed the calculus actuarial exam expect to make in the insurance industry?

Spokesperson: A graduate with no exams would start at about $20,000 and would have a tough time getting hired as an actuarial student. Having passed the calculus exam raises the salary expectations to the low/mid twenties and also makes the landing of a job much easier. Ultimately, a new Fellow who has passed all ten exams can expect to make $50,000 or more.

Random House: What would be some good courses other than calculus, probability, and statistics for students to take if they are interested in a career in the insurance industry?

Spokesperson: I firmly believe the best background is still a liberal arts education. The mathematics is critical but just as important is the ability to communicate the solutions. Obvious course work would include Economics and English.

12.3   Areas and Volumes

PURPOSE

We show how the definite integral can be used to find the area between two graphs in the plane. We then move to three dimensions and show how to find the volume of a solid of revolution by the method of disks.

We close by introducing the important economic concepts known as the consumers' surplus and the producers' surplus.

Introduction

Today, we sometimes think that without the integral calculus it would be impossible to find areas other than those of simple figures such as rectangles, triangles, and the like. This is not exactly true. Over 2000 years ago, the brilliant Greek mathematician Archimedes of Syracuse (287(?)–212 B.C.) proved that the area of a circle is the same as the area of a right triangle (a triangle with one right angle) with the length of one leg equal to the circumference of the circle and the length of the other leg equal to the radius of the circle. (See Figure 12.)

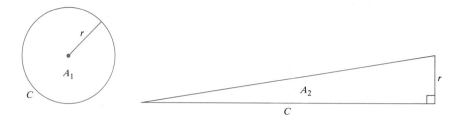

Figure 12
Archimedes' great discovery:
$A_1 = A_2$

This fact essentially proves the formula $A = \pi r^2$ for the area A of a circle in terms of its radius. However, to find the area of most regions in the plane, it was necessary to wait another 1900 years until the development of the integral calculus.

Area Between Graphs

We saw earlier how the integral of a nonnegative function f represented the area between the graph of $y = f(x)$ and the x-axis. We can generalize this concept to find the area between two graphs, since the x-axis is merely the graph $y = 0$. Consider finding the shaded area A shown in Figure 13 between the graphs $y = f(x)$ and $y = g(x)$ and the two vertical lines $x = a$ and $x = b$.

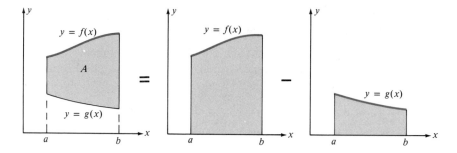

Figure 13
Area expressed as the difference between two areas

Since the area between the two graphs is clearly the area under the top graph $y = f(x)$ minus the area under the bottom graph $y = g(x)$, we can write

Area between $f(x)$ and $g(x)$ = Area under $f(x)$ − Area under $g(x)$

$$= \int_a^b f(x)\,dx - \int_a^b g(x)\,dx$$

$$= \int_a^b [f(x) - g(x)]\,dx$$

This gives rise to the general formula for finding areas between two graphs.

Area Between the Graphs of Two Functions

Let f and g be continuous functions defined on the interval $[a, b]$ with $f(x) \geq g(x)$ for all x in this interval. The area A between the graphs $y = f(x)$ and $y = g(x)$ over the interval $[a, b]$ is given by

$$A = \int_a^b [f(x) - g(x)]\, dx$$

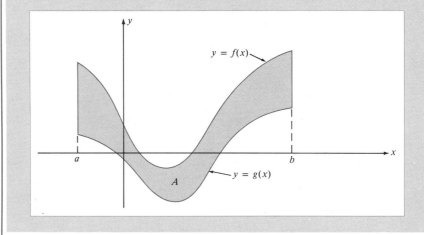

Example 1

Areas Between Graphs   Find the area between the curves $y = e^{-x}$ and $y = x^2$ on the interval $[1, 2]$.

Solution   It is always useful to sketch the graphs of the two functions to determine which graph is the upper boundary and which graph is the lower boundary. (See Figure 14.)

We see that $y = x^2$ is the upper boundary and $y = e^{-x}$ is the lower boundary. Hence we write

$$A = \int_1^2 (x^2 - e^{-x})\, dx$$

$$= \left(\frac{x^3}{3} + e^{-x} \right)\bigg|_1^2$$

$$= \left(\frac{8}{3} + e^{-2} \right) - \left(\frac{1}{3} + e^{-1} \right)$$

$$= \frac{7}{3} + e^{-2} - e^{-1}$$

$$\cong 2.101 \quad \text{square units} \qquad \square$$

Example 2

Area Between Graphs   Find the area between the graphs $f(x) = x^2 - 2$ and $g(x) = 2x + 1$.

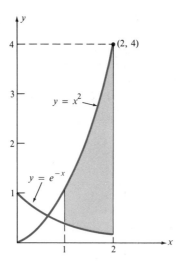

Figure 14
The area between two curves

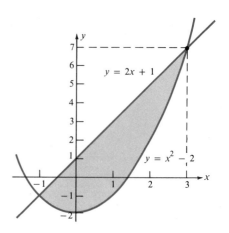

Figure 15
The area bounded by two graphs

Solution   The graphs of the functions are shown in Figure 15. To find the area of the region enclosed between the graphs, we must find the values of x where the graphs intersect. We find these values by setting $f(x)$ equal to $g(x)$ and solving for x. This will give us the endpoints of the interval needed as the limits of integration. Doing this, we get

$$x^2 - 2 = 2x + 1$$
$$x^2 - 2x - 3 = 0$$
$$(x - 3)(x + 1) = 0$$

Hence the limits of integration are

$$x = -1, 3$$

and the area of the region bounded by the two graphs is found by computing

$$A = \int_{-1}^{3} [\text{Upper boundary} - \text{Lower boundary}]\, dx$$

$$= \int_{-1}^{3} [(2x + 1) - (x^2 - 2)]\, dx$$

$$= \left[(x^2 + x) - \left(\frac{x^3}{3} - 2x \right) \right] \Big|_{-1}^{3}$$

$$= \frac{32}{3} \quad \text{square units} \qquad \qquad \square$$

We often wish to find the area between two graphs where the graphs cross each other somewhere in the interval of interest. The next example illustrates how to handle such problems.

Example 3 _____   Areas Between Intersecting Graphs   Find the area A of the region between $f(x) = x^2$ and $g(x) = \sqrt{x}$ over the interval $[0, 4]$.

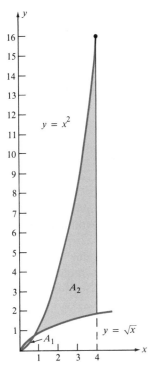

Figure 16
The area between two graphs

Solution   Again, these functions are easy to graph. (See Figure 16.)

This problem differs from the previous ones, since on the interval $[0, 4]$ neither graph always lies completely above the other. In problems like these, in which the two graphs "switch over" at some intermediate point, we subdivide the interval into subintervals and find the area over each subinterval. We then simply add up these areas to find the total area.

From Figure 16 we see that the square root function $\sqrt{x}$ lies above x^2 on the interval from 0 to 1. When x is equal to 1, the two graphs intersect; then for x greater than 1, the graph of x^2 lies above the graph of $\sqrt{x}$. Hence to find the two areas A_1 and A_2 shown in Figure 16, we compute

$$A_1 = \int_0^1 \left[\sqrt{x} - x^2\right] dx = \left(\frac{2}{3}x^{3/2} - \frac{x^3}{3}\right)\Bigg|_0^1 = \frac{1}{3}$$

$$A_2 = \int_1^4 \left[x^2 - \sqrt{x}\right] dx = \left(\frac{x^3}{3} - \frac{2}{3}x^{3/2}\right)\Bigg|_1^4 = \frac{49}{3}$$

Adding these two areas, we get the total area

$$A = A_1 + A_2$$

$$= \frac{1}{3} + \frac{49}{3}$$

$$= 16\frac{2}{3} \qquad \square$$

The Theory of Consumers' Surplus and Producers' Surplus

In a competitive market the price at which a commodity is sold is determined by the law of supply and demand. Suppose hot dogs are being sold by several vendors on the boardwalk in Atlantic City. From the point of view of the vendors the higher the price, the larger the number of vendors who will be willing to sell hot dogs. When the price is low, fewer vendors will be able to sell hot dogs at a profit, and hence the supply will be lower. The quantity q that producers are willing to supply at a price p defines the **supply curve**

$$q = S(p)$$

On the other hand, from the consumers' point of view the number q of hot dogs purchased depends on the price p of the hot dogs. The higher the price, the fewer hot dogs the consumers will buy. The lower the price, the more hot dogs the consumers will buy. The exact relationship between the quantity q demanded by the consumers and the price p of the hot dogs defines the **demand curve**

$$q = D(p)$$

Typical supply and demand functions are shown in Figure 17. We see that the supply curve is always increasing and the demand curve is always decreasing. Also, these two curves intersect at a point (p_0, q_0). The point at which the supply is equal to the demand is called the **equilibrium point**. In a competitive market the price p and quantity q will tend to stabilize at this point.

There are two economic indicators, known as the producers' surplus and the consumers' surplus, that can be found from the supply and demand curves.

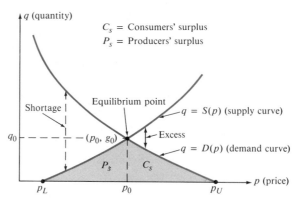

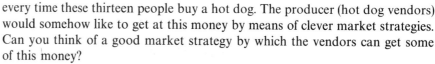

Figure 17
Typical supply and demand curves

To understand their meaning, suppose the going price p_0 (equilibrium price) of a hot dog in Atlantic City is $p_0 = \$0.50$. Of course, some people would be willing to pay more for a hot dog; hence in a certain sense they benefit from the price of $p_0 = \$0.50$. The total amount of money saved by these consumers is known by economists as the **consumers' surplus**. As a very simple example, if eight people were willing to pay $\$0.60$ for a hot dog and another five were willing to pay $\$0.70$, then these consumers are collectively saving a total of

$$\text{Consumers' surplus} = 8(\$0.60 - \$0.50) + 5(\$0.70 - \$0.50)$$
$$= \$1.80$$

every time these thirteen people buy a hot dog. The producer (hot dog vendors) would somehow like to get at this money by means of clever market strategies. Can you think of a good market strategy by which the vendors can get some of this money?

On the other hand, if the price of a hot dog is $p_0 = \$0.50$, there are probably several vendors who would be willing to sell hot dogs at a cheaper price. In a sense these vendors are earning "extra" money by being able to get the higher price of $\$0.50$. This extra money earned by these vendors is called the **producer's surplus**. For instance, if ten vendors were willing to sell hot dogs for $\$0.40$ and another seven were willing to sell hot dogs for $\$0.30$, then for each hot dog that these vendors sell (for $\$0.50$ each), they will collectively "earn" an extra

$$\text{Producers' surplus} = 10(\$0.50 - \$0.40) + 7(\$0.50 - \$0.30)$$
$$= \$2.40$$

The consumers would like to get this money by shopping intelligently.

The area C_s in Figure 17 represents the consumer's surplus, while the area P_s is the producers' surplus. They can be found in terms of the supply and demand curves by evaluating the two definite integrals

$$C_s = \int_{p_0}^{p_U} D(p)\, dp \qquad \text{(consumers' surplus)}$$

$$P_s = \int_{p_L}^{p_0} S(p)\, dp \qquad \text{(producers' surplus)}$$

The two numbers p_L and p_U in the above limits of integration are known as the **lower** and **upper price limits**, respectively. The lower price limit p_L is the lowest price at which any producer will be willing to supply the product, and the upper price limit p_U is the highest price at which any consumer is willing to buy the product.

Example 4 _____

Producers' and Consumers' Surpluses   Suppose the supply and demand curves for a given commodity are the following:

$$S(p) = 4p - 1 \qquad \text{(supply curve)}$$
$$D(p) = 4 - p^2 \qquad \text{(demand curve)}$$

where $S(p)$ and $D(p)$ are measured in millions of items and the price p is measured in dollars. What is the producers' surplus and what is the consumers' surplus for this commodity?

Solution   Graphs of the supply and demand curves are shown in Figure 18. The producers' surplus is given by

$$P_s = \int_{P_L}^{p_0} S(p)\,dp$$
$$= \int_{0.25}^{1} (4p - 1)\,dp$$
$$= 1.125 \quad \text{million dollars}$$

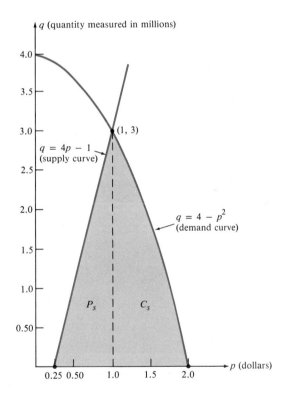

Figure 18 ·
Supply and demand curves

The consumers' surplus is given by

$$C_s = \int_{p_0}^{p_U} D(p)\,dp$$
$$= \int_{1}^{2} (4 - p^2)\,dp$$
$$= 1.667 \quad \text{million dollars} \qquad \square$$

Interpretation

A producers' surplus of 1.125 million dollars means that the producers have collectively "earned" an extra 1.125 million dollars in the sense that certain producers would have produced the product for less than the market price of p_0. It represents the amount of money out there that a smart entrepreneur might earn by buying from producers who are willing to produce at a cheaper price and then reselling the commodity at the going market price of p_0.

A consumers' surplus of 1.667 million dollars means that 1.667 million dollars was collectively "saved" by the consumers in the sense that certain consumers would have paid a higher price for the commodity (if the lower price were not available). If a smart entrepreneur could find these people, it would be possible to earn 1.667 million dollars by buying the commodity at the market price of p_0 and then reselling this commodity to these people at a higher price.

Volumes of Solids of Revolution

Besides finding areas under and between curves, the definite integral can also be used to find volumes of solid regions in three dimensions. The general method of attack is to subdivide the solid region into small subregions, each with known volume (such as cubes or cylinders) in much the same way that we used rectangles when finding areas in the plane. We then use a limiting process that is similar to the way we let the number of rectangles go to infinity in finding areas. The value of the limit will be the exact volume.

This section considers a particular type of solid region in three dimensions known as a **solid of revolution**. There are other solids that are *not* solids of revolution, but we can illustrate the idea of finding volumes by means of this particular type of region. To understand a solid of revolution, consider the graph of the function $y = f(x)$ for x between a and b as shown in Figure 19. If we revolve the region under the curve, called R, about the x-axis, this action will "sweep out" a three-dimensional solid. This solid is called the solid of revolution of R.

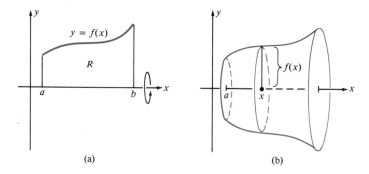

Figure 19
Solid of revolution of R

(a)   (b)

To find the volume of the solid of revolution, observe in Figure 20 that each small rectangle of width Δx in the region R will sweep out a disk (like a penny or nickel) when revolved about the x-axis. Because it is the functional value $f(x)$ that determines the radius of the disk at x, the disk located at x will have a volume of

$$\text{Volume of disk} = \pi[f(x)]^2 \Delta x$$

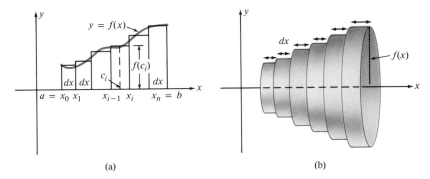

(a)   (b)

Figure 20
Volume V_n of n disks

We can now approximate the volume of the solid of revolution by summing the volume of n individual disks as demonstrated in Figure 20.

The sum of the volumes of these n disks is given by

$$V_n = \pi\big[f(x_1)^2 + f(x_2)^2 + \cdots + f(x_n)^2\big]\Delta x$$

If we now increase the number of disks where the width Δx of each individual disk becomes smaller and smaller, the approximation to the volume of the solid of revolution will become better and better. Taking the limiting value of this sum as the number of disks approaches infinity, we get the definite integral

$$\lim_{n \to \infty} V_n = \pi \int_a^b [f(x)]^2 \, dx$$

This gives the following result.

Volume of a Solid of Revolution

Let f be a nonnegative and continuous function on the interval $[a, b]$. Let a solid of revolution be defined by revolving the region bounded by $y = f(x)$ and the vertical lines $x = a$ and $x = b$ around the x-axis. The volume of this solid of revolution is given by

$$V = \pi \int_a^b [f(x)]^2 \, dx$$

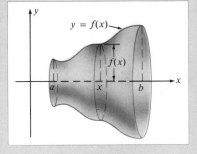

Example 5

Solid of Revolution   Find the volume of the solid of revolution generated by revolving $f(x) = \sqrt{x}$ around the x-axis over the interval $[0, 4]$.

Solution   The solid of revolution is shown in Figure 21. The volume of the solid of revolution is given by

$$V = \pi \int_0^4 [f(x)]^2\, dx$$

$$= \pi \int_0^4 x\, dx$$

$$= \pi \left[\frac{x^2}{2} \Big|_0^4 \right]$$

$$= \pi \frac{(16 - 0)}{2}$$

$$= 8\pi \quad \text{cubic units} \qquad \square$$

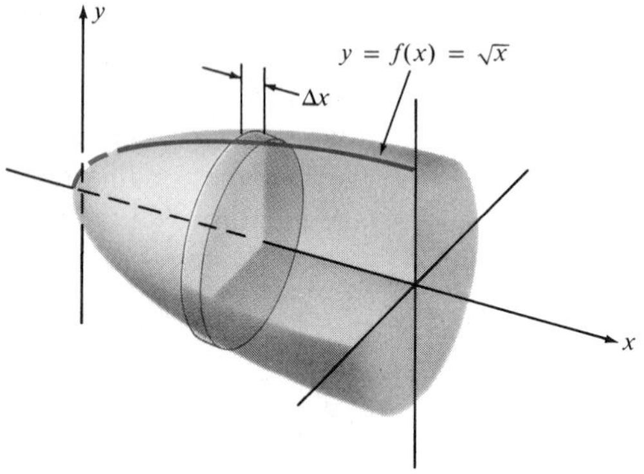

Figure 21
Solid of revolution of $y = \sqrt{x}$

The Paint Can Paradox

We close this section by presenting a puzzling result known as the **paint can paradox**. The paradox occurs when we revolve the curve $y = 1/x$ about the x-axis for $x \geq 1$. We will let you fall into this paradox by working the following example.

Example 6

The Paint Can Paradox   Find

(a) the area under the curve $y = 1/x$ for $x \geq 1$,

(b) the volume of the solid of revolution obtained by revolving $y = 1/x$ around the x-axis.

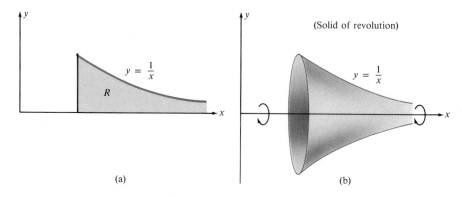

Figure 22
The region under $y = 1/x$ and its solid of revolution

(a)

(b)

Solution Both the curve $y = 1/x$ and the solid of revolution are shown in Figure 22.

(a) Area under the curve: The area A (if it exists) can be found by evaluating the improper integral

$$A = \int_1^\infty \frac{1}{x}\, dx$$
$$= \lim_{b \to \infty} \int_1^b \frac{1}{x}\, dx$$
$$= \lim_{b \to \infty} (\ln x)\Big|_1^b$$
$$= \lim_{b \to \infty} (\ln b - \ln 1)\Big|_1^b$$
$$= \lim_{b \to \infty} \ln b = \infty$$

In other words, the area under the curve will grow without bound as we move farther and farther to the right. We say that the area is infinite.

(b) Volume of the solid of revolution: The volume of the solid of revolution can also be found easily and is given by the improper integral

$$V = \pi \int_1^\infty \frac{1}{x^2}\, dx$$
$$= \pi \lim_{b \to \infty} \int_1^b \frac{1}{x^2}\, dx$$
$$= \pi \lim_{b \to \infty} \left[-x^{-1}\Big|_1^b \right]$$
$$= \pi \lim_{b \to \infty} \left(-\frac{1}{b} + 1 \right)$$
$$= \pi \quad \text{cubic units} \qquad \square$$

The Paradox

We have just seen that the area under the curve $y = 1/x$ is infinite, whereas the solid of revolution generated by this curve is finite ($\pi \cong 3.14$ cubic units). What this says in effect is that if the units of x and y were chosen as inches, then all the paint in the world couldn't paint the region under the curve $y = 1/x$, whereas the solid of revolution generated by this curve could be filled with roughly 3.14 cubic inches of paint (about a fourth of a cup). Believe it or not!

What is even more disturbing is that the surface area of the solid of revolution obtained by revolving $y = 1/x$ about the x-axis for $x \geq 1$ is also infinite. (We will not work this problem here.) This says in effect that we can fill the solid of revolution but cannot paint its surface. What is more amazing still is that if we filled the solid of revolution with paint, then the inside surface, which naturally has the same area as the outside surface, would appear to be covered with paint, yet the mathematics says that we cannot paint the surface. How can this be?

Interpretation of the Paint Can Paradox

Paradoxes in mathematics show that mathematics is relevant to the real world only when models accurately reflect real-world situations. In the paint can paradox the fact that a coat of paint actually has some thickness, and hence volume, has not been included in the discussion. There is a vast difference between painting with a paintbrush and "painting mathematically" in one's head.

Students of business management might think that such paradoxes as this are far afield from their course of study. This is not true. Recall from the interview with the head of mathematics division, Dr. McDonald of General Motors, that being able to argue quantitatively is one of the most important attributes of anyone in a management position. A successful decision maker must be able to find flaws in existing systems. The use of paradoxes teaches how and when to be constructively skeptical.

Problems

For Problems 1–14, find the area between the given curves over the indicated interval. Sketch the region between the two curves.

	Curve 1	Curve 2	Region
1.	$y = x^2 + 2$	$y = 1$	$[2, 5]$
2.	$y = 3x + 5$	$y = -2x + 4$	$[2, 4]$
3.	$y = 3x + 5$	$y = -2x + 4$	$[-2, -1/2]$
4.	$y = e^x$	$y = \ln x$	$[1, 2]$
5.	$y = \sqrt{x + 1}$	$y = 5x - 2$	$[2, 5]$
6.	$y = 3x^2 - 2$	$y = 3x^2 + 4$	$[-2, 1/2]$
7.	$y = xe^x$	$y = x \ln x$	$[1, 2]$
8.	$y = 3x^2 + 2$	$y = -x + 4$	$[2, 5]$
9.	$y = 3x^3$	$y = 3x + 2$	$[-1, 1]$
10.	$y = x^3 - 4$	$y = -5x - 1$	$[1, 3]$
11.	$y = 2x + 3$	$y = -x + 5$	$[1, 4]$
12.	$y = x^3$	$y = x^2$	$[0, 1]$
13.	$y = 2x^2 - 3$	$y = 5 - 2x^2$	$[-1, 1]$
14.	$y = e^x$	$y = e^{-x}$	$[1, \ln 8]$

For Problems 15–21, find the area between the curves. It will be necessary to find where the two curves intersect in order to find the limits of integration. Sketch the region whose area is being calculated.

	Curve 1	Curve 2
15.	$y = 2x^2 + 2$	$y = 18$
16.	$y = 3x^2 - 2$	$y = -2x^2 + 8$
17.	$y = x^3$	$y = x^2$
18.	$y = 3x^2 + 2$	$y = 7x$
19.	$y = 8 + 2x - x^2$	$y = x + 2$
20.	$y = \sqrt{x} \ (x \geq 0)$	$y = -x + 6$ and the y-axis
21.	$y = x^2 + 3$	$y = 12 - x^2$

For Problems 22–28, find the volume of the solids of revolution described. Sketch the volume being described.

22. $y = 3x$ on $[0, 2]$

23. $y = 2x^2 + 3$ on $[0, 3]$

24. $y = 6$ on $[1, 4]$

25. $y = \sqrt{1 - x^2}$ on $[0, 1]$

26. $y = \begin{cases} x & \text{on } [0,1 \,] \\ 2 - x & \text{on } [1, 2] \end{cases}$

27. $y = e^{-2x}$ on $[1, 4]$

28. $y = \ln x$ on $[e^2, e^4]$ (*Hint:* Use integration by parts.)

For Problems 29–37, find the consumers' and producers' surplus for the supply and demand curves.

	Supply	Demand
29.	$S(p) = p + 2$	$D(p) = -0.5p + 10$
30.	$S(p) = p + 1$	$D(p) = -2p + 5$
31.	$S(p) = 3p - 10$	$D(p) = -0.5p + 25$
32.	$S(p) = 5p - 50$	$D(p) = -p^2 + 100$
33.	$S(p) = p^2 + 2p + 5$	$D(p) = (p - 8)^2$
34.	$S(p) = p^2$	$D(p) = -p + 4$
35.	$S(p) = p^2$	$D(p) = -7p + 30$
36.	$S(p) = 0.01p^2$	$D(p) = -0.99p^2 + 400$
37.	$S(p) = 0.01p^2 - 1$	$D(p) = -0.99p^2 + 99$

Integrals in Business Management

38. Consumers' and Producers' Surpluses Suppose the supply and demand curves for a commodity are

$$S(p) = 5p - 2$$

$$D(p) = \frac{9}{p} - 1$$

where $S(p)$ and $D(p)$ are measured in millions of items and p is measured in dollars.

(a) Sketch the supply and demand curves.

(b) Find the equilibrium point (p_0, q_0).

(c) Find the consumers' and producers' surpluses, and interpret their values.

39. Consumer's and Producers' Surpluses An economic study was carried out on a given commodity and the supply and demand tables in Table 8 were constructed.

Supply Table		Demand Table	
Price p (Dollars)	Supply q (Millions of Items)	Price p (Dollars)	Demand q (Millions of Items)
0	0	0	∞
1	0	1	50
2	1	2	30
3	2	3	15
4	4	4	10
5	7	5	7
6	11	6	3
7	16	7	2
8	20	8	1
9	25	9	0
10	31	10	0

Plot the points in the pq-plane and determine the following:
(a) the equilibrium point (p_0, q_0),
(b) the total amount of money that exchanges hands from the consumers to the producers at this equilibrium point,
(c) the surplus (supply minus demand) of the commodity when the price of the commodity is $9,

(d) the shortage (demand minus supply) of the commodity when the price of the commodity is $3.

40. Consumers' and Producers' Surpluses Supply and demand curves are often approximated by linear curves. The general form for these functions is

$$S(p) = Ap - B \qquad \text{(supply curve)}$$
$$D(p) = C - Dp \qquad \text{(demand curve)}$$

where A, B, C, and D are nonnegative constants. (See Figure 23.) In terms of the general constants A, B, C, and D, find
(a) the lowest price p_L at which some producer will supply some of the product,
(b) the highest price p_U that some consumer will pay for the product,
(c) the equilibrium point (p_0, q_0),
(d) the consumers' surplus (leave in integral form),
(e) the producers' surplus (leave in integral form).

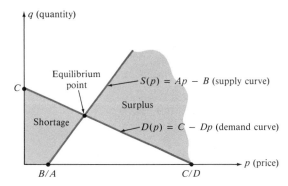

Figure 23
Graph for Problem 40

Numerical Integration (Calculus and Computers)

PURPOSE

We show how to approximate the value of a definite integral by

- the trapezoid rule and
- Simpson's Rule.

These methods are often used to approximate definite integrals that cannot be found exactly by the fundamental theorem of calculus. Because a great deal of computation is generally involved in carrying out these approximations, these methods are usually carried out on a computer or with the aid of a hand-held calculator.

Introduction

In our studies of the definite integral you may have the misguided impression that all definite integrals can be evaluated quickly and easily, either by a direct application of the fundamental theorem of calculus or by getting help from the method of substitution or by integration by parts. This is far from the truth. These are simple functions, such as

$$f(x) = e^{-x^2}$$

and others for which no elementary antiderivative exists. In such instances (as long as the function is continuous) the definite integral exists, and we may wish to evaluate it. This is where numerical integration is useful. Then too, it often happens in applied problems that the integrand is not provided as an algebraic expression at all but by a collection of data points such as Table 9. The y-values of these data points can often be thought of as the value of a function $y = f(x)$ at the points x_i for $1 \leq i \leq n$, and we may wish to integrate this function. Hence we again resort to numerical integration.

TABLE 9

Finding "Areas" Under Data Points

x	y
x_1	y_1
x_2	y_2
x_3	y_3
$\vdots$	$\vdots$
x_n	y_n

The Trapezoid Rule

We begin by presenting one of the simplest but most powerful techniques for approximating a definite integral, the trapezoid rule. Consider a function f whose graph $y = f(x)$ is shown in Figure 24(a) and whose area under the graph is represented by the definite integral

$$\int_a^b f(x)\, dx$$

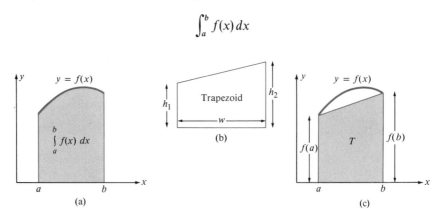

Figure 24
Single trapezoid approximation of a function

When we defined the definite integral we saw that the area under the graph of a function could be approximated by finding the area of approximating rectangles. We now see that the area under the graph can be approximated more effectively by approximating with trapezoids. It can be proven quite easily that the area T of a trapezoid, such as the one drawn in Figure 24(b) with width w and heights h_1 and h_2, is given by

$$T = w\frac{h_1 + h_2}{2}$$

Hence to approximate the integral

$$\int_a^b f(x)\,dx$$

we first draw the straight line that connects the endpoints $(a, f(a))$ and $(b, f(b))$. If we then compute the area of the resulting trapezoid as shown in Figure 24(c), we will get the trapezoid approximation to the integral

$$\int_a^b f(x)\,dx = (b - a)\frac{f(a) + f(b)}{2}$$

The power of the trapezoid rule, however, is to approximate the area under the curve not with one trapezoid, but with several. Consider again the integral

$$\int_a^b f(x)\,dx$$

but now subdivide the interval $[a, b]$ into four equal parts. On each interval, find the area of the shaded trapezoid as shown in Figure 25. Then add these areas to get an estimate of the integral on the entire interval $[a, b]$.

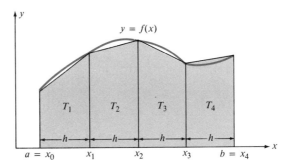

Figure 25
Trapezoid rule with four trapezoids

More generally, we introduce the following equally spaced points:

$$a = x_0 < x_1 < x_2 < x_3 < x_4 = b$$

where the length h between two adjacent points is computed by

$$h = \frac{b - a}{4}$$

Hence the approximation of the integral can be found by adding the areas T_1,

T_2, T_3, and T_4 of the four trapezoids shown in Figure 25. That is,

$$\int_a^b f(x)\,dx = \int_{x_0}^{x_1} f(x)\,dx + \int_{x_1}^{x_2} f(x)\,dx + \int_{x_2}^{x_3} f(x)\,dx + \int_{x_3}^{x_4} f(x)\,dx$$

$$\cong T_1 + T_2 + T_3 + T_4$$

$$= (x_1 - x_0)\frac{f(x_0) + f(x_1)}{2} + (x_2 - x_1)\frac{f(x_1) + f(x_2)}{2}$$

$$+ (x_3 - x_2)\frac{f(x_2) + f(x_3)}{2} + (x_4 - x_3)\frac{f(x_3) + f(x_4)}{2}$$

$$= \frac{h}{2}\{[f(x_0) + f(x_1)] + [f(x_1) + f(x_2)]$$

$$+ [f(x_2) + f(x_3)] + [f(x_3) + f(x_4)]\}$$

$$= \frac{h}{2}[f(x_0) + 2f(x_1) + 2f(x_2) + 2f(x_3) + f(x_4)]$$

This formula motivates the general trapezoid rule.

Trapezoid Rule for Numerical Integration

A definite integral can be approximated by the area of n trapezoids by the formula

$$\int_a^b f(x)\,dx \cong \frac{h}{2}[f(x_0) + 2f(x_1) + 2f(x_2) + \cdots + 2f(x_{n-1}) + f(x_n)]$$

where $h = \dfrac{b - a}{n}$ is the width of each subinterval and $x_j = a + jh$ for $j = 0, 1, 2, \ldots, n$.

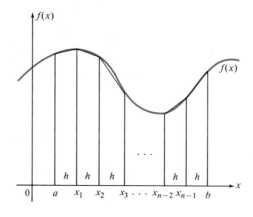

Note that the two endpoints a and b are denoted by x_0 and x_n, respectively, to give more uniformity in the notation.

Example 1

Trapezoid Rule   Approximate the definite integral

$$\int_1^3 \frac{dx}{x}$$

by the trapezoid rule using two trapezoids.

Solution   Here we are given

$$n = 2$$
$$a = x_0 = 1$$
$$b = x_2 = 3$$

We begin by computing the width h of each trapezoid. We have

$$h = \frac{b - a}{n} = \frac{3 - 1}{2} = 1$$

The endpoints of the interval are $x_0 = 1$, $x_2 = 3$, and there is one intermediate value x_1 given by

$$x_1 = a + jh \qquad (j = 1)$$
$$= 1 + 1$$
$$= 2$$

The value of $f(x)$ at this point is

$$f(x_1) = \frac{1}{x_1}$$
$$= 0.50$$

Substituting the above values into the formula for the trapezoid rule, we find

$$\int_1^3 \frac{dx}{x} \cong \frac{h}{2}[f(x_0) + 2f(x_1) + f(x_2)]$$
$$= 0.50[f(1) + 2f(2) + f(3)]$$
$$= 0.50[1 + 2(0.5) + 0.333]$$
$$= 1.167$$

Observation: If the above integral were evaluated by using the fundamental theorem of calculus, we would get

$$\int_1^3 \frac{dx}{x} = \ln x \Big|_1^3$$
$$= \ln 3 - \ln 1$$
$$= \ln 3$$
$$= 1.09861\ldots$$

The trapezoid rule approximation gives a value that is slightly larger than the real value. We can see from the graph of $y = 1/x$ in Figure 26 the reason for the discrepancy.   □

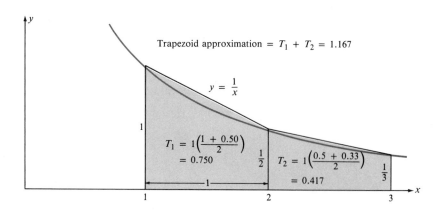

Figure 26
*Trapezoid approximation with
two trapezoids*

We now show how the approximation by the trapezoid rule changes if four subintervals are used instead of two. Intuitively, it is clear that approximating with four trapezoids should be better than with two trapezoids.

Example 2

Trapezoid Rule   Approximate the integral

$$\int_1^3 \frac{dx}{x}$$

using the trapezoid rule with four trapezoids.

Solution   Here we have

$$n = 4$$
$$a = x_0 = 1$$
$$b = x_4 = 3$$

Again, we first compute the width of the trapezoids. We have

$$h = \frac{b - a}{n} = \frac{3 - 1}{4} = 0.50$$

(See Figure 27.) The endpoints of the interval are $x_0 = 1$, $x_4 = 3$, and there are three intermediate values:

$$x_j = a + jh$$
$$= 1 + 0.50j \qquad (j = 1, 2, 3)$$

The heights of the trapezoids are

$$f(x_j) = \frac{1}{x_j} \qquad (j = 1, 2, 3)$$

These values are displayed in Table 10. We now substitute these values into the trapezoid rule, getting

$$\int_1^3 \frac{dx}{x} \cong \frac{h}{2} \left[f(x_0) + 2f(x_1) + 2f(x_2) + 2f(x_3) + f(x_4) \right]$$

$$= \frac{0.50}{2} \left[f(1) + 2f(1.5) + 2f(2) + 2f(2.5) + f(3) \right]$$

$$= 0.25 \left[1 + 2(0.6667) + 2(0.5) + 2(0.4) + 0.3333 \right]$$

$$= 1.1167$$

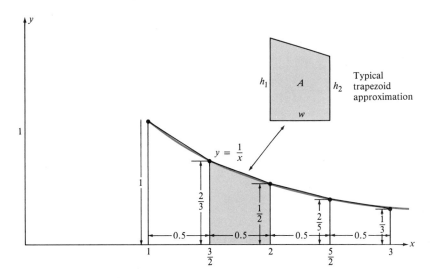

Figure 27
Use of the trapezoid rule

TABLE 10
Endpoints of Subintervals
and Heights of Trapezoids

	j	$x_j = 1 + 0.50j$	$f(x_j) = 1/x_j$
Endpoint ⟶ 0		$x_0 = 1 + 0\ \ = 1.0$	$f(1.0) = 1/1.0 = 1.000$
1		$x_1 = 1 + 0.5 = 1.5$	$f(1.5) = 1/1.5 = 0.6667$
2		$x_2 = 1 + 1.0 = 2.0$	$f(2.0) = 1/2.0 = 0.5000$
3		$x_3 = 1 + 1.5 = 2.5$	$f(2.5) = 1/2.5 = 0.4000$
Endpoint ⟶ 4		$x_4 = 1 + 2.0 = 3.0$	$f(3.0) = 1/3.0 = 0.3333$

   Table 11 shows how the trapezoid approximation approaches the real value
of the integral ln 3 = 1.09861 . . . as the number of trapezoids increases. Note
that the error goes to zero as the number of subintervals increases.

TABLE 11
Approximation of
$\int_1^3 \dfrac{dx}{x}$ by
the Trapezoid Rule

Number of Trapezoids	Trapezoid Approximation	Error (Approximation − ln 3)	
1	1.3333	0.2347	
2	1.1667	0.0680	⟵ *Example 1*
3	1.1301	0.0315	
4	1.1167	0.0180	⟵ *Example 2*
5	1.1103	0.0144	
6	1.1067	0.0081	
7	1.1046	0.0060	
8	1.1032	0.0046	
9	1.1022	0.0036	
10	1.1016	0.0030	
20	1.0993	0.0007	
30	1.0989	0.0003	
40	1.0988	0.0002	
50	1.0987	0.0001	

A good rule of thumb in estimating the accuracy of the trapezoid rule is to try different numbers of trapezoids and observe whether the approximations are converging to some value as the number of trapezoids increases. Given the values in Table 11, we would be tempted to say that the first three digits of the integral are 1.09. This in fact is correct. □

Figure 28 shows a flow diagram that illustrates the steps required in using the trapezoid rule.

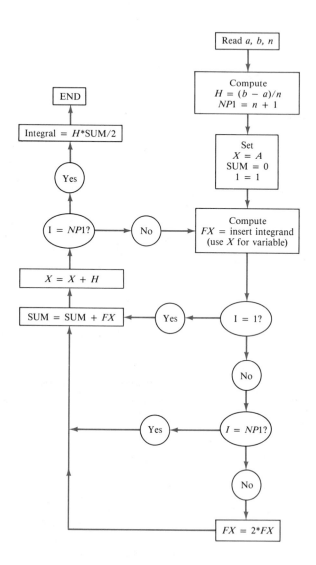

Figure 28
*Flow diagram for the trapezoid
rule*

The following BASIC program carries out the computations described in the flow diagram in Figure 28.

```
10   REM    TRAPEZOID RULE
20   PRINT  "ENTER LEFT AND RIGHT ENDPOINTS"
30   INPUT A, B
```

```
40   PRINT "ENTER NUMBER OF SUBINTERVALS"
50   INPUT N
60   H = (B − A)/N
70   SUM = 0
80   NP1 = N + 1
90   X = A
100  FOR I = 1 TO NP1
110     REM    INSERT INTEGRAND NEXT
120     FX = 1/X
130     IF I = 1    THEN GO TO 160
140     IF I = NP1  THEN GO TO 160
150     FX = 2*FX
160     SUM = SUM + FX
170     X = X + H
180  NEXT I
190  INTEGRAL = H*SUM/2
200  PRINT "INTEGRAL IS"; INTEGRAL
210  END
```

BASIC Computer Program for Trapezoid Rule

We mentioned earlier in this section that the integrand $f(x) = e^{-x^2}$ had no "elementary" antiderivative, and hence a definite integral of this function could not easily be found by using the fundamental theorem of calculus. We now approximate a definite integral of this function using the trapezoid rule.

Example 3

Trapezoid Rule   Approximate the definite integral

$$\int_0^1 e^{-x^2}\,dx$$

using the trapezoid rule with four trapezoids.

Solution   The graph of $f(x) = e^{-x^2}$ is shown in Figure 29. The basic values used in the trapezoid rule are

$$n = 4 \quad \text{(number of trapezoids)}$$
$$a = 0 \quad \text{(lower limit of integration)}$$
$$b = 1 \quad \text{(upper limit of integration)}$$

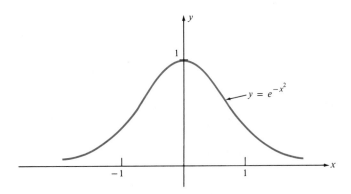

Figure 29
Graph of $f(x) = e^{-x^2}$

Hence we can compute

$$h = \frac{b - a}{n} \qquad \text{(width of trapezoids)}$$

$$= \frac{1 - 0}{4}$$

$$= 0.25$$

The intermediate points and heights of the trapezoids are found from the formulas in Table 12. Hence the trapezoid rule gives the approximation

$$\int_0^1 e^{-x^2}\,dx \cong \frac{h}{2}\left[f(x_0) + 2f(x_1) + 2f(x_2) + 2f(x_3) + f(x_4)\right]$$

$$= 0.125[1 + 2(0.9394) + 2(0.7788) + 2(0.5698) + 0.3679]$$

$$= 0.7430 \qquad\qquad\qquad \square$$

TABLE 12

Intermediate Points and Heights of Trapezoids

	j	$x_j = a + jh$	$f(x_j) = e^{-x_j^2}$
Endpoint ⟶	0	0	$e^{-(0)^2} \cong 1.0000$
	1	0.25	$e^{-(0.25)^2} \cong 0.9394$
	2	0.50	$e^{-(0.50)^2} \cong 0.7788$
	3	0.75	$e^{-(0.75)^2} \cong 0.5698$
Endpoint ⟶	4	1.00	$e^{-(1)^2} \cong 0.3679$

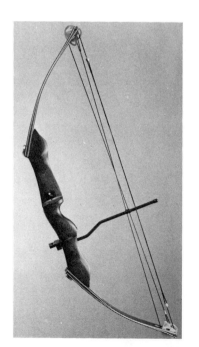

Using the Trapezoid Rule with Data Points

In the experimental sciences such as biology, psychology, and chemistry an experimenter often obtains observed values of two variables x and y. The values of y can often be interpreted as values of a function at specific values of x. It is often desirable to find the area under the curve defined by this function. The following example illustrates this idea.

Bows and arrows have changed since the good old days of Robin Hood. Today, the serious archer uses a compound bow. The compound bow is designed so that the force required to bend the bow is not directly proportional to the amount that the bow is bent. The force required to hold an arrow steady when the string is completely extended is actually much less than when the string is half-extended. This desirable property allows the archer to hold the arrow in the "cocked" position for long periods of time without tiring.

An experiment is conducted in which the string of a compound bow is slowly pulled until the bow is completely extended. As the string is pulled, the distance x the string is pulled and the force y applied to the string are periodically measured. The resulting observations are shown in Figure 30. A physical principle says that the area under this resulting curve of y versus x represents the energy stored in the bow and hence the energy transferred to the arrow upon release of the string.

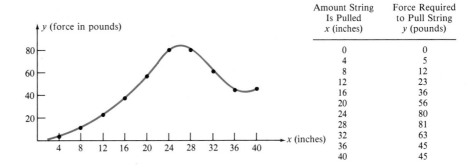

Amount String Is Pulled x (inches)	Force Required to Pull String y (pounds)
0	0
4	5
8	12
12	23
16	36
20	56
24	80
28	81
32	63
36	45
40	45

Figure 30
Force y required to pull a compound bow x inches

Example 4

Energy of a Compound Bow   Find the amount of energy stored by the compound bow represented by the observations in Figure 30.

Solution   The energy in inch-pounds stored by the bow is the area under the curve that approximates the data points in Figure 30. This force curve is shown in Figure 31.

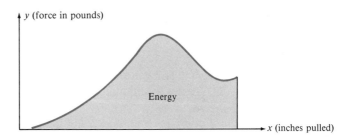

Figure 31
The area under the curve is energy

Since the observed force y applied to the string can be interpreted as values of a function $y = f(x)$ at 11 equally spaced points, we can use the trapezoid rule with

$$a = 0$$
$$b = 40$$
$$n = 10$$

Computing the width of each trapezoid, we have

$$h = \frac{b - a}{n} = 4$$

and so the trapezoid rule states that

$$\int_0^{40} f(x)\,dx \cong \frac{h}{2}\left[f(a) + 2f(x_1) + 2f(x_2) + \cdots + 2f(x_9) + f(b)\right]$$

$$= \frac{4}{2}\left[0 + 2(5) + 2(12) + \cdots + 2(45) + 45\right]$$

$$= 1694 \quad \text{inch-pounds}$$

Interpretation: The energy stored by the bow is 1694 inch-pounds or $1694/12 = 141$ foot-pounds of energy. To an archer studying compound bows the energy stored by a compound bow is an important statistic.   □

Simpson's Rule (Approximating by Parabolas)

Intuition suggests that the trapezoid rule can be improved by replacing trapezoids by parabolas (second-order polynomials). The general idea behind Simpson's Rule is quite simple. Since it is possible to find a unique parabola that will pass through any three points on a graph (other than points on a straight line), we simply replace the integrand of the definite integral by a parabola and integrate the parabola (which is easy to do). The value of the definite integral of the parabola will approximate the definite integral of the original function. We can also extend this idea by replacing the integrand by several parabolas and integrating each parabola separately. (See Figure 32.)

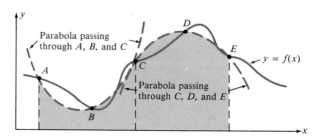

Figure 32
Approximating areas by areas under parabolas

To be more specific, it can be shown that the area under the parabola that passes through the three points $(a, f(a)), (a + h, f(a + h))$, and $(a + 2h, f(a + 2h))$ is

$$A = \frac{h}{3}\left[f(a) + 4f(a + h) + f(a + 2h)\right]$$

This is illustrated in Figure 33.

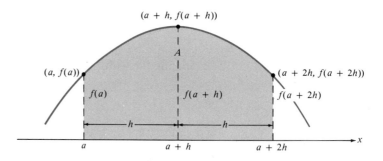

Figure 33
The area under a parabola

We can now approximate the integral

$$\int_a^b f(x)\,dx$$

by subdividing the entire interval $[a, b]$ into an even number (the number of subintervals must be $2, 4, 6, \ldots$) of subintervals and then applying the above parabola formula to successive pairs of subintervals. This line of reasoning will result in the following approximation.

Simpson's Rule for Approximating Integrals

Let f be a continuous function defined on the interval $[a, b]$ and subdivide the interval into an even number n of subintervals of equal length. An approximation to the integral of f over $[a, b]$ based on replacing the integrand $f(x)$ by a series of parabolas is called **Simpson's Rule** and is given by

$$\int_a^b f(x)\, dx \cong \frac{h}{3}\left[f(x_0) + 4f(x_1) + 2f(x_2) + 4f(x_3) + \cdots + 2f(x_{n-2}) \right.$$
$$\left. + 4f(x_{n-1}) + f(x_n) \right]$$

where

$$h = \frac{b - a}{n}$$

is the width of each subinterval and $x_j = a + jh$ for $j = 0, 1, 2, \ldots, n$.

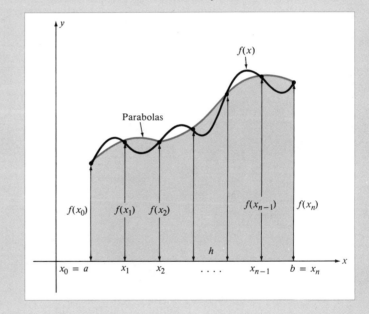

We have denoted the limits of integration as $a = x_0$ and $b = x_n$. Note that the coefficients in the formula for Simpson's Rule have the pattern

$$[1\ \ 4\ \ 2\ \ 4\ \ 2\ \ 4\ \ 2\ \cdots\ 2\ \ 4\ \ 2\ \ 4\ \ 2\ \ 4\ \ 2\ \ 4\ \ 1]$$

Example 5 Simpson's Rule   Approximate the definite integral

$$\int_1^3 \frac{dx}{x}$$

using Simpson's Rule with $n = 4$ subintervals.

Solution   We are given

$$a = 1$$
$$b = 3$$
$$n = 4$$

and so the length of each subinterval is

$$h = \frac{b - a}{n} = \frac{3 - 1}{4} = \frac{1}{2}$$

See Figure 34.

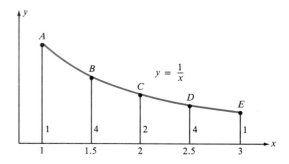

Figure 34
Simpson's Rule must have an even number of subintervals

TABLE 13
Relevant Points Used by Simpson's Rule

	j	$x_j = a + jh$	$f(x_j) = 1/x_j$
Endpoint ⟶	0	$x_0 = 1.0$	$f(1.0) = 1.000$
	1	$x_1 = 1.5$	$f(1.5) = 0.667$
	2	$x_2 = 2.0$	$f(2.0) = 0.500$
	3	$x_3 = 2.5$	$f(2.5) = 0.400$
Endpoint ⟶	4	$x_4 = 3.0$	$f(3.0) = 0.333$

Finding the intermediate points x_j and the values of $f(x_j)$, we compute the values in Table 13. Substituting these values into Simpson's Rule, we have

$$\int_1^3 \frac{dx}{x} \cong \frac{h}{3} \left[f(1) + 4f(1.5) + 2f(2.0) + 4f(2.5) + f(3.0) \right]$$

$$= 0.166[1 + 4(0.667) + 2(0.5) + 4(0.4) + 0.333]$$

$$= 1.099$$

Interpretation: The exact value of the above integral is ln 3 = 1.09861 . . . , which means that Simpson's Rule with four subintervals has an error of 0.00039. This compares with the trapezoid rule using four subintervals, which has a much larger error of 0.0180.   □

HISTORICAL NOTE

Thomas Simpson (1710–1761) was born to the wife of an English weaver and had no formal education. His interest in mathematics was aroused when he received an arithmetic book from a peddler. His first break in life came as a young man when he married his landlady, a marriage that allowed him to be partially financially independent.

In 1736, Simpson moved to London and published his first mathematical work in a periodical called *Ladies Diary*, of which he later became editor. In 1737 he published one of the all-time successful calculus textbooks. After that he concentrated on his textbook writing until one Robert Heath accused him of plagiarism. The publicity was great, and Simpson proceeded to dash off a series of popular textbooks that made him a wealthy man.

It is interesting to note that the rule that made Simpson famous was not discovered by Simpson at all, having been discovered in 1668 by the English mathematician James Gregory.

Problems

In Problems 1–15, calculate the definite integral using both the trapezoid method and Simpson's Rule with $n = 4$ subintervals. Compare the accuracy of these methods when the exact answer can be found by using the fundamental theorem of calculus.

1. $\int_0^4 2\,dx$

2. $\int_0^4 x\,dx$

3. $\int_1^5 (2x + 1)\,dx$

4. $\int_{-1}^3 x\,dx$

5. $\int_0^1 x^2\,dx$

6. $\int_1^3 (x^2 - 1)\,dx$

7. $\int_0^4 x^3\,dx$

8. $\int_{-2}^2 x^3\,dx$

9. $\int_{-2}^2 (x^3 + 2)\,dx$

10. $\int_0^8 \sqrt{x + 1}\,dx$

11. $\int_1^2 \frac{dx}{x}$

12. $\int_0^4 e^{-x}\,dx$

13. $\int_0^1 e^{x^2}\,dx$

14. $\int_0^1 xe^{-x^3}\,dx$

15. $\int_0^1 xe^x\,dx$

In Problems 16–22, approximate the definite integrals using both the trapezoid rule and Simpson's Rule with $n = 6$ subintervals.

16. $\int_0^1 \sqrt{x^3 + 1}\,dx$

17. $\int_{-1}^1 \sqrt{x^4 + 1}\,dx$

18. $\int_0^1 \sqrt{1 - x^2}\,dx$

19. $\int_0^2 \frac{1}{1 + x^2}\,dx$

20. $\int_1^7 \frac{1}{\sqrt{1 + x^2}}\,dx$

21. $\int_1^5 \frac{\ln x}{x}\,dx$

22. $\int_1^3 (x \ln x - x)\,dx$

23. Trapezoid Rule Approximate ln 3 by calculating the integral

$$\int_1^3 \frac{dx}{x}\,dx$$

using the trapezoid rule with $n = 6$ subintervals. Sketch the graph of the integrand showing the trapezoids used in the approximation.

24. Simpson's Rule Approximate ln 5 by evaluating the integral

$$\int_1^5 \frac{dx}{x}$$

using Simpson's Rule with $n = 8$ subintervals. Sketch the graph of the integrand showing the trapezoids used in the approximation.

25. Simpson's Rule Using a calculator or tables, approximate the value of the definite integral

$$\int_2^4 e^{-x^2}\,dx$$

using Simpson's Rule with $n = 10$ subintervals.

26. Trapezoid Rule A well-known identity in mathematics is

$$\int_0^1 \frac{dx}{1+x^2}\,dx = \frac{\pi}{4}$$

Approximate this integral using the trapezoid rule with $n = 10$ subintervals. Use this value to approximate the value of π.

27. Trapezoid Rule with Data Points The integral

$$S = \int_a^b v(t)\,dt$$

gives the distance traveled by an object moving at the velocity $v(t)$ during the time interval $[a, b]$. Suppose you are driving a car and you record the velocities in Table 14 every 30 minutes (0.5 hour). Use the trapezoid rule to approximate the total distance that you have traveled.

TABLE 14
Velocities for Problem 27

t (hours)	$v(t)$ (miles/hour)
0.0	0
0.5	45
1.0	52
1.5	60
2.0	35
2.5	45
3.0	53
3.5	48
4.0	37
4.5	49
5.0	53

28. Computing the Wood in a Tree The circumference of a tree trunk is measured at 3-foot intervals from the base of the tree to a height at which the tree is to be cut. Assuming the cross section of the tree to be circular, what is the estimate of the volume of wood in the tree described by the values in Table 15 when you use the trapezoid rule for approximation? *Hint:* The volume of a tree trunk with changing radius is given by

$$V = \pi \int_0^h r^2(h)\,dh = \frac{1}{4\pi}\int_0^h C^2(h)\,dh$$

where V is the volume of the wood, $r(h)$ is the radius of the tree at height h, and $C(h)$ is the circumference of the tree at height h.

TABLE 15
Values for Problem 28

h (feet)	$C(h)$ (feet)
0	10.5
3	9.8
6	9.4
9	9.0
12	8.4
15	7.5
18	6.8
21	6.0
24	5.3
27	4.6
30	4.0
33	3.3

Simpson's Rule with Data Points

Simpson's Rule can also be used to evaluate integrals whose integrands are defined by data points. Suppose we are given the equally spaced data points in Table 16. If we interpret these data points as points on the graph of a function $y = f(x)$, then Simpson's Rule for approximating the integral

$$\int_{x_1}^{x_n} f(x)\,dx$$

is

$$\int_{x_1}^{x_n} f(x)\,dx \cong \frac{h}{3}\left[y_1 + 4y_2 + 2y_3 + 4y_4 + \cdots + 2y_{n-2} + 4y_{n-1} + y_n\right]$$

TABLE 16
Data Points for Simpson's Rule

x_i	y_i	
x_1	y_1	
x_2	y_2	
x_3	y_3	$(h = x_i - x_{i-1})$
$\vdots$	$\vdots$	
x_n	y_n	

29. Simpson's Rule with Data Points Use Simpson's Rule to approximate the integral determined by the data points in Problem 27.

30. Simpson's Rule with Data Points Use Simpson's Rule to approximate the integral determined by the data points in Problem 28. Use the values $x = 3, 6, 9, \ldots, 30, 33$.

Numerical Analysis to Find Producers' and Consumers' Surpluses

31. Producers' Surplus The observations in Table 17 were taken by a marketing company to determine the supply and demand of a new product as a function of price p.

TABLE 17
Supply and Demand Data for Problems 31–36

Price, P	Supply, $q = S(p)$ (millions)	Demand, $q = D(p)$ (millions)
$1	0	20
$2	2	17
$3	4	14
$4	6	11
$5	8	8
$6	10	5
$7	12	2

These values are plotted in Figure 35. Using the trapezoid rule, use the necessary observations from Table 17 to

approximate the producers' surplus. The equilibrium point is $(p_0, q_0) = (5, 8)$.

32. Producers' Surplus Using Simpson's Rule, use the necessary observations from Table 17 to approximate the producers' surplus.

33. Accuracy of the Trapezoid Rule The observations of the supply and demand in Table 17 are in reality the values of the functions

$$S(p) = 2p - 2$$
$$D(p) = 23 - 3p$$

Find the exact value of the producers' surplus and compare it to the approximation found in Problem 31.

34. Consumers' Surplus Using the trapezoid rule, use the necessary observations in Table 17 to approximate the consumers' surplus with $p \leq 7$.

35. Consumer's Surplus Using Simpson's Rule, use the necessary observations in Table 17 to approximate the consumers' surplus with $p \leq 7$.

36. Accuracy of Simpson's Rule The observations of the supply and demand in Table 17 are in reality the value of the functions

$$S(p) = 2p - 2$$
$$D(p) = 23 - 3p$$

Find the exact consumers' surplus with $p \leq 7$ and compare it to the approximations found in Problems 34 and 35.

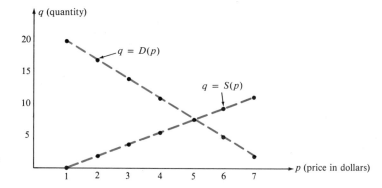

Figure 35
Supply and demand curves for Problem 31

12.5

Differential Equations (Growth and Decay)

PURPOSE

We introduce the basic concept of a differential equation and in particular the growth and decay equations. The important ideas introduced are

- the meaning of a differential equation,
- the general solution of a differential equation,
- the growth and decay equations.

Basics of Differential Equations

A **differential equation** is an equation that involves at least one derivative of an unknown function. Differential equations are important because they are able to describe so many physical phenomena. Often the derivatives of a function represent physical phenomena (such as velocity, acceleration, force, or friction); hence equations that relate a function and its derivatives often describe natural laws of science. Differential equations were originally used by astronomers to describe planetary motion, but nowadays they are used to describe phenomena in such diversified areas as biology, economics, ecology, and genetics. Sending vehicles to the moon and the further exploration of space would be impossible without the solution of differential equations to guide the spaceships.

Until now, when you have been asked to find the solution of an equation such as

$$x^2 - x - 2 = 0$$

the unknown x has always been a real number. Now, however, we will solve a differential equation in which the unknown is not a number but a function. For example, to solve the differential equation

$$\frac{dy}{dt} + 3y = 0$$

which relates an unknown function $y(t)$ with its derivative dy/dt, we seek to find the function $y(t)$ whose derivative plus three times itself is identically zero. Such a function is called the **solution** of the differential equation, and t (which generally stands for time) is called the **independent variable**. For this differential equation, it is a simple matter to verify that

$$y(t) = e^{-3t}$$

is a solution, since

$$\underbrace{\frac{d}{dt} e^{-3t}}_{y'} + \underbrace{3e^{-3t}}_{3y} = -3e^{-3t} + 3e^{-3t} = 0$$

Furthermore, we can verify that

$$y(t) = 2e^{-3t}$$
$$y(t) = -3e^{-3t}$$
$$y(t) = \frac{1}{2} e^{-3t}$$

are also solutions. In fact, any function of the form

$$y(t) = Ce^{-3t}$$

where C is any real number is a solution of the differential equation. We call this collection of solutions, as we allow C to "run through" all the real numbers, the **general solution** of the differential equation. It constitutes all of the solutions of the differential equation. The name *general solution* is often misinterpreted by beginning students of differential equations and is thought to consist of only one solution; however, it is an entire family or set of solutions. Specific solutions, such as $2e^{-3t}$ and $-5e^{-3t}$, attained by assigning specific values of C to the general solution Ce^{-3t}, are called **particular solutions** of the differential equation.

This section does not attempt to study all types of differential equations but an important class of differential equations called the growth and decay equations.

Growth and Decay Differential Equations

One of the simplest differential equations is the most important. It is called the **growth** and **decay equation**, and it is given by

$$\frac{dy}{dt} = ky \qquad \text{(growth/decay equation)}$$

More specifically, when $k > 0$ the equation is called the growth equation, and when $k < 0$ it is called the decay equation. In this equation, $y(t)$ is the dependent variable (which we denote simply by y) and t is the independent variable. Some phenomena that can be described by the growth and decay equations are described in Table 18. The graphs of a few of these phenomena are displayed in Figure 36.

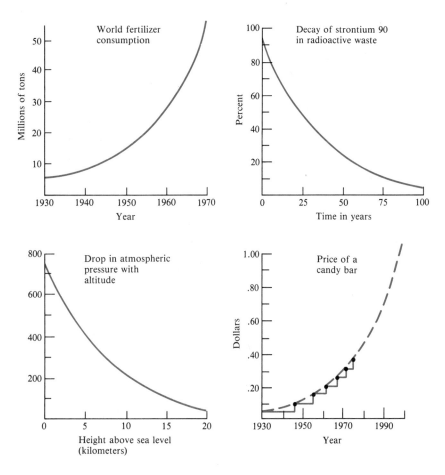

Figure 36
Typical growth and decay phenomena

TABLE 18
Common Phenomena
Described by the Growth
and Decay Equation

Growth Phenomena ($k > 0$)	**Decay Phenomena ($k < 0$)**
bank account paying compound interest	drug injected into bloodstream
unchecked biological growth	water temperature in a bathtub
short-term human and animal populations	radioactive decay
	intensity of x rays through the body
	intensity of light passing through water

Solution of the Growth and Decay Equation

Since the growth and decay equation

$$\frac{dy}{dt} = ky$$

is an important tool for describing natural phenomena, we show how its solutions can be found. Also, since y usually describes positive quantities, we only look for solutions y that satisfy $y(t) > 0$ for all t.

To find its solutions, we carry out the following steps.

Step 1: (Divide by y.)   Dividing each side of the equation by y, we get

$$\frac{1}{y}\frac{dy}{dt} = k$$

Step 2: (Find antiderivative.)   Setting the antiderivative of the left-hand side of the above equation (remember, $y(t)$ is a function of t) equal to the antiderivative of the right-hand side, we get

$$\int \left(\frac{1}{y(t)} \frac{dy}{dt} \right) dt = \int k\, dt$$

or

$$\ln |y(t)| = kt + C$$

Since we are only looking for functions that satisfy $y(t) > 0$ for all t, we can drop the absolute value sign and simply write

$$\ln y(t) = kt + C$$

Step 3: (Solve for y.)   We can solve for y in the above equation by writing

$$e^{\ln y(t)} = e^{(kt + C)}$$
$$= e^C e^{kt}$$
$$= A e^{kt}$$

Here A is merely the new arbitrary constant $A = e^C$. However, the left-hand side of the above equation is simply y. Hence, we have the solution

$$y(t) = A e^{kt}$$

This result is summarized below.

Solution of the Unlimited Growth and Decay Equation

For any constant k, the growth and decay equation

$$\frac{dy}{dt} = ky(t)$$

has an infinite number of solutions. They are

$$y(t) = Ae^{kt}$$

where A is any real number (the infinite number is a result of letting A be any real number). The constant k is positive for growth equations and negative for decay equations.

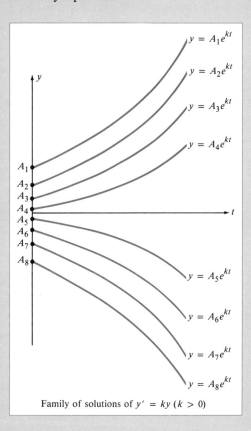

Family of solutions of $y' = ky$ $(k > 0)$

Check:

$$\frac{dy}{dt} = \frac{d}{dt}\left[Ae^{kt}\right] = Ake^{kt} = k\left[Ae^{kt}\right] = ky(t)$$

Unlimited Growth and Decay Curves

What do the growth of sunflowers, the growth of the total number of miles of railway tracks in the United States during the early 1900s, and the growth of money in a bank account in Lubbock, Texas, have in common? No, this is not a trick question; it is just that they can all be described by the exponential growth equation

$$y = Ae^{kt} \qquad (k > 0)$$

The graphs of these phenomena are displayed in Figure 37. Ready for another question? What does the decay of sunlight as one goes down into the ocean and the decay of the amount of alcohol in the blood after a person has stopped drinking have in common? Another trick question? No, it is just that these two phenomena can both be described by an exponential decay equation

$$y = Ae^{kt} \qquad (k < 0)$$

(See Figure 38.)

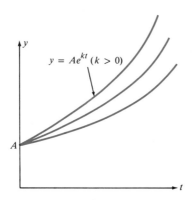

Figure 37
Typical growth curves

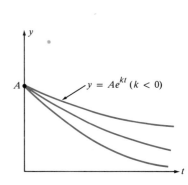

Figure 38
Typical decay curves

We illustrate some of the basic ideas of the growth and decay curves with the following examples.

Example 1

Solution of a Growth Equation   Verify that the growth curve $y = e^{0.05t}$ is a solution of the differential equation

$$\frac{dy}{dt} = 0.05y$$

Solution   We first calculate dy/dt, getting

$$\frac{dy}{dt} = 0.05e^{0.05t}$$

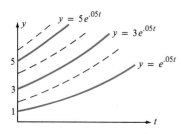

Figure 39
Family of curves $Ce^{0.05t}$ for $C > 0$

Since $y = e^{0.05t}$, we have

$$\frac{dy}{dt} = 0.05e^{0.05t} = 0.05y$$

Hence, $y = e^{0.05t}$ is a solution of the differential equation. By the same argument, any function of the form $Ce^{0.05t}$, where C is a constant, will satisfy the above differential equation. This entire family of solutions $Ce^{0.05t}$ is shown in Figure 39.   □

Example 2

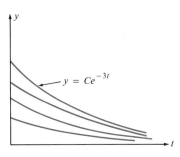

Figure 40
Typical particular solutions in the general solution

Solution of a Decay Equation   Find all the solutions of the decay equation

$$\frac{dy}{dt} = -3y$$

Solution   The solutions consist of the functions

$$y = Ce^{-3t}$$

where C is any constant. These decay curves are shown in Figure 40.   □

The Initial-Value Problem

Generally, when we solve the growth or the decay differential equation, our goal is not to find all the solutions, but that specific one that satisfies an additional initial condition $y(0) = y_0$. In order that the growth/decay curve $y = Ce^{kt}$ takes on a given value y_0 at $t = 0$, it is necessary that the constant C is y_0.
    These ideas introduce the initial-value problem.

Initial-Value Problem and Solution

The **initial-value problem** for the unlimited growth and decay equation consists of finding the function $y(t)$ that satisfies the two equations:

$$\frac{dy}{dt} = ky \qquad \text{(growth/decay equation)}$$

$$y(0) = y_0 \qquad \text{(initial condition)}$$

where y_0 is a given constant.
    The solution of this initial-value problem is the single function

$$y = y_0 e^{kt}$$

It is called a growth curve when $k > 0$ and a decay curve when $k < 0$.

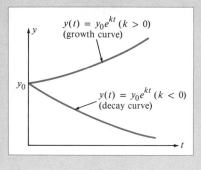

Example 3

Solution of Initial-Value Problem   Find the solution of the initial-value problem

$$\frac{dy}{dt} = -0.15y$$

$$y(0) = 100$$

Solution   We are given

$$k = -0.15$$
$$y_0 = 100$$

Hence, the solution is

$$y = y_0 e^{kt}$$
$$= 100 e^{-0.15t}$$

This decay curve is shown in Figure 41.   □

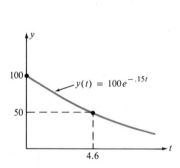

Figure 41
Solution of an initial-value problem

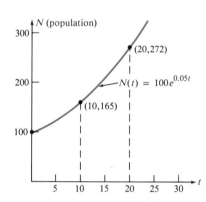

Figure 42
Growth curve with rate constant k = 0.05

Unrestricted Biological Growth

Consider the growth of a biological culture (such as a fungus, virus, bacteria, or cancer cells). We call the number of cells in the culture $N(t)$ and the number of cells when observations were begun N_0. If the culture is not contained in a crowded environment and has room to grow, then over the short run the rate of growth is often proportional to the number of cells present in the culture. This fact, along with the initial size N_0 of the culture, says that $N(t)$ must satisfy the initial value problem

$$\frac{dN}{dt} = kN(t) \qquad (k > 0)$$

$$N(0) = N_0 \qquad \text{(initial population)}$$

The growth constant k varies from population to population. The larger the rate of growth, the larger the constant. Figure 42 shows the growth curve $N(t)$ for the rate constant $k = 0.05$ with an initial population of $N_0 = 100$.

Example 4

Unrestricted Biological Growth   Assume the population of a certain culture of *Escherichia coli* grows at a rate proportional to its size N. A researcher has determined that every hour the size of the culture is 7% larger than it was the previous hour. Find the differential equation

$$\frac{dN}{dt} = kN(t)$$

that describes the size of this population.

Solution   The fact that the size is increasing by 7% every hour means that the population size after t hours will be

$$N(t) = N_0(1.07)^t$$

where N_0 is the initial population size. To determine the growth constant k, we use a property of exponentials to rewrite the above equation as

$$\begin{aligned} N(t) &= N_0(1.07)^t \\ &= N_0 e^{\ln(1.07)t} \\ &= N_0 e^{0.068t} \end{aligned}$$

Hence, the growth equation is

$$\frac{dN}{dt} = 0.068N$$

The growth curve in Figure 43 gives the number of *Escherichia coli* at any time t.   □

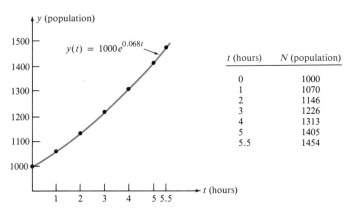

t (hours)	N (population)
0	1000
1	1070
2	1146
3	1226
4	1313
5	1405
5.5	1454

Figure 43
Growth of Escherichia coli

Example 5

Decay Problem in Biology   Suppose N_0 milligrams of a drug are injected into the bloodstream of a patient. The drug is carried by the oxygen in the blood to organs that consume and eliminate it from the body. Researchers have determined that the rate of decay of the drug in the bloodstream is proportional to the amount of the drug present. Suppose they have also determined that the amount of drug present decreases at a rate of 5% every hour. What is the initial-value problem that describes the amount of drug in the bloodstream?

Solution   Starting with N_0 milligrams of the drug in the bloodstream, the future amount is described by the decay curve

$$\begin{aligned} N(t) &= N_0(0.95)^t \\ &= N_0 e^{\ln(0.95)t} \\ &= N_0 e^{-0.051t} \end{aligned}$$

where t is measured in hours from the time the drug was injected.

Hence, the initial-value problem for $N(t)$ is given by

$$\frac{dN}{dt} = -0.051N \quad \text{(decay equation)}$$

$$N(0) = N_0 \quad \text{(initial condition)}$$

Interpretation   Note that in order to find the decay constant k, we had to convert the hourly decay rate of 5% to the continuous decay rate of 5.1%. Figure 44 shows the amount of drug that will be in the bloodstream at any time if 9 milligrams are initially injected. □

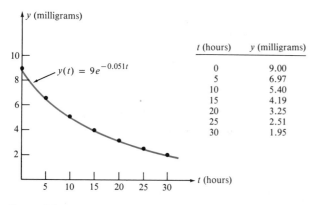

t (hours)	y (milligrams)
0	9.00
5	6.97
10	5.40
15	4.19
20	3.25
25	2.51
30	1.95

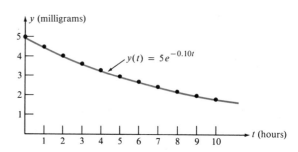

Figure 44
Decay of a drug in the bloodstream

Figure 45
Continuous decay of insulin in the bloodstream

Example 6

Typical Decay Phenomenon   A diabetic patient is injected with 5 milligrams of insulin directly into the bloodstream. From past experience, it is known that the insulin in the bloodstream decays exponentially and that the decay constant is $k = -0.10$.

- What initial-value problem describes the amount of insulin in the bloodstream?
- What is the solution of the initial-value problem?

Solution   Calling $N(t)$ the number of milligrams of insulin in the blood after t hours, the initial-value problem is given by

$$\frac{dN}{dt} = -0.10N \quad \text{(decay equation)}$$

$$N(0) = 5 \quad \text{(initial condition)}$$

The solution of this problem is given by

$$N(t) = 5e^{-0.10t}$$

Figure 45 displays this decay curve; Table 19 displays the decay of insulin in the bloodstream. □

TABLE 19
Decay of Drug in
Bloodstream

t (hours)	$N(t) =$ Amount of Drug (mg)
0	5.00
1	4.52
2	4.09
3	3.70
4	3.35
5	3.03
6	2.74
7	2.48
8	2.24
9	2.03
10	1.84
20	0.667
30	0.249
40	0.092
50	0.034
100	0.000227
250 (10.4 days)	0.00000000694

Money Growth by Continuous Compounding

When an initial amount of money P_0 is deposited in a bank that pays an annual interest of r percent compounded continuously, then the future value $P(t)$ of this account is known to grow exponentially according to the law

$$P(t) = P_0 e^{rt}$$

Another way of stating this exponential law is to say that the rate of growth dP/dt of the account is proportional to the amount present $P(t)$. Furthermore, the proportionality constant is the annual interest rate r. In other words,

$$\frac{dP}{dt} = rP$$

Hence, the future value $P(t)$ can be described by the initial-value problem

$$\frac{dP}{dt} = rP \qquad \text{(growth equation)}$$

$$P(0) = P_0 \qquad \text{(initial amount)}$$

Example 7

Compound Interest   An initial amount of $1000 is deposited in a bank that pays an annual interest of 12% compounded continuously.

· What initial-value problem describes the future value of this account?
· What is the future value of this account at time t?

Solution

▪ The initial-value problem is

$$\frac{dP}{dt} = 0.12P$$

$$P(0) = 1000$$

▪ The future value of the account is the solution of this initial-value problem, which is

$$P(t) = 1000e^{0.12t}$$

Table 20 shows the value of the above account for the first five years.

□

TABLE 20
Future Value of a Bank
Account that Pays an Annual
Interest of 12% Compounded
Continuously

t (years)	$P(t)$	
0	\$1000	← *initial deposit*
1	\$1127.50	
2	\$1271.25	
3	\$1433.33	
4	\$1616.07	
5	\$1822.12	

Problems

For Problems 1–10, find all the solutions of the following differential equations. Sketch the graphs of the solutions.

1. $\dfrac{dy}{dt} = 1$

2. $\dfrac{dy}{dt} = -1$

3. $\dfrac{dy}{dt} = 3y(t)$

4. $\dfrac{dy}{dt} = -3y(t)$

5. $\dfrac{dy}{dt} = y(t)$

6. $\dfrac{dN}{dt} = -0.01N(t)$

7. $\dfrac{dy}{dt} = 0.05y(t)$

8. $\dfrac{dy}{dt} = -y(t)$

9. $\dfrac{dy}{dt} = 10y(t)$

10. $\dfrac{dN}{dt} = 0.05N(t)$

For Problems 11–20, determine the exponential growth or decay curves that satisfy the following initial-value problems. Sketch the graph of the solution.

	Growth/Decay Equation	Initial Condition
11.	$\dfrac{dy}{dt} = y$	$y(0) = 100$
12.	$\dfrac{dN}{dt} = 2N$	$N(0) = 10$
13.	$\dfrac{dN}{dt} = -2N$	$N(0) = 50$
14.	$\dfrac{dy}{dt} = 0.08y$	$y(0) = 100$
15.	$\dfrac{dN}{dt} = -0.08N$	$N(0) = 50$
16.	$\dfrac{dN}{dt} = -0.05N$	$N(0) = 100$
17.	$\dfrac{dP}{dt} = 0.05P$	$P(0) = 1$

18. $\dfrac{dP}{dt} = 0.15P$ $P(0) = 50$

19. $\dfrac{dP}{dt} = 0.12P$ $P(0) = 1000$

20. $\dfrac{dP}{dt} = 0.50$ $P(0) = 50,000$

For Problems 21–25, the following derivatives dy/dt represent the velocity of a car along a road (miles per hour), where t is time measured in hours. The initial conditions $y(0) = y_0$ represent the initial position of the car when $t = 0$. Find the position $y(t)$ of the car at any time t by solving the initial-value problems.

Differential Equation	Initial Condition
21. $\dfrac{dy}{dt} = t^2$	$y(0) = 1$
22. $\dfrac{dy}{dt} = t - 1$	$y(0) = 1$
23. $\dfrac{dy}{dt} = t^2 + 2t + 1$	$y(0) = 3$
24. $\dfrac{dy}{dt} = t^2 - t$	$y(0) = 0$
25. $\dfrac{dy}{dt} = 1$	$y(0) = 0$

Relative Rate of Growth

Another way of stating the growth equation

$$\frac{dy}{dt} = ky$$

is to rewrite it as

$$\frac{1}{y}\frac{dy}{dt} = k$$

The left-hand side of this equation is called the relative rate of growth. This form of the growth equation says that the relative rate of growth is a constant. The relative rate is very useful because it does not depend on the amount present at a given time.

For Problems 26–31, find the populations $y(t)$ with the given relative rate of growth (RRG) and specified initial populations.

26. $RRG = 0.08$; $y(0) = 100$
27. $RRG = 0.05$; $y(0) = 1$
28. $RRG = 0.12$; $y(0) = 50$
29. $RRG = 0.00$; $y(0) = 100$
30. $RRG = 0.50$; $y(0) = 1$
31. $RRG = 0.50$; $y(0) = 1$

Finding the Rate Constant k

32. Yeast Growth   The size $P(t)$ of a yeast culture grows by 10% every week. What is the continuous growth constant k in the equation

$$P(t) = P_0 e^{kt}$$

if t is time measured in weeks and P_0 is the initial size of the yeast population?

33. Decline of the Whales   It has been shown that a certain whale population $N(t)$ is decreasing at a rate of 5% per year. What is the continuous decay constant k in the curve

$$N(t) = N_0 e^{kt}$$

if t is time measured in years and N_0 is the initial whale population?

34. Bacterial Growth   A bacteria grows in such a way that its growth rate at time t is equal to one-tenth its population. What is the initial-value problem that describes this phenomenon?

35. Yeast Growth   Yeast is growing in a medium at a rate equal to one-twentieth the amount present. If the initial amount present is 10 grams, determine the amount present at any time t.

36. Population Problem   Suppose the earth's human population is growing at the relative growth rate

$$\frac{1}{y}\frac{dy}{dt} = 0.02 \qquad \text{(two percent)}$$

per year.
   (a) What initial-value problem describes this population? Assume the population of the earth was 4.9 billion in in 1988 (set the time scale so $t = 0$ in 1988).
   (b) What will the population of the earth be at any time t provided this relative growth rate holds?

37. Thorium Decay   Thorium is used to date coral and other marine life. After the death of a coral, the amount of radioactive thorium present in the coral decays according to the differential equation

$$\frac{dy}{dt} = (-9.2 \times 10^{-8})y$$

where t is time measured in years. What is the half-life of radioactive thorium? (*Hint:* Remember that the half-life of a substance is the time it takes for the substance to decrease to one-half of the current amount.)

38. Flu Epidemic   Flu often spreads at a rate proportional to the number y of infected individuals. Suppose the relative rate of increase in the number of infected individuals is 5% per week.
   (a) What differential equation describes the number of infected individuals?
   (b) If there are presently 100 infected individuals, how many will be infected after one week?

39. Decay Rate The half-life of radioactive carbon-14 is approximately 5600 years.
   (a) What is the differential equation that describes the amount present at any time $t > 0$?
   (b) If the initial amount present is $y(0) = 10$ grams, how much will be present at any time $t > 0$?

40. Compound Interest The growth of a bank account is always equal to one-twentieth of the amount present.
   (a) What differential equation describes the future value of this account at any time $t > 0$?
   (b) If the initial deposit is $1, how much will be in the account after 10 years?

Epilogue: Population Curves in Biology

One area of mathematics that has been particularly useful in biology is the analysis of growth of cells, organisms, and other populations. We describe here some useful growth models.

Exponential Growth Curve

The **exponential growth curve** describes unrestricted growth and is often used to describe short-term populations of a whole host of phenomena such as multicellular organisms. (See Figure 46.) The main weakness of this model for long-term predictions of populations lies in the fact that the curve goes to infinity.

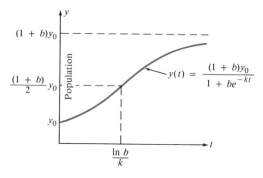

Figure 47
The logistic curve

The **Gompertz growth curve** applies to those phenomena in which the relative growth rate is not a constant but a linear function of time. (See Figure 48.) It was formulated in 1825 by the German mathematician of the same name for work in actuarial (life insurance) studies. In 1940, P. B. Medawar published a paper in which he theorized that the growth of an embryo chicken heart should follow this type of function. Since then, other biological organisms have been accurately described by the Gompertz curve.

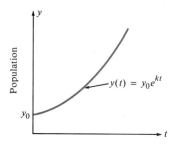

Figure 46
Unrestricted growth

The Logistics Growth Curve

The **logistics growth curve** overcomes the major criticism of the exponential growth curve; the logistics growth curve does not go to infinity. It describes populations that grow slowly at first, then grow rapidly, and then grow slowly. (See Figure 47.) The rate of growth in this model is proportional to both the current population and the difference between the current population and some asymptotic value of the population. It is a very popular model in biology.

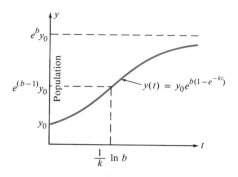

Figure 48
The Gompertz growth curve

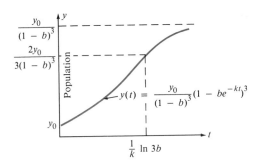

Figure 49
The von Bertalanffy growth curve

The von Bertalanffy Growth Curve

The **von Bertalanffy growth curve** is often used to describe the growth of a cell. The curve starts fast and reaches an upper limit. (See Figure 49.)

Key Ideas for Review

areas between curves, 776
consumers' surplus, 779
decay equation, 805
differential equations, 803
growth equation, 805

improper integrals, 757
initial-value problem, 809
integration by parts, 766
paint can paradox, 784
producers' surplus, 779

Simpson's Rule, 798
solid of revolution, 782
trapezoid rule, 790
trapezoid rule with data points, 796

Review Exercises

For Problems 1–5, find the antiderivative of the given functions.

1. $\int x e^{3x}\, dx$ **2.** $\int x \ln x\, dx$

3. $\int \ln (x^2)\, dx$ **4.** $\int x e^{-x}\, dx$

5. $\int x\sqrt{x + 4}\, dx$

For Problems 6–10, find the area of the regions bounded by the curve $y = f(x)$ and the x-axis for x defined on the indicated interval. Draw the indicated regions.

6. $y = x^2$ on $[-1, 1]$ **7.** $y = \sqrt{x}$ on $[0, 1]$
8. $y = x e^{-x}$ on $[0, 2]$ **9.** $y = e^x + e^{-x}$ on $[0, 1]$
10. $y = \ln (x^2)$ on $[1, 2]$

For Problems 11–15, find the area of the regions bounded by the curves $y = f(x)$ and $y = g(x)$. Draw the indicated regions.

11. $y = x^2,\ y = \sqrt{x}$ **12.** $y = x^3,\ y = x^2$
13. $y = x^2,\ y = 2x + 1$ **14.** $y = x^2,\ y = 2 - x$
15. $y = x^2,\ y = 1 - x^2$

For Problems 16–20, find the volumes of the solids of revolution formed by rotating the functions around the x-axis. Draw a rough picture of the function along with the solid of revolution.

16. $y = x^2$ on $[0, 1]$ **17.** $y = x$ on $[0, 1]$
18. $y = \sqrt{4 - x^2}$ on $[0, 2]$ **19.** $y = x^{1/3}$ on $[0, 1]$
20. $y = 1 - x$ on $[-1, 1]$

For Problems 21–24, use the given number of subintervals n to approximate the given integral by both the trapezoid method and Simpson's Rule.

21. $\int_0^1 x^2\, dx,\ n = 4$ **22.** $\int_0^1 e^{-x}\, dx,\ n = 4$

23. $\int_0^2 \dfrac{dx}{x + 1},\ n = 2$ **24.** $\int_0^4 \dfrac{dx}{\sqrt{x + 1}},\ n = 4$

25. Finding Distance from Velocity An automobile starting from rest accelerates for 10 seconds. Velocity readings (in feet per second) taken at one-second intervals are 2, 3, 4, 7, 11, 15, 20, 26, 32, 43. Estimate the distance traveled.

26. Present Value of an Ore Deposit   A coal mine is expected to earn an income of

$$p(t) = 100e^{-0.05t}$$

during the next ten years where p is measured in thousands of dollars and t is time measured in years. If money is discounted at 10% a year, what is the present value of the mine?

27. Producers' Surplus   If the supply and demand for a given commodity are given by

$$S(p) = p^2 - 4$$
$$D(p) = 10 - p$$

what is the producers' surplus?

28. Consumers' Surplus   For Problem 27, what is the consumers' surplus?

29. Finding the Continuous Rate Constant   Suppose the size of a bacteria culture is growing according to the exponential function

$$P(t) = P_0(1.15)^t$$

Rewrite this growth equation as

$$P(t) = P_0 e^{kt}$$

For Problems 30–35, evaluate the indicated improper integrals.

30. $\int_1^\infty \frac{1}{x} \, dx$ **31.** $\int_1^\infty \frac{1}{x^{3/2}} \, dx$

32. $\int_0^\infty e^{-2t} \, dt$ **33.** $\int_0^\infty xe^{-x^2} \, dx$

34. $\int_0^\infty \frac{dx}{x+1}$ **35.** $\int_{-\infty}^0 e^x \, dx$

For Problems 36–40, determine the exponential growth and decay curves that satisfy the given initial-value problem.

	Growth/Decay Equation	**Initial Condition**
36.	$\dfrac{dy}{dt} = 0.08y$	$y(0) = 1$
37.	$\dfrac{dP}{dt} = -0.08P$	$P(0) = 100$
38.	$\dfrac{dN}{dt} = 0.15N$	$N(0) = 50$
39.	$\dfrac{dP}{dt} = 0.25P$	$P(0) = 100$
40.	$\dfrac{dP}{dt} = 0.10P$	$P(0) = 15$

Chapter Test

1. Evaluate

$$\int_0^1 xe^{6x} \, dx$$

2. Approximate the following integral using Simpson's Rule with four subintervals

$$\int_0^4 x^2 \, dx$$

3. Find the area bounded by the curves $y = x^2 + x - 1$ and $y = x$.

4. Find the volume of the solid of revolution obtained by rotating the curve $y = x^2 + 1$ around the x-axis for $0 \le x \le 1$. Sketch both the graph of the function and the resulting solid of revolution.

5. Suppose the supply and demand curves for a commodity

are

$$S(p) = p - 5$$

$$D(p) = \frac{5}{p} - 1$$

What are the consumers' and producers' surpluses. Sketch the supply and demand curves.

6. Evaluate the improper integral

$$\int_0^\infty e^{-3x} \, dx$$

7. Solve the initial-value problem

$$\frac{dN}{dt} = 0.08N \qquad N(0) = 10$$

13

Multivariable
Calculus

In this final chapter we introduce the concept of a function of more than one variable and present some of the important problems related to such functions. In Section 13.1 we show how a function of two variables, $z = f(x, y)$, is graphed as a surface in three dimensions. In Section 13.2 we define the partial derivatives of a function of several variables that are the natural extensions of the ordinary derivative of a function of a single variable. In Sections 13.3 and 13.4 we find the maximum and minimum of functions of two variables and show how such optimization problems are useful in business management problems. Finally, in Section 13.5 we show how functions of two variables can be integrated as double integrals in much the same way that functions of one variable are integrated.

Functions of Several Variables

13.1

PURPOSE

We introduce functions of two or more variables and show how such functions arise in business and science. We also show how functions of two variables can be illustrated in three dimensions by means of surfaces.

Typical Functions of Two Variables

Until now we have considered only functions of one independent variable, say x, of the form $y = f(x)$. However, many situations arise in which a value depends on several variables. In biology, for example, the growth of an organism may depend on many variables, while in economics the price of a given commodity depends on many factors as well.

For instance, a company like the Owens Corning Fiberglass Corporation manufactures literally dozens of products that we use every day. Two of its products are roofing shingles and ceiling systems. Suppose the company earns a profit of $75 for each unit of a given type of roofing shingle sold and $350 for each unit of a given type of ceiling system. (We use hypothetical numbers for purposes of illustration. The precise values are not important for our discussion.) The total profit from the sale of x units of roofing shingles and y units of ceiling systems is a function of two variables $P(x, y)$. In this case it is the function

$$P(x, y) = 75x + 350y$$

Another company, the Ralston Purina Company of St. Louis, produces a wide variety of products ranging from pet foods to bakery goods. Two of its well-known products are *Chuck Wagon* dog food and *Cat Chow* cat food. Suppose the cost to produce a certain size bag of Chuck Wagon dog food is $1.40 and the cost to produce a given size bag of Cat Chow is $1.10. Suppose, too, that the plant that produces these two products has a fixed cost of $10,000 per week. Then if the plant produces x bags of Chuck Wagon per week and y bags of Cat Chow, the total weekly cost $C(x, y)$ to produce these products is given by the function of two variables:

$$C(x, y) = 1.4x + 1.1y + 10,000$$

In reality, of course, both the *Owens Corning Fiberglass Corporation* and the *Ralston Purina Company* produce dozens of products, and hence their revenues, costs, and profits will be functions of dozens of variables.

In this section we will consider only functions that depend on two variables. We begin by describing the cartesian coordinate system in three dimensions.

Three-Dimensional Coordinate Systems

For a function $y = f(x)$ of one variable a simple graph in the xy-plane provides an excellent way to visualize its properties. For a function that we can graph accurately, we can easily see where the function is increasing or decreasing, where the curve is concave upward or downward, and where the maximum and minimum points are located. However, to visualize properties of two variables

$$z = f(x, y)$$

we must use three-dimensional space.

In three dimensions we construct three coordinate axes, the x-, y-, and z-axes, each of which is perpendicular to each of the others, and the three axes intersect at a point. We make it our convention in this book to have the positive x-axis pointing towards us, the positive y-axis pointing to our right, and the positive z-axis pointing upward. (See Figure 1(a).)

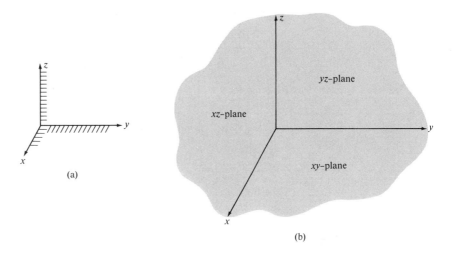

Figure 1
Basic ideas of the cartesian coordinate system

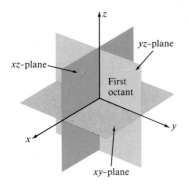

Figure 2
The eight octants of the three-dimensional coordinate system

These three coordinate axes form a three-dimensional rectangular or **cartesian coordinate system**. The point of intersection of the three axes is called the **origin** of the coordinate system. Each pair of axes determines a plane; these planes are called the **xy-plane**, the **xz-plane**, and the **yz-plane**, as shown in Figure 1(b).

For each point in three dimensions we assign a **triple** of numbers (a, b, c), called the **coordinates** of the point. For instance, to locate the point with coordinates $(3, 3, -2)$, start at the origin and go three units along the positive x-axis (out of the page). Then go three units in the positive y-direction (to the right) and finally two units in the negative z-direction (downward).

Another difference between two and three dimensions is that while the x- and y-axes in two dimensions divide the plane into four quadrants, the three coordinate planes in three dimensions divide three-dimensional space into eight **octants**. The points in three dimensions having all three of their coordinates positive form the first octant. (See Figure 2.)

Some of the important planes and lines in a three-dimensional cartesian coordinate system are described in Table 1.

TABLE 1
Important Lines and Planes in Three Dimensions

Region	Mathematical Description
x-axis	Points of the form $(x, 0, 0)$
y-axis	Points of the form $(0, y, 0)$
z-axis	Points of the form $(0, 0, z)$
xy-plane	Points of the form $(x, y, 0)$
xz-plane	Points of the form $(x, 0, z)$
yz-plane	Points of the form $(0, y, z)$

Distance Between Points in Three Dimensions

In two dimensions the distance D between two points (x_1, y_1) and (x_2, y_2) is determined by the Pythagorean theorem

$$D = \sqrt{(x_1 - x_2)^2 + (y_1 - y_2)^2}$$

To understand how to find the distance between two points in three dimensions, consider the three-dimensional box drawn in Figure 3. By use of the Pythagorean theorem the diagonal of the base of this box has the length

$$\text{Diagonal length of base} = \sqrt{a^2 + b^2}$$

Now note that the diagonal of the three-dimensional box is the hypotenuse of a right triangle; the triangle having one leg as the diagonal of the base and the

Figure 3
Finding distances in three dimensions

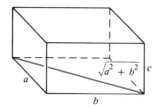 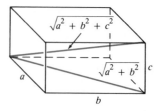

other leg as one of the vertical edges of the box. Hence we can apply the Pythagorean theorem again to find the length of the diagonal of the box. We have

$$(\text{Diagonal length of box})^2 = (\sqrt{a^2 + b^2})^2 + c^2$$

Hence we have

$$\text{Diagonal length of box} = \sqrt{a^2 + b^2 + c^2}$$

This discussion leads to the general formula for the distance between two points in three-dimensional space.

Distance in Three Dimensions

The distance D between any two points (x_1, y_1, z_1) and (x_2, y_2, z_2) in three-dimensional space is given by the generalized Pythagorean theorem

$$D = \sqrt{(x_1 - x_2)^2 + (y_1 - y_2)^2 + (z_1 - z_2)^2}$$

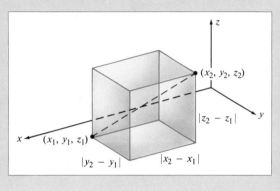

Example 1

Distance Between Points   Find the distance between the two points $(1, 2, 3)$ and $(1, 3, -4)$.

Solution

$$D = \sqrt{(1 - 1)^2 + (2 - 3)^2 + (3 + 4)^2}$$
$$= \sqrt{50} \qquad \square$$

Function of Two Variables

We say that an equation

$$z = f(x, y)$$

defines a function f of two variables x and y if a unique value of z is obtained for each value of x and y in a set, called the **domain** of the function. The domain of f is often taken as all points (x, y) where $f(x, y)$ makes mathematical sense. The **range** of a function is the set of values

$$\{f(x, y) \mid (x, y) \text{ is in the domain of } f\}$$

The variable z is called the dependent variable, and x and y are called the independent variables.

We have already seen a few functions of two variables. We list a few additional ones in Table 2.

TABLE 2
Functions of Two Variables

Subject	Function	Variables	
Area of a rectangle	$A(x, y) = xy$	$x =$ Width $y =$ Length	
Volume of a cylinder	$V(r, h) = \pi r^2 h$	$r =$ Radius $h =$ Height	
Future value of simple interest of $1	$F(r, t) = 1 + rt$	$r =$ Annual interest $t =$ Time period	
Future value of continuous compounding of $1	$F(r, t) = e^{rt}$	$r =$ Annual interest $t =$ Time period	
Revenue	$R(p, q) = pq$	$p =$ Price per item $q =$ Items sold	
Profit	$P(R, C) = R - C$	$R =$ Revenue $C =$ Cost	

Graphing Surfaces in Three Dimensions

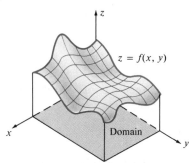

Figure 4
A function of two variables represented by a surface

The graph of a function of one variable $y = f(x)$ consists of the points $(x, f(x))$ in the xy-plane such that x belongs to the domain of f. To find the graph of a function, the usual strategy is to plot a few well-selected points, find the first and second derivatives at these points, and use all this information to draw the graph.

Analogously, the graph of a function of two variables $z = f(x, y)$ in three dimensions consists of the points $(x, y, f(x, y))$ such that (x, y) is a point in the domain of f. Geometrically, the graphs of functions of two variables represent surfaces in three-dimensional space. It should be noted that for every point (x, y) in the domain of the function the vertical line passing through the point (x, y) should intersect the surface exactly once. Figure 4 shows a typical surface drawn in three dimensions.

It is often difficult to draw the graph of a function $z = f(x, y)$. With the introduction of computer graphics packages the task has become much easier. Figure 5 shows computer drawings of several surfaces.

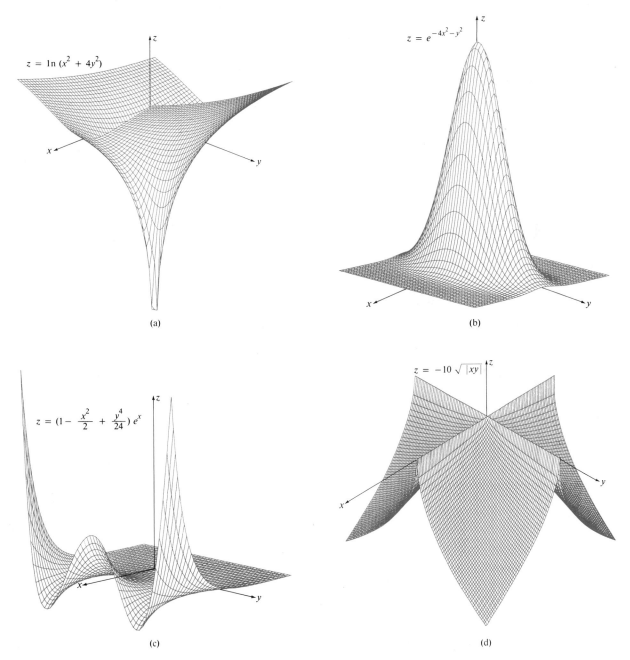

$z = \ln (x^2 + 4y^2)$

(a)

$z = e^{-4x^2 - y^2}$

(b)

$z = (1 - \dfrac{x^2}{2} + \dfrac{y^4}{24}) e^x$

(c)

$z = -10 \sqrt{|xy|}$

(d)

Figure 5
Computer-generated graphs of six surfaces

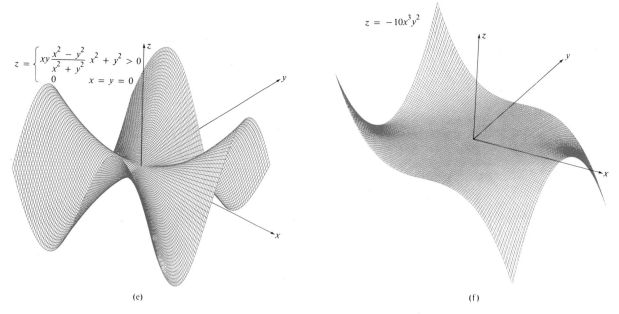

$$z = \begin{cases} xy\,\dfrac{x^2 - y^2}{x^2 + y^2} & x^2 + y^2 > 0 \\ 0 & x = y = 0 \end{cases}$$

$$z = -10x^3y^2$$

(e)

(f)

Figure 5
(*continued*)

Spheres in Three Dimensions

A **sphere** in three-dimensional space with **center** (a, b, c) and **radius** r consists of all points (x, y, z) at a distance of r units from the center (a, b, c). This fact can be stated by writing

$$\sqrt{(x - a)^2 + (y - b)^2 + (z - c)^2} = r$$

Squaring each side of this equation leads to the equation for the sphere.

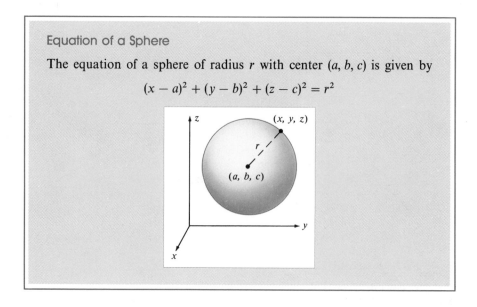

Equation of a Sphere

The equation of a sphere of radius r with center (a, b, c) is given by

$$(x - a)^2 + (y - b)^2 + (z - c)^2 = r^2$$

Note, however, that a sphere does not define a function, since vertical lines that intersect the sphere intersect the sphere at two points (with the exception of those vertical lines that intersect the sphere around the outer rim).

Example 2 _____   Typical Sphere   Sketch the surface described by the equation

$$(x - 1)^2 + (y - 5)^2 + (z + 2)^2 = 1$$

Solution   First observe that the surface is a sphere. The sphere is centered at $(1, 5, -2)$ and has radius 1. (See Figure 6.) □

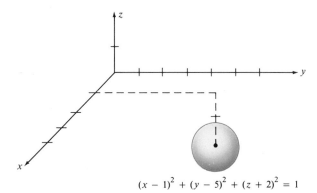

Figure 6
The equation of a sphere

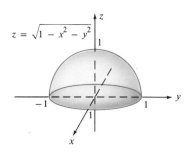

Figure 7
The upper part of a sphere

Example 3 _____   Graphing a Function of Two Variables   Sketch the surface described by the function

$$z = \sqrt{1 - x^2 - y^2} \qquad (x^2 + y^2 \leq 1)$$

Solution   We first note that in the domain of the function $x^2 + y^2 \leq 1$ we have $z \geq 0$. If we now square each side of the equation, we obtain the equation of a sphere

$$x^2 + y^2 + z^2 = 1$$

Hence the graph of the function is the upper half of the sphere shown in Figure 7. □

Planes in Three Dimensions

We learned in the Finite Mathematics Preliminaries that the general formula for a straight line in two dimensions is given by

$$ax + by = c$$

This equation generalizes to the equation of a plane in three dimensions.

Plane in Three Dimensions

Let a, b, c, and d be constants where not all a, b, and c are zero. The points (x, y, z) in three dimensions that satisfy the equation

$$ax + by + cz = d$$

define a plane.

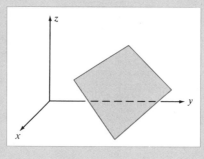

Example 4

Sample Plane   Graph

$$x + 2y + z = 2$$

Solution   A convenient way to graph planes in three dimensions is to find the points at which the plane crosses each of the x-, y-, and z-axes. By letting $x = y = 0$ we see that $z = 2$. Likewise, if we let $x = z = 0$, then we have $y = 1$. Finally, if we let $y = z = 0$, we get $x = 2$. Hence the plane passes through the three points $(0, 0, 2)$, $(0, 1, 0)$, and $(2, 0, 0)$. We can draw a portion of the plane by connecting these points in the manner shown in Figure 8.   □

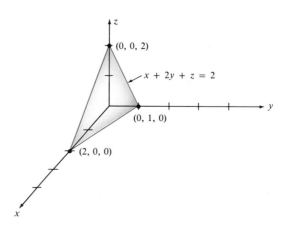

Figure 8

Graphing a plane in the first octant

Another graphing hint is the following.

> **Hint for Graphing Functions in Three Dimensions**
>
> It is often useful when graphing a surface $z = f(x, y)$ to let $x = 0$ and graph the intersection of $z = f(x, y)$ on the yz-plane. Then let $y = 0$ and graph the intersection of $z = f(x, y)$ on the xz-plane.

Example 5

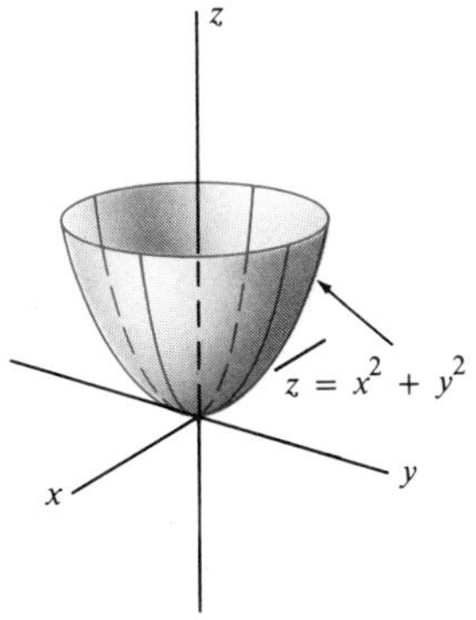

Figure 9
The graph of a paraboloid

Paraboloid   Graph the function

$$z = x^2 + y^2$$

Solution   Using the above hint, we begin by letting $x = 0$ and graph

$$z = 0^2 + y^2 = y^2$$

on the yz-plane. This is the equation that represents a parabola. We then let $y = 0$ and graph the parabola

$$z = x^2 + 0^2 = x^2$$

in the xz-plane. Both of these individual graphs drawn on the yz- and xz-planes are parabolas (see the Calculus Preliminaries). It can be shown that the surface $z = x^2 + y^2$ is the one that is formed when one rotates either of these two parabolas around the z-axis. This resulting cup-shaped surface is called a **paraboloid**. (See Figure 9.)   □

Problems

Problems 1–5 are concerned with the three common functions that occur in the theory of the firm (revenue, cost, and profit) where we assume that the firm produces two products. For each problem, evaluate the function at the indicated point.

1. $R(x, y) = 50x + 25y$ $R(10, 15)$
2. $C(x, y) = 10{,}000 + 5x + 7y$ $C(10, 20)$
3. $P(x, y) = 7x + 5y$ $P(100, 50)$
4. $C(x, y) = x^2 + xy + y^2$ $C(10, 5)$
5. $C(x, y) = 2x^2 + y^2$ $C(5, 10)$

For Problems 6–14, evaluate the given functions of two variables at the indicated points.

6. $f(x, y) = \sqrt{x^2 + y^2}$ $f(1, 2), f(3, 4)$

7. $f(x, y) = \dfrac{e^x + e^y}{e^x}$ $f(0, \ln 3), f(\ln 4, \ln 5)$

8. $f(x, y) = \ln(x - 2y + 4)$ $f(7, 1), f(11, 4)$

9. $f(x, y) = \dfrac{x + y}{x - y}$ $f(2, 4), f(3, 7)$

10. $f(x, y) = x^2 + y^2 - 2x + 4y + 1$ $f(-2, 3), f(4, -2)$

11. $f(x, y) = \dfrac{x}{x + y}$ $f(-3, 1), f(2, 5)$

12. $f(x, y) = \dfrac{2xy}{x^2 + y^2}$ $f(-3, 2), f(4, -3)$

13. $f(x, y) = xy + x^2 - 1$ $f(1, 3), f(5, 7)$

14. $f(x, y) = \log\left(\dfrac{x}{\sqrt{x^2 + y^2}}\right)$ $f(2, 3), f(5, 4)$

For Problems 15–19, calculate the distance between the given pairs of points.

15. $(3, 1, 5), (8, 2, -3)$ **16.** $(2, 2, 5), (-3, 2, 1)$
17. $(4, -3, 1), (2, 3, -2)$ **18.** $(3, 1, 5), (-1, -4, 2)$
19. $(0, 1, 3), (2, 7, -1)$

Identify the Sphere

For Problems 20–23, determine the center (a, b, c) and the radius r of the spheres. You may have to review the technique of completing the square as described in the Calculus Preliminaries.

20. $x^2 + (y - 1)^2 + (z + 1)^2 = 16$

21. $x^2 + y^2 - 2y + z^2 + 2z = 14$
22. $x^2 + 2x + y^2 + 4y + z^2 + 8z = 25$
23. $x^2 + y^2 - 2y + z^2 + 2z - 7 = 0$

Graphing Equations in Three Dimensions

For Problems 24–30, draw the surfaces in three dimensions described by the following equations. Systematically let $x = 0$ and $y = 0$ to determine what the surfaces look like in the yz- and xz-planes, respectively.

24. Plane $x + 2y + z = 2$ **25. Plane** $y = 3x$
26. Mystery Figure $x = 5$ **27. Plane** $x + y + 2z = 5$
28. Paraboloid $z = x^2 + y^2$
29. Paraboloid $z = x^2 + 4y^2$
30. Upper Hemisphere $z = \sqrt{1 - x^2 - y^2}$

Functions of Two Variables in the Publishing Industry

31. Cost to Produce Textbooks Suppose a certain publishing company produces a calculus textbook and a finite math textbook. Suppose the cost to produce a single calculus book is \$20, and the cost to produce a single finite math book is \$25. If we neglect fixed costs (electricity, taxes, general expenses, and so on), then the total cost in dollars of producing x calculus books and y finite math books is the function of two variables

$$C(x, y) = 20x + 25y$$

Find $C(500, 750)$.

32. Revenue for Calculus and Finite Math Texts A certain publishing company sells its calculus and finite math books to college bookstores for \$27 for each calculus book and \$30 for each finite math book. Hence the total revenue

in dollars taken in by the company will be the function of two variables

$$R(x, y) = 27x + 30y$$

Suppose the company sells 750 calculus textbooks and 1000 finite math textbooks to the University of Louisville. What is the revenue obtained from this sale?

33. Profit on Calculus and Finite Math Textbooks Based on the cost and revenue functions in Problems 31 and 32, the publishing company's profit in dollars for selling x calculus books and y finite math books will be

$$\begin{aligned} P(x, y) &= R(x, y) - C(x, y) \\ &= (27x + 30y) - (20x + 25y) \\ &= 7x + 5y \end{aligned}$$

If the company sells 500 calculus books and 750 finite math books to Florida State University, what will be the company's profit from this sale?

34. Author's Royalties Suppose the author of a certain calculus and finite math sequence earns \$2.25 for each calculus textbook sold and \$2.75 for each finite math textbook sold. If the publisher sells x calculus textbooks and y finite math textbooks, then the author's collective royalties will be

$$P(x, y) = 2.25x + 2.75y$$

Suppose the publisher sells the following numbers of textbooks to the following universities. How much are these sales worth to the author?

> University of Colorado:
> 1000 calculus, 750 finite math textbooks

> Colorado State University:
> 750 calculus, 900 finite math textbooks

35. Isoprofit Lines Suppose Figure 10 illustrates the number of calculus and finite math textbooks bought by some colleges and universities. We have seen that the profit from the sale of x calculus books and y finite math books is

$$P(x, y) = 7x + 5y$$

Draw the isoprofit lines in the xy-plane defined by

$$7x + 5y = c \qquad (x \geq 0, y \geq 0)$$

for $c = \$10,000$, \$15,000, and \$20,000. Each of these lines will define different combinations of sales x and y that yield profits of \$10,000, \$15,000, and \$20,000, respectively. Of course x and y are really integers, but the numbers x and y are large, and so the error in this approximation is small.

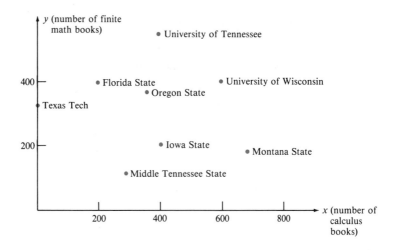

Figure 10
Textbook sales for Problem 35

Functions of Two Variables Are Everywhere

36. Cost of Producing Jogging and Racing Shoes Suppose a sporting goods manufacturer produces two types of running shoes, jogging shoes and racing shoes. If it costs $15 to produce each pair of jogging shoes and $20 to produce each pair of racing shoes, what is the general function $C(x, y)$ that describes the cost of producing x pairs of jogging shoes and y pairs of racing shoes? Neglect any fixed costs.

37. Cobb-Douglas Production Function The **Cobb-Douglas production function**

$$p(x, y) = Ax^k y^{1-k}$$

where $A > 0$ and $0 < k < 1$ can be used to predict the number p of units of a product produced as a function of x hours of labor and y dollars of capital expenses. By capital expenses we mean the cost of machinery, buildings, supplies, and so on. Suppose that it has been determined that the monthly production of automobiles in a given plant is described by the Cobb-Douglas function

$$p(x, y) = 1.2x^{2/3}y^{1/3}$$

where

  p = Number of cars produced (in thousands)

  x = Number of hours of labor (in thousands of hours)

  y = Capital expenses (in millions of dollars)

Find $p(10, 30)$. What does the value of this function mean?

38. How Big Are You? A biologist has constructed a mathematical model that predicts the surface area (measured in square feet) of a person's body. The model is given by

$$S(w, h) = 0.67w^{0.4}h^{0.7}$$

where w is the person's weight in pounds and h is the person's height in feet. What is the surface area of your body? How would you estimate the accuracy of this model?

39. Measuring Your Weight with a Ruler On the basis of years of collecting observations a physician has constructed a mathematical model that can predict a person's weight from the person's height and waist size. The formula is

$$W(h, w) = 6.4h + 4.2w - 450$$

where

  h = Person's measured height (in inches)

  w = Person's measured waist size (in inches)

  W = Person's predicted weight (in pounds)

On the basis of your own height and waist size, compute your predicted weight from this model. What is the error in this prediction?

40. Predicting Verbal Test Scores The U.S. Office of Education has conducted a study to determine the factors that contribute to verbal test scores of high school students. A mathematical model was constructed that depended on two variables. The model is

$$V(s, t) = 0.50s + 0.75t + 15$$

where

  s = Measure of the student's socioeconomic environment (ranges from -20 to $+20$)

  t = Measure of the teacher's verbal scores

  V = Predicted student's verbal scores

Evaluate $V(5, 20)$. What is the interpretation of this value?

41. What Is Your Cephalic Index?   A useful measurement for anthropologists is the ratio of width W in inches of the human skull to its length L in inches expressed as a percentage. This measurement is called the **cephalic index** and is given by

$$C(W, L) = 100\, \frac{W}{L}$$

A cephalic index of over 80 is called *brachycephalic*, and a measurement below 75 is called *dolicocephalic*. What is your cephalic index? What is the cephalic index of a person whose head has a width of 9 inches and a length of 12 inches?

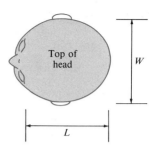

Figure 11
Head measurements for the cephalic index

13.2 Partial Derivatives

PURPOSE

We introduce the partial derivatives of a function of two variables. The major topics are

- the definition of a partial derivative and
- a geometric interpretation of partial derivatives.

We close this section by showing how partial derivatives can be used to study ideas in a multicommodity economy.

Introduction

We have seen that the derivative of a function of a single variable $y = f(x)$ gives the rate of change of the dependent variable y with respect to the independent variable x. When we study functions of two variables $z = f(x, y)$, we are interested in knowing how the function changes with respect to x (keeping y fixed) and how the function changes with respect to y (keeping x fixed). This brings us to the concept of the partial derivative.

Meaning of Partial Derivatives

Let f be a function of two variables x and y. If y is held constant, say $y = y_0$, then $f(x, y_0)$ can be viewed as a function of x alone. If this function is differentiable at $x = x_0$, then this derivative is called the **partial derivative of f with respect to x** at the point (x_0, y_0) and is denoted

$$f_x(x_0, y_0)$$

On the other hand, if x is held constant, say $x = x_0$, then $f(x_0, y)$ can be viewed as a function of y alone. If this function is differentiable at $y = y_0$, then this derivative is called the **partial derivative of f with respect to y** at the point (x_0, y_0) and is denoted by

$$f_y(x_0, y_0)$$

The partial derivatives $f_x(x_0, y_0)$ and $f_y(x_0, y_0)$ are generally found by first finding expressions for $f_x(x, y)$ and $f_y(x, y)$ at an arbitrary point (x, y) and then letting $x = x_0$ and $y = y_0$. To find $f_x(x, y)$ simply differentiate $f(x, y)$ with respect to x, treating y as a constant. To find $f_y(x, y)$, simply differentiate $f(x, y)$ with respect to y, treating x as a constant.

Example 1 ——————— Partial Derivatives Find $f_x(1, 3)$ and $f_y(0, 0)$ for

$$f(x, y) = x^2 y + y^2 + 3xy + 4$$

Solution Treating y as a constant and differentiating with respect to x, we obtain

$$f_x(x, y) = 2xy + 3y$$

Evaluating this partial derivative at the point $(1, 3)$, we have

$$f_x(1, 3) = 2(1)(3) + 3(3)$$
$$= 15$$

To find the partial derivative with respect to y, we treat x as a constant and differentiate with respect to y, getting

$$f_y(x, y) = x^2 + 2y + 3x$$

Hence the partial derivative $f_y(x, y)$ at the origin $(0, 0)$ is

$$f_y(0, 0) = (0)^2 + 2(0) + 3(0)$$
$$= 0 \qquad \qquad \square$$

Notation for Partial Derivatives

Table 3 lists some of the common ways for denoting the partial derivative of $z = f(x, y)$ with respect to x and y. Table 4 lists the common ways for denoting the partial derivatives of $z = f(x, y)$ evaluated at a point (x_0, y_0).

TABLE 3 ———————
Notation for Partial Derivatives

Partial Derivative of f (or z) with Respect to x	Partial Derivative of f (or z) with Respect to y
$f_x(x, y)$	$f_y(x, y)$
$\dfrac{\partial f(x, y)}{\partial x}$	$\dfrac{\partial f(x, y)}{\partial y}$
z_x	z_y
$\dfrac{\partial z}{\partial x}$	$\dfrac{\partial z}{\partial y}$

TABLE 4
Notation for Partial
Derivatives (x_0, y_0)

Partial Derivative of f (or z) with Respect to x Evaluated at (x_0, y_0)	Partial Derivative of f (or z) with Respect to y Evaluated at (x_0, y_0)
$f_x(x_0, y_0)$	$f_y(x_0, y_0)$
$\left.\dfrac{\partial f}{\partial x}\right|_{\substack{x=x_0 \\ y=y_0}}$	$\left.\dfrac{\partial f}{\partial y}\right|_{\substack{x=x_0 \\ y=y_0}}$
$z_x(x_0, y_0)$	$z_y(x_0, y_0)$
$\left.\dfrac{\partial z}{\partial x}\right|_{\substack{x=x_0 \\ y=y_0}}$	$\left.\dfrac{\partial z}{\partial y}\right|_{\substack{x=x_0 \\ y=y_0}}$

Example 2

Partial Derivatives   Find z_x and z_y for

$$z = x^2 + y^2 + 5xy + 1$$

Solution   To find z_x, we treat y as a constant and differentiate z with respect to x, getting

$$z_x = 2x + 5y$$

To find z_y, we treat x as a constant and differentiate z with respect to y, getting

$$z_y = 2y + 5x \qquad \square$$

Example 3

Partial Derivatives   Find f_x and f_y for

$$f(x, y) = x \ln y + ye^x$$

Solution   Just remember that when finding f_x, we treat y as a constant and differentiate with respect to x. This gives

$$f_x(x, y) = \ln y + ye^x$$

Likewise, we have

$$f_y(x, y) = \frac{x}{y} + e^x \qquad \square$$

Example 4

Partial Derivatives   Find f_x and f_y for

$$f(x, y) = \frac{x + 2y}{4x - 3y}$$

Solution   We use the quotient rule in each case, getting

$$f_x(x, y) = \frac{(4x - 3y)(1) - (x + 2y)(4)}{(4x - 3y)^2}$$

$$= \frac{-11y}{(4x - 3y)^2}$$

$$f_y(x, y) = \frac{(4x - 3y)(2) - (x + 2y)(-3)}{(4x - 3y)^2}$$

$$= \frac{11x}{(4x - 3y)^2} \qquad \square$$

Geometric Interpretation of Partial Derivatives

When we studied the derivative of a function of one variable, we learned that it represented the slope of the tangent line. We now see that partial derivatives also represent slopes of tangent lines.

Let P be a point on the surface

$$z = f(x, y)$$

If y is held constant, $y = y_0$, and x is allowed to vary, then the point P will move along the curve C_1, which is the intersection of the surface $z = f(x, y)$ and the vertical plane $y = y_0$. (See Figure 12a.) Hence the partial derivative $f_x(x_0, y_0)$ can be interpreted as the slope of the tangent line (the change in z per unit change in x) to the curve C_1 at the point (x_0, y_0). In other words, it tells the rate of change of z with respect to x.

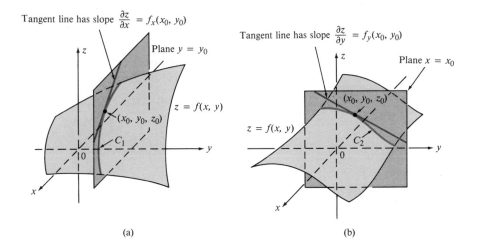

Figure 12
Geometric interpretation of the partial derivatives

(a)   (b)

On the other hand, if x is held constant, $x = x_0$, and y varies, then the point P moves along the curve C_2, which is the intersection of the surface $z = f(x, y)$ and the vertical plane $x = x_0$. (See Figure 12b.) Hence the partial derivative $f_y(x_0, y_0)$ can be interpreted as the slope of the tangent line (the change in z per unit change in y) to the curve C_2 at the point (x_0, y_0). In other words, it tells the rate of change in z with respect to y.

Higher-Order Partial Derivatives

Since the partial derivatives $\partial f/\partial x$ and $\partial f/\partial y$ are themselves functions of x and y, we can differentiate these functions and find higher-order partial derivatives. However, unlike the case of functions of one variable for which we had one second derivative, we now have four second-order partial derivatives.

Second-Order Derivatives

For $z = f(x, y)$ the four **second-order partial derivatives** are

$$z_{xx} = f_{xx} = \frac{\partial^2 f}{\partial x^2} = \frac{\partial}{\partial x}\left(\frac{\partial f}{\partial x}\right) \qquad \text{(second partial with respect to } x\text{)}$$

$$z_{yx} = f_{yx} = \frac{\partial^2 f}{\partial x\,\partial y} = \frac{\partial}{\partial x}\left(\frac{\partial f}{\partial y}\right) \qquad \begin{array}{l}\text{(cross partial, first with respect}\\ \text{to } y\text{, then with respect to } x\text{)}\end{array}$$

$$z_{xy} = f_{xy} = \frac{\partial^2 f}{\partial y\,\partial x} = \frac{\partial}{\partial y}\left(\frac{\partial f}{\partial x}\right) \qquad \begin{array}{l}\text{(cross partial, first with respect}\\ \text{to } x\text{, then with respect to } y\text{)}\end{array}$$

$$z_{yy} = f_{yy} = \frac{\partial^2 f}{\partial y^2} = \frac{\partial}{\partial y}\left(\frac{\partial f}{\partial y}\right) \qquad \text{(second partial with respect to } y\text{)}$$

Note that f_{xy} means that the partial derivative of f is taken with respect to x first, and then the resulting partial derivative f_x is differentiated with respect to y. This is in contrast to f_{yx}, where the partial derivative of f is taken with respect to y first, and the resulting partial derivative f_y is then differentiated with respect to x. You can remember which partial derivative to compute first by reading the x and y in the symbol f_{xy} as you would read a book (from left to right). Although it is not always true that f_{xy} is equal to f_{yx}, for all functions seen in this course the two partial derivatives will be the same.

Example 5

Higher-Order Partial Derivatives   Find the partial derivatives f_x, f_y, f_{xx}, f_{xy}, f_{yx}, and f_{yy} of

$$f(x, y) = x^2 + y \ln x + e^{xy}$$

Solution   We have

$$f_x = 2x + \frac{y}{x} + ye^{xy} \qquad \text{(treating } y \text{ as a constant)}$$

$$f_y = \ln x + xe^{xy} \qquad \text{(treating } x \text{ as a constant)}$$

$$f_{xx} = \frac{\partial}{\partial x}(f_x) = \frac{\partial}{\partial x}\left(2x + \frac{y}{x} + ye^{xy}\right)$$

$$= 2 - \frac{y}{x^2} + y^2 e^{xy}$$

$$f_{yx} = \frac{\partial}{\partial x}(f_y) = \frac{\partial}{\partial x}(\ln x + xe^{xy})$$

$$= \frac{1}{x} + e^{xy} + xye^{xy}$$

$$f_{xy} = \frac{\partial}{\partial y}(f_x) = \frac{\partial}{\partial y}\left(2x + \frac{y}{x} + ye^{xy}\right)$$

$$= \frac{1}{x} + e^{xy} + xye^{xy}$$

$$f_{yy} = \frac{\partial}{\partial y}(f_y) = \frac{\partial}{\partial y}(\ln x + xe^{xy})$$

$$= x^2 e^{xy} \qquad\qquad \square$$

Multicommodity Theory in Business

Until now we have studied the theory of the firm for companies that only produce one product. However, most companies produce several products, and for that reason a serious study of revenue, cost, and profit should consider multicommodity firms.

A typical multicommodity firm is the *A. H. Robbins Company* of Richmond, Virginia, which produces literally dozens of different products from pharmaceutical goods (Robitussin, Dimetapp, Chap Stick) to medical instruments and pet products. Although a serious study of revenue, cost, and profit conducted by a company such as this would involve many products, for mathematical simplicity we will focus our attention on only two products.

Multicommodity Costs

Suppose the A. H. Robbins Company has determined the cost $C(x, y)$ to produce x units of a given Item 1 and y units of a given Item 2. The two partial derivatives C_x and C_y are called the marginal costs of the two commodities. The interpretation of these partial derivatives is

$$C_x(x, y) = \text{Cost to make one additional unit of Item 1}$$
$$C_y(x, y) = \text{Cost to make one additional unit of Item 2}$$

when the production level is x units of Item 1 and y units of Item 2. For example, if

$$C_x(100, 50) = \$500$$
$$C_y(100, 50) = \$200$$

then at the production level of 100 units of Item 1 and 50 units of Item 2, it would cost the company \$500 to produce the next unit of Item 1 and \$200 to produce the next unit of Item 2.

Multicommodity Revenues

If the prices that the company charges for the two products are p_1 and p_2, respectively (and assuming that the company sells all that it produces), then the firm's total revenue is given by

$$R(x, y) = p_1 x + p_2 y$$

The two partial derivatives R_x and R_y are called the marginal revenues of the two products. These partial derivatives have the following interpretations:

$R_x(x, y)$ = Increase in revenue as a result of selling one more unit of Item 1

$R_y(x, y)$ = Increase in revenue as a result of selling one more unit of Item 2

when sales levels are at x units of Item 1 and y units of Item 2. Note that if the company charges \$100 and \$50 for each unit of Items 1 and 2, respectively, then the revenue function will be

$$R(x, y) = 100x + 50y$$

Hence the marginal revenues will be

$$R_x(x, y) = \$100$$
$$R_y(x, y) = \quad \$50$$

Ordinarily the marginal revenue functions represent the prices the company charges for the two products.

Multicommodity Profits

Finally, the profit the company makes when producing x and y units of Items 1 and 2, respectively, is the revenue minus the cost. That is,

$$P(x, y) = R(x, y) - C(x, y)$$
$$= p_1 x + p_2 y - C(x, y)$$

Just as there are marginal costs and revenues, there are marginal profits. The two partial derivatives P_x and P_y of the profit functions are called the marginal profits. They have the economic interpretation

$P_x(x, y)$ = Increase in profit from selling one more unit of Item 1

$P_y(x, y)$ = Increase in profit from selling one more unit of Item 2

when the sales level is x units of Item 1 and y units of Item 2.

Example 6

Multicommodity Firm *Cincinnati Milicron* of Cleveland, Ohio, manufactures different types of robots for industrial use. Suppose the monthly cost $C(r, s)$ of producing two models of robots at a certain plant is

$$C(r, s) = 20r^2 + 10rs + 10s^2 + 300{,}000$$

where C is measured in dollars and

r = Number of Model R robots made per month

s = Number of Model S robots made per month

Suppose the company charges

p_1 = \$5000 for each Model R robot

p_2 = \$8000 for each Model S robot

Find

(a) the monthly cost and marginal costs,

(b) the monthly revenue and marginal revenues, and

(c) the monthly profit and marginal profits,

when the monthly production of robots is

50 Model R robots

70 Model S robots

Solution

(a) Cost and marginal costs: The monthly cost to produce 50 Model R and 70 Model S robots is

$$C(50, 70) = 20 \cdot 50^2 + 10(50)(70) + 10 \cdot 70^2 + 300{,}000$$
$$= \$434{,}000$$

The marginal costs are

$$C_r(r, s) = 40r + 10s$$
$$C_s(r, s) = 10r + 20s$$

An economist would say that these partial derivatives give the "marginal costs" to produce more robots. For instance, if the company is currently producing 50 Model R robots and 70 Model S robots per month, the marginal costs are

$$C_r(50, 70) = 40(50) + 10(70) = \$2700$$
$$C_s(50, 70) = 10(50) + 20(70) = \$1900$$

In other words, it will cost the company approximately \$2700 to make the next Model R robot and \$1900 to make the next Model S robot. Note from the marginal revenue formulas that as the monthly production levels r and s rise, it costs more to make each of the two types of robots. This may be partly due to the fact that overtime wages must be paid and special ordering of materials is necessary.

(b) Revenue and marginal revenues: The firm's monthly revenue R is given by

$$R(r, s) = \$5000r + \$8000s$$

At the production level of 50 Model R robots and 70 Model S robots per month the firm will have a monthly revenue of

$$R(50, 70) = \$5000(50) + \$8000(70)$$
$$= \$810{,}000$$

The marginal revenues are the partial derivatives of the revenue function R. Hence

$$R_r(r, s) = \$5000$$
$$R_s(r, s) = \$8000$$

These values simply represent the prices charged for the two products.

(c) Profit and marginal profits: The firm's monthly profit is found by computing the money taken in minus the money expended in costs. That is,

$$P(r, s) = R(r, s) - C(r, s)$$
$$= 5000r + 8000s - (20r^2 + 10rs + 10s^2 + 300,000)$$

At the production level of 50 Model R robots and 70 Model S robots the company will earn monthly profit of

$$P(50, 70) = \$810,000 - \$434,000$$
$$= \$376,000$$

The marginal profits are given by

$$P_r(r, s) = 5000 - 40r - 10s$$
$$P_s(r, s) = 8000 - 10r - 20s$$

This means that when the production level is 50 Model R robots and 70 Model S robots per month, the company will make a profit on each additional Model R robot sold of

$$P_r(50, 70) = 5000 - 40(50) - 10(70)$$
$$= \$2300$$

while the profit on each additional Model S robot will be

$$P_s(50, 70) = 8000 - 10(50) - 20(70)$$
$$= \$6100 \qquad \square$$

Deeper Interpretation

As the production levels r and s increase, the revenue increases only linearly (first-order polynomial), whereas the cost increases quadratically (second-order polynomial). This means that costs will eventually grow faster than revenues (more money going out than coming in). When this happens, it makes sense that the company should not increase its level of production. This means that there will be an optimum production level that will maximize the company's profit. Of course, if the company can actually sell more than this optimum amount produced, it should consider a number of alternatives, such as raising prices or building a new factory to change the present cost structure.

PROBLEMS

For Problems 1–20, find $f_x(x, y)$ and $f_y(x, y)$ for the given functions.

1. $f(x, y) = 2x - xy + y + 3x^2$

2. $f(x, y) = (x - 3y)^2$

3. $f(x, y) = x/y$

4. $f(x, y) = xy^3 - yx^3$

5. $f(x, y) = \dfrac{x + y}{x - y}$

6. $f(x, y) = \sqrt{x^2 - 3y^2}$

7. $f(x, y) = \dfrac{x + y}{x^2 + y^2}$

8. $f(x, y) = \dfrac{3xy + y^2}{x^3 - y^3}$

9. $f(x, y) = e^{xy}$

10. $f(x, y) = xe^y + ye^x$

11. $f(x, y) = \dfrac{e^x + e^y}{e^x - e^y}$

12. $f(x, y) = e^y \ln x$

13. $f(x, y) = e^x \ln y + e^y \ln x$

14. $f(x, y) = xe^{-y/2} + xy + \dfrac{x}{y}$

15. $f(x, y) = e^{(x^2 + y^2)}$ **16.** $f(x, y) = \dfrac{(3x - 2y)^3}{y^2 - x^2}$

17. $f(x, y) = (x^2 e^y - 2xye^x + y^2 e^y)^{2/3}$

18. $f(x, y) = x^y + y^x$

19. $f(x, y) = \ln \sqrt{x^2 + 2xy + y^2}$ (Sometimes a little preliminary algebra will go a long, long way.)

20. $f(x, y) = \ln\left(\dfrac{x + y}{x - y}\right)$ (The same hint is true here, too.)

For Problems 21–25, evaluate the indicated partial derivatives.

21. $f(x, y) = 9 - y^2 - 5x^2$ $f_x(3, 1)$ $f_y(2, 3)$
22. $f(x, y) = y^2 + y^3 x$ $f_x(2, 3)$ $f_y(3, 2)$
23. $f(x, y) = e^{(2x + 3y)}$ $f_x(0, 1)$ $f_y(-3, 2)$

24. $f(x, y) = \dfrac{3xy}{x^2 + y^2}$ $f_x(-1, -1)$ $f_y(-1, 0)$

25. $f(x, y) = (x + 3y)(\ln x + e^y)$ $f_x(3, 4)$ $f_y(2, 5)$

For Problems 26–35, find $f_x(x, y)$, $f_y(x, y)$, $f_{xx}(x, y)$, $f_{xy}(x, y)$, and $f_{yy}(x, y)$ for the given functions.

26. $f(x, y) = 4xy - 5y + 4$
27. $f(x, y) = x^2 + 2xy - 3y^2 - 2y + 10x - 3$
28. $f(x, y) = e^{(x + y)}$
29. $f(x, y) = (x^2 + 2xy - y^2)^2$

30. $f(x, y) = \dfrac{xy}{x + y}$

31. $f(x, y) = (x + y) \ln (x + y)$
32. $f(x, y) = x^3 - x^2 y + yx^2 - y^3$
33. $f(x, y) = (x + 2y)(3x - y)$

34. $f(x, y) = \sqrt{x + y}$ **35.** $f(x, y) = \dfrac{(x + y)^2}{(x - y)^3}$

Geometric Interpretation of Partial Derivatives

For Problems 36–39, give rough approximations of the partial derivatives f_x and f_y of the functions $f(x, y)$ at the indicated points A, B, and C. Your answers do not have to be accurate to the nearest tenth, but you should at least be able to determine whether the partial derivatives are positive or negative at the points.

36.

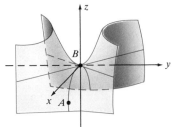

37.

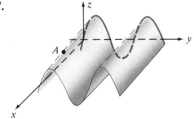

38.

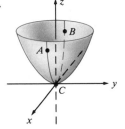

39.

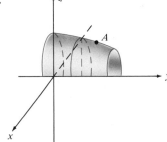

Marginal Functions in a Multicommodity Firm

The functions in Problems 40–48 represent common revenue, cost, and profit functions found in the theory of the multicommodity firm. For each function, find the two marginal functions. Evaluate all the marginal functions when $x = 10$, $y = 5$, and interpret the results.

40. $R(x, y) = 30x + 50y$
41. $R(x, y) = 5x + 10y - 0.05x^2$
42. $R(x, y) = 10x + 20y - 0.2x^2 - 0.3y^2$
43. $C(x, y) = x^2 + xy + y^2 + 10$
44. $C(x, y) = 10x + 20y + x^2 + xy + y^2 + 5000$

45. $C(x, y) = 100x + 50y + 5x^2 + xy + 10,000$
46. $P(x, y) = 100x + 200y - x^2 - xy - y^2 - 5000$
47. $P(x, y) = 50x + 200y - x^2 - 5000$
48. $P(x, y) = 2x + 2y - xy - 100$

Partial Derivatives in Mathematical Economics

49. Marginal Cost   Suppose the cost to a firm to produce x units of Product A and y units of Product B is given by

$$C(x, y) = 50x + 100y + x^2 + xy + y^2 + 10,000$$

Find $C_x(10, 20)$ and $C_y(10, 20)$. What is the interpretation of your answer?

50. Marginal Revenue   The revenue from the production of x units of Product A and y units of Product B is given by

$$R(x, y) = 50x + 100y - 0.01x^2 - 0.01y^2$$

Find $R_x(10, 20)$ and $R_y(10, 20)$. What is the interpretation of your answer?

51. Marginal Profit   A firm's profit $P(x, y)$ for producing x units of Product A and y units of Product B is given by

$$P(x, y) = 10x + 20y - x^2 + xy - 0.5y^2 - 10,000$$

Find $P_x(10, 20)$ and $P_y(10, 20)$. What is the interpretation of your answer?

52. Change in Present Value   The present value of an annuity for which R dollars are to be paid every year for t years is given by

$$\text{Present value} = \frac{R}{i}\left[1 - \left(\frac{1}{1+i}\right)^t\right]$$

where i is the annual rate of interest. If the length t of the annuity is fixed, how does the present value change as the annual rate of interest i changes? Evaluate this change when $R = \$750$ and $i = 0.10$. What is the interpretation of your answer?

53. Cobb-Douglas Production Function   The Cobb-Douglas production function describes the production output p of a firm in terms of the cost x of capital and the cost y of labor. It is given by

$$p(x, y) = Ax^k y^{1-k}$$

where $A > 0$ and $0 < k < 1$. Show that a Cobb-Douglas production function $p(x, y)$ satisfies the relations

$$\frac{p_x(x, y)}{p(x, y)} = \frac{k}{x}$$

$$\frac{p_y(x, y)}{p(x, y)} = \frac{1 - k}{y}$$

Cobb and Douglas were able to see the economic significance of these two equations. Can you interpret these two equations?

54. Marginal Productivity   The partial derivative $p_x(x, y)$ of the Cobb-Douglas production function is called the **marginal productivity of capital** and $p_y(x, y)$ is the **marginal productivity of labor.** Find the points (x, y) for which the marginal productivity of capital is equal to the marginal productivity of labor if the Cobb-Douglas function is given by

$$p(x, y) = 40x^{2/3}y^{1/3}$$

Problems 55–57 concern themselves with three important **partial differential equations** (equations that contain partial derivatives) that describe natural phenomena.

55. Partial Differential Equation   If

$$u(x, y) = \frac{2x + y}{x - y}$$

show that $u(x, y)$ satisfies the partial differential equation

$$xu_x + yu_y = 0$$

56. Laplace's Equation   If

$$u(x, y) = \ln(x^2 + y^2)$$

show that $u(x, y)$ satisfies Laplace's equation

$$u_{xx} + u_{yy} = 0$$

57. Cauchy-Riemann Equations   If

$$u(x, y) = x^2 - y^2$$
$$v(x, y) = 2xy$$

show that $u(x, y)$ and $v(x, y)$ satisfy the Cauchy-Riemann equations

$$u_x(x, y) = v_y(x, y)$$
$$u_y(x, y) = -v_x(x, y)$$

58. Just How Big Are You Getting?   A biologist has determined that the surface area in square inches of the human body can be reasonably approximated by

$$S(h, w) = 16h^{0.4}w^{0.7}$$

where h is a person's height in inches and w is a person's weight in pounds.
(a) Compute $S_w(64, 100)$. For a person who is 64 inches tall (5′4″) and weighs 100 pounds this partial derivative with respect to w will tell roughly how many square inches that person's surface area changes if height remains constant but weight increases by one pound. It will have units of square inches of surface area per pound of increase.

(b) Evaluate the partial derivative S_w at your own height and weight to estimate your own increase in surface area as a function of weight increase (assuming that you are not growing in height). If you gain x pounds (assuming that x is small), you can approximate your actual increase in surface area by using the differential approximation by multiplying the computed partial derivative by the weight increase x.

13.3

Unconstrained Optimization

PURPOSE

We show how the first- and second-partial derivatives can be used to find relative maximum and minimum points of a function of two variables. We do this by using the following two-step process.

- **Step 1.** Use the first-partial derivatives to find the critical points.
- **Step 2.** Use the second-partial derivatives to determine if the critical points are relative maximum or relative minimum points (or neither).

Introduction

One of the major applications of differential calculus is in finding maximum and minimum points of a function. Earlier, we used the first and second derivatives to find the maximum and minimum points of a function of a single variable. Now we will see how partial derivatives can be used to find maximum and minimum points of a function $z = f(x, y)$ of two variables.

We assume that the surface $z = f(x, y)$ is "continuous and smooth" as shown in Figure 13(a), which means it does not have any jumps, sharp corners, or edges of the type shown in Figure 13(b).

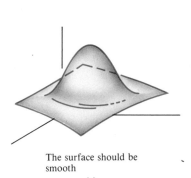

The surface should be smooth

(a)

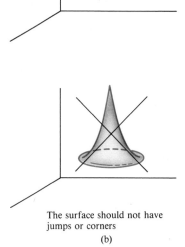

The surface should not have jumps or corners

(b)

Figure 13
Surfaces that we will and will not study

Relative Maximum and Minimum Points

What does it mean for a point (a, b) to be a relative maximum or minimum point of a function f of two variables? Roughly, it means that the surface $z = f(x, y)$ has a "high point" or a "low point" at (a, b) as shown in Figure 14.

If we were to state more precisely what our intuition tells us, we would say the following.

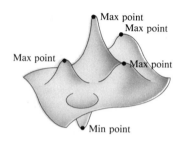

Figure 14
Relative maximum and minimum points

Definition of Relative Maximum and Minimum Points

A function f of two variables is said to have a **relative maximum** at a point (a, b) if there is a circle centered at (a, b) such that

$$f(a, b) \geq f(x, y)$$

for all points (x, y) inside the circle. Similarly, a function f of two variables has a **relative minimum** at a point (a, b) if there is a circle centered at (a, b) such that

$$f(a, b) \leq f(x, y)$$

for all points (x, y) inside the circle.

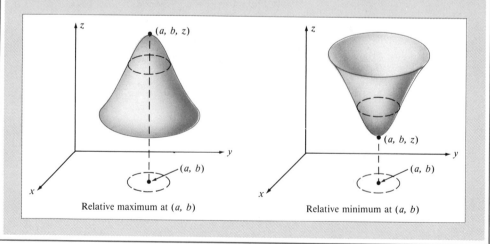

Relative maximum at (a, b) — Relative minimum at (a, b)

What is true about the partial derivatives $f_x(a, b)$ and $f_y(a, b)$ at a point (a, b) where $f(a, b)$ is a relative maximum or minimum point? The above diagram indicates that both partial derivatives $f_x(a, b)$ and $f_y(a, b)$ are zero, since the slopes of the tangent lines in the x- and y-directions are both zero. The following theorem indicates that our intuition is correct.

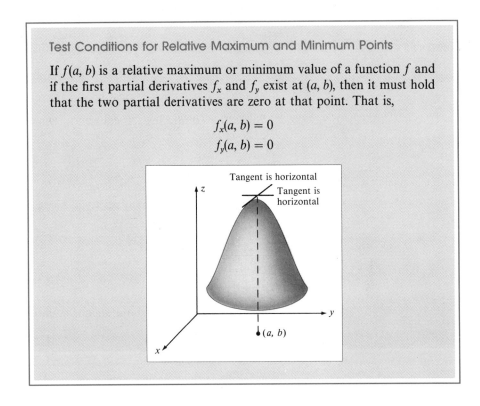

Test Conditions for Relative Maximum and Minimum Points

If $f(a, b)$ is a relative maximum or minimum value of a function f and if the first partial derivatives f_x and f_y exist at (a, b), then it must hold that the two partial derivatives are zero at that point. That is,

$$f_x(a, b) = 0$$
$$f_y(a, b) = 0$$

The points (a, b) where both partial derivatives f_x and f_y are zero are called **critical points**. It is important to understand that if (a, b) is a relative maximum or minimum point, then

$$f_x(a, b) = 0$$
$$f_y(a, b) = 0$$

but not vice versa. That is, just because $f_x(a, b)$ and $f_y(a, b)$ are both zero at (a, b), it is not necessarily true that $f(a, b)$ is a relative maximum or minimum value. Figure 15 shows an example of a function $f(x, y)$ that has the property

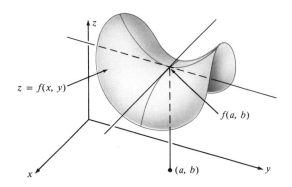

Figure 15
$f_x(a, b)$ and $f_y(a, b)$ are both zero, but (a, b) is not a relative maximum or minimum point

that both partial derivatives $f_x(a, b)$ and $f_y(a, b)$ are zero but $f(x, y)$ does not have a relative maximum or minimum value at (a, b). We call the point (a, b) a **saddle point**. The reason for the name is that the surface $z = f(x, y)$ at the point (a, b) is shaped somewhat like a saddle.

Example 1

Finding Critical Points　Find the critical points of the function

$$f(x, y) = x^2 - xy + y^2 + 3x + 10$$

and determine which critical points are relative maximum or minimum points.

Solution　To find the critical points, we look for solutions (a, b) of the system of equations

$$f_x(x, y) = 0$$
$$f_y(x, y) = 0$$

Setting the partial derivatives to zero, we get

$$f_x(x, y) = \frac{\partial}{\partial x}(x^2 - xy + y^2 + 3x + 10) = 2x - y + 3 = 0$$

$$f_y(x, y) = \frac{\partial}{\partial y}(x^2 - xy + y^2 + 3x + 10) = -x + 2y = 0$$

This gives us the two simultaneous equations

$$2x - y = -3$$
$$-x + 2y = 0$$

which have the unique solution

$$x = -2$$
$$y = -1$$

The above solution gives the single critical point $(-2, -1)$. Of course, we cannot be absolutely sure that this point is a relative maximum or minimum point. What we can say, however, is that every other point can be eliminated as a candidate for a relative maximum or minimum point. If we now evaluate the function $f(x, y)$ at the critical point $(-2, -1)$ and at several nearby points, we get a pretty good indication that $(-2, -1)$ is a relative maximum or minimum point. Figure 16 illustrates the value of $f(x, y)$ at the critical point $(-2, -1)$ and at four nearby points.

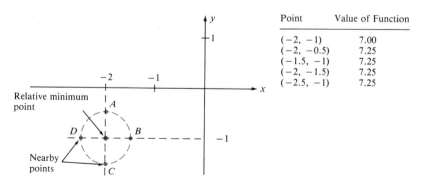

Point	Value of Function
$(-2, -1)$	7.00
$(-2, -0.5)$	7.25
$(-1.5, -1)$	7.25
$(-2, -1.5)$	7.25
$(-2.5, -1)$	7.25

Figure 16
Testing to determine a relative minimum point

A computer-driven plotter was also used by Professor Norton Starr of Amherst University to draw the surface that is shown in Figure 17. This surface also indicates that $(-2, -1)$ is a relative minimum point. Later, when we study the second-partials test, we will confirm this conclusion.   □

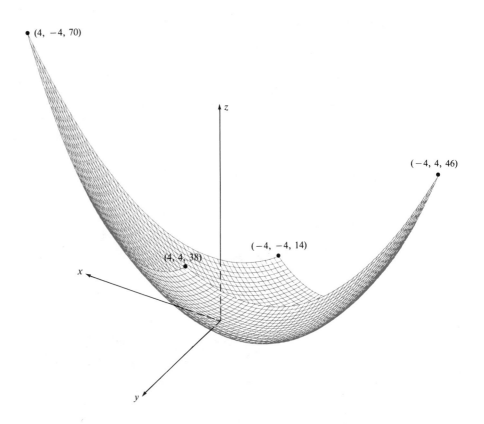

Figure 17
Graph of
$z = x^2 - xy + y^2 + 3x + 10$

Example 2

Finding Critical Points   Find the critical point(s) of

$$f(x, y) = 2x^2 + y^2 + 2$$

and determine whether they are relative maximum or minimum points.

Solution   Setting the partial derivatives f_x and f_y to zero, we have

$$f_x(x, y) = \frac{\partial}{\partial x}(2x^2 + y^2 + 2) = 4x = 0$$

$$f_y(x, y) = \frac{\partial}{\partial y}(2x^2 + y^2 + 2) = 2y = 0$$

Since $x = 0$ and $y = 0$ is the only solution of the system of equations, we conclude that the origin $(0, 0)$ is the only critical point. Hence this point is the only candidate for a relative maximum or minimum point. In this case it is clear that the value of the function

$$f(x, y) = 2x^2 + y^2 + 2$$

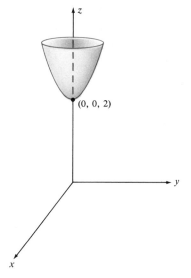

Figure 18
Relative and absolute minimum at (0, 0, 2)

is always greater than the value of the function at the origin. Hence we can conclude that $f(0, 0) = 2$ is a relative minimum value of the function. In fact, since $f(0, 0) = 2$ is smaller than any other value of $f(x, y)$, this means that $f(0, 0)$ is the absolute minimum value of the function. This surface is shown in Figure 18.

□

The Second-Partials Test

Knowing the critical points of a function of two variables narrows the search for the relative maximum and minimum points. However, the critical point may be a saddle point and not a relative maximum or minimum point. To determine whether a critical point is in fact a relative maximum or minimum point, there is a convenient test involving the second-partial derivatives f_{xx}, f_{xy}, and f_{yy}. The only price we pay for using this test is the computation of the second derivatives. We state the test here without proof.

Second-Partials Test for Relative Maximum and Minimum Points

Let $z = f(x, y)$ be a function of two variables with a critical point (a, b). That is, $f(x, y)$ satisfies

$$f_x(a, b) = 0$$
$$f_y(a, b) = 0$$

Define the **discriminant** D as

$$D = f_{xx}(a, b) \cdot f_{yy}(a, b) - f^2_{xy}(a, b)$$

The nature of the critical point (a, b) depends on the following values of D.

$$D = f_{xx}(a, b) \cdot f_{yy}(a, b) - f^2_{xy}(a, b)$$

$D < 0$	$D = 0$	$D > 0$	
Saddle point	Nothing can be said (test fails)	$f_{xx}(a, b) < 0$ Relative maximum point	$f_{xx}(a, b) > 0$ Relative minimum point

---

It might be pointed out here that when $D > 0$, it is impossible that the second partial derivative $f_{xx}(a, b)$ is zero. (This is why this possibility is not included in the table.)

Figure 19 shows the steps that should be taken in finding relative maximum and minimum points of a function.

Example 3 —————— Second-Partials Test   Use the second-partials test to find the relative maximum and minimum points of the function

$$f(x, y) = x^2 - xy + y^2 + 3x + 10$$

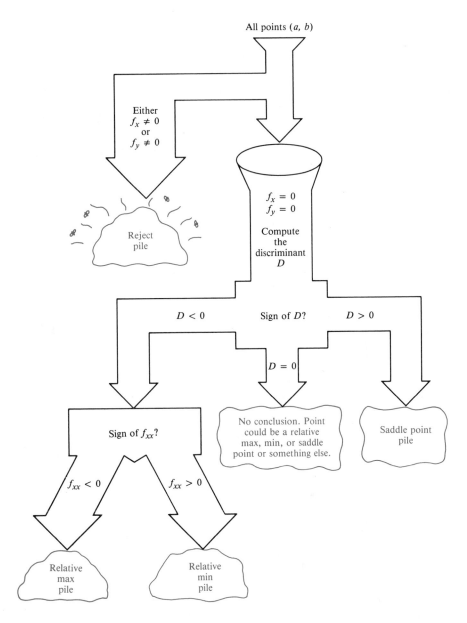

All points (a, b)

Either
$f_x \neq 0$
or
$f_y \neq 0$

Reject
pile

$f_x = 0$
$f_y = 0$

Compute
the
discriminant
D

$D < 0$ Sign of D? $D > 0$

$D = 0$

Sign of f_{xx}?

No conclusion. Point
could be a relative
max, min, or saddle
point or something else.

Saddle point
pile

$f_{xx} < 0$ $f_{xx} > 0$

Relative
max
pile

Relative
min
pile

Figure 19
The "classification machine"

Solution

- **Step 1** (find first-partial derivatives). Computing the first-partial derivatives, we get

$$f_x(x, y) = \frac{\partial}{\partial x}(x^2 - xy + y^2 + 3x + 10) = 2x - y + 3$$

$$f_y(x, y) = \frac{\partial}{\partial y}(x^2 - xy + y^2 + 3x + 10) = -x + 2y$$

- **Step 2** (find the critical points). Setting the first-partial derivatives to zero gives the simultaneous equations

$$f_x(x, y) = 2x - y + 3 = 0$$
$$f_y(x, y) = -x + 2y = 0$$

Solving for x and y, we have

$$x = -2$$
$$y = -1$$

Hence the only critical point is $(-2, -1)$. This provides one candidate for a relative maximum or minimum point.

- **Step 3** (compute the discriminant).   Computing $f_{xx}(-2, -1)$, $f_{yy}(-2, -1)$, and $f_{xy}(-2, -1)$, we get

$$f_{xx}(x, y) = \frac{\partial}{\partial x} f_x = \frac{\partial}{\partial x}(2x - y + 3) = 2$$

$$f_{xy}(x, y) = \frac{\partial}{\partial y} f_x = \frac{\partial}{\partial y}(2x - y + 3) = -1$$

$$f_{yy}(x, y) = \frac{\partial}{\partial y} f_y = \frac{\partial}{\partial y}(-x + 2y) = 2$$

Thus the discriminant is

$$\begin{aligned} D &= f_{xx}(-2, -1) \cdot f_{yy}(-2, -1) - f_{xy}^2(-2, -1) \\ &= (2)(2) - (-1)^2 \\ &= 3 \end{aligned}$$

- **Step 4** (apply the second-partials test).   Since the discriminant D is positive and $f_{xx}(-2, -1) = 2$ is positive, we conclude from the second-partials

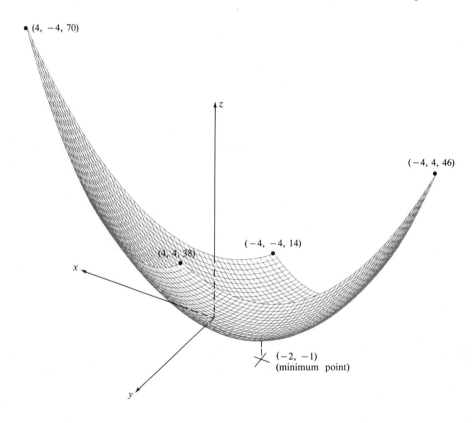

Figure 20
Minimum point of
$z = x^2 - xy + y^2 + 3x + 10$

test that the critical point $(-2, -1)$ is a relative minimum point. We would also say that the value $f(-2, -1) = -3$ is a relative minimum value.

The graph of

$$f(x, y) = x^2 - xy + y^2 + 3x + 10$$

was plotted and is shown in Figure 20.   □

Example 4

Second-Partials Test   Use the second-partials test to find the relative maximum and minimum points of the function

$$f(x, y) = x^3 + y^3 - 3xy + 10$$

Solution

· **Step 1** (find the first-partial derivatives).   Computing f_x and f_y, we have

$$f_x(x, y) = \frac{\partial}{\partial x}(x^3 + y^3 - 3xy + 10) = 3x^2 - 3y$$

$$f_y(x, y) = \frac{\partial}{\partial y}(x^3 + y^3 - 3xy + 10) = 3y^2 - 3x$$

· **Step 2** (find the critical points).   Setting f_x and f_y to zero gives

$$3x^2 - 3y = 0$$
$$3y^2 - 3x = 0$$

or

$$x^2 = y$$
$$y^2 = x$$

Substituting $y = x^2$ into the second equation gives

$$x^4 = x$$
$$x^4 - x = 0$$
$$x \cdot (x^3 - 1) = 0$$

The solutions of this equation are $x = 0$ and 1. Substituting these values into the equation for y in terms of x gives $x = 0, y = 0$, and $x = 1, y = 1$ as solutions. Therefore the critical points are $(0, 0)$ and $(1, 1)$.

· **Step 3** (compute the discriminant).   First we find $f_{xx}(x, y)$, $f_{xy}(x, y)$, and $f_{yy}(x, y)$:

$$f_{xx}(x, y) = \frac{\partial}{\partial x} f_x = \frac{\partial}{\partial x}(3x^2 - 3y) = 6x$$

$$f_{xy}(x, y) = \frac{\partial}{\partial y} f_x = \frac{\partial}{\partial y}(3x^2 - 3y) = -3$$

$$f_{yy}(x, y) = \frac{\partial}{\partial y} f_y = \frac{\partial}{\partial y}(3y^2 - 3x) = 6y$$

We then evaluate the above second-partial derivatives and the discriminant at each of the two critical points $(0, 0)$ and $(1, 1)$. We display these values in Table 5.

Critical Point	$f_{xx}(x, y)$	$f_{xy}(x, y)$	$f_{yy}(x, y)$	$D = f_{xx}f_{yy} - f_{xy}^2$
$(0, 0)$	0	-3	0	-9
$(1, 1)$	6	-3	6	27

- **Step 4** (apply the second-partials test).   Computing the discriminant D at the critical point $(0, 0)$, we get

$$D = f_{xx}(0, 0)f_{yy}(0, 0) - f_{xy}^2(0, 0)$$
$$= (0)(0) - (-3)^2$$
$$= -9$$

Since this value is negative, we conclude that the critical point $(0, 0)$ is a saddle point. On the other hand, since the discriminant is positive at the critical point $(1, 1)$ and $f_{xx}(1, 1)$ is positive, we conclude that $(1, 1)$ is a relative minimum point. The graph of

$$z = x^3 + y^3 - 3xy + 10$$

is shown in Figure 21.   □

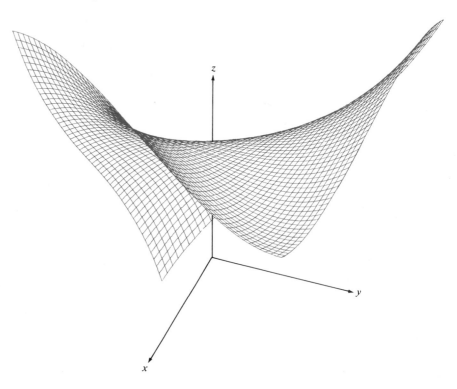

Figure 21
Graph of $z = x^3 + y^3 - 3xy + 10$

Profit Maximization in the Multicommodity Theory of the Firm

We saw in the previous section that the theory of the firm can be extended to multicommodity companies. We present an example that shows how production levels can be determined to maximize a company's profits.

Example 5

Profit Maximization   Suppose a company that produces x units of Product A per day and y units of Product B per day makes a daily profit (in dollars) of

$$P(x, y) = 200x + 300y - x^2 - 2xy - 2y^2 - 2000$$

Find the production levels x and y of the two products that will maximize the profit. What is the maximum profit that the company can earn?

Solution   To find the relative maximum value of the function

$$P(x, y) = 200x + 300y - x^2 - 2xy - 2y^2 - 2000$$

we carry out the following steps.

- **Step 1** (find the first-partial derivatives).   Computing P_x and P_y, we have

$$P_x(x, y) = 200 - 2x - 2y$$
$$P_y(x, y) = 300 - 2x - 4y$$

- **Step 2** (find the critical points).   Setting P_x and P_y to zero, we find

$$2x + 2y = 200$$
$$2x + 4y = 300$$

Solving for x and y, we get

$$x = 50$$
$$y = 50$$

This gives the critical point $(50, 50)$ as the sole candidate for a relative maximum point of $P(x, y)$.

- **Step 3** (find the discriminant).   Computing the second-order partial derivatives $P_{xx}(x, y)$, $P_{xy}(x, y)$, and $P_{yy}(x, y)$, we find

$$P_{xx}(x, y) = -2$$
$$P_{xy}(x, y) = -2$$
$$P_{yy}(x, y) = -4$$

Evaluating these partial derivatives at the critical point $(50, 50)$ gives the values

$$P_{xx}(50, 50) = -2$$
$$P_{xy}(50, 50) = -2$$
$$P_{yy}(50, 50) = -4$$

Hence the discriminant is given by

$$D = P_{xx}(50, 50) \cdot P_{yy}(50, 50) - P_{xy}^2(50, 50)$$
$$= (-2)(-4) - (-2)^2$$
$$= 4$$

· **Step 4** (apply the second-partials test). Since the discriminant is positive and $P_{xx}(50, 50)$ is negative, the second-partials test says that (50, 50) is a relative maximum point. The value of the profit function $P(x, y)$ at this point is

$$P(50, 50) = \$14{,}500$$

In Figure 22 we see a drawing of the profit surface $z = P(x, y)$ as a function of the production levels x and y. It is clear from this graph that \$14,500 is in fact the absolute maximum profit.

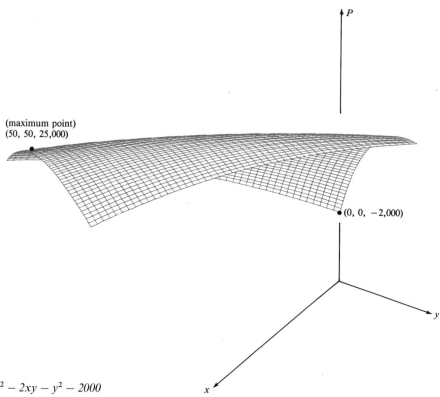

Figure 22
Graph of $P(x, y) = 200x + 300y - x^2 - 2xy - y^2 - 2000$

Interpretation: The optimal production level for this company is 50 units of Product A and 50 units of Product B per day. This production level results in a maximum daily profit of \$14,500.   □

Problems

For Problems 1–20, find all the critical points of the given functions.

1. $f(x, y) = x^2 - 2xy + 2y^2 - 3y - x$
2. $f(x, y) = 3 + 2x - 3x^2 - y^2$
3. $f(x, y) = x + 2y + x^2 - 3y^2$

4. $f(x, y) = -2x + xy^3 + 2x^3$
5. $f(x, y) = -2x + 4y + 3xy$
6. $f(x, y) = 3x - 7xy + x^2 + y^2$
7. $f(x, y) = 5x^2 + 2y^3 - 12y - 3y^2$
8. $f(x, y) = 6 + 3x - 4y + x^2 - 2y^2$

9. $f(x, y) = 7 + 6x - 3y + x^3 - y^2$
10. $f(x, y) = 6xy - 2x^2 - 3y^2$
11. $f(x, y) = 4 + 5x - 6y + 3xy + 2x^2 - 3y^2$
12. $f(x, y) = -7 - 3x + 2y + 4xy + 3x^2 + 4y^2$
13. $f(x, y) = -3 + 5x + 4y + 3xy + y^2$
14. $f(x, y) = 5xy + x^2 - 3y^2$
15. $f(x, y) = -8 - 5x + 12y + x^3 - 4y^3$
16. $f(x, y) = 4xy - xy^2 + 3x^2y$
17. $f(x, y) = 5 - 12xy + 3x^3 + 2y^3$
18. $f(x, y) = 8xy + 2xy^2 - 3x^2y$
19. $f(x, y) = ye^x$ **20.** $f(x, y) = xe^y$

For Problems 21–40, use the second-partials test to classify the critical points as relative maximum or minimum points, saddle points, or unclassified.

21. $f(x, y) = x^2 - 2xy + 4y^2 + 6x$
22. $f(x, y) = 5 + 2xy - 6x + 4y - x^2 - 2y^2$
23. $f(x, y) = 7 + 24x - 3x^2 + 8y - 2y^2$
24. $f(x, y) = x^4 + y^4 - 4xy$
25. $f(x, y) = (x - 2)(y - 2)(x + y - 2)$
26. $f(x, y) = 120x + 120y - xy - x^2 - y^2$
27. $f(x, y) = y^2 - 3x^2y + 2x^4$
28. $f(x, y) = 3x^2 + 4xy + y^2 - 6y$
29. $f(x, y) = x^2 - 3xy + 2y^2 + 5x$
30. $f(x, y) = y^3 - 3xy + x^2 + x - 4$
31. $f(x, y) = 8x^3 + y^3 - 6x^2 - 6y^2 + 4$
32. $f(x, y) = 7 + 4x - 3y - x^2 + y^3$
33. $f(x, y) = -x^3 - y^3 + 9xy + 7$
34. $f(x, y) = xy - y^2 - x^2 - 3y + 4x$

35. $f(x, y) = xy + \dfrac{8}{y} + \dfrac{2}{x}$ **36.** $f(x, y) = 2xy + \dfrac{10}{y}$

37. $f(x, y) = x^3 - y^3 - 9x$
38. $f(x, y) = 2 + x^4 + y^4 - 32y$
39. $f(x, y) = e^x + e^{-y}$ **40.** $f(x, y) = xe^y$

Finding Minimum Distances

41. Closest Point to the Origin   Find the point on the plane

$$x + 2y + 2z = 4$$

that is the closest to the origin $(0, 0, 0)$. What is this distance? *Hint:* The square of the distance D is

$$D^2 = x^2 + y^2 + z^2$$

Solve for z in the above equation of the plane, substitute this value into the formula for D^2, and find the minimum of D^2. This strategy is valid, since D is minimized when D^2 is minimized.

42. Closest Point to the Origin   Using the strategy discussed in Problem 41, find the point on the plane

$$x + y + z = 1$$

that is the closest to the origin. What is this distance?

43. Closest Point to the Origin   Using the strategy discussed in Problem 41, find the point on the plane

$$x = 2$$

that is the closest to the origin. What is this distance? Is it the same distance that you suspected?

Maximizing Problems in a Multicommodity Economy

44. Pricing Candy Bars   A company produces two kinds of candy bars, Gold Blocks and Wildbars. It costs the company \$0.20 to make each Gold Block and \$0.25 for each Wildbar. The demand for Gold Blocks and Wildbars has been determined to be

$$D_G(g, w) = \frac{10{,}000}{g^2 w} \qquad \text{(demand for Gold Blocks)}$$

$$D_W(g, w) = \frac{50{,}000}{g w^2} \qquad \text{(demand for Wildbars)}$$

where g is the price for a Gold Block and w is the price of a Wildbar.
(a) Find the company's revenue

$$R(g, w) = g D_G(g, w) + w D_W(g, w)$$

in terms of the prices g and w.
(b) Find the cost

$$C(g, w) = 0.20 D_G(g, w) + 0.25 D_W(g, w)$$

in terms of the prices g and w.
(c) Find the profit $P(g, w)$ in terms of g and w.
(d) Find the price of each candy bar g and w that will maximize the profit.
(e) What is the maximum profit?

45. Profit Maximization   A company's revenue and cost function are given by

$$R(x, y) = 25x + 35y$$

$$C(x, y) = \frac{3}{2}x^2 - 3xy + \frac{5}{2}y^2$$

Find the company's profit-maximizing output level and its maximum profit.

46. General Formulation of Production Levels   A firm markets two products and charges p_1 for the first product and p_2 for the second product. Hence the revenue from the sale of x and y units of these two products will be

$$R(x, y) = p_1 x + p_2 y$$

Suppose the cost function to produce these products is given by

$$C(x, y) = 2x^2 + xy + 2y^2$$

Determine as a function of p_1 and p_2 the number of items the company should produce to maximize profits. How many items should the company produce when the prices are $p_1 = \$5$ and $p_2 = \$2$?

47. **Finding Maximum Profits from Demand Functions** A company markets two products. The demands for these two products x and y depend on the prices p_1 and p_2 that the company charges for the products. The two demand functions are

$$x = 40 - 2p_1 + p_2$$
$$y = 25 + p_1 - p_2$$

Suppose the cost to the company to produce x and y units of each product is

$$C(x, y) = x^2 + xy + y^2$$

(a) Solve the demand functions for p_1 and p_2 in terms of x and y.
(b) Find the revenue function $R(x, y) = xp_1 + yp_2$ in terms of x and y.
(c) Find the profit function $P(x, y)$ in terms of x and y.
(d) Find the production levels x and y that maximize the profit.

48. **Optimal Pricing of Candy Bars** The Mr. NiceBar Company produces two kinds of candy bars, Mr. NiceBar Jr.

and Mr. NiceBar Sr. The cost to produce each of these candy bars is \$0.15 for Mr. NiceBar Jr. and \$0.25 for Mr. NiceBar Sr. The weekly demands x and y (in thousands) for Mr. NiceBar Jr. and Mr. NiceBar Sr. are

$$x = 10(p_2 - p_1) \qquad \text{(Mr. NiceBar Jr.)}$$
$$y = 5 + 3p_1 - 5p_2 \qquad \text{(Mr. NiceBar Sr.)}$$

respectively, where p_1 and p_2 are the prices in cents for Mr. NiceBar Jr. and Mr. NiceBar Sr., respectively.
(a) Find the revenue function $R(p_1, p_2) = p_1 x + p_2 y$ as a function of the prices p_1 and p_2.
(b) Find the cost function $C(p_1, p_2) = 0.15x + 0.25y$ as a function of the prices p_1 and p_2.
(c) Find the profit $P(p_1, p_2)$ function as a function of the prices p_1 and p_2.
(d) Find the prices that maximize the profit.

49. **Spreading Fertilizer** An experiment measures the results of applying two fertilizers A and B to an artichoke field. The yield of artichokes in bushels per acre is given by

$$V(x, y) = 10x + 5y + 2x^2 + y^2 - 8xy + 10$$

where x is the number of pounds of fertilizer A used and y is the number of pounds of fertilizer B used. What is the maximum yield of artichokes that can be produced? How many pounds of each fertilizer should be used?

13.4   Constrained Optimization (Lagrange Multipliers)

PURPOSE

We show how to find the maximum and the minimum of a function $y = f(x, y)$ where x and y are restricted to satisfy an additional "side condition" or constraint equation $g(x, y) = 0$. The major topics studied are

- a comparison between constrained and unconstrained optimization problems and

- the Lagrange multiplier rule for transforming constrained optimization problems into unconstrained optimization problems.

We close the section by showing how constrained optimization problems can be applied to the economic topic of consumer or utility theory.

Geometry of Constrained Problems

Consider the two optimization problems in Table 6. Note that the **constrained maximization** problem has a smaller maximum value than the **unconstrained maximization** problem. In constrained optimization problems we seek the highest or lowest point on a surface such that the point also lies on a particular part of the surface. As one might imagine, optimization problems with constraints are generally much more difficult to solve than problems without constraints. One approach for attacking optimization problems with constraints goes back 200 years to a method devised by the brilliant mathematician Joseph Louis Lagrange. Today, the method is called the **method of Lagrange multipliers**.

TABLE 6
Unconstrained and
Constrained Optimization
Problems

Unconstrained Optimization Problem	Constrained Optimization Problem
Find the maximum point on the surface $$f(x, y) = \sqrt{4 - x^2 - y^2}$$	Find the maximum point on the surface $$f(x, y) = \sqrt{4 - x^2 - y^2}$$ where x and y are constrained to lie on the plane $$x + 2y = 1$$

$z = \sqrt{4 - x^2 - y^2}$ Unconstrained maximum

$z = \sqrt{4 - x^2 - y^2}$ Constrained maximum

Plane: $x + 2y = 1$

HISTORICAL NOTE

Joseph Louis Lagrange (1736–1813) was born in Turin, Italy, and is considered to be one of the two "great mathematicians" of the eighteenth century (the Swiss mathematician Leonhard Euler being the other). Lagrange was one of the first mathematicians to bring a rigorous level of precision to mathematics and was responsible for introducing our moden notation $f(x)$ to denote a function.

Constraints and the Lagrange Multiplier

A typical constrained optimization problem would be to find the minimum value of a function, called the **objective function**,

$$z = x^2 + y^2$$

subject to the **constraint**

$$x + y = 2$$

From a geometric point of view we seek to find the lowest point (smallest value of z) on the intersection of the plane $x + y = 2$ and the paraboloid $z = x^2 + y^2$ as shown in Figure 23.

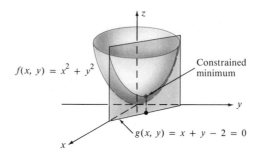

Figure 23
Constrained minimum problem

Finding a maximum or a minimum point of a constrained optimization problem is generally quite difficult, since the constraint equation makes it hard to describe the domain of the problem. However we now introduce a method known as the Lagrange multiplier method, which transforms a constrained optimization problem into an unconstrained optimization problem. In this way the maximum and minimum points of the original constrained problem can be found by finding the maximum and minimum points of the unconstrained problem. For example, consider minimizing the objective function.

$$z = x^2 + y^2$$

subject to the constraint

$$x + y = 2$$

The method of Lagrange multipliers uses the above functions to define a new function

$$L(x, y, \lambda) = x^2 + y^2 + \lambda(x + y - 2)$$

known as the **Lagrange function**. It is important that you understand a few things about this function.

· Observe how the constraint equation $x + y = 2$ enters into the Lagrange function.

· The Lagrange function $L(x, y, \lambda)$ is a function of three variables x, y, and λ. The variables x and y are the same variables as in the original constrained problem, but the variable λ is a new variable. We could just as well call them x, y, and z, but convention dictates that we call them x, y, and λ. The new variable λ is called the **Lagrange multiplier**.

The reason the Lagrange function $L(x, y, \lambda)$ is important is the interesting fact that when the three-dimensional point (x_0, y_0, λ_0) is a relative minimum point of the Lagrange function of three variables $L(x, y, \lambda)$, then (x_0, y_0) is a relative minimum point of the original constrained problem. Since minimizing a function without constraints is easier than minimizing a function with constraints, Lagrange in effect discovered a powerful rule for solving constrained optimization problems. His rule is stated here without proof.

Lagrange's Method for Solving Constrained Optimization Problems

If we define the Lagrange function as

$$L(x, y, \lambda) = f(x, y) + \lambda g(x, y)$$

then all the relative maximum and minimum points of $f(x, y)$ with x and y constrained to satisfy the equation $g(x, y) = 0$ will be among those points (x_0, y_0) for which (x_0, y_0, λ_0) is a maximum or minimum point of $L(x, y, \lambda)$. These points (x_0, y_0, λ_0) will be solutions of the system of simultaneous equations

$$L_x(x, y, \lambda) = 0$$
$$L_y(x, y, \lambda) = 0$$
$$L_\lambda(x, y, \lambda) = 0 \qquad \text{(this is just } g(x, y) = 0\text{)}$$

We assume that all indicated partial derivatives exist.

Example 1

Constrained Optimization Find the maximum and the minimum values of

$$z = x + 3y$$

where x and y satisfy the side condition

$$x^2 + y^2 = 1$$

Solution Geometrically, we seek to find the highest and lowest points on the plane

$$z = x + 3y$$

that also lies on the cylinder

$$x^2 + y^2 = 1$$

Think of slicing the inner cardboard core of a roll of paper towels at an angle. When the core is held vertically, the highest and lowest points on the core's cut will be analogous to the maximum and the minimum points. (See Figure 24.)

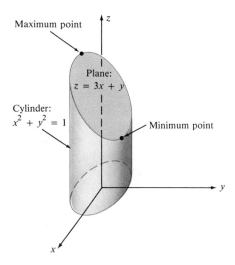

Figure 24
Constrained maximum and
minimum problem

· **Step 1** (define the Lagrange function).　We begin by writing the Lagrange function

$$L(x, y, \lambda) = f(x, y) + \lambda g(x, y)$$
$$= x + 3y + \lambda(x^2 + y^2 - 1)$$

Note that the constraint $x^2 + y^2 = 1$ is written in the Lagrange function as $x^2 + y^2 - 1$ (and not $x^2 + y^2$).

· **Step 2** (solve the equations $L_x = 0$, $L_y = 0$, $L_\lambda = 0$).　Computing the partial derivatives L_x, L_y, and L_λ, we have

$$L_x(x, y, \lambda) = \frac{\partial}{\partial x}\left[(x + 3y) + \lambda(x^2 + y^2 - 1)\right] = 1 + 2x\lambda$$

$$L_y(x, y, \lambda) = \frac{\partial}{\partial y}\left[(x + 3y) + \lambda(x^2 + y^2 - 1)\right] = 3 + 2y\lambda$$

$$L_\lambda(x, y, \lambda) = x^2 + y^2 - 1$$

Setting these partial derivatives to zero, we have

$$L_x(x, y, \lambda) = 1 + 2x\lambda = 0$$
$$L_y(x, y, \lambda) = 3 + 2y\lambda = 0$$
$$L_\lambda(x, y, \lambda) = x^2 + y^2 - 1 = 0 \qquad \text{(constraint equation)}$$

These equations are nonlinear algebraic equations and cannot be solved by using the usual Gauss-Jordan method. In this problem (and in many problems of this kind) the strategy is to solve for x and y in terms of the Lagrange multiplier λ in the two equations $L_x = 0$ and $L_y = 0$. Doing this, we find

$$x = -\frac{1}{2\lambda}$$

$$y = -\frac{3}{2\lambda}$$

Substituting these values into the constraint equation

$$x^2 + y^2 = 1$$

we obtain

$$\frac{1}{4\lambda^2} + \frac{9}{4\lambda^2} = 1$$

Solving for λ, we get

$$\lambda = \pm \frac{1}{2}\sqrt{10}$$

Now, by direct substitution of these values into the above equations for x and y, we get the two critical points of $L(x, y, \lambda)$:

$$\text{Critical point 1} = (x_0, y_0, \lambda_0) = \left(-\frac{1}{\sqrt{10}}, -\frac{3}{\sqrt{10}}, \frac{\sqrt{10}}{2} \right)$$

$$\text{Critical point 2} = (x_0, y_0, \lambda_0) = \left(\frac{1}{\sqrt{10}}, \frac{3}{\sqrt{10}}, -\frac{\sqrt{10}}{2} \right)$$

· **Step 3** (interpret the results). The values of the Lagrange multiplier λ_0 were needed so that the values of x_0 and y_0 could be found. However, once the values of x_0 and y_0 are known, the values of λ_0 can be ignored. We conclude that the two points

$$\left(-\frac{1}{\sqrt{10}}, -\frac{3}{\sqrt{10}} \right) \quad \text{and} \quad \left(\frac{1}{\sqrt{10}}, \frac{3}{\sqrt{10}} \right)$$

are the only candidates for relative maximum and minimum points of the constrained optimization problem. If we remember the physical analogy of slicing the inner core of a roll of paper towels (Figure 28), then it is clear that the maximum value of $z = x + 3y$ will occur at one of these points and the minimum value at the other point. By simply evaluating $z = x + 3y$ at these two points we find that the larger value occurs at the point

$$\left(\frac{1}{\sqrt{10}}, \frac{3}{\sqrt{10}} \right)$$

Hence we conclude that

$$\text{Maximum point} = \left(\frac{1}{\sqrt{10}}, \frac{3}{\sqrt{10}} \right) \qquad \text{Maximum value} = \frac{10}{\sqrt{10}}$$

$$\text{Minimum point} = \left(-\frac{1}{\sqrt{10}}, -\frac{3}{\sqrt{10}} \right) \qquad \text{Minimum value} = -\frac{10}{\sqrt{10}} \qquad \square$$

We summarize the steps that should be carried out when using the Lagrange multiplier method.

Steps in the Lagrange Method

To find candidates for the relative maximum and minimum points of $f(x, y)$ subject to the constraint

$$g(x, y) = 0$$

perform the following steps.

- **Step 1.** Write the Lagrange function

$$L(x, y, \lambda) = f(x, y) + \lambda g(x, y)$$

- **Step 2.** Solve the system of equations

$$L_x(x, y, \lambda) = 0$$
$$L_y(x, y, \lambda) = 0$$
$$L_\lambda(x, y, \lambda) = 0 \quad \text{(same as } g(x, y) = 0)$$

- **Step 3.** Interpret the results. The relative maximum and minimum points (x_0, y_0) for $z = f(x, y)$ subject to $g(x, y) = 0$ will be among the solutions (x_0, y_0, λ_0) of the three equations listed in Step 2 (among the critical points of the Lagrange function).

Comments on Lagrange's Method

Note that the Lagrange function $L(x, y, \lambda)$ is a function of *three* variables, and hence we cannot find its maximum and minimum points by using the discriminant function as we did in the previous section. (In that section we found maximum and minimum points of functions of *two* variables.) The power of Lagrange's method lies in its ability to find candidates for maximum and minimum points of constrained problems. It does not say, "This is a maximum point or this is a minimum point," as did the discriminant function of the previous section.

Sometimes, if the functions $f(x, y)$ and $g(x, y)$ are fairly simple, a rough graph might be drawn to determine whether a critical point is a relative maximum or minimum point. Also, by evaluating the function $f(x, y)$ at the points that are candidates for relative maximum and minimum points and at nearby points (nearby points that satisfy $g(x, y) = 0$) it is often possible to determine which points are relative maximum points and which points are relative minimum points.

Example 2

Lagrange Multiplier Method   Find the absolute maximum value of

$$f(x, y) = xy$$

subject to the constraint

$$x + y = 4$$

Solution

- **Step 1** (find the Lagrange function).   The Lagrange function is

$$L(x, y, \lambda) = f(x, y) + \lambda g(x, y)$$
$$= xy + \lambda(x + y - 4)$$

- **Step 2** (solve the equations $L_x = 0, L_y = 0, L_\lambda = 0$).   Computing the partial derivatives L_x, L_y, and L_λ and setting them to zero, we have

$$L_x(x, y, \lambda) = y + \lambda = 0$$
$$L_y(x, y, \lambda) = x + \lambda = 0$$
$$L_\lambda(x, y, \lambda) = x + y - 4 = 0$$

Solving these equations for x, y, and λ, we have

$$x = 2$$
$$y = 2$$
$$\lambda = -2$$

- **Step 3** (interpret the results).   The only critical point or candidate for a relative maximum or minimum point is $(2, 2)$. To determine whether $(2, 2)$ is a relative maximum or minimum point, we evaluate the objective function

$$z = xy$$

at this point and at several nearby points (points that satisfy $x + y = 4$). Table 7 shows the results of these computations. The values lead us to believe that $(2, 2)$ is the absolute maximum point and that $z = 4$ is the absolute maximum value.

TABLE 7
Value of an Objective
Function in a Neighborhood
of a Critical Point

Point on $x + y = 4$	Value of $z = xy$	
$(0, 4)$	0	
$(0.5, 3.5)$	1.75	
$(1, 3)$	3.00	
$(1.5, 2.5)$	3.75	
(2, 2)	**4.00**	⟵ *Maximum value*
$(2.5, 1.5)$	3.75	
$(3, 1)$	3.00	
$(3.5, 0.5)$	1.75	
$(4, 0)$	0	

Consumer or Utility Theory in Economics

Suppose that you are in a bakery that has just replenished its shelves with fresh doughnuts, cookies, cakes, and pies. The major problem for you is to decide what to buy. The secondary problem is that you have only $1.00 to spend. You, the consumer, are trying to maximize your pleasure (or, as economists say, utility) under given budgetary constraints. You are not the first person to face such a problem. In fact, millions of people try to solve some variation of this problem every day. No doubt you have tried to solve a similar problem already today.

In consumer theory, economists try to determine how consumers should spend a given amount of their money to maximize what economists call "**utility**." For instance, would you be happier (get more utility) spending your dollar on three doughnuts, or on two doughnuts and a cookie, or on one doughnut and two cookies, or some other possibility?

To make these ideas more precise, assume that a consumer has the option to buy two products, which we call Product A and Product B. We denote

$$x = \text{number of units of Product A purchased}$$
$$y = \text{number of units of Product B purchased}$$

Suppose that the cost of each product is given by

$$c_1 = \text{Cost per unit of Product A}$$
$$c_2 = \text{Cost per unit of Product B}$$

Under these circumstances, economists define a **utility function** $U(x, y)$ that gives a "measure" of the desirability of purchasing x and y units of the two products. Although a utility function is not as precise a measurement as other quantitative concepts, such as profit, revenue, and cost, general qualitative results can be obtained from an analysis of this function. Some typical utility functions of two variables that economists have found useful are the following:

$U(x, y) = \ln x + \ln y$	(logarithmic utility)
$U(x, y) = x + y$	(linear utility)
$U(x, y) = x + y - 0.05x^2 - 0.05y^2$	(quadratic utility)

Note that each of these utility functions will increase as the consumer buys more and more of the two products (although the quadratic utility will eventually start to decrease).

Quite often a consumer's utility will start to increase at a fast rate as the consumer starts buying but will grow at a slower rate for larger values of x and y. What this means is that although the consumer will always receive more "enjoyment" for larger purchases, the enjoyment is not proportional to the amount of goods bought. For instance, a consumer might get a great deal of enjoyment from eating the first doughnut, but not twice as much from eating two doughnuts, and certainly not ten times as much enjoyment from eating ten doughnuts (at least for most people). With this general concept of the utility function a typical problem in consumer or utility theory is to maximize a utility function subject to the budgetary constraint that the total amount of money spent by the consumer is some fixed amount. Mathematically, we can state the consumer problem as finding x and y in order to

$$\text{maximize} \quad U(x, y)$$

where x and y are subject to the budgetary constraint

$$c_1 x + c_2 y = M$$

with c_1 and c_2 defined as above and

$$M = \text{Total amount of money the consumer has to spend}$$

We illustrate this idea with the following example.

Example 3

Utility Theory   Hannah has \$50 to spend. She has decided to spend it on record albums and computer diskettes. She has determined that her utility from the purchase of x record albums and y boxes of computer diskettes is

$$U(x, y) = 3 \ln x + \ln y$$

where

$$\text{Cost per record album} = \$6$$
$$\text{Cost per box of computer diskettes} = \$4$$

How many record albums and boxes of computer diskettes should Hannah buy to maximize her utility while at the same time spending a total amount of \$50?

Solution   We must find the purchase levels x and y that maximize

$$U(x, y) = 3 \ln x + \ln y$$

subject to the spending constraint

$$6x + 4y = 50$$

- **Step 1** (write the Lagrange function).   The Lagrange function is

$$\begin{aligned} L(x, y, \lambda) &= U(x, y) + \lambda g(x, y) \\ &= 3 \ln x + \ln y + \lambda(6x + 4y - 50) \end{aligned}$$

- **Step 2** (solve the equations $L_x = 0$, $L_y = 0$, and $L_\lambda = 0$).   Computing the partial derivatives L_x, L_y, and L_λ and setting them to zero, we have

$$L_x(x, y, \lambda) = \frac{3}{x} + 6\lambda = 0$$

$$L_y(x, y, \lambda) = \frac{1}{y} + 4\lambda = 0$$

$$L_\lambda(x, y, \lambda) = 6x + 4y - 50 = 0$$

Solving for x and y in the first two equations in terms of λ, we find

$$x = -\frac{1}{2\lambda}$$

$$y = -\frac{1}{4\lambda}$$

Substituting these values into the third (constraint) equation

$$6x + 4y - 50 = 0$$

gives

$$-\frac{6}{2\lambda} - \frac{4}{4\lambda} = 50$$

or

$$\lambda_0 = -\frac{2}{25}$$

Finally, substituting this value of λ_0 back into the previous equations for x and y in terms of λ gives the values of x_0 and y_0:

$$x_0 = \frac{25}{4} = 6.250$$

and

$$y_0 = \frac{25}{8} = 3.125$$

Hence the candidate for a constrained maximum point is (6.250, 3.125).

· **Step 3** (interpret the results).   By evaluating the utility function $U(x, y)$ at nearby points on the line

$$6x + 4y = 50$$

we convince ourselves that the utility function attains its relative maximum value at the point (6.250, 3.125).

Interpretation: Hannah should purchase 6.25 record albums and 3.125 boxes of computer diskettes to maximize her utility while at the same time spending exactly \$50. (You can check to see that Hannah has spent exactly \$50.) Of course, it is impossible to purchase either a fractional number of record albums or a fractional number of boxes of computer diskettes. Hannah would have to resolve this dilemma by rounding off to some integer values (as long as she does not overspend her \$50). Generally, in utility theory the numbers are only meant to give a rough guide to the spending strategy and not meant to be read to the last decimal. Hannah might conclude that six record albums and three boxes of computer diskettes would be appropriate.

Also note that the value of the utility function at the relative maximum point is

$$U(6.25, 3.125) = 3 \ln (6.25) + \ln (3.125)$$
$$= 6.64$$

This relative maximum value of $U(x, y)$ is not significant in our analysis. The more important aspect of this problem to economists is the point (6.25, 3.125).

□

Problems

For Problems 1–10, find the critical point(s) for the functions such that the given constraint is satisfied.

Function	Constraint
1. $f(x, y) = x + 2xy$	$x - y = 3$
2. $f(x, y) = -3x - 2y$	$x + xy = 1$
3. $f(x, y) = 2x - y$	$x^2 - 2y^2 = 3$
4. $f(x, y) = 5x + 3y + 4$	$x^2 + y^2 = 1$
5. $f(x, y) = x^2 - y^2$	$x^2 - 2y^2 = 3$
6. $f(x, y) = 2x^2 - 3y^2 + 2xy$	$x + 3y = 5$
7. $f(x, y) = x^2 + y^2 + xy$	$x + 2y = 4$
8. $f(x, y) = x^2 + y^2 - 3x - 4y$	$x^2 + y^2 = 1$
9. $f(x, y) = 2x^2 - 3y^2 + 5x - 3y$	$3x - 5y = 2$
10. $f(x, y) = x^2 - y^2 - 3xy$	$5x - 3y = 2$

General Optimization Problems with Constraints

Problems 11–13 concern themselves with finding the optimal shapes of various solids under certain constraints.

11. Minimum Surface Area A box with a square base and a top with volume 125 cubic feet is to be constructed to minimize the surface area. What should be the dimensions of the box?

12. Maximum Volume A box with a square base and no top with surface area of 125 square feet is to be constructed to maximize the volume. What should be the dimensions of the box?

13. Minimum Surface Area A cylindrical tank with an open top has a volume of 256 cubic inches. What should its dimensions be to minimize the surface area? *Note:* Compare this problem to the cat food problem (Example 2, Section 9.5).

Constrained Optimization Problems in Business

14. Cobb-Douglas Production Function Maximize the Cobb-Douglas production function

$$u(x, y) = x^{1/2}y^{1/2}$$

in the case when the resources that are available satisfy the constraint

$$7x + 3y = 84$$

15. Minimum Cost The cost of producing x units of Product A and y units of Product B is given by

$$C(x, y) = 100 + 3x^2 + 5y^2$$

The production capacity for both products together is 75 units. How many units of each product should be produced to minimize costs?

16. Doughnuts Versus Cookies George is trying to decide how many doughnuts and cookies to buy. Each doughnut costs \$0.35, and each cookie costs \$0.25. He decides that his "index of enjoyment" or utility function is

$$U(x, y) = \ln x + \ln y$$

where

$$x = \text{Number of doughnuts he buys}$$
$$y = \text{Number of cookies he buys}$$

If George can spend \$1.50, how many doughnuts and cookies should he buy to maximize his pleasure?

17. Maximizing a Logarithmic Utility Function A consumer's utility function is given by

$$U(x, y) = 2 \ln x + \ln y$$

with budget constraints

$$2x + 4y = 50$$

Find the levels of x and y that this consumer should purchase to maximize the utility function, subject to the budgetary constraints.

18. Marginal Utility of Money In the consumer problem the value of the Lagrange multiplier λ is called by economists the **marginal utility of money**. Find the marginal utility of money for Problem 17. It can be shown that the marginal utility of money is the change in the optimal value of the utility with respect to the total amount of money spent. In other words, if the total amount spent of \$50 in Problem 17 were to be increased to \$51, then the change in the maximum utility would be approximately equal to the marginal utility of money (or the value of the Lagrange multiplier λ).

19. Consumer Problem A consumer's utility function is given by

$$U(x, y) = xy$$

with budgetary constraints

$$5x + 10y = 100$$

Find the values of x and y that maximize this utility function subject to the budgetary constraint. What is the marginal utility of money? (See Problem 18 for the definition of marginal utility.)

20. General Consumer Problem A consumer's utility function is given by

$$U(x, y) = xy$$

with budgetary constraints

$$c_1x + c_2y = M$$

Find the values of x and y that maximize this utility function subject to the general constraints. The answers, of course, will depend on c_1, c_2, and M.

Max/Min Problems Solved by the Lagrange Multiplier Technique

Back in Section 9.4, Optimization Problems (Max–Min Problems), we maximized (and minimized) functions of two variables, where the variables were related by a constraint equation. We solved those problems by solving the constraint equation for one variable and then substituting that value into the maximizing (or minimizing) function. In that way we reduced the maximizing (or minimizing), function to a function of only one variable. It is possible to solve those same problems by the method of Lagrange multipliers.

For Problems 21–32, solve the indicated problems in Section 9.4 using Lagrange multipliers.

21. Solve Problem 1 (Maximum Product).
22. Solve Problem 2 (Minimum Sum of Squares).
23. Solve Problem 3 (Constrained Maximum).
24. Solve Problem 6 (Optimal Pricing)
25. Solve Problem 14 (Minimal Cost Box).
26. Solve Problem 16 (Minimum Surface Area).
27. Solve Problem 18 (Maximum Volume).
28. Solve Problem 20 (Minimum Cost Box).
29. Solve Problem 21 (Wire Problem).
30. Solve Problem 26 (Maximum Area).
31. Solve Problem 27 (Telephone Lines).
32. Solve Problem 32 (Maximum Area of Field).

13.5 Double Integrals

PURPOSE

We introduce the concept of the double integral and show how to integrate functions $f(x, y)$ of two variables. The major topics discussed are

- the antiderivatives of functions of two variables,
- the iterated integral,
- the double integral, and
- the computation of volumes using double integrals.

We then show how a double integral can be used to compute the total photosynthetic production in a water column.

Introduction

Earlier in this chapter, we extended the concept of the derivative from functions of one variable to functions of two variables (three in the case of the Lagrange multiplier). We now see how the integral can be extended with the introduction of the double integral. We start by showing how to find antiderivatives of functions of two variables.

Antiderivatives of Functions of Two Variables

When we found a partial derivative of a function of two variables, we differentiated with respect to one of the two variables while treating the other variable as a constant. We can also find the antiderivative of a function of two variables in much the same way. That is, we take the antiderivative with respect to one of the two variables while treating the other variable as a constant. When we take the antiderivative or indefinite integral of a function $f(x, y)$ with respect to x (treating y as a constant), we denote this by

$$\int f(x, y)\, dx$$

Similarly, the antiderivative or indefinite integral of a function $f(x, y)$ with respect to y (treating x as a constant) is denoted by

$$\int f(x, y)\, dy$$

Example 1

Antiderivative with Respect to x Find the antiderivative of

$$f(x, y) = 2xy + 3y^2 + 3x^2$$

with respect to x.

Solution Taking the antiderivative with respect to x while treating y as a constant, we have

$$\int (2xy + 3y^2 + 3x^2)\, dx = x^2 y + 3xy^2 + x^3 + C(y)$$

The constant of integration $C(y)$ is not really a constant but can, in fact, be any function of y.

Check:

$$\frac{\partial}{\partial x}[x^2 y + 3xy^2 + x^3 + C(y)] = 2xy + 3y^2 + 3x^2$$

Note that any function $C(y)$ of y alone has a derivative of zero when differentiated with respect to x.   □

Example 2

Antiderivative with Respect to y Find the antiderivative

$$\int (x^2 + y^2)\, dy$$

Solution Integrating with respect to y while treating x as a constant, we get

$$\int (x^2 + y^2)\, dy = x^2 y + \frac{1}{3}y^3 + C(x)$$

Check:

$$\frac{\partial}{\partial y}\left[x^2 y + \frac{1}{3}y^3 + C(x)\right] = x^2 + y^2$$

Note that any function $C(x)$ of x alone has a derivative of zero when differentiated with respect to y.   □

Definite Integral of Functions of Two Variables

Just as the concept of the antiderivative or the indefinite integral can be extended to functions of two variables, so can the definite integral. Consider the two definite integrals of the form

$$\int_a^b f(x, y)\, dx \qquad \text{and} \qquad \int_a^b f(x, y)\, dy$$

The limits of integration refer to limits of integration for the variable for which integration is being performed. For example, in the integral involving dx the limits of integration refer to limits for x, while in the integral involving dy the limits of integration refer to limits for y.

Example 3

Definite Integrals   Evaluate the definite integral

$$\int_0^1 (x^2 + y^2)\, dy$$

Solution   In Example 2 we saw that

$$\int (x^2 + y^2)\, dy = x^2 y + \frac{1}{3} y^3 + C(x)$$

Hence we have

$$\int_0^1 (x^2 + y^2)\, dy = \left[x^2 y + \frac{1}{3} y^3 + C(x) \right]\Bigg|_{y=0}^{y=1}$$

$$= \left[x^2 + \frac{1}{3} + C(x) \right] - [0 + 0 + C(x)]$$

$$= x^2 + \frac{1}{3}$$

Note that the function $C(x)$ "canceled," as it always will. In the future we will choose it to be zero. □

Iterated Integrals

In Example 3 the definite integral with respect to y resulted in a function of x. On the other hand, if we integrated a function $f(x, y)$ with respect to x, we would get a function of y. In either case the resulting function is simply a continuous function of one variable and can be integrated a second time. This process of first integrating a function $f(x, y)$ with respect to one of the two variables and then integrating the resulting function with respect to the remaining variable is called **iterated integration**. The two successive integrals are called an **iterated integral**.

Example 4

Iterated Integration   Evaluate the iterated integral

$$\int_1^2 \left[\int_0^1 (x^2 + y^2)\, dy \right] dx$$

Solution   We first evaluate the inside or **inner integral**. We evaluated this integral in Example 3 and found

$$\int_0^1 (x^2 + y^2)\, dy = \left(x^2 y + \frac{1}{3} y^3 \right)\Bigg|_{y=0}^{y=1}$$

$$= x^2 + \frac{1}{3}$$

Hence this function now acts as the integrand in the second or **outer integral**. Hence we write

$$\int_1^2 \left[\int_0^1 (x^2 + y^2)\,dy\right]dx = \int_1^2 \left(x^2 + \frac{1}{3}\right)dx$$

$$= \left(\frac{1}{3}x^3 + \frac{x}{3}\right)\Bigg|_{x=1}^{x=2}$$

$$= \left(\frac{8}{3} + \frac{2}{3}\right) - \left(\frac{1}{3} + \frac{1}{3}\right)$$

$$= \left(\frac{1}{3} + \frac{7}{3}\right) - \left(\frac{0}{3} + \frac{0}{3}\right)$$

$$= \frac{8}{3}$$

In the above iterated integral, integration was performed first with respect to y (inner integral) and then with respect to x (outer integral). We wonder what would happen if we integrated the same function $f(x, y)$ in the opposite order. The following example gives the answer.   □

Example 5

Interchanging the Order of Integration   Evaluate the iterated integral in Example 4 in the opposite order. That is, instead of integrating with respect to y first and x second, integrate with respect to x first and y second as in the iterated integral

$$\int_0^1 \left[\int_1^2 (x^2 + y^2)\,dx\right]dy$$

Solution   Evaluating the inner integral, we get

$$\int_1^2 (x^2 + y^2)\,dx = \left(\frac{1}{3}x^3 + xy^2\right)\Bigg|_{x=1}^{x=2}$$

$$= y^2 + \frac{7}{3}$$

Substituting this value into the outer integral gives

$$\int_0^1 \left[\int_1^2 (x^2 + y^2)\,dx\right]dy = \int_0^1 \left(y^2 + \frac{7}{3}\right)dy$$

$$= \frac{1}{3}y^3 + \frac{7y}{3}\Bigg|_{y=0}^{y=1}$$

$$= \left(\frac{1}{3} + \frac{7}{3}\right) = \left(\frac{0}{3} + \frac{0}{3}\right)$$

$$= \frac{8}{3}$$

Note that when we changed the order of integration, we changed the order of the limits of integration. Note too that the two iterated integrals of Examples 4 and 5 have the same answer.   □

The Double Integral

The two previous examples lead us to believe that it makes no difference in which order we integrate a function of two variables. Although there are functions in which the order of integration will make a difference in the value of the integral, the functions studied in this book satisfy reasonable continuity properties and will have the same value no matter in what order the integration is performed. This leads to the following definition of the double integral.

Definition of the Double Integral

The **double integral** of a function $f(x, y)$ over the rectangle

$$R = \{(x, y): a \leq x \leq b, c \leq y \leq d\}$$

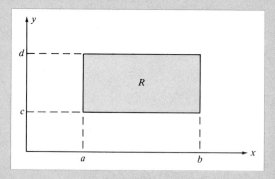

is denoted by

$$\iint\limits_{R} f(x, y)\, dx\, dy$$

and is defined by either (it makes no difference) of the two iterated integrals

$$\iint\limits_{R} f(x, y)\, dx\, dy = \int_{a}^{b}\left[\int_{c}^{d} f(x, y)\, dy\right] dx$$

$$= \int_{c}^{d}\left[\int_{a}^{b} f(x, y)\, dx\right] dy$$

The expression $dx\, dy$ in the double integral merely indicates that the double integral is taken over a two-dimensional region. The order in which $dx\, dy$

is written is not meant to imply any order in which the integration is to be performed.

Example 6

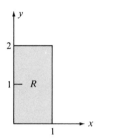

Figure 25
Domain of integration

Double Integral   Evaluate the double integral

$$\iint_R (e^x + e^y)\, dx\, dy$$

over the rectangle

$$R = \{(x, y): 0 \le x \le 1,\, 0 \le y \le 2\}$$

See Figure 25.

Solution   We can choose either order of integration. The final results will be the same in either case. We will integrate with respect to y first and x second. Writing the double integral as an iterated integral, we have

$$\iint_R (e^x + e^y)\, dx\, dy = \int_0^2 \left[\int_0^1 (e^x + e^y)\, dy \right] dx$$

Integrating the inner integral, we find

$$\int_0^1 (e^x + e^y)\, dy = ye^x + e^y \Big|_{y=0}^{y=1}$$
$$= (e^x + e) - (0 + 1)$$
$$= e^x + e - 1$$

Substituting this resulting integral into the outer integral, we obtain

$$\int_0^2 \left[\int_0^1 (e^x + e^y)\, dy \right] dx = \int_0^2 (e^x + e - 1)\, dx$$
$$= (e^x + ex - x) \Big|_{x=1}^{x=2}$$
$$= (e^2 + 2e - 2) - (e + e - 1)$$
$$= e^2 - 1$$

You can perform the integration in the opposite order, and you will get the same results. (Make sure the limits of integration are in the correct order for the other order of integration.)   □

Double Integrals as Volumes

In Chapter 12 we computed the volume of a surface of revolution. Now by using the double integral we can find the volume of other types of solids. For functions $f(x, y)$ with $f(x, y) \ge 0$ for all (x, y) in the region over which the integral is defined we can interpret the double integral as a volume in three dimensions.

Volume Under a Surface

Assume that the surface $z = f(x, y)$ lies above the xy-plane over a rectangular region

$$R = \{(x, y): a \le x \le b, c \le y \le d\}$$

The volume V of the solid under the surface over the rectangle R is given by the double integral

$$V = \iint_R f(x, y)\, dx\, dy$$

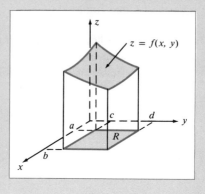

Example 7

Double Integrals as Volumes   Find the volume of the solid region under the plane

$$f(x, y) = 3x + 4y + 1$$

and over the square

$$R = \{(x, y): 0 \le x \le 1, 0 \le y \le 1\}$$

Solution   The solid region is shown in Figure 26. The volume is a double integral, which we write as an iterated integral with x integrated first. That is,

$$\iint_R (3x + 4y + 1)\, dx\, dy = \int_0^1 \left[\int_0^1 (3x + 4y + 1)\, dx \right] dy$$

Figure 26
Square domain of integration

Integrating the inner integral, we get

$$\int_0^1 (3x + 4y + 1)\,dx = \left(\frac{3}{2}x^2 + 4xy + x\right)\Bigg|_{x=0}^{x=1}$$

$$= \left(\frac{3}{2} + 4y + 1\right) - (0 + 0 + 0)$$

$$= 4y + \frac{5}{2}$$

Using this value as the integrand of the outer integral and integrating with respect to y, we have

$$\int_0^1 \left[\int_0^1 (3x + 4y + 1)\,dx\right]dy = \int_0^1 \left(4y + \frac{5}{2}\right)dy$$

$$= \left(2y^2 + \frac{5}{2}y\right)\Bigg|_{y=0}^{y=1}$$

$$= \left(2 + \frac{5}{2}\right) - (0 + 0)$$

$$= \frac{9}{2} \quad \text{cubic units} \qquad \square$$

Double Integrals in Marine Biology

Green plants consume carbon dioxide with the help of sunlight and produce sugars and carbohydrates by the process of photosynthesis. Much photosynthesis is carried out in an aquatic environment where the rate of photosynthetic action is directly proportional to the intensity of sunlight filtering into the water. It is important that a biologist know the total photosynthetic production in a column of water in a lake, sea, or region of ocean, since it represents the energy input into the biological energy chain. If we assume that the photosynthetic production $p(x, t)$ over a given cross section of water depends on both the depth x of the water and time t, then the total photosynthetic production in a water column over this cross section to a depth d over a time period $[t_1, t_2]$ is given by the double (or iterated) integral

$$\text{Total photosynthetic production} = \int_{t_1}^{t_2}\left[\int_0^d p(x, t)\,dx\right]dt$$

Example 8

Photosynthetic Production   A marine biologist has determined that in Lake Superior the photosynthetic production $p(x, t)$ that occurs in one cubic foot of water at a given depth and time of day is given by

$$p(x, t) = \frac{6 - t}{(x + 1)^2} \qquad (0 \le t \le 6, 0 \le x \le 50)$$

where x is the depth of the water in feet and t is time measured in hours with $t = 0$ being noon. What is the total photosynthetic production that takes place from noon to 6 P.M. in a column of water with a cross section of one square foot that has a depth of 50 feet?

Solution   The total photosynthetic production is given by the iterated integral

$$\text{Total photosynthetic production} = \int_0^6 \int_0^{50} \frac{6-t}{(x+1)^2}\, dx\, dt$$

To simplify this integral, we can factor out the term $6 - t$ from the inner integral involving dx, since $6 - t$ is treated as a constant in this integral. Hence we can write

$$\text{Total photosynthetic production} = \int_0^6 \int_0^{50} \frac{6-t}{(x+1)^2}\, dx\, dt$$

$$= \int_0^{50} \frac{1}{(x+1)^2}\, dx \int_0^6 (6-t)\, dt$$

$$= \left(\frac{50}{51}\right)(18)$$

$$= 17.64 \quad \text{units of energy} \qquad \square$$

Problems

In Problems 1–10, calculate the indicated antiderivative with respect to each of the two variables indicated. Check the answers by differentiating.

1. $\int (x^2 + xy - 2y + y^2)\, dx \qquad \int (x^2 + xy - 2y + y^2)\, dy$

2. $\int (x^3 - y^3 + 3xy)\, dx \qquad \int (x^3 - y^3 + 3xy)\, dy$

3. $\int xy\sqrt{x^2 + y^2}\, dx \qquad \int xy\sqrt{x^2 + y^2}\, dy$

4. $\int (x + xy - 2x^2 y)\, dx \qquad \int (x + xy - 2x^2 y)\, dy$

5. $\int (2x^2 y + 1)\, dx \qquad \int (2x^2 y + 1)\, dy$

6. $\int r(r + s)\, dr \qquad \int r(r + s)\, ds$

7. $\int x^2 y^2\, dx \qquad \int x^2 y^2\, dy$

8. $\int \frac{2t}{(s+t)^2}\, ds \qquad \int \frac{2t}{(s+t)^2}\, dt$

9. $\int e^x \ln y\, dx \qquad \int e^x \ln y\, dy$

10. $\int xe^{xy}\, dx \qquad \int xe^{xy}\, dy$

In Problems 11–25, calculate the value of the iterated integrals.

11. $\int_0^1 \left[\int_0^1 4\, dx\right] dy$

12. $\int_0^1 \left[\int_0^1 dy\right] dx$

13. $\int_{-2}^1 \left[\int_{-1}^2 (x + y)^2\, dy\right] dx$

14. $\int_1^3 \left[\int_0^3 xy\, dx\right] dy$

15. $\int_0^3 \left[\int_1^3 xy\, dy\right] dx$ (see Problem 14)

16. $\int_0^{\ln 2} \left[\int_0^1 xe^y\, dx\right] dy$

17. $\int_0^1 \left[\int_0^{\ln 2} xe^y\, dy\right] dx$ (see Problem 16)

18. $\int_0^1 \left[\int_0^1 \frac{x^2 y}{x^3 + 3}\, dx\right] dy$

19. $\int_{-2}^1 \left[\int_{-1}^2 (x + y)^3\, dx\right] dy$

20. $\int_{-1}^1 \left[\int_0^1 (x^2 + y^2)xy\, dx\right] dy$

21. $\int_0^1 \left[\int_{-1}^1 (x^2 + y^2)xy\, dy\right] dx$ (see Problem 20)

22. $\int_0^1 \left[\int_0^1 \frac{x}{(xy + 1)^2}\, dy\right] dx$

23. $\int_0^{\ln 2} \left[\int_0^1 xye^{y^2 x}\, dy\right] dx$

24. $\int_0^{\ln 2} \left[\int_0^{\ln 3} e^{(x+y)}\, dx\right] dy$

25. $\int_{-1}^1 \left[\int_0^1 (x^4 y + y^2)\, dy\right] dx$

In Problems 26–35, find the volume under the surface above the region in the xy-plane indicated by the inequalities.

Surface	Region
26. $f(x, y) = x + y$,	$R = \{(x, y): 0 \le x \le 2, 0 \le y \le 3\}$
27. $f(x, y) = e^{(x+y)}$,	$R = \{(x, y): 0 \le x \le 1, 1 \le y \le \ln 2\}$

28. $f(x, y) = (x + y)^2$, $R = \{(x, y): -1 \leq x \leq 1, 2 \leq y \leq 4\}$
29. $f(x, y) = xe^y$, $R = \{(x, y): 0 \leq x \leq 1, 1 \leq y \leq e\}$
30. $f(x, y) = x + y^2$, $R = \{(x, y): 1 \leq x \leq 3, 2 \leq y \leq 4\}$
31. $f(x, y) = x^2 y$, $R = \{(x, y): 2 \leq x \leq 4, 0 \leq y \leq 1\}$

32. $f(x, y) = \dfrac{x}{x^2 + y}$, $R = \{(x, y): 3 \leq x \leq 5, 1 \leq y \leq 2\}$

33. $f(x, y) = xy\sqrt{x^2 + y^2}$, $R = \{(x, y): 1 \leq x \leq 3, 0 \leq y \leq 3\}$
34. $f(x, y) = xy^2$, $R = \{(x, y): 2 \leq x \leq 4, 1 \leq y \leq 4\}$
35. $f(x, y) = 2x^2 y + 3xy^2$, $R = \{(x, y): 2 \leq x \leq 4, 0 \leq y \leq 2\}$

Double Integrals in Business

A company's production (say per day) in goods or services often can be estimated from the number of employees x and the company's invested capital y according to the Cobb-Douglas production function

$$P(x, y) = Cx^k y^{1-k} \qquad (0 < k < 1)$$

where C and k depend on the company. If over a given period of time the amount of labor x varies more or less uniformly between a and b while capital y varies between c and d, then the average daily production is given by the double integral

$$\text{Average production} = \frac{1}{A} \iint_R P(x, y)\, dx\, dy$$

where R is the rectangular region

$$R = \{(x, y): a \leq x \leq b, c \leq y \leq d\}$$

and A is the area of R. Use the above formula to find the average production in units per day for the Cobb-Douglas functions in Problems 36–38.

36. $P(x, y) = 100x^{0.2} y^{0.8}$,
$R = \{(x, y): 0 \leq x \leq 10, 0 \leq y \leq 10\}$
37. $P(x, y) = 50x^{0.5} y^{0.5}$,
$R = \{(x, y): 0 \leq x \leq 25, 5 \leq y \leq 20\}$
38. $P(x, y) = 10x^{0.9} y^{0.1}$,
$R = \{(x, y): 0 \leq x \leq 20, 20 \leq y \leq 30\}$

Epilogue: Calculus Crossword Puzzle*

* Answers are given in the Chapter Review.

Across

2. Something that gives the rate of change of a function.

3. The point at which a function changes its concavity, spelled backward.

5. The rule for differentiating compositions.

8. The number 2 is a _____ of the equation $x^2 - 3x + 2 = 0$

11. The number 1 is a _____ of the equation $x^2 - 3x + 2 = 0$.

12. To find the derivative dy/dx given the expression $x^3 + y^3 = 0$, we use _____ differentiation.

15. The logarithm to base 10 is called the _____ logarithm.

17. Initials of the founder of analytical geometry.

18. Probably the single most important concept in the calculus is the _____.

20. If $f'(a) = 0$ and $f''(a) < 0$, then the point a is a relative _____ point.

24. If $f'(a) > 0$ then f is _____ at a.

25. To find areas of unbounded regions, we use the _____ integral.

28. The line $y = 0$ is a horizontal _____ of the curve $y = e^{-x}$.

29. The lines $x = 0$ and $y = 0$ are called _____ for the cartesian plane.

31. Functions whose graphs can be drawn without lifting the pencil from the paper are _____ functions.

35. If $f'(a) = 0$, then a is called a _____ point.

36. The kinds of problems that involve finding the maximum or minimum of functions are called _____ problems.

39. Many growth phenomena are described mathematically by _____ functions.

40. If $f'(a) = 0$ and $f''(a) < 0$, then a is a relative _____ point.

42. The integral gives the _____ of a nonnegative function.

44. The method of _____ allows us to find the volume of solids of revolution.

45. The name of the mathematician whose name is associated with the numerical method for approximating integrals.

46. The initials of one of the founders of the calculus.

48. If $f''(a) > 0$, then the function f is _____ up at a.

50. The Swiss mathematician who introduced the symbol e.

51. The integral often represents the _____ under the graph of a function.

53. Initials of a famous Swiss mathematician.

54. When we wish to approximate the real change in the functional value, we can use the _____.

Down

1. The derivative of the velocity is the _____.

4. One of the two major operations of the calculus.

6. One of the two founders of the calculus.

7. When the tangent line is horizontal, the derivative is _____.

9. An interval without its endpoints is called an _____ interval.

10. The initials of one of the founders of the calculus.

13. A famous limiting sum in calculus.

14. The initials of the supposed discoverer of a numerical method for evaluating integrals.

16. An interval without its endpoints is called an _____ interval.

19. A point a is an absolute _____ point if the function is greater at this point than at any other point in the domain of the function.

22. When we integrate functions such as $f(x) = xe^x$, it is convenient to use integration by _____.

23. The German mathematician who cofounded the calculus.

26. The curve $y(t) = Ae^{-kt}$ decribes _____ phenomena.

28. A type of tree.

30. The equation $x^2 + 1 = 0$ does not have a real _____.

32. If a function is differentiable, the graph will appear _____.

33. If $f'(a) = 0$ and $f''(a) < 0$, then a is a relative _____ point.

36. Lagrange multipliers help one to solve _____ optimization problems.

37. If a function is greater at a certain point than at neighboring points, then the point is called a _____ maximum point.

38. The initials of the inventor of the logarithm.

39. An oblong circle is called an _____.

41. The derivative of the revenue is the _____ revenue.

43. The number 2 is the only _____ of the equation $3x - 4 = 2$.

48. An interval without its endpoints is called an _____ interval.

49. The integral is often used to find an _____.

52. The ratio of the circumference of a circle to the diameter is the number _____.

Key Ideas for Review

constrained optimization, 856
consumer theory, 863
critical points, 845
discriminant, 848
double integral, 872
iterated integral, 870
Lagrange multiplier method, 858

partial derivatives
  definition, 832
   economic applications, 837
   geometric interpretation, 835
relative maximum points, 844
relative minimum points, 844

saddle point, 846
second-partials test, 848
three dimensions, graphing, 824
unconstrained optimization, 843
utility theory, 863
volumes with double integrals, 873

Review Exercises

For Problems 1–12, find the indicated partial derivatives.

1. $f(x, y) = x^2 y + y$ $f_x(x, y)$
2. $f(x, y) = \ln x + x$ $f_{xx}(x, y)$
3. $f(x, y) = \log_{10} x + \ln x$ $f_x(x, y)$
4. $f(x, y) = x^2 + 1$ $f_{xy}(x, y)$
5. $f(x, y) = \sqrt{x^2 + y}$ $f_{xx}(x, y)$
6. $f(x, y) = e^{xy}$ $f_{yy}(x, y)$
7. $f(x, y) = x^2 y + y$ $f_{xy}(x, y)$
8. $f(x, y) = x^2 y$ $f_{xy}(x, y)$
9. $f(x, y) = (x^2 + y^2)$ $f_x(x, y)$
10. $f(x, y) = \ln(x + y)$ $f_y(x, y)$
11. $f(x, y) = x^y$ $f_x(x, y)$
12. $f(x, y) = x^y$ $f_y(x, y)$

13. Price-Earnings Ratio The price-earnings ratio of a stock is given by

$$R(P, E) = \frac{P}{E}$$

where P is the price per share of the stock and E is the annual earnings per share of the stock. Draw the level curves for this function in the PE-plane. That is, draw

curves of the form

$$R(P, E) = k$$

for different positive values of k. The following is a list of recent earnings and stock prices for five companies. Plot these companies in the PE-plane and compute their price-earnings ratio.

Company	Recent Price	Earnings per Share
A	$20	$4
B	$45	$6
C	$81	$8
D	$34	$2
E	$146	$49

For Problems 14–22, find the critical points and determine if they are relative maximum or minimum points.

14. $f(x, y) = 2xy - x^3 - y^2$ **15.** $f(x, y) = x^2 - y^2$

16. $f(x, y) = x^2 + y^2 - 2$

17. $f(x, y) = y^2 - 4x^2$

18. $f(x, y) = x^3 + y^3 - 3xy$

19. $f(x, y) = x + 2y + 1$

20. $f(x, y) = 5$

21. $f(x, y) = xy$

22. $f(x, y) = x$

23. Maximum Profits A company makes two kinds of baseball gloves. Model A gloves sell for \$15 each, and model B gloves sell for \$25 each. The total revenue in thousands of dollars from the sale of x thousand model A gloves and y thousand model B gloves will be given by

$$R(x, y) = 15x + 25y$$

The company determines that the total cost, in thousands of dollars, to produce x thousand model A gloves and y thousand model B gloves is given by

$$C(x, y) = x^2 + y^2 + 3x + 9y + 50$$

Find the amount of each type of glove that must be produced and sold to maximize their profit.

For Problems 24–37, find the critical points of the following functions and use the second-partials test to classify each of the critical points. Use any information you know about the surface to justify your answers.

24. $f(x, y) = x + y + 1$

25. $f(x, y) = -2x + y + xy$

26. $f(x, y) = xy$

27. $f(x, y) = x^2 + y^2$

28. $f(x, y) = 1 - x^2 - y^2$

29. $f(x, y) = x - x^2$

30. $f(x, y) = 10$

31. $f(x, y) = 120x + 60y - xy - x^2 - y^2$

32. $f(x, y) = \dfrac{1}{x^2} + \dfrac{1}{y^2}$

33. $f(x, y) = \dfrac{1}{x^2 + y^2}$

34. $f(x, y) = \ln(x^2 + y^2)$

35. $f(x, y) = e^{xy}$

36. $f(x, y) = e^{-(x^2 + y^2)}$

37. $f(x, y) = \ln x + y$

For Problems 38–47, find the point(s), if any, at which the functions attain their relative maximum and minimum values subject to the given constraints. Try working the problems using what you know about the graphs of the functions before resorting to the method of Lagrange multipliers.

Function	Constraint
38. $f(x, y) = x + y$	$x - y = 1$
39. $f(x, y) = x + y$	$x = 0$
40. $f(x, y) = x^2 + y^2$	$x = 0$
41. $f(x, y) = x^2 + y^2$	$x + y = 1$
42. $f(x, y) = x^2 + y^2$	$x^2 + y^2 = 1$
43. $f(x, y) = x^2 + y^2 + x$	$x + y = 1$
44. $f(x, y) = x^2 + y^2 - 3x$	$x + y = 1$
45. $f(x, y) = x^2 + y^2 + x + y$	$x^2 + y^2 = 1$
46. $f(x, y) = 5$	$x \ln x + e^y = 1$
47. $f(x, y) = x^2 + y^2$	$x - 3y = 0$

A			D	E	R	I	V	A	T	I	V	E		N	O	I	T	C	E	L	F	N	I
C																							N
C	H	A	I	N	R	U	L	E							Z			R	O	O	T		
E			E					W						Z	E	R	O		P			E	
L			W		I	M	P	L	I	C	I	T			R				E			G	
E			T				N						C	O	M	M	O	N				R	
R	D		O		L	I	M	I	T						O				P			A	
A		M	I	N			A		E			C		P			E			L			
T				L		X		G		I	N	C	R	E	A	S	I	N	G				
I	M	P	R	O	P	E	R		R			I		R				D					
O				I			A	S	Y	M	P	T	O	T	E		A	X	E	S			
N			B		R		L		E		I		S					C					
	C	O	N	T	I	N	U	O	U	S		W		C			M		A				
C			I		O		M			S	T	A	T	I	O	N	A	R	Y				
O	P	T	I	M	I	Z	A	T	I	O	N		R		L			X					
N							O			R		E			J		I						
S		E	X	P	O	N	E	N	T	I	A	L		M	I	N	I	M	U	M			
T		L			H			A	R	E	A			U		A							
R		L						T		R				M		R							
A	D	I	S	K	S		S	I	M	P	S	O	N			G							
I		P					V		O					I									
N		S		C	O	N	C	A	V	E		T		A		N							
E	U	L	E	R		P						A	R	E	A								
D			E	P								E		L	E								
	D	I	F	F	E	R	E	N	T	I	A	L			A								

For Problems 48–54, find the volumes under the given surfaces above the region R in the xy-plane.

Function	**Region**
48. $f(x, y) = 1$	$R = \{(x, y): 0 \le x \le 1, 0 \le y \le 5\}$
49. $f(x, y) = xy$	$R = \{(x, y): 0 \le x \le 2, 0 \le y \le 2\}$

50. $f(x, y) = e^x + e^y$ $\quad R = \{(x, y): 0 \le x \le 1, 0 \le y \le 1\}$
51. $f(x, y) = e^y$ $\quad R = \{(x, y): 0 \le x \le 1, 0 \le y \le 1\}$
52. $f(x, y) = xe^y$ $\quad R = \{(x, y): 0 \le x \le 1, -1 \le y \le 1\}$
53. $f(x, y) = 1$ $\quad R = \{(x, y): 5 \le x \le 7, -5 \le y \le 5\}$
54. $f(x, y) = (x + 2y)^2$ $\quad R = \{(x, y): 1 \le x \le 2, 1 \le y \le 2\}$

Chapter Test

1. For the function

$$f(x, y) = \ln (x + y)$$

find
(a) $f_x(x, y)$
(b) $f_y(x, y)$
(c) $f_{xx}(x, y)$
(d) $f_{xy}(x, y)$
(e) $f_{yy}(x, y)$

2. Find the critical points of

$$f(x, y) = x^3 + y^3 + 3xy$$

and use the second-partials test to classify them.

3. Of all numbers whose sum is 142, find the two that have a maximum product. Use the method of Lagrange multipliers to solve this problem.

4. Find the minimum value of

$$f(x, y) = x^2 + y^2$$

where x and y satisfy

$$x + y = 1$$

5. Evaluate the double integral

$$\iint_R (xy^2 + yx^2) \, dx \, dy$$

where

$$R = \{(x, y): 0 \le x \le 2, 0 \le y \le 1\}$$

Appendix A: Enriched Linear Programming: Mixed Constraints and the Two-Phase Method

PURPOSE

We show how the simplex method can be modified to solve linear programming problems that are different from the maximizing problems with all $\leq$ constraints or the minimizing problems with all $\geq$ constraints. We show how to solve maximizing and minimizing linear programming problems that can contain two or even all three of the types of constraints $\leq$, $\geq$, and $=$. We solve these mixed constraint problems by introducing artificial variables and using the two-phase method. To do this, we study

- how to change a minimizing objective function to a maximizing objective function,
- how artificial variables "enlarge" the feasible set,
- the "cousins" of slack variables—surplus variables, and
- the steps of the two-phase method.

Introduction

In Chapter 2 we used the simplex method to maximize the objective function of a linear programming problem in which all the constraints were of the form $\leq$. We also learned how to change a linear programming problem with a minimizing objective function and all $\geq$ constraints to the standard linear programming problem with a maximizing objective function and all $\leq$ constraints by finding the dual problem. We then solved the dual problem, using the simplex method. However, there are many linear programming problems that are not maximizing problems with all $\leq$ constraints or minimizing problems with all $\geq$ constraints.

It is often the case that a linear programming problem will contain some equality constraints ($=$), or even a mixture of $\leq$, $\geq$, and equality constraints. These **mixed constraint** problems cannot be solved by the simplex method as previously described, since the simplex method always starts at the origin (feasible corner-point). However, in mixed constraint problems the origin is often not a feasible point; hence the previously described simplex method fails. The *two-phase method*, using what are called **artificial variables**, allows us to start at the origin outside of the feasible set.

Before studying problems with mixed constraints we show how to change problems with minimizing objective functions to problems with maximizing objective functions. The reader should remember that the simplex method described in Chapter 2 solved only maximizing problems. Hence minimizing problems must be changed to maximizing problems if the simplex method is to be used.

How to Change a Min Problem to a Max Problem

The strategy for changing "min problems" to "max problems" is quite simple once the relationship between the graph of a function f and the graph of its negative $-f$ is seen. (See Figure 1.) From the graphs we can make the following observations.

Important Relationships Between f and $-f$

The point m where f attains its minimum value is the same point where $-f$ attains its maximum value. Also, the minimum value of f is the negative of the maximum value of $-f$.

The above observations can be restated in the language of linear programming.

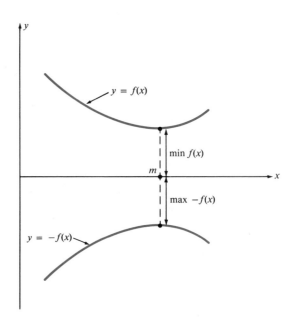

Figure 1
Geometric verification that
min f(x) = −max[−f(x)]

> ### Changing a Minimizing Problem to a Maximizing Problem
>
> The point $(x_1{}^*, x_2{}^*)$ that maximizes the objective function
>
> $$f(x_1, x_2) = c_1 x_1 + c_2 x_2$$
>
> is the same point that minimizes its negative
>
> $$-f(x_1, x_2) = -(c_1 x_1 + c_2 x_2)$$

We now show how the above principle can be applied to a linear programming problem.

Example 1

Changing a Min Problem to a Max Problem   Change the minimizing objective function

$$\min (2x_1 + x_2)$$

to a maximizing objective function.

Solution   We simply replace the above objective function by the new maximizing objective function

$$\max (-2x_1 - x_2)$$

By making this simple change in the problem the optimal feasible point $(x_1{}^*, x_2{}^*)$ of the new maximizing problem will be the same point as the optimal feasible point of the original minimizing problem. The actual minimum and maximum values of the respective functions, however, have opposite signs. For instance, if the optimal feasible point of the original minimizing problem were $(2, 3)$ with a minimum objective function of

$$\min (2x_1 + x_2) = 7$$

then the optimal feasible point of the corresponding maximizing problem would be $(2, 3)$. However, the maximum of the objective function would be

$$\max (-2x_1 - x_2) = -7 \qquad \qquad \square$$

We now study the problem of solving linear programming problems with mixed constraints. This brings us to the two-phase method and artificial variables.

The Two-Phase Method and Artificial Variables (Overview)

The simplex method was designed to start at the origin and move from corner-point to corner-point of the feasible set. The difficulty, however, in solving many problems (especially with $\geq$ and equality constraints) is that the origin may not be feasible. Hence the simplex method as described in Chapter 2 will fail, since we have no place to start.

However, there is a way to circumvent these difficulties by introducing what are called **artificial variables**. Artificial variables are defined in such a way that the simplex method "thinks" that the origin is a feasible corner-point and hence a valid starting point.

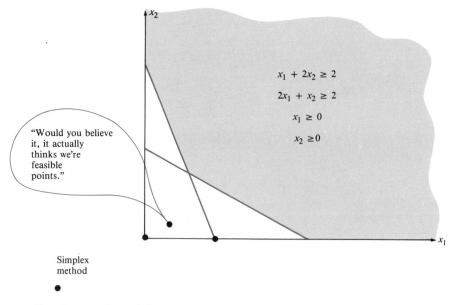

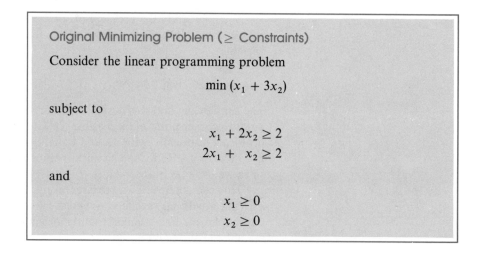

Starting at the origin, we can then solve linear programming problems using what is called the **two-phase method**. The first phase consists of starting at the origin (outside the feasible set) and searching for the feasible set. The artificial variables tell us in what direction to go and when we have arrived at the feasible set. Once we arrive at the feasible set, we have no more need for the artificial variables (they are thrown out), and we begin with Phase 2. Phase 2 simply consists of moving from corner-point to corner-point of the feasible set by manipulation of the simplex tableau as was done in Chapter 2.

We now illustrate the two-phase method by means of an example.

Using the Two-Phase Method and Artificial Variables

We start with the following minimizing linear programming problem with all $\geq$ constraints (which could be solved by solving its dual problem). We use this specific problem because it illustrates the method very clearly.

Original Minimizing Problem ($\geq$ Constraints)

Consider the linear programming problem

$$\min (x_1 + 3x_2)$$

subject to

$$x_1 + 2x_2 \geq 2$$
$$2x_1 + x_2 \geq 2$$

and

$$x_1 \geq 0$$
$$x_2 \geq 0$$

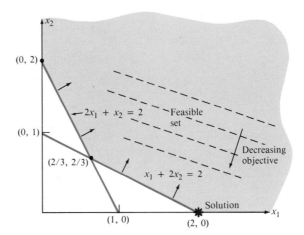

Figure 2
*Feasible set of linear
programming problem*

It is useful to show the solution of this problem before it is worked. This is shown in Figure 2.

To solve this problem using the two-phase method and artificial variables, we start by subtracting two variables s_1 and s_2 called **surplus variables** from the left-hand sides of the above inequalities. Use of these variables is analogous to the way we added slack variables to the $\leq$ constraints. By subtracting these variables we can rewrite the above $\geq$ inequalities as equalities, thus arriving at the new (but equivalent) form of the problem.

Equivalent Problem with Surplus Variables (Equality Constraints)

$$\min (x_1 + 3x_2)$$

subject to

$$x_1 + 2x_2 - s_1 \qquad = 2$$
$$2x_1 + \ x_2 \qquad - s_2 = 2$$

and

$$x_1 \geq 0$$
$$x_2 \geq 0$$
$$s_1 \geq 0$$
$$s_2 \geq 0$$

We interrupt our solution of the problem to make a few useful comments about surplus variables.

Notes on Surplus Variables

The subtracting of the two surplus variables in the above $\geq$ inequalities is reminiscent of adding slack variables to $\leq$ constraints as we did in Chapter 2. Slack and surplus variables differ only in that slack variables are added to $\leq$ constraints, whereas surplus variables are subtracted from $\geq$ constraints. They

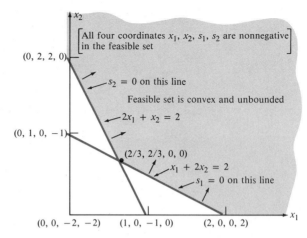

Figure 3

Illustration of problem showing surplus variables

are both denoted by the same notation $s_1, s_2, \ldots, s_m$; hence their meaning must be determined by the context of the problem.

The purpose of subtracting surplus variables is analogous to the reason for adding slack variables. They allow inequalities to be rewritten as equalities and thus be easier to manipulate. Figure 3 shows how points are labeled by using the four coordinates (x_1, x_2, s_1, s_2).

Back to the Problem

Now that we have changed the inequalities to equalities, we continue. However, we still face one major obstacle. We would like to begin the search for the solution by starting at the origin $(0, 0)$. The problem is that Dantzig designed the simplex method to move from feasible corner-point to feasible corner-point. Since the origin is clearly not feasible in this problem (see Figure 2), the simplex method as previously described will fail. Here is where artificial variables come to the rescue. We modify the above linear programming problem one more time by introducing the two new variables A_1 and A_2 in the following manner.

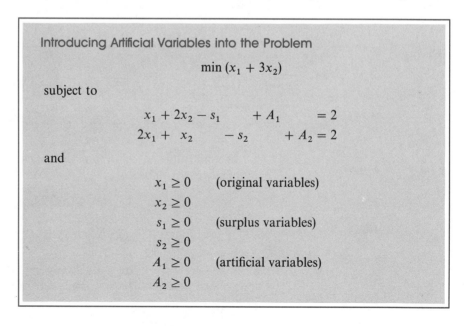

Introducing Artificial Variables into the Problem

$$\min (x_1 + 3x_2)$$

subject to

$$x_1 + 2x_2 - s_1 \qquad + A_1 \qquad = 2$$
$$2x_1 + x_2 \qquad - s_2 \qquad + A_2 = 2$$

and

$$x_1 \geq 0 \quad \text{(original variables)}$$
$$x_2 \geq 0$$
$$s_1 \geq 0 \quad \text{(surplus variables)}$$
$$s_2 \geq 0$$
$$A_1 \geq 0 \quad \text{(artificial variables)}$$
$$A_2 \geq 0$$

By introducing the two new variables A_1 and A_2, which are called **artificial variables**, we are able to, as we say, "trick" the simplex method into thinking that the origin is a feasible corner-point. To understand the reasoning behind artificial variables, we make another slight digression.

Labeling Points with Artificial Variables

The surplus and the artificial variables are what we might call the "inside" and "outside" variables of the feasible set. If a point (x_1, x_2) lies inside the feasible set (satisfies the original constraints of the problem), then the artificial variables A_1 and A_2 will be zero. On the other hand, if a point (x_1, x_2) lies outside of the feasible set, the surplus variables s_1 and s_2 will be zero. If we label each point in the plane with all six values x_1, x_2, s_1, s_2, A_1, and A_2, the simplex method will "think" that the origin $(0, 0)$ is a feasible corner-point. The following example shows how this labeling is performed.

Example 2

Labeling Points with Surplus and Artificial Variables   For the feasible set

Constraint 1: $x_1 + 2x_2 \geq 2$
Constraint 2: $2x_1 + x_2 \geq 2$
                          $x_1 \geq 0$
                          $x_2 \geq 0$

label the points

(a) $(x_1, x_2) = (0, 0)$
(b) $(x_1, x_2) = (1, 0)$
(c) $(x_1, x_2) = (2, 0)$

with the six "coordinates" $(x_1, x_2, s_1, s_2, A_1, A_2)$.

Solution   For reference we have drawn the feasible set in Figure 4.
We first rewrite the inequalities in equality form:

Constraint equation 1: $x_1 + 2x_2 - s_1 \quad\quad + A_1 \quad\quad = 2$
Constraint equation 2: $2x_1 + x_2 \quad\quad - s_2 \quad\quad + A_2 = 2$

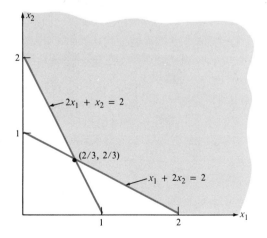

Figure 4
Feasible set for Example 2

(a) Since the origin $(x_1, x_2) = (0, 0)$ does not satisfy constraint 1, we set the first surplus variable $s_1 = 0$ and solve for the artificial variable A_1 using the constraint equation 1. Solving for A_1 gives

$$\begin{aligned}
A_1 &= 2 - x_1 - 2x_2 + s_1 \\
&= 2 - 0 - 2(0) + 0 \\
&= 2
\end{aligned}$$

Since the origin $(x_1, x_2) = (0, 0)$ does not satisfy constraint 2, we set the second surplus variable $s_2 = 0$ and solve for A_2 using constraint equation 2. Solving for A_2 gives

$$\begin{aligned}
A_2 &= 2 - 2x_1 - x_2 + s_2 \\
&= 2 - 2(0) - 0 + 0 \\
&= 2
\end{aligned}$$

Hence the origin $(0, 0)$ has the augmented coordinates

$$(x_1, x_2, s_1, s_2, A_1, A_2) = (0, 0, 0, 0, 2, 2)$$

(b) Since the point $(x_1, x_2) = (1, 0)$ does not satisfy constraint 1, we set $s_1 = 0$ and solve for A_1 using constraint equation 1. This gives

$$\begin{aligned}
A_1 &= 2 - x_1 - 2x_2 + s_1 \\
&= 2 - 1 - 2(0) + 0 \\
&= 1
\end{aligned}$$

However, since $(x_1, x_2) = (1, 0)$ does satisfy constraint 2, we set the second artificial variable $A_2 = 0$ and solve for s_2 using constraint equation 2. This gives

$$\begin{aligned}
s_2 &= 2x_1 + x_2 + A_2 - 2 \\
&= 2(1) + 0 + 0 - 2 \\
&= 0
\end{aligned}$$

Hence the augmented coordinates of $(1, 0)$ are

$$(x_1, x_2, s_1, s_2, A_1, A_2) = (1, 0, 0, 0, 1, 0)$$

(c) Since the point $(2, 0)$ satisfies constraint 1, we set $A_1 = 0$ and solve for s_1 using constraint equation 1. This gives

$$\begin{aligned}
s_1 &= x_1 + 2x_2 + A_1 - 2 \\
&= 2 + 2(0) + 0 - 2 \\
&= 0
\end{aligned}$$

Since the point $(x_1, x_2) = (2, 0)$ satisfies constraint 2, we set $A_2 = 0$ and solve for s_2 using constraint equation 2. This gives

$$\begin{aligned}
s_2 &= 2x_1 + x_2 + A_2 - 2 \\
&= 2(2) + 0 + 0 - 2 \\
&= 2
\end{aligned}$$

Hence the augmented coordinates of $(2, 0)$ are

$$(x_1, x_2, s_1, s_2, A_1, A_2) = (2, 0, 0, 2, 0, 0)$$

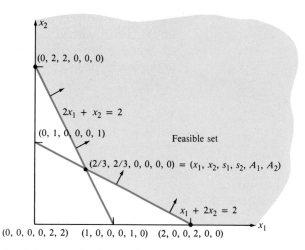

Figure 5
Points labeled with six coordinates
$(x_1, x_2, s_1, s_2, A_1, A_2)$

The labels of these and other points are shown in Figure 5.   □

Back to the Problem

Note that by labeling points in this way, all points in the first quadrant have all six coordinates nonnegative. From the point of view of the simplex method this says that every point in the first quadrant is a feasible point. Even the origin, which has the coordinates $(0, 0, 0, 0, 2, 2)$, is considered a feasible corner-point with zero (nonbasic) variables x_1, x_2, s_1, and s_2 and basic variables A_1 and A_2. Since we have now "tricked" the simplex method into thinking that the origin is a feasible corner-point, we can at last begin the two-phase method.

Phase 1 of the Two-Phase Method (in Search of the Feasible Set)

Phase 1 of the two-phase problem consists of solving the following problem.

> **Phase 1 Problem**
>
> $$\min (A_1 + A_2)$$
>
> subject to
>
> $$x_1 + 2x_2 - s_1 \qquad + A_1 \qquad = 2$$
> $$2x_1 + x_2 \qquad - s_2 \qquad + A_2 = 2$$
>
> and
>
> $$x_1 \geq 0 \qquad x_2 \geq 0$$
> $$s_1 \geq 0 \qquad s_2 \geq 0$$
> $$A_1 \geq 0 \qquad A_2 \geq 0$$

Note that the above objective function is chosen as the sum of the artificial variables. In this way the simplex method will keep moving in the direction such that the artificial variables will eventually become 0. This occurs when the feasible set is reached. So that we can use the simplex method as described in Chapter 2, however, we must change the above minimizing objective function to the corresponding maximizing function:

$$\max{(-A_1 - A_2)}$$

We can finally write the initial simplex tableau for Phase 1, as shown in Table 1.

TABLE 1
Initial Simplex Tableau for Phase 1 (Improper ROW 0)

BV	ROW	P	x_1	x_2	s_1	s_2	A_1	A_2	RHS
*	0	1	0	0	0	0	1	1	0
A_1	1	0	1	2	-1	0	1	0	2
A_2	2	0	2	1	0	-1	0	1	2

Unfortunately, after all this work, something is still slightly wrong with the tableau in Table 1. (Do you see what it is?) For the value of the objective function to lie in the RHS column of ROW 0 at every step, ROW 0 must contain zeros under the basic variables A_1 and A_2. We can do this very easily by simply replacing ROW 0 by ROW 0 − ROW 1 − ROW 2. This gives the new row

$$
\begin{array}{rrrrrrrr}
 & x_1 & x_2 & s_1 & s_2 & A_1 & A_2 & \text{RHS} \\
\text{Old ROW } 0 = \begin{bmatrix} & 0 & 0 & 0 & 0 & 1 & 1 & 0 \end{bmatrix} \\
- \text{ROW } 1 = -\begin{bmatrix} & 1 & 2 & -1 & 0 & 1 & 0 & 2 \end{bmatrix} \\
- \text{ROW } 2 = -\begin{bmatrix} & 2 & 1 & 0 & -1 & 0 & 1 & 2 \end{bmatrix} \\
\hline
\text{New ROW } 0 = \begin{bmatrix} -3 & -3 & 1 & 1 & 0 & 0 & -4 \end{bmatrix}
\end{array}
$$

This gives us the desired initial simplex tableau in Table 2.

TABLE 2
Initial Simplex Tableau for Phase 1 (Proper ROW 0)

BV	ROW	P	x_1	x_2	s_1	s_2	A_1	A_2	RHS
*	0	1	-3	-3	1	1	0	0	-4
A_1	1	0	1	2	-1	0	1	0	2
A_2	2	0	2	1	0	-1	0	1	2

We Are Ready to Begin

Using the simplex method, we change the tableau in Table 2 and find the tableau in Table 3. (Note the new basic variables A_1 and x_1.) The tableau in Table 3 corresponds to the point $(x_1, x_2) = (1, 0)$. Note that one of the artificial variables is now zero and one nonzero.

Continuing for one more step, we arrive at the new tableau in Table 4. (Note the new basic variables x_1 and x_2.) Since both artificial variables A_1 and A_2 are zero, we have arrived at the feasible set. We have arrived at the feasible set at the corner-point $(x_1, x_2) = (2/3, 2/3)$, which is shown in Figure 6.

TABLE 3
New Simplex Tableau
(Nonbasic Variables Circled,
Coefficients of Basic
Variables in Bold)

BV	ROW	P	x_1	$\widehat{x_2}$	$\widehat{s_1}$	$\widehat{s_2}$	A_1	$\widehat{A_2}$	ROW
*	0	1	0	$-3/2$	1	$-1/2$	0	$3/2$	-1
A_1	1	0	**0**	$3/2$	-1	$1/2$	**1**	$-1/2$	1
x_1	2	0	**1**	$1/2$	0	$-1/2$	**0**	$1/2$	1

TABLE 4
Final Tableau for Phase 1
(Zero Variables Circled)

BV	ROW	P	x_1	x_2	$\widehat{s_1}$	$\widehat{s_2}$	$\widehat{A_1}$	$\widehat{A_2}$	RHS
*	0	1	0	0	0	0	1	1	0
x_2	1	0	**0**	1	$-2/3$	$+1/3$	$2/3$	$-1/3$	$2/3$
x_1	2	0	**1**	0	$1/3$	$-2/3$	$-1/3$	$2/3$	$2/3$

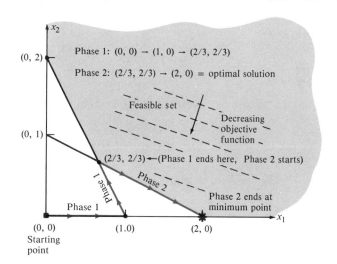

Phase 1: $(0, 0) \rightarrow (1, 0) \rightarrow (2/3, 2/3)$

Phase 2: $(2/3, 2/3) \rightarrow (2, 0) =$ optimal solution

Feasible set

Decreasing objective function

$(2/3, 2/3) \leftarrow$ (Phase 1 ends here, Phase 2 starts)

Phase 2 ends at minimum point

Phase 1

Phase 2

$(0, 2)$
$(0, 1)$
$(0, 0)$ Starting point
(1.0)
$(2, 0)$

Figure 6
*Path of Phase 1 and Phase 2 of
the two-phase method*

Phase 2 (We Are Now in the Feasible Set)

Since we now have no need for the artificial variables, we drop the two columns labeled A_1 and A_2 from the simplex tableau. We now take as the objective function the original minimizing function

$$\min (x_1 + 3x_2)$$

which we change to the equivalent maximizing function

$$\max -(x_1 + 3x_2)$$

This gives rise to the initial tableau for Phase 2, as shown in Table 5.

TABLE 5
Initial Tableau for Phase 2
(Improper ROW 0)

BV	ROW	P	x_1	x_2	$\widehat{s_1}$	$\widehat{s_2}$	RHS
*	0	1	1	3	0	0	0
x_2	1	0	0	1	$-2/3$	$1/3$	$2/3$
x_1	2	0	1	0	$1/3$	$-2/3$	$2/3$

Again, we must perform that little step of modifying ROW 0, since the coefficients of the basic variables x_1 and x_2 are not zero. Again, the process works about the same as before:

$$
\begin{array}{rrrrrr}
 & x_1 & x_2 & s_1 & s_2 & \text{RHS} \\
\text{Old ROW } 0 = & [\ 1 & 3 & 0 & 0 & 0\ ] \\
-3(\text{ROW } 1) = -3 & [\ 0 & 1 & -2/3 & 1/3 & 2/3] \\
-(\text{ROW } 2) = - & [\ 1 & 0 & 1/3 & -2/3 & 2/3] \\
\hline
\text{New ROW } 0 = & [\ 0 & 0 & 5/3 & -1/3 & -8/3]
\end{array}
$$

We now have the proper form for the initial tableau for phase 2, as shown in Table 6.

TABLE 6
Initial Simplex Tableau for Phase 2 (Proper ROW 0)

BV	ROW	P	x_1	x_2	(s_1)	(s_2)	RHS
*	0	1	0	0	5/3	-1/3	-8/3
x_2	1	0	0	1	-2/3	1/3	2/3
x_1	2	0	1	0	1/3	-2/3	2/3

Carrying out one more step, as you can do in Problem 14 at the end of the section, we will arrive at the final simplex tableau in Table 7.

TABLE 7
Final Simplex Tableau for Phase 2

BV	ROW	P	x_1	(x_2)	(s_1)	s_2	RHS
*	0	1	0	1	1	0	-2
s_2	1	0	0	3	-2	1	2
x_1	2	0	1	2	-1	0	2

The End (Summary)

Since ROW 0 has all nonnegative entries under the variables x_1, x_2, s_1, and s_2, we stop the process. The final tableau corresponds to the optimal feasible point $(x_1{}^*, x_2{}^*) = (2, 0)$ with surplus variables $(s_1, s_2) = (0, 2)$. The minimum value of the objective function is the negative of the value in ROW 0 in the RHS column. That is, $-(-2) = +2$.

Figure 6 shows the path that the two-phase method follows in finding the optimal feasible point.

Summary of the Two-Phase Method

We summarize the two-phase method here with a general problem in two variables and two constraints. Problems with more variables and constraints would follow along the same lines. Consider the problem

$$\min (c_1 x_1 + c_2 x_2)$$

subject to

$$a_{11}x_1 + a_{12}x_2 \geq b_1$$
$$a_{21}x_1 + a_{22}x_2 \geq b_2$$

and

$$x_1 \geq 0$$
$$x_2 \geq 0$$

Phase 1

Write the initial tableau for Phase 1 (ROW 0 in proper form), as shown in Table 8. The tableau is the general case of the tableau in Table 2. This tableau corresponds to the point

$$(x_1, x_2, s_1, s_2, A_1, A_2) = (0, 0, 0, 0, b_1, b_2)$$

TABLE 8

Initial Tableau for Phase 1 (ROW 0 in Proper Form)

BV	ROW	P	x_1	x_2	s_1	s_2	A_1	A_2	RHS
*	0	1	$-a_{11} - a_{21}$	$-a_{12} - a_{22}$	1	1	0	0	$-b_1 - b_2$
A_1	1	0	a_{11}	a_{12}	-1	0	1	0	b_1
A_2	2	0	a_{21}	a_{22}	0	-1	0	1	b_2

Carry out the steps of the simplex method until both artificial variables are removed from the BV column (in other words, until they are both zero). In choosing the pivot column simply pick in any order the columns under x_1, x_2, s_1, s_2. After choosing the columns under x_1 and x_2 both artificial variables will be "driven" from the basic variable set. When this happens, we have arrived at the feasible set.

Phase 2

First, change the final tableau found in Phase 1 by deleting the two columns labeled A_1 and A_2. It is difficult to write the general expression for the initial tableau for Phase 2, since we have no way of knowing in advance which of the variables x_1, x_2, s_1, and s_2 will be basic and which will be nonbasic. The best that we can say is to follow the steps that produced Tables 5 and 6. After the proper tableau has been found (make sure the basic variable columns have zeros in ROW 0), the optimal feasible solution is found by using the standard steps of the simplex method.

Problems with All Three Types of Constraints ($\leq$, $\geq$, and $=$)

The previous problem illustrated how to solve minimizing problems in which all the constraints were of the type $\geq$. We now show how to solve problems that have all three types of constraints: $\geq$, $\leq$, and $=$. Consider the following minimizing problem, a mixed constraint problem:

$$\min C = 2x_1 + 3x_2$$

subject to

$$x_1 + x_2 \leq 10$$
$$-x_1 + x_2 = 1$$
$$x_1 \qquad \geq 1$$

and

$$x_1 \geq 0$$
$$x_2 \geq 0$$

After solving this problem, you should be able to solve other problems with more constraints and more variables. We have restricted the problem to only two variables in order to graph the solution in the plane. The feasible set and solution to this problem are shown in Figure 7.

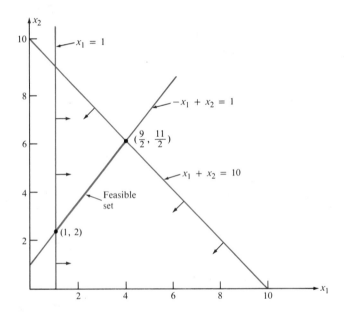

Figure 7
Feasible set: line segment between (1, 2) and (9/2, 11/2)

To solve this problem, we first rewrite the constraints in the proper form.

> **Rules for Slack, Surplus, and Artificial Variables**
>
> - Add a slack variable to each $\leq$ constraint.
> - Add an artificial variable to each $=$ constraint.
> - Subtract a surplus variable and add an artificial variable to each $\geq$ constraint.

Doing this, we get the following equivalent problem, a problem with slack, surplus, and artificial variables:

$$\min C = 2x_1 + 3x_2$$

subject to

$$
\begin{aligned}
x_1 + x_2 + s_1 &= 10 \\
-x_1 + x_2 + A_1 &= 1 \\
x_1 - s_2 + A_2 &= 1
\end{aligned}
$$

and

$$
\begin{aligned}
x_1 &\geq 0 & x_2 &\geq 0 \\
s_1 &\geq 0 & s_2 &\geq 0 \\
A_1 &\geq 0 & A_2 &\geq 0
\end{aligned}
$$

We are now ready to write the initial tableau for Phase 1. Since we always minimize the sum of the artificial variables in Phase 1 (or maximize $-A_1 - A_2$), the initial tableau has the form in Table 9.

TABLE 9
Initial Simplex Tableau for Phase 1 (Improper ROW 0)

BV	ROW	P	x_1	x_2	s_1	s_2	A_1	A_2	RHS
*	0	1	0	0	0	0	1	1	0
s_1	1	0	1	1	1	0	0	0	10
A_1	2	0	-1	1	0	0	1	0	1
A_2	3	0	1	0	0	-1	0	1	1

Note that the basic variables are chosen to be s_1, A_1, and A_2 for the three respective constraints. We now list the rules for choosing the initial basic variables.

> **Rules for Choosing Initial Basic Variables**
> - For a constraint of the type $\leq$ the initial basic variable is the slack variable that is added to that inequality.
> - For an equality constraint the initial basic variable is the artificial variable that is added to that equation.
> - For a constraint of the type $\geq$ the initial basic variable is the artificial variable that is added to that inequality.

Note in Table 9 that ROW 0 does not contain zeros under the basic variables A_1 and A_2. Since the simplex method requires zeros in these locations, we must replace ROW 0 by

$$\text{New ROW 0} = \text{Old Row 0} - \text{Row 2} - \text{Row 3}$$

This will give us the desired initial tableau for Phase 1, as shown in Table 10.

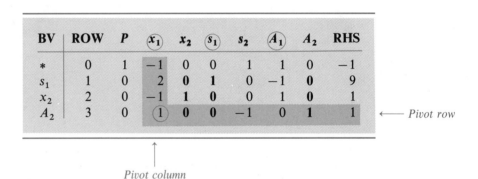

TABLE 10
Initial Tableau for Phase 1
(Note Zeros in ROW 0 Under
Basic Variables)

BV	ROW	P	x_1	x_2	s_1	s_2	A_1	A_2	RHS	
*	0	1	0	-1	0	1	0	0	-2	
s_1	1	0	1	1	1	0	0	0	10	
A_1	2	0	-1	1	0	0	1	0	1	← Pivot row
A_2	3	0	1	0	0	-1	0	1	1	

Pivot column

Table 11 shows the results of carrying out the next two steps. Note that after two more steps, both artificial variables have been removed from the BV column, and we have arrived at the feasible set.

TABLE 11
Two Steps in Phase 1

BV	ROW	P	x_1	x_2	s_1	s_2	A_1	A_2	RHS	
*	0	1	-1	0	0	1	1	0	-1	
s_1	1	0	2	0	1	0	-1	0	9	
x_2	2	0	-1	1	0	0	1	0	1	
A_2	3	0	1	0	0	-1	0	1	1	← Pivot row

Pivot column

BV	ROW	P	x_1	x_2	s_1	s_2	A_1	A_2	RHS
*	0	1	0	0	0	0	1	1	0
s_1	1	0	0	0	1	2	-1	-2	7
x_2	2	0	0	1	0	-1	1	1	2
x_1	3	0	1	0	0	-1	0	1	1

We have now completed Phase 1. To begin Phase 2, we delete the columns containing the artificial variables in Table 11 and replace ROW 0 above with the original objective function

$$P = 2x_1 + 3x_2$$

This gives Table 12.

TABLE 12
Initial Tableau for Phase 2
(Improper ROW 0)

BV	ROW	P	x_1	x_2	s_1	$\widehat{s_2}$	RHS
*	0	1	-2	-3	0	0	0
s_1	1	0	0	0	1	2	7
x_2	2	0	0	1	0	-1	2
x_1	3	0	1	0	0	-1	1

However, again ROW 0 does not have zeros under the two basic variables x_1 and x_2, and so we let

$$\text{New ROW 0} = \text{Old ROW 0} + 3(\text{ROW 2}) + 2(\text{ROW 3})$$

The new "proper tableau" for Phase 2 is shown in Table 13.

TABLE 13
Initial Tableau for Phase 2
(Proper ROW 0)

BV	ROW	P	x_1	x_2	s_1	$\widehat{s_2}$	RHS	
*	0	1	0	0	0	-5	8	
s_1	1	0	0	0	1	$\textcircled{2}$	7	← Pivot row
x_2	2	0	0	1	0	-1	2	
x_1	3	0	1	0	0	-1	1	

Pivot column

If we now carry out one more step, we will arrive at the final tableau, Table 14. The final tableau corresponds to the optimal feasible point

$$(x_1{}^*, x_2{}^*) = (9/2, 11/2)$$

with a minimum objective function of $P^* = 51/2$. Note that the values of the surplus and slack variables are

$$s_1 = 0 \qquad \text{(slack variable)}$$
$$s_2 = 7/2 \qquad \text{(surplus variable)}$$

TABLE 14
Final Tableau for Phase 2

BV	ROW	P	x_1	x_2	$\widehat{s_1}$	s_2	RHS
*	0	1	0	0	5/2	0	51/2
s_2	1	0	0	0	1/2	1	7/2
x_2	2	0	0	1	1/2	0	11/2
x_1	3	0	1	0	1/2	0	9/2

Figure 8 shows the path that the two-phase method takes to get the solution.

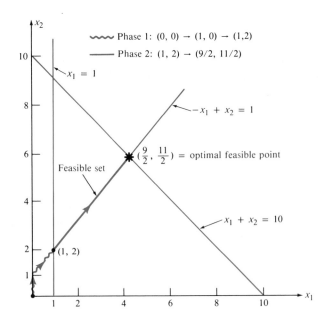

Figure 8
Path to the solution in the two-phase method

Problems

For Problems 1–5, find the minimum of the following functions f and the point where the minimum occurs by

1. Finding the point where $-f$ is a maximum.

2. Taking the negative of this maximum to get the minimum of f.

3. Graphing the functions f and $-f$ to illustrate the problems.

1. $f(x) = x^2$
2. $f(x) = x^2 + 1$
3. $f(x) = |x|$
4. $f(x) = |x| + 1$
5. $f(x) = x^2 - 2x + 2$

6. **Labeling Points**   Given the inequalities

$$x_1 + 3x_2 \geq 3$$
$$3x_1 + x_2 \geq 3$$
$$x_1 \geq 0$$
$$x_2 \geq 0$$

label the following points with the necessary slack, surplus, and artificial variables.

$$(x_1, x_2)$$

(a) (0, 0)
(b) (1, 0)
(c) (2, 0)
(d) (3, 0)
(e) (2, 2)

7. **Labeling Points**   Given the constraints

$$x_1 + x_2 \leq 10$$
$$x_1 - x_2 = 0$$
$$x_1 \geq 1$$
$$x_1 \geq 0$$
$$x_2 \geq 0$$

label the following points with the necessary slack, surplus, and artificial variables.

$$(x_1, x_2)$$

(a) (0, 0)
(b) (10, 0)
(c) (1, 0)
(d) (1, 1)
(e) (0, 10)

Problems 8–14 refer to the following problem:

$$\min (x_1 + 2x_2)$$

subject to
$$x_1 + x_2 \geq 2$$
$$x_1 + 2x_2 \geq 3$$

and

$$x_1 \geq 0$$
$$x_2 \geq 0$$

8. Draw the feasible set.
9. Rewrite the inequality constraints as equality constraints by introducing surplus and artificial variables.
10. State the Phase 1 linear programming problem.
11. Write the initial simplex tableau for the Phase 1 problem of the two-phase method. Be sure that Row 0 has zeros under the basic variables.
12. Carry out the steps of Phase 1 of the two-phase method.
13. Write the initial simplex tableau for Phase 2 of the two-phase method. Be sure that Row 0 has zeros under the basic variables.
14. Carry out the steps of Phase 2 of the two-phase method to find the optimal feasible solution.
15. **Carrying Out One Step** Starting with the simplex tableau in Table 2 in the text, carry out one step to get the tableau in Table 3.
16. **Carrying Out One Step** Starting with the simplex tableau in Table 6 in the text, carry out one step to get the tableau in Table 7.

Problems 17–21 refer to the following problem:

$$\min (5x_1 + 7x_2)$$

subject to

$$x_1 + x_2 \geq 10$$
$$x_1 + 4x_2 \geq 20$$

and

$$x_1 \geq 0$$
$$x_2 \geq 0$$

17. Draw the feasible solution and find the optimal feasible solution by the graphical method.
18. Find the simplex tableau for Phase 1. Make sure that there are zeros in Row 0 under the basic variables.
19. Solve the Phase 1 problem of the two-phase method.
20. Find the initial simplex tableau for Phase 2. Be sure that there are zeros in Row 0 under the basic variables.
21. Use the initial tableau found in Problem 20 to find the optimal solution.

Problems 22–26 refer to the following problem:

$$\min (x_1 + x_2)$$

subject to

$$x_1 + x_2 \geq 2$$
$$x_1 + 5x_2 \geq 5$$

and

$$x_1 \geq 0$$
$$x_2 \geq 0$$

22. Draw the feasible solution and solve the above problem graphically.
23. Write the initial simplex tableau for Phase 1. Be sure that there are zeros in Row 0 under the basic variables.
24. Use the initial tableau from Problem 23 to find a feasible point. Plot the path from $(0, 0)$ to the feasible point.
25. Write the initial simplex tableau for Phase 2. Be sure that there are zeros in Row 0 under the basic variables.
26. Use the initial tableau from Problem 25 to find the optimal solution.

Problems 27–32 refer to the following problem:

$$\min (2x_1 + 3x_2)$$

subject to

$$x_1 + x_2 \leq 10$$
$$-x_1 + x_2 = 1$$
$$x_1 \geq 1$$

and

$$x_1 \geq 0$$
$$x_2 \geq 0$$

27. Draw the feasible set and solve the problem by the graphical method.
28. Rewrite the problem introducing slack, surplus, and artificial variables.
29. Write the initial simplex tableau for Phase 1. Be sure that there are zeros in Row 0 under the basic variables.
30. Solve Phase 1 of the two-phase method.
31. Write the initial simplex tableau for Phase 2. Be sure that Row 0 has zeros under the basic variables.
32. Carry out Phase 2 of the two-phase method to find the optimal feasible solution.

Problems 33–37 refer to the following problem:

$$\max P = x_1 + x_2 + x_3$$

subject to

$$x_1 + 2x_2 + x_3 = 5$$
$$x_1 + x_2 + 2x_3 = 5$$

and

$$x_1 \geq 0$$
$$x_2 \geq 0$$
$$x_3 \geq 0$$

33. Rewrite the problem, introducing artificial variables.
34. Write the initial simplex tableau for Phase 1. Be sure that Row 0 has zeros under the basic variables.
35. Solve Phase 1 of the two-phase method.
36. Write the initial simplex tableau for Phase 2. Be sure that Row 0 has zeros under the basic variables.
37. Carry out Phase 2 of the two-phase method to find the optimal solution.

Applications

Describe the following problems by a linear programming problem and solve by the two-phase method.

38. **Advertising** A soft drink company is introducing a new diet cola and would like to place a total of 15 advertisements in three magazines, *Newsday*, *Newsmonth*, and *Newsyear*.

    ▪ *Newsday* charges $150 per ad and has a circulation of 1000.

    ▪ *Newsmonth* charges $200 per ad and has a circulation of 1500.

    ▪ *Newsyear* charges $200 per ad and has a circulation of 2000.

    Suppose the company want the ads to reach at least 20,000 people. How many ads should the company run in each magazine to meet its goals at minimum cost?

39. **Agriculture** A farmer can add three types of plant food to the corn crop, mixes A, B, and C. The contents (pounds) of nitrogen, phosphorus, and potash in each 100-pound bag of these mixes is given in Table 15. The farmer determines that at least the nutrients per acre shown in Table 16 are needed. If food A costs $2 per bag, food B costs $3 per bag, and food C costs $2.50 per bag, how many pounds of each should be added to each acre of corn to minimize the cost?

TABLE 15
Pounds per 100-Pound Bag

Ingredient	Mix A	Mix B	Mix C
Nitrogen	16	32	24
Phosphorus	32	16	24
Potash	32	16	32

TABLE 16
Minimum Needs (per acre)

500 lb nitrogen
200 lb phosphorus
200 lb potash

40. **Diet Problem** Find the number of ounces of Crispies (C) and Milk (M) that minimize the following total cost:

$$\text{Total cost} = 4C + 5M$$

subject to the nutritional requirements

$$2C + M \geq 10 \quad \text{(vitamin A)}$$
$$C + 2M \geq 10 \quad \text{(vitamin C)}$$

Appendix B: Tables

TABLE 1
Common Logarithms

N	0	1	2	3	4	5	6	7	8	9
1.0	0.0000	0.004321	0.008600	0.01284	0.01703	0.02119	0.02531	0.02938	0.03342	0.03743
1.1	0.04139	0.04532	0.04922	0.05308	0.05690	0.06070	0.06446	0.06819	0.07188	0.07555
1.2	0.07918	0.08279	0.08636	0.08991	0.09342	0.09691	0.1004	0.1038	0.1072	0.1106
1.3	0.1139	0.1173	0.1206	0.1239	0.1271	0.1303	0.1335	0.1367	0.1399	0.1430
1.4	0.1461	0.1492	0.1523	0.1553	0.1584	0.1614	0.1644	0.1673	0.1703	0.1732
1.5	0.1761	0.1790	0.1818	0.1847	0.1875	0.1903	0.1931	0.1959	0.1987	0.2014
1.6	0.2041	0.2068	0.2095	0.2122	0.2148	0.2175	0.2201	0.2227	0.2253	0.2279
1.7	0.2304	0.2330	0.2355	0.2380	0.2405	0.2430	0.2455	0.2480	0.2504	0.2529
1.8	0.2553	0.2577	0.2601	0.2625	0.2648	0.2673	0.2695	0.2718	0.2742	0.2765
1.9	0.2788	0.2810	0.2833	0.2856	0.2878	0.2900	0.2923	0.2945	0.2967	0.2989
2.0	0.3010	0.3032	0.3054	0.3075	0.3096	0.3118	0.3139	0.3160	0.3181	0.3201
2.1	0.3222	0.3243	0.3263	0.3284	0.3304	0.3324	0.3345	0.3365	0.3385	0.3404
2.2	0.3424	0.3444	0.3464	0.3483	0.3502	0.3522	0.3541	0.3560	0.3579	0.3598
2.3	0.3617	0.3636	0.3655	0.3674	0.3692	0.3711	0.3729	0.3747	0.3766	0.3784
2.4	0.3802	0.3820	0.3838	0.3856	0.3874	0.3892	0.3909	0.3927	0.3945	0.3962
2.5	0.3979	0.3997	0.4014	0.4031	0.4048	0.4065	0.4082	0.4099	0.4116	0.4133
2.6	0.4150	0.4166	0.4183	0.4200	0.4216	0.4232	0.4249	0.4265	0.4281	0.4298
2.7	0.4314	0.4330	0.4346	0.4362	0.4378	0.4393	0.4409	0.4425	0.4440	0.4456
2.8	0.4472	0.4487	0.4502	0.4518	0.4533	0.4548	0.4564	0.4579	0.4594	0.4609
2.9	0.4624	0.4639	0.4654	0.4669	0.4683	0.4698	0.4713	0.4728	0.4742	0.4757
3.0	0.4771	0.4786	0.4800	0.4814	0.4829	0.4843	0.4857	0.4871	0.4886	0.4900
3.1	0.4914	0.4928	0.4942	0.4955	0.4969	0.4983	0.4997	0.5011	0.5024	0.5038
3.2	0.5051	0.5065	0.5079	0.5092	0.5105	0.5119	0.5132	0.5145	0.5159	0.5172
3.3	0.5185	0.5198	0.5211	0.5224	0.5237	0.5250	0.5263	0.5276	0.5289	0.5302
3.4	0.5315	0.5328	0.5340	0.5353	0.5366	0.5378	0.5391	0.5403	0.5416	0.5428
3.5	0.5441	0.5453	0.5465	0.5478	0.5490	0.5502	0.5514	0.5527	0.5539	0.5551
3.6	0.5563	0.5575	0.5587	0.5599	0.5611	0.5623	0.5635	0.5647	0.5658	0.5670
3.7	0.5682	0.5694	0.5705	0.5717	0.5729	0.5740	0.5752	0.5763	0.5775	0.5786
3.8	0.5798	0.5809	0.5821	0.5832	0.5843	0.5855	0.5866	0.5877	0.5888	0.5899
3.9	0.5911	0.5922	0.5933	0.5944	0.5955	0.5966	0.5977	0.5988	0.5999	0.6010
4.0	0.6021	0.6031	0.6042	0.6053	0.6064	0.6075	0.6085	0.6096	0.6107	0.6117
4.1	0.6128	0.6138	0.6149	0.6160	0.6170	0.6180	0.6191	0.6201	0.6212	0.6222
4.2	0.6232	0.6243	0.6253	0.6263	0.6274	0.6284	0.6294	0.6304	0.6314	0.6325
4.3	0.6335	0.6345	0.6355	0.6365	0.6375	0.6385	0.6395	0.6405	0.6415	0.6425
4.4	0.6435	0.6444	0.6454	0.6464	0.6474	0.6484	0.6493	0.6503	0.6513	0.6522
4.5	0.6532	0.6542	0.6551	0.6561	0.6571	0.6580	0.6590	0.6599	0.6609	0.6618
4.6	0.6628	0.6637	0.6646	0.6656	0.6665	0.6675	0.6684	0.6693	0.6702	0.6712
4.7	0.6721	0.6730	0.6739	0.6749	0.6758	0.6767	0.6776	0.6785	0.6794	0.6803
4.8	0.6812	0.6821	0.6830	0.6839	0.6848	0.6857	0.6866	0.6875	0.6884	0.6893
4.9	0.6902	0.6911	0.6920	0.6928	0.6937	0.6946	0.6955	0.6964	0.6972	0.6981
5.0	0.6990	0.6998	0.7007	0.7016	0.7024	0.7033	0.7042	0.7050	0.7059	0.7067
5.1	0.7076	0.7084	0.7093	0.7101	0.7110	0.7118	0.7126	0.7135	0.7143	0.7152
5.2	0.7160	0.7168	0.7177	0.7185	0.7193	0.7202	0.7210	0.7218	0.7226	0.7235
5.3	0.7243	0.7251	0.7259	0.7267	0.7275	0.7284	0.7292	0.7300	0.7308	0.7316
5.4	0.7324	0.7332	0.7340	0.7348	0.7356	0.7364	0.7372	0.7380	0.7388	0.7396

TABLE 1
Common Logarithms (*continued*)

N	0	1	2	3	4	5	6	7	8	9
5.5	0.7404	0.7412	0.7419	0.7427	0.7435	0.7443	0.7451	0.7459	0.7466	0.7474
5.6	0.7482	0.7490	0.7497	0.7505	0.7513	0.7520	0.7528	0.7536	0.7543	0.7551
5.7	0.7559	0.7566	0.7574	0.7582	0.7589	0.7597	0.7604	0.7612	0.7619	0.7627
5.8	0.7634	0.7642	0.7649	0.7657	0.7664	0.7672	0.7679	0.7686	0.7694	0.7701
5.9	0.7709	0.7716	0.7723	0.7731	0.7738	0.7745	0.7752	0.7760	0.7767	0.7774
6.0	0.7782	0.7789	0.7796	0.7803	0.7810	0.7818	0.7825	0.7832	0.7839	0.7846
6.1	0.7853	0.7860	0.7868	0.7875	0.7882	0.7889	0.7896	0.7903	0.7910	0.7917
6.2	0.7924	0.7931	0.7938	0.7945	0.7952	0.7959	0.7966	0.7973	0.7980	0.7987
6.3	0.7993	0.8000	0.8007	0.8014	0.8021	0.8028	0.8035	0.8041	0.8048	0.8055
6.4	0.8062	0.8069	0.8075	0.8082	0.8089	0.8096	0.8102	0.8109	0.8116	0.8122
6.5	0.8129	0.8136	0.8142	0.8149	0.8156	0.8162	0.8169	0.8176	0.8182	0.8189
6.6	0.8195	0.8202	0.8209	0.8215	0.8222	0.8228	0.8235	0.8241	0.8248	0.8254
6.7	0.8261	0.8267	0.8274	0.8280	0.8287	0.8293	0.8299	0.8306	0.8312	0.8319
6.8	0.8325	0.8331	0.8338	0.8344	0.8351	0.8357	0.8363	0.8370	0.8376	0.8382
6.9	0.8388	0.8395	0.8401	0.8407	0.8414	0.8420	0.8426	0.8432	0.8439	0.8445
7.0	0.8451	0.8457	0.8463	0.8470	0.8476	0.8482	0.8488	0.8494	0.8500	0.8506
7.1	0.8513	0.8519	0.8525	0.8531	0.8537	0.8543	0.8549	0.8555	0.8561	0.8567
7.2	0.8573	0.8579	0.8585	0.8591	0.8597	0.8603	0.8609	0.8615	0.8621	0.8627
7.3	0.8633	0.8639	0.8645	0.8651	0.8657	0.8663	0.8669	0.8675	0.8681	0.8686
7.4	0.8692	0.8698	0.8704	0.8710	0.8716	0.8722	0.8727	0.8733	0.8739	0.8745
7.5	0.8751	0.8756	0.8762	0.8768	0.8774	0.8779	0.8785	0.8791	0.8797	0.8802
7.6	0.8808	0.8814	0.8820	0.8825	0.8831	0.8837	0.8842	0.8848	0.8854	0.8859
7.7	0.8865	0.8871	0.8876	0.8882	0.8887	0.8893	0.8899	0.8904	0.8910	0.8915
7.8	0.8921	0.8927	0.8932	0.8938	0.8943	0.8949	0.8954	0.8960	0.8965	0.8971
7.9	0.8976	0.8982	0.8987	0.8993	0.8998	0.9004	0.9009	0.9015	0.9020	0.9025
8.0	0.9031	0.9036	0.9042	0.9047	0.9053	0.9058	0.9063	0.9069	0.9074	0.9079
8.1	0.9085	0.9090	0.9096	0.9101	0.9106	0.9112	0.9117	0.9122	0.9128	0.9133
8.2	0.9138	0.9143	0.9149	0.9154	0.9159	0.9165	0.9170	0.9175	0.9180	0.9186
8.3	0.9191	0.9196	0.9201	0.9206	0.9212	0.9217	0.9222	0.9227	0.9232	0.9238
8.4	0.9243	0.9248	0.9253	0.9258	0.9263	0.9269	0.9274	0.9279	0.9284	0.9289
8.5	0.9294	0.9299	0.9304	0.9309	0.9315	0.9320	0.9325	0.9330	0.9335	0.9340
8.6	0.9345	0.9350	0.9355	0.9360	0.9365	0.9370	0.9375	0.9380	0.9385	0.9390
8.7	0.9395	0.9400	0.9405	0.9410	0.9415	0.9420	0.9425	0.9430	0.9435	0.9440
8.8	0.9445	0.9450	0.9455	0.9460	0.9465	0.9469	0.9474	0.9479	0.9484	0.9489
8.9	0.9494	0.9499	0.9504	0.9509	0.9513	0.9518	0.9523	0.9528	0.9533	0.9538
9.0	0.9542	0.9547	0.9552	0.9557	0.9562	0.9566	0.9571	0.9576	0.9581	0.9586
9.1	0.9590	0.9595	0.9600	0.9605	0.9609	0.9614	0.9619	0.9624	0.9628	0.9633
9.2	0.9638	0.9643	0.9647	0.9652	0.9657	0.9661	0.9666	0.9671	0.9675	0.9680
9.3	0.9685	0.9689	0.9694	0.9699	0.9703	0.9708	0.9713	0.9717	0.9722	0.9727
9.4	0.9731	0.9736	0.9741	0.9745	0.9750	0.9754	0.9759	0.9763	0.9768	0.9773
9.5	0.9777	0.9782	0.9786	0.9791	0.9795	0.9800	0.9805	0.9809	0.9814	0.9818
9.6	0.9823	0.9827	0.9832	0.9836	0.9841	0.9845	0.9850	0.9854	0.9859	0.9863
9.7	0.9868	0.9872	0.9877	0.9881	0.9886	0.9890	0.9894	0.9899	0.9903	0.9908
9.8	0.9912	0.9917	0.9921	0.9926	0.9930	0.9934	0.9939	0.9943	0.9948	0.9952
9.9	0.9956	0.9961	0.9965	0.9969	0.9974	0.9978	0.9983	0.9987	0.9991	0.9996

TABLE 2

The Natural Logarithm
Function: $\ln x = \log_e x$

x	$\ln x$	x	$\ln x$	x	$\ln x$
0.0	—	4.5	1.5041	9.0	2.1972
0.1	−2.3026	4.6	1.5261	9.1	2.2083
0.2	−1.6094	4.7	1.5476	9.2	2.2192
0.3	−1.2040	4.8	1.5686	9.3	2.2300
0.4	−0.9163	4.9	1.5892	9.4	2.2407
0.5	−0.6931	5.0	1.6094	9.5	2.2513
0.6	−0.5108	5.1	1.6292	9.6	2.2618
0.7	−0.3567	5.2	1.6487	9.7	2.2721
0.8	−0.2231	5.3	1.6677	9.8	2.2824
0.9	−0.1054	5.4	1.6864	9.9	2.2925
1.0	0.0000	5.5	1.7047	10	2.3026
1.1	0.0953	5.6	1.7228	11	2.3979
1.2	0.1823	5.7	1.7405	12	2.4849
1.3	0.2624	5.8	1.7579	13	2.5649
1.4	0.3365	5.9	1.7750	14	2.6391
1.5	0.4055	6.0	1.7918	15	2.7081
1.6	0.4700	6.1	1.8083	16	2.7726
1.7	0.5306	6.2	1.8245	17	2.8332
1.8	0.5878	6.3	1.8405	18	2.8904
1.9	0.6419	6.4	1.8563	19	2.9444
2.0	0.6931	6.5	1.8718	20	2.9957
2.1	0.7419	6.6	1.8871	25	3.2189
2.2	0.7885	6.7	1.9021	30	3.4012
2.3	0.8329	6.8	1.9169	35	3.5553
2.4	0.8755	6.9	1.9315	40	3.6889
2.5	0.9163	7.0	1.9459	45	3.8067
2.6	0.9555	7.1	1.9601	50	3.9120
2.7	0.9933	7.2	1.9741	55	4.0073
2.8	1.0296	7.3	1.9879	60	4.0943
2.9	1.0647	7.4	2.0015	65	4.1744
3.0	1.0986	7.5	2.0149	70	4.2485
3.1	1.1314	7.6	2.0281	75	4.3175
3.2	1.1632	7.7	2.0142	80	4.3820
3.3	1.1939	7.8	2.0541	85	4.4427
3.4	1.2238	7.9	2.0669	90	4.4998
3.5	1.2528	8.0	2.0794	95	4.5539
3.6	1.2809	8.1	2.0919	100	4.6052
3.7	1.3083	8.2	2.1041	200	5.2983
3.8	1.3350	8.3	2.1163	300	5.7038
3.9	1.3610	8.4	2.1282	400	5.9915
4.0	1.3863	8.5	2.1401	500	6.2146
4.1	1.4110	8.6	2.1518	600	6.3069
4.2	1.4351	8.7	2.1633	700	6.5511
4.3	1.4586	8.8	2.1748	800	6.6846
4.4	1.4816	8.9	2.1861	900	6.8024

TABLE 3
The Exponential Function: e^x

x	e^x	e^{-x}	x	e^x	e^{-x}
0.00	1.0000	1.0000	3.0	20.086	0.0498
0.05	1.0513	0.9512	3.1	22.198	0.0450
0.10	1.1052	0.9048	3.2	24.533	0.0408
0.15	1.1618	0.8607	3.3	27.113	0.0369
0.20	1.2214	0.8187	3.4	29.964	0.0334
0.25	1.2840	0.7788	3.5	33.115	0.0302
0.30	1.3499	0.7408	3.6	36.598	0.0273
0.35	1.4191	0.7047	3.7	40.447	0.0247
0.40	1.4918	0.6703	3.8	44.701	0.0224
0.45	1.5683	0.6376	3.9	49.402	0.0202
0.50	1.6487	0.6065	4.0	54.598	0.0183
0.55	1.7333	0.5769	4.1	60.340	0.0166
0.60	1.8221	0.5488	4.2	66.686	0.0150
0.65	1.9155	0.5220	4.3	73.700	0.0136
0.70	2.0138	0.4966	4.4	81.451	0.0123
0.75	2.1170	0.4724	4.5	90.017	0.0111
0.80	2.2255	0.4493	4.6	99.484	0.0101
0.85	2.3396	0.4274	4.7	109.95	0.0091
0.90	2.4596	0.4066	4.8	121.51	0.0082
0.95	2.5857	0.3867	4.9	134.29	0.0074
1.0	2.7183	0.3679	5.0	148.41	0.0067
1.1	3.0042	0.3329	5.1	164.02	0.0061
1.2	3.3201	0.3012	5.2	181.27	0.0055
1.3	3.6693	0.2725	5.3	200.34	0.0050
1.4	4.0552	0.2466	5.4	221.41	0.0045
1.5	4.4817	0.2231	5.5	244.69	0.0041
1.6	4.9530	0.2019	5.6	270.43	0.0037
1.7	5.4739	0.1827	5.7	298.87	0.0033
1.8	6.0496	0.1653	5.8	330.30	0.0030
1.9	6.6859	0.1496	5.9	365.04	0.0027
2.0	7.3891	0.1353	6.0	403.43	0.0025
2.1	8.1662	0.1225	6.5	665.14	0.0015
2.2	9.0250	0.1108	7.0	1096.6	0.0009
2.3	9.9742	0.1003	7.5	1808.0	0.0006
2.4	11.023	0.0907	8.0	2981.0	0.0003
2.5	12.182	0.0821	8.5	4914.8	0.0002
2.6	13.464	0.0743	9.0	8103.1	0.0001
2.7	14.880	0.0672	9.5	13,360	0.00007
2.8	16.445	0.0608	10.0	22,026	0.00004
2.9	18.174	0.0550			

TABLE 4

Future Value of an Annuity: $s_{\overline{n}|i} = \dfrac{(1 + i)^n - 1}{i}$

n \ i	$\frac{1}{2}\%$ 0.005	1% 0.010	$1\frac{1}{2}\%$ 0.015	2% 0.020	$2\frac{1}{2}\%$ 0.025	3% 0.030	$3\frac{1}{2}\%$ 0.035	4% 0.040
1	1.000000	1.000000	1.000000	1.000000	1.000000	1.000000	1.000000	1.000000
2	2.005000	2.010000	2.015000	2.020000	2.025000	2.030000	2.035000	2.040000
3	3.015025	3.030100	3.045225	3.060400	3.075625	3.090900	3.106225	3.121600
4	4.030100	4.060401	4.090903	4.121608	4.152516	4.183627	4.214943	4.246464
5	5.050251	5.101005	5.152267	5.204040	5.256329	5.309136	5.362466	5.416323
6	6.075502	6.152015	6.229551	6.308121	6.387737	6.468410	6.550152	6.632975
7	7.105879	7.213535	7.322994	7.434283	7.547430	7.662462	7.779408	7.898294
8	8.141409	8.285671	8.432839	8.582969	8.736116	8.892336	9.051687	9.214226
9	9.182116	9.368527	9.559332	9.754628	9.954519	10.159106	10.368496	10.582795
10	10.228026	10.462213	10.702722	10.949721	11.203382	11.463879	11.731393	12.006107
11	11.279167	11.566835	11.863262	12.168715	12.483466	12.807796	13.141992	13.486351
12	12.335562	12.682503	13.041211	13.412090	13.795553	14.192030	14.601962	15.025805
13	13.397240	13.809328	14.236830	14.680332	15.140442	15.617790	16.113030	16.626838
14	14.464226	14.947421	15.450382	15.973938	16.518953	17.086324	17.676986	18.291911
15	15.536548	16.096896	16.682138	17.293417	17.931927	18.598914	19.295681	20.023588
16	16.614230	17.257864	17.932370	18.639285	19.380225	20.156881	20.971030	21.824531
17	17.697301	18.430443	19.201355	20.012071	20.864730	21.761588	22.705016	23.697512
18	18.785788	19.614748	20.489376	21.412312	22.386349	23.414435	24.499691	25.645413
19	19.879717	20.810895	21.796716	22.840559	23.946007	25.116868	26.357180	27.671229
20	20.979115	22.019004	23.123667	24.297370	25.544658	26.870374	28.279682	29.778079
21	22.084011	23.239194	24.470522	25.783317	27.183274	28.676486	30.269471	31.969202
22	23.194431	24.471586	25.837580	27.298984	28.862856	30.536780	32.328902	34.247970
23	24.310403	25.716302	27.225144	28.844963	30.584427	32.452884	34.460414	36.617889
24	25.431955	26.973465	28.633521	30.421862	32.349038	34.426470	36.666528	39.082604
25	26.559115	28.243200	30.063024	32.030300	34.157764	36.459264	38.949857	41.645908
26	27.691911	29.525631	31.513969	33.670906	36.011708	38.553042	41.313102	44.311745
27	28.830370	30.820888	32.986678	35.344324	37.912001	40.709634	43.759060	47.084214
28	29.974522	32.129097	34.481479	37.051210	39.859801	42.930923	46.290627	49.967583
29	31.124395	33.450388	35.998701	38.792235	41.856296	45.218850	48.910799	52.966286
30	32.280017	34.784892	37.538681	40.568079	43.902703	47.575416	51.622677	56.084938
31	33.441417	36.132740	39.101762	42.379441	46.000271	50.002678	54.429471	59.328335
32	34.608624	37.494068	40.688288	44.227030	48.150278	52.502759	57.334502	62.701469
33	35.781667	38.869009	42.298612	46.111570	50.354034	55.077841	60.341210	66.209527
34	36.960575	40.257699	43.933092	48.033802	52.612885	57.730177	63.453152	69.857909
35	38.145378	41.660276	45.592088	49.994478	54.928207	60.462082	66.674013	73.652225
36	39.336105	43.076878	47.275969	51.994367	57.301413	62.275944	70.007603	77.598314
37	40.532785	44.507647	48.985109	54.034255	59.733948	66.174223	73.457869	81.702246
38	41.735449	45.952724	50.719885	56.114940	62.227297	69.159449	77.028895	85.970336
39	42.944127	47.412251	52.480684	58.237238	64.782979	72.234233	80.724906	90.409150
40	44.158847	48.886373	54.267894	60.401983	67.402554	75.401260	84.550278	95.025516
41	45.379642	50.375237	56.081912	62.610023	70.087617	78.663298	88.509537	99.826536
42	46.606540	51.878989	57.923141	64.862223	72.839808	82.023196	92.607371	104.819598
43	47.839572	53.397779	59.791988	67.159468	75.660803	85.483892	96.848629	110.012382
44	49.078770	54.931757	61.688868	69.502657	78.552323	89.048409	101.238331	115.412877
45	50.324164	56.481075	63.614201	71.892710	81.516131	92.719861	105.781673	121.029392
46	51.575785	58.045885	65.568414	74.330564	84.554034	96.501457	110.484031	126.870568
47	52.833664	59.626344	67.551940	76.817176	87.667885	100.396501	115.350973	132.945391
48	54.097832	61.222608	69.565219	79.353519	90.859582	104.408396	120.388257	139.263206
49	55.368321	62.834834	71.608698	81.940590	94.131072	108.540648	125.601846	145.833735
50	56.645163	64.463182	73.682828	84.579401	97.484349	112.796867	130.997910	152.667084

TABLE 4
Future Value of an Annuity (*continued*)

n	5% 0.050	6% 0.060	7% 0.070	8% 0.080	9% 0.090	10% 0.100	12% 0.120	15% 0.150	18% 0.180
1	1.000000	1.000000	1.000000	1.000000	1.000000	1.000000	1.000000	1.000000	1.000000
2	2.050000	2.060000	2.070000	2.080000	2.090000	2.100000	2.120000	2.150000	2.180000
3	3.152500	3.183600	3.214900	3.246400	3.278100	3.310000	3.374400	3.472500	3.572400
4	4.310125	4.374616	4.439943	4.506112	4.573129	4.641000	4.779328	4.993375	5.215432
5	5.525631	5.637093	5.750739	5.866601	5.984711	6.105100	6.352847	6.742381	7.154210
6	6.801913	6.975319	7.153291	7.335929	7.523335	7.715610	8.115189	8.753738	9.441968
7	8.142008	8.393838	8.654021	8.922803	9.200435	9.487171	10.089012	11.066799	12.141522
8	9.549109	9.897468	10.259803	10.636628	11.028474	11.435888	12.299693	13.726819	15.326996
9	11.026564	11.491316	11.977989	12.487558	13.021036	13.579477	14.775656	16.785842	19.085855
10	12.577893	13.180795	13.816448	14.486562	15.192930	15.937425	17.548735	20.303718	23.521309
11	14.206787	14.971643	15.783599	16.645487	17.560293	18.531167	20.654583	24.349276	28.755144
12	15.917127	16.869941	17.888451	18.977126	20.140720	21.384284	24.133133	29.001667	34.931070
13	17.712983	18.882138	20.140643	21.495297	22.953385	24.522712	28.029109	34.351917	42.218663
14	19.598632	21.015066	22.550488	24.214920	26.019189	27.974983	32.392602	40.504705	50.818022
15	21.578564	23.275970	25.129022	27.152114	29.360916	31.772482	37.279715	47.580411	60.965266
16	23.657492	25.672528	27.888054	30.324283	33.003399	35.949730	42.753280	55.717472	72.939014
17	25.840366	28.212880	30.840217	33.750226	36.973705	40.544703	48.883674	65.075093	87.068036
18	28.132385	30.905653	33.999033	37.450244	41.301338	45.599173	55.749715	75.836357	103.740283
19	30.539004	33.759992	37.378965	41.446263	46.018458	51.159090	63.439681	88.211811	123.413534
20	33.065954	36.785591	40.995492	45.761964	51.160120	57.274999	72.052442	102.443583	146.627970
21	35.719252	39.992727	44.865177	50.422921	56.764530	64.002499	81.698736	118.810120	174.021006
22	38.505214	43.392290	49.005739	55.456755	62.873338	71.402749	92.502584	137.631638	206.344786
23	41.430475	46.995828	53.436141	60.893296	69.531939	79.543024	104.602894	159.276384	244.486849
24	44.501999	50.815577	58.176671	66.764759	76.789813	88.497327	118.155241	184.167843	289.494480
25	47.727099	54.864512	63.249038	73.105940	84.700896	98.347059	133.333870	212.793018	342.603489
26	51.113454	59.156383	68.676470	79.954415	93.323977	109.181765	150.333935	245.711969	405.272114
27	54.669126	63.705766	74.483823	87.350768	102.723135	121.099942	169.374008	283.568768	479.221092
28	58.402583	68.528112	80.697691	95.338830	112.968217	134.209936	190.698889	327.104080	566.480888
29	62.322712	73.639798	87.346529	103.965936	124.135356	148.630930	214.582754	377.169693	669.447449
30	66.438848	79.058186	94.460786	113.283211	136.307541	164.494024	241.332684	434.745148	790.947990
31	70.760790	84.801677	102.073041	123.345868	149.575218	181.943426	271.292606	500.956921	934.318634
32	75.298829	90.889778	110.218154	134.213537	164.036987	201.137768	304.847721	577.100456	1103.495987
33	80.063771	97.343165	118.933425	145.950621	179.800316	222.251545	342.429447	664.665527	1303.125259
34	85.066959	104.183755	128.258764	158.626671	196.982344	245.476700	384.520981	765.365349	1538.687805
35	90.320307	111.434780	138.236877	172.316805	215.710756	271.024368	431.663494	881.170158	1816.651611
36	95.836323	119.120867	148.913460	187.102148	236.124723	299.126804	484.463116	1014.345680	2144.648895
37	101.628139	127.268119	160.337402	203.070320	258.375950	330.039490	543.598686	1167.497543	2531.685699
38	107.709546	135.904205	172.561020	220.315947	282.629784	364.043434	609.830536	1343.622162	2988.389130
39	114.095023	145.058458	185.640291	238.941221	309.066463	401.447781	684.010201	1546.165482	3527.299164
40	120.799774	154.761967	199.635113	259.056519	337.882442	442.592556	767.091423	1779.090302	4163.213013
41	127.839763	165.047684	214.609570	280.781040	369.291862	487.851810	860.142395	2046.953857	4913.591370
42	135.231752	175.950546	230.632240	304.243523	403.528133	537.636993	964.359482	2354.996918	5799.037842
43	142.993340	187.507578	247.776497	329.583004	440.845665	592.400696	1081.082611	2709.246460	6843.864624
44	151.143005	199.758032	266.120853	356.949650	481.521774	652.640762	1211.812531	3116.633453	8076.760254
45	159.700155	212.743513	285.749313	386.505615	525.858734	718.904839	1358.230026	3585.128448	9531.577148
46	168.685163	226.508125	306.751762	418.426067	574.186028	791.795319	1522.217636	4123.897705	11248.260986
47	178.119423	241.098612	329.224388	452.900150	626.862762	871.974854	1705.883759	4743.482361	13273.947876
48	188.025394	256.564529	353.270092	490.132164	684.280411	960.172333	1911.589813	5456.004761	15664.258545
49	198.426664	272.958401	378.999001	530.342735	746.865646	1057.189575	2141.980591	6275.405457	18484.825195
50	209.347996	290.335903	406.528927	573.770157	815.083557	1163.908524	2400.018250	7217.716309	21813.093750

TABLE 5

Present Value of an Ordinary Annuity: $a_{\overline{n}|i} = \dfrac{1 - (1 + i)^{-n}}{i}$

n \ i	$\frac{1}{2}\%$ 0.005	1% 0.010	$1\frac{1}{2}\%$ 0.015	2% 0.020	$2\frac{1}{2}\%$ 0.025	3% 0.030	$3\frac{1}{2}\%$ 0.035	4% 0.040
1	0.995025	0.990099	0.985222	0.980392	0.975610	0.970874	0.966184	0.961538
2	1.985099	1.970395	1.955883	1.941561	1.927424	1.913470	1.899694	1.886095
3	2.970248	2.940985	2.912200	2.883883	2.856024	2.828611	2.801637	2.775091
4	3.950496	3.901966	3.854385	3.807729	3.761974	3.717098	3.673079	3.629895
5	4.925866	4.853431	4.782645	4.713460	4.645828	4.579707	4.515052	4.451822
6	5.896384	5.795476	5.697187	5.601431	5.508125	5.417191	5.328553	5.242137
7	6.862074	6.728195	6.598214	6.471991	6.349391	6.230283	6.114544	6.002055
8	7.822959	7.651678	7.485925	7.325481	7.170137	7.019692	6.873956	6.732745
9	8.779064	8.566018	8.360517	8.162237	7.970866	7.786109	7.607687	7.435332
10	9.730412	9.471305	9.222185	8.982585	8.752064	8.530203	8.316605	8.110896
11	10.677027	10.367628	10.071118	9.786848	9.514209	9.252624	9.001551	8.760477
12	11.618932	11.255077	10.907505	10.575341	10.257765	9.954004	9.663334	9.385074
13	12.556151	12.133740	11.731532	11.348374	10.983185	10.634955	10.302738	9.985648
14	13.488708	13.003703	12.543382	12.106249	11.690912	11.296073	10.920520	10.563123
15	14.416625	13.865053	13.343233	12.849264	12.381378	11.937935	11.517411	11.118387
16	15.339925	14.717874	14.131264	13.577709	13.055003	12.561102	12.094117	11.652296
17	16.258632	15.562251	14.907649	14.291872	13.712198	13.166118	12.651321	12.165669
18	17.172768	16.398269	15.672561	14.992031	14.353364	13.753513	13.189682	12.659297
19	18.082356	17.226008	16.426168	15.678462	14.978891	14.323799	13.709837	13.133939
20	18.987419	18.045553	17.168639	16.351433	15.589162	14.877475	14.212403	13.590326
21	19.887979	18.856983	17.900137	17.011209	16.184549	15.415024	14.697974	14.029160
22	20.784059	19.660379	18.620824	17.658048	16.765413	15.936917	15.167125	14.451115
23	21.675681	20.455821	19.330861	18.292204	17.332110	16.443608	15.620410	14.856842
24	22.562866	21.243387	20.030405	18.913926	17.884986	16.935542	16.058368	15.246963
25	23.445638	22.023156	20.719611	19.523456	18.424376	17.413148	16.481515	15.622080
26	24.324018	22.795204	21.398632	20.121036	18.950611	17.876842	16.890352	15.982769
27	25.198028	23.559608	22.067617	20.706898	19.464011	18.327031	17.285365	16.329586
28	26.067689	24.316443	22.726717	21.281272	19.964889	18.764108	17.667019	16.663063
29	26.933024	25.065785	23.376076	21.844385	20.453550	19.188455	18.035767	16.983715
30	27.794054	25.807708	24.015838	22.396456	20.930293	19.600441	18.392045	17.292033
31	28.650800	26.542285	24.646146	22.937702	21.395407	20.000428	18.736276	17.588494
32	29.503284	27.269589	25.267139	23.468335	21.849178	20.388766	19.068865	17.873551
33	30.351526	27.989693	25.878954	23.988564	22.291881	20.765792	19.390208	18.147646
34	31.195548	28.702666	26.481728	24.498592	22.723786	21.131837	19.700684	18.411198
35	32.035371	29.408580	27.075595	24.998619	23.145157	21.487220	20.000661	18.664613
36	32.871016	30.107505	27.660684	25.488842	23.556251	21.832252	20.290494	18.908282
37	33.702504	30.799510	28.237127	25.969453	23.957318	22.167235	20.570525	19.142579
38	34.529854	31.484663	28.805052	26.440641	24.348603	22.492462	20.841087	19.367864
39	35.353089	32.163033	29.364583	26.902589	24.730344	22.808215	21.102500	19.584485
40	36.172228	32.834686	29.915845	27.355479	25.102775	23.114772	21.355072	19.792774
41	36.987291	33.499689	30.458961	27.799489	25.466122	23.412400	21.599104	19.993052
42	37.798300	34.158108	30.994050	28.234794	25.820607	23.701359	21.834883	20.185627
43	38.605274	34.810008	31.521232	28.661562	26.166446	23.981902	22.062689	20.370795
44	39.408232	35.455454	32.040622	29.079963	26.503849	24.254274	22.282791	20.548841
45	40.207196	36.094508	32.552337	29.490160	26.833024	24.518713	22.495450	20.720040
46	41.002185	36.727236	33.056490	29.892314	27.154170	24.775449	22.700918	20.884654
47	41.793219	37.353699	33.553192	30.286582	27.467483	25.024708	22.899438	21.042936
48	42.580318	37.973959	34.042554	30.673120	27.773154	25.266707	23.091244	21.195131
49	43.363500	38.588079	34.524683	31.052078	28.071369	25.501657	23.276564	21.341472
50	44.142786	39.196118	34.999688	31.423606	28.362312	25.729764	23.455618	21.482185

TABLE 5
Present Value of an Ordinary Annuity (*continued*)

i n	5% 0.050	6% 0.060	7% 0.070	8% 0.080	9% 0.090	10% 0.100	12% 0.120	15% 0.150	18% 0.180
1	0.952381	0.943396	0.934579	0.925926	0.917431	0.909091	0.892857	0.869565	0.847458
2	1.859410	1.833393	1.808018	1.783265	1.759111	1.735537	1.690051	1.625709	1.565642
3	2.723248	2.673012	2.624316	2.577097	2.531295	2.486852	2.401831	2.283225	2.174273
4	3.545951	3.465106	3.387211	3.312127	3.239720	3.169865	3.037349	2.854978	2.690062
5	4.329477	4.212364	4.100197	3.992710	3.889651	3.790787	3.604776	3.352155	3.127171
6	5.075692	4.917324	4.766540	4.622880	4.485919	4.355261	4.111407	3.784483	3.497603
7	5.786373	5.582381	5.389289	5.206370	5.032953	4.868419	4.563757	4.160420	3.811528
8	6.463213	6.209794	5.971299	5.746639	5.534819	5.334926	4.967640	4.487322	4.077566
9	7.107822	6.801692	6.515232	6.246888	5.995247	5.759024	5.328250	4.771584	4.303022
10	7.721735	7.360087	7.023582	6.710081	6.417658	6.144567	5.650223	5.018769	4.494086
11	8.306414	7.886875	7.498674	7.138964	6.805191	6.495061	5.937699	5.233712	4.656005
12	8.863252	8.383844	7.942686	7.536078	7.160725	6.813692	6.194374	5.420619	4.793225
13	9.393573	8.852683	8.357651	7.903776	7.486904	7.103356	6.423548	5.583147	4.909513
14	9.898641	9.294984	8.745468	8.244237	7.786150	7.366687	6.628168	5.724476	5.008062
15	10.379658	9.712249	9.107914	8.559479	8.060688	7.606080	6.810864	5.847370	5.091578
16	10.837770	10.105895	9.446649	8.851369	8.312558	7.823709	6.973986	5.954235	5.162354
17	11.274066	10.477260	9.763223	9.121638	8.543631	8.021553	7.119630	6.047161	5.222334
18	11.689587	10.827603	10.059087	9.371887	8.755625	8.201412	7.249670	6.127966	5.273164
19	12.085321	11.158116	10.335595	9.603599	8.950115	8.364920	7.365777	6.198231	5.316241
20	12.462210	11.469921	10.594014	9.818147	9.128546	8.513564	7.469444	6.259331	5.352746
21	12.821153	11.764077	10.835527	10.016803	9.292244	8.648694	7.562003	6.312462	5.383683
22	13.163003	12.041582	11.061240	10.200744	9.442425	8.771540	7.644646	6.358663	5.409901
23	13.488574	12.303379	11.272187	10.371059	9.580207	8.883218	7.718434	6.398837	5.432120
24	13.798642	12.550358	11.469334	10.528758	9.706612	8.984744	7.784316	6.433771	5.450949
25	14.093945	12.783356	11.653583	10.674776	9.822580	9.077040	7.843139	6.464149	5.466906
26	14.375185	13.003166	11.825779	10.809978	9.928972	9.160945	7.895660	6.490564	5.480429
27	14.643034	13.210534	11.986709	10.935165	10.026580	9.237223	7.942554	6.513534	5.491889
28	14.898127	13.406164	12.137111	11.051078	10.116128	9.306567	7.984423	6.533508	5.501601
29	15.141074	13.590721	12.277674	11.158406	10.198283	9.369606	8.021806	6.550877	5.509831
30	15.372451	13.764831	12.409041	11.257783	10.273654	9.426914	8.055184	6.565980	5.516806
31	15.592811	13.929086	12.531814	11.349799	10.342802	9.479013	8.084986	6.579113	5.522717
32	15.802677	14.084043	12.646555	11.434999	10.406240	9.526376	8.111594	6.590533	5.527726
33	16.002549	14.230230	12.753790	11.513888	10.464441	9.569432	8.135352	6.600463	5.531971
34	16.192904	14.368141	12.854009	11.586934	10.517835	9.608575	8.156564	6.609099	5.535569
35	16.374194	14.498246	12.947672	11.654568	10.566821	9.644159	8.175504	6.616607	5.538618
36	16.546852	14.620987	13.035208	11.717193	10.611763	9.676508	8.192414	6.623137	5.541201
37	16.711287	14.736780	13.117017	11.775179	10.652993	9.705917	8.207513	6.628815	5.543391
38	16.867893	14.846019	13.193473	11.828869	10.690820	9.732651	8.220993	6.633752	5.545247
39	17.017041	14.949075	13.264928	11.878582	10.725523	9.756956	8.233030	6.638045	5.546819
40	17.159086	15.046297	13.331709	11.924613	10.757360	9.779051	8.243777	6.641778	5.548152
41	17.294368	15.138016	13.394120	11.967235	10.786569	9.799137	8.253372	6.645025	5.549281
42	17.423208	15.224543	13.452449	12.006699	10.813366	9.817397	8.261939	6.647848	5.550238
43	17.545912	15.306173	13.506962	12.043240	10.837950	9.833998	8.269589	6.650302	5.551049
44	17.662773	15.383182	13.557908	12.077074	10.860505	9.849089	8.276418	6.652437	5.551737
45	17.774070	15.455832	13.605522	12.108402	10.881197	9.862808	8.282516	6.654293	5.552319
46	17.880066	15.524370	13.650020	12.137409	10.900181	9.875280	8.287961	6.655907	5.552813
47	17.981016	15.589028	13.691608	12.164267	10.917597	9.886618	8.292822	6.657310	5.553231
48	18.077158	15.650027	13.730474	12.189136	10.933575	9.896926	8.297163	6.658531	5.553586
49	18.168722	15.707572	13.766799	12.212163	10.948234	9.906296	8.301038	6.659592	5.553886
50	18.255925	15.761861	13.800746	12.233485	10.961683	9.914814	8.304498	6.660515	5.554141

TABLE 6
Compound Interest: $(1 + i)^n$

n \ i	$\frac{1}{2}\%$ 0.005	1% 0.010	$1\frac{1}{2}\%$ 0.015	2% 0.020	$2\frac{1}{2}\%$ 0.025	3% 0.030	$3\frac{1}{2}\%$ 0.035	4% 0.040
1	1.005000	1.010000	1.015000	1.020000	1.025000	1.030000	1.035000	1.040000
2	1.010025	1.020100	1.030225	1.040400	1.050625	1.060900	1.071225	1.081600
3	1.015075	1.030301	1.045678	1.061208	1.076891	1.092727	1.108718	1.124864
4	1.020151	1.040604	1.061364	1.082432	1.103813	1.125509	1.147523	1.169859
5	1.025251	1.051010	1.077284	1.104081	1.131408	1.159274	1.187686	1.216653
6	1.030378	1.061520	1.093443	1.126162	1.159693	1.194052	1.229255	1.265319
7	1.035529	1.072135	1.109845	1.148686	1.188686	1.229874	1.272279	1.315932
8	1.040707	1.082857	1.126493	1.171659	1.218403	1.266770	1.316809	1.368569
9	1.045911	1.093685	1.143390	1.195093	1.248863	1.304773	1.362897	1.423312
10	1.051140	1.104622	1.160541	1.218994	1.280085	1.343916	1.410599	1.480244
11	1.056396	1.115668	1.177949	1.243374	1.312087	1.384234	1.459970	1.539454
12	1.061678	1.126825	1.195618	1.268242	1.344889	1.425761	1.511069	1.601032
13	1.066986	1.138093	1.213552	1.293607	1.378511	1.468534	1.563956	1.665074
14	1.072321	1.149474	1.231756	1.319479	1.412974	1.512590	1.618695	1.731676
15	1.077683	1.160969	1.250232	1.345868	1.448298	1.557967	1.675349	1.800944
16	1.083071	1.172579	1.268986	1.372786	1.484506	1.604706	1.733986	1.872981
17	1.088487	1.184304	1.288020	1.400241	1.521618	1.652848	1.794676	1.947900
18	1.093929	1.196147	1.307341	1.428246	1.559659	1.702433	1.857489	2.025817
19	1.099399	1.208109	1.326951	1.456811	1.598650	1.753506	1.922501	2.106849
20	1.104896	1.220190	1.346855	1.485947	1.638616	1.806111	1.989789	2.191123
21	1.110420	1.232392	1.367058	1.515666	1.679582	1.860295	2.059431	2.278768
22	1.115972	1.244716	1.387564	1.545980	1.721571	1.916103	2.131512	2.369919
23	1.121552	1.257163	1.408377	1.576899	1.764611	1.973587	2.206114	2.464716
24	1.127160	1.269735	1.429503	1.608437	1.808726	2.032794	2.283328	2.563304
25	1.132796	1.282432	1.450945	1.640606	1.853944	2.093778	2.363245	2.665836
26	1.138460	1.295256	1.472710	1.673418	1.900293	2.156591	2.445959	2.772470
27	1.144152	1.308209	1.494800	1.706886	1.947800	2.221289	2.531567	2.883369
28	1.149873	1.321291	1.517222	1.741024	1.996495	2.287928	2.620172	2.998703
29	1.155622	1.334504	1.539981	1.775845	2.046407	2.356566	2.711878	3.118651
30	1.161400	1.347849	1.563080	1.811362	2.097568	2.427262	2.806794	3.243398
31	1.167207	1.361327	1.586526	1.847589	2.150007	2.500080	2.905031	3.373133
32	1.173043	1.374941	1.610324	1.884541	2.203757	2.575083	3.006708	3.508059
33	1.178908	1.388690	1.634479	1.922231	2.258851	2.652335	3.111942	3.648381
34	1.184803	1.402577	1.658996	1.960676	2.315322	2.731905	3.220860	3.794316
35	1.190727	1.416603	1.683881	1.999890	2.373205	2.813862	3.333590	3.946089
36	1.196681	1.430769	1.709140	2.039887	2.432535	2.898278	3.450266	4.103933
37	1.202664	1.445076	1.734777	2.080685	2.493349	2.985227	3.571025	4.268090
38	1.208677	1.459527	1.760798	2.122299	2.555682	3.074783	3.696011	4.438813
39	1.214721	1.474123	1.787210	2.164745	2.619574	3.167027	3.825372	4.616366
40	1.220794	1.488864	1.814018	2.208040	2.685064	3.262038	3.959260	4.801021
41	1.226898	1.503752	1.841229	2.252200	2.752190	3.359899	4.097834	4.993061
42	1.233033	1.518790	1.868847	2.297244	2.820995	3.460696	4.241258	5.192784
43	1.239198	1.533978	1.896880	2.343189	2.891520	3.564517	4.389702	5.400495
44	1.245394	1.549318	1.925333	2.390053	2.963808	3.671452	4.543342	5.616515
45	1.251621	1.564811	1.954213	2.437854	3.037903	3.781596	4.702359	5.841176
46	1.257879	1.580459	1.983526	2.486611	3.113851	3.895044	4.866941	6.074823
47	1.264168	1.596263	2.013279	2.536344	3.191697	4.011895	5.037284	6.317816
48	1.270489	1.612226	2.043478	2.587070	3.271490	4.132252	5.213589	6.570528
49	1.276842	1.628348	2.074130	2.638812	3.353277	4.256219	5.396065	6.833349
50	1.283226	1.644632	2.105242	2.691588	3.437109	4.383906	5.584927	7.106683

TABLE 6
Compound Interest (*continued*)

n \ i	5% 0.050	6% 0.060	7% 0.070	8% 0.080	9% 0.090	10% 0.100	12% 0.120	15% 0.150	18% 0.180
1	1.050000	1.060000	1.070000	1.080000	1.090000	1.100000	1.120000	1.150000	1.180000
2	1.102500	1.123600	1.144900	1.166400	1.188100	1.210000	1.254400	1.322500	1.392400
3	1.157625	1.191016	1.225043	1.259712	1.295029	1.331000	1.404928	1.520875	1.643032
4	1.215506	1.262477	1.310796	1.360489	1.411582	1.464100	1.573519	1.749006	1.938778
5	1.276282	1.338226	1.402552	1.469328	1.538624	1.610510	1.762342	2.011357	2.287758
6	1.340096	1.418519	1.500730	1.586874	1.677100	1.771561	1.973823	2.313061	2.699554
7	1.407100	1.503630	1.605781	1.713824	1.828039	1.948717	2.210681	2.660020	3.185474
8	1.477455	1.593848	1.718186	1.850930	1.992563	2.143589	2.475963	3.059023	3.758859
9	1.551328	1.689479	1.838459	1.999005	2.171893	2.357948	2.773079	3.517876	4.435454
10	1.628895	1.790848	1.967151	2.158925	2.367364	2.593742	3.105848	4.045558	5.233836
11	1.710339	1.898299	2.104852	2.331639	2.580426	2.853117	3.478550	4.652391	6.175926
12	1.795856	2.012196	2.252192	2.518170	2.812665	3.138428	3.895976	5.350250	7.287593
13	1.885649	2.132928	2.409845	2.719624	3.065805	3.452271	4.363493	6.152788	8.599359
14	1.979932	2.260904	2.578534	2.937194	3.341727	3.797498	4.887112	7.075706	10.147244
15	2.078928	2.396558	2.759032	3.172169	3.642482	4.177248	5.473566	8.137062	11.973748
16	2.182875	2.540352	2.952164	3.425943	3.970306	4.594973	6.130394	9.357621	14.129023
17	2.292018	2.692773	3.158815	3.700018	4.327633	5.054470	6.866041	10.761264	16.672247
18	2.406619	2.854339	3.379932	3.996019	4.717120	5.559917	7.689966	12.375454	19.673251
19	2.526950	3.025600	3.616528	4.315701	5.141661	6.115909	8.612762	14.231772	23.214436
20	2.653298	3.207135	3.869684	4.660957	5.604411	6.727500	9.646293	16.366537	27.393035
21	2.785963	3.399564	4.140562	5.033834	6.108808	7.400250	10.803848	18.821518	32.323781
22	2.925261	3.603537	4.430402	5.436540	6.658600	8.140275	12.100310	21.644746	38.142061
23	3.071524	3.819750	4.740530	5.871464	7.257874	8.954302	13.552347	24.891458	45.007632
24	3.225100	4.048935	5.072367	6.341181	7.911083	9.849733	15.178629	28.625176	53.109006
25	3.386355	4.291871	5.427433	6.848475	8.623081	10.834706	17.000064	32.918953	62.668627
26	3.555673	4.549383	5.807353	7.396353	9.399158	11.918177	19.040072	37.856796	73.948980
27	3.733456	4.822346	6.213868	7.988061	10.245082	13.109994	21.324881	43.535315	87.259797
28	3.920129	5.111687	6.648838	8.627106	11.167140	14.420994	23.883866	50.065612	102.966560
29	4.116136	5.418388	7.114257	9.317275	12.172182	15.863093	26.749930	57.575454	121.500541
30	4.321942	5.743491	7.612255	10.062657	13.267678	17.449402	29.959922	66.211772	143.370638
31	4.538039	6.088101	8.145113	10.867669	14.461770	19.194342	33.555113	76.143538	169.177355
32	4.764941	6.453387	8.715271	11.737083	15.763329	21.113777	37.581726	87.565068	199.629278
33	5.003189	6.840590	9.325340	12.676050	17.182028	23.225154	42.091533	100.699829	235.562548
34	5.253348	7.251025	9.978114	13.690134	18.728411	25.547670	47.142517	115.804803	277.963802
35	5.516015	7.686087	10.676581	14.785344	20.413968	28.102437	52.799620	133.175524	327.997292
36	5.791816	8.147252	11.423942	15.968172	22.251225	30.912681	59.135574	153.151852	387.036804
37	6.081407	8.636087	12.223618	17.245626	24.253835	34.003949	66.231843	176.124630	456.703426
38	6.385477	9.154252	13.079271	18.625276	26.436680	37.404343	74.179664	202.543324	538.910049
39	6.704751	9.703507	13.994820	20.115298	28.815982	41.144778	83.081224	232.924824	635.913857
40	7.039989	10.285718	14.974458	21.724521	31.409420	45.259256	93.050970	267.863544	750.378342
41	7.391988	10.902861	16.022670	23.462483	34.236268	49.785181	104.217087	308.043079	885.446449
42	7.761588	11.557033	17.144257	25.339482	37.317532	54.763699	116.723137	354.249538	1044.826813
43	8.149667	12.250455	18.344355	27.366640	40.676110	60.240069	130.729914	407.386971	1232.895630
44	8.557150	12.985482	19.628460	29.555972	44.336960	66.264076	146.417503	468.495014	1454.816849
45	8.985008	13.764611	21.002452	31.920449	48.327286	72.890484	163.987604	538.769272	1716.683868
46	9.434258	14.590487	22.472623	34.474085	52.676742	80.179532	183.666117	619.584656	2025.686981
47	9.905971	15.465917	24.045707	37.232012	57.417649	88.197485	205.706051	712.522362	2390.310638
48	10.401270	16.393872	25.728907	40.210573	62.585237	97.017234	230.390776	819.400711	2820.566559
49	10.921333	17.377504	27.529930	43.427419	68.217908	106.718957	258.037670	942.310814	3328.268524
50	11.467400	18.420154	29.457025	46.901613	74.357520	117.390853	289.002193	1083.657440	3927.356873

Appendix B

TABLE 7

Present Value: $(1 + i)^{-n}$

n \ i	$\frac{1}{2}\%$ 0.005	1% 0.010	$1\frac{1}{2}\%$ 0.015	2% 0.020	$2\frac{1}{2}\%$ 0.025	3% 0.030	$3\frac{1}{2}\%$ 0.035	4% 0.040
1	0.995025	0.990099	0.985222	0.980392	0.975610	0.970874	0.966184	0.961538
2	0.990075	0.980296	0.970662	0.961169	0.951814	0.942596	0.933511	0.924556
3	0.985149	0.970590	0.956317	0.942322	0.928599	0.915142	0.901943	0.888996
4	0.980248	0.960980	0.942184	0.923845	0.905951	0.888487	0.871442	0.854804
5	0.975371	0.951466	0.928260	0.905731	0.883854	0.862609	0.841973	0.821927
6	0.970518	0.942045	0.914542	0.887971	0.862297	0.837484	0.813501	0.790315
7	0.965690	0.932718	0.901027	0.870560	0.841265	0.813092	0.785991	0.759918
8	0.960885	0.923483	0.887711	0.853490	0.820747	0.789409	0.759412	0.730690
9	0.956105	0.914340	0.874592	0.836755	0.800728	0.766417	0.733731	0.702587
10	0.951348	0.905287	0.861667	0.820348	0.781198	0.744094	0.708919	0.675564
11	0.946615	0.896324	0.848933	0.804263	0.762145	0.722421	0.684946	0.649581
12	0.941905	0.887449	0.836387	0.788493	0.743556	0.701380	0.661783	0.624597
13	0.937219	0.878663	0.824027	0.773033	0.725420	0.680951	0.639404	0.600574
14	0.932556	0.869963	0.811849	0.757875	0.707727	0.661118	0.617782	0.577475
15	0.927917	0.861349	0.799852	0.743015	0.690466	0.641862	0.596891	0.555265
16	0.923300	0.852821	0.788031	0.728446	0.673625	0.623167	0.576706	0.533908
17	0.918707	0.844377	0.776385	0.714163	0.657195	0.605016	0.557204	0.513373
18	0.914136	0.836017	0.764912	0.700159	0.641166	0.587395	0.538361	0.493628
19	0.909588	0.827740	0.753607	0.686431	0.625528	0.570286	0.520156	0.474642
20	0.905063	0.819544	0.742470	0.672971	0.610271	0.553676	0.502566	0.456387
21	0.900560	0.811430	0.731498	0.659776	0.595386	0.537549	0.485571	0.438834
22	0.896080	0.803396	0.720688	0.646839	0.580865	0.521893	0.469151	0.421955
23	0.891622	0.795442	0.710037	0.634156	0.566697	0.506692	0.453286	0.405726
24	0.887186	0.787566	0.699544	0.621721	0.552875	0.491934	0.437957	0.390121
25	0.882772	0.779768	0.689206	0.609531	0.539391	0.477606	0.423147	0.375117
26	0.878380	0.772048	0.679021	0.597579	0.526235	0.463695	0.408838	0.360689
27	0.874010	0.764404	0.668986	0.585862	0.513400	0.450189	0.395012	0.346817
28	0.869662	0.756836	0.659099	0.574375	0.500878	0.437077	0.381654	0.333477
29	0.865335	0.749342	0.649359	0.563112	0.488661	0.424346	0.368748	0.320651
30	0.861030	0.741923	0.639762	0.552071	0.476743	0.411987	0.356278	0.308319
31	0.856746	0.734577	0.630308	0.541246	0.465115	0.399987	0.344230	0.296460
32	0.852484	0.727304	0.620993	0.530633	0.453771	0.388337	0.332590	0.285058
33	0.848242	0.720103	0.611816	0.520229	0.442703	0.377026	0.321343	0.274094
34	0.844022	0.712973	0.602774	0.510028	0.431905	0.366045	0.310476	0.263552
35	0.839823	0.705914	0.593866	0.500028	0.421371	0.355383	0.299977	0.253415
36	0.835645	0.698925	0.585090	0.490223	0.411094	0.345032	0.289833	0.243669
37	0.831487	0.692005	0.576443	0.480611	0.401067	0.334983	0.280032	0.234297
38	0.827351	0.685153	0.567924	0.471187	0.391285	0.325226	0.270562	0.225285
39	0.823235	0.678370	0.559531	0.461948	0.381741	0.315754	0.261413	0.216621
40	0.819139	0.671653	0.551262	0.452890	0.372431	0.306557	0.252572	0.208289
41	0.815064	0.665003	0.543116	0.444010	0.363347	0.297628	0.244031	0.200278
42	0.811009	0.658419	0.535089	0.435304	0.354485	0.288959	0.235779	0.192575
43	0.806974	0.651900	0.527182	0.426769	0.345839	0.280543	0.227806	0.185168
44	0.802959	0.645445	0.519391	0.418401	0.337404	0.272372	0.220102	0.178046
45	0.798964	0.639055	0.511715	0.410197	0.329174	0.264439	0.212659	0.171198
46	0.794989	0.632728	0.504153	0.402154	0.321146	0.256737	0.205468	0.164614
47	0.791034	0.626463	0.496702	0.394268	0.313313	0.249259	0.198520	0.158283
48	0.787098	0.620260	0.489362	0.386538	0.305671	0.241999	0.191806	0.152195
49	0.783182	0.614119	0.482130	0.378958	0.298216	0.234950	0.185320	0.146341
50	0.779286	0.608039	0.475005	0.371528	0.290942	0.228107	0.179053	0.140713

TABLE 7
Present Value (*continued*)

n \ i	5% 0.050	6% 0.060	7% 0.070	8% 0.080	9% 0.090	10% 0.100	12% 0.120	15% 0.150	18% 0.180
1	0.952381	0.943396	0.934579	0.925926	0.917431	0.909091	0.892857	0.869565	0.847458
2	0.907029	0.889996	0.873439	0.857339	0.841680	0.826446	0.797194	0.756144	0.718184
3	0.863838	0.839619	0.816298	0.793832	0.772183	0.751315	0.711780	0.657516	0.608631
4	0.822702	0.792094	0.762895	0.735030	0.708425	0.683013	0.635518	0.571753	0.515789
5	0.783526	0.747258	0.712986	0.680583	0.649931	0.620921	0.567427	0.497177	0.437109
6	0.746215	0.704961	0.666342	0.630170	0.596267	0.564474	0.506631	0.432328	0.370432
7	0.710681	0.665057	0.622750	0.583490	0.547034	0.513158	0.452349	0.375937	0.313925
8	0.676839	0.627412	0.582009	0.540269	0.501866	0.466507	0.403883	0.326902	0.266038
9	0.644609	0.591898	0.543934	0.500249	0.460428	0.424098	0.360610	0.284262	0.225456
10	0.613913	0.558395	0.508349	0.463193	0.422411	0.385543	0.321973	0.247185	0.191064
11	0.584679	0.526788	0.475093	0.428883	0.387533	0.350494	0.287476	0.214943	0.161919
12	0.556837	0.496969	0.444012	0.397114	0.355535	0.318631	0.256675	0.186907	0.137220
13	0.530321	0.468839	0.414964	0.367698	0.326179	0.289664	0.229174	0.162528	0.116288
14	0.505068	0.442301	0.387817	0.340461	0.299246	0.263331	0.204620	0.141329	0.098549
15	0.481017	0.417265	0.362446	0.315242	0.274538	0.239392	0.182696	0.122894	0.083516
16	0.458112	0.393646	0.338735	0.291890	0.251870	0.217629	0.163122	0.106865	0.070776
17	0.436297	0.371364	0.316574	0.270269	0.231073	0.197845	0.145644	0.092926	0.059980
18	0.415521	0.350344	0.295864	0.250249	0.211994	0.179859	0.130040	0.080805	0.050830
19	0.395734	0.330513	0.276508	0.231712	0.194490	0.163508	0.116107	0.070265	0.043077
20	0.376889	0.311805	0.258419	0.214548	0.178431	0.148644	0.103667	0.061100	0.036506
21	0.358942	0.294155	0.241513	0.198656	0.163698	0.135131	0.092560	0.053131	0.030937
22	0.341850	0.277505	0.225713	0.183941	0.150182	0.122846	0.082643	0.046201	0.026218
23	0.325571	0.261797	0.210947	0.170315	0.137781	0.111678	0.073788	0.040174	0.022218
24	0.310068	0.246979	0.197147	0.157699	0.126405	0.101526	0.065882	0.034934	0.018829
25	0.295303	0.232999	0.184249	0.146018	0.115968	0.092296	0.058823	0.030378	0.015957
26	0.281241	0.219810	0.172195	0.135202	0.106393	0.083905	0.052521	0.026415	0.013523
27	0.267848	0.207368	0.160930	0.125187	0.097608	0.076278	0.046894	0.022970	0.011460
28	0.255094	0.195630	0.150402	0.115914	0.089548	0.069343	0.041869	0.019974	0.009712
29	0.242946	0.184557	0.140563	0.107328	0.082155	0.063039	0.037383	0.017369	0.008230
30	0.231377	0.174110	0.131367	0.099377	0.075371	0.057309	0.033378	0.015103	0.006975
31	0.220359	0.164255	0.122773	0.092016	0.069148	0.052099	0.029802	0.013133	0.005911
32	0.209866	0.154957	0.114741	0.085200	0.063438	0.047362	0.026609	0.011420	0.005009
33	0.199873	0.146186	0.107235	0.078889	0.058200	0.043057	0.023758	0.009931	0.004245
34	0.190355	0.137912	0.100219	0.073045	0.053395	0.039143	0.021212	0.008635	0.003598
35	0.181290	0.130105	0.093663	0.067635	0.048986	0.035584	0.018940	0.007509	0.003049
36	0.172657	0.122741	0.087535	0.062625	0.044941	0.032349	0.016910	0.006529	0.002584
37	0.164436	0.115793	0.081809	0.057986	0.041231	0.029408	0.015098	0.005678	0.002190
38	0.156605	0.109239	0.076457	0.053690	0.037826	0.026735	0.013481	0.004937	0.001856
39	0.149148	0.103056	0.071455	0.049713	0.034703	0.024304	0.012036	0.004293	0.001573
40	0.142046	0.097222	0.066780	0.046031	0.031838	0.022095	0.010747	0.003733	0.001333
41	0.135282	0.091719	0.062412	0.042621	0.029209	0.020086	0.009595	0.003246	0.001129
42	0.128840	0.086527	0.058329	0.039464	0.026797	0.018260	0.008567	0.002823	0.000957
43	0.122704	0.081630	0.054513	0.036541	0.024584	0.016600	0.007649	0.002455	0.000811
44	0.116861	0.077009	0.050946	0.033834	0.022555	0.015091	0.006830	0.002134	0.000687
45	0.111297	0.072650	0.047613	0.031328	0.020692	0.013719	0.006098	0.001856	0.000583
46	0.105997	0.068538	0.044499	0.029007	0.018984	0.012472	0.005445	0.001614	0.000494
47	0.100949	0.064658	0.041587	0.026859	0.017416	0.011338	0.004861	0.001403	0.000418
48	0.096142	0.060998	0.038867	0.024869	0.015978	0.010307	0.004340	0.001220	0.000355
49	0.091564	0.057546	0.036324	0.023027	0.014659	0.009370	0.003875	0.001061	0.000300
50	0.087204	0.054288	0.033948	0.021321	0.013449	0.008519	0.003460	0.000923	0.000255

TABLE 8
Binomial Probabilities: $b(k; n, p)$

							p					
n	k	0.05	0.1	0.2	0.3	0.4	0.5	0.6	0.7	0.8	0.9	0.95
2	0	0.902	0.810	0.640	0.490	0.360	0.250	0.160	0.090	0.040	0.010	0.002
	1	0.095	0.180	0.320	0.420	0.480	0.500	0.480	0.420	0.320	0.180	0.095
	2	0.002	0.010	0.040	0.090	0.160	0.250	0.360	0.490	0.640	0.810	0.902
3	0	0.857	0.729	0.512	0.343	0.216	0.125	0.064	0.027	0.008	0.001	
	1	0.135	0.243	0.384	0.441	0.432	0.375	0.288	0.189	0.096	0.027	0.007
	2	0.007	0.027	0.096	0.189	0.288	0.375	0.432	0.441	0.384	0.243	0.135
	3		0.001	0.008	0.027	0.064	0.125	0.216	0.343	0.512	0.729	0.857
4	0	0.815	0.656	0.410	0.240	0.130	0.062	0.026	0.008	0.002		
	1	0.171	0.292	0.410	0.412	0.346	0.250	0.154	0.076	0.026	0.004	
	2	0.014	0.049	0.154	0.265	0.346	0.375	0.346	0.265	0.154	0.049	0.014
	3		0.004	0.026	0.076	0.154	0.250	0.346	0.412	0.410	0.292	0.171
	4			0.002	0.008	0.026	0.062	0.130	0.240	0.410	0.656	0.815
5	0	0.774	0.590	0.328	0.168	0.078	0.031	0.010	0.002			
	1	0.204	0.328	0.410	0.360	0.259	0.156	0.077	0.028	0.006		
	2	0.021	0.073	0.205	0.309	0.346	0.312	0.230	0.132	0.051	0.008	0.001
	3	0.001	0.008	0.051	0.132	0.230	0.312	0.346	0.309	0.205	0.073	0.021
	4			0.006	0.028	0.077	0.156	0.259	0.360	0.410	0.328	0.204
	5				0.002	0.010	0.031	0.078	0.168	0.328	0.590	0.774
6	0	0.735	0.531	0.262	0.118	0.047	0.016	0.004	0.001			
	1	0.232	0.354	0.393	0.303	0.187	0.094	0.037	0.010	0.002		
	2	0.031	0.098	0.246	0.324	0.311	0.234	0.138	0.060	0.015	0.001	
	3	0.002	0.015	0.082	0.185	0.276	0.312	0.276	0.185	0.082	0.015	0.002
	4		0.001	0.015	0.060	0.138	0.234	0.311	0.324	0.246	0.098	0.031
	5			0.002	0.010	0.037	0.094	0.187	0.303	0.393	0.354	0.232
	6				0.001	0.004	0.016	0.047	0.188	0.262	0.531	0.735
7	0	0.698	0.478	0.210	0.082	0.028	0.008	0.002				
	1	0.257	0.372	0.367	0.247	0.131	0.055	0.017	0.004			
	2	0.041	0.124	0.275	0.318	0.261	0.164	0.077	0.025	0.004		
	3	0.004	0.023	0.115	0.227	0.290	0.273	0.194	0.097	0.029	0.003	
	4		0.003	0.029	0.097	0.194	0.273	0.290	0.227	0.115	0.023	0.004
	5			0.004	0.025	0.077	0.164	0.261	0.318	0.275	0.124	0.041

TABLE 8
Binomial Probabilities (*continued*)

n	k	0.05	0.1	0.2	0.3	0.4	0.5	0.6	0.7	0.8	0.9	0.95
7	6				0.004	0.017	0.055	0.131	0.247	0.367	0.372	0.257
	7					0.002	0.008	0.028	0.082	0.210	0.478	0.698
8	0	0.663	0.430	0.168	0.058	0.017	0.004	0.001				
	1	0.279	0.383	0.336	0.198	0.090	0.031	0.008	0.001			
	2	0.051	0.149	0.294	0.296	0.209	0.109	0.041	0.010	0.001		
	3	0.005	0.033	0.147	0.254	0.279	0.219	0.124	0.047	0.009		
	4		0.005	0.046	0.136	0.232	0.273	0.232	0.136	0.046	0.005	
	5			0.009	0.047	0.124	0.219	0.279	0.254	0.147	0.033	0.005
	6			0.001	0.010	0.041	0.109	0.209	0.296	0.294	0.149	0.051
	7				0.001	0.008	0.031	0.090	0.198	0.336	0.383	0.279
	8					0.001	0.004	0.017	0.058	0.168	0.430	0.663
9	0	0.630	0.387	0.134	0.040	0.010	0.002					
	1	0.299	0.387	0.302	0.156	0.060	0.018	0.004				
	2	0.063	0.172	0.302	0.267	0.161	0.070	0.021	0.004			
	3	0.008	0.045	0.176	0.267	0.251	0.164	0.074	0.021	0.003		
	4	0.001	0.007	0.066	0.172	0.251	0.246	0.167	0.074	0.017	0.001	
	5		0.001	0.017	0.074	0.167	0.246	0.251	0.172	0.066	0.007	0.001
	6			0.003	0.021	0.074	0.164	0.251	0.267	0.176	0.045	0.008
	7				0.004	0.021	0.070	0.161	0.267	0.302	0.172	0.063
	8					0.004	0.018	0.060	0.156	0.302	0.387	0.299
	9						0.002	0.010	0.040	0.134	0.387	0.630
10	0	0.599	0.349	0.107	0.028	0.006	0.001					
	1	0.315	0.387	0.268	0.121	0.040	0.010	0.002				
	2	0.075	0.194	0.302	0.233	0.121	0.044	0.011	0.001			
	3	0.010	0.057	0.201	0.267	0.215	0.117	0.042	0.009	0.001		
	4	0.001	0.011	0.088	0.200	0.251	0.205	0.111	0.037	0.006		
	5		0.001	0.026	0.103	0.201	0.246	0.201	0.103	0.026	0.001	
	6			0.006	0.037	0.111	0.205	0.251	0.200	0.088	0.011	0.001
	7			0.001	0.009	0.042	0.117	0.215	0.267	0.201	0.057	0.010
	8				0.001	0.011	0.044	0.121	0.233	0.302	0.194	0.075
	9					0.002	0.010	0.040	0.121	0.268	0.387	0.315
	10						0.001	0.006	0.028	0.107	0.349	0.599

TABLE 8
Binomial Probabilities (*continued*)

							p					
n	*k*	0.05	0.1	0.2	0.3	0.4	0.5	0.6	0.7	0.8	0.9	0.95
11	0	0.569	0.314	0.086	0.020	0.004						
	1	0.329	0.384	0.236	0.093	0.027	0.005	0.001				
	2	0.087	0.213	0.295	0.200	0.089	0.027	0.005	0.001			
	3	0.014	0.071	0.221	0.257	0.177	0.081	0.023	0.004			
	4	0.001	0.016	0.111	0.220	0.236	0.161	0.070	0.017	0.002		
	5		0.002	0.039	0.132	0.221	0.226	0.147	0.057	0.010		
	6			0.010	0.057	0.147	0.226	0.221	0.132	0.039	0.002	
	7			0.002	0.017	0.070	0.161	0.236	0.220	0.111	0.016	0.001
	8				0.004	0.023	0.081	0.177	0.257	0.221	0.071	0.014
	9				0.001	0.005	0.027	0.089	0.200	0.295	0.213	0.087
	10					0.001	0.005	0.027	0.093	0.236	0.384	0.329
	11							0.004	0.020	0.086	0.314	0.569
12	0	0.540	0.282	0.069	0.014	0.002						
	1	0.341	0.377	0.206	0.071	0.017	0.003					
	2	0.099	0.230	0.283	0.168	0.064	0.016	0.002				
	3	0.017	0.085	0.236	0.240	0.142	0.054	0.012	0.001			
	4	0.002	0.021	0.133	0.231	0.213	0.121	0.042	0.008	0.001		
	5		0.004	0.053	0.158	0.227	0.193	0.101	0.029	0.003		
	6			0.016	0.079	0.177	0.226	0.177	0.079	0.016		
	7			0.003	0.029	0.101	0.193	0.227	0.158	0.053	0.004	
	8			0.001	0.008	0.042	0.121	0.213	0.231	0.133	0.021	0.002
	9				0.001	0.012	0.054	0.142	0.240	0.236	0.085	0.017
	10					0.002	0.016	0.064	0.168	0.283	0.230	0.099
	11						0.003	0.017	0.071	0.206	0.377	0.341
	12							0.002	0.014	0.069	0.282	0.540
13	0	0.513	0.254	0.055	0.010	0.001						
	1	0.351	0.367	0.179	0.054	0.011	0.002					
	2	0.111	0.245	0.268	0.139	0.045	0.010	0.001				
	3	0.021	0.100	0.246	0.218	0.111	0.035	0.006	0.001			
	4	0.003	0.028	0.154	0.234	0.184	0.087	0.024	0.003			
	5		0.006	0.069	0.180	0.221	0.157	0.066	0.014	0.001		
	6		0.001	0.023	0.103	0.197	0.209	0.131	0.044	0.006		
	7			0.006	0.044	0.131	0.209	0.197	0.103	0.023	0.001	
	8			0.001	0.014	0.066	0.157	0.221	0.180	0.069	0.006	
	9				0.003	0.024	0.087	0.184	0.234	0.154	0.028	0.003

TABLE 8
Binomial Probabilities (*continued*)

n	k	0.05	0.1	0.2	0.3	0.4	0.5	0.6	0.7	0.8	0.9	0.95
13	10				0.001	0.006	0.035	0.111	0.218	0.246	0.100	0.021
	11					0.001	0.010	0.045	0.139	0.268	0.245	0.111
	12						0.002	0.011	0.054	0.179	0.367	0.351
	13							0.001	0.010	0.055	0.254	0.513
14	0	0.488	0.229	0.044	0.007	0.001						
	1	0.359	0.356	0.154	0.041	0.007	0.001					
	2	0.123	0.257	0.250	0.113	0.032	0.006	0.001				
	3	0.026	0.114	0.250	0.194	0.085	0.022	0.003				
	4	0.004	0.035	0.172	0.229	0.155	0.061	0.014	0.001			
	5		0.008	0.086	0.196	0.207	0.122	0.041	0.007			
	6		0.001	0.032	0.126	0.207	0.183	0.092	0.023	0.002		
	7			0.009	0.062	0.157	0.209	0.157	0.062	0.009		
	8			0.002	0.023	0.092	0.183	0.207	0.126	0.032	0.001	
	9				0.007	0.041	0.122	0.207	0.196	0.086	0.008	
	10				0.001	0.014	0.061	0.155	0.229	0.172	0.035	0.004
	11					0.003	0.022	0.085	0.194	0.250	0.114	0.026
	12					0.001	0.006	0.032	0.113	0.250	0.257	0.123
	13						0.001	0.007	0.041	0.154	0.356	0.359
	14							0.001	0.007	0.044	0.229	0.488
15	0	0.463	0.206	0.035	0.005							
	1	0.366	0.343	0.132	0.031	0.005						
	2	0.135	0.267	0.231	0.092	0.022	0.003					
	3	0.031	0.129	0.250	0.170	0.063	0.014	0.002				
	4	0.005	0.043	0.188	0.219	0.127	0.042	0.007	0.001			
	5	0.001	0.010	0.103	0.206	0.186	0.092	0.024	0.003			
	6		0.002	0.043	0.147	0.207	0.153	0.061	0.012	0.001		
	7			0.014	0.081	0.177	0.196	0.118	0.035	0.003		
	8			0.003	0.035	0.118	0.196	0.177	0.081	0.014		
	9			0.001	0.012	0.061	0.153	0.207	0.147	0.043	0.002	
	10				0.003	0.024	0.092	0.186	0.206	0.103	0.010	0.001
	11				0.001	0.007	0.042	0.127	0.219	0.188	0.043	0.005
	12					0.002	0.014	0.063	0.170	0.250	0.129	0.031
	13						0.003	0.022	0.092	0.231	0.267	0.135
	14							0.005	0.031	0.132	0.343	0.366
	15								0.005	0.035	0.206	0.463

Selected Answers

FINITE MATHEMATICS PRELIMINARIES

Problems I

1. 125   **3.** -1 **5.** $\dfrac{5}{2}$ **7.** Undefined   **9.** $\dfrac{-1}{16}$ **11.** $\dfrac{-1}{16}$ **13.** -1 **15.** 1024   **17.** 0.0816327

19. 1.6446318   **21.** $\dfrac{1}{25}$ **23.** 10^8 **25.** ab^4 **27.** $\dfrac{1}{(8m)}$ **29.** $\dfrac{y}{x}$ **31.** 2   **33.** $2^{22} = 4{,}194{,}304$, yes, 222

Problems II

1. 2   **3.** 5   **5.** 3   **7.** -5 **9.** 2   **11.** Units are not the same (the old apples and oranges argument)

Problems III

1. $5x\sqrt{x}$ **3.** $3\sqrt{x}$ **5.** $\dfrac{\sqrt{xy}}{xy}$ **7.** $z^3\sqrt{z}$ **9.** uv^2 **11.** $\sqrt{ab} = \sqrt{a}\sqrt{b}$ only if $a > 0$ and $b > 0$

Problems IV

1. $\ln 5 \cong 1.6094379$ **3.** $x = \dfrac{\ln 50}{5} \cong 0.7824046$ **5.** $\ln 4 \cong 1.3862944$ **7.** $\dfrac{\ln 10}{0.2} \cong 11.512925$

Problems V

1. $\dfrac{13}{3}$ **3.** $\dfrac{11}{13}$ **5.** $\dfrac{2}{5}$ **7.** -4 **9.** -1 **11.** $\dfrac{3}{4}$ **13.** -3

Problems VI

1. $(-4, \infty)$ **3.** $\left(-\infty, -\dfrac{1}{2}\right]$ **5.** $\left(-\infty, -\dfrac{5}{8}\right]$

Problems VII

1. 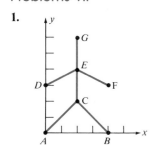   **3.** 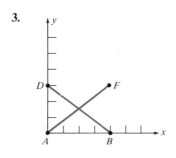   **5.**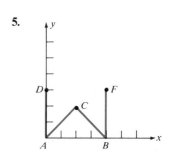

7.

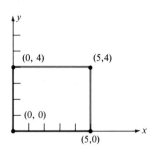

9.

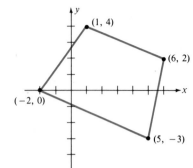

11.

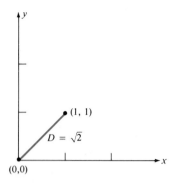

13.

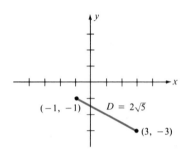

15. $D[(1, 1), (2, 3)] = \sqrt{1^2 + 2^2} = \sqrt{5}$
$D[(2, 3), (4, -1)] = \sqrt{2^2 + (-4)^2} = 2\sqrt{5}$
$D[(4, -1), (1, 1)] = \sqrt{(-3)^2 + 2^2} = \sqrt{13}$
Total distance $= 3\sqrt{5} + \sqrt{13}$

Problems VIII

1.

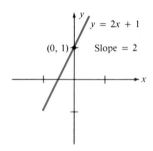

3.

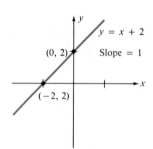

5.

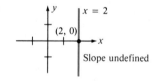

7. Slope $= 2$, y-intercept $= 0$

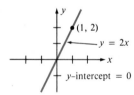

9. Slope $= 3$, y-intercept $= 5$

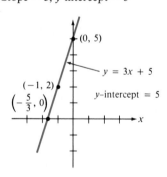

11. Slope $= 1$, y-intercept $= 0$

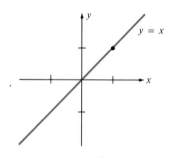

13. Slope $= -1$, y-intercept $= 0$

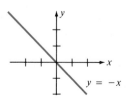

15. Slope $= -1$, y-intercept $= 4$

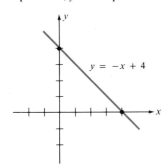

$y = -x + 4$

17.

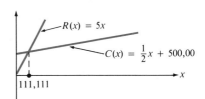

$R(x) = 5x$

$C(x) = \frac{1}{2}x + 500{,}00$

111,111

19.

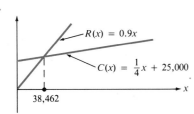

$R(x) = 0.9x$

$C(x) = \frac{1}{4}x + 25{,}000$

38,462

21.

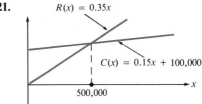

$R(x) = 0.35x$

$C(x) = 0.15x + 100{,}000$

500,000

23. (a) 50   **(b)** Fixed costs are generally larger than $200

25. (a)

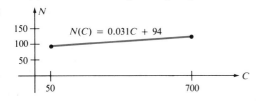

$N(C) = 0.031C + 94$

C	N
50	95.5
700	115.7

(b) 97.1   **27.** 3 min, 49.92 sec

Problems IX

1. -1 **3.** -2 **5.** $x^2 - 2$ **7.** $x^2 - 1$ **9.** $4x - 2$ **11.** 1   **13.** Undefined   **15.** $\dfrac{1}{x^2}$ **17.** $\dfrac{1 + x^2}{x^2}$

19. $\dfrac{1 + 4x^2}{2x}$ **21.** 1   **23.** 0   **25.** x^4 **27.** $x^4 + 1$ **29.** $4x^2 + 2x$

Practice Test

1. (a) 16   **(b)** $\dfrac{1}{9}$ **(c)** $\dfrac{4}{9}$ **(d)** $-\dfrac{1}{16}$ **(e)** 1   **2. (a)** 4   **(b)** 2   **(c)** Undefined   **(d)** -5 **(e)** $\dfrac{2}{3}$

3. (a) $6x$ **(b)** $3\sqrt{x}$ **4.** $\dfrac{1}{2} \ln 10$ **5.** $\dfrac{-5}{3}$ **6.** $(6, \infty)$ **7.** $\sqrt{13}$ **8.** Slope $= 3$, y-intercept $= -1$

9. $y = 2x + 1$ **10.** $y = -x + 2$ **11. (a)** $f(g(x)) = 4x^2 + 4x + 1$ **(b)** $g(f(x)) = 2x^2 + 1$

CHAPTER 1

Section 1.2

1. (a) No   **(b)** No   **(c)** Yes   **3. (a)** No   **(b)** No   **(c)** Yes

5. Solution is $(1, 0)$

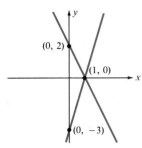

$3x - y = 3$	
x	y
0	-3
1	0

$2x + y = 2$	
x	y
0	2
1	0

7. Solution is $(1, 0)$

$2x - y = 2$	
x	y
0	-2
1	0

$4x + 2y = 4$	
x	y
0	2
1	0

9. $x = 2$, $y = 2$ **11.** $x = -\dfrac{33}{5}$, $y = \dfrac{17}{5}$ **13.** All numbers of the form $\left(\dfrac{5 + 3y}{2}, y\right)$ for y any real number are solutions.

15. No solution

17. (a) $ad - bc \neq 0$ **(b)** $a = ct$, $b = dt$, $r = st$ for some real number t **(c)** $a = ct$, $b = dt$ for some real number t, but $r \neq st$

19. If $k \neq 6$, the system has no solution.   **21.** $\begin{aligned} 5n + 10d + 25q &= 1000 \\ n + d + q &= 52 \\ n - 3q &= 0 \end{aligned}$

23. (a) $D(p) = -3p + 45$ **(b)** $S(p) = 2p$ **(c)**

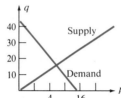

(d) $(p_0, q_0) = (9, 18)$ **(e)** Shortage $= 5$

(f) Surplus $= 15$ **25.**

$q = -2p + 7$

p	q
0	7
$\dfrac{7}{2}$	0

27. (a) \$4.50   **(b)** \$4.00   **(c)** \$5.00   **(d)** $\left(\dfrac{9}{2}, \infty\right)$

(e) $\left(0, \dfrac{9}{2}\right)$ **29.** \$35,715 in stocks, \$14,286 in bonds   **31.** Daughter 1 $= 25$, Daughter 2 $= 15$

33. Food 1 $= 18$ oz, Food 2 $= 16$ oz   **35.** $s = 1$, $t = 4$ **37.** $s = 9$, $t = 18$

Section 1.3

1. Last system will be

$$E_1:\quad x - \frac{1}{2}y = \frac{5}{2}$$
$$E_2:\quad \frac{1}{2}x + y = \frac{3}{2}$$

3. Last system will be

$$E_1:\quad x + 2y = -1$$
$$E_2:\quad -9y = 5$$

5. Last system will be

$$E_1:\quad x - \frac{1}{2}y = \frac{3}{2}$$
$$E_2:\quad -y = -4$$

7. Last system will be

$$E_1:\quad 4x + y = 1$$
$$E_2:\quad x - y - z = -1$$
$$E_3:\quad 5x + \frac{5}{2}y + z = 3$$

9. $x = \dfrac{13}{5}$, $y = -\dfrac{2}{5}$ **11.** $x = 0$, $y = \dfrac{1}{2}$ **13.** $x = \dfrac{7}{4}$, $y = -\dfrac{5}{4}$, $z = \dfrac{1}{2}$

15. $x = \dfrac{1}{11}$, $y = \dfrac{7}{11}$, $z = \dfrac{28}{11}$ **17.** $x = \dfrac{19}{3}$, $y = \dfrac{7}{3}$, $z = -\dfrac{5}{3}$

19. Growth stocks = \$71,429, Dividend-producing stocks = \$0, Tax-free bonds = \$178,571
21. No portfolio meets these requirements   **23.** Special 1 = 200   **25.** 1000 pounds of Ore 1   **27.** 10 oz of Food I
                                                  Special 2 = 100   500 pounds of Ore 2   10 oz of Food II
                                                  Special 3 = 100
29. $r = 0, s = 8, t = 1$ **31.** $r = 0, s = 2, t = -1$ (impossible)   **33.** $r = -8, s = 20, t = -14$ (impossible)
35. B: $x_1 + x_2 = 700 + 200$
       C: $x_2 + x_3 = 300 + 400$
       D: $x_3 + x_4 = 200 + 900$
37. $x_4 = 1100$ The solution with the largest x_4 is (200, 700, 0, 1100)
39. $x_1 = 900 - x_2$
       $x_3 = 700 - x_2$
       $x_4 = 400 + x_2$
       The value of x_2 must not be larger than 700. The solution with the largest x_2 is (200, 700, 0, 1100)

Section 1.4

1. $x + 3y = 4$ **3.** $x + 3y = 4$ **5.** $x = 2$ **7.** $x - \dfrac{3}{2}y + \dfrac{7}{2}z = 1$ **9.** $x + 3y + 4z + 6t = 3$
        $y = 1$ $y = 2$ $z = 1$ $2z = 1$
                                                $3z = 4$ $y - \dfrac{1}{4}z = -\dfrac{3}{2}$ $z = 2$
                                                                                                                         $t = 5$
                                                                  $z = 6$

11. $x = 3, y = 2, z = 1$ **13.** $x = -3 - t$ **15.** $x = 1, y = 1$ **17.** $x = -\dfrac{15}{39}, y = \dfrac{2}{13}$ **19.** $x = 8, y = 3, z = -6$
                                        $y = 2$
                                        $z = 3 + t$
                                        t any real number

21. $x = -\dfrac{3}{2}z, y = -\dfrac{1}{2}z$, z arbitrary, $t = 1$ **23.** $x = 1, y = 0$ **25.** No solution

27. $x = -1 + 3z - 2t, y = 2 - 2z$, z arbitrary, t arbitrary   **29.** $x = -\dfrac{7}{5} + \dfrac{1}{5}z, y = \dfrac{4}{5} - \dfrac{2}{5}z$, z arbitrary

31. $x = 3 + y$, y arbitrary   **33.** $u = -2 - 5w, v = 1 + 3w$, w arbitrary   **35.** $x = \dfrac{5}{3} - \dfrac{4}{3}z, y = \dfrac{7}{3} + \dfrac{1}{3}z$, z arbitrary

37. $a = \dfrac{8}{5} - \dfrac{4}{5}c + \dfrac{1}{5}d, b = \dfrac{9}{5} + \dfrac{3}{5}c - \dfrac{7}{5}d$, c and d arbitrary   **39.** $x = 4, y = 3$ **41.** $x = 1, y = 1, z = 1$

43. $x = -y$, y arbitrary, $z = 1$ **45.** $x = 0, y = 50, z = 0$ **47.** $x = 100, y = 200, z = 0$
49. $x = 520, y = -760, z = 280$ (no solution with these requirements)
51. generic I = 2, generic II = 3, generic III = 2   **53.** 105 tons in Eugene, 5 tons in Corvallis, 55 tons in Monroe

Section 1.5

1. (a) $2 \times 2, 2 \times 3, 2 \times 3$ **(b)** $B + C = \begin{bmatrix} 1 & 3 & 1 \\ 4 & 5 & 4 \end{bmatrix}$

(c) $3A = \begin{bmatrix} -6 & 3 \\ 6 & -3 \end{bmatrix}$, $-2B = \begin{bmatrix} -4 & -2 & -6 \\ -2 & -8 & -4 \end{bmatrix}$, $2A - B$ is not compatible, $-C = \begin{bmatrix} 1 & -2 & 2 \\ -3 & -1 & -2 \end{bmatrix}$

3. $w = \dfrac{7}{3}, x = 0, y = 4, z = 3$ **5.** $x = -2, y = -3$ **7.** $x = 0, y = -1$ **9.** Direct computation

11. (a) 2×2 **(b)** Product not defined **(c)** 3×5 **(d)** Product not defined **(e)** Product not defined

(f) 3×3 **(g)** 4×1 **(h)** 2×3 **13.** $\begin{bmatrix} 4 & 9 \end{bmatrix}$ **15.** $\begin{bmatrix} 20 & 40 & 24 \\ 2 & 5 & 2 \\ 11 & 24 & 13 \end{bmatrix}$ **17.** Both are $\begin{bmatrix} 0 & 2 \\ 20 & 18 \\ 10 & 11 \end{bmatrix}$

19. $\begin{bmatrix} 3 & -9 & 20 \\ 29 & -93 & 15 \\ -1 & -5 & -92 \end{bmatrix}$ **21.** Direct computation **23.** $\begin{bmatrix} 3 & 3 & 2 & 1 \\ 1 & -4 & 0 & -1 \\ 0 & 1 & 3 & 0 \\ 6 & 1 & 0 & 1 \end{bmatrix} \begin{bmatrix} w \\ x \\ y \\ z \end{bmatrix} = \begin{bmatrix} 1 \\ 0 \\ 2 \\ 3 \end{bmatrix}$

25. $\begin{bmatrix} 1 & 4 & 1 \\ 1 & -1 & 3 \end{bmatrix} \begin{bmatrix} u \\ v \\ w \end{bmatrix} = \begin{bmatrix} 1 \\ 0 \end{bmatrix}$ **27. (a)** 100 **(b)** $\begin{bmatrix} 7.56 & 3.24 & 15.12 & 4.32 \end{bmatrix}$ **(c)** $\begin{bmatrix} 8.16 & 3.50 & 16.33 & 4.67 \end{bmatrix}$
(d) Total assets year 1: 30.24 **(e)** Year 2: 32.66

29. $\begin{bmatrix} 3{,}050 \\ 2{,}125 \\ 2{,}150 \\ 1{,}300 \\ 10{,}700 \end{bmatrix}$ **31. (a)** 3000 **(b)** 105,000 **33.** 90 cats, 136 dogs, 46 mice

35. (a) 2700 **(b)** 450 **(c)** 375 **(d)** The total nutritional value contributed by Grains A and B to each mix

37. $\begin{bmatrix} 0.16 & 0.08 & 0.14 & 0.15 \\ 0.17 & 0.07 & 0.14 & 0.10 \\ 0.08 & 0.03 & 0.07 & 0.05 \\ 0.04 & 0.02 & 0.04 & 0.05 \\ 0.03 & 0.00 & 0.03 & 0.00 \end{bmatrix}$ **39.** $\begin{bmatrix} 18{,}840 \\ 15{,}840 \\ 17{,}640 \end{bmatrix}$ **41.** $\begin{bmatrix} 20{,}185 \\ 18{,}408 \\ 18{,}279 \end{bmatrix}$ **43.** $\begin{bmatrix} b_1 & b_2 & b_3 & b_4 \\ s_1 & 0 & 0 & 0 \\ 0 & s_2 & 0 & 0 \\ 0 & 0 & s_3 & 0 \end{bmatrix}$

45. 2, Mary–Peter–Carol, Mary–Ann–Carol

Section 1.6

1. $\begin{bmatrix} -2 & 1 \\ \frac{3}{2} & -\frac{1}{2} \end{bmatrix}$ **3.** $\begin{bmatrix} -2 & \frac{3}{2} \\ 1 & -\frac{1}{2} \end{bmatrix}$ **5.** $\begin{bmatrix} 5 & -3 \\ -3 & 2 \end{bmatrix}$ **7.** $\begin{bmatrix} \frac{7}{2} & -3 & \frac{1}{2} \\ -\frac{1}{2} & 0 & \frac{1}{2} \\ -\frac{1}{2} & 1 & -\frac{1}{2} \end{bmatrix}$ **9.** $\begin{bmatrix} 1 & 2 & -3 \\ 2 & 0 & -1 \\ -3 & -1 & 3 \end{bmatrix}$

11. $\begin{bmatrix} \frac{1}{4} & -\frac{5}{4} & \frac{3}{4} \\ \frac{1}{2} & -\frac{3}{2} & \frac{1}{2} \\ -2 & 9 & -4 \end{bmatrix}$ **13.** $\begin{bmatrix} -1 & 1 & 0 \\ 6 & -2 & -3 \\ 1 & 0 & 1 \end{bmatrix}$ **15.** $\begin{bmatrix} 1 & -2 & 3 \\ 0 & 1 & -3 \\ 0 & 0 & 1 \end{bmatrix}$ **17.** $\begin{bmatrix} -2 & 1 \\ 3 & -1 \end{bmatrix}$ **19.** $\begin{bmatrix} \frac{3}{8} & -\frac{1}{4} \\ -\frac{1}{4} & \frac{1}{2} \end{bmatrix}$

21. $\begin{bmatrix} 0 & 1 \\ \frac{1}{5} & -1 \end{bmatrix}$ **23.** $x = \frac{15}{39}, y = -\frac{18}{39}$ **25.** $x = 2, y = 0$ **27.** $x = 19, y = -4, z = -3$ **29.** $x = -9, y = 3, z = 2$

31. 30,000 lb high-grade ore, 40,000 lb low-grade ore

33. 37–116–22–52–26–62–29–62–6–14–47–148–29–88–38–112–32–118–5–12–18–46 **35.** $\frac{1}{2}\begin{bmatrix} 4 & -1 \\ -2 & 1 \end{bmatrix}$ HI

37. 111–93–97–51–118–64–25–15–215–175–169–137–92–56–109–67
39. 43–36–9–62–35–15–42–15–1–37–28–1–54–41–14–29–21–1–19–18–5–32–29–20–73–46–19

Section 1.7

1.

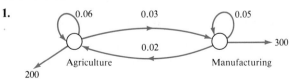

3. $\begin{bmatrix} 0.06 & 0.03 \\ 0.02 & 0.05 \end{bmatrix}$ **5.** $\begin{bmatrix} 1.064545 & 0.0336172 \\ 0.022411 & 1.0533393 \end{bmatrix}\begin{bmatrix} 200 \\ 300 \end{bmatrix} = \begin{bmatrix} 222.99416 \\ 320.4839 \end{bmatrix}$

7. To produce \$1 of output by Sector 1 requires \$0.05 worth of output from Sector 1 and \$0.04 worth of output from Sector 2. To produce \$1 worth of output by Sector 2 requires \$0.06 worth of output from Sector 1 and \$0.08 worth of output from Sector 2.

9. $\begin{bmatrix} \$114.73 \\ \$224.87 \end{bmatrix}$ **11.** $\begin{bmatrix} \$169.80 \\ \$282.81 \end{bmatrix}$ **13.** $\begin{bmatrix} \frac{2}{3} & -\frac{1}{2} & 0 \\ -\frac{1}{3} & \frac{2}{3} & -\frac{1}{6} \\ -\frac{1}{4} & 0 & \frac{5}{6} \end{bmatrix}$ **15.** $\begin{bmatrix} \$1450 \\ \$1405 \\ \$1030 \end{bmatrix}$ **17.** $\begin{bmatrix} \$66.67 \\ \$175.00 \\ \$450.00 \end{bmatrix}$

19. $\begin{bmatrix} 40.6 \\ 232.6 \\ 181.9 \end{bmatrix}\begin{bmatrix} 102.05 \\ 290.30 \\ 231.95 \end{bmatrix}\begin{bmatrix} 200.70 \\ 671.20 \\ 505.40 \end{bmatrix}$ **21.** \$0.13   **23.**

	Computer	Chip	Drive
Computer	0	0	0
Chip	25	0	0
Drive	2	0	0

25. 500 computers, 15,000 chips, 5000 disk drives   **27.** $\begin{bmatrix} 1 & 0 & 0 \\ -2 & 1 & 0 \\ 0 & -50 & 1 \end{bmatrix}$ **29.** 50 bicycles, 175 wheels, 9750 spokes

31.

Control system   Component

Chapter Review

1. (a) No **(b)** Yes **(c)** No   **3.** $x = \frac{7}{9}, y = \frac{2}{3}$ **5.** $x = \frac{3}{5}, y = \frac{4}{5}$ **7.** $x = -\frac{1}{7}, y = \frac{12}{7}$

9. $x = 1 - \frac{5}{7}z, y = -\frac{1}{7}z, z$ arbitrary   **11.** $x = 3, y = 0, z = 1$ **13.** $\begin{array}{rl} 2x + 3y - z = 4 & E_1 \\ 4x - 2y + 3z = 5 & E_2 \\ -3x - 5y - z = 0 & E_3 \end{array}$

15. $\begin{array}{rl} 10x - y + 5z = 14 & E_1 \\ 4x - 2y + 3z = 5 & E_2 \\ x + y - 3z = 8 & E_3 \end{array}$ **17.** $x = \frac{10}{7}, y = -\frac{6}{7}, z = \frac{5}{7}$ **19.** $\left[\begin{array}{ccc|c} 1 & 0 & 0 & 1 \\ 0 & 1 & 0 & 3 \\ 0 & 0 & 1 & 2 \end{array}\right]\begin{array}{c} 1 \\ 2 \\ 1 \end{array}$ **21.** $\left[\begin{array}{ccc|c} 1 & -\frac{1}{2} & \frac{3}{2} & 1 \\ 0 & \frac{1}{2} & -\frac{9}{2} & -2 \\ 0 & 5 & -7 & -3 \end{array}\right]$

23. $x = -3, y = -2, z$ arbitrary   **25.** $x = 1 - z, y = $ arbitrary, z arbitrary   **27.** $x = -5, y = 4, z = -11$

29. $x = 1, y = -1, z = 1$ **31.** $x = \frac{1}{6}, y = -\frac{1}{6}$ **33.** $x = -\frac{3}{4}, y = 3, z = \frac{11}{4}$ **35.** $\begin{bmatrix} 4 & 0 & 5 \\ 4 & 2 & 4 \\ 2 & 3 & 3 \end{bmatrix}$

37. $\begin{bmatrix} -6 & -2 & -10 \end{bmatrix}$ **39.** $\begin{bmatrix} 10 & 3 & -5 \\ 8 & 3 & -4 \\ 9 & 5 & -4 \end{bmatrix}$ **41.** Undefined   **43.** $\begin{bmatrix} 16 & 14 & 39 \\ 12 & 11 & 30 \\ 11 & 11 & 29 \end{bmatrix}$ **45.** $\begin{bmatrix} 15 & 12 & 39 \\ 16 & 10 & 30 \\ 13 & 10 & 30 \end{bmatrix}$

47. $\begin{bmatrix} 1 & 2 & -3 \\ 4 & -3 & 1 \\ -1 & 2 & -4 \end{bmatrix}\begin{bmatrix} x \\ y \\ z \end{bmatrix} = \begin{bmatrix} 5 \\ 2 \\ 6 \end{bmatrix}$ **49.** $\begin{bmatrix} \$1498.75 \\ \$1226.25 \\ \$1690.00 \\ \$2105.00 \end{bmatrix}$ **51.** $\begin{bmatrix} \$787.50 \\ \$1112.50 \end{bmatrix}$ **53.** Show $AB = I$ **55.** $\begin{bmatrix} 1 & -2 \\ -1 & 3 \end{bmatrix}$

57. $\begin{bmatrix} 2 & -1 & 0 \\ \dfrac{1}{3} & 0 & -\dfrac{2}{3} \\ -\dfrac{5}{3} & 1 & \dfrac{1}{3} \end{bmatrix}$ **59.** $\begin{aligned} x &= 0 \\ y &= \dfrac{4}{3} \\ z &= 1 \end{aligned}$ **61.** $\begin{bmatrix} 369.23 \\ 507.69 \end{bmatrix}$

Chapter Test

1. $\begin{aligned} x &= 0 \\ y &= 4 \end{aligned}$ **2.** $\begin{aligned} x &= 0 \\ y &= 4 \end{aligned}$ **3.** $A = \dfrac{1}{3}\begin{bmatrix} -1 & 1 \\ 4 & -1 \end{bmatrix}\begin{bmatrix} 4 \\ 4 \end{bmatrix} = \begin{bmatrix} 0 \\ 4 \end{bmatrix}$

4. $\begin{bmatrix} 3 & 6 & 9 & | & 1 \\ 2 & 3 & 5 & | & 2 \\ 4 & 1 & 0 & | & 0 \end{bmatrix}$

5. (a) $\begin{aligned} x &= 3 \\ y \text{ arbitrary} \end{aligned}$ **(b)** $\begin{aligned} x &= 3 \\ y &= 4 \\ z &= 0 \end{aligned}$ **(c)** No solution   **(d)** $\begin{aligned} x &= 0 \\ y &= 1 \\ z \text{ arbitrary} \end{aligned}$ **6.** $D = \begin{bmatrix} 26.77 \\ 25.20 \end{bmatrix}$

$\left(\dfrac{1}{3}\right)R_1 \longrightarrow R_1$

$\begin{bmatrix} 1 & 2 & 3 & | & \dfrac{1}{3} \\ 2 & 3 & 5 & | & 2 \\ 4 & 1 & 0 & | & 0 \end{bmatrix}$

$R_2 + (-2)R_1 \longrightarrow R_2$

$\begin{bmatrix} 1 & 2 & 3 & | & \dfrac{1}{3} \\ 0 & -1 & -1 & | & \dfrac{4}{3} \\ 4 & 1 & 0 & | & 0 \end{bmatrix}$

$R_3 + (-4)R_1 \longrightarrow R_3$

$\begin{bmatrix} 1 & 2 & 3 & | & \dfrac{1}{3} \\ 0 & -1 & -1 & | & \dfrac{4}{3} \\ 0 & -7 & -12 & | & -\dfrac{4}{3} \end{bmatrix}$

CHAPTER 2

Section 2.1

1.

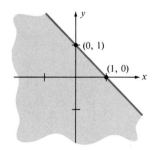

3.

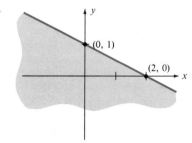

5.

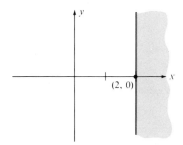

7.

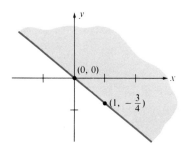

9.

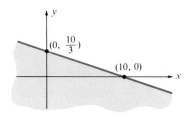

11.

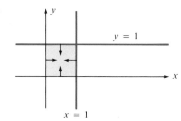

13.

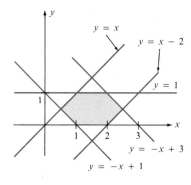

15.

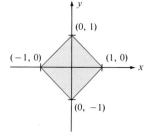

17.

19.

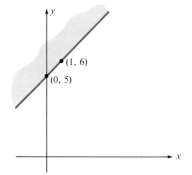

21. $(0, 0)$ satisfies $x + y \le 1$ **23.** $\left(\dfrac{1}{2}, \dfrac{1}{2}\right)$ satisfies $2x + 2y \le 2$

25.

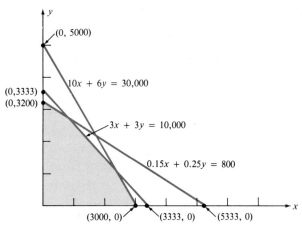

27.

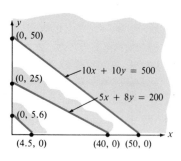

29. $8P + 4F + 2E \leq 5000$
$4P + 4F + 8E \leq 2000$
$2P + 4F + 2E \leq 3000$
$2P + 4F + 4E \leq 1000$

31.

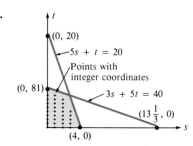

33. $15b \qquad \geq 25$
$2b + 5t \geq 15$
$\qquad 5t \geq 10$
$b \geq 0$
$t \geq 0$

35. $10f \qquad \geq 15$
$7t \geq 7$
$5t \geq 10$
$f \geq 0$
$t \geq 0$

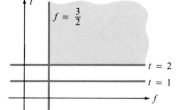

37. $12m + 7v \qquad \geq 30$
$8m + 2v + 7t \geq 15$
$10m + \qquad 5t \geq 20$
$m \geq 0$
$v \geq 0$
$t \geq 0$

39. $10f \qquad \geq 20$
$7t \qquad \geq 15$
$5t + 5a \geq 10$
$f \geq 0$
$t \geq 0$
$a \geq 0$

Section 2.2

1.

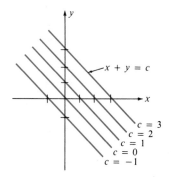

3.

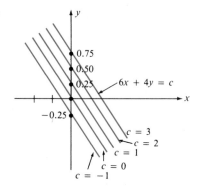

5.

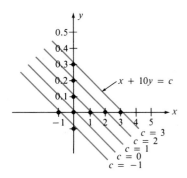

7.

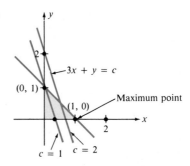

9.

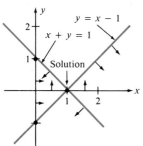

11.

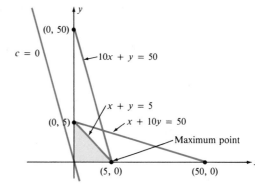

13.

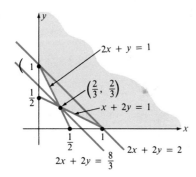

15.

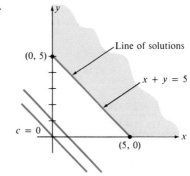

17. Optimal point = (1, 0), maximum value = 3

19. Optimal point = (1, 0), maximum value = 3

21. Optimal point = (5, 0), maximum value = 10

23. Optimal point = $\left(\frac{2}{3}, \frac{2}{3}\right)$, minimum value = $\frac{8}{3}$

25. Optimal point = (5, 0), minimum value = 35

27. yes   **29.** empty feasible set

31. Yes

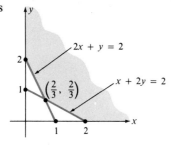

33. $333,334 in AAA bonds
    $166,666 in AA bonds

35. (a) 1 can of Brand A and 2.5 cans of Brand B   **(b)** $1.50

37. (a) 600 units of Product A and 0 units of Product B   **(b)** $6000

39. $\frac{5}{3}$ bread exchanges and 2 meat exchanges

41. 1.5 fruit exchanges and 2 meat exchanges

Section 2.3

1. (a)

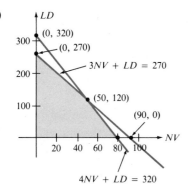

3NV + LD = 270

4NV + LD = 320

(b) $50NV$ and $120LD$ **(c)** 590   **(d)** Yes   **(e)** 30

3. (a) Max

$$P = 0.4x + 0.55y$$

(b) $x = 12,000$ **(c)** Nothing is left over
$y = 4,000$

subject to

$$0.35x + 0.2y \le 5000$$
$$0.10x + 0.2y \le 2000$$
$$0.05x + 0.1y \le 1000$$
$$x \ge 0$$
$$y \ge 0$$

(d) The optimal point will not change, but the profit will increase by \$1200.

5. (a) $\dfrac{80}{3}$ ounces of Food 2   **(b)** \$5.33   **(c)** Carbohydrates

7. (a) Min

$$C = 175x_{11} + 100x_{12} + 125x_{13} + 200x_{21} + 150x_{22} + 100x_{23}$$

subject to

$$x_{11} + x_{12} + x_{13} = 75,000$$
$$x_{21} + x_{22} + x_{23} = 25,000$$
$$x_{11} + x_{21} = 40,000$$
$$x_{12} + x_{22} = 30,000$$
$$x_{13} + x_{23} = 30,000$$
$$x_{11} \ge 0, \qquad x_{12} \ge 0, \qquad x_{13} \ge 0$$
$$x_{21} \ge 0, \qquad x_{22} \ge 0, \qquad x_{23} \ge 0$$

(b) 6   **(c)** 11

(d) All of the functional constraints are equality constraints.

9. (a) 142 tables and 0 chairs (rounded)   **(b)** \$9940.00   **(c)** There is a surplus of about 90 hours of labor.

11. (a) Max

$$C = 10,000x + 20,000y$$

subject to

$$5x + 15y \ge 100$$
$$50x + 75y \ge 400$$
$$500x + 500y \ge 1000$$
$$x \ge 0$$
$$y \ge 0$$

(b) Operate Mine 2 for 6.66 days and leave Mine 1 idle. The daily cost will be \$133,200.

13. The maximum reward is \$79. This is achieved by putting 11 squares and 7 triangles in the pot.

Section 2.4

1.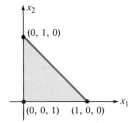

3.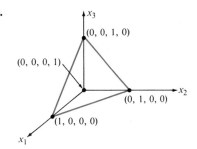

5.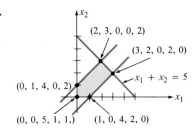

7.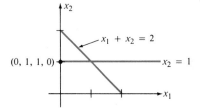

9.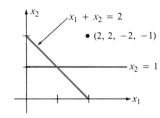

11.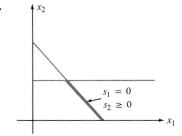

13. $A = (0, 0, 3, 2)$ **15.** $A = (0, 0, 2, 3)$
$B = (0, 3, 0, 2)$ $B = (0, 2, 0, 1)$
$C = (2, 1, 0, 0)$ $C = \left(\dfrac{1}{2}, \dfrac{5}{2}, 0, 0\right)$
$D = (2, 0, 1, 0)$
                        $D = (0, 3, -1, 0)$

17. **(a)** $x_1 + 2x_2 + s_1 = 2$
       $2x_1 + x_2 + s_2 = 2$

(b)

BV	x_1	x_2	s_1	s_2	RHS
s_1	1	2	1	0	2
s_2	2	1	0	1	2

(c)

BV	x_1	x_2	s_1	s_2	RHS
x_1	1	2	1	0	2
s_2	0	-3	-2	1	-2

(d) $(x_1, x_2, s_1, s_2) = (2, 0, 0, -2)$ **(e)**

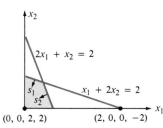

19. (a)
$$x_1 + x_2 + s_1 \qquad = 2$$
$$\qquad x_2 \qquad + s_2 = 1$$

(b)

BV	x_1	x_2	s_1	s_2	RHS
s_1	1	1	1	0	2
s_2	0	1	0	1	1

(c)

BV	x_1	x_2	s_1	s_2	RHS
x_1	1	1	1	0	2
s_2	0	1	0	1	1

(d) $(x_1, x_2, s_1, s_2) = (2, 0, 0, 1)$ **(e)**

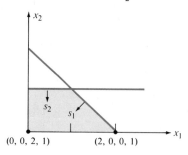

$(0, 0, 2, 1)$ $(2, 0, 0, 1)$

21. $A = (P, P, P, P)$
$B = (P, P, N, P)$
$C = (P, P, P, N)$
$D = (P, P, N, N)$
$E = (P, N, P, P)$
$F = (P, N, P, N)$
$G = (P, N, N, N)$

23. (a) s_1, s_2 basic; x_1, x_2 nonbasic   **(b)**

BV	x_1	x_2	s_1	s_2	RHS
x_2	1	1	1	0	2
s_2	-2	0	-3	1	-5

(c) $(x_1, x_2, s_1, s_2) = (0, 2, 0, -5)$

25. (a) s_1, s_2 basic; x_1, x_2 nonbasic   **(b)**

BV	x_1	x_2	s_1	s_2	RHS
x_2	0	1	1	0	2
s_2	1	0	-1	1	2

(c) $(x_1, x_2, s_1, s_2) = (0, 2, 0, 2)$

27.
$$x_1 + x_2 + s_1 \qquad = 2$$
$$x_1 + 3x_2 \qquad + s_2 = 1$$

29.
$$x_2 + s_1 \qquad = 2$$
$$x_1 + x_2 \qquad + s_2 = 4$$

31. (a)

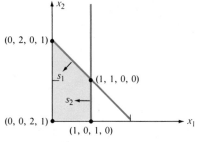

$(0, 2, 0, 1)$

s_1

$(1, 1, 0, 0)$

s_2

$(0, 0, 2, 1)$

$(1, 0, 1, 0)$

33. (a) s_1, s_2, s_3 basic; x_1, x_2 nonbasic   **(b)** $(0, 0)$ **(c)** 3

(b) Max
$$P = 5x_1 + 4x_2$$
subject to
$$x_1 + x_2 + s_1 \qquad = 2$$
$$x_1 \qquad + s_2 = 1$$
$$x_1 \geq 0, \qquad x_2 \geq 0$$
$$s_1 \geq 0, \qquad s_2 \geq 0$$

35. (a) x_1, x_2, s_3 basic; s_1, s_2 nonbasic   **(b)** $(2, 1)$ **(c)** 3   **37. (a)** x_1, s_1, s_2 basic; x_2, s_3 nonbasic   **(b)** $(1, 0)$ **(c)** 3

39. Points on the line $x_1 + 2x_2 = 2$ for $0 \leq x_1 \leq \dfrac{2}{3}$ **41.** $\{(x_1, x_2): 0 \leq x_1 \leq 1 \text{ and } x_2 = 0\}$

Section 2.5

1. Optimal feasible solution $= (0, 2)$
Optimal value $= 2$

3. Optimal feasible solution $= \left(\dfrac{5}{2}, \dfrac{3}{2}\right)$

Optimal value $= \dfrac{13}{2}$

5. Optimal feasible solution $= (2, 0, 0)$
Optimal value $= 4$

7. Optimal feasible point $= (2, 0)$
Optimal value $= 8$

9. Optimal feasible point $= \left(0, \dfrac{5}{3}, \dfrac{5}{3}\right)$

Optimal value $= \dfrac{25}{3}$

11. Optimal feasible point $= (1, 0, 0)$
Optimal value $= 1$

13. (a) Max   **(b)** Yes, the initial basic variables are s_1, s_2, and s_3.   **(c)** s_1, s_2, s_3 **(d)** x_1, x_2

$$P = 3x_1 + 2x_2$$
subject to
$$
\begin{aligned}
x_1 + 2x_2 + s_1 \qquad\qquad &= 2 \\
2x_1 + \ x_2 \qquad + s_2 \qquad &= 2 \\
2x_1 + 2x_2 \qquad\qquad + s_3 &= 3 \\
x_1 \ge 0, \qquad x_2 \ge 0 \\
s_1 \ge 0, \qquad s_2 \ge 0, \qquad s_3 \ge 0
\end{aligned}
$$

(e) 3   **(f)** 2   **(g)** $(0, 0, 2, 2, 3)$; yes, since all variables are greater than or equal to zero.   **(h)** x_1 **(i)** 2   **(j)** 2   **(k)** No
(l) The next tableau is not optimal; it is

BV	ROW	P	x_1	x_2	s_1	s_2	s_3	RHS
*	0	1	0	$-\dfrac{1}{2}$	0	$\dfrac{3}{2}$	0	3
s_1	1	0	0	$\dfrac{3}{2}$	1	$-\dfrac{1}{2}$	0	1
x_1	2	0	1	$\dfrac{1}{2}$	0	$\dfrac{1}{2}$	0	1
s_3	3	0	0	1	0	-1	1	1

15. (a) Max   **(b)** s_1, s_2, x_1 **(c)** x_2, s_3 **(d)** No   **(e)** x_2 **(f)** 2   **(g)** 7.33

$$P = 5x_2 - 2s_3 + 2$$
subject to
$$
\begin{aligned}
5.67x_2 + s_1 \qquad\qquad - 1.33s_3 &= 4.67 \\
7.33x_2 \qquad + s_2 - 2.67s_3 &= 5.33 \\
x_1 - 1.33x_2 \qquad\quad + 0.67s_3 &= 0.67 \\
x_1 \ge 0, \qquad x_2 \ge 0 \\
s_1 \ge 0, \qquad s_2 \ge 0, \qquad s_3 \ge 0
\end{aligned}
$$

(h)

BV	ROW	P	x_1	x_2	s_1	s_2	s_3	RHS
*	0	1	0	0	0	$\dfrac{15}{22}$	$\dfrac{2}{11}$	$\dfrac{62}{11}$
s_1	1	0	0	0	1	$-\dfrac{17}{22}$	$\dfrac{8}{11}$	$\dfrac{6}{11}$
x_2	2	0	0	1	0	$\dfrac{3}{22}$	$-\dfrac{4}{11}$	$\dfrac{8}{11}$
x_1	3	0	1	0	0	$\dfrac{2}{11}$	$\dfrac{2}{11}$	$\dfrac{18}{11}$

(i) $\left(\dfrac{18}{11}, \dfrac{8}{11}, \dfrac{6}{11}, 0, 0\right)$ **(j)** $\dfrac{62}{11}$; the objective has been increased by $\dfrac{62}{11} - 2$.

Section 2.6

1. Min
$$C = 2y_1 + 2y_2$$
subject to
$$2y_1 + y_2 \geq 1$$
$$y_1 + 2y_2 \geq 1$$
$$y_1 \geq 0, \quad y_2 \geq 0$$

3. Min
$$C = 2y_1 + 4y_2 + y_3$$
subject to
$$2y_1 + y_2 - 3y_3 \geq 4$$
$$y_1 - y_2 - y_3 \geq 5$$
$$y_1 \geq 0, \quad y_2 \geq 0, \quad y_3 \geq 0$$

5. Min
$$C = 5y_1$$
subject to
$$6y_1 \geq 1$$
$$7y_1 \geq 5$$
$$-3y_1 \geq 9$$
$$y_1 \geq 0$$

7. Max
$$P = y_1$$
subject to
$$y_1 \leq 1$$
$$4y_1 \leq 4$$
$$3y_1 \leq 0$$
$$y_1 \geq 0, \quad y_2 \geq 0, \quad y_3 \geq 0$$

9. Min
$$C = 5y_1 + 5y_2$$
subject to
$$-y_1 + y_2 \geq 7$$
$$y_1 + y_2 \geq 3$$
$$y_1 \geq 0, \quad y_2 \geq 0$$

11. The second resource

13. (a) $(2, 2)$ **(b)** x_1, x_2 **(c)** s_1, s_2 **(d)** $\left(\frac{2}{5}, \frac{3}{5}\right)$ **(e)** $\frac{2}{5}$ of a unit **(f)** $\frac{3}{5}$ of a unit

15. (a) 75 chairs, 0 tables, 0 cabinets **(b)** \$375 **(c)** Nothing **(d)** Nothing **(e)** \$2.50 **(f)** A gain of \$25 profit

(g) Nothing **17.** $(x_1, x_2, x_3) = \left(\frac{1}{2}, \frac{1}{2}, 0\right)$ **19.** $(x_1, x_2, x_3, x_4) = \left(0, 0, 0, \frac{1}{4}\right)$

Minimum value $= \dfrac{13}{2}$ Minimum value $= \dfrac{1}{4}$

21. Optimal feasible point $= \left(\dfrac{52}{15}, \dfrac{49}{15}\right)$ **23.** Optimal feasible point $= (10, 0)$

Optimal value $= \dfrac{101}{15}$ Optimal value $= 10$

Optimal value increases by 1/15   Optimal value increases by 3.33

25. It is useful to know which resources affect the profit the most so that these resources will always be available.

Chapter Review

1.

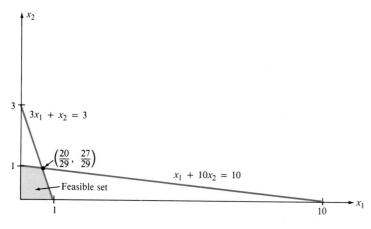

$3x_1 + x_2 = 3$

$\left(\dfrac{20}{29}, \dfrac{27}{29}\right)$

$x_1 + 10x_2 = 10$

Feasible set

3.

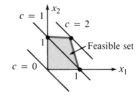

$c = 1$ $c = 2$

Feasible set

$c = 0$

5. Max
$$P = x_1 + x_2$$
subject to
$$x_1 + 10x_2 + s_1 \quad = 10$$
$$3x_1 + \quad x_2 \quad + s_2 = 3$$
$$x_1 \geq 0, \quad x_2 \geq 0, \quad s_1 \geq 0, \quad s_2 \geq 0$$

7.

BV	ROW	P	x_1	x_2	s_1	s_2	RHS
*	0	1	−1	−1	0	0	0
s_1	1	0	1	10	1	0	10
s_2	2	0	3	1	0	1	3

9.

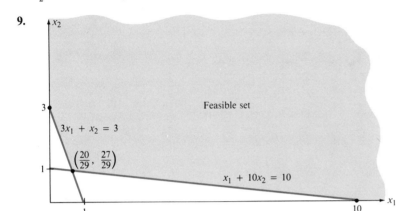

11. Optimal feasible solution $= \left(\dfrac{20}{29}, \dfrac{27}{29}\right)$

13. Max
$$P = 10y_1 + 3y_2$$
subject to
$$y_1 + 3y_2 \geq 1$$
$$10y_1 + \quad y_2 \geq 1$$
$$y_1 \geq 0$$
$$y_2 \geq 0$$

15. Min
$$C = 2y_1 + 2y_2$$
subject to
$$2y_1 + \quad y_2 \geq 3 \quad \text{(dual problem)}$$
$$y_1 + 2y_2 \geq 4$$
$$y_1 \geq 0$$
$$y_2 \geq 0$$

Optimal solution of dual $= \left(\dfrac{2}{3}, \dfrac{5}{3}\right)$

Optimal solution of primal $= \left(\dfrac{2}{3}, \dfrac{2}{3}\right)$

Optimal value of primal and dual $= \dfrac{14}{3}$

17. $\dfrac{14}{3}$

19. It increases by $\left(\dfrac{2}{3}\right)3 = 2.$

21. The optimal point becomes $(1, 0)$ and the optimal value 10.

23. 50 gm   **25.** 20 gm   **27.** Only enough resources to make five bracelets at the most   **29.** $70

31. $2 for each pound of chemical I and $3 for each pound of chemical II

33. 3 ounces   **35.** 2 ounces   **37.** 0 ounces   **39.** 15 ounces

Chapter Test

1.

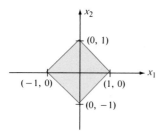

2.

BV	ROW	P	x_1	x_2	x_3	x_4	s_1	RHS
*	0	1	-3	-5	-7	-2	0	0
s_1	1	0	1	1	-1	2	1	4

3. Optimal feasible point of dual problem $= \left(\dfrac{1}{3}, \dfrac{1}{3}\right)$

Maximum value of primal problem $= \dfrac{2}{3}$

Optimal feasible point of dual problem $= \left(\dfrac{1}{3}, \dfrac{1}{3}\right)$

Minimum value of dual problem $= \dfrac{2}{3}$

4. No   **5.** $(2, 0, 0, 6)$ **6.** x_1, s_2 basic; x_2, s_1 nonbasic   **7.** 6   **8.** 1   **9.** x_2 **10.** $\dfrac{1}{2}$

11.

BV	ROW	P	x_1	x_2	s_1	s_2	RHS
*	0	1	2	0	$\dfrac{4}{3}$	0	10
x_2	1	0	2	1	$\dfrac{2}{3}$	0	4
s_2	2	0	$-\dfrac{4}{3}$	0	$-\dfrac{7}{36}$	1	$\dfrac{5}{4}$

CHAPTER 3

Section 3.1

1. $93.75   **3.** $1100   **5.** $12,500   **7.** $45.21   **9.** $170.14   **11.** 20%   **13.** $13\dfrac{1}{3}\%$

15. **(a)** After 2 years: $475   **(b)** After 4 years: $950   **(c)** After 5 years: $1187.50   **(d)** After 10 years: $2375
(e) After 10.52 years it will be worth $5000.   **17.** 12.5 years to double, 25 years to triple   **19.** $75
21. Discount = $30   **23.** Discount = $750
  Proceeds = $470   Proceeds = $1750
  Effective rate of interest = 8.51%   Effective rate of interest = 17.14%

25. (a) $15   **(b)** $485   **(c)** $500   **(d)** 6.18%   **27.** Proceeds = $1800
                                                                   Maturity value = $2000
                                                                   Effective rate of interest = 11.11%

Section 3.2

1. Future value after 2 years = $11,698.59   **3.** Future value after 2 years = $5940.50
Interest earned after 2 years = $1698.59   Interest earned after 2 years = $940.50
Future value after 5 years = $14,802.44   Future value after 5 years = $7693.12
Interest earned after 5 years = $4802.44   Interest earned after 5 years = $2693.19
5. Future value after 2 years = $11,051.71   **7.** 69.66 months or 5.8 years   **9.** 3.736 six-month periods or 1.868 years
Interest earned after 2 years = $1051.71
Future value after 5 years = $12,840.25
Interest earned after 5 years = $2840.25
11. 34.52 years   **13.** $2349.23   **15.** $30,420.67   **17.** $117.59   **19.** Better 9% compounded semiannually
21. Better 11.5% compounded daily   **23.** $10,000 now   **25.** $50,000 now   **27.** $2067.69   **29.** 18.3%
31. (a) 7.713%   **(b)** 7.763%   **(c)** 7.782%   **(d)** 7.787%   **(e)** 7.788%   **33.** $15,415.93
35. First quarter = $1.02   **37.** First year = $1.08   **39.** 117.19 million bottles
Second quarter = $1.0404   Second year = $1.1664
Third quarter = $1.061208   Third year = $1.259712
Fourth quarter = $1.0824322   Fourth year = $1.360489
                                         Fifth year = $1.469328

Section 3.3

1. $40,568.08   **3.** $144,865.63   **5.** $3052.55   **7.** $362,164.06   **9.** $78,058.08   **11.** $43,076.88   **13.** $14,714.28
15. $596.28   **17.** $1581.05   **19.** $304.79   **21.** $1003.55
23. Future value of sinking fund = $10,288.18 They will not be able to afford a house selling for more than $51,440.90
25. $1677.52   **27.** $94,569.63   **29.** $104,550.78   **31.** $8175.72   **33.** $10,678.69
35. Present value = $341.24%, total value = $391.24   **37.** Annual payments for 6.5% compounded monthly are $4888.39
                                                            Annual payments for 7% compounded quarterly are $4656.74
                                                            Better deal is 7% compounded quarterly

39. Better to make payments

Section 3.4

1. $1769.84   **3.** $396.53   **5.** $5009.13   **7. (a)** $125.47   **(b)** $26.01   **(c)** $148.99
9. (a) 0   **(b)** $245.10   **(c)** $4.90   **11. (a)** $16,859.21   **(b)** $390.02   **(c)** $344.98
13. 16.47 payments   **15.** 14.2 payments   **17.** 21.54 payments   **19.** $164.76   **21.** $1113.27
23. $29.06 for 30 payments, $19.02 for 48 payments
25. Payments = $924.68   **27.** Payments = $809.75   **29.** Payments = $681.53
Owed = $4116.51   Owed = $79,836.34   Payment against loan on first payment = $56.53
Paid = $3383.49   Paid = $10,163.06   Payment to interest on first payment = $625.00
31. 51.33 months   **33. (a)** $187   **(b)** 1000   **(c)** Interest is paid on the balance of the loan and not the total loan amount.

Chapter Review

1. $85.33   **3.** 7.22%   **5.** $10,500   **7.** $4850   **9.** $9500   **11.** $700.93
13. $46,319.35 should be invested, $53,680.65 earned   **15.** 5.64%; interest or 10.5% if it takes 11 years to triple in growth.
17. $13,754.45   **19.** 19.56%   **21.** $6954.07   **23.** $34,379.31   **25.** 50.64 months   **27.** Use computer program in text
29. Monthly payments = $155.69   **31.** Balance due = $78,874.73
Amount due = $3407.87
33. The payments are so small that the interest accumulates faster than the principle decreases.

Chapter Test

1. $83.33   **2.** 0.5714 years   **3.** $2819.72   **4.** 9.38%   **5.** $9818.15   **6.** $27,441.91   **7.** $526.61

CHAPTER 4

Section 4.1

1. {2, 4, 6, 8, 10, 12, 14, 18, 20}

3. ϕ, {1}, {2}, {3}, {4}, {1, 2}, {1, 3}, {1, 4}, {2, 3}, {2, 4}, {3, 4}, {1, 2, 3}, {1, 2, 4}, {1, 3, 4}, {2, 3, 4}, {1, 2, 3, 4}

5. {(x, y):(x, y) are both integers}   **7.** $\left\{ \begin{bmatrix} a & b & c \\ d & e & f \\ g & h & i \end{bmatrix} : a = e = i = 1, \text{ all other entries integers} \right\}$

9.

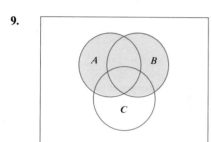

$A \cup B$

11.

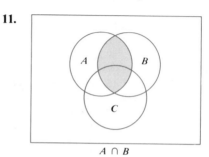

$A \cap B$

13.

$A' \cap B' \cap C'$

15.

$A \cup (A \cap C)'$

17.

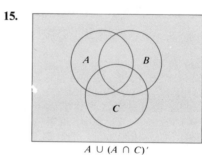

$A \cap (B \cup C)$

19.

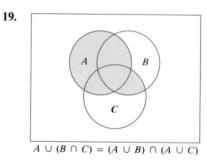

$A \cup (B \cap C) = (A \cup B) \cap (A \cup C)$

21. F   **23.** T   **25.** T   **27.** T   **29.** T   **31.** F   **33.** T   **35.** F   **37.** T   **39.** T   **41.** F   **43.** F   **45.** F

Section 4.2

1. (a) 2500   **(b)** 1000   **(c)** 5500   **(d)** 29986   **(e)** 3500

3.

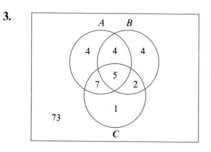

5. $2 \cdot 26 \cdot 26 \cdot 27 = 36{,}504$ **7.** 6 (any one face can be down)
9. For the reader   **11.** $3 \cdot 5 = 15$
13. $10^5 = 100{,}000$ Zip Codes   **15.** $36^5 = 60{,}466{,}176$ (no)
17.

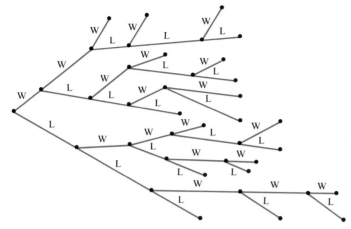

19. $6 \cdot 2 = 12$ **21.** $26 \cdot 10^3 = 26{,}000$

Section 4.3

1. $P(5, 3) = 60$ **3.** $P(3, 3) = 6$ **5.** $C(4, 1) = 4$ **7.** $C(10, 8) = 45$ **9.** $C(9, 2) = 36$
11. (a) $10{,}000$ **(b)** $n(n + 1)!$ **(c)** n **13.** $P(10, 3) = 720$ **15.**

17. $P(4, 4) = 24$ **19.** $P(2, 2) = 2$ **21.** 3   **23.** 60   **25.** 180   **27.** 1260   **29.** 3360   **31.** 3780
33. There are 16 people in the picture; hence $P(16, 3) = 3360$.   **35.** Poker hands: $C(52, 5) = 2{,}598{,}960$
37. Straight flush: $10 \cdot 4 = 40$ **39.** Full house: $C(4, 3)C(4, 2)C(13, 1)C(12, 1) = 3744$
41. Straight: $C(10, 1)C(4, 1)^5 - C(10, 1)C(4, 1) = 10{,}200$ **43.** Two pairs: $C(4, 2)C(4, 2)C(4, 1)C(13, 2)C(11, 1) = 123{,}552$

45. $C(12, 5) = 792$ **47.** $\dfrac{10!}{9!1!} = 10$ **49.** 56   **51.** $\dfrac{1}{2}C(4, 2) = 3$ **53.** $\dfrac{1}{2}C(8, 4) = 35$ **55.** $C(25, 2) = 300$

57. $C(25, 2)C(23, 3)C(20, 2) = 100{,}947{,}000$ **59.** $C(25, 2)C(25, 4) = 3{,}795{,}000$ **61.** $C(25, 2)^4 = 1{,}296{,}000{,}000{,}000$
63. $C(5, 2) \cdot C(5, 2) = 100$

Chapter Review

1. {MJS, MSJ, SMJ, SJM, JSM, JMS}   **3.** $\{x : 1 \le x \le 13\}$ **5.** $\{-1\}$ **7.** $\{1, -1\}$ **9.** 15   **11.** 12
13. 20   **15.** 10   **17.** The set of women (410 of them)   **19.** People who did not like the ship very much (450 of them)
21. People who disliked or greatly disliked the ship or liked it very much (400 of them)
23. Men who disliked or greatly disliked the ship or liked the ship very much (240 of them)

25. U.S. male citizens who liked the ship very much (300 of them)

27. Women who disliked or greatly disliked the ship (10 of them)   **29.** 1/50   **31.** 3/10

Problems 33–40 can be answered from the following Venn diagram.

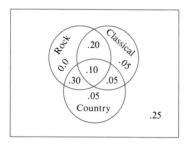

41. For the reader   **43.** $P(1000, 999) = 1000!$ **45.** $C(20, 4) = 4845$ **47.** $C(9, 1) = 9$

49. $\dfrac{(m + 3)!}{(m - 1)!} = (m + 3)(m + 2)(m + 1)m$ **51.** Same, since $C(5, 3) = C(5, 2)$

Chapter Test

1. There cannot be 15 girls who are B students and play sports if there are only 14 students who are B students and play sports.   **2.** 9

3. $\{a, b\}$
$\{a, c\}$ $\{b, c\}$
$\{a, d\}$ $\{b, d\}$ $\{c, d\}$
$\{a, e\}$ $\{b, e\}$ $\{c, e\}$ $\{d, e\}$

4.

ab	ba	ca	da	ea
ac	bc	cb	db	eb
ad	bd	cd	dc	ec
ae	be	ce	de	ed

5.

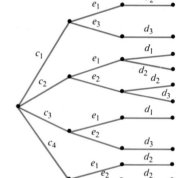

CHAPTER 5

Section 5.1

1. $\{HHH, HHT, HTH, THH, HTT, THT, TTH, TTT\}$ **3.** $\{6, 9, 12, 15\}$ **5.** $\{3, -3, 9, -9\}$

7.

Red \ Green	1	2	3	4	5	6
1	2	3	4	5	6	7
2	3	4	5	6	7	8
3	4	5	6	7	8	9
4	5	6	7	8	9	10
5	6	7	8	9	10	11
6	7	8	9	10	11	12

9.

Red \ Green	1	2	3	4	5	6
1	2	5	10	17	26	37
2	5	8	13	20	29	40
3	10	13	18	25	34	45
4	17	20	25	32	41	52
5	26	29	34	41	50	61
6	37	40	45	52	61	72

11.

Red \ Green	1	2	3	4	5	6
1	0	-1	-2	-3	-4	-5
2	1	0	-1	-2	-3	-4
3	2	1	0	-1	-2	-3
4	3	2	1	0	-1	-2
5	4	3	2	1	0	-1
6	5	4	3	2	1	0

13. 4

15. 36 (number of ways in which the lowest card can be chosen (9) times the number of different suits (4)

17. $\{(o, m, y) : o, m,$ and y can be MALE or FEMALE$\}$

19. $\{x : x$ is a real number$\}$. Negative incomes need be considered. Not able to put an exact range on positive and negative values that occur.

21. $\{x : 0 \le x \le 100\}$ **23.** $\{x : 0 \le x \le 100\}$ **25.** $\{x : x$ is a brand ice cream$\}$

27. $\{x : x$ is a real number$\}$. Weight gain and loss are possible, so positive and negative numbers must be considered.

29. $\{(H, 1), (H, 2), (H, 3), (H, 4), (H, 5), (H, 6)\}$ **31.** $\{(H, 3), (T, 3), (H, 6), (T, 6)\}$

33. $\{(1, 5), (2, 4), (3, 3), (4, 2), (5, 4), (4, 5), (6, 3), (3, 6)\}$

35. $(2, 1)$ $(2, 2)$ $(2, 3)$ $(2, 4)$ $(2, 5)$ $(2, 6)$ $(3, 1)$ $(3, 2)$ $(3, 3)$ $(3, 4)$ $(3, 5)$ $(3, 6)$

37. $\{(1, 1)\}$ **39.** $\{HHT, HTH\}$ **41.** $\{HHH\}$ **43.** $\varnothing$ **45.** $\varnothing$ **47.** $\{TTT, TTH, THT, HTT, HHH\}$

49. $\{TTT, TTH, THT\}$

Section 5.2

1. $\dfrac{1}{16}$ **3.** $\dfrac{3}{8}$ **5.** $\dfrac{1}{16}$ **7.** $\dfrac{1}{2}$ **9.** $\dfrac{1}{13}, 1:12$ **11.** $\dfrac{4}{13}, 4:9$ **13.** $\dfrac{7}{13}, 7:6$ **15.** $\dfrac{1}{216}$ **17.** $\dfrac{1}{216}$ **19.** $\dfrac{1}{108}$

21. $\dfrac{1}{2}$ **23.** 0.4   **25.** 0.8   **27.** 0.5   **29.** 0.4   **31.** 0.9   **33.** $\dfrac{1}{32}, \dfrac{5}{32}, \dfrac{10}{32}, \dfrac{10}{32}, \dfrac{5}{32}, \dfrac{1}{32}$ **35.** $\dfrac{11}{32}$ **37.** 4

39. $1 - \dfrac{48}{52}\dfrac{47}{51}\dfrac{46}{50} = 0.2520664$ **41.** $1 - \dfrac{5}{100}\dfrac{5}{100} = \dfrac{9975}{10,000} = 0.9975$ **43.** Same answer as Problem 42 (0.89394)

45. 3:5   **47.** 4:9   **49.** $\dfrac{5}{6} = 0.83333$ **53.** $\dfrac{150}{750} = \dfrac{1}{5}$ **55.** $\dfrac{250}{750} = \dfrac{1}{3}$

51.

NCAA	NBA
$\dfrac{5}{6}$	$\dfrac{7}{8}$
$\dfrac{2}{3}$	$\dfrac{2}{3}$
$\dfrac{1}{2}$	$\dfrac{5}{9}$

Section 5.3

1. $\dfrac{26}{51}$ **3.** $\dfrac{4}{51}$ **5.** 0.20   **7.** 0.25   **9.** 0.15   **11.** $\dfrac{5}{8}$

13.

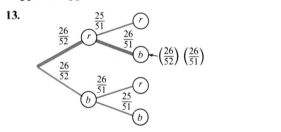

15.

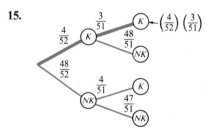

17.

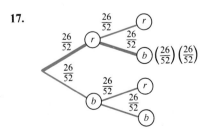

19.

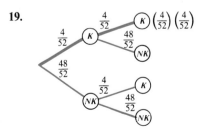

Venn diagram for Problems 21, 23:   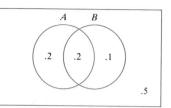   **21.** $\dfrac{2}{3}$ **23.** $\dfrac{2}{7}$ **25.** Dependent

Venn diagram for Problems 27, 29:   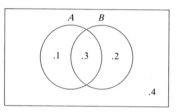   **27.** $\dfrac{3}{5}$ **29.** $\dfrac{1}{5}$ **31.** P (At least one) $= 0.99$

P (Making both) $= 0.81$

33. $\dfrac{1}{6}$ **35.** $\dfrac{1}{36}$ **37.** $\dfrac{1}{12}$ **39.** $\dfrac{7}{12}$ **41.** $\dfrac{5}{12}$ **43.** 5% of the population cheats on their taxes.   **45.** $\left(\dfrac{1}{2}\right)^8 = \dfrac{1}{256}$

47. $(0.350)(0.275)(0.300) \cong 0.029$ **49.** Left for the reader

Section 5.4

1. $\{0, 1, 2\}$ **3.** $\dfrac{5}{6}$ **5.** Continuous   **7.** Finite discrete   **9.** 0.90   **11.** \$.25 (or $-$\$.75 net gain)   **13.** 1

15. $P(1) = \dfrac{1}{36}$ $P(12) = \dfrac{4}{36}$ **17.** $12\dfrac{1}{4}$ **19.** 0.4   **21.** 2   **23.** 6   **25.** $\left(\dfrac{18}{37}\right)^{10} = 0.0007425$

$P(2) = \dfrac{2}{36}$ $P(15) = \dfrac{2}{36}$

$P(3) = \dfrac{2}{36}$ $P(16) = \dfrac{1}{36}$

$P(4) = \dfrac{3}{36}$ $P(18) = \dfrac{2}{36}$

$P(5) = \dfrac{2}{36}$ $P(20) = \dfrac{2}{36}$

$P(6) = \dfrac{4}{36}$ $P(24) = \dfrac{2}{36}$

$P(8) = \dfrac{2}{36}$ $P(25) = \dfrac{1}{36}$

$P(9) = \dfrac{1}{36}$ $P(30) = \dfrac{2}{36}$

$P(10) = \dfrac{2}{36}$ $P(36) = \dfrac{1}{36}$

27. $500(5)(10)(10)(1/37) = \$6756.76$ **29.** $P(5000) = 0.99,\ P(-245{,}000) = 0.01$ **31.** \$2500   **33.** 30.33   **35.** \$0.25
37. \$25   **39.** \$0.25   **41.** \$12,500   **43.** \$0.25

Chapter Review

The sample space for Problems 1–5 is as follows.

Die 2

	1	*2*	*3*	*4*	*5*	*6*	
	(1, 1)	(1, 2)	(1, 3)	(1, 4)	(1, 5)	(1, 6)	*1*
	(2, 1)	(2, 2)	(2, 3)	(2, 4)	(2, 5)	(2, 6)	*2*
Die 1	(3, 1)	(3, 2)	(3, 3)	(3, 4)	(3, 5)	(3, 6)	*3*
	(4, 1)	(4, 2)	(4, 3)	(4, 4)	(4, 5)	(4, 6)	*4*
	(5, 1)	(5, 2)	(5, 3)	(5, 4)	(5, 5)	(5, 6)	*5*
	(6, 1)	(6, 2)	(6, 3)	(6, 4)	(6, 5)	(6, 6)	*6*

1. $\dfrac{1}{36}$ **3.** $\dfrac{2}{9}$ **5.** 0 **7.** $\dfrac{2}{3}$ **9.** $\dfrac{1}{3}$ **11.**

2: $\dfrac{20}{1000} = 0.02$ versus $\dfrac{1}{36} = 0.027778$

3: $\dfrac{35}{1000} = 0.035$ versus $\dfrac{2}{36} = 0.055555$

4: $\dfrac{45}{1000} = 0.045$ versus $\dfrac{3}{36} = 0.083333$

5: $\dfrac{120}{1000} = 0.12$ versus $\dfrac{4}{36} = 0.111111$

6: $\dfrac{150}{1000} = 0.15$ versus $\dfrac{5}{36} = 0.138889$

7: $\dfrac{200}{1000} = 0.20$ versus $\dfrac{6}{36} = 0.166667$

8: $\dfrac{140}{1000} = 0.14$ versus $\dfrac{5}{36} = 0.138889$

9: $\dfrac{125}{1000} = 0.125$ versus $\dfrac{4}{36} = 0.111111$

10: $\dfrac{80}{1000} = 0.08$ versus $\dfrac{3}{36} = 0.083333$

11: $\dfrac{60}{1000} = 0.06$ versus $\dfrac{2}{36} = 0.055555$

12: $\dfrac{25}{1000} = 0.025$ versus $\dfrac{1}{36} = 0.027778$

13. 0.2215336 **15.** $\{(r, g, b) : 1 \le r, g, b \le 6\}$ **17.** $\dfrac{3}{216}$ **19.** $\dfrac{10}{216}$ **21.** $\dfrac{6}{216}$ **23.** $\left(\dfrac{1}{2}\right)^{28} = 3.73 \times 10^{-9}$

25. $S = \{(a, b, c), (a, c, b), (b, a, c), (b, c, a), (c, a, b), (c, b, a)\}$ $P(\text{All letters go to the right place}) = \dfrac{1}{6}$

27. $\left(\dfrac{18}{37}\right)^{28} = 0.00000000173$ **29.** $S = \{(x, y, z, t) : x, y, z, t \text{ are either BOY or GIRL}\}$ **31.** $\dfrac{1}{2}$ **33.** $\dfrac{0.2}{0.5} = 0.4$

35. $1 - 0.5 = 0.5$ **37.** 0.3 **39.** 0.8 **41.** $\dfrac{4}{52}\dfrac{48}{51} = 0.0724$ **43.**

	C	*M*	*L*
GPA ≥ 2.5	0.075	0.175	0.300
GPA < 2.5	0.125	0.150	0.175

45. 0.150 **47.** $P(\text{GPA below 2.5}) = 0.45$ **49.** $P(\text{below 2.5}) = 0.45$ **51.** $4\dfrac{1}{6}$
$\qquad\qquad\qquad P(C) = 0.20$ $\qquad\qquad\qquad P(L) = 0.475$
$\qquad\qquad\qquad P(\text{GPA below 2.5 and } C) = 0.125$ $\qquad P(\text{below 2.5 and } L) = 0.175$
$\qquad\qquad\qquad$ The events are dependent $\qquad\qquad$ The events are dependent

53. Since $A \cap B$ is a subset of $A \cup B$ **55.** $\dfrac{1}{2}$

Chapter Test

1. Expected value $= 1$

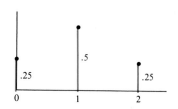

2.

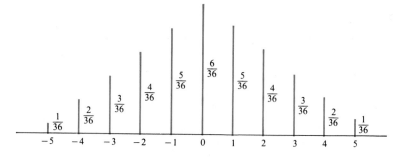

3.

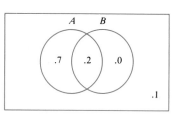

4. (a) $\dfrac{4}{51}$ (b) $\dfrac{3}{51}$

5. 0.24; we cannot find $P(\text{rain tomorrow})$ from the given information.   **6.** $S = \{0, 1, 2\}$

$$P(0) = \frac{3}{10}$$

$$P(1) = \frac{6}{10}$$

$$P(2) = \frac{1}{10}$$

$$E[BM] = \frac{4}{5}$$

CHAPTER 6

Section 6.1

1. Yes   **3.** Yes   **5.** Yes   **7.** No   **9.** No   **11.** Yes   **13.** No   **15.** No   **17.** No   **19.** Yes

21. $P_1 = \begin{bmatrix} \frac{1}{2} & \frac{1}{2} \end{bmatrix}$, $P_2 = \begin{bmatrix} \frac{1}{2} & \frac{1}{2} \end{bmatrix}$, $P_3 = \begin{bmatrix} \frac{1}{2} & \frac{1}{2} \end{bmatrix}$

23. $P_1 = \begin{bmatrix} \frac{3}{4} & \frac{1}{4} \end{bmatrix}$, $P_2 = \begin{bmatrix} \frac{7}{8} & \frac{1}{8} \end{bmatrix}$, $P_3 = \begin{bmatrix} \frac{15}{16} & \frac{1}{16} \end{bmatrix}$

25. $P_1 = \begin{bmatrix} \frac{1}{2} & \frac{1}{2} \end{bmatrix}$, $P_2 = \begin{bmatrix} \frac{5}{16} & \frac{11}{16} \end{bmatrix}$, $P_3 = \begin{bmatrix} \frac{49}{128} & \frac{79}{128} \end{bmatrix}$

27. $P_1 = \begin{bmatrix} 0 & 1 & 0 \end{bmatrix}$, $P_2 = \begin{bmatrix} 1 & 0 & 0 \end{bmatrix}$, $P_3 = \begin{bmatrix} 0 & 1 & 0 \end{bmatrix}$

29. $P_1 = \begin{bmatrix} 1 & 0 & 0 \end{bmatrix}$, $P_2 = \begin{bmatrix} 1 & 0 & 0 \end{bmatrix}$, $P_3 = \begin{bmatrix} 1 & 0 & 0 \end{bmatrix}$

31.

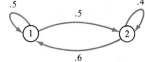

33.

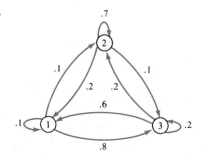

35. $\begin{bmatrix} \frac{1}{2} & \frac{1}{2} \\ \frac{1}{2} & \frac{1}{2} \end{bmatrix}$ **37.** Student project   **39.** $[0.4 \quad 0.6]$ **41.** $\begin{bmatrix} \frac{2}{3} & \frac{1}{3} \end{bmatrix}$ **43.** $[1 \quad 0 \quad 0]$ **45.** $\begin{bmatrix} \frac{1}{7} & \frac{6}{7} \end{bmatrix}$

47. $P_1 = \begin{bmatrix} \frac{9}{16} & \frac{7}{16} \end{bmatrix}$

$P_2 = \begin{bmatrix} \frac{41}{64} & \frac{23}{64} \end{bmatrix}$

$P_3 = \begin{bmatrix} \frac{169}{256} & \frac{87}{256} \end{bmatrix}$

Steady state vector $= \begin{bmatrix} \frac{2}{3} & \frac{1}{3} \end{bmatrix}$

49. Student project   **51.** $1.10   **53.** $3.60   **55.** $P_1 = \begin{bmatrix} \frac{3}{8} & \frac{4}{8} & \frac{1}{8} \end{bmatrix}$

$P_2 = \begin{bmatrix} \frac{5}{16} & \frac{8}{16} & \frac{3}{16} \end{bmatrix}$

$P_3 = \begin{bmatrix} \frac{9}{32} & \frac{16}{32} & \frac{7}{32} \end{bmatrix}$

Steady state $= [0.28 \quad 0.50 \quad 0.22]$

57. All-Star Game: If we can assume that the chance of one league's winning in a given year depends only on the outcome of the previous year, then we can say that this process is a Markov chain. This is true to some extent. If we model this process by a Markov chain, then the transition matrix will be

$$\begin{array}{cc} & \begin{array}{cc} \text{AL} & \text{NL} \end{array} \\ \begin{array}{c} AL \\ NL \end{array} & \begin{bmatrix} \frac{1}{4} & \frac{3}{4} \\ \frac{22}{25} & \frac{3}{25} \end{bmatrix} \end{array}$$

Section 6.2

1. $P(H) = \frac{1}{2}$ **3.** $R(H|R) = \frac{9}{11}$ **5.** $P(H|B) = \frac{9}{13}$ **7.** $P(H_1) = \frac{1}{4}$ **9.** $P(H_2|H_1) = \frac{12}{51}$ **11.** $P(H_2) = \frac{1}{4}$

$P(T) = \frac{1}{2}$

$P(R) = \frac{1}{2}$

$P(B) = \frac{1}{2}$

13. $P(H_1|H_2) = \frac{12}{51}$ **15.** $P(A|F) = 0.0748$ **17.** $P(A|F') = 0.0022396$ **19.** $P(R|C) = 0.087$

21. 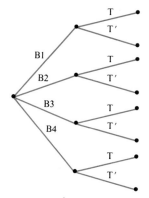   **23.** $P(B2|T') = \frac{6}{23}$ **25.** $P(B4|T') = \frac{9}{23}$ **27.** $P(B2|T) = \frac{6}{25}$

29. $P(B4|T) = \frac{3}{23}$ **31.** $P(Th|N) = 0.476$

Section 6.3

1. Yes   **3.** No   **5.** No   **7.** 0.0576   **9.** 0.0009766   **11.** 0.656   **13.** 0.0039   **15.** 1   **17.** 0.539   **19.** 0.132

21. 0.055   **23.** 0.999   **25.** $P(B = 0) = 0.037$ **27.** 0.578   **29.** 0.262   **31.** $P(X = 0) = 0.316$ **33.** 0.255

$$P(B = 1) = 0.222$$
$$P(B = 2) = 0.444$$
$$P(B = 3) = 0.296$$

$$P(X = 1) = 0.422$$
$$P(X = 2) = 0.211$$
$$P(X = 3) = 0.047$$
$$P(X = 4) = 0.004$$

35. 0.9775   **37.** $P(X \geq 7) = C(10, 7)(0.85)^7(0.15)^3 + C(10, 8)(0.85)^8(0.15)^2 + C(10, 9)(0.85)^9(0.15) + C(10, 10)(0.85)^{10}(0.15)^0$

39. $P(X \geq 1) = 0.167$ **41.** $P(0) = \dfrac{1}{4}$ **43.** $P(0) = \dfrac{1}{16}$ **45.** $P(0) = \dfrac{27}{64}$ **47.** $P(0) = \dfrac{1}{16}$ **49.** $P(0) = \dfrac{1}{32}$

$$P(1) = \frac{1}{2} \qquad P(1) = \frac{3}{8} \qquad P(1) = \frac{27}{64} \qquad P(1) = \frac{1}{4} \qquad P(1) = \frac{5}{32}$$

$$P(2) = \frac{1}{4} \qquad P(2) = \frac{9}{16} \qquad P(2) = \frac{9}{64} \qquad P(2) = \frac{3}{8} \qquad P(2) = \frac{10}{32}$$

$$P(3) = \frac{1}{64} \qquad P(3) = \frac{1}{4} \qquad P(3) = \frac{10}{32}$$

$$P(4) = \frac{1}{16} \qquad P(4) = \frac{5}{32}$$

$$P(5) = \frac{1}{32}$$

51. $P(0) = \dfrac{1}{256}$ **53.** $P(0) = \dfrac{1}{2048}$

$$P(1) = \frac{8}{256} \qquad\qquad P(1) = \frac{11}{2048}$$

$$P(2) = \frac{28}{256} \qquad\qquad P(2) = \frac{55}{2048}$$

$$P(3) = \frac{56}{256} \qquad\qquad P(3) = \frac{165}{2048}$$

$$P(4) = \frac{70}{256} \qquad\qquad P(4) = \frac{330}{2048}$$

$$P(5) = \frac{56}{256} \qquad\qquad P(5) = \frac{462}{2048}$$

$$P(6) = \frac{28}{256} \qquad\qquad P(6) = \frac{462}{2048}$$

$$P(7) = \frac{8}{256} \qquad\qquad P(7) = \frac{330}{2048}$$

$$P(8) = \frac{1}{256} \qquad\qquad P(8) = \frac{165}{2048}$$

$$P(9) = \frac{55}{2048}$$

$$P(10) = \frac{11}{2048}$$

$$P(11) = \frac{1}{2048}$$

Section 6.4

1. THHTTHHTTT

3. $0 < x < 0.25$ means clubs
$0.25 \le x < 0.50$ means diamonds
$0.50 \le x < 0.75$ means hearts
$0.75 \le x < 1$ means spades

5. $0 < x < 0.10$ walk
$0.10 \le x < 0.75$ out
$0.75 \le x < 0.90$ single
$0.90 \le x < 0.95$ double
$0.95 \le x < 0.98$ triple
$0.98 \le x < 1$ home run

Beginning with 0.045343, we get

Walk	Walk
Walk	Single
Out	Double
Out	Out
Out	Single

7. For the student

9. $P(0) = 0.3164$
$P(1) = 0.4219$
$P(2) = 0.2109$
$P(3) = 0.0469$
$P(4) = 0.0039$

11.

n	r_n	$17r_n + 41$	$\dfrac{17r_n + 41}{100}$	r_{n+1}	$\dfrac{r_{n+1}}{100}$
0	30	551	5.51	51	0.51
1	51	908	9.08	8	0.08
2	8	177	1.77	77	0.77
3	77	1350	13.50	50	0.50
4	50	891	8.91	91	0.91
5	91	1588	15.88	88	0.88
6	88	1537	15.37	37	0.37
7	37	670	6.70	70	0.70
8	70	1231	12.31	31	0.31
9	31	568	5.68	68	0.68
10	68	1197	11.97	97	0.97

13. Student project   **15.** Student project   **17.** Student project
19. Divide the random numbers between 0 and 1 into six equal-sized intervals and number the consecutive intervals with the values 1, 2, 3, 4, 5, and 6, respectively.
21. Consider a trial a success if the random number is between 0 and 2.5. Otherwise, consider the trial a failure.
23. Divide the interval $[0, 1]$ into the subintervals

$[0, 0.100)$	walk
$[0.1, 0.300)$	single
$[0.3, 0.350)$	double
$[0.35, 0.375)$	triple
$[0.375, 0.450)$	home run
$[0.450, 0.650)$	strike out
$[0.650, 0.850)$	fly out
$[0.850, 1.00)$	ground out

The random number will fall into one of these intervals. Associate the baseball event described with the random number.

Chapter Review

1.

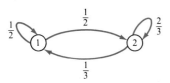

3.

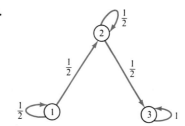

5. $\begin{bmatrix} 0 & \frac{2}{3} & \frac{1}{3} \\ 0 & 0 & 1 \\ \frac{2}{5} & \frac{1}{5} & \frac{2}{5} \end{bmatrix}$
7.

	Blue	White
Blue	$\frac{1}{2}$	$\frac{1}{2}$
White	$\frac{3}{5}$	$\frac{2}{5}$

9. Not a state vector   **11.** Not a state vector

13. State vector   **15.** Not a transition matrix   **17.** Not a transition matrix   **19.** Student project

21. $P_1 = \left(\frac{1}{2} \quad \frac{1}{2}\right)$ **23.** $P_1 = \left(\frac{1}{2} \quad \frac{1}{2}\right)$ **25.** $P_1 = \left(\frac{1}{2} \quad \frac{1}{2}\right)$

$P_2 = \left(\frac{1}{2} \quad \frac{1}{2}\right)$ $P_2 = \left(\frac{1}{2} \quad \frac{1}{2}\right)$ $P_2 = \left(\frac{1}{2} \quad \frac{1}{2}\right)$

$P_3 = \left(\frac{1}{2} \quad \frac{1}{2}\right)$ $P_3 = \left(\frac{1}{2} \quad \frac{1}{2}\right)$ $P_3 = \left(\frac{1}{2} \quad \frac{1}{2}\right)$

$P_4 = \left(\frac{1}{2} \quad \frac{1}{2}\right)$ $P_4 = \left(\frac{1}{2} \quad \frac{1}{2}\right)$ $P_4 = \left(\frac{1}{2} \quad \frac{1}{2}\right)$

27. $P_1 = (0 \quad 0.50 \quad 0.50)$
$P_2 = (0 \quad 0.35 \quad 0.65)$
$P_3 = (0 \quad 0.305 \quad 0.695)$
$P_4 = (0 \quad 0.2915 \quad 0.7085)$

It appears that the steady state vector is approximately
$$P_{ss} = (0 \quad 0.3 \quad 0.7)$$

29. $P_1 = (0.111 \quad 0.455 \quad 0.433)$
$P_2 = (0.037 \quad 0.389 \quad 0.574)$
$P_3 = (0.012 \quad 0.311 \quad 0.654)$
$P_4 = (0.004 \quad 0.306 \quad 0.689)$

It appears that the steady state vector is approximately
$$P_{ss} = (0 \quad 0.3 \quad 0.7)$$

31. $P^2 = \begin{bmatrix} 1 & 0 \\ \frac{3}{4} & \frac{1}{4} \end{bmatrix}$, $P^3 = \begin{bmatrix} 1 & 0 \\ \frac{7}{8} & \frac{1}{8} \end{bmatrix}$, $P^4 = \begin{bmatrix} 1 & 0 \\ \frac{15}{16} & \frac{1}{16} \end{bmatrix}$

Hence

$$P_1 = P_0 P = \begin{bmatrix} \frac{1}{2} & \frac{1}{2} \end{bmatrix} \begin{bmatrix} 1 & 0 \\ \frac{1}{2} & \frac{1}{2} \end{bmatrix} = \begin{bmatrix} \frac{3}{4} & \frac{1}{4} \end{bmatrix}$$

$$P_2 = P_0 P^2 = \begin{bmatrix} \frac{1}{2} & \frac{1}{2} \end{bmatrix} \begin{bmatrix} 1 & 0 \\ \frac{3}{4} & \frac{1}{4} \end{bmatrix} = \begin{bmatrix} \frac{7}{8} & \frac{1}{8} \end{bmatrix}$$

$$P_3 = P_0 P^3 = \begin{bmatrix} \frac{1}{2} & \frac{1}{2} \end{bmatrix} \begin{bmatrix} 1 & 0 \\ \frac{7}{8} & \frac{1}{8} \end{bmatrix} = \begin{bmatrix} \frac{15}{16} & \frac{1}{16} \end{bmatrix}$$

$$P_4 = P_0 P^4 = \begin{bmatrix} \frac{1}{2} & \frac{1}{2} \end{bmatrix} \begin{bmatrix} 1 & 0 \\ \frac{15}{16} & \frac{1}{16} \end{bmatrix} = \begin{bmatrix} \frac{31}{32} & \frac{1}{32} \end{bmatrix}$$

33. $P_1 = \begin{bmatrix} 1 & 0 \end{bmatrix} \begin{bmatrix} 0.999 & 0.001 \\ 0.0001 & 0.9999 \end{bmatrix} = [0.999 \quad 0.001]$

35. $P(\text{Stomach gas} \mid \text{Test negative}) = 0.874$ **37.** $P(\text{Gallstone} \mid \text{Test negative}) = 0.0097$ **39.** $C(10, 7)\left(\frac{3}{4}\right)^7\left(\frac{1}{4}\right)^3 = 0.2503$

41. $C(10, 9)\left(\dfrac{3}{4}\right)^9\left(\dfrac{1}{4}\right) = 0.1877$ **43.**

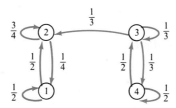

45. $b(0; 3, 0.5) = 0.125$
$b(1; 3, 0.5) = 0.375$
$b(2; 3, 0.5) = 0.375$
$b(3; 3, 0.5) = 0.125$

47. Toss a coin 3 times. If HHH, die = 1; HHT, die = 2; HTH, die = 3; HTT, die = 4; THH, die = 5; THT, die = 6; otherwise toss again 3 times.

49. Toss a coin 2 times.   **51.** Toss a coin 4 times. X = number of heads.
    TT means region 1
    TH means region 2
    HT means region 3
    HH means region 4

53. Student project

Chapter Test

1. (a) $\left(\dfrac{3}{4}\right)^2\left(\dfrac{1}{4}\right)^2 = 0.035$ **(b)** $C(4, 2)\left(\dfrac{3}{4}\right)^2\left(\dfrac{1}{4}\right)^2 = 0.211$ **2.** $C(10, 9)\left(\dfrac{1}{5}\right)^9\left(\dfrac{4}{5}\right) = 0.0000041$

3.

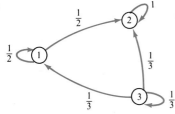

4. $\left[c/(b + c)\ b/(b + c)\right]\begin{bmatrix} a & b \\ c & d \end{bmatrix} = \left[c/(b + c)\ b/(b + c)\right]$

5.
$$\begin{array}{cc} & A \quad\ B \\ \begin{matrix} A \\ B \end{matrix} & \begin{bmatrix} 0.4 & 0.6 \\ 0.7 & 0.3 \end{bmatrix} \end{array}$$

Since $b = 0.6$ and $c = 0.7$, we have the steady state probabilities

$$\frac{c}{b + c} = \frac{0.7}{1.3} = 0.538$$

$$\frac{b}{b + c} = \frac{0.6}{1.3} = 0.462$$

or $P_{ss} = [0.538 \quad 0.462]$.

6. (a) Toss coins separately 3 times each. TTT: die = 1, TTH: die = 2, ... HTH: die = 6. Discard HHT, HHH.
   (b) Toss the pair of coins twice. Success = HH.

7. $P(A\mid B) = 0.1$. This is the probability that the student who got an A was a business major.

CHAPTER 7

Section 7.1

1. Student project   **3.** Student project   **5.** Student project
7. Summarizing rat data

(a)

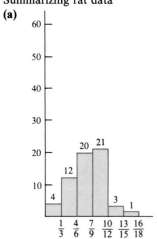 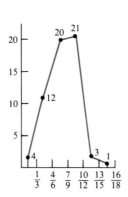

(b)

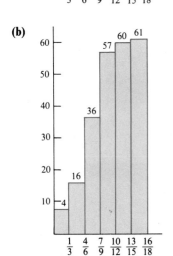 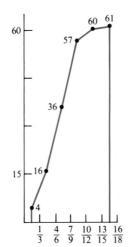

Section 7.2

1. Mean: 4	**3.** Mean: 3.57	**5.** Mean: 60.44	**7.** Mean 1.63	**9.** Mean: 27.78
Median: 4	Median: 3	Median: 5	Median: 1.7	Median: 50
Mode: none	Mode: 1, 7	Mode: 5	Mode: none	Mode: 50

11. It may have a numerical value that gives a false impression of the distribution of values.
13. The value of the mode is not likely to tell very much about the distribution of values since it relates only to the most repeated values.
15. 0, 1, 8, 8, 9; mean < mode ≤ median
                    5.2   8   8
17. 0, 7, 8, 8, 60; median ≤ mode < mean
                    8   8   16.6
19. −1, −1, 4, 5, 6; mode < mean < median
                  −1   2.6   4

21. Mean: 81   **23.** Mean: 605   **25.** Mean: 313.33
Median: 85   Median: 460   Median: 320
Mode: 95   Mode: none   Mode: 340, 350

27. Range: 0   **29.** Range: 1   **31.** Range: 9
Standard deviation: 0   Standard deviation: 0.527   Standard deviation: 3.0

33. Range: 6   **35.** Range: 10
Standard deviation: 1.85   Standard deviation: 3.81

37. Standard deviation of zero means that all the values are the same, but there is no indication of how many values are involved.

39. World Series time   **41.** Population problem
  (a) Range: 7   **(a)** 32.40
  (b) Mean: .48   **(b)** 11.82
  (c) Standard deviation: 1.1   **(c)** 27.67
                                         (d) 17.28

Section 7.3

1. (a) Sample: population too hard to poll
  (b) Population: all unemployed and employed workers
  (c) Parameter: fraction of population unemployed
  (d) Statistic: fraction of sample unemployed
3. (a) Population: every car is driven off the assembly line
  (b) Population: all cars manufactured in a given time period
  (c) Parameter: fraction of the population with engines that start
  (d) Statistic: fraction of the sample tested that have engines that start
5. (a) Sample: census would require too much effort
  (b) Population: all the wheat being shipped to the Soviet Union in some time period
  (c) Parameter: fraction of the population that is weed seed
  (d) Statistic: fraction of the sample that is weed seed
7. (a) Sample: observations are destructive
  (b) Population: all birds of a given species in a given area
  (c) Parameter: average amount of insects consumed per day by birds in the population
  (d) Statistic: average amount of insects consumed per day by birds in the sample
9. (a) Sample: life test is destructive
  (b) Population: all batteries imported from Korea in a given time period
  (c) Parameter: fraction of batteries in population with a given lifetime
  (d) Statistic: fraction of batteries in the sample that have a given lifetime
11. Fruit flies. There is a 95% chance that the fraction of fruit flies in the population with the trait is between 74% and 80%.
13. $n = 400$ **15.** [0.475, 0.525]   **17.** [0.35, 0.45]   **19.** [0.387, 0.487]   **21.** [0.36, 0.76]   **23.** [0.227, 0.284]
25. [0.210, 0.268]   **27.** [0.181, 0.237]   **29.** [0.177, 0.233]   **31.** [0.173, 0.229]   **33.** No   **35.** No
37. $n = 40,000$ **39.** $n = 400$

Chapter Review

1. Student project   **3.** $s = 5.296 \times 10^5$ **5.** Median $= 224,523$ **7.** Student project
9. Mean: 0   **11.** Mean: 2   **13.** Mean: 8.285   **15.** Mean: 18.33   **17.** Mean: 3.6
Median: 0   Median: 0   Median: 8   Median: 20   Range: 4
Mode: 1, −1   Mode: 0   Mode: 7, 8, 9   Mode: 20   Standard deviation: 1.67
19. Mean: 6   **21.** Mean: 3
Range: 10   Range: 0
Standard deviation: 5.48   Standard deviation: 0
23. Population: all households in the U.S.
Sample: 100 households interviewed
25. Population: all possible core samples that could be taken
Sample: the core samples actually taken

27. Length of 95% interval $= 0.10$
95% interval $= [0.25, 0.35]$

29. The pollster can be 95% sure that the fraction of individuals who prefer the soda is between 68.45% and 70.45%
31. Increasing the sample size by a factor of 4 decreases the length of the confidence interval by 2.

Chapter Test

1. 10, 4, 3, 1, 1   **2.** 5, 5, 5, 1, 2   **3.** 10, 1, 1, 1, 1   **4.** It is not possible.
Mean: 3.8   Mean: 3.6   Range: 9
Median: 3   Median: 5   Standard deviation: 4.02
Mode: 1   Mode: 5
5. Mean: 4.54
Standard deviation: 2.00
6. Statistic: fraction of people interviewed who oppose public transportation $= 0.214$
Population parameter: fraction of people in the city who oppose public transportation
95% confidence interval $= [15.1\%, 24.6\%]$
7. $n = 10,000$
New confidence interval $= [0.204, 0.224]$

CALCULUS PRELIMINARIES

Problem Set I

1. 5.88×10^{12} **3.** 1.58×10^{18} **5.** 6.0×10^{11} **7.** 5.0×10^{-3} **9.** 1.0×10^{-2}
11. 2,400,000,000,000,000,000   **13.** 100,000,000,000,000,000,000   **15.** Too large to write
17. 0.00000001

Problem Set II

1. 0   **3.** 7   **5.** Yes   **7.** Yes   **9.** Both have the same value

Problem III

1. 2   **3.** 4   **5.** 1.4142136   **7.** 2

Problem IV

1. $4\sqrt{2}$ **3.** $2\sqrt{5} + 5\sqrt{45} = 2\sqrt{5} + 15\sqrt{5} = 17\sqrt{5}$ **5.** $4x^2\sqrt{2x}$ **7.** $3x^2y^2\sqrt{10x}$ **9.** $xy^2\sqrt[3]{3x}$ **11.** $xyz^2\sqrt{2y}$
13. $4x^2w^4\sqrt{5xw}$ **15.** $x + x^2$

Problem Set V

1. Multiplication: $3x^3 + 6x$ **3.** Multiplication: $q^4 + q^3 - 4q^2 + 11q - 3$
Division: $\dfrac{x}{3} + \dfrac{2}{3x}$ Division: 1 with remainder $5q - 4$

5. Multiplication: $x^6 - x^4 + x^2 - 1$ **7.** Multiplication: $p^5 + 2p^4 + 2p^3 + 2p^2 + 2p + 1$
Division: 0 with remainder $x^2 - 1$ Division: $p^3 + p + \dfrac{1}{p + 1}$

Problem Set VI

1. $x^{3/2} + 2x^{1/2} - 1$ **3.** $x^2 + 2x^{3/2} + x^{1/2} + 2$ **5.** 1   **7.** Student project

Problem Set VII

1. $(x - 8)(x - 2)$ **3.** $(2x - 3y)(x + 5y)$ **5.** $(2x - 5)(4x^2 + 10x + 25)$ **7.** $(x^2 + a^2)(x + a)(x - a)$
9. $(a - b)(a^2 + ab + b^2)(a + b)(a^2 - ab + b^2)$ **11.** $(2x + 5)(4x^2 - 10x + 25)$ **13.** $(3 - x)(3 + x)$ **15.** $(y - x)(y + x)$
17. $x(1 - x) + (1 - x)^2 = (1 - x)(x + 1 - x) = 1 - x$ **19.** $(mn - 6)(mn - 4)$

Problem Set VIII

1. $-7, 2$ **3.** $3, -3$ **5.** $\dfrac{-3}{4}, \dfrac{3}{4}$ **7.** $\dfrac{3}{5}, -1$ **9.** $4, 2$ **11.** $1, -1, 2$ **13.** $\dfrac{-1 + \sqrt{31}}{3}, \dfrac{-1 - \sqrt{31}}{3}$

15. $\dfrac{3}{4}, \dfrac{2}{3}$ **17.** $-4 + \sqrt{14}, -4 - \sqrt{14}$ **19.** $2m + \sqrt{4m^2 - m}, 2m - \sqrt{4m^2 - m}$ **21.** $\dfrac{1 + \sqrt{13}}{6}, \dfrac{1 - \sqrt{13}}{6}$

Problem Set IX

1. $\dfrac{2 + x^2}{2x}$ **3.** $\dfrac{-x^2 + 2x + 2}{(x - 1)(x + 2)}$ **5.** $\dfrac{x - 1}{x - 2}$ **7.** $\dfrac{x^{3/2} + x - x^{1/2} + 1}{x(1 - \sqrt{x})}$

Problem Set X

1. $\dfrac{\sqrt{x} - \sqrt{y}}{x - y}$ **3.** $\dfrac{\sqrt{x + a} + \sqrt{x}}{a}$ **5.** $\dfrac{\sqrt{x + 2} - \sqrt{x - 2}}{4}$

Problem Set XI

1. $-3 \le x \le 3$ **3.** $-1 < x < 4$ **5.** The inequality has no real solution   **7.** $x > -1$ or $x < -2$

Problem Set XII

1. $x < -2$ or $0 < x < 4$ **3.** $x < -3$ or $x > 2$ **5.** $-2 < x < 0$ or $x > 4$

Problem Set XIII

1. 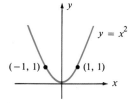   **3.** 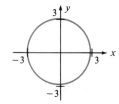   **5.** 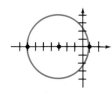   **7.**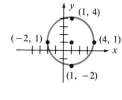

9. 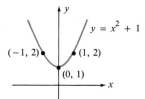   **11.** 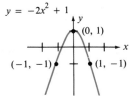   **13.**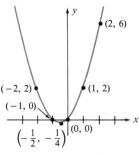

15. 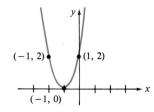   **17.** 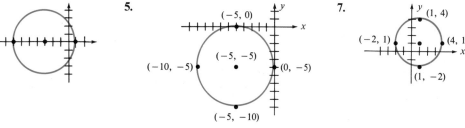   **19.**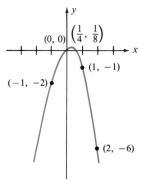

Problem Set XIV

1.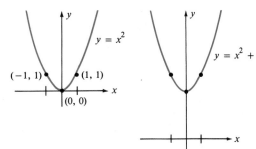

3.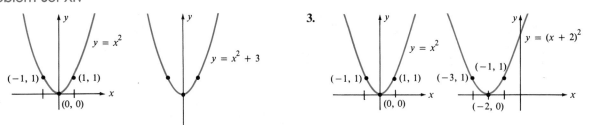

5.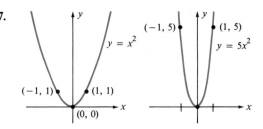

7.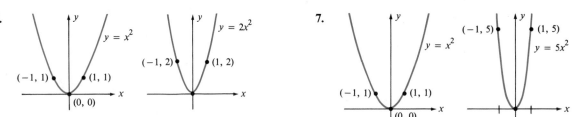

9.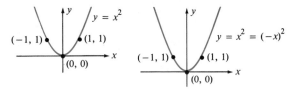

11.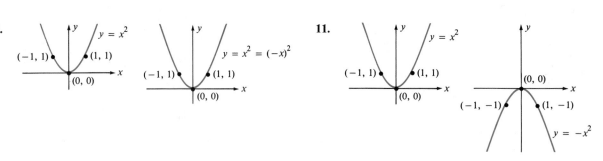

Problem Set XV

1. $f[g(x)] = 3(x^2 + 2x + 1)$ **3.** $g[f(x)] = 9x^2 + 6x + 1$ **5.** $h[f(x)] = 1 + \dfrac{1}{3x}$

Calculus Practice Test

1. (a) 2   **(b)** $\dfrac{1}{2}\left(\dfrac{1}{2}\right)^5$ **(c)** 4   **(d)** Does not exist, since the fourth powers are positive   **(e)** 9

2. (a) $4x$ **(b)** $ab + c$ **(c)** $2xy^2$ **3. (a)** $x^3 - 2x^2 + 6x - 4$ **(b)** $2x^3 + 5x^2 - 3x$

4. (a) $(x + 3)(x - 3)$ **(b)** $(x - 4)(x + 3)$ **5.** $\dfrac{-5 + \sqrt{33}}{2}, \dfrac{-5 - \sqrt{33}}{2}$ **6.** $\dfrac{x - 3}{(x + 1)(x + 3)}$ **7.** $\dfrac{\sqrt{x} + \sqrt{y}}{x - y}$

8. $x < \dfrac{1}{2}$ **9.** Center: $(-3, 0)$ **10.** Center: $(-1, 2)$ **11. (a)** $f[g(x)] = (2x + 3)^2$ **(b)** $g[f(x)] = 2x^2 + 3$
              Radius: 4   Radius: 2

CHAPTER 8

Section 8.2

1. 6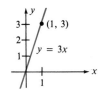

3. $\dfrac{1}{3}$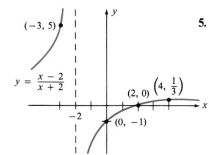

5. $\sqrt{5}$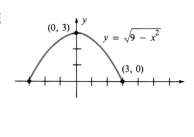

7. $\dfrac{1}{4}$

9. $\dfrac{1}{2}$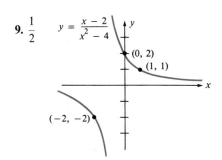

11. 2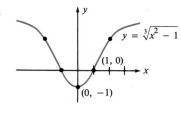

13. Does not exist **15.** Undefined **17.** 0 **19.** 6 **21.** 10 **23.** 12 **25.** Discontinuous **27.** Continuous
29. Discontinuous **31.** Discontinuous **33.** $(-\infty, \infty)$ **35.** $(-\infty, \infty)$ **37.** Discontinuous at $x = -1$
39. $(-\infty, \infty)$ **41.** Discontinuous at $-1, 3$ **43.** $-3, -1, 2$ **45.** -1

47. $f(x) = \begin{cases} 22 & 0 < x < 1 \\ 22 + 17n & n < x \le n + 1 \ (n = 1, 2, 3, \dots 11) \end{cases}$

Discontinuous at each integer value of x for $0 < x < 12$

Section 8.3

1. 8 **3.** 3 **5.** 3 **7.** 3 **9.** 1 **11.** 8 **13.** -4 **15.** 2 **17.** 0 **19.** 2 **21.** -2 **23.** 144 ft

25. 96 ft/sec **27.** 2.5 sec **29.** 1 **31.** $2t$ **33.** 2π **35.** $-\dfrac{128}{5}$ **37.** $\dfrac{-81}{10}$ **39.** -5.1 **41.** -2.9

43. -2.2 **45.** $\dfrac{-1}{2}$ **47.** $\dfrac{-9}{10}$ **49.** $-0.00234 \ (-0.234\%)$ **51.** $-0.189 \dfrac{\text{sec}}{\text{year}}$ **53.** 10.5 **54.** 330

Section 8.4

1. $y = 3x - 1$ **3.** $2 + h$ **5.** $f'(x) = 1$ **7.** $f'(x) = 2$ **9.** $f'(x) = 2x$ **11.** $f'(x) = 2x + 1$ **13.** $f'(x) = 14x + 5$
$\qquad\qquad\qquad\qquad\qquad\qquad\quad f'(0) = 1 \qquad\quad f'(0) = 2 \qquad\quad f'(0) = 0 \qquad\quad f'(0) = 1 \qquad\qquad f'(0) = 5$
$\qquad\qquad\qquad\qquad\qquad\qquad\quad f'(1) = 1 \qquad\quad f'(1) = 2 \qquad\quad f'(1) = 2 \qquad\quad f'(1) = 3 \qquad\qquad f'(1) = 19$

15. $f'(x) = -1/x^2$ **17.** $f'(x) = -2/x^3$ **19.** $f'(x) = \dfrac{1}{2\sqrt{x}} + 1$ **21.** $f'(x) = 2x$
$\quad f'(0)$ undefined $\qquad f'(0)$ undefined $\qquad\qquad\qquad\qquad\qquad\qquad\quad f'(2) = 4$
$\quad f'(1) = -1 \qquad\qquad f'(1) = -2 \qquad\quad f'(0)$ undefined $\qquad\qquad y = 4x - 4$

$\qquad\qquad\qquad\qquad\qquad\qquad\qquad\qquad\qquad\qquad\quad f'(1) = \dfrac{3}{2}$

23. $f'(x) = -2/x^3$ **25.** $f'(0) = 0$ **27.** $f'(-1) = -2$ **29.** $\left.\dfrac{dy}{dx}\right|_{x-4} = 8$ **31.** $\left.\dfrac{dy}{dx}\right|_{x-16} = \dfrac{1}{8}$
$f'(1) = -2$
$y = -2x + 3$

33. $y = 3x + 5$ **35.** $y = 6x - 4$

37. (a)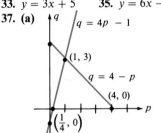

39. (a)

41. (a)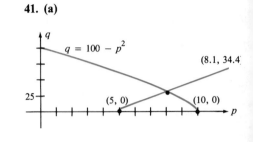

(b) $\left(\dfrac{1}{4}, 4\right)$

(b) $\left(\dfrac{1}{4}, 2\right)$

(b) $(5, 10)$
(c) $(8.10, 34.4)$
(d) $S'(p) = 2p - 2,\ D'(p) = -2p$
$S'(7.5) = 13,\quad D'(7.5) = -15$

(c) $(1, 3)$
(d) $S'(p) = 4,\quad D'(p) = -1$
$S'(0.5) = 4,\ D'(0.5) = -1$

(c) $(1, 3)$
(d) $S'(p) = 4,\ D'(p) = -2p$
$S'(1) = 4,\ D'(1) = -2$

45. $x = 0$

47. $x = 2$

43. (a) $D'(p) = -2p$

$$E(p) = \dfrac{2p^2}{16 - p^2}$$

(b) $E(2) = \dfrac{2}{3}$

Section 8.5

1. $f'(x) = 0$ **3.** $f'(x) = 2.5x^{1.5}$ **5.** $f'(x) = -2x^{-3}$ **7.** $f'(x) = \dfrac{3}{2}x^{1/2}$ **9.** $f'(x) = -3x^{-4}$ **11.** $f'(x) = 0$

13. $f'(x) = 0$ **15.** $f'(x) = 2$ **17.** $f'(x) = 2(x + 2)$ **19.** $f'(x) = 2ax + b$ **21.** $f'(x) = \dfrac{3}{2}x^{1/2}$ **23.** $f'(x) = 2x - \dfrac{1}{x^2}$

25. $f'(x) = -40x^{-5} + 6x + 2$ **27.** $s'(t) = 32t$ **29.** $h'(t) = 12 - 32t$ **31.** $M'(t) = 6t + \dfrac{1}{2}$ **33.** $p_1(t) = 100t - 20$
$\qquad\qquad\qquad\qquad\qquad\qquad\quad s''(t) = 32$ $\qquad h'(1) = -20$ $\qquad\qquad\qquad\qquad\quad p_1(4) = 380$
$\qquad\qquad\qquad\qquad\qquad\qquad\quad s''(1) = 32$ $\qquad\qquad\qquad\qquad\qquad\qquad\quad M'(1) = \dfrac{13}{2}$ $p_1(6) = 580$
$\qquad\qquad\qquad\qquad\qquad\qquad\quad s''(2) = 32$ $\qquad h'(t) = 0$ if $t = \dfrac{3}{8}$

35. $p_1'(t) + p_2'(t) = 160t - 100$ **37.** $y = 3x + 3$ **39.** $s'(t) = 4t + 4$ $M'(2) = \dfrac{25}{2}$
$\qquad\qquad\qquad\qquad\qquad\qquad\qquad\qquad\qquad\qquad\qquad\qquad\quad s'(10) = 44$
$\qquad\qquad\qquad\qquad\qquad\qquad\qquad\qquad\qquad\qquad\qquad\qquad\quad s'(20) = 84$

41. (a) $P(q) = -0.01q^2 + 5q - 5000$ **(b)** $R'(q) = 10 - 0.02q,\ C'(q) = 5,\ P'(q) = -0.02q + 5$ **(c)** $P'(q) = 0$ if $q = 250$

43. $R(q) = \dfrac{q^2 + 10q}{4}$ **45.** $C'(q) = 6q - 3$
$\qquad\qquad\qquad\qquad\qquad\quad C'(100) = 597$

Section 8.6

1. $f'(x) = 4x^3 + 3x^2 + 6x$ **3.** $f'(x) = 4x^3$ **5.** $f'(t) = 4t^3$ **7.** $f'(u) = 4u^3 + 3u^2 + 1$ **9.** $f'(x) = -2x^{-3}$

11. $f'(x) = \dfrac{-15x^2}{(x^3 + 5)^2}$ **13.** $h'(x) = \dfrac{x^2 + 4x + 1}{(x + 2)^2}$ **15.** $f'(x) = \dfrac{2x(x + 1)^2 - 2}{(x + 1)^2}$ **17.** $y' = \dfrac{-2x^4 + 4x - 6}{(2x + 1)^2(x^2 + 3)^2}$

19. $y = -\dfrac{1}{2}x^{-3/2}$ **21.** $\dfrac{d}{dx}[f(x)g(x)h(x)] = f(x)g(x)h'(x) + f(x)g'(x)h(x) + f'(x)g(x)h(x)$ **23.** $R'(x) = P(x) + xP'(x)$

25. $C'(t) = \dfrac{100{,}000}{(t+2)^2}$ **27.** $s'(t) = \dfrac{-100}{(t+1)^2}$ **29.** $R'(x) = 60 - 6x$ **31.** $C'(x) = \dfrac{50}{\sqrt{x}}$

$C'(1) = -11{,}111$ $\qquad s'(1) = -25$ $\qquad\qquad\qquad\qquad C'(100) = 5$
$C'(2) = -6{,}250$
$C'(3) = -4{,}000$ $\qquad s'(2) = -11\dfrac{1}{9}$

$\qquad\qquad\qquad\qquad s'(3) = -6\dfrac{1}{4}$

33. $AC(x) = \dfrac{100\sqrt{x} + 100{,}000}{x}$ **35.** $AC(10{,}000) = 11$ **37.** $\dfrac{d}{dx}\left[\dfrac{C(x)}{x}\right] = \dfrac{-50\sqrt{x} - 100{,}000}{x^2}$

39. $\left.\dfrac{d}{dr}R(r)\right|_{r=0.2} = \dfrac{-kL}{0.00008}$ **41.** $y' = \dfrac{(x^2+3)^2(x^2+1) + 2x^2(x^2+3)^2 - 4x^2(x^2+1)(x^2+3)}{(x^2+3)^4}$

43. $W(t) = (0.5t^2 + t + 1000)(0.3t^2 + 2t + 1)$
$W'(t) = (t+1)(0.3t^2 + 2t + 1) + (0.5t^2 + t + 1000)(0.6t + 2)$
$W'(1) = 2610.5$
$W'(2) = 3231.4$
45. Student project

Section 8.7

1. $f'(x) = 4(x^2 + 3x + 1)^3(2x + 3)$ **3.** $f'(x) = \dfrac{2x + 3}{2\sqrt{x^2 + 3x + 1}}$ **5.** $f'(x) = -\dfrac{3}{2}(2x + 1)(x^2 + x + 1)^{-5/2}$

7. $f'(x) = 3\left(x^2 + \dfrac{1}{x}\right)\left(2x - \dfrac{1}{x^2}\right)$ **9.** $f'(x) = 3[(x+1)^2 + x]^2[2(x+1) + 1]$ **11.** $f'(x) = \dfrac{-20x}{(x^2 + 1)^{11}}$

13. $f'(x) = \dfrac{(x+5)^4}{(x+4)^4}(2x + 5)$ **15.** $g[f(x)] = 3(x^2 + 1)^3 - 4$ **17.** $\dfrac{d}{dx}g[f(x)] = 18x(x^2 + 1)^2$ **19.** $\left.\dfrac{d}{dx}g[f(x)]\right|_{x=1} = 72$

21. $y = 24x - 16$ **23.** $y = 81x - 54$ **25.** $y' = 10x(x^2 + 1)^4$ **27.** $y' = \dfrac{1}{\sqrt{2x + 1}}$ **29.** $R'(q) = -4\left(2 - \dfrac{q}{500}\right)$

31. $N'(t) = \dfrac{25}{\sqrt{t + 4}}$ **33.** $y' = \dfrac{1}{3}k(x - a)^{-2/3}$ **35.** $y' = \dfrac{7}{2}k(x - a)^{5/2}$

37. $\dfrac{dB}{dt} = 2\dfrac{dM}{dt}, \dfrac{dM}{dt} = 3\dfrac{dJ}{dt}$ then $\dfrac{dB}{dt} = 2\left(3\dfrac{dJ}{dt}\right) = 6\dfrac{dJ}{dt}$ **39.** $\dfrac{dy}{dx} = 24x(3x^2 - 7)$ **41.** $P'(q) = \dfrac{1200 - 3q^2}{2(1200q - q^3)^{1/2}}$

$\qquad\qquad\qquad\qquad\qquad\qquad\qquad\qquad\qquad\qquad\qquad\qquad\qquad\qquad\qquad\qquad\qquad P'(10) = \dfrac{450}{\sqrt{11{,}000}}$

$\qquad\qquad\qquad\qquad\qquad\qquad\qquad\qquad\qquad\qquad\qquad\qquad\qquad\qquad\qquad\qquad\qquad P'(20) = 0$

Chapter Review

1. 4 **3.** $\sqrt{2} - 1$ **5.** $\dfrac{11}{24}$ **7.** 0 **9.** $\dfrac{1}{6}$ **11.** $2 - \dfrac{1}{x^2}$ **13.** $f'(x) = 18x(3x^2 + 4)^2$ **15.** $f'(t) = \dfrac{3}{2\sqrt{t}} - \dfrac{1}{2t^{3/2}}$

17. $f'(z) = \dfrac{1}{2}(z + \sqrt{z})^{-1/2}\left(1 + \dfrac{1}{2\sqrt{z}}\right)$ **19.** $f'(r) = 8r$ **21.** 5 **23.** $\dfrac{-2}{9}$ **25.** $32t$ **27.** $12(2t + 1)$ **29.** $\dfrac{1}{4}$

31. $2x[(x^2 - 1)(x^2 + 1) + (x^2 + 1)(x^2 + 2) + (x^2 + 2)(x^2 - 1)]$ **33.** Use the product rule to differentiate.

$$S(t) = \pi\left(3t + \frac{1}{t}\right)^2 (2t^2 + 3t - 1)$$

$$S'(1) = 176\pi$$

35. $R'(x) = 9$ **37.** $P(x) = -x^2 + 8x - 5$ **39.** $P'(x) = 7.20 - 0.72x$
$\qquad\qquad\qquad P'(x) = -2x + 8$
$\qquad\qquad\qquad P'(x) = 0, x = 4$

Chapter Test

1. 0   **2.** $\dfrac{-1}{(x-3)^2}$ **3.** $\dfrac{1}{4\sqrt{x}\sqrt{x^{1/2}+1}}$ **4.** $15x^2 - 2x + 5$ **5.** $f'(x) = \dfrac{4}{3}x^{1/3} + \dfrac{3}{4}x^{-1/4}$ **6.** $s'(t) = -4 - 10t$
$\qquad\qquad\qquad\qquad\qquad\qquad\qquad\qquad\qquad\qquad\qquad\qquad\qquad\qquad\qquad\qquad\qquad\qquad s'(1) = -14$
$\qquad\qquad\qquad\qquad\qquad\qquad\qquad\qquad\qquad\qquad\qquad\qquad\qquad\qquad\qquad\qquad\qquad\qquad s'(2) = -24$

7. Does not exist   **8.** Yes, since $\lim\limits_{x \to a} f(x) = f(a)$ for all a in $(0, 1)$

9. Yes, since $\lim\limits_{x \to a} f(x) = f(a)$ for all a in $(1, 2)$ **10.** No (there is a discontinuity at 1)

CHAPTER 9
Section 9.1

1. $(a, b), (c, d), (e, f), (g, h), (i, j)$
3. $b, c, d, e, f, g, h, i, j$
5. b, c, d, e, f, h
7.

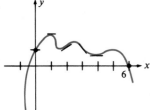

9.

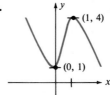

11. (a) +   **(b)** −   **(c)** +   **(d)** 0
13. f has no critical nor stationary points
$\qquad f$ is increasing on $(-\infty, \infty)$

15. f has a critical and stationary point at 0
$\qquad f$ is decreasing on $(-\infty, 0)$
$\qquad f$ is increasing on $(0, \infty)$

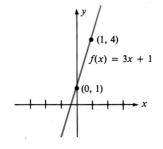

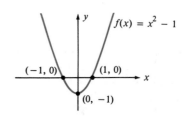

17. f has a critical and stationary point at 4
f is decreasing on $(-\infty, 4)$
f is increasing on $(4, \infty)$

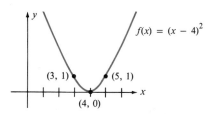

19. f has a critical and stationary point at 0
f is increasing on $(-\infty, 0)$ and $(0, \infty)$

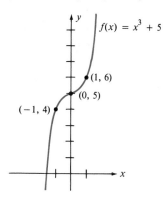

21. f has a critical and stationary point at 3
f is increasing on $(-\infty, 3)$ and $(3, \infty)$

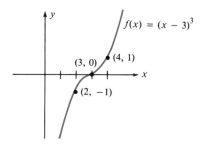

23. 0 is a critical point, $2^{1/3}$ is a critical and stationary point
f is increasing on $(-\infty, 0)$ and $(2^{1/3}, \infty)$
f is decreasing on $(0, 2^{1/3})$

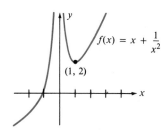

25. 0 is a critical and stationary point
f is decreasing on $(-\infty, 0)$
f is increasing on $(0, \infty)$

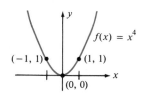

27. -2 is a critical and stationary point
f is decreasing on $(-\infty, -2)$
f is increasing on $(-2, \infty)$

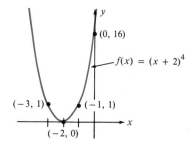

29. 0 is a critical point
f is increasing on $(0, \infty)$

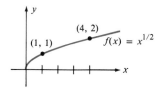

31. 0 is a critical point
f is increasing on $(-\infty, \infty)$

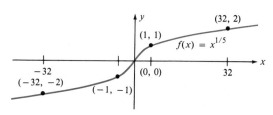

33. 4 is a critical point
   f is decreasing on $(-\infty, 4)$
   f is increasing on $(4, \infty)$

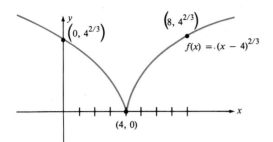

35. 0 is a critical point
   f is increasing on $(-\infty, 0)$
   f is decreasing on $(0, \infty)$

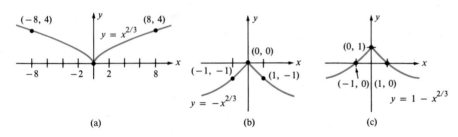

(a)   (b)   (c)

37. Increasing: $(0, \infty)$
Decreasing: $(-\infty, 0)$
Critical points: $x = 0$
Stationary points: $x = 0$

39. Increasing: $(-\infty, -1), (-1, 0)$
Decreasing: $(0, 1), (1, \infty)$
Critical points: $x = -1, 0, 1$
Stationary points: $x = 0$

41. Increasing: $(-\infty, -1), (3, \infty)$
Decreasing: $(-1, 3)$
Critical points: $x = -1, 3$
Stationary points: $x = -1, 3$

43. The slope is positive from $x = -4$ to $x = 0$. At $x = 0$ the slope is 0. From $x = 0$ to $x = 4$ the slope is negative. The slope increases from $x = -4$ to $x = 0$ and decreases from $x = 0$ to $x = 4$.

45. The slope is positive for all x different from zero. The slope increases from $x = -\infty$ to $x = 0$ and decreases from $x = 0$ to $x = \infty$.

47. $R(p) = p(9 - p^2)$

49. $P(q) = q\sqrt{36 - q} - 5 - 0.5q$

51. $R'(q) = \sqrt{64 - q} - \dfrac{q}{2\sqrt{64 - q}}$

$$P'(q) = q\sqrt{64 - q} - \frac{q}{2\sqrt{64 - q}} - 0.5 - 0.0024q$$

Section 9.2

1. $f''(x) = 6x + 2$
$f''(0) = 2$
$f''(1) = 8$
$f''(2) = 14$

3. $f''(x) = 2a$
$f''(0) = 2a$
$f''(1) = 2a$
$f''(2) = 2a$

5. $f''(x) = -\dfrac{1}{4}x^{-3/2}$

$f''(0)$ undefined

$f''(1) = -\dfrac{1}{4}$

$f''(2) = -\dfrac{1}{4}(2)^{-3/2}$

7. $f''(x) = \dfrac{-4}{(x+1)^3}$

$f''(0) = -4$

$f''(1)$ undefined, since $f(1)$ is undefined

$f''(2) = \dfrac{-4}{27}$

9. $f''(x) = 2$
$f''(0) = 2$
$f''(1) = 2$
$f''(2) = 2$

11. $f''(x) = 6x + 4$
$f''(0) = 4$
$f''(1) = 10$
$f''(2) = 16$

13. $f''(x) = 18x + 8$
$f'''(x) = 18$
$f^{(4)} = 0$

15. $f''(x) = n(n-1)x^{n-2}$
$f'''(x) = n(n-1)(n-2)x^{n-3}$
$f^{(4)}(x) = n(n-1)(n-2)(n-3)x^{n-4}$

17. $f''(x) = 12x^{-5} - \dfrac{1}{4}x^{-3/2}$

$f'''(x) = -60x^{-6} + \dfrac{3}{8}x^{-5/2}$

$f^{(4)}(x) = 360x^{-7} - \dfrac{15}{16}x^{-7/2}$

19. $f'(x) = 2x$
$f''(x) = 2$
   (a) Critical points: 0
   (b) Stationary points: 0
   (c) Increasing: $(0, \infty)$
   (d) Decreasing: $(-\infty, 0)$
   (e) Inflection points: none
   (f) Concave up: $(-\infty, \infty)$
   (g) Concave down: never

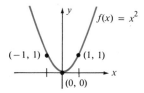

21. $f'(x) = 2x - 4$
$f''(x) = 2$
   (a) Critical points: 2
   (b) Stationary points: 2
   (c) Increasing: $(2, \infty)$
   (d) Decreasing: $(-\infty, 2)$
   (e) Inflection points: none
   (f) Concave up: $(-\infty, \infty)$
   (g) Concave down: never

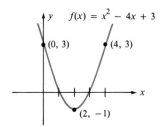

23. $f'(x) = 3x^2$
$f''(x) = 6x$
   (a) Critical points: 0
   (b) Stationary points: 0
   (c) Increasing: $(-\infty, 0), (0, \infty)$
   (d) Decreasing: never
   (e) Inflection points: 0
   (f) Concave up: $(0, \infty)$
   (g) Concave down: $(-\infty, 0)$

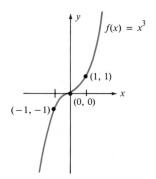

25. $f'(x) = 3x^2 - 1$
$f''(x) = 6x$
   (a) Critical points: $\pm\sqrt{1/3}$
   (b) Stationary points: $\pm\sqrt{1/3}$
   (c) Increasing: $(-\infty, -\sqrt{1/3}), (\sqrt{1/3}, \infty)$
   (d) Decreasing: $(-\sqrt{1/3}, \sqrt{1/3})$
   (e) Inflection points: 0
   (f) Concave up: $(0, \infty)$
   (g) Concave down: $(-\infty, 0)$

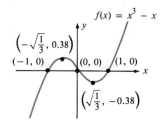

27. $f'(x) = 1 - \dfrac{2}{x^3}$

$f''(x) = \dfrac{6}{x^4}$

(a) Critical points: $0, 2^{1/3}$
(b) Stationary points: $2^{1/3}$
(c) Increasing: $(-\infty, 0), (2^{1/3}, \infty)$
(d) Decreasing: $(0, 2^{1/3})$
(e) Inflection points: none
(f) Concave up: $(-\infty, 0), (0, \infty)$
(g) Concave down: never

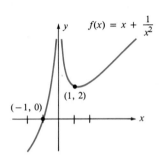

29. $f'(x) = 3(x - 2)^2$
$f''(x) = 6(x - 2)$
(a) Critical points: 2
(b) Stationary points: 2
(c) Increasing: $(-\infty, \infty)$
(d) Decreasing: never
(e) Inflection points: 2
(f) Concave up: $(2, \infty)$
(g) Concave down: $(-\infty, 2)$

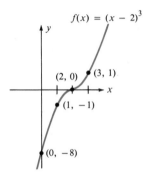

31. $f'(x) = 4(x + 2)^3$
$f''(x) = 12(x + 2)^2$
(a) Critical points: -2
(b) Stationary points: -2
(c) Increasing: $(-2, \infty)$
(d) Decreasing: $(-\infty, -2)$
(e) Inflection points: none
(f) Concave up: $(-\infty, \infty)$
(g) Concave down: never

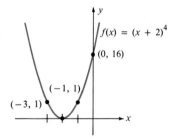

33. $f'(x) = \dfrac{1}{5} x^{-4/5}$

$f''(x) = -\dfrac{4}{25} x^{-9/5}$

(a) Critical points: 0
(b) Stationary points: none
(c) Increasing: $(-\infty, 0), (0, \infty)$
(d) Decreasing: never
(e) Inflection points: 0
(f) Concave up: $(-\infty, 0)$
(g) Concave down: $(0, \infty)$

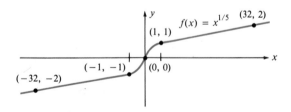

35. Concave down: $(-\infty, 1)$
Inflection point: none

37. Concave up: $(0, \infty)$
Concave down: $(-\infty, 0)$.
Inflection point: 0

39. Concave up: $(1, \infty)$
Concave down: $(-\infty, 1)$
Inflection point: 1

41. Plus infinity: 3
Minus infinity: 3
Horizontal asymptote: $y = 3$

43. Plus infinity: 3
Minus infinity: 3
Horizontal asymptote: $y = 3$

45. Plus infinity: 0
Minus infinity: 0
Horizontal asymptote: $y = 0$

47. Plus infinity: 9
Minus infinity: 9
Horizontal asymptote: $y = 9$

49. Plus infinity: ∞
Minus infinity: $-\infty$
No horizontal asymptote

51. Vertical asymptote: $x = -1$
Horizontal asymptote: $y = 0$

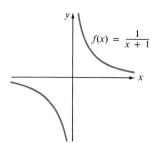

53. Vertical asymptote: $x = -1$
Horizontal asymptote: $y = 1$

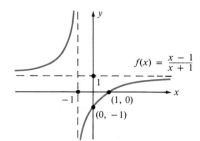

55. Vertical asymptote: $x = 1, -1$
Horizontal asymptote: $y = 0$

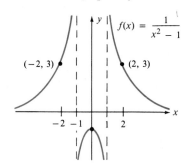

57. Vertical asymptote: $x = 2$
Horizontal asymptote: $y = 1$

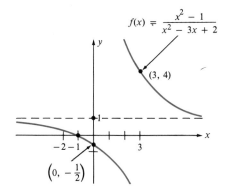

59. Vertical asymptote: $x = 1, 2, 3$
Horizontal asymptote: $y = 0$

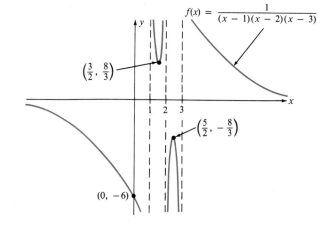

Section 9.3

1. $(a, b), (c, d), (e, f), (g, h)$ **3.** g, i **5.** c, e, g
7. e **9.** i **11.**

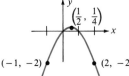

13.

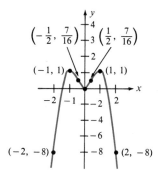

15. No relative max or min points   **17.** Relative minimum: $x = \dfrac{1}{2}$

19. Relative maximum: $-\sqrt{\dfrac{1}{3}}$ **21.** Relative minimum: $x = 12$ **23.** Relative maximum: $x = -2$

Relative minimum: $\sqrt{\dfrac{1}{3}}$ Relative minimum: $x = \dfrac{1}{2}$

25. First derivative test does not apply at $x = 0$; however, $x = 0$ is a relative minimum
27. No relative maximum or minimum
29. Relative minimum at $x = 0$ **31.** Second derivative test does not apply, since $f'(x)$ is never zero.
33. Critical point $x = 0$ is a relative minimum   **35.** Critical point $x = 1$ is a relative minimum
37. Critical point $x = 1$ (relative minimum)   **39.** Critical point $x = \sqrt{2}$ (relative maximum)
Critical point $x = -2$ (relative minimum)   Critical point $x = -\sqrt{2}$ (relative minimum)
Critical point $x = -1$ (relative maximum)
41. Critical point $x = 0$ (second derivative does not apply)   **43.** Critical point $x = 3$ (relative minimum)
Critical point $x = 1$ (relative maximum)
45. Absolute maximum: $x = 2$ **47.** Absolute maximum: $x = 2$ **49.** Absolute maximum: $x = 0$
Absolute minimum: $x = -1$ Absolute minimum: $x = 5$ Absolute minimum: $x = 3$
51. Absolute maximum: $x = 1$ **53.** Absolute maximum: $x = 4$ **55.** Absolute maximum: $x = 2$
Absolute minimum: $x = -\dfrac{1}{2}$ Absolute minimum: $x = 2$ Absolute minimum: $x = 3$

57. Absolute maximum: $x = -1, 1$ **59. (a)** 79 (approximately)   **(b)** Evaluate the cost at the nearest integer values.
Absolute minimum: $x = 0$

61. 6   **63.** Maximum at $q = \dfrac{98}{3}$ **65.** Maximum profit $= \dfrac{25}{8}$, $q = \dfrac{5}{4}$

Maximum profit $= \$17.179$

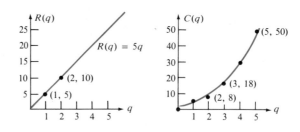

67. The function has no maximum on the open interval $(0, 1)$.   **69.** Student project

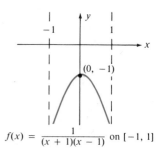

$f(x) = \dfrac{1}{(x + 1)(x - 1)}$ on $[-1, 1]$

Section 9.4

1. $x = y = 7.5$ **3.** $x = 25$ **5.** 70.7   **7.** $50   **9.** 1250   **11.** $1750   **13.** 25 trees
                          $y = 50$
15. $x = 10$ feet   **17.** 10 by 10 by 20 feet   **19.** $x = \dfrac{32 + \sqrt{304}}{24} = 2.05$ feet
$y = 10$ feet

$y = \dfrac{32 - \sqrt{304}}{24} = 0.607$ feet, $V(.607) = 4.104$

21. $x = \dfrac{25}{4 + \pi} \cong 3.5$ inches, $4x \cong 14$ inches   **23.** $x = \sqrt{160/3}$ inches (7.3 inches)   **25.** 18 square inches

$r = \dfrac{25}{8 + 2\pi} \cong 1.75$ inches, $2\pi r \cong 11$ inches $y = 80\sqrt{3/160}$ inches (10.95 inches)
(page is 11.3×16.95 inches)

27. $x = \dfrac{14}{\sqrt{5}}$ (x as given in figure)   **29.** \$7.50   **31.** 500 by 1000 feet   **33.** $t = 5\sqrt{2}$ weeks
Maximum size $= 250\sqrt{2} \cong 353.6$

Section 9.5

1. $\dfrac{dy}{dx} = 2x$ **3.** $\dfrac{dy}{dx} = -\dfrac{2y}{x}$ **5.** $\dfrac{dy}{dx} = -\dfrac{1+y}{x}$ **7.** $\dfrac{dy}{dx} = \dfrac{1 - y - 3x^2}{x}$ **9.** $\dfrac{dy}{dx} = \dfrac{1-y}{x + 2y}$ **11.** $\dfrac{dy}{dx} = -\dfrac{9x}{16y}$

13. $\dfrac{dy}{dx} = \dfrac{x-y}{x+y}$ **15.** $\dfrac{dy}{dx} = \dfrac{-x}{y(x^2 - 1)^2}$ **17.** $\dfrac{dy}{dx} = \dfrac{3x^2 + 4x - 2xy - y^2}{2xy + x^2}$ **19.** $\dfrac{dy}{dx} = \dfrac{-2xy}{x^2 + 1}, 0$ **21.** $\dfrac{dy}{dx} = \dfrac{-2y}{x}, -2$

23. $\dfrac{dy}{dx} = \dfrac{2x - 3y}{3x}, \dfrac{7}{6}$ **25.** $\dfrac{dy}{dx} = -\dfrac{\sqrt{y}}{\sqrt{x}} \dfrac{(\sqrt{y}+1)}{(\sqrt{x}+1)}, -3$ **27.** $y = -2x + 3$ **29.** $y = x - 2$ **31.** $y = 3$

33. $y = -3x + 5$ **35.** $y = \dfrac{2x}{\sqrt{3}} - \dfrac{1}{\sqrt{3}}$ **37.** $\dfrac{dy}{dt} = 4$ **39.** $\dfrac{1}{\pi} = 0.32 \dfrac{\text{inch}}{\text{sec}}$ **41.** $\dfrac{50\pi}{3} = 52.4 \dfrac{\text{cubic feet}}{\text{min}}$

43. (a) $t = 12$ min   **(b)** Minimum distance $= 10$ miles   **45.** $\dfrac{95}{3} = 31.67$ weeks

Section 9.6

1. $dy = 0$ **3.** $dy = 0.02$ **5.** $dy = -0.1$ **7.** $dy = -0.09$ **9.** $dy = 0.0375$ **11.** 0.3 million   **13.** 1.96%
15. 0.04   **17.** 0.03   **19.** 0.06   **21.** 1.29   **23.** 0   **25.** 2500.7   **27.** 0.55   **29.** 0.91   **31.** Profit = \$2000
Relative error = 0.01

33. Revenue = \$100,000   **35.** Actual change: $I(r_0 + dr) - I(r_0)$ **37.** $\dfrac{2x}{k_2 - x}$%
Relative error = 0.0075   Approximate change: $dI = 2cr_0\, dr$

Chapter Review

1. $x = -\dfrac{2}{3}$ (relative max)   **3.** $x = \sqrt{2}$ (relative max)   **5.** $x = \dfrac{1}{2}$ (relative min)   **7.** $x = 3$ (relative min)
$x = -\sqrt{2}$ (relative min) $x = -1$ (relative max)
$x = 2$ (relative min)
9. $x = 1$ (stationary point)   **11.** $x = 0$ (stationary point)   **13.** Absolute max: $f(0)$
Second derivative test inconclusive   Second derivative test inconclusive   Absolute min: $f(3)$ and $f(1)$
15. Absolute max: $f(0), f(4)$ **17.** Absolute max: $f(1)$ **19.** Absolute max: $f(4)$ **21.** Absolute max: $f(1)$
Absolute min: $f(2)$ Absolute min: $f(0)$ Absolute min: $f(1)$ Absolute min: $f(0)$

23. $x = -b/3a$ **25.** $\dfrac{dy}{dx} = -\dfrac{y}{x}$ **27.** $\dfrac{dy}{dx} = -2\dfrac{y}{x}$ **29.** $\dfrac{dy}{dx} = 1$ **31.** $\dfrac{dy}{dx} = \dfrac{2\sqrt{xy} - y}{x}$ **33.** $\dfrac{dA}{dt} = 2\pi r \dfrac{dr}{dt}$
35. shadow is shortening by 40/3 ft/sec   **37.** 95   **39.** 1000.00333   **41.** 1.009   **43.** 0.01001   **45.** 0.03
47. $x = y = 5$

Chapter Test

1. (a) Increasing: $\left(\dfrac{1}{2}, \infty\right)$,   **(b)** $x = \dfrac{1}{2}$ **(c)** $x = \dfrac{1}{2}$ **(d)** Relative min: $x = \dfrac{1}{2}$ **(e)** No inflection points
Decreasing: $\left(-\infty, \dfrac{1}{2}\right)$

(f) Concave up: $(-\infty, \infty)$ **(g)** Absolute max: $x = 5$ **2.** $\dfrac{dy}{dx} = -\dfrac{2y^{3/2}}{2xy^{1/2} + 1}$ **3.** $\dfrac{dy}{dx} = 3x^2 \dfrac{dx}{dt}$ **4.** 50 and 50
5. 0.10

CHAPTER 10

Section 10.1

1. 1, 3, 9, 27, 81, 243   **3.** $\frac{1}{2}$, 2, 8, 32, 128, 512   **5.** 2, 6, 18, 54, 162, 486   **7.** 1, 0.1, 0.01, 0.001, 0.0001, 0.00001

9. 3, −9, 27, −81, 243, −729   **11.** Yes, $\frac{226}{199} = 1.13$ **13.** 1990: 283.61 million
                                                          2000: 319.17 million

15.
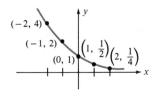

17.
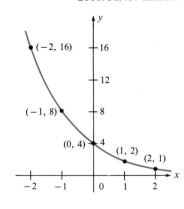

19.
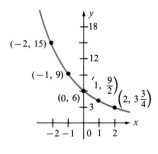

21. −1   **23.** 1   **25.** 0, $\frac{5}{3}$ **27.** −1, 2   **29.** \$4202 billion   **31.** Japan: \$2336 billion
                                                                     United States: \$5927 billion

33. Six months: \$105   **35.** 4.95, 4.90, 4.85, 4.80, 4.75 billion   **37.** $G(t) = 500(0.8)^t$ **39.** $V(t) = 20(0.9)^t$
     One year: \$110.25 $= 20e^{-0.105t}$
     18 months: \$115.76 $V(60) = 3.6 \times 10^{-2}$ mile/hr
     Two years: \$121.55

41. $E(1) = 511.7$ **43.** $P(1) = 800$ **45.** $P(t) = \$500e^{0.09t}$ **47.** $P(t) = \$10e^{0.05t}$ **49.** Week 1: 9.5 units
     $E(2) = 523.6$ $P(2) = 1600$ $P(1) = \$547.09$ $P(75) = \$425.21$ Week 2: 18.1 units
     $E(3) = 535.9$ $P(4) = 6400$ $P(1.5) = \$572.27$ Week 3: 25.9 units
     $E(4) = 548.4$ $P(10) = 409,600$ Week 4: 32.9 units
     $E(10) = 630.0$ $P(20) = 4.19 \times 10^8$ Week 5: 39.3 units

51. Present: 1 g   **53.** $y(-1) = 1.54$ **55.** $500(250)^{25}$ (huge)
     1000 years: 0.83 g $y(-0.5) = 1.13$
     2000 years: 0.70 g $y(0) = 1$
     4000 years: 0.48 g $y(0.5) = 1.13$
     8000 years: 0.23 g $y(1) = 1.54$

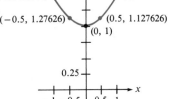

Section 10.2

1. 1.55   **3.** 0.15   **5.** −0.7   **7.** 7.05   **9.** 2.1   **11.** 4.075   **13.** 1.00   **15.** $\frac{3}{4}$ **17.** 9

19. $3 \log e - \log a$ **21.** $2^3 \cdot 3^2$ **23.** 2^{14} **25.** −1   **27.** 3   **29.** $\pm\frac{3}{2}$ **31.** $\frac{4}{3}$ **33.** $\frac{1}{2 \log_2 10 - 1}$

35. $\frac{\log_5 6}{\log_5 6 - 3}$ **37.** 6.94 years   **39.** 6.60 years   **41.** Double: 4.6 years   **43.** 9.78%   **45.** 1.58 hours
                                                                 Triple: 7.3 years

47. 12.88 days   **49.** 3960 units   **51.** (a) 5.38   (b) 5.10   (c) 7.22   (d) 3.20   **53.** $10^{4.5}$ **55.** 30 decibels

Section 10.3

1. $e^{x \ln 2}$ **3.** $e^{x \ln 4}$ **5.** $e^{x \ln 1.5}$ **7.** $e^{x \ln 0.9}$ **9.** $e^{x \ln 100}$ **11.** $y' = 6e^{3x}$ **13.** $y' = 5e^{5x}$ **15.** $y' = 2xe^{x^2}$

17. $y' = \dfrac{e^{\sqrt{x}}}{2\sqrt{x}}$ **19.** $y' = \dfrac{xe^x - e^x}{x^2}$ **21.** $y' = (2x + 1)e^{x^2 + x}$ **23.** $y' = e^x - e^{-x}$ **25.** $y' = 2 - 4e^{4x} + 6xe^{x^2}$

27. $y' = (\ln 5)\,5^x$ **29.** $y' = (\ln 2)\,2^x$ **31.** $y' = (2x \ln 2)2^{x^2}$ **33.** $y' = (2 \ln 9)\,9^{(2x + 5)}$ **35.** $y'' = e^x > 0$ for all x
(y concave up for all x)

37. $y' = (\ln 2)e^{x \ln 2} > 0$ for all x **39.** $y = x$ **41.** 9.8 years   **43.** 14.8 years
$y'' = (\ln 2)^2 e^{x \ln 2}$ (y concave up for all x)

45. Now (put money in bank)   **47.** $\dfrac{dw}{dt} = \dfrac{(500)49ke^{-kt}}{(1 + 49ke^{-kt})^2}$ **49.** $\dfrac{dN}{dt} = 3bkN_0 e^{-kt}(1 - be^{-kt})^2$

Section 10.4

1. $\dfrac{dy}{dx} = \dfrac{3}{3x + 5}$ **3.** $\dfrac{dy}{dx} = \dfrac{-x}{4 - x^2}$ **5.** $\dfrac{dy}{dx} = \dfrac{1}{x + 1}$

7. $\dfrac{dy}{dx} = \dfrac{e^x - e^{-x}}{e^x + e^{-x}}$ **9.** $\dfrac{dy}{dx} = \dfrac{-2}{x}$ **11.** $\dfrac{dy}{dx} = \dfrac{2(1 - \ln x^2)}{x^3}$ **13.** $\dfrac{dy}{dx} = \ln x$ **15.** $\dfrac{dy}{dx} = e^x \ln (2x) + \dfrac{e^x}{x}$ **17.** $\dfrac{dy}{dx} = \dfrac{\ln x}{\ln 5}$

19. $\dfrac{dy}{dx} = x$ **21.** $\dfrac{dy}{dx} = \dfrac{\ln (x^2 + 3x + 5)}{\ln 10}$ **23.** $\dfrac{dy}{dx} = \dfrac{\ln (e^x + 1)}{\ln 2}$ **25.** $\dfrac{dy}{dx} = \dfrac{\left(\dfrac{\ln x}{\ln 10}\right) + 1}{\ln 2}$ **27.** $\dfrac{dy}{dx} = \dfrac{1}{\ln 10(x + 1)}$

29. $\dfrac{dy}{dx} = \dfrac{2x}{\ln 10(x^2 + 1)}$ **31.** $\dfrac{dy}{dx} = \dfrac{-1}{x \ln 10}$ **33.** $\dfrac{dy}{dx} = \dfrac{2x + 3}{\ln 10(x^2 + 3x + 1)}$ **35.** $\dfrac{dy}{dx} = \dfrac{-1}{x \ln 2}$ **37.** Yes, $y' = y$

39. e^2 **41.** $y' = 1/x$ (x different from 0)   **43.** Elastic: $100 < p < 200$ **45.** $E(p) = 3$
Inelastic: $0 < p < 100$ Always elastic

47. Elastic: $533.33 < p < 800$ **49.** (a) $E(p) = 2$ (b) $R'(p) = -5000/p^2$ (c) Yes; as p decreases, revenue increases.
Inelastic: $0 < p < 533.33$

51. (a) $E(p) = \dfrac{p}{2(8000 - p)}$ (b) Inelastic   (c) Inelastic   (d) Raised   (e) Lowered   **53.** $\dfrac{y'}{y} = r$

55.

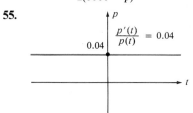

57. $y' = \left(\dfrac{3}{x} - \dfrac{2x}{x^2 + 1} - \dfrac{3}{3x - 4}\right)\dfrac{x^3}{(x^2 + 1)(3x - 4)}$

Chapter Review

1. 2   **3.** -2 **5.** 2   **7.** 4   **9.** e^{-1} **11.** 0   **13.** $f'(x) = 2e^{(2x + 1)}$ **15.** $f'(x) = \dfrac{e^x}{\ln 10}\left[\ln x + \dfrac{1}{x}\right]$

17. $f'(x) = \dfrac{1}{2x}$ **19.** $f'(x) = 1$ **21.** $f'(x) = \dfrac{e^x - e^{-x}}{2}$ **23.** $f'(x) = \dfrac{1}{x \ln 2}$ **25.** $f'(x) = \dfrac{-4x}{(x^2 + 1)(x^2 - 1)}$

27. $y' = -2xe^{-x^2}$, $y'' = -2e^{-x^2}[1 - 2x^2]$ **29.** $R'(100) = 0$ **31.** After 1.56 years   **33.** 3.16%   **35.** $r = \ln (1 + i)$

37. (a) $E(p) = \dfrac{2p}{-2p + 110}$ (b) Elastic: $p > 27.50$ (c) $R(p) = -2p^2 + 110p$ (d) $R'(p) = -4p + 110$
Inelastic: $0 < p < 27.50$

39. $P(t) = 65e^{t \ln (0.85)}$ **41.** 13,055 years   **43.** 15.6 times stronger

Chapter Test

1. $\dfrac{dy}{dx} = 2x \ln(x^2 + 1) + \dfrac{2x^3}{x^2 + 1}$ **2.** $\dfrac{dy}{dx} = \dfrac{1}{2 \ln 10}\left[\dfrac{x^2 - 1}{x(x^2 + 2x + 1)}\right]$ **3.** $P(t) = P_0 e^{t \ln(1.05)}$ **4.** 27.73 months

5. $x = e$ **6.** 11.61%

CHAPTER 11

Section 11.1

1. $y(x) = \dfrac{x^2}{2} + 3x + C$ **3.** $y(x) = \dfrac{1}{8}x^4 - \dfrac{2}{9}x^3 + \dfrac{x}{5} + C$ **5.** $y(x) = x^4 + 3x^{-1} + \dfrac{x^2}{2} - 3x + C$

7. $y(t) = \dfrac{1}{2}e^{2t} - e^{-3t} + C$ **9.** $y(x) = \dfrac{14}{3}x^{3/2} - e^{-2x} + C$ **11.** $y(t) = -\dfrac{1}{3}e^{-3t} + C$ **13.** $y(t) = -2e^{-0.5t} + C$

15. $F(x) = \dfrac{2}{3}(x^3 + 2x + 3)^{3/2} + C$ **17.** $F(x) = \dfrac{2}{3}x^3 + \dfrac{5}{4}x^4 + \dfrac{1}{5}x^5 + C$ **19.** $F(t) = \dfrac{4}{3}t^3 + 6t^2 + 9t + C$

21. $F(x) = -\dfrac{1}{2}(x^6 - 3)^{-2} + C$ **23.** $F(x) = \dfrac{1}{5}e^{5x} + C$ **25.** $F(t) = -25\,e^{-0.04t} + C$

27. $F(x) = \dfrac{2}{3}(3x + 2)^{3/2} + C$ **29.** $F(x) = \dfrac{1}{4}(x^2 + 1)^4 + C$ **31.** $f(x) = 1$ **33.** $f(s) = -s + 1$

35. $f(x) = \dfrac{3}{2}x^2 + 4x - 8$ **37.** $f(u) = \dfrac{1}{3}(u - 2)^3 - \dfrac{2}{3}$ **39.** $f(u) = u^2 - u^{-1} - 1$ **41.** $f(x) = \dfrac{3}{4}x^{4/3} + 48x^{1/3} - 96.6$

43. $f(t) = \dfrac{t^2}{2} + \dfrac{2}{3}t^{3/2} - \dfrac{34}{3}$ **45.** $f(t) = -100e^{-0.01t} + 101$ **47.** $f(x) = \dfrac{2}{3}(x^2 - 9)^{3/2} - \dfrac{113}{3}$

49. $f(x) = \dfrac{(x^2 + 1)^4}{4} - \dfrac{1}{4}$ **51.** $C(q) = 20q^2 - \dfrac{0.01}{3}q^3 + 100q + 100$ **53.** $C(q) = q^2 - \dfrac{0.02}{3}q^3 + 500$

55. $C(q) = 50e^{0.15q} + 30$ **57.** $P(q) = 10q - \dfrac{1}{4}q^2$ **59.** $R(q) = 10q - 0.04q^2 - \dfrac{0.10}{3}q^3$

61. $C(100) = \$9166.66$ **63.** $p(t) = -50e^{-0.02t} + 50$ **65.** $s(1) = 50$ miles   **67.** $s(5) = 400$ feet   **69.** $f(x) = e^x + 1$
$C(200) = \$48,333.33$
$C(300) = \$745,833.33$

Section 11.2

1. Left endpoint: $A = 16$
Right endpoint: $A = 80$
Midpoint: $A = 32$

3. Left endpoint: $A = 30$
Right endpoint: $A = 46$
Midpoint: $A = 37$

5. Left endpoint: $A = \dfrac{189}{8}$
Right endpoint: $A = \dfrac{81}{8}$
Midpoint: $A = \dfrac{297}{16}$

7. Left endpoint: $A = 0$
Right endpoint: $A = 8$
Midpoint: $A = 4$

9. Left endpoint: $A = 3$
Right endpoint: $A = 5$
Midpoint: $A = 4$

11. 2 subintervals: $A = 2$
4 subintervals: $A = 2$

13. 2 subintervals: $A = 16$
4 subintervals: $A = 16$

15. 2 subintervals: $A = \dfrac{113}{2}$
4 subintervals: $A = 56.625$

17. 2 subintervals: $A = \dfrac{24}{35}$
4 subintervals: $A = 0.6912$

19. 15   **21.** $\dfrac{1}{3}$ **23.** 31

Section 11.3

1. 10   **3.** 19.5   **5.** 1016.75   **7.** 12   **9.** 60   **11.** 29.6   **13.** 384   **15.** $\dfrac{e^2}{2} - \dfrac{1}{e} + \dfrac{1}{2}$ **17.** $\dfrac{65}{12}$ **19.** $\dfrac{33}{10}$

21. $4[\sqrt{33} - \sqrt{12}]$ **23.** $\dfrac{1}{12}$ **25.** $\dfrac{(a + b + c)^2}{2} - \dfrac{c^2}{2}$ **27.** 24   **29.** 12   **31.** 6   **33.** $\dfrac{200}{3}$ **35.** $\dfrac{67 - 16\sqrt{2}}{12}$

37. $e - 1$ **39.** $-15e^{(-\ln 2)/5} + 15$ **41.** $3 - e^{-2}$ **43.** $\dfrac{1}{2}(e - 1)$ **45.** $3e^2 - 3e + 1$ **47.** Revenue increase: $1587.50

49. Increase in profits: 880 units   **51.** Total revenue: $1,036,800   **53.** 336 feet

Section 11.4

1. $F(x) = \dfrac{1}{24}(3x^2 - 5)^4 + C$ **3.** $F(w) = \dfrac{(w^2 + 3)^3}{3} + C$ **5.** $F(x) = \dfrac{6}{5}(2x - 5)^{5/2} + C$ **7.** $F(t) = \ln|t + 3| + C$

9. $F(x) = 2\ln|x - 3| + C$ **11.** $F(x) = \dfrac{(\sqrt{x} + 2)^6}{3} + C$ **13.** $F(x) = \dfrac{2}{3}(x^3 + 12x^2 + 3)^{3/2} + C$

15. $F(x) = \dfrac{2}{3}(e^x + 1)^{3/2} + C$ **17.** $F(u) = \dfrac{1}{2}\ln|e^{2u} + e^{-2u}| + C$ **19.** $F(x) = \dfrac{[\ln(x^2 + 4)]^2}{2} + C$

21. $F(x) = -\dfrac{1}{2}e^{-x^2} + C$ **23.** $F(x) = x - 2\ln|x + 2| + C$ **25.** $F(u) = \dfrac{2}{5}(u - 1)^{5/2} - \dfrac{2}{3}(u - 1)^{3/2} + C$

27. $\dfrac{1}{12}[7^4 - 4^4]$ **29.** $\dfrac{3}{2}\ln 2$ **31.** $\dfrac{1}{2}\ln 11 - \dfrac{1}{2}\ln 5$ **33.** $\dfrac{2}{3}$ **35.** 9   **33.** $\dfrac{2}{9}[56\sqrt{7} - 64]$ **39.** $\dfrac{1}{15}[1281^{3/2} - 6^{3/2}]$

41. $\dfrac{2}{3}[8(17)^{3/4} - 24^{3/4}]$ **43.** $3(2 - \sqrt{3})$ **45.** $\dfrac{1}{2}$ **47.** $\dfrac{7}{24}$ **49.** Direct computation   **51.** Student project

Chapter Review

1. $F(x) = 7x + C$ **3.** $F(u) = \dfrac{2}{7}u^{7/2} + \dfrac{2}{3}u^{3/2} + C$ **5.** $F(x) = e^x - e^{-x} + C$ **7.** $f(t) = t^2 + 5t$ **9.** $f(x) = e^x$

11. $f(u) = \dfrac{1}{3}(u^2 - 9)^{3/2} - \dfrac{64}{3}$ **13.** $C(q) = \dfrac{45}{2}q^2 - \dfrac{2}{3}q^3 + 10,000$ **15.** $C(q) = 10q + 4e^{-0.5q} + 996$ **17.** $\dfrac{4}{3}$

19. $\dfrac{3}{2} - e^{-1}$ **21.** e

23. $\displaystyle\int_0^5 (x + 1)\,dx$

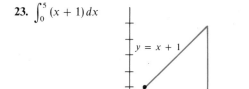

$y = x + 1$

25. $\displaystyle\int_0^2 (4 - x^2)\,dx$

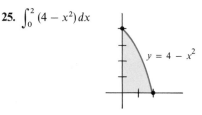

$y = 4 - x^2$

27. $\int_{-1}^{1} (e^x + e^{-x})\,dx$

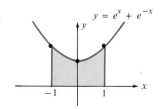

29. $\int_{1}^{2} x^{-n}\,dx$

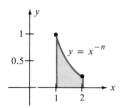

31. $F(x) = \dfrac{1}{8}(x^2 - 1)^4 + C$ **33.** $F(u) = \dfrac{(u+1)^7}{7} - \dfrac{(u+1)^6}{6} + C$ **35.** $F(x) = \dfrac{5(x+1)^{11}}{11} + C$

37. $F(x) = 5 \ln|x - 1| + C$ **39.** $F(t) = \ln|t + 1| + C$ **41.** $\dfrac{1}{3}17^{3/2} - \dfrac{1}{3}$ **43.** $\dfrac{1}{3}\left[128 - 2(11)^{3/2}\right]$

45. $R(q) = 100q - \dfrac{5}{2}q^2$ **47.** $5000

49. (a) 1.9% **(b)** 27.5% **(c)** $L(x)$ is increasing   **(d)** $L(x)$ is concave up   **(e)** $CI = 0.3$

Chapter Test

1. $F(t) = \dfrac{1}{2}t^2 + t + \ln|t| + C$ **2.** $\dfrac{3}{16}$ **3.** $F(u) = -\dfrac{2}{9}(1 - u^3)^{3/2} + C$ **4.** $10[10 + 2(e^{-1} - e^{-2})]$ **5.** $-\$29,500$ (loss)

CHAPTER 12

Section 12.1

1. ∞ **3.** ∞ **5.** ∞ **7.** ∞ **9.** $-\dfrac{1}{2}$ **11.** ∞ **13.** $-\infty$ **15.** $-\infty$ **17.** 0 **19.** ∞ **21.** $\dfrac{1}{k}$

23. $-\dfrac{1}{2}$ **25.** ∞ **27.** $\dfrac{15,000}{0.07} = \$357,142.86$ **29.** $\dfrac{100,000}{.06}(1 - e^{-3}) = \$1,583,686.22$ **31.** $\dfrac{10,000}{0.13} = \$76,923$

33. $125   **35.** Scenario 1: $7500
             Scenario 2: $1000

Section 12.2

1. $F(x) = \dfrac{1}{3}xe^{3x} - \dfrac{1}{9}e^{3x} + C$ **3.** $F(x) = x(\ln x)^2 - 2(x \ln x - x) + C$ **5.** $F(x) = \dfrac{x^2}{3}(x^2 + 2)^{3/2} - \dfrac{1}{3}\left[\dfrac{2}{5}(x^2 + 2)^{5/2}\right] + C$

7. $F(x) = \dfrac{x^2}{2}(\ln x)^2 - \dfrac{x^2}{2}\ln x + \dfrac{x^2}{4} + C$ **9.** $F(x) = -x^2\sqrt{1 - x^2} - \dfrac{2}{3}(1 - x^2)^{3/2} + C$ **11.** $F(x) = -\dfrac{ax}{b}e^{-bx} - \dfrac{a}{b^2}e^{-bx} + C$

13. $F(x) = \dfrac{ax^2 \ln(bx)}{2} - \dfrac{ax^2}{4} + C$ **15.** $F(x) = -\dfrac{ax^2}{b}e^{-bx} + \dfrac{2}{b}\left[-\dfrac{ax}{b}e^{-bx} - \dfrac{a}{b^2}e^{-bx}\right] + C$ **17.** $\dfrac{1}{4}(e^2 + 1)$

19. $10.25e^{10} - 0.25e^2$ **21.** $e^e(e - 1)$ **23.** $12,500 + 25,000e^{-10}$ **25.** $\dfrac{12,500}{e} = 4598$ hits in 50 hours

Batting average: $\dfrac{1}{e} = 0.368$ (not bad)

27. $500 million dollars (*Hint*: $ne^{-0.1n}$ approaches 0 as n gets large.)   **29.** No, $39,075

Section 12.3

1. Area: 42

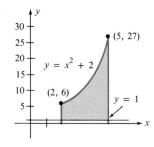

3. Area: $7\frac{7}{8}$

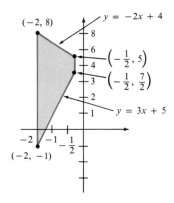

5. Area: $\dfrac{93}{2} - \dfrac{2}{3}\left(6^{3/2} - 3^{3/2}\right)$

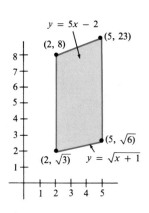

7. Area: $e^2 - 2\ln 2 + \dfrac{3}{4}$

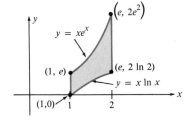

9. Area: 4

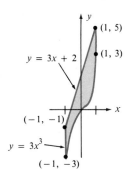

11. Area: 16.5

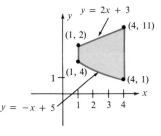

13. Area: $\dfrac{40}{3}$

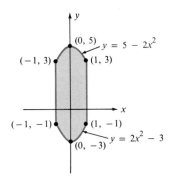

15. $\dfrac{128\sqrt{2}}{3}$

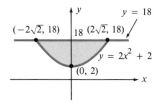

17. $\dfrac{1}{12}$

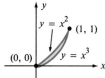

19. $\dfrac{125}{6}$

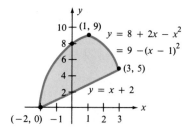

21. $\dfrac{36}{\sqrt{2}}$

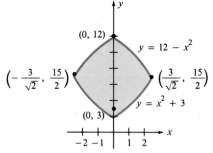

23. $\dfrac{1647}{5}\pi$

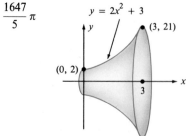

25. $\dfrac{2}{3}\pi$

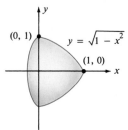

27. $\dfrac{1}{4}e^{-4}(1 - e^{-12})\pi$

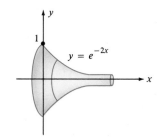

29. $P_S = \dfrac{224}{9} = 24.88$

$C_S = \dfrac{484}{9} = 53.78$

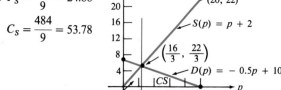

31. $P_S = \dfrac{200}{3}$

$= 66.67$

$C_S = 400$

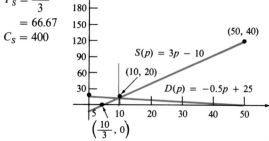

33. $P_S = 38.9$

$C_S = 35.1$

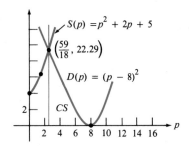

35. $P_S = 9$

$C_S = 5.78$

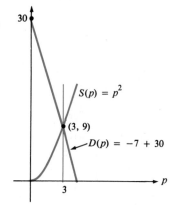

37. $P_S = 0$
$C_S = 0$

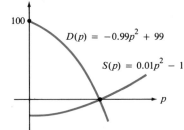

39. (a) $(5, 7)$
(b) 7 million dollars
(c) 25
(d) 13

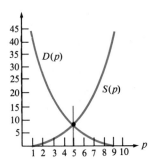

Section 12.4

1. Exact value: 8
Trapezoid rule: 8
Simpson's rule: 8

3. Exact value: 28
Trapezoid rule: 28
Simpson's rule: 28

5. Exact value: $\dfrac{1}{3}$
Trapezoid rule: 0.34375
Simpson's rule: $\dfrac{1}{3}$

7. Exact value: 64
Trapezoid rule: 68
Simpson's rule: 64

9. Exact value: 8
Trapezoid rule: 8
Simpson's rule: 8

11. Exact value: $\ln 2 \cong 0.6931472$
Trapezoid rule: 0.8845
Simpson's rule: 0.8183

13. Exact value: We cannot integrate
Trapezoid rule: 1.4907
Simpson's rule: 1.4637

15. Exact value: 1
Trapezoid rule: 1.023
Simpson's rule: 1.001

17. Trapezoid rule: 2.205
Simpson's rule: 2.179

19. Trapezoid rule: 1.106
Simpson's rule: 1.107

21. Trapezoid rule: 1.260
Simpson's rule: 1.290

23. 1.106

25. 0.004152

27. 225.25 **29.** 232.5 **31.** P_S (trapezoid rule) = 16 **33.** P_S (exact) = 16 **35.** C_S (Simpson's rule) = 10

Section 12.5

1. $y(t) = t + C$

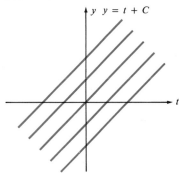

3. $y(t) = Ce^{3t}$

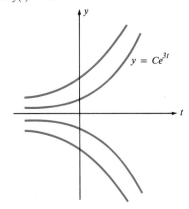

5. $y(t) = Ce^{t}$

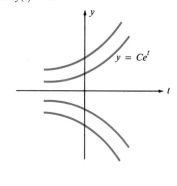

7. $y(t) = Ce^{0.05t}$

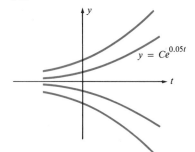

9. $y(t) = Ce^{10t}$

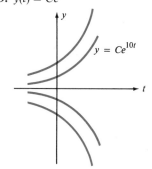

11. $y(t) = 100e^{t}$

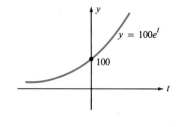

13. $N(t) = 50e^{-2t}$

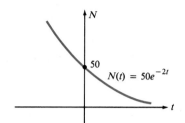

15. $N(t) = 50e^{-0.08t}$

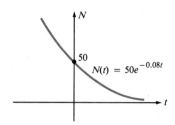

17. $P(t) = e^{0.05t}$

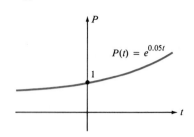

19. $P(t) = 1000e^{0.12t}$

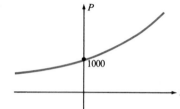

21. $y(t) = \dfrac{t^3}{3} + 1$ **23.** $y(t) = \dfrac{t^3}{3} + t^2 + t + 3$ **25.** $y(t) = t$ **27.** $y(t) = e^{0.05t}$ **29.** $y(t) = 100$

31. $y(t) = e^{0.5t}$ **33.** $k = \ln(.95) = -0.051$ **35.** $P(t) = P_0(1.05)^t$ **37.** $t_h = \dfrac{\ln 2}{9.2} \times 10^8 = 7{,}534{,}208$ years
$$= P_0 e^{0.0488t}$$

39. (a) $P(t) = P_0 e^{-0.0001237t}$ **(b)** $P(t) = 10e^{-0.0001237t}$

Chapter Review

1. $F(x) = \dfrac{x}{3} e^{3x} - \dfrac{1}{9} e^{3x} + C$ **3.** $F(x) = 2x \ln x - 2x + C$ **5.** $F(x) = \dfrac{2}{5}(x+4)^{5/2} - (x+4)^{3/2} + C$

7. $\dfrac{2}{3}$

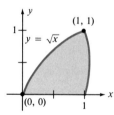

9. $e - \dfrac{1}{e}$

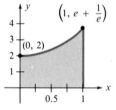

11. $\dfrac{1}{3}$

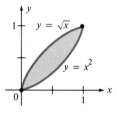

13. $\dfrac{8}{3}\sqrt{2}$

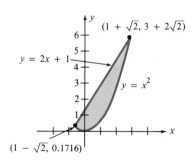

15. $\dfrac{4}{3\sqrt{2}}$

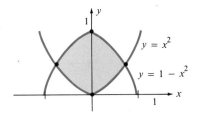

17. $\dfrac{\pi}{3}$

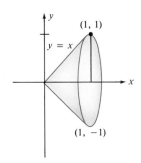

19. $\dfrac{3}{5}\pi$

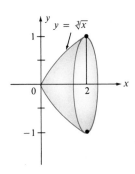

21. Trapezoid rule: 0.3437   **23.** Trapezoid rule: 1.1667   **25.** 140.5 feet   **27.** Producer's surplus = $3.94
    Simpson's rule: 0.3229   Simpson's rule: 1.1111

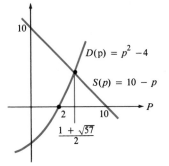

29. $P(t) = P_0 e^{t \ln (1.15)}$ **31.** 2   **33.** $\dfrac{1}{2}$ **35.** 1   **37.** $y(t) = e^{0.08t}$ **39.** $N(t) = 50e^{0.15t}$ **41.** $P(t) = 15e^{0.10t}$

Chapter Test

1. $\dfrac{1}{36}(1 + 5e^6)$ **2.** $\dfrac{65}{3}$ **3.** $\dfrac{4}{3}$ **4.** $\dfrac{28}{15}$ **5.** The consumers' and producers' surplus are both zero.   **6.** $\dfrac{1}{3}$

7. $N(t) = 10e^{0.08t}$

CHAPTER 13

Section 13.1

1. 875   **3.** 950   **5.** 150   **7.** $f(0, \ln 3) = 4$ **9.** $f(2, 4) = -3$ **11.** $f(-3, 1) = \dfrac{3}{2}$ **13.** $f(1, 3) = 3$ **15.** $3\sqrt{10}$
                                $f(\ln 4, \ln 5) = \dfrac{9}{4}$ $f(3, 7) = -\dfrac{5}{2}$ $f(2, 5) = \dfrac{2}{7}$ $f(5, 7) = 59$

17. 7   **19.** $2\sqrt{14}$ **21.** Center: $(0, 1, -1)$ **23.** Center: $(0, 1, -1)$
                                        Radius: 4   Radius: 3

25.

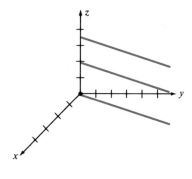

27.

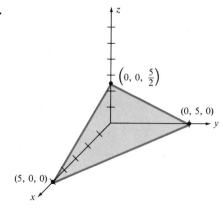

29.

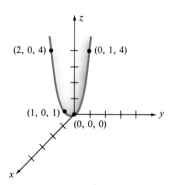

31. $C(500, 70) = \$11,750$ **33.** $P(500, 750) = \$7250$

35. Isoprofit lines: $7x + 5y = c$ **37.** $p(10, 30) = 17.31$ thousand cars   **39.** Student project   **41.** $C(9, 12) = 75$

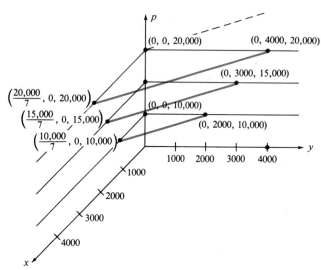

Section 13.2

1. $f_x(x, y) = 2 - y + 6x$

$f_y(x, y) = -x + 1$

3. $f_x(x, y) = \dfrac{1}{y}$

$f_y(x, y) = -\dfrac{x}{y^2}$

5. $f_x(x, y) = -\dfrac{2y}{(x-y)^2}$

$f_y(x, y) = \dfrac{2x}{(x-y)^2}$

7. $f_x(x, y) = \dfrac{y^2 - x^2 - 2xy}{(x^2 + y^2)^2}$

$f_y(x, y) = \dfrac{x^2 - y^2 - 2xy}{(x^2 + y^2)^2}$

9. $f_x(x, y) = ye^{xy}$

$f_y(x, y) = xe^{xy}$

11. $f_x(x, y) = -\dfrac{2e^{x+y}}{(e^x - e^y)^2}$

$f_y(x, y) = \dfrac{2e^{x+y}}{(e^x - e^y)^2}$

13. $f_x(x, y) = e^x \ln y + \dfrac{e^y}{x}$

$f_y(x, y) = \dfrac{e^x}{y} + e^y \ln x$

15. $f_x(x, y) = 2xe^{(x^2 + y^2)}$

$f_y(x, y) = 2ye^{(x^2 + y^2)}$

17. $f_x(x, y) = \dfrac{2}{3}(2xe^y - 2ye^x - 2xye^x)(x^2e^y - 2xye^x + y^2e^y)^{-1/3}$

$f_y(x, y) = \dfrac{2}{3}(x^2e^y - 2xye^x + y^2e^y)^{-1/3}(x^2e^y - 2xe^x + 2ye^y + y^2e^y)$

19. $f_x(x, y) = \dfrac{1}{x+y}$

$f_y(x, y) = \dfrac{1}{x+y}$

21. $f_x(3, 1) = -30$

$f_y(2, 3) = -6$

23. $f_x(0, 1) = 2e^3$

$f_y(-3, 2) = 3$

25. $f_x(3, 4) = \ln 3 + e^4 + 5$

$f_y(2, 5) = 3 \ln 2 + 20e^5$

27. $f_x(x, y) = 2x + 2y + 10$

$f_y(x, y) = 2x - 6y - 2$

$f_{xx}(x, y) = 2$

$f_{xy}(x, y) = 2$

$f_{yy}(x, y) = -6$

29. $f_x(x, y) = 4(x + y)(x^2 + 2xy - y^2)$

$f_y(x, y) = 4(x - y)(x^2 + 2xy - y^2)$

$f_{xx}(x, y) = 4(3x^2 + 6xy + y^2)$

$f_{xy}(x, y) = 4(3x^2 + 2xy - 3y^2)$

$f_{yy}(x, y) = 4(x^2 - 6xy + 3y^2)$

31. $f_x(x + y) = \ln(x + y) + 1$

$f_y(x, y) = \ln(x + y) + 1$

$f_{xx}(x, y) = \dfrac{1}{x+y}$

$f_{xy}(x, y) = \dfrac{1}{x+y}$

$f_{yy}(x, y) = \dfrac{1}{x+y}$

33. $f_x(x, y) = 6x + 5y$

$f_y(x, y) = 5x - 4y$

$f_{xx}(x, y) = 6$

$f_{xy}(x, y) = 5$

$f_{yy}(x, y) = -4$

35. $f_x(x, y) = -\dfrac{(x + y)(x + 5y)}{(x - y)^4}$

$f_y(x, y) = \dfrac{(x + y)(5x + y)}{(x - y)^4}$

$f_{xx}(x, y) = -\dfrac{2(x^2 + 10xy + 13y^2)}{(x - y)^5}$

$f_{xy}(x, y) = -\dfrac{2(5x^2 + 14xy + 5y^2)}{(x - y)^5}$

$f_{yy}(x, y) = \dfrac{2(13x^2 + 8xy + 3y^2)}{(x - y)^5}$

37. $f_x(A) = 0$

$f_y(A) = 0$

39. $f_x(A) = 0$

$f_y(A) < 0$

41. $R_x(10, 5) = 4$

$R_y(10, 5) = 10$

43. $C_x(10, 5) = 25$

$C_y(10, 5) = 20$

45. $C_x(10, 5) = 205$

$C_y(10, 5) = 60$

47. $P_x(10, 5) = 30$

$P_y(10, 5) = 200$

49. $C_x(10, 20) = 90$

$C_y(10, 20) = 150$

51. $P_x(10, 20) = 10$

$P_y(10, 20) = 10$

53. Direct computation   **55.** Direct computation

57. Direct computation

Section 13.3

1. $\left(\dfrac{5}{2}, 2\right)$ **3.** $\left(-\dfrac{1}{2}, \dfrac{1}{3}\right)$ **5.** $\left(-\dfrac{4}{3}, \dfrac{2}{3}\right)$ **7.** $(0, 2), (0, -1)$ **9.** No critical points   **11.** $\left(-\dfrac{4}{11}, -\dfrac{13}{11}\right)$

13. $\left(-\dfrac{2}{9}, -\dfrac{5}{3}\right)$ **15.** $\left(\sqrt{\dfrac{5}{3}}, 1\right), \left(\sqrt{\dfrac{5}{3}}, -1\right), \left(-\sqrt{\dfrac{5}{3}}, 1\right), \left(-\sqrt{\dfrac{5}{3}}, -1\right)$ **17.** $(0, 0), \left(2\left(\dfrac{4}{9}\right)^{1/3}, 2\left(\dfrac{2}{3}\right)^{1/3}\right)$

19. No critical points   **21.** Relative minimum: $(-4, -1)$ **23.** Relative maximum: $(4, 2)$

25. Saddle points: $(2, 2), (2, 0), (0, 2)$ **27.** Test fails at $(0, 0)$ **29.** Saddle point: $(20, 15)$

Relative maximum: $\left(\dfrac{4}{3}, \dfrac{4}{3}\right)$

31. Saddle points: $(0, 4)$, $\left(\dfrac{1}{2}, 0\right)$ **33.** Saddle point: $(0, 0)$ **35.** Relative minimum: $(2^{-1/3}, 2^{5/3})$

Relative minimum: $\left(\dfrac{1}{2}, 4\right)$ Relative maximum: $(3, 3)$

Relative maximum: $(0, 0)$

37. Test fails at $(\sqrt{3}, 0), (-\sqrt{3}, 0)$ **39.** Test fails at $(0, 0)$ **41.** $\left(\dfrac{4}{9}, \dfrac{8}{9}\right)$, distance $= \sqrt{\dfrac{80}{81}}$ **43.** $(2, 0, 0)$, distance $= 2$

45. Output level: $\left(\dfrac{115}{3}, 30\right)$

Maximum profit: \$1,004.17

47. (a) $p_1(x, y) = 65 - x - y$ (b) $R(x, y) = 65x - x^2 - 2xy + 90y - 2y^2$ (c) $P(x, y) = 65x - 2x^2 - 3xy + 90y - 3y^2$
$p_2(x, y) = 90 - x - 2y$

(d) Maximum production levels: $(8, 11)$ **49.** Pounds A $= \dfrac{15}{14}$

Pounds B $= \dfrac{25}{14}$

Section 13.4

1. $\left(\dfrac{5}{4}, -\dfrac{7}{4}\right)$ **3.** $x = -\sqrt{\dfrac{24}{7}}, y = -\dfrac{1}{4}\sqrt{\dfrac{24}{7}}$ **5.** $(0, 0)$ **7.** $(0, 2)$ **9.** $\left(-\dfrac{58}{23}, -\dfrac{44}{23}\right)$

$x = \sqrt{\dfrac{24}{7}}, y = \dfrac{1}{4}\sqrt{\dfrac{24}{7}}$

11. Side and height both 5 feet   **13.** Diameter and height both $\left(\dfrac{256}{\pi}\right)^{1/3} \cong 4.33$ feet   **15.** $x = \dfrac{375}{8}, y = \dfrac{225}{8}$

17. $x = \dfrac{50}{3}, y = \dfrac{25}{6}$ **19.** $x = 10, y = 5$ **21.** Maximum product: minimize   **23.** Constrained minimum
$L(x, y, \lambda) = xy + \lambda(x + y - 15)$ $L(x, y, \lambda) = xy^2 + \lambda(x + y - 75)$

25. Minimum cost box   **27.** Maximum volume
$L(x, y, \lambda) = 6x^2 + 4xy + \lambda(x^2 y - 1600)$ $L(x, y, \lambda) = x^2 y + \lambda(y + 4x - 108)$
29. Wire problem   **31.** Telephone lines
$L(x, y, \lambda) = 2000y + 3000\sqrt{49 + x^2} + \lambda(x + y - 12)$

$L(x, y, \lambda) = x^2 + \pi\left(\dfrac{y}{2\pi}\right)^2 + \lambda(4x + 2y - 25)$

Section 13.5

1. x-antiderivative: $\dfrac{x^3}{3} + \dfrac{x^2 y}{2} - 2xy + xy^2 + C(y)$ **3.** x-antiderivative: $\dfrac{y}{3}(x^2 + y^2)^{3/2} + C(y)$

y-antiderivative: $x^2 y + \dfrac{xy^2}{2} - y^2 + \dfrac{y^3}{3} + C(x)$ y-antiderivative: $\dfrac{x}{3}(x^2 + y^2)^{3/2} + C(x)$

5. x-antiderivative: $\dfrac{2}{3}x^3 y + x + C(y)$ **7.** x-antiderivative: $\dfrac{x^3}{3} y^2 + C(y)$

y-antiderivative: $x^2 y^2 + y + C(x)$ y-antiderivative: $x^2 \dfrac{y^3}{3} + C(x)$

9. x-antiderivative: $e^x \ln y + C(y)$ **11.** 4   **13.** $\dfrac{27}{2}$ **15.** 18   **17.** $\dfrac{1}{2}$ **19.** 0   **21.** 0   **23.** $\dfrac{1}{2}(1 - \ln 2)$
y-antiderivative: $e^x(y \ln y - y) + C(x)$

25. $\dfrac{13}{15}$ **27.** $3e - e^2 - 2$ **29.** $\dfrac{1}{2}(e^e - e)$ **31.** $\dfrac{28}{3}$ **33.** $\dfrac{1}{15}[18^{5/2} - 10^{5/2} - 9^{5/2} + 1]$ (roughly 54.5)   **35.** $\dfrac{368}{3}$

37. $\dfrac{200}{27}[20^{3/2} - 5^{3/2}]$ (roughly 579.4)

Chapter Review

1. $f_x(x, y) = 2xy$ **3.** $f_x(x, y) = \dfrac{1}{x}\left(\dfrac{\ln 10 + 1}{\ln 10}\right)$ **5.** $f_{xx}(x, y) = \dfrac{y}{(x^2 + y)^{3/2}}$ **7.** $f_{xy}(x, y) = 2x$ **9.** $f_x(x, y) = 4x(x^2 + y^2)$

11. $f_x(x, y) = yx^{y-1}$ **13.**   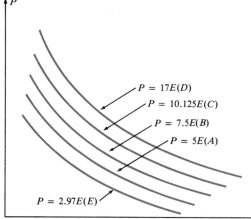   **15.** Saddle point: $(0, 0)$

17. Saddle point: $(0, 0)$ **19.** No critical points   **21.** Saddle point: $(0, 0)$ **23.** Maximum profit: $x = 6, y = 8$
25. Saddle point: $(-1, 2)$ **27.** Relative minimum: $(0, 0)$ **29.** No critical point   **31.** Relative maximum: $(60, 0)$
33. No critical points   **35.** Saddle point: $(0, 0)$ **37.** No critical points

39. $f(x, y) = y$ on the constraint surface (no max or min)   **41.** Relative minimum: $\left(\dfrac{1}{2}, \dfrac{1}{2}\right)$ **43.** Relative minimum: $\left(\dfrac{1}{4}, \dfrac{3}{4}\right)$

45. $\left(\dfrac{\sqrt{2}}{2}, \dfrac{\sqrt{2}}{2}\right)$ is the maximum point   **47.** $(0, 0)$ is the minimum point   **49.** 4   **51.** $e - 1$ **53.** 20

$\left(-\dfrac{\sqrt{2}}{2}, \dfrac{\sqrt{2}}{2}\right)$ is the minimum point

Chapter Test

1. **(a)** $f_x(x, y) = (x + y)^{-1}$ **(b)** $f_y(x, y) = (x + y)^{-1}$ **(c)** $f_{xx}(x, y) = -(x + y)^{-2}$ **(d)** $f_{xy}(x, y) = -(x + y)^{-2}$
(e) $f_{yy}(x, y) = -(x + y)^{-2}$

2. Saddle point: $(0, 0)$ **3.** $x = y = 71$ **4.** $\left(\dfrac{1}{2}, \dfrac{1}{2}\right)$ **5.** 2
    Relative maximum point $(-1, -1)$

APPENDIX A: ENRICHED LINEAR PROGRAMMING

1. The function $-f(x) = -x^2$ has a maximum value of 0 at $x = 0$, and so x^2 has a minimum value of $-0 = 0$ at $x = 0$.

3. The function $-f(x) = -|x|$ has a maximum value of 0 at $x = 0$, and so $|x|$ has a minimum value of $-0 = 0$ at $x = 0$.

5. The function $-f(x) = -x^2 + 2x - 2$ has a maximum value of -1 at $x = 1$, and so $x^2 - 2x + 2$ has a minimum value of $-(-1) = 1$ at $x = 1$.

$(x_1, x_2, s_1, s_2, A_1, A_2)$

7. (a) $(0, 0, 10, 0, 10, 1)$
   (b) $(10, 0, 0, 9, -10, 0)$
   (c) $(1, 0, 9, 0, -1, 0)$
   (d) $(1, 1, 8, 0, 0, 0)$
   (e) $(2, 0, 8, 0, -2, -1)$

9. $\begin{aligned} x_1 + x_2 - s_1 \quad\quad + A_1 \quad\quad &= 2 \\ x_1 + 2x_2 \quad\quad - s_2 \quad\quad + A_2 &= 3 \\ x_1 \geq 0 \quad s_1 \geq 0 \quad A_1 \geq 0 & \\ x_2 \geq 0 \quad s_2 \geq 0 \quad A_2 \geq 0 & \end{aligned}$

11.

BV	ROW	P	x_1	x_2	s_1	s_2	A_1	A_2	RHS
*	0	1	-2	-3	1	1	0	0	-5
A_1	1	0	1	1	-1	0	1	0	2
A_2	2	0	1	2	0	-1	0	1	3

(Note zeros under A_1 and A_2 in ROW 0.)

13.

BV	ROW	P	x_1	x_2	s_1	s_2	RHS
*	0	1	0	0	0	1	-3
x_1	1	0	1	0	-2	1	1
x_2	2	0	0	1	1	-1	1

(Note zeros in ROW 0 under basic variables x_1 and x_2.)

15. Student project

17.

(x_1, x_2)	$C = 5x_1 + 7x_2$
$(0, 10)$	70
$(20, 0)$	100
$\left(\dfrac{20}{3}, \dfrac{10}{3}\right)$	$\dfrac{170}{3}$ ⟵ min

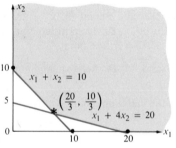

19.

BV	ROW	P	x_1	x_2	s_1	s_2	A_1	A_2	RHS	
*	0	1	-2	-5	1	1	0	0	-30	*Initial*
A_1	1	0	1	1	-1	0	1	0	10	*tableau*
A_2	2	0	1	4	0	-1	0	1	20	
*	0	1	0	-3	-1	1	2	0	-10	*Second*
x_1	1	0	1	1	-1	0	1	0	10	*tableau*
A_2	2	0	0	3	1	-1	-1	1	10	
*	0	1	0	0	0	0	1	1	0	
x_1	1	0	1	0	$-\dfrac{4}{3}$	$\dfrac{1}{3}$	$\dfrac{4}{3}$	$-\dfrac{1}{3}$	$\dfrac{20}{3}$	*Final* *tableau*
x_2	2	0	0	1	$\dfrac{1}{3}$	$-\dfrac{1}{3}$	$-\dfrac{1}{3}$	$\dfrac{1}{3}$	$\dfrac{10}{3}$	

(The first feasible point is $(x_1, x_2) = \left(\dfrac{20}{3}, \dfrac{10}{3}\right).$)

21. $(x_1{}^*, x_2{}^*) = \left(\dfrac{20}{3}, \dfrac{10}{3}\right)$

Minimum value $= 10$

23.

BV	ROW	P	x_1	x_2	s_1	s_2	A_1	A_2	RHS
*	0	1	-2	-6	1	1	0	0	-7
A_1	1	0	1	1	-1	0	1	0	2
A_2	2	0	1	5	0	-1	0	1	5

(Note the zeros in ROW 0 under A_1 and A_2.)

25. Student project   **27.**

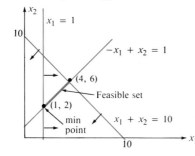

29.

BV	ROW	P	x_1	x_2	s_1	s_2	A_1	A_2	RHS
*	0	1	0	-1	0	1	0	0	-2
s_1	1	0	1	1	1	0	0	0	10
A_1	2	0	-1	1	0	0	1	0	1
A_2	3	0	1	0	0	-1	0	1	1

(Note the zeros in ROW 0 under A_1 and A_2.)

31.

BV	ROW	P	x_1	x_2	s_1	s_2	RHS
*	0	1	0	0	0	5	-8
s_1	1	0	0	0	1	2	7
x_2	2	0	0	1	0	-1	2
x_1	3	0	1	0	0	-1	1

(Note the zeros in ROW 0 under s_1, x_2, and x_1.)

33. $x_1 + 2x_2 + \ x_3 + A_1 = 5$
$x_1 + \ x_2 + 2x_3 + A_2 = 5$
$x_1 \geq 0 \qquad A_1 \geq 0$
$x_2 \geq 0 \qquad A_2 \geq 0$
$x_3 \geq 0$

35. Student project   **37.** Student project

39. Solve the linear programming problem

$$\min (2x_1 + 3x_2 + 2.5x_3)$$

subject to

$$16x_1 + 32x_2 + 24x_3 \geq 500$$
$$32x_1 + 16x_2 + 24x_3 \geq 200$$
$$32x_1 + 16x_2 + 32x_3 \geq 200$$

and

$$x_1 \geq 0$$
$$x_2 \geq 0$$
$$x_3 > 0$$

Credits

Calculus Preliminaries: page 461, The Bettmann Archive; page 463, Sidney Harris; page 474, The Bettmann Archive; page 483, Sidney Harris.

Chapter 8: page 493, Dr. Harold Edgerton, MIT, Cambridge, MA; pages 496 and 497, North Wind Picture Archives; page 528, Boston Public Library.

Chapter 9: page 567, George Gardner/Stock, Boston; page 579, courtesy of Carnation; page 620, courtesy of General Motors Corporation.

Chapter 10: page 647, Don W. Fawcett/Photo Researchers; page 656, author photo; page 666, The Bettmann Archive; page 684, Karen Preuss/Taurus Photos.

Chapter 11: page 707, Ellis Herwig/Stock, Boston.

Chapter 12: page 755, Jean-Claude Lejeune/Stock, Boston; page 774, courtesy of Chevron Corporation; page 780, John Coletti/Stock, Boston; page 796, photo by Bruce Iverson for Random House.

Chapter 13: page 819, Mark Antman/The Image Works; page 830, Herb Levar/Photo Researchers; page 857, North Wind Picture Archives.

Index

Definition of the Derivative

The derivative of a function $f(x)$ is

$$f'(x) = \frac{d}{dx} f(x) = \lim_{h \to 0} \frac{f(x + h) - f(x)}{h}$$

Rules for the Derivative

Constant rule

If $f(x) = c$, where c is a constant, then

$$f'(x) = 0$$

Power rule

If $f(x) = x^n$, for any real number n, then

$$f'(x) = nx^{n-1}$$

Constant times a function

If $y = cf(x)$, where the factor c is a constant, then

$$y' = cf'(x)$$

Sum and difference rule

If $y = f(x) \pm g(x)$, then

$$y' = f'(x) \pm g'(x)$$

Product rule

If $y = f(x)g(x)$, then

$$y' = f'(x)g(x) + g'(x)f(x)$$

Quotient rule

If $y = \dfrac{f(x)}{g(x)}$, where $g(x) \neq 0$, then

$$y' = \frac{g(x)f'(x) - f(x)g'(x)}{g^2(x)}$$

Chain rule

If y is a function of u, say $y = f(u)$, and u is a function of x, say $u = g(x)$, then $y = f(u) = f[g(x)]$, and

$$\frac{dy}{dx} = \frac{dy}{du} \frac{du}{dx}$$

Chain rule (alternate form)

If $y = f[g(x)]$, then

$$y' = f'[g(x)] \cdot g'(x)$$

General power rule

If $y = u^n$, where $u = u(x)$ is a differentiable function of x, then

$$y' = n \cdot u^{n-1} \cdot u'(x)$$

Derivative of logarithm functions

If $y = \log |u(x)|$, then

$$y' = \frac{u'(x)}{u(x)}$$

General exponent function

If $y = e^u$, where $u(x)$ is a differentiable function of x, then

$$y' = e^u \cdot u'(x)$$